Markus J Aschwanden

Physics of the
Solar Corona

An Introduction with Problems and Solutions

 Springer

Published in association with
Praxis Publishing
Chichester, UK

Dr Markus J Aschwanden
Lockheed Martin Advanced Technology Center
Solar and Astrophysics Laboratory
Palo Alto
California
USA

SPRINGER–PRAXIS BOOKS IN ASTRONOMY AND PLANETARY SCIENCES
SUBJECT *ADVISORY EDITOR*: Dr. Philippe Blondel, C.Geol., F.G.S., Ph.D., M.Sc., Senior Scientist, Department of Physics, University of Bath, Bath, UK; John Mason B.Sc., M.Sc., Ph.D.

ISBN 3-540-30765-6 Springer-Verlag Berlin Heidelberg New York

Springer is part of Springer-Science + Business Media (springeronline.com)
Bibliographic information published by Die Deutsche Bibliothek
Die Deutsche Bibliothek lists this publication in the Deutsche Nationalbibliografie; detailed bibliographic data are available from the Internet at http://dnb.ddb.de

Library of Congress Control Number: 2005937065

Cover design: Jim Wilkie
Completed in LaTex: EDV-Beratung, Germany

Printed on acid-free paper

To Susanna

Contents

Preface

In this book we provide a comprehensive introduction into the basic physics of phenomena in the solar corona. Solar physics has evolved over three distinctly different phases using progressively more sophisticated observing tools. The first phase of naked-eye observations that dates back over several thousands of years has been mainly concerned with observations and reports of solar eclipses and the role of the Sun in celestial mechanics. In the second phase that lasted about a century before the beginning of the space age, ground-based solar-dedicated telescopes, spectrometers, coronagraphs, and radio telescopes were built and quantitative measurements of solar phenomena developed, which probed the basic geometric and physical parameters of the solar corona. During the third phase, that started with the beginning of the space age around 1950, we launched solar-dedicated spacecraft that explored the Sun in all possible wavelengths, conveying to us high-resolution images and spectral measurements that permitted us to conduct quantitative physical modeling of solar phenomena, supported by numerical simulations using the theories of magneto-hydrodynamics and plasma particle physics. This book focuses on these new physical insights that have mostly been obtained from the last two decades of space missions, such as from soft X-ray observations with *Yohkoh*, extreme-ultraviolet (EUV) observations with *SoHO* and *TRACE*, and hard X-ray observations with *Compton* and the recently launched *RHESSI*. The last decade (1992−2002) has been the most exciting era in the exploration of the solar corona in all wavelengths, producing unprecedented stunning pictures, movies, and high-precision spectral and temporal data.

There are a lot of new insights into the detailed structure of the solar corona, mass flows, magnetic field interactions, coronal heating processes, and magnetic instabilities that lead to flares, accelerated particles, and coronal mass ejections, which have not been covered in previous textbooks. The philosophy and approach of this textbook is to convey a physics introduction course to a selected topic of astrophysics, rather than to provide a review of observational material. The material is therefore not described in historical, phenomenological, or morphological order, but rather structured by physical principles. In each chapter we outline the basic physics of relevant physical models that explain coronal structures, flare processes, and coronal mass ejections. We include in each chapter analytical and numerical model calculations that have been applied to these solar phenomena, and we show comprehensive data material that illustrate the models and physical interpretations. This textbook is aimed to be an up-to-date introduction into the physics of the solar corona, suitable for graduate students and researchers. The 17 chapters build up the physical concepts in a systematic way, starting

Table 1: Selection of textbooks and monographs on solar physics.

Author	Year	Book title
Sun: general		
Bruzek & Durrant	1977	*Illustrated Glossary for Solar and Solar-Terrestrial Physics*
Zirin	1988	*Astrophysics of the Sun*
Foukal	1990	*Solar Astrophysics*
Phillips	1992	*Guide to the Sun*
Lang	1995	*Sun, Earth, and Sky*
Lang	2000	*The Sun from Space*
Murdin	2000	*Encyclopedia of Astronomy and Astrophysics*
Lang	2001	*The Cambridge Encyclopedia of the Sun*
Golub & Pasachoff	2001	*Nearest Star: The Surprising Science of Our Sun*
Stix	2002	*The Sun*
Zirker	2002	*Journey from the Center of the Sun*
Photosphere		
Stenflo	1994	*Solar Magnetic Fields: Polarized Radiation Diagnostics*
Schüssler & Schmidt	1994	*Solar Magnetic Fields*
Transition region		
Mariska	1992	*The Solar Transitions Region*
Corona, flares, plasma physics		
Kundu	1965	*Solar Radio Astronomy*
Zheleznyakov	1970	*Radio Emission of the Sun and Planets*
Svestka	1976	*Solar Flares*
Krueger	1979	*Introduction to Solar Radio Astronomy and Radio Physics*
Melrose	1980	*Plasma Astrophysics (Vol. 1 & 2):*
		Nonthermal Processes in Diffuse Magnetized Plasmas.
Priest	1982	*Solar Magnetohyrdodynamics*
McLean & Labrum	1985	*Solar Radiophysics*
Melrose	1986	*Instabilities in Space and Laboratory Plasmas*
Bray et al.	1991	*Plasma Loops in the Solar Corona*
Benz	1993	*Plasma Astrophysics. Kinetic Processes in Solar and Stellar Coronae*
Sturrock	1994	*Plasma Physics. An Introduction to the Theory of*
		Astrophysical, Geophysical and Laboratory Plasmas
Kirk, Melrose, & Priest	1994	*Plasma Astrophysics*
Tandberg–Hanssen	1995	*The Nature of Solar Prominences*
Golub & Pasachoff	1997	*The Solar Corona*
Strong et al.	1999	*The Many Faces of the Sun:*
		A Summary of the Results from NASA's Solar Maximum Mission
Schrijver & Zwaan	2000	*Solar and Stellar Magnetic Activity*
Priest & Forbes	2000	*Magnetic Reconnection (MHD Theory and Applications)*
Tajima & Shibata	2002	*Plasma Astrophysics*
Aschwanden	2002	*Particle Acceleraion and Kinematics in Solar Flares*
Heliosphere and interplanetary		
Russell, Priest, Lee	1990	*Physics of Magnetic Flux Ropes*
Schwenn & Marsch	1991	*Physics of the Inner Heliosphere:*
		Vol. 1: Large-Scale Phenomena
		Vol. 2: Particles, Waves and Turbulence
Kivelson & Russell	1995	*Introduction to Space Physics*
Crooker et al.	1997	*CMEs*
Song et al.	2001	*Space Weather*
Balogh, Marsden, & Smith	2001	*The Heliosphere near Solar Minimum — The Ulysses Perspective*
Carlowicz & Lopez	2002	*Storms from the Sun — The Emerging Science of Space Weather*

with introductions into the basic concepts (§1−2), magneto-hydrodynamics (MHD) of the coronal plasma (§3−6), MHD of oscillations and waves (§7−8), coronal heating (§9), magnetic reconnection processes (§10), particle acceleration and kinematics (§11−12), flare dynamics and emission in various wavelengths (§13−16), and coronal mass ejection (CME) phenomena (§17). So the first-half of the book is concerned with phenomena of the quiet Sun (§1−9), while the second-half focuses on eruptive phenomena such as flares and CMEs (§10−17). The scope of the book is restricted to the solar corona, while other parts of the Sun, such as the solar interior, the photosphere, the chromosphere, or the heliosphere, are referred to in other textbooks, of which a selection is given in Table 1. Extensive literature (reviews and theoretical or modeling studies) are referenced preferentially at the beginning or end of the chapters and subsections. The material of this textbook is based on some 10,000 original papers published in solar physics, which grows at a current rate of about 700 papers per year. The physical units used in this textbook are the *cgs units*, as they are most frequently used in the original publications in solar physics literature. Conversion into *SI units* can be done easily by using the conversion factors given in Appendix B.

The author is most indebted to invaluable discussions, comments, reviews, and graphics provisions from colleagues and friends, which are listed in alphabetical order: Loren Acton, Spiro Antiochos, Tim Bastian, Arnold Benz, John Brown, Ineke DeMoortel, Brian Dennis, Marc DeRosa, George Doschek, Robertus (von Fay-Siebenburgen) Erdélyi, Terry Forbes, Allen Gary, Costis Gontikakis, Marcel Goossens, Joe Gurman, Iain Hannah, Jack Harvey, Eberhard Haug, Jean Heyvaerts, Gordon Holman, Joe Huba, Marion Karlický, Jim Klimchuk, Boon Chye Low, John Mariska, Eckart Marsch, Don Melrose, Zoran Mikić, Jim Miller, Valery Nakariakov, Aake Nordlund, Leon Ofman, Eugene Parker, Spyridon Patsourakos, Alex Pevtsov, Ken Phillips, Eric Priest, Reza Rezaei, Bernie Roberts, Jim Ryan, Salvatore Serio, Karel Schrijver, Gerry Share, Kazunari Shibata, Daniele Spadaro, Barbara Thompson, Alan Title, Kanaris Tsinganos, Tong-Jiang Wang, Harry Warren, Stephen White, Thomas Wiegelmann, David Williams, Amy Winebarger, Takaaki Yokoyama, and Paolo Zlobec.

The author wishes to acknowledge the efficient and most helpful support provided by *PRAXIS Publishing Ltd.*, in particular for reviewing by John Mason, copy-editing by Neil Shuttlewood (*Originator*), and cover design by Jim Wilkie. LATEX support was kindly provided by Stephen Webb, image scanning by David Schiff (*Lockheed Martin*), *SolarSoft* software by Sam Freeland (*Lockheed Martin*), and archive support by the *NASA Astrophysics Data System (ADS)*. Most gratefully I wish to thank the publisher Clive Horwood, who had the vision to produce this textbook, and who invited me to undertake this project, provided untiring encouragement and patience during the 3-year production period.

Special thanks goes to my entire family, my late wife Susanna to whom this book is dedicated with eternal love, my children Pascal Dominique and Alexander Julian, and to my friend Carol J. Kersten, who all supported this work with an enthusiastic spirit.

Palo Alto, California, June 2004   *Markus J. Aschwanden*

Preface to 2nd Edition

The second edition (in paperback) reproduces the entire body of the first edition (with hardbound cover) with identical text, except for minor corrections of typographical misprints, but contains in addition a section with some 170 *Problems* and *Solutions* at the end of the main text body (p. 739−788), suitable as exercises for students and researchers to obtain a deeper understanding of the text. The analytical and numerical solutions provided for these specific problems allows the reader to verify whether he/she understands practical applications to solar observations, which are not always exemplified in detail in the main body of the text. The author is most indebted to the following friends and colleagues who double-checked and proof-read the sections of the Problems and Solutions, as well as parts or the first edition text: Henry Aurass, Mitchell Berger, Jeff Brosius, Amir Caspi, Steven Christe, Ineke DeMoortel, Robertus Erdélyi, Gregory Fleishman, Andre Fludra, Dale Gary, Sarah Gibson, Holly Gilbert, Costis Gontikakis, Iain Hannah, Leon Kocharov, Bernhard Kliem, Eduard Kontar, Enrico Landi, Yuri Litvinenko, Scott McIntosh, Thomas Metcalf, Thomas Neukirch, Hardi Peter, Gordon Petrie, John Raymond, Pete Riley, Julia Saba, David Smith, Youra Taroyan, David Tsiklauri, Aad VanBallegooijen, Gary Verth, Angelos Vourlidas, Harry Warren, Thomas Wiegelmann, and Jie Zhang.

Due to the short interval between the first and second edition we refrained from updating the text and references. However, we are happy to report that significant new science results have been produced as a consequence of the flawless continued performance of the major solar space missions, such as SoHO, TRACE, and RHESSI. The reader may consult the official NASA mission websites or the author's website on solar literature:

http://www.lmsal.com/∼aschwand/publications/index.html

for recent updates of SoHO, TRACE, and RHESSI publications. New solar missions such as STEREO, Solar-B, and SDO are planned for launch shortly after this second edition comes out in print. We are looking forward to another exciting decade of vibrant solar research.

Palo Alto, California, December 2005   *Markus J. Aschwanden*

Chapter 1

Introduction

1.1 History of Solar Corona Observations

What we see with our eye from our Sun or from the billions of stars in our galaxy is optical radiation that is emitted at the surface of the star, in the so-called photosphere. The optical emission produced by Thomson scattering in the much more tenuous atmosphere or corona above, is many orders of magnitude less intense and thus can only be seen when the solar surface is occulted (e.g., by the moon during a total solar eclipse). The first observations of the solar corona thus date back to ancient eclipse observations, which have been reported from Indian, Babylonian, or Chinese sources. A detailed account of historical eclipses can be found in the book of Guillermier & Koutchmy (1999), which mentions, for example Chinese solar eclipses as early as 2800 BC, the failure of a prediction by the two luckless Chinese royal astrologers Hsi and Ho around 2000 BC, the successful prediction of the solar eclipse of 28 May 585 BC by the Greek mathematician and philosopher Thales, or the eclipse of 1919 May 29 in Sobral (Brazil) and Principe (West Africa), which has been observed by two expeditions of the British astronomer Arthur Stanley Eddington to prove Einsteins theory of relativity.

   Regular observations of solar eclipses and prominences started with the eclipse of 1842, which was observed by experienced astronomers like Airy, Arago, Baily, Littrow, and Struve. Photographic records start since the 1851 eclipse in Norway and Sweden, when the professional photographer Berkowski succeded to produce a daguerrotype of prominences and the inner corona. Visual and spectroscopic observations of prominence loops were carried out by Pietro Angelo Secchi in Italy and by Charles Augustus Young of Princeton University during the 19th century. The element helium was discovered in the solar corona by Jules Janssen in 1868. George Ellery Hale constructed a spectroheliograph in 1892 and observed coronal lines during eclipses. Bernard Lyot built his first coronagraph at the Pic-du-Midi Observatory in 1930, an instrument that occults the bright solar disk and thus allows for routine coronal observations, without need to wait for one of the rare total eclipse events. In 1942, Edlén identified forbidden lines of highly ionized atoms and in this way established for the first time the

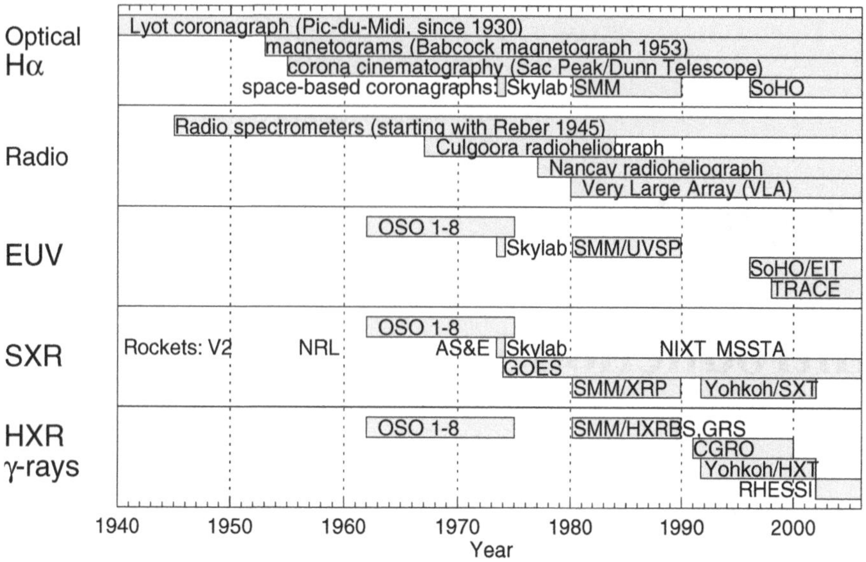

Figure 1.1: The operation periods of major instruments and space missions that provided unique observations of the solar corona are shown in historical order, sorted in different wavelength regimes.

million-degree temperature of the corona. A historical chronology of these early coronal observations is given in Bray et al. (1991), while coronal observations in the space age are reviewed in the encyclopedia article of Alexander & Acton (2002).

If we observe the Sun or the stars in other wavelengths, for instance in soft X-rays, hard X-rays, or radio wavelengths, the brightest emission comes from the corona, while the photosphere becomes invisible, just opposite to optical wavelengths. A breakthrough of coronal observations therefore started with the space era of rocket flights and spacecraft missions, which enabled soft X-ray and extreme ultraviolet (EUV) observations above the absorbing Earth's atmosphere (Fig. 1.1). Early Areobee rocket flights were conducted by the U.S. *Naval Research Laboratory (NRL)* in 1946 and 1952, recording spectrograms at EUV wavelengths down to 190 nm and Lyman-α emission of H I at 121.6 nm. In 1974, G. Brueckner and J.D. Bartoe achieved a resolution of $\approx 4''$ in EUV lines of He I, He II, O IV, O V, and Ne VII. The first crude X-ray photograph of the Sun was obtained by Friedman in 1963 with a pinhole camera from an Areobee rocket of *NRL* on 1960 April 19. These rocket flights last typically about 7 minutes, and thus allow only for short glimpses of coronal observations. Long-term observations were faciliated with the satellite series *Orbiting Solar Observatory (OSO-1* to *OSO-8)*, which were launched into orbit during $1962-1975$, equipped with non-imaging EUV, soft X-ray, and hard X-ray spectrometers and spectroheliographs. Finally, the launch of *Skylab*, which operated from 1973 May 14 to 1974 February 8 initiated a new era of multi-wavelength solar observations from space. *Skylab* carried a white-light coronagraph, two grazing-incidence X-ray telescopes, EUV spectroheliometers/spectroheliographs, and an UV spectrograph, recording $\approx 32,000$ pho-

Figure 1.2: The *Solar Maximum Mission (SMM)* satellite was operated during 1980−1989. Some instrument failures occurred early in the mission, such as the position encoder of *FCS* and several gyroscopes. Repairs of the satellite were performed by "Pinky" Nelson and Dick Scobee from the Space Shuttle Challenger in April 1984. The approach of the maneouvering astronaut in his flying armchair to the ailing spacecraft is documented in this figure (courtesy of *NASA*).

tographs during its mission. A comprehensive account of Skylab observations and its scientific results on cool and hot active region loops, as well as on flare loops, is described in the book *Plasma Loops in the Solar Corona* by Bray et al. (1991).

The first solar-dedicated space mission that operated a full solar cycle was the *Solar Maximum Mission (SMM)*, launched on 1980 February 14 and lasting until orbit decay on 1989 December 2 (Fig. 1.2). The *SMM* instrumental package contained a *Gamma-Ray Spectrometer (GRS)* with an energy range of 10 keV−160 MeV, a *Hard X-Ray Burst Spectrometer (HXRBS)* in the energy range of 20−300 keV, a *Hard X-ray Imaging Spectrometer (HXIS)* with moderate spatial resolution (8″ pixels) in the energy range of 3.5−30 keV, two soft X-ray spectrometers [*Bent Crystal Spectrometer (BCS)*, 1.7−3.2 Å, and a *Flat Crystal Spectrometer (FCS)*, 1.4−22 Å], an *Ultraviolet Spectrometer/Polarimeter (UVSP)* in the 1150−3600 Å wavelength range, a white-light *Coronagraph/Polarimeter (CP)* in the 4448−6543 Å wavelength range, and an *Active Cavity Radiometer Irradiance Monitor (ACRIM)* in ultraviolet and infrared wavelengths. *SMM* made a number of scientific discoveries in observing some 12,000 solar flares and several hundred *coronal mass ejections (CME)*, which are lucidly summarized in the book *The Many Faces of the Sun: A Summary of the Results from NASA's Solar Maximum Mission* by Strong et al. (1999). There was also a Japanese mission (*Hinotori*) flown with similar instrumentation, from 1981 February 21 to 1982 October 11 (see instrumental descriptions in Makishima, 1982 and Takakura et al. 1983a).

Figure 1.3: Deployment of the *Compton Gamma-Ray Observatory (CGRO)* in 1991 April, photographed from the space shuttle *Atlantis*. The 8 *Large-Area Detectors (LAD)* of the *Burst and Transient Source Experiment (BATSE)* are located at the 8 corners of the spacecraft. The cylindrical detectors on the top are *COMPTEL* and *EGRET*, while *OSSE* is on the far side. The *BATSE* detectors provided for the first time sufficient photon statistics to measure electron time-of-flight delays in solar flares (courtesy of *NASA*).

In April 1991, the *Compton Gamma-Ray Observatory (CGRO)* was deployed (Fig. 1.3), which was designed to detect gamma-ray bursts from astrophysical and cosmological objects, but actually recorded more high-energy γ-ray and hard X-ray photons from solar flares than from the rest of the universe. In particular the *Burst And Transient Source Experiment (BATSE)* with its high sensitivity (with a collecting area of 2,000 cm^2 in each of the 8 detectors) in the energy range of $25-300$ keV (Fishman et al. 1989) delivered unprecedented photon statistics, so that energy-dependent electron time-of-flight delays could be determined in solar flares down to accuracies of a few milliseconds, and thus crucially contributed to a precise localization of particle acceleration sources in solar flares. Also the *Oriented Scintillation Spectrometer Experiment (OSSE)* (Kurfess et al. 1998) provided crucial measurements of γ-ray lines in the energy range of 50 keV to 10 MeV in a number of large X-class flares. *CGRO* recorded a total of some 8000 solar flares during its lifetime. The truck-heavy spacecraft was de-orbited by *NASA* in May 2000 because of gyroscope malfunctions.

The great breakthrough in soft X-ray imaging of the solar corona and flares came with the Japanese mission *Yohkoh* (Ogawara et al. 1991), which contained four instruments (Fig. 1.4): the *Hard X-ray Telescope (HXT)* with four energy channels ($14-23$

Figure 1.4: On 1991 August 30, the *Yohkoh* satellite was launched into space from the *Kagoshima Space Center (KSC)* in Southern Japan. The *Yohkoh* satellite, meaning "Sunbeam" in Japanese, is a project of the Japanese Institute of Space and Astronautical Science (ISAS) (courtesy of ISAS).

keV, 23−33 keV, 33−53 keV, and 53−93 keV) and a spatial resolution of ≈ 8″ (Kosugi et al. 1991); the *Soft X-ray Telescope (SXT)* with mulitple filters sensitive to temperatures of $T \gtrsim 1.5$ MK, having a spatial resolution of 5″ in full-disk images and 2.5″ in partial frames (Tsuneta et al. 1991); a *Wide-Band Spectrometer (WBS)* containing soft X-ray, hard X-ray, and γ-ray spectrometers (Yoshimori et al. 1991); and a *Bent Crystal Spectrometer (BCS)* (Culhane et al. 1991). *Yohkoh* provided a full decade of soft X-ray images from the solar disk and from flares, revealing for the first time the geometry and topology of large-scale magnetic field reconfigurations and magnetic reconnection processes in flares. The satellite lost its pointing during a solar eclipse in December 2001 and reentered in September 2005.

The next major solar mission to observe the Sun from inside out to 30 solar radii in the heliosphere was the *ESA/NASA* jointly-built spacecraft *Solar and Heliospheric Observatory (SoHO)*, launched on 1995 December 2, and still being fully operational at the time of writing. Twelve instrument teams with over 200 co-investigators participate in this mission. The *SoHO* spacecraft includes the following 12 instrument packages: 3 instruments for helioseismology (*GOLF, VIRGO, SOI/MDI*), 5 instruments for observing the solar atmosphere (*SUMER, CDS, EIT, UVCS, LASCO*), and 4 particle detector instruments that monitor the solar wind (*CELIAS, COSTEP, ERNE, SWAN*). Observations of the solar corona are performed by the *Solar Ultraviolet Measurements*

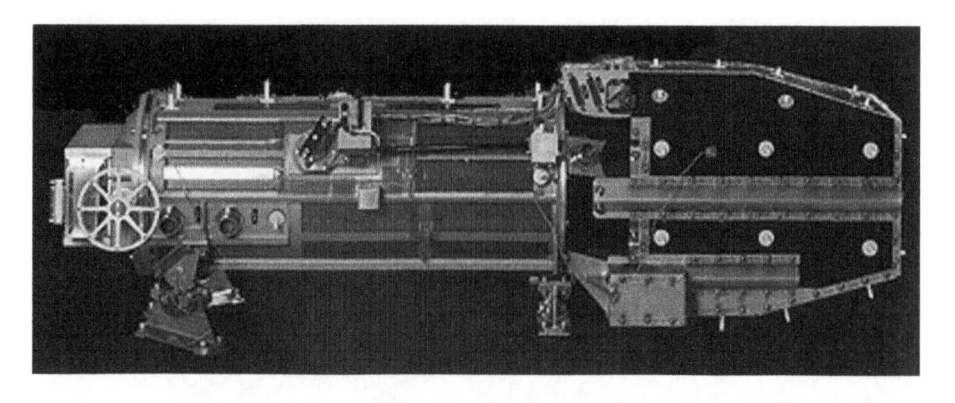

Figure 1.5: The *Extreme-ultraviolet Imaging Telescope (EIT)* on board the *Solar and Helio-spheric Observatory (SoHO)* is a normal-incidence, multi-layer telescope (Delaboudinière et al. 1995). It records since 1996 full-disk solar images in multiple filters, sensitive to coronal temperatures of $1-2$ MK, with a spatial resolution of $2.5''$ and a cadence of a few hours (courtesy of *EIT* team).

of Emitted Radiation (SUMER) telescope and spectrometer in the $500-1610$ Å wavelength range (Wilhelm et al. 1995), the *Coronal Diagnostic Spectrometer (CDS)* at $150-800$ Å (Harrison et al. 1995), the *Extreme-ultraviolet Imaging Telescope (EIT)* (Fig. 1.5) which takes full-disk images in Fe IX (171 Å), Fe XII (195 Å), Fe XV (284 Å), and He II (304 Å) (Delaboudinière et al. 1995), the *Ultraviolet Coronagraph Spectrometer (UVCS)* which provides spectral line diagnostics between 1 and 3 solar radii, and the *Large Angle Spectroscopic COronagraph (LASCO)* which images the heliosphere from 1.1 to 30 solar radii (Brueckner et al. 1995). Instrument descriptions and first results can be found in Fleck et al. (1995) and Fleck & Svestka (1997). The *SoHO* observatory is a highly successful mission that provided a wealth of data on dynamic processes in the solar corona, which will be described in more detail in the following chapters.

The next solar mission was designed to provide coronal observations with unprecedented high spatial resolution of $\approx 1''$ (with $0.5''$ pixel size), built as a spacecraft with a single telescope, called the *Transition Region And Coronal Explorer (TRACE)* (Fig. 1.6). The *TRACE* telescope was mated on a Pegasus launch vehicle, which launched in 1998 April and is still successfully operating at the time of writing. An instrumental description is provided in Handy et al. (1999). The instrument features a 30-cm Cassegrain telescope with a *field of view (FOV)* of 8.5×8.5 arc minutes and operates in three coronal EUV wavelengths (Fe IX/X, 171 Å; Fe XII/XXIV, 195 Å; and Fe XV, 284 Å), as well as in H I Lyman-α (1216 Å), C IV (1550 Å), UV continuum (1600 Å), and white light (5000 Å). This wavelength set covers temperatures from 6000 K to 10 MK, with the main sensitivity in the $1-2$ MK range for the EUV filters. *TRACE* has provided us with stunning high-resolution images that reveal intriguing details about coronal plasma dynamics, coronal heating and cooling, and magnetic reconnection processes.

The latest solar mission was launched on 2002 February 5, − a *NASA small ex-*

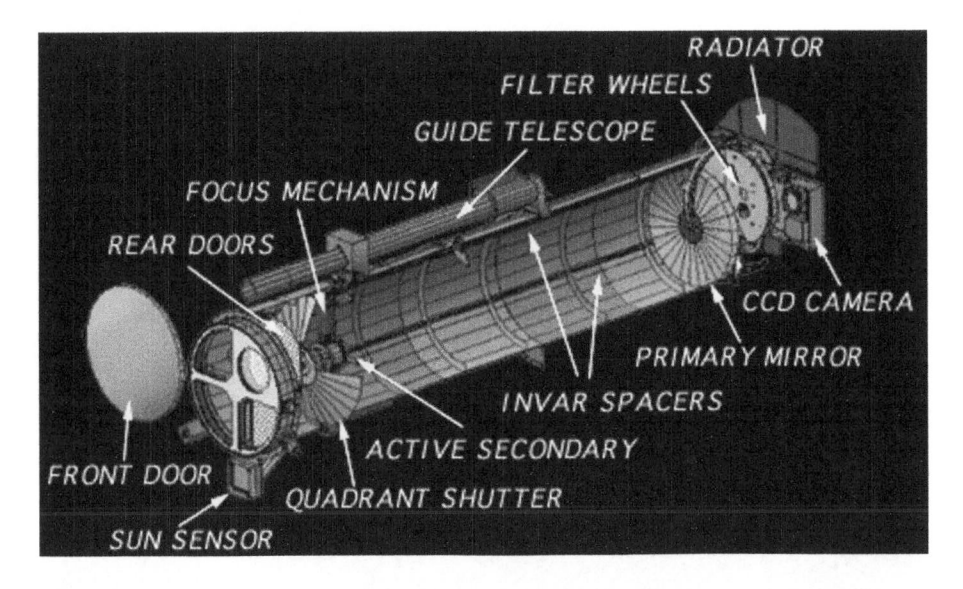

Figure 1.6: A cut-away view of the *Transition Region And Coronal Explorer (TRACE)* telescope is shown. The 30-cm aperture *TRACE* telescope uses four normal-incidence coatings for the EUV and UV on quadrants of the primary and secondary mirrors. The segmented coatings on solid mirrors form identically sized and perfectly co-aligned images. Pointing is internally stabilized to 0.1 arc second against spacecraft jitter. A 1024×1024 CCD detector collects images over an 8.5 x 8.5 arc minute *field-of-view* (FOV). A powerful data handling computer enables very flexible use of the CCD array including adaptive target selection, data compression, and fast operation for a limited FOV (courtesy of *TRACE* team).

plorer mission (SMEX) called the (Reuven) *Ramaty High Energy Solar Spectroscopic Imager (RHESSI)* (Fig. 1.7), designed to explore the basic physics of particle acceleration and explosive energy release in solar flares. This instrument provides hard X-ray images with the highest ever achieved spatial resolution of $\approx 2.3''$, as well as with the highest energy resolution (achieved by germanium-cooled detectors). An instrumental description of the *RHESSI* spacecraft can be found in Lin et al. (1998b, 2002). Although this mission was launched late in solar cycle XXIII, it has already recorded 7500 solar flares during the first year.

Progress in the physical understanding of the structure and dynamics of the solar corona is mainly driven by space-based observations in EUV, soft X-rays, and hard X-rays, obtained by the high-resolution imagers that were flown over the last decade. However, very valuable and complementary observations have also been produced by ground-based observatories. There are essentially two wavelength regimes that are observable from ground and provide information on the solar corona (i.e., optical and radio wavelengths). Imaging radio observations of the solar corona have mainly been accomplished by the *Culgoora radioheliograph* in Australia (1967–1984), the *Nançay radioheliograph* in France (since 1977), the *Very Large Array (VLA)* in New Mexico (since 1980), the *Owens Valley Radio Observatory (OVRO)* in California (since 1978),

Figure 1.7: The *Ramaty High Energy Solar Spectroscopic Imager (RHESSI)* is a spinning spacecraft that contains a single instrument, a rotation-modulated collimator with 9 germanium-cooled detectors (courtesy of *RHESSI* Team).

RATAN-600 in Russia (since 1972), and the *Nobeyama Radio Observatory* in Japan (since 1992). There is also a larger number of radio spectrometers distributed all around the world, which complement imaging radiotelescopes with dynamic spectra of high temporal and spectral resolution.

1.2   Nomenclature of Coronal Phenomena

We start now with a tour of coronal phenomena, describing them first by their morphological appearance in observed images, giving a more physical definition and quantitative characterization in the following chapters of the book. For an illustrated glossary of phenomena in solar and solar-terrestrial physics see also Bruzek & Durrant (1977).

It is customary to subdivide the solar corona into three zones, which all vary their size during the solar cycle: (1) active regions, (2) quiet Sun regions, and (3) coronal holes.

Active regions: Like cities on the Earth globe, active regions on the solar surface harbor most of the activity, but make up only a small fraction of the total surface area. Active regions are located in areas of strong magnetic field concentrations, visible as sunspot groups in optical wavelengths or magnetograms. Sunspot groups typically exhibit a strongly concentrated leading magnetic polarity, followed by a more fragmented trailing group of opposite polarity. Because of this bipolar nature active regions are mainly made up of closed magnetic field lines. Due to the permanent magnetic activity in terms of magnetic flux emergence, flux cancellation, magnetic reconfigurations, and magnetic reconnection processes, a number of dynamic processes such as plasma heating, flares, and coronal mass ejections occur in active regions. A consequence of plasma heating in the chromosphere are upflows into coronal loops, which give active regions the familiar appearance of numerous filled loops, which are hotter and denser than the background corona, producing bright emission in soft X-rays and extreme ultraviolet (EUV) wavelengths. In the *Yohkoh* soft X-ray image shown in Fig. 1.8, active regions appear in white. The heliographic position of active regions is typically confined within latitudes of $\pm 40°$ from the solar equator.

Quiet Sun: Historically, the remaining areas outside of active regions were dubbed *quiet Sun* regions. Today, however, many dynamic processes have been discovered all over the solar surface, so that the term *quiet Sun* is considered as a misnomer, only justified in relative terms. Dynamic processes in the quiet Sun range from small-scale phenomena such as network heating events, nanoflares, explosive events, bright points, and soft X-ray jets, to large-scale structures, such as transequatorial loops or coronal arches. The distinction between active regions and quiet Sun regions becomes more and more blurred because most of the large-scale structures that overarch quiet Sun regions are rooted in active regions. A good working definition is that quiet Sun regions encompass all closed magnetic field regions (excluding active regions), clearly demarcating the quiet Sun territory from coronal holes, which encompass open magnetic field regions.

Coronal holes: The northern and southern polar zones of the solar globe have generally been found to be darker than the equatorial zones during solar eclipses. Max Waldmeier thus coined those zones as "Koronale Löcher" (in german, i.e., *coronal holes*). Today it is fairly clear that these zones are dominated by open magnetic field lines, that act as efficient conduits for flushing heated plasma from the corona into the solar wind, if there are any chromospheric upflows at their footpoints. Because of this efficient transport mechanism, coronal holes are empty of plasma most of the time, and thus appear much darker than the quiet Sun, where heated plasma upflowing from the chromosphere remains trapped until it cools down and precipitates back to the chromosphere.

Like our Earth atmosphere displays a large variety of cloud shapes, from bulky stratocumuli to fine-structured cirrus clouds, the solar corona exhibits an equally rich menagery of loop morphologies, which can reveal important clues about the underlying magnetic reconnection and reconfiguration processes. Pointed and cusp-shaped structures may pinpoint coronal nullpoints of X-type magnetic reconnection points, while circular geometries may indicate relaxed, near-dipolar magnetic field geometries. In Fig. 1.9 we show some characteristic coronal structures that are typically seen in soft X-ray images (Acton et al. 1992).

Helmet streamers: Streamers (e.g., Fig. 1.9[A]) are huge, long-lived, radially oriented structures that extend from the base of the corona out to several solar radii. A

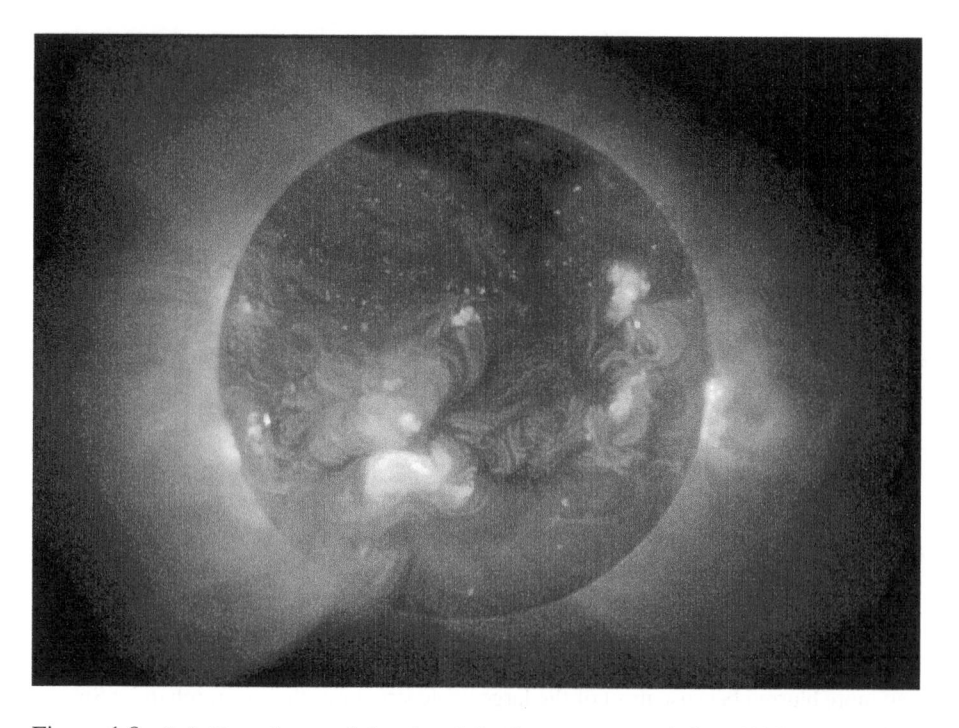

Figure 1.8: Soft X-ray image of the extended solar corona recorded on 1992 August 26 by the *Yohkoh Soft X-ray Telescope (SXT)*. The image was made up from two pointings of the spacecraft, one to the east and one to the west, to capture the distant corona far above the Sun's limb (courtesy of *Yohkoh* Team).

subclass are *helmet streamers*, which are connected with active regions and are centered over prominences (above the limb) or filaments (on the disk). The lower part contains closed field lines crossing a neutral line, while the upper part turns into a cusp-shaped geometry, giving the appearance of a "helmet". Above the helmet, a long, straight, near-radial stalk continues outward into the heliosphere, containing plasma that leaks out from the top of the helmet where the thermal pressure starts to overcome the magnetic confinement (plasma-β parameter $\gtrsim 1$).

Loop arcades: Regions of opposite magnetic polarity can sometimes have a quite large lateral extent, so that dipolar loops can be found lined up perpendicularly to a neutral line over a large distance, with their footpoints anchored in the two ribbons of opposite magnetic polarity on both sides of the neutral line. Such a loop arcade is shown end-on in Fig. 1.9(B), which illustrates that the cross section of an arcade can appear like a single loop.

Soft X-ray jets: If heated plasma flows along an open field line, it is called a *soft X-ray jet*, which can have the form of a linear structure, sometimes slightly bent. Such a jet feature is shown in Fig. 1.9(C), which grew with a velocity of 30 km s^{-1} to a length of 200 Mm. Such jet features are visible until the flow fades out or the structure erupts.

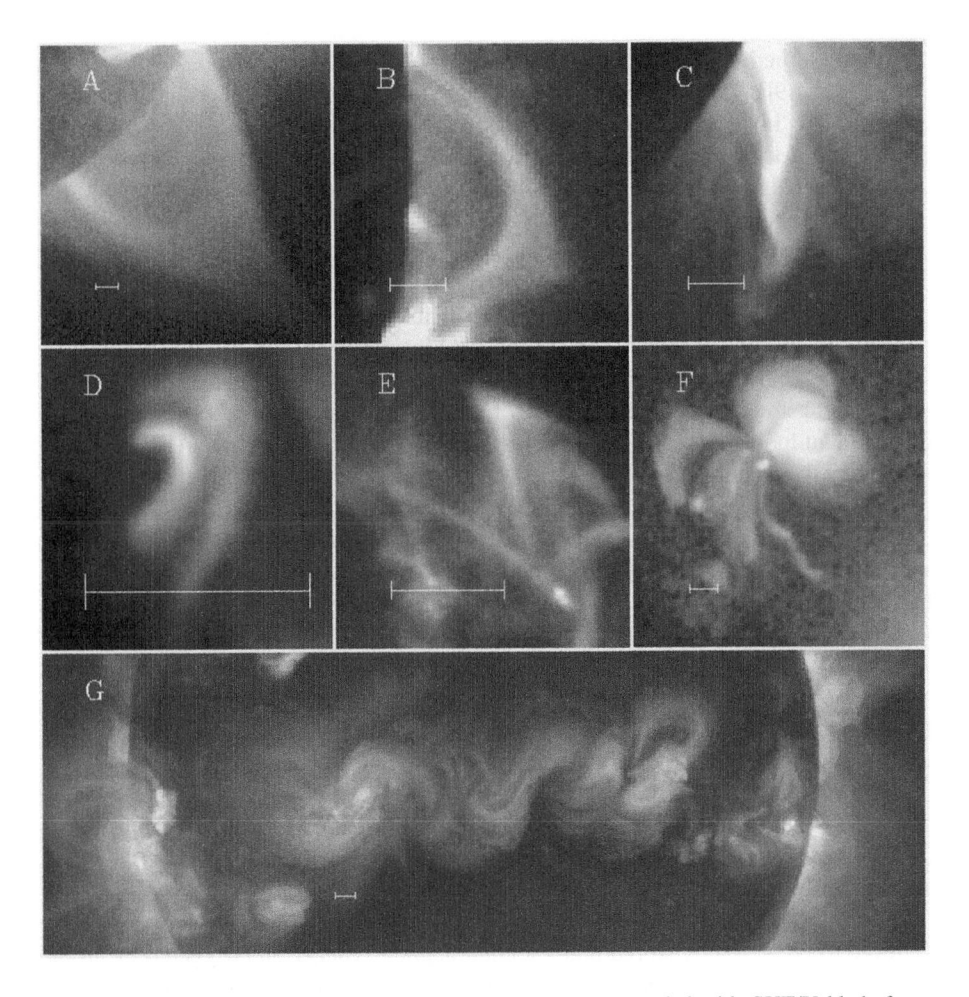

Figure 1.9: A collection of soft X-ray structures is shown, recorded with *SXT/Yohkoh*, from 1991 October 3 to 1992 January 25, described in Acton et al. (1992). The images exemplify the variety of solar coronal features seen in soft X-rays: A) Large helmet-type structure; B) arcade of X-ray loops seen end-on; C) dynamic eruptive feature which grew at a velocity of about 30 km s^{-1}; D) a pair of small symmetrical flaring loops; E) two cusped loops with heating in the northern loop; F) a tightly beamed X-ray jet towards the southwest at 200 km s^{-1}; and G) the sinuous magnetic connection between active regions (courtesy of *Yohkoh* Team).

Postflare loops: Flare loops become bright in soft X-rays after they become filled up by upflowing heated plasma, a process that is dubbed *chromospheric evaporation*, driven by intense chromospheric heating at the loop footpoints from precipitating non-thermal particles or thermal conduction fronts. Flare (or postflare) loops trace out dipole-like magnetic field lines, after relaxation from flare-related magnetic reconnection processes. A pair of flaring loops are shown in Fig. 1.9(D), where the smaller flaring loop has a height of 10 Mm and a footpoint separation of 18 Mm.

Cusp-shaped loops: Loop segments with a pointed shape at the top, are also called *cusps* (Fig. 1.9[E]). Such cusps represent deviations from dipole-like magnetic field lines and thus indicate some dynamic processes. Cusps are expected in X-type reconnection geometries, and thus may occur right after a magnetic reconnection process, if the chromospheric plasma filling process happens sufficiently fast before the loop relaxes into a dipole-like (near-circular) geometry. The cusp of the northern loop shown in Fig. 1.9(E) is particularly bright, and thus may even indicate local heating at the cusp.

Multiple arcades: It is not uncommon that multiple neutral lines occur in active regions, which organize the magnetic field into multiple arcades side-by-side, as shown in the "bow-tie" structure in Fig. 1.9(F). Multiple arcades can lead to quadrupolar magnetic structures, which play a key role in eruptive processes, because a coronal magnetic nullpoint occurs above such neighboring magnetic flux systems, which aids in opening up the field by reconnecting away the overlying closed field lines (according to the so-called *magnetic breakout model* (Antiochos et al. 1999b).

Sigmoid structures: If a dipole field is sheared, the field lines deform into S-shaped geometries, which are also called *sigmoids*. A string of sigmoid-shaped field line bundles are shown in Fig. 1.9(G), stringing together a series of sheared active regions along the solar equator. Sigmoids can imply nonpotential magnetic fields that have an excess energy over the potential field configuration, and thus may contain free energy to spawn a flare or coronal mass ejection. It is not clear yet whether the eruptive phase is initiated by a kink instability of the twisted field lines or by another loss-of-equilibrium process.

The foregoing inventory describes quasi-stationary or slowly varying coronal structures. Rapidly varying processes, which all result from a loss of equilibrium, are also called eruptive processes, such as flares, coronal mass ejections, or small-scale variability phenomena.

Flares: A flare process is associated with a rapid energy release in the solar corona, believed to be driven by stored nonpotential magnetic energy and triggered by an instability in the magnetic configuration. Such an energy release process results in acceleration of nonthermal particles and in heating of coronal/chromospheric plasma. These processes emit radiation in almost all wavelengths: radio, white light, EUV, soft X-rays, hard X-rays, and even γ-rays during large flares.

Microflares and nanoflares: The energy range of flares extends over many orders of magnitude. Small flares that have an energy content of 10^{-6} to 10^{-9} of the largest flares fall into the categories of microflares and *nanoflares*, which are observed not only in active regions, but also in quiet Sun regions. Some of the microflares and nanoflares have been localized to occur above the photospheric network, and are thus also dubbed *network flares* or *network heating events*. There is also a number of small-scale phenomena with rapid time variability for which it is not clear whether they represent miniature flare processes, such as *active region transients, explosive events, blinkers,* etc. It is conceivable that some are related to *photospheric* or *chromospheric* magnetic reconnection processes, in contrast to flares that always involve *coronal* magnetic reconnection processes.

Coronal mass ejections (CME): Large flares are generally accompanied by eruptions of mass into interplanetary space, the so-called *coronal mass ejections (CME)*.

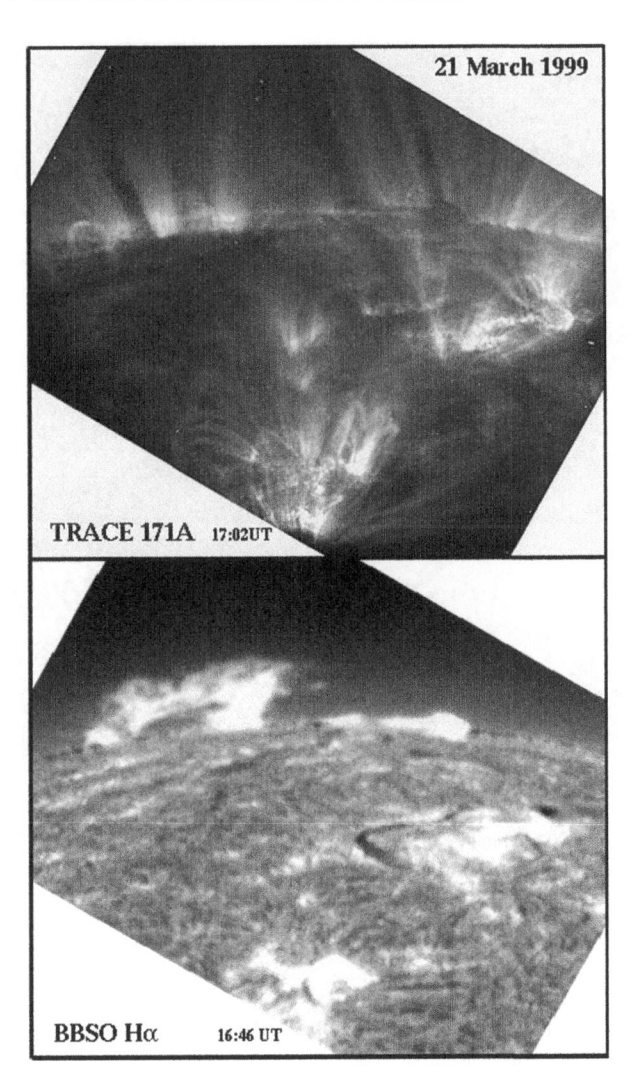

Figure 1.10: These images (taken on 1999 March 21) compare the corona seen in EUV (top panel: *TRACE*, 171 Å, T=1 MK) and the chromosphere seen in Hα (bottom panel: *Big Bear Solar Observatory (BBSO)*, T=10,000 K). The cool filaments (on the disk) and prominences (above the limb) show up as bright structures in Hα (bottom frame), but as dark, absorbing features in EUV (top frame) (courtesy of *TRACE* and *BBSO*).

Flares and CMEs are two aspects of a large-scale magnetic energy release, but the two terms evolved historically from two different observational manifestations (i.e., *flares* denoting mainly the emission in hard X-rays, soft X-rays, and radio, while *CMEs* being referred to the white-light emission of the erupting mass in the outer corona and heliosphere). Recent studies, however, clearly established the co-evolution of both processes triggered by a common magnetic instability.

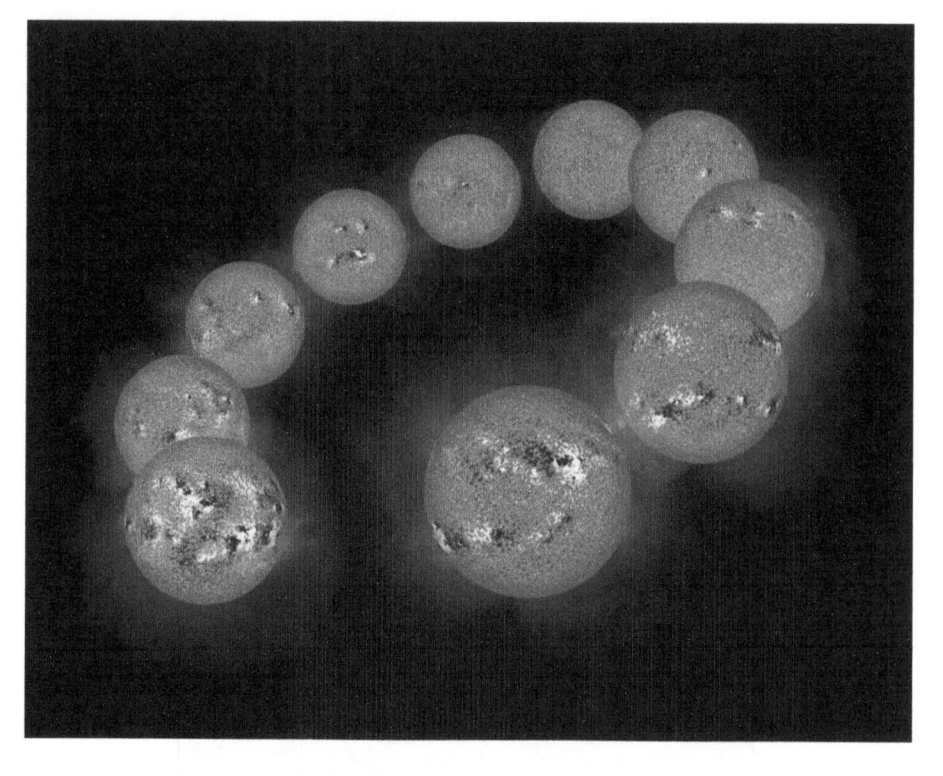

Figure 1.11: Ten maps of photospheric magnetograms are shown, spanning a period of almost a solar cycle, from 1992 January 8 to 1999 July 25, recorded with the Vacuum Telescope of the *National Solar Observatory (NSO)* at *Kitt Peak National Observatory (KPNO)*, Tucson, Arizona. White and black colors indicate positive and negative magnetic polarity (of the longitudinal magnetic field component $B_{\parallel}$), while grey indicates the zero field (courtesy of KPNO).

Filaments and prominences: A filament is a current system above a magnetic neutral line that builds up gradually over days and erupts during a flare or CME process. Historically, filaments were first detected in Hα on the solar disk, but later they were also discovered in He 10,830 Å and in other wavelengths. On the other side, erupting structures above the limb seen in Hα and radio wavelengths were called *prominences*. Today, both phenomena are unified as an identical structure, with the only significant difference being that they are observable on the disk in absorption (filaments) and above the limb in emission (prominences), (e.g., see Fig. 1.10).

1.3   The Solar Magnetic Cycle

The solar magnetic cycle of about 11 years, during which the magnetic polarity of the global solar magnetic field is reversed, modulates the total radiation output in many wavelengths in a dramatic way. A full cycle of 22 years, after which the original magnetic configuration is restored, is called a *Hale cycle*. The total magnetic flux

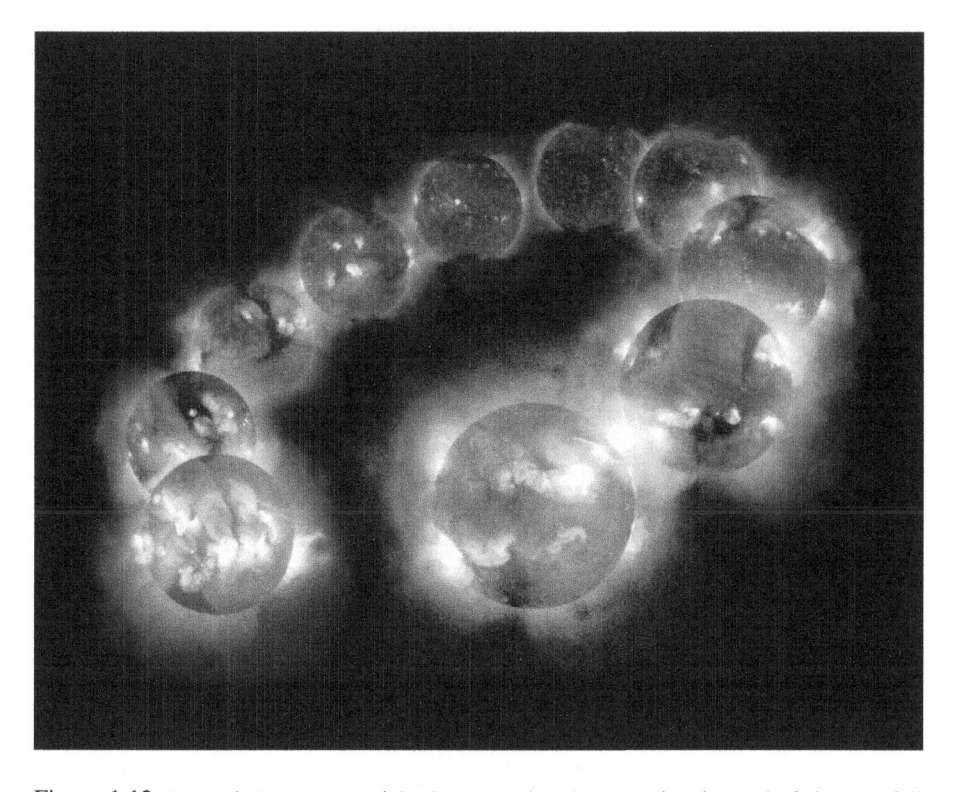

Figure 1.12: Ten soft X-ray maps of the Sun, covering the same time interval of almost a full solar cycle as the magnetograms shown in Fig. 1.11, observed with the *Solar X-ray telescope (SXT)* on the *Yohkoh* spacecraft (courtesy of Yohkoh Team).

reaches a maximum during the peak of a cycle and drops to a low level during the minimum of the cycle (see magnetograms in Fig. 1.11). During the solar cycle, the magnetic flux varies by a factor of ≈ 8 in active regions, by a factor of ≈ 2 in ephemeral regions, and even in anti-phase in small regions (Hagenaar et al. 2003). Since many radiation mechanisms are directly coupled to the dissipation of magnetic energy and related plasma heating, the radiation output in these wavelengths is correspondingly modulated from solar maximum to minimum (e.g., in soft X-rays, hard X-rays, and radio wavelengths). This can clearly be seen from the sequence of Yohkoh soft X-ray images shown in Fig. 1.12.

Although the modulation of optical emission during a cycle is much less dramatic than in X-rays, the magnetic solar cycle was discovered long ago, based on the increase and decrease of sunspots. The east-west orientation of the magnetic field in active regions was found to be opposite in the northern and southern hemispheres, switching for every 11-year cycle, known as *Hale's polarity law*. During a cycle, the active regions migrate from high latitudes ($\approx 40°$) towards lower latitudes ($\approx 10°$) near the equator, leading to the famous *butterfly diagram of sunspots* (Spörer's law), when their latitudinal position is plotted as a function of time.

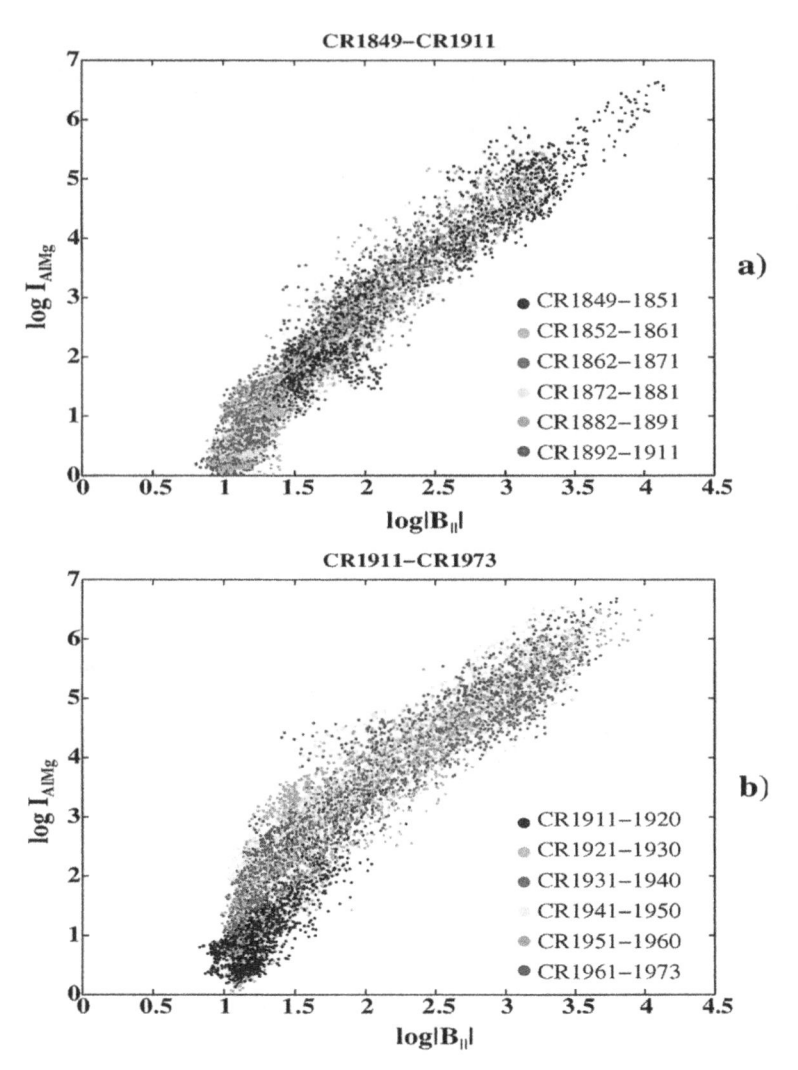

Figure 1.13: Scatter plots of the soft X-ray intensity from *SXT* (Yohkoh) data in the AlMg filter as a function of the magnetic flux (from the Kitt Peak Observatory) in the natural logarithmic scale for the latitudinal zone from $-55°$ to $+55°$ for two periods: (a) 1991 November 11 to 1996 July 25, and (b) 1996 June 28 to 2001 March 13. Different subperiods are marked by different grey tones. The color-coding in terms of Carrington rotation numbers is shown in the figures (Benevolenskaya et al. 2002).

The cyclic evolution of the sunspots can be understood in terms of a reversal of the global solar magnetic field, which evolves from an initial poloidal field towards a toroidal field under the influence of differential rotation, according to a model by Horace W. Babcock proposed in 1961. In recent versions of this scenario, the cause of the solar cycle is a dynamo-like process that is driven by the internal magnetic field in the tachocline at the bottom of the convection zone. There a strong magnetic field in the order of 100,000 G is periodically strengthened and weakened, from which occasionally buoyant magnetic fluxtubes arise and emerge at the photospheric surface, visible as bipolar sunspot pairs. The differential surface rotation shears the new emerging fields gradually into a more toroidal field, until surface diffusion by granular convection breaks up the field and meridional flows transport the fragments towards the poles. The surface diffusion neutralizes the toroidal field component increasingly during the decay of the cycle, so that a weak poloidal global field is left at the cycle minimum. When the internal dynamo strengthens the tachocline field again, the rate of buoyant fluxtubes increases and the cycle starts over.

The coupling of the photospheric magnetic field (Fig. 1.11) to the coronal soft X-ray emission (Fig. 1.12) has been quantitatively investigated (e.g., Fisher et al. 1998 and Benevolenskaya et al. 2002), resulting in an unambiguous correlation between the two quantities, when sampled over a solar cycle (Fig. 1.13). It was found that the soft X-ray intensity, I_{SXR}, scales approximately with the square of the magnetic field, B^2, and thus the soft X-ray intensity can be considered as a good proxy of the magnetic energy ($\varepsilon_m = B^2/8\pi$). Slightly different powerlaw values have been found for the solar maximum and minimum (Benevolenskaya et al. 2002),

$$I_{SXR} \propto < |B_\parallel| >^n \left\{ \begin{array}{ll} n = 1.6 - 1.8 & \text{solar maximum} \\ n = 2.0 - 2.2 & \text{solar minimum .} \end{array} \right. \tag{1.3.1}$$

The soft X-ray flux is an approximate measure of the energy rate (E_H) that is deposited into heating of coronal plasma according to this positive correlation. On the other hand, global modeling of magnetic fields and soft X-ray flux from the Sun and cool stars yields a more linear relationship, $E_H \propto B^{1.0\pm0.5}$ (e.g., Schrijver & Aschwanden, 2002). It is conceivable that the positive correlation applies more generally to all energy dissipation processes in the solar corona (i.e., that the processes of coronal heating, flare plasma heating, and particle acceleration are ultimately controlled by the amount of magnetic energy that emerges through the solar surface, modulated by about two orders of magnitude during a solar cycle).

1.4 Magnetic Field of the Solar Corona

The solar magnetic field controls the dynamics and topology of all coronal phenomena. Heated plasma flows along magnetic field lines and energetic particles can only propagate along magnetic field lines. Coronal loops are nothing other than conduits filled with heated plasma, shaped by the geometry of the coronal magnetic field, where cross-field diffusion is strongly inhibited. Magnetic field lines take on the same role for coronal phenomena as do highways for street traffic. There are two different magnetic zones in the solar corona that have fundamentally different properties: open-field and closed-field regions. Open-field regions (white zones above the limb in Fig. 1.14), which always exist in the polar regions, and sometimes extend towards the equator, connect the solar surface with the interplanetary field and are the source of the *fast solar wind* (≈ 800 km s^{-1}). A consequence of the open-field configuration is efficient plasma transport out into the heliosphere, whenever chromospheric plasma is heated at the footpoints. Closed-field regions (grey zones in Fig. 1.14), in contrast, contain

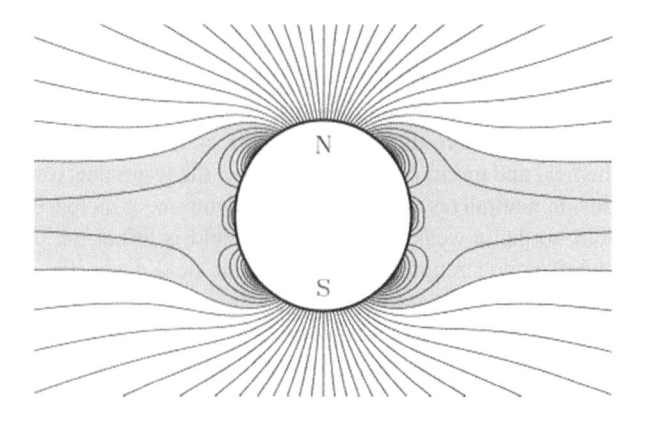

Figure 1.14: Depiction of lines of magnetic force in the semi-empirical multipole-current sheet coronal model of Banaszkiewicz et al. (1998). The high-speed solar wind fills the unshaded volume above the solar surface (Cranmer, 2001).

mostly closed field lines in the corona up to heights of about one solar radius, which open up at higher altitudes and connect eventually to the heliosphere, but produce a *slow solar wind* component of $\approx$ 400 km s^{-1}. It is the closed-field regions that contain all the bright and overdense coronal loops, produced by filling with chromospheric plasma that stays trapped in these closed field lines. For loops reaching altitudes higher than about one solar radius, plasma confinement starts to become leaky, because the thermal plasma pressure exceeds the weak magnetic field pressure which decreases with height (plasma-β parameter > 1).

The magnetic field on the solar surface is very inhomogeneous. The strongest magnetic field regions are in sunspots, reaching field strengths of $B = 2000 - 3000$ G. Sunspot groups are dipolar, oriented in an east-west direction (with the leading spot slightly closer to the equator) and with opposite leading polarity in both hemispheres, reversing for every 11-year cycle (*Hale's laws*). Active regions and their plages comprise a larger area around sunspots, with average photospheric fields of $B \approx 100 - 300$ G (see Fig. 1.11), containing small-scale pores with typical fields of $B \approx 1100$ G. The background magnetic field in the quiet Sun and in coronal holes has a net field of $B \approx 0.1 - 0.5$ G, while the absolute field strengths in resolved elements amount to $B = 10 - 50$ G. Our knowledge of the solar magnetic field is mainly based on measurements of Zeeman splitting in spectral lines, while the coronal magnetic field is reconstructed by extrapolation from magnetograms at the lower boundary, using a potential or force-free field model. The extrapolation through the chromosphere and transition region is, however, uncertain due to unknown currents and non-force-free conditions. The fact that coronal loops exhibit generally much less expansion with height than potential-field models underscores the inadequacy of potential-field extrapolations. Direct measurements of the magnetic field in coronal heights are still in their infancy.

An empirical compilation of coronal magnetic field measurements is given in the paper of Dulk & McLean (1978), reproduced in Fig. 1.15. This compilation combines

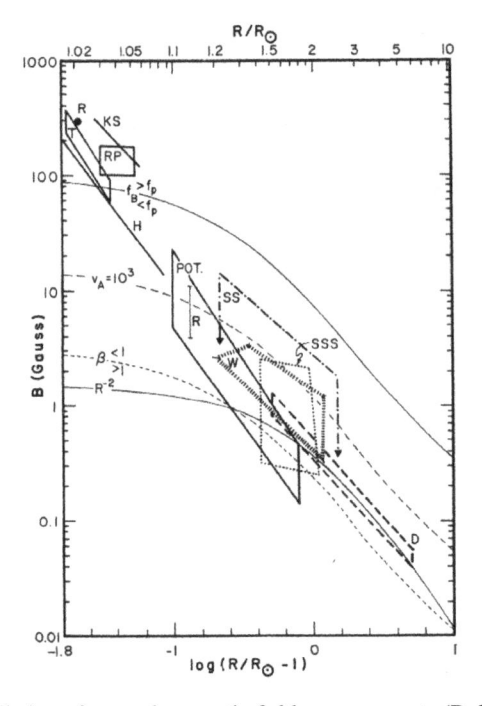

Figure 1.15: Compilation of coronal magnetic field measurements (Dulk & McLean, 1978).

the following measurement methods of the magnetic field: (1) in situ measurements by spacecraft at $\gtrsim 0.5$ AU; (2) Zeeman effect in active region prominences; (3) extrapolations from photospheric magnetograms; (4) microwave radio bursts with gyroresonance emission; (5) decimetric bursts that involve a frequency drift related to the Alfvén speed; (6) metric type II radio bursts, which have a shock speed related to the Alfvénic Mach number; and (7) type III bursts that show a circular polarization, which depends on the refractive index and local magnetic field. Although the magnetic field strength varies $1 - 2$ orders of magnitude at any given height (see Fig. 1.15), Dulk & McLean (1978) derived an empirical formula that approximately renders the average decrease of the magnetic field with height between 1.02 and 10 solar radii,

$$B(R) = 0.5 \left(\frac{R}{R_\odot} - 1 \right)^{-1.5} \quad \mathrm{G} \quad (1.02 \lesssim R/R_\odot \lesssim 10) . \tag{1.4.1}$$

Of course, the variation of the magnetic field strength by $1 - 2$ orders of magnitude is mainly caused by the solar cycle (see Fig. 1.13). Alternative methods to measure the coronal magnetic field directly employ the effects of Faraday rotation, the polarization of free-free emission, Hanle effect in H Lyman-α, or Stokes polarimetry in infrared lines.

In order to give a practical example of how the magnetic field varies in typical active region loops, we show the height dependence $B(h)$ of calculated potential-field lines (using *MDI/SoHO* magnetograms) that are co-spatial with some 30 active regions

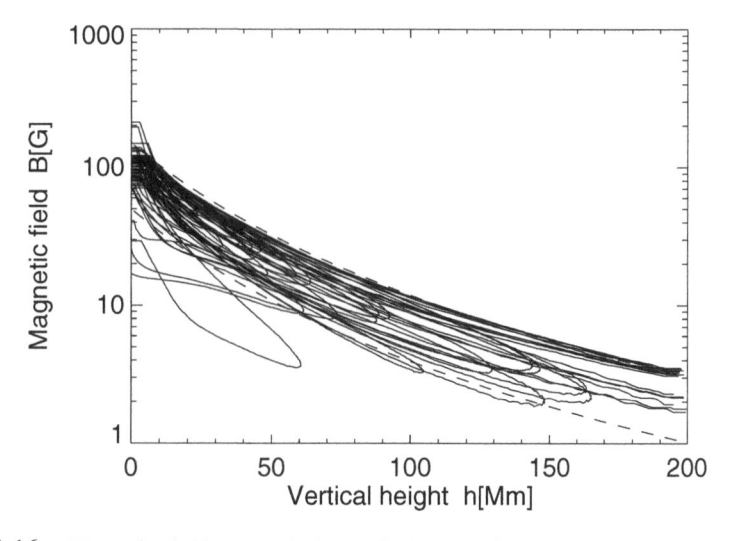

Figure 1.16: Magnetic field extrapolations of 30 loops in an active region based on a *MDI/SoHO* magnetogram (Aschwanden et al. 1999a).

loops traced from *EIT/SoHO* 171 Å images (Aschwanden et al. 1999a). These active region loops reach heights of $h \approx 50 - 200$ Mm. The photospheric magnetic field strength at their footpoints varies in the range of $B_{foot} \approx 20 - 200$ G (averaged over a pixel of the MDI magnetogram), but drops below $B \lesssim 10$ G for heights of $h \gtrsim 100$ Mm. The height dependence of the magnetic field can be approximated with a dipole model (dashed lines in Fig. 1.16),

$$B(h) = B_{foot} \left(1 + \frac{h}{h_D} \right)^{-3}, \qquad (1.4.2)$$

with a mean dipole depth of $h_D \approx 75$ Mm. Of course, the dipole approximation is contingent on the potential-field model. Alternatively, if one assumes a constant magnetic flux $\Phi(s) = B(s)A(s)$ along the loop, as the almost constant cross-sectional area $A(s)$ of observed loops suggests, one might infer in the corona an almost constant field with height, $B(h) \approx const$, while the largest magnetic field decrease occurs in the chromospheric segment, which expands like a canopy above the photospheric footpoint. It is therefore imperative to develop direct measurement methods of the coronal field to decide between such diverse models.

1.5   Geometric Concepts of the Solar Corona

Solar or stellar atmospheres are generally characterized in lowest order by spherical shells, with a decreasing density as a function of the radial distance from the surface. The spherical structure is, of course, a result of the gravitational stratification, because the gravitational potential $U(r)$ is only a function of the radial distance r from the mass center of the object. The gravitational stratification, which just makes the simplest assumption of pressure equilibrium and homogeneity, is a useful basic concept of the average radial density structure for portions of the atmosphere that are horizontally quasi-homogeneous.

The solar corona, however, is highly inhomogeneous, due to the structuring of the magnetic field. The decisive parameter is the ratio of the thermal pressure p_{th} to the magnetic pressure p_{mag}, also called the plasma-β parameter (see §1.8). In the major part of the solar corona, the value of the plasma-β parameter is less than unity, which constitutes a rigorous topological constraint, inasmuch as the thermal pressure is insufficient to warrant horizontal stratification across the magnetic field. This inhibition of cross-field transport has the natural consequence that every plasma that streams from the chromosphere to the corona, traces out bundles of magnetic field lines, with cross sections that are roughly determined by the geometric area where the chromospheric upflows pass the boundary of $\beta \lesssim 1$, which is generally at heights of the transition region. This topological structuring of the corona has the nice property that the radiating coronal plasma can be used to delineate the 3D coronal magnetic field $\mathbf{B}(x, y, z)$, but it also produces a highly inhomogeneous density structure, which is more difficult to model than a homogeneous atmosphere.

The evolution of our perception of the topological structure of the solar corona is depicted in Fig. 1.17. While the concept of gravitationally stratified spheres seemed to be adequate to model the average density structure of the solar corona in the 1950s, the concept of magnetic structuring into horizontally separated fluxtubes, which essentially represent isolated "mini-atmospheres", was introduced in the 1980s. In the last decade before 2000, the high-resolution images from spacecraft observations revealed a spatially highly inhomogeneous and temporally dynamic corona, which is constantly stirred up by dynamic processes such as heated chromospheric upflows, cooling downflows, magnetic reconfigurations, and interactions with waves. The trend is clear, our picture evolved from stationary and spherical symmetry away towards dynamic and highly inhomogeneous topologies.

So we can conceive the topology of the solar corona by structures that are aligned with the magnetic field. There are two types of magnetic field lines: *closed field lines* that start and end at the solar surface, and *open field lines* which start at the solar surface and end somewhere in interplanetary space. Also for theoretical reasons, open field lines close over very large distances, to satisfy Maxwell's equation $\nabla \mathbf{B} = 0$, which states that there are no static sources of magnetic field in the absence of magnetic monopoles. The appearance of closed magnetic field lines loaded with over-dense plasma, with respect to the ambient density, has led to the familiar phenomenon of *coronal loops*, which are ubiquitous, and most conspicuously seen in active regions, but can also be discerned everywhere in quiet Sun regions, except in coronal holes near the poles.

A loop structure can essentially be parametrized in a 1D coordinate system given by the central magnetic field line. However, the extent of the transverse cross section is less well defined. What a telescope perceives as a cross-sectional area, depends very much on the angular resolution. Theoretically, a loop can be as thin as an ion gyroradius, because cross-field transport is inhibited in a low-β plasma. The instrument-limited fine structure appears as shown in Fig. 1.18: an apparently compact loop seen at low resolution may reveal a bundle of fine threads at high resolution.

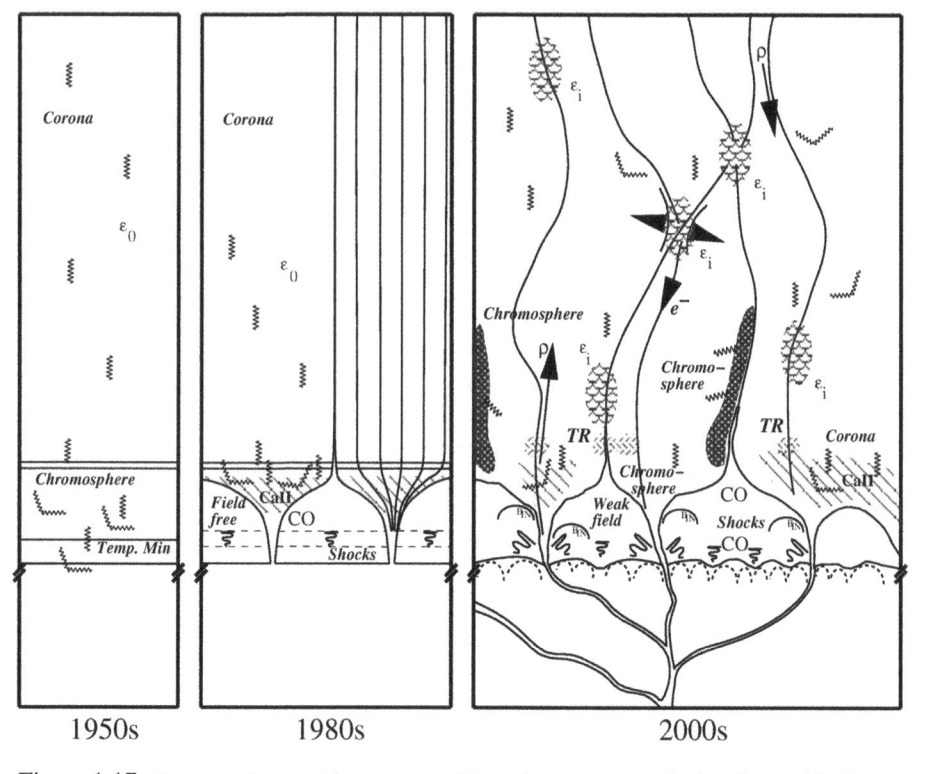

1950s   1980s   2000s

Figure 1.17: Cartoon of geometric concepts of the solar corona: gravitationally stratified layers in the 1950s (*left*), vertical fluxtubes with chromospheric canopies in the 1980s (*middle*), and a fully inhomogeneous mixing of photospheric, chromospheric, and coronal zones by dynamic processes such as heated upflows, cooling downflows, intermittent heating (ε), nonthermal electron beams (e), field line motions and reconnections, emission from hot plasma, absorption and scattering in cool plasma, acoustic waves, and shocks (*right*) (Schrijver, 2001b).

Given these geometrical concepts, we can partition the solar corona into open-field and closed-field regions, as shown in Fig. 1.14. Because the 3D magnetic field is space-filling, every location can be associated with a particular magnetic field line. Depending on the desired spatial resolution of a geometric model, each domain of the corona can further be subdivided into magnetic fluxtubes with a certain cross-sectional area, each one representing an isolated "mini-atmosphere", having its own gravitational stratification and hydrostatic pressure balance, constrained by different densities and temperatures at the lower boundary. This breakdown of the inhomogeneous atmosphere into separate fluxtubes simplifies the magneto-hydrostatics into 1D transport processes. Measurements of the flux in EUV or soft X-rays, which is an optically thin emission, however, involves various contributions from different fluxtubes along a line-of-sight, requiring the knowledge of the statistical distribution of fluxtubes. EUV and soft X-ray data can therefore only be modeled in terms of *multi-fluxtube* or *multi-temperature* concepts.

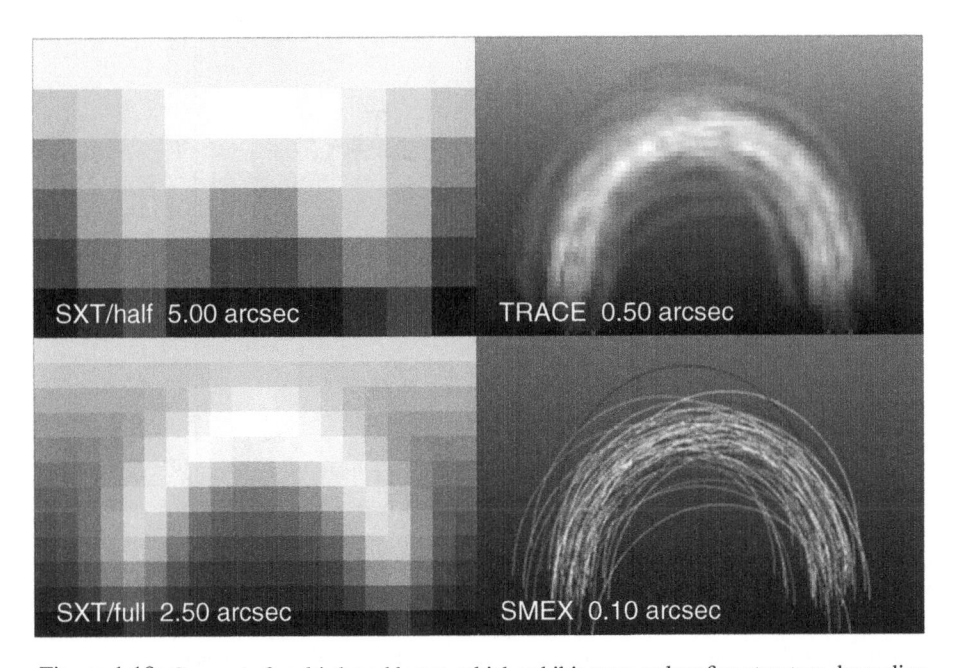

Figure 1.18: Concept of multi-thread loops, which exhibit more or less fine structure depending on the angular resolution of the observing instrument. Four different representations of the same bundle of bright magnetic field lines are shown, observed with angular resolutions of $5''$ and $2.5''$ (*SXT/Yohkoh, EIT/SoHO*), $0.5''$ (*TRACE*), and with $0.1''$ (a proposed future *SMEX* mission).

1.6 Density Structure of the Solar Corona

The particle density in the corona, and even in the chromosphere, is much lower than the best vacuum that can be generated in any laboratory on Earth. Electron densities in the solar corona range from $\approx 10^6$ cm^{-3} in the upper corona (at a height of one solar radius) to $\approx 10^9$ cm^{-3} at the base in quiet regions , and can increase up to $\approx 10^{11}$ cm^{-3} in flare loops. The transition region at the base of the corona demarcates an abrupt boundary where the chromospheric density increases several orders of magnitude higher than coronal values (Fig. 1.19) and the temperature drops below 11,000 K, the ionization temperature of hydrogen. The chromospheric plasma is therefore only partially ionized, while the coronal plasma is fully ionized. Chromospheric density models have been calculated in great detail, based on ion abundance measurements from a large number of EUV lines, constrained by hydrostatic equilibrium, radiation transfer (Vernazza et al. 1973, 1976, 1981), and ambipolar diffusion (Fontenla et al. 1990, 1991).

The density models shown in Fig. 1.19 represent an average 1D model for a gravitationally stratified vertical fluxtube. There are, however, a lot of dynamic processes that heat up the chromospheric plasma, which is then driven by the overpressure upward into the corona, producing over-dense structures with densities in excess of the ambient quiet corona. In order to give a feeling for the resulting density variations in

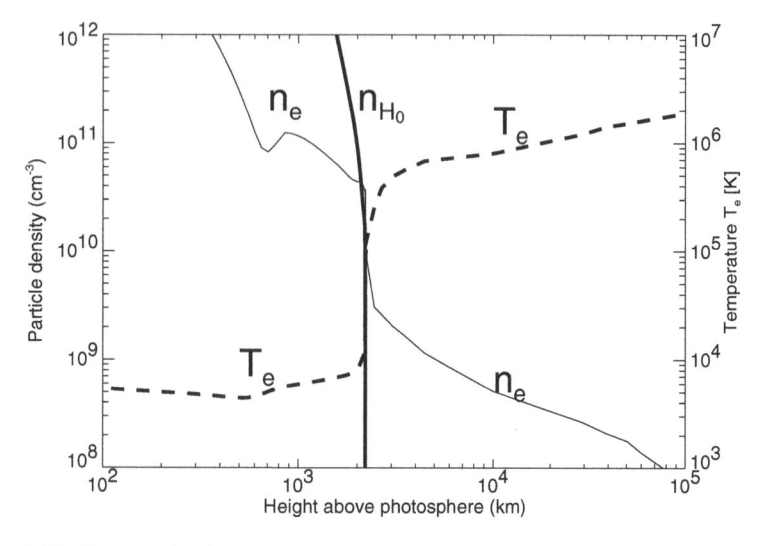

Figure 1.19: Electron density and temperature model of the chromosphere (Fontenla et al. 1990; Model FAL-C) and lower corona (Gabriel, 1976). The plasma becomes fully ionized at the sharp transition from chromospheric to coronal temperatures. In the chromosphere, the plasma is only partially ionized: n_e indicates the electron density, n_{H_0} the neutral hydrogen density.

the highly inhomogeneous corona we show a compilation of density measurements in Fig. 1.20, sampled in coronal holes, quiet Sun regions, coronal streamers, and active regions. At the base of the corona, say at a nominal height of ≈ 2500 km above the photosphere, the density is lowest in coronal holes, typically around $\approx (0.5-1.0) \times 10^8$ cm^{-3}. In quiet Sun regions the base density is generally higher (i.e., $\approx (1-2) \times 10^8$ cm^{-3}). At the base of coronal streamers, the density climbs to $\approx (3-5) \times 10^8$ cm^{-3}, and in active regions it is highest at $\approx 2 \times 10^8 - 2 \times 10^9$ cm^{-3}. In the upper corona, say at heights larger than 1 solar radius, the density drops below $10^6 - 10^7$ cm^{-3}.

Coronal densities were first measured from white-light (polarized brightness) data using a Van de Hulst (1950a,b) inversion, assuming that the polarized brightness of white light is produced by Thomson scattering and is proportional to the line-of-sight integrated coronal electron density. Another ground-based method uses the frequency of radio bursts that propagate through the corona, assuming that their emission frequency corresponds to the fundamental or harmonic plasmafrequency, which is a direct function of the electron density. During the last several decades, space-borne observations in EUV and soft X-rays provided another method, based on the emission measure, which is proportional to the squared density integrated along the column depth, for optically thin radiation. This latter method can be performed in different spectral lines so that the densities can be measured independently for plasmas with different temperatures. However, since every instrument has a limited spatial resolution that may not adequately resolve individual plasma structures, the inferred densities place only lower limits on the effective densities. Absolute densities can be measured from some density sensitive lines in the $1 - 20$ Å range, at relatively high densities of $n_e \gtrsim 10^{12}$ cm^{-3} (e.g., the Fe XXI and Fe XXII lines; Phillips et al. 1996).

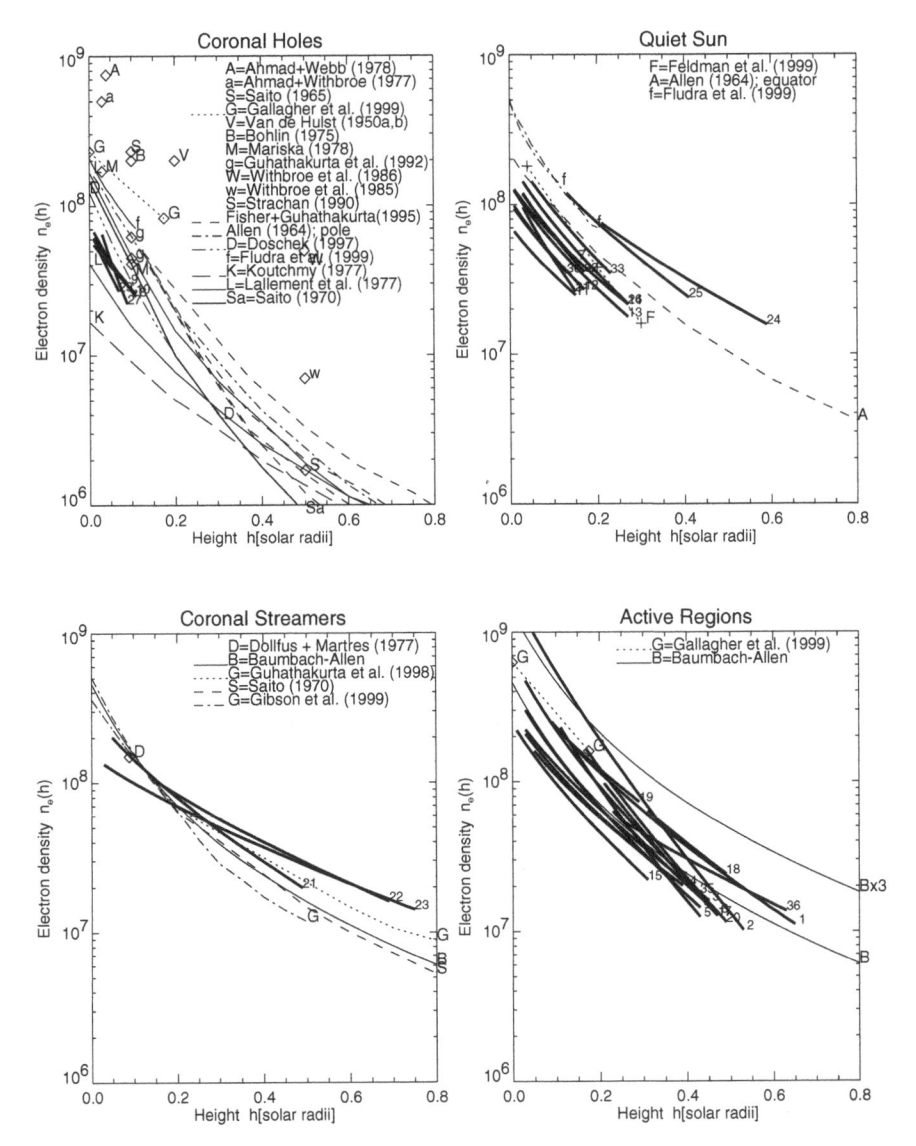

Figure 1.20: Compilation of electron density measurements in the height range of $0.003 < h < 0.8$ solar radii or $2500 < h < 560,000$ km: in coronal holes (top left), quiet Sun regions (top right), coronal streamers (bottom left), and in active regions (bottom right). Note that the electron density at at any given height varies $1-2$ orders of magnitude over the entire corona. The numbered curves (with thick linestyle) represent soft X-ray measurements from Aschwanden & Acton (2001).

From white-light observations during solar eclipses, the corona was characterized by three components: (1) the K-corona, made of partially polarized continuum emission from photospheric light scattered at free electrons (dominating at $h \lesssim 0.3$ solar radius);

(2) the L-corona, consisting of spectral line emission from highly ionized atoms (dominating at $h \lesssim 0.5$ solar radius); and (3) the F corona, which presents absorption lines of the photospheric Fraunhofer spectrum caused by diffraction from interplanetary dust (dominating at $h \gtrsim 0.5$ solar radius). The line-of-sight integrated density profiles of these three components can be approximated by a powerlaw function, each leading to an average density profile known as the *Baumbach–Allen formula* (e.g., Cox, 2000),

$$n_e(R) = 10^8 \left[2.99 \left(\frac{R}{R_\odot} \right)^{-16} + 1.55 \left(\frac{R}{R_\odot} \right)^{-6} + 0.036 \left(\frac{R}{R_\odot} \right)^{-1.5} \right] \quad \mathrm{cm}^{-3}$$

$$(1.6.1)$$

parametrized by the distance $R = R_\odot + h$ from the Sun center. Density measurements from eclipse observations can be performed out to distances of a few solar radii. Further out in interplanetary space, heliospheric densities can be measured by scintillations of radio sources (Erickson, 1964),

$$n_e(R) \approx 7.2 \times 10^5 \left(\frac{R}{R_\odot} \right)^{-2} \quad \mathrm{cm}^{-3} , \qquad R \gg R_\odot \qquad (1.6.2)$$

or from the plasmafrequency or interplanetary radio bursts.

1.7 Temperature Structure of the Solar Corona

The temperature of the solar corona was first asserted only 60 years ago. In 1940, Bengt Edlén analyzed spectral lines following the diagrams of Walter Grotrian and established that coronal emission arises from highly ionized elements, at temperatures of $\gtrsim 1$ MK (1 MegaKelvin = 1,000,000 Kelvin). The physical understanding of this high temperature in the solar corona is still a fundamental problem in astrophysics, because it seems to violate the second thermodynamic law, given the much cooler photospheric boundary, which has an average temperature of $T = 5785$ K (and drops to $T \approx 4500$ K in sunspots). The rapid temperature rise from the chromosphere to the corona is shown in Fig. 1.19 (dashed curve). Further out in the heliosphere, the coronal temperature drops slowly to a value of $T \approx 10^5$ K at 1 AU.

The temperature structure of the solar corona is far from homogeneous. The optically thin emission from the corona in soft X-rays (Fig. 1.8) or in EUV implies overdense structures that are filled with heated plasma. The only available reservoir is the chromosphere, which apparently needs to be heated at many locations, to supply coronal loops with hot upflowing plasma. Because the thermal pressure is generally smaller than the magnetic pressure (plasma-β parameter< 1), plasma transport occurs only in one dimension along the magnetic field lines, while cross-field diffusion is strongly inhibited. This has the consequence that every coronal loop represents a thermally isolated system, having only the tiny chromospheric footpoints as valves for rapid heat and mass exchange. Because the heating of the footpoints of coronal loops seems to be spatially and temporally intermittent, every loop ends up with a different energy input and settles into a different temperature, when a quasi-steady heating rate is obtained. Consequently, we expect that the corona is made up of many filled loops with different

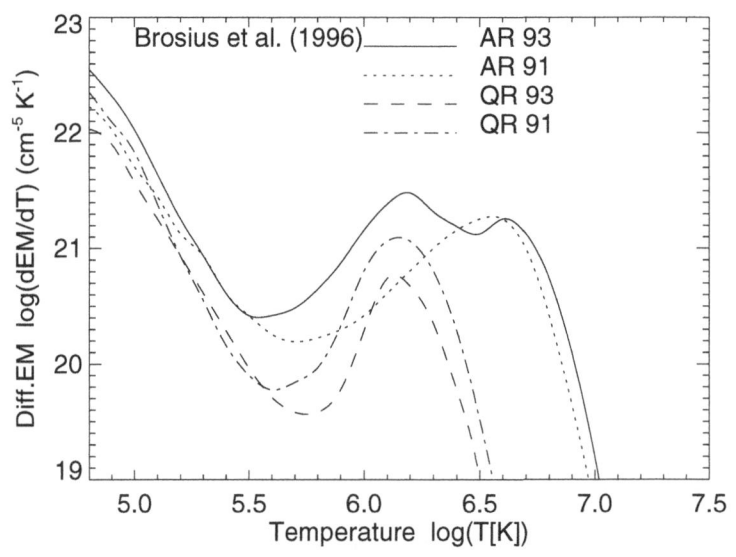

Figure 1.21: Differential emission measure distribution $dEM(T)/dT$ of two active regions (AR 93, AR 91) and two quiet Sun regions (QR 93, QR 91) measured by Brosius et al. (1996) with *SERTS* data.

temperatures, the cooler ones visible in EUV and the hotter ones shining in soft X-rays. Such a multi-temperature picture of the solar corona is shown in Plate 1, composed of three images taken with different temperature filters with *EIT/SoHO*: a blue image at 1.0 MK (171 Å), a green image at 1.5 MK (195 Å), and a red image at 2.0 MK (284 Å). Following the color coding, one can clearly see in Plate 1 that the northern coronal hole contains the coolest regions and that the temperature seems to increase with altitude above the limb. In §3 we will learn that this apparent temperature increase does not reflect a positive temperature gradient with height along individual field lines, but merely results from the relative density weighting of cool and hot (hydrostatic) temperature scale heights. Another effect that can be seen in Plate 1 is the law of additive color mixing according to Isaac Newton: If blue, green, and red are mixed with equal weighting, white results. The white color-coding in active regions seen in Plate 1 can therefore be taken as evidence that active regions contain comparable temperature contributions from 1.0, 1.5, and 2.0 MK temperature loops.

The multi-temperature distribution of the corona can quantitatively be expressed by the so-called *differential emission measure distribution* $dEM(T)/dT$, which is a measure of the squared density $n_e(T)$ integrated over the column depth along the line-of-sight for any given temperature,

$$\frac{dEM(T)}{dT}dT = \int n_e^2(T,z)dz \ . \tag{1.7.1}$$

This quantity can be measured with a broad range of EUV and soft X-ray lines at any location (or line-of-sight) on the Sun. Such differential emission measure distributions obtained from 4 different locations are shown in Fig. 1.21, two in quiet Sun regions and two in active regions, obtained from He II, C IV, Mg V, Ca VII, Mg VI, Mg VIII, Fe X, Si IX, Fe XI, Al X, Si XI, Fe XV, Fe XVI, and Ni XVIII lines with the *Solar EUV Research Telescope and Spectrograph (SERTS)* on sounding rocket flights during 1991 and 1993 (Brosius et al. 1996). The set of EUV emission lines used is sensitive in the temperature range of $log(T) = 4.8 - 6.5$ (T = 63,000 K $-$ 3.2 MK). The DEM distribution in Fig. 1.21 clearly shows a temperature peak around $log(T) = 6.0 - 6.3$ for the quiet Sun regions, which corresponds to the $T = 1 - 2$ MK range, which is coincident with the blue-red color range in Plate 1. In active regions, the DEM distributions in Fig. 1.21 show a comparable amount of emission measure in the temperature range of $log(T) = 6.3 - 6.8$, which corresponds to temperatures of $T = 2.0 - 6.3$ MK.

This coarse temperature characterization of the corona already shows an interesting trend. Open-field regions, such as coronal holes (Fig. 1.14), have the coolest temperatures of $T \lesssim 1$ MK; closed-field regions, such as the quiet Sun, have intermediate temperatures of $T \approx 1 - 2$ MK; while active regions exhibit the hottest temperatures of $T \approx 2 - 6$ MK. Open-field regions seem to be cooler because plasma transport is very efficient, while closed-field regions seem to be hotter because the heated plasma is trapped and cannot easily flow away. The temperature difference between quiet Sun and active regions is a consequence of different magnetic flux emergence rates, heating rates, conductive loss rates, radiative loss rates, and solar wind loss rates.

1.8   Plasma-β Parameter of the Solar Corona

The magnetic field B exerts a Lorentz force on the charged particles of the coronal plasma, consisting of electrons and ions, guiding them in a spiraling gyromotion along the magnetic field lines. Only when the kinetic energy exceeds the magnetic energy (i.e., at high temperatures and low magnetic fields), can particles escape their gyro-orbits and diffuse across the magnetic field. The critical parameter between these two confinement regimes is the plasma-β parameter, defined as the ratio of the thermal plasma pressure p_{th} to the magnetic pressure p_m,

$$\beta = \frac{p_{th}}{p_m} = \frac{2\xi n_e k_B T_e}{B^2/8\pi} \approx 0.07 \, \xi \, n_9 \, T_6/B_1^2 \, , \tag{1.8.1}$$

where $\xi = 1$ is the ionization fraction for the corona (and $\xi = 0.5$ in the photosphere), $k_B = 1.38 \times 10^{-16}$ erg K^{-1} the Boltzmann constant, $B_1 = B/10$ G the magnetic field strength, $n_9 = n_e/10^9$ cm^{-3} the electron density, and $T_6 = T/10^6$ K the electron temperature. In Table 1.1 we list some typical parameters for the chromosphere, and cool, hot, and outer parts of the corona.

We see that most parts of the corona have a plasma-β parameter of $\beta < 1$, but are sandwiched between the higher values $\beta > 1$ in the chromosphere and outer corona. Therefore, most parts of the corona are magnetically confined. However, hot regions with low magnetic fields (e.g., in the magnetic cusps above streamers), can easily have values of $\beta > 1$, so that plasma can leak out across the cusped magnetic field lines. Most magnetic field extrapolation codes do not or cannot take into account the full range of β and essentially assume $\beta \ll 1$. The inference of the plasma-β parameter in

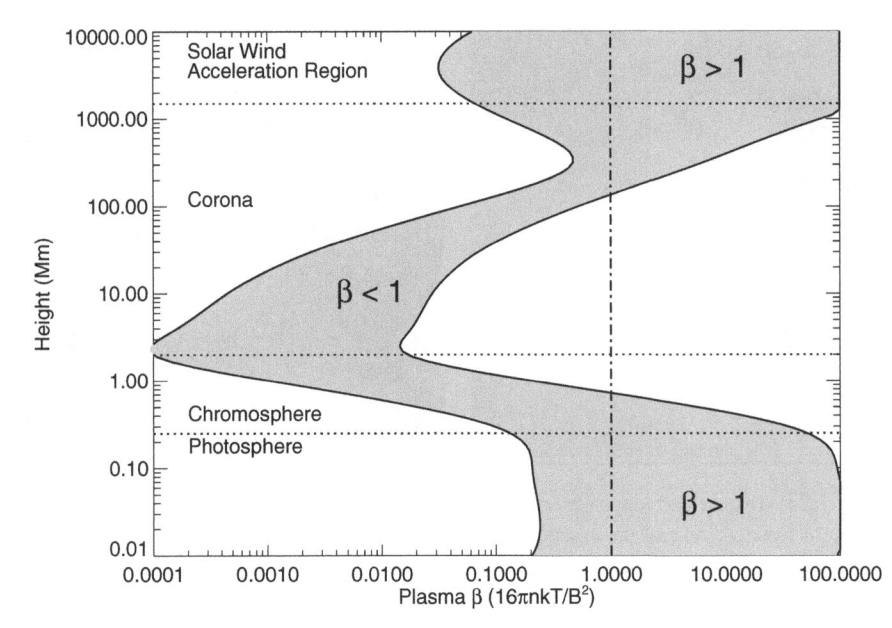

Figure 1.22: Plasma β in the solar atmosphere for two assumed field strengths, 100 G and 2500 G. In the inner corona ($R \lesssim 0.2 R_\odot$), magnetic pressure generally dominates static gas pressure. As with all plots of physical quantities against height, a broad spatial and temporal average is implied (Gary, 2001).

Table 1.1: The plasma-β parameter in the solar atmosphere.

Parameter	Photosphere	Cool corona	Hot corona	Outer corona
Electron density n_e (cm^{-3})	2×10^{17}	1×10^9	1×10^9	1×10^7
Temperature T (K)	5×10^3	1×10^6	3×10^6	1×10^6
Pressure p (dyne cm^{-2})	1.4×10^5	0.3	0.9	0.02
Magnetic field B (G)	500	10	10	0.1
Plasma-β parameter	14	0.07	0.2	7

different locations of the solar corona thus strongly depends on the employed magnetic field model, in particular because the magnetic field strength B is the least known physical parameter in the corona, while the density n_e and temperature T_e can readily be measured in EUV and soft X-rays for structures with good contrast to the coronal background. A comprehensive model of the plasma-β parameter has been built by Gary (2001), using a large number of physical parameters quoted in the literature, resulting in a well-constrained range of β-values for any given height, $\beta(h)$, shown as a grey zone in Fig. 1.22. One conclusion of Gary (2001) is that even in coronal heights of $h \gtrsim 0.2 R_\odot$ high β-values above unity can occur, which might be responsible for the dynamics of cusp regions (Fig. 1.23) or overpressure near the apices of large loops seen with *TRACE*.

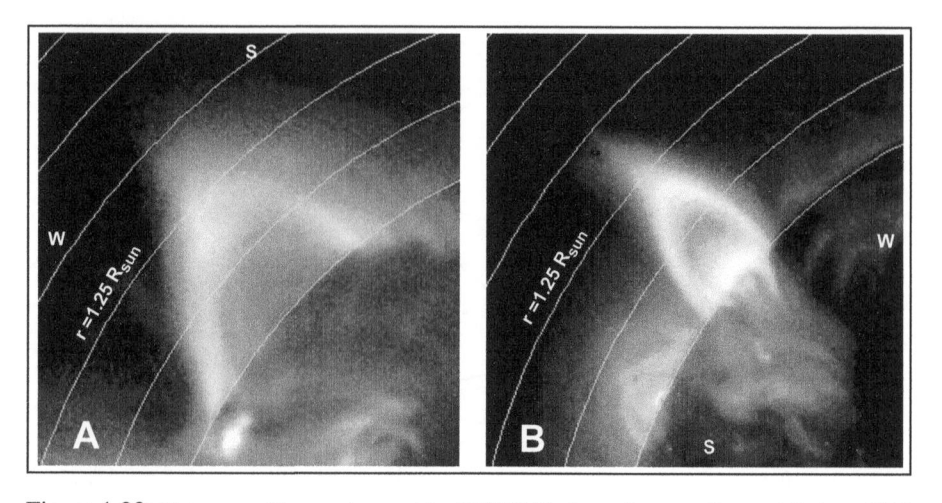

Figure 1.23: Two cusped loops observed by *SXT/Yohkoh* are shown at the west limb on 1992 January 24 (rotated, A) and the east limb on 1999 March 18 (B). The cusps are at a height of $r = 0.23R_\odot$ and $r = 0.25R_\odot$. The pressure and inertial forces of the solar wind dominate over the magnetic forces at the cusp and distend the field outwards. This implies $\beta \gtrsim 1$ at these heights (Gary, 2001).

1.9   Chemical Composition of the Solar Corona

Atomic spectroscopy that led to a first quantitative analysis of the chemical composition of the photosphere was pioneered by Henry Noris Russell in 1929. The chemical composition of the Sun has been measured most accurately in the photosphere, where line emission is brightest, while measurements of elemental abundances in the corona are much less sensitive, due to the corona being many orders of magnitude fainter. The chemical composition in the photosphere and corona are largely similar for most of the elements, and are also consistent with cosmic abundances, as they have been measured from chemical analyses of meteorites. The most recent comparison of photospheric and meteorite elemental abundances can be found, for example in Grevesse & Sauval (2001), where significant differences are listed only for the element of lithium. From the same source we list elemental abundances measured in the photosphere and corona (Table 1.2), and show the logarithmic abundances in Fig. 1.24. It can be seen that coronal elements have only currently been detected up to an atomic number of $Z \leq 30$, essentially elements down to a fraction of $\gtrsim 10^{-6}$ of the hydrogen abundance, while the sensitivity limit in the photosphere reaches $\approx 10^{-12}$.

There are a few special elements. The abundance of helium cannot be measured in the photosphere, simply because no helium line (neutral or single-ionized species) falls within the wavelength range covered by the photospheric spectrum, in the (photospheric) temperature range around 5000 K. Also for the same reason, the noble gases neon and argon cannot be measured in the photosphere. Theoretical stellar models that fit the observed age, mass, diameter, and luminosity of the Sun yield a helium abundance of $A_{He}/A_H = 9.8 \pm 0.4\%$ for the protosolar cloud. A slightly lower value is

Table 1.2: Elemental abundances in the solar photosphere and corona.

Element	Abundance[1] Photosphere	Abundance[1] Corona	FIP [eV]	Element	Abundance[1] Photosphere
1 H	12.00	12.00	13.6	42 Mo	1.92 ± 0.05
2 He	-	10.93 ± 0.004	24.6	44 Ru	1.84 ± 0.07
3 Li	1.10 ± 0.10^2	-	5.4	45 Rh	1.12 ± 0.12
4 Be	1.40 ± 0.09	-	9.3	46 Pd	1.69 ± 0.04
5 B	(2.55 ± 0.30)	-	8.3	47 Ag	(0.94 ± 0.25)
6 C	8.52 ± 0.06	8.5	11.3	48 Cd	1.77 ± 0.11
7 N	7.92 ± 0.06	7.9	14.5	49 In	(1.66 ± 0.15)
8 O	8.83 ± 0.06^2	8.8	13.6	50 Sn	2.0 ± 0.3
9 F	4.56 ± 0.30^2	-	17.4	51 Sb	1.0 ± 0.3
10 Ne	-	8.08 ± 0.06	21.6	52 Te	-
11 Na	6.33 ± 0.03	6.3^3	5.2	53 I	-
12 Mg	7.58 ± 0.05	7.6^3	7.6	54 Xe	-
13 Al	6.47 ± 0.07	6.4^3	6.0	55 Cs	-
14 Si	7.55 ± 0.05	7.6^3	8.1	56 Ba	2.13 ± 0.05
15 P	5.45 ± 0.04	5.5	10.5	57 La	1.17 ± 0.07
16 S	7.33 ± 0.11	7.2	10.3	58 Ce	1.58 ± 0.09
17 Cl	5.50 ± 0.30^2	5.8	13.0	59 Pr	0.71 ± 0.08
18 Ar	-	6.40 ± 0.06^3	15.8	60 Nd	1.50 ± 0.06
19 K	5.12 ± 0.13	-	4.3	62 Sm	1.01 ± 0.06
20 Ca	6.36 ± 0.02	6.3^3	6.1	63 Eu	0.51 ± 0.08
21 Sc	3.17 ± 0.10	-		64 Gd	1.12 ± 0.04
22 Ti	5.02 ± 0.06	-		65 Tb	(-0.1 ± 0.3)
23 V	4.00 ± 0.02	-		66 Dy	1.14 ± 0.08
24 Cr	5.67 ± 0.03	-	6.8	67 Ho	(0.26 ± 0.16)
25 Mn	5.39 ± 0.03	-	7.4	68 Er	0.93 ± 0.06
26 Fe	7.50 ± 0.05	7.6^3	7.9	69 Tm	(0.00 ± 0.15)
27 Co	4.92 ± 0.04	5.0		70 Yb	1.08 ± 0.15
28 Ni	6.25 ± 0.04	6.3^3	7.6	71 Lu	0.06 ± 0.10
29 Cu	4.21 ± 0.04	4.1		72 Hf	0.88 ± 0.08
30 Zn	4.60 ± 0.08	4.4		73 Ta	-
31 Ga	2.88 ± 0.10^2	-	6.0	74 W	(1.11 ± 0.15)
32 Ge	3.41 ± 0.14^2	-		75 Re	-
33 As	-	-		76 Os	1.45 ± 0.10
34 Se	-	-		77 Ir	1.35 ± 0.10
35 Br	-	-		78 Pt	1.8 ± 0.3
36 Kr	-	-	14.0	79 Au	(1.01 ± 0.15)
37 Rb	2.60 ± 0.15^2	-	4.2	80 Hg	-
38 Sb	2.97 ± 0.07	-		81 Tl	(0.9 ± 0.2)
39 Y	2.24 ± 0.03	-		82 Pb	1.95 ± 0.08
40 Zr	2.60 ± 0.02	-		83 Bi	-
41 Nb	1.42 ± 0.06	-		90 Th	-
				92 U	$< -0.47)$

[1]) abundances are given on a logarithmic scale, $12.0 + ^{10} log(A/A_H)$
[2]) derived in addition or only from sunspots
[3]) abundance could be a factor of ≈ 3 times higher in corona and solar wind (low-FIP)
(...) Values between parentheses are less accurate

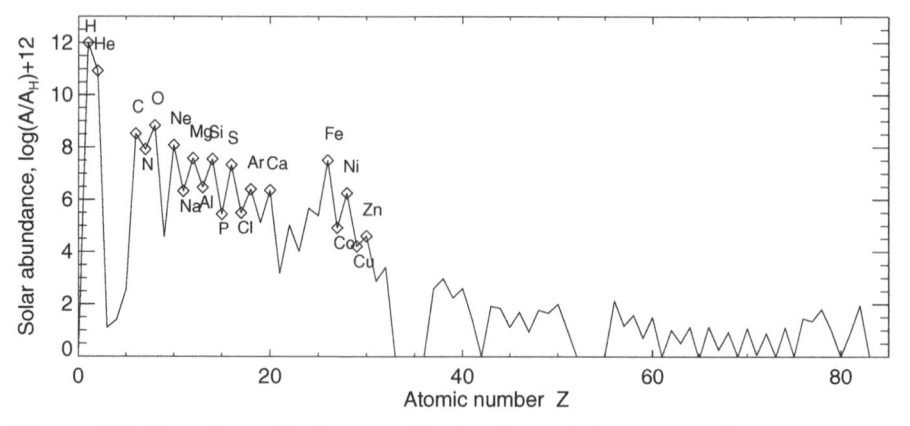

Figure 1.24: Logarithmic distribution of the abundance of the elements relative to hydrogen, normalized to $12 + log(A/A_H[Z = 1])$, as a function of the atomic number, Z. Elements that have been detected in the corona are labeled.

obtained from inversion of helioseismic data. Other exceptions are the light elements of lithium, beryllium, and boron, which have a strikingly low abundance as a consequence of their low *nuclear binding energy*, so that these nuclei are destroyed by proton collisions at solar temperatures of a few million Kelvins.

Since the solar corona is a highly dynamic and inhomogeneous place, subtle variations in the chemical composition have been observed in various coronal structures. A fractionation seems to occur in the lower chromospheric layers, where elements with a low *first ionization potential (FIP)* (≤ 10 keV) show abundances that are locally and temporarily larger than in the photosphere. Such a FIP fractionation effect can change the relative abundances typically by a factor of ≈ 3 (e.g., Na, Mg, Al, Si, Ca, Ni), or up to a factor of 10 for iron (Fe), between the chromosphere and the solar wind. Unfortunately the relative abundances cannot directly be measured with respect to hydrogen, because hydrogen is completely ionized in the corona, so that it has no emission lines that can be used for absolute abundance determinations. Instead, only relative abundances of "heavy" elements (metals) can be measured, so that there is an ambiguity whether the FIP fractionation effect causes an enhancement of low-FIP elements into the corona, or a depletion of high-FIP elements out of the corona (to the chromosphere).

1.10 Radiation Spectrum of the Solar Corona

The atmosphere of our planet Earth filters out emission from the Sun and stars in many wavelengths, except for two spectral windows at optical and radio wavelengths. The major progress in solar physics achieved over the last decade thus involves a number of space missions, floating aloft our absorbing atmosphere. These space missions provide us unprecedented information over the entire wavelength spectrum, covering gamma-rays, hard X-rays, soft X-rays, X-ray ultraviolet (XUV), extreme ultraviolet (EUV), and ultraviolet (UV). It has opened our eyes about the physical processes that govern solar and stellar atmospheres that we could not anticipate without space data. An overview of the solar radiation spectrum is shown in Fig. 1.25, covering a wavelength range over 14 orders of magnitude.

The physical units of the spectrum are generally quantified in terms of the wavelength λ (bottom axis of Fig. 1.25), having units of meters (m) in the SI system or centimeters (cm) in the cgs system. However, astronomers often prefer Ångstrøm (Å) units, defined by,

$$1 \text{ Å} = 10^{-8} \text{ cm} = 10^{-10} \text{ m} . \tag{1.10.1}$$

In the radio domain, the frequency unit ν is used (top axis in Fig. 1.25), which has the unit (Hz) or (s^{-1}) in both the SI or cgs system and can be calculated from the basic dispersion relation of electromagnetic waves in a vacuum,

$$\nu = \frac{c}{\lambda} , \tag{1.10.2}$$

with $c = 2.9979 \times 10^{10}$ cm s^{-1} being the speed of light. For relativistic particles, as we observe in hard X-rays and gamma-rays in the solar corona and chromosphere, a practical unit is eV. An electron-Volt (1 eV) is defined by the energy gained by an electron (e) if it is accelerated by a potential difference of 1 Volt (V), which is in SI and cgs units,

$$1 \text{ eV} = 1.60 \times 10^{-19} \text{ C} \times 1 \text{ V} = 1.60 \times 10^{-19} \text{ J} = 1.60 \times 10^{-12} \text{ erg} . \tag{1.10.3}$$

Equating the particle energy ε with the photon quantum energy $\epsilon = h\nu$, we then have (with Eq.1.10.2) a relation to the wavelength λ,

$$\epsilon = h\nu = h\frac{c}{\lambda} = 12.4 \left(\frac{1 \text{ Å}}{\lambda}\right) \text{ keV} . \tag{1.10.4}$$

For thermal plasmas, which are observed in the solar corona in EUV, XUV, and in soft X-rays, it is customary to associate a temperature T with a wavelength λ, using the definition of the quantum energy of a photon (and the dispersion relation 1.10.2),

$$\varepsilon_{th} = k_B T = \epsilon = h\nu = h\frac{c}{\lambda} . \tag{1.10.5}$$

This yields (with the Boltzmann constant $k_B = 1.38 \times 10^{-16}$ erg K^{-1} and the Planck constant $h = 6.63 \times 10^{-27}$ erg s) the conversion formula,

$$T = \frac{hc}{k_B \lambda} = 1.44 \left(\frac{1 \text{ cm}}{\lambda}\right) \text{ K} . \tag{1.10.6}$$

With these relations we can conveniently convert any physical unit used in the context of a wavelength spectrum. The solar irradiance spectrum from gamma-rays to radio waves is shown in Fig.1.25, given in units of energy per time, area, and wavelength (erg s^{-1} cm^{-2} μm^{-1}).

Let us tour quickly from the shortest to the longest wavelength regime, summarizing in each one the essential physical emission processes that can be detected with space-borne instruments and ground-based telescopes.

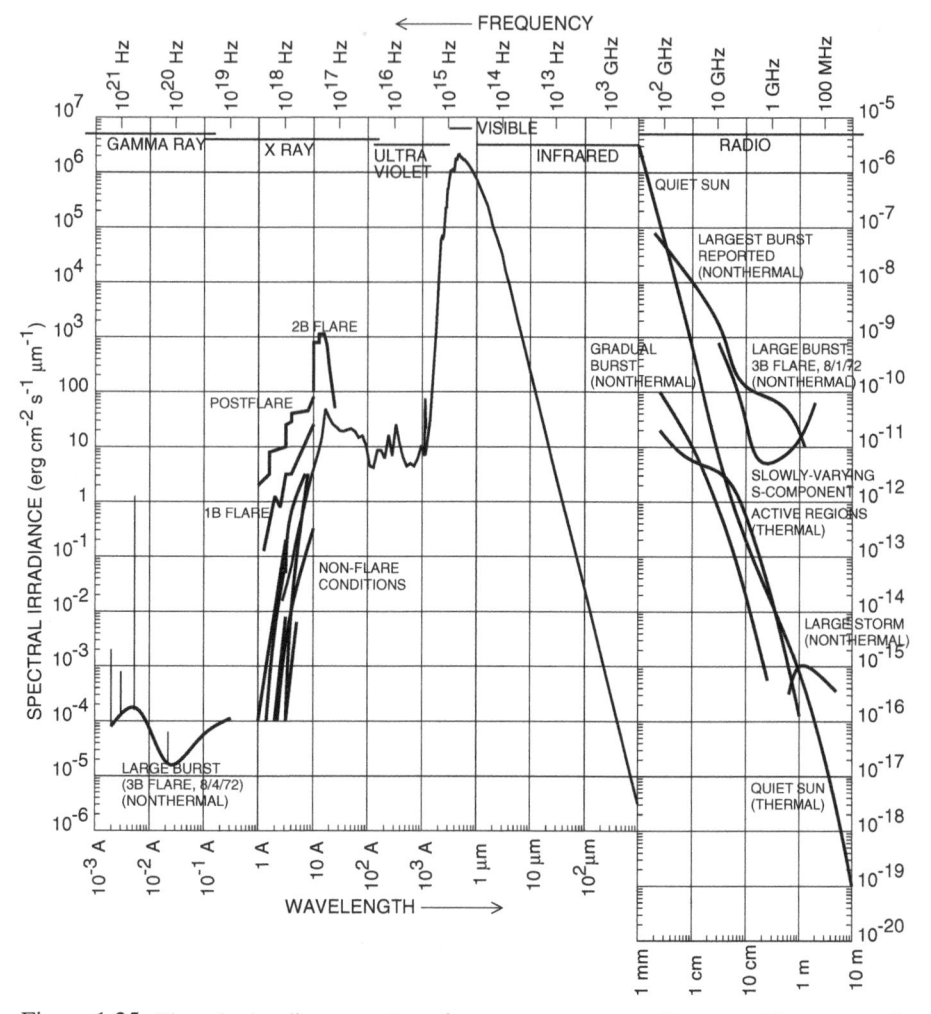

Figure 1.25: The solar irradiance spectrum from gamma-rays to radio waves. The spectrum is shifted by 12 orders of magnitude in the vertical axis at $\lambda = 1$ mm to accommodate for the large dynamic range in spectral irradiance (after Zombeck, 1990 and Foukal, 1990).

The shortest wavelength regime, which is emitted during only the most energetic processes in large flares, is the *gamma-ray* (γ-ray) regime, in a wavelength range of $\lambda \approx 10^{-3} - 10^{-1}$ Å, which corresponds to energies of $\epsilon \approx 100$ keV-10 MeV. Obviously, such high-energy radiation is only emitted from the Sun when particles are accelerated in flares to sufficiently high energies, so that they can interact with atomic nuclei. Solar gamma-ray emission is produced when particles, accelerated to high energies in the collisionless corona, precipitate to the chromosphere, where a number of nuclear processes produce gamma-rays (e.g., continuum emission by primary electron bremsstrahlung, nuclear de-excitation lines, neutron capture lines, positron annihilation radiation, or pion decay radiation).

The next wavelength regime is the *hard X-rays* regime, corresponding to wavelengths of $\lambda \approx 0.1 - 1$ Å or energies of $\epsilon \approx 10 - 100$ keV. This energy range corresponds to mild-to-medium relativistic electrons (with a rest mass of $m_e c^2 = 511$ keV and a relativistic speed of $\beta = v/c \approx 0.2 - 0.5$), and thus requires electrons that are accelerated in a collisionless plasma in the corona. When these nonthermal electrons enter the high-density transition region and chromosphere, they lose their energy by collisions and emit electron (thick-target) bremsstrahlung, which is seen at hard X-ray wavelengths.

The next wavelength regime is called *soft X-rays*, at wavelengths of $\lambda \approx 1 - 100$ Å. This soft X-ray wavelength range corresponds to thermal energies of $\varepsilon_{th} \approx 0.1 - 10$ keV and plasma temperatures of $T \approx 1.5 - 150$ MK. Such plasma temperatures are found in active regions ($T \approx 1.5 - 10$ MK) and in flare loops ($T \approx 10 - 40$ MK). This entire wavelength regime is well observed with the *Soft X-Ray Telescope (SXT)* on board *Yohkoh*. The longer wavelength decade at $\lambda \approx 10 - 100$ Å is also called *X-ray ultraviolet (XUV)*. Soft X-rays are produced by free-free emission of electrons that are scattered off highly-ionized ions in the solar corona.

The next wavelength decade is *extreme ultraviolet (EUV)*, covering wavelengths of $\lambda \approx 100 - 1000$ Å or temperatures of $T \approx 0.15 - 1.5$ MK. As it can be seen from the differential emission measure distribution (Fig.1.21), most of the plasma in the quiet Sun has temperatures around $T \approx 1 - 3$ MK, and thus the quiet Sun is best probed with EUV lines. Both EUV imagers, the *EIT/SoHO* and *TRACE*, have their primary passbands in this wavelength regime, i.e., 171 Å ($T \approx 1.0$ MK), 195 Å ($T \approx 1.5$ MK), and 284 Å ($T \approx 2.0$ MK) wavebands. EUV emission is also produced by free-free emission like for soft X-rays, but by scattering off ions with lower temperatures, Fe IX to Fe XV.

The peak of the solar irradiance spectrum is, of course, in the optical wavelength regime of $\lambda \approx 3000 - 7000$ Å that is visible to the eye. It is probably an evolutionary adaption process that our eyes developed the highest sensitivity where our Sun provides our life-supporting daylight. Optical emission from the Sun originates mostly from continuum emission in the photosphere, so that we do not gain much information from the corona in visible light. It is only during total solar eclipses, when the bright light from the solar disk is occulted by the moon, that faint scattered light reveals us some structures from the tenuous corona.

The next three decades of the wavelength spectrum is called *infrared*, extending over $\lambda \approx 1$ μm - 1 mm. Infrared emission does not contain many lines, the most prominent one longward of 1 μm is the He 10,830 Å line. Most of the infrared is dominated by low-excitation species: neutral atoms and molecules (e.g., carbon monoxide, hydroxil, water vapor). Infrared emission is thus only produced in quiescent gas not strongly heated, conditions that can be found in supergranulation cell interiors and in sunspot umbrae.

Finally, the longest wavelength regime is occupied by *radio wavelengths*, from $\lambda \approx 1$ mm to 10 m, corresponding to frequencies of $\nu \approx 300$ GHz to 30 MHz. In radio there are a variety of emission mechanisms that radiate at characteristic frequencies: gyrosynchrotron emission produced by high-relativistic electrons during flares (microwaves: dm, cm, mm wavelengths), gyroresonance emission from mildly relativistic electrons in strong magnetic field regions, such as above sunspots (in microwaves), free-free emission from heated plasma in flares and active regions (dm and cm), plasma emission excited by propagating electron beams (m and dm), and a number of other radiation mechanisms produced by kinetic plasma instabilities. The reason why solar radio emission is so multi-faceted is because the gyrofrequency of coronal magnetic fields as well as the plasma frequency of coronal densities coincide with the radio wavelength range. A large number of ground-based radio telescopes and interferometers have studied this highly interesting spectral window over the last five decades.

1.11   Summary

Observations of the solar corona started with solar eclipses by the moon, which was the only natural means to suppress the strong contrast caused by the six orders of magnitude brighter solar disk in optical wavelengths. Eclipse observations provided the first crude density models of the corona and the discovery of coronal holes. Only the space age, however, allowed a systematic exploration of the other more important wavelength regimes in EUV, soft X-rays, hard X-rays, and gamma-rays. Decade-long spacecraft missions such as SMM, Yohkoh, CGRO, SoHO, and TRACE opened up multi-wavelength observations that were instrumental for diagnostics of coronal plasma physics and high-energy particle physics in flare processes. We note that even the most basic physical properties, such as the density, temperature, and magnetic field structure of solar and stellar coronae are far more complex than any planetary magnetosphere. The coronal magnetic field is driven by an internal dynamo that modulates the surface magnetic field, the solar activity, the coronal brightness, the flare and CME frequency, and the solar wind with an eleven-year cycle, waxing and waning in complexity and nonlinear behavior. The density and temperature structure is far from the initially conceived hydrostatic, uniform, and gravitationally stratified atmospheric models, but rather displays a highly dynamic, restless, intricate, and global system that is hidden to us in stellar atmospheres by the concealment of distance. A key parameter that distinguishes the plasma dynamics in the solar corona is the plasma-β parameter, which indicates that the magnetic pressure rather than the thermal pressure governs the coronal plasma dynamics, in contrast to the chromosphere, the solar interior, or the heliosphere. The chemical composition is another peculiarity of the corona, which shows anomalies as a function of the first ionization potential energy (FIP-effect) with respect to the universal chemical composition that is found in the photosphere or in meteorites. The radiation spectrum of the solar corona spans over at least 14 orders of magnitude, from the longest wavelengths in radio over infrared, visible, ultraviolet, soft X-rays, hard X-rays, to gamma-rays, with each wavelength regime revealing different physical processes. Only simultaneous and complementary multi-wavelength observations allow us to establish coherent theoretical models of the fundamental plasma physics processes that operate in the solar corona and in other high-temperature astrophysical plasmas.

Chapter 2

Thermal Radiation

Thermal radiation from the Sun is emitted from temperatures as low as $T = 4400$ K in sunspot umbrae (during solar minimum) to as high as $T \approx 40$ MK in the superhot component of flare plasmas, a temperature range that covers about 4 orders of magnitude. Concentrating on the thermal emission from the corona, quiet Sun regions emit in the temperature range of $T \approx 1 - 3$ MK, active regions in the range of $T \approx 2 - 8$ MK, and typical flares in the range of $T \approx 10 - 40$ MK. Therefore, thermal emission from the solar corona falls mainly in the wavelength domain of EUV and soft X-rays. In this chapter we describe the basic physics of thermal radiation mechanisms in the EUV and soft X-ray wavelength domain, which will then be applied to data analysis from the *Yohkoh*, *SoHO/EIT*, *CDS*, and *TRACE* missions. For complementary literature the reader is referred to textbooks by Zirin (1988), Foukal (1990), Phillips (1992), Golub & Pasachoff (1997), or to *Astrophysical Formulae* by Lang (1980).

2.1 Radiation Transfer and Observed Brightness

First we have to define a few basic elements of radiation theory before we can go into discussion of the solar radiation in various wavelengths. The quantity we observe at Earth is the *specific intensity* I_ν, which is the energy (erg) radiated from a solar source at a frequency interval $\nu...\nu + d\nu$ (Hz^{-1}), per time interval dt (s^{-1}), from a unit area dA (cm^{-2}) of the source, which is foreshortened by a factor $\cos(\theta)$ for a line-of-sight with an angle θ to the solar vertical, emitted over a solid angle $d\Omega$ (ster^{-1}), so it has the physical (cgs) units of (erg s^{-1} cm^{-2} Hz^{-1}ster^{-1}). The source is assumed to contain both emitting and absorbing elements.

The definition of the *specific intensity* I_ν is illustrated in Fig. 2.1. The associated energy dW is thus

$$dW = I_\nu \, dt \, dA \, \cos(\theta) \, d\nu \, d\Omega \qquad \text{(erg)} \, . \tag{2.1.1}$$

The intensity along the path s through the solar source increases in every emitting volume element $dV = ds \, dA \, \cos(\theta)$ per path increment ds by

$$dW^{em}(s) = \epsilon_\nu(s) \, dt \, dV \, d\nu \, d\Omega \, , \tag{2.1.2}$$

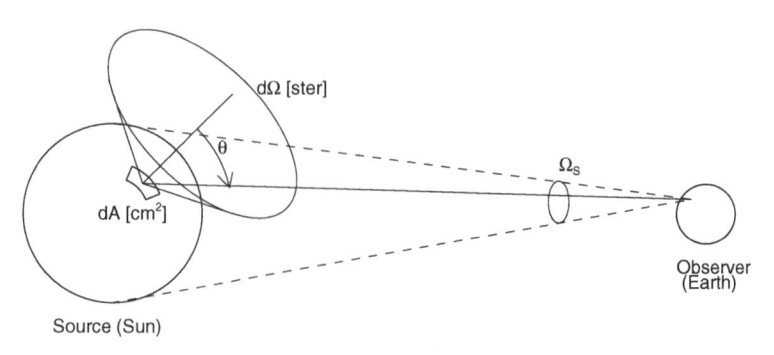

Figure 2.1: Definition of *specific intensity* I_ν emitted per unit area dA and solid angle $d\Omega$ at an angle θ. The source appears under a solid angle Ω_s from the observer's position.

where $\epsilon_\nu(s)$ is the local *emission coefficient* at position s. Equating Eqs. (2.1.1) and (2.1.2) we have a relation between the intensity increase dI_ν^{em} and the emission coefficient ϵ_ν,

$$dI_\nu^{em}(s) = \epsilon_\nu(s)ds \ . \tag{2.1.3}$$

Along the same path s, there may be absorbing material (e.g. cool prominence material absorbs EUV and soft X-ray emission) with an *absorption coefficient* $\alpha_\nu(s)$, so that the intensity decreases by an amount

$$dI_\nu^{ab}(s) = -\alpha_\nu(s)I_\nu(s)ds \ . \tag{2.1.4}$$

Note that the emission and absorption coefficients are not defined by the same physical units. For absorption the intensity change is proportional to the incident intensity, while the emission coefficient is independent of it.

Combining the intensity changes from emission (Eq. 2.1.3) and absorption (Eq. 2.1.4) leads to the *radiative transfer equation*, $dI_\nu(s) = dI_\nu^{ab}(s) + dI_\nu^{em}(s)$, or

$$\frac{dI_\nu(s)}{ds} = -\alpha_\nu(s)I_\nu(s) + \epsilon_\nu(s) \ , \tag{2.1.5}$$

which is a linear differential equation of first order. It is more convenient to solve the radiative transfer equation by expressing it in terms of the so-called *source function* S_ν, which is defined as the ratio of the emission coefficient to the absorption coefficient,

$$S_\nu(s) = \frac{\epsilon_\nu(s)}{\alpha_\nu(s)} \ . \tag{2.1.6}$$

Furthermore, we define the *optical depth* $\tau_\nu(s)$ at location s, which includes the integrated absorption over the path segment $(s - s_0)$,

$$\tau_\nu(s) = \int_{s_0}^{s} \alpha_\nu(s')ds' \ , \tag{2.1.7}$$

where the path starts at the remotest source location s_0 and moves along s towards the observer. Inserting the source function $S_\nu(s)$ (Eq. 2.1.6) into the radiative transfer equation (2.1.5) and substituting the path variable ds by the opacity variable $d\tau_\nu = \alpha_\nu ds$ (Eq. 2.1.7), the radiative transfer equation can easily be integrated by multiplying both sides of the equation with the factor $e^{-\tau_\nu}$ (e.g., see Rybicki,G.B. & Lightman,A.P. 1979, p.13; Zirin 1988, p.57), leading to the solution,

$$I_\nu(\tau_\nu) = I_\nu(0)e^{-\tau_\nu} + \int_0^{\tau_\nu} e^{-(\tau_\nu - \tau_\nu')} S_\nu(\tau_\nu') \, d\tau_\nu' \qquad (2.1.8)$$

Let us now apply the radiative transfer equation to a simple case, for instance to a source that is observed through an absorbing cloud with a constant absorption coefficient $\alpha_\nu(s) = const$ and length L (yielding an opacity of $\tau_\nu = \int_0^L \alpha_\nu ds = \alpha_\nu L$), and with a constant emission coefficient $\epsilon_\nu(s) = const$. The limit of the optically thick ($\tau_\nu \gg 1$) and optically thin case ($\tau_\nu \ll 1$) is then

$$I_\nu(\tau_\nu) = I_0 e^{-\tau_\nu} + S_\nu \left[1 - \exp(-\tau_\nu) \right] \approx \begin{cases} S_\nu & \text{if } \tau_\nu \gg 1 \\ I_0(1 - \tau_\nu) + S_\nu \tau_\nu & \text{if } \tau_\nu \ll 1 \end{cases}.$$
$$(2.1.9)$$

This application is especially important at radio wavelengths, where the observed brightness temperature T_B is proportional to the brightness $B_\nu(T)$ [Rayleigh−Jeans approximation, see Eq. (2.2.5) in §2.2] or source intensity $I_\nu(T)$. For an optically thick source, the observed brightness temperature is then identical to the electron temperature of the source (i.e., $T_B = T_e$), which is usually the case for long wavelengths (dm and m in the corona), while the observed brightness temperature decreases at higher frequencies (cm, mm) in the optically thin case (i.e., $T_B = T_e \tau_\nu \ll T_e$). In the optically thick case we see essentially the temperature at a surface layer where the source becomes optically thick ($\tau_\nu \approx 1$), while the interior of the source is hidden and has no effect on the observed brightness temperature, regardless how hot or cold it is. In the opposite case, optically thin emission is proportional to the opacity τ_ν, which is proportional to the depth of the source for a constant absorption coefficient α_ν. Optically thin emission, therefore, always shows a center-to-limb effect, because the optical path through the solar corona is increasing from disk center to the limb by a factor of $1/\cos(l - l_0)$ as a function of the longitude l (relative to the longitude l_0 at the disk center), and is jumping by a factor of two from inside to outside of the disk, when the coronal segment behind the disk is added to the segment in front of the disk. This center-to-limb effect is observable at short radio wavelengths as well as in EUV and in soft X-rays.

2.2 Black-Body Thermal Emission

The concept of electromagnetic radiation in thermodynamic equilibrium was introduced by Gustav Kirchhoff in 1860, who envisioned a radiating gas that is confined in a so-called *black body* and reaches an equilibrium between emission and absorption. He found that the ratio of the emission coefficient ϵ_ν to the absorption coefficient α_ν is given by a universal brightness function $B_\nu(T)$ that depends only on the temperature T and frequency ν, the so-called *Kirchhoff law*,

$$\epsilon_\nu = n_\nu^2 \alpha_\nu B_\nu(T) . \qquad (2.2.1)$$

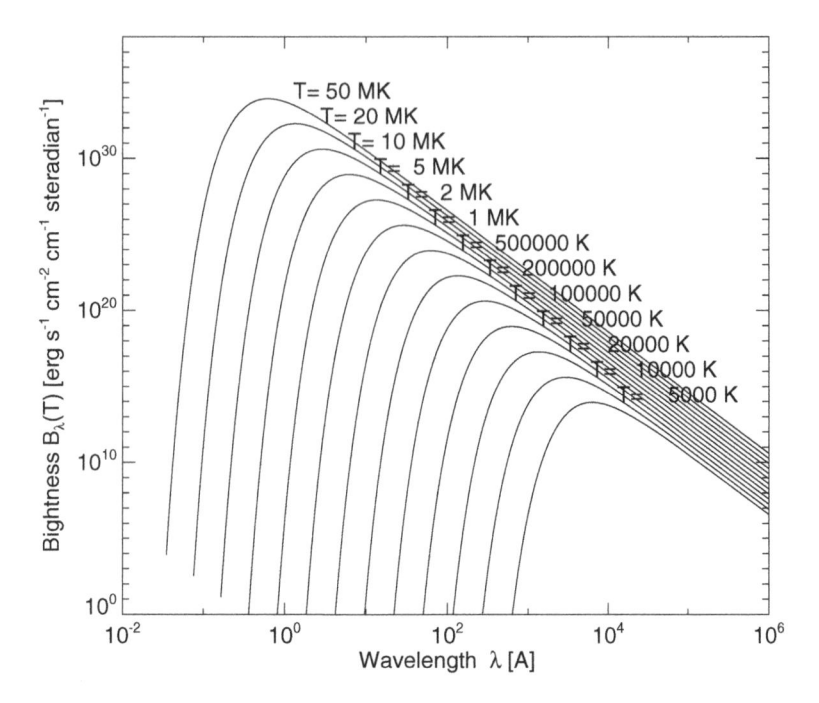

Figure 2.2: The brightness function $B_\nu(T)$ of a black-body radiator as a function of wavelength λ (Å) for solar temperatures, ranging from $T \approx 5000$ K in sunspots to $T = 50$ MK in superhot flares. Note that the brightness peaks in the $\lambda \approx 0.5 - 20$ Å range for flare temperatures.

where n_ν is the refractive index of the medium. The calculation of the universal brightness function $B_\nu(T)$ was achieved by Max Planck in 1900, after he introduced the quantum mechanical theory, where a photon with energy $h\nu$ is represented by a harmonic oscillator with a mean thermal energy of $k_B T$ [with $h = 6.63 \times 10^{-27}$ (erg s) being the Planck constant and $k_B = 1.38 \times 10^{-16}$ (erg K^{-1}) the Boltzmann constant]. *Planck's law* of the brightness distribution as a function of frequency ν (Planck 1901, 1913) is,

$$B_\nu(T) = \frac{2h\nu^3 n_\nu^2}{c^2} \frac{1}{[\exp(h\nu/k_B T) - 1]} , \qquad (2.2.2)$$

or, in terms of wavelength λ,

$$B_\lambda(T) = \frac{2hc^2 n_\nu^2}{\lambda^5} \frac{1}{[\exp(hc/\lambda k_B T) - 1]} . \qquad (2.2.3)$$

The function $B_\lambda(T)$ is plotted in Fig. 2.2 for a range of temperatures. *Planck's function* $B_\nu(T)$ can be simplified to *Wien's law* in the short-wavelength approximation ($h\nu \gg k_B T$) (Wien 1893, 1894),

$$B_\nu(T) = \frac{2h\nu^3 n_\nu^2}{c^2} \exp\left(-\frac{h\nu}{k_B T}\right) , \qquad (2.2.4)$$

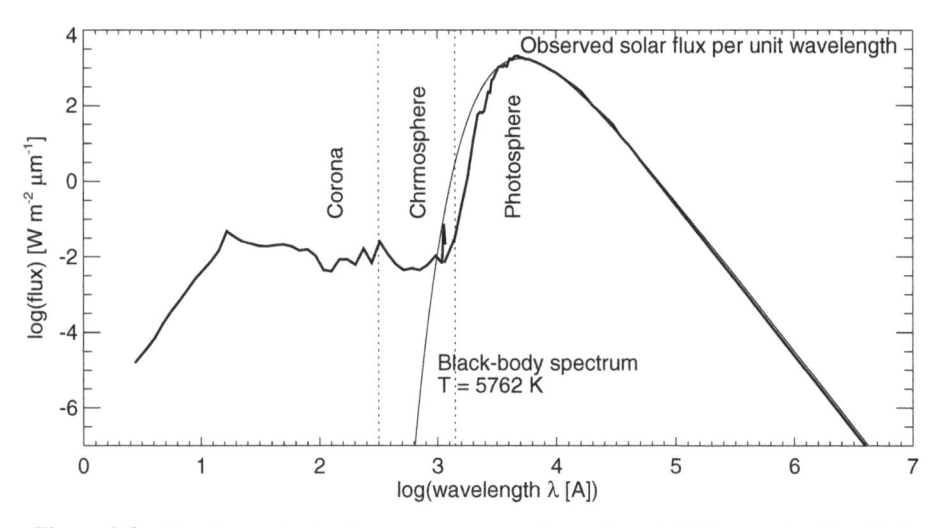

Figure 2.3: The observed solar flux spectrum per unit wavelength (thick curve) with a black-body spectrum of temperature $T = 5762$ K.

or to the *Rayleigh–Jeans law* in the long-wavelength approximation ($h\nu \ll k_B T$) (Rayleigh 1900, 1905; Jeans 1905, 1909),

$$B_\nu(T) = \frac{2\nu^2 k_B T n_\nu^2}{c^2} . \qquad (2.2.5)$$

The Rayleigh–Jeans law is an important approximation used in radio wavelengths (see Fig. 1.25). For solar temperatures, $T \gtrsim 5 \times 10^3$ K, the Rayleigh–Jeans approximation is good at all frequencies $\nu \ll k_B T/h \lesssim 10^{14}$ Hz, in infrared as well as in radio.

The wavelength λ_{max} where the Planck function $B_\lambda(T)$ has the maximum is only a function of the temperature $T(K)$ and can be found from the derivative $dB_\lambda(T)/d\lambda = 0$, and is

$$\lambda_{max} = \frac{0.2898}{T(\mathrm{K})} \quad (\mathrm{cm}) . \qquad (2.2.6)$$

This relation is also called the *Wien displacement law*. When the Planck function $B_\lambda(T)$ is integrated over all frequencies, the total flux over the visible hemisphere, πB, of the black body is obtained (Stefan 1879; Boltzmann 1884; Milne 1930), the so-called *Stefan–Boltzmann law*,

$$\pi B(T) = \pi \int_0^\infty B_\lambda(T) d\lambda = n_\nu^2 \, \sigma T^4 , \qquad (2.2.7)$$

with the Stefan–Boltzmann constant $\sigma = 2\pi^5 k_B^4/(15c^2 h^3) = 5.6774 \times 10^{-5}$ (erg s^{-1} cm^{-2} K^{-4}). The total output of the photosphere $\pi B(T)$ defines with Eq. 2.2.7 the *effective temperature* of the photosphere, which is $T = 5762$ K.

We plot the Planck brightness distribution $B_\nu(T)$ in Fig. 2.2 for various temperatures relevant to the solar chromosphere and corona (from $T = 5000$ K to $T = 50$ MK), in units of (erg s^{-1} cm^{-2} cm^{-1} ster^{-1}). In Fig. 2.3 we overplot the Planck brightness function $B_\nu(T)$ for an effective temperature of $T = 5762$ K to the observed solar spectrum (for a complete solar spectrum see Fig. 1.25). The flux scale is fully determined by the source size for a black body. The source size is given by the solar disk with a radius of $R_\odot = 6.96 \times 10^5$ km, while a steradian unit corresponds to the angular area of D^2 subtended by the Sun–Earth distance, $D = 1$ AU $= 1.50 \times 10^8$ km. The angular area of the Sun in steradian units is

$$\Omega_\odot = \left(\frac{\pi R_\odot^2}{D^2}\right) = 6.76 \times 10^{-5} \qquad \text{(ster)} . \tag{2.2.8}$$

The solar irradiance spectrum I_λ is given in flux intensity units (e.g., tabulated for the optical wavelength range of $\lambda = 0.115, ..., 50$ μm in Foukal (1990), quantified in SI units of [W m^{-2} μm^{-1}]). So the relation between the flux intensity I_λ and the brightness function B_λ is given by

$$I_\lambda(T) = B_\lambda(T)\Omega_S \tag{2.2.9}$$

where Ω_S is the solid angle subtended by the source (e.g., the solar disk here, $\Omega_S = \Omega_\odot$). A effective temperature of $T = 5762$ K fits the solar irradiance spectrum quite well over the optical, infrared, and shorter radio wavelength range (Fig. 2.3). The wavelength range of $\lambda = 0.18 - 10$ μm (1800-10^5 Å) over which a good fit is found to the black-body spectrum, comprises 99.9% of the solar irradiance output.

At shorter wavelengths, such as in EUV and soft X-rays, the solar irradiance spectrum cannot be modeled by black-body spectra for a number of reasons: (1) the tenuous corona is not optically thick at EUV and soft X-rays, so that the brightness is drastically reduced by the small opacity τ, $I_\lambda = \tau B_\lambda(T)$ with $\tau \ll 1$; (2) *line blanketing* effects occur in UV due to the increasing number of Fraunhofer absorption lines; (3) the EUV and SXR spectrum is dominated by thin emission lines from highly ionized metals (e.g. O, Ne, Mg, Al, Si, Ca, Fe). At the longer radio wavelengths (dm, m), the solar spectrum can, in the absence of other radio emission mechanisms, again be modeled with a black-body spectrum, because the corona becomes optically thick.

2.3   Thermal Bremsstrahlung (Free–Free Emission)

The corona is a fully ionized gas, a *plasma*, so that electrons and ions move freely, interacting with each other through their electrostatic charge. Electrons and atomic nuclei are separated in the coronal plasma and can undergo manyfold interactions. We describe in the following the most important atomic processes that contribute to coronal emission in EUV, soft X-rays, and radio wavelengths. The most common and least interfering interaction (with the atomic structure) is when a free electron is scattered in the Coulomb field of an ion. Essentially a free electron of the coronal plasma is elastically scattered off an ion and escapes as a free electron, which is called a *free-free transition*. A photon is emitted with an energy corresponding to the difference of the outgoing to the incoming kinetic energy of the electron (i.e., $h\nu = \epsilon' - \epsilon$), according to the principle of energy conservation. Because the energy of the emitted photon is only positive when the electron loses energy, this *free-free emission* is also called

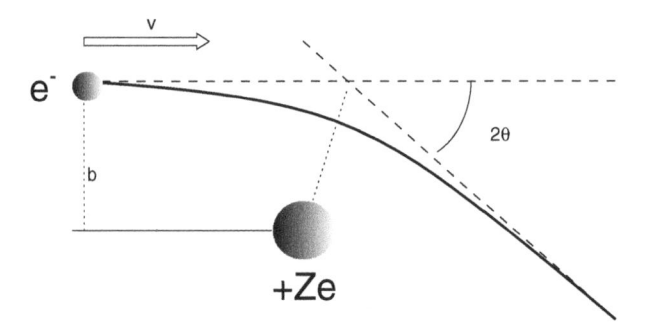

Figure 2.4: Elastic scattering of an electron (e^-) off a positively charged ion ($+Ze$). The electron moves with velocity v on a path with impact parameter b and is deflected by an angle of 2ϑ, with $\tan \vartheta/2 = Ze^2/(mv^2b)$, according to Rutherford (1911).

bremsstrahlung, the german word for *braking radiation* introduced by Bohr, Bethe, and Heitler. We provide a brief derivation of the thermal bremsstrahlung spectrum in the following, while a more detailed treatment can be found in Jackson (1962, §13, 15) and Lang (1980).

When a nonrelativistic electron with mass m, charge e, and velocity v passes the Coulomb field of an atom, molecule, or ion with charge $+Ze$, considered stationary, the angular deflection ϑ of the particle is given by (Rutherford, 1911),

$$tan\left(\frac{\vartheta}{2}\right) = \frac{Ze^2}{mv^2b} \, , \qquad (2.3.1)$$

where the impact parameter b designates the perpendicular distance from the charge Ze to the original trajectory of the incoming electron (Fig. 2.4). The angular deflection of the electron on its hyperbolic trajectory corresponds to an acceleration in a direction perpendicular to its original path by an amount of Δv. The total power of the electromagnetic radiation (emitted per unit time in all directions) of an accelerated point charge is

$$P = \frac{2}{3}\frac{q^2}{c^3}\left(\frac{dv}{dt}\right)^2 . \qquad (2.3.2)$$

(see Larmor 1897; for derivation, see Jackson 1962, §14). Thus, the total power radiated by such a deflected electron is, by combining Eqs. (2.3.1) and (2.3.2), using the approximation $\Delta v \approx v \times \tan(\vartheta) \approx v \times 2\tan(\vartheta/2)$ that results from the momentum transfer equation for elastic scattering,

$$W(\nu)d\nu \approx \frac{2e^2}{3\pi c^3}|\Delta v|^2 d\nu \approx \frac{8}{3\pi}\left(\frac{e^2}{mc^2}\right)^2 \frac{Z^2e^2}{c}\left(\frac{c}{vb}\right)^2 d\nu \, , \qquad \text{for } \nu < \frac{v}{b} \qquad (2.3.3)$$

where Δv is the change in electron velocity caused by the collision.

The total rate of encounters between an electron and a volume density n_i of ions in the parameter range of $b...b + db$ (which is a cylindric tube with radius b around each

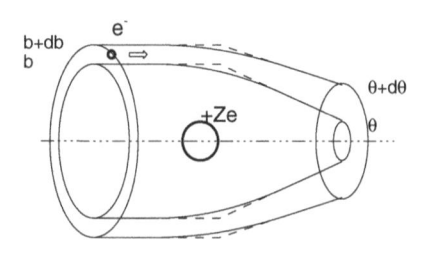

Figure 2.5: Definition of *differential cross section* $d\sigma_s/d\Omega$, which quantifies the solid angle $d\Omega = 2\pi \sin \vartheta d\vartheta$ that corresponds to scattering angles ϑ from the impact parameter range $b, ..., b + db$.

ion, see Fig. 2.5), is

$$n_i v 2\pi \, b \, db = n_i \, \text{v} \, 2\pi \, \frac{d\sigma_s}{d\Omega} \sin(\vartheta) \, d\vartheta \tag{2.3.4}$$

where the *differential scattering cross section* $d\sigma_s/d\Omega$ can be calculated from Eq. (2.3.4) and Rutherford's expression (Eq. 2.3.1),

$$\frac{d\sigma_s}{d\Omega} = \frac{b}{\sin(\vartheta)} \left| \frac{db}{d\vartheta} \right| = \frac{Z^2}{4} \left(\frac{e^2}{mv^2} \right)^2 \frac{1}{\sin^4(\vartheta/2)} . \tag{2.3.5}$$

The *total bremsstrahlung power* $P_i(\text{v}, \nu)d\nu$ radiated by a single electron in the frequency interval $\nu, ..., \nu + d\nu$ in collisions with n_i ions is

$$P_i(\text{v}, \nu)d\nu = n_i \text{v} Q_r(\text{v}, \nu)d\nu , \tag{2.3.6}$$

where the *radiation cross section* $Q_r(\text{v}, \nu)$ is given by

$$Q_r(\text{v}, \nu) = \frac{16}{3} \frac{Z^2 e^6}{m^2 c^3 \text{v}^2} \int_{b_{min}}^{b_{max}} \frac{db}{b} \quad (\text{cm}^2 \, \text{erg} \, \text{Hz}^{-1}) . \tag{2.3.7}$$

The integral in the radiation cross section is also called *Coulomb integral* $\ln \Lambda$, or *Gaunt factor* $g(\nu, T)$ if multiplied with the constant $\sqrt{3}/\pi$,

$$g(\nu, T) = \frac{\sqrt{3}}{\pi} \ln \Lambda = \frac{\sqrt{3}}{\pi} \int_{b_{min}}^{b_{max}} \frac{db}{b} = \frac{\sqrt{3}}{\pi} \ln \left(\frac{b_{max}}{b_{min}} \right) . \tag{2.3.8}$$

The *total bremsstrahlung power* radiated from a thermal plasma per unit volume, unit frequency, and unit solid angle, is called the *volume emissivity* ϵ_ν and is obtained by integrating the *total bremsstrahlung power* of a single electron (Eq. 2.3.6) over the thermal distribution $f(\text{v})$,

$$\epsilon_\nu = n_\nu \frac{n_e}{4\pi} \int P_i(\text{v}, \nu) f(\text{v}) d\text{v} . \tag{2.3.9}$$

We insert the Maxwell–Boltzmann distribution of a plasma with temperature T,

$$f(\mathrm{v})d\mathrm{v} = \left(\frac{2}{\pi}\right)^{1/2}\left(\frac{m}{k_BT}\right)^{3/2}\mathrm{v}^2\exp\left(-\frac{m\mathrm{v}^2}{2k_BT}\right)d\mathrm{v}\ . \tag{2.3.10}$$

Inserting the *total bremsstrahlung power* $P_i(\mathrm{v},\nu)$ (Eq. 2.3.6) of a single electron into Eq. (2.3.9) and integrating over the Maxwell–Boltzmann distribution $f(\mathrm{v})$ Eq. (2.3.10), in the classical bremsstrahlung treatment ($h\nu = \frac{1}{2}m\mathrm{v}^2$) yields then

$$\epsilon_\nu d\nu = \frac{8}{3}\left(\frac{2\pi}{3}\right)^{1/2}n_\nu\frac{Z^2e^6}{m^2c^3}\left(\frac{m}{k_BT}\right)^{1/2}n_in_e\ g(\nu,T)\ \exp\left(-\frac{h\nu}{k_BT}\right)d\nu$$

$$\approx 5.4\times10^{-39}Z^2n_\nu\frac{n_in_e}{T^{1/2}}g(\nu,T)\exp\left(-\frac{h\nu}{k_BT}\right)d\nu \quad \left(\mathrm{erg\ s^{-1}\ cm^{-3}\ Hz^{-1}\ rad^{-2}}\right)\ .$$

$$\tag{2.3.11}$$

The observable energy flux F at Earth can be derived from the emissivity $\epsilon_\nu = d\epsilon/(dt\ dV\ d\nu\ d\Omega)$ according to Eq. (2.3.11) by the following dimensional relation

$$dF = \frac{d\epsilon}{dt\ dA\ d(h\nu)} = \frac{d\epsilon}{dt\ (R^2d\Omega)\ h\ d\nu} = \frac{1}{R^2h}\frac{d\epsilon}{dt\ d\Omega\ d\nu} = \frac{dV}{R^2h}\epsilon_\nu\ , \tag{2.3.12}$$

where the factor $1/(R^2h) \approx 1.5$ in cgs units, based on the Earth–Sun distance of $R = 1.5\times10^{13}$ cm and $h = 6.63\times10^{-27}$ erg s. Thus the observed bremsstrahlung flux at Earth can be expressed as a volume integral over the source (with Eqs. 2.3.11 and 2.3.12),

$$Fd\nu = \int dV\frac{1}{R^2h}\epsilon_\nu d\nu$$

$$\approx 8.1\times10^{-39}Z^2n_\nu\int_V\frac{n_in_e}{T^{1/2}}g(\nu,T)\exp\left(-\frac{h\nu}{k_BT}\right)dV\ d\nu \quad \left(\mathrm{erg\ s^{-1}\ cm^{-2}\ erg^{-1}}\right)\ .$$

$$\tag{2.3.13}$$

A standard expression for the bremsstrahlung spectrum $F(\epsilon)$ is, as a function of the photon energy $\epsilon = h\nu$, setting the coronal electron density equal to the ion density ($n = n_i = n_e$), and neglecting factors of order unity, such as from the Gaunt factor, $g(\nu,T)$, and the ion charge number, $Z\approx1$, yielding (e.g. Brown 1974; Dulk & Dennis 1982),

$$F(\epsilon) \approx 8.1\times10^{-39}\int_V\frac{\exp\left(-\epsilon/k_BT\right)}{T^{1/2}}n^2\ dV \quad \left(\mathrm{keV\ s^{-1}\ cm^{-2}\ keV^{-1}}\right)\ .$$

$$\tag{2.3.14}$$

For an isothermal plasma, the spectral shape thus drops off exponentially with photon energy ϵ, which is the main way of discriminating between thermal emission from nonthermal emission, the latter generally exhibiting powerlaw spectra through a power-law energy distribution of energetic electrons at the impulsive stage of flares. We show an example of thermal bremsstrahlung emission in Fig. 2.6, recorded during the flare of 1980-Jun-27 with balloon-borne detectors (Lin et al. 1981a). Fitting Eq. (2.3.14) to

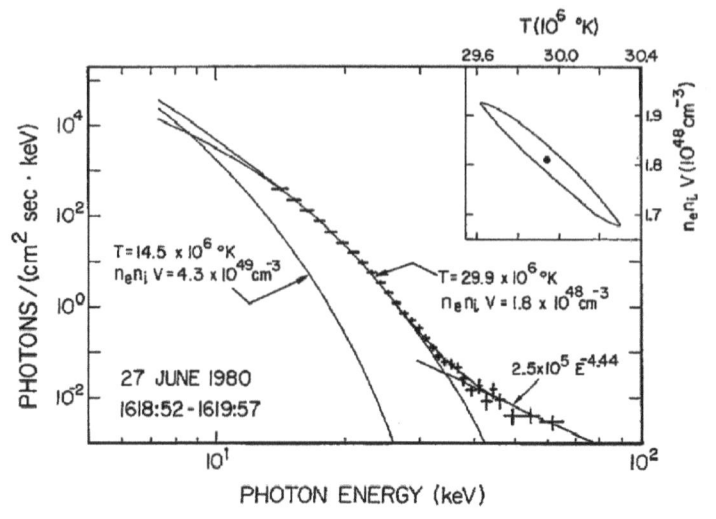

Figure 2.6: Example of a thermal spectrum, observed during the flare on 1980-June-27, 16:16 UT, by balloon-borne cooled germanium detectors. The thermal spectrum extends up to $\lesssim 35$ keV and corresponds to a temperature of $T = 29.9$ MK and a total emission measure of $EM = 1.8 \times 10^{48}$ cm^{-3} (Lin et al., 1981a).

the observed spectrum yields a temperature of $T \approx 30$ MK and an emission measure of $EM \approx n^2 V = 1.8 \times 10^{48}$ cm^{-3}. Since typical flare densities amount to $n \approx 10^{11}$ cm^{-3}, the observed emission measure suggests a flare volume of $V \approx 1.8 \times 10^{26}$ cm$^3 \approx (5600$ km$)^3$ for this so-called *superhot* flare plasma.

For radiation transfer calculations it is useful to define the absorption coefficient α_ν. This absorption coefficient per unit length for electron-ion bremsstrahlung is according to the Kirchhoff law (Eq. 2.2.1) and the Planck function $B_\nu(T)$ for a black body (Eq. 2.2.2),

$$\alpha_\nu = \frac{\epsilon_\nu}{n_\nu^2 B_\nu(T)} = \frac{\epsilon_\nu}{n_\nu^2} \frac{c^2}{2h\nu^3} \left[\exp\left(\frac{h\nu}{k_B T}\right) - 1 \right]. \tag{2.3.15}$$

For frequencies $\nu \ll k_B T/h$, which is $\nu \lesssim 10^{14}$ Hz in the corona ($T \gtrsim 1$ MK), the Rayleigh–Jeans approximation (Eq. 2.2.5) can be used, which leads to a simple form of the absorption coefficient after inserting the emissivity ϵ_ν from Eq. (2.3.11) into Eq. (2.3.15),

$$\alpha_\nu = \frac{1}{n_\nu} \frac{n_e n_i}{(2\pi\nu)^2} \left[\frac{32\pi^2 Z^2 e^6}{3(2\pi)^{1/2} m^3 c} \right] \left(\frac{m}{k_B T} \right)^{3/2} \ln \Lambda \approx \frac{9.786 \times 10^{-3}}{\nu^2} \frac{n_e n_i}{T^{3/2}} \ln \Lambda, \tag{2.3.16}$$

(in units of cm^{-1}), where the Coulomb logarithm $\ln \Lambda$ amounts to

$$\ln \Lambda = \ln \left[4.7 \times 10^{10} \left(\frac{T}{\nu} \right) \right] \tag{2.3.17}$$

for coronal temperatures $T > 3.16 \times 10^5$ [K] (i.e., $\ln \Lambda \approx 20$ for coronal conditions). The free-free opacity τ_{ff} of the quiet corona is about unity at a radio frequency of

$\nu = 1$ GHz, according to Eq. (2.3.16),

$$\tau_{ff}(T, n, \nu) \approx \int \alpha_\nu ds =$$

$$0.9786 \left(\frac{\nu}{\text{GHz}}\right)^{-2} \left(\frac{n}{10^9 \text{ cm}^{-3}}\right)^2 \left(\frac{T}{\text{MK}}\right)^{-3/2} \left(\frac{\ln \Lambda}{20}\right) \left(\frac{\lambda_T}{5 \times 10^9 \text{ cm}}\right) , \quad (2.3.18)$$

where $\lambda_T \approx 50$ Mm corresponds to the thermal scale height of the 1 MK corona. For lower radio frequencies, the corona becomes optically thick to free-free emission, while it is generally optically thin for higher frequencies and higher temperatures, such as during flares.

2.4 Atomic Energy Levels

While *free-free transitions*, as discussed in the foregoing section, represent close encounters of an electron with an ion that do not change the atomic structure, more intrusive processes are *free-bound transition, excitation*, and *ionization*, where the incoming free electron becomes bound to the target ion, excites an electron, or hits an electron out of the atom, respectively. In order to understand such processes quantitatively we have to resort to the quantum-mechanical theory of the atomic structure.

The simplest atom is hydrogen (H), consisting of a nucleus with a (positively charged) proton, neutron, and an orbiting electron. In the 1 MK hot solar corona, hydrogen is completely ionized, so that we have a plasma of free electrons (e^-) and protons (positive H ions, H^+), with He nuclei. If an electron becomes captured by a hydrogen ion, quantum mechanics rules allow only discrete energy levels (or orbits in the semi-classical treatment), numbered as $n = 1, 2, 3, 4, ...$ (Fig. 2.7). The wavelengths of these transitions were found by Balmer (1885) to follow discrete values, which are called the *Balmer series*, denoted as Hα, Hβ, Hγ, Hδ in Fig. 2.7,

$$\lambda = 3645.6 \frac{n^2}{n^2 - 4} \text{ Å}, \qquad n = 3, 4, 5, ... \qquad (2.4.1)$$

Johannes Rydberg later (1906) found more such series of discrete wavelengths and generalized Balmer's formula to any transition from orbit n_1 to n_2, with $n_2 > n_1$,

$$\frac{1}{\lambda} = \frac{\nu}{c} = R_H \left[\frac{1}{n_1^2} - \frac{1}{n_2^2}\right] , \qquad (2.4.2)$$

with $R_H = 1.0974 \times 10^5$ cm^{-1} the *Rydberg constant*. Transitions to $n_1 = 1$ are called the *Lyman series* (Lyα, Lyβ, ...), to $n_1 = 2$ the *Balmer series* (Hα, Hβ, ...), to $n_1 = 3$ the *Paschen series* (Pα, Pβ, ...), to $n_1 = 4$ the *Brackett series*, to $n_1 = 5$ the *Pfund series*, and so forth. A diagram of the transitions and associated wavelengths in the hydrogen atom is shown in Fig. 2.8. We see that the shortest wavelength of transitions in the hydrogen atom is at the continuum limit of the Lyman series, at $\lambda = 912$ Å. This wavelength is in the ultraviolet regime (Fig. 1.25), and all other hydrogen lines are at longer wavelengths in the optical and infrared regime.

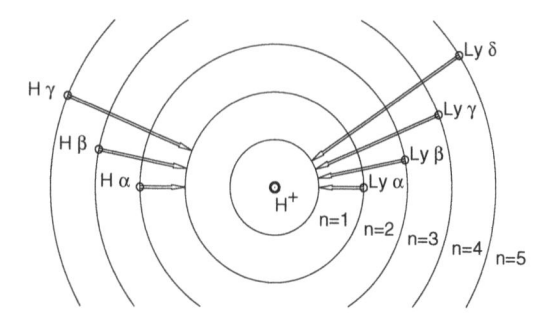

Figure 2.7: Bohr–Sommerfeld model of the hydrogen atom. The lowest quantum mechanical orbits $n = 1, 2, 3, 4, 5, ...$, are indicated, along with the transitions of the shortest wavelengths, the Lyman series (Lyα, Lyβ, ...) and the Balmer series (Hα, Hβ, ...), that are emitted when an electron makes the indicated transitions.

An electron transition from an inner orbit n_1 (with lower energy state) to an outer orbit n_2 (with a higher energy state) is called *excitation*, requiring energy input either by a collision with another particle or by absorption of a photon. The inverse process is *de-excitation*, when an electron jumps from a higher orbit n_2 to a lower one n_1 and emits a photon with the discrete wavelength λ given by the Rydberg formula (Eq. 2.4.2) or corresponding energy $\epsilon = h\nu = hc/\lambda$. If the electron is excited with an energy above 13.6 eV in the hydrogen atom, it can escape the hydrogen atom altogether, a process called *ionization*. The *photoelectric effect*, observed first by Hertz and explained by Einstein in 1900, is such an ionization process where an incoming photon hits an electron out of an atom.

A quantum-mechanical model of the atom was developed by Niels Bohr in 1914, called the *Bohr–Sommerfeld atomic model*. Bohr's semi-classical model assumes that: (1) electrons move in orbits around the nucleus in a Coulomb potential, $V = -Ze^2/r$; (2) the orbital angular momentum is quantized in integer multiples of the Planck constant, $L = mvr = n\hbar$, $n = 1, 2, 3, ...$; and (3) the electron energy remains constant except when it changes orbits by emission or absorption of a photon with quantized energies. These quantum-mechanical rules lead directly to a physical derivation of the Rydberg constant (Eq. 2.4.2), so that the Rydberg formula can be expressed in terms of physical constants,

$$\frac{1}{\lambda} = \frac{\nu}{c} = \frac{Z^2 m e^4}{4\pi\hbar^3 c}\left[\frac{1}{n_f^2} - \frac{1}{n_i^2}\right] , \qquad (2.4.3)$$

where the indices i and f indicate the initial and final state, respectively.

Going from the simple hydrogen atom to atoms with higher masses and multiple electrons, the quantum mechanical treatment becomes increasingly more complex, similar to the mechanical orbits in a multi-body system. However, the Rydberg formula (Eq. 2.4.3) still gives approximately the correct transition energies for special cases, such as for singly ionized atoms, if the correct charge number Z is used. If we take singly ionized helium, He II (with $Z = +2$), the first transition from $n_i = 2$ to $n_f = 1$

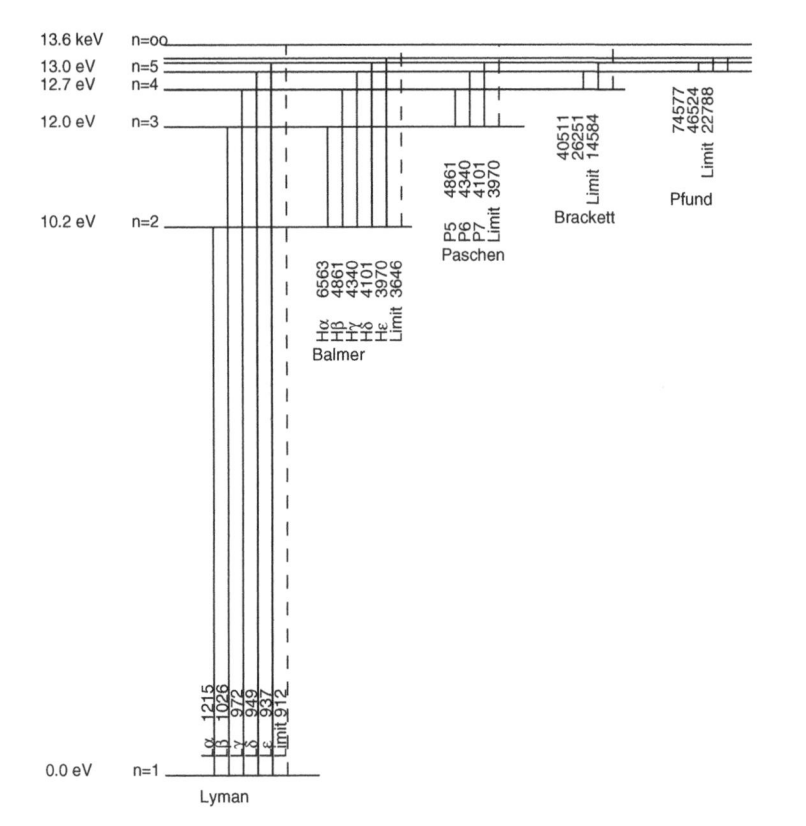

Figure 2.8: The principal hydrogen series are shown with the y-axis proportional to the photon energy of the emission. The wavelengths of the transitions are given in Ångstrøm units.

should have a factor of $Z^2 = 4$ higher energy difference according to Eq. (2.4.3), or a factor of $1/4$ smaller wavelength, than in hydrogen. Thus, the Lyman-α transition at $\lambda = 1215$ Å in hydrogen is expected to happen at $\lambda/4 = 304$ Å in singly ionized helium, which is indeed one of the strongest lines in the solar XUV spectrum.

The basic energy levels in hydrogen-like atoms are given by the *principal quantum number* $n = 1, 2, 3, \ldots$. The next quantum number is the *orbital angular momentum* $l = 0, \ldots, n - 1$, quantized in units of $\hbar$. Additional quantum numbers are the *projected angular momentum* m_l, also quantized in units of $\hbar$, and the *electron spin number* $m_s = -\hbar/2, +\hbar/2$. Thus, an energy state of an electron in an atom is specified by the 4 quantum numbers nlm_lm_s. In spectroscopic nomenclature, the principal quantum number is given in arabic numbers $n = 1, 2, 3, \ldots$, the orbital angular momentum l by the letters s, p, d, f, and the number of electrons occupying this state are indicated by a superscript. A few examples of this spectroscopic nomenclature of atomic configurations are listed in Table 2.1.

The ionization state of an atom (i.e., the number of electrons that have been removed), is conventionally denoted by an arabic superscript m with a plus sign (X^{+m}), indicating the positive charge of the ion (e.g., Fe^{+9} for a ninefold-ionized iron ion).

Table 2.1: Spectroscopic nomenclature for ground configurations of elements between Z=1 and Z=26 (adapted from Zirin 1988 and Golub & Pasachoff 1997).

Element	Symbol	Z	Ground configuration
Hydrogen	H	1	$1s$
Helium	He	2	$1s^2$
Lithium	Li	3	$1s^2 2s$
Beryllium	Be	4	$1s^2 2s^2$
Boron	B	5	$1s^2 2s^2 2p$
Carbon	C	6	$1s^2 2s^2 2p^2$
Nitrogen	N	7	$1s^2 2s^2 2p^3$
Oxygen	O	8	$1s^2 2s^2 2p^4$
Fluorine	F	9	$1s^2 2s^2 2p^5$
Neon	Ne	10	$1s^2 2s^2 2p^6$
Sodium	Na	11	$1s^2 2s^2 2p^6 3s$
Potassium	K	19	$1s^2 2s^2 2p^6 3s^2 3p^6 4s$
Calcium	Ca	20	$1s^2 2s^2 2p^6 3s^2 3p^6 4s^2$
Vanadium	V	23	$1s^2 2s^2 2p^6 3s^2 3p^6 4s^2 3d^3$
Chromium	Cr	24	$1s^2 2s^2 2p^6 3s^2 3p^6 4s 3d^5$
Iron	Fe	26	$1s^2 2s^2 2p^6 3s^2 3p^6 4s^2 4p^6$

The spectra from such ions are also labeled with roman numerals, but starting with the neutral ion (e.g., Fe I corresponds to the neutral iron atom, Fe II is a singly ionized iron ion, or Fe X is the ninefold-ionized iron ion Fe^{+9}).

2.5   Atomic Transition Probabilities

The solar spectrum in EUV and soft X-rays is dominated by a large number of emission lines (Fig. 1.25). The relative line strengths can only be understood if we know the relative transition probabilities between the quantized atomic energy levels in each chemical element. Obviously the emission of photons, and thus a particular line strength, must strongly depend on the temperature of the plasma. So we turn back to the blackbody concept introduced in §2.2.

Let us consider a plasma with a temperature T consisting of a large number of identical atoms of some chemical element. The electrons are found in the available quantized energy levels, while the emitted photons form an electromagnetic radiation field that continuously disturbs the electrons. If the atom absorbs a photon, resulting in an orbiting electron moving from a lower to a higher energy orbit, we call it *induced absorption*, while the reverse process, when an electron moves from a higher orbit to a lower one as a result of external radiation and emits a photon with energy $h\nu = \epsilon_2 - \epsilon_1$ is called *stimulated emission*. We denote the transition rates from level n to m with R_{nm}, and R_{mn} vice versa. Whatever the initial conditions and temperatures are, we expect that the system settles into an equilibrium with a stable temperature T after some time. According to Kirchhoff's law (Eq. 2.2.1) we know already that the electromagnetic radiation will settle into a black-body spectrum given by the Planck function (Eq. 2.2.2), when it reaches an equilibrium with temperature T. To calculate the transition probabilities P_{mn} in this equilibrium situation, Einstein made in 1917 the following additional assumptions: (1) the rate for *induced absorption* of photons is proportional to the energy density U_ν of the radiation field, (2) the distribution of atomic energy states is assumed to be in equilibrium, and (3) the statistical distribution of the energy level population is given by the Boltzmann probability distribution (Eq. 2.3.10).

For transitions from a high energy state ϵ_m to a lower state ϵ_n, Einstein introduced a coefficient A_{mn} for *spontaneous emission* (not affected by the external radiation field) and a coefficient B_{mn} for *stimulated emission* (proportional to the energy density U_ν of the radiation field), so that the probability per unit time is

$$P_{mn} = P_{spon} + P_{stim} = A_{mn} + B_{mn}U_\nu \ . \tag{2.5.1}$$

The opposite process, the excitation of electrons from lower energy states ϵ_n to higher ones ϵ_m by *induced absorption*, has a probability of

$$P_{nm} = P_{abs} = B_{nm}U_\nu \ . \tag{2.5.2}$$

The energy density U_ν of black-body radiation follows from the Planck function (Eq. 2.2.2),

$$U_\nu = \frac{4\pi}{c}B_\nu(T) = \frac{8\pi h\nu^3}{c^3}\frac{1}{[\exp(h\nu/k_BT)-1]} \ . \tag{2.5.3}$$

The Boltzmann equation for the population density of excited states at temperature T is (Boltzmann 1884)

$$N_m = \frac{g_m}{g_0}N_0\exp\left(-\frac{\chi_m}{k_BT}\right) , \tag{2.5.4}$$

where χ_m is the excitation energy (of energy level ϵ_m), N_0 is the number of molecules in ground state, and g_m is the statistical weight of the state m. Thus the relative population of the energy states ϵ_n and ϵ_m is

$$\frac{N_n}{g_n} = \frac{N_m}{g_m}\exp\left(-\frac{\epsilon_n - \epsilon_m}{k_BT}\right) , \tag{2.5.5}$$

Combining Einstein's rate equations for emission (Eq. 2.5.1) and absorption (Eq. 2.5.2) with the Boltzmann equation (2.5.5) for the population of energy levels ϵ_n and ϵ_m we find

$$\frac{g_n}{g_m}\exp\left[-\frac{\epsilon_n}{k_BT}\right]B_{nm}U_\nu = \exp\left[-\frac{\epsilon_m}{k_BT}\right](A_{mn} + B_{mn}U_\nu) \ . \tag{2.5.6}$$

Choosing a photon energy of $h\nu = \epsilon_m - \epsilon_n$ in the Planck function (Eq. 2.5.3) we obtain for the Einstein coefficients A_{mn} and B_{mn} the following relation, which provides the ratio of transition rates between *induced absorption* and *spontaneous emission*,

$$A_{mn} = \frac{8\pi h\nu^3}{c^3}B_{mn} , \tag{2.5.7}$$

Table 2.2: Excitation, ionization, and recombination processes in atoms. Note: ν=photon, e^-=electron, Z=atom, Z^+=ion, Z'=excited atom (adapted from Zirin 1988).

Process	Particle and Photon notation	Transition rate
Induced absorption	$\nu + Z \mapsto Z'$	$U_\nu B_{mn} N_m$
Stimulated emission	$\nu + Z' \mapsto 2\nu + Z$	$U_\nu B_{nm} N_n$
Spontaneous emission	$Z' \mapsto Z + \nu$	$N_{nm} A_{nm}$
Photo-ionization	$\nu + Z \mapsto Z^+ + e^-$	$U_\nu N_m B_{m\kappa}$
2-body (radiative) recombination	$e^- + Z^+ \mapsto \nu + Z$	$N_e N_i A_{\kappa m}$
Dielectronic recombination	$e^- + Z^+ \mapsto ... \mapsto \nu + Z', \nu + Z$	$N_e N_i \alpha_{diel}$
Auto-ionization	$Z'' \mapsto Z^+ + e^-$	$N_\nu U_\nu \kappa_{diel}$
Thomson scattering	$\nu + e^- \mapsto \nu + e^-$	$\sigma_T N_e$
Free-free emission (bremsstrahlung)	$e^- + Z^+ \mapsto e^- + Z^+ + \nu$	$N_e N_i \kappa \kappa'$
Free-free absorption	$\nu + e^- + Z^+ \mapsto \nu + e^- + Z^+$	$N_e N_i B_{\kappa'\kappa} U_\nu$
Collisional excitation	$e^- + Z \mapsto e^- + Z'$	$N_m N_e C_{mn}$
Collisional de-excitation	$e^- + Z' \mapsto e^- + Z$	$N_n N_e C_{nm}$
Collisional ionization	$e^- + Z \mapsto 2e^- + Z^+$	$N_m N_e C_{m\kappa}$
3-body recombination	$2e^- + Z^+ \mapsto e^- + Z$	$N_e^2 N_i C_{\kappa, m}$

and

$$g_m B_{mn} = g_n B_{nm} , \qquad (2.5.8)$$

for the statistical weights g_m and g_n of the energy levels ϵ_m and ϵ_n. We see that the temperature dependence dropped out in Eqs. (2.5.7) and (2.5.8), so the relation between the Einstein coefficients are of atomic nature and independent of the plasma temperature.

The Einstein probability coefficients have been calculated for the dipole and quadrupole moments of the atomic electromagnetic field. The highest transition probability for spontaneous emission results from the electric dipole moment ($A_{mn} \approx 10^9$ s^{-1}). The transition probability for spontaneous emission from a magnetic dipole is much lower ($A_{mn} \approx 10^4$ s^{-1}), and for the electric quadrupole moment is many orders of magnitude lower ($A_{mn} \approx 10$ s^{-1}). Although electric dipole transitions provide the strongest spectral lines, non-electric dipole transitions are important for providing electron density sensitive diagnostics.

2.6   Ionization and Recombination Processes

A selection of the most important atomic processes that contribute to the continuum and line emission of the solar corona in soft X-rays and EUV is enumerated in Table 2.2 and pictorially represented in Fig. 2.9, including a number of interactions between photons, electrons, atoms, and ions, such as absorption, emission, excitation, de-excitation, ionization, and recombination processes. All these processes can occur when a (incoming) photon or electron interacts with an atom or ion. We describe some of these processes briefly in the context of coronal conditions, in the same order as listed in Table 2.2 and shown in Fig. 2.9.

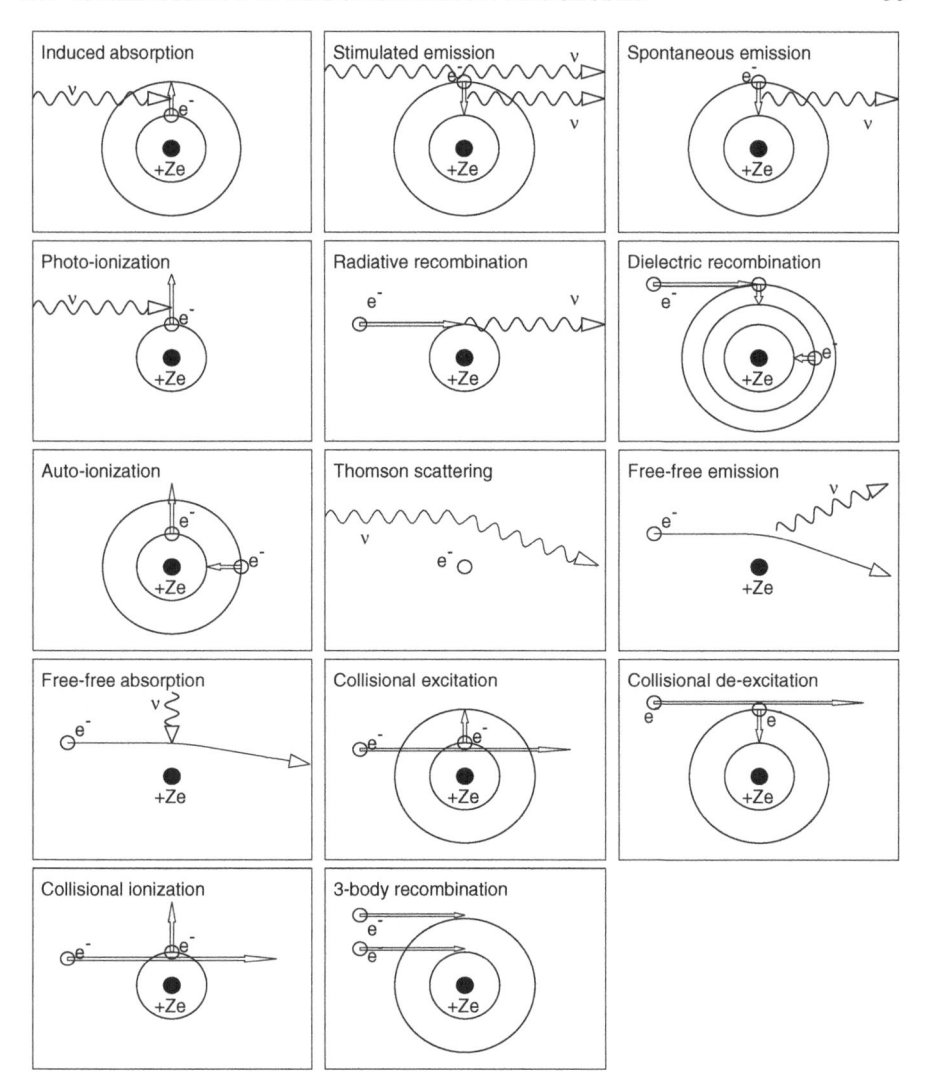

Figure 2.9: Diagrams of absorption, emission, ionization, recombination, excitation, and de-excitation processes, sorted in same order as in Table 2.2. Atoms and ions are marked with filled dots, electrons with open dots, electron orbits with circles, electron transitions with arrows, and photons with a wiggly arrow. Time is proceeding from left to right.

Induced absorption : An incoming photon can excite an electron in an atom to a higher energy state $\epsilon_n = \epsilon_m + h\nu$. This process occurs with a probability that is proportional to the occupation number N_m in state m and the energy density U_ν of the radiation field, where the transition probabilities are specified by the Einstein coefficients B_{mn} (Eq. 2.5.2), giving a transition rate $R = U_\nu B_{mn} N_m$. This process is of the type of *discrete bound-bound transitions*, which produce numerous absorption lines in the solar corona (e.g., the Fraunhofer lines at optical wavelength and UV wavelengths).

Stimulated emission : An electron in an excited atom at energy state ϵ_n is stimulated, by a passing photon ν from the ambient radiation field, to fall back into a lower state ϵ_m, emitting during this process a second photon with energy $h\nu = \epsilon_n - \epsilon_m$. The probability of this process is also proportional to the energy density U_ν of the radiation field (Eq. 2.5.1), as for induced absorption (i.e., with a rate of $R = U_\nu B_{nm} N_n$). This process is relevant to laser emission.

Spontaneous emission: In contrast to stimulated emission, no incoming photon is needed. An electron spontaneously falls from a higher energy state ϵ_n to a lower state ϵ_m by emission of a photon with $h\nu = \epsilon_n - \epsilon_m$. The probability is given by the Einstein coefficients A_{nm}, so the rate is $R = N_{nm} A_{nm}$.

Photo-ionization: This a *bound-free transition*, where the incoming photon has a higher energy than the ionization energy (i.e., > 13.6 eV for hydrogen), so that the bound electron becomes free from the atom. The wavelength of the incoming photon is $\lambda = hc/\Delta\epsilon$, with $\Delta\epsilon = \epsilon_i + \frac{1}{2} m_e v_e^2$, where ϵ_i is the ionization energy of the atom from the bound state ϵ_i of the electron, and $\frac{1}{2} m_e v_e^2$ is the kinetic energy of the free electron after escape.

Radiative (or 2-body) recombination: Recombination (also called *free-bound transition*) can occur by several processes, such as *radiative recombination, collisional recombination*, or *dielectronic recombination*. In *radiative recombination* a free electron is captured by an ion into one of the available energy states ϵ_i, while the excess energy is removed by emission of a photon with energy $h\nu = \frac{1}{2} m_e v^2 - \epsilon_i$, just as the time-reverse process of photo-ionization. This type of *bound-free transition* produces series limit continua such as the Balmer (3646 Å) and Lyman continua (912 Å) of hydrogen (Fig. 2.8). These wavelengths correspond to the ionization energies of $\epsilon_1 = 13.6$ eV from the ground state level $n = 1$ and $\epsilon_2 = 3.6$ eV from the first excited state $n = 2$ (Fig. 2.7).

Dielectronic recombination: The term *dielectronic recombination* indicates (since the Greek word "di" means "two"), that two electrons are involved in this process. A free electron e_1^- is captured by an ion, resulting in a double excitation of the ion: (1) the original free electron lands in an excited state, and (2) a bound electron of the ion also becomes excited. Dielectroni recombination is accomplished when the highly unstable doubly excited configuration subsequently stabilizes, with one or both excited electrons falling back into vacancies of the lowest available states.

Auto-ionization: An ion is initially in a doubly excited state Z'' and auto-ionizes (i.e., it spontaneously ionizes without induced particle or photon) to $Z^+ + e^-$, thus leaving an ion and a free electron. If an electron from a lower energy state is knocked off, an electron from a higher energy level has to fall back to the emptied lower state to stabilize the ion. This process is also called *auto-ionization*.

Thomson scattering: Photons are scattered off electrons of the coronal plasma, which produces the white-light corona as visible during total solar eclipses. This is a continuum scattering process with no wavelength dependence, and thus the name *K-corona* (from the german word "kontinuierlich") has been used for this part of the corona during eclipses. Without Thomson scattering, none of the solar photons would reach the Earth during blocking by the moon. The scattering rate is proportional to the electron density, which provided the first method of determining the electron density in the solar corona.

Free-free emission: This process, also called *bremsstrahlung* is discussed in §2.3. Electrons are non-elastically scattered off ions and emit photons that have energies corresponding to the energy difference of the incoming and outcoming electron (i.e., $h\nu = \epsilon' - \epsilon$). Free-free emission from the chromosphere and corona is responsible for much of the emission in microwaves and soft X-rays, and up to $\lesssim 20$ keV in hard X-rays during flares.

Collisional ionization: This occurs by collisions of ions with free electrons, when an orbiting electron of the ion (generally the outermost) is removed, and the ion is left in the next higher ionization state. This process is much more important than photo-ionization (when a photon incident on an ion results in removal of an orbiting electron).

"Forbidden line" transitions: Some lines that occur in the the solar corona are called *"forbidden lines"*, because the electron transitions involved are highly improbable by certain quantum-mechanical rules. For instance, the *green line* of Fe XIV at 5303 Å involves two transitions from 4d levels to two closely spaced 3p levels, for which the splitting corresponds to the 5303 Å line. In laboratory plasmas, this line would never be observed, because collisional de-excitation would occur within a time of $\approx 10^{-8}$ s. In the solar corona however, the temperature is so high and the density so low, that collisional de-excitation can take a day, giving rise to continuous *green line emission*. There are many other forbidden lines (mostly in Ca, Fe, and Ni), the most prominent ones being Ca XV 5694 Å (*yellow line*) and Fe X 6375 Å (*red line*), both visible in the optical continuum.

In the next section we will see that the relative transition rates of these processes can be constrained by a detailed equilibrium between the *direct* (e.g. ionization) and *reverse* (e.g., recombination) processes, at least in a closed system in equilibrium.

2.7 Ionization Equilibrium and Saha Equation

A standard method used to calculate the relative line strengths in the coronal plasma involves the assumption of *ionization equilibrium*. This assumption is made in the Boltzmann equilibrium relation (Eq. 2.5.5) for the relative population of two energy states ϵ_n and ϵ_m. The same assumption made for *bound-bound transitions* can also be extended to *bound-free transitions* to include ionization processes. While the energy difference is $h\nu = \epsilon_n - \epsilon_m$ for *bound-bound transitions*, we have an energy difference that consists of the ionization energy ϵ_i plus the kinetic energy of the free (ionized) electron for *bound-free transitions* (i.e., $h\nu = \epsilon_k + \frac{1}{2}m_e v^2$). Thus the Boltzmann equilibrium equation (2.5.5) for the population ratio N_k/N_0 of the continuum population N_k to the ground state population N_0 is

$$\frac{N_k}{N_0} = \frac{g_k}{g_0} \exp\left(-\frac{\epsilon_k + \frac{1}{2}m_e v^2}{k_B T}\right), \qquad (2.7.1)$$

where the statistical weight $g_k = g'_k \cdot g_e$ is now composed of the statistical weight of the ground state g'_k of the ionized atom and g_e of the free (ionized) electron. The statistical weight of the free electron is proportional to the number of quantum-mechanical phase space elements and inversely proportional to the electron density n_e,

$$g_e = \frac{8\pi m_e^3 v_e^2}{n_e h^3} \, . \tag{2.7.2}$$

Substituting the statistical weight g_e (Eq. 2.7.2) into the Boltzmann equilibrium equation (Eq. 2.7.1) and integrating over velocity space, which involves the integral,

$$\int_0^\infty \exp\left(-\frac{\frac{1}{2} m_e v^2}{k_B T}\right) 4\pi v^2 dv = \left(\frac{2\pi k_B T}{m_e}\right)^{3/2} , \tag{2.7.3}$$

leads to the *Saha equation*, first derived by M. Saha in 1920,

$$\frac{N_k}{N_0} = \frac{2}{n_e} \frac{(2\pi m_e k_B T)^{3/2}}{h^3} \frac{g_k'}{g_0} \exp\left(-\frac{\epsilon_k}{k_B T}\right) , \tag{2.7.4}$$

The *Saha equation*, which yields the population ratios relative to the ground state, N_0, can be generalized for the relative ratio between any two different ionization states k and $k + 1$,

$$\frac{N_{k+1}}{N_k} = \frac{2}{n_e} \frac{(2\pi m_e k_B T)^{3/2}}{h^3} \frac{g_{k+1}}{g_k} \exp\left(-\frac{\epsilon_{k+1} - \epsilon_k}{k_B T}\right) . \tag{2.7.5}$$

We see that the electron density n_e appears in the *Saha equation* (2.7.5), which tells us that soft X-ray line ratios of the same element are density sensitive, although the density does not occur in the Boltzmann equation (which depends only on the temperature).

Although the derivation of the Boltzmann and Saha equations rests on the assumption of ionization equilibrium, and thus strictly applies only to processes with photons (induced absorption, stimulated emission, and photo-ionization), a similar argument can be made for the equilibrium situation of collisional processes (collisional excitation, de-excitation, and ionization). This is primarily important in the chromosphere and transition region, where the collision rates are much higher than interactions between photons and atoms or ions. The ionization equilibrium in the corona is an equilibrium between collisional ionizations (including auto-ionizations) and radiative and dielectronic recombinations. At coronal densities, the radiative decay rates (spontaneous emission with Einstein coefficient A_{nm}), is so high compared with collisional excitation rates, that the population of higher excited states can be neglected relative to the ground state (though not always true, e.g., for Fe IX). The following approximation for the rate equation constrains the *collisional excitation coefficient* C_{mn},

$$A_{nm} N_n = C_{mn} N_e N_m \, . \tag{2.7.6}$$

The coefficient C_{mn} for collisions between electrons and ions can be derived by integrating the cross section $\sigma(v)$ over the Maxwell–Boltzmann distribution (Eq. 2.3.10), yielding the *effective collision strengths*:

$$C_{mn} = 8.63 \times 10^{-6} T_e^{-1/2} n_e \left(\frac{\Omega_{mn}}{g_m}\right) \quad (s^{-1}) , \tag{2.7.7}$$

where Ω_{mn} are the so-called *collision strengths* (see e.g., Lang 1980, p.101).

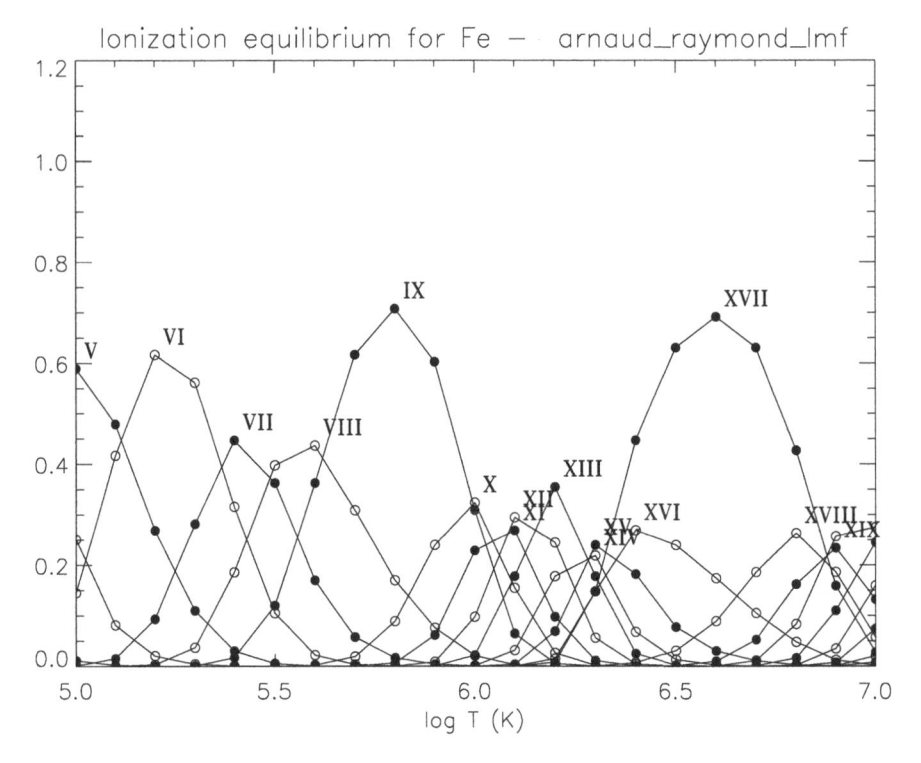

Figure 2.10: Ionization equilibrium calculations for Fe V to Fe XIX in the temperature range of $T = 10^5 - 10^7$ K, for the ionization ratios from Arnaud & Raymond (1992), using the CHIANTI code.

2.8 Emission Line Spectroscopy

To calculate optically thin synthetic spectra and to perform spectral analysis and plasma diagnostics of solar soft X-ray and EUV data, we need data from thousands of atomic energy levels. The most comprehensive and accurate atomic database used for the analysis of astrophysical and solar spectra is *CHIANTI* (Dere et al. 1997a, 2001; Young et al. 1998; Landi et al. 1999; see also URL http://wwwsolar.nrl.navy.mil/chianti.html), a database that contains atomic energy levels, wavelengths, radiative data, and electron excitation data for ions, covering the soft X-ray and EUV wavelength range of $\lambda \approx 1 - 1700$ Å. This database is widely used for the analysis of solar data, e.g. from *SoHO/CDS, SoHO/EIT, TRACE*, and *RHESSI*. In the following we define the physical quantities that are used in such spectroscopic data analysis.

The intensity $I(\lambda_{ij})$ of an optically thin spectral line of wavelength λ_{ij} (or frequency ν_{ij}), produced by photons via spontaneous transitions from electrons jumping from a higher energy level ϵ_j to a lower level ϵ_i is defined as,

$$I(\lambda_{ij}) = \frac{h\nu_{ij}}{4\pi} \int N_j(X^{+m})A_{ji}dz \qquad (\text{erg cm}^{-2} \text{ s}^{-1} \text{ ster}^{-1}), \qquad (2.8.1)$$

where A_{ji} is the Einstein coefficient of the spontaneous transition probability (i.e.,

$N_j(X^{+m})$ the number density of the upper level j of the emitting ion), and z is the line-of-sight through the emitting plasma. In the coronal plasma it is assumed that the collisional excitation processes are faster than the ionization and recombination time scales, so that collisional excitation determines the population of excited states. It allows treatment of low-level populations separately from ionization and recombination processes, the so-called *coronal approximation*. The population of level j is generally expressed as a chain of ratios,

$$N_j(X^{+m}) = \frac{N_j(X^{+m})}{N(X^{+m})} \frac{N(X^{+m})}{N(X)} \frac{N(X)}{N(H)} \frac{n(H)}{n_e} n_e , \qquad (2.8.2)$$

where $N(X^{+m})/N(X)$ is the ionization ratio of the ion X^{+m} relative to the total number density of element X (e.g. Arnaud & Raymond 1992; Mazzotta et al., 1998; see example for Fe in Fig. 2.10), $A_X = N(X)/N(H)$ is the elemental abundance relative to hydrogen (see Table 1.2), and $n(H)/n_e \approx 0.83$ is the hydrogen ratio to the free electron density (based on an abundance of H:He=10:1 with complete ionization). The level populations $N_j(X^{+m})/N(X^{+m})$ can be calculated with the *CHIANTI* code by solving the statistical equilibrium for a number of low levels of the ion including all important collisional and radiative excitation and de-excitation mechanisms. For allowed transitions the approximation $N_j(X^{+m})A_{ji} \propto n_e$ holds.

It is customary to combine all atomic physics parameters in a so-called *contribution function* $C(T, \lambda_{ij}, n_e)$, which is strongly peaked in temperature and is defined as (e.g. as used in the CHIANTI code),

$$C(T, \lambda_{ij}, n_e) = \frac{h\nu_{ij}}{4\pi} \frac{A_{ji}}{n_e} \frac{N_j(X^{+m})}{N(X^{+m})} \frac{N(X^{+m})}{N(X)} \qquad (\text{erg cm}^{-2} \text{ s}^{-1} \text{ ster}^{-1}) .$$
$$(2.8.3)$$

If we insert the population number $N_j(X^{+m})$ (Eq. 2.8.2) into the line intensity (Eq. 2.8.1) and make use of the contribution function $C(T, \lambda_{ij}, n_e)$ as defined in (Eq. 2.8.3), we have for the line intensity,

$$I(\lambda_{ij}) = A_X \int C(T, \lambda_{ij}, n_e) \, n_e n_H \, dz , \qquad (2.8.4)$$

where the abundance factor $A_X = N(X)/N(H)$ is not included in the contribution function in this particular definition. This means that the assumption has to be made that the abundances are constant along the line-of-sight, if Eq. (2.8.4) is used. There are alternative definitions of the *contribution function* in literature that include the abundance factor, which we refer to by

$$G(T, \lambda_{ij}, A_X, n_e) = A_X \, C(T, \lambda_{ij}, n_e) . \qquad (2.8.5)$$

There are further differences in the definition of the *contribution function* in the literature, sometimes the factor $1/4\pi$ is not included, or a fixed value of $N(H)/n_e = 0.83$ is assumed. An example of a synthetic spectrum with line intensities $I(\lambda_{ij})$ (Eq. 2.8.4) in the wavelength range that contains the strong Fe IX 171 Å and Fe XII 195 Å lines is shown in Fig. 2.11.

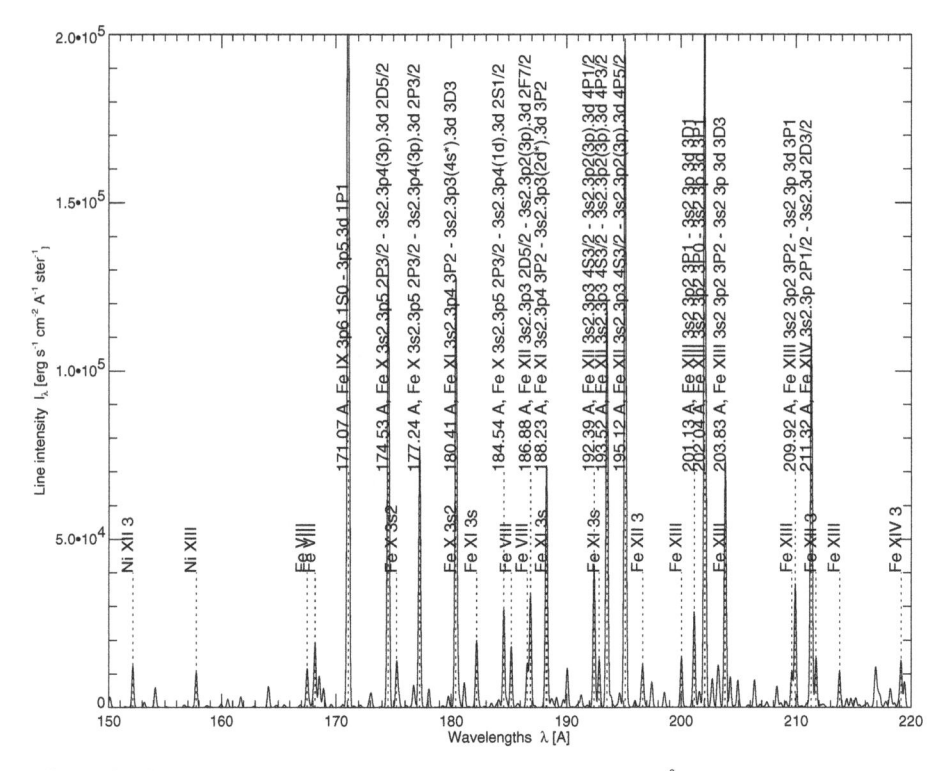

Figure 2.11: A synthetic spectrum calculated in the $\lambda = 150-220$ Å wavelength range with the CHIANTI code, using the Feldman (1992) abundances, the Arnaud & Raymond (1992) values for the ionization equilibrium, a DEM for a typical active region, and $n_eT = 10^{15}$ cm^{-3} K, plotted with a wavelength resolution of 0.2 Å. The strongest lines are identified, mostly Fe lines, and transitions calculated by CHIANTI are indicated. Note the prominent Fe IX at 171 Å and Fe XII at 195 Å that are frequently used in *SoHO/EIT* and *TRACE* images.

Another convenient quantity is the *differential emission measure* function $dEM(T)/dT$ (cm^{-5} K^{-1}), which is a measure of the amount of plasma along the line-of-sight that contributes to the emitted radiation in the temperature range $T, ..., T + dT$ (Craig & Brown 1976),

$$\frac{dEM(T)}{dT} = n_e n_H \frac{dz}{dT} , \qquad (2.8.6)$$

which allows us to express the line intensity (Eq. 2.8.4) in the simple form of an integral over the temperature range,

$$I(\lambda_{ij}) = A_X \int_T C(T, \lambda_{ij}, n_e) \frac{dEM(T)}{dT} dT \qquad \left(\text{erg cm}^{-2}\text{ s}^{-1}\text{ ster}^{-1}\right) . \qquad (2.8.7)$$

We note that quantitative analysis of spectral line fluxes requires three different inputs: (1) the density profile $n(z) = n_e(z) \approx n_H(z)$ of the source along the line-of-sight z in some range of temperatures (T), in order to define the differential emission measure $dEM(T)/dT$; (2) the atomic abundance A_X in the source, for which often standard

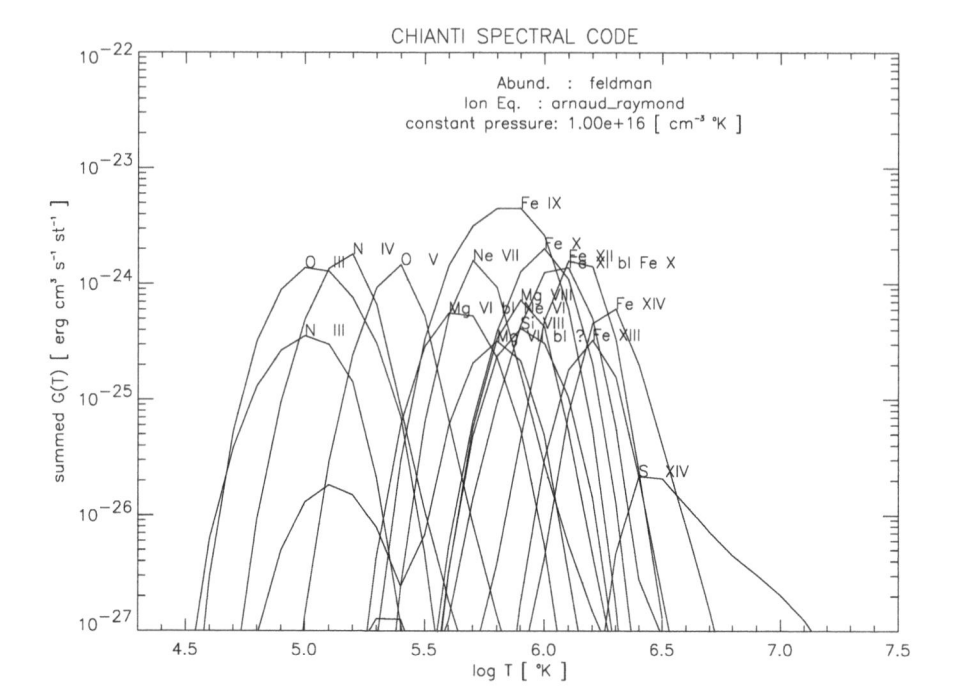

Figure 2.12: A set of contribution functions $G(T)$ from different lines is shown which contribute to an observed spectrum. The contribution functions were calculated with the CHIANTI code, using elemental abundances from Feldman (1992), ionization equilibrium values from Arnaud & Raymond (1992) and Mazzotta et al. (1998), and a constant pressure of $p = n_e T = 10^{16}$ [cm^{-3} K].

values of photospheric abundances are used (Table 1.2); and (3) atomic parameters, which are all included in the contribution function $C(T, \lambda_{ij}, n_e)$ (e.g., as provided by CHIANTI), but slightly depend on the source density n_e (because of the collisional excitation rates).

An example of a spectral data analysis is illustrated in Figs. 2.12 and 2.13. Observations (e.g., from *SoHO/CDS*), provide line intensities $I(\lambda_{ij})$ in a set of spectral lines. Using such a list of line intensities $I(\lambda_{ij})$ and wavelengths λ_{ij}, and choosing a predefined set of elemental abundances A_X (e.g., Meyer 1985; Feldman 1992) and ionization equilibrium data (e.g., Arnaud & Raymond 1992), the CHIANTI spectral code is able to calculate the contribution functions for each transition or line (Fig. 2.12) and to determine the best-fitting differential emission measure distribution $dEM(T)/dT$ (Fig. 2.13) that is consistent with the data. The solution of the $dEM(T)/dT = \int n_e n_I dz \approx n^2 z$ provides a measure of the squared average density n^2 multiplied with the depth z of the source along the line-of-sight. If observations allow for a geometric model (e.g., the column depth z of a flare loop can be equated to the transverse width w for a cylindrical loop geometry, $z \approx w$), then the average density $n_e(T) = \sqrt{[dEM(T)/dt]/w}$ can be estimated for a range of temperatures T.

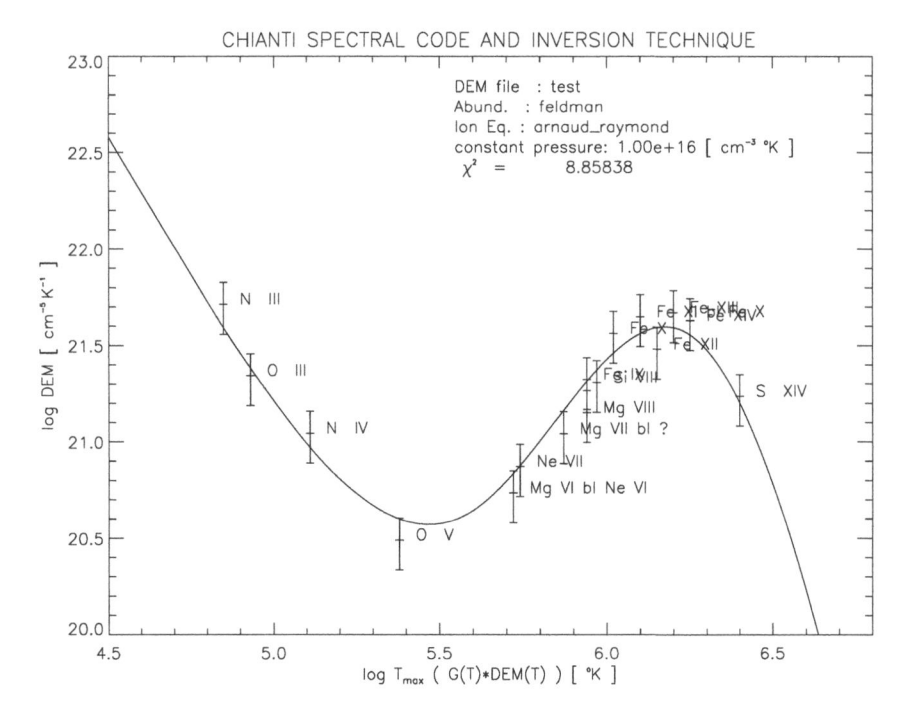

Figure 2.13: A differential emission measure distribution $DEM(T)$ (smooth curve) is reconstructed with the CHIANTI code based on the observed spectral line intensities, elemental abundances from Feldman (1992), ionization equilibrium values from Arnaud & Raymond (1992), and a constant pressure of $p = n_e T = 10^{16}$ (cm^{-3} K). The data points of the spectral line intensities with error bars are overplotted in DEM units. The polynomial fit (smooth curve) is only accurate whithin the range of measured spectral lines.

2.9 Radiative Loss Function

While the line intensities $I(\lambda_{ij})$ are important for the analysis of spectra, it is more practical for the analysis of soft X-ray and EUV images to sum all line intensities in a temperature interval $T, ..., T+dT$ to obtain an emission measure as a function of temperature (i.e., $EM(T)$). Because the excitation of atomic levels is a strong function of temperature, the spectral emissivity $P(\lambda, T)$ has to be calculated for each temperature T separately (e.g., as shown for the synthetic spectrum in Fig. 2.11). It is customary to define a *radiative loss rate* $E_R(T, n_e)$ that sums up all the line contributions radiating at temperature T with density n_e, as having the physical units of (erg cm^3 s^{-1}). This *radiative loss rate* E_R can be written as a product of densities and a temperature-dependent function, called the *radiative loss function* $\Lambda(T)$ having the physical units of (erg cm^{-3} s^{-1}),

$$E_R = n_e n_p \alpha_{FIP} \Lambda(T) \approx n_e^2 \Lambda(T) , \qquad (\text{erg cm}^{-3} \text{ s}^{-1}) \qquad (2.9.1)$$

for optically thin plasmas, where α_{FIP} is a correction factor for abundance enhancements due to the FIP effect (see §2.10). This *radiative loss function* $\Lambda(T)$ has been

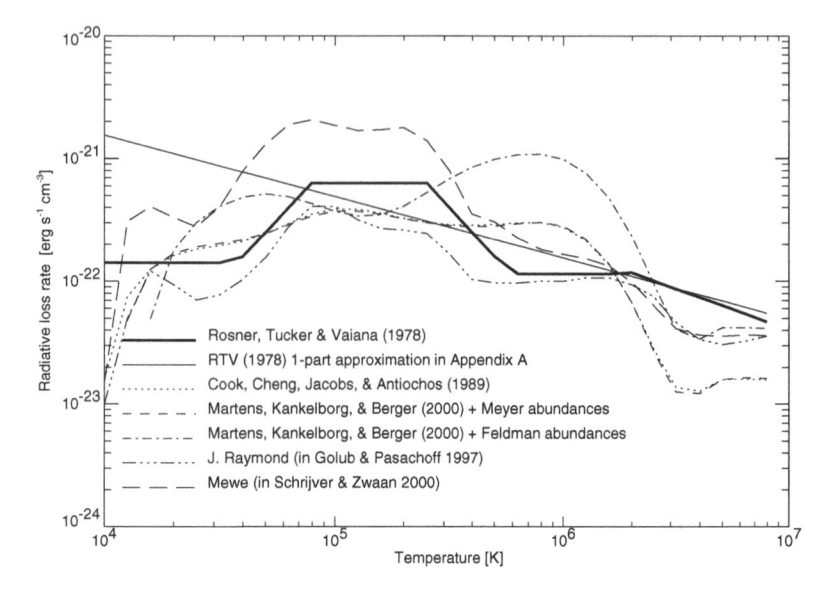

Figure 2.14: A compilation of radiative loss functions is shown. The differences mainly result from the assumptions of elemental abundances. Coronal abundances (e.g., Feldman 1992) have approximately a 3 times higher iron content than photospheric abundances (e.g., Meyer 1985), and thus increase the value of the radiative loss function by a factor of ≈ 10 at temperatures of $T \approx 0.5 - 2.0$ MK. The one-piece powerlaw approximation is used in the derivation of the RTV scaling law.

calculated by many authors (e.g., Tucker & Koren 1971; Rosner et al. 1978a; Mewe & Gronenschild 1981; Cook et al. 1989; Landini & Monsignori—Fossi 1990; Martens et al. 2000, etc.). A compilation of various calculations is shown in Fig. 2.14. As the contribution functions in Fig. 2.12 show, oxygen contributes mostly at $log(T) \approx 5.3$ and iron contributes mostly around $log(T) \approx 6.0$. In fact, the radiative loss function between $log(T) = 4.5$ and 7.0 is mostly dominated by contributions from the elements C, O, Si, and Fe, while continuum emission processes dominate at $log(T) \gtrsim 7.0$.

The differences in the various calculations shown in Fig. 2.14 result mainly from different assumptions in the elemental abundances. For instance, the two curves calculated by Martens et al. (2000) use the same code, but apply it to photospheric abundances (Meyer 1985) and coronal abundances (Feldman 1992). The radiative loss function for coronal abundances shows an enhancement over that of photospheric abundances in the temperature range of $log(T) \approx 5.2 - 6.2$, by up to an order of magnitude, which is mainly due to the enhanced coronal iron abundance. An iron enhancement of ≈ 3 is reflected in an emission measure increase of $EM \propto n_e^2 \approx 9$. This is compensated by the same factor in the *radiative loss function* $\Lambda(T)$, because a factor ≈ 3 times a lower density is then needed to explain the same observed radiative loss rate E_R according to Eq. (2.9.1).

For practical applications in hydrostatic loop models an analytical function for the *radiative loss function* $\Lambda(T)$ is needed. A widely used approximation was made by

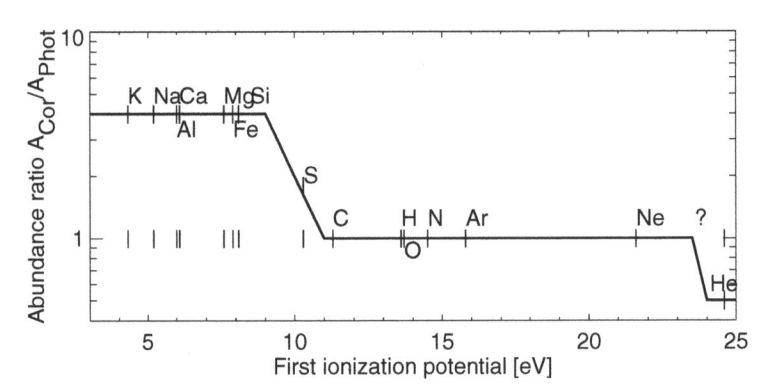

Figure 2.15: Ratio of coronal to photospheric abundances versus their *first ionization potential* (Feldman & Widing 2003).

a powerlaw parameterization in six pieces by Rosner et al. (1978a), shown as a thick solid line in Fig. 2.14. An even cruder approximation with a single powerlaw function has been used in Rosner et al. (1978a),

$$\Lambda(T) \approx 10^{-17.73} T^{-2/3} \qquad (6.3 < log\ T < 7.0) \qquad (2.9.2)$$

in the coronal temperature range of $T \approx 2 - 10$ MK to calculate a temperature-dependent function in the hydrostatic equilibrium. It turns out that this single powerlaw function (shown as a thin solid straight line in Fig. 2.14) is sufficiently accurate for most coronal loop models, because the emission measure contributions below temperatures of $T \lesssim 1$ MK are generally negligible due to the small extent of the transition layer compared with the coronal part of the loop length.

2.10 First Ionization Potential (FIP) Effect

In §1.9 and in Table 1.2 we quantified the chemical composition of the chromosphere and corona in terms of relative atomic abundances. Because most of the ionization processes are collisional and thus depend on the product of the electron density n_e with the ion density n_i, the line intensities are essentially proportional to the squared electron density, $n^2 \approx n_e n_i$, as well as to the abundance of the ion A_X (Eq. 2.8.4). The absolute abundance of an element could be determined if the absolute abundance of hydrogen is known, since the relative abundances to hydrogen; $A_X = n(X)/n(H)$ can readily be measured from line ratios of the same element with multiple ionization states. Ratios of the line/continuum fluxes (especially in soft X-rays) give the absolute abundance of the elements that emit the line. The absolute abundance of hydrogen, however, cannot be measured with spectroscopic line ratios, because hydrogen is completely ionized at coronal temperatures and produces no line emission. The absolute abundance of hydrogen in the solar corona is, therefore, still a big uncertainty factor.

A further uncertainty is the spatial and temporal variation of coronal abundances compared with chromospheric abundances. An enhancement of coronal abundances

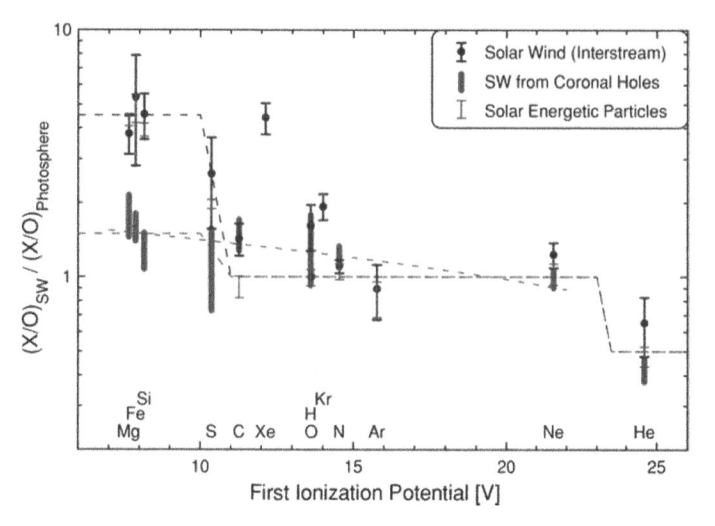

Figure 2.16: Abundances of elements measured in the solar wind relative to their photospheric abundances versus their *first ionization potential (FIP)* (Von Steiger 2001).

of Mg, Al, and Si by a factor of $\approx$ 3 over the chromospheric abundance was first noticed by Pottasch (1964a,b) from UV and XUV spectra recorded during sounding rocket flights. Extending abundance measurements to other elements it was found that the coronal enhancement is mainly a function of the *first ionization potential (FIP)* (see Table 1.2). Elements with *first ionization potentials* of $\leq$ 10 eV (K, Na, Al, Ca, Mg, Fe, Si), the so-called *low-FIP* elements, were found to have enhanced coronal abundances with respect to photospheric abundances, while the so-called *high-FIP* elements with *first ionization potentials* of $\gtrsim$ 10 eV (C, H, O, N, Ar, Ne) show no enhancement. The element S with a FIP of 10 eV is an "intermediate" FIP case. The situation is illustrated in Fig. 2.15, taken from a recent review by Feldman & Widing (2003), which also recounts the measurements of the FIP effect in more detail. A similar FIP effect is measured in the slow solar wind (Fig. 2.16)

Helium is also a special case. Essentially the helium abundance cannot be determined in the chromosphere because there are no helium lines available at chromospheric temperatures. Measurements of the 1216 Å H I resonance line and the 304 Å He II resonance line yielded He/H$\approx$ 0.07$\pm$0.01 (Gabriel et al. 1995), and an even lower ratio of He/H$\approx$ 0.052 $\pm$ 0.011 was measured by Feldman & Laming (2000), rendering coronal helium underabundant with respect to photospheric abundance (Fig. 2.15).

As mentioned above, the biggest uncertainty is still the absolute abundance of hydrogen, which also makes the reference level of the FIP diagram (Fig. 2.15) uncertain. In Fig. 2.15 it is assumed that hydrogen has the same coronal abundance as the photospheric abundance, so that the low-FIP elements seem to be enhanced in the corona. However, if hydrogen would have an enhanced coronal abundance, the FIP fractionation effect has to be interpreted in terms of a depletion of high-FIP elements out of the corona. Recently, spectroscopic EUV line analysis has been combined with free-free opacity measurements in radio wavelengths, which provides the absolute Fe/H

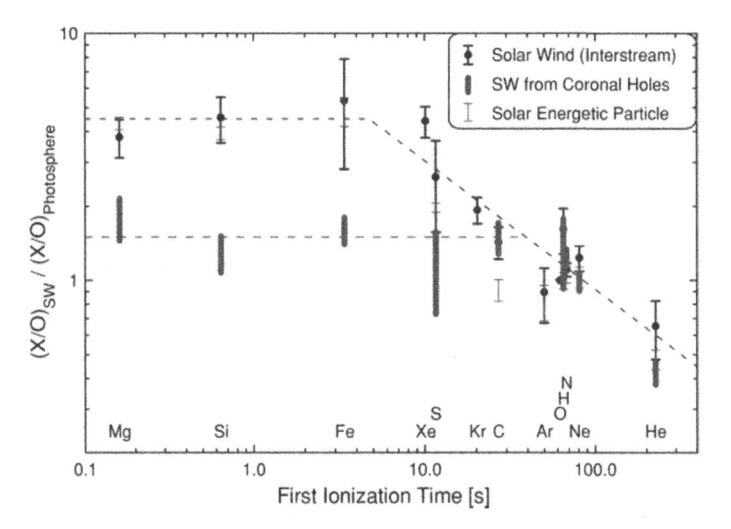

Figure 2.17: Abundances of elements measured in the solar wind relative to their photospheric abundances versus their *first ionization time (FIT)* (Von Steiger 2001).

abundance ratio (White et al. 2000). In this study it was found that coronal iron in an active region has an abundance of Fe/H= 1.56×10^{-4} (log(Fe/H)+12=8.2), which is enhanced by a factor of $10^{(8.2-7.5)} = 4.9$ above the photospheric abundance (Table 1.2; log(Fe$_{Phot}$/H)+12=7.5). This result implies that low-FIP elements such as Fe are indeed enhanced in the corona relative to photospheric values, rather than high-FIP elements being depleted in the corona.

The physical reason behind the FIP effect may be linked to the transport process of chromospheric plasma upward into coronal structures, or into the heliosphere in the case of open field lines in coronal holes. Several models have been proposed to explain the FIP fractionation by means of ion-neutral separation in the chromosphere, predominantly by diffusion perpendicular to the magnetic field (e.g., Geiss 1985). Alternative approaches involve the time scales of excitation and photo-ionization in the ion-neutral separation. They all rely on the fact that the low-FIP elements are partially ionized in the photosphere. Low-FIP elements have a shorter ionization time than high-FIP elements. Since transport from a highly collisional plasma (in the chromosphere), which warrants ionization equilibrium, to a near-collisionless state in the outer corona or heliosphere freezes the ionization ratios, the ratio of the transport time to the ionization time could be a crucial parameter in the relative coronal/chromospheric abundance ratios. It has therefore been suggested that the FIP effect should rather be considered as *first ionization time (FIT)* effect. In Fig. 2.17 the same abundance measurements from the solar wind shown in Fig. 2.16 are shown as a function of the *first ionization times (FIT)*, yielding a similar ordering as the FIP effect. The result suggests that any separation of ions from neutrals has to occur on short time scales of the order of $\lesssim 1$ minute.

2.11   Summary

Thermal radiation is produced by particles with a Maxwellian distribution function, which are found in collisional plasmas, such as in the chromosphere or in flare loops. The main type of thermal radiation is free-free (bremsstrahlung), which can be observed in hard X-rays as well as in microwaves. The equilibrium between absorption and emission processes is expressed with the radiative transfer equation (§2.1). A thermal equilibrium is realized in a black body according to Kirchhoff's law (§2.2). In a plasma, where electrons and ions are free (above the ionization temperature), free electrons become scattered in the Coulomb field of ambient free ions, which produces the so-called free-free bremsstrahlung emission (§2.3). In a thermal plasma there are abundant collisions between free electrons and free ions. Occasionally such collisions excite a bound electron of an ion into a higher quantum-mechanical energy level, from where it falls back into the ground state, emitting a photon at the same time to conserve energy. The transitions from excited to ground state levels are associated with specific energy differences, which can be calculated with the Rydberg formula (§2.4). There are a large number of atomic processes that are possible in a collisional plasma, including various ionization and recombination processes (§2.6). A local thermodynamic and ionization equilibrium defines the number of bound-bound transitions in a plasma, expressed by the Saha equation (§2.7). Quantitative analysis of fluxes in a particular EUV or soft X-ray wavelength involve the computation of many spectral lines (mostly produced by collisional excitation). A comprehensive code containing all known line transitions is the CHIANTI code. From these lines, the radiative loss rate can be computed for a coronal plasma with a given density and temperature (§2.9). The abundance of elements in the solar chromosphere is fairly consistent with cosmic abundances or abundances measured in meteorites. In the corona, however, some elements seem to have an enhanced abundance, in particular those elements with a low first ionization potential < 10 eV (K, Na, Al, Ca, Mg, Si, Fe), typically a factor of 3, but can amount up to a factor of 10 for the iron (Fe) element (§2.10). This FIP effect seems to be related to the first ionization time, which is $\lesssim 10$s for the low-FIP elements. In summary, thermal plasmas (in the solar corona and in flares) give rise to continuum (free-free bremsstrahlung) as well as to line emission (in soft X-ray and EUV wavelengths). Quantitative modeling of the soft X-ray and EUV fluxes provide a powerful temperature and density diagnostic of the coronal plasma.

Chapter 3

Hydrostatics

When you stand on a high mountain and look at the horizon, you can notice how the color of the sky blends from a pale blue into a dark blue with increasing elevation angle, a clear manifestation of the gravitational settling of the atmosphere. While the Earth atmosphere has a density or pressure scale height of about 4 km, a fact that forces most climbers of the Himalayan mountains to carry their own oxygen, the density scale height in the solar corona is much larger ($\gtrsim 50,000$ km). The larger scale height in the solar corona is a consequence of the much larger coronal temperature and lighter hydrogen gas. The physics of hydrostatics provides us with a description of the pressure and density variation with height, which strongly depends on the temperature of the coronal plasma. The physical parameters of temperature, density, and pressure represent the most basic quantities needed, if we want to study any phenomenon in the solar corona in a quantitative way. Without them we could not do any diagnostics from the observed EUV, soft X-ray, or radio emissions. Of course, hydrostatics is applicable only to static or quasi-stationary structures, strictly speaking. However, because dynamic phenomena in the solar corona often evolve into a stable equilibrium, most of the coronal structures are found near a stationary state most of the time. Later on, we will generalize the *hydrostatic models* by including flows, which lead to *hydrodynamic models* (§ 4), and by including magnetic fields (§ 5), leading to *magneto-hydrodynamics (MHD)* (§ 6).

3.1 Hydrostatic Scale Height

If the pressure is only balanced by the gravitational force, we call it *gravitational stratification, gravitational settling,* or *hydrostatic equilibrium*. The gravity force always points to the center of mass (i.e., the center of the Sun), so the *gravitational potential* ε_{grav} is a function of a 1D spatial coordinate (i.e., the distance r from Sun's center),

$$\varepsilon_{grav} = -\frac{GM_\odot m}{r} \, , \tag{3.1.1}$$

where $G = 6.67 \times 10^{-8}$ (dyne cm^2 g^{-2}) is the gravitational constant in Newton's gravitational force equation, $M_\odot = 1.989 \times 10^{33}$ g is the solar mass, and m is a particle mass in the solar corona. The *solar gravitation* $g_\odot$ (i.e., the acceleration constant due to solar gravity; in Newton's law $F = mg_\odot$), is defined by

$$g_\odot = \frac{GM_\odot}{R_\odot^2} = 2.74 \times 10^4 \; (\text{cm s}^{-2}) \tag{3.1.2}$$

where $R_\odot = 6.9551 \times 10^{10}$ cm is the solar radius. Thus, with Eqs. (3.1.1) and (3.1.2) we have the *gravitational potential*,

$$\varepsilon_{grav}(r) = -mg_\odot \left(\frac{R_\odot^2}{r} \right) . \tag{3.1.3}$$

The *gravity force* F_{grav} is derived from the gradient of the gravitational potential,

$$F_{grav}(r) = -\frac{d\varepsilon_{grav}(r)}{dr} = -mg_\odot \left(\frac{R_\odot^2}{r^2} \right) . \tag{3.1.4}$$

which is just $F_{grav}(r = R_\odot) = -mg_\odot$ at the solar surface, $r = R_\odot$. The pressure p is defined as a force F per area dA, having the physical dimension of (dyne cm^{-2}). A pressure gradient dp/dr thus has the physical dimension of a force per volume (i.e., $dp/dr = F/(dA \times dr) = F/dV$), which can be expressed as a product of the force with a particle volume density n (cm^{-3}), $dp/dr = F \times n$. We can, therefore, deduce the *pressure equilibrium* or *momentum equation* by multiplying the force $F_{grav}(r)$ in Eq. (3.1.4) with the particle density n,

$$\frac{dp}{dr}(r) = \frac{dp_{grav}(r)}{dr} = F_{grav}(r)n(r) = -mn(r)g_\odot \left(\frac{R_\odot^2}{r^2} \right) . \tag{3.1.5}$$

In plasma physics and fluid dynamics it is customary to define a *mass density* $\rho = mn$, where m (g) is the average particle mass and n (cm^{-3}) the particle density. For a fully ionized gas, such as the solar corona, the mass density is composed of the electron density n_e and ion density n_i,

$$\rho = mn = m_e n_e + m_i n_i \approx \mu m_H n_e , \tag{3.1.6}$$

where μ is the molecular (or mean atomic) weight of the ion, m_H the hydrogen mass, while the electron mass $m_e = m_H/1836$ can be neglected for a neutral plasma with $n_e \approx n_i$. It is sufficient to include only the most abundant two ions (i.e., hydrogen and helium), because all heavier elements have an abundance of many orders of magnitude lower (Table 1.2). In the solar corona, which consists of H:He=10:1, the molecular weight (with $\mu = 1$ for hydrogen ^{1}H and $\mu = 4$ for ^{4}He) is

$$\mu = \frac{10 \times 1 + 1 \times 4}{11} \approx 1.27 . \tag{3.1.7}$$

We can now write the *momentum equation* (Eq. 3.1.5) in terms of the coronal electron density $n_e(r)$ by inserting the definition of mn (Eq. 3.1.6),

$$\frac{dp}{dr}(r) = -\mu m_H n_e(r) g_\odot \left(\frac{R_\odot^2}{r^2} \right) . \tag{3.1.8}$$

To solve this momentum equation, we need a relation between the pressure $p(r)$ and the electron density $n_e(r)$, which is provided by a thermodynamic *equation of state* (e.g., the *ideal gas equation*, $p = nk_BT$). For the coronal pressure, we have to add the ion pressure p_i and electron pressure p_e, so the pressure is twice that of the electron gas (assuming $n_e \approx n_i$ for the corona),

$$p(r) = 2n_e(r)k_BT_e(r) , \qquad (3.1.9)$$

The electron density $n_e(r)$ can then be substituted into the momentum equation (3.1.8) from the ideal gas equation (Eq. 3.1.9),

$$\frac{dp}{dr}(r) = -p(r)\frac{\mu m_H g_\odot}{2k_BT_e(r)}\left(\frac{R_\odot^2}{r^2}\right) . \qquad (3.1.10)$$

We introduce the more practical height variable h above the solar surface,

$$h = r - R_\odot , \qquad (3.1.11)$$

so that we can express the momentum equation as a function of height h,

$$\frac{dp}{dh}(h) = -p(h)\frac{\mu m_H g_\odot}{2k_BT_e(h)}\left(1 + \frac{h}{R_\odot}\right)^{-2} . \qquad (3.1.12)$$

For near-isothermal coronal structures where the approximation $T_e(h) \approx const$ holds, we can separate the pressure variable $p(h)$ in the momentum equation (3.1.12),

$$\frac{dp}{p} = -dh\frac{\mu m_H g_\odot}{2k_BT_e}\left(1 + \frac{h}{R_\odot}\right)^{-2} = -\frac{dh}{\lambda_p(T_e)}\left(1 + \frac{h}{R_\odot}\right)^{-2} , \qquad (3.1.13)$$

and integrate both sides of the equation,

$$\int_{p_0}^{p}\frac{dp}{p} = -\int_{h_0}^{h}\frac{1}{\lambda_p(T_e)}\left(1 + \frac{h}{R_\odot}\right)^{-2} dh , \qquad (3.1.14)$$

which has a near-exponential solution,

$$p(h) = p_0 \exp\left[-\frac{(h - h_0)}{\lambda_p(T_e)(1 + \frac{h}{R_\odot})}\right], \qquad (3.1.15)$$

where the *pressure scale height* $\lambda_p(T)$ is defined by Eq. (3.1.12) as

$$\lambda_p(T_e) = \frac{2k_BT_e}{\mu m_H g_\odot} \approx 4.7 \times 10^9 \left(\frac{T_e}{1\ \mathrm{MK}}\right) \quad (\mathrm{cm}) . \qquad (3.1.16)$$

Near the solar surface (i.e., for $h \ll R_\odot$), the pressure drops off exponentially according to Eq. (3.1.15) in an isothermal atmosphere. For relatively small loops (i.e., for $h \ll \lambda_p(T_e)$), even an isobaric approximation ($p \approx const$) can be used. Since the *pressure scale height* or *hydrostatic scale height* λ_p scales linearly with temperature T_e, the isobaric assumption is a quite valid approximation for near-isothermal flare loops.

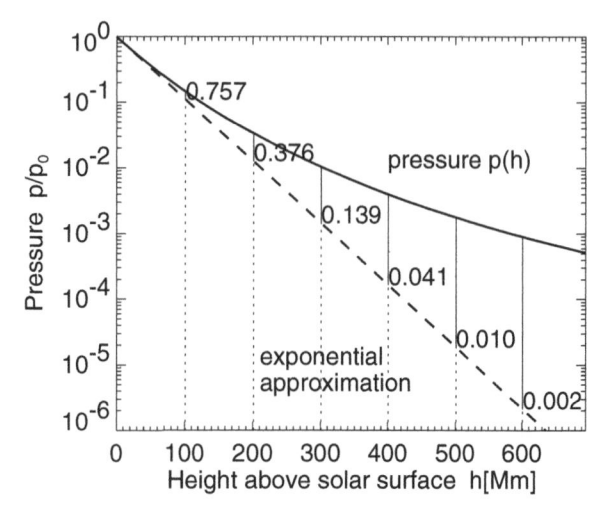

Figure 3.1: Comparison of exact hydrostatic solution (solid line) for the pressure $p(r)$ in an isothermal atmosphere ($T = 1$ MK) and an exponential approximation (dashed line).

The r-dependence is usually neglected in hydrostatic modeling, but we show the magnitude of the effect in Fig. 3.1. The exponential approximation underestimates a pressure of 76% at a height of $h = 100$ Mm, or 38% at a height of $h = 200$ Mm, a common altitude seen for active region loops in *TRACE* 171 Å images.

An observational test of the hydrostatic equilibrium can be performed by measuring the emission measure $EM(h)$ along a vertical loop segment, which yields the density profile $n_e(h) = \sqrt{EM(h)/w}$ for a loop with diameter w, which can then be fitted by an exponential density model in order to obtain the pressure scale height λ_p and the corresponding temperature T_λ (with Eq. 3.1.16). On the other hand, if the loop is observed with two different temperature filters, a filter-ratio temperature T_{filter} can be obtained from the flux ratio over the same segment (after appropriate background subtraction to avoid contamination from the background with different temperatures). Such measurements of the filter-ratio temperature T_{filter} are shown in Fig. 3.2, determined from a sample of some 60 loops measured from either *SoHO/EIT* 171/195 Å or 195/284 Å filter ratios, plotted versus the scale height temperature T_λ. There is obviously a good agreement between the two independently measured temperature values (within $\lesssim 20\%$) for this particular set of coronal loops observed in the active region NOAA 7986 on 1996 Aug 30 with *SoHO/EIT* (Aschwanden et al. 2000a). Because this was an old active region, less dynamic heating or passive cooling seems to occur, so that the hydrostatic equilibrium of these loops is little disturbed. Loops in hydrostatic equilibrium can easily be identified at the limb from their small height extent. The emission measure scale height is half the density (or pressure) scale height (i.e., $\lambda_{EM} = \lambda_p/2$), due to the dependence of $EM(h) \propto n_e^2(h)w$. Thus, hydrostatic loops at $T = 1$ MK, as seen in TRACE 171 Å images, have only an emission measure (or flux) scale height of $\lambda_{EM} \approx 23$ Mm, according to Eq. (3.1.16).

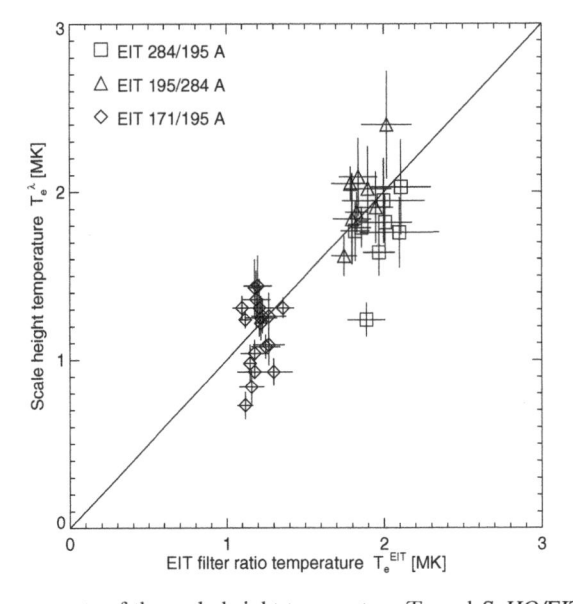

Figure 3.2: Measurements of the scale height temperature T_λ and *SoHO/EIT* filter-ratio temperatures T_{filter}, for 30 loops detected in 171 Å, 195 Å, and 284 Å, (Aschwanden et al. 1999a, 2000a).

3.2 Hydrostatic Weighting Bias

The fact that the plasma-β parameter is generally much lower than unity in the solar corona allows us to consider each magnetic fluxtube as an isolated mini-atmosphere, each one being in its own hydrostatic equilibrium, since cross-field transport by thermal conduction is inhibited. The isothermal assumption is often a good approximation for a single fluxtube, but *never* for a line-of-sight integrated quantity. If we measure the intensity I_ν in the optically thin soft X-ray or EUV wavelengths, there are always contributions from many fluxtubes with different temperatures along the line-of-sight. This multi-thermal effect is most conspicuous from the differential emission measure distributions (e.g., Fig. 1.21), which show a broad temperature range along any line-of-sight. The multi-temperature contributions are illustrated in Plate 1, where the temperature is color-coded.

We have seen in Eq. (3.1.16) that the pressure scale height λ_p is proportional to the temperature T, and is identical to the density scale height in the isothermal approximation (since $p(r) \propto n(r)$ for $T(r) = const$, according to Eq. 3.1.9). Therefore, hotter fluxtubes have a larger density scale height than cooler ones, if they are at or near hydrostatic equilibrium. If we average the brightness of a cool and a hot loop, say with equal base density $n_0 = n(h = 0)$, the relative density contributions from the hotter loop will be larger with increasing height, because the hot loop has a larger density (or pressure) scale height λ_p than the cooler one. This is illustrated in Fig. 3.3. Consequently, we have the squared weighting effect for the averaged emission measure, because $EM(h) \propto n_e(h)^2$. The observed soft X-ray or EUV flux is proportional to

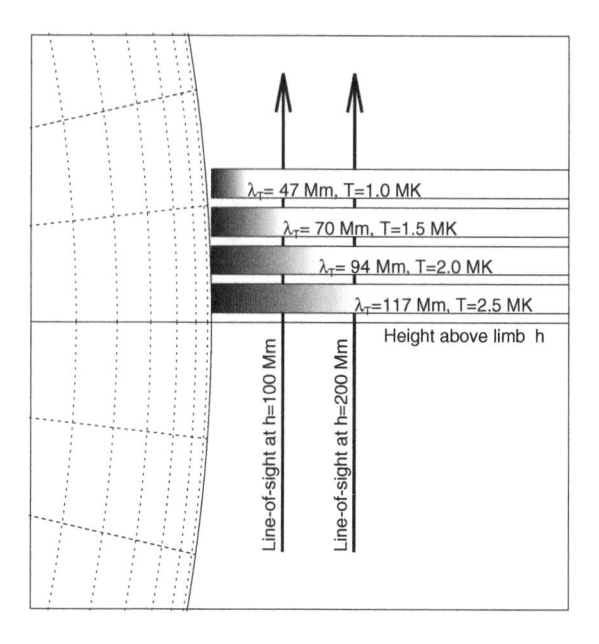

Figure 3.3: Illustrating scale height-weighted contributions of hydrostatic loops or open flux-tubes to the emission measure observed along two line-of-sights above the solar limb. The left line-of-sight at a height of $h = 100$ Mm above the limb samples significant emission from the 3 loops with temperatures of 1.5–2.5 MK. The right line-of-sight at a height of $h = 200$ Mm above the limb samples significant emission only from the hottest loop with $T = 2.5$ MK, (Aschwanden & Nitta 2000).

the emission measure, and thus any temperature analysis of the averaged flux is subject to this height-dependent density weighting, which we call the *hydrostatic weighting bias* (Aschwanden & Nitta 2000).

Let us quantify this *hydrostatic weighting bias* effect. In the isothermal approximation ($T_e(h) = const$, for the coronal segment of a single fluxtube), neglecting the variation of gravity with height ($h \ll R_\odot$), we obtain an exponential density profile $n_e(h)$ as a function of height h (from Eqs. 3.1.15 and 3.1.9),

$$n_e(h, T) = n_{e0} \exp\left(-\frac{h}{\lambda_p(T)}\right) . \tag{3.2.1}$$

The emission measure $EM(T)$ sums up all the density contributions $n_e^2(z)$ along a given line-of-sight z at a particular temperature T,

$$EM(T) = \int n_e^2(T, z) dz . \tag{3.2.2}$$

The flux F_i or intensity measured by a detector or filter i is given by the product of the differential emission measure distribution $dEM(T)/dT$ and the instrumental response

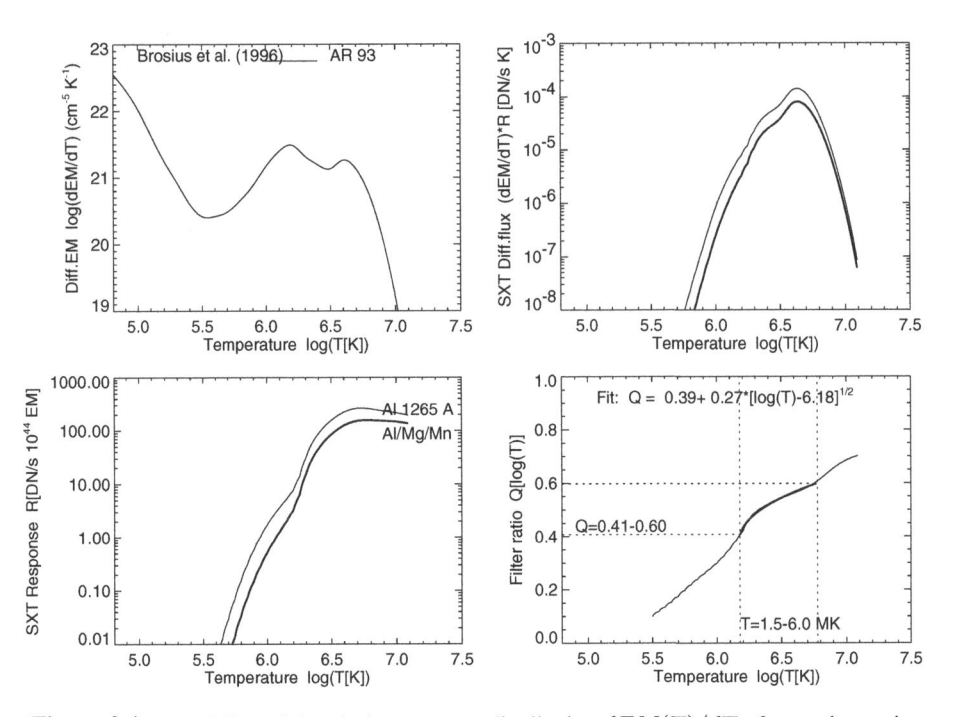

Figure 3.4: The differential emission measure distribution $dEM(T)/dT$ of an active region (AR93) measured by Brosius et al. (1996) with SERTS data is shown (top left panel), the Yohkoh/SXT response function for the two thinnest filters (bottom left panel), the contribution functions $dF(T)/dT = [dEM(T)/dT]*R(T)$ for the two SXT filters (thin and thick linestyles) (top right panel), and the filter ratio $Q(T) = R_2(T)/R_1(T)$ for the two Yohkoh/SXT filters and an analytical approximation in the range of $T = 1.5 - 6.0$ MK (bottom right panel).

function $R_i(T)$ (defined as a function of temperature),

$$F_i = \int \frac{dEM(T)}{dT} R_i(T) dT .$$ (3.2.3)

Combining Eqs. (3.2.1-3) leads to the following height dependence of the flux,

$$F_i(h) = \int \frac{dEM(T, h = 0)}{dT} \exp\left(-\frac{2h}{\lambda_p(T)}\right) R_i(T) dT .$$ (3.2.4)

In Fig. 3.4 we take an observed differential emission measure distribution from an active region (Brosius et al. 1996) and assign it to height $h = 0$, $dEM(T, h = 0)/dT$ (Fig. 3.4, top left). The instrumental response function R_i for the mostly used Yohkoh/SXT filters (i.e., the Al 1265 Å and Al/Mg/Mn filters), which we call $R_1(T)$ and $R_2(T)$ are shown in Fig. 3.4 (bottom left). The product of the differential emission measure distribution with the response function quantifies the contribution function, $C_i(T) = [dEM(T)/dT] \times R_i(T)$, shown in Fig. 3.4 (top right). Temperature analysis is usually done with a filter-ratio method. The temperature dependence of this filter

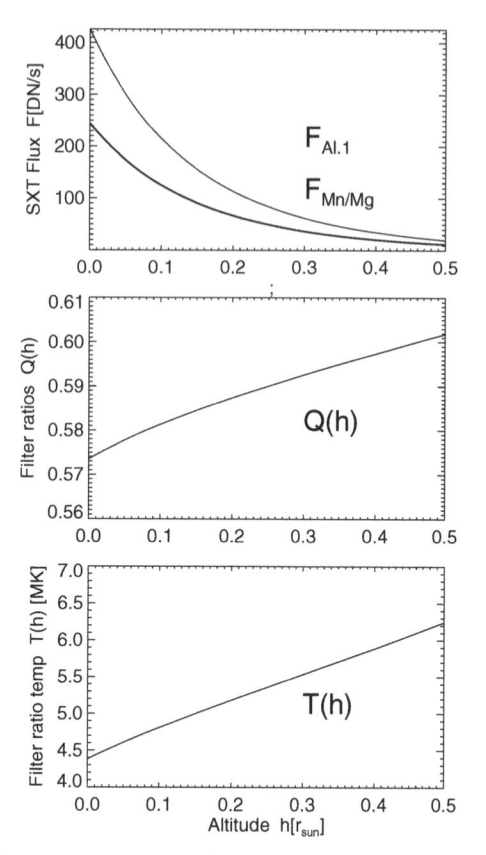

Figure 3.5: The height dependence of the observed SXT fluxes $F(h)$ for the two filters are shown with thin and thick linestyles (top panel), the resulting filter ratio $Q(h)$ (second panel), and the inferred filter-ratio temperatures (bottom panel). Note that the filter-ratio temperature $T(h)$ shows a systematic increase with height, although a model with isothermal loops was assumed.

ratio, $Q(T) = R_2(T)/R_1(T)$, is explicitly shown in Fig. 3.4 (bottom right), and can be approximated with the analytical function,

$$Q(T) = \frac{R_2(T)}{R_1(T)} \approx 0.39 + 0.27[\log(T) - 6.18]^{1/2} , \qquad 1.5 \text{ MK} < T < 6.0 \text{ MK} ,$$

(3.2.5)

This analytical approximation allows us to conveniently invert the filter-ratio temperature (in the range of $Q = 0.4 - 0.6$),

$$\log[T(Q)] = 6.18 + \left(\frac{Q - 0.39}{0.27}\right)^2 , \qquad 0.4 < Q < 0.6 .$$

(3.2.6)

We can now extrapolate the fluxes $F_1(h)$ and $F_2(h)$ as a function of height h by using the hydrostatic model of Eq. (3.2.4) for both filters, which are shown in Fig. 3.5

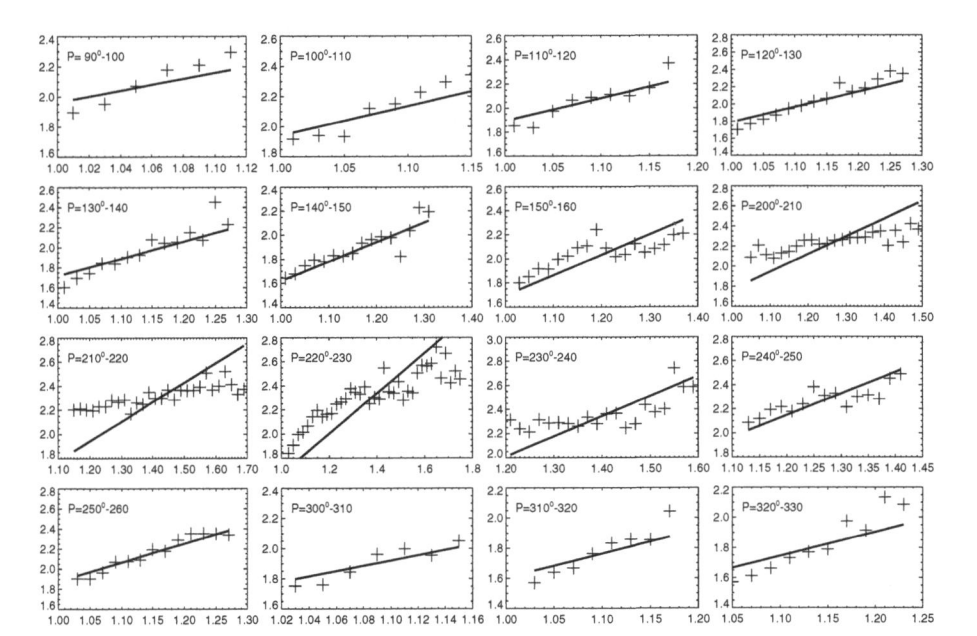

Figure 3.6: Filter-ratio temperature measurements $T(r)$ as functions of height $r = 1 + h/R_\odot$ in 16 radial sectors of the Sun, observed in quiet Sun and coronal hole regions on 1992-Aug-26 with *Yohkoh/SXT* (Plate 2). The crosses mark the measurements, while the thick lines represent the expected temperature gradients due to the *hydrostatic weighting bias* according to the approximation of Eq. (3.2.7). The y-axis indicates the temperature $T(MK)$ (Aschwanden & Acton 2001).

(top panel). The resulting flux ratio $Q(T) = F_2(T)/F_1(T)$, which varies in this case from $Q(h = 0) \approx 0.57$ at the coronal base to $Q(h = 0.5R_\odot) \approx 0.60$ at a height of a half solar radius (Fig. 3.5, middle), can then be inverted into a filter-ratio temperature $T(Q)$ according to Eq. (3.2.6), which varies from $T(h = 0) \approx 4.4$ MK at the coronal base to $T(h = 0.5R_\odot) \approx 6.3$ MK at a height of a half solar radius (Fig. 3.5 bottom). Thus the hydrostatic weighting bias results in a temperature gradient of $dT_{bias}/dh \approx 0.005$ K m^{-1}. Applying the same calculations to other active regions or quiet Sun regions, the typical temperature gradient $\Delta T_{bias}(h)$ due to the hydrostatic weighting bias has the following height dependence,

$$\Delta T_{bias}(h) \approx T_0 \left(\frac{h}{R_\odot} \right) . \tag{3.2.7}$$

From this calculation we predict quite a substantial positive temperature gradient with height in the solar corona. Such positive temperature gradients with height, $dT(h)/dh > 0$, have been reported in a number of studies, often using *Yohkoh/SXT* filter ratios (Mariska & Withbroe 1978; Kohl et al. 1980; Falconer 1994; Foley et al. 1996; Sturrock et al. 1996a,b; Wheatland et al. 1997; Fludra et al. 1999; Priest et al. 1999, 2000; Aschwanden & Acton 2001). Because the reported temperature gradients have approximately the magnitude predicted by Eq. (3.1.22), it is very plausible that they reflect the *hydrostatic weighting bias* (as derived here for isothermal fluxtubes), rather than a real temperature variation $T(h)$ along individual fluxtubes. A larger number of filter-ratio temperature measurements as functions of height are shown in Fig. 3.6 for quiet Sun and coronal hole regions, along with the estimated *hydrostatic weighting bias*, which predicts a fully compatible temperature gradient. In other words, although the filter-ratio temperatures show a positive height gradient, the temperature profile of each fluxtube is consistent with a near-isothermal model, $T(h) \approx const.$

3.3   Multi-Hydrostatic Corona

If we observe the most homogeneous parts of the quiet Sun corona, without any detectable features of loops or fluxtubes (e.g. as seen in the north-east quadrant above the limb in Plate 1), what is the simplest appropriate model to quantify the optically thin soft X-ray or EUV brightness? Obviously we cannot apply an isothermal model, because observed differential emission measure distributions reveal a broad temperature distribution at any location in the solar corona. Moreover, gravitational stratification in any thermally insulated structure is an important effect we have to include in a physical model. Let us characterize the differential emission measure distribution $dEM(T)/dT$ or the corresponding *differential density distribution* $dn_e(T)/dT$ with a simple Gaussian function of temperature (following Aschwanden & Acton, 2001),

$$\frac{dn_e(\mathbf{x}, T)}{dT} = \frac{n_e(\mathbf{x})}{\sqrt{2\pi}\sigma_T} \exp\left[-\frac{(T - T_0)^2}{2\sigma_T^2}\right] . \tag{3.3.1}$$

This temperature distribution is characterized by 3 parameters: the total electron density $n_e(\mathbf{x})$, the mean electron temperature T_0, and the Gaussian temperature distribution width σ_T. The differential electron density distribution is normalized so that the integral over the entire temperature range yields the total electron density $n_e(\mathbf{x})$ at a spatial location $\mathbf{x}$,

$$\int \left(\frac{dn_e(\mathbf{x}, T)}{dT}\right) dT = n_e(\mathbf{x}) . \tag{3.3.2}$$

We define the corresponding *differential emission measure (DEM)* distribution integrated along a line-of-sight z,

$$\frac{dEM(x, y, T)}{dT} = \int \left(\frac{dn_e(\mathbf{x}, T)}{dT}\right)^2 2\sqrt{\pi}\sigma_T \, dz = \int \frac{n_e^2(\mathbf{x})}{\sqrt{\pi}\sigma_T} \exp\left[-\frac{(T - T_0)^2}{\sigma_T^2}\right] dz , \tag{3.3.3}$$

which contains a normalization factor so that the total emission measure EM, obtained by integrating over the entire temperature range, meets the standard definition of column emission measure,

$$EM(x, y) = \int \left(\frac{dEM(x, y, T)}{dT}\right) dT = \int n_e^2(\mathbf{x}) \, dz . \tag{3.3.4}$$

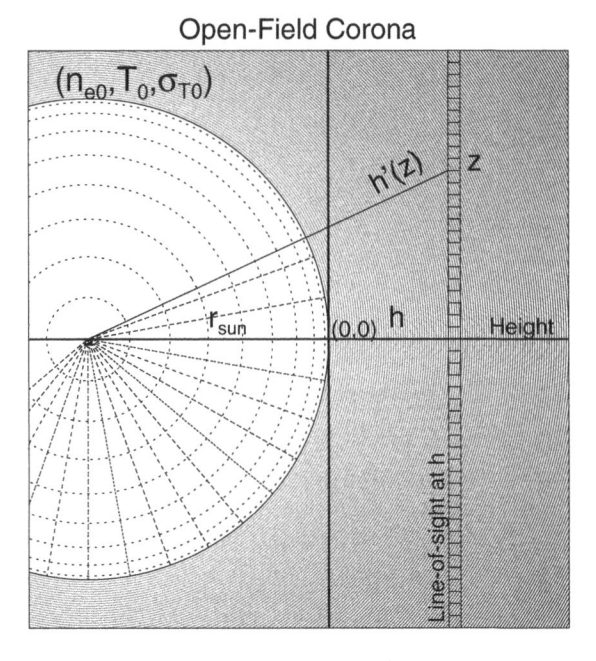

Figure 3.7: Geometric definitions are shown for a homogeneous corona in a plane along the line-of-sight. The height as a function of the line-of-sight distance z is denoted with $h'(z)$.

Implicitly, we assume here that the ion density is equal to the electron density, $n_i = n_e$, because the general definition of the DEM is $dEM(T)/dT \propto n_i n_e dV$.

We follow now the multi-hydrostatic concept outlined in § 3.2 and assume that the quiet corona is filled by many structures (closed loops or open fluxtubes) all being in hydrostatic equilibrium, each one at a different temperature, where the sum of all temperature contributions is constrained by the differential emission measure distribution $dEM(T)/dT$ at the base of the corona ($h = h_0$). Attributing a hydrostatic density profile $n_e(h)$ to each structure as defined by Eq. (3.2.1), the variation of the electron density $n_e(h)$ as a function of height h is then obtained by integrating over the temperature range (Eq. 3.3.2),

$$n_e(h) = \int \left(\frac{dn_e(h,T)}{dT}\right) dT = \int \frac{n_e(h_0)}{\sqrt{2\pi}\,\sigma_T} \exp\left[-\frac{[h-h_0]}{\lambda_p(T)} - \frac{(T-T_0)^2}{2\sigma_T^2}\right] dT\,,$$
(3.3.5)

which can be calculated by numerical integration. The density height profile $n_e(h)$ would be an exponential function exactly for an isothermal atmosphere ($\sigma_T \ll T_0$), but falls off less steeply with height for a multi-thermal atmosphere ($\sigma_T \approx T_0$).

Thus, to model the diffuse corona observed above the limb, we assume (in a spherical geometry) density distributions with a radial dependence according to the hydrostatic scaling $n_e(h,T)$ (Eq. 3.2.1), but having randomly distributed temperatures, characterized by the differential electron density distribution $dn_e(\mathbf{x},T)/dT$ (Eq. 3.3.1) at the coronal base ($h = h_0$). Denoting the direction of the line-of-sight with z, a simple

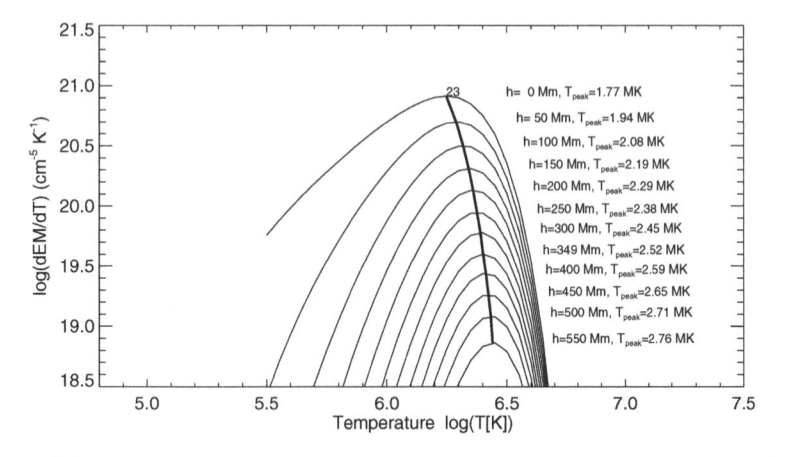

Figure 3.8: The multi-hydrostatic differential emission measure distributions $dEM(T, h)/dT$ are shown for different heights h, computed from parameters measured in Aschwanden & Acton (2001). Note the systematic increase of the peak temperature from $T_{EM} = 1.77$ MK at the coronal base ($h = 0$) to $T_{EM} = 2.76$ MK at a height of $h = 550$ Mm, which is a manifestation of the *hydrostatic weighting bias.*

geometric consideration yields the following height dependence $h'(z)$ for a point z on the line-of-sight axis that passes the limb at a lowest height h (at location $z = 0$),

$$h'(z) = \sqrt{(R_\odot + h)^2 + z^2} - R_\odot \qquad (3.3.6)$$

(see Fig. 3.7). Combining Eqs. (3.3.3−6) and Eq. (3.2.1) we obtain then the following differential emission measure $dEM(h, T)/dT$ for a position h above the limb, integrated along the line-of-sight z,

$$\frac{dEM(h, T)}{dT} = \int_{-\infty}^{\infty} \frac{n_e^2(h_0)}{\sqrt{\pi}\, \sigma_T} \exp\left[-\frac{2[h'(z) - h_0]}{\lambda_p(T)} - \frac{(T - T_0)^2}{\sigma_T^2} \right] dz . \qquad (3.3.7)$$

Examples of such $dEM(T)$ distributions inferred from *Yohkoh/SXT* data are shown for different heights h in Fig. 3.8. The peak of the $dEM(T)$ distributions shifts systematically to higher temperatures T_0 with increasing altitude h, due to the hydrostatic weighting bias.

Multiplying the differential emission measure distribution $dEM(h, T)/dT$ with the instrumental response function $R_W(T) = dF_W/dEM(T)$ of a filter with wavelength W and integrating over the temperature range we then directly obtain the flux $F_W(h)$ at a given height h,

$$F_W(h) = \int_0^{\infty} \frac{dF_W(h)}{dEM(h, T)} \frac{dEM(h, T)}{dT} dT = \int_0^{\infty} R_W(T) \frac{dEM(h, T)}{dT} dT .$$
$$(3.3.8)$$

It is instructive to visualize the column depth of a hydrostatically stratified atmosphere as a function of the distance to the Sun's center, because the electron density can then directly be estimated from this column depth and an observed emission measure. We

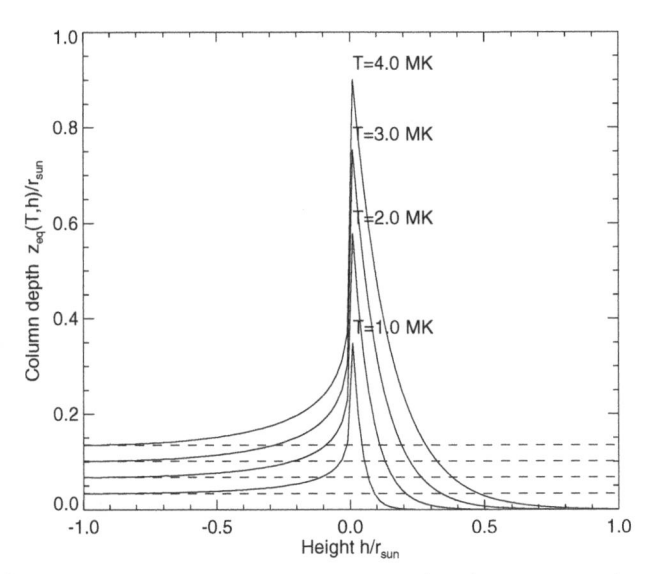

Figure 3.9: Calculated equivalent column depths $z_{eq}(T, h)$ (Eq. 3.3.10) for a gravitationally stratified atmosphere as a function of the limb distance h, for isothermal plasmas with temperatures of column depths corresponding to the emission measure scale heights (i.e. half the density scale height) for vertical line-of-sights at the Sun's center ($h/r_\odot = -1$), indicated with dashed lines.

define an *equivalent column depth along the line-of-sight* $z_{eq}(h, T)$ as a function of the height h above the limb and for a coronal temperature T_0 (in the isothermal approximation $\sigma_{T_0} \ll T_0$),

$$EM(h, T_0) = \int_{-\infty}^{\infty} n_e^2(h, z, T_0)dz = n_e^2(h_0, T_0)z_{eq}(h, T_0) . \qquad (3.3.9)$$

From Eqs. (3.3.4−7) the following relation follows for this equivalent column depth,

$$z_{eq}(h, T_0) = \int_{z_1}^{z_2} \exp\left(-\frac{2}{\lambda_p(T_0)}\left[\sqrt{(R_\odot + h)^2 + z^2} - R_\odot\right]\right) dz \quad \text{for } h \geq 0 .$$
$$(3.3.10)$$

with the integration limits $z_1 = -\infty$ and $z_2 = +\infty$. Inside the disk ($-1 \leq h \leq 0$), we have only to change the integration limit z_2 to,

$$z_2(h) = -\sqrt{R_\odot^2 - (R_\odot + h)^2} \quad \text{for } -1 < h < 0 . \qquad (3.3.11)$$

The column depths $z_{eq}(h, T_0)$ are shown in Fig. 3.9 for a height range from disk center ($h = -1$) to one solar radius outside the limb ($h = +1$), for temperatures in the range of $T = 1.0 - 4.0$ MK. At disk center ($h = -1$), the equivalent column depth matches the emission measure scale height, which is the half density scale height ($\lambda_{EM} = \lambda_p/2 = \lambda_p(T_0)/2$) (indicated with dashed lines in Fig. 3.8). At the limb

($h = 0$), there is in principle a discontinuous change by a factor of 2, which, however, is difficult to measure because of the extremely high instrumental resolution required to resolve this jump. Above the limb, the column depth drops quickly with height. This dependence of the column depth on the Sun center distance constrains the soft X-ray brightness across and outside the solar disk.

After we have derived the differential emission measure distribution $dEM(h, T)$ as a function of height (Eq. 3.3.7), which can predict the flux $F_W(h)$ for an arbitrary instrument, we test this model with *Yohkoh/SXT* data from 1992-Aug-26 (shown in Plate 2). We apply the model to the limb profiles in two filters, $F_{Al.1}(h)$ and $F_{Mn/Mg}(h)$, for 36 different position angles (as indicated with radial $10°$ sectors in Plate 2). We introduce an additional free parameter q_λ that allows for an arbitrary observed exponential scale height λ_{obs},

$$\lambda_{obs} = q_\lambda \, \lambda_p(T) \, , \tag{3.3.12}$$

so that we can actually probe deviations from hydrostatic equilibrium. The best-fit parameters are shown in Fig. 3.10: electron base densities of $n_e \approx 10^8 - 10^9$ cm^{-3}, electron base temperatures of $T_0 \approx 1.0 - 1.5$ MK, temperature widths of $\sigma_T \approx 1.0$ MK, and scale height factors of $q_\lambda \approx 1$. The last result represents a test of the *"hydrostaticity"* of the quiet corona. Besides a significant deviation of $q_\lambda \lesssim 2.3$ at the location of a coronal streamer (at position angle $P \approx 200° - 250°$, most parts of the corona are found to be near $q_\lambda \approx 1$, which means that the pressure scale height is nearly hydrostatic, $\lambda \approx \lambda_p(T)$ (Eq. 3.3.12). Of course, slight deviations are expected wherever inhomogeneities in the form of brighter loops or active regions occur.

The white-light spectrum of the solar corona is composed of three components: (1) the *K-corona*, made of partially polarized continuum emission from photospheric light scattered at free electrons (dominating at $h \lesssim 0.3 \, r_\odot$); (2) the *L-corona*, consisting of spectral line emission from highly ionized atoms (dominating at $h \lesssim 0.5 \, r_\odot$); and (3) the *F-corona*, which presents absorption lines of the photospheric Fraunhofer spectrum caused by diffraction from interplanetary dust (dominating at $h \gtrsim 0.5 \, r_\odot$). The line-of-sight integrated density profiles of these three components can each be approximated by a powerlaw, leading to an average density profile known as the *Baumbach−Allen formula*, given in Eq. (1.6.1). This density model $n_e(h)$ is shown in Fig. 3.11 (square symbols). We now fit each of these three components with our multi-hydrostatic (line-of-sight integrated DEM) model. Approximate fits are shown in Fig. 3.11. The coronal base densities n_0 are directly given by the Baumbach−Allen formula (Eq. 1.6.1) [i.e., $n_0 = 2.99 \times 10^8$ cm^{-3} (K-corona), $n_0 = 1.55 \times 10^8$ cm^{-3} (L-corona), and $n_0 = 3.6 \times 10^6$ cm^{-3} (F-corona)], where the densities of the K and L-corona (see fits in Fig. 3.11) are also fully consistent with other quiet Sun values we infer here. Also the DEM temperature ranges inferred from the best fits, $T = 0.75 - 1.25$ MK for the K-corona, and $T = 0.9 - 2.3$ MK for the F and L-corona, are fully consistent with quiet Sun values. Most interesting is that the best fits confirm that the K-corona is exactly hydrostatic (i.e., $q_\lambda = 1.0$; Fig. 3.11), while the L-corona has an extended scale height of $q_\lambda = 1.9$ times the hydrostatic scale height, a value that is typical for coronal streamers. We can, therefore, associate the L-corona dominantly to coronal streamer regions. Finally, the F-corona exhibits an extreme density scale height factor of $q_\lambda \approx 10$ times the hydrostatic scale height. Such an excessive scale height cannot be consistent with

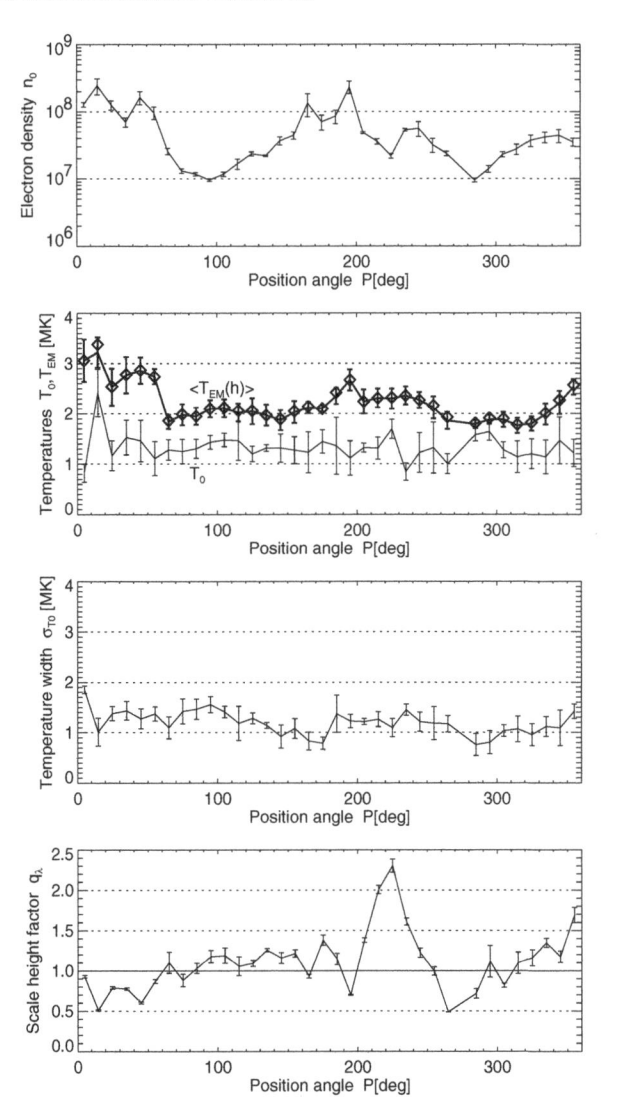

Figure 3.10: Best-fit parameters of the multi-hydrostatic coronal density model (Eqs. 3.1.1−9) in *Yohkoh/SXT* data (Plate 2) at 36 position angles around the disk: the base electron density $n_{e,0}$ (top panel), the electron temperature T_0 at the dEM(T) peak, along with the averaged temperature over height, $< T_{EM}(h) >$ (second panel), the Gaussian temperature width σ_T of the $dEM(T)$ distribution (third panel), and the ratio q_λ of the best-fit exponential scale height λ to the thermal scale height $\lambda_p(T)$ (from Aschwanden & Acton 2001).

gravitational stratification. It is, therefore, suggestive that the F-corona does not represent coronal plasma in hydrostatic equilibrium, but rather the density distribution of interplanetary dust that has its own dynamics escaping solar gravitation. Thus, our DEM model is not only capable of reproducing the Baumbach−Allen formula and

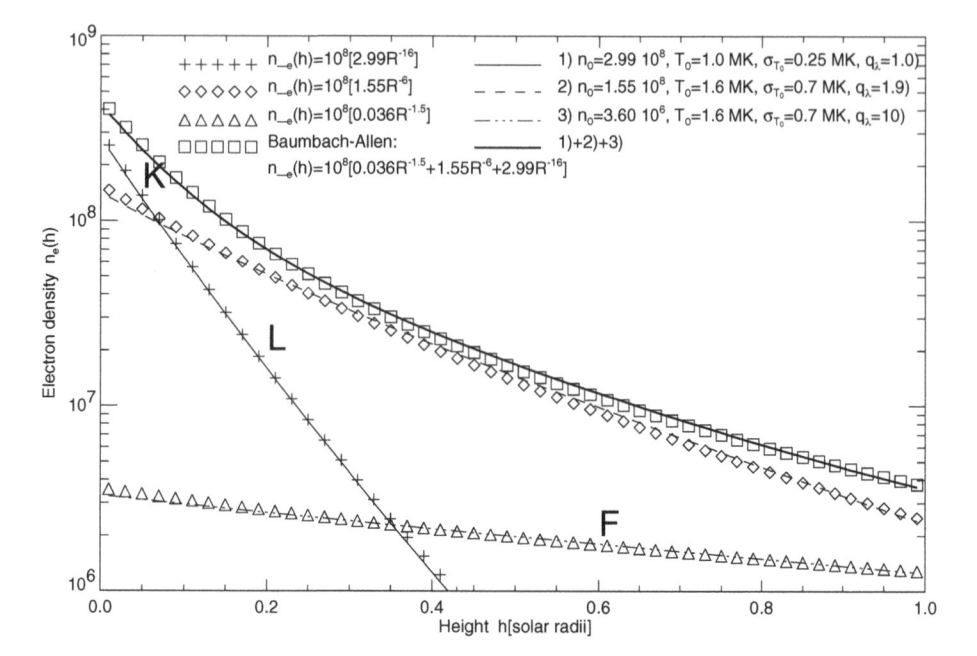

Figure 3.11: The Baumbach−Allen formula $n_e(h)$ (squares) and its powerlaw components, associated with the K-corona (crosses), the L-corona (diamonds), and the F-corona (triangles). Approximate fits to these 3 components and the sum (i.e., Baumbach−Allen formula) are shown with different linestyles, and the fit parameters $(n_0, T_0, \sigma_{T_0}, q_\lambda)$ are listed in the top part of the figure (Aschwanden & Acton 2001).

each of its three coronal components quantitatively in detail (see fits in Fig. 3.11), but it also provides us with additional physical information that is not explicitly used in the derivation of the Baumbach−Allen formula, namely the DEM temperature ranges and the density scale heights of the three coronal components. We can, therefore, relate the 6 coefficients of the semi-empirical Baumbach−Allen formula to 12 physical parameters in the framework of our DEM model (using Eq. 3.3.5),

$$
n_e(R) = \sum_{i=1}^{3} \int \frac{n_i}{\sqrt{2\pi}\,\sigma_{T_i}} \exp\left[-\frac{(R-1)R_\odot}{q_{\lambda,i}\lambda_0 T} - \frac{(T-T_i)^2}{2\sigma_{T_i}^2}\right] dT , \qquad (3.3.13)
$$

where the summation $i = 1, 2, 3$, represents the 3 coronal components $(K, L, F\text{-}corona)$, each one specified by a base density n_i, a DEM temperature range $T_i \pm \sigma_{T_i}$, and a scale height ratio $q_{\lambda,i} = \lambda_{n,i}/\lambda_{T_i}$.

Thus we found that the Baumbach−Allen formula is fully consistent with our DEM model, where every magnetic field line has a constant temperature with height, $T_e(h)$, while the height variation of the emission measure weighted temperatures $T_{EM}(h)$ can naturally be explained by the multi-scale height weighting effects of a broad DEM distribution.

3.4 Loop Geometry and Inclination

After we have modeled the homogeneous part of the corona, which is also called *quiet Sun corona* or *diffuse background corona*, we turn now to inhomogeneous structures. Inhomogeneous structures entail active regions, filaments, prominences, or any magnetic field topology that is filled with denser plasma. One of the foremost geometric topologies is that of a *coronal loop*. Many parts of the solar corona are made of dipole-like field lines, and thus can be represented by a set of coronal loops. In this section we deal with the elementary geometry of loop structures, which provides us with a framework for physical modeling.

In all previous sections we considered the hydrostatic equation as a function of height h or radial distance from the Sun's center, $r = R_\odot + h$. Such a parameterization works for vertical structures, such as fluxtubes along open field lines in coronal holes. Now we quantify the geometry for structures along closed field lines, such as active region loops. A convenient approximation of closed magnetic field lines is a *coplanar circular geometry*, where the center of the circle can be offset from the solar surface, and the loop plane can be inclined to the solar vertical.

3.4.1 Vertical Semi-Circular Loops

Let us start with the simplest geometry (i.e., a semi-circular loop with a vertical loop plane). The height $h(s)$ of a position specified by the loop length coordinate s (starting at one footpoint) is

$$h(s) = r(s) - R_\odot = \frac{2L}{\pi} \sin\left(\frac{\pi s}{2L}\right) , \tag{3.4.1}$$

where L is the *loop half length*. The *pressure balance* or *momentum equation* (Eq. 3.1.5) as a function of loop length s is then

$$\frac{dp}{ds}(s) = \frac{dp_{grav}(r)}{dr}\left(\frac{dr}{ds}\right) , \tag{3.4.2}$$

which involves the derivative $dr(s)/ds$, which is (from Eq. 3.4.1)

$$\left(\frac{dr}{ds}\right) = \left(\frac{dh}{ds}\right) = \cos\left(\frac{\pi s}{2L}\right) = \cos\varphi \tag{3.4.3}$$

where φ is the angle in the circular loop plane. If we want to specify the pressure $p(s)$ as a function of the loop length s, we can use the solution derived for h (Eq. 3.1.15) where we substitute the scaling $h(s)$ between the height h and loop length s for a specific geometric model (e.g., Eq. 3.4.1 for a semi-circular loop),

$$p(s) = p_0 \exp\left[-\frac{(h[s] - h_0)}{\lambda_p(T_e)(1 + \frac{h[s]}{R_\odot})}\right] . \tag{3.4.4}$$

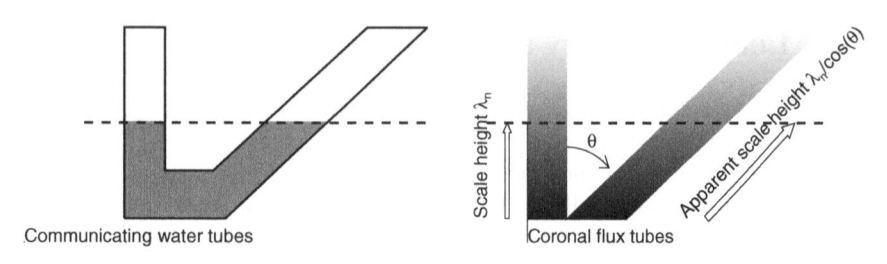

Figure 3.12: Vertical and apparent density scale heights in coronal loops (right) and analogy with communicating water tubes (left).

3.4.2 Inclined Semi-Circular Loops

In the next step we relax the assumption of a vertical loop plane and allow for an inclination angle θ. A large inclination angle of the loop plane has two primary effects: (1) the observed density scale height along a loop does not correspond to the hydrostatic scale height, which is strictly defined in vertical direction; and (2) the chance to observe complete large loops is much higher for inclined loop planes than for vertical ones, in particular if they have a size in excess of several density scale heights (e.g., Fig. 3.13). The hydrostatic scale height λ_p has always the same vertical extent, regardless of how much the loop is inclined, similar to the water level in communicating water tubes with different slopes (Fig. 3.12). The apparent density scale height as observed for an inclined loop is stretched by the cosine angle,

$$\lambda_p^{obs}(\theta) = \frac{\lambda_p}{\cos(\theta)} \; , \tag{3.4.5}$$

For solutions of the hydrostatic equations it is most convenient to keep the same coordinate system with s the loop coordinate in the (inclined) loop plane, but to correct for the loop inclination by modifying the gravity $g_\odot$ by an *effective gravity*,

$$g_{eff} = g_\odot \cos(\theta) \; , \tag{3.4.6}$$

in the definition of the pressure scale height λ_p (Eq. 3.1.16). The hydrostatic solutions can also be applied to other stars, in which case the *effective gravity* can be adjusted in the same way by replacing the solar gravity $g_\odot$ by the stellar gravity g_*,

$$g_{eff} = g_* \cos(\theta) \; . \tag{3.4.7}$$

Thus, we will use in the following the generalized definition of the *pressure scale height* λ_p in terms of g_{eff} (Eq. 3.4.6),

$$\lambda_p(T_e) = \frac{2k_B T_e}{\mu m_H g_{eff}} \approx 4.6 \times 10^9 \left(\frac{g_\odot}{g_{eff}} \right) \left(\frac{T_e}{1 \text{ MK}} \right) \text{ [cm]} \; . \tag{3.4.8}$$

In essence, inclined loops have larger pressure scale heights, which are equivalent to a lower effective gravity, but the functional form of the hydrostatic solutions remains the same.

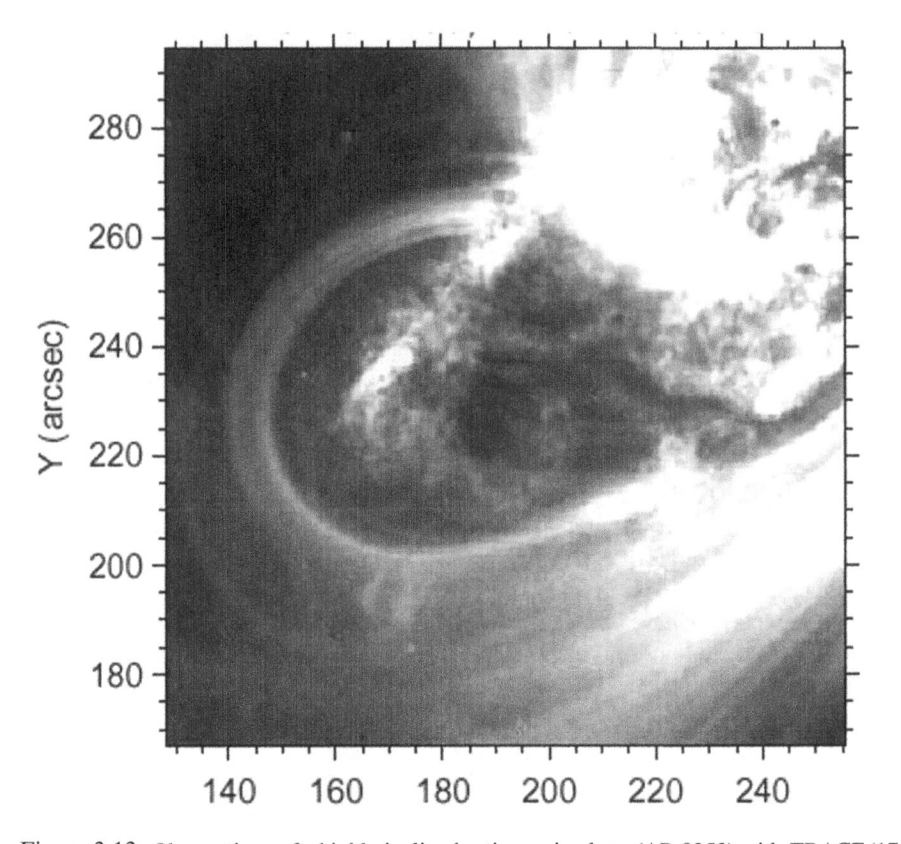

Figure 3.13: Observations of a highly inclined active region loop (AR 8253) with *TRACE* (171 Å) on 1998-Jun-26, 14:25 UT. The loop is estimated to have a length of $L \approx 100$ Mm and and an inclination angle of $\theta \approx 60°$ from the vertical (Reale et al. 2000a).

3.4.3 Diverging Semi-Circular Loops

For a 3D model we need in addition to the 1D parameterization along the loop (with coordinate s) also a 2D parameterization transverse to the loop coordinate s, which is generally characterized by a spherical cross-sectional area $A(s)$. The simplest assumption is a constant cross section $A(s) = const$, which is often fulfilled to a good approximation, and can also be justified on theoretical grounds for current-carrying loops (Klimchuk 2000). On the other side, if a magnetic potential field is assumed, loops are expected to have a diverging cross section with height. The cross-sectional variation $A(s)$ along the loop can then be estimated by two adjacent magnetic field lines of a buried line dipole, as shown in Fig. 3.14. In the geometry of Vesecky et al. (1979), the two defining field lines join at the center of the buried dipole (Fig. 3.14 left), and where the cross section expands by a factor Γ above the solar surface. A restriction of this geometry is that the expansion factor Γ is a strict function of the height of the curvature center with respect to the photospheric surface, which cannot always be matched with observations. For instance, semi-circular loops can only have

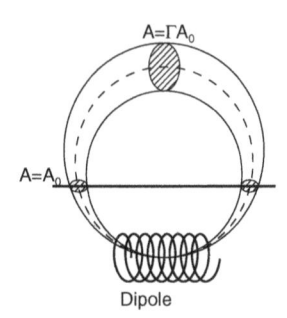
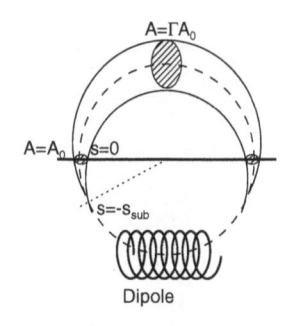

Figure 3.14: Geometry of expanding loops with a buried dipole: (a) circular loops with zero cross section at the lowest point and loop expansion factor Γ above the solar surface (Vesecky et al. 1979); (b) semi-circular loops with zero cross section at coordinate $s = -s_{sub}$ and loop expansion factor of Γ above the solar surface (Aschwanden & Schrijver 2002).

an expansion factor of $\Gamma = 2$ in the geometry of Vesecky et al. (1979). In order to allow for an arbitrary range of expansion factors Γ for semi-circular loops, Aschwanden & Schrijver (2002) introduced a generalized definition shown in Fig. 3.14 (right). A semi-circular loop can have any (positive) expansion factor above the solar surface, but the zero cross section point needs to be at a depth of $s = -s_{sub}$. Algebraically, the cross-sectional area depends on the loop coordinate s as,

$$A(s) = A_0 \Gamma \sin^2 \left(\frac{\pi}{2} \frac{s + s_{sub}}{L + s_{sub}} \right), \tag{3.4.9}$$

where the location of the zero cross section point depends on Γ as,

$$s_{sub}(\Gamma) = \frac{L}{[(\pi/2)/\arcsin(1/\Gamma^{1/2}) - 1]}. \tag{3.4.10}$$

This definition yields an area of $A(s = 0) = A_0$ at the solar surface, and an expanded area $A(s = L) = A_0\Gamma$ at the looptop position ($s = L$). Loops with a constant cross section require $\Gamma = 1$ and $s_{sub} = \infty$. Note, however, that this geometric model for loop divergence is only appropriate for the coronal parts of loops, above the canopy geometry in the transition region (Fig. 4.25).

3.4.4 Coordinate Transformations

There are three natural coordinate systems when it comes to analyze, model, or simulate loops: (1) the *observers coordinate system* (x, y, z); (2) the *heliographic coordinate system* (l, b) with longitude l and latitude b, which is invariant to solar rotation; and (3) the *loop plane coordinate system* (X, Y, Z) aligned with the loop plane. Because the Sun is rotating, the coordinate transformations are time-dependent, if non-simultaneous images are analyzed. In the following section we provide the most general coordinate transformations between the three coordinate systems (also described in Aschwanden et al. 1999a), which are particularly useful for analysis of stereoscopic observations or any type of 3D modeling.

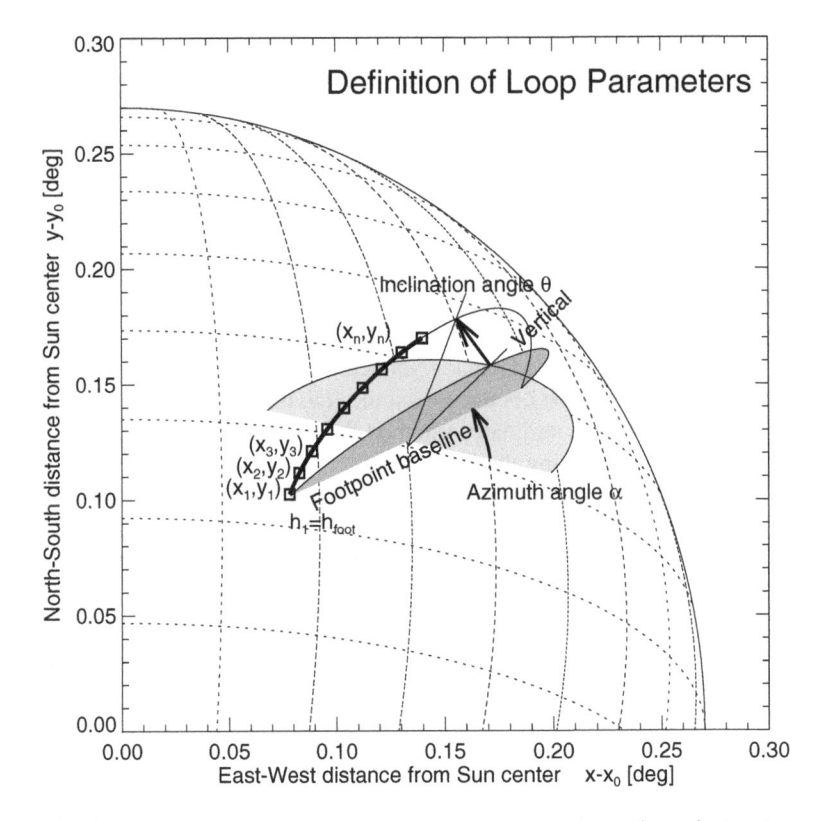

Figure 3.15: Definition of loop geometry parameters: loop positions (x_i, y_i), $i = 1, ..., n$, the azimuth angle α between the loop footpoint baseline and heliographic east-west direction, and the inclination angle θ of the loop plane with respect to the vertical on the solar surface.

Image (or Observer's) Coordinate System

(x, y, z): The (x, y) coordinates refer to the x-axis and y-axis of an observed image, while the coordinate (z) is orthogonal to the image, or parallel to the line-of-sight direction, defined as positive towards the observer. The origin $(x, y, z) = (0, 0, 0)$ of this coordinate system is most conveniently assumed at the Sun center position. A solar FITS image should contain the position of the Sun center in pixel units (i_{x0}, j_{y0}) (in FITS header *CRPIX1*, *CRPIX2* or *E_XCEN*, *E_YCEN*), the pixel size $(\Delta x, \Delta y)$ in units or arcseconds (in FITS header *CDELT1*, *CDELT2*), and the solar radius i_{r0} in pixel units (in FITS header *SOLAR_R*, or *E_XSMD*, *E_YSMD* if the semi-diameters of an ellipse are fitted). With this information, a pixel position (i, j) can then be converted into the coordinate system (x, y) by

$$x_i = \Delta x \left(i - i_{x0} \right) ,$$

$$y_i = \Delta y \left(j - j_{x0} \right) , \tag{3.4.11}$$

where Δx =arcseconds/pixel for (x, y) in units of arcseconds, or $\Delta x = R_\odot/i_{r0}$, with $R_\odot = 696$ Mm, if physical length units (Mm) are preferred.

Heliographic Coordinate System

(l, b, r): The heliographic coordinate system is co-rotating with the solar surface. A position on the solar surface is generally specified by heliographic longitude and latitude coordinates (l, b) (in units of heliographic degrees), with reference to the *Carrington rotation* grid. The heliographic longitude and latitude $[l_0(t), b_0(t)]$ of the Sun's center and the solar position angle $P(t)$ of the solar rotation axis are published as a function of time t in *The Astronomical Almanac* (Nautical Almanac Office, NRL, Washington DC), or can conveniently be obtained from the *IDL* routine *SUN.PRO* in the *SSW* package. The 2D spherical coordinate system (l, b) can be generalized into a 3D coordinate system by incorporating the height h above the solar surface, which can be expressed as a dimensionless distance to the Sun center (in units of solar radii),

$$r = \left(1 + \frac{h}{R_\odot}\right) . \tag{3.4.12}$$

The transformation from the 3D heliographic coordinate system (l, b, r) into image coordinates (x, y, z) can be accomplished by applying a series of 4 rotations to the (normalized) vector $(0, 0, r)$ (Loughhead et al. 1983; Aschwanden et al. 1999a),

$$\begin{pmatrix} x/R_\odot \\ y/R_\odot \\ z/R_\odot \end{pmatrix} = \begin{pmatrix} \cos(P + P_0) & -\sin(P + P_0) & 0 \\ \sin(P + P_0) & \cos(P + P_0) & 0 \\ 0 & 0 & 1 \end{pmatrix} \begin{pmatrix} 1 & 0 & 0 \\ 0 & \cos(b_0) & -\sin(b_0) \\ 0 & \sin(b_0) & \cos(b_0) \end{pmatrix}$$

$$\begin{pmatrix} \cos(l_0 - l) & 0 & -\sin(l_0 - l) \\ 0 & 1 & 0 \\ \sin(l_0 - l) & 0 & \cos(l_0 - l) \end{pmatrix} \begin{pmatrix} 1 & 0 & 0 \\ 0 & \cos(-b) & -\sin(-b) \\ 0 & \sin(-b) & \cos(-b) \end{pmatrix} \begin{pmatrix} 0 \\ 0 \\ r \end{pmatrix} ,$$

$$\tag{3.4.13}$$

where (l_0, b_0) are the heliographic longitude and latitude of the Sun center, P is the position angle of the solar rotation axis with respect to the north-south direction (defined positive towards east), and P_0 is the image rotation (roll) angle with respect to the north-south direction, $(P + P_0 = 0$ for images rotated to solar north). In stereoscopic correlations, only the longitude of the Sun center $l_0(t)$ is time-dependent in first order (according to the solar rotation rate), while $b_0(t)$ and $P(t)$ are slowly varying, and thus almost constant for short time intervals.

Loop Plane Coordinate System

(X, Y, Z): To parameterize coronal loops it is convenient to introduce a cartesian system that is aligned with the loop footpoint baseline (X-axis) and coincides with the loop plane (X–Z plane, $Y = 0$). For instance, a circular loop model defined in the X–Z plane is specified in Eqs. (3.4.1). The transformation of loop coordinates $(X, Y = 0, Z)$ into a cartesian coordinate system (X', Y', Z') that is aligned with the heliographic coordinate system (l, b, r), can simply be accomplished by two rotations,

$$
\begin{pmatrix} X' \\ Y' \\ Z' \end{pmatrix} = \begin{pmatrix} \cos(\alpha) & -\sin(\alpha) & 0 \\ \sin(\alpha) & \cos(\alpha) & 0 \\ 0 & 0 & 1 \end{pmatrix} \begin{pmatrix} 1 & 0 & 0 \\ 0 & \cos(\theta) & \sin(\theta) \\ 0 & -\sin(\theta) & \cos(\theta) \end{pmatrix} \begin{pmatrix} X \\ Y \\ Z \end{pmatrix} ,
$$

$$(3.4.14)$$

where the *azimuth angle* α denotes the angle between the loop footpoint baseline and the east-west direction, and θ represents the inclination or tilt angle between the loop plane and the vertical to the solar surface (Fig. 3.15). Placing the origin of the loop coordinate system $(X = 0, Y = 0, Z = 0)$ [which is also the origin of the rotated coordinate system $(X' = 0, Y' = 0, Z' = 0)$] at heliographic position (l_1, b_1) at an altitude h_{Foot} above the solar surface, the transformation into heliographic coordinates is given by

$$
\begin{pmatrix} l \\ b \\ r \end{pmatrix} = \begin{pmatrix} l_1 + \arctan[X'/(Z' + h_{Foot} + R_\odot)] \\ b_1 + \arctan[Y'/(Z' + h_{Foot} + R_\odot)] \\ \sqrt{[X'^2 + Y'^2 + (Z' + h_{Foot} + R_\odot)^2]}/R_\odot \end{pmatrix} .
$$

$$(3.4.15)$$

Column Depth of Loops with Constant Cross Section

In order to convert observed emission measures $EM(x, y) = \int n_e^2(x, y, z)dz$ into local electron densities $n_e(x, y, z)$ we need information on the column depth $\int dz$. An approximation that is often useful can be obtained from coronal loops with a constant cross section w, which can be measured from the FWHM as it appears perpendicular to the line-of-sight in the plane of the sky. For 3D models of loops parameterized by coordinates (x_i, y_i, z_i), the angle ψ between the line-of-sight and a loop segment can then directly be derived by the ratio of the projected to the effective length of a loop segment $[i, i + 1]$,

$$
\cos(\psi[x_i, y_i, z_i]) = \frac{\sqrt{(x_{i+1} - x_i)^2 + (y_{i+1} - y_i)^2}}{\sqrt{(x_{i+1} - x_i)^2 + (y_{i+1} - y_i)^2 + (z_{i+1} - z_i)^2}} ,
$$

$$(3.4.16)$$

yielding the column depth w_z along the line-of-sight axis z,

$$
w_z[x_i, y_i, z_i] = \frac{w}{\cos(\psi[x_i, y_i, z_i])} .
$$

$$(3.4.17)$$

3.5 Loop Line-of-Sight Integration

Whenever we analyze images in soft X-rays and EUV wavelengths, we deal with optically thin emission, which represents an integral along the line-of-sight. While the relevant theory is tersely defined by the *radiative transfer equation* (Eqs. 2.1.5−7), the application to real data is a kind of art that requires careful treatment to obtain correct results. Unfortunately, many papers have been published with over-simplified models of the line-of-sight integration and have produced incorrect results, in particular in studies on coronal loops. It is, therefore, important that we familiarize ourselves first with the subtle effects that matter in the line-of-sight integration, before we proceed

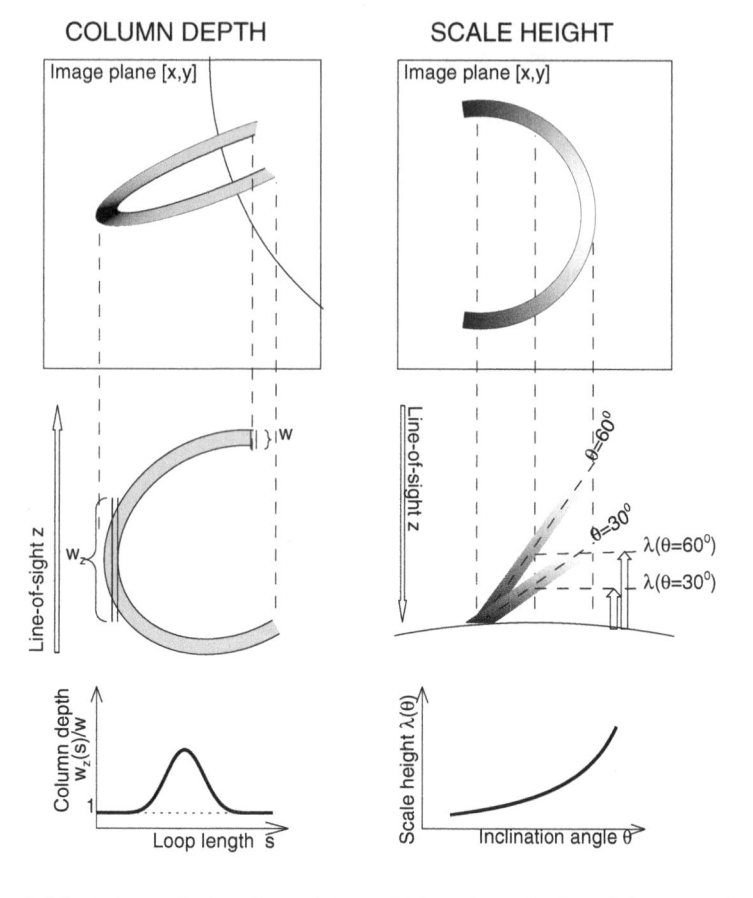

Figure 3.16: *Left panel:* the effect of the variable column depth $w_z(s)$ measured parallel to the line-of-sight z is illustrated as a function of the loop length parameter s, for a loop with a constant diameter w. *Right panel:* the effect of the inclination angle θ of the loop plane on the inferred density scale height $\lambda(\theta)$ is shown. Both effects have to be accounted for when determining the electron density $n_e(s)$ along the loop.

with physical loop models. In this section we discuss 3 cases: (1) an intense single loop, (2) a faint loop embedded in the background corona, and (3) a statistical distribution of loops embedded in the background corona. All cases are highly relevant to the analysis of *Yohkoh/SXT*, *SoHO/EIT*, *SoHO/CDS*, and *TRACE* images.

3.5.1 Bright Single Loop

What we mean with a bright loop is that the observed brightness of the loop structure has sufficient contrast to the background (at least 10:1) so that the background can be neglected and ignored in the data analysis. However, even if we manage to find an isolated bright loop that is most suitable for a quantitative analysis, there are still a number of effects that need to be considered: (1) projection effects, (2) loop plane

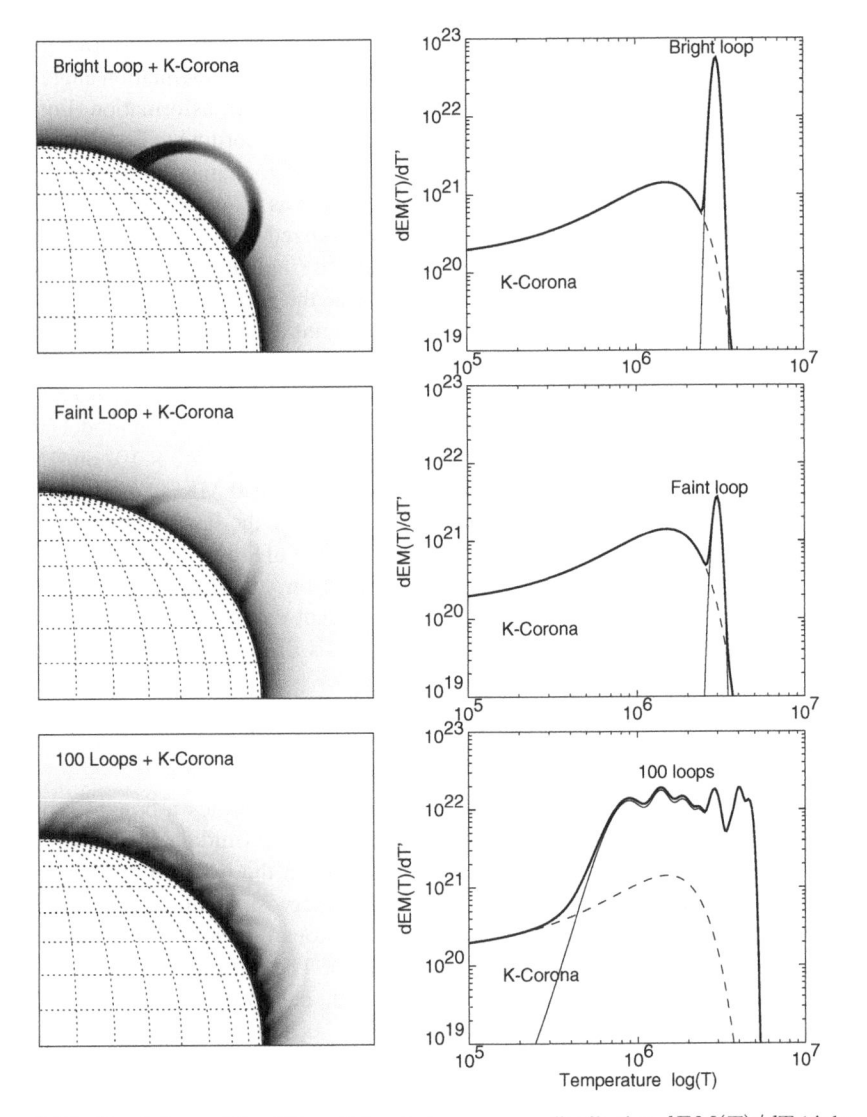

Figure 3.17: Simulations of the differential emission measure distribution $dEM(T)/dT$ (right panels) and emission measure maps $EM(x,y)$ (left panels) for a bright loop (top panels), a faint loop (middle panels), and a statistical distribution of 100 loops (bottom panels). The emission measure of the diffuse K-corona is also added (dashed line in right panels). The greyscale is logarithmic, with maximum contrast in the range of $EM = 0.8 - 1.0 \times 10^{28}$ cm^{-5}. The physical parameters are described in the text.

inclination, (3) cross-sectional variation, (4) instrumental temperature bandpass, etc. Ideally, the simplest case would be a coronal loop above the limb with a vertical loop plane, perpendicular to the line-of-sight, and a constant cross section. If the loop plane appears at an angle to the line-of-sight, the angle $\psi(x_i, y_i, z_i)$ of each loop segment

$s(x_i, y_i, z_i)$ varies along the loop according to Eq. (3.4.16), and the resulting projected column depth w_z is approximately given by Eq. (3.4.17). The inclination angle θ of the loop plane is taken into account by the proper coordinate transformation (Eq. 3.4.14) and affects the inference of the pressure scale height λ_p according to Eq. (3.4.5). Cross-sectional variations $A(s)$, require a geometric model, e.g. characterized with an expansion factor Γ (Eq. 3.4.9), and affect the column depth as $w_z(s) = \sqrt{A(s)}$. The effects of column depth and inclined scale height are visualized in Fig. 3.16.

In Fig. 3.17 we perform a simulation of a bright single loop embedded in a diffuse background corona. For the bright loop we choose the parameters: base density $n_e = 2 \times 10^9$ cm^{-3}, temperature $T_e = 3.0$ MK, footpoint distance $d_{foot} = 400$ Mm, and loop width $w = 50$ Mm. This bright loop has a total emission measure of $EM_{loop} \approx n_e^2 * w = 2 \times 10^{28}$ cm^{-5} at the footpoint. The background corona has been modeled from parameters that approximately reproduce the Baumbach–Allen model (which is dominated by the *K-corona* component): a base density of $n_e = 5 \times 10^8$ cm^{-3}, a peak temperature of $T_0 = 1.5$ MK, a Gaussian width $\sigma_T = 1.0$ MK, and a hydrostaticity factor of $q_\lambda = 1.9$. The equivalent column depth above the limb is about $z_{eq} \approx 10^{10}$ cm (Eq. 3.3.10 and Fig. 3.9), and the emission measure of the background corona above the limb is about $EM_{back} \approx n_e^2 z_{eq} = 2.5 \times 10^{27}$ cm^{-5}. Thus the contrast is about $EM_{loop}/E_{back} \approx 10$, so that there is no significant confusion with the background which can legitimately be neglected in the analysis of the physical loop parameters. However, such bright and isolated loops are rare.

3.5.2 Faint Single Loop

If we go to fainter loops, the background confusion becomes a crucial problem in the data analysis. In Fig. 3.17 we simulate a faint loop (middle panel) which has a density ($n_e = 5 \times 10^8$ cm^{-3}) four times lower than the bright loop (Fig. 3.17 top panel) defined in the previous section, but with otherwise identical parameters. This means that the emission measure contrast is a factor of 16 lower than for the bright loop, and comparable with the background corona. The differential emission measure distribution $dEM(T)/dT$ in Fig. 3.17 (middle right panel) shows that the loop has still some contrast above the background corona at $T \approx 2.5 - 3.5$ MK, so a background subtraction could be made in the data, provided that a narrowband filter is available in this temperature range (e.g., with *TRACE* 284 Å or with a *Yohkoh/SXT* filter). However, if an emission measure distribution $dEM^{loop+back}(T)/dT$ is reconstructed at a loop location, the non-loop contributions of the background corona along the line-of-sight cannot be subtracted out properly, because they are co-spatial. A separation of the loop from the background corona could only be accomplished by performing an emission measure distribution reconstruction of the background corona at adjacent positions to the loop (in the same altitudes), which would yield the dashed distribution $dEM^{back}(T)/dT$ in Fig. 3.17.

3.5.3 Statistical Distribution of Loops

Let us illustrate the line-of-sight problem in loop analysis in a more realistic way. In the third case shown in Fig. 3.17 (bottom panels) we simulated 100 loops with randomly chosen sizes, densities (in the range of $n_e = 1 - 5 \times 10^8$ cm^{-3}), and temperatures (in the range $T_e = 0.5 - 5.0$ MK), superimposed on the background corona. The differential emission measure distributions of both the 100 loops and the background corona are shown in the bottom right panel of Fig. 3.17. Obviously the forest of loops forms a diffuse inhomogeneous background that is hard to distinguish from the background corona. Also in the differential emission measure distribution, where the 100 loops dominate, none of the 100 loops can be separated properly. Multi-temperature analysis is done with a limited number of SXR and EUV lines [e.g., from *SERTS* (Brosius et al. 2000) or from *SoHO/CDS* (Schmelz et al. 2001)], which show a relatively broad differential emission measure distribution. Generally, the contrast between the brightest loop feature and the background corona is weaker the broader the temperature. This is expected for the situation shown in Fig. 3.17 (bottom panel). It is, therefore, recommended only to perform multi-temperature analysis on loops with strong contrast, or to use narrow-band filters (e.g., EIT or TRACE) to benefit from a better temperature discrimination (Aschwanden 2002a).

3.6 Hydrostatic Solutions and Scaling Laws

In the previous sections we described all coronal loops with an *isothermal approximation* to make the essential aspects of the multi-temperature structure of the corona transparent. However, while the isothermal approximation works fine for the coronal segments of loops, it breaks down at the footpoints of the loops, in the transition region and chromosphere. When we talk about *hydrostatic solutions*, we mean a stationary solution of the density $n_e(s)$ (or pressure $p_e(s)$) and temperature profile $T_e(s)$ that is in hydrostatic equilibrium and matches the chromospheric (at $s = 0$) and coronal boundary conditions (at $s = L$).

A *hydrostatic solution* has to fulfill both the *momentum equation* and the *energy equation*,

$$\frac{dp}{ds} - \frac{dp_{grav}}{dr}\left(\frac{dr}{ds}\right) = 0 \,, \tag{3.6.1}$$

$$E_H(s) - E_R(s) - \frac{1}{A(s)}\frac{d}{ds}A(s)F_C(s) = 0 \,. \tag{3.6.2}$$

where the energy equation (expressed in conservative form) contains a heating rate $E_H(s)$ and two loss terms, the radiative loss $E_R(s)$ (defined in Eq. 2.9.1) and the conductive loss term, which is expressed as the divergence of the conductive flux $F_C(s)$,

$$F_C(s) = \left[-\kappa T^{5/2}(s)\frac{dT(s)}{ds}\right] = -\frac{2}{7}\kappa\frac{d}{ds}\left[T^{7/2}(s)\right] \,, \tag{3.6.3}$$

with $\kappa = 9.2 \times 10^{-7}$ (erg s^{-1} cm^{-1} K$^{-7/2}$) the Spitzer conductivity. The least known term is the volumetric heating rate $E_H(s)$ along the loop, which crucially depends on assumptions for the physical heating mechanism. Many previous loop models assumed uniform heating, $E_H(s) = const$ (e.g. Rosner et al. 1978a), for the sake of simplicity.

A more general parameterization of the heating function involves two free parameters; the base heating rate E_{H0} and an *exponential heating scale length* s_H, which was introduced by Serio et al. (1981),

$$E_H(s) = E_0 \exp\left(-\frac{s}{s_H}\right) = E_{H0} \exp\left(-\frac{s - s_0}{s_H}\right) . \qquad (3.6.4)$$

where these authors defined the base heating rate E_0 at height $s = 0$, while we refer to the *coronal base heating rate* E_{H0} at some height $s = s_0$ above the chromosphere.

A hydrostatic solution is uniquely defined by three free parameters (i.e., the loop length L, the base heating rate E_{H0}, and the heating scale length s_H) plus the boundary conditions of vanishing conductive flux at both sides of the loop boundary,

$$F_C(s = s_0) = 0 , \qquad (3.6.5)$$

$$F_C(s = L) = 0 , \qquad (3.6.6)$$

as well as the temperature T_0 at the loop base,

$$T(s = s_0) = T_0 , \qquad (3.6.7)$$

which is usually associated with the temperature minimum region (i.e., $T_0 \approx 2 \times 10^4$ K at a height of $h_0 \approx 1300$ km). Alternatively, instead of the three independent parameters (L, s_H, E_{H0}), an equivalent set of (L, s_H, T_{max}) is used (e.g. in Serio et al. 1981), where T_{max} represents the loop maximum temperature, often located near the looptop.

The hydrostatic solutions of the momentum and energy equation (Eqs. 3.6.1−2) can be found numerically, either by explicit codes (e.g., a *shooting method* that starts from one boundary and varies the free parameters until the solutions meet the second boundary condition) or by implicit codes (keeping the parameters at the boundaries fixed and optimizing the solutions in between). No strict analytical solution can be derived, mainly because the energy equation contains a differential equation of second order (in temperature), and also because the radiative loss function is parameterized in arbitrary ways (e.g., with a 7-piece powerlaw function in Rosner et al. 1978a). In the following we describe how the hydrostatic solutions have been solved in historical order, first in the approximation of Rosner et al. (1978a) for the case of *uniform heating*, then in the generalization of Serio et al. (1981) for the case of *nonuniform heating*, and finally we provide the most accurate analytical approximations known today, derived by Aschwanden & Schrijver (2002).

3.6.1   Uniform Heating and RTV Scaling Laws

In their seminal paper, Rosner et al. (1978a) derived hydrostatic solutions under the assumption of constant pressure $p(s) = 2n_e(s)k_B T(s) = const$ and a constant heating rate $E_H(s) = const$. The assumption of a constant pressure can be justified for small loops with a length of less than a pressure scale height $L < \lambda_p$, or for hot loops, where the pressure scale height can be much longer than for cool loops. The assumption of a constant heating rate can be justified for some long-range heating mechanisms, but seems not to be consistent with the latest *TRACE* observations. The way Rosner et al. (1978a) solved the hydrostatic equations was by expressing the conductive flux $F_C(s)$ as a function of temperature $F_C(T)$ (using Eq. 3.6.3),

$$\frac{dF_C(s)}{ds} = \frac{dF_C(T)}{dT}\frac{dT(s)}{ds} = \frac{dF_C}{dT}\left(-\frac{F_C(T)}{\kappa T^{5/2}}\right) , \qquad (3.6.8)$$

which then allows us to write the energy equation (Eq. 3.6.2) as a temperature integral,

$$-\int F_C dF_C = -\frac{1}{2}F_C^2\Big|_{F_C(T_0)}^{F_C(T)} = \int_{T_0}^{T} \kappa T'^{5/2}E_H(T')dT' - \int_{T_0}^{T} \kappa T'^{5/2}E_R(T')dT' . \qquad (3.6.9)$$

Rosner et al. (1978a) then introduced two auxiliary functions $f_H(T)$ and $f_R(T)$ for the two temperature integrals,

$$f_H(T) = \int_{T_0}^{T} 2\kappa T'^{5/2}E_H(T')dT' , \qquad (3.6.10)$$

$$f_R(T) = \int_{T_0}^{T} 2\kappa T'^{5/2}E_R(T')dT' , \qquad (3.6.11)$$

leading to an equation that separates the length coordinate s and temperature profile T, and thus provides us with an explicit solution for the inverse temperature profile $s(T)$,

$$(s - s_0) = \int_{s_0}^{s} ds = \int_{T_0}^{T} -\frac{\kappa T'^{5/2}}{F_C(T')}dT' = \int_{T_0}^{T} \frac{-\kappa T'^{5/2}}{\sqrt{f_R(T') - f_H(T')}}dT' . \qquad (3.6.12)$$

Rosner et al. (1978a) made then the approximations of: (1) a single powerlaw function for the radiative loss function, $\Lambda(T) \approx \Lambda_0 T^{-1/2}$ (Fig. 2.14); (2) the neglect of the auxiliary function with the heating term $f_H(T) \lesssim f_R(T)$; (3) constant acceleration $g_{eff}(r) \approx g_\odot$; and (4) constant cross section $A(s) = const$, and derived from Eq. (3.6.12) an analytical expression for the inverse temperature profile $s(T)$,

$$s(T) = s(T_0) + 2.5 \times 10^5 p_0^{-1}$$

$$\times \left\{9.6 \times 10^{-16}T_{max}^3 \left[\arcsin\left(T/T_{max}\right) - (T/T_{max})(1 - (T/T_{max})^2)^{1/2}\right] + 1\right\} . \qquad (3.6.13)$$

This leads directly to the famous RTV scaling laws, which specify two relations between three independent parameters: the loop length L, the looptop temperature T_{max}, and the base pressure p_0 (assumed to be constant throughout the loop in the RTV approximation),

$$T_{max} \approx 1400(p_0 L)^{1/3} , \qquad (3.6.14)$$

$$E_{H0} \approx 0.98 \times 10^5 p_0^{7/6}L^{-5/6} = 0.95 \times 10^{-6}T_{max}^{7/2}L^{-2} . \qquad (3.6.15)$$

These *RTV scaling laws* have been applied mainly to soft X-ray loops over the last 20 years, which have a sufficiently high temperature to make the assumption of a constant loop pressure viable. Tests of the *RTV scaling laws* seemed to be roughly consistent with observations (at least when parameters are compared on a log-log scale), so that other assumptions in the RTV derivation were tacitly accepted also, such as that of a uniform heating rate. Only recently, when more accurate measurements with *TRACE* data were performed on cooler EUV loops did it become clear that the assumption of a uniform heating rate was untenable.

3.6.2 Nonuniform Heating and Gravity

Two important generalizations were introduced by Serio et al. (1981): (1) the application to a nonuniform heating rate, and (2) the consideration of gravity. (The effects of gravity and loop expansion was also considered at the same time by Wragg & Priest 1981, 1982). The most common parameterization of the heating rate is a concentration near the loop footpoints, which can be characterized by an exponential function with a heating scale height of s_H (Eq. 3.6.4), as originally introduced by Serio et al. (1981). The consideration of gravity involves the height dependence of the pressure (Eq. 3.1.15) in the momentum equation (Eq. 3.1.8), which is approximately given by the pressure scale height $\lambda_p \approx \lambda_p(T)$ (Eq. 3.1.16). The consideration of these two important effects leads to a significant modification of the RTV scaling laws, which read in Serio's version as,

$$T_{max} \approx 1400(p_0 L)^{1/3} \exp\left(-0.08\frac{L}{s_H} - 0.04\frac{L}{\lambda_p}\right), \qquad (3.6.16)$$

$$E_{H0} \approx 0.95 \times 10^{-6} T_{max}^{7/2} L^{-2} \exp\left(0.78\frac{L}{s_H} - 0.36\frac{L}{\lambda_p}\right). \qquad (3.6.17)$$

The deviations from the RTV scaling laws for uniform heating ($s_H = \infty$) are most prominent for short heating scale heights $s_H < L$. For instance, if the heating scale height is 10% of the loop length $L/s_H = 10$, then the loop base pressure increases by a factor of $\exp(-0.08 * 10)^{-3} \approx 11$, and the heating rate by a factor $\exp(0.78 * 10) \approx 2.4 \times 10^3$ for a hydrostatic solution. Essentially, the shorter the heating scale height, the higher are the base pressure p_0 and base heating rate E_{H0}. Note that the Serio et al. (1981) treatment, although it corrects for the largest physical effects, still contains a number of approximations: (1) the pressure scale height λ_p is defined with the approximation $p(s) \approx p_0 \exp\left(-s/\lambda_p\right)$, but neglects the temperature variation T_s along the loop, and thus the variation of the pressure scale height $\lambda_p(s)$ (see Eq. 3.4.4); (2) the height dependence in the pressure function is approximated by the semi-circular loop coordinate s, $p(h) \approx p(s)$; (3) the numerical coefficients (in Eqs. 3.6.16 and 3.6.17) were calculated for loops with an expansion factor of $\Gamma = 0$ (but with different weighting of chromospheric and coronal lines); and (4) the height variation of gravity is neglected, $g(r) \approx g_\odot$.

3.6.3 Analytical Approximations of Hydrostatic Solutions

Since no analytical solutions of the hydrostatic equations (Eqs. 3.6.1−2) are known, the solutions for the pressure $p(s)$, density $n_e(s)$, and temperature $T(s)$ have to be

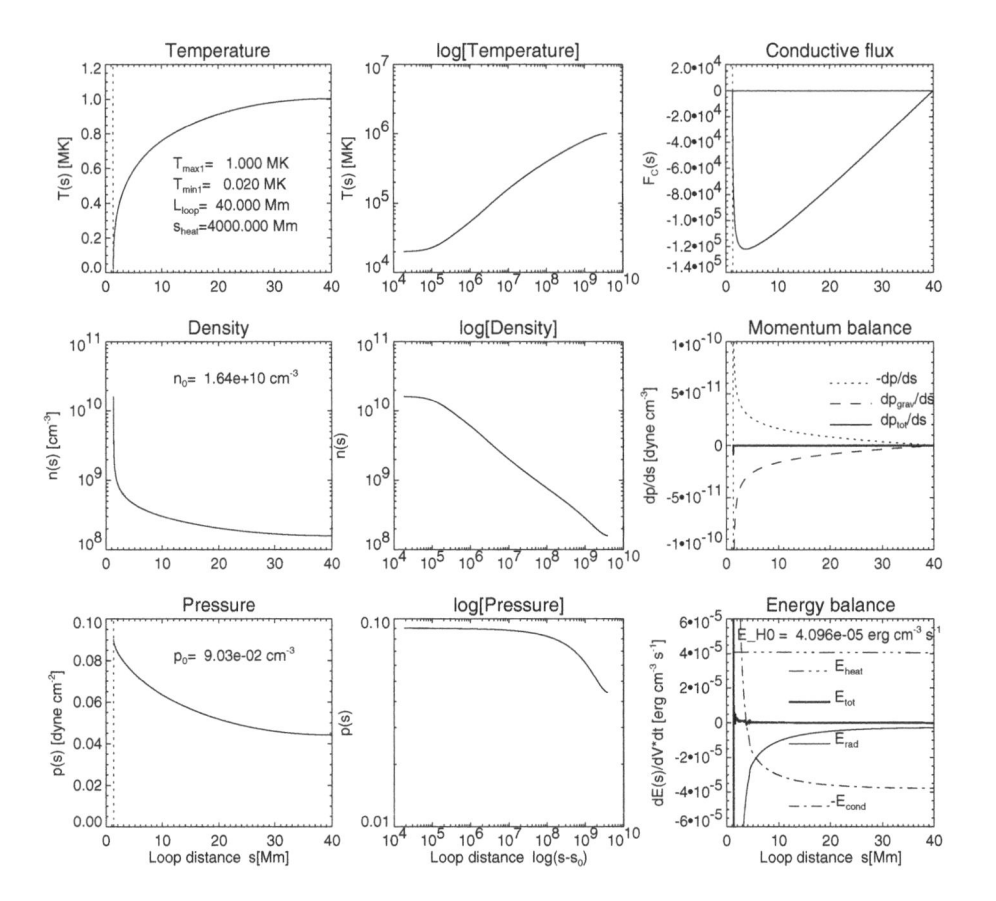

Figure 3.18: A hydrostatic solution of a uniformly heated loop with a looptop temperature of $T_{max} = 1.0$ MK and loop half length of $L = 40$ Mm, computed with a numeric code. Note that the transition region is adequately resolved with the numeric code and the boundary conditions of $T(s_0) = 2 \times 10^4$ MK and vanishing flux $dT/ds(s_0) = 0$ are accurately met (see temperature profile in middle top panel). The numeric solution is tested by the criteria that the momentum balance is zero along the loop (thick solid line in middle right panel) and that the energy balance is zero along the loop (thick solid line in bottom right panel) (Aschwanden & Schrijver 2002).

calculated by a numerical code (see example in Fig. 3.18). However, accurate approximations have been found by Aschwanden & Schrijver (2002), based on numerical solutions from a large parameter space. The analytical approximations are expressed for the temperature function $T(s)$ with

$$T(s) = T_{max} \left[1 - \left(\frac{L-s}{L-s_0} \right)^a \right]^b , \qquad (3.6.18)$$

and the pressure function $p(s)$ with

$$p(s) = p_0 \exp \left[-\frac{h(s) - h_0}{\lambda_p(s)} \right] , \qquad (3.6.19)$$

with the pressure scale height $\lambda_p(s)$,

$$\lambda_p(s) = \lambda_0 \left(\frac{T(s)}{10^6 K}\right) \left(1 + \frac{h(s)}{R_\odot}\right) q_\lambda(L, s_H), \tag{3.6.20}$$

and the density function $n_e(s)$,

$$n(s) = \frac{p(s)}{k_B T(s)}. \tag{3.6.21}$$

All coefficients are mainly functions of the ratio of the heating scale height s_H to the loop half length L, and depend only weakly on the temperature T_{max}. This dependence of the three coefficients a, b, and q_λ is characterized by,

$$a(L, s_H) = a_0 + a_1 \left(\frac{L}{s_H}\right)^{a_2}, \tag{3.6.22}$$

$$b(L, s_H) = b_0 + b_1 \left(\frac{L}{s_H}\right)^{b_2}, \tag{3.6.23}$$

$$q_\lambda(L, s_H) = c_0 + c_1 \left(\frac{L}{s_H}\right)^{c_2}, \tag{3.6.24}$$

and the subcoefficients a_i, b_i, and c_i are listed in Table 3.1 for a valid temperature range of $T = 1 - 10$ MK.

Also the RTV or Serio's scaling laws can be expressed with more accurate analytical approximations, with the coefficients d_i and e_i listed in Table 3.1,

$$T_{max} \approx (p_0 L)^{1/3} \times d_0 \left[\exp\left(d_1 \frac{L_0}{s_H} + d_2 \frac{L_0}{\lambda_p}\right) + d_3 \frac{L_0}{s_H} + d_4 \frac{L_0}{\lambda_p}\right], \tag{3.6.25}$$

$$E_{H0} \approx T_{max}^{7/2} L^{-2} \times e_0 \left[\exp\left(e_1 \frac{L_0}{s_H} + e_2 \frac{L_0}{\lambda_p}\right) + e_3 \frac{L_0}{s_H} + e_4 \frac{L_0}{\lambda_p}\right]. \tag{3.6.26}$$

These approximations have been tested in the parameter space of loop lengths $L = 4 - 400$ Mm, heating scale heights $s_H = 4 - 400$ Mm, and maximum temperatures $T_{max} = 1 - 10$ MK. The application is most straightforward by using the parameters $[L, s_H, T_{max}]$ as independent parameters. If the base heating rate E_{H0} is preferred as the independent parameter (i.e., $[L, s_H, E_{H0}]$), which is more physical from the point of view of an initial-value problem, an inversion of T_{max} is required, which is approximately,

$$T_{max}(L_0, s_H, E_{H0}) \approx 55.2 \left[E_{H0}^{0.977} L_0^2 \exp\left(-0.687 \frac{L_0}{s_H}\right)\right]^{2/7}. \tag{3.6.27}$$

A summary of the analytical formulae is given in Table 3.2, where the equations are listed in the order they need to be calculated (a numerical *IDL* code is available in the *SolarSoftWare SSW*). These analytical approximations also accommodate for: (1) a finite height s_0 of the coronal loop base, at which the base pressure p_0 and base heating

Table 3.1: Coefficients in analytical approximations of hydrostatic solutions

Coefficient	$T_{max} = 1$ MK	$T_{max} = 3$ MK	$T_{max} = 5$ MK	$T_{max} = 10$ MK
a_0	2.098	2.000	2.055	2.026
a_1	0.258	0.343	0.328	0.298
a_2	1.565	1.418	1.649	1.570
b_0	0.320	0.323	0.329	0.309
b_1	-0.009	-0.008	-0.009	-0.009
b_2	0.877	0.902	0.852	0.928
c_0	0.693	0.699	0.700	0.707
c_1	0.026	0.024	0.014	0.029
c_2	1.199	1.240	2.427	0.915
d_0	1452	1416	1428	1506
d_1	-0.074	-0.087	-0.064	-0.036
d_2	-0.030	-0.044	0.000	0.001
d_3	-0.001	-0.003	-0.023	-0.021
d_4	0.015	0.043	0.010	0.011
$e_0 \times 10^6$	0.686	0.831	0.808	0.707
e_1	0.558	0.848	0.847	0.685
e_2	-0.423	-0.707	-0.634	-0.403
e_3	0.548	0.057	-0.058	0.063
e_4	0.156	0.365	0.361	0.145

rate E_{H0} are defined (where the half loop length above the base is called $L_0 = L - s_0$); (2) for an inclination angle θ of the loop plane; (3) for loop expansion factors Γ; as well as for (4) extremely short heating scale heights $s_H \ll L$, which require a slightly modified parameterization of the temperature function $T(s)$ (Eq. 3.6.18).

The scaling laws (using the accurate approximations) are shown in Fig. 3.19 for the base pressure $p_0(L, s_H, T_{max})$ (Eq. 3.6.25 and Fig. 3.19), and for the base heating rate $E_{H0}(L, s_H, T_{max})$ (Eq. 3.6.26 and Fig. 3.20). One sees that the base pressure p_0 as well as the base heating rate E_{H0} is always lowest for uniform heating, whereas both parameters systematically increase with shorter heating scale heights s_H or with temperature T_{max}.

3.7 Heating Function

The heating function $E_H(s)$ is the least known term in the energy balance equation of hydrostatic solutions (Eq. 3.6.2), while the radiative loss function $E_R(s)$ is well determined for known elemental abundances (Eq. 2.9.1), and the conductive flux $F_C(s)$ is well-described in terms of Spitzer conductivity (Eq. 3.6.3). The formulation of thermal conductivity of ionized plasmas in the framework of Spitzer & Harm (1953) assumes that the electron mean free path is much less than the temperature scale height, which is generally assumed to be the case in the corona, otherwise models with non-local heat transport have to be employed (e.g., Ciaravella et al. 1991). The reason why the heating

Table 3.2: Summary of analytical formulae to calculate hydrostatic solutions and scaling laws.

Constants:	
— Height of loopbase:	$s_0 = 1.3 \times 10^8$ cm
— Temperature at loopbase:	$T_0 = 2.0 \times 10^4$ K
— Solar radius:	$R_\odot = 6.96 \times 10^{10}$ cm
— Solar gravity:	$g_\odot = 2.74 \times 10^4$ cm s^{-2}
— Spitzer conductivity:	$\kappa = 9.2 \times 10^{-7}$ erg s^{-1} cm^{-1} K$^{-7/2}$
Independent Variables:	
Loop half length	$L[cm]$
Heating scale length	$s_H[cm]$
Loop top temperature	$T_{max}[K]$
Base heating rate	$E_{H0}[erg\ cm^{-3}\ s^{-1}]$
Loop plane inclination angle	$\theta[deg]$
Loop expansion factor	$\Gamma \geq 1$
Choice 1: $[L, s_H, T_{max}, \theta, \Gamma]$	
Choice 2: $[L, s_H, E_{H0}, \theta, \Gamma]$	$\mapsto T_{max} \approx 55.2[E_{H0}^{0.977} L_0^2 \exp\left(-0.687 L_0/s_H^\Gamma\right)]^{2/7}$
Dependent Parameters:	
Half loop length above base	$L_0 = L - s_0$
Loop height	$h_1 = (2L/\pi)$
Subphotospheric zero point	$s_{sub} = L\left[(\pi/2)/\arcsin(1/\Gamma^{1/2}) - 1\right]^{-1}$
Equivalent heating scale length	$s_H^\Gamma = s_H \quad$ if $(\Gamma = 1)$
	$s_H^\Gamma = s_H/[1 + \ln\Gamma + 2\ln(\sin[(\pi/2)(s_H + s_{sub})/(2L + s_{sub})])]$
	if $\Gamma > 1$
Temperature index 1	$a = a_0 + a_1 \left(L_0/s_H^\Gamma\right)^{a_2}$
Temperature index 2	$b = b_0 + b_1 \left(L_0/s_H^\Gamma\right)^{b_2}$
Scale height factor	$q_\lambda = c_0 + c_1 \left(L_0/s_H^\Gamma\right)^{c_2}$
Effective gravity	$g_{eff} = g_\odot \cos\theta$
Effective scale height	$\lambda_0 = \left(2k_B 10^6[K]/\mu m_p g_{eff}\right) = 4.6 \times 10^9 \ (1/\cos\theta)$ cm
Serio scale height	$\lambda_p = \lambda_0 \left(T_{max}/10^6 K\right)$
Scaling law factor 1	$S_1 = d_0 \left[\exp\left(d_1 L_0/s_H^\Gamma + d_2 L_0/\lambda_p\right) + d_3 L_0/s_H^\Gamma + d_4 L_0/\lambda_p\right]$
Scaling law factor 2	$S_2 = e_0 \left[\exp\left(e_1 L_0/s_H^\Gamma + e_2 L_0/\lambda_p\right) + e_3 L_0/s_H^\Gamma + e_4 L_0/\lambda_p\right]$
Base heating rate (for Choice 1)	$E_{H0} = L_0^{-2} T_{max}^{7/2} S_2$
Base pressure	$p_0 = L_0^{-1}(T_{max}/S_1)^3$
Analytical Approximations:	
Normalized length coordinate	$z(s) = (L - s)/(L - s_0)$
Height (in loop plane)	$h'(s) = h_1 \sin(s/h_1)$
Loop cross-sectional area	$A(s) = \Gamma \sin^2\left[(\pi/2)(s + s_{sub})/(L + s_{sub})\right]$
Temperature (if $s_H^\Gamma/L > 0.3$)	$T(s) = T_{max}\left[1 - z^a\right]^b$
Temperature (if $s_H^\Gamma/L \leq 0.3$)	$T(s) = T_{max}[1 - z^a]^b \left[1 + 0.5\ ^{10}\log(L/s_H)(1 - z)z^5\right]$
Conductive flux	$F_C(s) = -\kappa T(s)^{5/2} dT(s)/ds$
Pressure scale height	$\lambda_p(s) = \lambda_0 \left(T(s)/10^6 K\right)(1 + h'(s)/R_\odot) q_\lambda$
Pressure	$p(s) = p_0 \exp\left[-[h'(s) - h'(s_0)]/\lambda_p(s)\right]$
Density	$n(s) = (p(s)/2k_B T(s))$

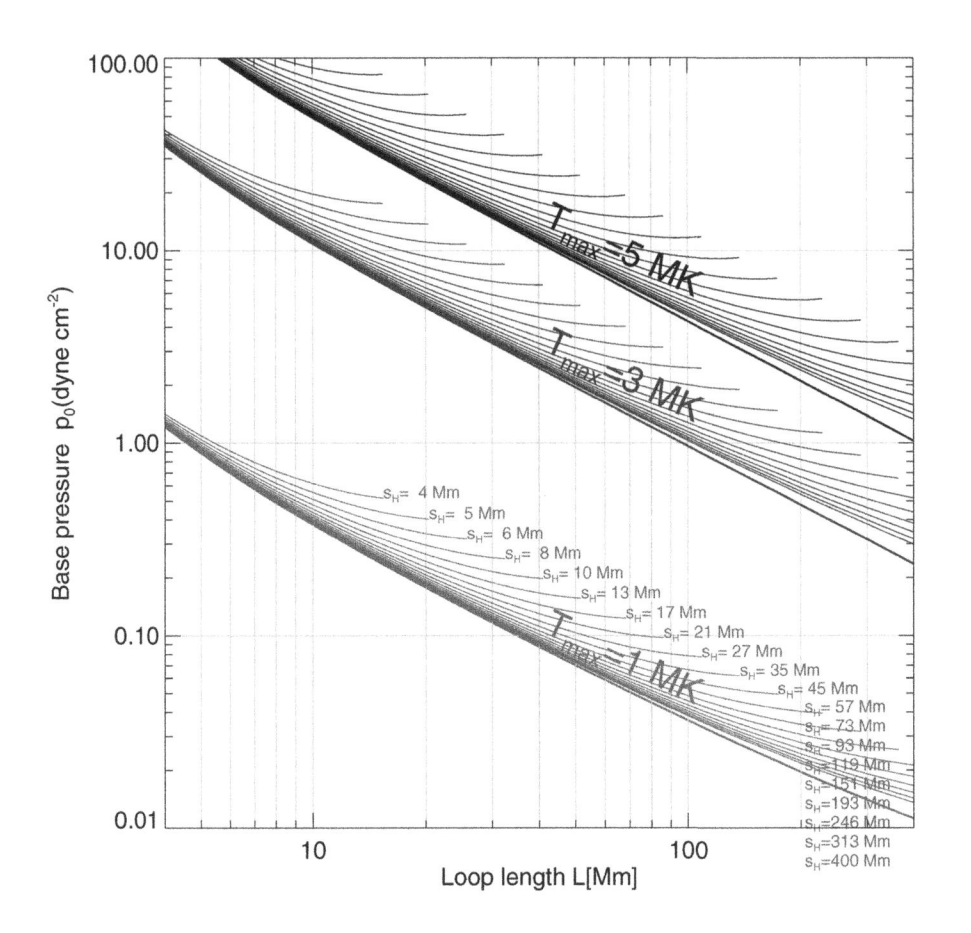

Figure 3.19: The scaling law for the base pressure $p_0(L, s_H, T_{max})$ according to the analytical approximation (Eq. 3.6.25) (Aschwanden & Schrijver 2002).

function cannot be quantified is simply because neither the dominant physical heating mechanism of coronal loops nor the spatial distribution function of the energy input is known. Energy could be conveyed by plasma heating processes in the corona (via magnetic reconnection) or in the chromosphere (via particle bombardment), or by means of waves and turbulence. We will discuss theoretical coronal heating mechanisms in more detail in § 9. Most of the theoretical heating mechanisms are insufficiently advanced to quantify their spatial distribution. At this point we are in a situation of trying arbitrary mathematical functions for the parameterization of the heating function $E_H(s)$, which we explore by fitting the resulting hydrostatic solutions to observations, hoping to find constraints. A number of heating functions have been proposed, e.g.

$$E_H(s) = E_{H0} \ (constant) \,, \tag{3.7.1}$$

$$E_H(s) = E_{H0} \exp\left(-\frac{s - s_0}{s_H}\right) \,, \tag{3.7.2}$$

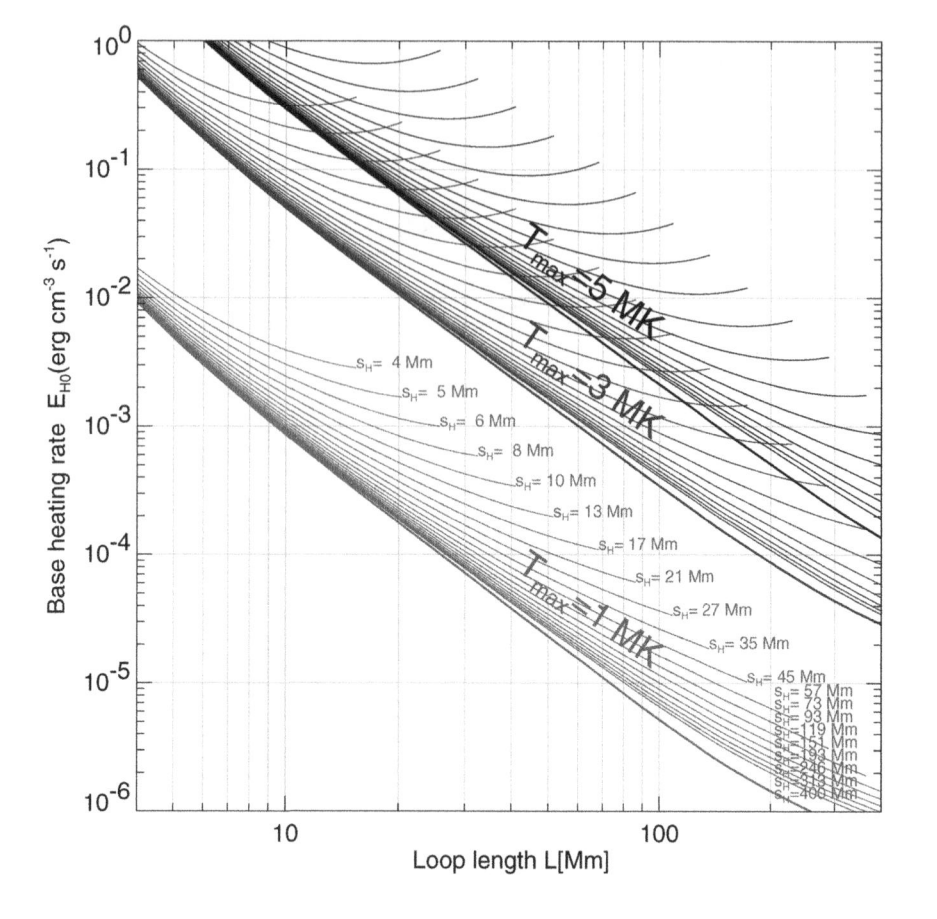

Figure 3.20: The scaling law for the base heating rate $E_{H0}(L, s_H, T_{max})$ according to our analytical approximation (Eq. 3.6.26), (Aschwanden & Schrijver 2002).

$$E_H(s) = E_{H0} \exp\left(\frac{s - L}{s_H}\right), \qquad (3.7.3)$$

including *uniform heating* (Eq. 3.7.1) or *nonuniform heating*, characterized by an exponential function, either concentrated at the footpoint $s \approx s_0$ (Eq. 3.7.2) or at the looptop $s = L$ (Eq. 3.7.3). The case of uniform heating was already treated by Rosner et al. (1978a), the case of footpoint heating by Serio et al. (1981), while the case of looptop heating was applied to data by Priest et al. (2000).

In Fig. 3.21 we compare a hydrostatic solution of a uniformly heated loop (with a heating scale height of $s_H \gg L$) with that of a footpoint heated loop (with a heating scale height of $s_H/L = 0.3$). The main effect in the energy balance along the loop (Fig. 3.21 bottom panels) is that the reduced heat input in the upper part of the loop requires less heat conduction, and thus less of a temperature gradient dT/ds in the loop segment above the heating scale height $s_H \lesssim s < L$. Consequently, footpoint heated

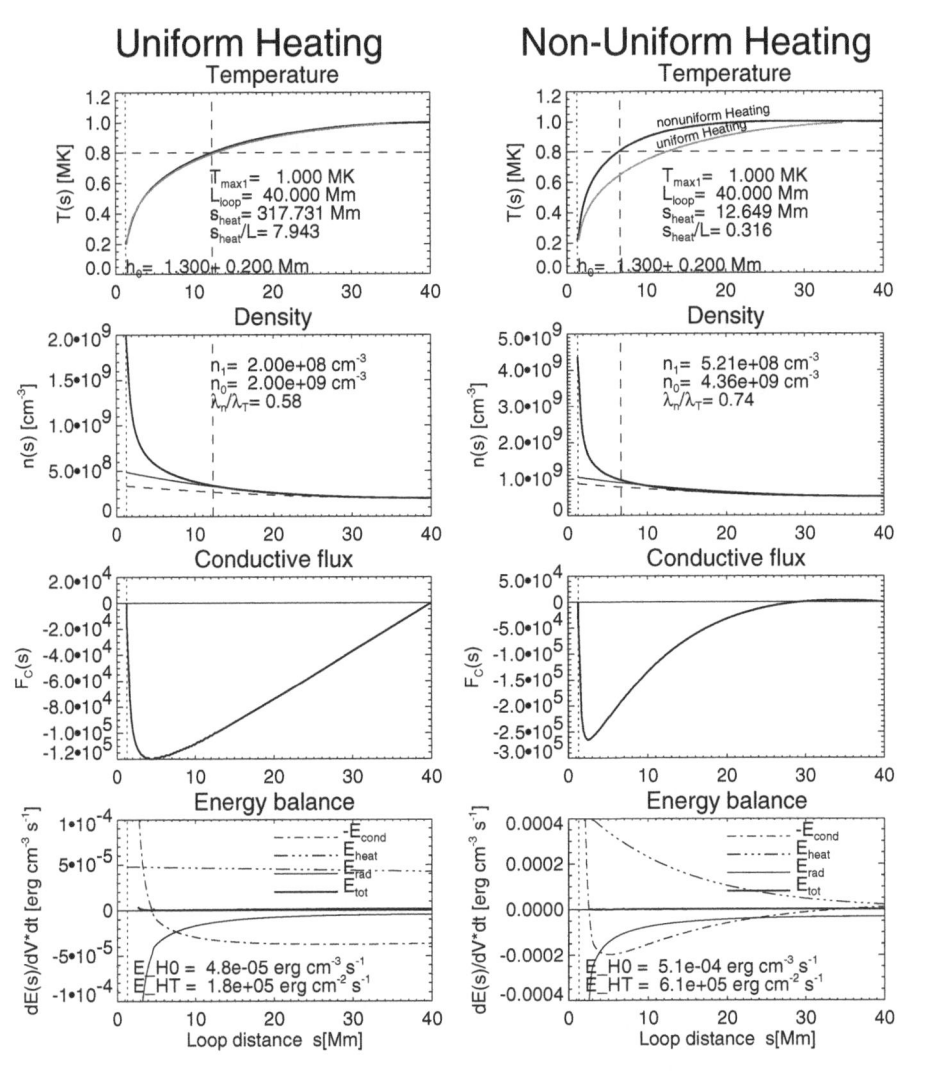

Figure 3.21: Hydrostatic solution for a loop with half length $L = 40$ Mm, looptop temperature of $T_{max} = 1.0$ MK, and uniform heating (left panels), and for nonuniform heating $s_H = L/3 = 13$ Mm (right panels). The four panels show the solution of the temperature profile $T(s)$ (top), the density profile $n_e(s)$ (second row), the conductive flux $F_C(s)$ (third row), and the energy balance (bottom row). Note that the footpoint heated loop (left) is more isothermal in the coronal segment $s > s_H$ (Aschwanden et al. 2001).

loops are more isothermal in their coronal segments than uniformly heated loops (see temperature profiles in Fig. 3.21 top). On the other hand, the temperature gradient and thus the conductive flux is larger near the footpoints, and thus requires a larger base heating rate E_{H0} than uniformly heated loops. Therefore, another characteristic of footpoint heated loops is their higher base heating rate E_{H0} and higher base pressure

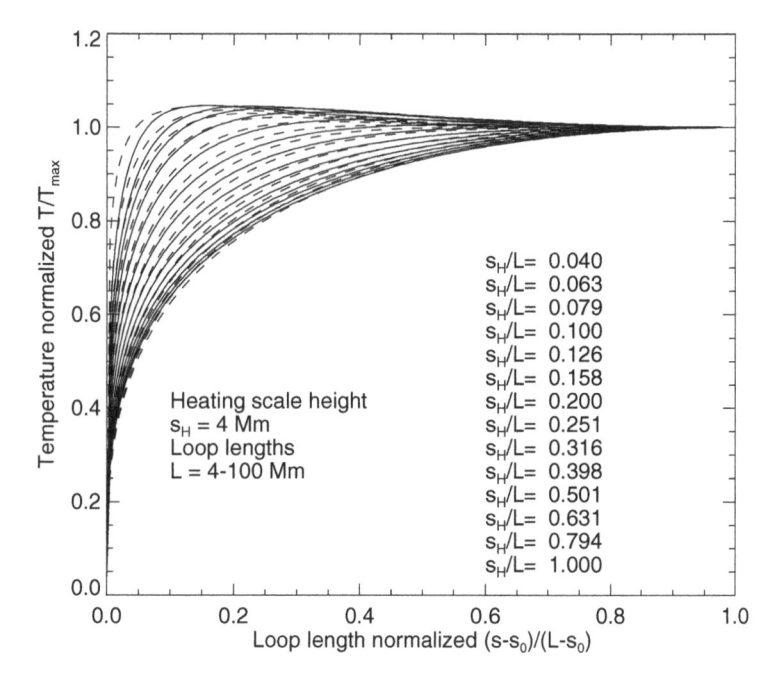

Figure 3.22: Hydrostatic solutions of the temperature profile $T(s)$ for near-uniform heating ($s = L$) and short heating scale heights ($s_H/L = 0.04, ..., 1.0$). The numeric solutions are shown in solid lines, the analytical approximations (Eqs. 3.6.18 and 3.7.4) in dashed lines (Aschwanden & Schrijver 2002).

p_0, which is also seen in the scaling laws (Figs. 3.19 and 3.20), and generally produces a higher density $n_e(s)$ along the entire loop than a uniformly heated loop.

The higher degree of *isothermality* of footpoint heated loops is illustrated in Fig. 3.22, where temperature profiles $T_e(s)$ are calculated for a range of short heating scale heights $s_H/L = 0.04, ..., 1.0$. For very short heating scale heights, say $s_H/L \lesssim 1/3$, we see also that the temperature maximum is not located at the loop apex anymore, but closer to the footpoints. This means that we have upward thermal conduction in the upper part of the loop to redistribute the excess of the footpoint heating rate that cannot be radiated away. The analytical approximation for $T(s)$ (Eq. 3.6.18) needs to be modified for such short heating scale heights, which can be approximated by,

$$T'(s) = T_{max}[1 - z^a]^b \left[1 + 0.5 \, ^{10}\log\left(\frac{L}{s_H}\right)(1 - z)z^5\right] , \qquad (3.7.4)$$

where $z = (s - s_0)/L - s_0)$ represents the normalized length coordinate, shown as dashed curves in Fig. 3.22. A comparison with observed temperature profiles, obtained from the 171/195 Å filter ratio of 41 loops observed with *TRACE*, indeed reveals that the temperature profiles are near isothermal in their coronal segments, say at $s/L \gtrsim 0.2$, and thus significantly flatter than in the RTV model with uniform heating (Fig. 3.23, top panel).

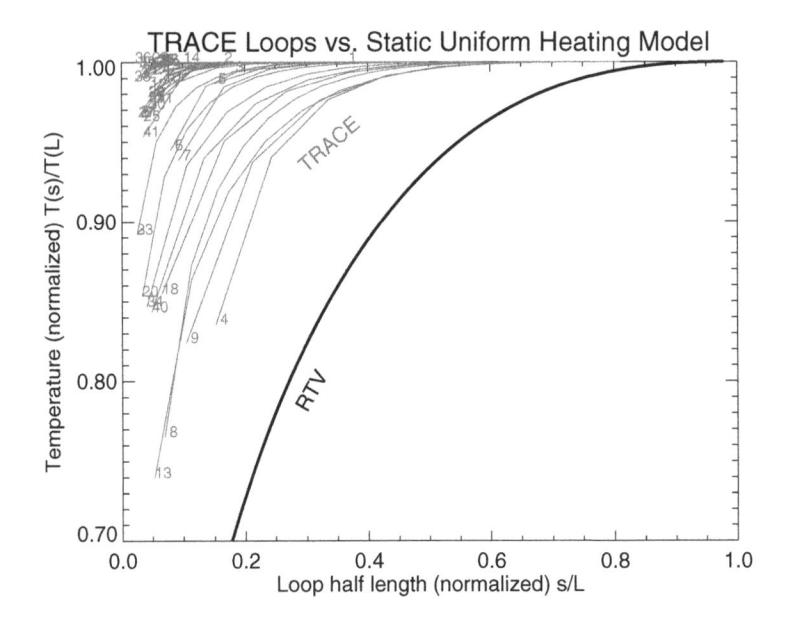

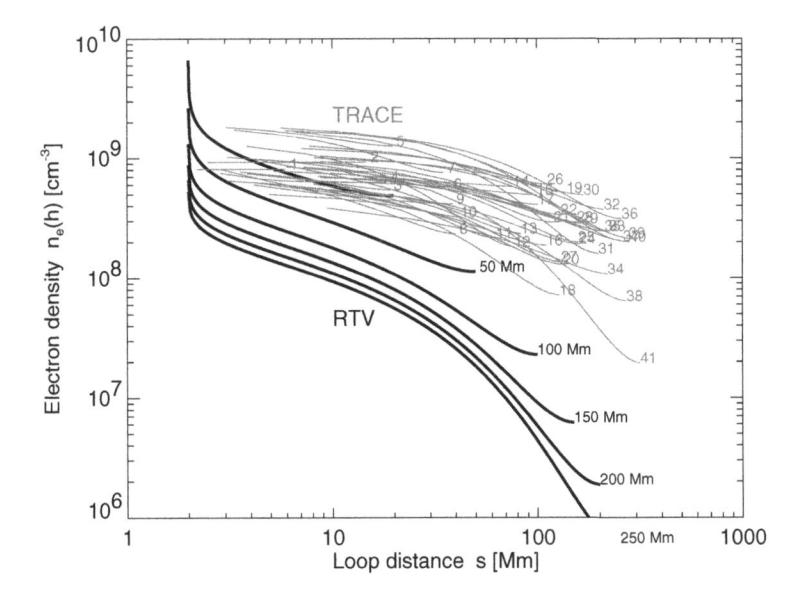

Figure 3.23: Normalized temperature profiles $T(s)/T_{max}$ as a function of the normalized loop lengths s/L are shown in the top panel, inferred from the *TRACE* filter ratio of 171 Å and 195 Å in 41 coronal loops. The inferred electron density profiles $n_e(s)$ are shown in the lower panel, compared with hydrostatic solutions of the RTV model with uniform heating. Note that the observed loops are more isothermal and have higher densities, as expected for a nonuniform (footpoint) heating function (Aschwanden et al. 2000d).

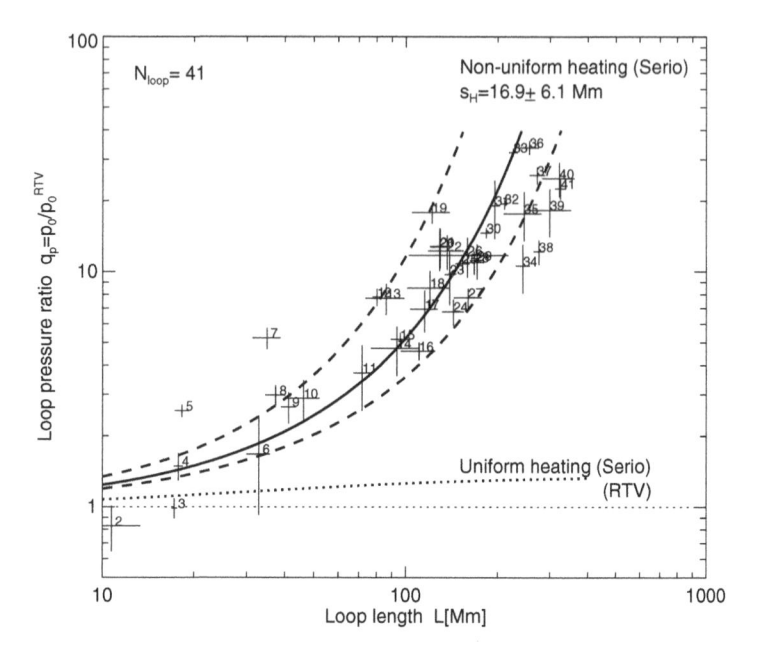

Figure 3.24: A fit of the nonuniform heating model of Serio et al. (1981), to the observed loop pressure factors $q_p = p_0/p_0^{RTV}$. The best fit yields a heating scale height of $s_H = 16.9 \pm 6.1$ Mm: the mean is shown with a thick solid line, the $\pm\sigma_{s_H}$ standard deviations with dashed lines. Note that the alternative model with uniform heating according to the RTV scaling law of Rosner et al. (1978a), or using the formalism of Serio et al. (1981) with $s_H = \infty$ are not consistent with the data (Aschwanden et al. 2000d).

The other hallmark of footpoint heated loops is their enhanced base pressure p_0, or equivalently, their higher density $n_e(s)$ throughout the loop. This characteristic is clearly confirmed in the observed density profiles, as shown from 41 loops observed with *TRACE* in Fig. 3.23 (bottom panel). The observed densities are typically an order of magnitude higher than predicted by the RTV model with uniform heating. The ratio of the base pressure p_0 of a footpoint heated loop to that p_0^{RTV} of a uniformly heated loop can be expressed from Eqs. (3.6.14) and (3.6.16),

$$\frac{p_0}{p_0^{RTV}} = \exp\left(0.24\frac{L}{s_H} + 0.12\frac{L}{\lambda_p}\right), \tag{3.7.5}$$

This overpressure ratio is shown in Fig. 3.24 for the same sample of 41 *TRACE* observed loops. The longer the loops are, the higher the overpressure ratio is. If we assume the same heating scale height s_H for all loops, we find a best fit of $s_H = 16.9 \pm 6.1$ Mm (Fig. 3.24). This is a remarkable result for two reasons: first, the heating scale height s_H seems to be independent of the loop length, showing a variation of only a factor 1.4, while the loop lengths vary over 2 orders of magnitude; second, the independence of the heating process on the loop length implies that the energy balance is accomplished locally, barely depending on the physical conditions at the upper

Table 3.3: Summary of measurements of heating scale heights s_H in coronal loops.

Data set	Instrument	Method	Heating scale height	Refs.
41 AR loops	TRACE	Scaling laws	$s_H = 17 \pm 6$ Mm	Aschwanden et al. 2000a
41 AR loops	TRACE	Hydrostatic solutions	$s_H = 12 \pm 5$ Mm	Aschwanden et al. 2001
1 limb loop	Yohkoh	Filter ratios	$s_H \approx \infty$	Priest et al. 1999, 2000
		Hydrostatic solutions	$s_H \ll L$	Mackay et al. 2000a
		Hydrostatic solutions	$s_H \approx 13 \pm 1$ Mm	Aschwanden 2001a
1 disk loop	Yohkoh	Hydrostatic solutions	$s_H \approx 8.4 \pm 2.5$	Aschwanden 2002c

boundary near the apex of the loop.

This result of the observationally inferred heating rate in the order of $s_H \approx 10 - 23$ Mm can be understood in theoretical terms. Since the coronal part of the loop is nearly isothermal, we can basically neglect the thermal conduction term $-\nabla F_C(s)$ in the energy equation (Eq. 3.6.2), at least in the coronal segment of the loop. The heating rate $E_H(s)$ is then mainly balanced by the radiative loss, which has a scale length that is half of the density or pressure scale height $\lambda \approx \lambda_p/2$, because the radiative loss is proportional to the squared density $E_{rad}(s) \propto n_e(s)^2$. The 41 loops observed with *TRACE* 171 Å all have a looptop temperature of $T_{max} \approx 1.0$ MK, which corresponds to a pressure scale height of $\lambda_p = 46$ Mm, yielding a scale height of $\lambda \approx \lambda_p/2 \approx 23$ Mm for the radiative loss or the balancing heating rate. This argument explains the observationally inferred heating scale heights of $s_H \lesssim 23$ Mm.

In Table 3.3 we give a summary of measurements of heating scale heights s_H in coronal loops. The first measurements were conducted by applying the scaling law for loop base pressure as expressed in Eq. 3.7.5 and shown in Fig. 3.24 (Aschwanden et al. 2000a). A second analysis of the same loop data was done by fitting the fluxes in two filters from calculated hydrostatic solutions, yielding $s_H = 12 \pm 5$ Mm (Aschwanden et al. 2001), corroborating the initial result. While these results were inferred from relatively cool EUV loops ($T \approx 1.0$ MK), the first attempt to determine the heating function in hotter loops seen by *Yohkoh/SXT* was performed by Priest et al. (1999, 2000). These authors found an opposite result with a best fit for uniform or looptop heating, but since filter ratios with a broadband temperature filter like *Yohkoh/SXT* are prone to the *hydrostatic weighting bias* (§ 3.2), it is likely that the analyzed faint loop contains contaminations from the background corona. The result of looptop heating was again criticized because the underlying hydrostatic solutions are not consistent with the observed location of the transition region at the limb (Mackay et al. 2000a). The same *Yohkoh* loop was also re-analyzed with a two-component model that includes the cooler background corona which accounts for the hydrostatic weighting bias in first order, finding a heating scale height of $s_H = 13 \pm 1$ Mm with this correction (Aschwanden 2001a). A further case of a hot loop observed by *Yohkoh/SXT* was analyzed near the disk center, where background confusion is minimal. This loop also showed a strong contrast, so that no background modeling was necessary, and a heating scale height of $s_H = 8.4 \pm 2.5$ Mm was found (Aschwanden 2002c). Thus, we conclude that all measurements that properly account for the hydrostatic weighting bias yield a fairly small range of heating scale heights of $s_H \approx 10 - 20$ Mm. The

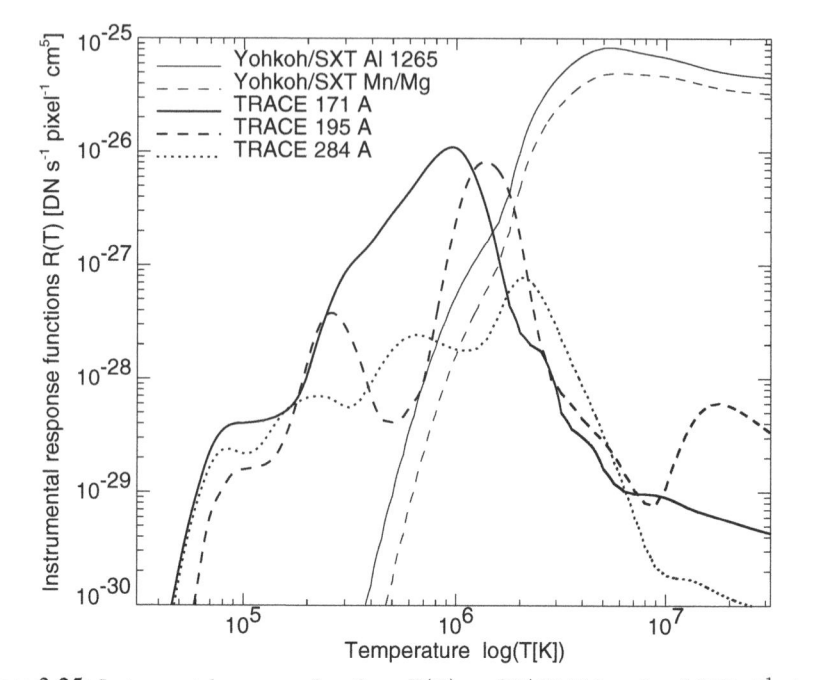

Figure 3.25: Instrumental response functions $R(T) = dF/dEM$ in units of (DN s^{-1} pixel^{-1} cm^5) of *TRACE* filters in 171 Å, 195 Å, 284 Å, and of *Yohkoh/SXT* filters Al 1265 Å and Mn/Mg.

dominant heating at the footpoints of loops seems to apply equally to cool EUV and hotter soft X-ray loops (Neupert et al. 1998; Lenz et al. 1999).

3.8 Instrumental Response Functions

With the knowledge of the hydrostatic solutions, which predict near-isothermal temperature profiles $T(s)$ in the coronal parts of the loops for short heating scale lengths, $s_H < L$, we expect to observe "complete loops" with narrowband temperature filters, such as those used on *SoHO/EIT* and *TRACE*. If the temperature variation $T(s)$ along a loop would change by more than a factor of ≈ 1.5, such narrowband filters would detect only a partial segment of coronal loops. As the loop temperature profiles $T(s)/T_{max}$ observed with *TRACE* show in Fig. 3.23 (top panel), almost "complete loops" with a length portion of $\approx 85\% - 95\%$ are observed, except the footpoint parts where the transition to chromospheric temperatures are unobservable with EUV filters.

The instrumental response functions $R(T)$ of the three *TRACE* coronal filters and two *Yohkoh/SXT* filters are shown in Fig. 3.25. The peak response of each filter essentially determines the temperature of the observed loops. Since the response functions are relatively narrowband for *TRACE*, there is not much ambiguity in the temperature determination. Temperature determinations with triple filters confirmed the unambiguity of double filter methods for most of the practical applications (Chae et al. 2002a).

However, caution is needed. Both the *TRACE* 171 Å and 195 Å filters have also secondary peaks at temperatures of $T \approx 0.2 - 0.3$ MK, and the 195 Å filter has a secondary peak at temperatures of $T \approx 20$ MK. If emission at $T \approx 0.25$ MK or $T \approx 20$ MK is $1 - 2$ orders of magnitude brighter than emission of the plasma at $1 - 2$ MK, it can contribute a competing brightness in the *TRACE* filters. Emission from 20 MK plasma has been observed in the 195 Å filter at the initial phases of flares (Warren & Reeves 2001).

Yohkoh/SXT has a broadband filter response, with an increasing sensitivity towards higher temperatures. Thus plasma at $T = 2.0$ MK is about an order of magnitude brighter than $T = 1.0$ MK plasma, or hot plasma with $T = 5.0$ MK is up to 2 orders of magnitude brighter in the same filter. Since *Yohkoh/SXT* does not have a temperature discrimination like narrowband filters, it always sees a forest of many loops with different temperatures, unequally weighted with preference for the highest temperatures. Loop modeling with *Yohkoh/SXT* can, therefore, only be done in terms of *differential emission measure* distributions $dEM(T)/dT$.

3.9 Observations of Hydrostatic Loops

Let us now have a look at observations of hydrostatic loops. Coronal loops in hydrostatic equilibrium are most likely to be found in older active regions that decay and do not show many signs of new emerging magnetic flux and sporadic heating. Such an active region was observed with *SoHO/EIT* on 1996-Aug-30, 0020:14 UT, as shown in 171 Å in Plate 3. This active region is located close to the disk center, displaying an obvious dipole-like magnetic field in the east-west direction. So, we see a bundle of dipolar loops from a top perspective, where most loops are seen incompletely. We observe only the segments that correspond to their lowest density scale heights, as is typical for hydrostatic loops. A high-pass filtered image (Plate 3 bottom) shows the outlines of the dipolar loops more clearly. As the loops obviously exhibit a wide range of inclination angles, a 3D reconstruction of their geometry is needed in order to properly measure their vertical density scale height (see § 3.4). This was accomplished with a method called *"dynamic stereoscopy"*, which reconstructs the inclination angle of each loop plane by measuring the stereoscopic parallax effect due to the solar rotation over $1 - 2$ days (Aschwanden et al. 1999a). The results of the reconstructed 3D geometries of the loops is shown in Fig. 3.26, where the loop outlines are shown from three orthogonal directions. A comparison image of the same active region 7.2 days earlier at the east limb (Fig. 3.26 top left) confirms the large range of loop plane inclination angles, but reveals a more active phase in the earlier life of the active region, where many loops (probably postflare loops) are filled with plasma all the way up to the looptop, far above the hydrostatic scale height.

The analysis of the density profiles is illustrated in Figs. 3.27 and 3.28. The loop coordinates $[x(s), y(s)]$ parameterized as functions of the 3D loop length s are traced in Fig. 3.26. The curved loop coordinates aligned with the loop axis s have then to be transformed into an orthogonal coordinate system (s, r) and the loop-associated fluxes have to be (bilinearly) interpolated, so that the cross-sectional flux profiles $f(r, s)$ can be obtained as shown in Fig. 3.27. A background flux has to be modeled to subtract

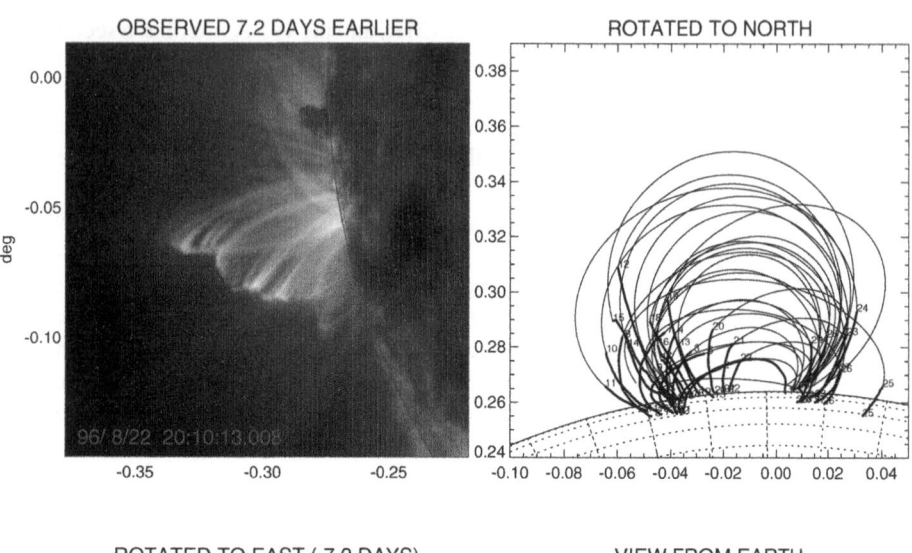

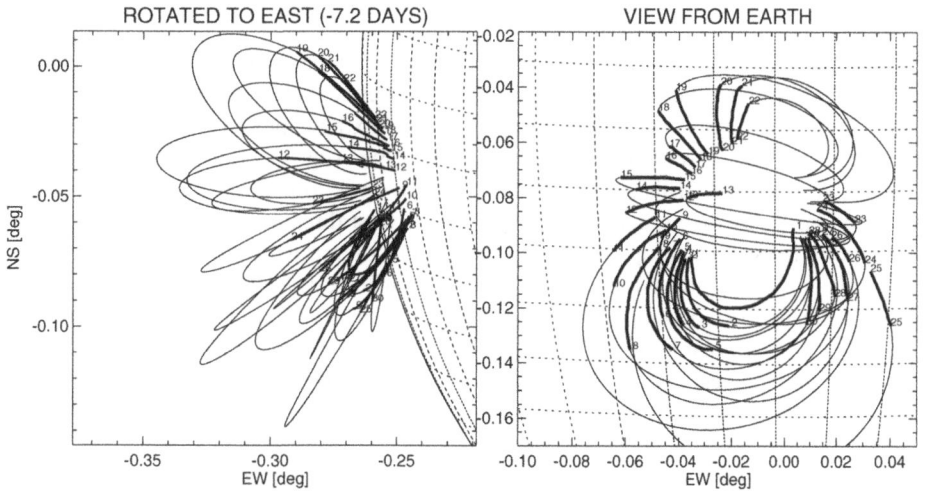

Figure 3.26: Three different projections of the stereoscopically reconstructed 30 loops of AR 7986. The loop segments that were traced from the 96-Aug-30, 171 Å image are marked with thick solid lines, while the extrapolated segments (marked with thin solid lines) represent circular geometries extrapolated from the traced segments. The three views are: (1) as observed from Earth with l_0, b_0 (bottom right panel); (2) rotated to north by $b'_0 = b_0 - 100°$ (top right panel); and (3) rotated to east by $l'_0 = l_0 + 97.2°$ (corresponding to -7.2 days of solar rotation; bottom left panel). An *EIT* 171 Å image observed at the same time (-7.2 days earlier) is shown for comparison (top left panel), illustrating a similar range of inclination angles and loop heights as found from stereoscopic correlations a week later. The heliographic grid has a spacing of 5^0 degrees or 60 Mm (Aschwanden et al. 1999a).

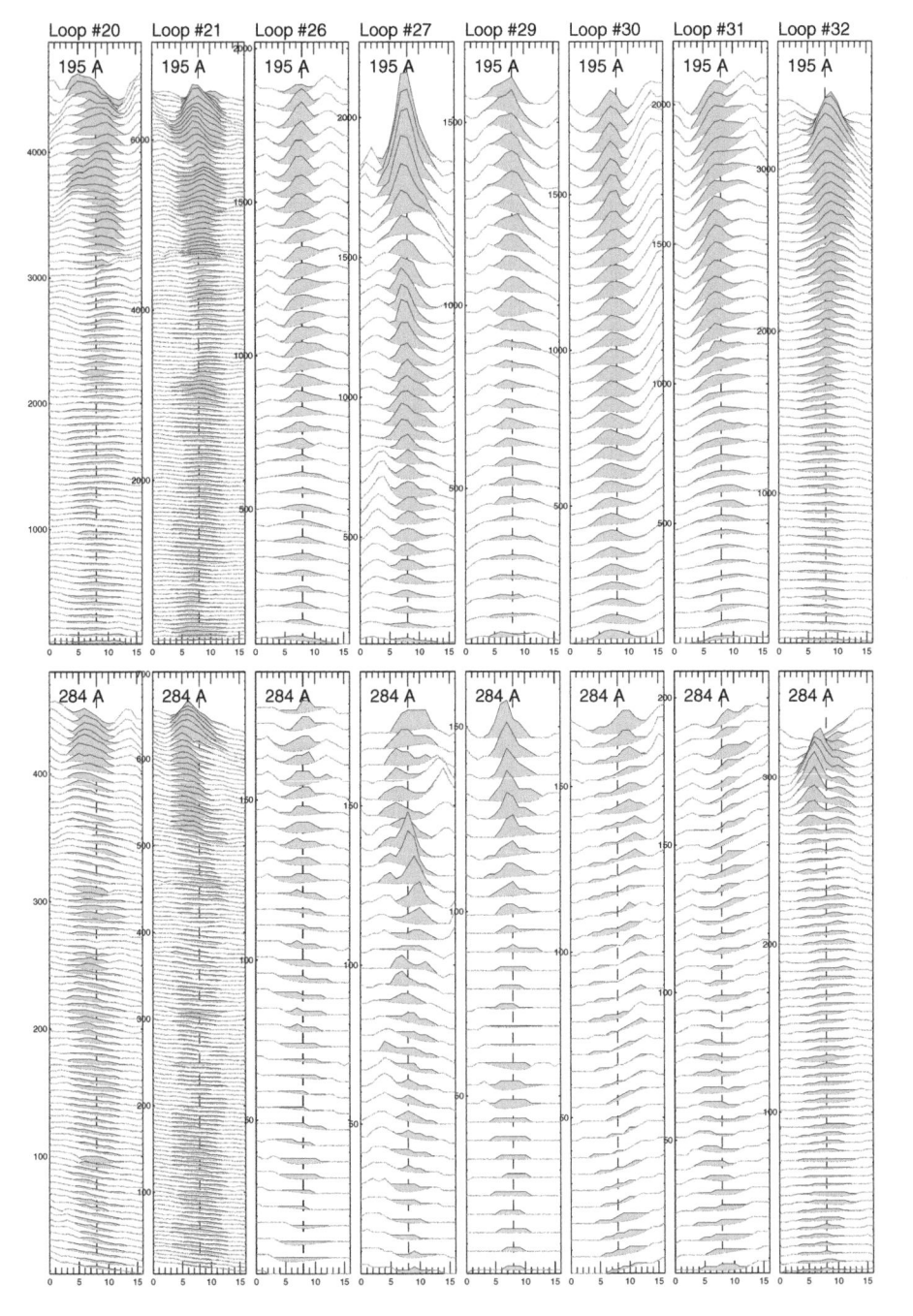

Figure 3.27: Cross-sectional flux profiles of coronal loops in AR 7986 observed with *SoHO/EIT* in 171 Å and 195 Å. The loop coordinate *s* has been stretched out in the vertical direction. The background flux profile across a loop cross section is modeled with a cubic spline fit, the grey areas mark the loop-associated EUV fluxes (Aschwanden et al. 2000a).

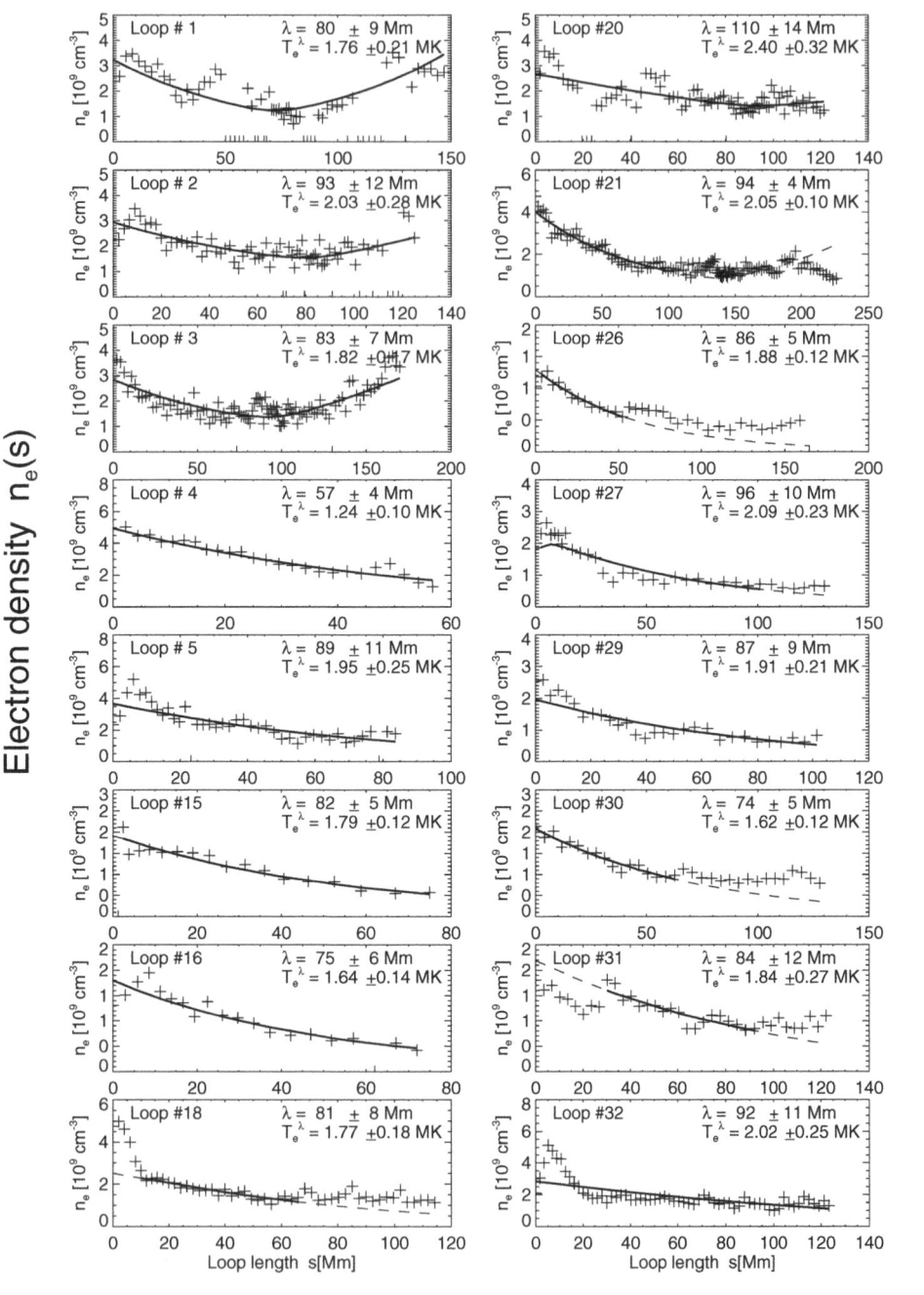

Figure 3.28: The density profiles $n_e(s)$ inferred from the loop-associated fluxes shown in Fig. 3.27 are shown with crosses, while an exponential density profile (solid line) is fitted to infer the density scale height λ_n and the associated scale height temperature T_e^λ (Aschwanden et al. 2000a).

out the loop non-associated EUV fluxes. Once the background-subtracted flux profiles are obtained, the loop widths $w(s)$ can be measured from the FWHM and the density $n_e(s) = \sqrt{EM(s)/w(s)}$ can be determined at locations s from the emission measures $EM(s)$ corresponding to the loop width-integrated fluxes $F(s) = \int f(r,s)dr$. A fit of an exponential density profile $n_e(h) = n_{e0} \exp(-h/\lambda)$ can then be performed with proper coordinate transformation $h(s) = h_{loop} \sin(\pi s/2L)\cos(\theta)$ (see § 3.4). The results are shown in Fig. 3.28, along with the inferred scale height temperatures T_e^λ (Eq. 3.1.16). The obtained scale height temperatures T_e^λ agree with the filter-ratio temperatures T_e^{filter} fairly well for this set of loops (see Fig. 3.2), which confirms that this particular set of loops is close to hydrostatic equilibrium.

3.10 Hydrostatic DEM Distributions

The temperature and density-dependent emission from an observed coronal region is most conveniently quantified in terms of a *differential emission measure (DEM) distribution* $dEM(T)/dT$, which yields a measure of the relative contributions from plasmas at different temperatures. Of course, such a DEM distribution cannot be obtained by a measurement in a single filter or wavelength, but has to be assembled from many filters covering the entire temperature range of interest (see, e.g., Pallavicini et al. 1981). Differential emission measure distributions are defined either integrated in one dimension along the line-of-sight,

$$\left(\frac{dEM(T)}{dT}\right)dT = \int_{-\infty}^{+\infty} dz \int_{T}^{T+dT} n_e^2(z,T)dT \qquad (\text{cm}^{-5}), \qquad (3.10.1)$$

or integrated over a unit area A (e.g., per cm^2 or per pixel area),

$$\left(\frac{dEM(T)}{dT}\right)dT = A \int_{-\infty}^{+\infty} dz \int_{T}^{T+dT} n_e(z,T)^2 dT \qquad (\text{cm}^{-3}). \qquad (3.10.2)$$

If we observe a quiet part of the solar corona above the limb, it is likely that it consists of a myriad of hydrostatic loops, all with different temperatures, which we call a *multi-hydrostatic corona* (§ 3.2). The corresponding DEM distribution is then composed of many elementary DEM distributions from individual hydrostatic loops. Thus, we can define a DEM of a loop by replacing the column depth dz at temperature T by the loop segment $ds(T)/dT$, i.e.

$$\frac{dEM(T)}{dT} = A(s)n_e^2(s[T])\frac{ds(T)}{dT} \qquad (\text{cm}^{-3}\,\text{K}^{-1}). \qquad (3.10.3)$$

Such elementary DEM distributions of individual loops are then straightforward to calculare from a hydrostatic solution, by using the density profile $n_e(s)$ and by inverting the temperature profile $T(s)$ to obtain $ds(T)/dT$. We show such hydrostatic DEM distributions for a variety of loop temperatures, $T_{max} = 1, 3, 5, 10$ MK and expansion factors $\Gamma = 1, 2, 5, 10$, for a loop length $L = 40$ Mm, for footpoint heating $s_H = 20$ Mm (Fig. 3.29, left), as well as for uniform heating (Fig. 3.29, right). The DEM distributions of hydrostatic loops are almost δ-functions at $T \lesssim T_{max}$ due to their near-isothermal coronal segments, with a faint tail to lower temperatures, resulting from

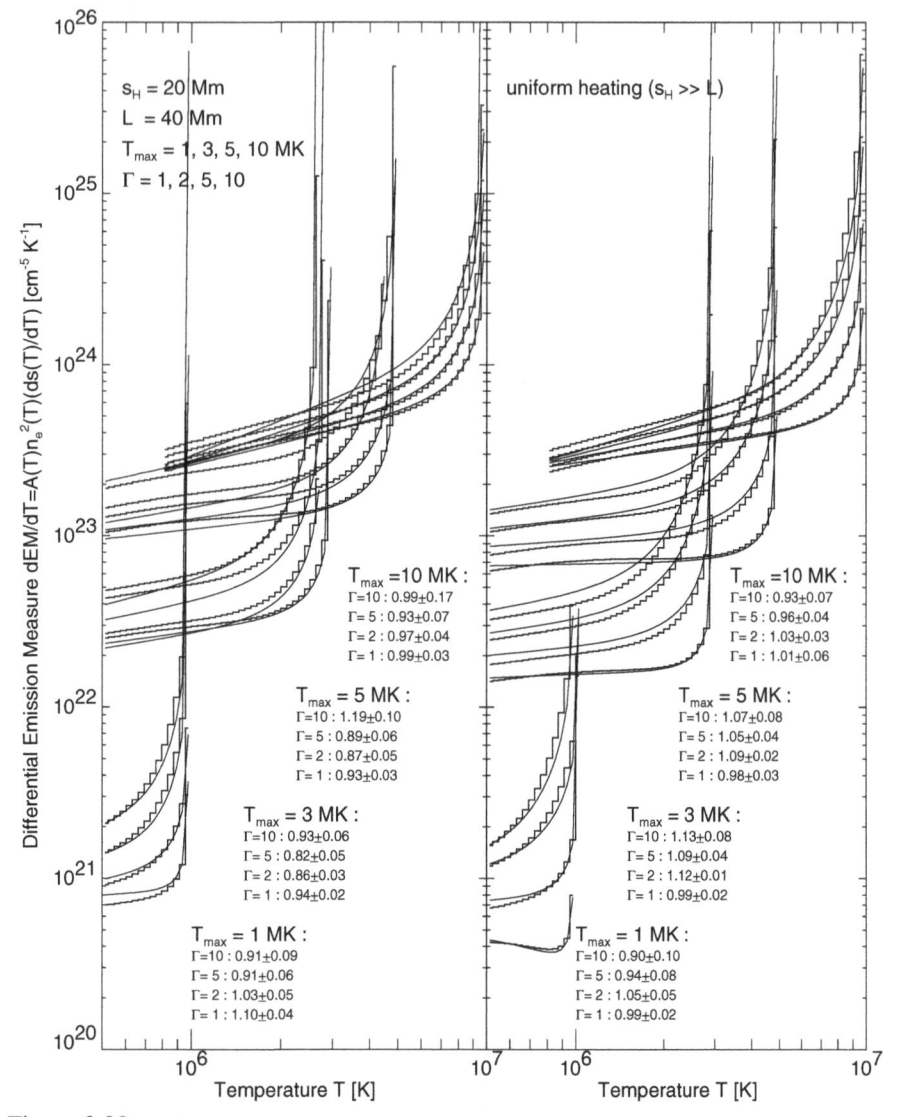

Figure 3.29: Differential emission measure (DEM) distributions of hydrostatic loops, calculated for a loop length of $L = 40$ Mm, a heating scale height of $s_H = 20$ Mm (left) and uniform heating (right), for looptop temperatures of $T = 1, 3, 5, 10$ Mm, and for expansion factors of $\Gamma = 1, 2, 5, 10$. The histograms represent the numerical hydrostatic solutions, the curves the analytical approximations (Aschwanden & Schrijver 2002).

the thin transition region segment. The absolute values of the loop DEMs shown in Fig. 3.29 illustrate the dramatically enhanced weight of hot hydrostatic loops in DEM distributions, which essentially reflects the fact that soft X-ray loops have significantly higher pressure and density than the cooler EUV loops (see also scaling laws of base pressure in Fig. 3.19). When we look at a typical DEM distribution of an active re-

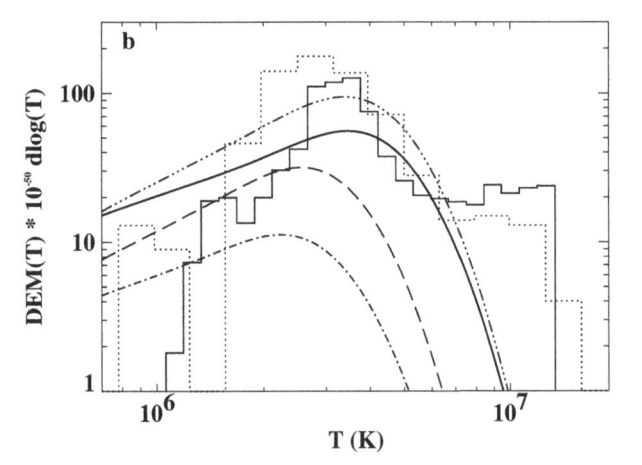

Figure 3.30: Stellar DEM distributions of χ^1 Ori (dotted histogram) and ζ Boo A (solid histogram) compared with various model DEM distributions: multi-hydrostatic atmosphere with uniform heating without flows (thick solid line) and with flows (dot-dashed line), and footpoint heating $s_H = 20$ Mm without flows (triple-dot dashed) and with flows (long-dashed line) (Schrijver & Aschwanden 2002).

gion, say those shown in Fig. 1.21 by Brosius et al. (1996), we clearly see a secondary peak at temperatures of $log(T) \approx 6.6 - 7.0$, which results from the hotter soft X-ray loops in active regions. If the majority of loops are in hydrostatic equilibrium, it takes a much smaller number of hot loops than cool loops to produce a comparable peak in the DEM distribution at a higher temperature than the myriad of cooler EUV loops that make up the primary DEM peak originating from the quiet background corona. The relative DEM scaling shown in Fig. 3.29 demonstrates that a hot hydrostatic loop with $T = 10$ MK produces a contribution to the DEM distribution that is several orders of magnitude larger than the contribution from a cool hydrostatic loop with $T = 1.0$ Mm.

All of what is known from the emission of a star can be characterized by the DEM distribution of the total star irradiance. Assuming that most of the loops in a stellar corona are in hydrostatic equilibrium, the resulting DEM distribution can be modeled with a small number of parameters, essentially the distribution of the independent parameters $[L, s_H, E_{H0}]$ that define the hydrostatic solutions. Such a stellar DEM model has been tackled in a recent study (Schrijver & Aschwanden 2002), where a prototype of a stellar atmosphere was composed by (1) a diffusion-controlled surface magnetic field, (2) potential field extrapolations of the coronal magnetic field, (3) heating rates of coronal loops constrained by the magnetic field at the photospheric loop footpoints, and (4) hydrostatic solutions of the density and temperature profile of the coronal loops. Some trial DEM distributions with this model are shown in Fig. 3.30. Global DEM distributions for the Sun have been calculated for different phases of the solar cycle and for different elemental abundances. The simulated DEMs fit the observed stellar DEMs (χ^1 Ori and ζ Boo A, Fig. 3.30) to some extent, but a major mismatch occurs at temperatures of $T \lesssim 1.0$ MK. This unexplained steepness of the DEM at the low-temperature side has not been successfully modeled so far, and thus provides a powerful constraint.

Factors that help to make the DEM steeper are short heating scale heights and loop expansion factors. One effect seems not to be sufficient but a combination of both could reconcile the theory with the data. Nevertheless, this example shows that hydrostatic models can even constrain stellar DEM distributions and provide information on stellar heating functions.

3.11   Summary

The solar corona consists of many isolated loops that have their own gravitational stratification, depending on their plasma temperature. A useful quantity is the hydrostatic scale height, which depends only on the temperature (§ 3.1). Observing the solar corona in soft X-rays or EUV, which are both optically thin emission, the line-of-sight integrated brightness intercepts many different scale heights, leading to a hydrostatic weighting bias towards systematically hotter temperatures in larger altitudes above the limb (§ 3.2). The observed height dependences of the density needs to be modeled with a statistical ensemble of multi-hydrostatic loops, which can be characterized with a height-dependent differential emission measure distribution (§ 3.3). Measuring a density scale height of a loop requires careful consideration of projection effects (§ 3.4), loop plane inclination angle, cross-sectional variation, line-of-sight integration (§ 3.5), and the instrumental response function (§ 3.8). Hydrostatic solutions have been computed from the energy balance between the heating rate, the radiative energy loss, and the conductive loss (§ 3.6). The major unknown quantity is the spatial heating function (§ 3.7), but analysis of loops in high-resolution images indicate that the heating function is concentrated near the footpoints, say at altitudes of $h \lesssim 20$ Mm (§ 3.8–3.9). Also stellar differential emission measure distributions are sensitive to the heating function (§ 3.10). Of course, a large number of coronal loops are found to be not in hydrostatic equilibrium, but nearly hydrostatic loops have been found preferentially in the quiet corona and in older dipolar active regions.

Chapter 4

Hydrodynamics

Although the Sun appears as lifeless and unchanging to our eyes, except for the monotonic rotation that we can trace from the sunspot motions, there are actually a lot of vibrant dynamic plasma processes continuously happening in the solar corona, which can be detected mainly in EUV and soft X-rays. There is currently a paradigm shift stating that most of the apparently static structures seen in the corona are probably controlled by plasma flows. It is, however, not easy to measure and track these flows with our remote sensing methods. Imagine you are sitting in an airplane and are looking down to the Earth's surface, where you see meandering rivers, but because of the distance cannot make out whether the water in the river bed is flowing or standing still. For slow flow speeds, so-called *laminar flows*, there is no feature to track and water at rest cannot be distinguished from laminar flows, unless we find a drifting boat or tree as tracer of the river flow. Turbulent flows may be easier to detect, because they produce whirls and vortices that can be tracked. A similar situation happens in today's solar physics. Occasionally we detect a moving plasma blob in a coronal loop that can be used as a tracer. Most of the flows in coronal loops seem to be subsonic (like laminar flows) and thus featureless. Occasionally we observe turbulent flows, which clearly reveal motion, especially when cool and hot plasma becomes mixed by turbulence and thus yields contrast by emission and absorption in a particular temperature filter. Another method that is available is the Doppler shift measurement, which however, measures only the flow component along the line-of-sight. At any rate, we are at the verge of detecting ubiquitous flows in the solar corona, so it is appropriate to consider the basic physics of *hydrodynamics* applied to the coronal plasma.

4.1 Hydrodynamic Equations

While we had a momentum equation (3.6.1) and energy equation (3.6.2) to describe *hydrostatic solutions*, the generalization to flows ($v \neq 0$) requires a third equation, the *continuity equation*, which expresses *particle number conservation*. For the time-dependent set of hydrodynamic equations, the operator for the total derivative, which contains an advective term that also takes the motion of the fluid into account, is

$$\frac{D}{Dt} = \left(\frac{\partial}{\partial t} + \mathbf{v} \cdot \nabla\right) = \left(\frac{\partial}{\partial t} + \mathbf{v} \cdot \frac{\partial}{\partial s}\right) , \tag{4.1.1}$$

where $v(s)$ is the velocity and s the length coordinate along a 1D loop. With this operator we can express the continuity, momentum, and energy equation in the most compact form,

$$\frac{D}{Dt}\rho = -\rho\nabla \cdot \mathbf{v} , \tag{4.1.2}$$

$$\rho\frac{D}{Dt}\mathbf{v} = -\nabla p + \nabla p_{grav} , \tag{4.1.3}$$

$$\rho T\frac{D}{Dt}S = E_H - E_R - \nabla F_C . \tag{4.1.4}$$

where S is the entropy per unit mass of the plasma. The left-hand side of the energy equation (4.1.4) describes the heat changes of the plasma while it moves in space through different energy sinks and sources, which are specified on the right-hand side of the energy equation.

The heat or energy equation occurs in many forms in literature, so we derive the most common forms here. For this purpose we summarize the definitions of a number of thermodynamic quantities (e.g., Priest 1982). The total number density n, mass density ρ (with mean mass m), and pressure p for a fully ionized gas (as it is the case in the corona, i.e., $n_p = n_e$) are, according to the *ideal gas law*,

$$n = n_p + n_e = 2n_e , \tag{4.1.5}$$

$$\rho = nm = n_p m_p + n_e m_e \approx n_p m_p = n_e m_p , \tag{4.1.6}$$

$$p = nk_B T = 2n_e k_B T . \tag{4.1.7}$$

When a plasma is thermally isolated, so that there is no heat exchange with the ambient plasma, the thermodynamic state is called *adiabatic*, (i.e., the *entropy S* is constant), defined by

$$S = c_v \log\left(p\rho^{-\gamma}\right) + S_0 , \tag{4.1.8}$$

where S_0 is a constant. The factor c_v is the *specific heat at constant volume*, relating to the *specific heat at constant pressure* c_p by the relation

$$c_p = c_v + \frac{k_B}{m} . \tag{4.1.9}$$

The hydrodynamics of a plasma is generally described with the thermodynamics of a polytropic gas, which can be characterized by the *polytropic index* γ,

$$\gamma = \frac{c_p}{c_v} \tag{4.1.10}$$

which expresses the ratio of the specific heats at constant pressure (c_p) and at constant volume (c_v). The *polytropic index* is theoretically $\gamma = (N_{free} + 2)/N_{free} = 5/3$ for a fully ionized gas, such as hydrogen in the corona, with $N_{free} = 3$ degrees of freedom, but the effects of partial ionization can make it as low as $\gamma \gtrsim 1$. From Eqs. (4.1.9–10) we can express the specific heats c_p and c_v as functions of the polytropic index γ,

$$c_p = \frac{\gamma}{(\gamma - 1)} \frac{k_B}{m} \qquad (4.1.11)$$

$$c_v = \frac{1}{(\gamma - 1)} \frac{k_B}{m} \qquad (4.1.12)$$

A commonly used thermodynamic quantity is the *internal energy e*, which is defined in an *ideal gas* by

$$e = c_v T = \frac{1}{(\gamma - 1)} \frac{k_B T}{m} = \frac{1}{(\gamma - 1)} \frac{p}{\rho}. \qquad (4.1.13)$$

A related thermodynamic quantity is the *enthalpy* ε_{enth}, which comprises the heat energy acquired (or lost) at constant volume, plus the work done against the pressure force when the volume changes, and is defined (per unit mass) by

$$\frac{\varepsilon_{enth}}{m} = c_p T = \frac{\gamma}{(\gamma - 1)} \frac{k_B T}{m} = \frac{\gamma}{(\gamma - 1)} \frac{p}{\rho} = \gamma e. \qquad (4.1.14)$$

With these definitions we can now express the energy or heat equation (4.1.4) in the various forms used in the literature. Inserting the expression for the entropy S, Eq. (4.1.8) leads to the form (Priest 1982, Eq. 2.28e):

$$\rho c_v T \frac{D}{Dt} \log \left(p \rho^{-\gamma} \right) = E_H - E_R - \nabla F_C. \qquad (4.1.15)$$

Taking the derivative of the logarithm and inserting the definition of the *internal energy e*, Eq. (4.1.13) eliminates the temperature T (Priest 1982, Eq. 2.28d; Bray et al. 1991, Eq. 5.53):

$$\frac{\rho^{\gamma}}{(\gamma - 1)} \frac{D}{Dt} \left(p \rho^{-\gamma} \right) = E_H - E_R - \nabla F_C. \qquad (4.1.16)$$

Expanding the derivatives on the left-hand side yields (e.g., Field 1965, Eq. 8)

$$\frac{1}{(\gamma - 1)} \frac{Dp}{Dt} - \frac{\gamma}{(\gamma - 1)} \frac{p}{\rho} \frac{D\rho}{Dt} = E_H - E_R - \nabla F_C, \qquad (4.1.17)$$

and using the continuity equation (4.1.2) eliminates the mass density ρ, so that the energy equation is expressed in terms of pressure p and velocity $\mathbf{v}$ (Bray et al. 1991, Eq. 5.54),

$$\frac{1}{(\gamma - 1)} \frac{Dp}{Dt} + \frac{\gamma}{(\gamma - 1)} p \nabla \mathbf{v} = E_H - E_R - \nabla F_C. \qquad (4.1.18)$$

Alternatively, the energy equation can be expressed in terms of pressure p and temperature T by using the ideal gas law (Eq. 4.1.7), (Bray et al. 1991, Eq. 5.55)

$$\frac{\gamma}{(\gamma - 1)} \frac{p}{T} \frac{DT}{Dt} - \frac{Dp}{Dt} = c_p \rho \frac{DT}{Dt} - \frac{Dp}{Dt} = E_H - E_R - \nabla F_C. \qquad (4.1.19)$$

Another form of the energy equation is to express it in terms of the internal energy e, which is proportional to the temperature T. We can express the pressure as a function of the internal energy with Eq. (4.1.13) by $p = (\gamma - 1)\rho e$. Inserting the pressure p into the energy equation (4.1.18), taking the total derivative, and using the continuity equation then yields

$$\rho \frac{De}{Dt} + p\nabla \mathbf{v} = E_H - E_R - \nabla F_C \,. \tag{4.1.20}$$

The hydrodynamic equations (4.1.2–4) for coronal loops can be written more specifically, in one dimension, with loop coordinate s, by inserting the operator D/Dt explicitly, using $\rho = mn$, and the form (Eq. 4.1.20) of the energy equation,

$$\frac{\partial n}{\partial t} + \frac{\partial}{\partial s}(n\mathbf{v}) = 0 \,, \tag{4.1.21}$$

$$mn\frac{\partial \mathbf{v}}{\partial t} + mn\mathbf{v}\frac{\partial \mathbf{v}}{\partial s} = -\frac{\partial p}{\partial s} + \frac{\partial p_{grav}}{\partial r}(\frac{\partial r}{\partial s}) \,, \tag{4.1.22}$$

$$mn\frac{\partial e}{\partial t} + mn\mathbf{v}\frac{\partial e}{\partial s} + p\frac{\partial \mathbf{v}}{\partial s} = E_H - E_R - \frac{\partial F_C}{\partial s} \,. \tag{4.1.23}$$

The hydrodynamic equations can be generalized for a variable cross-sectional area $A(s)$. The operator (4.1.1) for the total derivative can be transformed (in 3D) into a coordinate system that follows the loop coordinate s, with $\mathbf{v}(s)$ being the parallel velocity. Integrating the flux quantities over the perpendicular cross-sectional area $\int dA(s)$ and dividing the equations by the cross section $A(s)$ then yields

$$\frac{\partial n}{\partial t} + \frac{1}{A}\frac{\partial}{\partial s}(n\mathbf{v}A) = 0 \,, \tag{4.1.24}$$

$$mn\frac{\partial \mathbf{v}}{\partial t} + mn\mathbf{v}\frac{\partial \mathbf{v}}{\partial s} = -\frac{\partial p}{\partial s} + \frac{\partial p_{grav}}{\partial r}\left(\frac{\partial r}{\partial s}\right) \,, \tag{4.1.25}$$

$$mn\frac{\partial e}{\partial t} + \frac{mn\mathbf{v}}{A}\frac{\partial}{\partial s}(eA) + \frac{p}{A}\frac{\partial}{\partial s}(\mathbf{v}A) = E_H - E_R - \frac{1}{A}\frac{\partial}{\partial s}(F_C A) \,. \tag{4.1.26}$$

Note that the area dependence enters the advective terms of scalars (ρ, S, F_C), but cancels out for vectors $\mathbf{v}$ [e.g., in the momentum equation (4.1.25), if the nonlinear terms of pressure gradients due to curvature are neglected].

A physically more intuitive form of the hydrodynamic equations is to express them in *conservative form*, which means that the energy equation can be written in terms of energies that are conserved in their sum, so that the left-hand side contains the changes in enthalpy ε_{enth}, kinetic energy ε_{kin}, and gravitational energy ε_{grav}, which have to balance the heating input (E_H), radiative (E_R), and conductive losses $(-\nabla F_C)$ on the right-hand side of the energy equation. Note that the energy equation has the physical dimension of energy density per time (erg cm^{-3} s^{-1}) in all terms. The total energy is conserved in a closed system (i.e., when no external forces act on the system). This will be the case once the system settles into a steady-state equilibrium with no time dependence ($\partial/\partial t = 0$). The hydrodynamic equations of a 1D loop with variable cross section $A(s)$ can be written in *conservative form*,

$$\frac{1}{A}\frac{\partial}{\partial s}(nvA) = 0 , \tag{4.1.27}$$

$$mnv\frac{\partial v}{\partial s} = -\frac{\partial p}{\partial s} + \frac{\partial p_{grav}}{\partial r}(\frac{\partial r}{\partial s}) , \tag{4.1.28}$$

$$\frac{1}{A}\frac{\partial}{\partial s}(nvA[\varepsilon_{enth} + \varepsilon_{kin} + \varepsilon_{grav}] + AF_C) = E_H - E_R , \tag{4.1.29}$$

where the *kinetic energy* $\varepsilon_{kin}(s)$ is

$$\varepsilon_{kin}(s) = \frac{1}{2}mv^2(s) , \tag{4.1.30}$$

and the derivative of the gravitational potential $d\varepsilon_{grav}/ds$ is defined by Eqs. (3.1.3−5),

$$\frac{\partial \varepsilon_{grav}(s)}{\partial s} = -\frac{1}{n}\frac{\partial p_{grav}}{\partial s} . \tag{4.1.31}$$

We can verify that the energy equation in conservative form (Eq. 4.1.29) is equivalent to the standard form of the energy equation (4.1.26) with the following few steps. Ignoring the cross-sectional dependence (i.e., $A(s) = const$), the continuity equation (4.1.27) is ,

$$\frac{\partial}{\partial s}(nv) = 0 . \tag{4.1.32}$$

This implies that the total derivative d/ds in the energy equation (4.1.29) can be written as a partial derivative of the energy terms,

$$\frac{\partial}{\partial s}(nv\varepsilon) = \varepsilon\frac{\partial}{\partial s}(nv) + nv\frac{\partial \varepsilon}{\partial s} = nv\frac{\partial \varepsilon}{\partial s} . \tag{4.1.33}$$

In the same way, we can simplify the total derivative of the product (pv), because it contains the product (nv) after inserting the pressure $p = nk_BT$,

$$\frac{\partial}{\partial s}(pv) = \frac{\partial}{\partial s}(nvk_BT) = k_BT\frac{\partial}{\partial s}(nv) + nv\frac{\partial}{\partial s}(k_BT) = nvk_B\frac{\partial T}{\partial s} = mnv(\gamma-1)\frac{\partial e}{\partial s} , \tag{4.1.34}$$

where we expressed T in terms of the internal energy e from the definition of Eq. (4.1.13). The energy equation in conservative form (Eq. 4.1.29), after inserting the definitions of the enthalpy (Eq. 4.1.14), kinetic energy (Eq. 4.1.30), and gravity (Eq. 4.1.31), and using the relation of the total derivative (Eq. 4.1.33), reads

$$v(mn\gamma\frac{\partial e}{\partial s} + mnv\frac{\partial v}{\partial s} - \frac{\partial p_{grav}}{\partial s}) = E_H - E_R - \frac{\partial F_C}{\partial s} . \tag{4.1.35}$$

Substituting the momentum equation (Eq. 4.1.28),

$$mnv\frac{\partial v}{\partial s} - \frac{\partial p_{grav}}{\partial s} = -\frac{\partial p}{\partial s} \tag{4.1.36}$$

eliminates the term $mv(\partial v/\partial s)$ in Eq. (4.1.35) and simplifies the energy equation to

$$mnv\gamma\frac{\partial e}{\partial s} - v\frac{\partial p}{\partial s} = E_H - E_R - \frac{\partial F_C}{\partial s} \ . \tag{4.1.37}$$

We can now insert $v(\frac{\partial p}{\partial s}) = \frac{\partial}{\partial s}(pv) - p\frac{\partial v}{\partial s}$ and make use of Eq. (4.1.34),

$$v\frac{\partial p}{\partial s} = \frac{\partial}{\partial s}(pv) - p\frac{\partial v}{\partial s} = mnv(\gamma - 1)\frac{\partial e}{\partial s} - p\frac{\partial v}{\partial s} \ , \tag{4.1.38}$$

which inserted into the energy equation (4.1.37) yields the standard form of the (time-independent) energy equation

$$mnv\frac{\partial e}{\partial s} + p\frac{\partial v}{\partial s} = E_H - E_R - \frac{\partial F_C}{\partial s} \ , \tag{4.1.39}$$

corresponding to the notation given in Eq. (4.1.23). Thus, we have proven that the standard form of the time-independent hydrodynamic energy equation (4.1.23) is identical to the conservative form given in (Eq. 4.1.29), and that both describe the same thermodynamics. These are two of the most basic notations of the hydrodynamic energy equation, although a number of other variants can be found in the literature (e.g., see summaries in Priest 1982 or Bray et al. 1991). Solutions of the time-independent hydrodynamic equations are also called *steady-flow solutions* which are described in the next section.

4.2   Steady-Flow and Siphon-Flow Solutions

The hydrodynamic equations (4.1.21−23) describe relations between the four functions of density $n(s,t)$, pressure $p(s,t)$, temperature $T(s,t)$, and velocity $v(s,T)$, which are generally space and time-dependent. We expect that dynamic processes in loops will be smoothed out after a time scale that corresponds to the sound travel time through the loop length. Especially for loops with longer lifetimes, we expect that dynamic processes eventually settle into a near-stationary state, and it is therefore useful to consider time-independent solutions ($\partial/\partial t = 0$), i.e., steady-flow solutions for $n(s)$, $p(s)$, $T(s)$, and $v(s)$.

A first approach in solving the hydrostatic equations analytically is to ignore the energy equation Eq. (4.1.23) and instead to assume an adiabatic process (i.e., a constant entropy with $\partial S/\partial s = 0$ in Eq. 4.1.8), for which

$$p\rho^{-\gamma} = const \ , \qquad \frac{d}{ds}\left(p\rho^{-\gamma}\right) = 0 \ , \tag{4.2.1}$$

with γ being the polytropic index Eq. (4.1.10), the ratio of specific heats. Thus the spatial derivative d/ds of this quantity along the loop coordinate s vanishes in adiabatic or polytropic processes, and the expansion of the derivative yields the relation (using $\rho = mn$),

$$\frac{\partial p}{\partial s} = \left(\frac{\gamma p}{\rho}\right) \frac{\partial \rho}{\partial s} = c_s^2 \frac{\partial \rho}{\partial s} = mc_s^2 \frac{\partial n}{\partial s} , \tag{4.2.2}$$

where the *sound speed* c_s is defined by

$$c_s = \sqrt{\frac{\gamma p}{\rho}} . \tag{4.2.3}$$

Using the continuity equation (4.1.27) for variable cross sections $A(s)$,

$$\frac{d}{ds}(nvA) = 0 , \tag{4.2.4}$$

and the expansion of the continuity equation, the pressure gradient $\partial p / \partial s$ (Eq. 4.2.2) becomes

$$\frac{\partial p}{\partial s} = mc_s^2 \frac{\partial n}{\partial s} = -mc_s^2 \left(\frac{n}{v}\frac{\partial v}{\partial s} + \frac{n}{A}\frac{\partial A}{\partial s}\right) . \tag{4.2.5}$$

Using the approximation of constant gravity $\partial p_{grav}/\partial r \approx -mng_\odot$ (Eq. 3.1.5) and considering semi-circular loops, $\partial r/\partial s = \cos(\pi s/2L)$ (Eq. 3.4.3), the momentum equation (4.1.28) becomes,

$$mnv\frac{\partial v}{\partial s} = -\frac{\partial p}{\partial s} - mng_\odot \cos\left(\frac{\pi s}{2L}\right) , \tag{4.2.6}$$

Inserting the expression for the pressure gradient (4.2.5) then yields a differential equation for the flow speed $v(s)$,

$$\left(v - \frac{c_s^2}{v}\right)\frac{\partial v}{\partial s} = -g_\odot \cos\left(\frac{\pi s}{2L}\right) + \frac{c_s^2}{A}\frac{\partial A}{\partial s} . \tag{4.2.7}$$

This solution is described in Cargill & Priest (1980) and Noci (1981), and is similar to the solar wind solution of Parker (1958). The solutions of this differential equation are depicted in Fig. 4.1. Similar solutions are also studied in Noci & Zuccarello (1983).

The solutions for the flow speed are symmetric on both sides of the loop. If the initial flow speed at one footpoint is subsonic, the flow speed increases towards the looptop ($s = L$), because the continuity equation, $n(s)v(s) = const$, requires a reciprocal change to the density decrease with altitude. When the flow reaches sonic speed at the looptop (dashed line), it will become supersonic beyond the summit. But at some point a discontinuous jump must occur to connect the supersonic flow with the subsonic boundary condition at the secondary footpoint and the supersonic flow will be slowed down to subsonic speed. There are also a number of unphysical solutions shown in Fig. 4.1 (dotted lines). Solutions with subsonic speed throughout the loop are also called *siphon flow solutions*, which can be driven by a pressure difference between both loop footpoints.

While the approach of replacing the energy equation by an adiabatic process reveals the basic features of hydrodynamic solutions in terms of subsonic and supersonic flows

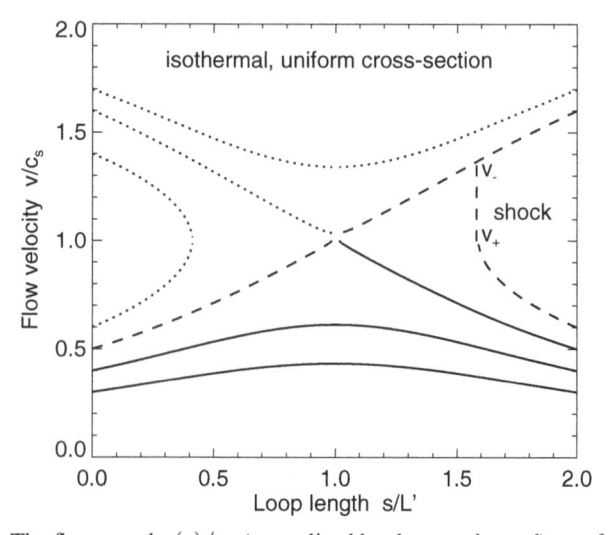

Figure 4.1: The flow speed $v(s)/c_s$ (normalized by the sound speed) as a function of the normalized loop length s/L according to the steady-flow solution of Eq. (4.2.7), for an isothermal loop with uniform loop cross section. Flows at the critical speed (dashed line) become supersonic beyond the loop summit (s=L) and are slowed down from v_- to v_+ by a shock wave. Unphysical or fully supersonic solutions are marked with dotted lines (after Cargill & Priest 1980).

$v(s)$, accurate steady-flow solutions for the temperature $T(s)$ and density $n(s)$ require a rigorous solution of the (time-independent) momentum and energy equations, using a numeric hydrodynamic code. Essentially, this involves finding a solution for the two functions $T(s)$ and $n(s)$ from the momentum equation (4.1.22) and energy equation (4.1.23),

$$mn(s)v(s)\frac{\partial v(s)}{\partial s} = -\frac{\partial p(s)}{\partial s} - mn(s)g_\odot \left(\frac{R_\odot}{r(s)}\right)^2 \left(\frac{\partial r(s)}{\partial s}\right), \qquad (4.2.8)$$

$$mn(s)v(s)\frac{\partial e(s)}{\partial s} + p(s)\frac{\partial v(s)}{\partial s} = E_H(s) - E_R(s) - \nabla F_C(s), \qquad (4.2.9)$$

where all terms can be expressed as a function of the temperature $T(s)$ and density $n(s)$, using the continuity equation (4.1.27) for v(s), the ideal gas equation (4.1.7) for the pressure $p(s)$, a semi-circular loop geometry (Eq. 3.4.1) for the height $r(s)$, the enthalpy per mass $e(s)$ (Eq. 4.1.13), a heating function of choice $E_H(s)$ (§3.7), the radiative loss function $E_R(s)$ (Eq. 2.9.1), the conductive loss $\nabla F_C(s)$ (Eq. 3.6.3), and an arbitrary cross section function $A(s)$ (§3.4). In numeric codes, the following standard definitions are often used,

$$v(s) = v_0 \frac{n_0 A_0}{n(s)A(s)}, \qquad (4.2.10)$$

$$p(s) = 2n_e(s)k_B T(s) , \tag{4.2.11}$$

$$r(s) = R_\odot + \frac{2L}{\pi} \sin\left(\frac{\pi s}{2L}\right) , \quad \frac{dr(s)}{ds} = \cos\left(\frac{\pi s}{2L}\right) , \tag{4.2.12}$$

$$e(s) = \frac{1}{(\gamma - 1)} \frac{k_B T(s)}{m} , \tag{4.2.13}$$

$$E_R(s) = n(s)^2 \Lambda[T(s)] , \tag{4.2.14}$$

$$\nabla F_C(s) = \frac{1}{A(s)} \frac{\partial}{\partial s} \left[-A(s)\kappa T^{5/2}(s) \frac{\partial T(s)}{\partial s} \right] , \tag{4.2.15}$$

while the heating function $E_H(s)$ and cross section function $A(s)$ can be arbitrarily chosen. From this full set of equations needed to calculate a steady-flow solution, we see that a solution depends on the following independent parameters: the loop half length L, base density n_0, base temperature T_0, base upflow speed v_0, base heating rate E_{H0}, and cross section A_0. In addition, a heating scale height s_H is needed for nonuniform heating, and a loop expansion factor Γ for expanding loops.

In the following we describe some results of steady-flow solutions that have been found with such hydrodynamic codes or from the asymptotic limit of a time-dependent simulation. For slow flow speeds, $v \lesssim c_s$, the velocity-dependent terms in the momentum and energy equation can be neglected, and the flow speed is approximately reciprocal to the density (for a constant cross section) according to the continuity equation,

$$v(s) = v_0 \frac{n_0}{n(s)} \approx v_0 \left(\frac{n_0}{n(s)}\right)_{static} \tag{4.2.16}$$

The solutions of the velocity $v(s)$ and density $n(s)$ are shown for subsonic speeds of ($v < c_s$) in Fig. 4.2, for a loop with a half length of $L = 200$ Mm and for a range of initial upflow speeds $v_0/c_s = 0.001, ..., 1$. This example illustrates that the solution for the density profile does not change much in the presence of flows, it is only a factor of ≈ 0.7 lower near the looptop or sonic point. The flow speed is always increasing with height, approximately reciprocal to the density for subsonic zones, but somewhat faster near the sonic point. The almost identical density and temperature profiles of hydrostatic loops and steady-flow loops makes it almost impossible to diagnose subsonic flow speeds in steady coronal loops (except with measurements of the Doppler shift).

The general case of subsonic siphon flows was studied by a number of authors (e.g. Mariska & Boris 1983; Craig & McClymont 1986; Orlando et al. 1995a,b). The most common situation are uni-directional siphon flows, where the overpressure at one footpoint of the loop drives an upflow, which flows along the entire loop and drains at the opposite side (Fig. 4.3). The flow may stay subsonic all the time (Fig. 4.3 left) or become supersonic near the looptop (Fig. 4.3 right). If all loop parameters are chosen symmetrically (spatial heating rate, loop cross section), the solutions in $n(s)$, $T(s)$, and $|v(s)|$ are also symmetric, except that the flow direction changes the sign, $v(s) = -v(2L - s)$. Flows driven by asymmetric heating or by asymmetric loop cross sections were considered by Craig & McClymont (1986). Asymmetric heating enhances thermal conduction into the direction of decreasing heating rate (Fig. 4.4), but does not change the temperature or velocity profile much. Asymmetric loop cross

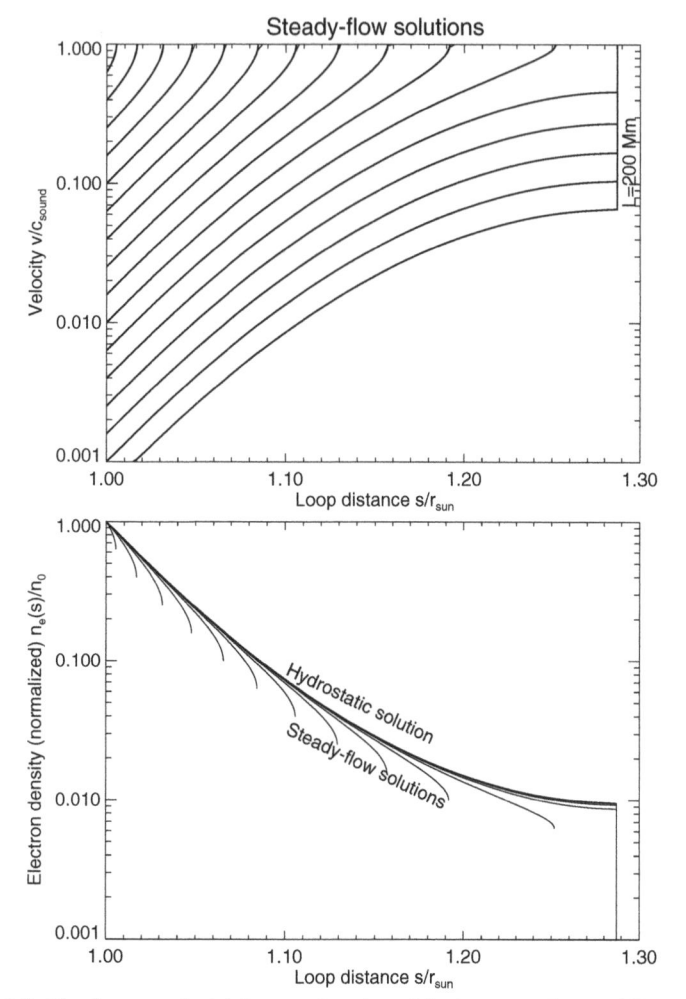

Figure 4.2: The flow speed $v(s)/c_s$ as a function of the loop coordinate $s/R_\odot + 1$ for subsonic steady-flow solutions, for a loop with a length of $L = 200$ Mm and a range of initial upflow speeds $v(s)/c_s = 0.001, ..., 1$. The asymptotic limit v=0 of the hydrostatic solution for the density $n(s)$ is shown with a thick solid line (bottom panel).

sections mainly affect the flow speed in a reciprocal way, as expected from the continuity equation. The solutions for $v(s)$, $T(s)$, and $n(s)$ calculated by Craig & McClymont (1986), shown in Fig. 4.4, demonstrate that the temperature $T(s)$ and density profiles $n(s)$ are not much different from the hydrostatic solutions in the presence of flows (see also Fig. 4.2), even in the presence of asymmetric drivers. Mariska & Boris (1983) and Klimchuk & Mariska (1988) also simulated flow solutions with asymmetric heating sources in the loops and found that the density and temperature profiles were rather insensitive to the location of the heating function. They found the highest flow speed $v(s)$ in locations that were bracketed by a localized coronal heating source or in converging

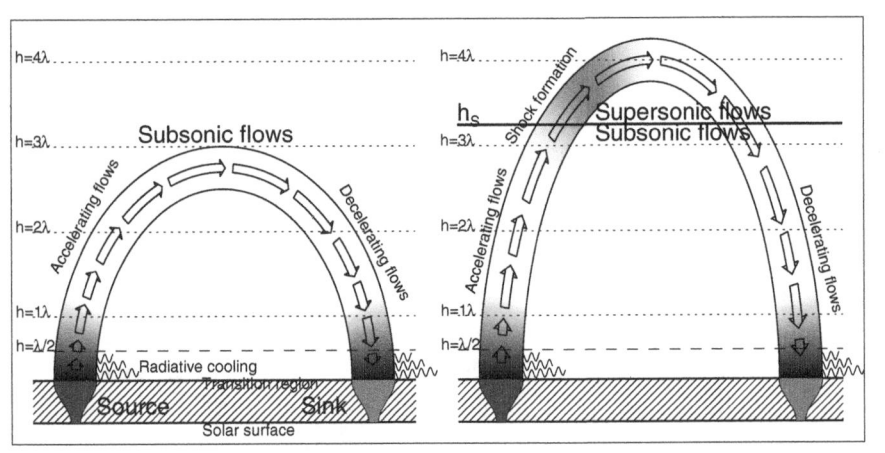

Figure 4.3: Scenario of a (uni-directional) siphon flow model, where an asymmetric pressure drives an upflow at the left-hand footpoint, which flows through the loop and drains at the opposite (right-hand) footpoint. The flow is accelerating with height due to the density decrease according to the continuity equation. The siphon flow may always be subsonic (left panel), or become supersonic if the initial speed is high or the loop extends over several scale heights (right panel). At the sonic point, the resulting shock wave produces a density compression that should also be observable as a brightening of the emission measure near the looptop.

footpoints, possibly explaining the observed redshifts ($v \lesssim 5$ km s^{-1}) in some spectral lines (C II, Si IV, C IV) formed in the network at transition region temperatures. Other effects have been investigated for siphon flow models, such as nonequilibrium ionization effects (Noci et al. 1989; Spadaro et al. 1990a,b, 1991, 1994; Peres et al. 1992), but the same conclusions were confirmed, namely that asymmetric heating cannot explain large redshifts. Models for stationary siphon flows with shocks were calculated by Orlando et al. (1995a,b), where they found that, (1) the shock position depends on the volumetric heating rate of the loop, and (2) there exists a range of volumetric heating rates that produce two alternative positions for shock formation. Thus, observed positions of shocks in coronal loops could potentially provide a diagnostic of heating rates and flow profiles.

4.3 Thermal Stability of Loops

We considered hydrostatic solutions of loops (§ 3) as well as steady-flow solutions (§4.2), which both assume an equilibrium state. An equilibrium state can be stable or unstable, depending on whether the system returns to the same equilibrium state after a disturbance or not. It is important to select the stable solutions amongst the mathematically possible equilibria solutions, because unstable solutions will never be observed in the real world. An introduction into instabilities in solar MHD plasmas is given in Priest (1982). Most of the instabilities occur at boundaries between two plasma layers with different physical parameters. Examples in classical mechanics are heavy fluids on top of lighter ones, or waves on water. Examples in plasma

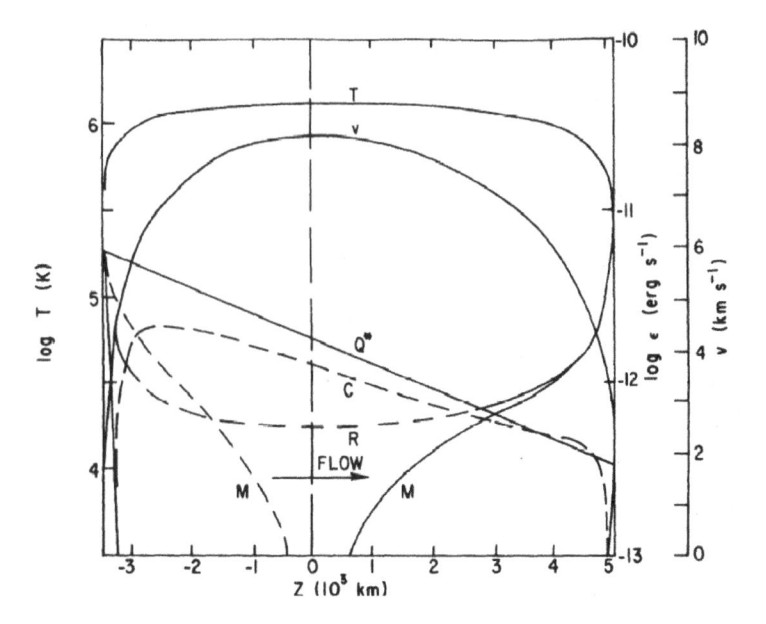

Figure 4.4: Siphon flow solutions of $T(s)$ and $v(s)$ in an asymmetrically heated loop [with heating rate $Q^*(s)$]. Radiative loss $R(s)$, conductive loss $C(s)$, and enthalpy $M(s)$ are also shown (Craig & McClymont 1986).

physics are the (hydrodynamic) *Rayleigh–Taylor instability*, where a boundary between two plasmas of different densities and pressures is disturbed by the gravity force; the *Kruskal–Schwarzschild instability* (or the hydromagnetic analog of the *Rayleigh–Taylor instability*), where the plasma boundary is supported by a magnetic field; the *Kelvin–Helmholtz instability*, where different flow speeds shear at both sides of the boundary; the *convective instability*, where convection cells form due to a large temperature gradient; or the *radiatively driven thermal instability*, where the radiative loss rate leads to a thermal instability in the case of insufficient thermal conduction.

4.3.1   Radiative Loss Instability

Parker (1953) already pointed out the thermal instability of coronal plasmas as a consequence of the dependence of the radiative loss function on the density and temperature. A simple derivation of the *radiatively driven thermal instability*, which occurs in the case of insufficient thermal conduction, is given in Priest (1982, p. 277). If we neglect thermal conduction (∇F_C) and flows ($v = 0$), and assume a constant pressure ($Dp/Dt = 0$) in the time-dependent energy equation (4.1.19), we have

$$c_p \rho \frac{\partial T}{\partial t} = E_H - E_R \,. \tag{4.3.1}$$

Defining a heating rate h per unit volume (that is proportional to the number density),

$$E_H = h\rho \,, \tag{4.3.2}$$

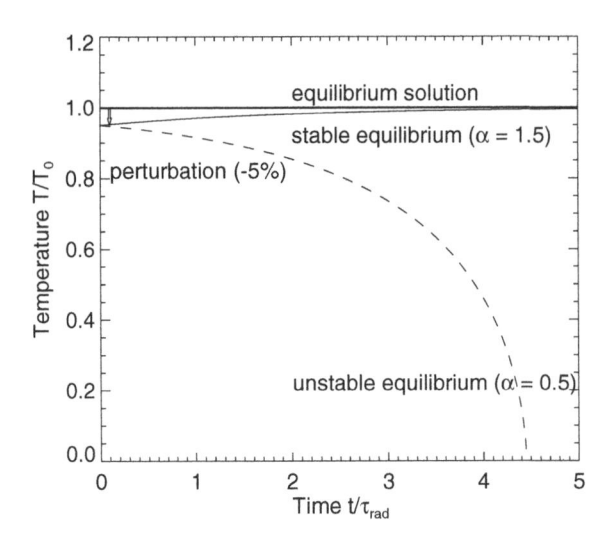

Figure 4.5: The temperature evolution $T(t)$ of a loop after a small perturbation by -5% from the equilibrium situation (thick line): for a stable loop (thin line: with a powerlaw slope of $a > 1$ in the radiative loss function), and an unstable loop (dashed line: with $a < 1$).

and approximating the radiative loss function $E_R(s)$ with a powerlaw function in temperature,

$$E_R = n^2\Lambda(T) \approx \chi\rho^2 T^\alpha \,, \tag{4.3.3}$$

the energy equation reads (Priest 1982, Eq. 7.69),

$$c_p\frac{\partial T}{\partial t} = h - \chi\rho T^\alpha \,. \tag{4.3.4}$$

Assuming a constant pressure, $p(s,t) \approx p_0 = const$, the mass density ρ can be substituted by the temperature T,

$$\rho = \frac{mp_0}{k_B T} \,. \tag{4.3.5}$$

The heating constant h can be constrained from the equilibrium situation (from Eq. 4.3.4) and $dT/dt = 0$,

$$0 = h - \chi\rho_0 T_0^\alpha(s) \,, \tag{4.3.6}$$

where the mass density is $\rho_0 = mp_0/k_B T_0$. Substituting h into Eq. (4.3.4) and ρ from Eq. (4.3.5) then yields an equation for the temperature evolution,

$$c_p\frac{\partial T}{\partial t} = \chi\rho_0 T_0^\alpha\left(1 - \frac{T^{\alpha-1}}{T_0^{\alpha-1}}\right) \,. \tag{4.3.7}$$

From this evolutionary equation, the stability conditions for a purely radiatively cooling loop can easily be seen. If the exponent $\alpha < 1$, a decrease in temperature ($T < T_0$) makes the right-hand side of Eq. (4.3.7) negative, and thus the cooling continues (with the same sign as $\partial T/\partial t$ on the left-hand side), so we have a *thermal instability*, while a loop with $\alpha > 1$ has a stable behavior (Fig. 4.5). A comparison with the radiative loss function (Fig. 2.14) shows that $\alpha < 1$ for temperatures of $T \gtrsim 10^5$ K, so most of the coronal loops would undergo such a *radiatively driven instability* in the case of insufficient thermal conduction. The *radiative cooling time* τ_{rad} follows from Eq. (4.3.7), using $\partial T / \partial t \approx T_0 / \tau_{rad}$,

$$\tau_{rad} = \frac{c_p}{\chi \rho_0 T_0^{\alpha - 1}} \; . \tag{4.3.8}$$

Some numerical values of radiative cooling times τ_{rad} for typical coronal conditions are given in Table 4.1. If cooling is dominated by thermal conduction, the energy equation could be written as

$$c_p \rho \frac{\partial T}{\partial t} = E_H - \nabla F_C = E_H + \frac{\partial}{\partial s}\left(\kappa T^{5/2} \frac{\partial T}{\partial s} \right) , \tag{4.3.9}$$

which yields a *conductive cooling time* τ_{cond} (with approximation $\partial T / \partial s \approx T_0 / L$),

$$\tau_{cond} = \frac{L^2 \rho_0 c_p}{\kappa T_0^{5/2}} \; . \tag{4.3.10}$$

Therefore, the *radiatively driven thermal instability* can be prevented if the *conductive cooling time* is shorter than the *radiative cooling time* (i.e., $\tau_{cond} < \tau_{rad}$), which yields with Eqs. (4.3.8) and (4.3.10) a condition for the maximum loop length (Priest 1978),

$$L < L_{max} = \left(\frac{\kappa T_0^{7/2 - \alpha}}{\chi \rho_0^2} \right)^{1/2} . \tag{4.3.11}$$

Thus, if a loop expands and exceeds this limit of the maximum length L_{max}, thermal instability will take place, cooling down until a new equilibrium is reached, with a cooler temperature and, consequently, higher density. This process is also called *condensation* and may be important for the formation of coronal loops and prominences.

A general stability analysis of a hydrodynamic system can be performed via two different methods: (1) a time-dependent perturbation equation is inserted into the hydrodynamic equations and the eigen values of the resulting combined differential equation are determined (with numerical methods) to determine the sign of the growth rate in the perturbation (*normal-mode method* or *eigen-mode analysis*); (2) a small perturbation is added to a hydrostatic solution in a time-dependent hydrodynamic numeric code, and the time evolution is simulated to see whether the perturbed state returns to the initial hydrostatic solution or diverges from it.

Thermal instabilities have been pioneered for astrophysical plasmas under general conditions by Parker (1953) and Field (1965). However, applications to coronal loops require more specific geometries. Antiochos (1979) examined the stability of hydrostatic solutions in coronal loops, which we describe here as an example of an eigenmode analysis. We use the continuity equation (4.1.21), the force equation (4.1.22), the energy equation in the form of Eq. (4.1.18), a polytropic index of $\gamma = 5/3$, neglect the height dependence of the gravity in the momentum equation (i.e., $\partial p_{grav} / \partial r \approx 0$,

Table 4.1: Radiative cooling times τ_{rad} for typical densities n_0 and temperatures T_0 in coronal loops (after Priest 1982).

	$T_0 = 0.1$ MK	$T_0 = 0.5$ MK	$T_0 = 1.0$ MK	$T_0 = 2.0$ MK
$n_0 = 10^8$ cm^{-3}	$\tau_{rad} = 440$ s	2200 s	3.2×10^4 s	1.3×10^5 s
$n_0 = 10^9$ cm^{-3}	$\tau_{rad} = 44$ s	220 s	3200 s	1.3×10^4 s
$n_0 = 10^{10}$ cm^{-3}	$\tau_{rad} = 4.4$ s	22 s	320 s	1300 s

and use a powerlaw parameterization for the heating rate, $E_H = E_{H0} n^{\gamma_1} T^{-\gamma_2}$, and for the radiative loss rate, $E_R = n^2 \Lambda_0 T^{-l}$,

$$\frac{\partial n}{\partial t} + \frac{\partial}{\partial s}(nv) = 0 , \tag{4.3.12}$$

$$mn\frac{\partial v}{\partial t} + mnv\frac{\partial v}{\partial s} = -\frac{\partial p}{\partial s} , \tag{4.3.13}$$

$$\frac{3}{2}\frac{\partial p}{\partial t} + \frac{3}{2}v\frac{\partial p}{\partial s} + \frac{5}{2}p\frac{\partial v}{\partial s} = E_{H0} n^{\gamma_1} T^{-\gamma_2} - n^2 \Lambda_0 T^{-l} - \frac{\partial}{\partial s}\left(\kappa T^{5/2}\frac{\partial T}{\partial s}\right) . \tag{4.3.14}$$

Inserting a time-independent hydrostatic solution $n_0(s)$, $T_0(s)$, $p_0(s)$, $v_0(s) = 0$, yields the equilibrium model,

$$0 = -\frac{\partial p_0}{\partial s} , \tag{4.3.15}$$

$$0 = E_{H0} n_0^{\gamma_1} T_0^{-\gamma_2} - n_0^2 \Lambda_0 T_0^{-l} - \frac{\partial}{\partial s}\left(\kappa T_0^{5/2}\frac{\partial T_0}{\partial s}\right) . \tag{4.3.16}$$

A perturbation can be described in each variable $n(s,t)$, $p(s,t)$, $T(s,t)$, and $v(s,t)$ by the general function,

$$f(s,t) = f_0(s) + f_1(s)\exp(+\nu t) , \tag{4.3.17}$$

which is composed of the stationary solution $f_0(s)$ and a time-dependent disturbance with a spatial function $f_1(s)$ that is initially exponentially growing or decreasing at a rate ν, for positive or negative values of ν, respectively. Explicitly, the perturbation terms for each of the four variables for a hydrostatic solution are ($v_0 = 0$),

$$n(s,t) = n_0(s) + n_1(s)\exp(+\nu t) , \tag{4.3.18}$$

$$p(s,t) = p_0(s) + p_1(s)\exp(+\nu t) , \tag{4.3.19}$$

$$T(s,t) = T_0(s) + T_1(s)\exp(+\nu t) , \tag{4.3.20}$$

$$v(s,t) = v_1(s)\exp(+\nu t) . \tag{4.3.21}$$

Of course, the pressure p is not independent of the disturbances in density n and temperature T, so the equation of state (e.g., $p = nk_B T$ for an ideal gas), yields the linearized relation, from the total derivative of $\partial p/\partial s = nk_B(\partial T/\partial s) + k_B T(\partial n/\partial s)$,

$$\frac{p_1}{p_0} = \frac{T_1}{T_0} + \frac{n_1}{n_0} . \tag{4.3.22}$$

Inserting the disturbance equations (4.3.18−21) into the time-dependent hydrodynamic equations (4.3.12−14), and subtracting the hydrostatic solutions (4.2.15−16), and linearizing the powerlaw functions for the heating and radiative cooling terms, yields the following first-order plasma equations (i.e., the linearized equations without second-order terms in the perturbed variables),

$$\nu n_1 + \frac{d}{ds}(n_0 v_1) = 0 \, , \tag{4.3.23}$$

$$\nu m n_0 v_1 = -\frac{dp_1}{ds} \, , \tag{4.3.24}$$

$$\frac{3}{2}\nu p_1 + \frac{5}{2}p_0\frac{dv_1}{ds} - \kappa\frac{5}{7}\frac{d}{ds^2}(T_0^{5/2}T_1)$$
$$= E_{H0}\gamma_1 n_0^{\gamma_1-1} n_1 T_0^{-\gamma_2} - E_{H0}\gamma_2 n_0^{\gamma_1} T_0^{-\gamma_2-1} T_1 - 2n_0 n_1 \Lambda_0 T_0^{-l} + n_0^2 \Lambda_0 l T_0^{-l-1} T_1 \, . \tag{4.3.25}$$

To study perturbations that are entirely located inside the coronal loop, the boundary conditions of vanishing perturbation pressure, velocity, and heat flux can be imposed. Using these boundary conditions and changing to dimensionless variables,

$$x = s/L \, , \tag{4.3.26}$$

$$y = \left(\frac{T_0}{T_{max}}\right)^{7/2} \, , \tag{4.3.27}$$

Antiochos (1979) derived from Eqs. (4.2.23−25) a single eigen-value equation of the Sturm−Liouville type,

$$\frac{\partial}{\partial x}\left(y\frac{\partial}{\partial x}\mu\right) + Q(x)\mu + \lambda\mu = 0 \, , \tag{4.3.28}$$

where the eigen function μ is

$$\mu = y^{5/14}\frac{\partial v_1}{\partial x} \, , \tag{4.3.29}$$

and the coefficients $Q(x)$ and λ are,

$$Q(x) = \frac{1}{14}\frac{\partial^2 y}{\partial x^2} - \frac{1}{y}\left(\frac{5}{14}\frac{\partial y}{\partial x}\right)^2 - y\left(\frac{\partial y}{\partial x}\right)^{-1}\frac{\partial^3 y}{\partial x^3} \, , \tag{4.3.30}$$

$$\lambda = -\nu\frac{5}{2}\frac{p_0 L^2}{\kappa T_{max}^{7/2}}, \tag{4.3.31}$$

with T_{max} the temperature at the looptop, and L the loop half length. Since neither the eigen functions $\mu(s)$ nor the the coefficient functions $Q(s)$ and $T(s)$ can be expressed in closed form, numerical algorithms are needed to determine the eigen functions $\mu(s)$ and the eigen values λ. The stability analysis comes down to determine the sign of the eigen value λ. If an eigen value λ comes out negative, the growth rate ν is positive according to definition (4.3.31), because all other parameters are positive in Eq. (4.3.31), and thus the perturbation function (4.3.17) will grow, meaning that the system is unstable. If an eigen value λ comes out positive, the growth rate ν is negative, which means that the perturbation will damp out, and the system will return to its hydrostatic equilibrium (i.e., it is a *stable* equilibrium). With this type of stability analysis, Antiochos (1979) found that hydrostatic models with vanishing heat flux at the loop base are unstable, and thus concluded that the transition region is intrinsically dynamic.

Analytical work with the eigen-mode method has been developed to quite sophisticated levels to find thermally stable regimes of coronal loops (for a review see, e.g., Bray et al. 1991, p. 312−320). It was found that the thermal stability of coronal loops depends critically on assumptions of the lower boundary conditions, in particular the heat flux $F_C(s = 0)$ across the loop footpoint and the chromospheric temperature $T(s = 0)$ (Chiuderi et al. 1981; McClymont & Craig 1985a; Antiochos et al. 1985), on the stabilizing effects of chromospheric evaporation (McClymont & Craig 1985b), but also on the stabilizing effects of gravity for large loops (Wragg & Priest 1982). Of course, the results from analytical stability studies have restricted validity, because a number of assumptions and approximations (e.g., the radiative loss function) have to be made to make the problem treatable. So, an obvious next step was to study the stability problem by means of numerical simulations, which do not suffer from analytical approximations. However, these could still produce unreliable results if the numerical resolution in the transition region is insufficient. Stable loop solutions were found in time-dependent simulations for a variety of lower boundary conditions (Craig et al. 1982; Peres et al. 1982; Mok et al. 1991). We will discuss numerical simulations of the dynamic evolution of loops in more detail in §4.7.

4.3.2 Heating Scale Instability

Another crucial factor in the question of loop stability is the effect of the heating function $E_H(s)$. Early work assumed a constant heating function ($E_H(s) = const$, e.g., Rosner et al. 1978a) or a heating rate that is proportional to the loop density ($E_H(s) \propto \rho(s)$, Eq. 4.3.2, e.g., Priest 1978). Serio et al. (1981) divided hydrostatic solutions of coronal loops into two classes, depending on whether the temperature maximum is at the looptop or below. For the second class (with a temperature maximum *below* the looptop), Serio et al. (1981) recognized that some solutions have a density inversion, and thus are unstable against the *Rayleigh−Taylor instability*, whenever $\nabla n \cdot \mathbf{g} < 0$. From the energy scaling laws Serio et al. (1981) derived a criterion for unstable loop solutions for heating scale heights s_H that are about three times shorter than the loop half length L, i.e.,

$$s_H \lesssim \frac{L}{3} \, . \tag{4.3.32}$$

Hydrostatic solutions were calculated for a large parameter space of heating scale heights (4 Mm < s_H < 400 Mm) and loop half lengths (4 Mm < L < 400 Mm) by Aschwanden et al. (2001), finding an approximate limit for unstable loops at $s_{H,Mm} \lesssim \sqrt{L_{Mm}}$. More accurate calculations with a higher resolution in the transition region (Aschwanden & Schrijver, 2002) revealed a limit that was closer to that given by Serio et al. (1981) (i.e., Eq. 4.3.32). Critical heating scale heights s_H were also independently

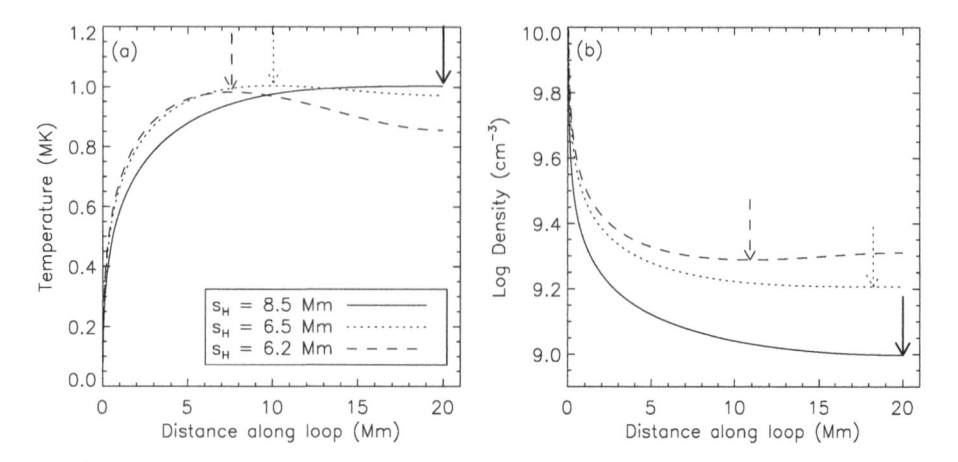

Figure 4.6: The temperature (a) and density (b) solutions for a $L = 20$ Mm loop with three different values around the critical heating scale height s_H. The locations of the maximum and minimum temperatures are shown with arrows (Winebarger et al. 2003b).

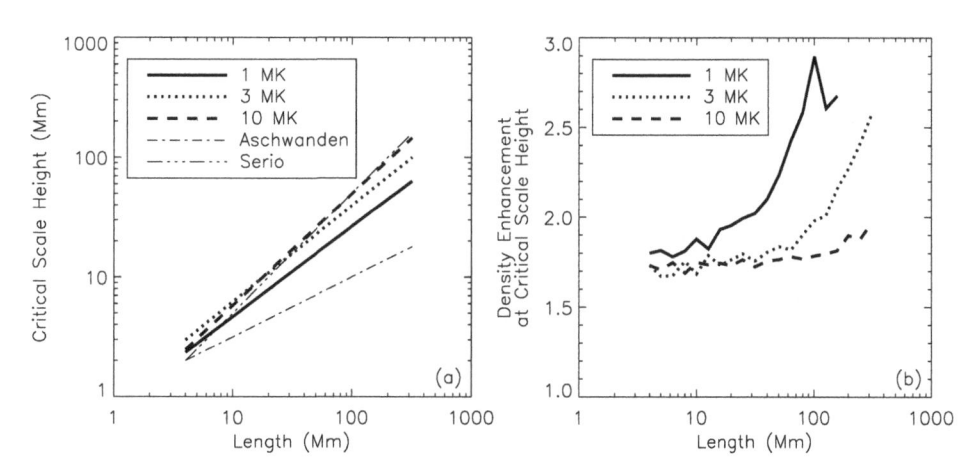

Figure 4.7: (a) The critical heating scale height $s_H(L)$ for loops with an apex temperature of $T = 1.0$ (Serio et al. 1981; Aschwanden et al. 2001) and for $T = 1.0, 3.0, 10.0$ MK (Winebarger et al. 2003b). The critical heating scale heights correspond to slightly different loop stability criteria in the different works. (b) The corresponding maximum density enhancements at the critical heating scale height are shown for $T = 1.0, 3.0, 10.0$ MK (Winebarger et al. 2003b).

determined with a hydrodynamic code in the same parameter space by Winebarger et al. (2003b), where the instability limit, defined by a temperature maximum $T_{max}(s)$ below the looptop ($s < L$), was found to have the following dependence on the loop length L and temperature T,

$$\left(\frac{s_H}{1 \text{ Mm}}\right) \lesssim A(T) \left(\frac{L}{1 \text{ Mm}}\right)^{\delta(T)} \qquad (4.3.33)$$

Table 4.2: Flow measurements in the corona ($T > 0.5$ MK).

Observer	Instrument	Wavelength	Temperature	Flow speed
Coronal holes:				
Cushman & Rense (1976)	rocket	Si XI, 303 Å	< 1.4 MK	13±5 km s^{-1}
		Mg X, 610 Å	1.1 MK	12±5 km s^{-1}
		Mg IX, 368 Å	1.0 MK	14±3 km s^{-1}
Rottman et al. (1982)	rocket	Mg X, 625 Å	1.4 MK	12 km s^{-1}
Orrall et al. (1983)	rocket	Mg X, 625 Å	1.4 MK	8 km s^{-1}
Quiet Sun regions:				
Hassler et al. (1991)	rocket	Ne VIII, 770 Å	0.6 MK	0±4 km s^{-1}
Mariska & Dowdy (1992)	Skylab	Ne VII, 465 Å	0.5 MK	0±18 km s^{-1}
Brekke et al. (1997b)		Ne VIII, 770 Å	0.6 MK	5 ± 1.5 km s^{-1}
		Ne VIII, 770 Å	0.6 MK	6 ± 3 km s^{-1}
		Mg X, 625 Å	1.1 MK	6 ± 1.5 km s^{-1}
Active regions, plages:				
Mariska & Dowdy (1992)	Skylab	Ne VII, 465 Å	0.5 MK	< 70 km s^{-1}
Brekke (1993)	HRTS	Fe XII, 1242 Å	1.3 MK	7±4 km s^{-1}
Above sunspots:				
Neupert et al. (1992)	SERTS	Mg IX, 368 Å	1.1 MK	14±3 km s^{-1}
Active region loops:				
Brekke et al. (1997a)	SoHO/CDS	Mg IX, 368 Å	1.0 MK	< 50 km s^{-1}
	SoHO/CDS	Mg X, 624 Å	1.0 MK	< 50 km s^{-1}
	SoHO/CDS	Si XII, 520 Å	1.9 MK	≈ 25 km s^{-1}
	SoHO/CDS	Fe XVI, 360 Å	2.7 MK	≈ 25 km s^{-1}
Winebarger et al. (2001)	TRACE	Fe IX/X, 171 Å	1.0 MK	5 − 20 km s^{-1}
Winebarger et al. (2002)	SUMER	Ne VIII, 770 Å	0.6 MK	40 km s^{-1}

with numerical parameters $A(T) = 0.83, 1.2, ..., 0.68$ and $\delta(T) = 0.75, 0.73, ..., 0.93$ in the temperature range of $T = 1, 2, ..., 10$ MK. An example of a temperature and density solution is shown in Fig. 4.6, while a comparison of the critical limits $s_H(L)$ from Serio et al. (1981), Aschwanden et al. (2001), and Winebarger et al. (2003b) are shown in Fig. 4.7. As a rule of thumb, we can say that loops that are heated over a scale height of less than a third of the loop half length are expected to be dynamically unstable.

4.4 Observations of Flows in Coronal Loops

Before the advent of SoHO, most of the flow measurements were reported from the chromosphere and transition region below temperatures of $T \lesssim 0.25$ MK (e.g., see Table II in Brekke et al. 1997b). In the hotter portions of the solar transition region, a few flow velocities were reported in coronal holes (Table 4.2) from rocket flights in the pre-SoHO era (Cushman & Rense 1976; Rottman et al. 1982; Orrall et al. 1983), of the order of v= $8 - 16$ km s^{-1}. In quiet Sun regions, no significant flow speeds were found (Hassler et al. 1991) or only very marginal ones in the order of ≈ 5 km s^{-1} (Brekke

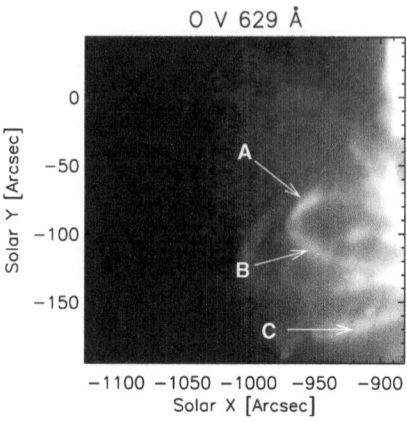

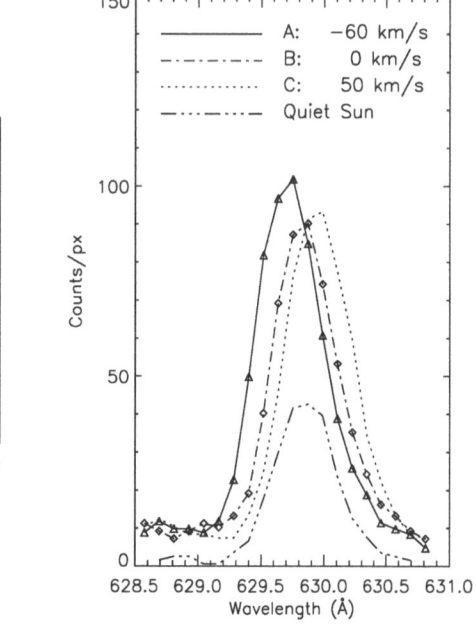

Figure 4.8: Active region loop system above the east limb observed with SoHO/CDS in O V on 1996 July 27, 10:00 UT. The Doppler-shifted line profiles (right frame) are measured at three different spatial positions (A,B,C, left frame) (Brekke et al. 1997a).

et al. 1997b). Significant flow speeds, in the order of $\approx 5 - 50$ km s^{-1}, were discovered in active region loops mainly with SoHO/CDS (Brekke et al. 1997a), but also with SoHO/SUMER (Winebarger et al. 2002) and with TRACE (Winebarger et al. 2001). Is is likely that flows exist in the majority of active region loops, but their measurement is difficult with every existing method, because: (1) if spectrographs (such as CDS or SUMER) are used, then the Doppler shift can only be measured along the line-of-sight and may largely cancel out in images with insufficient spatial resolution, and (2) if high-resolution imaging (such as TRACE) is used, then only inhomogeneities in flowing plasmas can be tracked, while laminar flows appear indifferent to static loops. In the following we describe some of the few existing coronal flow measurements.

SoHO/CDS measurements of high-speed velocities in active region loops were reported by Brekke et al. (1997a), displaying large Doppler shifts of the O V, 629 Å line at coronal locations in a loop system above the east limb (Fig. 4.8). At position A, a blueshifted velocity of v$\approx$ 60 km s^{-1} (towards the observer) was measured relative to the quiet Sun line profile (Fig. 4.8, right). At the opposite loop side (at position B) there is no evidence for significant flows. This asymmetry cannot be explained by a uni-directional siphon flow, but possibly indicates a one-sided catastrophic cooling with rapid downflow, or perhaps the positions A and B might even belong to two different loops that cannot be discriminated in the CDS image. In a loop nearby (at location C), a redshifted flow with a speed of v$\approx$ 25 km s^{-1} was detected, which cannot

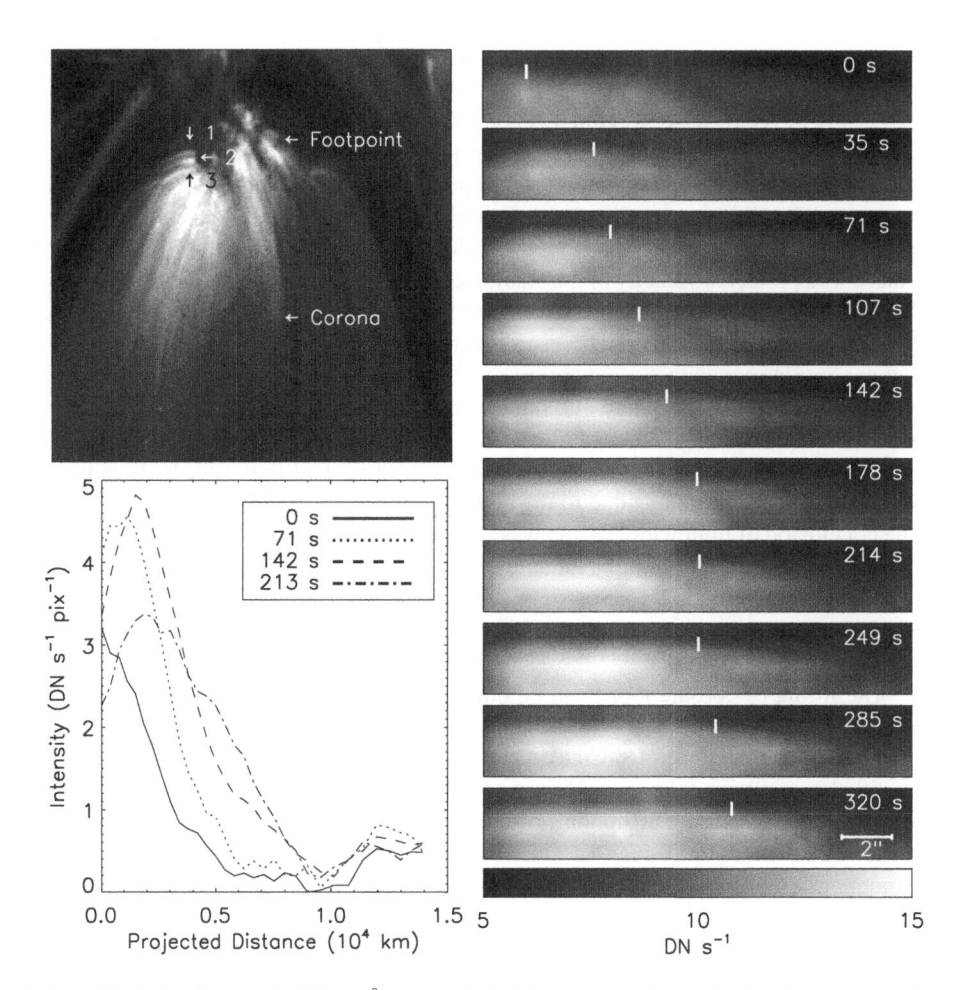

Figure 4.9: *Top left:* TRACE 171 Å image of fan-like loops in Active Region 8396 observed near the Sun center on 1998 Dec 1, 01:00:14 UT, shown with a field-of-view of 270 Mm. *Bottom left:* intensity profiles along loop #3 over a projected distance of 15 Mm taken at 4 different times. *Right side:* grey-scale representation of the flux along loop #3 (over a segment of 15 Mm) observed at 10 time intervals after 01:40:35 UT, with the footpoint of the fan-like loop on the left side (Winebarger et al. 2001).

be identified as downflow or upflow because of the unknown orientation of the loop plane. Brekke et al. (1997b) also measured high Doppler shifts in coronal lines such as Mg IX 368 Å and Mg X 624 Å, corresponding to plasma velocities of $v \approx 50$ km s^{-1}. However, those lines are contaminated by blends, which limits the accuracy of velocity measurements. While the largest speeds are measured in O V, which probably corresponds to cooling downflows (with a temperature of $T \approx 0.25$ MK), velocities in coronal temperatures ($T \gtrsim 1.0$ MK) are generally found to be subsonic ($v < 150$ km s^{-1}).

Apparent flows in active regions can also been measured by feature tracking. An example of such a measurement is shown in Fig. 4.9, observed with TRACE in Fe IX/Fe X (171 Å) in Active Region 8395 on 1998 December 1 (Winebarger et al. 2001). A bundle of fan-shaped loops (Fig. 4.9, top left panel) shows an increase in intensity that propagates along loop #3, which is interpreted as a tracer of apparent flows. A time sequence of 4 brightness profiles is shown in Fig. 4.9 (bottom left panel) and a sequence of 10 brightness maps of stripes extracted along loop #3 are shown in Fig. 4.9 (right panel). The projected velocity of the leading edge was measured to v= 15.9±3.1 (loop #1), 13.4 ± 1.6 (loop #2), 16.9 ± 2.1 (loop #3), and 4.6 ± 1.2 km s^{-1} (loop #4) for four different upflow events, each one lasting between 2 and 5 minutes. These dynamic features in plasmas with a temperature of $T \approx 1.0$ MK represent upflows of heated plasma from the chromosphere into the corona, because they could not be reproduced with quasi-static changes in hydrostatic loop models without flows. The measured velocities probably represent only lower limits to the actual flow speeds, because of projection effects as well as due to the convolution with spatial brightness variations. The conclusion of upflowing plasma was corroborated with SUMER Ne VIII Doppler shift measurements in a similar bundle of fan-shaped loops observed with TRACE 171 Å, which exhibited a line-of-sight flow speed of v $\lesssim$ 40 km s^{-1} (Winebarger et al. 2002). The same temperature of $10^{5.95\pm0.05}$ K was measured with the *SUMER* (Ne VIII, Ca X, Mg IX) lines and in the TRACE 171 Å passband.

The filling of coronal loops can be quite dynamic, especially for cool loops. An example is shown in Fig. 4.10, observed with TRACE in the C IV line, which has a formation temperature of $60, 000 - 250, 000$ K. A bundle of magnetic field lines are filled by upflows from one footpoint, which propagate to the opposite footpoint similar to a uni-directional siphon flow scenario (Fig. 4.3). Given the 30 s cadence in the 9 frames shown in Fig. 4.10 and the average loop length of $L \approx 40$ Mm, we estimate a speed of v$\approx$ 40,000 km/(8× 30 s) $\approx$ 170 km s^{-1} for the leading edge, which corresponds to a supersonic speed with a Mach number of about $M = \text{v}/\text{v}_{th} \approx$ 4. Comparing such highly dynamic images, where cool loops change on time scales shorter than 1 minute, with a CDS image as shown in Fig. 4.8, which took 17 minutes for a full raster scan, it is perhaps not surprising that most of the flows are not properly resolved in CDS images. Thus, the Doppler shift measurements reported by CDS represent only lower limits and averages over unresolved flows, which can even cancel out in the case of counterflows.

Although we have described a few analyses of flow measurements with high speeds in coronal loops, Doppler shift measurements with velocities of v$\approx$ $20 - 100$ km s^{-1} are more common in the transition region [i.e., at temperatures from 10,000 K (He I, 584 Å) to 500,000 K (Ne VI, 563 Å), while at coronal temperatures, e.g., in Mg IX (368 Å)] and particularly in Fe XVI (360 Å), large Doppler shifts are much less frequent (Fredvik et al. 2002). There are also more complicated flow patterns than uni-directional (or siphon) flows that have been observed in active region loops, such as helical or rotational flows in apparently sheared or twisted loops (Chae et al. 2000a), bi-directional or counter-streaming flows (Qiu et al. 1999), or highly fragmented down-flows in catastrophically cooling loops, also called *coronal rain* (Foukal 1987; Schrijver 2001a). Doppler-shifted emission, although it is a diagnostic of plasma motion, does not necessarily have its only explanation in terms of steady flows, it also can be pro-

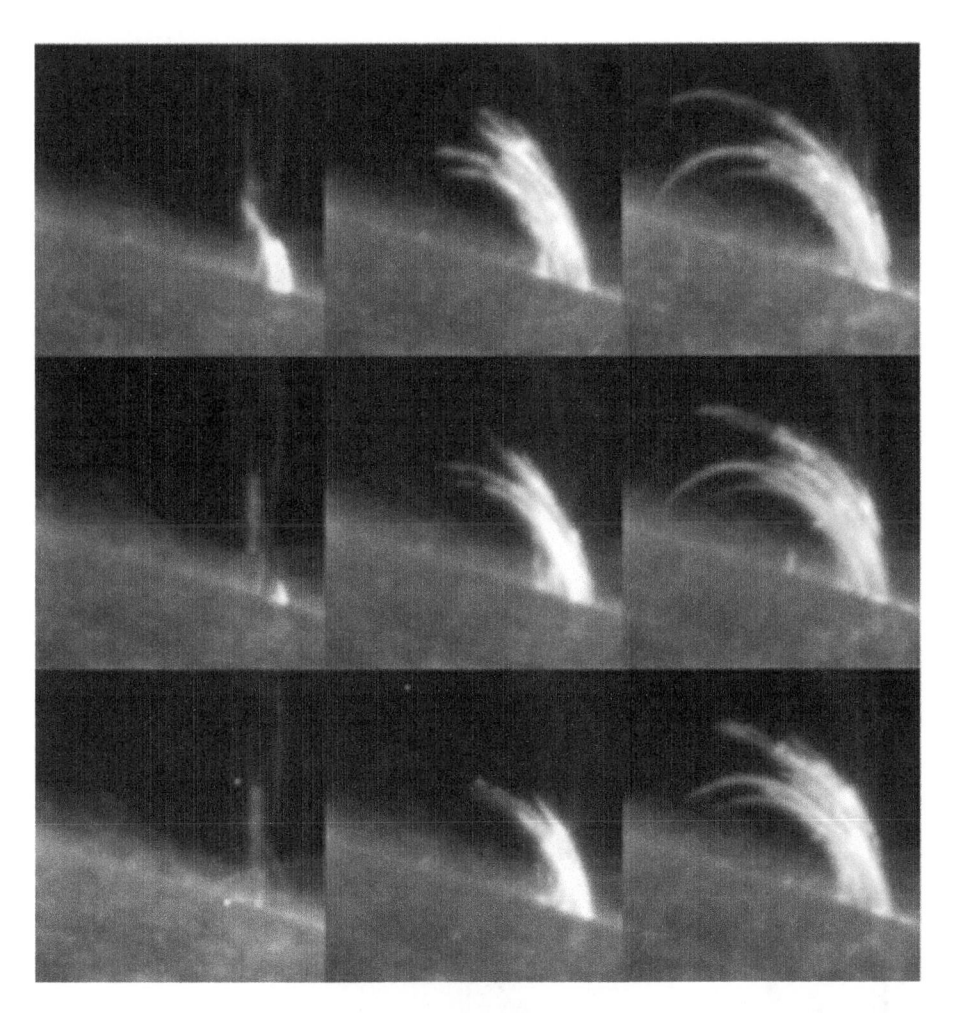

Figure 4.10: The filling of coronal field lines with cool chromospheric plasma in the temperature range of $60,000 - 250,000$ K observed with TRACE in the C IV, 1550 Å line. The sequence of images starts at the bottom left and continues in upward direction, each frame separated by ≈ 30 s. The field-of-view is 40 Mm, and rotated by $90°$ counterclockwise (courtesy of Paal Brekke).

duced by waves or oscillations. The latter, as we will see in § 7, can be distinguished from steady flows by the phase shift between velocity and intensity modulations.

4.5   Observations of Cooling Loops

Temperature changes that are observed in coronal loops are a result of the imbalance between heating and cooling rates. Multiple temperature filters are required to study such dynamic processes. Curiously, not much quantitative studies of these dynamic

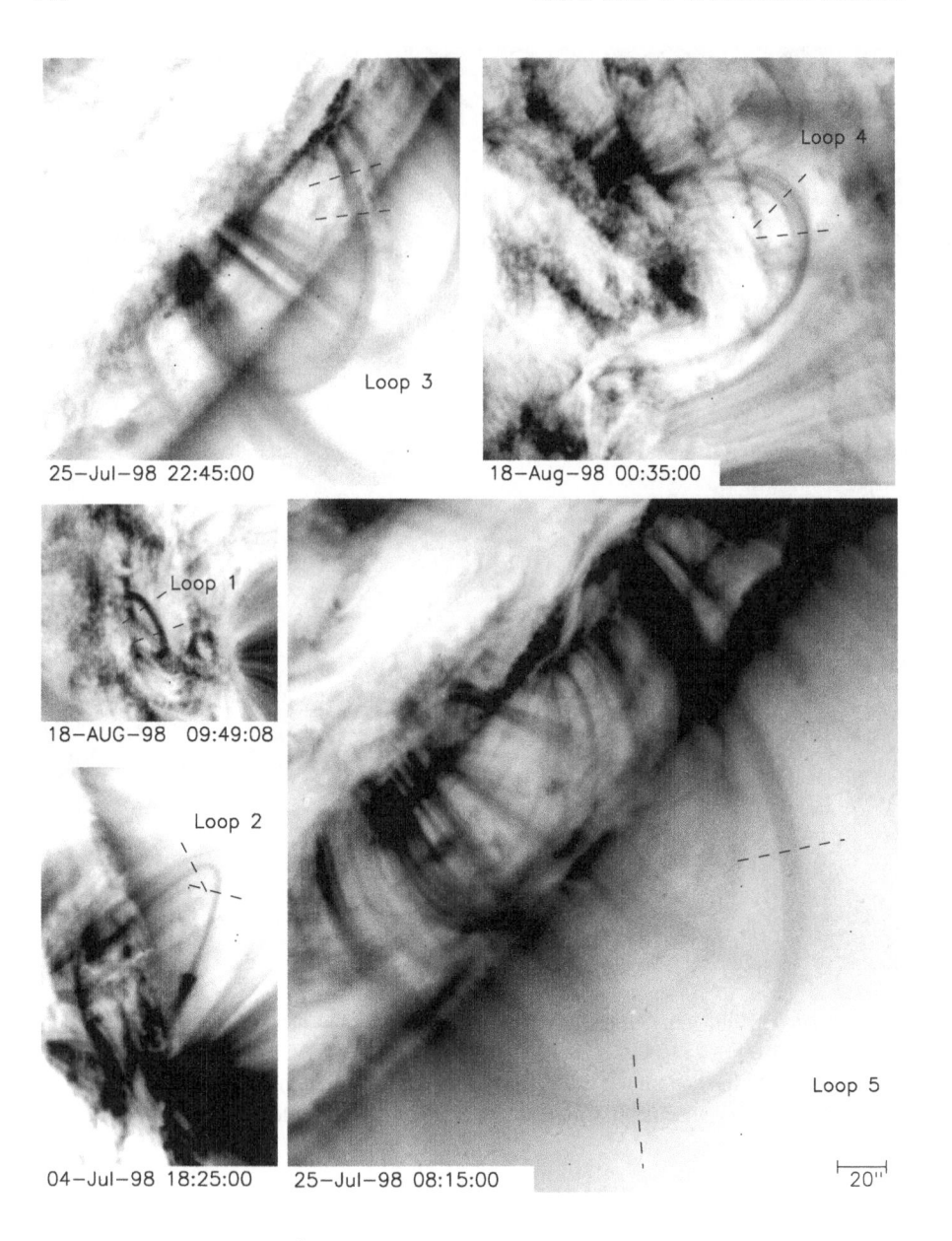

Figure 4.11: TRACE 171 Å images of 5 loops, outlined with some inner and outer curvature radii. The dashed segments, mostly located near the looptop, were used to measure the intensity as a function of time, as shown in Fig. 4.12 (Winebarger et al. 2003a).

processes have been performed, although multi-temperature filters with high spatial resolution have been available for a decade, on *Yohkoh*, SoHO, and TRACE. The first detailed quantitative studies appeared only recently, described below.

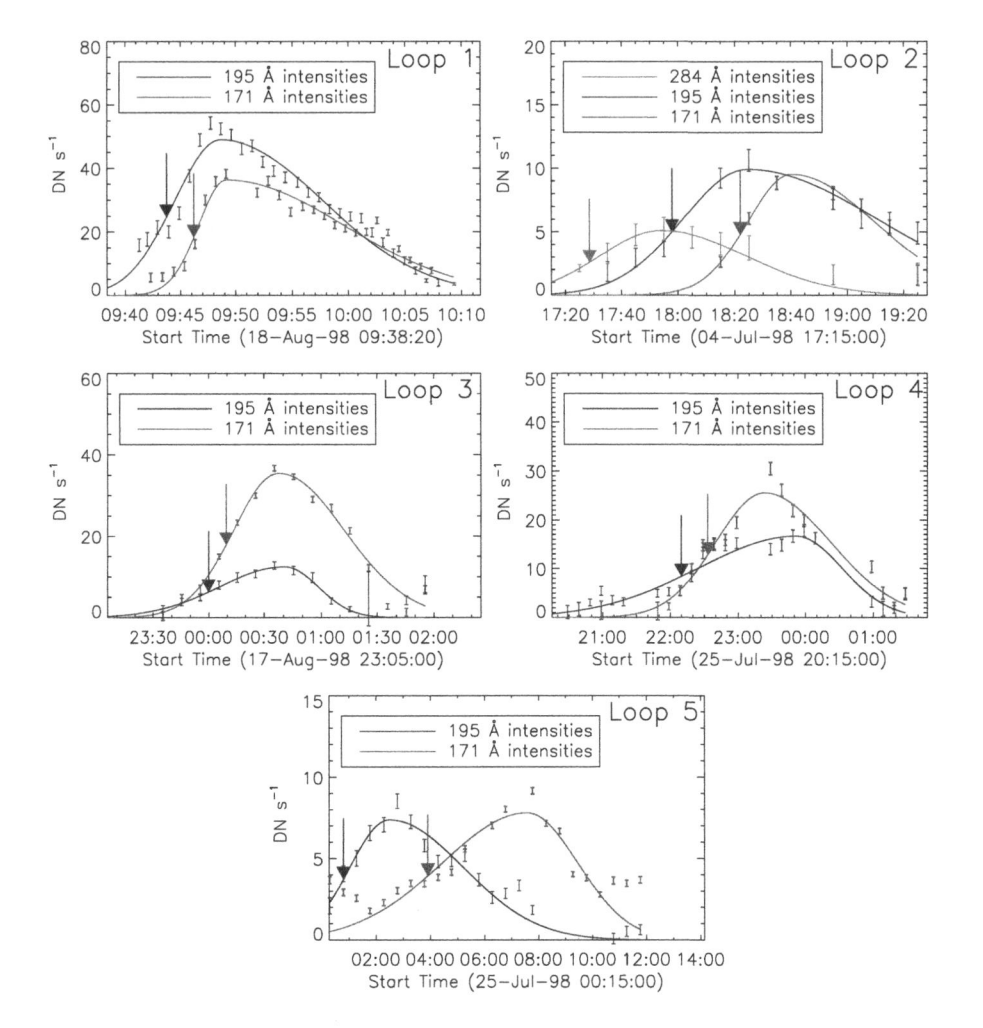

Figure 4.12: The intensities of the five loops in the 284 Å, 195 Å, and 171 Å filter as a function of time. The solid curves are fits of asymmetric Gaussians to each light curve and the arrows indicate the appearance time in each filter image, when the loop reaches half of the maximum flux (Winebarger et al. 2003a).

4.5.1 Cooling Delays

In the study of Winebarger et al. (2003a), a set of 5 active region loops was selected according to the following criteria: TRACE observations in multiple filters (171, 195, and 284 Å, if possible), at least 1-hour time coverage, no postflare loops, not obscured throughout their lifetime, and a somewhat secluded location to allow for accurate background subtraction. These 5 selected loops are shown in Fig. 4.11, exhibiting a range of sizes from $L \approx 26$ Mm to 356 Mm. All of these loops are found to show a temporal evolution with a rise and fall in flux during the observing periods. However, the rise and fall times are not simultaneous in different temperature filters. The rise time (indicated with arrows at FWHM in Fig. 4.12) always occurs first in the highest temperature filter and sequentially later in the cooler temperature filters. This is exactly the behavior expected for a cooling plasma. The cooling process is most convincingly shown for loop #2 (in Fig. 4.12), which was observed with 3 temperature filters. The delays of the rise times between the 195 Å and 171 Å filters were measured to be $\Delta t_{195-171} = 2.5, 23, 23, 10$, and 183 minutes for these 5 loops. The FWHM durations of the flux in each filter were compared with the expected lifetimes based on cooling of a single-temperature plasma, and it was found that the observed lifetimes of the loops were significantly longer than the estimated ones (in 4 out of the 5 cases), which was interpreted in terms of multiple loop strands with different temperatures and different cooling onset times. Furthermore, an approximate linear correlation was found between the cooling delay and the loop length (i.e., $\Delta t_{195-171} \propto L$).

4.5.2   Iron Abundance and Filling Factors

These observed properties of cooling times and their linear scaling to the loop lengths reveal physical relations that can be analytically derived and be used to determine the coronal iron abundance (Aschwanden et al. 2003a). Let us describe the plasma cooling through a narrow temperature range by an exponential function in the first approximation,

$$T_e(t) = T_1 \exp\left(-\frac{t}{\tau_{cool}}\right), \qquad (4.5.1)$$

where T_1 is the initial temperature at time $t = 0$ and τ_{cool} is the cooling time. Cooling over large temperature ranges would require the full consideration of the hydrodynamic equations, but the exponential approximation is fully justified for the narrow temperature range of $T_e \approx 1.0 - 1.4$ MK we are considering here, during the cooling of coronal loops through the two TRACE 171 and 195 Å filters. So, when a cooling plasma reaches the temperature T_1 of the peak response of the hotter filter ($T_1 = 1.4$ MK for TRACE 195 Å), the time delay Δt_{12} to cool down to the cooler filter T_2 (e.g., $T_2 = 0.96$ MK for TRACE 171 Å), can be expressed with Eq. (4.5.1) as

$$\Delta t_{12} = t_2 - t_1 = \tau_{cool} \ln\left(\frac{T_1}{T_2}\right). \qquad (4.5.2)$$

The cooling time scale could be dominated by thermal conduction losses in the initial phase, but is always dominated by radiative losses in the later phase. So, as a working hypothesis, we assume dominant radiative cooling, which is particularly justified near the almost isothermal looptops (where conductive cooling is very inefficient due to the small temperature gradient), and is also corroborated by the observational result found in Aschwanden et al. (2000a), where a median value of $\tau_{cool}/\tau_{rad} = 1.02$ was obtained from the statistics of 12 nanoflare loops observed with TRACE 171 and 195 Å. Thus we set the cooling time τ_{cool} equal to the radiative cooling time τ_{rad},

$$\tau_{cool} \approx \tau_{rad} = \frac{\varepsilon_{th}}{dE_R/dt} = \frac{3 n_e k_B T_e}{n_e n_H \alpha_{FIP} \Lambda(T_e)}, \qquad (4.5.3)$$

where n_e is the electron density, n_H the hydrogen density, T_e the electron temperature, k_B the Boltzmann constant, α_{FIP} the abundance enhancement factor for low *first ionization potential* elements, and $\Lambda(T_e)$ the radiative loss function, which can be approximated with a constant in the limited temperature range of $T_e \approx 0.5 - 2.0$ MK, according to the piecewise powerlaw approximation of Rosner et al. (1978a), see Fig. 2.14,

$$\Lambda(T) \approx \Lambda_0 = 10^{-21.94} \ (\text{erg s}^{-1} \text{ cm}^3), \text{ for } T \approx 0.5 - 2.0 \text{ MK} . \quad (4.5.4)$$

Inserting this radiative loss function into the time delay (Eq. 4.5.2) we find

$$\Delta t_{12} = \frac{3k_B T_e}{n_H \alpha_{FIP} \Lambda_0} \ln \left(\frac{T_1}{T_2} \right) . \quad (4.5.5)$$

The computation of the radiative loss function at a given temperature depends on the elemental atomic abundances, and thus we define a reference value $\Lambda_{0,ph}$ for photospheric abundances. Coronal abundances generally show a density enhancement for low first ionization potential (FIP) elements (§2.10), which we express with an enhancement factor α_{FIP}. Since the radiative loss scales with this FIP enhancement factor, we obtain for the coronal value of the radiative loss function $\Lambda_{0,Cor}$ (e.g., when iron (Fe) ions dominate),

$$\Lambda_{0,cor} = \Lambda_{0,ph} \times \alpha_{Fe} . \quad (4.5.6)$$

The cooling delay Δt_{12} as a function of the coronal iron abundance α_{Fe} is, assuming full ionization in the corona ($n_H = n_e$), thus,

$$\Delta t_{12} = \frac{3k_B T_e}{n_e \alpha_{Fe} \Lambda_{0,ph}} \ln \left(\frac{T_1}{T_2} \right) . \quad (4.5.7)$$

When a loop cools through a passband, the maximum of the flux $F(t)$ is detected at the time when the loop temperature matches the peak of the response function, $R_2 = R(T_2)$, so the peak flux F_2 of the light curve in the lower filter corresponds to the differential emission measure EM at the filter temperature T_2,

$$F_2 = EM \times R_2 = (n_e^2 w q_{fill}) \times R_2 \quad (4.5.8)$$

with the flux F_2 in units of DN/(pixel s), w is the loop width or diameter, q_{fill} is the linear filling factor in the case of unresolved substructures, and R_2 is the response function, which is $R_2 = 0.37 \times 10^{-26}$ cm^5 DN/(pixel s) for 171 Å and photospheric abundances (see Appendix A in Aschwanden et al. 2000c). Inserting the density from Eq. (4.5.8) into Eq. (4.5.7) we find the following expression for the iron abundance α_{Fe},

$$\alpha_{Fe} = \frac{3k_B T_2}{\Lambda_{0,ph} \Delta t_{12}} \sqrt{\frac{R_2 w q_{fill}}{F_2}} \ln \left(\frac{T_1}{T_2} \right)$$

$$\alpha_{Fe} = 4.17 \left(\frac{w \, q_{fill}}{1 \text{ Mm}} \right)^{1/2} \left(\frac{F_2}{10 \text{ DN/s}} \right)^{-1/2} \left(\frac{\Delta t_{12}}{1 \text{ min}} \right)^{-1} . \quad (4.5.9)$$

For a filling factor of unity ($q_{fill} = 1$), the iron enhancement factor can be determined with an accuracy of about $\lesssim 20\%$, because the observables w, F_2, and Δt_{12} can each

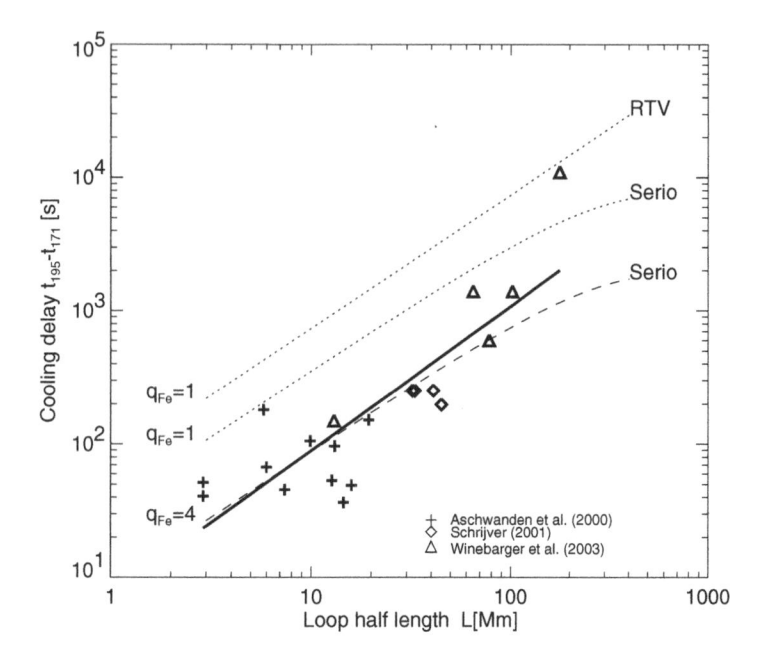

Figure 4.13: Cooling delays Δt_{12} are measured between the peak time in the TRACE 195 Å and 171 Å filters, as a function of the loop half length L, from three data sets with 11 nanoflare loops (crosses; Aschwanden et al. 2000c), 4 active region loops (diamonds; Schrijver 2001a), and 5 active region loops (triangles; Winebarger et al. 2003a). The thick line represents a linear regression fit with a slope of 0.94 ± 0.12. The theoretically predicted scaling laws (based on RTV and Serio et al. 1981, with $s_H = L/3$) are shown for an iron enhancement factor of $\alpha_{Fe} = 1.0$ (dotted) and $\alpha_{Fe} = 4.0$ (dashed) (Aschwanden et al. 2003a).

be measured better than $\lesssim 10\%$. In case of unresolved fine structures (i.e., filling factors of $q_{fill} < 1$), we obtain with Eq. (4.5.9) at least a firm upper limit for the iron enhancement. With this method, the following values were found in the study of Aschwanden et al. (2003a),

$$\begin{aligned} \alpha_{Fe} \quad &\leq 4.9 \pm 1.7 \qquad \text{for 11 nanoflare loops [Aschwanden et al. 2000c]} \\ &\leq 1.4 \pm 0.4 \qquad \text{for 5 coronal loops [Winebarger et al. 2003a] .} \end{aligned}$$
$$(4.5.10)$$

Interestingly, the iron enhancement was found to be higher in the short-lived (≈ 10 min) nanoflare loops than in the longer lived (few hours) coronal loops. The enhancement factors of order $\alpha_{Fe} \approx 4$ are typical for coronal abundances, as evidenced by other studies (e.g., Feldman 1992; White et al. 2000). So, this method independently corroborates the previous finding that low-FIP elements such as Fe are enhanced in the corona, at least initially after the filling of a coronal loop with heated chromospheric plasma. The fact that we find a lower value in longer lived loops, almost compatible with chromospheric abundances, could be interpreted in terms of a depletion process that occurs after the initial filling phase.

4.5.3 Scaling Law of Cooling Loops

By the same token, we can also explain the proportionality of the cooling delay Δt_{12} with the loop length L, found by Winebarger et al. (2003a). We can make use of the energy balance equation (§3.6), which is valid in a steady state before the cooling process, at the turning point from dominant heating to dominant cooling, or at the turning point from dominant conductive cooling to radiative cooling. Using the RTV scaling law of Rosner et al. (1978a),

$$T_{max} \approx 1400(p_0 L)^{1/3} \times q_{Serio} \qquad (4.5.11)$$

with p_0 the pressure and L the loop half length, generalized for gravity and nonuniform heating by Serio et al. (1981), with the correction factor

$$q_{Serio} = \exp\left(-0.08\frac{L}{s_H} - 0.04\frac{L}{\lambda_p}\right), \qquad (4.5.12)$$

where s_H is the heating scale length and $\lambda_p = 47,000 \times T_{MK}$ km is the pressure scale height. Using the ideal gas law we can eliminate the pressure p,

$$p = 2n_e k_B T_{max} = p_0 q_p, \qquad q_p = \exp\left(-\frac{h}{\lambda_p}\right) = \exp\left(-\frac{2L}{\pi\lambda_p}\right). \qquad (4.5.13)$$

in the RTV scaling law and find the following expression for the density n_e (with the understanding that the density is measured at the same location as T_{max}, which is generally at the looptop for RTV loops),

$$n_e = \frac{T_{max}^2 q_p}{2k_B L (1400\ q_{Serio})^3}. \qquad (4.5.14)$$

Inserting this density into the relation for the cooling delay (Eq. 4.5.7) we indeed find a proportional relation $\Delta t_{12} \propto L$,

$$\Delta t_{12} = L \times \left[\frac{6\ (1400\ q_{Serio})^3\ k_B^2}{T_{max}\Lambda_{0,ph}\alpha_{Fe}q_p} \ln\left(\frac{T_1}{T_2}\right)\right]. \qquad (4.5.15)$$

which should show up for cooling loops with similar maximum temperatures T_{max}. In Fig. 4.13 we show cooling times Δt_{12} versus loop lengths L observed in 20 different loops. A linear regression fit yields a slope of 0.94 ± 0.12, which is fully compatible with the theoretical prediction of a linear relationship. Also the absolute value of the linear regression fit yields a mean abundance enhancement of $\alpha_{Fe} \approx 4.0$, which is consistent with spectroscopic measurements (Feldman 1992).

4.5.4 Catastrophic Cooling Phase

The cooling process of coronal loops can be traced from coronal temperatures ($T \gtrsim 10^6$ K) in EUV wavelengths all the way down to transition region temperatures ($T \lesssim 10^4$ K) in Hα and Lyman-α wavelengths. Downflowing material observed in coronal loops in Hα wavelengths, with typical speeds of v$\approx$ 30 km s^{-1} (Brueckner 1981), many

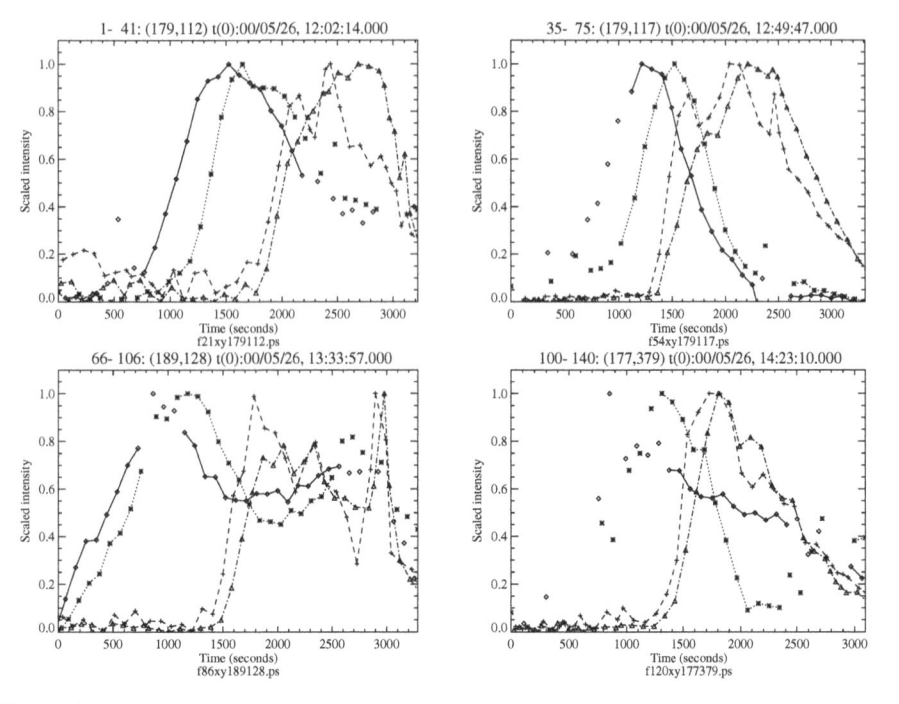

Figure 4.14: TRACE light curves of 4 cooling loops measured at the looptops, normalized to the minimum and maximum intensity during an interval of 3200 s. In each panel, 4 curves are shown taken with different filters: 195 Å (solid lines, diamonds), 171 Å (dotted lines, asterisks), 1600 Å (dashed lines, plus symbols), and 1216 Å (dashed-dotted lines, triangles). Note the similar but delayed evolution of the intensity rise in each filter, sequentially peaking in order of decreasing temperatures (Schrijver 2001a).

pressure scale heights above the gravitational scale height expected for cool Hα material, has been dubbed *coronal rain* (e.g., Foukal 1978, 1987). Recent observations with TRACE, taken with a cadence of 90 s, clearly show the entire cooling process, starting with dominant radiative cooling in EUV and progressing into a phase of *catastrophic cooling* followed by high-speed downflows with temperatures of $T \lesssim 100,000$ K (Schrijver 2001a). In Fig. 4.14 the light curves of 4 cooling loops are shown, observed in 195 Å (Fe XII, $T \approx 1.5$ MK), 171 Å (Fe IX, Fe X, $T \approx 1.0$ MK), 1600 Å (C IV, 0.1 MK), and 1216 Å (Lyman-α, C II, C IV; $T \approx 20,000 - 100,000$ K). The light curves peak in consecutively cooler temperatures, as expected for a cooling plasma, with typical cooling times of $\tau_{cool} \approx 10^3$ s. Once catastrophic cooling starts, cool material is observed to form clumps and to slide down on both sides of the loops with speeds up to 100 km s^{-1}, but the downward acceleration was found to be no more than 80 m s^{-2}, less than 1/3 of the surface gravity (Schrijver 2001a). All these cooling processes were observed in quiescent (nonflaring) loops in active regions, and are probably an inevitable consequence once heating stops and the loops become thermally unstable. Comprehensive hydrodynamic modeling of the spatio-temporal evolution of these heating and cooling processes is still lacking.

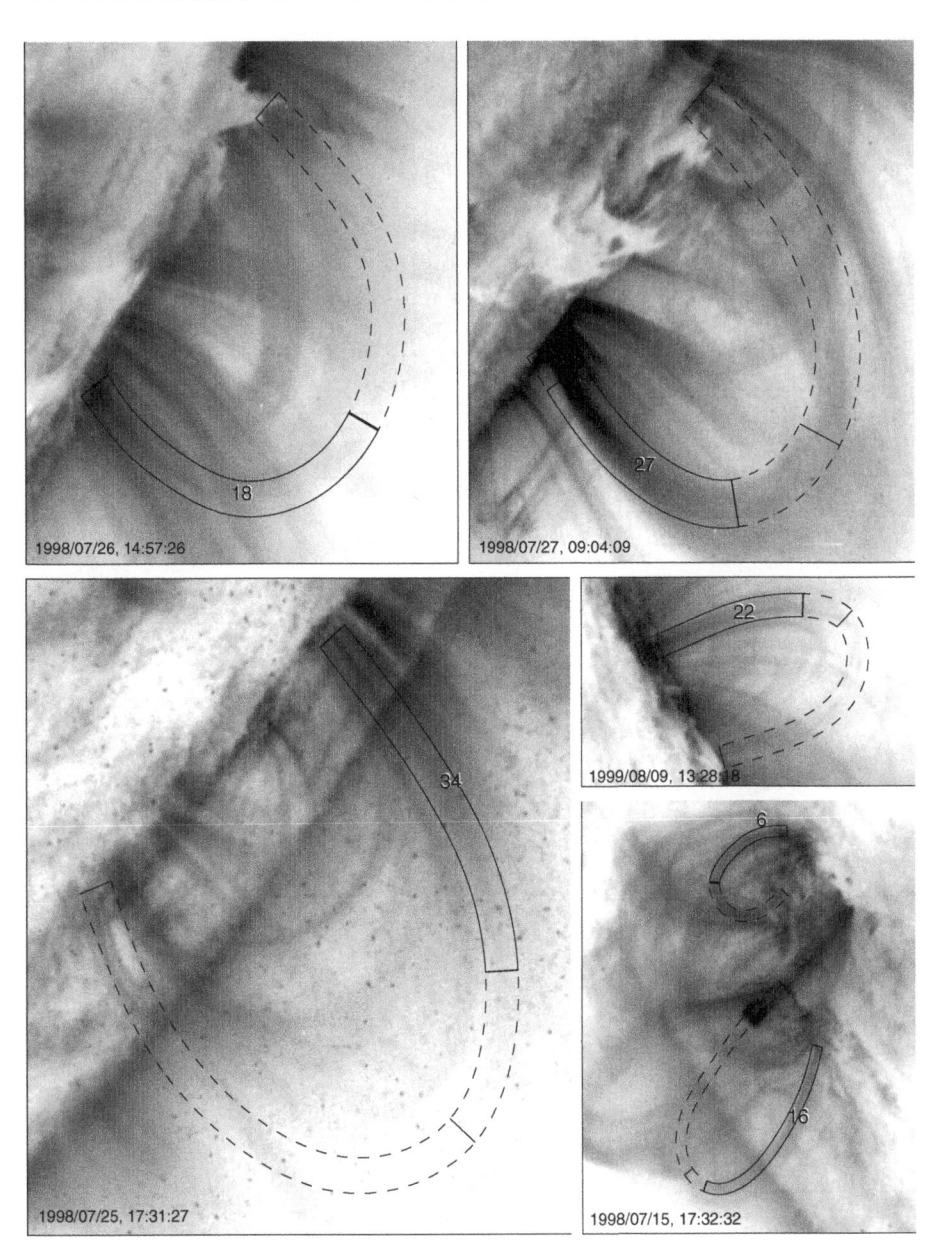

Figure 4.15: TRACE 171 Å images of selected active region loops. The selected loops are outlined with dashed lines, while the analyzed density profiles $n_e(s)$ were measured in the segments marked with solid linestyle. All partial images are shown on the same spatial scale, according to the scale bar at the bottom. The loops are numbered according to their length, with #1 being the smallest loop ($L = 4$ Mm) and #41 the largest loop ($L = 324$ Mm). Black color corresponds to high EUV intensity (Aschwanden et al. 2000d).

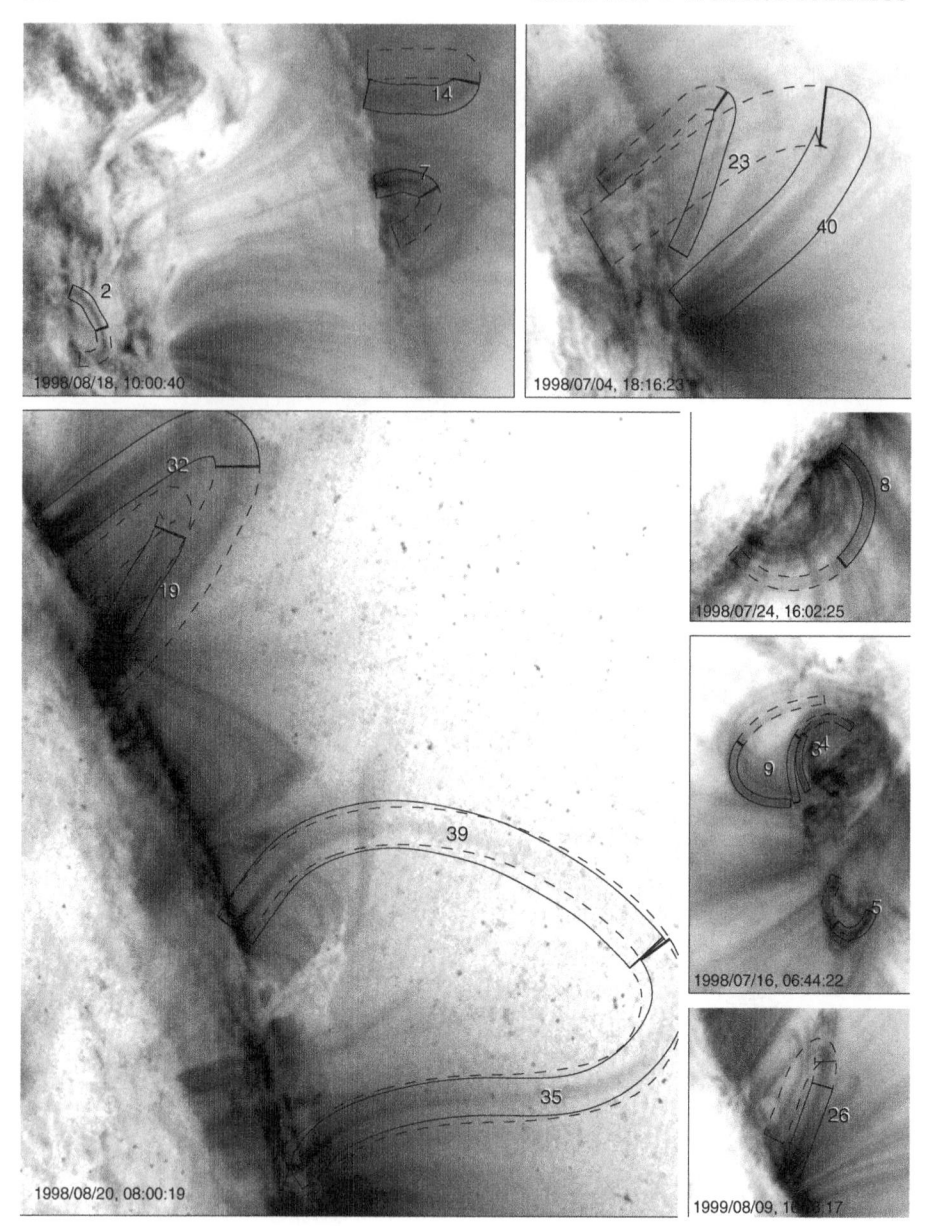

Figure 4.16: See description in Fig. 4.15.

4.6 Observations of Non-Hydrostatic Loops

In §3.9 we reported on observations of hydrostatic loops, based on the criterion that
their density profile as a function of height $n_e(h)$ corresponds roughly to the gravita-

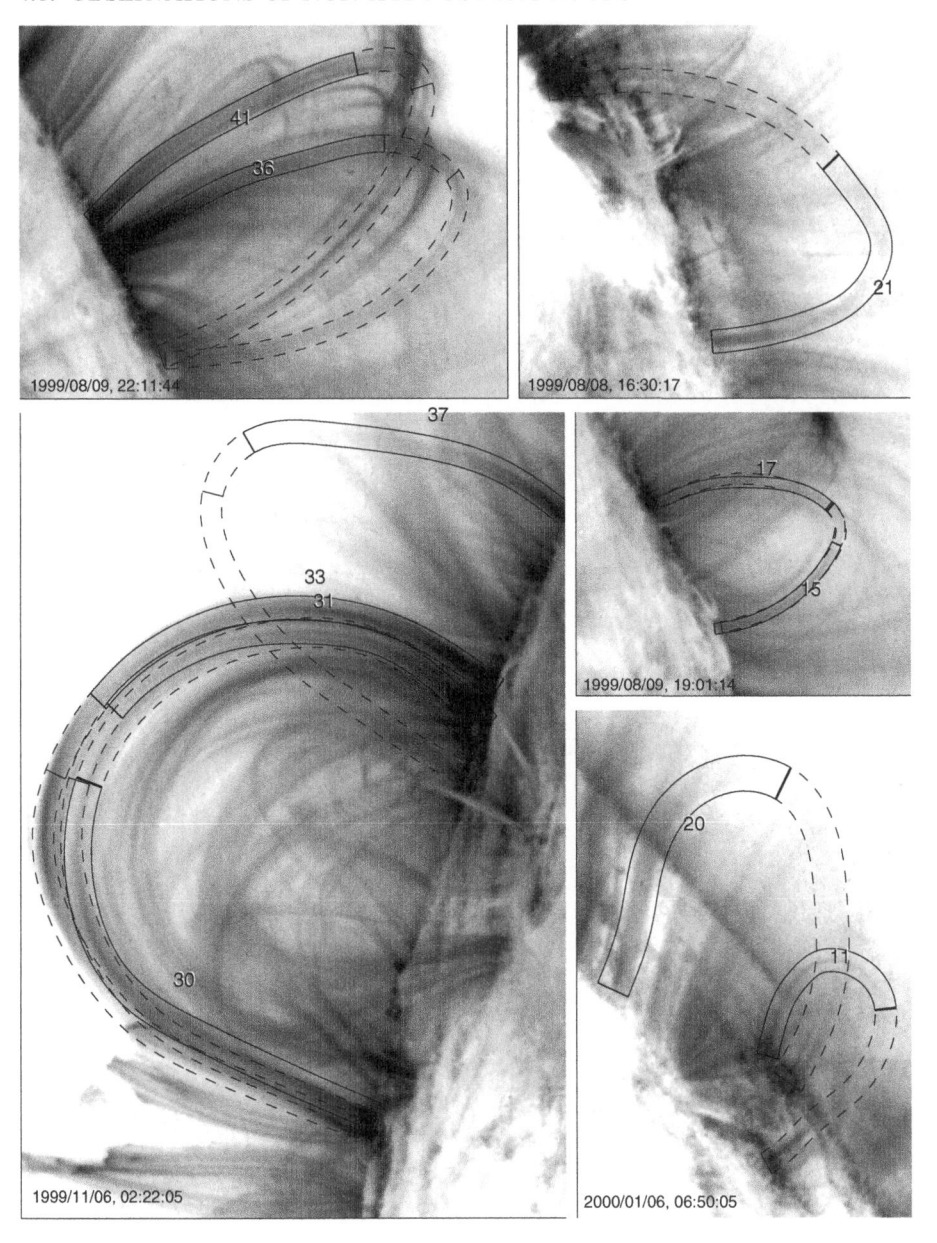

Figure 4.17: See description in Fig. 4.15.

tional scale height that is expected for a given loop temperature. Loops are expected to reach such a hydrostatic equilibrium in older active regions, where no impulsive heating or steady-state heating occurs, or in the initial phase of cooling after heating stopped, as long they are thermally stable.

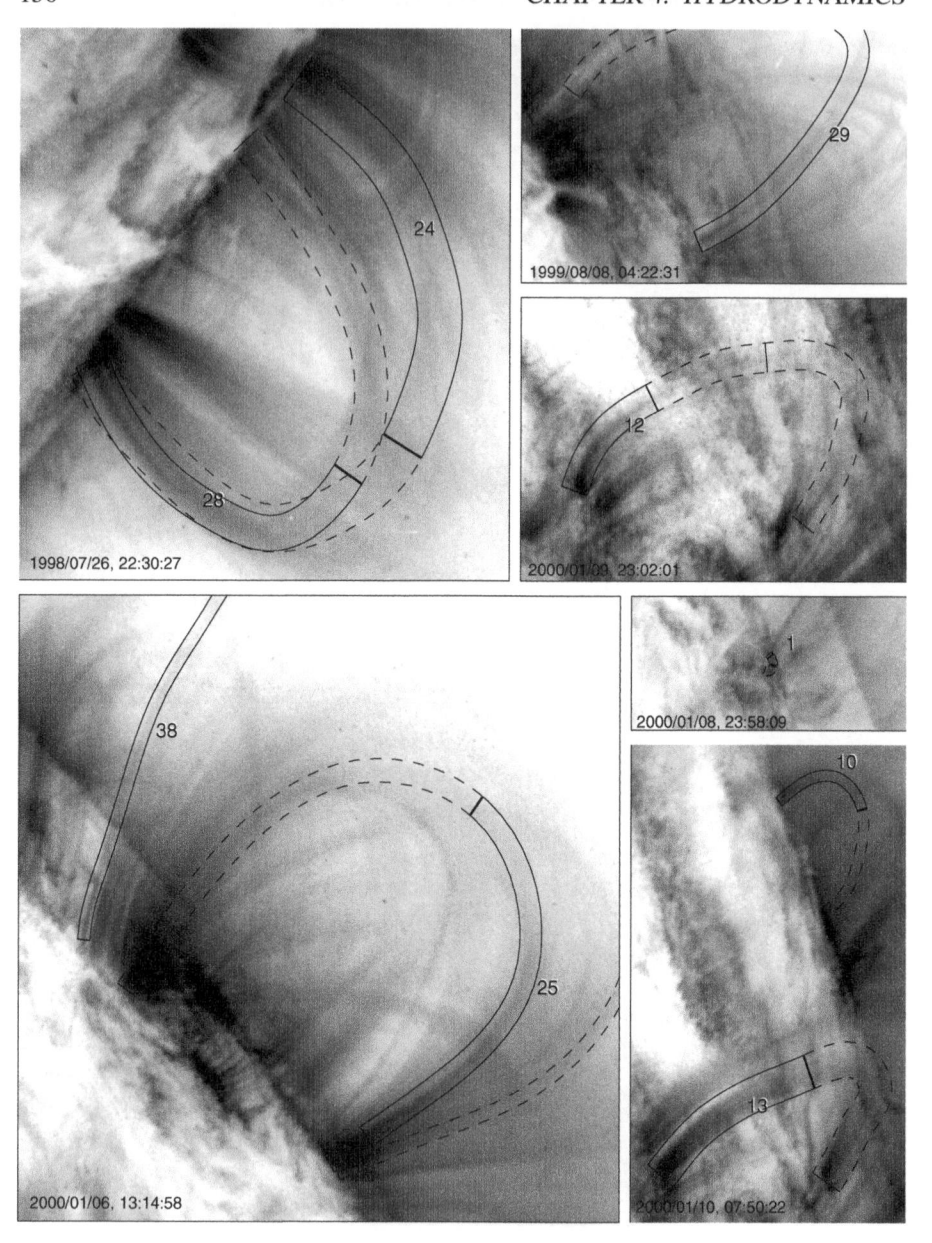

0   10 20 30 40 50 60 70 80 90 100 Mm

Figure 4.18: See description in Fig. 4.15.

4.6.1   TRACE Observations

However, recent TRACE observations now allow us to clearly distinguish between loops in hydrostatic equilibrium and those that are *not in hydrostatic equilibrium*, based on spatially resolved density profiles $n_e(h)$ measured as a function of height h. We

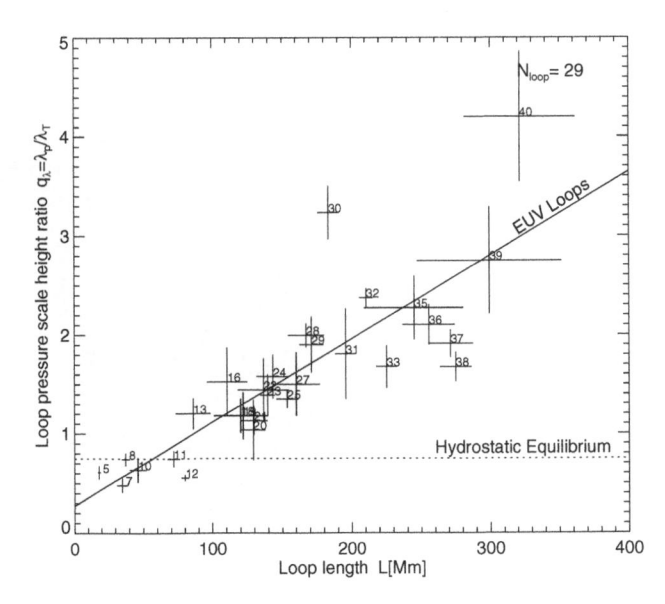

Figure 4.19: The ratio q_λ of scale heights λ_p (inferred from the best fits to the observations) to the hydrostatic scale height λ_{RTV} is shown as a function of the loop length L in the top panel. Note that short loops with a half length of $L \lesssim 80$ Mm have scale heights close to the hydrostatic equilibrium scale height, while they become systematically longer for larger loops, up to about $q_\lambda = 4$ hydrostatic scale heights. From the 41 loops shown in Figs. 4.15−18, only measurements from a subset of 29 loops with relatively small uncertainties ($\sigma_{q_\lambda}/q_\lambda \leq 0.7$) are shown (Aschwanden et al. 2000d).

describe here a data set of 41 observed EUV loops that turned out to be mostly *non-hydrostatic* (Aschwanden et al. 2000d). These loops were selected by the criterion of significant contrast above the background over the entire length of the loops, which obviously constitutes a bias for high densities at the looptop in the case of large loops that extend over several density scale heights, and thus prefers nonhydrostatic loops. The images of the selected 41 loops are shown in Figs. 4.15−18, having a large variety of sizes, with half lengths of $L = 4 - 324$ Mm, all shown on the same scale. Almost all of these loops were observed in active regions without previous flaring, except for the loops #6 and #16, where a C5.7 flare occurred 7 hours earlier, and the loops #30, #31, #33, #37, where a M3.0 flare occurred 8 hours earlier. Two of these loops (#2 and #23, both in Fig. 4.16 top panels) were also analyzed by Winebarger et al. (2003a), shown as cases #1 and #2 in Figs. 4.11 and 4.12, and were both identified to be in a cooling phase. The cooling process is clearly an indication that these loops do not have an energy balance between heating and cooling, and thus are not likely to be in hydrostatic equilibrium.

The nonhydrostaticity of the selected loops has been diagnosed from inferring the pressure scale height λ_p from their density $n_e(s)$ and temperature profiles $T_e(s)$, which was found to be in excess of the expected temperature scale height $\lambda_T = 47 \times T_{MK}$

The Non-Hydrostatic Sun   The Hydrostatic Sun

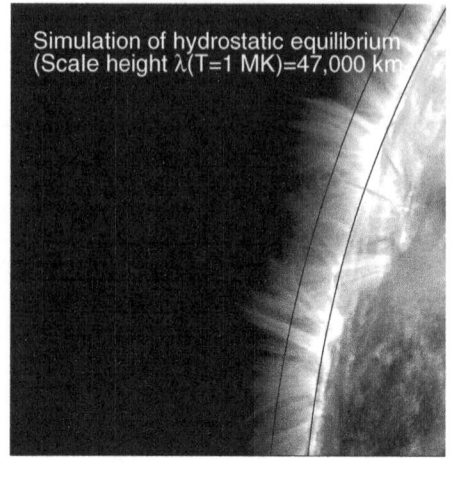

TRACE 171 A, 1999 Nov 6, 22:05 UT

Simulation of hydrostatic equilibrium (Scale height λ(T=1 MK)=47,000 km

Figure 4.20: An active region with many loops that have an extended scale height of $\lambda_p/\lambda_T \lesssim 3-4$ (left panel) has been scaled to the hydrostatic thermal scale height of $T = 1$ MK (right panel). The pressure scale height of the 1 MK plasma is $\lambda_T = 46$ Mm, but the observed flux is proportional to the emission measure $(F \mapsto EM \mapsto n_e^2)$, which has the half pressure scale height $\lambda_T/2 = 23$ Mm (Aschwanden et al. 2001).

Mm. The deviations in the pressure scale height were quantified by the ratio

$$q_\lambda = \frac{\lambda_p}{\lambda_T} , \qquad (4.6.1)$$

which are shown in Fig. 4.19. The scale height factor q_λ was found to be systematically larger for large loops (in $\approx 60\%$), up to factors of $q_\lambda \lesssim 4$. This of course reflects the bias of the detection threshold requirement. The largest loop (#41), with a half length of $L = 324$, Mm has a height of $h \approx 206$ Mm, which corresponds to $h/\lambda_T \approx 4.3$ thermal scale heights. The EUV flux or the emission measure is proportional to the squared density, so it would drop by a factor of $\exp(-4.3 \times 2) \approx 10^{-4}$ between the footpoint and the loop apex for the hydrostatic case, which is below the detection threshold. The scale height of the super-hydrostatic density profile is about 4 times larger, implying a density decrease of about a factor of $\exp(-2) \approx 0.1$ only, which is detectable. Nevertheless, although most of the small loops have a pressure scale height that is compatible with hydrostatic equilibrium, this particular selection demonstrates that there exists a considerable number of large loops evolving in a state far away from hydrostatic equilibrium, governed by hydrodynamic processes that have been little quantified and are poorly understood. The dynamic forces that are needed to suspend the enormous amount of plasma in such super-hydrostatic loops is illustrated in Fig. 4.20, which shows an image of super-hydrostatic loops observed on 1999-Nov-06, 22:05 UT (left panel), compared with a simulated image that mimics how the same loop constellation would appear in hydrostatic equilibrium (right panel).

4.6.2 Theoretical Models

What hydrodynamic processes could explain such super-hydrostatic loops? We discuss briefly three possible interpretations, all involving some dynamic scenarios: (1) accelerated flows, (2) impulsive heating of multiple loop strands with subsequent cooling, and (3) wave pressure that exceeds the hydrodynamic pressure.

Let us consider first a scenario with upflowing plasma that is accelerated, say with a constant vertical acceleration of $\partial v / \partial t = a \cos \theta$ over some time interval. This acceleration term has the same physical units as the gravitational pressure in the momentum equation (4.2.8), so that it effectively corresponds to a reduced gravity force,

$$mnv \frac{\partial v}{\partial s} = -\frac{\partial p}{\partial s} - mn(g_\odot - a) \cos \theta \qquad (4.6.2)$$

and has the same solution as the steady-flow case (Eq. 4.2.7),

$$\left(v - \frac{c_s^2}{v} \right) \frac{\partial v}{\partial s} = -(g_\odot - a) \cos \theta \qquad (4.6.3)$$

except for the reduced gravity, which yields an extended pressure scale height $\lambda_p = q_\lambda \lambda_T$ in the pressure solution, amounting to a factor of

$$q_\lambda = \frac{\lambda_p}{\lambda_T} = \frac{1}{1 - a/g_\odot} . \qquad (4.6.4)$$

In this scenario, the observed factors of $q_\lambda \lesssim 3$ are feasible with accelerations up to $a \lesssim (2/3) g_\odot$. The upflow speed at the loop base increases (for constant acceleration) with time as

$$v_0(t) = a(t - t_0) = g_\odot (1 - \frac{1}{q_\lambda})(t - t_0) . \qquad (4.6.5)$$

For a loop with a super-hydrostatic scale height of q_λ it takes a time interval of

$$\Delta t_s = \frac{c_s}{g_\odot} \frac{1}{1 - 1/q_\lambda} , \qquad (4.6.6)$$

to reach the sound speed, which is about $\Delta t_s \approx 14$ minutes for $q_\lambda \approx 3$ and a sound speed of $c_s = 150$ km s^{-1} (in a $T = 1.0$ MK plasma). After this time interval a supersonic point is reached and a shock has to develop. Therefore, subsonic flows with such extended scale heights can only be maintained over shorter acceleration intervals. Such a constraint conveys severe restrictions on the lifetime of loops with super-hydrostatic scale heights.

A second scenario was envisioned in terms of nonsteady heating, using time-dependent hydrodynamic simulations (Warren et al. 2002; Mendoza–Briceno et al. 2002). An active region was simulated by an ensemble of loop strands that have been impulsively heated and are cooling through the TRACE 171 Å (Fe IX/X) and 195 Å (Fe XII) bandpass. A heating function of the form

$$E_H(s, t) = E_0 + g(t) E_{H0} \exp \left[-\frac{(s - s_0)^2}{2\sigma_s^2} \right] , \qquad (4.6.7)$$

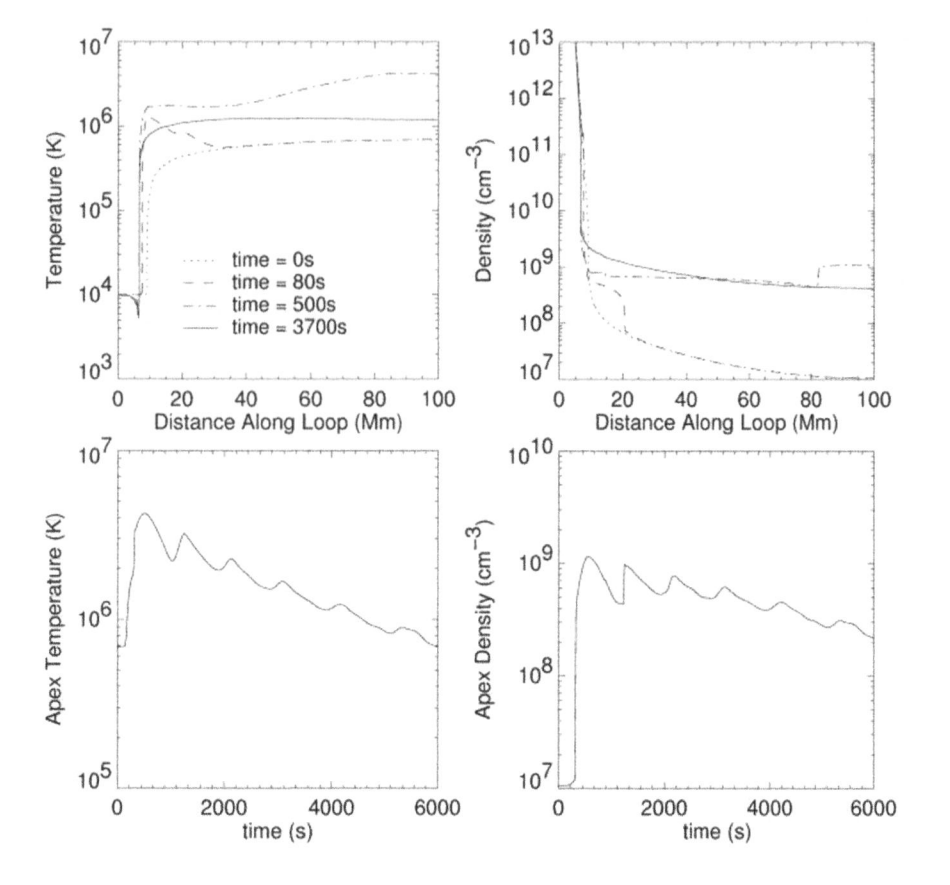

Figure 4.21: Evolution of an impulsively heated loop simulated with a time-dependent hydro-dynamic code, showing the temperature $T_e(s)$ and density profiles $n_e(s)$ at 4 different times (top panels), and the loop apex temperature $T_e(t)$ and apex density $n_e(t)$ (bottom panels) (Warren et al. 2002).

was assumed, where s_0 designates the location of the impulsive heating, σ_s is the spa-tial width of the heating, $g(t)$ a linearly decreasing time dependence, E_{H0} the peak heating rate, and E_0 a constant background heating rate. The results of the numer-ical simulation for $s_0 = 9.0$ Mm, $\sigma_s = 0.3$ Mm, $E_F = 0.5$ erg cm^{-3} s^{-1}, and a heating duration of 600 s are shown in Fig. 4.21, for temperature $T_e(s,t)$ and density profiles $n_e(s,t)$. After 800 s, the maximum temperature in the loop remains located near the loop apex, with temperature pulses propagating back down the loop after they passed the loop apex. The density increases almost two orders of magnitude, compared with the uniform heating solution (of the background heating) before impulsive heating starts. Moreover, the density profile clearly exhibits a larger scale height than the ther-mal scale height before heating, and thus can explain the observed super-hydrostatic loops. The simulations demonstrate that an impulsively heated loop cools faster than it drains, allowing for large departures from hydrostatic equilibrium as it cools through the TRACE bandpasses (Warren et al. 2002). Moreover, the flat filter ratios and the persistence of these loops for longer than the radiative cooling time imply substructures with different timing in their heating and cooling phases.

A third scenario to explain super-hydrostatic density scale-height enhancements was proposed by Litwin & Rosner (1998), where the presence of MHD waves exerts a ponderomotive force on a dielectric medium (such as the plasma), which opposes the force of gravity. The parallel momentum balance in steady state can then be expressed by

$$n_0 \nabla_\parallel (\Psi_g + \Psi_p) = -\nabla_\parallel p_0 , \qquad (4.6.8)$$

where Ψ_g is the gravitational potential, $\Psi_p = -mu^2/2$ is the ponderomotive potential, $u^2 = <v^2>$ is the average velocity fluctuation of the plasma shaken by MHD waves, n_0 is the mean particle number density, and $m = \rho_0/n_0$ is the ion mass. Similar to Eqs. (4.6.2−4), the reduced gravity leads to an extended pressure scale height. Essentially the presence of Alfvén waves produces a wave pressure in addition to the thermodynamic gas pressure, which in principle can explain the extended scale heights observed in nonhydrostatic loops.

In conclusion, although all three scenarios can produce extended or super-hydrostatic pressure scale heights, observational tests have not yet been performed to discriminate between the different theoretical possibilities. The difficulties are: (1) the requirements of flow measurements as a function of time, (2) fits of time-dependent temperature $T_e(s, t)$ and density profiles $n_e(s, t)$, and (3) the lack of diagnostics for wave pressure.

4.7 Hydrodynamic Numerical Simulations of Loops

The time-dependent evolution of physical parameters in coronal loops, such as the temperature $T_e(s, t)$ and density $n_e(s, t)$, is constrained by the hydrodynamic equations (4.1.21−23) and generally needs to be calculated numerically. If the evolution converges to a stable state, we call it a *hydrostatic solution* (§3.6), which is time-independent [i.e., $T_e(s)$ and $n_e(s)$], and which can also be calculated with a *shooting method*. Such hydrostatic solutions are of highest importance because they represent the most likely states observed in nature, while all other functions $T_e(s)$ and $n_e(s)$ are unstable and would be observable only in a transient stage. Thus, a number of numerical simulations have been conducted to find out into which hydrostatic state a loop system evolves, or whether initial conditions such as the unknown heating function can be constrained from the observed stationary states. A representative summary of such numerical simulations is given in Table 4.3, with the initial conditions and the final states listed.

Early time-dependent simulations essentially verified the hydrostatic solutions that have been derived analytically (e.g., Landini & Monsignori−Fossi 1975; Craig et al. 1978; Rosner et al. 1978a) or numerically (e.g., Serio et al. 1981). Vesecky et al. (1979) investigated hydrostatic solutions for uniform heating functions $E_H(s) = const$ and variable loop cross sections and concluded that the temperature maximum does not necessarily need to occur at the looptop as postulated in the RTV model (Rosner et al. 1978a). Variable heating and the subsequent coronal temperature and density

Table 4.3: Hydrodynamic time-dependent numerical simulations of coronal loops.

Reference	Initial conditions and external drivers	Temporal evolution and final state
Vesecky et al. (1979)	uniform heating rate E_H	hydrostatic solutions
Krall & Antiochos (1980)	heating increase/decrease	$T_e(t,s)$, $n_e(t,s)$ evolution
Peres et al. (1982)	perturbations in T_e, n_e, E_H	relaxation and stability
Craig et al. (1982)	perturbations in E_H	relaxation and stability
Mariska & Boris (1983)	asymmetric E_H, loop tapering	little steady flows
McClymont & Craig (1985a)	uniform E_H	thermal stability
McClymont & Craig (1985b)	footpoint E_H	unstable to antisym. perturb.
MacNeice (1986)	heating pulse at looptop	chromospheric ablation
Klimchuk et al. (1987)	hot low-lying (< 1 Mm) loops	unstable, cooling ($< 10^5$ K)
Klimchuk & Mariska (1988)	flows in cool loops	siphon flows v $\lesssim 20$ km s^{-1}
Mok et al. (1991)	short cool loops, free boundaries	unstable, chrom. expansion
Robb & Cally (1992)	steady, nonsteady siphon flows	subsonic flows, "surge" flows
Peres (1997)	asymmetric cold long loops	nonsteady siphon flows
Betta et al. (1999a,b)	asymmetric p, uniform E_H	unstable siphon flows
Reale et al. (2000a,b)	localized asymmetric heating	1998-Jun-26 event
Warren et al. (2002)	impulsive heating/cooling	super-hydrostatic
Winebarger et al. (2003b)	footpoint heating	unstable for $s_H/L < 1/3$
Spadaro et al. (2003)	asym., transient footpoint E_H	siphon flows, downflows

adjustments were simulated by Krall & Antiochos (1980), mimicking evaporation and condensation processes. Mariska (1987) verified the conjecture that areas of downflowing (upflowing) plasma correspond to loop bases with decreasing (increasing) heating rates. The hydrodynamic stability and uniqueness of hydrostatic solutions was verified by introducing finite-amplitude perturbations in temperature, density, and/or heating rates (Peres et al. 1982; Craig et al. 1982). Loops whose chromospheric footpoints are only marginally stable, however, were found to evolve dynamically away from the initial static configuration (Craig et al. 1982). Also loops with a heating rate that decreases with height, were found to be unstable to antisymmetric perturbations (McClymont & Craig 1985b). The consequences of a transient heating pulse released at the looptop was simulated with sufficient numerical resolution in the transition region by MacNeice (1986), showing the downward propagating conduction front, the downward compression of the transition region, and subsequent upward ablation (also called *chromospheric evaporation* in flare models). While large hot loops ($h \gtrsim 5$ Mm) are thermally stable, small-scale loops with heights of $h \lesssim 1$ Mm were found to be thermally unstable and to evolve into cool loops far below 10^5 K, forming an *"extended chromosphere"* (Klimchuk et al. 1987). The thermal stability has been scrutinized by changing the chromospheric boundaries from rigid to flow-through, and it was found that short cool loops are unstable to thermal chromospheric expansion (Mok et al. 1991). A more detailed discussion of loop dynamics based on these numerical simulations is given in Bray et al. (1991, p. 321).

The next generation of numerical hydrodynamic simulations mainly concentrated

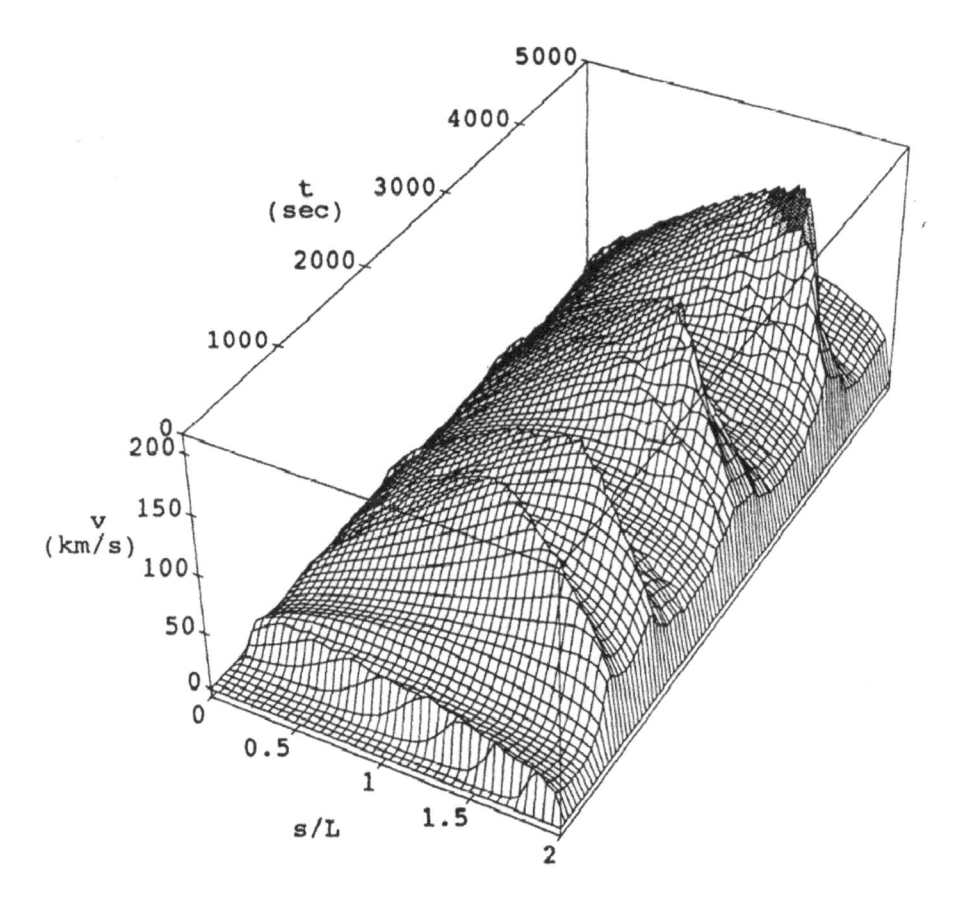

Figure 4.22: Flow velocity v as a function of position s/L ($0 \le s \le 2$) and time t for $L = 30$ Mm, $E_H = 10^{-4}$ W m^{-3}, and $p_2 = 0.02$ Pa. The second footpoint pressure p_2 is discontinuously dropped at $t = 0$ from 0.02 Pa by 15%, illustrating the phenomenon of surge flow (Robb & Cally 1992).

on flows in coronal loops. A central question was: how common are siphon flows? Mariska & Boris (1983) simulated asymmetric heating in straight and tapered flux-tubes and found little flows in the final steady phase. Robb & Cally (1992) found that siphon flows are inhibited by low heating rates, high pressures, short loop lengths, and turbulence, while small footpoint pressure asymmetries produce steady subsonic siphon flows. Time variations of the asymmetric heating rate were found to give rise to nonsteady *"surge flows"* without developing standing shocks (Fig. 4.22). In particular for cool (a few 10^5 K) long loops, no steady-state siphon flows were found, suggesting that cool loops (e.g., observed in O V), must be highly dynamic (Peres 1997). Siphon flows driven by a pressure difference between the two footpoints were found to be unstable for uniform heating, while siphon flows driven by asymmetric heat deposition were found to be stable (Betta et al. 1999a).

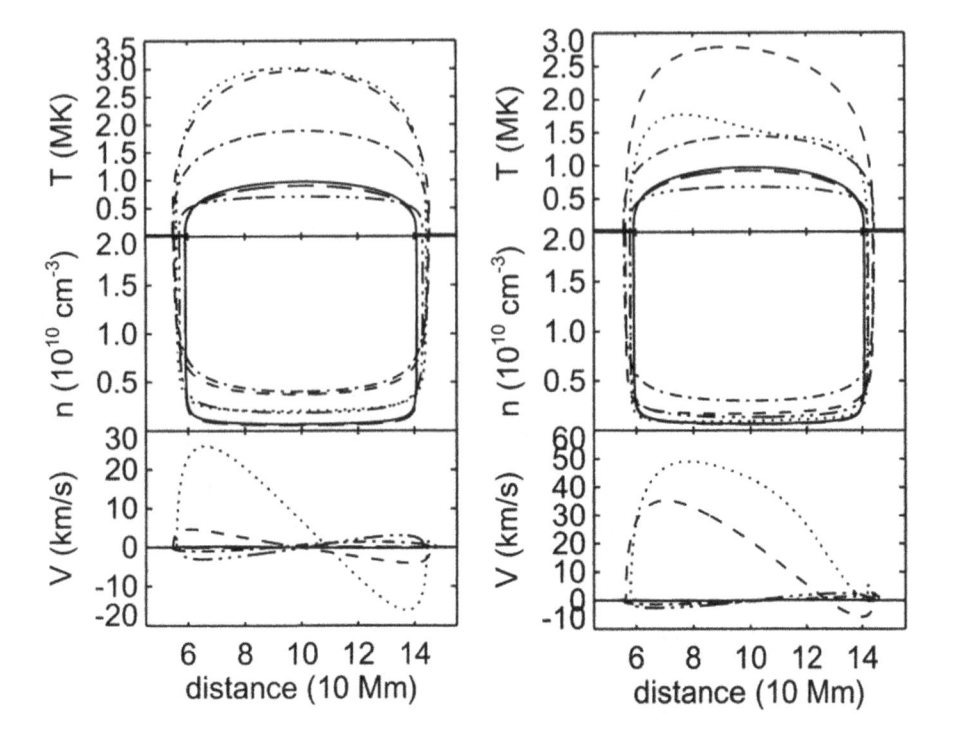

Figure 4.23: Temperature $T(s)$, electron density $n(s)$, and velocity v(s) for two loops with weakly asymmetric (left) and strongly asymmetric (right) footpoint heating are shown for 6 different times during the simulation. *Left:* $t = 0$ s (solid line), near maximum heating $t = 1504$ s (dotted line), near the end of transient heating $t = 2978$ s (dashed line), beginning of downflow $t = 5970$ s (dash-dotted line), maximum downflow $t = 14,958$ s (triple-dot-dashed line), and restored initial condition $t = 27,913$ s (long-dashed line). *Right:* $t = 0$ s (solid line), maximum siphon flow $t = 670$ s (dotted line), near maximum of transient heating $t = 1544$ s (dashed line), beginning of downflow $t = 7641$ s (dash-dotted line), maximum downflow $t = 16,641$ s (triple-dot-dashed line), and restored initial condition $t = 28,278$ s (long-dashed line) (Spadaro et al. 2003).

The most recent numerical simulations attempt to fit specific observations. For instance, the evolution of the brightening of a highly inclined loop observed with TRACE on 1998 June 26 (Fig. 3.13) was simulated with a localized, asymmetric heating function (Reale et al. 2000a,b). However, detailed quantitative fitting by hydrodynamic codes is still beyond current capabilities, and thus the obtained time-dependent solutions are only approximate and probably not unique. Warren et al. (2002) simulated impulsive heating, which could reproduce super-hydrostatic scale heights (Fig. 4.21). Winebarger et al. (2003b) varied the heating scale heights and confirmed the instability limit of Serio et al. (1981), namely that short heating scale heights $s_H/L < 1/3$ do not produce stable hydrostatic solutions (see also Figs. 4.6 and 4.7). Spadaro et al. (2003) simulated transient heating localized at footpoints (Fig. 4.23), as suggested by recent TRACE observations (§ 3.7; Table 3.3) and explained by a number of observed facts: (1) the higher density observed with TRACE than predicted by RTV, (2) persistent downflows in the transition region ($T \approx 0.1$ MK) giving rise to redshifted UV and EUV lines, and (3) strongly unequal heating at the two legs drives siphon flows (up to 50 km s^{-1}) as observed with CDS and SUMER. Their conclusions are that coronal heating is (1) transient in nature, (2) localized near the chromosphere, and (3) asymmetric about the loop midpoint. The localization and asymmetry are important to explain the observed flows, including siphon flows that would not occur for looptop heating (Spadaro et al. 2003).

Future numerical hydrodynamic simulations might also include the nonlinear evolution of instabilities, which can lead to limit cycles around stable solutions (Kuin & Martens 1982), when the time scale of chromosphere-corona coupling (evaporation) competes with the time scale of cooling (condensation). Gomez et al. (1990) studied the *Hopf bifurcation point* where the stability of the static equilibrium is lost. One predicted consequence is an excess width of chromospheric EUV lines.

4.8 Hydrodynamics of the Transition Region

The transition region between the cool chromosphere ($T \approx 10^4$ K) and hot corona ($T \gtrsim 10^6$ K) represents the most important boundary condition to all coronal structures. It acts as an energy source (for upflowing plasma and upwardly propagating waves) as well as an energy sink (for cooling and draining of coronal plasma). As discussed in §1.5, the geometric concept of this lower coronal boundary evolved from stratified layers, to canopies of vertical fluxtubes, and most recently to a rather inhomogeneous and rugged interface (Fig. 1.17). Not only has the spatial complexity increased in theoretical concepts of the transition region (Table 4.4), but also the temporal characterization: stationary models in hydrostatic equilibrium became increasingly criticized in favor of more dynamic pictures, whose time averages are not equivalent to hydrostatic equilibrium situations. Additional modeling complications also result from the fact that the transition region is not only characterized by a steep temperature and density gradient (Fig. 1.19), but also by time-dependent effects of the ionization equilibrium, abundance variations, transitions from partial to full ionization, from a high to a low plasma β-parameter, and from optically thick to optically thin regimes. Here in this section we discuss mainly aspects that are important for hydrodynamic modeling (geometry, density, temperature, flow speeds), while other physical properties (magnetic field, MHD dynamics, transient phenomena) are treated in later chapters. Complementary information can be found in the textbook by Mariska (1992) or in reviews by Withbroe & Noyes (1977), Mariska (1986), Schrijver et al. (1999), Brekke (1999), or Hansteen (2001).

In early models of the transition region (e.g., Reimers 1971a), only (downward) heat conduction $\nabla F_C(h)$ was included in the energy equation (4.1.4), balanced by an unknown heating source $E_H(h)$ constrained from temperature profiles $T(h)$ inferred from EUV and radio data (Reimers 1971b),

$$E_H = \nabla F_C = \frac{\partial}{\partial h}\left(\kappa T^{5/2}\frac{\partial T}{\partial h}\right) = \frac{2\kappa}{7}\frac{\partial T^{7/2}}{\partial h} . \tag{4.8.1}$$

Table 4.4: Hydrodynamic models of the transition region.

Reference	Physical model
Reimers (1971a)	heat conduction
Moore & Fung (1972)	heat conduction and radiative loss
Lantos (1972)	heat conduction, radiative loss, and enthalpy
Gabriel (1976)	"fluxtube canopy" expanding over size of supergranule
Pneuman & Kopp (1977)	energy equation with enthalpy flux of spicular downflows
Wallenhorst (1982)	energy equation with enthalpy flux of downflows
Antiochos (1984)	flare-like impulsive heating in high-pressure fluxtubes
Dowdy et al. (1985)	heat conduction inhibited by magnetic constriction
Dowdy et al. (1986)	"junkyard model" of mixed small and large-scale loops
Antiochos & Noci (1986)	DEM distribution of hydrostatic cool and hot loops
McClymont & Craig (1987)	fast downflows in steady-flow solutions of cool loops
McClymont (1989)	redshifts from steady-flow solutions of cool loops
Athay (1990)	heat conduction with an inhomogeneous temperature boundary
Woods et al. (1990)	opacity effects of hydrogen Lyman-α cooling
Woods & Holzer (1991)	multi-component plasma with downflows
Rabin (1991)	effects of canopy geometry on energy balance
Fontenla et al. (1990)	FAL I model: non-LTE, ambipolar diffusion
Fontenla et al. (1991)	FAL II model: diffusion of H atoms and ions
Fontenla et al. (1993)	FAL III model: He emission with diffusion
Chae et al. (1997)	non-LTE radiative transfer in H and He
Fontenla et al. (2002)	FAL IV model: H and He mass flows with diffusion

If all the heating occurs above the transition region ($E_H(s) = 0$ for $s < s_1$), this leads to a temperature dependence of $T(s) \propto (s - s_0)^{2/5}$. For a constant heating rate this leads to the solution $T(s) \propto (s - s_0)^{2/7}$, a functional dependence that approximately dominates most of the hydrostatic solutions in the transition region.

In the next step, the radiative loss term $E_R(h) \propto n(h)^2 \Lambda[T(h)]$ was included (e.g., Moore & Fung 1972), because EUV observations (e.g., Noyes et al. 1970) provided constraints on the radiative flux, but no local heating or mechanical flux was assumed in the transition region,

$$0 = -n^2 \Lambda(T) - \frac{\partial}{\partial h} \left(\kappa T^{5/2} \frac{\partial T}{\partial h} \right) . \qquad (4.8.2)$$

This model predicts a dependence of $\Delta log(p) \propto \Delta log(F_C)$ between the pressure and conductive flux, which was found not to be consistent with the EUV data of Withbroe & Gurman (1973). These strictly hydrostatic models could be substantially modified by mass flows v(h) [e.g., caused by upflows of spicular material, which was modeled by including the enthalpy term of Eq.4.1.14 (Lantos 1972)], using constraints from radio observations,

$$\frac{\partial}{\partial h} \left(\frac{5}{2} n v k_B T \right) = -n^2 \Lambda(T) - \frac{\partial}{\partial h} \left(\kappa T^{5/2} \frac{\partial T}{\partial h} \right) . \qquad (4.8.3)$$

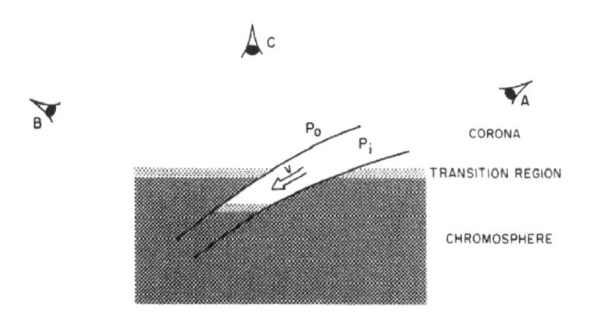

Figure 4.24: Geometry of the base of a fluxtube with a higher gas pressure than in the surrounding transition region. Observer A would see redshifts, observer B's view is fully obscured, and observer C would detect some redshift but also some absorption by neutral hydrogen (Antiochos 1984).

This model, which was applied to quiet Sun interspicular regions, produced relatively large upflow velocities of v $\lesssim$ 40 km s^{-1} in the $T = 10,000 - 300,000$ K temperature regime (Lantos 1972). However, observations from rocket flights, OSO-8 (Lites et al. 1976), and Skylab (Doschek et al. 1976) indicated clear evidence for downflows of spicular material of the order of v $\lesssim$ 5 − 15 km s^{-1}. Thus, Pneuman & Kopp (1977) suggested that downflows of spicular material which returns to the chromosphere, after being heated to coronal temperatures, needs to be included in the energy equation (4.1.29),

$$\frac{1}{A}\frac{\partial}{\partial h}\left(nvA\left[\varepsilon_{enth} + \varepsilon_{kin} + \varepsilon_{grav}\right] + AF_C\right) = E_H - E_R . \qquad (4.8.4)$$

Pneuman & Kopp (1977) estimated that the enthalpy flux from the spicular downflows was 10 − 100 times larger than the conductive flux. The dynamics of downflows from coronal condensations with observed speeds of v $\gtrsim$ 45 km s^{-1} was considered as a main contributor to the EUV emission seen in the entire temperature range of $T = 7000$ K−3 MK (Foukal 1978). Consequently, conduction and radiation-dominated models were abandoned in favor of enthalpy-dominated models (Wallenhorst 1982). However, although the enthalpy flux was found to be much larger than the conductive flux, the divergence of the enthalpy flux was found to be smaller than the divergence of the conductive flux, so the enthalpy term seems not to modify the temperature solution $T(h)$ of the energy equation drastically (Wallenhorst 1982). Nevertheless, siphon-flow models were explored in more detail and it was found that steady-flow hydrodynamic solutions with fast downflows (v$\approx$ 5 − 10 km s^{-1}) could be reproduced for cool ($T \lesssim 10^5$ K) loops, but not for hot coronal loops (McClymont & Craig 1987; McClymont 1989), which is consistent with observations.

Antiochos (1984) pointed out that the ubiquitous redshift of EUV emission observed in the temperature regime $10^{4.3} \lesssim T \lesssim 10^{5.3}$ (e.g. Gebbie et al. 1981) is unlikely to be caused by falling spicular material, because this model could not explain the observed absence of a center-limb variation of the redshifts. Also steady-state siphon flow models would not work because they produce an equal amount of blueshifts and red-

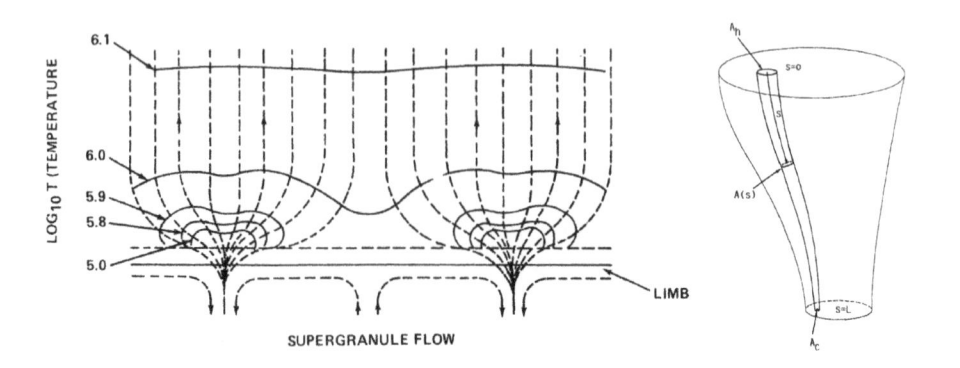

Figure 4.25: *Left:* The standard 2D geometry of the magnetic field in the transition region according to Gabriel (1976). The magnetic field emerges from the network boundaries, where they are concentrated by supergranulation flow and from where they diverge rapidly with height until they are uniform and vertical in the corona. *Right:* geometric expansion of an elementary fluxtube across the transition region (Dowdy et al. 1985).

shifts, which was not observed. Antiochos (1984) concluded that only fluxtubes with a higher gas pressure than the ambient transition region could explain the direction-independent redshift, because they possess a transition region that lies below the ambient height level (Fig. 4.24), and downflows in the lowered fluxtube transition regions would only be observable for line-of-sights that are aligned with the magnetic field lines. This model of immersed fluxtubes explains also the continuum absorption observed above quiet Sun transition regions (Schmahl & Orrall 1979) and displacements between hot and cool loop segments (Dere 1982). The required overpressure in this fluxtube necessarily leads to a dynamic fluxtube model with flare-like impulsive heating and condensation cycles, constituting a dynamic model for parts of the transition region (Antiochos 1984).

The geometry of coronal fluxtubes is most simplified in 1D models, where a constant cross section is assumed. Such an approximation can be reasonable for coronal fluxtubes, although there is some controversy whether the majority of coronal loops expand with height like a dipole field or whether they have a rather constant cross section due to current-induced twisting. In the transition region, however, such 1D models clearly represent an unacceptable oversimplification, because vertical magnetic field lines are believed to fan out from the network boundaries in the photosphere to the size of a supergranule in the corona, forming a *canopy structure*, according to a widely accepted model by Gabriel (1976), as shown in Fig. 4.25 (left). This geometry leads to a varying cross section of every fluxtube across the height of the transition region (Fig. 4.25, right), an effect that produces a strong inhibition of the conductive heat flow from the corona down to the chromosphere (Dowdy et al. 1985). In the steady-state case, the inhibition factor of the heat flow $\Phi = AF_C$ through a constricted fluxtube is simply the ratio of the harmonic mean area of the constricted tube to the area of the unconstricted fluxtube (Dowdy et al. 1985). For a cone-like magnetic fluxtube (with a

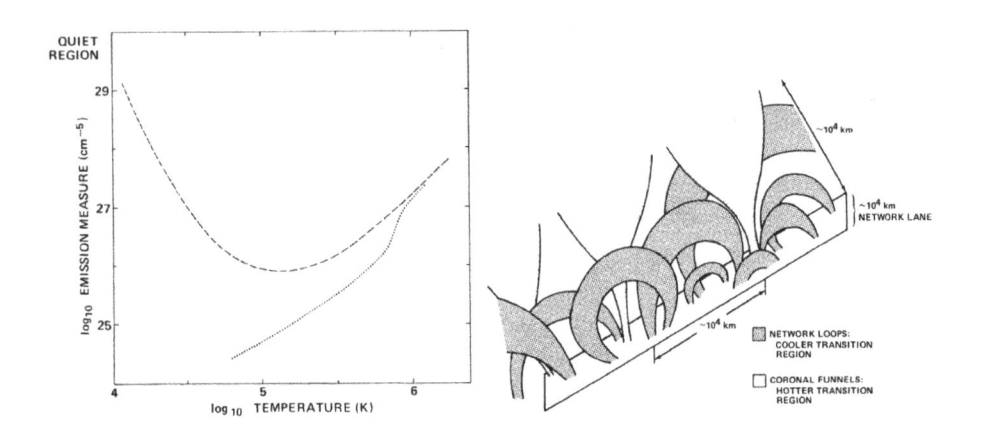

Figure 4.26: *Left:* The emission measure distribution in quiet Sun regions, derived from observations (dashed line; Rabin & Moore 1984). A theoretical model of hydrostatic energy balance in canopy-like coronal fluxtubes is shown for comparison, explaining the hotter part of the observed distribution (dotted line; Athay 1982). *Right:* The *"magnetic junkyard" model* of Dowdy et al. (1986) also invokes in addition to the hotter coronal funnels cooler small-scale network loops (grey structures), which account for the cooler part of the observed emission measure distribution.

hot looptop cross section A_h and cool bottom area A_c) the *inhibition factor* is simply

$$\frac{\Phi}{\Phi_\parallel} = \Gamma^{-1/2} \, , \qquad (4.8.5)$$

where $\Phi_\parallel$ is the heat flow AF_C through a parallel fluxtube and $\Gamma = A_h/A_c$ is the geometric expansion or constriction factor (e.g., Fig. 3.14 or Fig. 4.25). This inhibition factor was found to vary by no more than a factor of 2 for other geometries with different, but monotonic taper functions (Dowdy et al. 1985). A larger variety of funnel-shaped canopies was explored by Rabin (1991), finding that magnetic constriction does not greatly alter the overall energy budget of coronal loops for conduction-dominated models, with the heat flux varying by no more than a factor of 2 for constriction ratios up to 100. For a flow-dominated model only, Rabin (1991) found that constriction can reduce the total energy requirement by a factor of 5 in extreme cases, but it does not influence the differential emission measure distribution.

The canopy-like footpoint geometry of coronal loops has the consequence of a much smaller total emission measure at temperatures of $T \lesssim 0.25$ MK, far less than the observed emission measure distributions (Athay 1982), as shown in Fig. 4.26 (left). A solution to this dilemma was offered by Dowdy et al. (1986), by invoking a large number of cool small-scale loops that are anchored in the network but are disconnected from the hotter large-scale loops that reach up into the corona, the so-called *magnetic junkyard model* (Fig. 4.26, right). This second population of cool low-lying loops with lengths $\lesssim 10$ Mm make up most of the volume of the cooler transition region, while the pointed footpoints of the larger coronal loops occupy only a small filling factor in

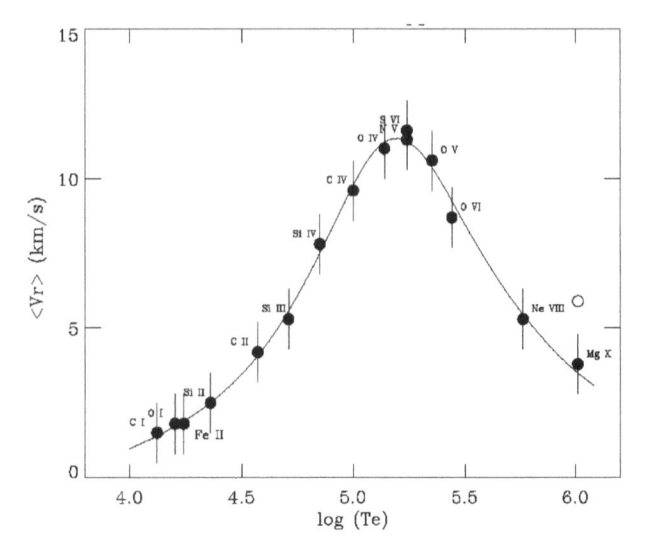

Figure 4.27: Spatially average Doppler shifts of various ions measured with SoHO/SUMER. The smooth curve represents an analytical fit of a model with steady downward flow with constant pressure and varying cross section as specified in Eq. (4.8.6) (Chae et al. 1998c).

the transition region. It was shown that hydrostatic solutions of these two populations can add up to the observed differential emission measure distribution (Antiochos & Noci 1986). However, the same differential emission measure distribution can also be reproduced with a model that includes only heat conduction, but arranged on an inhomogeneous temperature surface (Athay 1990).

Downflows have been measured in the transition region since Skylab. Downflows were generally found to increase with height in the temperature regime of $10^4 \lesssim T \lesssim 2 \times 10^5$ K, and to decrease at higher temperatures $2 \times 10^5 \lesssim T \lesssim 10^6$ K (e.g. Achour et al. 1995; Brekke et al. 1997b, 1999; Peter 2001). One of the more recent measurements is shown in Fig. 4.27, obtained from SoHO/SUMER (Chae et al. 1998c). The average downflow velocity shows a peak value of v= 11 km s^{-1} at a temperature near $T = 0.23$ MK, and decreases systematically at both sides of this peak temperature. Chae et al. (1998c) interpreted the velocity decrease with height above $T = 0.23$ K with the divergence of a fluxtube, which was found to have the analytical form

$$\frac{A(T)}{A(T_h)} = \frac{[1 + (\Gamma^2 - 1)(T/T_h)^p]^{1/2}}{\Gamma} \qquad (4.8.6)$$

with the numerical values $T_h = 10^6$ K, $\Gamma = 30$, and $p = 3.6$.

So far, all previous transition region models assumed full ionization, ionization equilibrium, and optically thin emission. Optically thin emission implicitly assumes that the collisional excitation rate of a particular atomic level can be set equal to the radiative de-excitation rate for that level, and that the radiated photon leaves the plasma without further interaction. McClymont & Canfield (1983a), however, suggested that

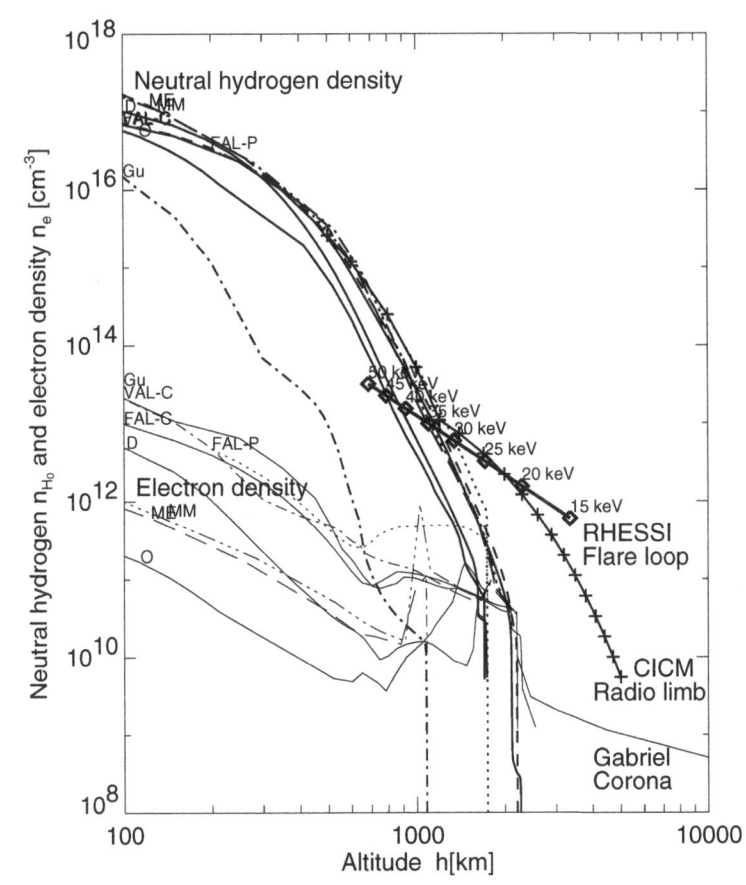

Figure 4.28: A compilation of transition region density models: VAL-C = Vernazza et al. (1981), model C; FAL-C = Fontenla et al. (1990), model C; FAL-P = Fontenla et al. (1990), model P; G = Gu et al. (1997); MM = Maltby et al. (1986), model M; ME = Maltby et al. (1986), model E; D = Ding & Fang (1989); O = Obridko & Staude (1988); Gabriel = Gabriel (1976), coronal model; CICM = Caltech Irreference Chromospheric Model, radio sub-millimeter limb observations (Ewell et al. 1993), and RHESSI flare loop (Aschwanden et al. 2002b).

optical depth effects drastically modify the radiative loss function $\Lambda(T)$ (Fig. 2.14) below temperatures of $T \lesssim 10^5$ K. In particular, radiative loss in hydrogen Lyman-α, which has a peak at $log(T) \approx 4.2$ in the radiative loss function $\Lambda(T)$, leads to singularities in hydrodynamic solutions of the force balance and energy balance equations (Woods et al. 1990; Kuin & Poland 1991). In further refinements of transition region models, a multi-component plasma (e.g. electrons, protons, ionized helium, minor ions) was implemented, so that each species could have different temperatures and abundance enhancements (Woods & Holzer 1991). A series of transition region models was developed by Fontenla et al. (1990, 1991, 1993, 2002), which assume hydrostatic equilibrium, but include beside the standard terms in the energy equation (e.g., Eq. 4.1.34) also hydrogen ionization energy flow due to ambipolar diffusion,

$$\frac{\partial}{\partial h}\left[v\left(\frac{5}{2}p - \varepsilon_I n_H\right) - \left(\frac{5}{2}k_B T + \varepsilon_I\right)\frac{n_p n_H}{(n_p + n_H)}v_{ambi} + F_C\right] - v\frac{\partial p}{\partial h} = -E_R$$

(4.8.7)

where ε_I and n_H are the ionization energy and the number density of hydrogen atoms, respectively, n_p is the proton density, and v_{ambi} is the ambipolar diffusion velocity. To treat the Lyman-α emission properly, which is partially optically thick, the radiative loss term E_R involves the radiative transfer equation (2.1.5),

$$E_R = \int d\Omega \int (\varepsilon_\nu - \alpha_\nu I_\nu)d\nu$$

(4.8.8)

where $\varepsilon_\nu, \alpha_\nu$, and I_ν are the emissivity, the absorption coefficient, and the radiation intensity, respectively, all at frequency ν, and $d\Omega$ is the solid angle element. In further models, the diffusion of helium was also included and the resulting He I 10,830 Å emission was compared with observations (Fontenla et al. 1993, 2002). Flow velocities also have important effects on H and He line intensities, radiative losses, and ionization rates, and thus on the ionization energy fluxes. These effects were integrated in the solution of the statistical equilibrium and non-LTE radiative transfer equation in an energy balance model by Chae et al. (1997). This effect was found to lead to an order of magnitude increase in the differential emission measure at $T \lesssim 2.5 \times 10^4$ K compared to an optically thin approximation with complete ionization.

A compilation of a number of chromospheric density models is shown in Fig. 4.28. Chromospheric and coronal density models are sharply divided by a thin transition region where the density drops sharply and the temperature rises reciprocally. Below this transition region, the plasma is only partially ionized, and thus the electron density n_e is lower than the hydrogen density n_H. At the transition region, where complete ionization sets in, the neutral hydrogen density n_{H^0} then drops to a very small value, at an altitude of approximately $h \approx 2.0$ Mm. Above the transition region the plasma is fully ionized and the electron density is almost equal to the ion density (i.e., $n_e \approx n_i$), as it is usually assumed in coronal density models. Chromospheric density models have been calculated in great detail based on ion abundance measurements from a larger number of EUV lines, constrained by hydrostatic equilibrium and radiation transfer assumptions (e.g., Vernazza et al. 1973, 1976, 1981; VAL models, Fig. 4.28), and ambipolar diffusion (Fontenla et al. 1990, 1991; FAL models; Fig. 4.28). Newer developments include sunspot umbral models (Maltby et al. 1986; Obridko & Staude 1988), sunspot penumbral models (Ding & Fang 1989), or stochastic multi-component models with hot fluxtubes randomly embedded in a cool medium (Gu et al. 1997), all shown in Fig. 4.28. At coronal heights, i.e., at $h \gtrsim 2.0$ Mm, an average quiet Sun density model was computed by Gabriel (1976), based on the expansion of the magnetic field of coronal fluxtubes that line up with the boundaries of the supergranule convection cells. The geometric expansion factor and the densities at the lower boundary computed from chromospheric models then constrain coronal densities in hydrostatic models. This yields electron densities of $n_e \approx 10^9$ cm^{-3} at the coronal base in quiet Sun regions (Gabriel 1976), as shown in Fig. 4.28. However, these hydrostatic models in the lower corona have been criticized because of indications of unresolved spatial fine structure and unresolved dynamic phenomena, which contribute in the statistical average to an extended chromosphere. Observational evidence for an extended chromosphere comes from spicules observed in EUV or radio (e.g., from submillimeter observations during a total eclipse; Ewell et al. 1993; CICM model in Fig. 4.28). These radio limb measurements yield electron densities that are 1-2 orders of magnitude higher in the height range of $500 - 5000$ km than predicted by hydrostatic models (VAL, FAL, Gabriel 1976), which have been interpreted in terms of the dynamic nature of spiculae (Ewell et al. 1993). Chromospheric density models have also recently been inferred from hard X-ray measurements with *RHESSI*, which mainly probe the total neutral and ionized hydrogen density via bremsstrahlung, and the total bound and free electron density from the collisional energy losses. The *RHESSI* measurements also confirm the presence of an elevated transition region significantly higher than hydrostatic models predict. The 1-MK interface between hot (> 1 MK) coronal loops and their cooler (< 1 MK) footpoints in the transition region is prominently seen in *TRACE* 171 Å images and is termed *"moss"* (Berger et al. 1999; DePontieu et al. 1999; Fletcher & DePontieu 1999; Martens et al. 2000). We will discuss some dynamic aspects of transition region models in § 6 in terms of MHD processes.

4.9 Hydrodynamics of Coronal Holes

Coronal holes are regions of low-density plasma (Fig. 4.29), located in *open magnetic field regions* where the unipolar field lines connect from the solar surface directly to interplanetary space (Fig. 1.14). This open-field configuration provides efficient conduits to convey upflowing plasma and accelerated particles into interplanetary space, and is thus the source of the *fast* ($v \approx 400 - 800$ km s^{-1}) *solar wind*. For reviews see Zirker et al. (1977) and Cranmer (2001, 2002a,b). Here we discuss some aspects of hydrodynamic modeling in coronal holes, while other effects (magnetic field, MHD processes, transient phenomena) are treated later on.

Since coronal holes entirely consist of open magnetic field lines, and the plasma β-parameter is below unity, the topology of coronal holes can essentially be represented by bundles of near-parallel radial fluxtubes, which, however, can have quite large radial expansion factors. For strictly radial fluxtubes, the expansion factor $A(h)$ would be proportional to the square of the distance $r = R_\odot + h$ from the Sun center,

$$A(h) = A(0) \left(1 + \frac{h}{R_\odot}\right)^2 f(r) , \qquad (4.9.1)$$

but observations and models (e.g., Fig. 1.14) show rather a super-radial expansion factor $f(r)$ for the magnetic field near the poles (Suess et al. 1998; Dobrzycka et al. 1999; DeForest et al. 2001).

For hydrodynamic modeling, the main differences to coronal loops are: (1) the open upper boundary condition and (2) the (super-radial) expansion $A(s)$ with height. The generalized versions of the hydrodynamic equations with consideration of variable loop cross sections $A(s)$ are given in Eqs. (4.1.24−26). For hydrostatic solutions of fluxtubes in coronal holes, the time dependence can be ignored ($d/dt = 0$) and a radial direction can be chosen ($s = r$). Thus we have three equations (4.1.24−26) for the five

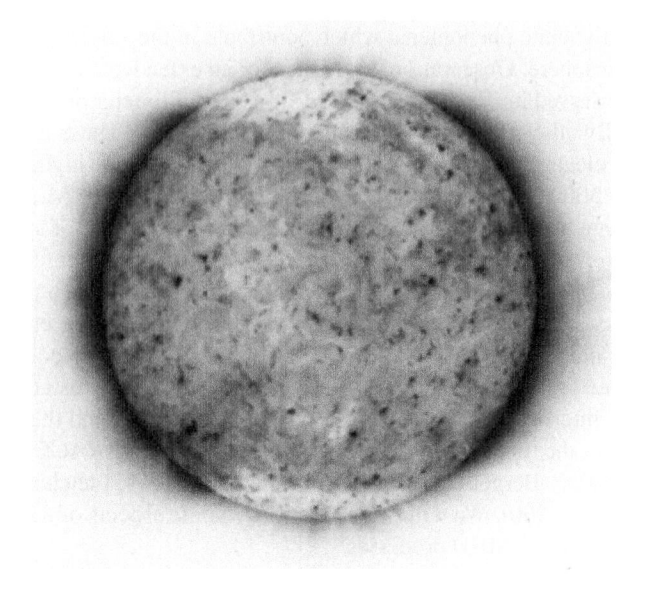

Figure 4.29: EUV image of the quiet Sun during solar minimum, 1996-May-23, taken by SoHO/EIT at a wavelength of 195 Å ($T \approx 1.5$ MK). Black color indicates enhanced emission. Note the absence of active regions over the entire solar disk and the deficit of EUV emission in the coronal holes at the northern and southern polar caps (Cranmer 2002a).

independent functions $n(r)$, $T(r)$, $v(r)$, $A(r)$, and $E_H(r)$, which need to be additionally constrained by observations or model assumptions. Observational constraints can be obtained from the emission measure $EM \propto \int n(r)^2 dz$ of optically thin EUV lines, as well as from radio brightness temperatures $T_B(r) \propto T(r)[1 - exp(-\tau_\nu)]$ in the case of free-free emission, which scales with $\tau_{ff} \propto n(r)^2 T(r)^{-3/2}$ (Eq. 2.3.18). Such methods have been applied in several studies (e.g., Drago 1974; Rosner & Vaiana 1977; Chiuderi−Drago et al. 1977, 1999; Chiuderi−Drago & Poletto 1977; Dulk et al. 1977) to invert the run of temperature $T(r)$ and density $n(r)$ in coronal holes, either locally (e.g., in a *plume* structure or in an *inter-plume region*) or averaged over a larger coronal hole region. A compilation of temperature measurements in coronal holes is shown in Fig. 4.30, being significantly lower ($T \approx 0.78−0.93$ MK between $r = 1.02−1.07\,R_\odot$) in coronal holes than in the surrounding quiet Sun regions ($T \approx 0.94 − 1.2$ MK), according to Habbal et al. (1993). A notorious problem in the determination of the true temperature in coronal hole areas are contaminations from hotter plasma in the foreground and background. Typical density profiles in coronal holes are shown in Fig. 1.20 and in Fig. 4.31, based on recent SoHO/SUMER measurements (Wilhelm et al. 1998).

Coronal holes were first detected during solar eclipses (Waldmeier 1950), because the lower density produces less scattered light than the high-density equatorial or streamer regions. Coronal holes have also been detected in the green line of Fe XIV, 5303 Å (e.g., Fisher & Musman 1975; Guhathakurta et al. 1996). Although the average elec-

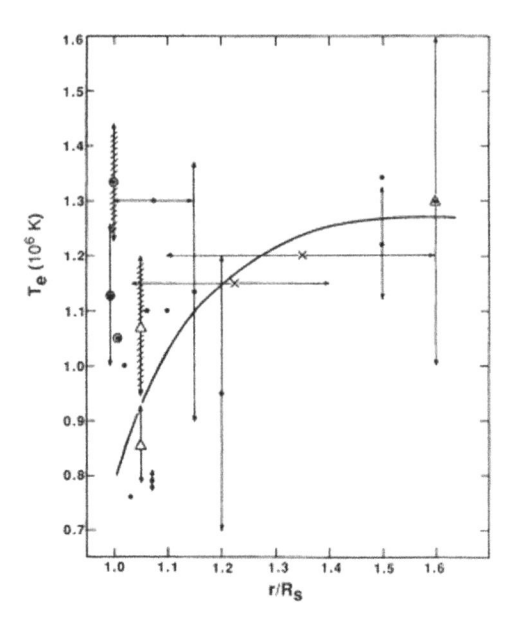

Figure 4.30: Average temperature run $T(h)$ in coronal holes compiled from a number of measurements listed in Habbal et al. (1993). The hatched arrows indicate measurements that are severely contaminated by plasma outside the boundaries of coronal holes (Habbal et al. 1993).

tron density profile $n(h)$ in coronal holes can be inverted from polarized brightness measurements, optical observations do not provide independent temperature information. However, temperature profiles $T(h)$ can be estimated in terms of the scale height λ_T from density gradients (i.e., $\partial n(h)/\partial h \approx n(h)/\lambda_T(h)$), in the case of hydrostatic equilibrium (e.g., Guhathakurta & Fisher 1998).

In EUV emission, densities and temperatures in coronal holes have been studied with *Skylab* (Huber et al. 1974; Doschek & Feldman 1977), OSO-7 (Wagner 1975), SoHO/SUMER, CDS, and UVCS (Warren et al. 1997; Doschek et al. 1997, 1998, 2001; Wilhelm et al. 1998, 2000; Warren & Hassler 1999; Kohl et al. 1999; Doyle et al. 1999; Dobrzycka et al. 1999; DelZanna & Bromage 1999; Stucki et al. 2000, 2002; Zangrilli et al. 2002). The measurement of EUV line intensities in different temperature bands allows for a characterization of the differential emission measure distribution $dEM(T)/dT$ in coronal holes. The inversion of the temperature profile $T(h)$ can in principle be achieved from multiple EUV lines observed above the limb, but the ambiguities introduced by the inhomogeneities due to the multi-temperature corona are better handled by forward-modeling (see §3.3). For hydrodynamic modeling, another important parameter is the flow speed measurement. Outflows in coronal holes, with typical velocities of v$\approx$ 10 − 15 km s^{-1} were measured by Cushman & Rense (1976), Rottman et al. (1982), Orrall et al. (1983), Hassler et al. (1991, 1999), Dupree et al. (1996), Patsourakos & Vial (2000), and Strachan et al. (2000), see also

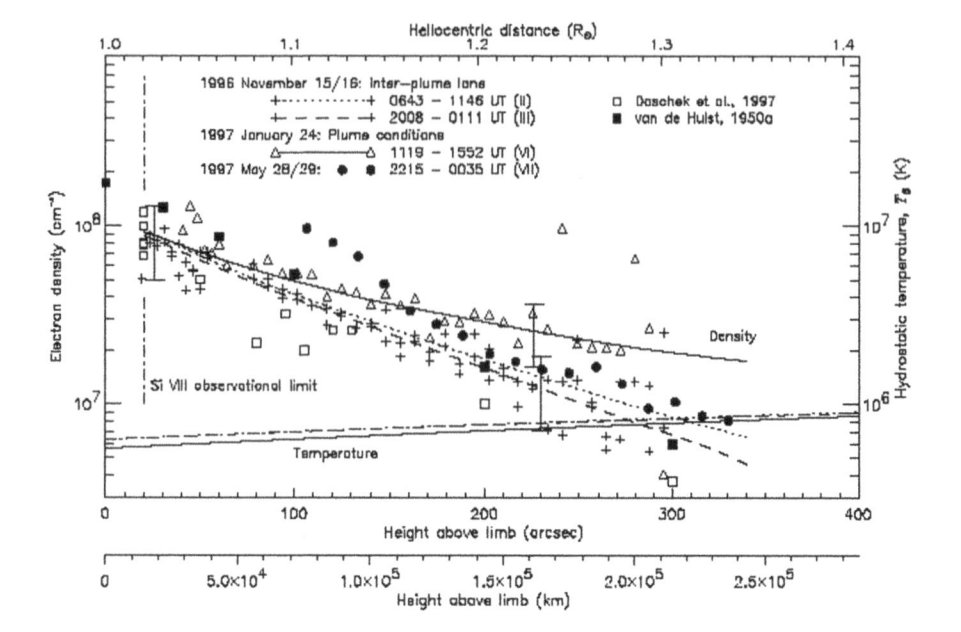

Figure 4.31: Electron density measurements in coronal holes based on Si VIII line ratios measured with SoHO/SUMER, compared with values found in literature. The hydrostatic temperature T_S used for the fits of the Si VIII (1445 Å) line is indicated on the right (Wilhelm et al. 1998).

Table 4.2 for measurements of coronal temperatures.

The boundaries of coronal holes are most pronounced in soft X-rays, because of the absence of hot plasma with $T \gtrsim 1.5$ MK (Fig. 4.29). Thus, the contrast between coronal holes and quiet Sun regions is probably best in soft X-rays. This makes it very suitable to study the contours and topology of coronal holes in soft X-rays, but density and temperature measurements are more difficult than in EUV because of the high sensitivity to contaminations from hotter plasma from the foreground and background. In soft X-ray emission, densities and temperatures of coronal holes have been investigated with OSO-7 and Skylab (Timothy et al. 1975), and Yohkoh/SXT (Hara et al. 1994, 1996; Hara 1997; Watari et al. 1995; Foley et al. 1997; Aschwanden & Acton 2001). Comparisons of density profiles $n(h)$ inferred from soft X-rays in coronal holes with other regions can be seen in Fig. 1.20

Because the free-free opacity is highest for low temperatures ($\tau_{ff} \propto T^{-3/2}$), the cool plasma in coronal holes is also clearly detectable in contrast to the hotter plasma of the quiet Sun in radio wavelengths, where a large number of observations have been made (Drago 1974; Lantos & Avignon 1975; Fürst & Hirt 1975; Kundu & Liu 1976; Chiuderi−Drago et al. 1977; Chiuderi−Drago & Poletto 1977; Dulk et al. 1977; Papagiannis & Baker 1982; Wang et al. 1987; Kundu et al. 1989a; Bogod & Grebinskij 1997; Chiuderi−Drago et al. 1999; Gopalswamy et al. 1999; Moran et al. 2001). Most of the coronal hole radio observations (e.g., Fig. 4.32), however, are of morphological

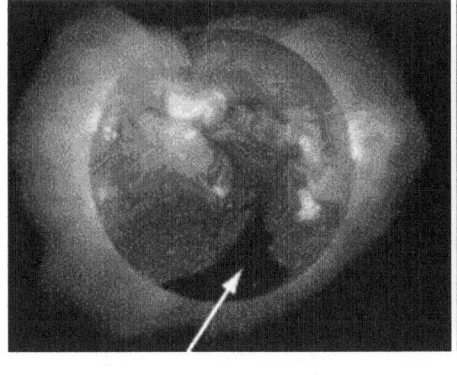 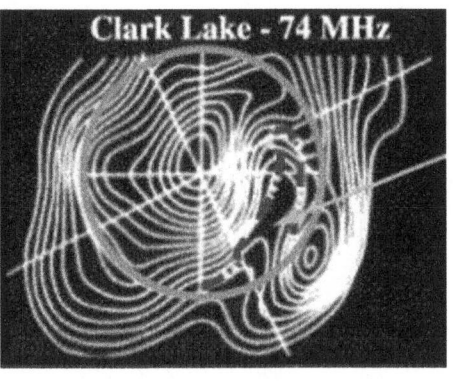

Figure 4.32: *Left:* A Yohkoh/SXT image shows a large coronal hole in the southern polar cap that extends all the way to the equator. *Right:* A radio image recorded with the *Clark Lake Radioheliograph* at a frequency of 74 MHz shows a deficit of free-free emission at the same location of the coronal hole (courtesy of Nat Gopalswamy).

nature.

The deconvolution of density $n(h)$ and temperature profiles $T(h)$ from the free-free opacity τ_{ff}, requires many frequencies and involves an unknown height scaling of the opacity with height $\tau_{ff}(h)$. Inversion procedures have been developed and applied to coronal holes (Bogod & Grebinskij 1997), now called *frequency tomography* (Aschwanden et al. 2004b). The ambiguity in the height scaling $\tau_{ff}(h)$, however, requires additional constraints from other observations (such as EUV) or additional model assumptions.

Despite the numerous observations in all wavelengths, relatively few attempts of hydrodynamic modeling in coronal holes have been made. A pioneering study was performed by Rosner & Vaiana (1977), but their best hydrodynamic solution constrained by extensive soft X-ray, EUV, and radio data, yields only a small flow speed of the order v $\lesssim$ 1 km s^{-1}, far below the required solar wind speed of v$\approx$ 300 $-$ 800 km s^{-1}. Early hydrodynamic models were not able to reconcile multi-wavelength data (Chiuderi$-$Drago & Poletto 1977). Other early hydrodynamic models attempted to model coronal holes with constant cross section fluxtubes and could not reproduce the differential emission measure distribution $dEM(T)/dT$ from network or cell centers in coronal holes (Raymond & Doyle 1981). The super-radial expansion $A(h)$ of fluxtubes in coronal holes has been determined from several observations (Suess et al. 1998; Dobrzycka et al. 1999; DeForest et al. 2001), but has not yet been applied to hydrodynamic models. In a parametric study it was found that hydrodynamic models of coronal holes that are most consistent with the data yield only a very low speed of v$\approx$10 km s^{-1} at an Earth's distance, while an MHD model that includes acceleration by Alfvén waves is capable of reproducing the observed velocities of v$\approx$ 630 km s^{-1} (Tziotziou et al. 1998). This is a strong indication that MHD effects are important and must be included in physical models of coronal hole structures and solar wind. Recent observations with SoHO/UVCS in Lyman-α and O VI (1032, 1037 Å) have demonstrated anisotropies of hydrogen atoms H^0 and oxygen ions O^{5+} that are differ-

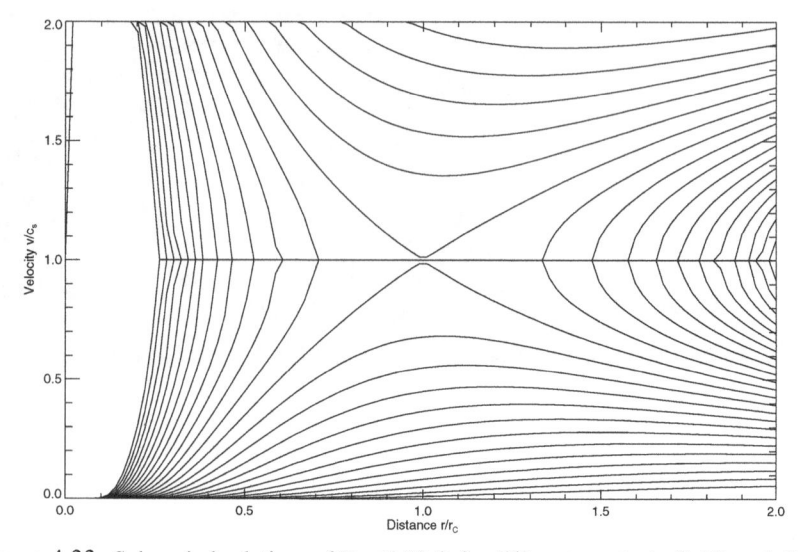

Figure 4.33: Solar wind solutions of Eq. (4.10.6) for different constants C. The solutions are shown for the normalized parameters v/c_s and r/r_c. The middle horizontal line represents the sonic limit, curves below this limit are subsonic solutions.

ent from the electron velocity distributions, leading to empirical (multi-species) models of coronal holes (Cranmer et al. 1999; Miralles et al. 2001; Cranmer 2001, 2002a,b). The empirical models of coronal holes deduced from *SoHO/UVCS* observations imply acceleration mainly at low heliocentric distances, different behavior in the H I and O VI heating, and anisotropies in the velocity distribution of the emitting ions, which all constitute constraints in the development of new and more accurate models of the solar wind plasma heating (see also §9.4.6) and acceleration in coronal holes.

4.10 Hydrodynamics of the Solar Wind

It was Parker's (1958) ingenious idea that the open corona cannot be in hydrostatic equilibrium and thus needs to be described by a dynamic solution that includes the continuous expansion in outward direction, with a vanishing pressure at a large distance. He found the simplest steady-state solution ($d/dt = 0$) by assuming: (1) a radial expansion, $A(r) \propto r^2$; (2) the ideal gas law, $p(r) = n(r)k_BT$; and (3) a constant temperature, $T(r) = const$. The assumption of a constant temperature decouples the momentum equation from the energy equation, so that a hydrodynamic solution can simply be calculated by solving only the continuity equation (4.1.24) and momentum equation (4.1.25), which then reads (using Eq. 3.1.5)

$$4\pi r^2 n v = const ,   (4.10.1)$$

$$mnv\frac{\partial v}{\partial r} = -\frac{\partial p}{\partial r} - mng_\odot \frac{R_\odot^2}{r^2} .   (4.10.2)$$

Inserting the ideal gass law, $p(r) = n(r)k_B T$, and the sound speed, $c_s = \sqrt{(k_B T/m)}$, eliminating the density, and using the continuity equation (4.10.1), $n(\mathrm{v}, r) \propto 1/(r^2 \mathrm{v})$, yields for the pressure gradient,

$$\frac{\partial p}{\partial r} = \frac{\partial}{\partial r}(nmc_s^2) = mc_s^2 \left(\frac{\partial n}{\partial r} + \frac{\partial n}{\partial \mathrm{v}} \frac{\partial \mathrm{v}}{\partial r} \right) = mc_s^2 \left(-\frac{2n}{r} - \frac{n}{\mathrm{v}} \frac{\partial \mathrm{v}}{\partial r} \right) . \qquad (4.10.3)$$

Inserting this pressure gradient $\partial p/\partial r$ into the momentum equation (4.10.2) yields a differential equation between the solar wind speed v and distance r,

$$\left(\mathrm{v} - \frac{c_s^2}{\mathrm{v}} \right) \frac{\partial \mathrm{v}}{\partial r} = \frac{2c_s^2}{r} - g_\odot \frac{R_\odot^2}{r^2} . \qquad (4.10.4)$$

This differential equation is formally similar to the siphon-flow solution given in Eq. (4.2.7), except that a quadratically expanding cross section is specified, $A(r) \propto r^2$, and the height dependence of the gravity, $g(r) \propto g_\odot/r^2$, is additionally included. This differential equation can simply be integrated by variable separation

$$\int \left(\mathrm{v} - \frac{c_s^2}{\mathrm{v}} \right) d\mathrm{v} = \int \left(\frac{2c_s^2}{r} - g_\odot \frac{R_\odot^2}{r^2} \right) dr , \qquad (4.10.5)$$

which has the solution

$$\frac{\mathrm{v}^2}{c_s^2} - \ln \left(\frac{\mathrm{v}^2}{c_s^2} \right) = 4 \ln \left(\frac{r}{r_c} \right) + \frac{2g_\odot R_\odot^2}{c_s^2 r} + C , \qquad (4.10.6)$$

The solutions have a critical point where $\partial \mathrm{v}/\partial r$ is not defined, occurring at the sonic point $\mathrm{v} = c_s$ with $r = r_c = g_\odot R_\odot^2/2c_s^2$ (from setting both sides of Eq. 4.10.4 to zero). The solutions are shown in Fig. 4.33 for a range of values for the constant C.

Acceptable solar wind solutions are only those which are always subsonic (referred to as *solar breezes*) or that go through the critical point (requiring $C = -3$ in Eq. 4.10.6), which has become the standard *solar wind solution*. Of course, this solution characterizes the observed solar wind only approximately, since the underlying assumptions of radial expansion and isothermality are not exactly met in reality.

Hydrostatic models that include the energy equation have been applied to the inner solar wind out to 3 solar radii (Kopp & Orrall 1976) as well as all the way to 1 AU (Withbroe 1988). Subsequent models have been made more realistic by including a number of additional effects, such as two-fluid models with different electron and proton temperatures, or MHD models that include the magnetic field (of the Parker spiral). A discussion of some effects can be found in Priest (1982; § 12), Kivelson & Russell (1995; § 4 by A.Hundhausen), and Baumjohann & Treumann (1997; § 8.1).

4.11 Summary

In contrast to the previous chapter (§ 3), where the solar corona was modeled with hydrostatic solutions, we go a step further in the direction of hydrodynamic models by including flows. We derived the 1D hydrodynamic equations in time-dependent and conservative form, for constant and variable loop or fluxtube cross

sections (§4.1). We derived time-independent steady-flow or siphon flow solutions, which do not differ much from hydrostatic solutions in the subsonic regime (§4.2). Thermal stability is an important criterion to constrain the parameter regime in which coronal loops are most likely to be observed. Loops at coronal temperatures are thermally unstable when the radiative cooling time is shorter than the conductive cooling time, or when the heating scale height falls below one-third of a loop half length (§4.3). Recent observations show ample evidence of the presence of flows in coronal loops (§4.4), as well as evidence for impulsive heating with subsequent cooling, rather than the prevalence of a long-lasting hydrostatic equilibrium (§4.5). High-resolution observations of coronal loops reveal that many loops have a super-hydrostatic density scale height, far in excess of hydrostatic equilibrium solutions (§4.6). Time-dependent hydrodynamic simulations are still in a very exploratory phase (§4.7) and hydrodynamic modeling of the transition region (§4.8), coronal holes (§4.9), and the solar wind (§4.10) remains challenging due to the number of effects that can not easily be quantified by observations, such as unresolved geometries, inhomogeneities, time-dependent dynamics, and MHD effects.

Chapter 5

Magnetic Fields

While the magnetic field of our planet Earth barely amounts to field strengths of the order $B \lesssim 1$ Gauss [G], it produces a significant shielding effect at the bow shock to the solar wind and channels energetic particles from the solar wind towards the polar cusps, where they precipitate in unpopulated polar regions. The magnetic field thus plays an important role in protecting life on Earth from high-energy particles.

Much more gigantic magnetic fields are found on our vital Sun, amounting to several 1000 G in sunspots, and probably harboring field strengths of $B \approx 10^5$ G in the tachocline at the bottom of the convection zone. While the thermal pressure exceeds the magnetic pressure ($\beta = p_{th}/p_m \gg 1$) inside the Sun, in the chromosphere, and in the heliosphere, the corona demarcates a special zone where the opposite is true ($\beta \ll 1$). This special physical property has the far-reaching consequence that all plasma flows in the corona are governed by the magnetic field, which creates an extremely inhomogeneous medium that is filled with myriads of thin fluxtubes with different densities and temperatures. Essentially, plasma transport is only allowed in a 1D direction along magnetic field lines, while cross-field transport is strongly inhibited and can be neglected on all time scales of interest. The physical structure of the coronal plasma is therefore drastically simplified by the 1D geometry of individual fluxtubes on the one hand, but the topology and diversity of the multitude of fluxtubes that make up the corona represents a complication on the other hand. Nevertheless, the beauty of the manifold magnetic fluxtubes that can be observed in the solar corona is their role as detailed tracers of the magnetic field, which fingerprint the coronal magnetic field with remarkable crispiness in EUV images (e.g., as provided by TRACE in 171 Å). The challenge to match up these observed EUV tracers with theoretically calculated 3D magnetic fields has just begun. While our simplest concepts to characterize these magnetic fields with potential fields has some success in the lowest order, detailed observations clearly reveal to us the nonpotentiality of the coronal magnetic field in almost all places, providing us essential clues about the underlying currents. In this section we consider idealized concepts of coronal magnetic fields that neglect the effects of the hydrodynamic plasma, which are then described in terms of magneto-hydrodynamics (MHD) in the next chapter (§ 6).

5.1 Electromagnetic Equations

5.1.1 Maxwell's Equations

Classical electrodynamics (e.g., Jackson 1962) relates the magnetic field **B** to the electric field **E** by Maxwell's equations (here in cgs units),

$$\nabla \cdot \mathbf{E} = 4\pi \rho_E \tag{5.1.1}$$

$$\nabla \cdot \mathbf{B} = 0 \tag{5.1.2}$$

$$\nabla \times \mathbf{E} = -\frac{1}{c}\frac{\partial \mathbf{B}}{\partial t} \tag{5.1.3}$$

$$\nabla \times \mathbf{B} = \frac{1}{c}\frac{\partial \mathbf{E}}{\partial t} + 4\pi \mathbf{j} \tag{5.1.4}$$

where ρ_E is the electric charge density, $\mathbf{j}$ the electric current density, and c the speed of light. In an astrophysical context, the *magnetic induction* **B** is referred to as the magnetic field, while the standard definition of the magnetic field is $\mathbf{H} = \mathbf{B}/\mu$ (i.e., the *magnetic induction* **B** divided by the *magnetic permeability* μ). The *electric field* **E** is related to the *electric displacement*, $\mathbf{D} = \epsilon \mathbf{E}$, by the *permittivity of free space* ϵ. For astrophysical plasmas, these constants have values close to that in a vacuum, which is near unity in Gaussian (cgs) units (i.e., $\mu \approx 1$ and $\epsilon \approx 1$). Note also that the term for the current density is $(4\pi)\mathbf{j}$ in Maxwell's and related equations if the current density is measured in *electromagnetic units (emu)*, but amounts to $(4\pi/c)\mathbf{j}$ if measured in *electrostatic units (esu)* (see Appendix C).

5.1.2 Ampère's Law

A fundamental assumption in magneto-hydrodynamics is the nonrelativistic approximation, in the sense that plasma motions with speed v_0 are much slower than the speed of light c,

$$v_0 \ll c . \tag{5.1.5}$$

This nonrelativistic approximation allows us to neglect the term $(1/c)(d\mathbf{E}/dt)$ in Maxwell's equation (5.1.4), because it is much smaller than the term $(\nabla \times \mathbf{B})$,

$$\frac{1}{c}\frac{\partial \mathbf{E}}{\partial t} \approx \frac{1}{c}\frac{E_0}{t_0} \approx \left(\frac{v_0}{c}\right)\frac{E_0}{l_0} \ll \frac{B_0}{l_0} \approx (\nabla \times \mathbf{B}) , \tag{5.1.6}$$

where we attributed the plasma speed $v_0 \approx l_0/t_0$ to a typical length scale l_0 and time scale t_0, that is given by the curl-operator ($\nabla = d/dl \approx 1/l_0$), and assumed that the typical electric E_0 and magnetic field B_0 are of the same order. Thus the simplified Maxwell equation (5.1.4) yields the following definition for the current density in the nonrelativistic approximation,

$$\mathbf{j} = \frac{1}{4\pi}(\nabla \times \mathbf{B}) \tag{5.1.7}$$

The integral equivalent of this definition of the current density **j** is called *Ampère's law*,

$$\int_S (\nabla \times \mathbf{B}) \cdot \mathbf{n} \, dS = 4\pi \int_S \mathbf{j} \cdot \mathbf{n} \, dS, \tag{5.1.8}$$

which integrates the current density **j** that flows perpendicular (in normal direction **n**) through a surface S. The surface integral $\int_S$ can be transformed into a contour integral $\int_C$ with *Green's theorem*, where the total integrated current I flowing through the surface area S can be computed from the integration of the magnetic field **B** along the contour line C that encompasses the area S,

$$\int_C \mathbf{B} \cdot d\mathbf{l} = 4\pi \int_S \mathbf{j} \cdot \mathbf{n} \, dS = 4\pi I \,. \tag{5.1.9}$$

5.1.3 Ohm's Law

Plasma moving at a nonrelativistic speed in a magnetic field is subject to an additional electric field component $(1/c)(\mathbf{v} \times \mathbf{B})$, besides the direct electric field **E** it sees at rest. *Ohm's law* defines an *electric conductivity* constant σ by setting the current density **j** proportional to the *total electric field* $\mathbf{E}'$,

$$\mathbf{j} = \sigma \mathbf{E}' = \sigma (\mathbf{E} + \frac{1}{c} \mathbf{v} \times \mathbf{B}) \,, \tag{5.1.10}$$

where $\mathbf{E}'$ is the *total electric field* in the frame of reference moving with the plasma.

5.1.4 Induction Equation

A most convenient description of the magnetic field **B** in a plasma is to eliminate the electric field **E** and the current density **j**. The electric field **E** can be eliminated in Maxwell's equation (5.1.3) by inserting the electric field **E** from Ohm's law (5.1.10),

$$\frac{d\mathbf{B}}{dt} = -c \, \nabla \times \mathbf{E} = \nabla \times (\mathbf{v} \times \mathbf{B}) - \frac{c}{\sigma} (\nabla \times \mathbf{j}) \,. \tag{5.1.11}$$

In addition, the current density **j** can be eliminated by inserting the nonrelativistic approximation (5.1.7) from Ampère's law,

$$\frac{d\mathbf{B}}{dt} = \nabla \times (\mathbf{v} \times \mathbf{B}) - \frac{c^2}{4\pi\sigma} \nabla \times (\nabla \times \mathbf{B}) \,. \tag{5.1.12}$$

so that we obtain an equation that contains only the magnetic field **B** and velocity **v**. The constant $\eta = c^2/4\pi\sigma$ is called the *magnetic diffusivity*. Making use of the vector identity,

$$\nabla \times (\nabla \times \mathbf{B}) = \nabla (\nabla \cdot \mathbf{B}) - (\nabla \cdot \nabla)\mathbf{B} \,, \tag{5.1.13}$$

yields, after inserting Maxwell's equation (5.1.2), $\nabla \cdot \mathbf{B} = 0$, the following form of the induction equation,

$$\frac{d\mathbf{B}}{dt} = \nabla \times (\mathbf{v} \times \mathbf{B}) + \eta \nabla^2 \mathbf{B} \,. \tag{5.1.14}$$

The first term on the right-hand side is called the *convective term*, while the second term is called the *diffusive term*. Depending on the value of the *Reynolds number* R_m, which gives the ratio of the convective term ($\propto v_0 B_0 / l_0$) to the diffusive term ($\propto \eta B_0 / l_0^2$),

$$R_m = \frac{l_0 v_0}{\eta} ,\qquad (5.1.15)$$

the induction equation can be approximated in the two limits by

$$\frac{d\mathbf{B}}{dt} \approx \nabla \times (\mathbf{v} \times \mathbf{B}) \qquad \text{for } R_m \gg 1 \qquad (5.1.16)$$

$$\frac{d\mathbf{B}}{dt} \approx \eta \nabla^2 \mathbf{B} \qquad \text{for } R_m \ll 1 \qquad (5.1.17)$$

The plasma in the solar corona is close to a *perfectly conducting medium* (with a high Reynolds number $R_m \approx 10^8 - 10^{12}$), so that approximation (5.1.16) with $R_m \gg 1$ applies, while the diffusive limit (5.1.17) with $R_m \ll 1$ is not relevant in the corona.

5.2 Potential Fields

A force field $\mathbf{F}(\mathbf{r})$ is called a *potential field*, as generally defined in physics, when the force field can be expressed by a gradient of a potential scalar function $\phi(\mathbf{r})$,

$$\mathbf{F}(\mathbf{r}) = \nabla \phi(\mathbf{r}) ,\qquad (5.2.1)$$

for instance the gravitational force field $\mathbf{F}_{grav}(\mathbf{r}) = \nabla \phi_{grav}(\mathbf{r})$ with $\phi_{grav}(\mathbf{r}) = GMm\mathbf{r}/r^2$, or the electric force field $\mathbf{E}(\mathbf{r}) = \nabla \phi_{el}(\mathbf{r})$ with $\phi_{el}(\mathbf{r}) = Qq\mathbf{r}/r^2$.

In the same way we can define a magnetic potential field $B(\mathbf{r})$ by a *magnetic scalar potential function* $\phi(\mathbf{r})$,

$$\mathbf{B}(\mathbf{r}) = \nabla \phi(\mathbf{r}) ,\qquad (5.2.2)$$

or expressed in cartesian coordinates,

$$(B_x, B_y, B_z) = \left(\frac{d}{dx}, \frac{d}{dy}, \frac{d}{dz} \right) \phi(x, y, z) .\qquad (5.2.3)$$

Because of Maxwell's equation (5.1.2), which states that the magnetic field is divergence free, $\nabla \cdot \mathbf{B} = 0$, it follows that the potential magnetic field also satisfies the Laplace equation,

$$\nabla \cdot \mathbf{B} = \nabla(\nabla \phi) = \nabla^2 \phi = 0 .\qquad (5.2.4)$$

Inserting the magnetic potential $\mathbf{B} = \nabla \phi$ into the nonrelativistic approximation of the current density $\mathbf{j}$ (Eq. 5.1.7),

$$\mathbf{j} = \frac{1}{4\pi}(\nabla \times \mathbf{B}) = \frac{1}{4\pi}(\nabla \times \nabla \phi) = 0 \qquad (5.2.5)$$

immediately shows that the current density $\mathbf{j} = 0$ vanishes in a magnetic potential field ϕ, because the vector product of the two parallel vectors ∇ and $\nabla \phi$ (in Eq. 5.2.5) is zero by definition. So, a *magnetic potential field* ϕ is equivalent to a "*current-free*" field,

$$\mathbf{j} = 0 .\qquad (5.2.6)$$

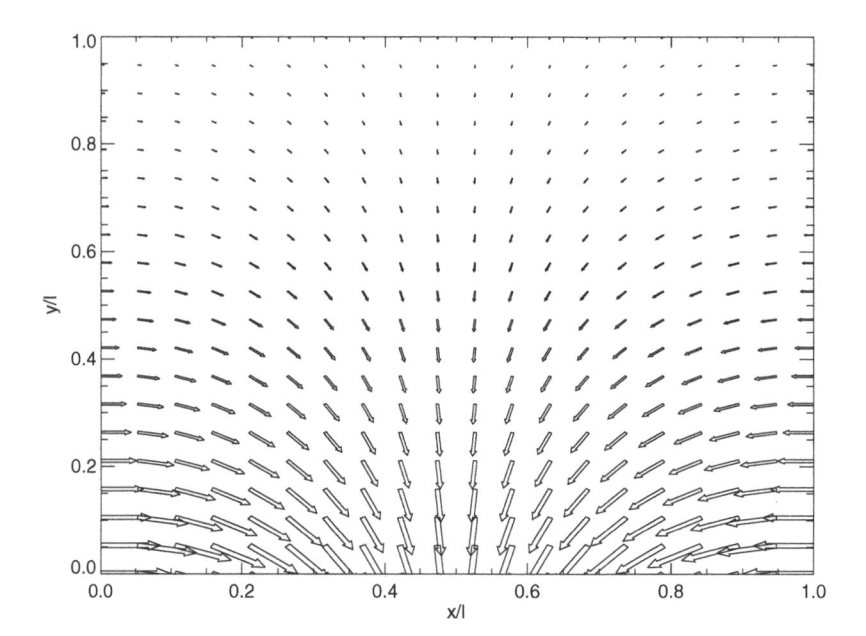

Figure 5.1: Potential vector field, $\mathbf{B}(x,y) = [B_x(x,y), B_y(x,y)]$, of a symmetric unipolar field, a first-order model of a sunspot. The potential scalar function $\phi(x,y)$ is given in Eq. (5.2.12), and the vector components, $B_x(x,y) = d\phi(x,y)/dx$ and $B_y(x,y) = d\phi(x,y)/dy$, in Eq. (5.2.13).

5.2.1 Unipolar Field

As a simple example we model a symmetric unipolar potential field (Fig. 5.1), which in first order approximates the magnetic field of a symmetric sunspot. We introduce a coordinate system with x representing horizontal direction and y representing vertical direction in a 2D coordinate system. The magnetic field of a sunspot is often nearly symmetrical to the central vertical axis and decreases with height with a vanishing field at large distance. So, for a sunspot placed in the middle of a rectangular box with length l we can impose the following boundary conditions for a magnetic potential $\phi(x,y)$,

$$\begin{aligned}
\phi(x,0) &= \text{symmetric} \\
\phi(x,\infty) &= 0 \\
\phi(0,y) &= \phi(l,y) = 0
\end{aligned} \tag{5.2.7}.$$

We attempt a separation ansatz [i.e., a representation of the potential scalar function $\phi(x,y)$ by a product of two separate functions $X(x)$ and $Y(y)$],

$$\phi(x,y) = X(x)Y(y) \tag{5.2.8}$$

A symmetric function in x that fulfills the boundary conditions $X(0) = X(l) = 0$ is for instance the sine function,

$$X(x) = X_0 \sin\left(\frac{\pi x}{l}\right). \tag{5.2.9}$$

For a function that decreases with height we can choose an exponential function, because it fulfills the boundary condition $Y(\infty) = 0$,

$$Y(y) = Y_0 \exp\left(-\frac{y}{y_0}\right), \qquad (5.2.10)$$

Inserting these expressions (Eqs. 5.2.8–10) into the Laplace equation (5.2.4) we obtain

$$\nabla^2 \phi = \frac{\partial^2 \phi}{\partial x^2} + \frac{\partial^2 \phi}{\partial y^2} = \frac{\partial^2 X}{\partial x^2} Y + X \frac{\partial^2 Y}{\partial y^2} = \left(-\frac{\pi^2}{l^2} + \frac{1}{y_0^2}\right) \phi = 0 \qquad (5.2.11)$$

which requires $y_0 = l/\pi$ to make the expression zero. Thus a solution for the *magnetic potential function* ϕ is

$$\phi(x, y) = \phi_0 \sin\left(\frac{\pi x}{l}\right) \exp\left(-\frac{\pi y}{l}\right). \qquad (5.2.12)$$

The vector magnetic field $\mathbf{B}(x, y)$ can now be calculated via Eq. (5.2.3),

$$\begin{aligned}
B_x(x, y) &= \partial \phi / \partial x &= B_0 \cos(\pi x/l) \exp(-\pi y/l) \\
B_y(x, y) &= \partial \phi / \partial y &= -B_0 \sin(\pi x/l) \exp(-\pi y/l)
\end{aligned} \qquad (5.2.13)$$

with the constant $B_0 = \phi_0 \pi / l$. The vector magnetic field $\mathbf{B}(x, y)$ is visualized in Fig. 5.1, where the local field strength $B(x, y)$,

$$B(x, y) = \sqrt{(B_x^2 + B_y^2)} = B_0 \exp\left(-\frac{\pi y}{l}\right), \qquad (5.2.14)$$

is represented by the length of the vector arrows, and the direction of the field $\mathbf{B}(x, y)$ is shown by the orientation of the field vectors.

5.2.2   Dipole Field

Many coronal loops have a semi-circular shape, with conjugate magnetic polarities at the opposite photospheric footpoints, and thus can approximately be described by a magnetic dipole field, generated by a dipole coil buried below the photosphere. Let us assume a buried coil with its axis parallel to the surface (in horizontal direction at location $(0, 0)$ in Fig. 5.2). The coil axis is aligned to the z-axis of a spherical coordinate system. The distance from the dipole center (at $x = 0$ and $y = 0$ in Fig. 5.2) is denoted with r, the zenith angle from the dipole axis with θ, and the azimuth angle with φ. The (r, θ)-plane is shown in Fig. 5.2. A general derivation of the dipole field in spherical coordinates is given in Jackson (1962), which in the far-field approximation $(r \gg a)$ takes the simple analytical form,

$$B_r(r, \theta) = \frac{2m \cos \theta}{r^3},$$

$$B_\theta(r, \theta) = \frac{m \sin \theta}{r^3},$$

$$B_\varphi(r, \theta) = 0. \qquad (5.2.15)$$

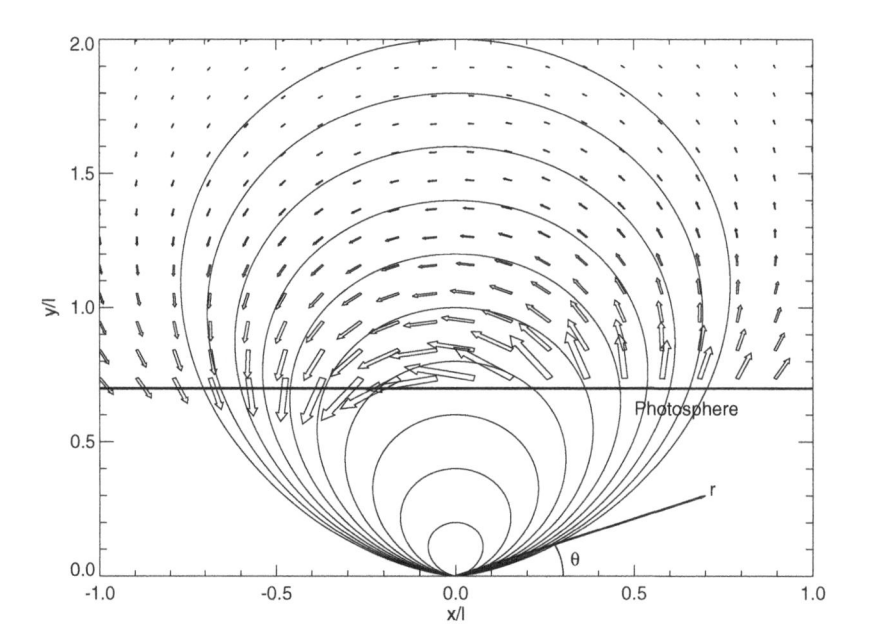

Figure 5.2: Potential vector field, $\mathbf{B}(x, y) = [B_x(x, y), B_y(x, y)]$, of a dipole model in the far-field approximation, according to Eqs. (5.2.15) and (5.2.18). The magnetic field vectors are drawn as arrows at equispaced mesh points above the photosphere ($y \geq 0.7$), while the dipole is buried in depth $y = 0.0$. Magnetic field lines are given by the analytic equation (5.2.22).

with $m = \pi a^2 I/c$ the magnetic moment induced by the ring current I with a coil radius a. Expressing the gradient ∇ of the magnetic potential scalar function $\phi(\mathbf{r})$ from (Eq. 5.2.2) in spherical coordinates,

$$B(\mathbf{r}) = (B_r, B_\theta, B_\varphi) = \nabla \phi(\mathbf{r}) = \left(\frac{\partial \phi}{\partial r}, \frac{1}{r} \frac{\partial \phi}{\partial \theta}, \frac{1}{r \sin \theta} \frac{\partial \phi}{\partial \varphi} \right) , \qquad (5.2.16)$$

we can easily find a potential function $\phi(r, \theta, \varphi)$ that satisfies the magnetic field components specified in Eqs. (5.2.15) as derived with the gradient given in Eq. (5.2.16), namely

$$\phi(r, \theta, \varphi) = -\frac{m \cos \theta}{r^2} . \qquad (5.2.17)$$

This is the potential scalar function of a dipole, which is the far-field approximation of a small coil. We plot the magnetic field vectors $\mathbf{B}(x, y)$ in cartesian coordinates in Fig. 5.2, which have the components

$$B_x(x, y) = B_r(r, \theta) \cos \theta - B_\theta(r, \theta) \sin \theta ,$$

$$B_y(x, y) = B_r(r, \theta) \sin \theta + B_\theta(r, \theta) \cos \theta . \qquad (5.2.18)$$

It is also illustrative to draw closed field lines of a dipole field. If the magnetic field $\mathbf{B}(x, y)$ is known as a function of position, then the magnetic field lines in a cartesian coordinate system can simply be derived from the proportionality relations,

$$\frac{dx}{B_x} = \frac{dy}{B_y} = \frac{dz}{B_z} = \frac{ds}{B} , \qquad (5.2.19)$$

where s is the distance in the direction of the field line. In spherical coordinates, the proportionality relations are according to Eq. (5.2.16),

$$\frac{dr}{B_r} = \frac{r \, d\theta}{B_\theta} = \frac{r \sin\theta \, d\varphi}{B_\varphi} = \frac{ds}{B} . \qquad (5.2.20)$$

These proportionality relations lead for our dipole field components [i.e., (B_r, B_θ) given in Eq. 5.2.15], to the following differential equation (after variable separation),

$$\frac{dr}{r} = \frac{B_r}{B_\theta} d\theta = \frac{2\cos\theta}{\sin\theta} d\theta , \qquad (5.2.21)$$

which can be easily integrated. To avoid the singularity at the limit $\theta = 0$, we integrate relative to the lower limit $\theta_1 = \pi/2$ and r_1 and find the functional relation $r(r_1, \theta)$,

$$r(r_1, \theta) = r_1 \, \exp\left[2 \, log(\sin\theta)\right] = r_1 \, \sin^2\theta . \qquad (5.2.22)$$

This parameterization allows us to directly plot the magnetic dipole field lines, with their distance r from the dipole center as a function of the zenith angle θ, for a set of constants r_1. The constants r_1 correspond to the distance of the dipole field line in the y-direction (at $\theta = \pi/2$). The dipole field lines plotted in Fig. 5.2 are chosen for the constants $r_1 = 0.2, 0.4, ..., 2.0$. We see that the magnetic field vectors B(r) from Eq. (5.2.15), marked with arrows in Fig. 5.2, are parallel to the parameterized field lines $r(\theta)$ from Eq. (5.2.22) at every location, both being different representations of the same dipole potential field.

5.2.3   Potential-Field Calculation Methods

At present, the solar vector magnetic field is routinely measured from the Zeeman effect in the photosphere, so we have to rely on field extrapolation methods to infer properties of the coronal field. The inference of a potential magnetic field **B** in the corona is thus only constrained by the values at the photospheric boundary, and thus, mathematically, represents a boundary problem. Classical potential theory yields a unique solution for the potential scalar function $\phi(\mathbf{r})$, either when the value of $\phi(\mathbf{r})$ is specified at the boundary $\mathbf{r} = \mathbf{r}_B$ (*Dirichlet problem*), or when its derivative normal to the boundary is specified (Neumann problem), which is here the normal component of the magnetic field, $B_n = d\phi(\mathbf{r})/dr_n$. To solve the Laplace equation (5.2.4) for this boundary problem, the *Green's function method* or the *eigen function expansion methods* are used in solar physics.

Green's Function Methods

The method of using the *Green's function* to calculate the solar magnetic field was first used by Schmidt (1964) and was further developed by Sakurai (1982), leading to the widely used *Sakurai code* for solar potential field extrapolations. In these codes, besides the lower boundary $r = a$ at the solar surface, where the magnetic field is constrained by observed magnetograms, an upper boundary at a distance of a few solar radii $r = b$ is also specified, the so-called *source surface*, where the magnetic field is assumed to become radial due to the solar wind.

Let us describe first the *Green's function* method for the simplest geometry, when the line-of-sight is perpendicular to the boundary surfaces (*classical Schmidt method*), as it is the case for an observation at the center of the solar disk. We define Cartesian coordinates with the solar surface parallel to the (x, y)-plane at $z = 0$, and the z-axis towards the observer. A magnetogram then yields the normal field component on the boundary, $B_n(x, y) = B_z(x, y, z = 0)$. The potential magnetic field in the volume $z > 0$ is obtained from the potential scalar function $\phi(\mathbf{r})$, which first satisfies the Laplace equation,

$$\nabla^2 \phi = 0 \qquad (z > 0) \,, \tag{5.2.23}$$

second, the lower boundary condition (the so-called *Neumann boundary condition*), with $\mathbf{n}$ being the unit vector in z-direction,

$$-\mathbf{n} \cdot \nabla \phi = B_n \qquad (z = 0) \,, \tag{5.2.24}$$

and third, the upper boundary condition

$$\lim_{r \mapsto \infty} \phi(\mathbf{r}) = 0 \qquad (z > 0) \,. \tag{5.2.25}$$

In analogy to an electrostatic potential (e.g., Jackson 1962), a 1/r-potential function can be chosen as a simple *Green's function*, with a "magnetic charge" envisioned at location $\mathbf{r}'$,

$$G_n(\mathbf{r}, \mathbf{r}') = \frac{1}{2\pi |\mathbf{r} - \mathbf{r}'|} \,. \tag{5.2.26}$$

The photospheric boundary with coordinates $\mathbf{r}' = (x', y', 0)$, which contains the normal magnetic components $B_n(\mathbf{r}')$, now contributes to the potential $\phi(\mathbf{r})$ at an arbitrary coronal location $\mathbf{r}$ via the Green's function $G_n(\mathbf{r}, \mathbf{r}')$. The potential function $\phi(\mathbf{r})$ can thus be calculated by integrating over all contributions from the photospheric surface (x', y'),

$$\phi(\mathbf{r}) = \int \int B_n(\mathbf{r}') G_n(\mathbf{r}, \mathbf{r}') \, dx' \, dy' \,. \tag{5.2.27}$$

This potential scalar function $\phi(\mathbf{r})$ satisfies the Laplace equation (5.2.23) and the two boundary conditions (Eqs. 5.2.24−25), as it can be verified mathematically (see, e.g., Jackson 1962; Sakurai 1982).

For a practical application, one has to consider the discreteness of the meshpoints (x', y') over which $B_n(x', y')$ is measured and the potential is calculated. If the mesh consists of pixels with a size Δ, the continuous integral (5.2.27) transforms into a summation over all positions $\mathbf{r}'_{ij} = (x'_i, y'_j, 0)$,

$$\phi(\mathbf{r}) = \sum_{ij'} B_n(\mathbf{r}'_{ij}) G_n(\mathbf{r}, \mathbf{r}'_{ij}) \Delta^2 \,. \tag{5.2.28}$$

This leads to a modified Green's function (Sakurai 1982) that is adjusted for the discreteness Δ of the mesh points in such a way that it still fulfills the Laplace equation (5.2.23) and the two boundary conditions (Eqs. 5.2.24–25),

$$G_n(\mathbf{r}, \mathbf{r}') = \frac{1}{2\pi|\mathbf{r} - \mathbf{r}' + (\Delta/\sqrt{2\pi})\mathbf{n}|} \,. \qquad (5.2.29)$$

A next generalization is for an arbitrary line-of-sight direction (*oblique Schmidt method*), so that the magnetic field can be reconstructed at any position away from the Sun's center. The normal vector $\mathbf{n}$ has now be replaced by a unit vector $\mathbf{l}$ that points towards the observer, but is not normal to the solar surface anymore. This changes the Neumann boundary condition (5.2.24) to

$$-\mathbf{l} \cdot \nabla \phi = B_l \qquad (z = 0) \,, \qquad (5.2.30)$$

with B_l the longitudinal magnetic field component along the line-of-sight. The magnetic potential scalar function (5.2.27) becomes

$$\phi(\mathbf{r}) = \int \int B_l(\mathbf{r}') G_l(\mathbf{r}, \mathbf{r}') \, dx' \, dy' \,, \qquad (5.2.31)$$

where the Green's function (5.2.26) can be expressed in terms of the direction vectors $\mathbf{n}$ and $\mathbf{l}$ as (Sakurai 1982)

$$G_n(\mathbf{r}, \mathbf{r}') = \frac{1}{2\pi} \left[\frac{\mathbf{n} \cdot \mathbf{l}}{R} + \frac{\mathbf{m} \cdot \mathbf{R}}{R(R + \mathbf{l} \cdot \mathbf{R})} \right] \,, \qquad (5.2.32)$$

$$\mathbf{R} = (\mathbf{r} - \mathbf{r}') \,, \qquad \mathbf{m} = \mathbf{l} \times (\mathbf{n} \times \mathbf{l}) \,. \qquad (5.2.33).$$

The oblique Schmidt method works only for small areas of the solar surface. A more general approach is to include the curvature of the solar surface, which is required for magnetic field extrapolations over larger areas of the Sun or for global magnetic field models. Generalizations of potential field line extrapolation codes have been developed with Green's function methods, the so-called *spherical Schmidt method* (e.g., Sakurai 1982), as well as with *eigen function (spherical harmonic) expansion* methods (e.g., Altschuler & Newkirk 1969).

Eigenfunction Expansion Methods

The natural coordinates to represent the curved solar surface are spherical coordinates (r, θ, φ), with the photosphere at $r = a$ and the z-axis oriented towards the observer in the l-direction. The Neumann boundary condition (5.2.30) now becomes,

$$-\mathbf{l} \cdot \nabla \phi = B_l(\theta, \varphi) \qquad (r = a) \,. \qquad (5.2.34)$$

The expansion into spherical harmonics was first applied to solar magnetic fields by Newkirk et al. (1968) and Altschuler & Newkirk (1969). The magnetic potential function $\phi(r, \theta, \varphi)$, (e.g., Rudenko 2001; Stix 2002),

$$\phi(r, \theta, \varphi) = R_\odot \sum_{l=1}^{N} \sum_{m=0}^{l} f_l(r) \, P_l^m(\theta)(g_l^m \cos m\varphi + h_l^m \sin m\varphi) \,, \qquad (5.2.35)$$

Figure 5.3: Potential field models of the global corona calculated from a source surface model during 6 different phases of the solar cycle. Each display shows 500 potential magnetic field lines, computed from simulated magnetograms that mimic the flux dispersion during the solar cycle (Schrijver & Aschwanden 2002).

is expressed in terms of the Legendre polynomials $P_l^m(\theta)$, and the radial dependence is

$$f_l(r) = \frac{(r_w/r)^{l+1} - (r/r_w)^l}{(r_w/R_\odot)^{l+1} - (R_\odot/r_w)^l} \; . \tag{5.2.36}$$

where r_w demarcates the location of the outer boundary, the so-called *source surface* where $f_l(r_w) = 0$. This means the magnetic field at $r = r_w$ points strictly in the radial direction, as it is assumed to happen due to the forces of the solar wind, at a typical distance of $r_w = 2.6$ solar radii. The functions f_l are normalized to $f_l(r = R_\odot) = 1$. The potential function expressed in terms of spherical harmonics (Eq. 5.2.35) essentially corresponds to a multipole expansion, where the weight of different multipoles is given by the expansion coefficients g_l^m and h_l^m, which have to be numerically fitted until the potential function $\phi(r, \theta, \varphi)$ satisfies the Neumann boundary condition (5.2.34) specified by the observed magnetic field $B_l(\theta, \varphi)$. Typical numeric codes calculate expansions up to orders of $N = 25, ..., 100$. More sophisticated extrapolation codes fit non-spherical source surface models (Levine et al. 1982).

A example of potential field extrapolations of the global magnetic field of the corona is shown in Fig. 5.3, based on a synthetic data set of magnetograms of solar observations covering several sunspot cycles (Schrijver & Aschwanden 2002). The time evolution of the global magnetic field was simulated by magnetic flux dispersion subjected to random walk, induced by the evolution of the supergranular network. The potential field lines were computed by integrating along the vector field that is the sum of fields from a set of $\approx 100,000$ point charges simulated in the surface diffusion model. Only a subset of 500 field lines is displayed in each image, selected from the photospheric mesh points with the highest magnetic flux concentrations.

5.3   Force-Free Fields

If n particles with electric charge q are moving with velocity $\mathbf{v}$ in a magnetic field $\mathbf{B}$, they experience the well-known *Lorentz force*

$$\mathbf{F} = \frac{q}{c} n (\mathbf{v} \times \mathbf{B}) \,. \tag{5.3.1}$$

Since the current density $\mathbf{j}$ associated with a moving electric charge q is defined as

$$\mathbf{j} = \frac{q}{c} n \mathbf{v} \,, \tag{5.3.2}$$

the Lorentz force is generally called the $\mathbf{j} \times \mathbf{B}$ force in magneto-hydrodynamics (MHD),

$$\mathbf{F} = \mathbf{j} \times \mathbf{B} \,, \tag{5.3.3}$$

and will show up as an additional force term in the momentum equation (4.1.3) when we include magnetic fields in the MHD equations (§ 6).

In magneto-statics, a magnetic field is said to be *force-free*, when the *"self-force"* or Lorentz force is zero,

$$\mathbf{j} \times \mathbf{B} = 0 \,. \tag{5.3.4}$$

As we already derived a simplified Maxwell equation (5.1.4) in the nonrelativistic approximation [i.e., $\mathbf{j} = (1/4\pi)(\nabla \times \mathbf{B})$ in Eq. (5.1.7)], we can insert this expression for the current density $\mathbf{j}$ into Eq. (5.3.4) and obtain a condition for force-free magnetic fields that is entirely expressed by the magnetic field $\mathbf{B}$,

$$(\nabla \times \mathbf{B}) \times \mathbf{B} = 0 \,. \tag{5.3.5}$$

The general calculation of force-free magnetic fields is non-trivial, because the equation (5.3.5) is nonlinear, since it contains terms of order B^2. Therefore, the computation of force-free fields subdivides the problem into the mathematically simpler case of *linear force-free fields*, which covers only a subset of the possible solutions, while the more general solutions are referred to as *nonlinear force-free fields*.

5.3.1 Linear Force-Free Fields

The general force-free condition (5.3.5) can be turned into a linear problem. A linear solution of Eq. (5.3.4) can be defined by

$$(\nabla \times \mathbf{B}) = 4\pi\mathbf{j} = \alpha(\mathbf{r})\mathbf{B} , \qquad (5.3.6)$$

where $\alpha(\mathbf{r})$ is a scalar function as a function of position (and may additionally depend on time). So, $\alpha(\mathbf{r}) \neq 0$ characterizes a *nonpotential field*, while the special case of $\alpha(\mathbf{r}) = 0$ yields $(\nabla \times \mathbf{B}) = 0$ and corresponds to the *potential field* case (Eq. 5.2.5).

The choice of the scalar function $\alpha(\mathbf{r})$ is not completely arbitrary, because it must also satisfy the divergence-free Maxwell equation (5.1.2), $\nabla \cdot \mathbf{B} = 0$, and the vector identity $\nabla \cdot (\nabla \times \mathbf{B}) = 0$. These two conditions yield

$$\nabla \cdot (\nabla \times \mathbf{B}) = \nabla \cdot (\alpha\mathbf{B}) = \alpha(\nabla \cdot \mathbf{B}) + \mathbf{B} \cdot \nabla\alpha = \mathbf{B} \cdot \nabla\alpha = 0 . \qquad (5.3.7)$$

This condition is fulfilled if $\alpha(\mathbf{r})$ is constant and does not vary along a magnetic field line, so that $\nabla\alpha = 0$. This means that $\alpha(\mathbf{r})$ is not a scalar function anymore, but a simple constant, and Eq. (5.3.6) becomes,

$$(\nabla \times \mathbf{B}) = \alpha\mathbf{B} . \qquad (5.3.8)$$

If α is constant everywhere, for each field line in a given volume, we obtain with (Eq. 5.3.8) for the curl of the current density $(\nabla \times \mathbf{B})$,

$$\nabla \times (\nabla \times \mathbf{B}) = \nabla \times \alpha\mathbf{B} = \alpha(\nabla \times \mathbf{B}) = \alpha^2\mathbf{B} . \qquad (5.3.9)$$

On the other hand we have the vector identity,

$$\nabla \times (\nabla \times \mathbf{B}) = \nabla(\nabla \cdot \mathbf{B}) - \nabla^2\mathbf{B} . \qquad (5.3.10)$$

Comparing these two expressions (Eq. 5.3.9) and (Eq. 5.3.10) and making use of the divergence-free Maxwell equation $(\nabla \cdot \mathbf{B} = 0)$ leads to the *Helmholtz equation*,

$$\nabla^2\mathbf{B} + \alpha^2\mathbf{B} = 0 . \qquad (5.3.11)$$

Numerical methods to calculate *linear force-free fields (lff)* have been developed by using Fourier series (Nakagawa & Raadu 1972; Alissandrakis 1981; Démoulin et al. 1997b; Gary 1989), using Green's functions (Chiu & Hilton 1977; Seehafer 1978; Semel 1988), spherical harmonics (Newkirk et al. 1969; Altschuler & Newkirk 1969), or a superposition of discrete flux sources (Lothian & Browning, 1995). A generalization of linear force-free fields that includes MHD terms, such as the thermal pressure (∇p) and gravity (∇p_{grav}) have been studied analytically by Low (1985, 1991, 1993a, 1993b) and Neukirch (1995). Numerical schemes to compute such linear force-free fields with MHD terms have been devised by using Fourier–Bessel series (Low 1992), Green's functions (Neukirch & Rastätter 1999; Petrie & Neukirch 2000), spherical solutions (Bogdan & Low 1986; Neukirch 1995; Neukirch & Rastätter 1999), or optimization methods (Wiegelmann & Neukirch 2003).

More generally, if $\alpha(\mathbf{r})$ is a function of position $\mathbf{r}$, which is called the *nonlinear force-free field (nfff)*, then the curl of the current density $(\nabla \times \mathbf{B})$ is,

$$\nabla \times (\nabla \times \mathbf{B}) = \nabla \times \alpha \mathbf{B} = \alpha(\nabla \times \mathbf{B}) + \nabla\alpha \times \mathbf{B} = \alpha^2 \mathbf{B} + \nabla\alpha \times \mathbf{B} . \quad (5.3.12)$$

Together with the vector identity (Eq. 5.3.10) and (Eq. 5.3.7) we have therefore two coupled equations for $\mathbf{B}$ and $\alpha(\mathbf{r})$ that need to be solved together to obtain a solution for a linear (α=constant) or nonlinear ($\alpha(\mathbf{r})$) force-free field,

$$\nabla^2 \mathbf{B} + \alpha^2 \mathbf{B} = \mathbf{B} \times \nabla\alpha(\mathbf{r}) ,$$

$$\mathbf{B} \cdot \nabla\alpha(\mathbf{r}) = 0 . \quad (5.3.13)$$

5.3.2   Sheared Arcade

As an example of a linear force-free field we model a *sheared arcade* (Priest 1982; Sturrock 1994). A *loop arcade* consists of a sequence of equal-sized loops which have a common axis of curvature, which usually coincides closely with a neutral line on the solar surface. A *sheared arcade* can be generated by shifting the footpoints parallel to the neutral line on one side along the solar surface. The longer the shearing motion is applied, the higher is the shearing angle of the arcade. With the following simple analytical model we will see that the shear angle is directly related to the nonpotentiality of the magnetic field of the loop arcade.

A simple setup is to characterize the horizontal component of the magnetic field with a periodic function that decreases exponentially with height,

$$B_x = B_{x0} \sin(kx) \exp(-lz) ,$$

$$B_y = B_{y0} \sin(kx) \exp(-lz) ,$$

$$B_z = B_0 \cos(kx) \exp(-lz) . \quad (5.3.14)$$

The components of the vector product $(\nabla \times \mathbf{B})$ are then,

$$(\nabla \times \mathbf{B})_x = \frac{\partial B_z}{\partial y} - \frac{\partial B_y}{\partial z} = lB_{y0} \sin(kx) \exp(-lz) ,$$

$$(\nabla \times \mathbf{B})_y = \frac{\partial B_x}{\partial z} - \frac{\partial B_z}{\partial x} = (-lB_{x0} + kB_0) \sin(kx) \exp(-lz) ,$$

$$(\nabla \times \mathbf{B})_z = \frac{\partial B_y}{\partial x} - \frac{\partial B_x}{\partial y} = kB_{y0} \cos(kx) \exp(-lz) . \quad (5.3.15)$$

Using the constant-α condition for a force-free field (Eq. 5.3.8), we obtain three equations,

$$lB_{y0} = \alpha B_{x0} ,$$

$$(-lB_{x0} + kB_0) = \alpha B_{y0} ,$$

$$kB_{y0} = \alpha B_0 . \quad (5.3.16)$$

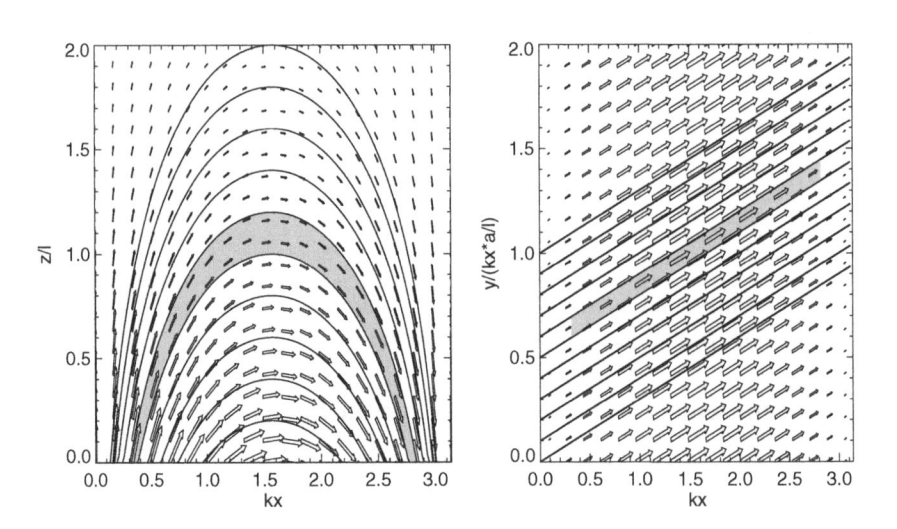

Figure 5.4: Force-free field of a sheared arcade, with the components (B_x, B_z) shown in the vertical plane (left), and the components (B_x, B_y) shown in the horizontal plane (right). The shear angle θ corresponds to the slope in the right panel. Magnetic field lines are given by the analytic equation (5.3.19).

These three equations yield the field components B_{x0} and B_{y0} and a condition between the parameters l, k, and α,

$$B_{x0} = \frac{l}{k} B_0 \ ,$$

$$B_{y0} = \frac{\alpha}{k} B_0 \ ,$$

$$k^2 - l^2 - \alpha^2 = 0 \ . \tag{5.3.17}$$

The solutions are plotted in Fig. 5.4, projected into the (x, z)-plane (Fig. 5.4 left) and into the (x, y)-plane (Fig. 5.4 right). Obviously each field line lies in a vertical plane and has a shear angle θ of

$$\tan \theta = \frac{B_y}{B_x} = \frac{B_{y0}}{B_{x0}} = \frac{\alpha}{l} \ . \tag{5.3.18}$$

Thus the α-parameter is related to the shear angle θ, and is even proportional for small values. The higher the shear angle θ, the higher the nonpotential parameter α. For the case of no shear, $\theta = 0$, we have the potential case with $\alpha = 0$. Thus, the shear angle of a loop arcade is a sensible indicator of the nonpotentiality of the magnetic field (i.e., the difference between the force-free field and the potential field).

This simple analytical model has a mathematical limitation. The maximum shear angle is $\theta = \pi/2$. The third condition in Eq. (5.3.17) implies a maximum value for α, namely $\alpha \leq \sqrt{k^2 - l^2}$, thus $\alpha < k$, since $l > 0$ (see also Sturrock 1994).

An explicit parameterization of the nonpotential field lines (Eq. 5.3.14) can be calculated with the proportionality relations given in (Eq. 5.2.19), by variable separation and integration. We find the following explicit parameterization for individual field lines,

$$y(x) = \frac{\alpha}{l}x + y_0 \ ,$$

$$z(x) = \frac{k}{l} \log\left[\sin\left(kx\right)\right] + z_0 \ . \tag{5.3.19}$$

The field lines shown in Fig. 5.4 are plotted with this explicit parameterization. One of the sheared loops is outlined with a shaded area in Fig. 5.4.

5.3.3   Nonlinear Force-Free Field Calculation Methods

After we have given the general definition of the *nonlinear force-free field* in §5.3.1, which requires a solution of the coupled equation system

$$(\nabla \times \mathbf{B}) = \alpha\mathbf{B} \ ,$$

$$\mathbf{B} \cdot \nabla\alpha = 0 \ . \tag{5.3.20}$$

(which is a different from of Eq. 5.3.13) we turn now to some numerical methods that have been developed to calculate such magnetic field extrapolations. The necessity of nonlinear force-free extrapolation methods is in particular motivated by the fact that a direct comparison of extrapolated fields and observed coronal structures in images shows that the potential and linear force-free extrapolations are often inadequate in representing coronal structures. Critical reviews on nonlinear force-free extrapolation methods with inter-comparisons and discussions of their problems can be found in Démoulin et al. (1997b), Amari et al. (1997), and McClymont et al. (1997), all contained in the Special Issue of *Solar Physics* **174** on the Workshop *Measurements and Analyses of the 3D Solar Magnetic Fields.*

The Vertical Integration Method

The most straightforward method is called the method of *vertical integration*. The photospheric boundary condition at height $z = 0$ is given by the measured magnetic field components B_x, B_y, B_z, which also define the α-parameter at the photospheric boundary (e.g., from the z-component of Eq. 5.3.6),

$$\alpha(x, y, z = 0) = \frac{1}{B_z}\left(\frac{\partial B_y}{\partial x} - \frac{\partial B_x}{\partial y}\right) \ . \tag{5.3.21}$$

Using the x and y-component of Eq. (5.3.6), the divergence-free condition $\nabla \cdot \mathbf{B}$ (Eq. 5.1.2), and the vector identity condition $\mathbf{B} \cdot \nabla\alpha = 0$ (5.3.7), all vertical derivatives of B_x, B_y, B_z, and α can be written as a function of horizontal derivatives,

$$\frac{\partial B_x}{\partial z} = \alpha B_y + \frac{\partial B_z}{\partial x} \ , \tag{5.3.22}$$

$$\frac{\partial B_y}{\partial z} = -\alpha B_x + \frac{\partial B_z}{\partial y}\,, \tag{5.3.23}$$

$$\frac{\partial B_z}{\partial z} = -\frac{\partial B_x}{\partial x} - \frac{\partial B_y}{\partial y}\,, \tag{5.3.24}$$

$$\frac{\partial \alpha}{\partial z} = -\frac{1}{B_z}\left(B_x \frac{\partial \alpha}{\partial x} + B_y \frac{\partial \alpha}{\partial y}\right)\,. \tag{5.3.25}$$

Integration of this system of four equations (5.3.22−25) in the vertical direction provides the extrapolated magnetic field at height $z + dz$, which can be continued in the upward direction to a desired upper boundary of a coronal volume. This apparently straightforward method (e.g., Wu et al. 1990) was numerically tested by Démoulin et al. (1992). Although the numeric computation is straightforward, it is a mathematically ill-posed problem, even for a potential field ($\alpha = 0$), because of the nature of the elliptical equations. The ill-posedness of the problem is manifested by the fact that a small variation of the base boundary conditions can change the solutions dramatically, as well as the fact that the numerical errors of integration make the integration unstable and the deviations from the solution are rapidly growing. This has been shown by a Fourier analysis for the linear force-free case (Alissandrakis 1981; Démoulin et al. 1992). The Fourier transforms of the magnetic field components with respect to the spatial coordinates (x, y) are defined in Fourier space (u, v) as,

$$\hat{\mathbf{B}}(u, v, z) = \hat{\mathbf{B}}(u, v, 0)\exp\left(-kz\right)\,, \tag{5.3.26}$$

where solutions of the Fourier components $\hat{\mathbf{B}}$ are considered that decrease exponentially with height z. Taking the Fourier transform of the linear force-free equation (5.3.20) and using the derivative theorem of the Fourier transform leads to a specific condition for the wave number k (Alissandrakis 1981),

$$k = \pm\sqrt{(4\pi^2(u^2 + v^2) - \alpha^2)}\,, \tag{5.3.27}$$

where the negative sign $(-)$ in Eq. (5.3.27) indicates wave numbers with decreasing modes, and the positive sign $(+)$ with growing modes. The growing modes lead to unphysical solutions, because the magnetic energy diverges to infinity with height, $E_B \mapsto \infty$. The growing modes grow faster for high spatial frequencies and seem to appear due to lack of boundary conditions. Various modifications have been developed to minimize the algorithm-dependent computational errors, for instance by spatial averaging of α at each step of the vertical integration (Cuperman et al. 1990), or by spatial averaging of both the magnetic field $\mathbf{B}$ and α separately (Démoulin et al. 1992). The increased numerical accuracy of the averaging approach was tested in Démoulin et al. (1992) with an analytical model of a weakly sheared nonlinear force-free field solution (Low 1982). In an alternative approach, growing modes were suppressed through Fourier Transform during the extrapolation (Amari et al. 1997). A further drawback of the vertical integration method is that the derivative $d\alpha/dz$ (Eq. 5.3.25) becomes singular at the neutral line, where $B_z = 0$. This obstacle has been circumvented by spatial smoothing methods (Wu et al. 1990), by imposing a maximum value for α (Cuperman et al. 1990) and B_z (Démoulin et al. 1992). Another difficulty with the vertical integration method is that a constant α-value along a field line cannot strictly be en-sured with noisy magnetogram data, as it is required by the condition $\mathbf{B} \cdot \nabla \alpha = 0$ (Eq. 5.3.20). While the vertical integration method is the most straightforward ap-proach, the numerical stability seems to represent a major obstacle to achieve accurate nonlinear force-free solutions.

The Boundary Integral Method

The previously described *vertical integration method* is based on techniques of solving differential equations (5.3.22−25) that have ill-posed features. Alternatively, if the same physical problem can be formulated by integral equations, one can benefit from the advantages of integral techniques (e.g., they are more robust because the residual errors cancel out better in integral summations).

Such an integral equation representation for force-free fields was developed and demonstrated in Yan (1995), Yan et al. (1995), and Yan & Sakurai (1997, 2000). In order to apply *Green's theorem*, two functions are required, for which the magnetic field $B(\mathbf{r})$ and the reference function $Y(\mathbf{r})$ are chosen, where

$$Y(\mathbf{r}) = \frac{\cos (\lambda r)}{4\pi r} \, , \tag{5.3.28}$$

with $r = |\mathbf{r} - \mathbf{r}_i|$ being the distance of an arbitrary location $\mathbf{r}$ to a fixed point at $\mathbf{r}_i$, and $\lambda(\mathbf{r}_i)$ the parameter that depends only on the fixed location $\mathbf{r}_i$. The reference function $Y(\mathbf{r})$ satisfies the *Helmholtz equation* (5.3.11)

$$\nabla^2 Y + \lambda^2 Y = \delta_i \, , \tag{5.3.29}$$

where δ_i is the Dirac function defined at point i. Because there is a singularity at point i, a small volume around point i needs to be excluded in the field extrapolation. The two functions B and Y satisfy the Green's second identity and the Helmholtz equation (5.3.29), so that we have

$$\int_V (Y \nabla^2 B - B \nabla^2 Y) dV = \int_S \left(Y \frac{dB}{dn} - B \frac{dY}{dn} \right) dS \, , \tag{5.3.30}$$

with $\int_V$ being a volume integral and $\int_S$ the surface surrounding the volume, consist-ing of all sides of the extrapolation box. The surface integral can be calculated for the chosen reference function Y (Eq. 5.3.28) and we obtain for the volume integral (5.3.30),

$$\begin{aligned} \int_V (Y \nabla^2 B - B \nabla^2 Y) dV &= \mathbf{B}(\mathbf{r}) + \int_V Y \left[\lambda^2 \mathbf{B} - \alpha^2 \mathbf{B} - (\nabla \alpha \times \mathbf{B}) \right] dV \\ &= \mathbf{B}(\mathbf{r}) + \psi_i(\lambda, \mathbf{r}) \, . \end{aligned} \tag{5.3.31}$$

Since the volume integral has to vanish at infinity, we require $\psi_i(\lambda) = 0$ at any point in the extrapolation box, which constitutes an implicit definition for the *pseudo force-free parameter* λ, which approximately equals the *force-free parameter* α. For each compo-nent of $\mathbf{B}$ a corresponding reference function Y is needed, as defined in Eq. (5.3.28), so we can express it as a diagonal matrix $\mathbf{Y} = diag(Y_x, Y_y, Y_z)$ with parameters

$(\lambda_x, \lambda_y, \lambda_z)$. This then finally leads to a boundary integral representation for the solution of the nonlinear force-free field problem,

$$\mathbf{B}(\mathbf{r_i}) = \int_S \left(\mathbf{Y}(\mathbf{r_i}, \mathbf{r}) \frac{d\mathbf{B}(\mathbf{r})}{dn} - \frac{d\mathbf{Y}(\mathbf{r}_i, \mathbf{r})}{dn} \mathbf{B}(\mathbf{r}) \right) dS \qquad (5.3.32)$$

This equation indicates that the values of $\mathbf{B}$ and the gradients $d\mathbf{B}/dn$ are required over the photospheric surface S. With this input one can then determine the magnetic field $\mathbf{B}$ at any location in the volume V by integration of the products of the reference function $\mathbf{Y}$ and the boundary values $\mathbf{B}(x, y, z = 0)$ and gradients $d\mathbf{B}(x, y, z = 0)/dn$.

Drawbacks of the boundary integral method are oversampling of photospheric boundary conditions and the difficulty in setting the *pseudo force-free parameter* $\lambda(\mathbf{r}_i)$. The method is well adapted for weakly nonlinear force-free cases, for which it has been successfully tested (Yan 1995; Yan & Sakurai 2000).

The Euler Potential Method

The magnetic field $\mathbf{B}(\mathbf{r})$ can also be defined in terms of *Euler potentials* (Stern 1966; Parker 1979),

$$\mathbf{B} = \nabla\alpha \times \nabla\beta \qquad (5.3.33)$$

where the scalar functions $\alpha(\mathbf{r})$ and $\beta(\mathbf{r})$ are also called *Clebsch variables* (Lamb 1963). This is equivalent to express the Euler potentials in terms of a vector potential $\mathbf{A}$,

$$\mathbf{A} = \alpha \cdot \nabla\beta \qquad (5.3.34)$$

since the curl of this vector potential $\mathbf{A}$ yields

$$\mathbf{B} = \nabla \times \mathbf{A} = \nabla \times (\alpha \cdot \nabla\beta) = \nabla\alpha \times \nabla\beta \qquad (5.3.35)$$

according to the vector identity $\nabla \times (\alpha \cdot \nabla\beta) = \nabla\alpha \times \nabla\beta + \alpha(\nabla \times \nabla\beta)$ and the fact that the curl of a gradient vanishes, $(\nabla \times \nabla\beta) = 0$. We can also see that the Maxwell equation (5.1.2) of the divergence-free condition, $\nabla \cdot \mathbf{B} = 0$, is automatically fulfilled by using the vector identity $\nabla \cdot (\nabla\alpha \times \nabla\beta) = \nabla\beta \cdot (\nabla \times \nabla\alpha)$ and $\nabla \times \nabla\alpha = 0$.

According to the vector product (5.3.33), both gradients $\nabla\alpha$ and $\nabla\beta$ are perpendicular to the vector $\mathbf{B}$, so that their scalar product is zero,

$$\mathbf{B} \cdot \nabla\alpha = 0 \,,$$

$$\mathbf{B} \cdot \nabla\beta = 0 \,, \qquad (5.3.36)$$

and thus the functions $\alpha(\mathbf{r})$ and $\beta(\mathbf{r})$ are constant along $\mathbf{B}(\mathbf{r})$. The surfaces $\alpha = const$ and $\beta = const$ are orthogonal to their gradients and tangential to $\mathbf{B}$. Therefore, the magnetic field lines lie on the intersection of the surfaces $\alpha = const$ and $\beta = const$ and each field line can uniquely be labeled by a pair of variables α and β.

Euler potentials have been applied to solar magnetic field problems in various contexts. Sakurai (1979) derived a formulation of the force-free field using Euler potentials and the variational principle, Low (1991) presented the 3D component equations, Uchida (1997a,b) reviews the theory of force-free fields in terms of Euler potentials,

Yang et al. (1986) and Porter et al. (1992) use them in their 2D magneto-frictional method to calculate force-free fields, Fiedler (1992) presents a numeric method, Emonet & Moreno–Insertis (1996) model rising fluxtubes, and Hennig & Cally (2001) discuss the boundary conditions and develop a new fast numeric method with a multi-grid technique.

The Euler potential method is mathematically a well-posed problem. A field line is defined by (α, β) and constrains the connectivity at the boundary ($z = 0$), but the connectivities are hard to obtain from observations. Strategies of numeric codes are therefore to evolve towards observed fields and relax into an equilibrium. The advection at the boundary ($z = 0$) is prescribed by

$$\frac{d\alpha(x, y, z = 0, t)}{dt} + \mathbf{v}(x, y, z = 0, t) \cdot \nabla \alpha = 0 \,,$$

$$\frac{d\beta(x, y, z = 0, t)}{dt} + \mathbf{v}(x, y, z = 0, t) \cdot \nabla \beta = 0 \,, \qquad (5.3.37)$$

while relaxation in the coronal volume ($z > 0$) is achieved by solving the momentum equation with magnetic induction, the Lorentz force, and viscous terms (Yang et al. 1986), or by iterative minimization of the magnetic energy E_B (Sakurai 1979). The main problem with the Euler potential method are the free boundaries and the related unconstraintness of the magnetic topology (Antiochos 1986). It is also impossible to superpose potentials from multiple sources (Stern 1966), which makes it difficult to obtain analytical solutions.

Full MHD Method

The most complete method to calculate the coronal magnetic field is to use the full set of *magneto-hydrodynamic (MHD)* equations, which we will discuss in more detail in § 6. The first numeric code that solves the full MHD equations has been developed by Mikić et al. (1990). Generally, the MHD equations can describe the time-dependent evolution of the coronal plasma, but for a magnetic field description we are mostly interested in stable equilibria solutions. Such stable equilibria are expected to be force-free ($\mathbf{j} \times \mathbf{B} = 0$), where the plasma density (and the associated pressure p and gravity p_{grav}) have no importance. The solutions of the force-free magnetic field are thus mainly governed by the interaction of the magnetic field with the plasma flows, which is expressed in the equations of motion (generalized from Eq. 4.1.3) and induction (Eq. 5.1.14),

$$\rho \frac{D}{Dt} \mathbf{v} \approx (\mathbf{j} \times \mathbf{B}) + \nu_{visc} \rho \nabla^2 \mathbf{v} \,, \qquad (5.3.38)$$

$$\frac{d\mathbf{B}}{dt} = \nabla \times (\mathbf{v} \times \mathbf{B}) + \eta \nabla^2 \mathbf{B} \,. \qquad (5.3.39)$$

where ν_{visc} is the *kinematic viscosity* and η the finite plasma resistivity, which both contribute to the numerical stability of MHD codes. Required boundary conditions for an MHD code are the magnetic field $\mathbf{B} = (B_x, B_y, B_z)$ and the velocity field $\mathbf{v} = (v_x, v_y, v_z = 0)$ at the photospheric surface.

The MHD method has proven to be efficient in calculations of the solar magnetic field. Mikić et al. (1990) study the dynamic evolution of twisted magnetic fluxtubes, following them until they become kink unstable. Mikić & McClymont (1994) and Mc-Clymont & Mikić (1994) use the MHD code to derive horizontal photospheric flows which form from the observed currents. Amari et al. (1997) simulate the 3D MHD evolution of a quadrupolar configuration and compare it with other extrapolation methods. Jiao et al. (1997) use an MHD code to model the nonpotential field of active region loops observed with Yohkoh/SXT. Difficulties with the MHD method are the same as they are typical for numerical simulations, such as heavy computing demand, convergence problems, insufficient spatial resolution to handle discontinuities, numerical resistivity and viscosity, too small Reynolds numbers for coronal conditions, and line-tying (continuous slippage of magnetic field lines).

Potential Vector Grad–Rubin Method

Another method to solve the nonlinear equation system (5.3.20) for force-free fields is to express it as an iterative sequence of linear equations

$$\mathbf{B}^n \cdot \nabla \alpha^n = 0 , \tag{5.3.40}$$

$$\nabla \times \mathbf{B}^{n+1} = \alpha^n \mathbf{B}^n , \tag{5.3.41}$$

with n the iteration number. This method was introduced by Grad & Rubin (1958) and is also described in Aly (1989) and Amari et al. (1997). The required boundary conditions are the vertical magnetic field B_z and the vertical current density $j_z \propto (\nabla \times \mathbf{B})_z$ or α. One can start with a potential field as an initial guess, and then successively iterate the sequence of hyperbolic equations (5.3.40) and elliptic equations (5.3.41). The problem is mathematically well-posed and the boundary conditions (B_z and α in one polarity, either for $B_z < 0$ or $B_z > 0$) are not oversampled. Applications to solar observations have been performed by Sakurai (1981). Magnetic field regions with high currents can lead to twisting and kink-unstable behavior of this code, so that this method has presently been applied only to observations with low magnetic shear (Sakurai et al. 1985).

Evolutionary Methods

An alternative technique of a nonlinear force-free calculation method is the so-called *Stress-and-relax method* (Roumeliotis 1996), which iteratively adjusts the vector potential field by stressing and relaxing phases until the transverse field at the lower boundary is optimally matched.

A variant of an integral technique is also given in Wheatland et al. (2000), where the following quantity is minimized with an evolutionary procedure,

$$L = \int_V \left[B^{-2} |(\nabla \times \mathbf{B}) \times \mathbf{B}|^2 + |\nabla \cdot \mathbf{B}|^2 \right] \, dV . \tag{5.3.42}$$

If the residual $L \mapsto 0$ converges, then obviously both conditions $(\nabla \times \mathbf{B}) \times \mathbf{B} = 0$ and $\nabla \cdot \mathbf{B} = 0$ are satisfied. A similar method minimizes the differences of the modeled to the observed tangent magnetic field on the surface (Wheatland 1999).

5.4 Magnetic Field in Active Region Corona

The theoretical models described in the previous sections provide us quantitative 3D geometries of magnetic field lines in the solar corona. From the theory it is already clear that there is no unique solution of the coronal field for any boundary condition given at the photosphere. So, the ultimate arbitors of the usefulness of these magnetic field extrapolation models are direct observations of the coronal field. Fortunately, heating of the corona is localized at the footpoints of many individual magnetic field lines, which causes upflows of heated plasma, which in turn trace out myriads of loop threads or strands, each one illuminating a set of individual magnetic field lines. The aim of this section is a quantitative comparison of such illuminated field lines, as it can be observed in EUV and soft X-rays, with theoretical field extrapolations, such as potential fields (§5.2) or force-free fields (§5.3).

5.4.1 3D Stereoscopy of Active Region Loops

If one wants to compare theoretical with observed field lines, the first problem is to overcome the limitation that observations provide only 2D projections of coronal field lines, while theoretical extrapolations provide 3D coordinates. An observed 2D projection is, therefore, ambiguous, because there is an infinite number of possible 3D curves that produce the same 2D projection along a given line-of-sight. The missing information of this 2D degeneracy can, however, be retrieved by stereoscopic observations. While truly stereoscopic observations will be available from the *STEREO* mission, which is planned to be launched in 2006, we can currently only use pseudo-stereoscopic techniques that make use of the solar rotation to vary the aspect angle. The solar rotation in heliographic longitude l, which has a differential rotation rate as a function of the latitude b, is (Allen 1973),

$$l'(t_2) = l(t_1) + \left[13.45° - 3° \sin^2(b)\right] \frac{(t_2 - t_1)}{1 \text{ day}} . \qquad (5.4.1)$$

Thus the aspect angle (from Earth) of a coronal loop varies about $13°$ per day, which is sufficient to introduce an appreciable parallax effect to constrain the 3D coordinates of its magnetic field line. The only shortcoming of such a solar rotation-based stereoscopy technique is the requirement of static structures, which seems to be a prohibitive obstacle at first glance, given the fact that the observed loop cooling times are of the order of hours (Fig. 4.12; Winebarger et al. 2003a). Fortunately, the global magnetic field configuration of an active region changes over much longer time scales (i.e., in the order of days or weeks). Therefore, the slowly varying 3D geometry of the coronal magnetic field represents a quasi-stationary system of conduits that is flushed through by heated and cooling plasma on much faster time scales. A logical consequence is therefore the concept of *dynamic stereoscopy* (Aschwanden et al. 1999a, 2000b), which makes the only assumption that the magnetic field in the neighborhood of a traced loop is quasi-stationary during the time interval of a stereoscopic correlation, but allows dynamic processes of plasma flows, heating, and cooling along the field line. The principle of dynamic stereoscopy is illustrated in Fig. 5.5.

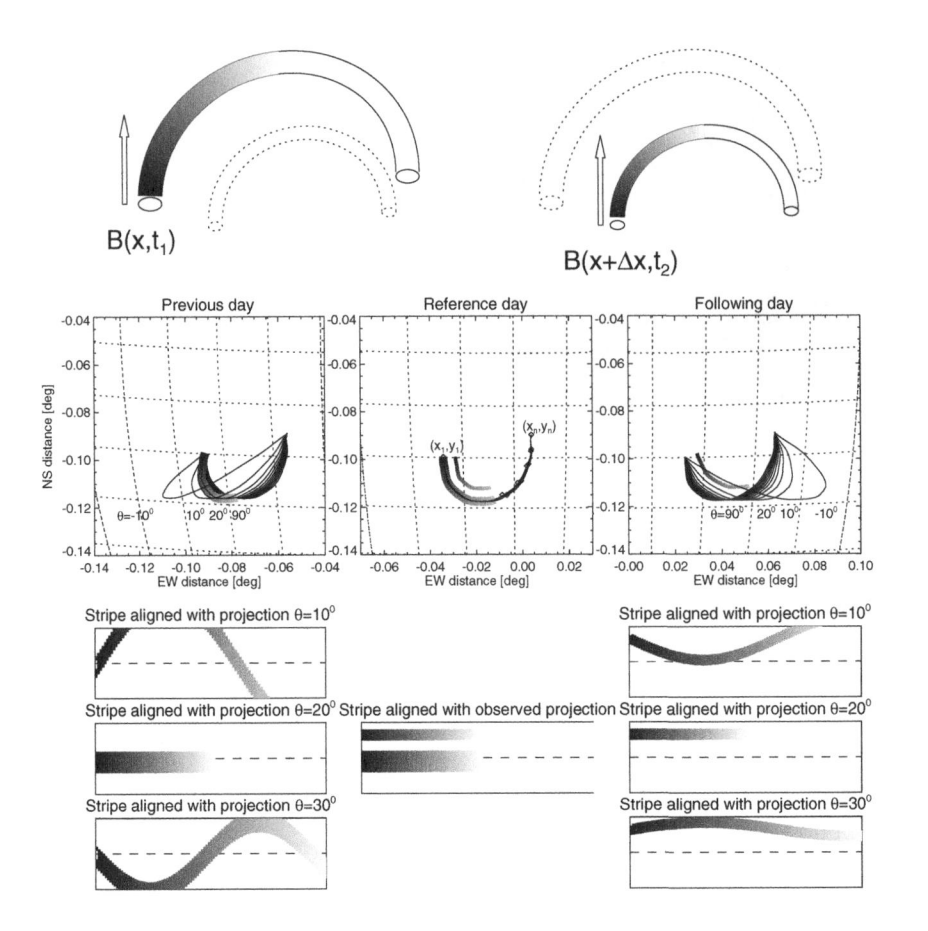

Figure 5.5: The principle of *dynamic stereoscopy* is illustrated here with an example of two adjacent loops, where a thicker loop is bright at time t_1, while a thinner loop is brightest at time t_2. From the loop positions (x_i, y_i) measured at an intermediate reference time t (i.e., $t_1 < t < t_2$; middle panel in middle row), projections are calculated for the previous and following days for different inclination angles θ of the loop plane (left and right panel in middle row). By extracting stripes parallel to the calculated projections $\theta = 10°, 20°, 30°$ (panels in bottom part), it can be seen that both loops appear only co-aligned with the stripe axis for the correct projection angle $\theta = 20°$, regardless of the footpoint displacement Δx between the two loops. The coalignment criterion can, therefore, be used to constrain the correct inclination angle θ, even for dynamically changing loops (Aschwanden et al. 1999a).

The technique of dynamic stereoscopy has been applied to 30 loops observed in 171 Å ($T \approx 1.0$ MK) with *Soho/EIT* (Aschwanden et al. 1999a), and to a set of 30 hotter loops observed in 195 Å ($T \approx 1.5$ MK) and in 284 Å ($T \approx 2.0$ MK) (Aschwanden et al. 2000b). The 3D coordinates of the traced loops were parameterized with a coplanar circular geometry, which can be expressed with 6 free parameters for each loop (the heliographic coordinates l_1, b_1 of a footpoint reference position, the height offset z_0 of

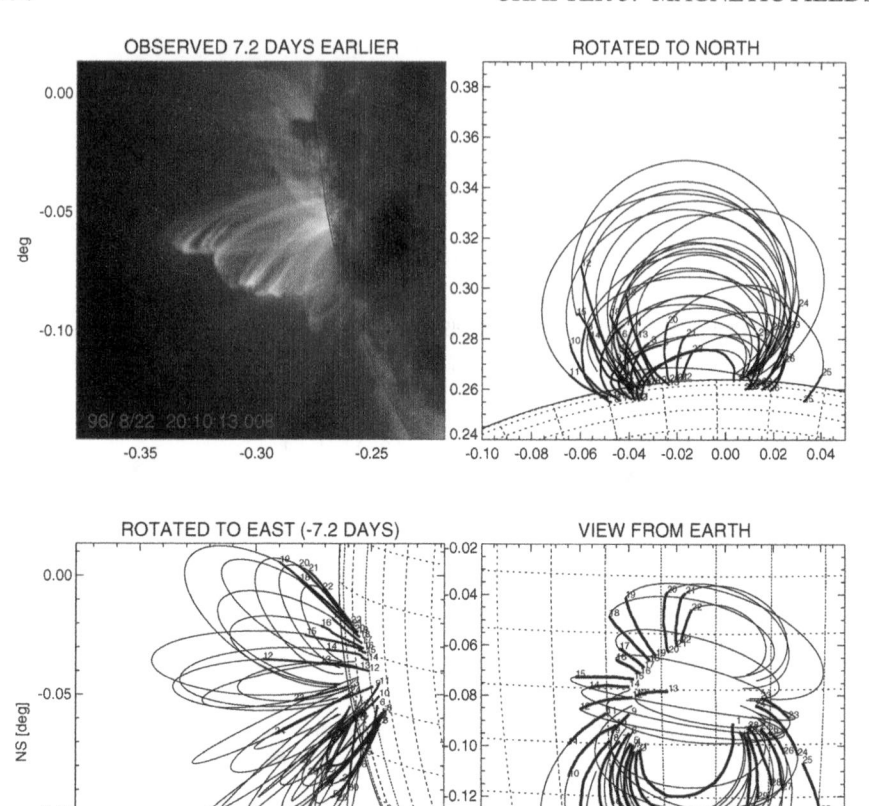

Figure 5.6: Three different projections of the stereoscopically reconstructed 30 loops of AR 7986 are shown. The loop segments that were traced from the 96-Aug-30, 171 Å image are marked with thick solid lines, while the extrapolated segments (thin solid lines) represent circular geometries extrapolated from the traced segments. The three views are: (1) as observed from Earth with l_0, b_0 (bottom right panel), (2) rotated to north by $b'_0 = b_0 - 100°$ (top right panel), and (3) rotated to east by $l'_0 = l_0 + 97.2°$ (corresponding to -7.2 days of solar rotation; bottom left panel). An *EIT* 171 Å image observed at the same time (-7.2 days earlier) is shown for comparison (top left panel), illustrating a similar range of inclination angles and loop heights as found from stereoscopic correlations a week later. The heliographic grid has a spacing of 5^0 degrees or 60 Mm (Aschwanden et al. 1999a).

the loop curvature center, the loop radius r_0, the azimuth angle α of the loop baseline, and the inclination angle θ of the loop plane) as defined in Fig. 3.15. The inclination angle θ can then be measured by stereoscopic correlation of the projected positions at two different times, using the coordinate transformations given in §3.4.4. The loops

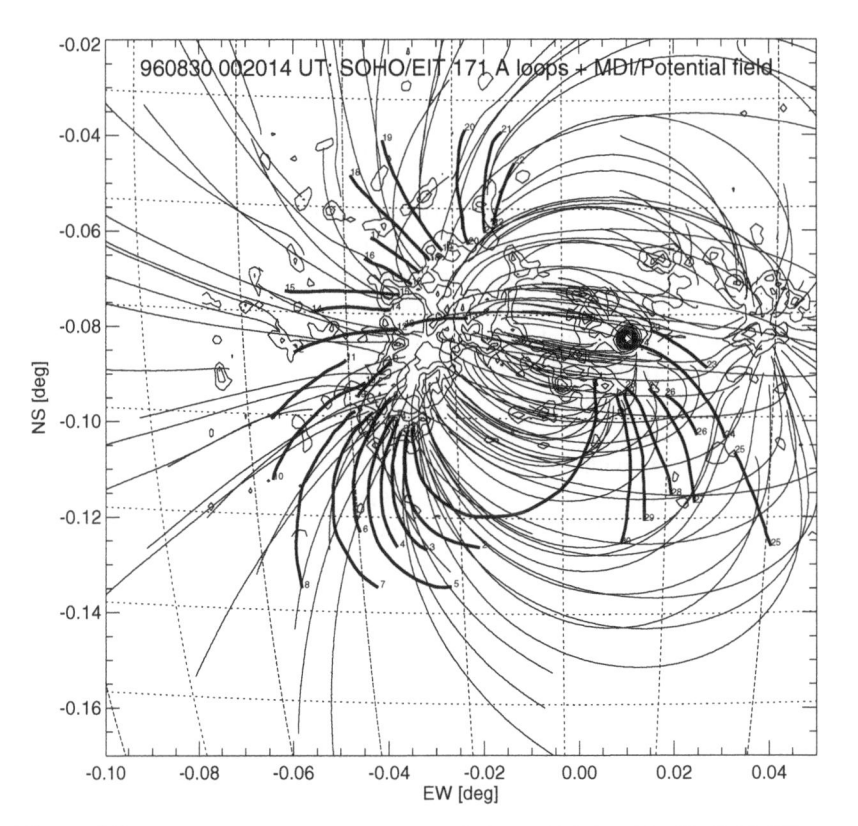

Figure 5.7: SoHO/MDI magnetogram recorded on 1996-Aug-30, 20:48:00 UT, rotated to the time of the analyzed EIT image (1996-Aug-30, 00:20:14 UT), with contour levels at $B = -350, -250, ..., +1150$ G (in steps of 100 G). Magnetic field lines calculated from a potential field model are overlaid (thin lines) onto the 30 loops (thick lines) traced from the SoHO/EIT image (Aschwanden et al. 1999a).

have been traced from a highpass-filtered image as shown in Plate 3. An example of such a stereoscopic reconstruction is shown in Fig. 5.6, which displays the solutions of the 3D coordinates in three different projections. A sanity test of the procedure can be made by rotating the stereoscopic solutions evaluated at disk center by 90 degrees to the east limb, and by comparing it with an image observed 7 days earlier at the same location (Fig. 5.6, top left panel). Apparently the long-lived loop system in this active region displayed a similar distribution of loop inclination angles, so the stereoscopic determination of loop inclination angles seems to be correct in the first order.

After we have retrieved the 3D coordinates of a set of observed coronal loops, we are primarily interested in whether they line up with a theoretical model of the 3D magnetic field. We first apply a potential field model, which has been calculated from a near-simultaneous *SoHO/Michelson Doppler Imager (MDI)* magnetogram, using the Green's function method described in §5.2.3 (Sakurai 1982). The active region displays a typical dipole field, with a group of stronger leading sunspots in the west and a trailing

region of weaker magnetic field in the east. A subset of potential field lines is shown
in Fig. 5.7 (thin lines), selected above a threshold value for the photospheric magnetic
field. From the 30 loops reconstructed in Fig. 5.6, we display only the loop segments
that actually have been traced in the EIT images (thick lines in Fig. 5.7), which roughly
correspond to the lowest density scale height up to height of $h \approx 50$ Mm, while the
upper parts of the loops could not be observed due to insufficient contrast, as expected
in a hydrostatic atmosphere (§3.1). Nevertheless, the traced loop segments clearly
reveal a mismatch to the calculated potential field in Fig. 5.7, although the overall
dipolar structure is roughly matched. The same mismatch is also present in EUV loops
traced at different temperatures, such as in 195 Å ($T = 1.5$ MK) and in 284 Å ($T = 2.0$
MK), see Plate 4. This brings us to the important conclusion that this active region has
a significant nonpotential characteristic, and therefore contains current-carrying loops.

A next step towards reality is to model the observed loop geometries with a nonpotential field model (e.g., with a force-free field, see § 5.3). This task has been conducted for the same set of field lines by Wiegelmann & Neukirch (2002), using the
stereoscopic 3D coordinates shown in Figs. 5.6 and 5.7 as geometric input. For each
of the observed EIT loops the 3D coordinates $\mathbf{r}$ have been measured, which can be
parameterized by the loop length coordinate s (i.e., $\mathbf{r}(s) = [x(s), y(s), z(s)]$) with
$0 \leq s \leq L$, where L is the full length of the loop here. In the simplest realization of
a force-free field, a magnetic field line can be parameterized by a single free parameter α (Eq. 5.3.8). Wiegelmann & Neukirch (2002) varied the parameter α for each
field line until the best match with the observed EIT loop was achieved. They explored
two different criteria to quantify the goodness of the fit. One criterion was to minimize the difference of the secondary footpoint position, at the force-free loop length
$s = s_{max}(\alpha)$ and at the observed loop length L,

$$min[\Delta_1(\alpha)] = |\mathbf{r}_{ff}(s = s_{max}, \alpha) - \mathbf{r}_{obs}(s = L)| , \qquad (5.4.2)$$

which works only when the footpoints of the observed loops $\mathbf{r}_{obs}(s = 0)$ and $\mathbf{r}_{obs}(s = L)$ are exactly known. In the example shown in Fig. 5.7, the full loop coordinates
could only be measured for the shortest loop, while the secondary footpoint of all
other 29 loops could only be estimated with poor accuracy. A more robust criterion
to fit the data is therefore to minimize the average least-square deviation between the
loop coordinates, which was quantified by the integral over the entire loop length by
Wiegelmann & Neukirch (2002),

$$min[\Delta_2(\alpha)] = \frac{1}{s_{max}^2} \int_0^{s_{max}} \sqrt{[\mathbf{r}_{obs}(s) - \mathbf{r}_{ff}(s, \alpha)]^2} \, ds . \qquad (5.4.3)$$

Since the parameter $\Delta_2(\alpha)$ is normalized, it yields a measure of the average deviation in units of the loop length. Wiegelmann & Neukirch (2002) obtained a fit with a
goodness of $\Delta_2(\alpha) \approx 3\%$ for the 4 best-fitting force-free field lines (Fig. 5.8), which
were the smallest loops, since those could be measured with highest accuracy with our
pseudo-stereoscopic method. A much higher accuracy cannot be expected at this point,
since the 3D coordinates of the pseudo-stereoscopic EIT loop observations were also
subject to the restriction of a circular and coplanar geometric model. A higher accuracy is likely to be achieved with true stereoscopy with the planned *STEREO* mission,

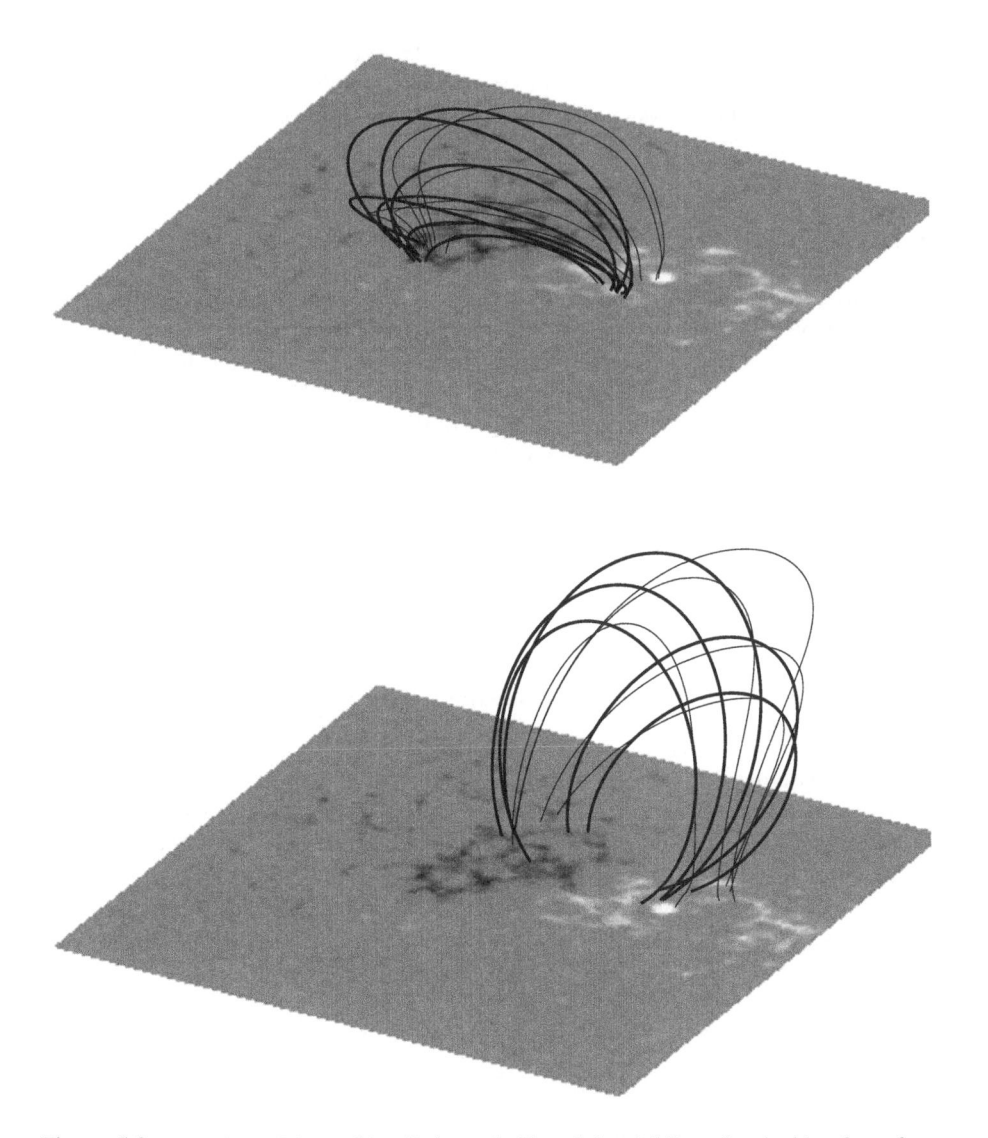

Figure 5.8: *Top:* the EIT loops #1 − 7 shown in Figs. 5.6 and 5.7 are fitted with a force-free magnetic field model for $\alpha = 2.5$. *Bottom:* the EIT loops #13, 16, 17, 20, and 21 of Figs. 5.6 and 5.7 are fitted with a force-free parameter $\alpha = -2.0$. The thin lines represent the force-free field, while the EIT loops are marked with thick lines (Wiegelmann & Neukirch 2002).

which will feature two spacecraft with an aspect angle that increases with time when the two spacecraft separate in opposite directions along the ecliptic plane.

The resulting force-free parameter was found to be $\alpha \approx 2.5$ for the group of loops with the best fits (Fig. 5.8, top), and $\alpha \approx -2.0$ for another group on the opposite side of the dipole axis (Fig. 5.8 bottom). The linear force-free field in each group was solved according to the method of Seehafer (1978), where the values of the force-free parameter α are normalized to the range of $-\sqrt{2}\pi < \alpha < \sqrt{2}\pi$. Given this limit of $\sqrt{2}\pi = 4.44$, we see that this active region exhibits a significant nonpotential field and that a force-free field provides a reasonably good approximation for the well-measured field lines.

5.4.2   Alfvén Velocity in Active Regions

For many dynamic phenomena in active regions, or in the solar corona in general, we need to know the Alfvén velocity, for instance to identify a propagating wave or an oscillation wave mode (§ 8). The Alfvén velocity v_A at a particular location $\mathbf{r}$ is a function of the local electron density $n_e(\mathbf{r})$ and magnetic field $B(\mathbf{r})$. In particular it is useful to know the height dependence $v_A(h)$ of the Alfvén velocity, because both the density and magnetic field vary strongest along the vertical direction of a magnetic field line. For the set of 30 active region loops analyzed from *SoHO/EIT* and *SoHO/MDI* data (Aschwanden et al. 1999a) we know the geometry $\mathbf{r}(h)$ from stereoscopic corre-lations (Fig. 5.6), the density $n_e(h)$ which agrees with hydrostatic models (Fig. 3.2, §3.9), and the approximate magnetic field $B(h)$ from a potential field model (Fig. 5.7). The electron density $n_e(h)$ decreases near-exponentially with height (e.g., Fig. 3.28),

$$n_e(h) \approx n_0 \exp\left(-\frac{h}{\lambda_T}\right), \tag{5.4.4}$$

where the base densities were measured as $n_0 \approx 2.0 \pm 0.5$ cm^{-3} for this set of 30 active region loops (Aschwanden et al. 1999a) and the thermal scale height is $\lambda_T = 47$ Mm $\times (T_e/1$ MK$)$. The vertical projection of the magnetic field $B(h)$ obtained from the potential field extrapolation using the *SoHO/MDI* data is shown in Fig. 5.9 (top), along with the inferred plasma β-parameter (Fig. 5.9, middle), and the Alfvén velocity v_A (Fig. 5.9. bottom). The magnetic field can be approximated with a dipole field (Eq. 5.2.15), with the far-field decreasing with the third power,

$$B(h) \approx B_0 \left(1 + \frac{h}{h_D}\right)^{-3}, \tag{5.4.5}$$

where h_D is defined as the *dipole depth*, demarcating the location of the buried mag-netic coil where the magnetic field becomes singular, $B(h = -h_D) \mapsto \infty$. The po-tential field lines shown in Fig. 5.9 (top) can closely be fitted with such a dipole model (Eq. 5.4.5), with footpoint magnetic fields in the range of $B_0 = 20, ..., 230$ G and a dipole depth of $h_D = 75$ Mm (Aschwanden et al. 1999a). If we plug the hydrostatic density model $n_e(h)$ (Eq. 5.4.4) and the magnetic dipole model $B(h)$ (Eq. 5.4.5) into the definition of the Alfvén velocity v_A (in cgs units),

$$v_A(h) = 2.18 \times 10^{11} \, \frac{B(h)}{\sqrt{n_e(h)}}, \tag{5.4.6}$$

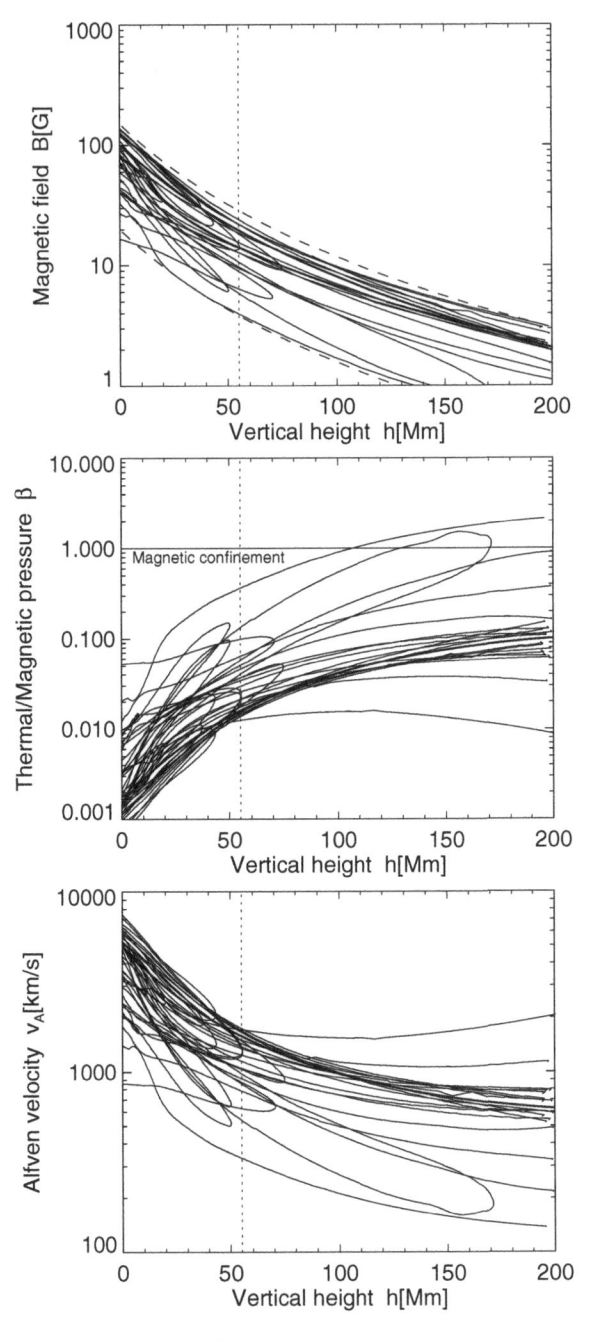

Figure 5.9: The magnetic field $B(h)$ (top), the plasma β-parameter or ratio of thermal to magnetic pressure, $\beta(h)$ (middle), and the Alfvén velocity $v_A(h)$ (bottom), determined as a function of height h for the 30 analyzed EIT loops. The magnetic field $B(h)$ is taken from the nearest potential field line (see Fig. 5.7). The vertical density scale height $\lambda = 55$ Mm is marked with a dotted line. A potential field model is indicated with dashed curves (top) (Aschwanden et al. 1999a).

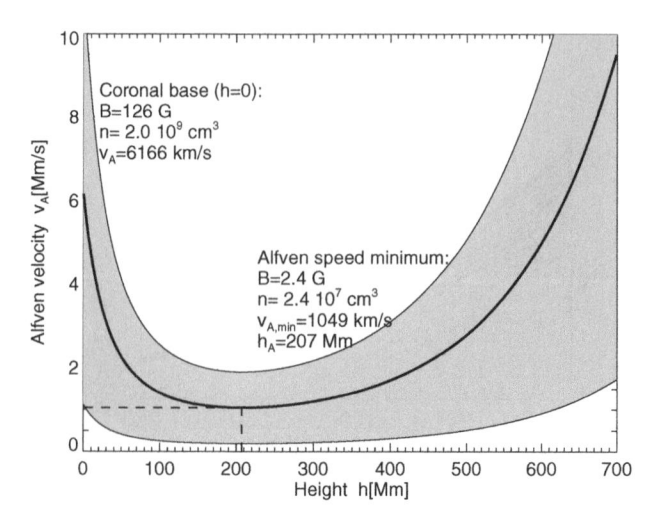

Figure 5.10: Run of the Alfvén velocity $v_A(h)$ as a function of height h, for the range of magnetic fields $B = 20 - 230$ G and density $n_e = 2.0\ 10^9$ cm^{-3} measured in 30 active region loops with SoHO/EIT (Aschwanden et al. 1999a). The average Alfvén velocity has a maximum of $v_{A0} = 6166$ km s^{-1} at the base and reaches a minimum of $v_{A,min} = 1049$ km s^{-1} at a height of $h_A = 207$ Mm.

we obtain the following height dependence

$$v_A(h) = v_{A0}\ \exp\left(\frac{h}{2\lambda_T}\right)\left(1 + \frac{h}{h_D}\right)^{-3}, \qquad (5.4.7)$$

with the constant v_{A0} at the coronal base $h \approx h_0$,

$$v_{A0} = 2180\ \left(\frac{B_0}{100\ \mathrm{G}}\right)\left(\frac{n_0}{10^{10}\ \mathrm{cm}^{-3}}\right)^{-1/2}\ \mathrm{km\ s}^{-1}. \qquad (5.4.8)$$

Both the density and magnetic field decrease with height, but the density drops faster in the lower corona, while the magnetic field drops faster in the outer corona. Interestingly, the Alfvén velocity reaches a minimum in the middle corona, which can be found from the derivative $dv_A(h)/dh = 0$ of Eq. (5.4.7). The corresponding height h_A of the minimum Alfvén velocity $v_{A,min}$ is

$$h_A = 6\lambda_T - h_D, \qquad (5.4.9)$$

which is found at $h_A = 6 \times 47 - 75 = 207$ Mm for our active region with $h_D = 75$ Mm and $\lambda_T = 47$ Mm at a temperature of $T = 1.0$ MK. The minimum value of the Alfvén velocity $v_{A,min}$ is

$$v_{A,min} = v_{A0}\ \exp\left(\frac{h_A}{2\lambda_T}\right)\left(1 + \frac{h_A}{h_D}\right)^{-3} = v_{A0}\ \exp\left(3 - \frac{h_D}{2\lambda_T}\right)\left(\frac{6\lambda_T}{h_D}\right)^{-3}.$$
$$(5.4.10)$$

Table 5.1: Alfvén velocity measurements in the lower corona (adapted from Schmelz et al. 1994).

Temperature T_e [MK]	Density $n_e/10^9$ cm^{-3}	Magnetic field $B[G]$	Alfvén speed v_A [km s^{-1}]	Reference
3.0	4.2 ± 2.0	250	7700	Nitta et al. (1991)
2.0	1.0	55	3500	Schmelz et al. (1992)
1.9	1.0	133	9800	Schmelz et al. (1992)
1.9	1.0	180	13,000	Schmelz et al. (1992)
3.1	2.6 ± 0.6	150	6000	Schmelz et al. (1994)
2.9	1.2 ± 0.9	178	10,000	Brosius et al. (1992)
2.5	1.0	$164-297$	$10-19,000$	Brosius et al. (1992)
		583	37,000	Brosius et al. (1992)
3.0	$3-8$	410	$10-16,000$	Webb et al. (1987)
$2.3-2.9$	1.0	$30-60$	$1900-3800$	Brosius et al. (1993)
1.0	2.0 ± 0.5	126	6200	Aschwanden et al. (1999a)

We show the height dependence of the Alfvén velocity $v_A(h)$ in Fig. 5.10. The plateau of the minimum Alfvén velocity extends over a quite large coronal height range. The Alfvén velocity varies less than a factor of 2 (e.g., $v_{A,min} < v_A < 2v_{A,min}$), over a height range of $0.28h_A < h < 2.14h_A$, or 57 Mm $< h <$ 443 Mm in our example. Therefore, there is a good justification to approximate the Alfvén velocity with a constant over large ranges of the solar corona. However, it should be noticed, that the Alfvén velocity tends to be significantly higher in the lower corona, which is often neglected in theoretical models. In our active region we see that the Alfvén velocity drops from 6166 km s^{-1} at the coronal base to 1049 km s^{-1} in the middle corona. In Table 5.1 we show a compilation of Alfvén velocity measurements in the lower corona, mostly inferred from the circular polarization of free-free emission in radio wavelengths (see §5.7.1).

5.4.3 Non-Potentiality of Soft X-Ray Loops

We reconstructed the 3D geometry of magnetic field lines with a stereoscopic method in §5.4.1 and found significant nonpotentiality in all investigated EUV loops of an active region. EUV loops are most suitable for this type of 3D reconstruction, because they appear as numerous thin loop strands, with the most "crispy" contrast at a temperature of $T \approx 1.0$ MK, as seen in 171 Å. Hotter active region loops are more difficult to disentangle, because the most-used instrument that is sensitive to hot plasma with temperatures of $T > 2.0$ MK is *Yohkoh/SXT*, which has very broadband temperature filters with a sensitivity that increases with temperature. Consequently, *Yohkoh/SXT* images tend to show dense, fat, and diffuse bundles of loops with many temperatures above $T \gtrsim 2.0$ MK, which are less suitable for stereoscopic 3D reconstruction. Nevertheless, because hotter plasma requires a higher heating rate than the cooler EUV loops, it is of interest to investigate the nonpotentiality of these soft X-ray loops, because this could reveal the underlying currents responsible for plasma heating. Although

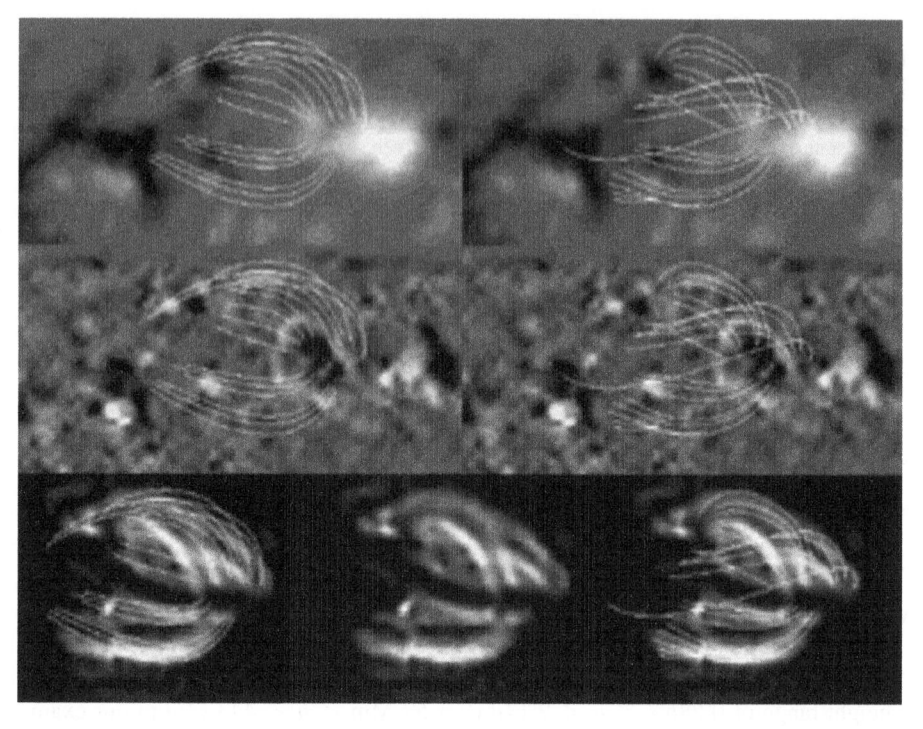

Figure 5.11: *Top:* a potential coronal field is calculated (left panels), which evolves into a force-free field (right panels). The field lines of the calculations are superimposed on a magnetogram map of positive (white) and negative (black) magnetic polarity. *Middle:* a map of vertical current density (J_z) flowing into (white) and out (black) of the corona. *Bottom:* the picture of soft X-ray loops over the same active region, taken by the Yohkoh satellite. The loops are shown with the potential field superimposed (left), no superposition (middle), and with the force-free field superimposed (right) (Jiao et al. 1997).

3D reconstruction might not be feasible, the nonpotentiality can at least be shown by matching the calculated magnetic field lines with the contours of soft X-ray loops in 2D projection.

Magnetic modeling of AR 7220/7222, observed with *Yohkoh/SXT* on 1992 July 13, was performed by Jiao et al. (1997) using vector magnetograms from Mees Observatory and longitudinal magnetograms from Kitt Peak Observatory (Fig. 5.11). The magnetic field varies between $B = -880$ G and $B = +2370$ G in this active region. The Yohkoh/SXT images shown in Fig. 5.11 were processed with a pixon method to remove the point spread function and to enhance the contrast of individual loops. A force-free magnetic field was computed with the *evolutionary MHD technique* described in Mikić & Clymont (1994) and McClymont et al. (1997), similar to the *stress-and-relax method* of Roumeliotis (1996), where the time-dependent evolution of the resistive and viscous MHD equations evolves in response to the changing boundary conditions from an initial current-free (potential) to a force-free (nonpotential) solution. The initial potential field is shown in the left panels of Fig. 5.11, clearly exhibiting a mismatch

Figure 5.12: (a) contains a Yohkoh/SXT image of a quadrupolar active region observed on 1994 Jan 4. (b) shows the line-of-sight magnetogram with positive/negative magnetic field colored in white/black. (c) displays a potential field calculation, and (d) shows a projection of the same field lines on the west limb (Gary 1997).

between the dipolar field lines and the east-west oriented soft X-ray loops. Once the time-dependent magnetic field code evolves into a force-free state as shown in the right panels of Fig. 5.11, all (selected) field lines match up with the soft X-ray loops. The current densities range from $J_z = -18$ to $+20$ mA m^{-2}. Thus, this method provides a sensitive diagnostic to measure the currents that may play a crucial role in the heating of these soft X-ray emitting loops.

A similar exercise was performed by Schmieder et al. (1996), who calculated a linear force-free magnetic field and matched it up with soft X-ray images from a *NIXT* flight on 1991 July 11 (Golub 1997). Based on the fit between the observed soft X-ray loops and the linear force-free field a differential magnetic field shear was inferred

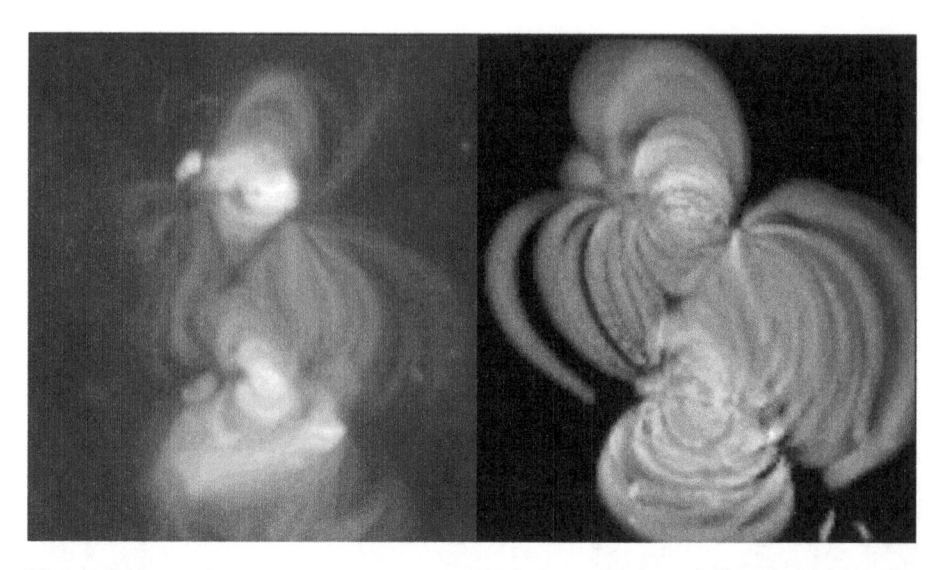

Figure 5.13: The left panel shows the same observed Yohkoh/SXT image of Fig. 5.12a, while the right panel displays an artificial image produced by rendering of calculated magnetic field lines with filling of optically thin plasma (Gary 1997).

that decreases from the low-lying inner loops towards the higher lying outer loops, suggesting a continual relaxation of the magnetic field to a lower energy state in the progressively older portions of the active region. Therefore, a linear force-free field (with a constant α) cannot describe the entire active region, the parameter α seems to decrease with height.

Another approach of comparing magnetic field models with soft X-ray loops was developed by Gary (1997), a *3D loop rendering technique*. Gary (1997) calculated a 3D potential field of the transequatorial AR 7645/7646/7647 based on a Kitt Peak magnetogram [Fig. 5.12(b)]. The quadrupolar potential field is shown in Fig. 5.12(c), as well as in a different projection rotated by $90°$ to the west limb [Fig. 5.12(d)]. A near-simultaneous soft X-ray image observed with *Yohkoh/SXT* on 1994 Jan 4, 07:35 UT, in the AlMg filter is shown in Fig. 5.12(a), covering a field-of-view of $16' \times 16'$. Gary (1997) developed a geometric code that combines bundles of magnetic field lines to fluxtubes and fills them with hot plasma of suitable density and temperature. The filling of the fluxtubes with hot plasma takes into account the exponential hydrostatic density scale height, the height dependence of gravity, and the scaling laws of Serio et al. (1981) between looptop temperature, loop length, and heating rate (see hydrostatic solutions in §3.6). The geometric code then iterates the fluxtube rendering with a non-negative least-square minimization between the synthetic soft X-ray brightness image and the observed *Yohkoh/HXT* image. The result of the *3D loop rendering technique* is shown in Fig. 5.13, which shows the line-of-sight integrated emission measure distribution of optically thin hot plasma loops that best matches the observed soft X-ray image. This forward-fitting technique allows modeling of the physical parameters within a given magnetic field model without requiring the disentangling of individ-

Table 5.2: Observed Loop width expansion factors $q_w = w_{top}/w_{foot}$.

Loop expansion q_w	Loop widths w [Mm]	Spatial resolution	Number of loops	Instrument	Refs.
1.13	4.3±0.3	4.9"	10	Yohkoh/SXT	1
1.3	4.9±2.2	2.45"	43	Yohkoh/SXT	2
1.1	7.1±0.8	2.62"	30	EIT/SOHO	3
1.20 (non-flare)	2.2±1.5	0.5"	15	TRACE	4
1.16 (postflare)	1.3±0.8	0.5"	9	TRACE	4

References: [1]) Klimchuk et al. (1992); [2]) Klimchuk (2000); [3]) Aschwanden et al. (1999a), [4]) Watko & Klimchuk (2000).

ual fluxtubes. The same method can also be used to vary the free parameters of the magnetic field model to constrain the nonpotentiality of observed loops.

5.4.4 The Width Variation of Coronal Loops

Based on the fact that coronal loops are governed by a low plasma-β parameter (Fig. 1.22), plasma can only flow along the loops, and the cross-sectional variation provides an accurate measure of whether neighboring magnetic field lines are parallel, expand, or converge. Since different magnetic field models predict specific geometries, the observed width variation of coronal loops bears a sensitive test of magnetic field models and current distributions in the solar corona.

In Fig. 5.14 we calculate the width variation that is expected for different magnetic field models: (a) for a potential field of a dipole according to the analytical model of §5.2.2, (b) for a force-free field of a sheared arcade according to the analytical model of §5.3.2, and (c) for a force-free field of a helically twisted fluxtube as modeled in §5.5.1, which has a constant cross section by definition. The loop width $w(s)$ can be calculated straightforwardly from the shortest distance between two neighboring field lines, parameterized by cartesian coordinates $z(x)$ [i.e., using Eq.(5.3.19)] for a force-free sheared arcade loop, or Eq. (5.2.22) for a potential-field dipole loop. We plot in Fig. 5.14 the calculated magnetic field lines for loops that have a width of 10% of the curvature radius and aspect ratios or $q_A = 0.5, 1.0$, and 1.5 (i.e., the ratio of the vertical to the horizontal curvature radius). The free parameters in each model are adjusted in such a way that the different magnetic field models coincide at the inner curvature radius at the footpoints and looptops. A convenient characterization of the loop width variation is the so-called *loop width expansion ratio* q_w (i.e., the ratio of the loop diameter at the top w_{top} to that at the footpoint w_{foot}),

$$q_w = \frac{w_{top}}{w_{foot}} . \qquad (5.4.11)$$

In the bottom frame of Fig. 5.14 we plot this loop width expansion factor $q_w(q_A)$ as a function of the loop aspect ratio q_A, computed numerically from the three analytical magnetic field models. The diameter of a helically twisted force-free fluxtube is constant by definition. However, a force-free fluxtube of a sheared arcade exhibits the

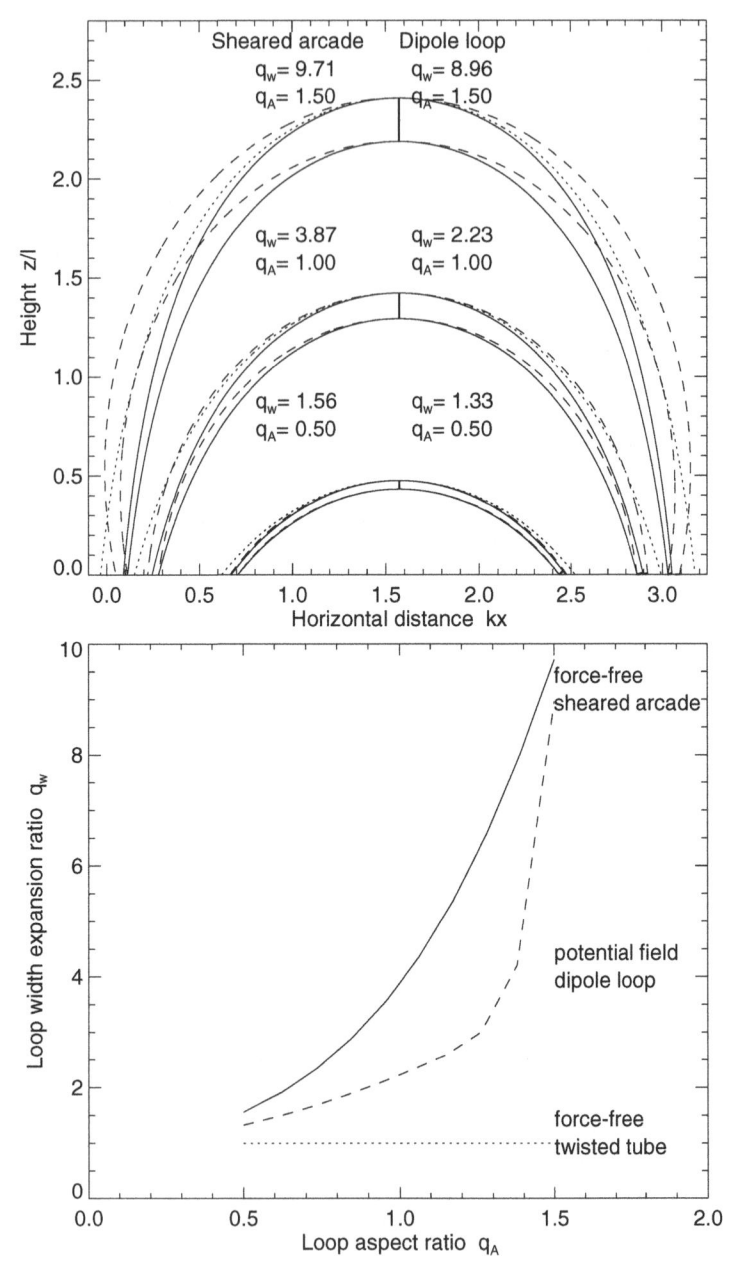

Figure 5.14: Thickness variations of coronal loops computed for three different magnetic field models: (a) potential field of dipole loop (dashed lines), (b) force-free field of sheared arcade loop (solid lines), and (c) force-free field of helically twisted loop (dotted line). *Top:* field lines for loop with a ratio of $q_r = 1.1$ between the outer and inner curvature radius. The variation of the width ratio $q_w = w_{top}/w_{foot}$ between the looptop and footpoint is listed. *Bottom:* the loop width ratio q_w as function of the geometric aspect ratio $q_A = r_2/r_1$ (i.e., the ratio of the vertical r_2 to the horizontal curvature radius r_1).

Figure 5.15: Post-flare loop system observed with TRACE 171 Å on 1998 Aug 25, 00:13:13 UT. Loop expansion factors q_w have been measured for 3 loops by Watko & Klimchuk (2000). Note the near-constancy of the loop diameters in the arcade.

largest width variation (solid line in Fig. 5.14), up to factors of $q_w \lesssim 10$ at $q_A = 1.5$. The potential field of the dipole loop shows some intermediate expansion factor. Thus we learn from these three examples that large loop expansion factors q_w occur systematically for loops with large vertical/horizontal aspect ratios $q_A > 1$, but can equally be produced by force-free as well as by potential fields, depending on the model.

Let us look now at observational measurements, which are compiled in Table 5.2. Klimchuk et al. (1992) investigated a set of 10 soft X-ray loops observed with *Yohkoh/ SXT* with half resolution (4.9") and a measured expansion factor of $q_w = 1.13$ between the looptop and footpoint, which indicates an almost constant diameter. A second study with *Yohkoh/SXT* full resolution (2.45") revealed a similar small number for the loop expansion factor, of the order $q_w \approx 1.3$ (Klimchuk 2000; Fig.17 therein). Yan & Sakurai (1997) calculated the expansion of loops from a force-free magnetic field model and obtained loop expansion factors up to $q_w \approx 4 - 6$, while the observed Yohkoh/SXT loops in arcades showed a much smaller ratio of $q_w \approx 1.0 - 1.5$. Another study on cooler EUV loops observed with *SoHO/EIT* with a resolution of 2.62" revealed a significant expansion only for 2 out of 30 loops, while the overall range was $q_w \approx 1.1\pm0.2$ (Fig. 8 in Aschwanden et al. 1999a). However, because the cooler EUV loops have a shorter hydrostatic scale height than the hotter soft X-ray loops, the expansion could only be measured over the lowest vertical scale height, which thus constitutes a lower limit for the expansion factor at the looptop. Measurements of the loop expansion factors with the highest spatial resolution have then been made with TRACE having a pixel size of 0.5" (Watko & Klimchuk 2000). An example of such an analyzed postflare loop system in shown in Fig. 5.15, where the near-constancy of loop cross sections can be seen clearly. The effective loop widths were determined by correcting also for the TRACE point spread function with a FWHM of 1.25" (Golub et al. 1999). However, no large loop expansion factors were measured, either for postflare loops ($q_w = 1.20$) or for non-flare loops ($q_w = 1.16$). Although longer loops are expected theoretically to have greater expansion than shorter loops (Klimchuk 2000), Watko & Klimchuk (2000) found no such correlation between the loop expansion factor q_w and loop length L.

Given the consistent observational result that loop expansion factors hover in the neighborhood of $q_w \approx 1.1 - 1.2$ (Table 5.2), which is significantly less than predicted for a dipolar field ($q_w \approx 2$ for a semi-circular loop) or for sheared arcade loops ($q_w \approx 4$), we are left with the interpretation in terms of current-carrying, helically twisted fluxtubes, for which force-free solutions exist with constant loop cross sections. Ampère's law (Eq. 5.1.9) implies that loops with constant cross sections $A(s)$ have a constant magnetic field $B(s)$, since the magnetic flux $\Phi = B(s) \cdot A(s)$ is conserved along a loop.

There exist analytical models and numerical simulations on the effect of twist on the structure of straight axis-symmetric fluxtubes (Parker 1977, 1979; Browning & Priest 1983; Zweibel & Boozer 1985; Craig & Sneyd 1986; Steinolfson & Tajima 1987; Browning 1988; Browning & Hood 1989; Lothian & Hood 1989; Mikić et al. 1990; Robertson et al. 1992; Bellan 2003), which can explain some twist-induced constriction of fluxtubes, but in many analytical models the assumption of an initial straight fluxtube is made, which defeats the purpose to explain its existence. Twisted magnetic fluxtubes embedded within a much larger untwisted dipole configuration have been numerically calculated (Klimchuk et al. 2000), which indeed reproduced a reduction of the expansion factor for side views, but produced increased expansion for a view from above (Fig. 5.16). The two effects almost compensate, leading to no significant change of the average expansion factor. However, an important systematic bias that comes into play is that the constriction occurs strongest near the footpoints, which are obscured for views from above (e.g., Fig. 5.15), and may even be outside the sensitivity range of a given narrowband filter, due to the temperature decrease between the lower corona and the transition region. Therefore, the instrumental response function (§3.8) has to be included in comparisons of theoretical models with observations.

Another property that has been observed is that the loop widths do not vary periodically as conceivable for a twisted 2D ribbon stripe, but vary monotonically, indicative of a circular shape of the loop envelope. This geometric property speaks against coronal heating in 2D sheets, as expected from magnetic reconnection models (e.g., Galsgaard & Nordlund, 1996). Therefore, it was concluded that loops with constant cross sections require axial-symmetric heating, dissipation, or transport mechanisms (Klimchuk 2000). A natural mechanism that fulfills this condition is upflow of heated plasma from

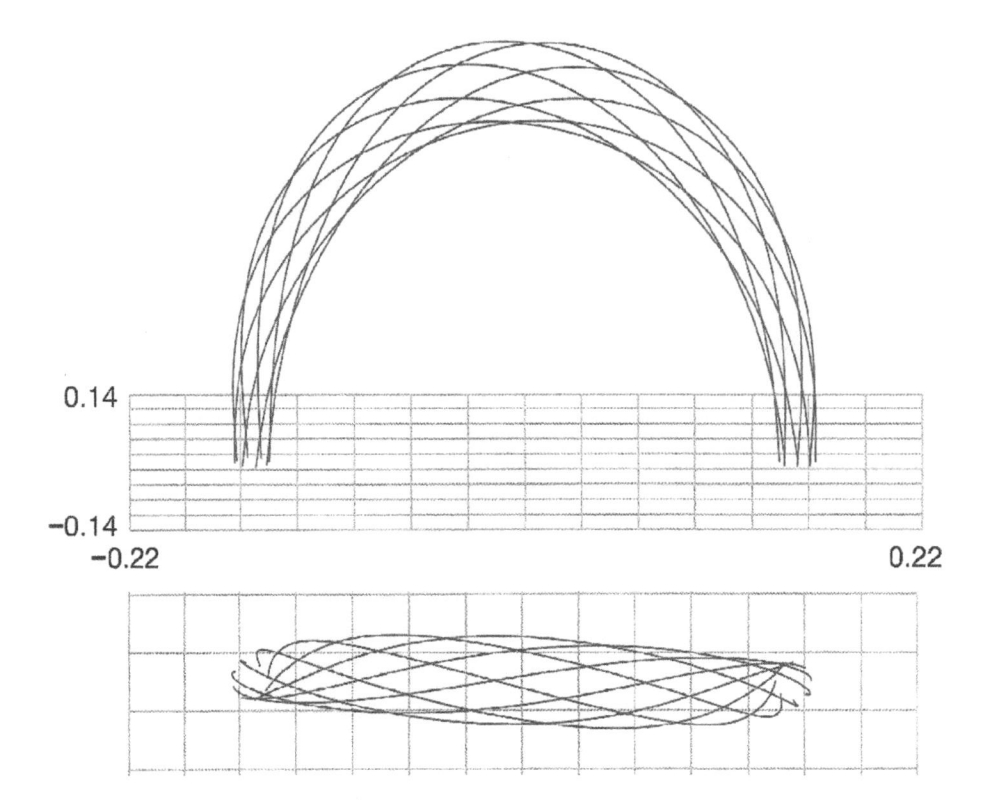

Figure 5.16: A coronal loop embedded within a large dipole configuration is twisted by an angle of 2π, viewed from the side (at 15° elevation angle) and from the top. Note that the loop expansion with height is more pronounced in side views than in top views (Klimchuk et al. 2000).

below the magnetic canopy (located at the height of the transition region) as envisioned in siphon-flow models (Fig. 4.3), where the canopy acts as a nozzle and distributes the upflowing plasma axis-symmetrically over an expanded cross section of the order of a granulation scale ($\gtrsim 1$ Mm).

5.5 Magnetic Helicity

Magnetic helicity is a measure of the topological structure of the magnetic field, particularly suited to characterize helically twisted, sheared, (inter-)linked, and braided field lines. The concept of magnetic helicity became a focus recently because of: (1) the helicity conservation (especially since magnetic helicity conserves better than the magnetic energy), (2) the importance of magnetic helicity for the effectiveness of the solar dynamo (since non-helical dynamos are generally not very effective), and (3) the role of helicity in magnetic reconnection and stability (since co-helicity and counter-helicity reconnection have different reconnection rates and energy outputs). In a plasma with high electric conductivity (σ) or high Reynolds number R_m (Eq. 5.1.15), magnetic he-

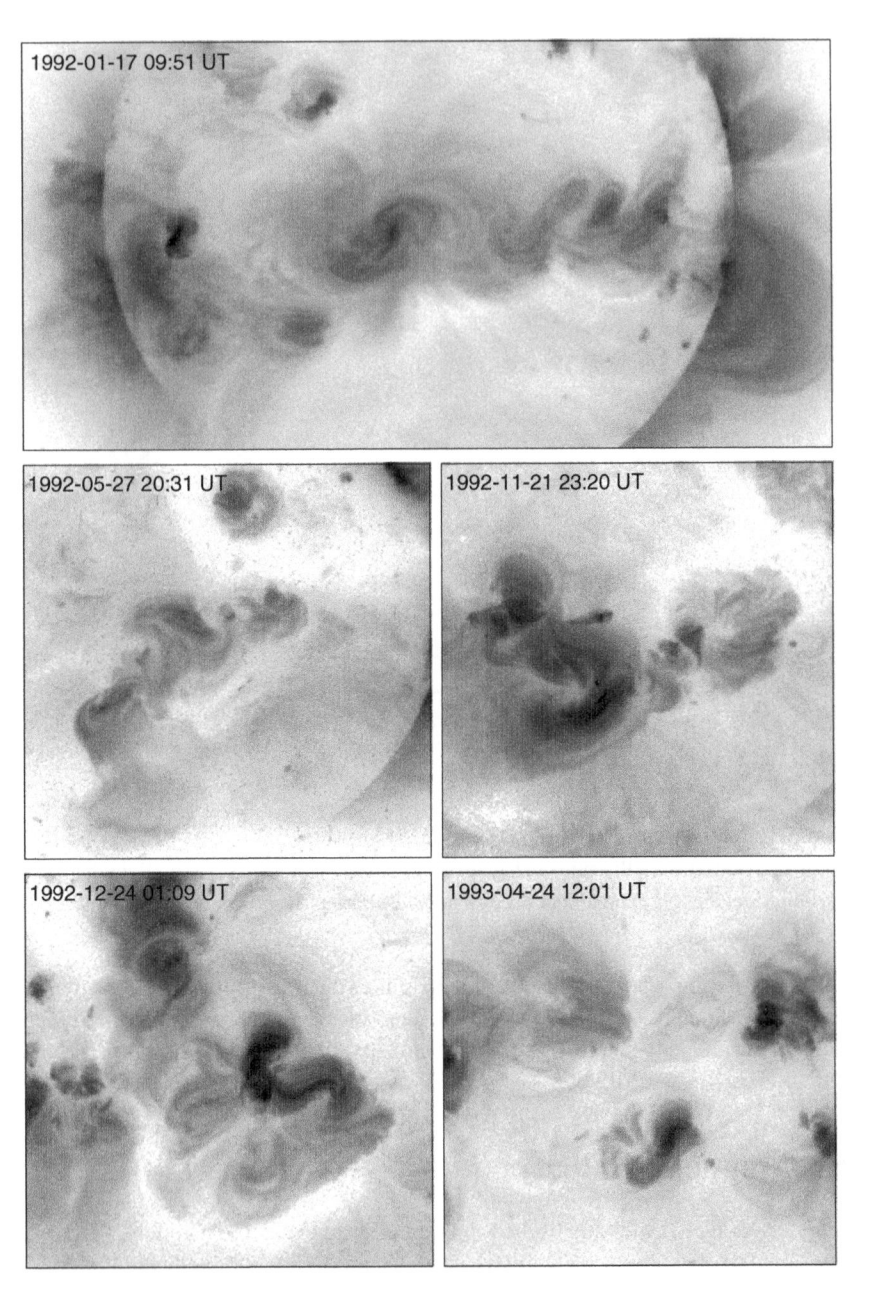

Figure 5.17: Sigmoid-shaped loop structures observed with *Yohkoh/SXT*: on 1992 Jan 17, 09:51 UT (top), 1992 May 27, 20:31 UT (middle left), 1992 Nov 21, 23:20 UT (middle right), 1992 Dec 24, 01:09 UT (bottom left), and 1993 Apr 24, 12:01 UT (bottom right). All images are plotted on the same spatial scale, to the scale of the solar diameter visible in the top frame. The greyscale is on a logarithmic scale and inverted (dark for bright emission). Note that many of the sigmoid structures consist of segmented loops that do not extend over the entire sigmoid length, but fit the overall curvature of the sigmoid shape.

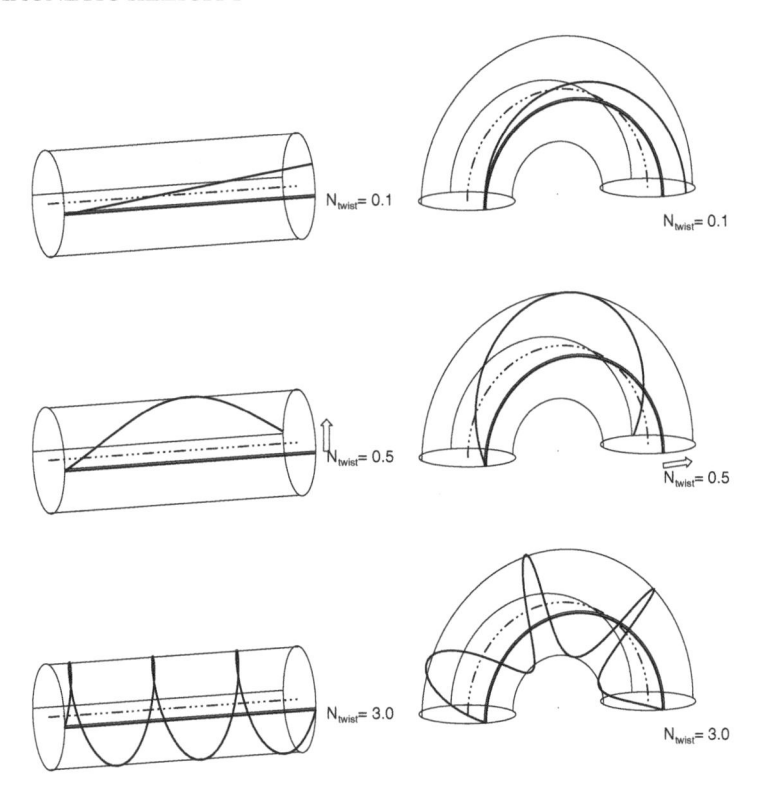

Figure 5.18: The geometry of twisted fluxtubes is visualized for straight fluxtubes (left) and for semi-circular fluxtubes (right). Twisted magnetic field lines (thick linestyle) are shown on the surface of the cylindrical fluxtubes for a small twist by $N_{twist} = 0.1$ (top panels), moderate twist by $N_{twist} = 0.5$, and large twist by $N_{twist} = 3.0$ turns. The twist is a measure of the helicity of magnetic field lines.

licity is conserved to a high degree, even during magnetic reconnection processes. The magnetic helicity could clearly be measured from the (forward or backward) *S-shaped loops* seen in *Yohkoh/SXT* images (Fig. 5.17), which are also called *sigmoids*, due to their similarity with the greek letter σ. Thematic publications on magnetic helicity can be found, for examplie, in Brown et al. (1999) and Buechner & Pevtsov (2003).

5.5.1 Uniformly Twisted Cylindrical Force-Free Fluxtubes

A common geometrical concept is to characterize coronal loops with cylindrical flux-tubes. For thin fluxtubes, the curvature of coronal loops and the related forces can be neglected, so that a cylindrical geometry can be applied. Because the footpoints of coronal loops are anchored in the photosphere, where a random velocity field creates vortical motion on the coronal fluxtubes, they are generally twisted. We consider now such twisted fluxtubes (Fig. 5.18) and derive a relation between the helical twist and the force-free parameter α (Eq. 5.3.6) (see also Sturrock 1994, § 13.7, or Boyd &

Sanderson 2003, § 4.3.4).

We consider a straight cylinder where a uniform twist is applied (Fig. 5.18 left), so that an initially straight field line $\mathbf{B} = (0, 0, B_z)$ is rotated by a number N_{twist} of turns over the cylinder length l, yielding an azimuthal field component B_φ,

$$\frac{B_\varphi}{B_z} = \frac{r\partial\varphi}{\partial z} = \frac{r 2\pi N_{twist}}{l} = br \ . \tag{5.5.1}$$

The fluxtube can be considered as a sequence of cylinders with radii r, each one twisted by the same shear angle $\partial\varphi/\partial z = 2\pi N_{twist}/l$. Obviously, the magnetic components depend only on the radius r, but not on the length coordinate z or azimuth angle φ. Thus, the functional dependence is

$$\mathbf{B} = [B_r, B_\varphi, B_z] = [0, B_\varphi(r), B_z(r)] \ . \tag{5.5.2}$$

Consequently, the general expression of $\nabla \times \mathbf{B}$ in cylindrical coordinates,

$$\nabla \times \mathbf{B} = \left[\frac{1}{r}\frac{\partial B_z}{\partial \varphi} - \frac{\partial B_\varphi}{\partial z}, \frac{\partial B_r}{\partial z} - \frac{\partial B_z}{\partial r}, \frac{1}{r}\left(\frac{\partial}{\partial r}(rB_\varphi) - \frac{\partial B_r}{\partial \varphi} \right) \right] , \tag{5.5.3}$$

is simplified (with Eq. 5.5.2), yielding a force-free current density $\mathbf{j}$ of

$$\mathbf{j} = [j_r, j_\varphi, j_z] = \frac{1}{4\pi}(\nabla \times \mathbf{B}) = \frac{1}{4\pi}\left[0, -\frac{\partial B_z}{\partial r}, \frac{1}{r}\left(\frac{\partial}{\partial r}(rB_\varphi) \right) \right] . \tag{5.5.4}$$

Requiring that the Lorentz force is zero, $\mathbf{F} = 0$, we obtain a single non-zero component in the r-direction, since $j_r = 0$ for the two other components,

$$\mathbf{F} = \mathbf{j} \times \mathbf{B} = [B_z j_\varphi - B_\varphi j_z, 0, 0] \ , \tag{5.5.5}$$

yielding a single differential equation for B_z and B_φ,

$$B_z \frac{dB_z}{dr} + B_\varphi \frac{1}{r}\frac{d}{dr}(rB_\varphi) = 0 \ . \tag{5.5.6}$$

By substituting $B_\varphi = brB_z$ from Eq. (5.5.1) this simplifies to

$$\frac{d}{dr}\left[(1 + b^2 r^2)B_z \right] = 0 \ . \tag{5.5.7}$$

A solution is found by making the expression inside the derivative a constant, which yields B_z, and B_φ with Eq. (5.5.1),

$$\mathbf{B} = [B_r, B_\varphi, B_z] = \left[0, \frac{B_0 \, br}{1 + b^2 r^2}, 0, \frac{B_0}{1 + b^2 r^2} \right] . \tag{5.5.8}$$

With the definition of the force-free α-parameter (Eq. 5.3.6) we can now verify that the α-parameter for a uniformly twisted fluxtube is,

$$\alpha(r) = \frac{2b}{(1 + b^2 r^2)} \ , \qquad b = \frac{2\pi N_{twist}}{l} \tag{5.5.9}$$

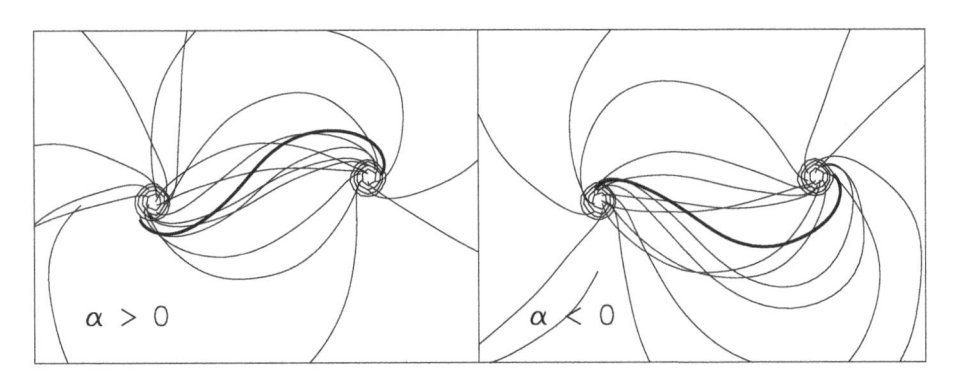

Figure 5.19: A model of a bipolar linear (constant-α) force-free magnetic field, computed for positive (left frame) and negative α (right frame). Contours show the vertical magnetic field strength B_z. An arbitrary field line near the central part of the dipole is marked (thick line) to visualize the sense of twist, i.e., a *forward S-shape* for $\alpha > 0$ and a *backward S-shape* for $\alpha < 0$ (Pevtsov et al. 1997).

For instance, for one of the two twisted fluxtubes shown in Fig. 5.18 with $N_{twist} = 0.5$ turn, we obtain, for a length $l = 10^9$ cm and tube radius $r = 0.25 \times 10^9$ cm, a numerical value of $b = 3.1 \times 10^{-9}$ cm^{-1} and a force-free parameter of $\alpha = 3.9 \times 10^{-9}$ cm^{-1}. The shear angle between the untwisted and the twisted field line is $\tan\theta = B_\varphi/B_z = br = 2\pi N_{twist} r/l = 45^0$. This way, the geometric shear angle θ, which can observationally be measured from twisted coronal loops, can be used to estimate the force-free α-parameter (Fig. 5.19). Note that we derived a similar relation for the shear angle θ of a sheared arcade, where we found (Eq. 5.3.18) that the shear angle θ of the arcade is proportional to the force-free parameter α for small angles (i.e., $\tan\theta = B_y/B_x = \alpha/l$).

5.5.2 Observations of Sigmoid Loops

In principle, the appearance of a sinuous-shaped magnetic field line does not rule out a magnetic potential field. An offset pair of dipoles with an oblique angle with respect to the mid-axis are connected by sinuous shaped field lines, as shown in Figs. 5.19 and 5.20. Such sinuous field lines can thus be produced by a current-free potential field. However, given the ubiquitous nonpotentiality of the coronal magnetic field we encountered already in most of the EUV (§ 5.4.1) and soft X-ray loops (§ 5.4.2), a more likely interpretation is that coronal structures are twisted because of current systems of sub-photospheric origin, which enter the corona as vertical currents in the photosphere. These vertical currents j_z,

$$j_z \propto (\nabla \times \mathbf{B})_z = \left(\frac{dB_y}{dx} - \frac{dB_x}{dy} \right) \propto \alpha B_z \,, \tag{5.5.10}$$

can then be measured from the observed horizontal magnetic field components (B_x, B_y), specified by an α.

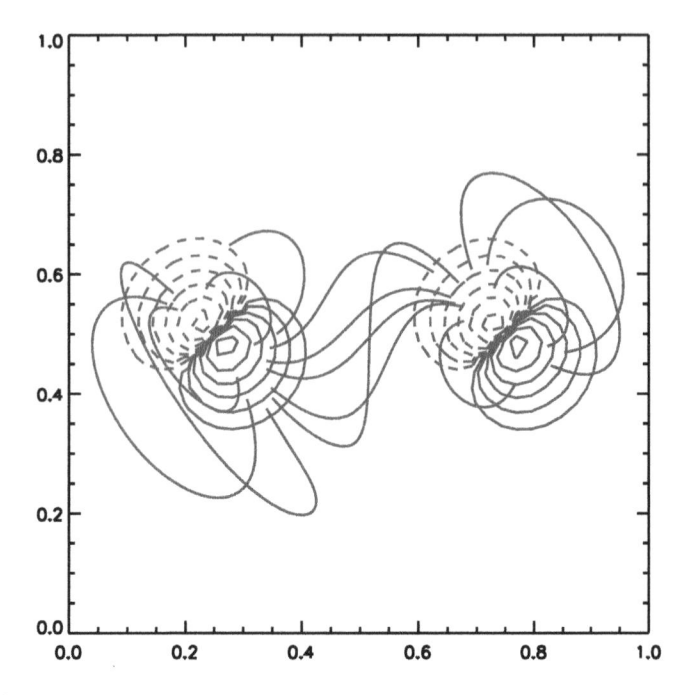

Figure 5.20: Magnetic potential field of an offset pair of dipoles. Contours show vertical magnetic field. Field lines connecting dipoles show distinct forward-S shapes, even for a current-free potential field (Pevtsov et al. 1997).

The interpretation of sheared and twisted field lines in terms of a current-carrying force-free field can be proven by testing the proportionality between the geometric shear angle θ and the force-free parameter α. This test was performed in the study of Pevtsov et al. (1997), where the force-free parameter α was measured for photospheric magnetograms in 140 active regions, and was compared with the geometric shear angle θ of individual soft X-ray loops observed with *Yohkoh/SXT*. A strong correlation was found between α and θ in active regions with a dominant α-map (Fig. 5.21), which led to the conclusion that coronal electric currents extend down to the photosphere, and probably to subphotospheric currents according to other studies (Lites et al. 1995; Leka et al. 1996).

5.5.3   Conservation of Helicity

Helically wounded coils provide a magnetic inductance. The energy of n circuits carrying currents $I_1, ..., I_n$ is proportional to $\sum_{i=1}^{n} \sum_{j=1}^{n} M_{ij} I_i I_j$, where M_{ii} is the self-inductance and M_{ij} is the mutual inductance. By analogy we can define the helicity H between fluxtubes that carry magnetic fluxes Φ_i, ($i = 1, ..., n$), and are interlinked by a number of L_{ij} turns,

$$H = \sum_{i=1}^{n} \sum_{j=1}^{n} L_{ij} \Phi_i \Phi_j \,, \qquad (5.5.11)$$

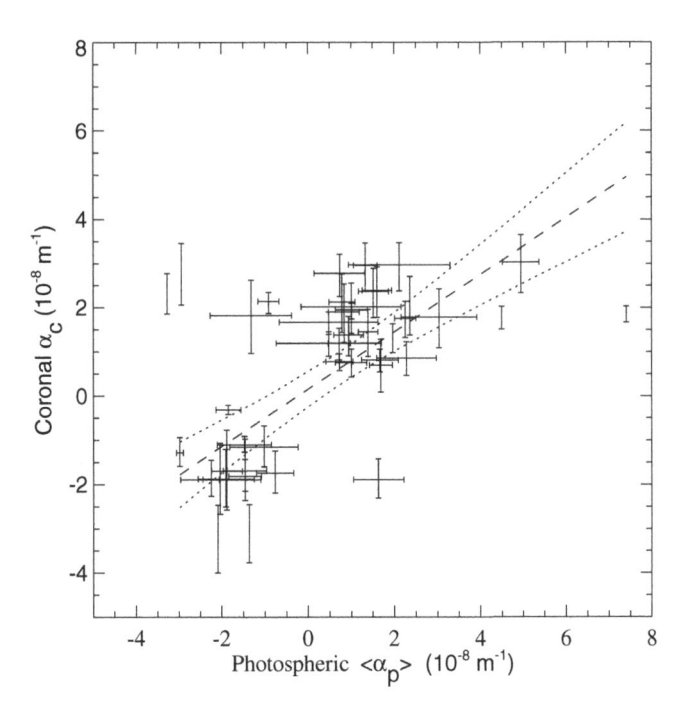

Figure 5.21: Observed force-free field parameter α_p from photospheric vector magnetograms and shear angle α_c (corresponding to the angle θ in Eq. 5.4.12) of coronal loops for 44 active regions with a dominant sign of α_p. A linear correlation is indicated with a dashed line (as predicted by Eq. 5.4.12 (i.e., $\tan\theta \propto \alpha$), and the 2-$\sigma$ error band is marked with dotted lines (Pevtsov et al. 1997).

where L_{ii} ("*self-helicity*") represents the number of helical twists within the same flux-tube i, and L_{ij} ("*mutual helicity*") represents the number of helical turns of fluxtube i around another fluxtube j. In the limit of $n \mapsto \infty$ the discrete double-summation in the helicity H can be transformed into a double-integral, provided that the volume V is bounded by a magnetic surface (with $\mathbf{n} \cdot \mathbf{B} = 0$),

$$H = \int_V \mathbf{A} \cdot \mathbf{B} \, dV \, , \tag{5.5.12}$$

where $\mathbf{A}$ is the vector potential ,

$$\mathbf{A}(\mathbf{x}) = -\frac{1}{4\pi} \int_V \frac{(\mathbf{x} - \mathbf{x}') \times \mathbf{B}(\mathbf{x}')}{|\mathbf{x}' - \mathbf{x}|^3} dV' \, , \tag{5.5.13}$$

satisfying

$$\nabla \times \mathbf{A} = \mathbf{B} \, , \tag{5.5.14}$$

$$\nabla \cdot \mathbf{A} = 0 \, . \tag{5.5.15}$$

The vector potential is defined only to within a gauge transformation,

$$\mathbf{A} \mapsto \mathbf{A}' = \mathbf{A} + \nabla\Phi \, , \tag{5.5.16}$$

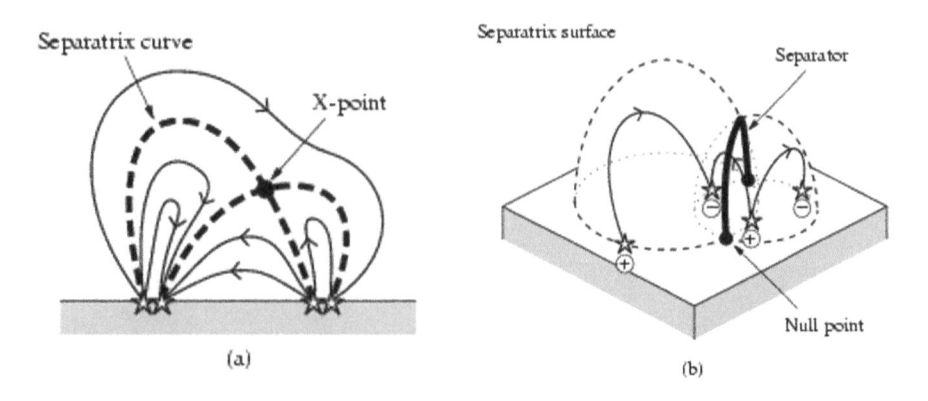

Figure 5.22: The definition of X-points or nullpoints, separatrix surfaces, and separator lines in the context of 2D concepts (left) and 3D magnetic topologies (courtesy of Eric Priest).

which does not change the helicity H, if the volume V is bounded by a magnetic surface (see derivation in Sturrock 1994, §13.8).

From the analytical model of a uniformly twisted fluxtube we have seen that the force-free parameter α is uniquely related to the number of twists N_{twist} (Eq. 5.5.9), which is a measure of the helicity. Therefore, we can conclude that the helicity H is conserved for a linear force-free field (with constant parameter α) for an expanding fluxtube with anchored footpoints. A more formal derivation can be seen in Sturrock (1994; § 13.9) or in Bellan (2002; § 3), also known as *Woltjer's theorem* (1958). The conservation of helicity is an important invariant during the evolution of coronal structures, which can be applied to active region loops (e.g., Liu et al. 2002; Kusano et al. 2002), flare loops (e.g., Pevtsov et al. 1996; Moon et al. 2002b), filaments (e.g., Chae 2000; Pevtsov 2002), prominences (e.g., Shrivastava & Ambastha 1998; DeVore & Antiochos 2000), magnetic fluxropes and interplanetary magnetic clouds (e.g., Kumar & Rust 1996; Cid et al. 2001), etc. We will deal with such dynamic phenomena in the sections on MHD (§ 6) and magnetic field changes during flares (§ 9).

5.6 Magnetic Nullpoints and Separators

5.6.1 Topological Definitions

Magnetic nullpoints are single-point locations where the magnetic field vanishes, $\mathbf{B} = (0, 0, 0)$. Magnetic topologies with such singular points are common in the presence of multiple magnetic sources. The magnetic field of a buried dipole as shown in Fig. 5.2 has no such singular point in the corona, but as soon as a second dipole with a different orientation is brought into proximity or emerges out of the surface, a coronal nullpoint will occur between the two dipoles where the oppositely directed field components cancel (Fig. 5.22, left). In the neighborhood of the X-point or nullpoint one sees that the magnetic field lines connect to different topological domains, which are separated by *separator curves* in 2D (Fig. 5.22, left) or by *separatrix surfaces* in 3D

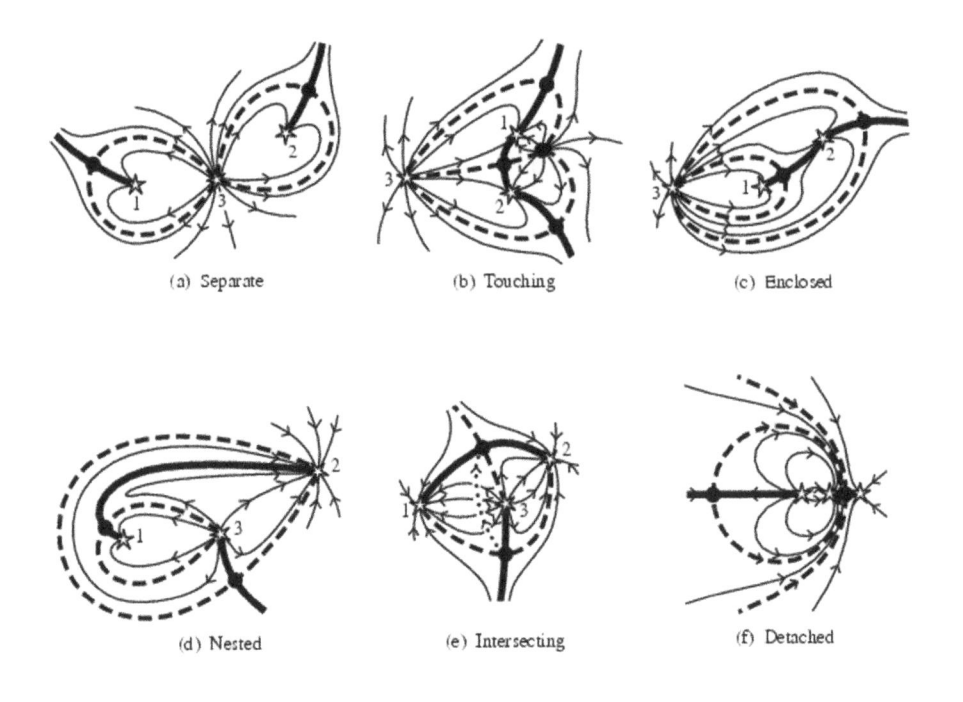

Figure 5.23: Complex 3D magnetic topologies created by three sources. Some of the eight structurally stable configurations that can be formed are shown, with magnetic sources marked with stars and separatrix curves with dashed lines (Priest et al. 1997).

(Fig. 5.22 right). A topological domain is defined by a family of magnetic field lines that connect to the same conjugate pair of magnetic polarities. In the quadrupolar configuration shown in Fig. 5.22 we have a pair of two dipoles, which we can number as 1 and 2, each one having a positive (+) and a negative (−) pole. The four topological domains confine families of field lines which either connect (a) 1+ with 1−, (b) 1+ with 2−, (c) 2+ with 2−, or (d) 2+ with 1−. In the presence of three sources, eight structurally stable configurations can be formed (Priest et al. 1997), of which a subset is shown in Fig. 5.23. The intersection of two separatrix surfaces is called a *separator* line, which is marked with thick linestyle in the examples of Figs. 5.22 and 5.23. Separators can also be found as magnetic field lines that connect two nullpoints. More extensive discussions of topologies with magnetic nullpoints can be found in Priest & Forbes (2002), Priest & Schrijver (1999), Brown & Priest (1999, 2001), or Longcope & Klapper (2002).

There are two different types of magnetic neutral points, one at the intersection of separatrix curves (X-point), and one at the center of a magnetic island (O-point). A simple mathematical model of the magnetic field at such neutral points is given in Priest & Forbes (2002, §1.3.1), with a vector potential **A** expressed as a quadratic function of

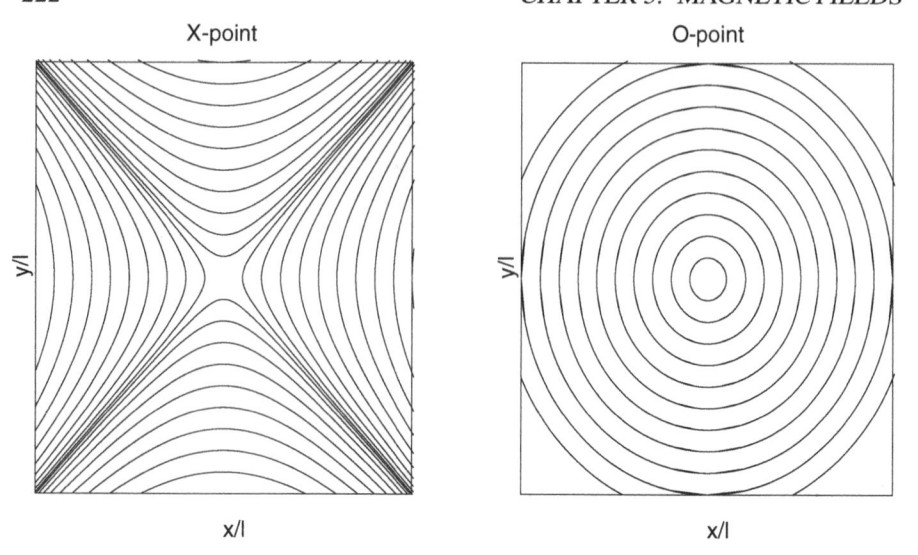

Figure 5.24: 2D X-point (left) and O-point (right). Iso-contours of the magnetic field B are calculated from $A = 0.1, ..., 0.9$ with $a = \pm 1$ with Eqs. (5.6.3).

the x and y-coordinates,

$$\mathbf{A} = (0, 0, A_z) = \left[0, 0, \frac{B_0}{2L_0}(y^2 - ax^2)\right] , \qquad (5.6.1)$$

which has elliptic functions as solutions for O-points (if $a < 0$), and hyperbolic functions as solutions for X-points (if $a > 0$). The resulting magnetic field $\mathbf{B}$ follows from this vector potential $\mathbf{A}$ as

$$\mathbf{B} = (B_x, B_y, B_z) = \nabla \times \mathbf{A} = (\frac{dA_z}{dy}, -\frac{dA_z}{dx}, 0) , \qquad (5.6.2)$$

and has the explicit solutions

$$\mathbf{B} = (B_x, B_y, B_z) = (B_0 \frac{y}{L_0}, B_0 a \frac{x}{L_0}, 0) , \qquad (5.6.3)$$

With L_0 being the length scale over which the magnetic field varies. We see that the magnetic field vanishes at the position $(x, y) = (0, 0)$. Magnetic field lines are defined by $A = constant$, plotted for $A = 0.1, ..., 0.9$ and $a = \pm 1$ in Fig. 5.24.

5.6.2 Observations of Coronal Nullpoints

We expect to observe coronal magnetic nullpoints in quadrupolar configurations, in the midpoint region between the four magnetic poles. Such an almost symmetric situation is shown in Fig. 5.12, observed with Yohkoh/SXT on 1994 Jan 4, and modeled with a 3D potential field in Fig. 5.13. Other examples are given in Tsuneta (1996d), or in Fig. 5.25, where a bundle of loops emanating from one magnetic polarity group in the

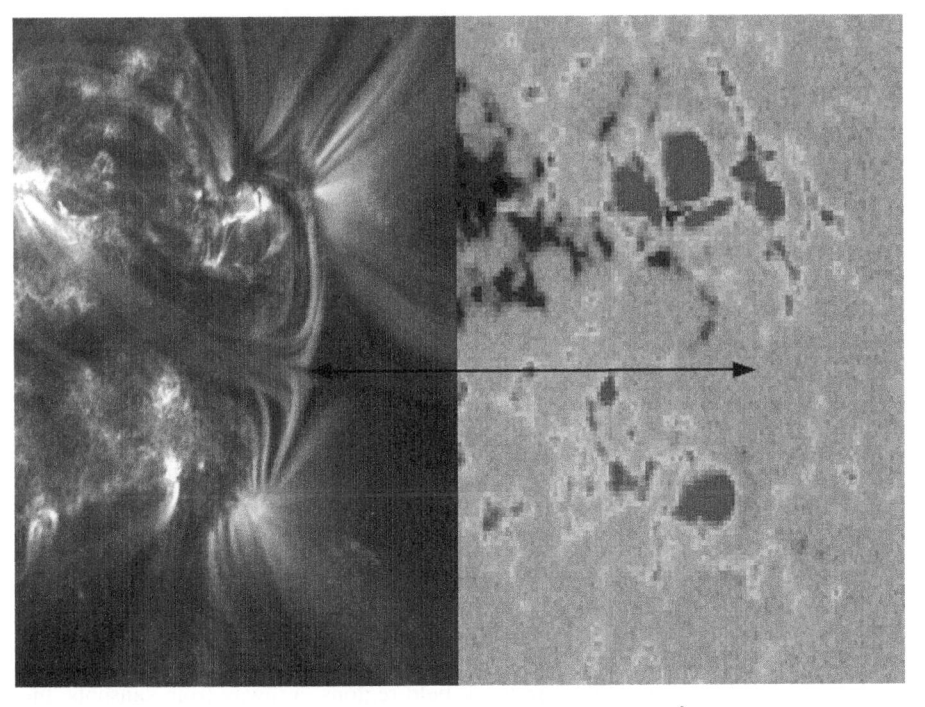

Figure 5.25: Coronal magnetic nullpoint observed by TRACE 171 Å (left) and SoHO/MDI magnetogram (right). The 171 Å image, taken on 2000-Sep-4, 10:17 UT, shows the corona between two sunspots of equal polarity in active regions 9149 (north) and 9147 (south). Between the spots, the loops meet and are deflected sideways, forming a so-called X-point in the magnetic field. The leftmost half of the field shows up clearly, but the rightmost half has a different temperature and is only vaguely visible.

east bifurcates and connects two sunspots, a northern and southern, almost symmetric pair, with equal magnetic polarity and similar field strength in the west. A separator surface can be imagined at the bifurcation line between the northern and southern bundle of loops. Probably, the complete magnetic configuration is even quadrupolar, but field lines on the west side of the magnetic nullpoint (marked with an arrow in Fig. 5.25) are not illuminated with heated plasma at the right temperature to be detected with TRACE 171 Å. We see that the magnetic field lines are hyperbolically curved near the X-point, similar to the simple mathematical model visualized in Fig. 5.24 (left).

What role do magnetic nullpoints play in the solar corona? Usually, an X-point is considered as a necessary condition for magnetic field line reconnection in flares. However, bifurcating structures with hyperbolic groups of field lines on each side, which are indicative of magnetic 3D null points, have been found to be in a quasi-stable configuration, such as the saddle-like structure observed with *SoHO/EIT* in active region NOAA 8113 from 1997-Dec-6 to 1997-Dec-10 (Filippov 1999). The saddle-like structure was visible without any signs of additional heating and was even darker than the loops of the neighboring arcade. Apparently, magnetic nullpoints are a necessary but not sufficient condition for fast reconnection processes that occur in flares. There probably exists a vast number of magnetic nullpoints in active regions, wherever multiple strong dipoles co-exist in proximity, but their detection requires cotemporaneous heating of hyperbolic field lines that surround a nullpoint from topologically disconnected domains.

5.7 Magnetic Field Measurements in Radio

Active regions harbor the strongest magnetic fields on the solar surface, which amounts up to several kiloGauss (kG) in sunspots. We used a simple analytical model for a unipolar magnetic field in terms of a potential field in § 5.2.1, which characterizes the coronal magnetic field above a symmetric sunspot to first order. While in all other coronal regions the relatively weak magnetic field can only be inferred from photospheric extrapolation methods, the strong fields above and around sunspots provide direct measurement methods at radio wavelengths, making use of the magneto-ionic mode splitting in the circular polarization of free-free emission (§ 5.7.1), gyroresonance emission (§ 5.7.2), gyroresonance stereoscopy (§ 5.7.3), and nonpotential field modeling of gyroresonance emission (§ 5.7.4). Above active regions, the magnetic field can sometimes be inferred from the change in sign and amplitude of the circular polarization when the microwave radiation passes through a *quasi-transverse (QT)* region (e.g., Ryabov et al. 1999). In weak-field regions far away from sunspots, the coronal field can also be probed by measuring the Faraday rotation (Alissandrakis & Chiuderi−Drago 1995). Alternative methods to measure the coronal magnetic field have also been demonstrated by using the Fe XIII line in the infrared (Lin et al. 2000) or the Hanle effect (Stenflo 1994).

5.7.1 Magnetic Fields Measured from Free-Free Emission

We derived the free-free absorption coefficient and free-free opacity for electron bremsstrahlung (or free-free emission) in an unmagnetized plasma in § 2.3. In the presence of a magnetic field **B**, the plasma becomes magneto-active, in the sense that it becomes anisotropic and gyrotropic. The anisotropy shows up in the direction of the particle motion along magnetic field lines as well as in the direction of wave propagation (refractive index) and in the gyrotropy of the circular polarization. The dispersion relation and refractive indices of a magnetized plasma are derived in a number of standard textbooks on plasma physics (e.g., Ratcliffe 1969; Ginzburg 1961; Chen 1974; Schmidt 1979; Lang 1980; Benz 1993; Sturrock 1994). The complex refractive index ϵ_ν for propagating electromagnetic (transverse) waves in a magnetized plasma is (Lang 1980),

$$\epsilon_{\nu,\sigma}^2 = [n_\nu - iq_\nu]^2 = 1 - \frac{X}{\frac{1-iZ-\frac{1}{2}Y_T^2}{1-X-iZ} + \sigma\left[\frac{\frac{1}{4}Y_T^4}{(1-X-iZ)^2} + Y_L^2\right]^{1/2}} , \qquad (5.7.1)$$

where the dimensionless frequency ratios are defined by

$$X = \left(\frac{\nu_p}{\nu}\right)^2 , \qquad (5.7.2)$$

$$Y = \left(\frac{\nu_B}{\nu}\right) , \tag{5.7.3}$$

$$Z = \left(\frac{\nu_{eff}}{2\pi\nu}\right) , \tag{5.7.4}$$

$$Y_T = Y \sin\psi , \tag{5.7.5}$$

$$Y_L = Y \cos\psi , \tag{5.7.6}$$

with ν being the observed frequency of the electromagnetic wave, ν_p the plasma frequency, ν_B the gyrofrequency, ν_{eff} the effective collision frequency, ψ the angle between the propagation vector $\mathbf{k}$ of the electromagnetic wave and the magnetic field $\mathbf{B}$, and σ denoting the two magneto-ionic modes [i.e., the ordinary mode ($\sigma = +1$) and extraordinary mode ($\sigma = -1$)], representing the sense of rotation of the gyrating electrons with respect to the electromagnetic wave vector. In the cold-plasma approximation, where the thermal motion or pressure is neglected, only the real part of the refractive index (5.7.1) is considered n_ν (i.e., the imaginary part is set to zero, $q_\nu = 0$, $Z = 0$, or $\nu_{eff} \ll \nu$). The refractive index $n_{\nu,\sigma}$ for this cool and collisionless plasma is then

$$n_{\nu,\sigma}^2 = \frac{k^2 c^2}{\omega^2} = 1 - \frac{X}{\frac{1-\frac{1}{2}Y_T^2}{1-X} + \sigma\left[\frac{\frac{1}{4}Y_T^4}{(1-X)^2} + Y_L^2\right]^{1/2}} , \tag{5.7.7}$$

also known as the Appleton–Hartree expression, which can be written by inserting Eqs. (5.7.2−6) (e.g., Brosius & Holman 1988),

$$n_{\nu,\sigma}^2 = 1 + \frac{2\nu_p^2(\nu_p^2 - \nu^2)}{\sigma\sqrt{\nu^4\nu_B^4\sin^4\theta + 4\nu^2\nu_B^2(\nu_p^2 - \nu^2)\cos^2\theta} - 2\nu^2(\nu_p^2 - \nu^2) - \nu^2\nu_B^2\sin^2\theta} , \tag{5.7.8}$$

where the plasma frequency ν_p and gyrofrequency ν_B are (in cgs units),

$$\nu_p = \left(\frac{e^2 n_e}{\pi m_e}\right)^{1/2} = 8979\sqrt{n_e} , \tag{5.7.9}$$

$$\nu_B = \left(\frac{eB}{2\pi m_e c}\right) = 2.80 \times 10^6 \, B . \tag{5.7.10}$$

The free-free absorption coefficient α_ν, which we derived for a vacuum in § 2.3 (with refractive index $n_\nu = 1$), is now generalized for a magnetized plasma by including the refractive index $n_{\nu,\sigma}$ (Eq. 5.7.7) for the two magneto-ionic modes (σ) in Eqs. (2.3.16−17),

$$\alpha_{\nu,\sigma}^{ff} = \frac{9.786 \times 10^{-3}}{n_{\nu,\sigma}\nu^2} \frac{n_e n_i}{T^{3/2}} \ln\left[4.7 \times 10^{10}\left(\frac{T}{\nu}\right)\right] \quad [\text{cm}^{-1}] \tag{5.7.11}$$

The free-free opacity $\tau_{\nu,\sigma}^{ff}$ in a magnetized plasma is then

$$\tau_{\nu,\sigma}^{ff} \approx \int \alpha_{\nu,\sigma}^{ff}(z) \, dz . \tag{5.7.12}$$

When free-free emission is observed at a radio frequency ν, the corresponding radio brightness temperature is then (according to Eq. 2.1.9) in the Rayleigh−Jeans approximation (Eq. 2.2.5), depending on the magneto-ionic mode (σ),

$$T_B(\nu, \sigma) = \int T(z) \exp^{-\tau_{\nu,\sigma}^{ff}(z)} \alpha_{\nu,\sigma}^{ff}(z) \, dz \; . \qquad (5.7.13)$$

Since the free-free opacities are different for the two magneto-ionic modes, we obtain a circular polarization, also called *Stokes V parameter*, which is defined by the difference of the brightness temperature between the extraordinary mode (X) and ordinary mode (O), normalized by their sum,

$$V = \frac{T_B(\nu, X) - T_B(\nu, O)}{T_B(\nu, X) + T_B(\nu, O)} \; . \qquad (5.7.14)$$

Therefore, the coronal magnetic field B can be inferred by measuring the circular polarization V at radio wavelengths. Neglecting the magnetic field, which can be justified at sufficiently high radio frequencies $\nu \gg \nu_B$ or for sufficiently weak magnetic fields, reduces the Appleton−Hartree expression to the simple dispersion relation for plasma emission (by setting $Y = 0$ in Eq. 5.7.7),

$$n_\nu^2 = \frac{k^2 c^2}{\omega^2} = 1 - X = 1 - \left(\frac{\nu_p}{\nu}\right)^2 \qquad (5.7.15)$$

yielding no circular polarization for Stokes parameter V in Eq. (5.7.14).

The application of such a model of free-free emission to observed brightness temperature maps $T_B(x, y)$, as they can readily be obtained in radio wavelengths, obviously involves the integration of the free-free opacity over a temperature model $T(z)$ and density model $n_e(z)$ along the line-of-sight. Brosius & Holman (1988) explored this method for a coronal loop by assuming a constant temperature T and constant density n_e over an appropriate length scale L of the coronal plasma along the line-of-sight, using radio maps obtained with the *Very Large Array (VLA)* (Webb et al. 1987). A *COronal Magnetic Structures Observing Campaign (COMSTOC)* was organized to measure the magnetic field in active regions, using *VLA* radio observations (Nitta et al. 1991; Schmelz et al. 1992; 1994; Brosius et al. 1992). Further studies with the same method were continued by Brosius et al. (1993, 1997, 2002) and Alissandrakis et al. (1996). A summary of the obtained temperatures T_e, densities n_e, magnetic fields B, and Alfvén velocities v_A is compiled in Table 5.1 (§ 5.4.2). The density and temperature models were parameterized only with a single or two-component plasma along the line-of-sight, which, of course, cannot do justice to the highly inhomogeneous structure of the solar corona, as evidenced by the relatively broad differential emission measure distributions observed by multi-temperature instruments (see discussion on line-of-sight integration in § 3.5). Consequently, significant discrepancies arose between the plasma temperatures inferred from the COMSTOC radio observations and soft X-ray or EUV observations, which could only be reconciled with realistic inhomogeneous models. Nevertheless, although this method allows only for very crude density and magnetic field models in the case of single radio frequencies, it has the potential of *tomographic reconstruction* (Aschwanden et al. 2003b), when radio maps will become available at many frequencies, such as anticipated with the *Frequency-Agile Solar Radio telescope (FASR)* (White et al. 2003).

5.7.2 Gyroresonance Emission

The *free-free emission* described in the previous section originates from the collisional energy loss when nonthermal electrons become slowed down in a dense plasma, also called *bremsstrahlung*. At the same time, electrons spiral along magnetic field lines and emit *gyroresonance emission* due to the Lorentz force that an electron experiences on its gyroorbit with Larmor radius, also called *magneto-bremsstrahlung* or *cyclotron emission*. In both cases, the radiated power $dP/d\Omega$ per stereo-angle is a function of the acceleration $\dot{v}$ of a moving charge e, according to Larmor (e.g., Jackson 1962, p. 663),

$$\frac{dP}{d\Omega} = \frac{e^2 \dot{v}^2}{4\pi c^3} \frac{\sin^2 \theta'}{(1 - \beta \cos \theta')^5} , \tag{5.7.16}$$

where e is the electric charge, $\dot{v}$ the acceleration, $\beta = v/c$ the relativistic speed, and θ' the angle between the electromagnetic wave vector and the direction of acceleration. The calculation of the gyroresonance emissivity involves the circular gyromotion of the electron and the integration over the velocity distribution of the electrons. The gyroemissivity peaks at harmonic frequencies $\nu = s\nu_B$ of the gyrofrequency ν_B and has a strong angular dependence. For a thermal (Maxwellian) velocity distribution of kinetic electron energies the gyroemission absorption coefficient per unit length is (derivations and discussions can be found in, e.g., Takakura 1960; Wild et al. 1963; Zheleznyakov 1970; Melrose 1980a; Krueger 1979; Lang 1980; Dulk 1985),

$$\alpha_{\nu,s,\sigma}^{gr}(\theta) = \frac{4\pi^{3/2}}{c} \frac{\nu_p^2}{\nu} \frac{(\frac{s}{2})^{2s}}{s!} \frac{(\sin \theta)^{2s-2}(1 - \sigma |\cos \theta|)^2}{\cos \theta} \beta_0^{2s-3} \exp \left[-\frac{(1 - \frac{s\nu_B}{\nu})^2}{\beta_0^2 \cos^2 \theta} \right] , \tag{5.7.17}$$

where the thermal velocity of the electron is defined by $\beta_0 = \sqrt{k_B T/m_e c^2}$, and the angle θ is defined between the directions of wave propagation and the magnetic field. Because of the exponential factor, the absorption coefficient $\alpha_{\nu,s,\sigma}^{gr}$ drops quickly at frequencies differing from the resonance frequencies $\nu = s\nu_B$. The integral of the exponential factor over a resonance line is

$$\int_{-\infty}^{+\infty} \exp \left[-\frac{(1 - \frac{s\nu_B}{\nu})^2}{\beta_0^2 \cos^2 \theta} \right] \frac{d\nu}{\nu_B} = \left(\frac{\pi}{2}\right)^{1/2} \frac{\beta_0 \cos \theta}{\sqrt{2}} . \tag{5.7.18}$$

It is therefore convenient to define a gyroresonance absorption coefficient averaged over a resonance line, which yields with Eqs. (5.7.17−18),

$$\langle \alpha_{\nu,s,\sigma}^{gr}(\theta) \rangle = \left(\frac{\pi}{2}\right)^2 \frac{2}{c} \frac{\nu_p^2}{\nu} \frac{s^2}{s!} \left(\frac{s^2 \beta_0^2 \sin^2 \theta}{2}\right)^{s-1} (1 - \sigma |\cos \theta|)^2 . \tag{5.7.19}$$

The line-of-sight depth of a gyroresonant layer can be estimated from the magnetic scale height $L_B = B/\nabla B$. With this length scale L_B we can calculate the gyroresonance opacity $\tau_{\nu,s,\sigma}$ of a resonance layer with harmonic s,

$$\tau_{\nu,s,\sigma}(\theta) = \int_0^\infty \langle \alpha_{\nu,s,\sigma}^{gr}(\theta) \rangle dz \approx \langle \alpha_{\nu,s,\sigma}^{gr}(\theta) \rangle L_B \tag{5.7.20}$$

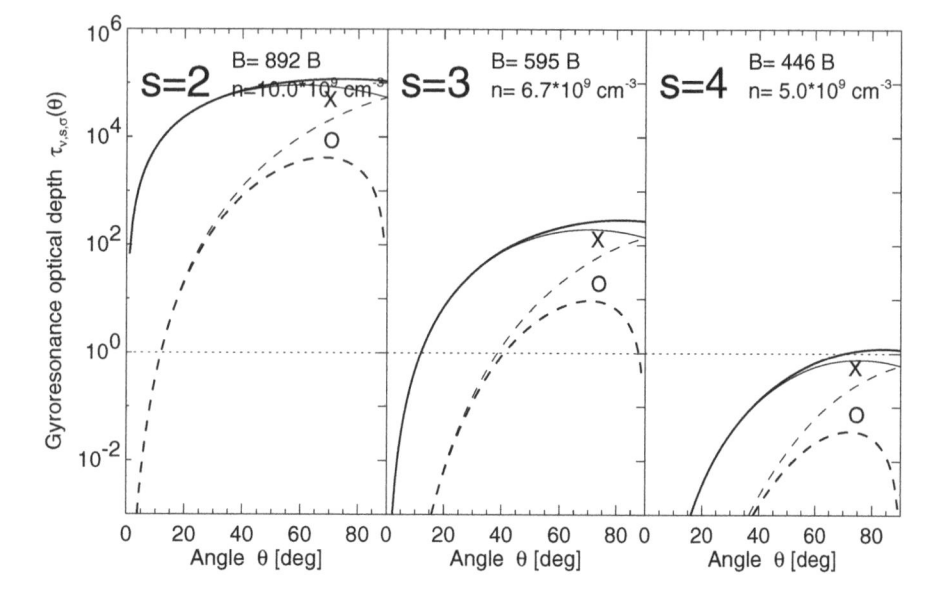

Figure 5.26: The line-of-sight integrated optical depth of the $s = 2, 3, 4$ gyroresonance layers at a radio frequency of $\nu = 5.0$ GHz as a function of the angle θ between the line-of-sight and the magnetic field direction. The plasma temperature is $T = 3.0$ MK and the magnetic scale height $L_B = 10^9$ cm. The magneto-ionic X-mode is represented with solid linestyle and the O-mode with dashed lines. The thin lines show the approximation of Eq. (5.7.21), while the thick lines show the more accurate calculation based on Zlotnik (1968) (adapted from White & Kundu 1997).

which with Eq. (5.7.19), after inserting the plasma frequency ν_p (Eq. 5.7.9), yields the simplified expression (White & Kundu 1997),

$$\tau_{\nu,s,\sigma}(\theta) = 0.0133 \times \frac{n_e L_B}{\nu} \frac{s^2}{s!} \left(\frac{s^2 \beta_0^2 \sin^2 \theta}{2} \right)^{s-1} (1 - \sigma |\cos \theta|)^2 . \qquad (5.7.21)$$

The expected radio brightness temperature can then be calculated with Eq. (5.7.13) in both circular polarizations $\sigma = \pm 1$. We show a typical example of the gyroresonance opacity in Fig. 5.26, calculated for a plasma temperature of $T = 3 \times 10^6$ K, a magnetic scale height of $L_B = 10^9$ cm, and at a frequency of $\nu = 5.0$ GHz, for the harmonics $s = 2, 3, 4$. Note that the lowest harmonic $s = 2$ is optically thick ($\tau_\nu \gg 1$) at most intermediate angles in both magneto-ionic modes, while the highest harmonic $s = 4$ is optically thin $\tau_\nu < 1$ in both polarizations. The expression of the gyroresonance opacity given in Eq. (5.7.21) represents an approximation for perfect circular polarization, while more general expressions are calculated by Zlotnik (1968), also shown in Fig. 5.26.

Gyroresonance emission is competing with free-free emission at radio wavelengths. In strong-field regions above sunspots, gyroresonance emission is always dominant

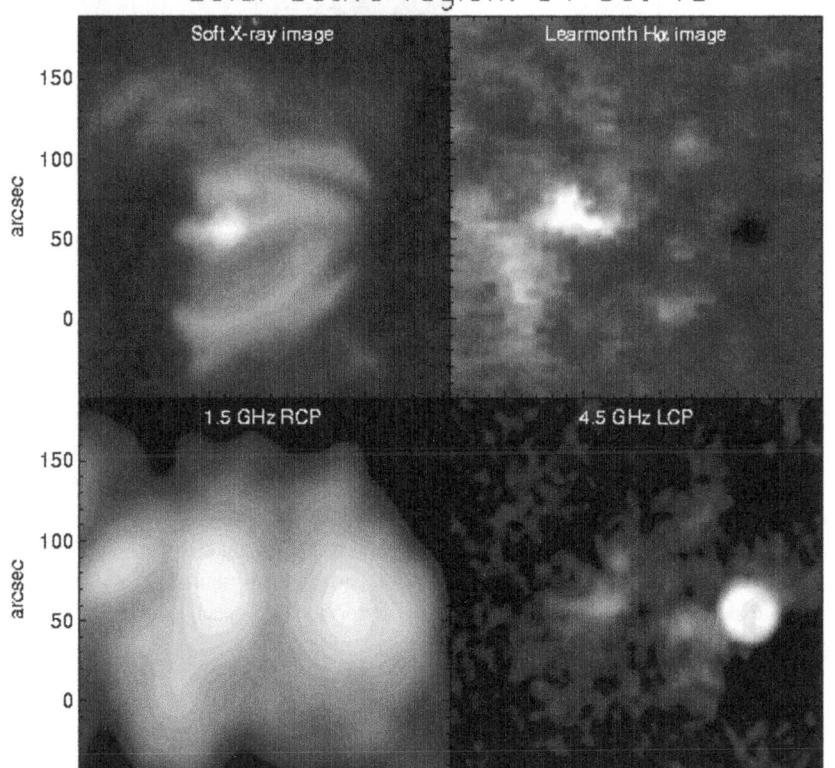

Figure 5.27: Free-free emission and gyroresonance emission observed in an active region on 1994 Oct 15. *Top left:* soft X-ray image obtained by *Yohkoh/SXT*, which outlines the hot ($T \gtrsim 2$ MK) plasma loops of the dipolar region. *Right top:* H-α image from Learmonth Observatory, which shows the dominant sunspot in the west as a dark feature. *Bottom left:* radio map made with the *VLA* at $\nu = 1.5$ GHz which shows free-free emission from the plasma envelope at temperatures of $T \gtrsim 1$ MK that become optically thick above the hotter active region seen in soft X-rays. *Bottom right:* radio map made with the *VLA* at $\nu = 4.5$ GHz, which shows bright gyroresonance emission above the western sunspot and faint free-free emission spread over the entire active region (courtesy of Stephen White).

at frequencies above a few GHz (Fig. 5.27, bottom right), while free-free emission is generally dominant in the weak-field regions in plages of active regions, almost always at frequencies of $\nu \lesssim 2$ GHz (Fig. 5.27, bottom left). Measuring the circular polarization at radio frequencies of $\nu \gtrsim 1$ GHz, however, provides information on the magnetic field strengths for both emission mechanisms. It is very fortunate that coronal magnetic fields, which have field strengths of up to $B \approx 2500$ G, fall in the range of microwave frequencies ($\nu = 1\text{-}20$ GHz), for the lower harmonics $s = 2, 3, 4$, which can be readily mapped with high spatial resolution (e.g., White & Kundu 1997). The

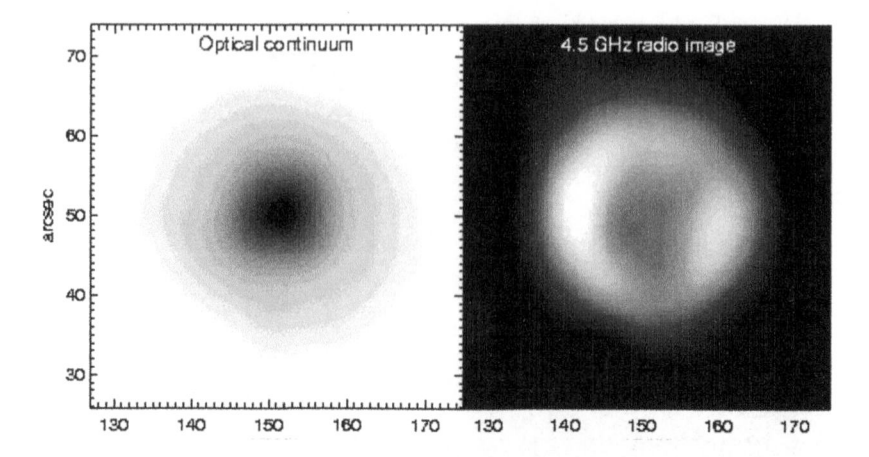

Figure 5.28: *Left:* white-light continuum image of a sunspot, showing the darkest emission in the central umbra, where the temperature is coolest ($T \approx 4500°$ K) and the magnetic fields are strongest ($B \approx 1 - 2$ kG). *Right:* radio map of the same sunspot observed with the *VLA* at a frequency of $\nu = 4.5$ GHz. The central part of the umbra, where the angle of the line-of-sight to the magnetic field is small, is darker than the peripheral parts in the penumbra, where the angles are larger, as expected from the angle dependence of gyroresonance opacity shown in Fig. 5.26 (courtesy of Stephen White).

magnetic field B, scales with the observed radio frequency ν and harmonic number as (according to Eq. 5.7.10),

$$B = 357 \left(\frac{1}{s}\right) \left(\frac{\nu}{1 \text{ GHz}}\right) \quad \text{(G)} . \tag{5.7.22}$$

For gyroresonance emission, a number of modeling studies have been conducted to infer the magnetic fields above sunspots (e.g., Alissandrakis et al. 1980; Krueger et al. 1985; Brosius & Holman 1989; Lee et al. 1993a,b; Vourlidas et al. 1997). A brightness temperature map of gyroresonance emission of a nearly symmetric sunspot is shown in Fig. 5.28. Generally, radio images at multiple microwave frequencies are required to disentangle the contributions from different gyroharmonics ($s = 2, 3, 4$) for inference of a magnetic field model. When only one single frequency is available, one can estimate the maximum field strength in the corona from the extent of the gyroresonance-emitting region. Maximum field strengths up to $B \approx 1800$ G have been measured in the corona at a frequency of $\nu = 15$ GHz (White et al. 1991), and up to $B \approx 2000$ G at $\nu = 17$ GHz with the Nobeyama heliograph (Shibasaki et al. 1994).

5.7.3   Gyroresonance Stereoscopy

All previously described models based on free-free or gyroresonance emission at radio wavelengths are scale-free, because they provide essentially a relation between the magnetic field as a function of the opacity $B(\tau)$, while the spatial scaling of the opacity

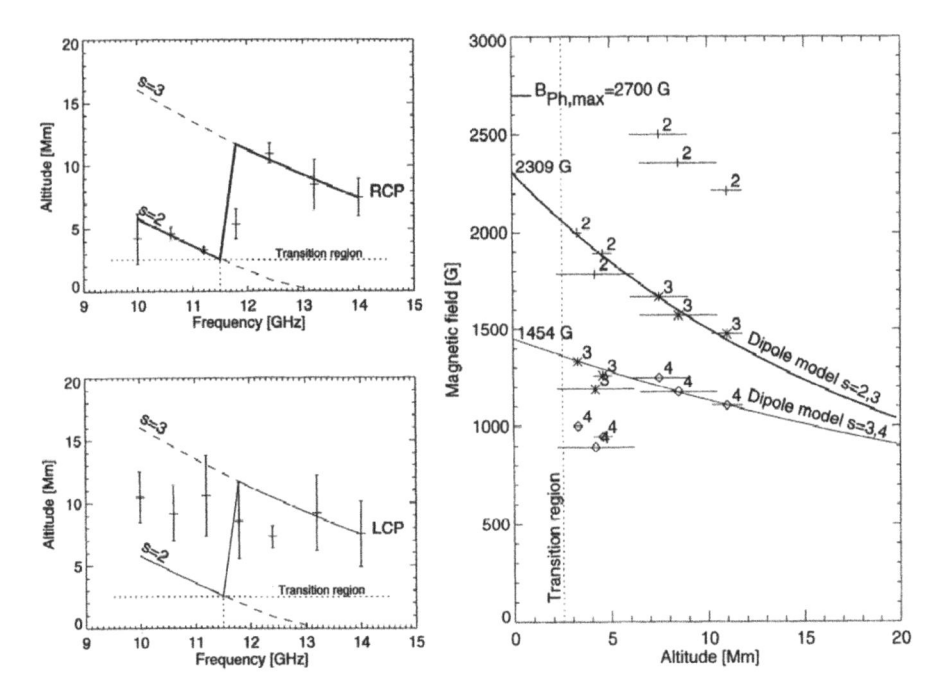

Figure 5.29: *Left:* stereoscopic height measurements $h(\nu)$ of a gyroresonance source above a sunspot as a function of radio frequency ν, observed with the *Owens Valley Radio Array (OVRA)*, showing the brightness temperature in right-hand-circular (RCP) and left-hand-circular (LCP) polarization. The dashed curves represent fits of a dipole-field model for the 2nd and 3rd harmonic. *Right:* magnetic field measurements $B(h)$ as a function of height h, reconstructed from the same stereoscopic correlations. The upper curve represents a dipole-field model involving the 2nd and 3rd harmonic (Aschwanden et al. 1995a).

$\tau(z)$ as a function of the line-of-sight depth z cannot be directly measured. One possible method to infer an absolute scaling of the coronal magnetic field $B(z)$ as a function of height z is *solar-rotation stereoscopy*, which works relatively well for gyroresonance emission above stationary sunspots. Such an experiment was conducted over four days (1992-Apr-13 to 16) of sunspot observations with the *Owens Valley Radio Observatory (OVRO)* in active region NOAA 7128 (Aschwanden et al. 1995a). Optical observations of the sunspot provided a photospheric rotation rate of $dL/dt = +0.240°$ day^{-1}. Radio images were reconstructed in both RCP and LCP polarizations and the rotation rates of the centroids of gyroresonance emission were measured at 7 different frequencies ($\nu = 10 - 14$ GHz). This yielded an average altitude $h(\nu)$ of the dominant gyroresonance layers at each frequency, depending on the parallax effect between the sunspot motion and the radio source motion. The obtained altitudes are shown in Fig. 5.29 for both circular polarizations. The altitude $h(\nu)$ of the stronger RCP polarization (X-mode) clearly shows a pattern to decrease with higher frequencies, as expected for an individual harmonic s in the case of gyroresonance emission. The situation is only complicated by the fact that multiple harmonics are involved and that the

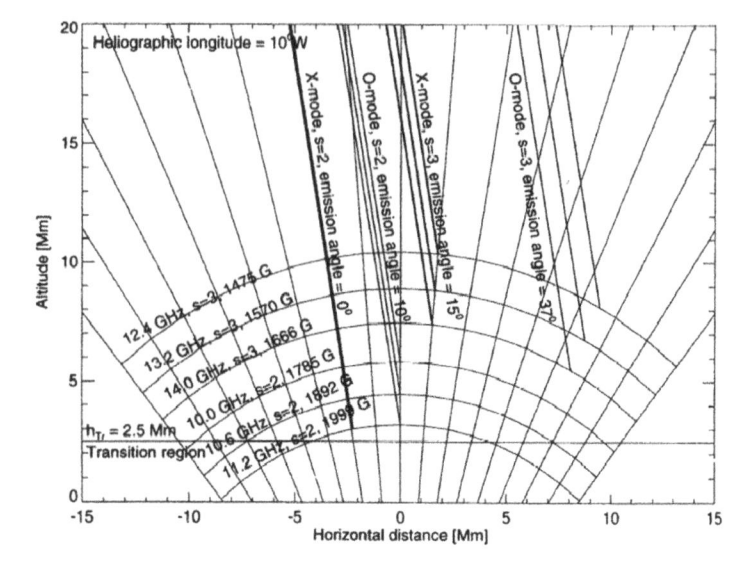

Figure 5.30: Spatial model of gyroresonance layers above a sunspot, constrained by stereo-scopic measurements at 7 frequencies between $\nu = 10$ GHz and 14 GHz (Aschwanden et al. 1995a).

harmonic number is not known a priori. The altitudes could be fitted by a dipole model in this case (Fig. 5.29, right),

$$B(h) = B_0 \left(1 + \frac{h}{h_D} \right)^{-3}, \qquad (5.7.23)$$

yielding a photospheric field strength of $B_0 = 2309$ G for a combination of the two harmonics $s = 2, 3$, or alternatively $B_0 = 1454$ G for the combination of $s = 3, 4$. Since the photospheric field could be independently measured in white light to $B_{Ph} \approx 2500$ from an (unsaturated) Mount Wilson magnetogram, the ambiguity of the two solutions could be resolved, indicating that the combination of harmonics with $s = 2, 3$ is the more consistent solution. The corresponding dipole depth was found to be $h_D = 65$ Mm, similar to the active region characterized in Fig. 5.9, where an average value of $h_D = 75$ Mm was found (Aschwanden et al. 1999a). Based on this model the height dependence of the gyroresonance source at an observed frequency ν is (with Eqs. 5.7.22–23),

$$h(\nu, s) = -h_D + \left[h_D \left(\frac{B_0}{357} \right)^{1/3} \right] \left(\frac{\nu}{s} \right)^{-1/3}. \qquad (5.7.24)$$

This dipole model also yields an estimate of the local magnetic scale height $L_B(h)$,

$$L_B(h) = -\frac{B}{\nabla B} = \frac{(h_D + h)}{3}, \qquad (5.7.25)$$

which yields values of $L_B \approx 24$ Mm for the lower corona. The magnetic scale height L_B is required for gyro-opacity models (Eq. 5.7.21). Additional knowledge of the geometry (as visualized with a symmetric model in Fig. 5.30), which fixes the aspect angle θ between the line-of-sight and the local magnetic field for every position, then allows together with brightness temperature measurements $T_B(x, y, h)$, reconstruction of the 3D density $n_e(x, y, z)$ and temperature $T_e(x, y, z)$ in the coronal region above the sunspots, using the relations for the gyro-opacity (Eqs. 5.7.13 and 5.7.21). Of course, this method works better the more frequencies that are available, because each frequency provides information about a different height (specified by the gyroresonance layer of the dominant gyroharmonics). Future solar-dedicated multi-frequency imaging instruments, such as FASR, will allow us to apply such 3D reconstruction methods to the coronal magnetic field, density, and temperature in great detail.

5.7.4 Non-Potential Field Modeling of Gyroresonance Emission

While we used a simple potential field to model the height dependence of the magnetic field $B(h)$ using stereoscopic methods (§ 5.7.3), a more sophisticated approach has been advanced by Lee et al. (1997, 1998, 1999) by employing nonlinear force-free field models. Earlier it was shown that potential-field (i.e., current-free) extrapolations of photospheric magnetic fields failed to predict the (maximum) magnetic field strength in some particular regions to explain the observed gyroresonance emission (Alissandrakis et al. 1980; Pallavicini et al. 1981; Schmahl et al. 1982; Alissandrakis & Kundu 1984; Lee et al. 1997). Thus, a convincing argument could be made that nonpotential fields are necessary to explain the coronal magnetic field in such locations.

In the study of Lee et al. (1998), the gyroresonance emission above a sunspot group was modeled with a nonlinear force-free code (Plate 5) and 6 different models for the plasma temperature: (1) $T_e \propto h$ (Fig. 5.31, top row); (2) $T_e \propto B$ (Fig. 5.31 second row); (3) $T_e \propto \alpha$, the force-free parameter (Fig. 5.31 third row); (4) $T_e \propto j$, the current density; (5) $T_e \propto j(l/L)$, mimicking an RTV loop; and (6) with a magnetic field correction. It was found that only a nonlinear force-free field model could be reconciled with the source morphology seen in radio and Hα maps, and that the scaling law of $T_e \propto B$ yields a good agreement in low-α regions, while $T_e \propto j$ seemed to work better in high-α regions. Although no unique answer was found, this type of forward-modeling has the potential to constrain the currents in coronal regions.

An alternative method to test coronal magnetic field models makes use of the spatial correlations expected for temporal variations of plasma temperatures (Lee et al. 1999). For a given 3D magnetic field model, plasma heating and cooling should produce correlated changes at two altitudes of the same field line, mapped at two radio frequencies, because plasma transport in a low-β corona is only possible along field lines (§ 1.8). This method successfully demonstrated that in a strongly sheared region only a nonlinear force-free field could explain the correlations at the right locations, while a potential-field model failed (Lee et al. 1999).

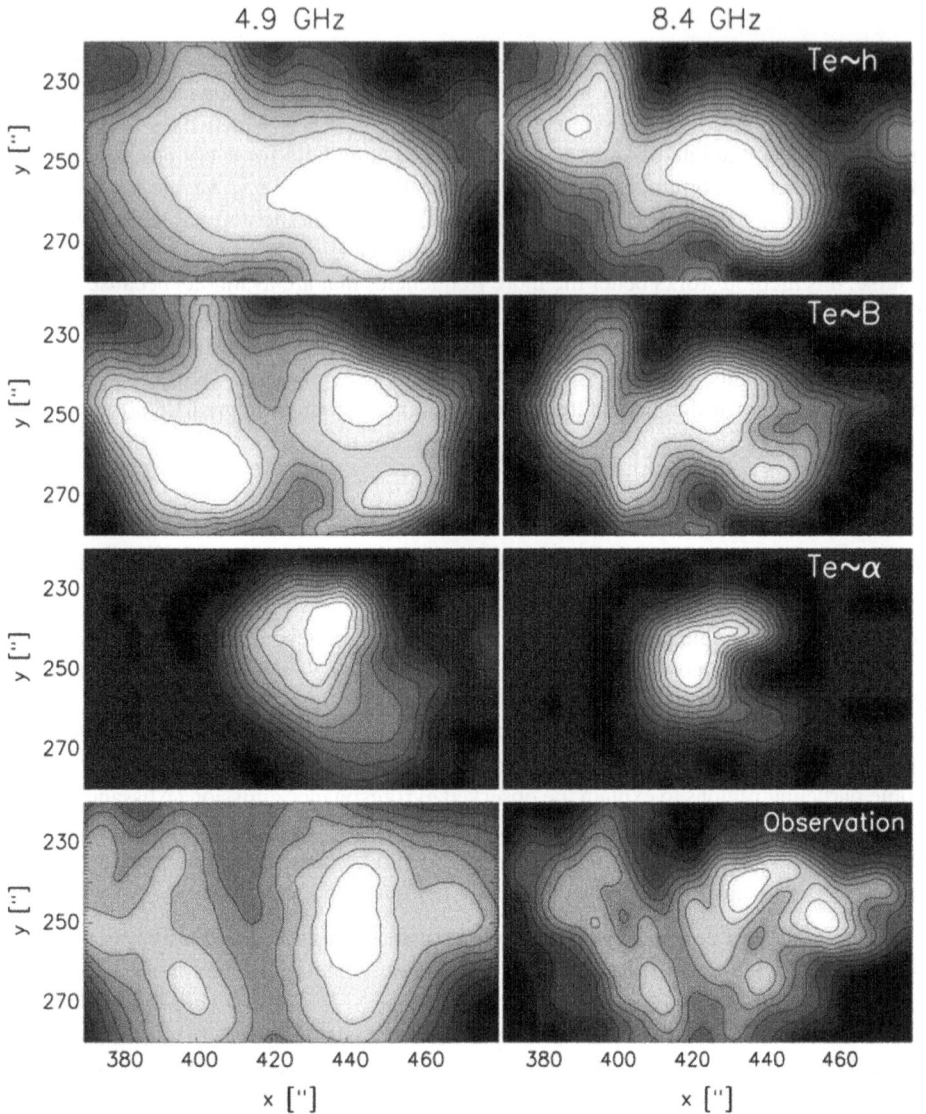

Figure 5.31: Radio brightness temperature maps of a sunspot region, observed with the *VLA* at 4.9 GHz (left bottom) and 8.4 GHz (right bottom), modeled with three different models of temperature scaling: $T_e \propto h$ (top row), $T_e \propto B$ (second row), and $T_e \propto \alpha$ (third row). (Lee et al. 1998).

5.8   Magnetic Field in the Transition Region

The transition region represents a boundary in many physical parameter regimes: it demarcates (1) a temperature jump from $T = 10^4$K to 10^6 K, (2) a density jump from $\gtrsim 10^{11}$ cm^{-3} to 10^9 cm^{-3} (Fig. 1.19), (3) a transition from a high-β parameter to a low-β regime (Fig. 1.22), and (4) a transition from a non-force-free to a force-free mag-

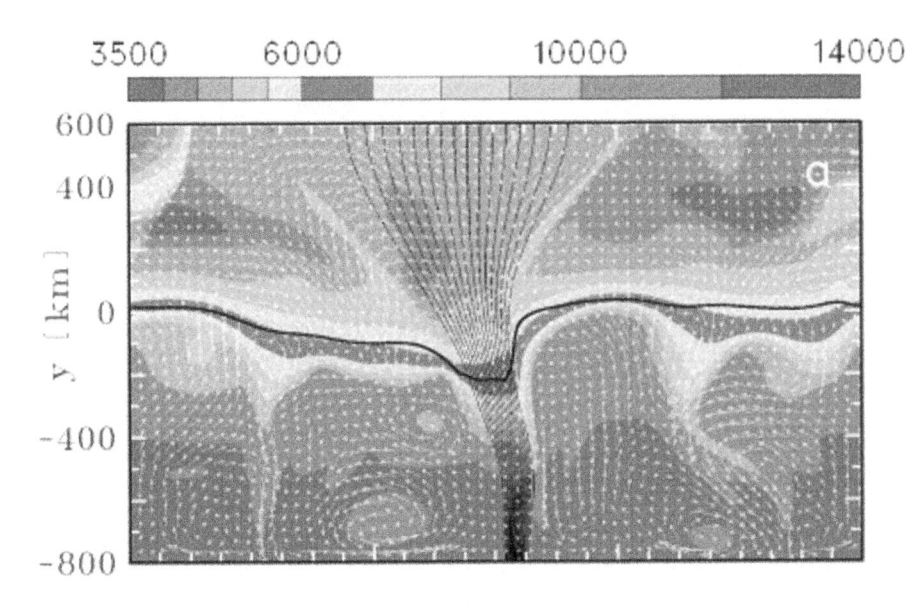

Figure 5.32: Numerical simulation of a magneto-hydrodynamic fluxtube. The temperature distribution is shown in greyscale, the velocity field with white arrows, and the magnetic field with black lines. Note the canopy-like expansion in the upper chromosphere (Steiner et al. 1998).

netic field. It is therefore not surprising that also the geometry of magnetic fluxtubes is subject to a drastic change at this boundary. Fluxtubes generally exhibit a rapid expansion with height, the so-called *magnetic canopy*, as inferred from magnetograms in chromospheric spectral lines or from theoretical models of magneto-hydrostatic extrapolations (e.g., Fig. 5.32; Steiner et al. 1998). In the following we describe a few aspects of the coronal magnetic field structure at this lower boundary, while more extensive descriptions of the magnetic field in the photosphere, chromosphere, and transition region can be found in other solar textbooks (e.g., Foukal 1987; Zirin 1988; Schüssler & Schmidt 1994; Stenflo 1994; Schrijver & Zwaan 2000; Stix 2002) or recent encyclopedia articles (e.g., Schüssler 2001; Lites 2001; Stenflo 2001a,b; Solanki 2001a,b,c; Roberts 2001; Steiner 2001; Keppens 2001).

5.8.1 The Magnetic Canopy Structure

The organizing pattern that structures the magnetic field in the transition region are the horizontal flows inside a supergranule, which transport emerging magnetic flux to the boundaries of the supergranulation pattern, i.e., the photospheric network, causing a congestion of magnetic flux in the network (Fig. 5.33). The concentration of the magnetic flux in the network causes field-free regions in the internetwork cell (of the supergranule). If we consider the total (thermal and magnetic) pressure in a fluxtube rooted in a high-field region in the network, it has to be balanced against the total

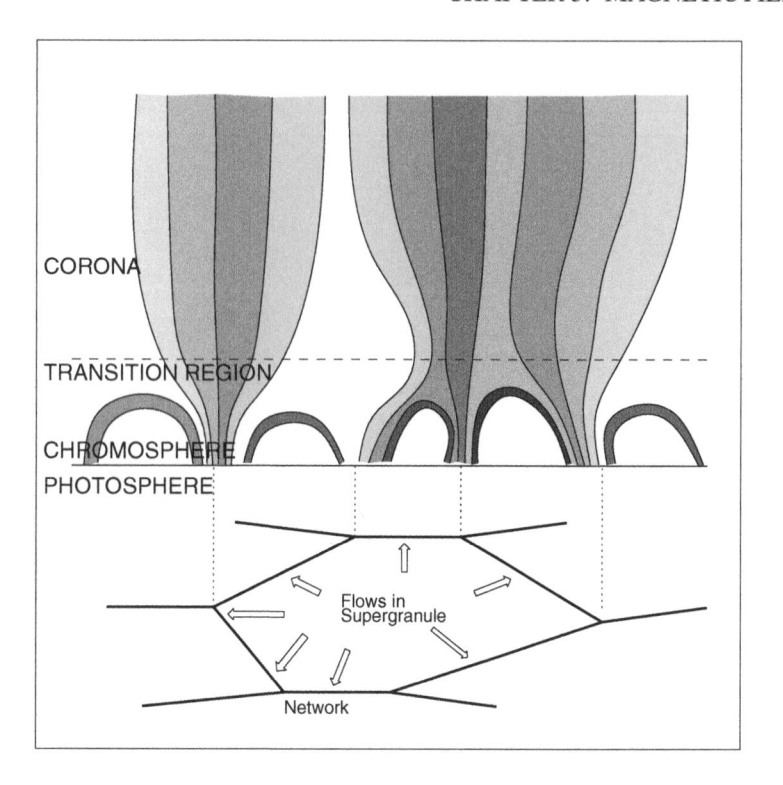

Figure 5.33: Cartoon of canopy structure of magnetic field in the chromosphere and transition region to the corona. The flows inside a supergranule transport the footpoints of coronal field lines into the network (arrows), which causes a bundling of the footpoint of vertical field lines. The expansion into the corona occurs in fluxtubes that have a higher magnetic pressure than the surrounding coronal plasma.

pressure of the external plasma,

$$p_{int}(h) + \frac{B_{int}^2(h)}{8\pi} = p_{ext}(h) + \frac{B_{ext}^2(h)}{8\pi} \ . \tag{5.8.1}$$

In the lower chromosphere, where the plasma-β parameter is high ($\beta = p_{th}/p_{magn} \gg 1$; Fig. 1.22), the difference in magnetic field pressure inside and outside the fluxtube does not matter much in the pressure balance, because $p_{th,int} \approx p_{th,ext} \gg p_{magn}$. However, above a height where the plasma-β parameter drops below unity, the pressures become comparable and the total internal pressure starts to dominate over the total external pressure. In the strong-field limit of $B_{int} \gg B_{ext}$, the pressure balance equation (5.8.1) becomes,

$$\frac{B_{int}^2(h)}{8\pi} \approx p_{ext}(h) - p_{int}(h) \ . \tag{5.8.2}$$

As a consequence of the decreasing external thermal pressure with height, the magnetic fluxtube has to expand, while keeping the total magnetic flux $\Phi(h) = B(h)A(h)$ constant. This geometric expansion of the fluxtube with height is called *magnetic canopy* (Fig. 5.33). This simplified model explains the physical connection between the canopy height and the critical height where the plasma-β parameter becomes unity.

The magnetic canopy structure was used by Gabriel (1976) to model the lower boundary of coronal magnetic fluxtubes. Using standard hydrostatic atmospheric models and magnetogram measurements near the solar limb, canopy heights of $h \approx 600 - 1000$ km have been inferred. Fluxtubes with large temperature differences between the internal and external plasma produce lower lying canopies. This applies in particular to sunspots, where the horizontal field of the sunspot canopy extends far beyond the penumbra (Plate 7). The canopy structure also explains the transition from a very low filling factor ($\approx 10^{-3}, ..., 10^{-4}$) of the photospheric magnetic flux (outside sunspots) to a high filling factor ($\lesssim 1$) of the coronal magnetic flux.

The dynamic picture of the chromosphere is far more complex. It is thought that the magnetic flux migration causes myriads of magnetic separatrix surfaces above the network, where magnetic energy is dissipated like in the interaction regions of geophysical plate tectonics (Priest et al. 2002). The texture of these small-scale magnetic fields (Plate 6) is also called *magnetic carpet* (Title & Schrijver 1998), involving the processes of magnetic flux emergence, cancellation, coalescence, and fragmentation (Parnell 2001). The energy dissipation of the magnetic separatrix surfaces is thought to be a key player of the coronal heating process.

5.8.2 Force-Freeness of the Chromospheric Magnetic Field

For extrapolations of the magnetic field to coronal heights it is important to establish where the field becomes force-free. The magnetic field is expected to be force-free in the corona wherever the plasma β-parameter is smaller than unity, and for the same reason the magnetic field is expected to be non-force-free in the photosphere due to the high value of the plasma β-parameter ($\beta \gg 1$). Where is the transition?

An observational measurement of the magnetic field in the chromosphere was made by using the Na I $\lambda 5896$ Å spectral line with the Stokes Polarimeter at the *Mees Solar Observatory* in active region NOAA 7216 (Metcalf et al. 1995). The magnetic field was measured at 6 wavelengths within the Na I spectral line, using densities from the VAL-F model atmosphere (Vernazza et al. 1981). A necessary condition for a magnetic field to be force-free is (Low 1984a),

$$F_x \ll F_0 , \quad F_y \ll F_0 , \quad F_z \ll F_0 , \tag{5.8.3}$$

where F_x, F_y, and F_z are the components of the net Lorentz force,

$$F_x = \frac{1}{4\pi} \int_{z=0} B_x B_z \, dxdy ,$$

$$F_y = \frac{1}{4\pi} \int_{z=0} B_y B_z \, dxdy ,$$

$$F_z = \frac{1}{8\pi} \int_{z=0} (B_z^2 - B_x^2 - B_y^2) \, dxdy , \tag{5.8.4}$$

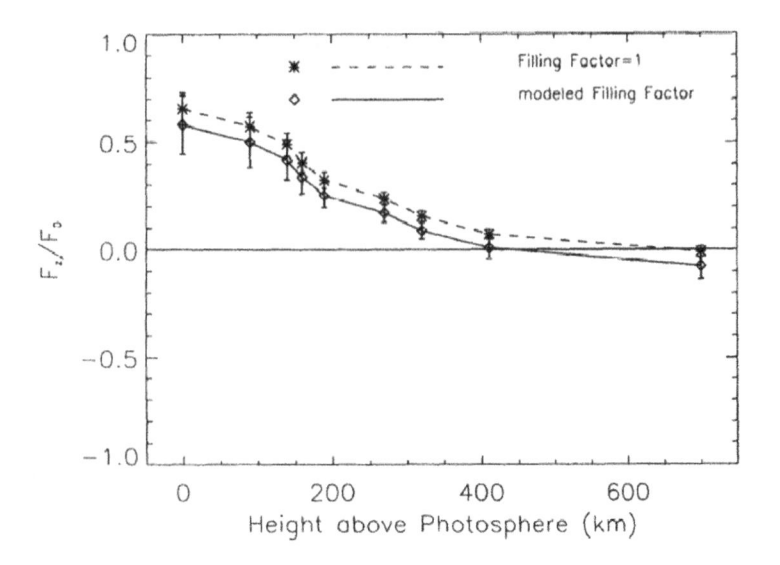

Figure 5.34: Ratio of vertical net Lorentz force component F_z to total Lorentz force F_0 as a function of height h, measured in AR 7216. Note that the magnetic field is not-force-free in the photosphere, but becomes force-free above an altitude of $h \gtrsim 400$ km in the chromosphere (Metcalf et al. 1995).

and F_0 is a characteristic magnitude of the total Lorentz force in a non-force-free atmosphere, which was set equal to the magnetic pressure force by Low (1984),

$$F_0 = \frac{1}{8\pi} \int_{z=0} (B_x^2 + B_y^2 + B_y^2) \, dx dy \, . \tag{5.8.5}$$

To measure the force-freeness condition (Eq. 5.8.3) the magnetic structure must be isolated, which was verified by checking the imbalance between the upward and downward vertical magnetic flux, found to be $\lesssim 0.5\%$ for the investigated active region (Metcalf et al. 1995). After correcting for the filling factor in the data and inversion of the wavelengths using the contribution functions from the VLA-F atmospheric model, Metcalf et al. (1995) inferred the net Lorentz force as a function of height, shown for F_z/F_0 in Fig. 5.34. Clearly the Figure reveals that the magnetic field becomes force-free above a height of $h \gtrsim 400$ km. This critical height approximately corresponds to the canopy height, which is generally found at altitudes of $h \approx 600 - 1000$ km from limb measurements (Steiner 2001). This confirms the theoretical expectation that the canopy height coincides with the value of the plasma-β parameter $\beta = 1$ and the transition from non-force-free to force-free fields. A more extended study with 12 active regions was conducted by Moon et al. (2002a), and somewhat smaller values were found for $|F_z/F_0| = 0.06, ..., 0.32$ than in the study of Metcalf et al. (1995), with $|F_z/F_0| \approx 0.4$ in the photosphere, as well as a dependence of the photospheric force-freeness on the evolutionary status of the active region.

5.9 Summary

The magnetic field controls virtually all physical processes and observed phenomena in the solar corona. It structures the corona into a highly inhomogeneous ensemble of isolated mini-atmospheres and drives many dynamical processes such as upflows and heating of coronal plasma, the evolution of filaments, prominences, flares, and coronal mass ejections. The coronal magnetic field and its associated currents can be described in the framework of Maxwell's equations ($\S$ 5.1). The simplest 3D models of magnetic fields can be quantified in terms of a potential field, which characterizes to first order unipolar fields in sunspots, dipolar fields in active regions, and is often used to compute the global coronal field with a source-surface model ($\S$ 5.2). A physically better motivated approach is the framework of force-free fields, where the Lorentz force on a magnetic field structure vanishes, as it can be justified in the low-β corona ($\S$ 5.3). A number of linear and nonlinear force-free codes have been developed to calculate the coronal magnetic field from photospheric magnetograms ($\S$ 5.3.3). Independent measurements of the 3D magnetic field have been attempted with stereoscopic methods, which constrain the force-free parameter α ($\S$ 5.4.1). Generally, most of the EUV and soft X-ray structures traced in the corona show significant deviations from potential fields ($\S$ 5.4.3) and reveal near-constant cross sections ($\S$ 5.4.4). A special subset of highly non-potential field structures are helically twisted loops, also called sigmoids, which most conspicuously indicate current systems in the corona ($\S$ 5.5). Singularities in the coronal fields are nullpoints and separatrix surfaces, which are now more and more frequently observed ($\S$ 5.6). One of the few methods of measuring the magnetic field strength in the corona directly is based on radio observations, utilizing the polarization of free-free emission or gyroresonance emission ($\S$ 5.7). The lower boundaries of coronal loops are generally formed by canopy-like structures in the transition region, which demarcate a transition from a high to a low plasma-β parameter, as well as a transition from non-force-free to force-free magnetic fields ($\S$ 5.8).

Chapter 6

Magneto-Hydrodynamics (MHD)

We started to study the structure of the coronal plasma first in terms of simple fluid mechanics, called *hydrostatics (HS)* (§ 3), then we moved on to fluid dynamics by adding flows, called *hydrodynamics (HD)* (§ 4), then we studied the coronal magnetic field separately (§ 5), and now we are going to combine these physical concepts into a single unified framework, called *magneto-hydrodynamics (MHD)*. The coronal magnetic field has many effects on the hydrodynamics of the plasma. It can play a passive role in the sense that the magnetic geometry does not change (e.g., by channeling particles, plasma flows, heat flows, and waves along its field lines, or by maintaining a thermal insulation between the plasmas of neighboring loops or fluxtubes). On the other hand, the magnetic field can play an active role (where the magnetic geometry changes), such as exertion of a Lorentz force on the plasma, build-up and storage of nonpotential energy, triggering an instability, changing the topology (by various types of magnetic reconnection), and accelerating plasma structures [filaments, prominences, coronal mass ejections (CMEs)]. To understand and quantify all these phenomena we need the tools of magneto-hydrodynamics. The study of magneto-hydrodynamic processes in the solar corona is particularly unique because it represents the only laboratory of astrophysical plasmas where we can spatially resolve the structures of interest on the one hand, whilst exhibiting more powerful plasma processes than can be created in any terrestrial laboratory on the other hand. General introductions into MHD can be found in various textbooks (Priest 1982, 1994; Bray et al. 1991; Heyvaerts 2000; Somov 2000; Davidson 2001; Boyd & Sanderson 2003; Goossens 2003).

6.1 MHD Equations

Magneto-hydrodynamics (MHD) is a *fluid theory*, expressed in terms of macroscopic parameters, such as density, pressure, temperature, and flow speed of the plasma. This plasma reacts to (macroscopic) electric and magnetic forces as described by the Maxwell equations. However, particle motion in a plasma can also be described by microscopic physics, called *kinetic theory*, such as in terms of the *Boltzmann equation*, or *Vlasov equation*. The fluid approach of MHD can be derived from kinetic theory by defining appropriate statistical (average) quantities.

6.1.1   Particle Conservation

A particle distribution of species α (e.g., electrons, protons, or ions) has a statistical distribution in space $\mathbf{x}$, time t, and (microscopic) velocity $\mathbf{v}'$,

$$f_\alpha(\mathbf{x}, \mathbf{v}', t) \, d^3x \, d^3v' \,. \tag{6.1.1}$$

The total time-dependent changes of the distribution f_α contains, besides time-dependent variations $\partial/\partial t$, also advection (which we included in the total time derivative D/Dt defined in Eq. 4.1.1), and the action of forces $\mathbf{F}$,

$$\left(\frac{\partial f_\alpha}{\partial t} + \frac{\partial \mathbf{x}}{\partial t} \frac{\partial f_\alpha}{\partial \mathbf{x}} + \frac{\partial \mathbf{v}'}{\partial t} \frac{\partial f_\alpha}{\partial \mathbf{v}'} \right) = \left(\frac{\partial f_\alpha}{\partial t} + \mathbf{v}' \nabla f_\alpha + \frac{\mathbf{F}}{m_\alpha} \frac{\partial f_\alpha}{\partial \mathbf{v}'} \right)$$

$$= \left(\frac{D f_\alpha}{Dt} + \frac{\mathbf{F}}{m_\alpha} \frac{\partial f_\alpha}{\partial \mathbf{v}'} \right) \tag{6.1.2}$$

where the force $\mathbf{F}$ may include acceleration by an electric field $\mathbf{E}$, and the Lorentz force due to a magnetic field $\mathbf{B}$ (plus forces due to microscopic electromagnetic fields $\Delta \mathbf{E}$ and $\Delta \mathbf{B}$, omitted here),

$$\mathbf{F} = q \left[\mathbf{E} + \frac{1}{c} (\mathbf{v}' \times \mathbf{B}) \right] \,. \tag{6.1.3}$$

In the statistical description of Boltzmann, the total time-dependent changes of a particle distribution f_α is balanced by the change in the particle collision rate,

$$\frac{\partial f_\alpha}{\partial t} + \mathbf{v}' \nabla f_\alpha + \frac{q}{m_\alpha} \left[\mathbf{E} + \frac{1}{c} (\mathbf{v}' \times \mathbf{B}) \right] \frac{\partial f_\alpha}{\partial \mathbf{v}'} = \left(\frac{\partial f_\alpha}{\partial t} \right)_{coll} \,, \tag{6.1.4}$$

while the right-hand term vanishes in a collisionless plasma, which is referred to as the *Vlasov equation*,

$$\frac{\partial f_\alpha}{\partial t} + \mathbf{v}' \nabla f_\alpha + \frac{q}{m_\alpha} \left[\mathbf{E} + \frac{1}{c} (\mathbf{v}' \times \mathbf{B}) \right] \frac{\partial f_\alpha}{\partial \mathbf{v}'} = 0 \,. \tag{6.1.5}$$

We now define the macroscopic quantities in terms of statistical averages over microscopic particle distributions by integration over velocity space, such as the average *particle density* n per unit volume $[\text{cm}^{-3}]$,

$$n(\mathbf{x}, t) = \int f_\alpha(\mathbf{x}, \mathbf{v}', t) \, d^3v' \,, \tag{6.1.6}$$

or the *average velocity* $\mathbf{v}$ $[\text{cm s}^{-1}]$, which is the bulk velocity of the macroscopic fluid,

$$\mathbf{v}(\mathbf{x}, t) = \frac{\int \mathbf{v}' \cdot f_\alpha(\mathbf{x}, \mathbf{v}', t) \, d^3v'}{\int f_\alpha(\mathbf{x}, \mathbf{v}', t) \, d^3v'} \,. \tag{6.1.7}$$

Integrating the *Boltzmann equation* (6.1.4) in velocity space, using the definitions of the macroscopic quantities n and $\mathbf{v}$ (Eq. 6.1.6–7), and summing over the particle species α, we can directly derive the *equation of continuity* or *equation of mass conservation* (Eq. 4.1.2) (see e.g., Benz 1993, p. 51; Sturrock 1994, p. 169; Golub & Pasachoff 1997, p. 212; Somov 2000, p. 167; Heyvaerts 2000),

$$\frac{\partial n}{\partial t} + \frac{\partial}{\partial \mathbf{x}}(n\mathbf{v}) = 0 \qquad (6.1.8)$$

or, using the mass density $\rho = mn$ (Eq. 4.1.6), we have the same form as Eq. (4.1.2),

$$\frac{D}{Dt}\rho = -\rho \, \nabla \cdot \mathbf{v} \,. \qquad (6.1.9)$$

6.1.2 Momentum or Force Equation

In order to derive the momentum equation we have to multiply the Boltzmann equation (6.1.4) (with velocity components v_i) with the momentum $m_\alpha v_j$, and integrate over the velocity space d^3v,

$$\int m_\alpha v'_j \frac{\partial f_\alpha}{\partial t} d^3 v' + \int m_\alpha v'_j v'_i \frac{\partial f_\alpha}{\partial x_i} d^3 v' + \int v'_j F_i \frac{\partial f_\alpha}{\partial v_i} d^3 v'$$

$$= \int m_\alpha v'_j \left(\frac{\partial f_\alpha}{\partial t}\right)_{coll,i} d^3 v' \,. \qquad (6.1.10)$$

The first term corresponds to the temporal change of the mean momentum, while the second term involves a second moment of the distribution function and is generally expressed in terms of the pressure tensor p_{ij},

$$p_{ij} = \int m_\alpha \, v'_i v'_j f_\alpha \, d^3 v' \,. \qquad (6.1.11)$$

The right-hand side of Eq. (6.1.10) includes the momentum transfer between electrons and ions due to collisions, which cancels out as a net effect and does not contribute to the momentum of the combined plasma in the single-fluid description. Using the definitions of the statistical averages for the density n (Eq. 6.1.6), velocity $\mathbf{v}$ (Eq. 6.1.7), and force $\mathbf{F}$, the momentum equation can be expressed in terms of macroscopic parameters for the combined mass $m = \sum m_\alpha$ (see derivation in, e.g., Benz 1993, p. 51; Sturrock 1994, p. 169; Golub & Pasachoff 1997, p. 212; Somov 2000, p. 167; Heyvaerts 2000),

$$\frac{\partial}{\partial t}(nm\mathbf{v}) = -\nabla p + n\mathbf{F} \,, \qquad (6.1.12)$$

The force term may include (ignoring electric forces $\mathbf{F}_{El}$ parallel to the magnetic field) the Lorentz force $\mathbf{F}_{Lor}$ from magnetic fields, the gravity force $\mathbf{F}_{grav}$ (Eq. 3.1.4), and viscous forces $\mathbf{F}_{visc}$,

$$\mathbf{F} = \mathbf{F}_{Lor} + \mathbf{F}_{grav} + \mathbf{F}_{visc} \,, \qquad (6.1.13)$$

where viscous forces $\mathbf{F}_{visc}$ are defined by

$$\mathbf{F}_{visc} = \nu_{visc} mn \left[\nabla^2 \mathbf{v} + \frac{1}{3} \nabla(\nabla \cdot \mathbf{v}) \right] , \qquad (6.1.14)$$

with ν_{visc} being the coefficient of *kinematic viscosity*. Using the conservation of mass, $d(nm)/dt = 0$, the definition of the mass density $\rho = mn$, and expressing the Lorentz force $\mathbf{F}_{Lor} = \mathbf{j} \times \mathbf{B}$ with the current density $\mathbf{j}$ (Eq. 5.3.3), Eq. (6.1.12) can be written in the standard form of the MHD momentum equation,

$$\rho \frac{D\mathbf{v}}{Dt} = -\nabla p - \rho \mathbf{g} + (\mathbf{j} \times \mathbf{B}) + \mathbf{F}_{visc} , \qquad (6.1.15)$$

which corresponds to the hydrodynamic momentum equation (4.1.3), except that the Lorentz force and viscosity force is added.

6.1.3   Ideal MHD

Many astrophysical plasmas are characterized by a set of equations that is called *ideal MHD equations* and includes the MHD continuity equation (6.1.9), the momentum equation (6.1.15), Maxwell's equations (5.1.1−4), Ohm's law (5.1.10), and a specialized equation of state for energy conservation (e.g., incompressible, isothermal, or adiabatic). Thus, a full set of *ideal MHD equations* includes (for an adiabatic equation of state):

$$\frac{D}{Dt}\rho = -\rho \, \nabla \cdot \mathbf{v} , \qquad (6.1.16)$$

$$\rho \frac{D\mathbf{v}}{Dt} = -\nabla p - \rho \mathbf{g} + (\mathbf{j} \times \mathbf{B}) , \qquad (6.1.17)$$

$$\frac{D}{Dt}(p\rho^{-\gamma}) = 0 , \qquad (6.1.18)$$

$$\nabla \times \mathbf{B} = 4\pi \mathbf{j} , \qquad (6.1.19)$$

$$\nabla \times \mathbf{E} = -\frac{1}{c}\frac{\partial \mathbf{B}}{\partial t} , \qquad (6.1.20)$$

$$\nabla \cdot \mathbf{B} = 0 , \qquad (6.1.21)$$

$$\mathbf{E} = -\frac{1}{c}(\mathbf{v} \times \mathbf{B}) . \qquad (6.1.22)$$

The usage of such a set of ideal MHD equations involves a number of implicit approximations: (1) the plasma is charge-neutral, $\rho_E = 0$, which yields the Maxwell equations $\nabla \cdot \mathbf{E} = 0$ and $\nabla \cdot \mathbf{j} = 0$; (2) the plasma has a very large magnetic Reynolds number, which yields the electric field $\mathbf{E} = -(1/c)(\mathbf{v} \times \mathbf{B})$ from Ohm's law (Eq. 5.1.10); (3) the nonrelativistic approximation, $v \ll c$, which also implies that MHD time scales are much longer than electron or ion gyroperiods; (4) a highly collisional plasma, implying the MHD time scales are much longer than collisional time scales; (5) isotropic pressure, where the pressure tensor (Eq. 6.1.11) simplifies to the diagonal elements, $p_{ij} = \delta_{ij}p_\alpha$, which for an ideal gas is $p_\alpha = n_\alpha k_B T_\alpha$; (6) the total pressure is the sum of the partial pressures $p = \sum_\alpha p_\alpha$, which yields (with Eq. 4.1.5) $p = (n_e + n_i)k_B T \approx 2n_e k_B T_e$; and (7) adiabatic gas with energy equation of state $p\rho^{-\gamma} = const$. Approximations in such sets of ideal MHD equations (6.1.16–22) are discussed in more detail elsewhere (see, e.g., Priest 1982, p. 73; Benz 1993, p. 56; Boyd & Sanderson 2003, p. 61).

6.1.4 Energy Equation

The MHD energy equation can be derived from the Boltzmann equation (6.1.4) by multiplication with the kinetic energy $(1/2)m_\alpha v_j^2$ and integration over the velocity space d^3v,

$$\int \frac{m_\alpha}{2}v_j'^2 \frac{\partial f_\alpha}{\partial t}d^3v' + \int \frac{m_\alpha}{2}v_j'^2 v_i' \frac{\partial f_\alpha}{\partial x_i}d^3v' + \int v_j'^2 \frac{F_i}{2}\frac{\partial f_\alpha}{\partial v_i}d^3v'$$

$$= \int \frac{m_\alpha}{2}v_j'^2 \left(\frac{\partial f_\alpha}{\partial t}\right)_{coll,i} d^3v' . \tag{6.1.23}$$

The first term contains the temporal change of the mean kinetic energy (Eq. 4.1.30), defined as a macroscopic average over the microscopic kinetic energies with distribution f_α,

$$\varepsilon_{kin}n_\alpha = \frac{1}{2}m_\alpha n_\alpha < v'^2 > = \int \frac{1}{2}m_\alpha v'^2 f_\alpha d^3v' . \tag{6.1.24}$$

The second term of Eq. (6.1.23) contains the divergence of the *heat flux density* or *conductive flux* F_C, which we introduced in Eq. (6.1.23),

$$\nabla F_C = \nabla \left(\frac{1}{2}m_\alpha n_\alpha < v_j'^2 v_i' >\right) = \int \frac{1}{2}m_\alpha v_j'^2 v_i' \frac{\partial f_\alpha}{\partial x_i}d^3v' . \tag{6.1.25}$$

The third term of Eq. (6.1.24) represents the work done by the force **F**, which can include acceleration by an electric field **E**, work by the gravitational force $\mathbf{F}_{grav}$, emission of radiation, or heat input. The Lorentz force, $\mathbf{F} = \mathbf{v} \times \mathbf{B}$, of course, cannot do any work, and is thus excluded in this term. Since this force **F** is not dependent on the velocity, it can be taken out of the integral in the third term of Eq. (6.1.23) and partial integration yields

$$n_\alpha < F_i \cdot v_i' > = \int v_j'^2 \frac{F_i}{2}\frac{\partial f_\alpha}{\partial v_i'}d^3v' . \tag{6.1.26}$$

Using Ohm's law for the electric field, $\mathbf{j} = \sigma\mathbf{E}$ (Eq. 5.1.10), the definition of the current density, $\mathbf{j} = qn_\alpha\mathbf{v}/c$ (Eq. 5.3.2), or expressed with the magnetic field by $\mathbf{j} = (1/4\pi)(\nabla \times \mathbf{B})$ (Eq. 5.1.7), Eq. (6.1.26) can be expressed as

$$n_\alpha < F_i \cdot v_i > = n_\alpha q E_i v_i = \frac{n_\alpha q}{\sigma}j_i v_i = \frac{c}{\sigma}j_i^2 = \frac{c}{(4\pi)^2\sigma}(\nabla \times \mathbf{B})^2 . \tag{6.1.27}$$

Adding in the work done by other forces, such as work against gravitational force F_{grav}, emission by radiation with a rate of E_R per time and unit volume, and heating with a rate of E_H per time and unit volume, the complete energy term of work done by the forces is,

$$n_\alpha < F_i \cdot v_i > = \frac{c}{(4\pi)^2 \sigma} (\nabla \times \mathbf{B})^2 + n_\alpha F_{grav} v_i + E_H - E_R . \qquad (6.1.28)$$

Finally, the right-hand term of the energy equation (6.1.23), which describes the energy changes due to collisions between electrons and ions, is conserved when summed over all particle species, and thus the net effect is zero on the summed energy equation,

$$\sum_\alpha \int \frac{m_\alpha}{2} v_j^2 \left(\frac{\partial f_\alpha}{\partial t} \right)_{coll,i} d^3 v = 0 \qquad (6.1.29)$$

Thus, if we insert the macroscopic quantities defined in Eqs. (6.1.24−29) in the energy equation (6.1.23) and sum over all species, we obtain the energy equation for a single-fluid plasma in conservative form,

$$\frac{\partial}{\partial t} (n_\alpha \varepsilon_{kin}) - \nabla F_C + \frac{c}{(4\pi)^2 \sigma} (\nabla \times \mathbf{B})^2 + n_\alpha F_{grav} v_i + E_H - E_R = 0 . \quad (6.1.30)$$

which we can arrange in the same order as the energy equation in conservative form given in Eq. (4.1.29),

$$\frac{\partial}{\partial t} (n\varepsilon_{kin}) + n v \frac{\partial}{\partial x} (\varepsilon_{grav}) + \nabla F_C + \frac{c}{(4\pi)^2 \sigma} (\nabla \times \mathbf{B})^2 = E_H - E_R , \quad (6.1.31)$$

or in the form of Eq. (4.1.18),

$$\frac{1}{(\gamma - 1)} \frac{Dp}{Dt} + \frac{\gamma}{(\gamma - 1)} p \nabla v + \frac{c}{(4\pi)^2 \sigma} (\nabla \times \mathbf{B})^2 = E_H - E_R - \nabla F_C . \quad (6.1.32)$$

commonly used in the framework of *resistive MHD*.

6.1.5   Resistive MHD

In the ideal MHD approximation it is assumed that the time scales of MHD processes are much longer than collisional processes, which guarantees that all species stay close to a Maxwellian distribution all the time. A plasma with a local Maxwellian distribution has zero viscosity and heat conduction, and thus the viscosity term F_{visc} and heat conduction ∇F_C does not appear in the ideal MHD approximation. This leaves the electrical resistivity as the only remaining dissipative mechanism. We see in the energy equation (6.1.31) that a perfect electric conductor ($\sigma \mapsto \infty$) makes the electric dissipation term vanish, $c/\sigma (\nabla \times \mathbf{B})^2 \mapsto 0$. Thus, only some finite resistivity $\sigma \ll \infty$ allows for electric energy dissipation. The set of MHD equations (mass, momentum, and energy conservation) with finite electrical resistivity σ is called *resistive MHD*, consisting of the equations (as a function of ρ, $\mathbf{v}$, p, and $\mathbf{B}$):

$$\frac{D}{Dt}\rho = -\rho\nabla\cdot\mathbf{v}\,,\tag{6.1.33}$$

$$\rho\frac{D\mathbf{v}}{Dt} = -\nabla p - \rho\mathbf{g} + (\mathbf{j}\times\mathbf{B})\,,\tag{6.1.34}$$

$$\frac{1}{(\gamma-1)}\frac{Dp}{Dt} + \frac{\gamma}{(\gamma-1)}p\nabla\cdot\mathbf{v} + \frac{c}{(4\pi)^2\sigma}(\nabla\times\mathbf{B})^2 = E_H - E_R - \nabla F_C\,.\tag{6.1.35}$$

$$\frac{\partial\mathbf{B}}{\partial t} = \nabla\times(\mathbf{v}\times\mathbf{B}) - \frac{c}{4\pi\sigma}(\nabla\cdot\nabla)\mathbf{B}\,,\tag{6.1.36}$$

$$\nabla\cdot\mathbf{B} = 0\,,\tag{6.1.37}$$

$$\mathbf{j} = \frac{1}{4\pi}(\nabla\times\mathbf{B})\,,\tag{6.1.38}$$

$$\mathbf{E} = \frac{1}{4\pi\sigma}(\nabla\times\mathbf{B}) - \frac{1}{c}(\mathbf{v}\times\mathbf{B})\,,\tag{6.1.39}$$

Many of the same implicit approximations are made in resistive MHD as in ideal MHD, except for perfect conduction and adiabatic equation of state: (1) charge-neutrality, (2) nonrelativistic speeds, (3) highly collisional plasma, and (4) isotropic pressure. For further discussion see, for example, Boyd & Sanderson (2003, p. 59).

6.2 MHD of Coronal Loops

We discuss now MHD effects in coronal loops, such as magneto-statics of vertical fluxtubes (§ 6.2.1), MHD effects near magnetic nullpoints (§ 6.2.2), in curved coronal fluxtubes (§ 6.2.3), in twisted fluxtubes (§ 6.2.4), the MHD of emerging fluxtubes (§ 6.2.5), the MHD dynamics of coronal loops based on numeric simulations (§ 6.2.6), and the MHD stability in coronal loops (§ 6.3).

6.2.1 Magneto-Statics in Vertical Fluxtubes

Let us consider one of the simplest applications of the MHD equations, namely the case of a static equilibrium, where the time-dependence vanishes ($\partial/\partial t = 0$) and flows are constant ($\mathbf{v} = const$). Thus the left-hand side of the ideal MHD momentum equation (6.1.17), also called *Euler's equation*, vanishes and we have

$$0 = -\nabla p - \rho\mathbf{g} + \mathbf{j}\times\mathbf{B}\,.\tag{6.2.1}$$

Neglecting gravity $\mathbf{g} = (0, 0, g_\odot)$, which is exactly zero when we consider the horizontal pressure balance, and inserting the current $\mathbf{j} = (1/4\pi)(\nabla\times\mathbf{B})$ from Maxwell's equation (6.1.19), we obtain the equation of *magneto-statics*,

$$-\nabla p - \frac{1}{4\pi}\mathbf{B}\times(\nabla\times\mathbf{B}) = 0\,.\tag{6.2.2}$$

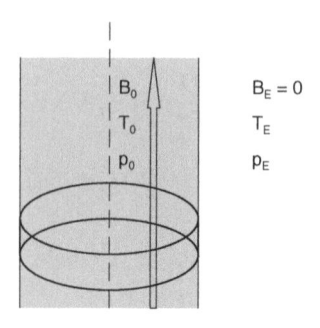

Figure 6.1: Concept of vertical fluxtube with internal parameters $(B_0 > 0, T_0, p_0)$ and external parameters $(B_E = 0, T_E, p_E)$, as it can be used to calculate the horizontal pressure balance at the boundary of a sunspot.

Using the vector identity $\nabla(\mathbf{a} \cdot \mathbf{b}) = (\mathbf{a} \cdot \nabla)\mathbf{b} + (\mathbf{b} \cdot \nabla)\mathbf{a} + \mathbf{a} \times (\nabla \times \mathbf{b}) + \mathbf{b} \times (\nabla \times \mathbf{a})$, which simplifies for $\mathbf{a} = \mathbf{b}$ to $\nabla(\mathbf{b}^2) = 2(\mathbf{b} \cdot \nabla)\mathbf{b} + 2\mathbf{b} \times (\nabla \times \mathbf{b})$, we obtain

$$-\nabla(p + \frac{B^2}{8\pi}) + \frac{1}{4\pi}(\mathbf{B} \cdot \nabla)\mathbf{B} = 0 \,, \tag{6.2.3}$$

where the first term represents the gradient of the total pressure, which is the sum of the thermal and magnetic pressure, while the second term represents the magnetic tension. For vertical fluxtubes, which are not bent and thus have no magnetic tension, we can neglect the second term. For the horizontal pressure balance we thus obtain the simple relation that the total pressure is constant,

$$-\nabla(p + \frac{B^2}{8\pi}) = 0 \,. \tag{6.2.4}$$

This simple magneto-static model is most suitable to describe the horizontal pressure balance in a sunspot, which contains a strong magnetic field (B_0) inside, so that the field outside can be neglected, $B_E = 0$ (Fig. 6.1). Denoting the pressure inside and outside with p_0 and p_E, the total pressure inside thus has to balance the thermal pressure outside,

$$p_E = p_0 + \frac{B_0^2}{8\pi} \,. \tag{6.2.5}$$

Inserting the pressures $p_0 = 2n_0 k_B T_0$ and $p_E = 2n_E k_B T_E$ and assuming equal densities inside and outside, $n = n_0 = n_E$, we obtain a relation between the magnetic field strength B_0 and temperature difference $(T_E - T_0)$,

$$B_0^2 = 16\pi n k_B (T_E - T_0) \,. \tag{6.2.6}$$

This immediately proves that every magnetic field $B_0 > 0$ implies a positive temperature difference (i.e., that the temperature inside the sunspot is cooler than outside). For a typical sunspot the temperatures in the photosphere are $T_E \approx 6000°$ K outside and $T_0 \approx 4500°$ K inside, while the density at photospheric height is about $n \approx 10^{18}$

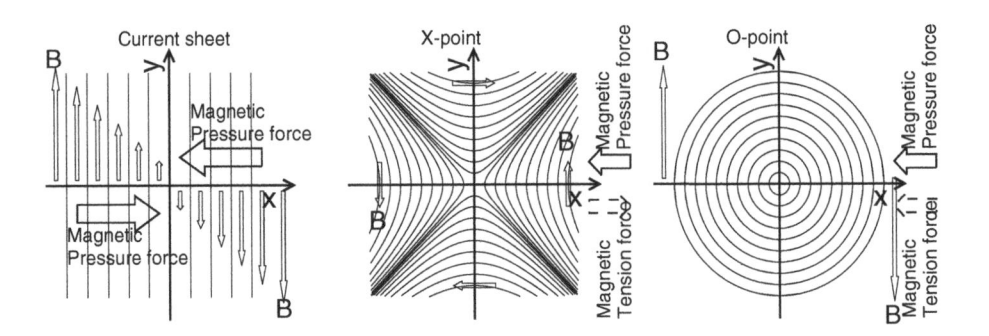

Figure 6.2: Magnetic pressure force (large solid arrow) and magnetic tension force (large dashed arrow) in three different magnetic field configurations: current sheet (left), X-point (middle), and O-point (right).

cm^{-3}, which allows according to Eq. (6.2.6) to balance a magnetic field of $B_0 = 3150$ G.

Higher up in the corona, where the density drops to $n \lesssim 10^9$ cm^{-3}, the same magnetic field could not be balanced by any observed temperature difference, which has the important consequence that the magnetic field has to fan out in the chromosphere (§ 5.8.1) so that it occupies the corona in a space-filling manner without field-free regions. In the corona, the plasma-β parameter is much lower than unity, so the magnetic pressure dominates over thermal pressure. The horizontal pressure balance (Eq. 6.2.4) can be expressed in terms of the internal (β_0) and external (β_E) plasma-β parameters,

$$\frac{B_0}{B_E} = \left(\frac{1 + \beta_E}{1 + \beta_0} \right)^{1/2} , \qquad (6.2.7)$$

which proves that for coronal conditions ($\beta_E \ll 1, \beta_0 \ll 1$) the difference between the magnetic field inside (B_0) and outside (B_E) a fluxtube is always small. For instance, for a typical coronal plasma with $B_E = 100$ G, $n_E = 10^9$ cm^{-3}, $T_E = 10^6$ K the plasma-β parameter is $\beta_E = 0.00035$. In a coronal fluxtube where the density is enhanced by 2 orders of magnitude ($n_0 = 10^{11}$ cm^{-3}), the plasma-β parameter inside the fluxtube becomes $\beta_0 = 0.035$, which requires only a decrease of the interior magnetic field by $B_0/B_E = 0.983$ to balance the thermal pressure, a mere 2%. Since the magnetic flux is conserved over the cross section of a fluxtube, the fluxtube has only to expand by a fraction of 1% to decrease the magnetic field inside the fluxtube by 2%. Such small loop expansions are not measurable.

6.2.2 Lorentz Force near Magnetic Nullpoints

In the previous example of a vertical fluxtube we assumed a uniform magnetic field inside and outside of the fluxtube, which has no magnetic tension ($\nabla \cdot \mathbf{B} = 0$) and thus yields a very simple relation for the pressure balance (Eq. 6.2.4). However, as soon as a magnetic field gradient is present in curved field lines $(\mathbf{B} \cdot \nabla)\mathbf{B} \neq 0$, a magnetic tension force occurs. We illustrate how this magnetic tension force compares with the magnetic pressure force in three simple examples of magnetic nullpoint geometries, where a gradient in the magnetic field is obviously present: in a current sheet (Fig. 6.2, left), in an X-point (Fig. 6.2, middle), and in an O-point (Fig. 6.2, right) (see also discussion in Priest 1994, p. 21).

Let us characterize a current sheet in two dimensions (x,y), with the magnetic field in the y-direction, varying linearly in field strength as a function of the distance from the center of the current sheet (Fig. 6.2, left),

$$\mathbf{B} = (B_x, B_y, B_z) = B_0(0, \frac{x}{l}, 0) , \qquad (6.2.8)$$

where l is a magnetic scale height. Applying the momentum equation (6.2.3) we find then that only the first term (i.e., the magnetic pressure) yields a non-zero contribution, while the second term (the magnetic tension force) vanishes ($dB_x/dx = 0, dB_y/dy = dx/dy = 0, dB_z/dz = 0$), so that we have,

$$-\nabla(p + \frac{B^2}{8\pi}) + \frac{1}{4\pi}(\mathbf{B} \cdot \nabla)\mathbf{B} = -\nabla(p + \frac{B_0}{8\pi}\frac{x^2}{l^2}) = -\nabla p - \frac{B_0}{4\pi}(\frac{x}{l}, 0, 0) , \quad (6.2.9)$$

which corresponds to an inward directed magnetic pressure (in the x-direction on the left side, and in anti-x-direction on the right side, see Fig. 6.2),

$$\nabla p = -\frac{B_0}{4\pi} \left(\frac{x}{l}, 0, 0 \right) . \qquad (6.2.10)$$

So there is no magnetic tension force, which is expected for straight magnetic field lines, and the magnetic pressure force points on both sides of the current sheet from a higher magnetic field towards the lower-field region, which is also expected just from the definition of the magnetic pressure ($p_m \propto B^2$). This behavior has the important implication that the magnetic pressure drives a lateral inflow of plasma into the current sheet, unless the lateral magnetic pressure is balanced by the thermal pressure from hotter plasma inside the current sheet.

Next let us turn to the magnetic field geometry of an X-point, as defined in Eq. (5.6.3), but here omit the asymmetry factor a for simplicity,

$$\mathbf{B} = (B_x, B_y, B_z) = B_0 \left(\frac{y}{l}, \frac{x}{l}, 0 \right) . \qquad (6.2.11)$$

The magnetic pressure force for such a magnetic field configuration is

$$-\nabla \left(\frac{B^2}{8\pi} \right) = -\nabla \left(\frac{B_0}{8\pi} \frac{(y^2 + x^2)}{l^2} \right) = \frac{B_0^2}{4\pi l} \left(-\frac{x}{l}, -\frac{y}{l}, 0 \right) , \qquad (6.2.12)$$

while the magnetic tension force is

$$\frac{B_0^2}{4\pi}(\mathbf{B} \cdot \nabla)\mathbf{B} = \frac{B_0^2}{4\pi} \left(\frac{y}{l}\frac{\partial}{\partial x} + \frac{x}{l}\frac{\partial}{\partial y} \right) \left(\frac{y}{l}, \frac{x}{l}, 0 \right) = \frac{B_0^2}{4\pi l} \left(\frac{x}{l}, \frac{y}{l}, 0 \right) , \qquad (6.2.13)$$

which turns out to be exactly the same magnitude as the magnetic pressure force, but with opposite sign. Thus, in such an X-point geometry the inward-directed magnetic pressure force (pointing towards the lower magnetic field) and the outward-directed tension force (which tries to reduce the curvature), exactly cancel out and no net pressure is exerted on the configuration (Fig. 6.2, middle). This implies that such an X-type point geometry is in pressure equilibrium and may be observable in the quiet corona (see, e.g., the TRACE image in Fig. 5.25), although we have not yet investigated its stability against disturbances. Equivalently, the current in this X-point configuration turns out to be zero,

$$\mathbf{j} = \frac{1}{4\pi}(\nabla \times \mathbf{B}) = \frac{1}{4\pi}\left(0, 0, \frac{\partial B_y}{\partial x} - \frac{\partial B_x}{\partial y}\right) = (0, 0, 0) \,, \qquad (6.2.14)$$

and implies that the Lorentz force is zero, corresponding to a pressure equilibrium between the magnetic pressure force and the magnetic tension force. An application of this X-type geometry to coronal loops can be made in postflare loops beneath a cusp, which may relax from an initial cusp-shaped geometry after reconnection through a transition of hyperbolic shapes into a near-circular geometry as indicated in Fig. 6.2 (middle panel). According to our understanding of the force balance obtained in this example, the downward-directed magnetic tension force would be stronger than the upward-directed magnetic pressure force (pointing to the lower field strength in the cusp) during the relaxation phase in the cusp after reconnection, until they balance out in the final, relaxed postflare position.

Let us finally proceed to the case of a magnetic O-point, which is mathematically related to a magnetic X-point (§ 5.6.1), with the difference that the magnetic field lines form ellipses (or circles) around the nullpoint instead of hyperbolae. The magnetic field around an O-point can be defined as (see Eqs. 5.6.1−3),

$$\mathbf{B} = (B_x, B_y, B_z) = B_0\left(\frac{y}{l}, -\frac{x}{l}, 0\right) \,, \qquad (6.2.15)$$

similar to the field around an X-point, except for an opposite sign in the y-coordinate. We can calculate the magnetic pressure force and magnetic tension force the same way as in Eqs. (6.2.12−14) with cartesian coordinates, or alternatively use a cylindrical coordinate system (see, e.g., Priest 1994, p. 24). We will obtain the same magnitude and sign for the magnetic pressure force as for the X-point (because B^2 is identical in both cases), but an opposite sign for the magnetic tension force. Thus, both the magnetic pressure force and the magnetic tension force will exert a force inward towards the center of the magnetic island (Fig. 6.2). There is a non-zero current associated with the magnetic island, which according to (Eq. 6.2.15) points into the anti-z-direction,

$$\mathbf{j} = \frac{1}{4\pi}(\nabla \times \mathbf{B}) = \frac{1}{4\pi}\left(0, 0, \frac{dB_y}{dx} - \frac{dB_x}{dy}\right) = \frac{B_0}{4\pi}(0, 0, -2) \qquad (6.2.16)$$

so such a magnetic island is not force-free. We will come back to the dynamical behavior of magnetic islands in the context of the tearing-mode instability during flares (§ 10.2.1).

6.2.3   Lorentz Force in Curved Fluxtubes

So far we have neglected the loop curvature in hydrodynamic models (§ 3 and 4). However, the loop curvature introduces a non-vanishing tension force term in the Lorentz force (Eqs. 6.2.1−3), which can be broken down into a magnetic pressure force and a magnetic tension force,

$$\mathbf{j} \times \mathbf{B} = -\nabla \left(\frac{B^2}{8\pi} \right) + \frac{1}{4\pi} (\mathbf{B} \cdot \nabla) \mathbf{B} . \qquad (6.2.17)$$

The effect of the loop curvature can be seen most clearly when we express the tension force in a coordinate system with unit vectors parallel ($\mathbf{e}_s$) and perpendicular ($\mathbf{e}_n$) to the magnetic field line (e.g., Priest 1982, p. 102; Priest 1994, p. 22),

$$\frac{B}{4\pi} \frac{d}{ds} (B\mathbf{e}_s) = \frac{B}{4\pi} \frac{dB}{ds} \mathbf{e}_s + \frac{B^2}{4\pi} \frac{d\mathbf{e}_s}{ds} = \frac{d}{ds} \left(\frac{B^2}{8\pi} \right) \mathbf{e}_s + \frac{B^2}{4\pi} \frac{1}{r_{curv}} \mathbf{e}_n , \qquad (6.2.18)$$

where r_{curv} is the curvature radius of the loop. The first term on the right-hand side cancels with the component of $-\nabla(B^2/8\pi)$ parallel to the magnetic field line. So, if the magnetic field has only a parallel component, $\mathbf{B} \parallel \mathbf{e}_s$, the Lorentz force is simply,

$$(\mathbf{j} \times \mathbf{B})_{curv} = \frac{B^2}{4\pi} \frac{1}{r_{curv}} \mathbf{e}_n , \qquad (6.2.19)$$

which is perpendicular (i.e., normal $\mathbf{e}_n$) to the magnetic field line and is stronger the smaller the curvature radius r_{curv}. Thus the Lorentz force is directed towards the curvature center and tries to reduce the curvature, unless a magnetic pressure gradient or thermal pressure gradient is present across the loop that compensates this curvature force. In the force-free configuration of an X-point shown in Fig. 6.2, the magnetic field was defined in such a way (Eq. 6.2.11) that the pressure force exactly compensates the magnetic tension or curvature force. We might ask how important is this Lorentz force due to the loop curvature. If we compare the Lorentz force $(\mathbf{j} \times \mathbf{B})_{curv}$ resulting from the curvature (Eq. 6.2.19) with the dominant term in the momentum equation (Eq. 6.2.1), i.e., the pressure gradient in the same (perpendicular) direction, $(-\nabla_\perp p)$, we find that the ratio scales with the perpendicular pressure gradient $\lambda_\perp$ and plasma β-parameter as

$$\frac{(\mathbf{j} \times \mathbf{B})_{curv}}{\nabla_\perp p} = \frac{(1/r_{curv})B^2/4\pi}{\nabla_\perp p} \approx \frac{\lambda_\perp}{r_{curv}} \frac{p_m}{p_{th}} = \frac{\lambda_\perp}{r_{curv}} \frac{1}{\beta} . \qquad (6.2.20)$$

Since the plasma β-parameter is generally small in the solar corona ($\beta \ll 1$, see Fig. 1.22), the curvature force in coronal MHDs can be neglected for slender loops with a transverse pressure scale height of $\lambda_T \ll \beta r_{curv} \ll r_{curv}$. If the magnetic field is force-free, the Lorentz force is exactly zero, which means that the tension force due to the loop curvature is exactly balanced by the perpendicular magnetic pressure force.

Studies that include the Lorentz force due to the loop curvature explicitly express the momentum equation with field-aligned coordinates (also called *Frenet vectors*; e.g., Garren et al. 1993; Petrie & Neukirch 1999, 2003), but without exact knowledge of the 3D magnetic field $\mathbf{B}$ it cannot be decided whether the curvature force in a static loop is balanced by a magnetic pressure gradient or by a thermal pressure gradient.

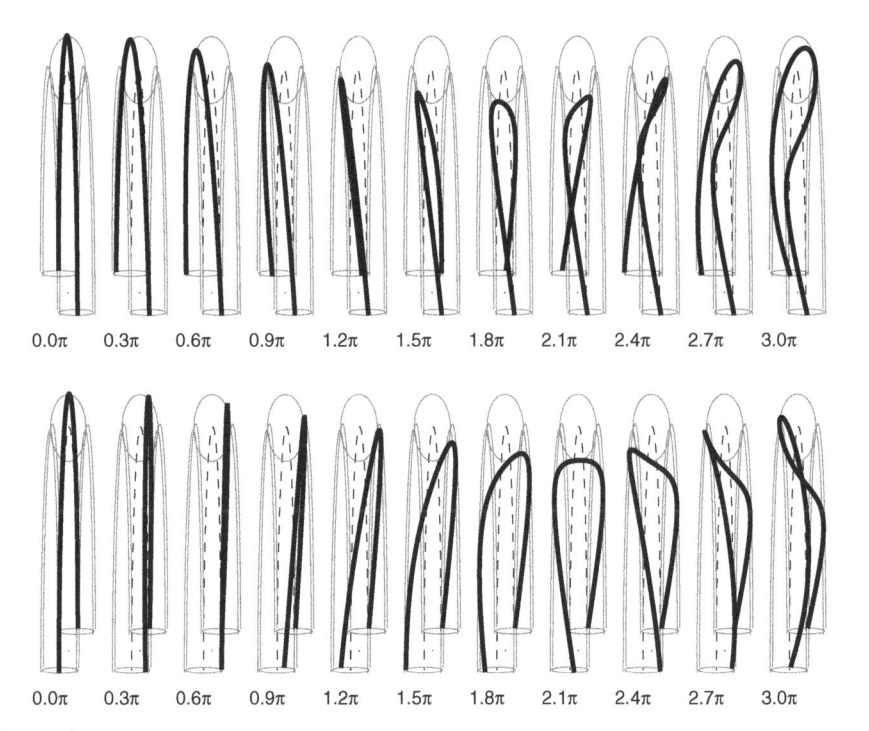

Figure 6.3: A series of loops is shown with twist angles of 0, 0.3π, ..., 3.0π, lying on a torus with a width-to-length ratio of 1:10. The projection angle is $+4°$ between the line-of-sight and the loop planes (top) and $-4°$ (bottom). Note the similarity of the sling-shot loop geometries with observed loop systems in Fig. 6.4.

6.2.4 Dynamics of Twisted Fluxtubes

The magnetic field enters the MHD momentum equation (6.1.17) as a $\mathbf{j} \times \mathbf{B}$ term. In § 5.5.1 we learned that a twisted fluxtube can be characterized with an azimuthal magnetic field component B_φ besides the parallel magnetic field component B_z (in cylindrical coordinates). The resulting Lorentz force in the momentum equation is then (Eq. 5.5.5),

$$\mathbf{F} = \mathbf{j} \times \mathbf{B} = [B_z j_\varphi - B_\varphi j_z, 0, 0] \ . \qquad (6.2.21)$$

The ratio of the azimuthal to the parallel magnetic field component can essentially be inferred in coronal loops from the twist angle or measured number of twists N_{twist} along the loop length $l = \pi r_{curv}$ (Eq. 5.5.1),

$$\frac{B_\varphi}{B_z} = \frac{r d\varphi}{dz} = \frac{r 2\pi N_{twist}}{l} = br \ . \qquad (6.2.22)$$

Thus, an important diagnostic for evaluation of the Lorentz force term is the measurement of twists, and a temporal change in the number of twists may provide a diagnostic on the dynamic evolution of the Lorentz force on the loop.

In Fig. 6.3 we show projections of twisted loop shapes, for a range of twist angles $N_{twist} \times 2\pi = 0, ..., 3\pi$. Obviously, the measurement of the number of twists seems

Figure 6.4: TRACE 171 Å observations of a postflare loop system on 1999-Sep-14, 08:13:18 UT (left frame) and 09:03:24 UT. The flare started around 06:34 UT and occurred near the limb. The loop planes are oriented nearly along the line-of-sight. This perspective is most favorable to display the twist of non-coplanar loops. The system seems to be highly twisted during the first time interval (left frame) and is more dipolar (coplanar) in the second configuration (right frame). Compare with theoretical models of twisted loop geometries with a similar projection in Fig. 6.3.

to be easiest for projections where the loop plane has almost the same direction as the line-of-sight. In the representation shown in Fig. 6.3 we have chosen a geometric model where the twisted field line lies on the surface of a semi-circular torus with annular radius r and curvature radius r_{curv}, with a ratio of $r/r_{curv} = 0.1$. The 3D coordinates of a twisted field line are then (with $X - Z$ the loop plane), parameterized as a function of the angle $\varphi = 0, ..., \pi$ along the loop length coordinate $s = r_{curv}\varphi$,

$$X(\varphi) = [r_{curv} + r\cos(\varphi N_{twist})]\cos(\varphi)$$

$$Y(\varphi) = r\sin(\varphi N_{twist})$$

$$Z(\varphi) = [r_{curv} + r\cos(\varphi N_{twist})]\sin(\varphi) , \qquad (6.2.23)$$

while an arbitrary projection into an observed image plane can be calculated with the coordinate transformation from loop coordinates (X, Y, Z) into image coordinates (x, y, z) given in Eqs. (3.4.13−15).

Observations of twisted loops are shown in Fig. 6.4, where a postflare loop system near the limb is shown as observed with TRACE in 171 Å. A peculiar feature of this observation is that the loop planes are oriented almost along the line-of-sight, so that the non-coplanarity of the loops is clearly seen and the number of twists can be measured from individual loops by fitting the projected geometries (Fig. 6.3) of our twist model given in Eq. (6.2.23). A qualitative comparison of Fig. 6.4 (left frame) with Fig. 6.3 suggests that many loops have twist angles of $N_{twist} \approx 1, ..., 1.5$. Thus, with Eq. (6.2.22) we estimate the azimuthal magnetic field to be of the order $B_\varphi/B_z = N_{twist}(2r/r_{curv}) \approx 0.2, ..., 0.3$, for $r/r_{curv} = 0.1$. If these loops were force-free, the current ratio would scale proportionally according to Eq. (6.2.21), $j_\varphi/j_z \approx B_\varphi/B_z \approx 0.2 - 0.3$. Vertical currents j_z have been measured (in other observations) from $\nabla \times \mathbf{B}$, using vector magnetographs, i.e.,

$$j_z(x, y) = \frac{1}{4\pi} \left(\frac{\partial B_y}{\partial x} - \frac{\partial B_x}{\partial y} \right) , \qquad (6.2.24)$$

with typical values of $j_z \approx 2 - 10 \, \mathrm{mA \, m^{-2}}$ (see Gary & Démoulin 1995 and references therein). However, the observations here show that these highly twisted loops are not stable, but rather relax into nearly untwisted dipolar geometries during the next hour (Fig. 6.4, right frame). This evolution suggests that the initial state was not force-free and that the Lorentz force tries to untwist the loops until they become force-free. Such a relaxation of a highly twisted postflare loop system into a near-dipolar configuration was also observed by Sakurai et al. (1992). There are also quantitative measurements of the untwisting of fluxtubes as observed with EIT over several days (Portier–Fozzani et al. 2001).

The MHD evolution of sheared or twisted fluxtubes or coronal loops is quite complex, but has virtually not yet been touched with quantitative data analysis of observations at all. Recent models of current-carrying loops motivated by laboratory experiments actually predict that the $\mathbf{j} \times \mathbf{B}$-force creates a pinch effect that gives a twisted fluxtube an axially uniform cross section also (Bellan 2002, 2003). Excessive twist could lead to a thermal instability and could force the core of the loop to cool faster than the outer envelope (Priest 1978). Instabilities in the current sheets between the sheared cylindrical layers of a fluxtube tend to relax the shear (Priest 1978), while rapid flows along the sheared layers tend to play a stabilizing role (Glencross 1980). If a fluxtube is thought as a bundle of filamentary structures, shear between adjacent filament channels is introduced by braiding due to random footpoint motions, which creates current sheets between the filaments (Mikić et al. 1989). The fluxtube evolves quasi-statically through sequences of equilibria with increasing twist, but it becomes linearly unstable to an ideal MHD kink mode when the twist exceeds a twist angle of about 2.5π or more, depending on the field structure (Hood & Priest 1979b, 1981). Twisted loops with local irregularities in the twist can develop an inflexional instability, which subsequently can relax into braided patterns, form a hammock configuration, or trigger a kink instability (Ricca 1997). The property of the magnetic field line twist can also affect the global dynamics of the solar corona. The emergence of a large-scale twisted magnetic flux from the convection zone is thought to lead to a kinked alignment of neighboring active regions (Matsumoto et al. 1996). Applying the photospheric random shear on a grander scale to the entire corona leads to eruption of part of the magnetic field (Steinolfson 1991).

A direct observation of twist-related dynamics has recently been made with the SoHO/SUMER instrument by measuring the spectral line profiles of active region

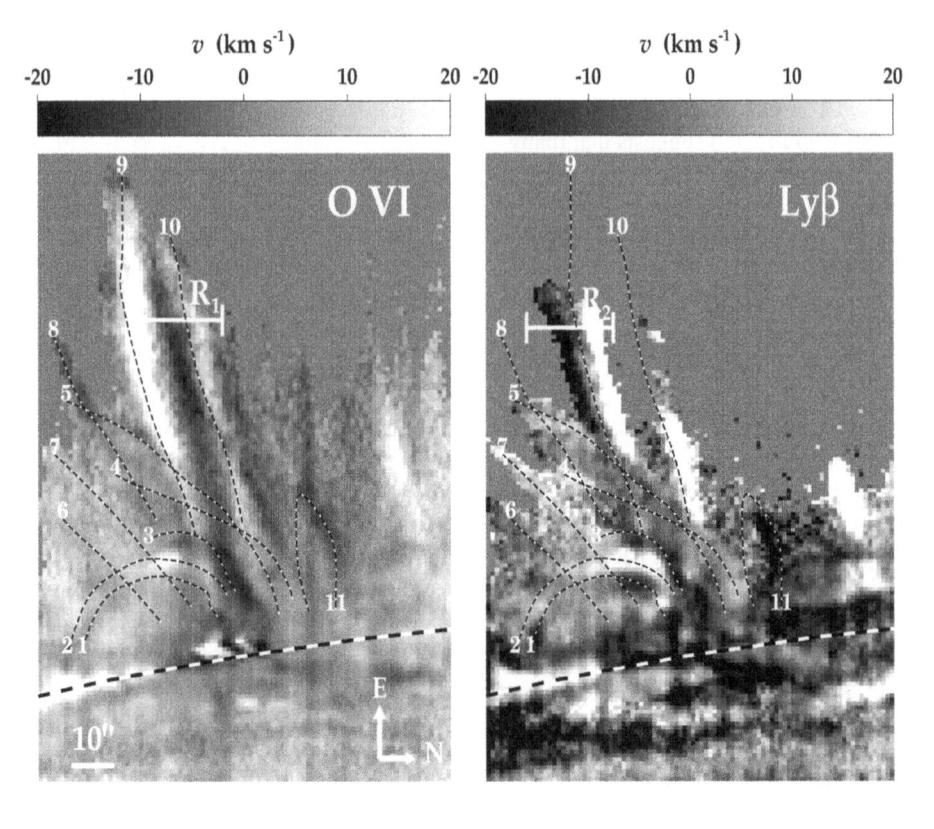

Figure 6.5: Doppler velocity maps of the active region in the O VI line (left) and the Ly$-\beta$ line (right). The bright features are redshifted and the dark features blueshifted. Note the large velocity shears at the cuts R_1 and R_2 (Chae et al. 2000a).

loops, which showed evidence of rotational motion of twisted loops (Chae et al. 2000a). If a coronal loop is twisting-up or untwisting, a rotational motion is expected, which should be observable as a Doppler blueshift at one lateral edge of the loop and as a redshift on the other edge. Such a Doppler signature has been observed in a number of active region loops in O VI ($T \gtrsim 10^5$ K) and in Lyβ ($T \lesssim 10^5$ K), with redshifts and blueshifts of up to v$\approx 20 - 30$ km s^{-1} (Fig. 6.5). This observation of rotational motion has interesting consequences for the underlying MHD dynamics. A static loop cannot twist-up or untwist without changing the position of the footpoints of the twisted magnetic field line. However, if the loop expands, the number of twists stays constant for fixed footpoints, but an observer will see a rotation because the twisted field line will rotate in azimuth at a fixed height. The same is true when a fluxtube erupts into an open field line, while helicity stays conserved to first order. The rotational velocity (v_{rot}) is essentially determined by the expansion velocity along the loop (v_{exp}) and by the twist angle (θ),

$$v_{rot} \approx v_{exp} \tan \theta \ . \tag{6.2.25}$$

Thus the observed rotation velocities of $v_{rot} \lesssim 30$ km s^{-1} would require quite rapid expansion speeds of $v_{exp} \geq v_{rot}$ for twist angles of $\theta \leq 45^0$. Therefore, such observations of rotating loops imply highly dynamic twisted fluxtubes that expand rapidly. Such a rapid expansion of a twisted fluxrope (e.g., with a vertical expansion speed of $v_{exp} \approx 400$ km s^{-1} between 2 and 5 solar radii), was observed and modeled by Mouschovias & Poland (1978). The interpretation in terms of vertical expansion would also predict that the twist angle θ reduces during expansion.

6.2.5 MHD Simulations of Emerging Fluxtubes

The formation process of coronal loops consists of two physically distinct steps: (1) the emergence of a buoyant subphotospheric magnetic fluxtube into the corona, which forms a dipolar magnetic field structure; and (2) and filling of coronal fluxtubes with heated plasma, which increases the density and emission measure of coronal loops. Generally it is thought that the two processes occur simultaneously. Therefore, the formation process of a coronal loop cannot be restricted to the corona itself, but rather includes the evolution of an MHD process starting at the bottom of the convection zone and propagating upward through the photosphere, chromosphere, transition region, and corona (Fig. 6.6), covering quite different regimes of plasma parameters (convection, diffusion, thermal conductivity, resistivity, and viscosity). The process of emergence of fluxtubes has therefore mainly been studied with numerical 3D MHD codes (see resistive MHD equations in § 6.1.5) that can handle these various plasma regimes.

A typical 3D MHD simulation of an emerging magnetic fluxtube is performed in Matsumoto et al. (1993). Initially, a horizontal magnetic flux sheet or tube is embedded at the bottom of a 3D box with photospheric/chromospheric conditions. The magnetic flux sheet or tube is unstable against undular modes ($\mathbf{k} \parallel \mathbf{B}$) of the *magnetic buoyancy instability* and starts to rise. The ascendance of the fluxtubes is subject to pinching in the longitudinal direction, as well as to fragmentation in the perpendicular direction due to the *interchange instability* ($\mathbf{k} \perp \mathbf{B}$). The interchange modes help to produce a fine fiber flux structure perpendicular to the magnetic field direction, while the undular modes determine the overall buoyant loop structure (Matsumoto et al. 1993). The shear flow at the surface of a buoyant fluxtube can be unstable to the Kelvin−Helmholtz instability (Tsinganos 1980). Starting the simulation at the base of the convection zone requires magnetic field strengths of typically $B \approx 30,000 - 100,000$ G to match the observed field strengths in the photosphere. However, simple buoyancy models of straight fluxtubes show a fragmentation into two parallel tubes with opposite senses and fluid circulation (Fig. 6.7), where the counter-rotating elements move apart from each other horizontally and eventually stop rising (Longcope et al. 1996). This simple model cannot explain the emergence into the corona. However, 3D MHD simulations that start with twisted fluxtubes show the evolution into buoyant helical structures through kink instability (Fig. 6.8 and Plate 8) which rise all the way through the convection zone and finally emerge into the corona, forming a sequence of strongly sheared magnetic loops (Matsumoto et al. 1998). The inclusion of twist into the buoyant fluxtubes is also important to explain the observed current in emerging active regions (Longcope & Klapper 1997; Fan et al. 1999), as has been observed by Van Driel−Gesztelyi et al. (1994), Leka et al. (1996), see Fig. 6.9, and Pevtsov et al. (1997). Once an Ω-fluxtube emerges through the photosphere, the fragmented fluxtube features eventually

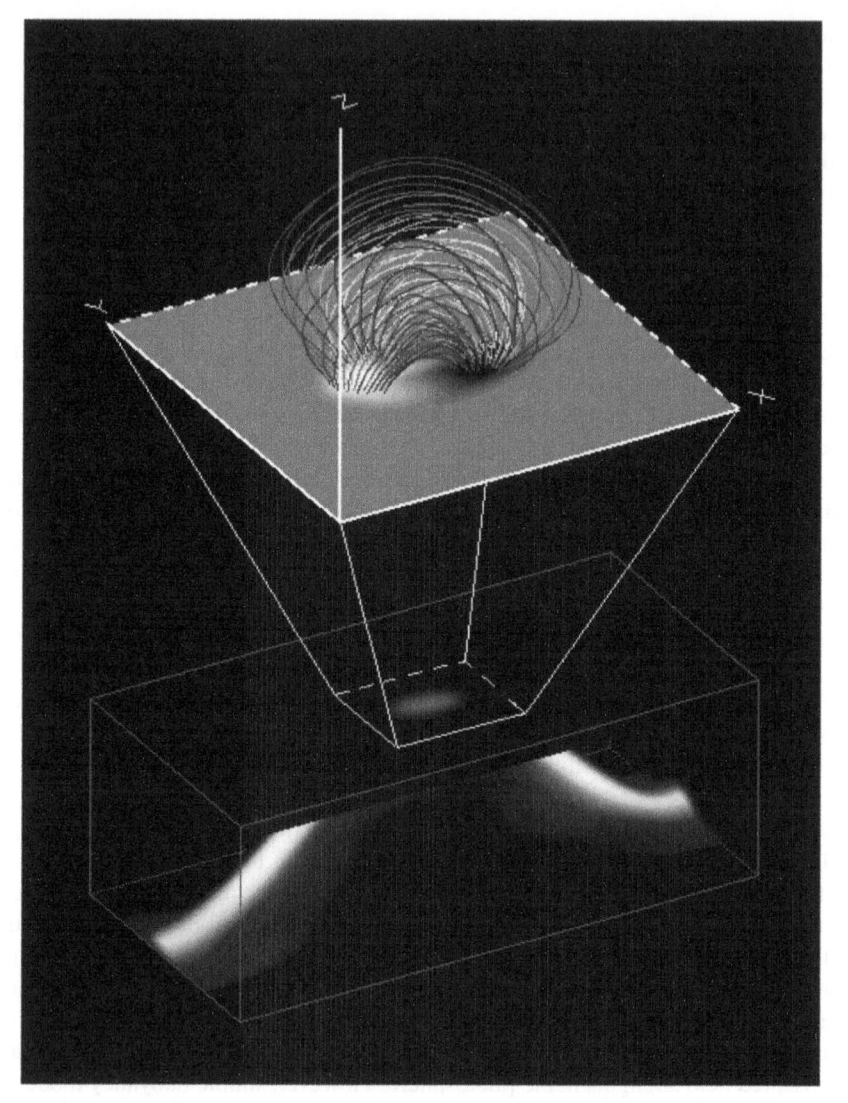

Figure 6.6: Snapshot of an emerging Ω-loop simulated with the ZEUS-3D/ANMHD code. The bottom rectangular box shows a volume rendering of $|\mathbf{B}|$, while the coronal magnetic field above the photosphere (on the $X - Y$ plane) is the top part. The magnetogram is shown with a greyscale (XY-plane), with the colors (white/black) indicating the (positive/negative) magnetic polarity (Abbett & Fisher 2003).

coalesce (Abbett et al. 2000). The complete evolution of a rising fluxtube from the base of the convection zone all the way up into the corona has been simulated by combining two 3D MHD codes, ANMHD (for the convection zone) and ZEUS-3D (for photosphere, chromosphere, and corona), see, e.g., Fig. 6.6 (Abbett & Fisher 2003).

Numerical simulations of emerging fluxtubes can explain many observed properties

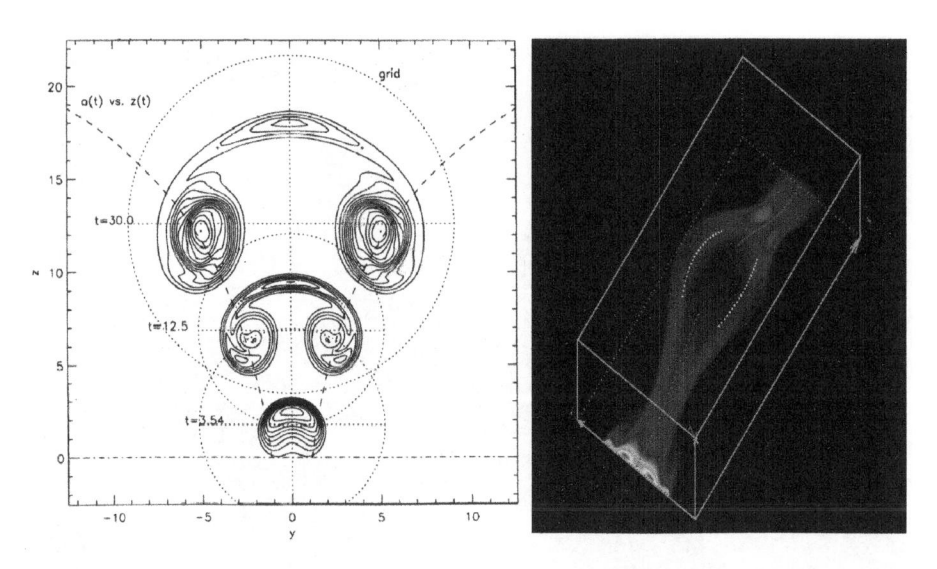

Figure 6.7: *Left:* contours of the relative underdensity $d\rho/\rho_0$ of an emerging magnetic flux-tubes at three different times. Note the fragmentation process into two structures (Longcope et al. 1996). *Right:* volume rendering of $|\mathbf{B}|$ during the emergence of an Ω-loop that has not fully fragmented (Abbett et al. 2000).

of bipolar regions, such as the asymmetry between leading and following spots (Fan et al. 1993), Joy's law of bipolar tilt angle (D'Silva & Choudhuri 1993), the asymmetric proper motion of sunspots (Caligiari et al. 1995), a relation between dipole tilt angle and net flux (Fisher et al. 1995), and a relation between the hemispheric helicity (with opposite chirality in both hemispheres) and helical turbulence in the convection zone (Σ-effect, Longcope et al. 1998). The 3D MHD simulations of Matsumoto et al. (1998) or Dorch et al. (1999) also reproduce the kinked alignment of S-shaped solar active regions along the same latitude, as can be seen in Yohkoh images (e.g., Fig. 5.17 or Fig. 1.9 bottom frame).

6.2.6 MHD Simulations of Coronal Loops

We summarized a number of hydrodynamic (HD) simulations on coronal loops in § 4.7, where the hydrodynamic processes are simulated without taking the role of the magnetic field into account, except for the assumption that plasma transport is restricted to one dimension along the field line, while cross-field transport is inhibited by the low value of the plasma β-parameter, generally assumed for coronal conditions. Including the magnetic field, however, using the full set of resistive MHD equations (§ 6.1.5), opens up a whole new dimension, where the Lorentz force plays an active role and can change the geometry and topology of the loops (by reducing the curvature, twisting, braiding, filamenting) and provides additional energy sources and sinks (in the form of nonpotential energy storage, dissipation by heating, acceleration, and disruption). In the following we describe a few numeric MHD simulations that exemplify the typical

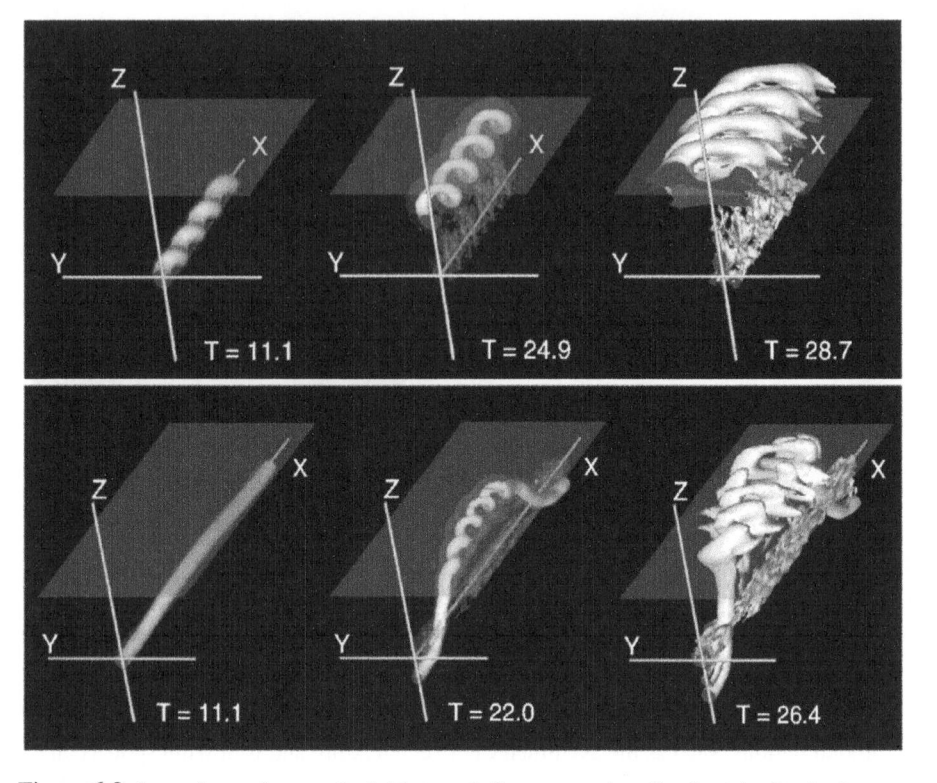

Figure 6.8: Isosurfaces of magnetic field strength for an emerging slim fluxtube for short wavelength perturbations (top) and long wavelength perturbations (Matsumoto et al. 1998).

MHD behavior of coronal loop simulations in general, while more special aspects in terms of coronal heating will be treated in § 9.

It is generally agreed that the dynamical evolution of coronal loops is driven by photospheric footpoint motion, where the coronal magnetic field lines are *line-tied*. The large photospheric density and the small value of the plasma resistivity warrant a *frozen flow* that moves the footpoints of coronal magnetic field lines around without *slippage*. Because the photospheric flows are organized in granular and super-granular cells, it drives a random walk of the footpoints of coronal magnetic field lines. The dynamical consequences determine the evolution and history of coronal fluxtubes, including the formation, equilibrium, linear instability, and nonlinear behavior. A time-dependent 3D MHD simulation performed by Mikić et al. (1990) starts with an initial uniform background magnetic field, where a twisted fluxtube is created by application of slow, localized photospheric vortex flows. For twists beyond a critical threshold (of $\gtrsim 4.8\pi$ according to Mikić et al. 1990), fluxtubes become linearly unstable to ideal or resistive modes. The nonlinear evolution of kink instabilities generate concentrated current filaments and their resistive dissipation provides a heating source for the corona. The 3D MHD simulation by Mikić et al. (1990) demonstrates the basic dynamic evolution of coronal fluxtubes in terms of twisting, current filamentation, and kink instability

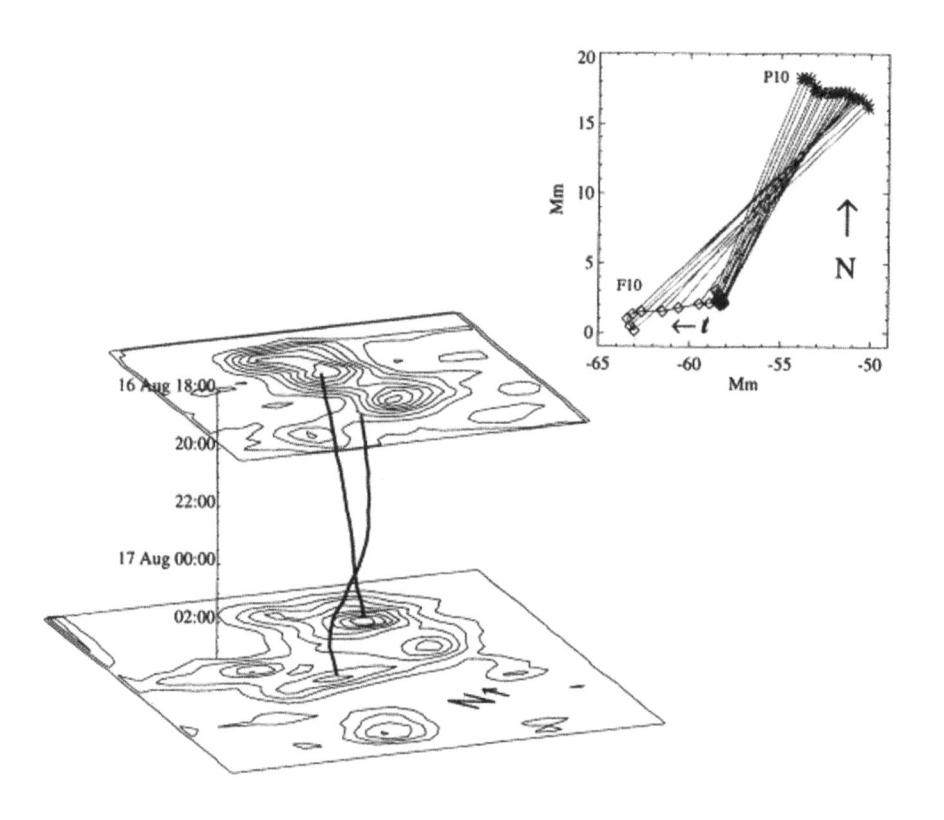

Figure 6.9: Proper-motion tracks for emerging dipoles from *Mees* CCD imaging spectrograph (MCCD) over a total elapsed time of $\approx$ 9 hours. The bottom left part shows two magnetograms at the beginning and end of the observation period, while the top right insert indicates the tracked position of two conjugate magnetic footpoints (P10 and F10), connected with a line for each time step. Note the twisting motion during the emergence process, which provides evidence that emerging loops are current-carrying (Leka et al. 1996).

in response to photospheric footpoint motion, but the energetics of the plasma in the fluxtube is not considered (no thermal conduction, radiative loss, or gravitational stratification).

One of the most realistic 3D MHD simulations of coronal loops to date has been performed by Gudiksen & Nordlund (2002), who simulate a typical scaled-down active region (Fig. 6.10 top) by constraining the coronal magnetic field with a potential field extrapolated from an observed magnetogram from SoHO/MDI (Fig. 6.10, bottom) and subject to random footpoint motion. The photospheric horizontal velocity pattern is generated from a velocity potential with randomly phased 2D Fourier components, with amplitudes $a(k)$, velocity power spectrum $P(k)$, typical velocities $v(k)$ at scales $1/k$, and corresponding turnover times $\tau(k)$ that follow the following power laws,

$$a(k) \propto k^{-p} \, ,$$

$$P(k) \propto k^{3-2p} \, ,$$

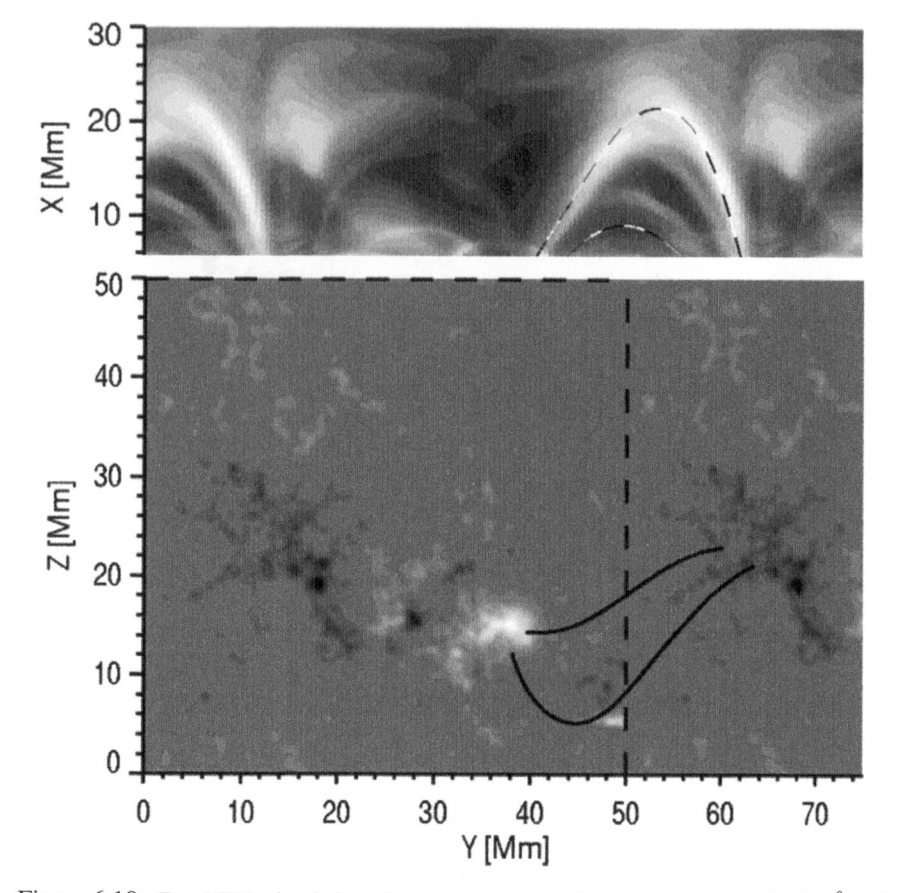

Figure 6.10: *Top:* MHD simulation of coronal loops, synthesized into a TRACE 171 Å emission measure image, averaged over the z-direction (line-of-sight) of the 3D simulation box. The greyscale is raised to the power of 0.5 to soften the contrast. *Bottom:* SOHO/MDI magnetogram of AR 9914, which has been used as the initial condition of the photospheric boundary of the simulated 3D box. One and a half box width is shown. Note the locations of the most prominent loops, marked with solid lines in the bottom panel and with dashed lines in the top panel (Gudiksen & Nordlund 2002).

$$v(k) \propto [kP(k)]^{1/2} \propto k^{2-p} \,,$$

$$\tau_k = \frac{1}{kv(k)} \propto k^{p-3} \,, \tag{6.2.26}$$

where $p = 1$ is chosen to be consistent with observed turnover times: ≈ 30 hr at supergranulation scales of ≈ 30 Mm, and ≈ 1000 s at granulation scales of ≈ 3 Mm. The 3D simulation box spans $30 \times 50 \times 50$ Mm^3, giving a resolution of 0.3 Mm vertically (including the transition region and corona) and 0.5 Mm horizontally. The lower boundary at the photosphere is kept at a constant temperature of 8000 K, while the upper coronal boundary is kept at an initial temperature of 10^6 K. The vertical extent of the chromosphere is realized with a thickness of 4 Mm, and the coronal plasma is initiated with pressure balance in a gravitationally stratified atmosphere, where the MHD code also includes the thermal conduction and radiative loss in the energy equation. The magnetic field was computed from an observed magnetogram of AR 9114. After an initial start-up phase, the simulation evolves towards a quasi-stationary configuration with a hot tenuous corona with temperatures of the order of a few MK, where slender loops are automatically created by the DC heating process from footpoint-driven braiding of magnetic field lines. The slender loops are visualized in Fig. 6.10 (top panel) by displaying the resulting emission measure in the temperature range ($T \approx 0.8 - 1.2$ MK) of the TRACE 171 Å filter. This simulation demonstrates a number of interesting physical effects. The average Joule dissipation, which balances the sum of thermal conduction losses and radiation losses decreases monotonically with height, as expected in a gravitationally stratified atmosphere (§ 3.6), and is in agreement with the footpoint-concentrated heating function measured observationally ($s_h \approx 17 \pm 6$ Mm, Aschwanden et al. 2000d; $s_h \approx 12 \pm 5$ Mm, Aschwanden et al. 2001). The average heating rate found in this simulation is $E_H \approx 2 \times 10^6$ erg cm^{-2} s^{-1} for plasma at temperatures of $T \gtrsim 1.0$ MK. Another interesting finding is that the heated coronal loops are not in hydrostatic equilibrium, exhibiting a nonhydrostatic stratification with a larger pressure scale height, and thus reflect the internal dynamics of the intermittent heating process. The loops are apparently formed by short duration excess heating episodes, which cause upflows that increase the density along those magnetic field lines that are subjected to the excess heating. The observed coronal loop structures therefore represent a selection effect of maximum density structures that are by definition not in hydrostatic equilibrium, consistent with the analysis of observed nonhydrostatic loops (§ 4.6). The 3D MHD simulations by Gudiksen & Nordlund (2002) represent a major breakthrough in our physical understanding of the formation and evolution of coronal loops, which could only be realized by a realistic driver in the photosphere and a realistic physical representation of the entire chromospheric transition region and coronal domain.

6.3 MHD Instabilities in Coronal Loops

We discussed the thermal stability of loops in § 4.3, which was mainly found to depend on the temperature gradient along a 1D fluxtube. When the transverse structure of a fluxtube or coronal loop is taken into account, the magnetic field comes into play and a whole new world of macroscopic (MHD) and microscopic (kinetic) processes become possible. The extra degree of freedom opens up a variety of instabilities. In Table 6.1 we give an overview of macroscopic hydrodynamic (HD) and magneto-hydrodynamic (MHD) plasma instabilities thought to be relevant in coronal loops.

The stability of a plasma structure such as a coronal loop can be investigated using two basic methods, either by analytical calculations of the eigen values of the linear growth rate in a perturbed system (*normal-mode method*), or by time-dependent numerical simulations starting with an initial perturbation of the system. An explicit example of the normal-mode method is given in § 4.3.1 for the case of the *radiative loss instability*, and applications to other instabilities are also given in Priest (1982,

Table 6.1: Overview of HD and MHD instabilities in coronal loops.

Instability	Unstable condition
1) Interchange or Pressure-Driven Instabilities:	
1.1. Rayleigh–Taylor instability:	
1.1.1 Hydrodynamic:	$\mathbf{g} \cdot \nabla n_0 < 0$
1.1.2 Hydromagnetic (Kruskal–Schwarzschild):	$\mathbf{k} \cdot \mathbf{B} = 0$
1.1.3 Hydromagnetic (Parker instability):	$\mathbf{k} \cdot \mathbf{B} \neq 0$
1.2) Kelvin–Helmholtz instability:	
1.2.1 Hydromagnetic:	$\mathrm{v}_1 > \mathrm{v}_{A,2}$
1.3) Ballooning instability:	$\mathbf{j} \times \mathbf{B} > \rho\mathbf{g}$
2) Thermal Instabilities:	
2.1) Convective instabilities:	$(dT/dz)_{crit}$
2.2) Radiatively-driven thermal instabilities:	$\tau_{cond} > \tau_{rad}$
2.3) Heating-driven thermal instabilities:	$s_H/L < 1/3$
3) Resistive Instabilities:	
3.1. Gravitational mode:	$F_{grav} > (\mathbf{j} \times \mathbf{B})$
3.2. Rippling mode:	$F_{adv} > (\mathbf{j} \times \mathbf{B})$
3.3. Tearing mode:	$(dB/dx)_{crit}$
4) Current Pinch Instabilities:	
4.1. Cylindrical pinch:	
4.1.1 Kink mode:	$B_{0\varphi}^2 \ln(L/a) > B_{0z}^2$
4.1.2 Sausage mode:	$B_{0\varphi}^2 > 2B_{0z}^2$
4.1.3 Helical/torsional mode:	$B_{0\varphi} > (2\pi a/L)B_{0z}$
4.2. Current sheet:	

§ 7). In the following we summarize the instabilities that can occur in coronal loops. An overview diagram is provided in Fig. 6.11 and more detailed descriptions of the instabilities can be found in reviews (Priest 1978; Sakurai 1989) and textbooks (Priest 1982, § 7; Bray et al. 1991, § 5.6.2; Sturrock 1994, § 15).

6.3.1   Rayleigh–Taylor Instability

The class of _interchange instabilities_ (Jeffrey & Taniuti 1966) includes instabilities that occur at the interface between two fluid layers, where the equilibrium is perturbed by rippling or meandering of the boundary layer, which can be thought of as an "interchange" of neighboring fluid elements. The classical hydrodynamic example is a pair of two horizontal fluid layers with different densities, ρ_1 and ρ_2, in a vertical gravitational field. A ripple in the horizontal interface implies that upper fluid elements are pushed downward and lose potential energy, while lower fluid elements are pushed upward and gain potential energy. It is intuitively clear that such rippling is only stable when the lighter fluid lies above the denser fluid (e.g. air above water waves), while an instability occurs when the lighter fluid lies below the heavier fluid ($\rho_2 > \rho_1$), the so-called *Rayleigh–Taylor instability* (Fig. 6.11-[1.1.1]). This instability condition is given when the density gradient of the unperturbed plasma (∇n_0) has an opposite direction to gravity (i.e., $\mathbf{g} \cdot \nabla n_0 < 0$).

6.3.2 Kruskal–Schwarzschild Instability

Kruskal & Schwarzschild (1954) showed that an analogous Rayleigh–Taylor instability occurs in a plasma supported by a magnetic field against gravity (Priest 1982, § 7.5.2; Boyd & Sanderson 2003, § 4.7.1). A ripple or disturbance along a longitudinal magnetic field will be restored by the curvature force of the bent magnetic field line, so that no instability arises in the longitudinal direction. In the perpendicular direction ($\mathbf{k} \cdot \mathbf{B} = 0$), however, there is no stabilizing curvature force for disturbances and longitudinal ripples can form, driven by perpendicular pressure differences (Fig. 6.11-[1.1.2]). Such longitudinal ripples do not involve any compression or rarefaction of the plasma, they just correspond to a redistribution (or interchange) of longitudinal plasma "pleats" in the perpendicular direction to the magnetic field, similar to the motion of curtain pleats in the wind. Applied to a coronal loop, such longitudinal ripples could possibly form along the surface of cylindrical loops, driven for instance by a pressure difference between a cool loop core plasma and a hotter surrounding sheet plasma (Priest 1978). However, such motions of ripples perpendicular to the (longitudinal) magnetic field are strongly inhibited by the line-tying of the magnetic field lines at the chromospheric/photospheric footpoints at both endpoints of the loop, and thus are unlikely to occur in active region loops (Priest 1978).

The *Parker instability* (Parker 1966, 1969, 1979; Shibata et al. 1989a,b) is a related kind of ideal MHD instability, occurring for long-wavelength perturbations with $\mathbf{k} \parallel \mathbf{B}$ (i.e., undular mode), while the *Kruskal–Schwarzschild instability* occurs for perturbations with $\mathbf{k} \perp \mathbf{B}$ and $\mathbf{k} \perp \mathbf{g}$. Numerical simulations of the Parker instability for an isolated horizontal magnetic flux sheet embedded in a two-temperature layer are shown in Shibata et al. (1989a,b).

6.3.3 Kelvin–Helmholtz Instability

The classical *Kelvin–Helmholtz instability* (Fig. 6.11-[1.2]) occurs at the interface between two fluids with different parallel flow speeds. Stable or laminar flows occur for small velocity differences, while the interface becomes unstable or turbulent when the Reynolds number exceeds a critical value. For plasmas, a longitudinal magnetic field has a stabilizing influence, as well as an azimuthal magnetic field component under certain conditions (Chandrasekhar 1961; Priest 1982, § 7.5.4). For the solar corona, such shear flows leading to a Kelvin–Helmholtz instability could potentially arise between adjacent fluxtubes with upflows and downflows. However, assuming comparable speeds, densities, and magnetic fields in the two counter-streaming fluxtubes, Priest (1978) estimates that the flow speed v_1 would have to exceed the Alfvén speed $v_{A,2}$ to produce a Kelvin–Helmholtz instability (i.e., $v_1 > v_{A,2}$), which is unlikely in active region loops.

6.3.4 Ballooning Instability

In 3D toroidal plasmas, and therefore in coronal loops, stability is often balanced by the inner and outer curvature radii, where inner curvatures are more favorable for stability and outer curvatures are less stable and can cause a *ballooning instability* locally (Fig. 6.11-[1.3]). The stability of cylindrical coronal loops and arcades against ballooning modes has been quantified by Hood (1986) and Hardie et al. (1991) for line-tying conditions at the coronal footpoints. Obviously, the ballooning instability is driven by the gas pressure gradient, which is balanced by the magnetic tension force in force-free magnetic fields. Thus, the ballooning instability is only an issue in non-force-free fields. In large-scale flare loops, the centrifugal or Lorentz force $\mathbf{j} \times \mathbf{B}$ could possibly exceed the gravity force $\rho\mathbf{g}$ in high-β regions (which may occur at altitudes of $h \gtrsim 0.25 R_{\odot}$, see § 1.8), and thus produce the right conditions for a ballooning instability with subsequent disruption of the flare loop (Shibasaki 2001).

6.3.5 Convective Thermal Instability

Thermal instabilities are driven by a temperature gradient. A chief example is the convective instability (Fig. 6.11-[2.1]), where a horizontal layer of viscous, thermally conducting fluid is heated from below and becomes unstable when the temperature difference between the lower and upper boundary becomes too large. Convection cells form in hexagonal patterns, also known as *Bénard cells*. Such convection dynamics occurs in the solar interior (i.e., in the convection zone at a radius of $r \gtrsim 0.7 R_{\odot}$), as well as on a finer scale in the photospheric granulation pattern. The type of convective instability depends on the thermal diffusivity and magnetic diffusivity (Priest 1982, § 7.5.6; Stix 2002, § 6). In the solar corona, plasma flows are essentially 1D along the magnetic field lines because of the low value of the plasma β-parameter, and thus convection is inhibited along a 1D coronal loop, even when a critical temperature gradient occurs along a loop. Unstable temperature gradients can only be stabilized by (parallel) thermal conduction or by uni-directional flows along the loops.

6.3.6 Radiatively-Driven Thermal Instability

Although all coronal loops have a large temperature difference between the coronal looptop and the chromospheric footpoints, stationary solutions exist where an energy balance is achieved between heating, thermal conduction, and radiative loss (see hydrostatic solutions in § 3.5 and hydrodynamic solutions in § 4.1 and 4.2). Under certain conditions, however, thermal conduction is not efficient enough to balance the radiative loss, leading to a so-called *radiatively-driven thermal instability* (Fig. 6.11-[2.2]) which we discussed in § 4.3.1 (see also Field 1965; Priest 1978; Priest 1982, § 7.5.7; Hood & Priest 1979a; Roberts & Frankenthal 1980). Numerical simulations of the radiatively-driven thermal stability of loops were performed by Klimchuk et al. (1987) and Cally & Robb (1991), which demonstrated that low-lying compact hot loops are generally thermally unstable, while large-scale hot loops are quite stable even to large-amplitude disturbances. Dahlburg et al. (1987) studied the nonlinear evolution of radiation-driven thermally unstable (unmagnetized) fluids, finding a turbulent contraction of the con-

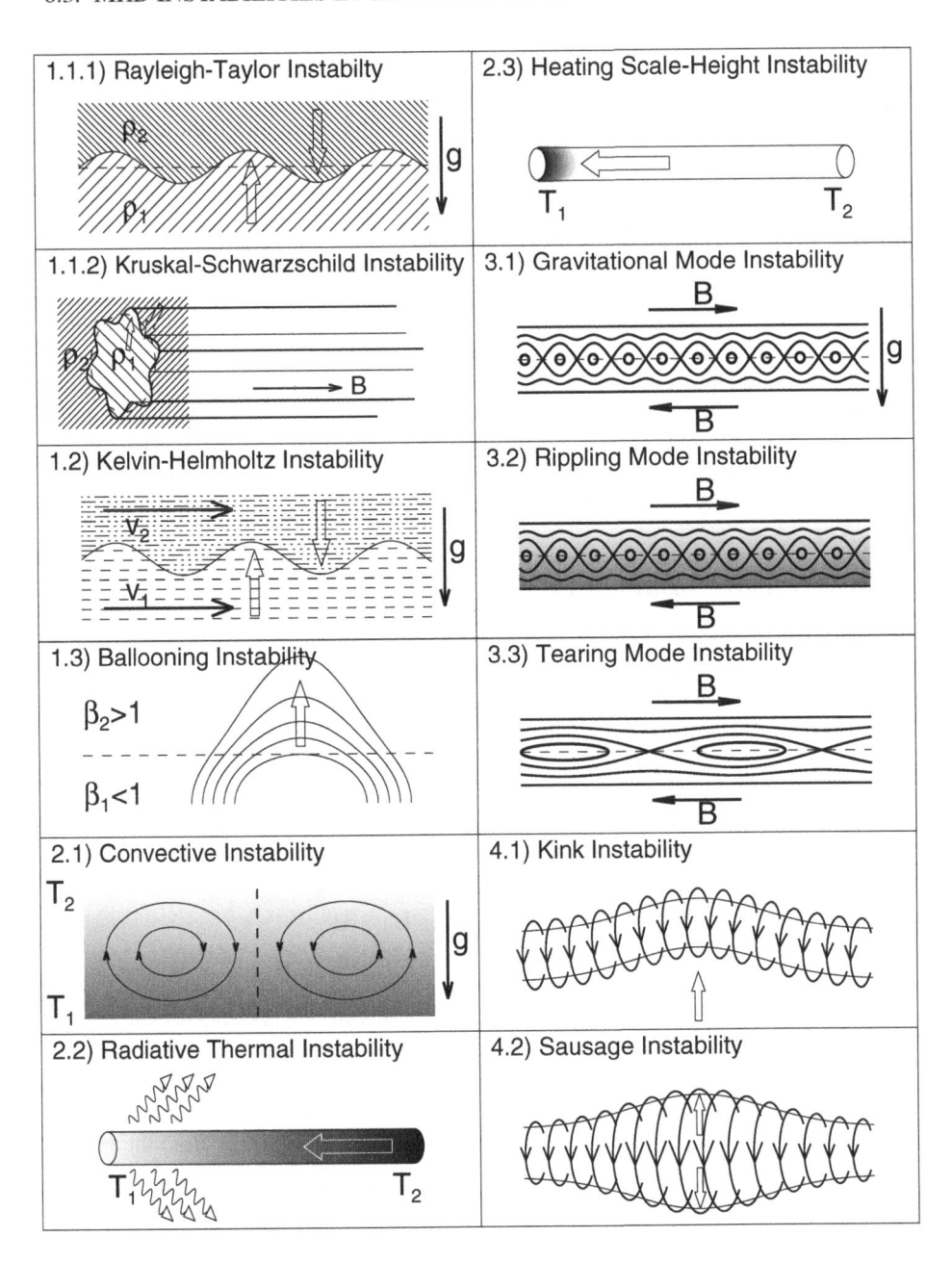

Figure 6.11: Hydrodynamic and MHD instabilities that can occur in coronal loop plasmas are illustrated (in the same order as in Table 6.1). Different densities (ρ_1, ρ_2) are rendered with hatched linestyle, different velocities (v_1, v_2) with dashed linestyle, temperature gradients (T_1, T_2) with greyscales, longitudinal magnetic field lines (B_0) with thin solid lines, azimuthal magnetic field components (B_φ) with thick solid lines, and radiation with wiggly lines. The directions of the disturbances that lead to an instability are indicated with thick white arrows.

densed region with subsequent increased radiative loss. The observational signature of coronal loops that undergo a radiatively-driven thermal instability occurs as *coronal rain* visible in Hα and UV (Schrijver 2001a) and has been called the *catastrophic cooling phase* (see § 4.5.4).

6.3.7 Heating Scale-Height Instability

Another thermal imbalance can occur when the heating function is very localized (Fig. 6.11-[2.3]), in particular if the exponential heating scale height at the footpoint is smaller than about one third of the loop half length (i.e., $s_H \lesssim L/3$). Numerical simulations have shown that the arising temperature gradients cannot be stabilized by radiative loss and thermal conduction (Winebarger et al. 2002), so that such a nonuniformly heated loop undergoes a dynamic cooling phase (see § 4.3). The heating is insufficient to maintain any stable temperature at the looptop, and thus the loop will cool down and finally enter a catastrophic cooling phase.

6.3.8 Resistive Instabilities

There is also a class of *resistive instabilities*, which are driven by a finite resistivity σ that converts magnetic energy into Ohmic heating. Since the magnetic diffusion time scales $t_D = l^2/\eta$ (with l the width of a current sheet and $\eta = 1/\sigma$ the magnetic diffusivity) are extremely long in coronal conditions, such resistive effects are only significant for very small spatial scales l (e.g., in thin current sheets). A random disturbance of a thin current sheet could occur in the form of a lateral inflow with speed $\mathbf{v}$, which causes a current density $\mathbf{j} \approx (\sigma/c)(\mathbf{v} \times \mathbf{B})$ (see Ohm's law, Eq. 5.1.10) and a Lorentz force which opposes the flow,

$$\mathbf{F}_L = \mathbf{j} \times \mathbf{B} \approx \frac{\sigma}{c}(\mathbf{v} \times \mathbf{B}) \times \mathbf{B} \,. \tag{6.3.1}$$

A resistive instability occurs then in the current sheet when the driving force of the inflow exceeds the opposing Lorentz force. There are different possible driving forces (Furth et al. 1963), such as : (1) gravity, leading to the *gravitational mode* (Fig. 6.11-[3.1]); (2) a spatial variation across the current sheet in magnetic diffusivity, e.g. caused by a temperature gradient, leading to the *rippling mode* (Fig. 6.11-[3.2]); or (3) a sheared magnetic field, leading to the *tearing mode* (Fig. 6.11-[3.3]). Both the gravitational and rippling-mode instability occur on wavelengths that are shorter than the width l of the current sheet, while the tearing mode occurs at longer wavelengths ($kl < 1$). The fastest growth rate is given by the longest wavelength, which has a value corresponding to the geometric mean of the magnetic diffusion time τ_D and Alfvén crossing time τ_A,

$$\tau_{tear} \approx \sqrt{\tau_D \tau_A} \,. \tag{6.3.2}$$

Since sheared magnetic fields have been often been observed in active regions before a flare, the tearing-mode instability plays a major role in triggering solar flares (see § 10 on magnetic reconnection).

6.3.9   Kink Instability (m=1)

Current pinch instabilities have been studied in laboratories, where cylindrical plasma columns are *pinched* (i.e., confined) by azimuthal magnetic field components B_φ and the related currents j_φ. Due to the cylindrical geometry of coronal loops, current pinch instabilities represent an important class of MHD instabilities in the solar corona, such as the kink mode, sausage mode, or helical/torsional mode. Considering modes that are periodic in axial direction z or azimuth angle φ, axially symmetric perturbations $\xi(\mathbf{x})$ can be written as,

$$\xi(\mathbf{x}) = \xi(r)e^{i(kz+m\varphi)} , \qquad (6.3.3)$$

where $m = 0$ is called the *sausage mode* (Fig. 6.11-[4.2]) and $m = 1$ the *kink mode* (Fig. 6.11-[4.1]). We see that $m = 0$ is independent of the azimuth angle φ and thus represents a purely radial oscillation, while $m = 1$ involves an azimuthal asymmetry (i.e., a sinusoidal oscillation in a particular azimuthal plane φ).

Let us consider the *kink instability*, in which a sinuous displacement grows along a fluxtube (Fig. 6.11-[4.1]). The kink instability can occur in fluxtubes that have azimuthal magnetic fields $B_{0\varphi}$ (or associated azimuthal currents j_φ) above some critical threshold. The basic physical effect can easily be understood from the diagram shown in Fig. 6.11-[4.1]. If a kink-like displacement deforms a straight fluxtube, the azimuthal magnetic field lines move closer together at the inner side of the kink than at the outer side, which creates a magnetic pressure difference $\nabla(B_\varphi^2/8\pi)$ towards the direction of the lower field (which is the outer side of the kink), and thus acts as a force in the same direction as the kink displacement, and therefore makes it grow further. A disturbance (i.e., the kink displacement) that is not stabilized by the resulting forces (i.e., the magnetic pressure difference) leads by definition to an instability. Of course, there are other forces present that may contribute to stabilization, such as the longitudinal magnetic pressure force $\nabla(B_z^2/8\pi)$, the gas pressure gradient ∇p, or line-tying at the ends of the fluxtube anchored in the photosphere.

The kink instability has been originally studied by Kruskal et al. (1958), Shafranov (1957), and Suydam (1958), while special applications to coronal loops were considered in Anzer (1968), and Hood & Priest (1979b). For force-free magnetic fields of uniform twist, a critical twist of $\varphi_{twist} \gtrsim 3.3\pi$ (or 1.65 turns) was found to lead to kink instability, while the critical value ranges between 2π and 6π for other types of magnetic fields. The effect of line-tying was included by Raadu (1972) and Velli et al. (1990), that of pressure gradients by Giachetti et al. (1977), and both combined by Hood & Priest (1979b) or Lothian & Hood (1992). The nonlinear development of the kink instability was studied by Sakurai (1976). Analytical evaluations of instability thresholds, of course, involve highly simplified geometries, while numerical MHD simulations provide more realistic conditions. Numerical MHD simulations of an increasingly twisted loop system demonstrated linear instability of the ideal MHD kink mode for twist angles in excess of $\approx 4.8\pi$ (or 2.4 turns) (Mikić et al. 1990). The kink instability is believed to be an important trigger for filament eruption, flare initiation, and CMEs (§ 17.3.2).

6.3.10 Sausage Instability (m=0)

In a cylindrical plasma column, the radially inwards directed $\mathbf{j} \times \mathbf{B}$ force is balanced by the outward pressure gradient ∇p. However, in the absence of a longitudinal magnetic field B_{0z}, the plasma column is unstable at locations where the confining field is concave, leading to the *sausage instability* (Fig. 6.11-[4.2]). On the other side, the pinched plasma column can be stabilized against the sausage instability with a sufficiently large longitudinal field B_{0z}. Stability is warranted for $B_{0z}^2 > (1/2)B_\varphi^2$. An analytical derivation of the stability analysis of the sausage mode can be found in Sturrock (1994, § 15), and general discussions of the sausage instability are given in Hasegawa (1975, § 3.5), Priest (1982, § 7.5.3), and Bray et al. (1991, § 5.6.2). Although sausage-type waves (e.g., Berghmans et al. 1996) and sausage-mode oscillations (Roberts et al. 1984) have been applied to coronal loops, no imaging observation of a sausage instability in coronal loops has been reported so far, so that it seems that the sausage stability criterion is generally met in coronal magnetic fields (i.e., the azimuthal magnetic field of a twisted coronal loop seems never to reach the critical limit of $|B_\varphi| > 1.4|B_{0z}|$).

6.4   MHD of Quiescent Filaments and Prominences

Some horizontal magnetic field lines overlying a neutral line (i.e., the magnetic polarity inversion line) of an active region are found to be filled with cool gas (of chromospheric temperature), embedded in the much hotter tenuous coronal plasma. On the solar disk, these cool dense features appear dark in $H\alpha$ or EUV images, in absorption against the bright background, and are called *filaments* (Fig. 6.12, bottom), while the same structures appear bright above the limb, in emission against the dark sky background, where they are called *prominences* (Fig. 6.12, top). Thus, *filaments* and *prominences* are identical structures physically, while their dual name just reflects a different observed location (inside or outside the disk). A further distinction is made regarding their dynamic nature: *quiescent filaments/prominences* are long-lived stable structures that can last for several months, while *eruptive filaments/prominences* are usually associated with flares and coronal mass ejections (CMEs). Typical parameters of prominences are given in Table 6.2.

Reviews on filaments/prominences can be found in Hirayama (1985), Martin (1990, 1998), Kucera & Antiochos (1999), Anzer (2002), Patsourakos & Vial (2002), in the encyclopedia articles of Gaizauskas (2001), Zirker (2001), Moore (2001), Engvold (2001a), Martin (2001), Van Ballegooijen (2001), Oliver (2001a), Rust (2001), Tandberg−Hanssen (2001), as well as in the textbooks of Svestka (1976), Priest (1982), Zirin (1988), Foukal (1990), Tandberg−Hanssen (1974, 1995), Schrijver & Zwaan (2000), Stix (2002), and proceedings edited by Jensen et al. (1979), Poland (1986), Ballester & Priest (1988), Priest (1989), Webb et al. (1998), and Kaldeich−Schurmann & Vial (1999). Despite this extensive literature there are still a number of unresolved theoretical problems in prominences, as pointed out by Forbes (1997).

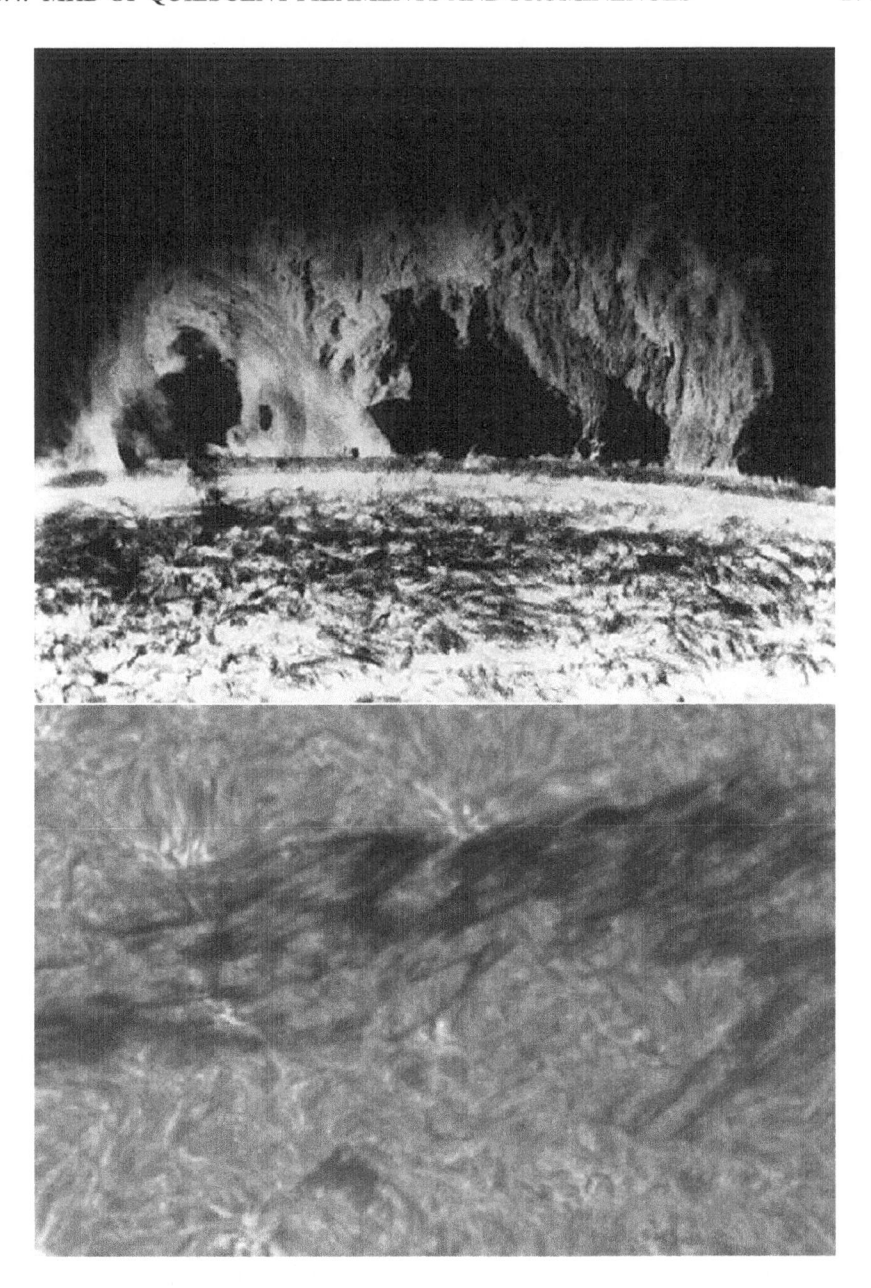

Figure 6.12: *Top:* A small "hedgerow-shaped" prominence above the limb (observed by the Big Bear Observatory, Zirin 1988), which appears bright in Hα against the black sky background. The fine structure of the prominence consists of many vertical threads. *Bottom:* A large polar crown prominence on the disk (observed with the Swedish Vacuum Solar Telescope on La Palma on 1998-Jun-19, Engvold 2001a), appearing dark in Hα against the bright chromospheric background. Note the orientation angle of the prominence threads by $\approx 20° - 30°$ relative to the major prominence axis.

Table 6.2: Typical physical parameters of prominences inferred from SoHO measurements (Patsourakos & Vial 2002), Hα (Engvold 2001a), and other sources.

Physical parameter	Range of values
Electron density (at $T = 10^5$ K)	$n_e = 1.3 \times 10^9 - 3 \times 10^{11}$ cm^{-3}
Neutral hydrogen column density	$n_H = 10^{16} - 10^{19}$ cm^{-2}
Electron temperature	$T_e = 5000 - 15,000$ K
Gas pressure (at $T = 10^5$ K)	$p = 0.03 - 0.38$ dyne cm^{-2}
Length of prominence	$L \approx 60,000 - 600,000$ km
Height of prominence	$h \approx 10,000 - 100,000$ km
Width of prominence	$w \approx 4000 - 30,000$ km
Number of threads	$N_{thread} = 15 - 20$
Lengths of threads	$L_{thread} = 5000 - 35,000$ km
Widths of threads	$d_{thread} = 200 - 400$ km
Filling factor	$q_{fill} = 0.001 - 0.1$
Lifetime of threads	$\tau_{thread} = 1 - 10$ min
Oscillation frequencies	$T = 3 - 5, 6 - 12, > 40$ min
Vertical and horizontal flow velocity (at $T = 10^5$ K)	v$= 2 - 13$ km s^{-1}
Nonthermal velocity (at $T = 10^5$ K)	Δv$= 26$ km s^{-1}
Magnetic field density	$B = 4 - 40$ G

6.4.1 Magnetic Field Configuration

Magnetic field measurements (e.g., Querfeld et al. 1985; Bommier et al. 1986a,b) with the Zeeman and Hanle effect have shown that the magnetic field in prominences has field strengths in the order of $B = 4 - 20$ G and that the field is mainly directed along the length of the prominence. On average, the magnetic field is inclined to the prominence axis by an angle of $\approx 25°$ (Leroy 1989), see example given in Fig. 6.12 (bottom). Quiescent prominences often have an inverse polarity with respect to the overlying loop arcade. These observations point to a model where the cool prominence material is suspended by a highly twisted fluxtube inside a loop arcade (Fig. 6.14). The high twist explains the inclined angle of $\approx 25°$, and the location of the prominence material in the lower trough of the interior fluxtube explains the opposite field direction with respect to the overlying arcade (Fig. 6.14). Viewing this magnetic model along the prominence axis, this scenario with opposite magnetic polarity corresponds to the Kuperus−Raadu (1974) model (Fig. 6.13, right), while the Kippenhahn−Schlüter (1957) model (Fig. 6.13, left), which proposes trapping of prominence material in dips near the apex of a loop arcade, predicts the same magnetic polarity in the prominence and overlying arcade. The magnetic dip on the top of the field lines preserves the stability of the quiescent prominence (Pikel'ner 1971). Prominence models based on a dip in the horizontal magnetic field have been developed, e.g., by Antiochos et al. (1994).

Magnetic field models of quiescent prominences inside large, twisted fluxtubes have been developed by a number of authors (e.g., Priest et al. 1989; Ridgway et al. 1991a,b). According to these models, the interior of a large loop arcade contains a highly twisted fluxtube (Priest et al. 1989b) anchored at both ends near the neutral line, having a helical twist of one or two turns in the coronal portion of the tube (Fig. 6.14).

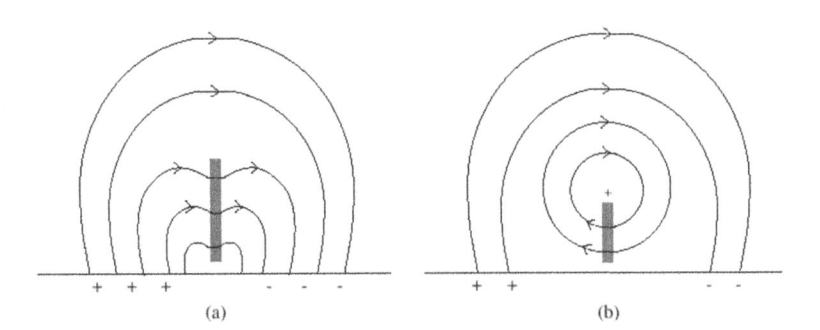

Figure 6.13: Two magnetic field models to support a prominence: (a) Kippenhahn—Schlüter model; (b) Kuperus—Raadu model. The figures show the field lines projected perpendicularly to the long axis of a prominence (shaded region) (Van Ballegooijen 2001).

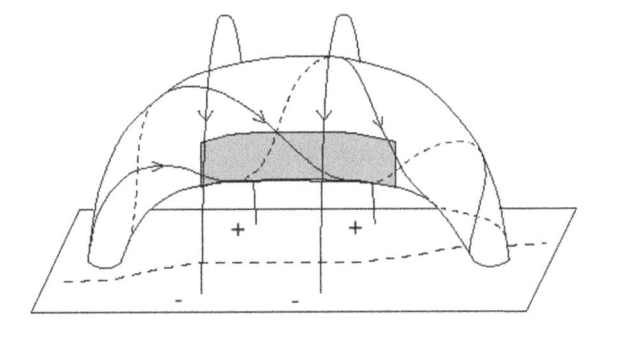

Figure 6.14: Side view of a twisted fluxtube model (Priest et al. 1989) for a solar prominence. The cool prominence material (shaded region) is suspended inside of the twisted fluxtube (Van Ballegooijen 2001).

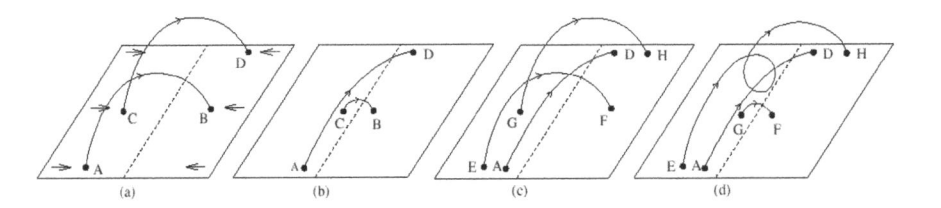

Figure 6.15: Model of the formation process of a prominence by flux cancellation in a sheared coronal arcade. The neutral line is indicated with a dashed line on the photospheric surface (rectangle). (a) Converging flows shear the initial field. (b) Magnetic reconnection between the sheared loops AB and CD exchanges the connectivity, producing a long loop AD and a short loop BC, which submerges subsequently. (c) Overlying loops EF and GH are pushed to the neutral line. (d) Reconnection produces the helical loop EH and a short loop GF, which again submerges (Van Ballegooijen & Martens 1989).

The overlying arcade straps down the twisted fluxtube and probably is essential for the overall stability and equilibrium (as simulated in a laboratory experiment by Hansen & Bellan 2001). The cool prominence material is suspended at the lower, upwardly con-cave formed trough of the twisted fluxtube, which in addition might have some dips in longitudinal direction to trap the prominence material. A possible scenario to produce such helical fields is outlined in Fig. 6.15, where shearing and reconnection forms heli-cal field lines above the neutral line (Priest et al. 1996). Alternative models assume that twisting occurs in the convection zone (Fig. 6.8) and that the fluxtube emerges through the photosphere already twisted (Rust & Kumar 1994; McKaig 2001). Observations also show that prominences tend to form in regions where opposite-polarity flux is can-celled at the neutral line, which can be explained by reconnection just above the neutral line (Van Ballegooijen & Martens 1989; Litvinenko & Martin 1999), as illustrated in Fig. 6.15. A consequence of this model is that reconnection also produces longer and more sheared field lines above the neutral line, gradually enlarging the prominence until it eventually becomes unstable and erupts. A more realistic rendering of the 3D magnetic field in a *filament channel* is illustrated in Fig. 6.16, computed with a magneto-hydrostatic code by Aulanier & Schmieder (2002).

The twist of magnetic field lines in prominences has been found to have a pre-ferred chirality or handedness in the northern and southern solar hemispheres (Rust 1967; Leroy et al. 1984). The majority of quiescent filaments are either left-handed (*sinistral*) or right-handed (*dextral*) (Martin et al. 1994). The chirality can be deter-mined either from the direction of *barbs* (lateral appendices) on both sides of a fila-ment *spine*, because they trace the flow field and thus the magnetic shear, or from the crossing of bright over dark (or vice versa) filament threads (Chae 2000). Most of the dextral filaments are found in the northern hemisphere (80%), while most sinis-trals are in the southern hemisphere (85%; Pevtsov et al. 2003), a pattern that remains unchanged from solar cycle to cycle, although the absolute direction of the axial mag-netic field in east-west direction reverses from cycle to cycle. A number of models has been developed to explain the chirality of filaments, either based on emergence of subphotospheric twisted fluxtubes (Rust & Kumar 1994), continuous shearing and reconnection (Priest et al. 1996), or based on twisting by meridional flows and other surface motions (Van Ballegooijen et al. 1998), all models also being affected by differ-ential rotation, but none of the models can explain all observed patterns (see review by Zirker 2001). Numerical simulations based on surface flow fields could reproduce the formation of filaments (Mackay et al. 1997; Gaizauskas et al. 1997), predict the correct chirality (Mackay et al. 2000b), but the cycle-dependent hemispheric pattern could not be reproduced (Van Ballegooijen et al. 2000; Mackay & Van Ballegooijen 2001). 3D magnetic modeling is able to simulate the formation process (Galsgaard & Longbottom 1999; DeVore & Antiochos 2000), to mimic structural details such as the *lateral feet*, *barbs*, *bald patches*, and *dips* of filaments (Aulanier & Démoulin 1998; Aulanier et al. 1998a, 1998b, 1999, 2000a, 2002; Kucera et al. 1999; Lionello et al. 2002), as well as the magnetic field strength and orientation (Aulanier & Schmieder 2002; Aulanier & Démoulin 2003; Démoulin 2003).

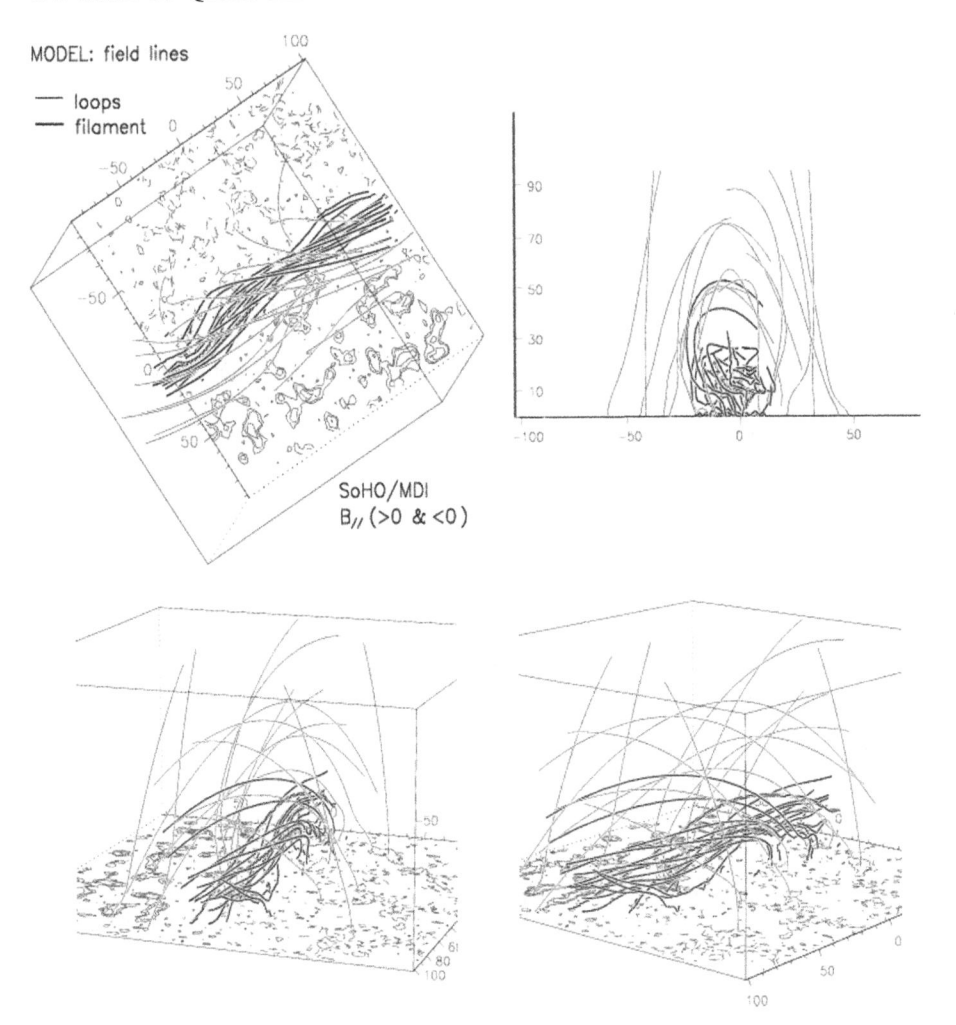

Figure 6.16: Magnetic field lines of a filament channel, computed with a linear magneto-hydrostatic (LMHS) code. The field lines passing through the Hα filament body and feet (resp. surrounding it) are drawn with thick (resp. thin) lines. The images show views in projection from Earth (top left), along the filament channel (top right), and in projections (bottom), (Aulanier & Schmieder 2002).

6.4.2 Equilibrium Models

The longevity observed in many quiescent filaments and prominences requires a stable equilibrium of the dense prominence gas against the gravity force. Since the partially ionized gas can only move along the magnetic field lines in the low-β corona, it will naturally slide to the lowest points along a field line, and thus many prominence models envision that the dense prominence gas is trapped in dips of the nearly horizontal field lines. Kippenhahn & Schlüter (1957) pioneered such a model of the equilibrium and stability of prominence plasma in a magnetic configuration with dipped field lines (Fig. 6.13 left). This standard model is summarized in many texts (e.g. Priest 1982, § 11.2; Stix 2002, § 9.3.4; Van Ballegooijen 2001). The stationary ($\partial/\partial t = 0$) momentum equation (6.1.17) implies an equilibrium between the gradient of the gas pressure ∇p, gravity $\rho \mathbf{g}$, and the Lorentz force ($\mathbf{j} \times \mathbf{B}$),

$$-\nabla p - \rho \mathbf{g} + (\mathbf{j} \times \mathbf{B}) = 0 \,. \tag{6.4.1}$$

Kippenhahn & Schlüter (1957) modeled the geometry of the prominence as a thin vertical current sheet in which the pressure $p(x)$, density $\rho(x)$, and vertical magnetic field $B_z(x)$ depend only on the horizontal coordinate x perpendicular to the sheet. The temperature T and horizontal magnetic field components B_x and B_y are assumed to be constant. The x and z-components of Eq. (6.4.1) then reduce to (see analogous derivation in § 6.2.1),

$$-\frac{\partial}{\partial x}\left(p + \frac{B^2}{8\pi}\right) = 0 \,, \tag{6.4.2}$$

$$-\rho g + \frac{B_x}{4\pi}\frac{\partial B_z}{\partial x} = 0 \,. \tag{6.4.3}$$

Defining the boundary conditions far away from the prominence sheet,

$$p(x \mapsto \pm\infty) = 0 \,, \qquad B_z(x \mapsto \pm\infty) = \pm B_{z\infty} \,, \tag{6.4.4}$$

and integrating Eq. (6.4.2) over the range $[x, \infty]$ yields,

$$p(x) = \frac{B_{z\infty}^2 - B_z^2(x)}{8\pi} \,. \tag{6.4.5}$$

Using now the equation of state for the ideal gas ($p = 2n_e k_B T_e$, Eq. 3.1.9), the mass density ($\rho = mn = \mu m_H n_e$, Eq. 3.1.6), the definition of the pressure scale height ($\lambda_p = 2k_B T_e/\mu m_H g$, Eq. 3.1.6), and inserting the pressure $p(x)$ from Eq. (6.4.5), equation (6.4.3) becomes

$$-\frac{B_{z\infty}^2 - B_z^2(x)}{2\lambda_p} + B_x \frac{\partial B_z(x)}{\partial x} = 0 \,, \tag{6.4.6}$$

which has the following analytical solution for the magnetic field $B_z(x)$,

$$B_z(x) = B_{z\infty} \tanh\left(\frac{B_{z\infty}}{B_x}\frac{x}{2\lambda_p}\right) \,, \tag{6.4.7}$$

and for the pressure $p(x)$, after inserting Eq. (6.4.7) into Eq. (6.4.5),

$$p(x) = \frac{B_{z\infty}^2}{8\pi}\left[\mathrm{sech}\left(\frac{B_{z\infty}}{B_x}\frac{x}{2\lambda_p}\right)\right]^2 = \frac{B_{z\infty}^2}{8\pi}\left[\cosh\left(\frac{B_{z\infty}}{B_x}\frac{x}{2\lambda_p}\right)\right]^{-2} \,. \tag{6.4.8}$$

The pressure scale height $\lambda_p = 46,000$ km $\times (T_e/1 \text{ MK})$ (Eq. 3.1.16) is only $230 - 690$ km for filaments and prominences (with $T_e = 5000 - 15,000$ K). The fact that the observed height extent of prominence threads spans over several 1000 km implies a

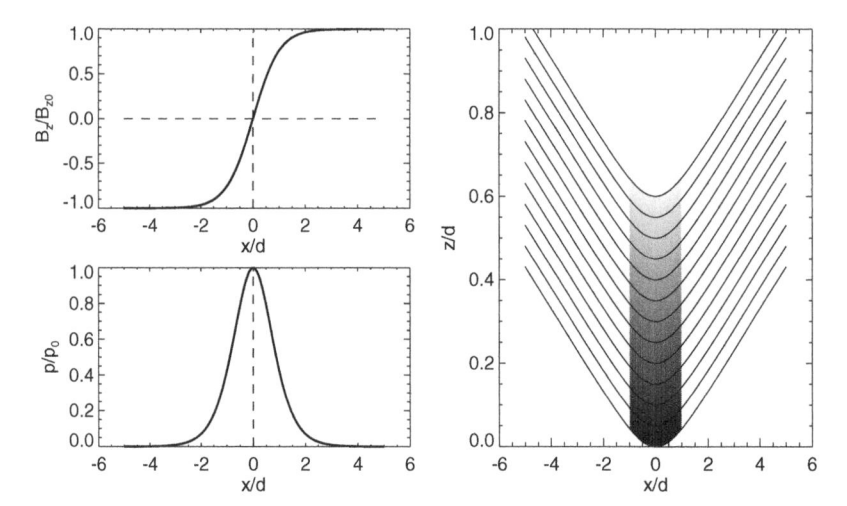

Figure 6.17: Vertical magnetic field strength $B_z(x)$ (top left), pressure $p(x)$ or density profile $n_e(x)$ (bottom left), and magnetic field lines $B(x, z)$ for the Kippenhahn−Schlüter (1957) prominence model, according to the analytical solutions specified in Eqs. (6.4.7−10). The prominence material (grey zone) is suspended by upwardly concave field lines in dips of loop arcades (Fig. 6.13 left).

highly nonhydrostatic and dynamic nature. The width of the sheet can be estimated from the pressure scale height and is about $w \approx 4(B_x/B_{z\infty})\lambda_p$. The observed widths of quiescent prominences (≈ 8000 km) can be reproduced with $B_{z\infty} \approx 0.1 B_z$. This implies that only a minor variation of the horizontal magnetic field by $\approx 10\%$ can support a prominence. Since the temperature is assumed to be constant, the density $n_e(x)$ has the same transverse dependence as the pressure $p(x)$ (Eq. 6.4.8), shown in Fig. 6.17. The magnetic field lines of the Kippenhahn & Schlüter (1957) model can be computed according to Eq. (5.2.19) using $dx/B_x = dz/B_z$,

$$\int \frac{B_{z\infty}}{B_{x\infty}} \tanh\left(\frac{B_{z\infty}}{B_{x\infty}} \frac{x}{2\lambda_p}\right) dx = z + c, \tag{6.4.9}$$

which can be integrated and yields,

$$2\lambda_p \ln\left[\cosh\left(\frac{B_{z\infty}}{B_{x\infty}} \frac{x}{2\lambda_p}\right)\right] = z + c. \tag{6.4.10}$$

So the analytical solutions (Fig. 6.17) indicate magnetic field lines that are bent upwardly concave. Since the magnetic field $B(z)$ generally decreases with height, the magnetic pressure force points in an upward direction (towards the lower magnetic field). Moreover, the magnetic tension force also points in an upward direction at the center, and towards the center location of the prominence plasma, due to the concave curvature. So, both the magnetic pressure force and the magnetic tension force oppose gravity and compress the prominence plasma.

The Kippenhahn−Schlüter (1957) model has been generalized in a number of ways, by including heat conduction and radiative loss in the energy balance equation

(6.1.35) (Orrall & Zirker 1961; Low 1975a, 1975b; Lerche & Low 1977; Heasley & Mihalas 1976; Milne et al. 1979). A prominence model with more realistic boundary conditions was modeled (Anzer 1972; Hood & Anzer 1990) by including the external magnetic field (in the form of a potential field). The MHD stability of the Kippenhahn−Schlüter (1957) prominence model was investigated by Anzer (1969), and that of a sheared field configuration by Nakagawa (1970).

In the alternative prominence model of Kuperus & Raadu (1974) where helical fields in the interior of a loop arcade suspend the prominence (Fig. 6.13, right frame, and Fig. 6.14), support of the prominence against gravity is provided by the azimuthal magnetic field B_φ, which is required to be

$$\frac{B_\varphi}{4\pi h} \approx (\rho_P - \rho_0)g \, ,   (6.4.11)$$

and can easily be satisfied within the observed parameters (Table 6.2). Equilibrium models of prominences supported by helical fields, which form after reconnection in vertical neutral sheets, have been considered by Kuperus & Raadu (1974), Van Tend & Kuperus (1978), Kuperus & Van Tend (1981), Lerche & Low (1980), Low (1981), and Malherbe & Priest (1983).

6.4.3   Formation and Evolution

How are prominences or filaments created? This question consists of two parts: first, how is an appropriate magnetic field configured so that it can support a prominence (as we discussed in § 6.4.1 for the Kippenhahn & Schlüter model), and second, how is the magnetic field structure filled with cool prominence material? The observations show that prominences always form in so-called *filament channels* (Fig. 6.18, top), regions where *chromospheric fibrils* (thread-like fine structures) are aligned with the neutral line (Fig. 6.18, bottom). The filament channels themselves form along boundaries of large-scale convection cells, typically occurring in mid-latitudes of $\approx 10° − 40°$, separated by about $\approx 45°$ in longitude (Schröter & Wöhl 1976; Stix 2002, § 6.6). For the question of how prominences acquire their mass there are three different scenarios (Fig. 6.19): (1) cooling and condensation of plasma from the surrounding corona (Pneuman 1972), (2) injection by chromospheric upflows, and (3) footpoint heating triggering condensation. Although filaments and prominences appear to be static over longer time intervals, there is a lot of observational evidence that the formation is a continuous process, where mass is continuously entering and exiting the filament magnetic field throughout its lifetime. Moreover, the continuous process of mass transport in filament spines and barbs has been found to consist of bi-directional streams (Zirker et al. 1998).

Coronal Condensation

The first group of scenarios for the formation of filaments or prominences envisions a coronal origin. For instance, Pneuman (1972) proposed that quiescent prominences with a high-density form at the base of *helmet streamers* inside a low-density cavity at a special location where conductive transport is inefficient and the radiating volume is

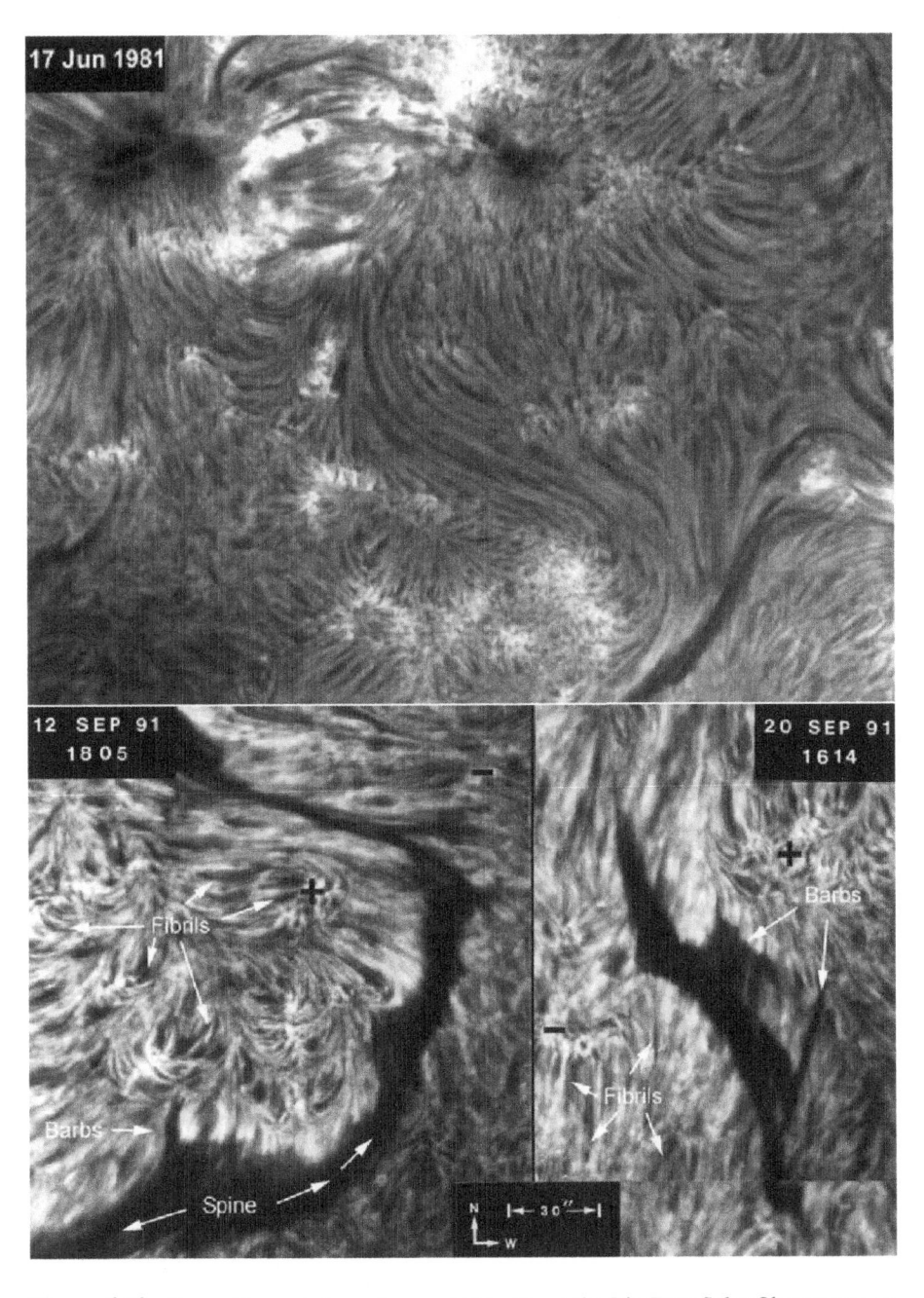

Figure 6.18: *Top:* a filament channel, recorded in Hα at the Big Bear Solar Observatory on 17 June 1981. The S-shaped filament channel consists of chromospheric fibrils and connects to opposite magnetic polarities (a sunspot in the north with a filament in the south) and will become dark once it is filled with cool material that forms a filament. *Bottom:* example of a sinistral filament (left) and a dextral filament (right). The fine structure of the filament barbs are aligned with the chromospheric fibrils (Martin 2001).

(1) Coronal
Condensation

(2) Footpoint
Injection

(3) Footpoint
Heating

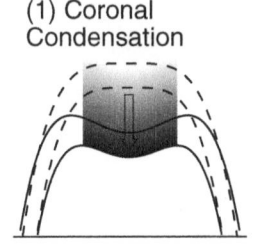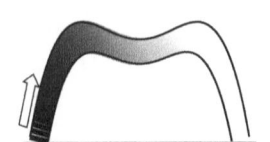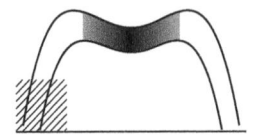

Figure 6.19: Three different model concepts for the formation of filaments or prominences: (1) condensation and cooling in a coronal loop, (2) upflows of cool prominence material from the chromosphere, and (3) footpoint heating that causes condensation in the coronal segments of the loop. Grey zones demarcate cool gas or plasma.

small. Similarly, Hood & Priest (1979a), and Priest & Smith (1979) modeled prominences as stretched-out loops in which thermal conduction becomes ineffective, so that radiative cooling dominates and forms cool prominence plasma. Prominence models of coronal origin, however, generally face the problem of insufficient mass supply, unless the mass flow circuit is closed through the chromosphere. The mass of a large quiescent prominence is estimated to be (with a density $n_p \approx 3 \times 10^{11}$ cm^{-3} and a volume of $V_{prom} = 5$ Mm $\times$ 50 Mm $\times$ 400 Mm $= 10^{29}$ cm^3),

$$M_{prom} \approx n_p m_p V_{prom} \approx 5 \times 10^{16} \text{ g} , \qquad (6.4.12)$$

which makes up about $\approx 10\%$ of the entire mass of the corona (with $n_p \approx 3 \times 10^8$ cm^{-3} and a density scale height of $\lambda_p \approx 100$ Mm),

$$M_{cor} \approx n_p m_p (4\pi R_\odot^2 \lambda_p) \approx 3 \times 10^{17} \text{ g} . \qquad (6.4.12)$$

Thus the corona would be completely depleted after a dozen prominence eruptions, and thus inevitably a dynamic process is required to replenish prominence mass from the much denser chromosphere (Tandberg—Hanssen 1986). There is, of course, a lot of observational evidence for cooling in coronal loops (§ 4.5), which eventually evolves into *catastrophic cooling* below temperatures of $T \lesssim 1.0$ MK followed by high-speed downflows at $T \approx 0.1$ MK (Schrijver 2001a), but this cooling or condensation process apparently does not lead to the formation of filaments or prominences.

Chromospheric Injection

The second group of filament/prominence formation models envisions injection or upflow of chromospheric material, which can be driven in various ways. Siphon flows can drive material upwards if there is a large enough positive pressure difference between the chromosphere and a dip near the apex of a coronal loop (Pikel'ner 1971; Engvold & Jensen 1977). In other models, a unspecified mechanism is assumed that produces a ballistic injection with some initial velocity, which may explain surge-like upflows, such as in spicules or soft X-ray jets.

Footpoint Heating

Poland & Mariska (1986) employ an asymmetric heating mechanism at both footpoints to drive the upflows. A one-sided reduction of the heating rate by $\approx 1\%$ was found to be sufficient to cause condensation at the looptop and to establish the pressure difference to the footpoints to drive a siphon flow. Once the condensation at the looptop progresses, it acts like a "second chromosphere" and stabilizes the prominence against fluctuations of the footpoint heating rate. Numerical simulations demonstrated that heating functions strongly concentrated at the footpoints (as independently measured by TRACE observations; Aschwanden et al. 2000a, 2001a) lead to condensations of coronal plasma in dips in the near-horizontal segments of coronal loops (Antiochos & Klimchuk 1991; Dahlburg et al. 1998; Antiochos et al. 1999a), or even in loops *without dips* (Karpen et al. 2001). Although dips were generally considered to be a necessary condition for the stability of prominences, Karpen et al. (2001) argue that an equilibrium situation is not necessary given the dynamic nature of prominences as manifested by the observations of counter-streaming flows (Zirker et al. 1998). Long-term simulations demonstrate that the prominence plasma is never in equilibrium, but rather subject to continuous oscillatory cycles of formation and destruction of condensation knots (Antiochos et al. 2000a). The footpoint-concentrated heating with subsequent mass upflows is thought to be ultimately driven by a coronal reconnection process (Kuperus & Tandberg–Hanssen 1967) or by chromospheric reconnection processes (Van Ballegooijen & Martens 1989, Priest et al. 1996; Litvinenko & Martin 1999), as evidenced by the observed correlation of magnetic cancellation features and upflows into filaments (Martin 1998; Chae et al. 2000b). Prominence formation in coronal reconnection geometries have been simulated for bipolar (Cheng & Choe 1998) and quadrupolar current sheets (Sakai et al. 1987).

Disappearance of Filaments/Prominences

The lifetime of a prominence is determined by the balance between heating and cooling. Several possibilities can provide a heating source: (1) thermal conduction from the hot corona, which is not efficient in the low-temperature prominence plasma; (2) heating by dissipation of MHD waves (Jensen 1986); or (3) heating by absorption of ultraviolet radiation (Lyman and Balmer lines). The latter mechanism has a more efficient penetration depth for threaded prominences (Heasley & Mihalas 1976; Anzer & Heinzel 1999; Heinzel & Anzer 2001).

The final stage of a filament or prominence is called *sudden disappearance* (or *disparition brusque* in french), which manifests itself as an instability with a subsequent eruption into the corona and interplanetary space, often accompanied by a flare or coronal mass ejection. We will discuss these phenomena of *eruptive filaments* or *eruptive prominences* in the section on flares (§ 10) and CMEs (§ 17), (see also models of Hirayama 1974; Kopp & Pneuman 1976; Martens & Kuin 1989; Smith et al. 1992).

6.5   Summary

**Magneto-hydrodynamics (MHD) is one of the key tools to understand the hydro-
dynamics of fully magnetized plasmas. It concerns virtually all phenomena observed in the solar corona: coronal loops, filaments, prominences, flares, coronal
mass ejections, etc. In this section we derived the basic MHD equations ($\S$ 6.1),
which includes particle conservation ($\S$ 6.1.1), the momentum or force equation
($\S$ 6.1.2), and the energy equation ($\S$ 6.1.4). This set of MHD equations, together
with the Maxwell electrodynamic equations, constitute the framework of "ideal
MHD" ($\S$ 6.1.3), or "resistive MHD" ($\S$ 6.1.4), if we include finite resistivity, respectively. On an introductory level we apply the basic MHD equations to some structures in the quiet corona, such as to coronal loops ($\S$ 6.2) and filaments and prominences ($\S$ 6.4). We familiarize ourselves with the application of MHD equations
to coronal loops by investigating the pressure equilibrium of magneto-static fluxtubes ($\S$ 6.2.1), the Lorentz force near magnetic nullpoints ($\S$ 6.2.2), the Lorentz
force in curved loops ($\S$ 6.2.3), and the dynamics of twisted fluxtubes ($\S$ 6.2.4).
More complicated situations require numerical MHD simulations, which became
feasible and popular over the last decade, for instance to reproduce the emergence
of magnetic fluxtubes from the convection zone to the corona ($\S$ 6.2.5), or to simulate the heating of coronal loops by random shuffling of their footpoints ($\S$ 6.2.6).
The evolution of coronal structures changes dramatically when the plasma becomes locally unstable (e.g., by a pressure-driven, thermal, resistive, or current-
pinch instability, $\S$ 6.3). Another fascinating application of MHD physics are filaments and prominences, which consist of cool chromospheric gas embedded in
the hot coronal plasma ($\S$ 6.4). The most challenging unsolved questions concern
their magnetic field configuration ($\S$ 6.4.1), their equilibrium ($\S$ 6.4.2), their mass
supply, and evolution ($\S$ 6.4.3). All these applications deal with quiescent, quasi-
stationary phenomena in the solar corona, while their unstable evolution during
flares and coronal mass ejections require, in addition to the MHD framework, also
a description in terms of kinetic plasma physics.**

Chapter 7

MHD Oscillations

Just as a music orchestra contains many instruments with different sounds, it was recently discovered that the solar corona also contains an impressively large ensemble of plasma structures that are capable of producing sound waves and harmonic oscillations. Global oscillations from the solar interior were discovered four decades ago, and are mainly pressure-driven (p-mode) oscillations at fundamental periods of 5 minutes, which display spatial nodes on the solar surface according to the spherical harmonics functions, similar to the nodes on the skin of a vibrating drum. The systematic study of these global oscillations created the discipline of *helioseismology*.

Thanks to the high spatial resolution, image contrast, and time cadence capabilities of the *SoHO* and *TRACE* spacecraft, oscillating loops, prominences, or sunspots, and propagating waves have been identified and localized in the corona and transition region, and studied in detail since 1999. These new discoveries established a new discipline that became known as *coronal seismology*. The theory of coronal MHD oscillations and waves was developed two decades ago and was ready for applications, but had to await the high-resolution EUV imaging capabilities that can only be obtained from space. One of the most exciting benefits of coronal seismology is the probing of physical parameters such as Alfvén velocities and magnetic field strengths, which are very difficult to measure in the solar corona by other means. Initial research in the new field of coronal seismology concentrates on measurements of oscillation periods, spatial displacements, damping times, and temperature and density diagnostics of oscillating structures. Future exploration is likely to reveal additional aspects on the fundamental physics of wave excitation, wave propagation, waveguides, wave damping, phase mixing, and resonant absorption.

Theoretical reviews on coronal oscillations can be found in Roberts (1984, 1985, 1991a, 2000, 2001, 2002), Bray et al. (1991), Goossens (1991), Poedts (1999), Goossens et al. (2002b), Roberts & Nakariakov (2003), in the Proceedings of the INTAS workshop on *MHD Waves in Astrophysical Plasmas* edited by Ballester & Roberts (2001), while observational reviews are covered in Aschwanden (1987a, 2003) Nakariakov (2003), and Wang (2004).

7.1 Dispersion Relation of MHD Waves

To understand the various oscillations and waves we observe in the coronal plasma we have to find wave solutions of the MHD equations (§6.1). The basic theory of hydromagnetic waves in plasmas can be found in many textbooks (e.g., Jackson 1962, § 10.7; Chen 1974, § 4.19; Cowling 1976; Schmidt 1979, § 4.4; Priest 1982, § 4; Benz 1993, § 3.2; Sturrock 1994, § 14).

The existence of wave solutions is generally derived by introducing a perturbation of the physical parameters of the plasma, using a spatio-temporal function that contains periodic solutions, like $\exp(i[\mathbf{kx} - \omega t])$, and to derive dispersion relations $\omega(\mathbf{k})$, which tell us either the group velocity $\mathbf{v}_g = d\omega(\mathbf{k})/d\mathbf{k}$ or phase speed $\mathbf{v}_{ph} = \omega/k$ of a wave.

We start from the ideal MHD equations as given in Eqs. (6.1.16−22), where we insert the current density $\mathbf{j}$ (Eq. 6.1.19) and electric field $\mathbf{E}$ (Eq. 6.1.22), so that the ideal MHD equations are expressed only in terms of the variables $\rho, p, \mathbf{v}$, and $\mathbf{B}$,

$$\frac{D}{Dt}\rho = -\rho \, \nabla \cdot \mathbf{v} \,, \tag{7.1.1}$$

$$\rho\frac{D\mathbf{v}}{Dt} = -\nabla p - \rho\mathbf{g} + \frac{1}{4\pi}[(\nabla \times \mathbf{B}) \times \mathbf{B}] \,, \tag{7.1.2}$$

$$\frac{D}{Dt}(p\rho^{-\gamma}) = 0 \,, \tag{7.1.3}$$

$$\nabla \times (\mathbf{v} \times \mathbf{B}) = \frac{\partial \mathbf{B}}{\partial t} \,, \tag{7.1.4}$$

$$\nabla \cdot \mathbf{B} = 0 \,. \tag{7.1.5}$$

For adiabatic processes we can eliminate the pressure term in Eq. (7.1.2), using the relation $\nabla p = c_s^2 \nabla \rho$ (Eq. 4.2.2) and the definition of the sound speed $c_s^2 = \gamma p/\rho$ (Eq. 4.2.3). Furthermore, using $\nabla \mathbf{B} = 0$ (Eq. 7.1.5), and inserting vector identities for $\nabla \times (\mathbf{v} \times \mathbf{B})$ in Eq.(7.1.4) and for $(\nabla \times \mathbf{B}) \times \mathbf{B}$ in Eq.(7.1.2), as shown in Eqs. (6.2.2−3), the ideal MHD equations can be reduced to three equations for the variables $\rho, \mathbf{v}$, and $\mathbf{B}$:

$$\frac{D}{Dt}\rho = -\rho \, \nabla \cdot \mathbf{v} \,, \tag{7.1.6}$$

$$\rho\frac{D\mathbf{v}}{Dt} = -c_s^2\nabla\rho - \rho\mathbf{g} + \frac{1}{4\pi}\left[-\frac{1}{2}\nabla B^2 + (\mathbf{B} \cdot \nabla)\mathbf{B}\right] \,, \tag{7.1.7}$$

$$\frac{\partial \mathbf{B}}{\partial t} = -\mathbf{B}(\nabla \cdot \mathbf{v}) + (\mathbf{B} \cdot \nabla)\mathbf{v} - (\mathbf{v} \cdot \nabla)\mathbf{B} \,. \tag{7.1.8}$$

Now we assume that there exists an equilibrium solution ($\partial/\partial t = 0$), with no flows ($\mathbf{v}_0 = 0$), so that the density and magnetic field have only a spatial dependence. We denote this equilibrium solution with the subscript 0: $\rho_0(\mathbf{x}), \mathbf{B}_0(\mathbf{x})$. Both sides of the equations (7.1.6) and (7.1.8) then vanish and there is only one equation left (Eq. 7.1.7) that defines the equilibrium solution,

$$-c_s^2 \nabla \rho_0 - \rho_0 \mathbf{g} + \frac{1}{4\pi} \left[-\frac{1}{2} \nabla B_0{}^2 + (\mathbf{B}_0 \cdot \nabla) \mathbf{B}_0 \right] = 0 . \qquad (7.1.9)$$

7.1.1 Unbounded Homogeneous Medium

We now introduce a small perturbation (in density, velocity, and magnetic field) from the equilibrium, where the perturbation amplitude is denoted with the subscript 1,

$$\rho(\mathbf{x}, t) = \rho_0 + \rho_1(\mathbf{x}, t) ,$$

$$\mathbf{v}(\mathbf{x}, t) = \qquad \mathbf{v}_1(\mathbf{x}, t) ,$$

$$\mathbf{B}(\mathbf{x}, t) = \mathbf{B}_0 + \mathbf{B}_1(\mathbf{x}, t) . \qquad (7.1.10)$$

In the simplest concept we can consider the perturbation as a local phenomenon and neglect the large-scale gradients of macroscopic parameters. Therefore, we neglect the gravity term $(-\rho_0 \mathbf{g})$ and the spatial variation of the stationary solution $\rho_0(x)$ and $B_0(x)$, so we assume constants ρ_0 and B_0 in the perturbation ansatz (Eq. 7.1.10), so $\nabla \rho_0 = 0$ and $\partial \mathbf{B}_0 / \partial \mathbf{x} = 0$. Plugging these perturbation functions (Eq. 7.1.10) into the ideal MHD equations (7.1.6-8) and subtracting the equilibrium solution (Eq. 7.1.9), we obtain the following linear terms (dropping quadratic terms of order $[\rho_1 \mathbf{v}_1]$, $[\rho_1 \mathbf{B}_1]$, or $[\mathbf{v}_1 \mathbf{B}_1]$),

$$\frac{\partial \rho_1}{\partial t} = -\rho_0 \nabla \cdot \mathbf{v}_1 , \qquad (7.1.11)$$

$$\rho_0 \frac{\partial \mathbf{v}_1}{\partial t} = -c_s^2 \nabla \rho_1 + \frac{1}{4\pi} [-\nabla(\mathbf{B}_0 \cdot \mathbf{B}_1) + (\mathbf{B}_0 \cdot \nabla) \mathbf{B}_1] , \qquad (7.1.12)$$

$$\frac{\partial \mathbf{B}_1}{\partial t} = -\mathbf{B}_0 (\nabla \cdot \mathbf{v}_1) + (\mathbf{B}_0 \cdot \nabla) \mathbf{v}_1 . \qquad (7.1.13)$$

Without loss of generality we can consider a homogeneous magnetic field in the z-direction (with unit vector $\mathbf{e}_z$), $\mathbf{B}_0 = (0, 0, B_0)$, and the associated Alfvén speed v_A,

$$v_A = \frac{B_0}{\sqrt{4\pi \rho_0}} . \qquad (7.1.14)$$

The momentum equation (7.1.12) then becomes

$$\frac{\partial \mathbf{v}_1}{\partial t} = -c_s^2 \frac{\nabla \rho_1}{\rho_0} + \frac{v_A^2}{B_0} \left[-\nabla(\mathbf{e}_z \mathbf{B}_1) + \frac{\partial \mathbf{B}_1}{\partial z} \right] . \qquad (7.1.15)$$

Taking the time derivative $\partial / \partial t$ of the momentum equation (7.1.15),

$$\frac{\partial^2 \mathbf{v}_1}{\partial t^2} = -\frac{c_s^2}{\rho_0} \nabla \left(\frac{\partial \rho_1}{\partial t} \right) + \frac{v_A^2}{B_0} \left[-\nabla (\mathbf{e}_z \frac{\partial \mathbf{B}_1}{\partial t}) + \frac{\partial}{\partial z} \left(\frac{\partial \mathbf{B}_1}{\partial t} \right) \right] , \qquad (7.1.16)$$

allows us to insert $(\partial\rho_1/\partial t)$ from the continuity equation (7.1.11) and $(\partial\mathbf{B}_1/\partial t)$ from the induction equation (7.1.13) and to obtain a single equation for the velocity perturbation $\mathbf{v}_1$,

$$\frac{\partial^2\mathbf{v}_1}{\partial t^2} = c_s^2\nabla(\nabla\mathbf{v}_1) + \mathrm{v}_A^2\left[-\nabla\left(\frac{\partial v_z}{\partial z} - \nabla\mathbf{v}_1\right) + \frac{\partial}{\partial z}\left(\frac{\partial\mathbf{v}_1}{\partial z} - \mathbf{e}_z\nabla\mathbf{v}_1\right)\right] .$$
(7.1.17)

For convenience we introduce the variables

$$\Delta = \nabla\cdot\mathbf{v}_1 ,$$
(7.1.18)

$$\Gamma = \frac{\partial v_z}{\partial z} ,$$
(7.1.19)

so that Eq. (7.1.17) reads,

$$\frac{\partial^2\mathbf{v}_1}{\partial t^2} = c_s^2\nabla\Delta + \mathrm{v}_A^2\left[-\nabla(\Gamma - \Delta) + \frac{\partial}{\partial z}\left(\frac{\partial\mathbf{v}_1}{\partial z} - \mathbf{e}_z\Delta\right)\right] .$$
(7.1.20)

Taking the z-component and the divergence of Eq. (7.1.18) then leads to the standard notation for the linear equations of MHD modes in an unbounded homogeneous medium (e.g., Cowling 1976; Roberts 1981a),

$$\frac{\partial^2 v_z}{\partial t^2} = c_s^2\frac{\partial\Delta}{\partial z} ,$$
(7.1.21)

$$\frac{\partial^2\Delta}{\partial t^2} = (c_s^2 + \mathrm{v}_A^2)\nabla^2\Delta - \mathrm{v}_A^2\nabla^2\Gamma .$$
(7.1.22)

The two equations (7.1.21) and (7.1.22) can be combined by taking the 4th time derivative,

$$\frac{\partial^4\Delta}{\partial t^4} - (c_s^2 + \mathrm{v}_A^2)\frac{\partial^2}{\partial t^2}\nabla^2\Delta + c_s^2\mathrm{v}_A^2\frac{\partial^2}{\partial z^2}\nabla^2\Delta = 0 .$$
(7.1.23)

To obtain the dispersion relation, one expresses the disturbance in terms of Fourier components for frequency ω and wave numbers (k_x, k_y, k_z) in the x, y, and z-direction, respectively,

$$\Delta \propto e^{i\omega t + ik_x x + ik_y y + ik_z z} .$$
(7.1.24)

Inserting this Fourier form in Eq. (7.1.23), where the temporal derivatives yield terms of $\partial/\partial t \mapsto i\omega$ and the spatial derivatives yield terms of $\nabla \mapsto i\mathbf{k}$, we can express Eq. (7.1.23),

$$k_x^2 + k_y^2 + m_0^2 = 0 ,$$
(7.1.25)

where

$$m_0^2 = \frac{(k_z^2 c_s^2 - \omega^2)(k_z^2\mathrm{v}_A^2 - \omega^2)}{(c_s^2 + \mathrm{v}_A^2)(k_z^2 c_T^2 - \omega^2)}$$
(7.1.26)

and c_T is defined as the *tube speed*,

$$c_T = \frac{c_s\mathrm{v}_A}{\sqrt{c_s^2 + \mathrm{v}_A^2}} ,$$
(7.1.27)

yielding the *dispersion relation for fast and slow magneto-acoustic waves*

$$\omega^4 - k^2(c_s^2 + v_A^2)\omega^2 + k_z^2 k^2 c_s^2 v_A^2 = 0 . \tag{7.1.28}$$

The wave propagation vector $\mathbf{k} = (k_x, k_y, k_z)$ has an absolute value of $k = |\mathbf{k}| = (k_x^2 + k_y^2 + k_z^2)^{1/2}$. The cosine of the wave vector in the direction to the magnetic field is k_z, so the *propagation angle* θ is

$$\cos\theta = \frac{k_z}{k} , \tag{7.1.29}$$

and the phase speed v_{ph} and group speed v_{gr} of the wave are defined as

$$v_{ph} = \frac{\omega}{k} , \tag{7.1.30}$$

$$\mathbf{v}_{gr} = \frac{\partial\omega(\mathbf{k})}{\partial\mathbf{k}} = \left(\frac{\partial\omega}{\partial k_x}, \frac{\partial\omega}{\partial k_y}, \frac{\partial\omega}{\partial k_z}\right) . \tag{7.1.31}$$

When the phase speed v_{ph} varies as a function of the wavelength $\lambda = 2\pi/k$ or wave vector k, the wave is called *dispersive*. For *non-dispersive* waves, phase speed and group speed are identical. The dispersion relation $\omega(k)$ can easily be expressed in explicit form from the quadratic equation (of ω^2) in Eq. (7.1.28). Dividing the dispersion relation (7.1.28) by k^4, we can express it as an implicit function of the phase speed v_{ph} and propagation angle θ,

$$v_{ph}^4 - v_{ph}^2(c_s^2 + v_A^2) + c_s^2 v_A^2 \cos^2(\theta) = 0 . \tag{7.1.32}$$

This phase speed diagram $v_{ph}(\theta)$ is shown as polar plot in Fig. 7.1 for a specific ratio c_s/v_A of the sound speed to the Alfvén speed.

From the phase speed relation (Eq. 7.1.32) we can see several special cases. For a vanishing magnetic field $v_A \propto B_0 = 0$, the relation degenerates to the simple solution of a wave with the phase speed equal to the sound speed, without direction dependence on the angle θ,

$$v_{ph} = c_s , \qquad \text{for} \quad v_A = 0 . \tag{7.1.33}$$

From the continuity equation (7.1.11) and the Fourier form (Eq. 7.1.24) it follows that $\omega\rho_1 = -\rho_0 k v_1$, or $c_s\rho_1 = -\rho_0 v_1$, which means that the pressure gradient is the only restoring force, where a velocity disturbance v_1 is restored by a proportional density change ρ_1. So, the *sound wave* or *acoustic wave* is a *non-dispersive* ($v_{gr} = v_{ph} = c_s$), *compressional* ($\mathbf{k} \cdot \mathbf{v}_1 \neq 0$), and *longitudinal* ($\mathbf{k} \parallel \mathbf{v}$) wave.

For a non-vanishing magnetic field, but perpendicular propagation direction ($\theta = 90°$), the phase speed becomes,

$$v_{ph} = \sqrt{c_s^2 + v_A^2} , \qquad \text{for} \quad \theta = 90° . \tag{7.1.34}$$

For parallel direction ($\theta = 0$) the phase speed is [using the standard solution for the variable v_{ph}^2 in the quadratic equation (7.1.32)],

$$v_{ph}^2 = \frac{(c_s^2 + v_A^2) \pm (c_s^2 - v_A^2)}{2} , \qquad \text{for} \quad \theta = 0° , \tag{7.1.35}$$

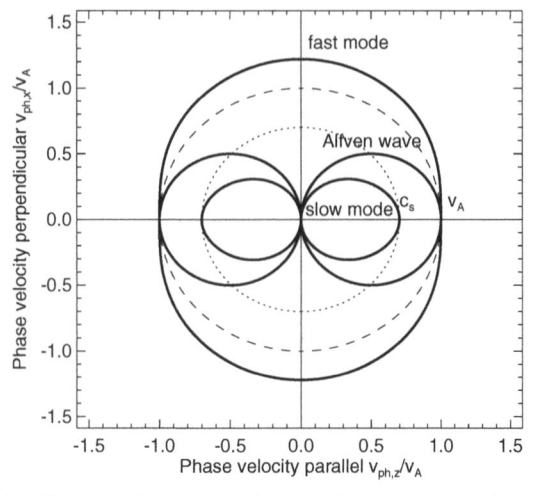

Figure 7.1: Polar diagram of phase speeds v_{ph} of magneto-acoustic waves, shown for a ratio of $c_s/v_A = 0.7$. The sound speed c_s is marked with a dotted circle, and the Alfvén speed v_A with a dashed line. The slow and fast magneto-sonic modes, and the Alfvén wave (also called "intermediate mode" in this context) are shown with thick curves.

which has the two asymptotic limits of the Alfvén speed v_A and the sound speed c_s,

$$v_{ph} \mapsto \pm v_A , \qquad \text{for} \quad v_A \gg c_s . \tag{7.1.36}$$

Since the dispersion relation (7.1.28) is a quadratic equation for ω^2, there are two solutions $\omega^2(\mathbf{k})$, which are called the *slow* and *fast modes*. If the perturbation is *incompressible* [i.e., the wave direction is perpendicular to the velocity disturbance ($\mathbf{k}\mathbf{v}_1 = 0$ or $\nabla \mathbf{v}_1 = 0$)], the velocity disturbance $\mathbf{v}_1$ is restored by a magnetic field change $\mathbf{B}_1$. This can easily be derived from the induction equation (7.1.13), where $\nabla \mathbf{v}_1 = 0$ leads to $\omega \mathbf{B}_1 = B_0 k_z \mathbf{v}_1$. Combining this condition with the momentum equation (7.1.12) leads then directly to the definition of the Alfvén speed (Eq. 7.1.14). This incompressible wave is called a *shear Alfvén wave* and falls into the category of *transverse waves*. Because it is an incompressible wave, no density or pressure changes are associated with it. The driving force of an Alfvén wave is the magnetic tension force alone. Using the constraint $k = k_z$ (i.e., that an Alfvén wave propagates parallel to the magnetic field), the dispersion relation (7.1.28) yields the following solution for the shear Alfvén wave,

$$\omega = v_A k_z = v_A k \cos \theta . \tag{7.1.37}$$

Thus the shear Alfvén wave propagates with speed v_A in a parallel direction, but cannot propagate in a perpendicular direction ($v_{ph} = \omega/k = 0$ for $\theta = 90°$). This solution is shown in the polar diagram (Fig. 7.1) as Alfvén wave (or *intermediate mode*).

So, for every propagation angle θ, there are generally three solutions, which are called the *slow, intermediate,* or *fast magneto-acoustic mode*. The order in the phase speed, however, depends whether the sound speed c_s is larger or smaller than the Alfvén speed v_A. For coronal conditions the Alfvén speed is generally larger ($v_A \approx$

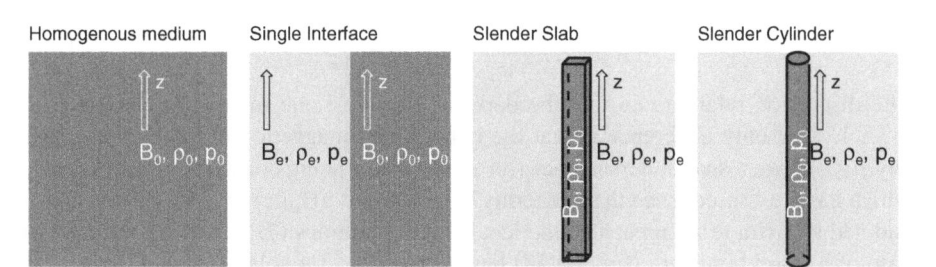

Figure 7.2: Geometries used in the derivation of the dispersion relation of magneto-acoustic waves: (1) unbounded homogeneous medium, (2) single interface, (3) slender slab, and (4) slender cylinder. The curvature of coronal loops is neglected in all models.

1000 km s^{-1}) than the sound speed ($c_s \approx 200$ km s^{-1} for a $T = 1.5$ MK plasma), say $c_s/v_A \approx 0.2$. In Fig. 7.1 we show the phase speeds of magneto-acoustic waves for a ratio of $c_s/v_A = 0.7$, but coronal ratios are typically lower, so that $v_{ph} \gtrsim v_A$. From the polar diagram shown in Fig. 7.1 we can see that the *slow mode* has a phase speed in the range of $0 \leq v_{ph} \leq min(c_s, v_A)$, having a maximum for propagation along the magnetic field and cannot propagate perpendicularly to the magnetic field. The *fast mode* has a phase speed in the range of $max(c_s, v_A) \leq v_{ph} \leq (c_s^2 + v_A^2)^{1/2}$, with the fastest mode propagating perpendicularly to the magnetic field.

7.1.2 Single Magnetic Interface

Since the solar corona is a highly inhomogeneous medium, this introduces a modification in the dispersion relation of waves and oscillations that can be supported in coronal structures. A first step towards inhomogeneous structures is the concept of two different plasma zones that are separated by a discontinuity interface. Such boundaries occur at the surface of overdense coronal loops, at the edge of sunspots, or at the boundaries of coronal holes. The dispersion relations for magneto-acoustic waves have been generalized for a magnetically structured atmosphere by Roberts (1981a), for a continuously changing magnetic field $\mathbf{B}_0(\mathbf{x}) = [0, 0, B_0(x)]$, as well as for a sharply structured medium with a step function at the discontinuity boundary $x = 0$ (Fig. 7.2, second left),

$$B_z(x), \rho_0(x), p_0(x) = \begin{cases} B_e, \rho_e, p_e , & x > 0 \\ B_0, \rho_0, p_0 , & x < 0 \end{cases} \qquad (7.1.38)$$

The physical parameters on the two sides are related by the pressure balance (Eq. 6.2.4) across the interface,

$$p_e + \frac{B_e^2}{8\pi} = p_0 + \frac{B_0^2}{8\pi} , \qquad (7.1.39)$$

with the corresponding definitions of the sound speeds (c_s, c_{Se}) and Alfvén speeds (v_A, v_{Ae}),

$$c_s = \sqrt{\frac{\gamma p_0}{\rho_0}} , \quad c_{Se} = \sqrt{\frac{\gamma p_e}{\rho_e}} , \qquad (7.1.40)$$

$$\mathrm{v}_A = \frac{B_0}{\sqrt{4\pi\rho_0}}, \qquad \mathrm{v}_{Ae} = \frac{B_e}{\sqrt{4\pi\rho_e}}. \qquad (7.1.41)$$

The dispersion relation can then be derived with the same method as described in § 7.1.1. The only difference is that the unperturbed magnetic field $\mathbf{B}_0(x)$ and density $\rho_0(x)$ have a spatial dependence (on x) instead of being constants (in Eq. 7.1.11), which has the consequence that the terms $\nabla\rho_0 \neq 0$ and $d\mathbf{B}_0(x)/dx \neq 0$ do not vanish and add an extra term in each of the ideal MHD equations (7.1.11−13). The special case of a sharp boundary (Eq. 7.1.38) has been treated for solar applications by Ionson (1978), Wentzel (1979), and Roberts (1981a). Each of the two separated media can support magneto-acoustic waves in their domain separately, with dispersion relations as calculated for the unbounded single medium (§ 7.1.1), requiring $m_0^2 < 0$ in Eq. (7.1.25), which are called *body waves*. A new phenomenon that occurs thanks to the presence of a boundary, is a mode or modes that are termed *surface waves*, with $m_0^2 > 0$ in Eq. (7.1.25). The obtained *dispersion relation for surface waves at a single magnetic interface*, for a wave vector $\mathbf{k} = (0, k_y, k_z)$ is found to be (Wentzel 1979; Roberts 1981a),

$$\frac{\omega^2}{k_z^2} = \mathrm{v}_A^2 - \frac{R}{R+1}(\mathrm{v}_A^2 - \mathrm{v}_{Ae}^2) = \mathrm{v}_{Ae}^2 + \frac{1}{R+1}(\mathrm{v}_A^2 - \mathrm{v}_{Ae}^2), \qquad (7.1.42)$$

$$R = \frac{\rho_e}{\rho_0}\left(\frac{m_0^2 + k_y^2}{m_e^2 + k_y^2}\right)^{1/2} > 0, \qquad (7.1.43)$$

$$m_e^2 = \frac{(k_z^2 c_{Se}^2 - \omega^2)(k_z^2 \mathrm{v}_{Ae}^2 - \omega^2)}{k_z^2 c_{Se}^2 \mathrm{v}_{Ae}^2 - (c^2 + \mathrm{v}_{Ae}^2)\omega^2}. \qquad (7.1.44)$$

The phase speed of this surface wave is between v_{Ae} and v_A. Since Eq. (7.1.42) is transcendental, there may be more than one surface mode (Roberts 1981a). For certain choices of sound and Alfvén speeds there are two surface waves; for other choices only one surface wave exists.

A special case is a *shear wave* that propagates along the y-axis of the surface. The corresponding dispersion relation is obtained for the limit $k_y^2 \gg k_z^2$ and $k_y^2 \gg m_0^2, m_e^2$, leading to $R \approx \rho_e/\rho_0$ and the simplified dispersion relation

$$\frac{\omega^2}{k_z^2} = \frac{\rho_0 \mathrm{v}_A^2 + \rho_e \mathrm{v}_{Ae}^2}{\rho_0 + \rho_e}. \qquad (7.1.45)$$

This special case is equivalent to the *Alfvén surface wave* in the incompressible fluid limit ($\gamma \mapsto \infty.$). The same dispersion relation (Eq. 7.1.45) is also obtained in the special case when the plasma β-parameter is small on both sides of the magnetic interface (i.e., $\beta_0 \ll 1$ and $\beta_e \ll 1$), which is highly relevant for the solar corona. There are actually at least two modes of surface waves: one with a phase speed of $\mathrm{v}_{ph} < min(c_{Se}, c_T)$, called a *slow surface wave*, and one with $c_s < \mathrm{v}_{ph} < min(c_{Se}, \mathrm{v}_A)$, called a *fast surface wave* (Roberts 1981a). If one side is field-free, then the interface supports a slow wave. Further, if the gas inside the magnetic field is cooler than the field-free medium, then the interface can support a fast surface wave. This shows that the introduction of a boundary between two media creates a new branch of *surface wave modes*, on top of the *body wave modes* that exist in each medium separately.

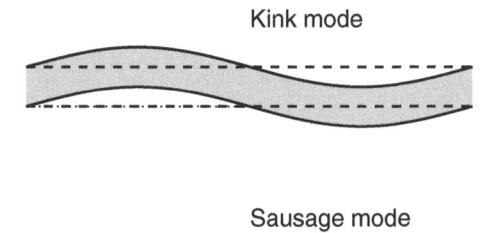

Kink mode

Sausage mode

Figure 7.3: Two types of fast magneto-acoustic modes: (1) the kink mode (top) and (2) the sausage mode (bottom). The half width x_0 of the slab is in the vertical direction and the wavelength $\lambda = 2\pi/k$ is in the horizontal direction.

7.1.3 Slender Slab Geometry

The next step towards a more realistic geometry of coronal structures is the consideration of an isolated magnetic slab, having a step-function transverse profile (for mathematical simplicity) (see Fig. 7.2, second-right panel),

$$B_0(x), \rho_0(x), p_0(x) = \left\{ \begin{array}{ll} B_0, \rho_0, p_0 \,, & |x| < x_0 \\ B_e, \rho_e, p_e \,, & |x| > x_0 \end{array} \right. \tag{7.1.46}$$

For the special case of an isolated slab in a field-free environment ($B_e = 0$), the following general *dispersion relation for surface waves* (for wave vector $\mathbf{k} = (0, 0, k_z)$) is found (Roberts 1981b),

$$(k^2 \mathrm{v}_A^2 - \omega^2) m_e = \left(\frac{\rho_e}{\rho_0} \right) \omega^2 m_0 \left(\begin{array}{c} \tanh \\ \coth \end{array} \right) m_0 x_0 \,, \tag{7.1.47}$$

where two wave modes are supported, a *sausage mode* for the *odd* tanh-function, and a *kink mode* for the *even* coth-function. In the sausage mode, opposite sides of the slab oscillate in anti-phase and the central axis remains undisturbed, while in the kink mode, opposite sides as well as the central axis oscillate in phase (Fig. 7.3). The more general case of a slab in a magnetic environment ($B_e \neq 0$) is treated in Edwin & Roberts (1982), yielding a similar dispersion relation as Eq. (7.1.47).

Special cases are the limits of *slender slabs* (where the wavelength is much larger than the width of the slab, $kx_0 \ll 1$) and *wide slabs* ($kx_0 \gg 1$). For slender slabs we have the simplification $\tanh(m_0 x) \approx m_0 x_0$ for $kx_0 \ll 1$. The resulting variety of longitudinal) slow modes and (transverse) fast magneto-acoustic modes (with the two manifestations of sausage and kink type modes) are discussed in Roberts (1981b).

7.1.4 Cylindrical Geometry

Transforming the cartesian slab geometry into a polar cylindrical geometry brings us to a suitable representation for magnetic fluxtubes and coronal loops (save for neglect-

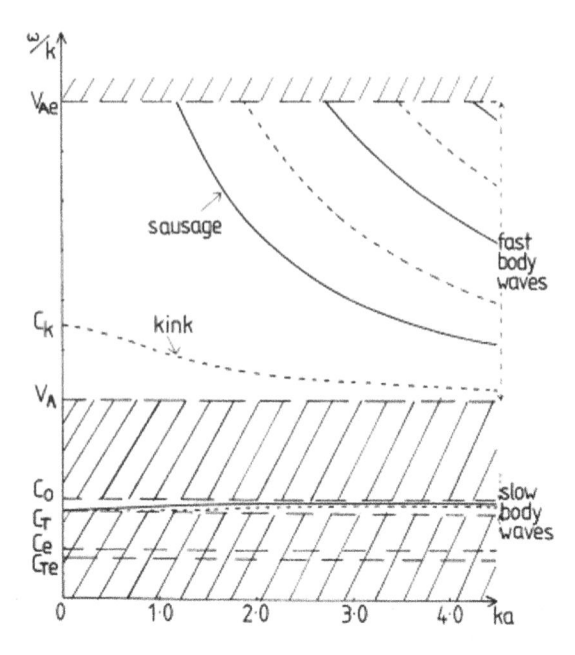

Figure 7.4: The phase speed ω/k is shown for magneto-acoustic waves in a cylindrical fluxtube (with radius a), as a function of the longitudinal wave number ka, for coronal conditions $v_{Ae} > v_A > c_T > c_s$. The sausage modes are indicated with solid lines, the kink modes with dashed lines. The notation in this figure corresponds to $k = k_z$ (Edwin & Roberts 1983).

ing the curvature). A straight cylinder is defined in cylindrical coordinates (r, θ, z), enclosed by a surface boundary at a radius of $r = a$ (Fig. 7.2, right),

$$B_0(r), \rho_0(r), p_0(r) = \begin{cases} B_0, \rho_0, p_0\,, & r < a \\ B_e, \rho_e, p_e\,, & r > a \end{cases}. \tag{7.1.48}$$

The derivation of the dispersion relation of fast and slow waves in such a cylindrical tube is given in Edwin & Roberts (1983) and is reviewed in Roberts & Nakariakov (2003). The analytical derivation is similar to the case of the cartesian slab: external and internal solutions of the MHD equations need to be matched at the boundary by the continuity of pressure and perpendicular velocity, and the wave energy has to vanish at a large distance from the tube. The Fourier form of the velocity disturbance $\mathbf{v}_1$ in cylindrical coordinates is

$$\mathbf{v}_1 = \mathbf{v}_1(r) \exp\left[i(\omega t + n\theta - k_z z)\right], \tag{7.1.49}$$

where n is an integer ($n = 0, 1, 2, ...$) which describes the azimuthal behavior of the oscillating tube.

The *dispersion relation for magneto-acoustic waves in a cylindrical magnetic flux-tube* is found to be (McKenzie 1970; Spruit 1982; Edwin & Roberts 1983; Cally 1986),

$$\rho_e(\omega^2 - k_z^2 v_{Ae}^2)m_0\frac{I_n'(m_0 a)}{I_n(m_0 a)} + \rho_0(k_z^2 v_A^2 - \omega^2)m_e\frac{K_n'(m_e a)}{K_n(m_e a)} = 0\,, \tag{7.1.50}$$

where m_0 and m_e are defined by m_α (with $\alpha = 0$ or $\alpha = e$),

$$m_\alpha^2 = \frac{(k_z^2 c_\alpha^2 - \omega^2)(k_z^2 v_{A\alpha}^2 - \omega^2)}{(c_\alpha^2 + v_{A\alpha}^2)(k_z^2 c_{T\alpha}^2 - \omega^2)} . \tag{7.1.51}$$

and $I_n(x)$ and $K_n(x)$ are modified Bessel functions of order n, with I_n' and K_n' being the derivatives with respect to the argument x. This dispersion relation (7.1.50) describes both surface (for $m_0^2 > 0$) and body waves (for $m_0^2 < 0$). Fig. 7.4 visualizes a typical dispersion diagram for fluxtubes under coronal conditions, $c_{Se} < c_s < v_A < v_{Ae}$, (Edwin & Roberts 1983). There are two branches of phase speeds: (1) a *fast-mode* branch with phase speeds $v_A < v_{ph} < v_{Ae}$, containing the kink ($n = 1$) and sausage mode ($n = 0$); and (2) a *slow-mode* branch with phase speeds $c_T < v_{ph} < c_s$. Higher modes with $n = 2, 3, ...$, are called the *fluting modes*. Under coronal conditions ($v_A \gg c_s$), the tube speed c_{T0} (Eq. 7.1.27) is almost equal to the sound speed c_s, making the slow mode only weakly dispersive. In contrast, the fast modes are dispersive, especially in the long wavelength part of the spectrum ($ka \lesssim 2\pi$).

7.2 Fast Kink-Mode Oscillations

7.2.1 Kink-Mode Period

Magneto-acoustic oscillations of the fast kink mode have now been directly observed in EUV wavelengths with TRACE (Aschwanden et al. 1999b, 2002a; Nakariakov et al. 1999; Schrijver et al. 1999, 2002). The geometric ratio of the loop width $w = 2a$ to the loop half length L is typically $w/L \approx 0.05 - 0.1$ (Aschwanden et al. 2002a), so the dimensionless wave number is $ka = 2\pi a/\lambda = (\pi/4)(w/L) \approx 0.04 - 0.08$, where the wavelength is $\lambda = 4L$ and the loop radius $a = w/2$. Therefore, the observed kink-mode oscillations are in the long-wavelength regime of $ka \ll 1$, where the phase speed of the kink mode is practically equal to the *kink speed* c_k (see Fig. 7.4), given in Eq. (7.1.45) and also discussed in earlier studies (e.g., Wilson 1980; Spruit 1981; Roberts et al. 1984),

$$c_k = \left(\frac{\rho_0 v_A^2 + \rho_e v_{Ae}^2}{\rho_0 + \rho_e} \right)^{1/2} . \tag{7.2.1}$$

In the low-β plasma of the corona, where the magnetic field is almost equal inside and outside of the loop ($B_e \approx B_0$), it can be approximated by

$$c_k \approx v_A \left(\frac{2}{1 + (\rho_e/\rho_0)} \right)^{1/2} . \tag{7.2.2}$$

Thus, for typical coronal density ratios around oscillating loops (i.e., $\rho_e/\rho_0 \approx 0.1 - 0.5$) (Aschwanden 2001b; Aschwanden et al. 2003b), the kink speed amounts to $c_k \approx (1.15 - 1.35)v_A$. As the solution for the kink mode for long wavelengths ($ka \ll 1$) shows in Fig. 7.4, the kink wave is essentially non-dispersive and has a phase speed equal to the kink speed,

$$v_{ph} = \frac{\omega}{k_z} \approx c_k \gtrsim v_A . \tag{7.2.3}$$

Table 7.1: Inference of the magnetic field in 11 oscillations events (based on parameters obtained in Aschwanden et al. 2002a, 2003b).

Observation date and start time [UT]	Loop Length $l[Mm]$	Oscillation Period $P[s]$	Damping time $\tau_D[s]$	Electron density n_0 cm^{-3}	Density ratio n_e/n_0	Magnetic field $B[G]$
1998-Jul-14, 12:59:57	168	261	870	15.9 10^8	0.20	20
1998-Jul-14, 12:57:38	72	265	300	23.2 10^8	0.26	11
1998-Jul-14, 12:57:38	174	316	500	11.2 10^8	0.27	15
1998-Jul-14, 12:56:32	204	277	400	4.9 10^8	0.42	14
1998-Jul-14, 13:02:26	162	272	849	14.2 10^8	0.44	20
1998-Nov-23, 06:35:57	390	522	1200	9.6 10^8	0.32	19
1999-Jul-04, 08:33:17	258	435	600	20.0 10^8	0.14	21
1999-Oct-25, 06:28:56	166	143	200	22.3 10^8	0.18	43
2001-Mar-21, 02:32:44	406	423	800	20.8 10^8	0.07	33
2001-May-15, 02:57:00	192	185	200	4.7 10^8	0.32	19
2001-Jun-15. 06:32:29	146	396	400	9.5 10^8	0.66	11

Standing waves occur when the kink wave is reflected at both ends of the loop, so that the nodes of the standing wave coincide with the footpoints of the loops, which are rigidly anchored in the chromosphere (by line-tying). If we denote the full loop length with l, the wavelength of the fundamental standing wave is the double loop length (due to the forward and backward propagation) (i.e., $\lambda = 2l$), and thus the wave number of the fundamental mode (N=1) is $k_z = 2\pi/\lambda = \pi/l$, while higher harmonics ($N = 2, ...$) would have wave numbers $k_z = N(\pi/l)$. Then the time period P_{kink} of a kink-mode oscillation at the fundamental harmonic is (with Eq. 7.2.2),

$$P_{kink} \approx \frac{2l}{c_k} = \frac{2l}{v_A} \left(\frac{1 + (\rho_e/\rho_0)}{2} \right)^{1/2} . \qquad (7.2.4)$$

Coronal loops observed with TRACE have full loop lengths of $l \approx 60 - 600$ Mm (Aschwanden et al. 2002a), so for Alfvén speeds of order $v_A \approx 1$ Mm s^{-1} we therefore expect kink periods in the range of $P_{kink} \approx 2l/v_A = 2 - 20$ minutes.

7.2.2   Magnetic Field Strength and Coronal Seismology

The observation of kink modes in the corona provides us with a valuable tool for determining the magnetic field strength B in coronal loops, a capability that has been dubbed *MHD coronal seismology* (Roberts et al. 1984; Roberts 2000, 2002; Roberts & Nakariakov 2003) and has been applied to (TRACE) images of oscillating loops (Nakariakov & Ofman 2001). The kink-mode period P_{kink} is a function of the Alfvén speed $v_A = B/\sqrt{4\pi\rho_0}$, and thus of the mean magnetic field strength B, which can be explicitly expressed from Eq. (7.2.4),

$$B = \frac{l}{P_{kink}} \sqrt{8\pi\rho_0(1 + \rho_e/\rho_0)} . \qquad (7.2.5)$$

Table 7.2: Parameters measured for loop oscillations during the 1998-Jul-14, 12:55 UT, flare (Aschwanden et al. 1999b).

Loop #	Length l[Mm]	Amplitude A_1[Mm]	Ratio A_1/l	Period T[s]	Phase t_0/T	Number of pulses
4	90	2.0 ± 0.2	0.022	258	0.79	2.1
6	120	4.6 ± 0.2	0.038	269 ± 6	0.67 ± 0.01	5.4
7	130	3.9 ± 0.4	0.030	320 ± 10	0.87 ± 0.07	2.8
8	150	4.4 ± 0.3	0.029	258 ± 10	0.68 ± 0.09	4.5
9	160	5.6 ± 0.1	0.035	278 ± 5	0.91 ± 0.01	2.4
	130 ± 30	4.1 ± 1.3	0.031 ± 0.006	276 ± 25	0.78 ± 0.10	3.4 ± 1.4

For coronal abundances (H:He=10:1) we have $\mu \approx 1.27$ (Eq. 3.16) in the mass density $\rho_0 = \mu m_H n_0$ (Eq. 3.1.6) and can express Eq. (7.2.5) in practical units ($l_8 = l/10^8$ cm^{-3}, $n_8 = n_e/10^8$ cm^{-3}, P_{kink} [s]), giving the field strength B in Gauss:

$$B = 7.3 \times \frac{l_8}{P_{kink}} \sqrt{n_8 \left(1 + n_e/n_0\right)} \qquad (G) . \qquad (7.2.6)$$

For instance, for an oscillating loop observed on 1998-Jul-14, 12:11 UT, Nakariakov & Ofman (2001) measured a loop length of $l = 1.3 \times 10^{10}$ cm, a density of $n_e = 10^{9.3 \pm 0.3}$ cm^{-3}, and an oscillation period of $P = 256$ s, and using an estimate of the density contrast of $n_e/n_0 \approx 0.1$ they determined a magnetic field strength of $B = 13 \pm 9$ G, an Alfvén speed of $v_A = 756 \pm 100$ km s^{-1}, and a kink speed of $c_k = 1020 \pm 132$ km s^{-1}. The largest uncertainty of this method is probably the density measurement n_0, which can be uncertain up to a factor of 4 due to unknown iron enhancement; This affects the magnetic field determination according to Eq. (7.2.6) by a factor of 2.

We apply this coronal magnetic field measurement technique, using Eq. (7.2.6), to 11 coronal loops, for which all necessary parameters have been measured [i.e. the full loop length l (Mm), the oscillation period P (s), the internal electron density n_0 (cm^{-3}), and the ratio n_e/n_0 of the external to the internal density], obtained in Aschwanden et al. (2003b) and listed in Table 7.1. The resulting magnetic field determinations fall into the range $B = 11, ..., 43$ G, with a median value of $B \approx 20$ G. The loop densities have been determined from the background-subtracted emission measure and could be subject to a filling factor, which would increase the inferred value of the magnetic field strength. Also, the additional uncertainty from the unknown coronal iron abundance could double the estimated values for the magnetic field strengths.

7.2.3 Observations of Kink-Mode Oscillations

The first *spatial oscillations* of coronal loops were discovered in extreme-ultraviolet wavelengths (171 Å) with the *Transition Region And Coronal Explorer (TRACE)*, in the temperature range of $T_e \approx 1.0 - 1.5$ MK (Aschwanden et al. 1999b). The observed loop oscillations occurred during a flare which began at 1998 July 14, 12:55 UT (Plate 9), and were most prominent during the first 20 minutes. The oscillating loops connected the penumbra of the leading sunspot to the flare site in the trailing portion.

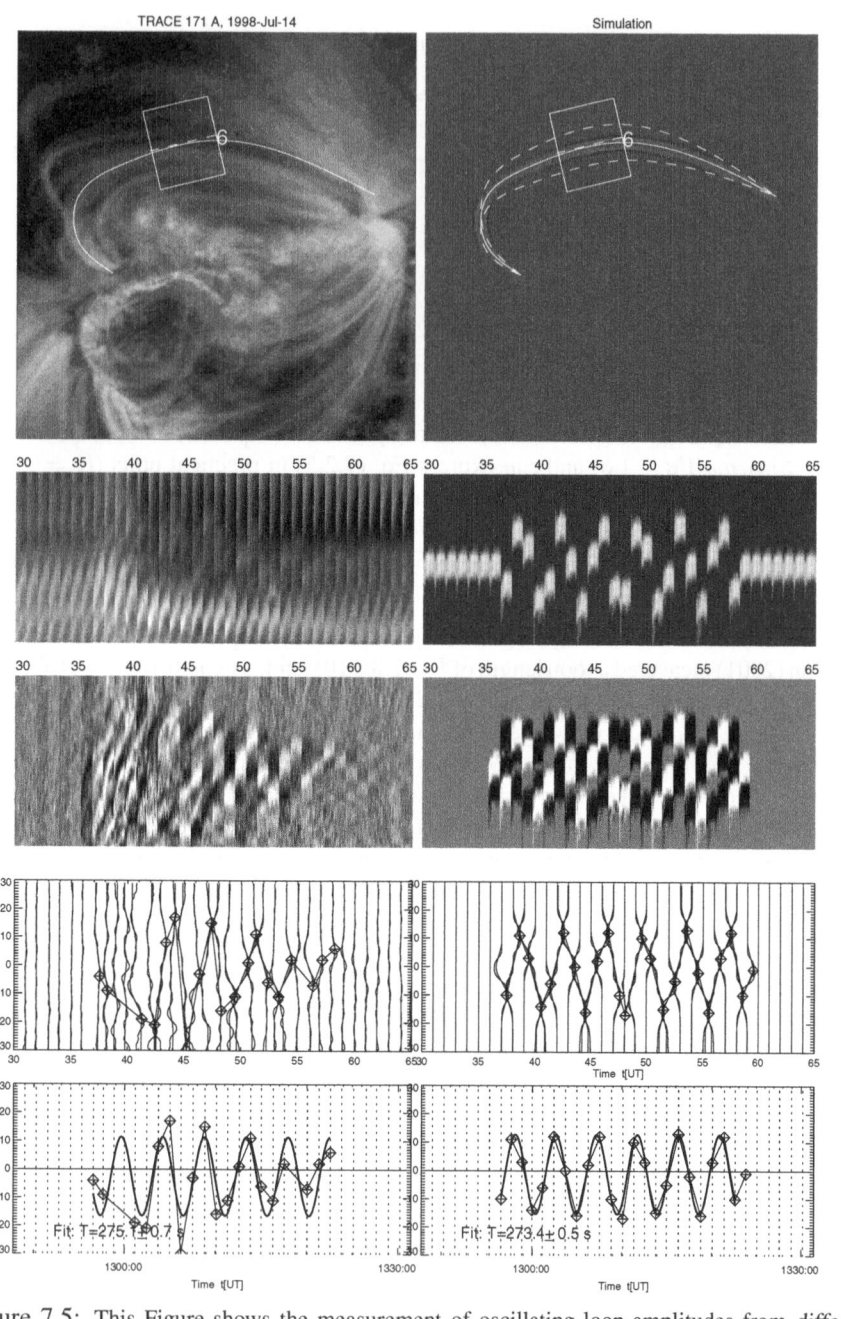

Figure 7.5: This Figure shows the measurement of oscillating loop amplitudes from *difference images* (left column) and a corresponding simulation (right column). The location of the analyzed loop (Loop #6) is outlined with a thin white curve (top map), where data slices are extracted perpendicular to the loop at the apex (second row), from which a *running difference* is taken to filter out the oscillating loop (third row), and the amplitudes are measured (forth row), fitted by a sine function (bottom row) (Aschwanden et al. 1999b).

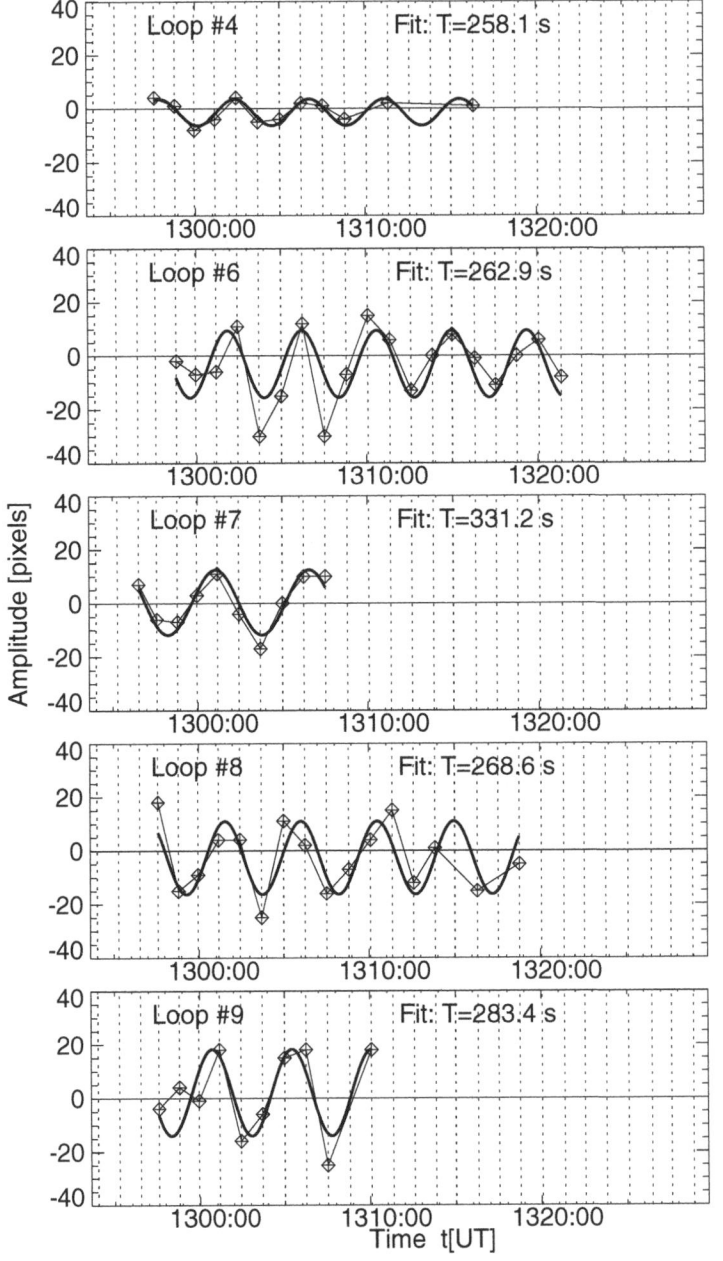

Figure 7.6: The spatial displacements of the 5 analyzed oscillating loops (Loop #4, 6−9) are shown (diamonds) and fitted with a sine function (thick solid curve) according to the procedure illustrated in Fig. 7.5. The spatial displacements are all measured in a perpendicular direction to the loops. Note that all periods and relative phases are similar, indicative of a common trigger (Aschwanden et al. 1999b).

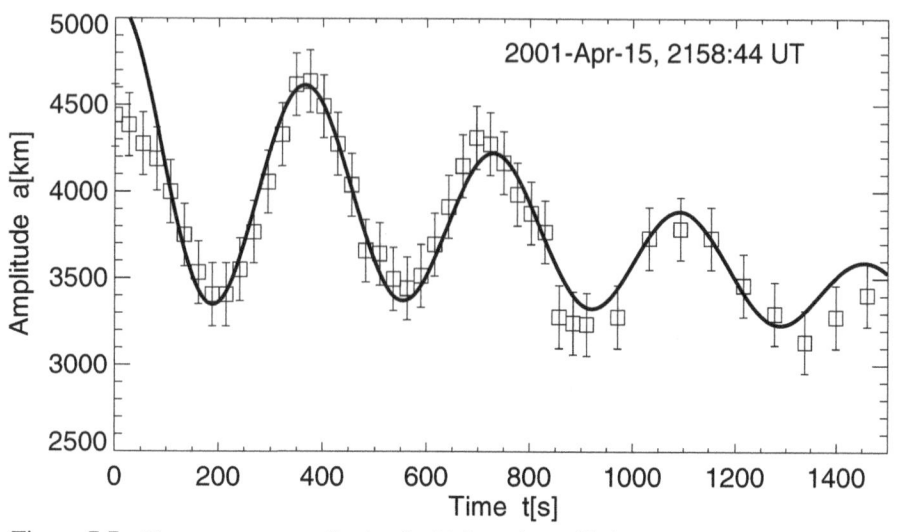

Figure 7.7: The transverse amplitude of a kink-mode oscillation measured in one loop of a postflare loop arcade observed with TRACE on 2001-Apr-15, 21:58:44 UT (see front cover of book). The amplitudes are fitted by a damped sine plus a linear function, $a(t) = a_0 + a_1 \sin\left(2\pi * (t - t_0)/P\right) \exp\left(-t/\tau_D\right) + a_2 * t$, with a period of $P = 365$ and a damping time of $\tau_D = 1000$ s (courtesy of Ed DeLuca and Joseph Shoer).

Table 7.3: Average and ranges of physical parameters of 26 loops exhibiting MHD fast kink-mode oscillations. All Alfvén speeds and times are calculated for a magnetic field of $B = 30$ G. The most extreme period of $P = 2004$ s is excluded in the statistics. Only the 10 most reliable decay times τ_D are included (Aschwanden et al. 2002a).

Parameter	Average	Range
Loop half length L	110 ± 53 Mm	$37 - 291$ Mm
Loop width w	8.7 ± 2.8 Mm	$5.5 - 16.8$ Mm
Oscillation period P	321 ± 140 s	$137 - 694$ s ^{2})
	5.4 ± 2.3 min	$2.3 - 10.8$ min^2)
Decay time τ_D	580 ± 385 s	$191 - 1246^3$) s
	9.7 ± 6.4 min	$3.2 - 20.8$ min^3)
Oscillation duration d	1392 ± 1080 s	$400 - 5388$ s
	23 ± 18 min	$6.7 - 90$ min
Oscillation amplitude A	2200 ± 2800 km	$100 - 8800$ km
Number of periods	4.0 ± 1.8	$1.3 - 8.7$
Electron density of loop n_{loop}	$(6.0 \pm 3.3)10^8$ cm^{-3}	$(1.3 - 17.1)10^8$ cm^{-3}
Maximum transverse speed v_{max}	42 ± 53 km/s	$3.6 - 229$ km/s
Loop Alfvén speed v_A	2900 ± 800 km/s	$1600 - 5600$ km/s
Mach factor v_{max}/v_{sound}	0.28 ± 0.35	$0.02 - 1.53$
Alfvén transit time τ_A	150 ± 64 s	$60 - 311$ s
Duration/Alfvén transit d/τ_A	9.8 ± 5.7	$1.5 - 26.0$
Decay/Alfvén transit τ_D/τ_A	4.1 ± 2.3	$1.7 - 9.6^3$)
Period/Alfvén transit P/τ_A	2.4 ± 1.2	$0.9 - 5.4^2$)

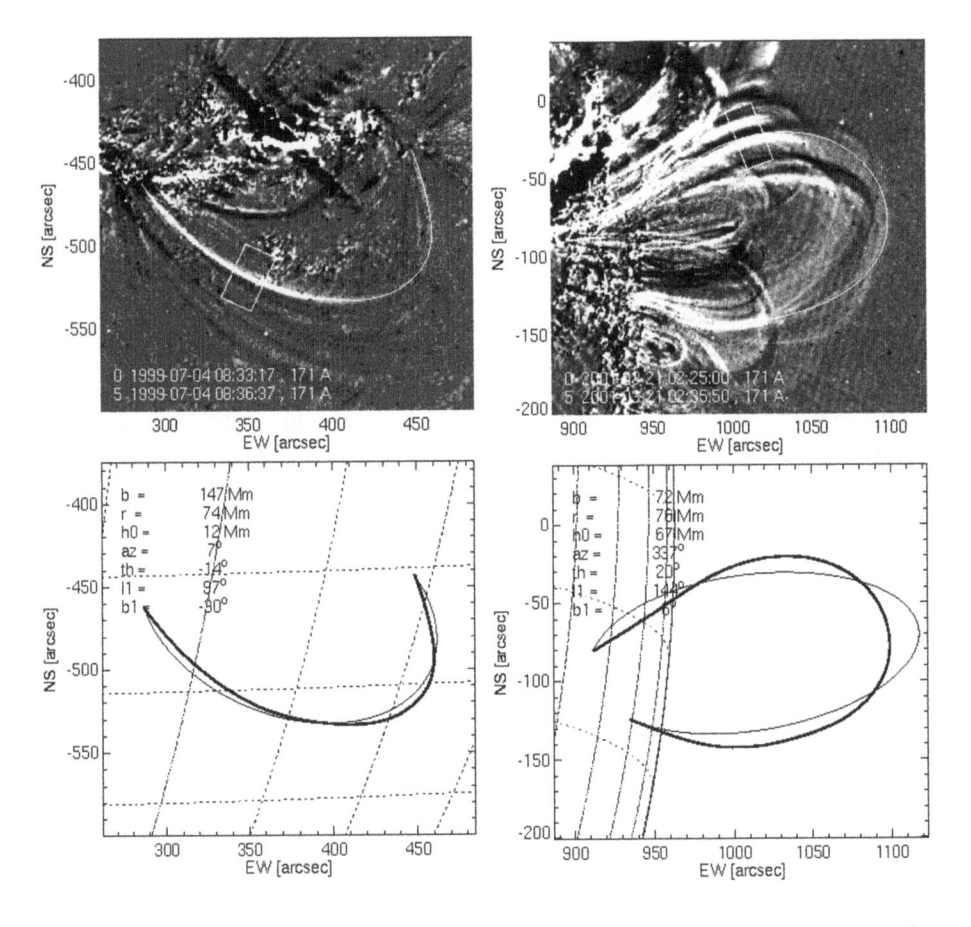

Figure 7.8: The displacement of loops during MHD fast kink-mode oscillations is shown for the 1998-Jul-4, 08:33 UT (left), and 2001-Mar-21, 02:32 UT (right) events. The top images show difference images of TRACE 171 Å, where white/black structures demarcate the new/old location of oscillating loops. Note that in the 1998-Jul-4 event only the left segment of a single loop oscillates, while during the 2001-Mar-21 flare all loops in the active region become displaced (Aschwanden et al. 2002a).

Five oscillating loops (Table 7.2) with an average length of $L = 130 \pm 30$ Mm were detected. The transverse amplitude of the oscillations is $A = 4.1 \pm 1.3$ Mm and the mean period is $T = 276 \pm 25$ s. The oscillation mode is consistent with a standing wave in the kink mode (with fixed nodes at the footpoints), based on the measured lateral displacements and the observed oscillation periods.

The steps in the analysis procedure of a loop oscillation are shown in Fig. 7.5 for a particular loop, and the results of the measured amplitudes of 5 loops are shown in Fig. 7.6. The parameters of the measured loop length l (Mm), transverse oscillation amplitude A_1 (Mm), ratio A_1/l, oscillation period P (s), phase t_0/T, and number of observed pulses are given in Table 7.2. The spatial location of the loops can be

Table 7.4: Coronal oscillations observed with periods in the range of 1−10 min, the domain in which MHD fast kink-mode oscillations and photospheric p-mode oscillations (3 to 5 min) occur.

Observer	N	Frequency or wavelength	Oscillation period P[s]	Spatial scale or instrument
Radio:				
Simon & Shimabukuro (1971)	1	86, 91 GHz	180	Kitt Peak
Trottet et al. (1979)	1	169 MHz	60	6′ (Nançay)
Strauss et al. (1980)	1	22 GHz	336	Itapetinga
Aurass & Mann (1987)	1	23−40 MHz	44−234	Tremsdorf
Chernov et al. (1998)	1	164−407 MHz	180	5′ -7′ (Nançay)
Gelfreikh et al. (1999)	4	17 GHz	120−220	0.2′ (Nobeyama)
Nindos et al. (2002)	2	5, 8 GHz	157, 202	$1.1'' - 5.7''$ (VLA)
Optical and Hα:				
Koutchmy et al. (1983)	2	5303 Å	43, 80, 300	Sac Peak
Jain & Tripathy (1998)	2	Hα	180-300	Udaipur
Ofman et al. (1997)	2	White light	360	SOHO/UVCS
Ofman et al. (2000a)	7	White light	400, 625	SOHO/UVCS
EUV:				
Chapman et al. (1972)	1	304,315,368 Å	300	OSO-7
Antonucci et al. (1984a)	1	554,625,1335 Å	141,117	Skylab
Aschwanden et al. (1999b)	5	171, 195 Å	276±25	TRACE
Nakariakov et al. (1999)	1	171 Å	256	TRACE
Nakariakov & Ofman (2001)	2	171, 195 Å	256, 360	TRACE
Aschwanden et al. (2002a)	26	171, 195 Å	120−1980	TRACE
Soft X-rays:				
Jakimiec & Jakimiec (1974)	2	1−8 Å	200−900	SOLRAD 9
Hard X-rays:				
Lipa (1978)	26	14−111 keV	10−100	OSO-5
Terekhov et al. (2002)	1	8−60 keV	143.2±0.8	GRANAT

identified from the numbers (#4−9) on Plate 9. Note that the oscillations can typically only be detected over a small number of $\approx 2 - 5$ periods.

The onset of the loop oscillations seem to be triggered in the core region of the flare, transmitted by a radially propagating disturbance with a speed of ≈ 700 km s^{-1}. Interestingly, there is an intriguing selection effect with which loops are seen to oscillate: sometimes only a few loops oscillate, while in strong flare events all loops of an active region become displaced, with only a subset of them continuing to oscillate after an initial displacement (Figs. 7.5−7.8). Apparently the oscillating loops need to fulfill a special selection criterion, which we will discuss further in § 7.5 on damping mechanisms.

Imaging observations of MHD fast kink-mode oscillations have mainly be made in EUV (Figs. 7.5−7.8) and the statistics of physical parameters of 26 events are provided in Table 7.3. A more comprehensive compilation of reports on oscillatory events with periods of $P \approx 1 - 10$ minutes is given in Table 7.4, which besides EUV observations

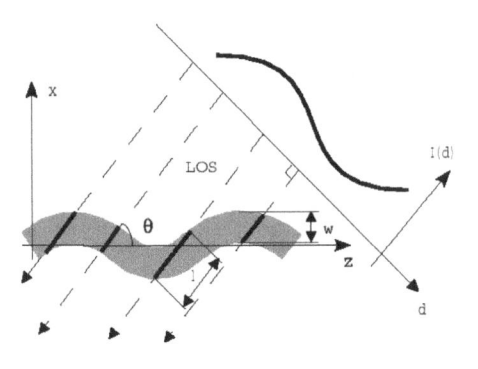

Figure 7.9: A snapshot of perturbation of a segment of a coronal loop by a harmonic kink wave, which produces a variation of the optically thin emission measure integrated along the line-of-sight, due to the variable column depth (Cooper et al. 2003).

encompasses also occasional reports in radio, optical, Hα, soft X-ray, and hard X-ray wavelengths. However, most of these other observations have only restricted spatial information, and thus only a subset are candidates for the interpretation of MHD fast kink-mode standing waves, based on the plausibility of the observed oscillation period. There is also the question of how the observed emission is modulated by the kink mode, which does not perturb the density in the first order. One possibility is that line-of-sight effects of an oscillating curved loop vary the projected column depth (Fig. 7.9), and this way produce a modulation of optically thin emission such as in EUV and soft X-rays (Cooper et al. 2003). Radio emission with high directivity, such as gyrosynchrotron emission (see § 5.7.2) or beam-driven plasma emission, could also be modulated by the changing orientation of oscillating curved loops.

Alternatively, gyroresonance emission (particularly above sunspots) may also be modulated by upward propagating sound waves driven by the well-known photospheric 3-min oscillations in sunspots. This has been clearly identified by high-resolution imaging of sunspot oscillations with the *Nobeyama Radioheliograph* (Gelfreikh et al. 1999) or with the *VLA* (Nindos et al. 2002). Upward propagating waves are also thought to cause density fluctuations and to modulate the white-light emission both in plumes and further out in the solar wind (Ofman et al. 1997, 2000a).

7.2.4 Prominence Oscillations

Quiescent prominences or filaments are made of bundles of closed magnetic field lines, and thus can exhibit similar fast kink-mode oscillations as ordinary coronal loops described in the previous section. What is different, however, is the temperature and density structure of the prominences. While ordinary coronal loops have an almost uniform temperature and density along their coronal top segment (save for the gravitational stratification), filaments and prominences are made of horizontally stretched field lines that have dipped segments filled with cool plasma. This nonuniform density and temperature structure prevents a pure kink-mode oscillation at the fundamental

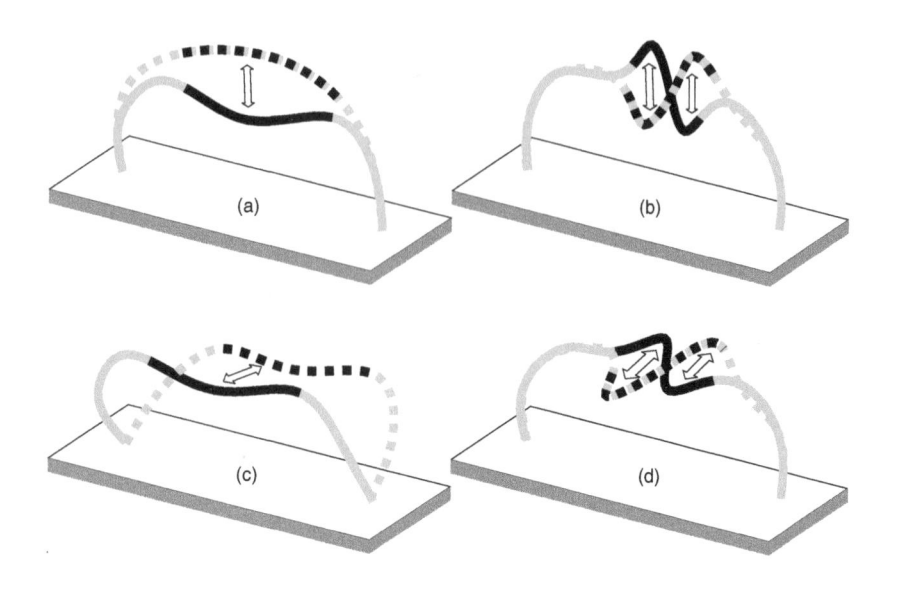

Figure 7.10: Various kink-type oscillation modes for 2D prominence fibrils: (a) a string-type magnetic Love mode, (b) an internal magnetic Love mode, (c) an Alfvén string mode, and (d) an internal Alfvén mode (after Joarder et al. 1997).

harmonic with nodes of the standing wave at the chromospheric footpoints. Instead, kink-type displacements have a different inertia in the most massive field line segments that make up the cool prominence material. Prominence oscillations have therefore been modeled analogously to a stretched elastic string with nonuniform density, either as a slab with a skewed field (e.g., Joarder & Roberts 1992; Díaz et al. 2001), or with a two-component model that consists of a cool dense segment in the middle (mimicking the prominence gas) and hotter less dense ends on both sides (mimicking the evacuated parts of the prominence fibril that contain coronal plasma) (e.g., Joarder et al. 1997; Díaz et al. 2002, 2003). The fundamental modes of such a two-component model are illustrated in Fig. 7.10, categorized as vertical kink-type oscillations (called magnetic Love modes in Joarder et al. 1997) and horizontal kink-type oscillations (called Alfvén modes in Joarder et al. 1997), both having either two nodes (string modes) or three nodes (internal modes). Since prominences are made of many fibrils, complex interactions can occur between neighboring fibrils (e.g., wave leakage) and a single oscillating fibril can excite oscillations in a neighboring one (Díaz et al. 2001). The oscillations are damped, for example, by radiative cooling (Terradas et al. 2001). Very long periods (of the order of hours and more) are believed to be due to *magneto-acoustic slow string modes* (see § 7.4). Those parts of prominences that show fast periods of $P \approx 3 - 5$ min are believed to be coupled to the photospheric global p-mode oscillations (e.g, Balthasar et al. 1988). Alternative prominence oscillation models invoke formation in current sheets with nonlinear oscillations during compression (Sakai et al. 1987), or induction-driven radial oscillations in current-carrying loops (Cargill et al. 1994).

Observations of prominence oscillations are compiled in Table 7.5. There are three

Table 7.5: Periods of prominence oscillations and size scale (adapted from Oliver & Ballester 2002).

Observer	Period $P[min]$	Size $l[Mm]$	Wavelength instrument
Ramsey & Smith (1966)	6−40	...	Hα; Lockheed
Harvey (1969)	1−17	...	Hα; HAO, Climax
Landman et al. (1977)	22	...	Hβ; He³, Ca⁺; Maui
Bashkirtsev et al. (1983)	77, 82	...	Hβ; Sayan
Bashkirtsev & Mashnich (1984)	42−82	...	Hβ; Sayan
Wiehr et al. (1984)	3−5, 50, 60, 64	...	Hα; Locarno
Tsubaki & Takeuchi (1986)	2.7, 3.5	20	Ca II; Hida
Balthasar et al. (1986)	3−6, 48	...	Hα; Locarno
Balthasar et al. (1988)	3−5, 60	...	Hα, Ca⁺, Hε; Tenerife
Suematsu et al. (1990)	≈ 60	2.8	Ca II K, Hβ; Hida
Thompson & Schmieder (1991)	4.4	84	Hα; Meudon
Yi, Engvold, & Keil (1991)	5.3, 8.6, 15.8	21	He I; NSO, Sac Peak
	5.3, 7.9, 10.6	7	He I; NSO, Sac Peak
Balthasar et al. (1993)	0.5−20	1−37	Hβ, Ca⁺, HeD₃; Tenerife
Balthasar & Wiehr (1994)	5.3−60	1−2	He 3889 A, H8, Ca⁺; Tenerife
Sütterlin et al. (1997)	3.5−62	2.1−7.7	Ca⁺ 8542; Locarno
Molowny−Horas et al. (1997)	4−12	1.4−7.3	Hβ; Tenerife
Blanco et al. (1999)	1−6	...	O IV, Si IV; SUMER
Régnier et al. (2001)	5, 6−20, 65	...	He I; SUMER
Bocchialini et al. (2001)	6−12	...	O IV, Si IV; SUMER
Terradas et al. (2002)	71	30	Hβ; NSO Sac Peak
Lin (2002)	5−23	...	Hα; NSO

major groups (see also Engvold 2001b): short periods with $P \approx 3 - 5$ min, longer periods with $P \approx 12 - 20$ min, and very long periods with $P \approx 40 - 90$ min, which respectively reflect the photospheric p-mode driven domain, the fast magneto-acoustic domain, and the slow magneto-acoustic domain. Typical Alfvén speeds in prominences are about $v_A \approx 170$ km s^{-1}, based on estimated magnetic fields of $B \approx 8$ G and densities of $n_e \approx 10^{10}$ cm^{-3}. In general, prominence oscillations have been observed in Doppler shifts (e.g., v $\approx 2 - 13$ km s^{-1}, Bocchialini et al. 2001) rather than in intensity, as expected for kink-type oscillations, which are non-compressional transverse displacements. However, there is some confusion in the interpretation of Doppler shift measurements, because there are also siginificant flows present along the filaments (Zirker et al. 1998) besides the kink-type displacement motion. The exciter of a large-scale prominence oscillation is generally a flare (Ramsey & Smith 1966), as for the kink-mode oscillations in active region loops (§ 7.2.3). Spatial observations allow us to distinguish between large-amplitude oscillations, triggered by a wave front from a flare (*winking filaments*), and small-amplitude oscillations that affect only parts (threads) of a prominence. The longest periods seem to involve entire prominences, whereas periods shorter than $\lesssim 20$ min appear to be tied to thread-like small-scale structures

(Engvold 2001b). Recent reviews on oscillations in prominences and filaments can be found in Roberts (1991b), Roberts & Joarder (1994), Oliver (2001a,b), Engvold (2001a,b), and Oliver & Ballester (2002).

7.3 Fast Sausage-Mode Oscillations

7.3.1 The Wave Number Cutoff

The dispersion relation for magneto-sonic waves in cylindrical magnetic fluxtubes (Eq. 7.1.50) has multiple types of solutions in the fast-mode branch, sausage modes ($n = 0$), kink modes ($n = 1$), and fluting modes ($n = 2, 3, ...$). The solutions are depicted in Fig. 7.4, showing that the kink-mode solution extends all the way to $ka \mapsto 0$, while the fast sausage mode has a cutoff at a phase speed of $v_{ph} = v_{Ae}$, which has no solution for small wave numbers $ka \lesssim 1$. In the following we ignore harmonic structures along the tube, we set $N = 1$. The propagation cutoff for the sausage mode occurs at the cutoff wave number k_c (Roberts et al. 1984),

$$k = k_c = \left[\frac{(c_s^2 + v_A^2)(v_{Ae}^2 - c_T^2)}{(v_{Ae}^2 - v_A^2)(v_{Ae}^2 - c_s^2)} \right]^{1/2} \left(\frac{j_{0,s}}{a} \right), \quad s = 1, 2, 3, ... \qquad (7.3.1)$$

where $j_{0,s} = (2.40, 5.52, ...)$ are the s-zeros of the Bessel function J_0. The cutoff frequency ω at the cutoff is $k_c v_{Ae} = \omega_c$. Let us consider this cutoff wave number for coronal conditions. Most of the active region loops have temperatures of $T \approx 1-3$ MK (e.g., see differential emission measure distribution by Brosius et al. 1996, illustrated in Fig. 3.4), so the sound speed typically amounts to $c_s = 147 \times \sqrt{T_{MK}} \approx 150 - 260$ km s^{-1}, which is much smaller than the typical Alfvén speed ($v_A \approx 1000$ km s^{-1}) in the corona (Fig. 5.10), so we have the coronal condition $c_s \ll v_A$. In this case the tube speed $c_T = c_s v_A / \sqrt{c_s^2 + v_A^2}$ (Eq. 7.1.27) is close to the sound speed, $c_T \lesssim c_s$, and the expression for the cutoff wave number k_c (Eq. 7.3.1) simplifies to

$$k_c \approx \left(\frac{j_{0,s}}{a} \right) \left[\frac{1}{(v_{Ae}/v_A)^2 - 1} \right]^{1/2} . \qquad (7.3.2)$$

In the low-β corona the thermal pressure is much smaller than the magnetic pressure, and thus we can assume almost identical magnetic field strengths inside and outside coronal loops (Eq. 6.2.6) (i.e., $B_0 \approx B_e$), so that the ratio of external to internal Alfvén speed v_{Ae}/v_A is essentially the density ratio $\sqrt{n_0/n_e}$. Then

$$k_c \approx \left(\frac{j_{0,s}}{a} \right) \left[\frac{1}{n_0/n_e - 1} \right]^{1/2} . \qquad (7.3.3)$$

From this expression we see that for typical density ratios inferred in the solar corona (e.g., $n_e/n_0 \approx 0.1-0.5$, see Table 7.1), the dimensionless cutoff wave number $k_c a$ (see Fig. 7.4) is expected to fall into the range $0.8 \lesssim k_c a \lesssim 2.4$. Therefore, we would expect that the global sausage-mode oscillation is completely suppressed for the slender loops for which kink-mode oscillations have been observed, which have wave numbers of

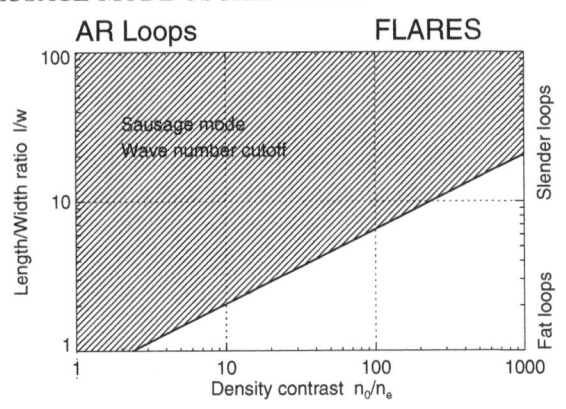

Figure 7.11: The wave number cutoff k_c for sausage-mode oscillations expressed as a require-ment of the loop length-to-width ratio l/w as a function of the density contrast n_0/n_e between the external and internal loop densities (see Eq. 7.3.5). Note that sausage mode oscillations in slender loops can occur only for flare conditions.

$ka = 2\pi a/\lambda = (\pi/4)(w/l) \approx 0.04 - 0.08$ (§ 7.2.1). The occurrence of sausage-mode oscillations therefore requires special conditions (e.g., very high-density contrast n_0/n_e and relatively thick (fat) loops). The high-density ratio (i.e., $n_0/n_e \gg 1$ or $v_{Ae} \gg v_A$), yields the following simple expressions for the cutoff wave number k_c (Eqs. 7.3.1−3),

$$k_c a \approx j_{0,s}\left(\frac{v_A}{v_{Ae}}\right) = j_{0,s}\left(\frac{n_e}{n_0}\right)^{1/2}. \qquad (7.3.4)$$

Since the wavelength of the fundamental eigen mode ($s = 1$) corresponds to the double loop length, so that the wave number relates to the loop length by $k = 2\pi/\lambda = \pi/l$, the cutoff wave number condition $k > k_c$ implies a constraint between the loop geometry ratio w/l and the density contrast n_e/n_0,

$$\frac{l}{w} = \frac{\pi}{2ak} < \frac{\pi}{2ak_c} = \frac{\pi}{2j_{0,s}}\sqrt{\frac{n_0}{n_e}} = 0.6545\sqrt{\frac{n_0}{n_e}}. \qquad (7.3.5)$$

The numerical factor 0.65 applies to the fundamental ($s = 1$) sausage mode. Since geometric parameters (such as l and w) can be measured easier than densities, we might turn the cutoff condition around and formulate it as a density contrast requirement for a given loop aspect ratio,

$$\frac{n_0}{n_e} > \left(\frac{1}{0.6545}\frac{l}{w}\right)^2 = 2.334\left(\frac{l}{w}\right)^2. \qquad (7.3.6)$$

This clearly indicates that slender loops with a high length-to-width ratio $l/w \gg 1$ would require extremely high-density contrast n_0/n_e. Typical active region loops, which have only a moderate density contrast in the order of $n_0/n_e \approx 2, ..., 10$, would be required to be extremely bulgy and fat, with width-to-length ratios of $l/w \approx 1 - 2$. The density contrast is much higher for flare loops or postflare loops, up to $n_0/n_e \approx 10^2, ..., 10^3$. In this case, a length-to-width ratio of $l/w \approx 6 - 20$ would be allowed for sausage-mode oscillations. This brings us to the conclusion that global sausage type oscillations are only expected in fat and dense loops, basically only in flare and postflare loops, a prediction that was not appreciated until recently (Nakariakov et al. 2004). The restriction of the sausage mode wave number cutoff is visualized in Fig. 7.11, where the permitted range of geometric loop aspect ratios (l/w) is shown as a function of the density contrast (n_0/n_e). Note that all these considerations apply to the fundamental harmonic $N = 1$ along the fluxtube.

7.3.2 Sausage-Mode Period

Since the sausage mode is highly dispersive (see Fig. 7.4), the phase speed v_{ph} is a strong function of the dimensionless wave number ka. The phase speed diagram (Fig. 7.4) shows that the phase speed equals the external Alfvén speed above the long-wavelength cutoff, $v_{ph}(k = k_c) = v_{Ae}$, and tends to approach the internal Alfvén speed in the short-wavelength limit, $v_{ph}(ka \gg 1) \mapsto v_A$. For coronal conditions, both the phase speed of the fast kink and fast sausage mode are bound by the internal and external Alfvén velocities,

$$v_A \leq v_{ph} = \frac{\omega}{k} \leq v_{Ae} \; . \tag{7.3.7}$$

Therefore, the period $(P = 2l/v_{ph})$ of the standing sausage mode is also bound by these two limits,

$$\frac{2l}{v_{Ae}} < P_{saus} = \frac{2l}{v_{ph}} < \frac{2l}{v_A} \; . \tag{7.3.8}$$

At the lower limit we can derive a simple relation for the sausage-mode period from the long-wavelength cutoff, where $v_{ph}(k = k_c) = v_{Ae}$, using Eq. (7.3.5),

$$P_{saus} = \frac{2l}{v_{ph}} = \frac{2\pi}{k v_{ph}} < \frac{2\pi}{k_c v_{ph}} \approx \frac{2\pi a}{j_{0,s} v_{Ae}} \left(\frac{v_{Ae}}{v_A} \right) = \frac{2\pi a}{j_{0,s} v_A} = \frac{2.62 \, a}{v_A} \; , \tag{7.3.9}$$

an approximation that was derived in Roberts et al. (1984), which agrees with a simplified solution of the wave equation in Rosenberg (1970). However, one should be aware that this relation represents only a lower limit that applies at $k = k_c$, while for all other valid wave numbers $(k > k_c)$ the sausage node length is shorter $(l = \pi/k)$, and thus the sausage-mode period is shorter, although the phase speed becomes lower, $v_{ph} < v_{Ae}$ for $k > k_c$ (Fig. 7.4). Therefore, we should use this relation (Eq. 7.3.9) only as an inequality, while the actual resonance frequency is determined by

$$P_{saus}(k) = \frac{2l}{v_{ph}(k)} \; , \tag{7.3.10}$$

provided that the inequality (Eq. 7.3.9) is satisfied (Nakariakov et al. 2004), which is equivalent to the density requirement (Eq. 7.3.5 or 7.3.6). Provided that sufficiently fat and overdense loops exist according to the requirement of Eq. (7.3.6), we expect, for loops with radii of $a \approx 500, ..., 10,000$ km and a typical Alfvén speed of $v_A \approx 1000$ km s^{-1} the following range of sausage-mode periods: $P \approx 1.3 - 26$ s, which is about 2 orders of magnitude shorter than kink-mode periods.

Table 7.6: Imaging observations of oscillations in the period range 0.5 s $< P < 60$ s, a period range in which MHD fast sausage oscillations are possible.

Observer	N	Frequency or wavelength	Oscillation period P[s]	Spatial scale or instrument
Radio:				
Kai & Takayanagi (1973)	1	160 MHz	<1.0	17' (Nobeyama)
Pick & Trottet (1978)	1	169 MHz	0.37, 1.7	5'-7' (Nançay)
Trottet et al. (1981)	1	140−259 MHz	1.7±0.5	5' (Nançay)
Sastry et al. (1981)	1	25−35 MHz	2−5	30' (Gauribidanur)
Kattenberg & Kuperus (1983)	1	5 GHz	1.5	0.15' (Westerbork)
Zlobec et al. (1992)	1	333 MHz	9.8−14.2	0.7' − 1.5' (VLA)
Aschwanden et al. (1992a)	1	1.5 GHz	8.8	0.2' − 0.9' (VLA, OVRO)
Baranov & Tsvetkov (1994)	1	8.5−15 GHz	22, 30, 34	3.6' − 6.0' (Crimea)
Makhmutov et al. (1998)	1	48 GHz	2.5−4.5	1.9' (Itapetinga)
Chernov et al. (1998)	1	164−407 MHz	0.2	5' − 7' (Nançay)
Asai et al. (2001)	1	17 GHz	6.6	10'' (Nobeyama)
Nakariakov et al. (2003)	1	17 GHz	8−11, 14−17	10'' (Nobeyama)
Optical and Hα:				
Pasachoff & Landman (1984)	1	5303 Å	0.5−2 (?)	2.5'' (Hydrabad eclipse)
Pasachoff & Ladd (1987)	1	5303 Å	0.5−4 (?)	2.5'' (East Java eclipse)
Williams et al. (2001, 2002)	1	5303 Å	6	4.0'' (Bulgaria eclipse)
Soft X-rays:				
Thomas et al. (1987)	1	2−8, 8−16 Å	1.6	20'' (OSO-7)
McKenzie & Mullan (1997)	16	3−45 Å	9.6−61.6	2.4'' − 4.7'' (Yohkoh/SXT)
Hard X-rays:				
Asai et al. (2001)	1	23−53 keV	6.6	5.0'' (Yohkoh/HXT)

7.3.3 Imaging Observations of Sausage-Mode Oscillations

Imaging observations (Table 7.6) should, in principle, provide the period and geometric dimensions of the oscillating sources, so that the hypothesis of MHD fast sausage oscillations could be tested (e.g., the relation $P < 2.62 \, a/\text{v}_A$, Eq. 7.3.9). However, most of the early imaging observations were made with 1D interferometers (e.g., Fig. 7.12), with poor spatial resolution ($> 1'$), and radio images are also subject to substantial wave scattering (Bastian 1994), so that the widths ($w = 2a$) of oscillating loops generally could not be properly resolved. One fast oscillation event has been observed with high spatial resolution ($\approx 4''$) in optical wavelengths during a solar eclipse (Williams et al. 2001, 2002), which was interpreted as a propagating magneto-acoustic wave (§ 8), rather than a standing wave. McKenzie & Mullan (1997) conducted a systematic search for oscillatory signals in *Yohkoh* soft X-ray images and found significant periods in 16 out of 544 power spectra. They interpreted the periods (9.6 s $< P <$ 61.6 s) in terms of kink-mode oscillations, but since the kink mode does not modulate the density (and thus the optically thin soft X-ray flux) to first order, an interpretation in terms of sausage-mode oscillations would be more plausible.

Asai et al. (2001) observed a fast oscillation with a period of $P = 6.6$ s simulta-

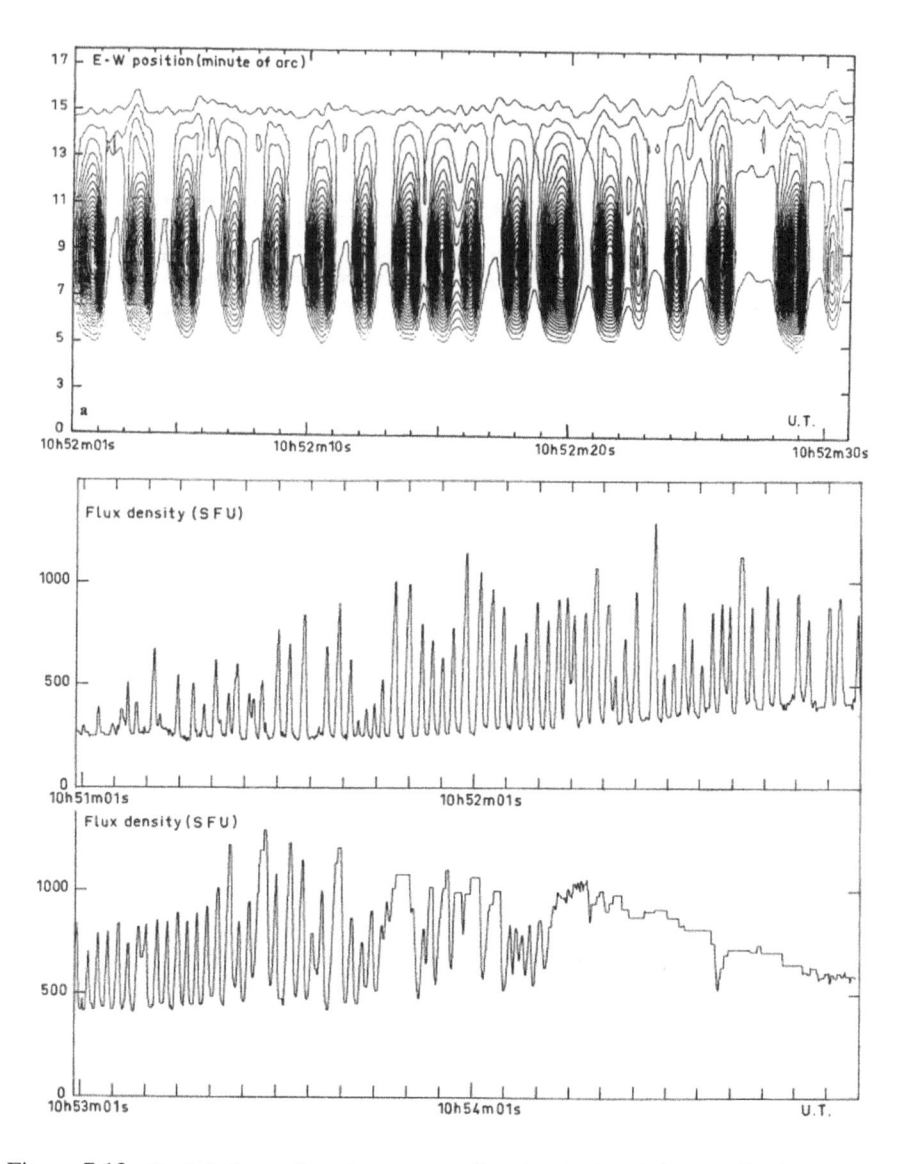

Figure 7.12: Spatial observations in east-west direction (top) and time profile (bottom) of a pulsating source ($P = 1.7 \pm 0.5$ s) observed with the *Nançay Radioheliograph* at 169 MHz. Note the slight deviations from strict periodicity and the transition from emission to absorption (Trottet et al. 1981).

neously in soft X-rays and radio during a flare. From their derived parameters (loop length $l = 16$ Mm, loop width $w = 6$ Mm, density $n_0 = 4.5 \times 10^{10}$ cm^{-3}) we estimate that the density contrast for the sausage mode wave number cutoff is $n_0/n_e \approx 17$ (Eq. 7.3.6), requiring an ambient density of $n_e < n_0/17 = 2.6 \times 10^9$, which is very likely to be satisfied in active regions. Therefore, an MHD fast sausage-mode oscilla-

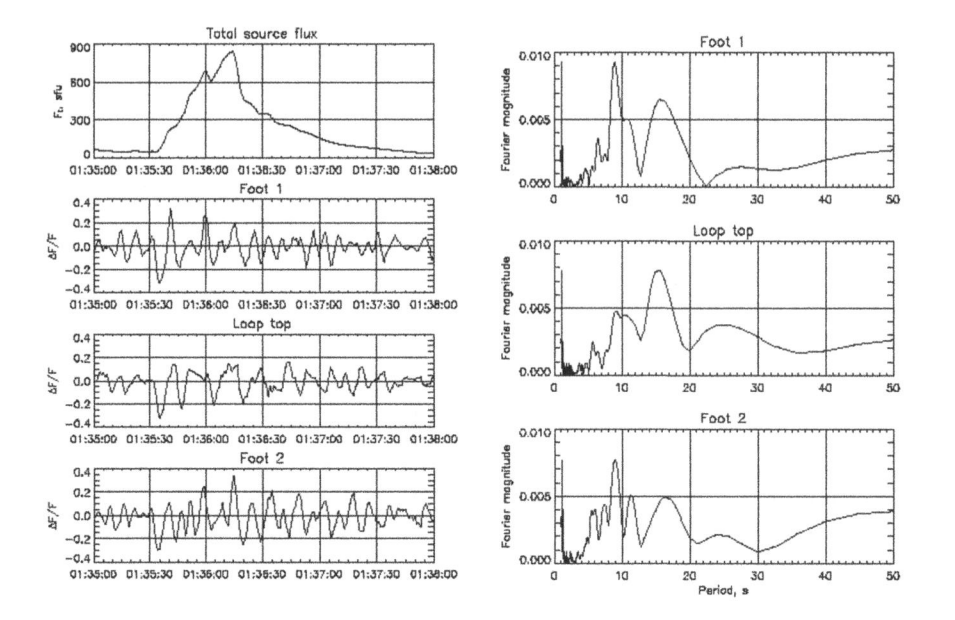

Figure 7.13: Observations of an oscillation event with the Nobeyama Radio Heliograph (NRH) on 2000-Jan-12 at 17 GHz, showing the time profiles (left) and Fourier power spectra (right). The total flux (top) integrated over the entire source does not reveal oscillations, but selected $10'' \times 10''$ regions at the footpoints and looptop of a flare loop show oscillating flux amplitudes, which are shown in normalized units $\Delta F/F = [F(t) - F_0]/F_0$ (left). The Fourier power spectra (right panels) reveal periods in the range of $P_1 = 14 - 17$ s and $P_2 = 8 - 11$ s in all three locations (Nakariakov et al. 2003).

tion is possible above the long-wavelength cutoff $k > k_c$ (Eq. 7.3.2).

A recent observation of an oscillating source (Fig. 7.12) interpreted in terms of sausage mode is presented in Nakariakov et al. (2003). The oscillatory flare loop was estimated to have a length of $l = 25$ Mm, a width of $w = 6$ Mm, and a density of $n_0 \approx 10^{11}$ cm^{-3}. If we apply the wave number cutoff criterion (Eq. 7.3.6) we find that a density contrast of $n_0/n_e > 2.4(l/w)^2 \approx 40$ is necessary to satisfy the sausage mode cutoff. Given the flare loop density of $n_0 = 10^{11}$ cm^{-3}, the density of the background corona around the flare loop is required to be lower than $n_e < 2.5 \times 10^9$ cm^{-3}, which is a possible value. Therefore, the necessary condition for the sausage mode is satisfied.

An important difference to kink-mode oscillations is that the sausage mode produces density variations, while the kink mode does not. The oscillatory density variations can easily modulate the gyrosynchrotron emissivity observed in microwaves, because the gyrosynchrotron radio flux is proportional to the electron number density (see, e.g., Rosenberg 1970). The fluctuations shown in Fig. 7.12 demonstrate that the radio flux is modulated locally up to $\lesssim 20\%$, which implies an oscillation amplitude of $\approx 10\%$ for the sausage mode.

Table 7.7: Non-imaging observations of oscillations in the period range 0.5 s $< P <$ 1 min, a period range in which fast MHD sausage oscillations are possible. N is the number of observed pulsation events.

Observer	N	Frequency or wavelength	Oscillation period $P(s)$	Instrument
Radio:				
Dröge (1967)	> 18	240, 460 MHz	0.2−1.2	Kiel
Janssens & White (1969)	1	2.7−15.4 GHz	17, 23	Sagamore Hill
Parks & Winckler (1969)	1	15.4 GHz	16	Sagamore Hill
Abrami (1970, 1972)	3	239 MHz	1.7−3.1	Trieste
Gotwols (1972)	1	565−1000 MHz	0.5	Silver Spring
Rosenberg (1970)	1	220−320 MHz	1.0−3.0	Utrecht
De Groot (1970)	> 25	250−320 MHz	2.2−3.5	Utrecht
McLean et al. (1971)	1	100−200 MHz	2.5−2.7	Culgoora
Rosenberg (1972)	1	220−320 MHz	0.7−0.8	Utrecht
McLean & Sheridan (1973)	1	200−300 MHz	4.28±0.01	Culgoora
Janssens et al. (1973)	7	3.0 GHz	11−35	Norco
Achong (1974)	1	18−28 MHz	4.0	Kingston
Tapping (1978)	14	140 MHz	0.06−5	Cranleigh
Elgaroy (1980)	8	310−340 MHz	1.1	Oslo
Bernold (1980)	> 13	100−1000 MHz	0.5−5	Zurich
Slottje (1981)	> 40	160−320 MHz	0.2−5.5	Dwingeloo
Zodi et al. (1984)	1	22, 44 GHz	1.5	Itapetinga
Zaitsev et al. (1984)	23	45−230 MHz	0.3−5	Izmiran
Wiehl et al. (1985)	1	300−1000 MHz	1−2	Zurich
Aschwanden (1986, 1987b)	10,60	300−1100 MHz	0.4−1.4	Zurich
Aschwanden & Benz (1986)	10	237, 650 MHz	0.5−1.5	Zurich
Aurass et al. (1987)	1	234 MHz	0.25−2	Tremsdorf
Correia & Kaufmann (1987)	1	30, 90 GHz	1−3	Itapetinga
Kurths & Herzel (1987)	1	480−800 MHz	1.0, 3.5	Tremsdorf
Kurths & Karlický (1989)	1	234 MHz	1.3, 1.5	Tremsdorf
Chernov & Kurths (1990)	10	224−245 MHz	0.35−1.3	Izmiran
Zhao et al. (1990)	1	2.84 GHz	1.5	Bejing
Fu et al. (1990)	1	1,2,3.8,9.4 GHz	0.18, 4	Yunnan
Kurths et al. (1991)	25	234−914 MHz	0.07−5.0	Zurich
Stepanov et al. (1992)	1	37 GHz	5.2	Metsähovi
Aschwanden et al. (1994a,b)	1,1	300−650 MHz	1.15, 1.8	Zurich
Qin & Huang (1994)	1	9.375 GHz	1.0−3.0	Nanjing
Qin et al. (1996)	1	2.84, 9.375, 15 GHz	1.5, 40	Nanjing
Wang & Xie (1997)	1	1.42, 2.0 GHz	44, 47	Yunnan
Kliem et al. (2000)	1	600−2000 MHz	0.5−3.0	Zurich
Hard X-rays:				
Parks & Winckler (1969)	1	> 20 keV	16	Balloon
Lipa (1978)	26	14−111 keV	10−100	OSO-5
Takakura et al. (1983b)	1	30−40 keV	0.3	Hinotori
Kiplinger et al. (1982)	30	> 25 keV	0.4, 0.8	SMM/HXRBS
Kiplinger et al. (1983)	1	> 25 keV	8.2	SMM/HXRBS
Nakajima et al. (1983)	2	> 25 keV	8.2	SMM/HXRBS
Desai et al. (1987)	3	> 30 keV	2−7	Venera

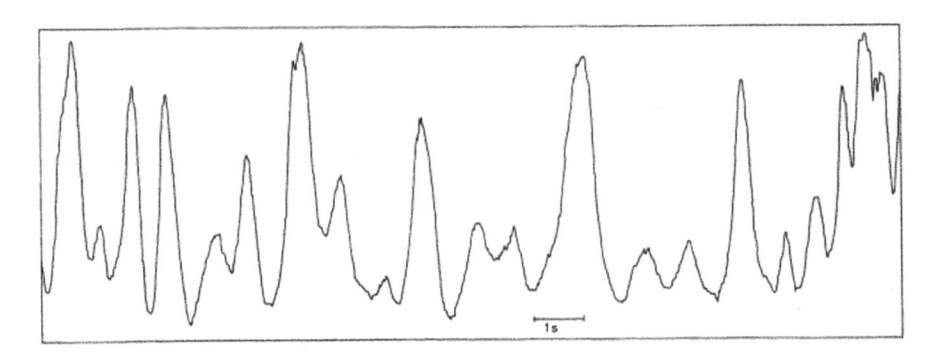

Figure 7.14: Oscillating time profile observed during a flare on 1969-Feb-25, 09:55 UT with the Utrecht spectrograph in the radio wavelength range of 160-320 MHz. Large pulses with a period of $P_1 = 3$ s and small pulses with a subperiod of $P_2 = 1.0$ s were interpreted as harmonics of the MHD fast sausage oscillation mode (Rosenberg 1970).

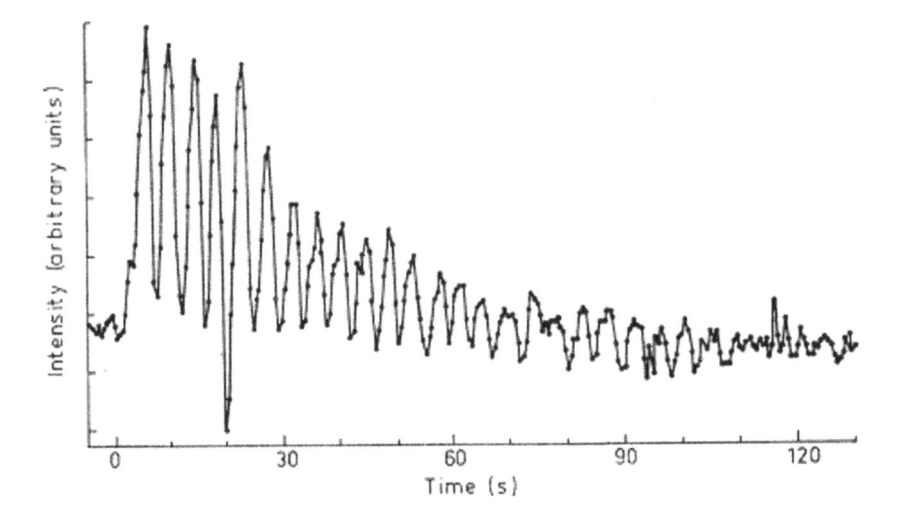

Figure 7.15: A damped oscillation observed with the *Culgoora radio spectrograph* on 1972-May-16, 03:07 UT, at a frequency of 230 MHz. The deep minimum at $t = 20$ is produced by an instrumental time marker. Note the strict periodicity, which yields a mean period of $P = 4.28 \pm 0.01$ s (McLean & Sheridan 1973).

7.3.4 Non-Imaging Observations of Sausage-Mode Oscillations

There are numerous observations of coronal oscillations in the period regime of $P \approx 0.5 - 5$ s that are without imaging information, mostly reported in radio wavelengths (Table 7.7 and Figs. 7.14−16). Although we have no direct imaging information on the oscillating loops, there are a number of reasons that support an interpretation in terms of MHD fast sausage mode oscillations: (1) the observed pulse periods $P \approx 0.5 - 5$ s

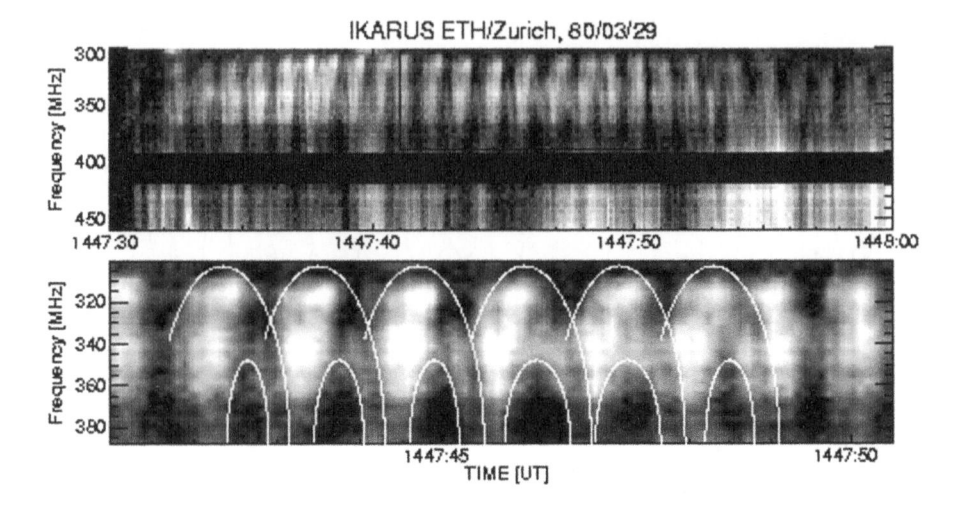

Figure 7.16: Periodic radio emission observed during the 1980-Mar-29, 14:47 UT, flare with the Zurich radio spectrometer *IKARUS*. The enlarged spectrogram with a frequency range of $310-385$ MHz and a time duration of 10 s shows a periodic sequence of *inverted U-bursts* with a mean period of $P = 1.15$ s, indicating that the radio emission originates within coronal loops with an apex density of $n_0 \approx 10^9$ cm^{-3}, corresponding to the turnover frequency $\nu \approx 310$ MHz (Aschwanden et al. 1994b).

coincide with the theoretically expected range of periods ($P \approx 1.3 - 26$ s) for typical loop widths ($a \approx 0.5 - 10$ Mm) and coronal Alfvén speeds ($v_A \approx 1000$ km s^{-1}), see § 7.3.2; (2) most of the reported oscillation events occurred during flares (in particular those made in decimetric, microwave, and hard X-ray wavelengths), which match the theoretical requirement of a high-density contrast $n_0/n_e \approx 10^2 - 10^3$ imposed by the long-wavelength cutoff $k > k_c$ of the MHD fast sausage mode (see § 7.3.1 and Fig. 7.11); (3) those observations that exhibit a regular periodicity can naturally be explained by an eigen-mode standing wave; and (4) the MHD fast sausage mode modulates the loop cross section, density, and magnetic field, which provides a natural mechanism to modulate plasma emission, gyroresonance, and gyrosynchrotron emission observed at radio wavelengths. These are just plausibility arguments for an interpretation of the observed oscillation events (Table 7.6) in terms of MHD fast sausage-mode oscillations, but it does not rule out that some events may be caused by alternative mechanisms (e.g., by limit cycles of nonlinear dissipative systems of wave-particle interactions). Nevertheless, even more arguments have been brought forward in support of MHD fast sausage mode oscillations for some particular events. For instance, a subharmonic period of $P_2 = 1.0$ was observed besides the fundamental period of $P_1 = 3.0$ s (Fig. 7.14), which matched the theoretically expected ratio derived for a radial MHD sausage oscillation expressed in terms of the first-order Bessel function (i.e., $5.3/1.8 = 2.94$, Rosenberg 1970). An important characteristic of MHD eigen modes is their regular periodicity. A classic example is that of McLean & Sheridan (1973), who observed a damped wave train with about 28 pulses with an extremely stable peri-

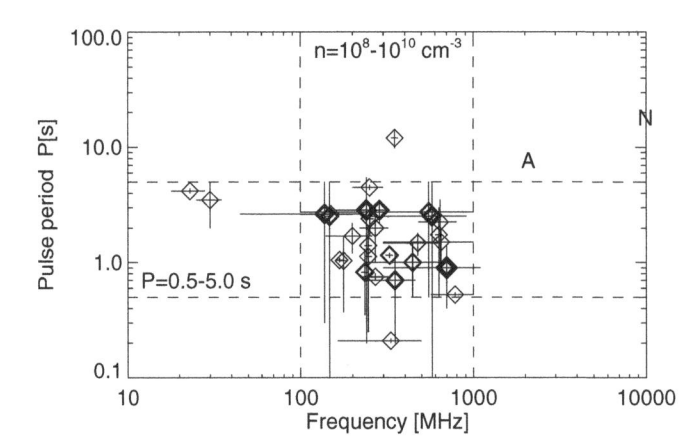

Figure 7.17: Observed oscillation periods as a function of the radio frequency, according to the list given in Tables 7.6 and 7.7. Horizontal and vertical bars indicate the ranges in each observation, diamonds represent the mean values, and thick diamonds indicate reports on multiple oscillation events. Note that most of the events fall in the range of $P \approx 0.5 - 5.0$ s and $n_0 = 10^8 - 10^{10}$ cm^{-3}, if one assumes that the radio frequency is near the plasma frequency, $\nu \approx \nu_p$. The letters A and N refer to the observations of Asai et al. (2001) and Nakariakov et al. (2003).

odicity of $P = 4.28 \pm 0.01$ and a damping time of $\tau_D \approx 35$ s (Fig. 7.15). There is also (non-imaging) evidence from radio dynamic spectra that the pulsed emission originates in coronal loops, because beam-driven radio emission caused by exciters along closed magnetic field lines show up as *inverted U-bursts* in dynamic spectra (e.g., Fig. 7.16).

The long-wavelength wave number cutoff k_c of the MHD fast sausage oscillation mode (Fig. 7.4 and Eqs. 7.3.1−4) has an important consequence on the density requirement ($n_0/n_e \approx 10^2 - 10^3$) that can only be met in flare loops or postflare loops. This requirement can be tested for a pulsed radio emission generated near the plasma frequency (e.g., radio type III bursts), because the plasma frequency ν_p provides a direct density diagnostic,

$$\nu \approx \nu_p = 8980 \sqrt{n_e [\mathrm{cm}^{-3}]} \quad [\mathrm{Hz}] \ . \tag{7.3.11}$$

According to a standard density model of the corona (e.g., the Baumbach−Allen model, Cox 2000), the density $n_e(h)$ varies as a function of the normalized height $R = 1 + h/R_\odot$ (Fig. 3.11 and Eq. 1.6.1),

$$n_e(h) = 10^8 \left[\frac{2.99}{R^{16}} + \frac{1.55}{R^6} + \frac{0.036}{R^{1.5}} \right] \approx \frac{4 \times 10^8}{R^9} \ [\mathrm{cm}^{-3}] \tag{7.3.12}$$

where the approximation on the right-hand side holds for $h \lesssim R_\odot/2$. This density model $n_e(h)$ is shown in Fig. 7.18 (top). Assuming that the fatest possible loops have a width-to-length ratio of $q_w \approx 1/4$, the minimum required density ratio is $n_{0,min}(h)/n_e(h) = 2.4/q_w^2 \approx 40$ (Eq. 7.3.6) for loops oscillating in the MHD fast

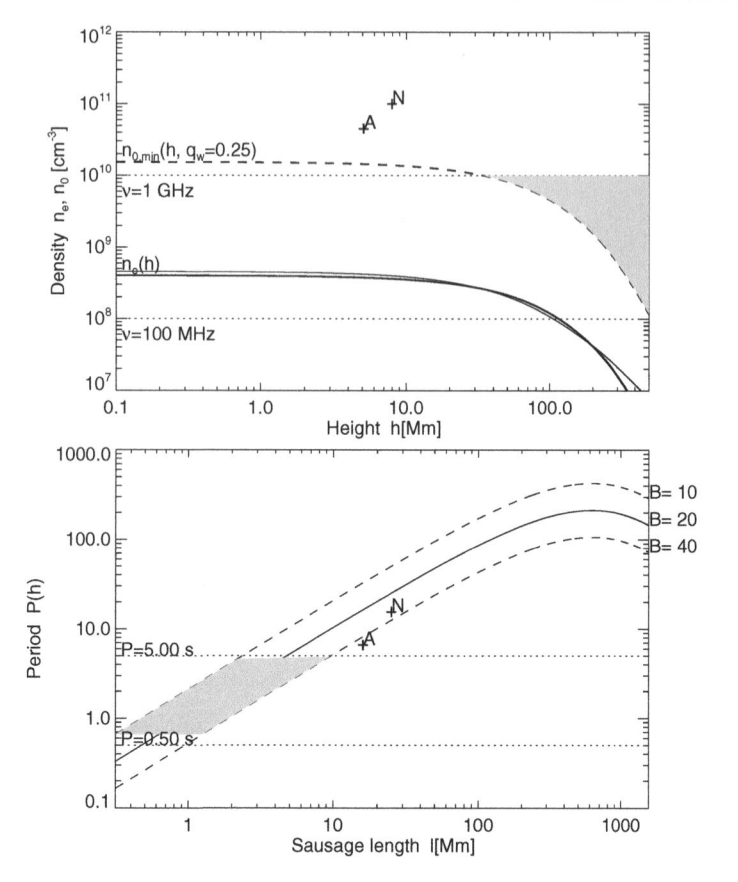

Figure 7.18: *Top:* the Baumbach−Allen density is shown as a model of the average coronal background (thin curve), together with the approximation $n_e(h) \approx 4 \times 10^8 (1 + h/R_\odot)^9$ (thick curve), and the minimum loop density $n_{0,min}(h)$ required for loops that have a width-to-length ratio of $q_w = w/l = 0.25$. The height and density range corresponding to a plasma frequency range of 100 MHz−1 GHz is indicated by the shaded area. *Bottom:* the MHD fast sausage-mode period $P(l_{saus})$ is shown as function of the sausage length l_{saus} for magnetic fields of $B = 20$ G within a factor of 2. The regime of physical solutions for periods $P = 0.5 - 5$ s and $B = 10 - 40$ G is indicated by the shaded area. A and N indicate the values inferred for the observations of Asai et al. (2001) and Nakariakov et al. (2003).

sausage mode due to the long-wavelength cutoff criterion (Eq. 7.3.1). This cutoff criterion implies the constraint that physical solutions are only possible for heights $h \gtrsim 50$ Mm, for oscillations observed in the plasma frequency range of $\nu_p \approx 1$ GHz (grey area in Fig. 7.18, top). The two events observed by Asai et al. (2001) and Nakariakov et al. (2003) do not fall within this range, because their density inferred from soft X-rays is higher and would correspond to plasma frequencies of 2 and 3 GHz. The important conclusion that follows, based on the two assumptions that the Baumbach−Allen model represents a realistic density model of the background corona and that oscillating loops are not fatter than a quarter of their length, is that physical solutions for the

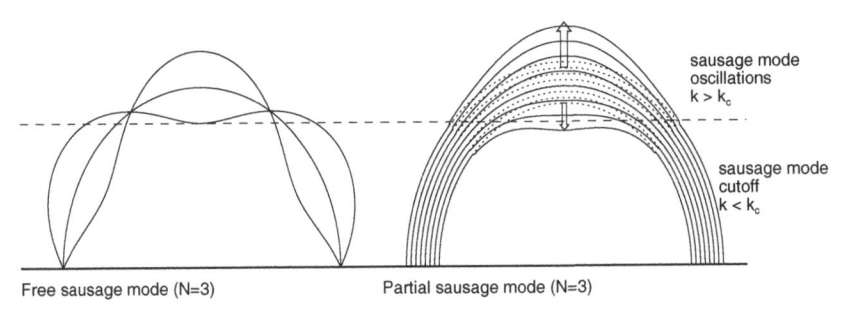

Free sausage mode (N=3)   Partial sausage mode (N=3)

Figure 7.19: Sketch of MHD fast sausage-mode oscillation in partial loop segment (right), where the cutoff condition $k > k_c$ is satisfied. For a given width-to-length ratio q_w, oscillations occur in segments where $n_0/n_e > 2.4/q_w^2$ (Eq. 7.3.6). The oscillating segment may correspond to a higher harmonic node [e.g., to $N = 3$ here (left)], (Aschwanden et al. 2004a).

MHD fast sausage mode are only possible for coronal heights of $h \gtrsim 50$ Mm in the observed frequency range of $\nu \leq 1$ GHz, where most of the fast oscillations have been observed (Fig. 7.17).

Next we investigate the constraints introduced by the pulse periods, which are found in the range of $P \approx 0.5 - 5.0$ s (Fig. 7.17). If we denote the length of the loop segment that participates in the MHD fast sausage mode oscillation with l_{saus}, we obtain an upper limit from the phase speed at the wave number cutoff k_c (Eqs. 7.3.8 and 7.3.10),

$$l_{saus} = \frac{P_{saus}}{2} \mathrm{v}_{ph}(k) \leq \frac{P_{saus}}{2} \mathrm{v}_{Ae} , \qquad (7.3.13)$$

where the Alfvén speed v_{Ae} is given by the background density $n_e(h)$ (Eq. 7.3.12),

$$\mathrm{v}_{Ae}(h) = \frac{B}{\sqrt{4\pi\rho_0}} = \frac{B}{\sqrt{4\pi\mu m_H n_e(h)}} = 1210 \left(\frac{B}{20\,\mathrm{G}}\right) \left(\frac{n_e(h)}{10^9\,\mathrm{cm}^{-3}}\right)^{-1/2} \,\,[\mathrm{km/s}] , \qquad (7.3.14)$$

where we used a mean molecular weight of $\mu = 1.27$ for a H:He=10:1 coronal abundance. We plot the sausage-mode period P_{saus} as a function of the length l_{saus} in Fig. 7.18 (bottom), for a mean coronal magnetic field of $B = 20$ G (solid line), as well as for values a factor of 2 smaller or larger ($B = 10 - 40$ G). We find that the corresponding lengths of the oscillating loop segment have values in the range of 0.3 Mm $< l_{saus} < 10$ Mm (Fig. 7.18 bottom, shaded area) for pulse periods of $0.5 < P_{saus} < 5.0$ s, which is much smaller than the full loop lengths where physical solutions are possible for the sausage mode ($h = \pi l > 120$ Mm, see Fig. 7.18, top, shaded area). Because physical solutions are only possible at heights $h > 50$ Mm, the oscillating loop segment has to be located near the top of the loop. Thus we conclude that only a small segment of the loop is oscillating in the fast MHD sausage mode, which can be a segment bound by a higher harmonic node N (see Fig. 7.19). Note that the measurements of Asai et al. (2001) and Nakariakov et al. (2003) have longer periods, and actually allow for global sausage-mode oscillations of the entire loop length for reasonable magnetic fields in the order of $B \approx 40$ G (marked with crosses in Fig. 7.17, bottom).

Alternatively, fast pulsations can also be explained in terms of propagating fast-mode waves, rather than by standing MHD sausage-mode oscillations. Such scenarios have been developed by Roberts et al. (1984), called *"impulsively generated MHD waves"* therein, and have been applied to the optical (eclipse) observations of fast pulses ($P = 6$ s), clearly showing propagating wave trains (Nakariakov et al. 2003). A hallmark of such propagating periodic wave trains would be parallel ridges of drifting burst structures in radio dynamic spectra, a fine structure known as *"fibers"* (e.g., Rosenberg 1972; Bernold 1980; Slottje 1981).

7.4 Slow-Mode (Acoustic) Oscillations

7.4.1 Slow-Mode Oscillation Period

The dispersion relation for magneto-acoustic waves in a cylindrical fluxtube (Fig. 7.4) shows two domains of solutions, a slow branch with acoustic phase speeds ($c_T \leq v_{ph} = \omega/k \leq c_s$), and a fast branch with Alfvén phase speeds ($v_A \leq v_{ph} \leq v_{Ae}$). Here we consider the slow-mode branch, which is only weakly dispersive in the low-β corona. The tube speed c_T (Eq. 7.1.27) is close to the sound speed c_s,

$$c_T = \frac{c_s v_A}{\sqrt{c_s^2 + v_A^2}} \approx c_s, \quad \text{for } c_s \ll v_A . \tag{7.4.1}$$

Thus, the phase speeds of slow-mode (acoustic) waves are given to a good approximation by $v_{ph} = \omega(k)/k \approx c_s$. As discussed in § 7.1.1, the slow-mode magneto-acoustic waves are essentially sound waves, non-dispersive ($v_{gr} \approx v_{ph} \approx c_s$), compressional ($kv_1 \neq 0$), and longitudinal ($\mathbf{k} \parallel \mathbf{v}$).

Let us consider the simplest solution of slow-mode acoustic oscillations, a standing wave in the fundamental mode, which has the two endpoints of a fluxtube as fixed nodes. Neglecting energy dissipation and the magnetic field, the linearized MHD equations yield (from Eqs. 7.1.11 and 7.1.12)

$$\frac{\partial \rho_1}{\partial t} = -\rho_0 \nabla \cdot \mathbf{v}_1 , \tag{7.4.2}$$

$$\rho_0 \frac{\partial \mathbf{v}_1}{\partial t} = -c_s^2 \nabla \rho_1 . \tag{7.4.3}$$

Taking the time derivative of the momentum equation (7.4.3) and inserting the continuity equation (7.4.2) yields,

$$\frac{\partial^2 \mathbf{v}_1}{\partial t^2} = -c_s^2 \rho_0 \nabla \left(\frac{\partial \rho_1}{\partial t} \right) = c_s^2 \nabla^2 \mathbf{v}_1 \tag{7.4.4}$$

which is just the equation for a harmonic oscillator and has the general solution $v(z, t) \propto \exp(\mathbf{k}z - \omega t)$, or

$$v(z, t) = v_1 \cos \left(\frac{\pi}{2l} z \right) \sin \left(\frac{\pi c_s}{2l} t \right) , \quad v_1 = c_s \frac{\rho_1}{\rho_0} , \tag{7.4.5}$$

Table 7.8: Average and ranges of physical parameters of 54 oscillating events observed with SUMER. The speed of phase propagation was only measured in cases with a clear signal (Wang et al. 2003b).

Parameter	Average	Range
Oscillation period P	17.6 ± 5.4 min	$7.1 - 31.1$ min
Decay time τ_D	14.6 ± 6.9 min	$5.5 - 37.3$ min
Doppler oscillation amplitude v_D	98 ± 75 km/s	$13 - 315$ km/s
Maximum Doppler amplitude v_m	75 ± 53 km/s	$11 - 234$ km/s
Derived displacement amplitude A	12.5 ± 9.9 Mm	$1.7 - 43.7$ Mm
Ratio of decay time to period τ_D/P	0.84 ± 0.34	$0.33 - 2.13$
Number of periods N_p	2.3 ± 0.7	$1.5 - 5$
Time lag of intensity peak ΔT_{IV}	8.5 ± 13.1 min	$-2.5 - 52.5$ min
Time lag of line width peak ΔT_{WV}	1.0 ± 3.0 min	$-4.1 - 9.1$ min
Intensity peak duration ΔT_I	36.2 ± 27.0 min	$10 - 141$ min
Number of intensity peak N_I	1.5 ± 0.7	$1 - 3$
Speed of phase propagation	43 ± 25 km/s	$8 - 102$ km/s
Spatial extent of oscillation along slit ΔY	35 ± 21 Mm	$7 - 87$ Mm

$$\rho(z,t) = \rho_1 \sin\left(\frac{\pi}{2l}z\right) \cos\left(\frac{\pi c_s}{2l}t\right) \qquad (7.4.6)$$

so the density perturbation is out of phase with respect to the velocity perturbation by $\pi/2$. Since the wave is longitudinal, the standing wave corresponds to a compression back and forth in a coronal loop, essentially the plasma is "sloshing" back and forth like the air in Scottish bagpipes.

The period of a standing wave in the MHD slow acoustic mode is obviously the sound travel time back and forth in a coronal loop for the fundamental harmonic ($j = 1$),

$$P_{slow} = \frac{2l}{jc_T} \approx \frac{2l}{jc_s} \qquad (7.4.7)$$

where the sound speed c_s in a fully ionized plasma is defined as

$$c_s = \sqrt{\frac{\gamma p}{\rho}} = \sqrt{\frac{\gamma 2 k_B T}{\mu m_p}} = 147\sqrt{\frac{T_e}{1 \text{ MK}}} \quad [\text{km/s}] \,, \qquad (7.4.8)$$

with $\gamma = 5/3$ being the adiabatic index (Eq. 4.1.10), $\mu = 1.27$ the mean molecular weight (Eq. 3.1.7), and m_p the proton mass. For typical coronal loops, say with a temperature of $T \approx 1$ MK and lengths between $l = 10$ Mm and 100 Mm, the period of MHD slow-mode oscillations, therefore, is expected to be in the range of $P_{slow} \approx 140 - 1400$ s $\approx 2 - 20$ min.

7.4.2 Observations of Slow-Mode Oscillations

Slow-mode oscillations in coronal loops have been imaged for the first time with the *SUMER* spectrograph on *SoHO*, by measuring the Doppler shift and intensity fluctuations of EUV lines, mainly in the Fe XIX and Fe XXI lines with a formation tempera-

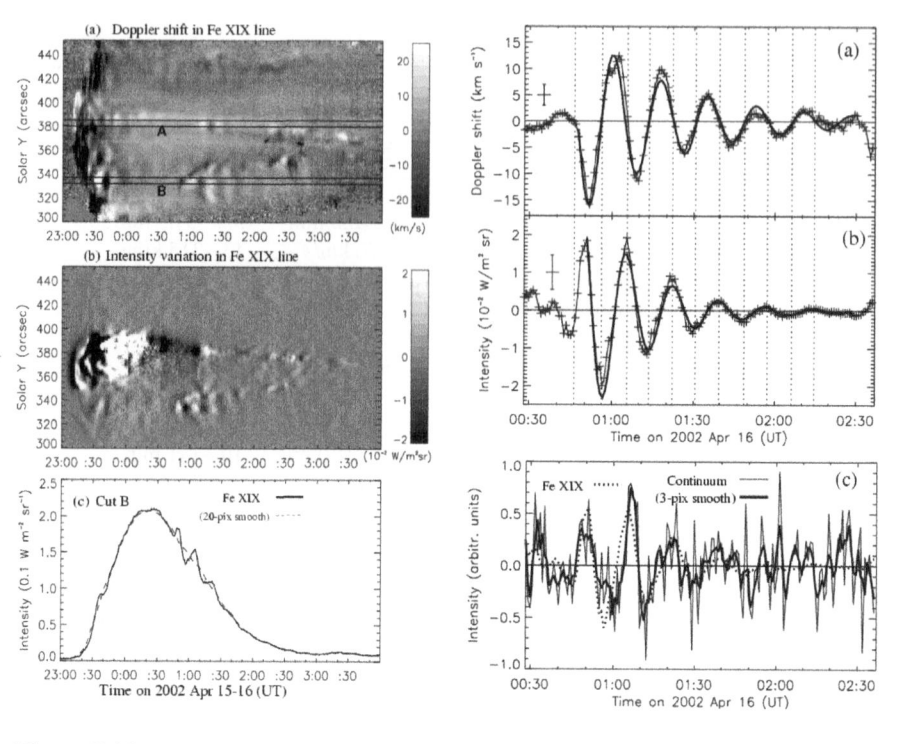

Figure 7.20: SUMER observations of MHD slow-mode standing waves. *Top left:* Doppler shift in Fe XIX line ($T \approx 6.3$ MK) shown across a slit (in the vertical direction on the Sun) as a function of the time. *Middle left:* background-subtracted intensity variation in the Fe XIX line. *Bottom left:* time profile of total intensity on position B marked in top left frame. *Top right:* time profile of Doppler shift at position B, with the fit of a damped sine function. *Middle right:* background-subtracted time profile of the Fe XIX intensity, also with the fit of a damped sine function. *Bottom right:* time profile of continuum intensity in the wings of S III ($T = 0.03-0.06$ MK) and Ca X ($T = 0.7$ MK) lines at position B (Wang et al. 2003a).

ture of $T \gtrsim 6$ MK (Wang et al. 2002, 2003a,b; Kliem et al. 2002). *SUMER* records only a 1D slit, generally located at a position $50'' - 100''$ above the limb, but the oscillating loops could clearly be located in cotemporaneous EUV images from SoHO/EIT and TRACE. In the statistical paper of Wang et al. (2003b), the authors analyzed a sample of 54 Doppler shift oscillations in 27 flare-like events. They found oscillation periods in the range of $P = 7 - 31$ min, with decay times of $\tau_D = 5.5 - 37.3$ min, and initial large Doppler shift velocities up to v$= 200$ km s^{-1}. The evidence for MHD slow-mode (standing wave) oscillations is based on the following facts: (1) the phase speed derived from the observed period and loop lengths roughly agrees with the expected sound speed; (2) the intensity fluctuation lags behind the Doppler shift by a quarter period (like the cosine and sine functions in Eqs. 7.4.5−6); (3) the scaling of the dissipation time (modeled as thermal conduction cooling time) with period agrees with the observed scaling in 90% of the cases. The slow-mode oscillations seem to be trig-

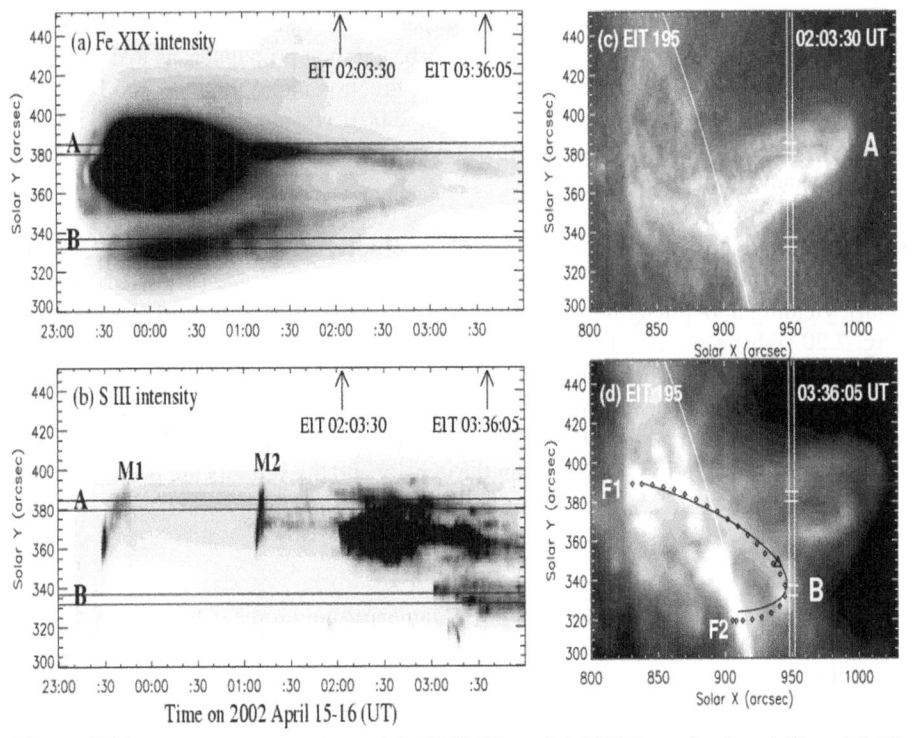

Figure 7.21: SUMER observations of Fe XIX ($T = 6.3$ MK) intensity (*top left*) and S III ($T = 0.03 - 0.06$ MK) intensity (*bottom left*) of M1.2 flare on 2002-Apr-15, 23 UT, and EIT 195 Å ($T \approx 1.5$ MK) intensity images (right panels) of the same event as displayed in Fig. 7.20 (Wang et al. 2003a).

Table 7.9: Observations of oscillations in the period range $P > 10$ min, a period range in which MHD slow-mode (acoustic) oscillations are expected.

Observer	N	Frequency or wavelength	Oscillation period P[s]	Instrument
Radio:				
Kaufmann (1972)	1	7.0 GHz	41 min	Brazil
Kobrin and Korshunov (1972)	3	9.67, 9.87 GHz	30−60 min	Gorky
Optical:				
Ofman et al. (1997)	3	White light	20−50 min	SOHO/UVCS
Ofman et al. (2000a)	9	White light	7−10 min	SOHO/UVCS
EUV:				
DeForest and Gurman (1998)	1	171 Å	10−15 min	SoHO/EIT
Soft X-rays:				
Jakimiec & Jakimiec (1974)	2	1−8 Å	3−15 min	SOLRAD 9
Harrison (1987)	1	3.5−5.5 keV	24 min	SMM/HXIS
Svestka (1994)	1	0.5−4, 1−8 Å	20 min	GOES
Wang et al. (2002a,b)	54	Fe XIX, 1118 Å	7−31 min	SoHO/SUMER

gered by a pressure disturbance near one footpoint (e.g., caused by a microflare, flare, or destabilizing filament). The pressure disturbance then propagates as a slow-mode magneto-sonic (acoustic) wave in longitudinal direction along the loop and becomes reflected at the opposite side (Nakariakov et al. 2000). The rapid damping of the propagating wave seems to be caused by thermal conduction, as simulated with an MHD code (Ofman & Wang 2002).

Interestingly, the oscillation is always more clearly detected in the Doppler shift signal than in intensity. An example of such an observation is given in Figs. 7.20 and 7.21. The geometry of the loop was inferred from an EIT image (Fig. 7.21, bottom right), yielding a loop length of $l = 191$ Mm. From the fit of a damped sine function (Fig. 7.20 right) a period of $P = 17.1 \pm 0.1$ min was found with a decay time of $\tau_D = 21.0 \pm 1.6$ min. The formation of the temperature of the Fe XIX line is $T = 6.3$ MK, yielding a sound speed of $c_s = 370$ km s^{-1} (with Eq. 7.4.8), and thus an expected slow-mode oscillation period of $P_{slow} = 2l/c_s = 1032$ s $= 17.2$ min. From Eq. (7.4.5) it follows that the relative amplitudes in the density and the velocity are proportional,

$$\frac{\rho_1}{\rho_0} = \frac{v_1}{c_s} . \tag{7.4.9}$$

For this event the authors measure an Fe XIX intensity amplitude of $I_1/I_0 \approx 0.19/2 = 0.09$, which roughly agrees with the ratio of the (line-of-sight corrected) Doppler shift divided by the sound speed, $v_1/c_s = 18.6/370 = 0.05$. The authors analyzed a total of 54 Doppler shift oscillations in 27 events, the statistical parameters are given in Table 7.8.

There are other reports of oscillations with periods that are commensurate with MHD slow-mode oscillations. A compilation of such observations with periods $P > 10$ min in different wavelengths (radio, optical, EUV, soft X-rays) is given in Table 7.9. Clearly most of these MHD slow-mode acoustic oscillations have been recorded by SUMER. However, since MHD slow-mode acoustic oscillations modulate the density along a loop, there is the possibility of a modulation of the scattered light in optical wavelengths, of free-free emission detectable in soft X-rays, or in radio wavelengths at $\nu \approx 1 - 5$ GHz. Alternatively, long-period oscillations could also be related to prominence oscillations (Kaufmann 1972).

7.5 Damping of MHD Oscillations

In contrast to p-mode oscillations driven from the solar interior, which seem to be sustained for indefinite time durations, almost all MHD oscillations in the corona are strongly damped, many having an exponential decay time of only about one oscillation period. This observational fact might have some important physical consequences, once we have identified the correct damping mechanism for each MHD mode. Currently, however, there are at least five, more or less competing mechanisms, which we will discuss in turn: non-ideal MHD effects (§ 7.5.1), lateral wave leakage (§ 7.5.2), footpoint wave leakage (§ 7.5.3), phase mixing (§ 7.5.4), and resonant absorption (§ 7.5.5).

7.5.1 Non-Ideal MHD Effects

In the derivation of the dispersion relation of MHD waves (§ 7.1) we used the adiabatic equation of state, and thus neglected heating, thermal conduction, and radiative loss in the energy equation. Also, using the set of *ideal MHD* equations (§ 6.1.3), we neglected non-ideal effects such as resistivity, viscosity, and Ohmic dissipation, which might contribute to damping of wave motion. Also the curvature of loops is neglected in straight slab and cylinder geometries.

Let us first investigate non-adiabatic effects, such as plasma heating and cooling times. There are two major cooling effects, thermal conduction and radiative loss. The thermal conduction time (Eq. 4.3.10) is generally estimated by dividing the thermal energy by the thermal conduction rate, which is the divergence of the conductive flux (∇F_C, Eq. 3.6.3),

$$\tau_{cond} = \frac{\varepsilon_{th}}{\nabla F_C} \approx \frac{3n_e k_B T_e}{(2/7)\kappa T^{7/2}/L^2} = \frac{21 k_B}{2\kappa} n_e T^{-5/2} L^2 = 1.6 \times 10^{-9} n_e T^{-5/2} L^2 \ .$$

$$(7.5.1)$$

Thus for active region loops ($n_e \approx 10^9$ cm^{-3}, $T_e \approx 1.0$ MK, $L = 10^9 - 10^{10}$ cm = $10 - 100$ Mm) we estimate cooling times due to thermal conduction from the corona to the chromosphere to be of the order $\tau_{cond} \approx 0.4 - 40$ hours), while for postflare loops ($n \approx 10^{10}$ cm^{-3}, $T_e \approx 10$ MK, $L = 1 - 5 \times 10^9$ cm = 10-50 Mm) the conductive cooling times shorten to $\tau_{cond} \approx 1 - 20$ min. Therefore, thermal conduction is definitely important for MHD slow-mode oscillations, which have comparable damping times ($\tau_D \approx 5 - 37$ min, Table 7.8) and flare-like temperatures. This is probably unimportant for MHD fast sausage-mode oscillations, which have flare-like densities but have the shortest oscillation periods ($P \approx 1 - 10$ s), and is definitely unimportant for MHD fast kink-mode oscillations, which have active region loop densities and damping times ($\tau_D \approx 3 - 20$ min) much shorter than the expected thermal conduction times ($\tau_{cond} \approx 0.4 - 40$ hours). Therefore, thermal conduction is considered as the main physical mechanism to damp MHD slow-mode oscillations, as observed by SUMER. Ofman & Wang (2002) include the thermal conduction term and kinematic viscosity (Eq. 6.1.14) in the MHD equations and perform a numeric simulation of an oscillating loop according to the physical parameters observed by SUMER. The damping time τ_D of the velocity disturbance v_1 by the viscosity force is then approximately (with Eqs. 6.1.14−15 and neglecting other force terms),

$$\frac{v_1}{\tau_D} \approx \frac{D\mathbf{v}}{Dt} \approx \frac{\mathbf{F}_{visc}}{\rho} = \nu_{visc} \left[\nabla^2 \mathbf{v} + \frac{1}{3} \nabla(\nabla \mathbf{v}) \right] \ . \tag{7.5.2}$$

Ofman & Wang (2002) find that the resulting damping times τ_D are consistent with the observed ones, and moreover that the theoretically expected scaling law of the dissipation time with period agrees with the observed one by SUMER (i.e., $\tau_D \propto P^{1.07\pm0.16}$).

Let us turn now to the other cooling process (i.e., radiative cooling in optically thin plasma). The radiative cooling time (Eq. 4.5.3) is generally defined by the ratio of the thermal energy divided by the radiative cooling rate,

$$\tau_{rad} = \frac{\varepsilon_{th}}{dE_R/dt} = \frac{3n_e k_B T_e}{n_e n_H \Lambda(T_e)} \approx \frac{3k_B T_e}{n_e \Lambda(T_e)} \ , \tag{7.5.3}$$

which yields, for active region loops, typical radiative cooling times of $\tau_{rad} \approx 1$ hour, and for postflare loops $\tau_{rad} \approx 10$ min. Therefore, we conclude that the radiative cooling time could contribute to the observed damping time in MHD slow-mode oscillations, but seems to be unimportant for the MHD fast sausage-mode oscillations (which have periods of $P \approx 1 - 10$ s) or for the MHD fast kink-mode oscillations (which have damping times of $\tau_D \approx 3 - 20$ min, much shorter than the radiative cooling times of active region loops with $\tau_{rad} \approx 1$ hour). For MHD slow-mode oscillations, the temperature decrease $T(t)$ due to radiative cooling would also decrease the sound speed $c_s(T[t])$ in the oscillating loops and prolong the oscillation periods $P(t) \propto 1/c_s(T[t])$ with time, an effect that has not yet been tested.

Regarding other adiabatic time scales, such as heating, we do not have much information in oscillating loops, except that the observed decay times were found to be significantly longer than the radiative cooling times (§ 4.5.1), a fact that requires continuous or intermittent heating (Winebarger et al. 2003a). The effects of heating time scales on the damping time of loop oscillations have not yet been explored much because of the unknown heating function. However, whether heating or cooling occurs in the oscillating loops, a most conspicuous consequence is that the intensity or brightness evolution of the loops is convolved with the instrumental temperature response function (§ 3.8). If a loop cools out of the passband of an instrument, the observed or detected damping time of a loop oscillation appears to be shorter than for the same loop with a constant temperature. Thus, the observed damping times of MHD slow-mode oscillations might be a lower limit to the effective damping times, which can only be corrected if the cooling time is known from multi-filter data.

Non-ideal MHD effects on damping of magneto-acoustic waves have also been considered for slab geometries by Van den Linden & Goossens (1991), Laing & Edwin (1995), and are discussed in Roberts (2000), suggesting that damping might be significant over $N_P \approx 20 - 100$ periods, and thus too slow to explain the observed damping of MHD fast-mode oscillations.

7.5.2 Lateral Wave Leakage

Consider a perfect waveguide, such as an optical fiber, which has no losses out of the fiber because of the phenomenon of total reflection, which guarantees 100% reflection along the internal sides as long as the light waves hit the side walls with a sufficiently small angle. The small angle becomes larger at the outer surface when you bend the fiber glass, and total reflection might not be guaranteed anymore, so the optical fiber becomes "leaky". By analogy, curvature in coronal loops may lead to leakage of MHD body or surface waves. The effect of lateral wave leakage occurs in curved loops because the fast kink and sausage modes become slightly coupled, so that energy can be transferred from one mode to the other and can be radiated away to infinity (Roberts 2000). For slender fluxtubes, this effect of lateral wave leakage is estimated to be very small, of order $(\tau_D/P) \approx (l/a)^2$ (Spruit 1982; Cally 1986, 2003; Roberts 2000), which is about 10^4 wave periods for a slender loop with $a/l = 0.01$ So, it cannot explain the observed damping of order $\tau_D/P \approx 1$.

7.5.3 Footpoint Wave Leakage

Now consider a laser, where a resonant cavity builds up wave energy coherently when the trapped waves are 100% reflected at the side walls that coincide with the wave nodes. If the reflection is less than 100% we have leakage. Similarly, waves in coronal loops can leak out of the chromospheric footpoints when the reflection coefficient at the coronal/chromospheric interface is less than 100%, due to ion-neutral collisions in the chromosphere. Leakage of fast waves from loop footpoints was considered by Berghmans & De Bruyne (1995), who estimated a decay time of about 10^2 periods. The energy reflection coefficient R between an incoming (downward propagating) and outgoing (upward propagating) coronal Alfvén wave at the coronal/chromospheric interface is given by Davila (1991),

$$R = \left| \frac{v_A^{phot} - v_A^{cor}}{v_A^{phot} + v_A^{cor}} \right| , \qquad (7.5.4)$$

which amounts to about $R \approx 0.98$ for typical coronal and chromospheric Alfvén velocities. The corresponding damping time τ_D of loop oscillations resulting from wave leakage at the footpoints is (Hollweg 1984a),

$$\tau_D = \frac{l}{(2 - R_1 - R_2)v_A^{cor}} = \frac{\tau_A}{2 - R_1 - R_2} , \qquad (7.5.5)$$

where τ_A is the Alfvén crossing time from one footpoint to the other, and R_1 and R_2 are the reflection coefficients at both footpoints. The high reflection coefficient given in Eq. (7.5.4), however, is derived from the assumption of a step function, while a more realistic treatment of the chromosphere with a finite scale height (i.e., $v_A \propto e^{h/2\lambda}$), leads to a lower reflection coefficient (De Pontieu et al. 2001),

$$R^2 = \frac{J_0^2 + Y_0^2 + J_1^2 + Y_1^2 - 4/(\pi\alpha)}{J_0^2 + Y_0^2 + J_1^2 + Y_1^2 + 4/(\pi\alpha)} \approx (1 - \pi\alpha)^2 , \qquad (7.5.6)$$

where the argument of the Bessel functions (J_0, J_1, Y_0, Y_1) is $\alpha = 2\lambda\omega/v_A^{cor}$ with $\omega = 2\pi/P$. For solar applications, the chromospheric scale height is $\lambda \approx 500$ km, $v_A^{cor} \approx 1000$ km s^{-1}, and thus $\alpha \ll 1$, leading to the approximation on the right-hand side of Eq. (7.5.6). The damping time τ_D due to wave leakage then becomes,

$$\tau_D \approx \frac{\tau_A}{\pi\alpha} = \frac{LP}{4\pi^2\lambda} . \qquad (7.5.7)$$

Applying this relation to the observed damping times τ_D, periods P, and loop lengths l to the MHD fast kink-mode oscillations observed in Aschwanden et al. (2002a), one finds that chromospheric scale heights between $\lambda = 400$ km and 2400 km are required. These values are somewhat higher than the scale height derived in chromospheric models with hydrostatic equilibrium (VAL and FAL models), yielding $\lambda \approx 500$ km, but can be consistent with more dynamic models (as manifested by the ubiquitous spicules) and with more recent measurements of the chromospheric scale height in microwaves and in hard X-rays with RHESSI (see Fig. 4.28). In contrast, chromospheric leakage was also studied with a nonlinear visco-resistive MHD code, where the damping time due chromospheric leakage was found to be about 5 times longer than the observed decay time for one MHD fast kink-mode oscillation event observed with TRACE (Ofman 2002). Thus the importance of chromospheric wave leakage is presently unresolved, awaiting more statistics of damping times and more accurate local measurements of chromospheric density scale heights.

7.5.4   Phase Mixing

Magneto-acoustic waves in a homogeneous medium (in density and magnetic field) propagate undamped and undisturbed because they are in perfect resonance with the magnetic field. However, if the medium has large gradients in the Alfvén velocity, then shear Alfvén waves suffer intense *phase mixing*, during which the oscillations of neighboring field lines become rapidly out of phase, which leads to enhanced viscous and ohmic dissipation (Heyvaerts & Priest 1983). The amount of wave damping due to phase mixing in a coronal loop with length l is derived in Heyvaerts & Priest (1983) and summarized in Roberts (2000), and depends on the Alfvén transit time $\tau_A = l/v_A$, the scale of inhomogeneity l_{inh}, and the coronal viscosity ν_{visc}; the damping time is

$$\tau_D = \left[\frac{6l^2 l_{inh}^2}{\nu_{visc} \pi^2 v_A^2} \right]^{1/3} . \tag{7.5.8}$$

The first application of this damping formula to an oscillating loop was presented in Nakariakov et al. (1999), where a coronal loop with MHD fast kink-mode oscillations was observed with TRACE during the 1999-Jul-14, 12:55 UT, flare (Plate 9) with the following parameters: full loop length $l = 130 \pm 6$ Mm, oscillation period $P_{kink} = 256$ s, and damping time $\tau_D = 870 \pm 162$ s. For the fundamental kink mode this corresponds to a phase speed of $c_k = 2l/P_{kink} = 1040 \pm 50$ km s^{-1}, and assuming a density ratio of $n_e/n_i \approx 0.1$, this yields (with Eq. 7.2.4) an Alfvén speed of $v_A = 770 \pm 40$ km s^{-1} or an Alfvén crossing time of $\tau_A = 1.3$ s. The full width of the oscillating loop was measured to $w = 7.2$ Mm in Aschwanden et al. (2002a). Whatever the fine structure of the oscillating loop is, for instance a bundle of unresolved threads, an upper limit on the spatial scale l_{inh} over which the Alfvén velocity changes drastically is a loop radius (i.e., $a = w/2 = 3.6$ Mm). Therefore, we can derive from Eq. (7.5.8) an upper limit for the coronal viscosity,

$$\nu_{visc} = \frac{6l^2 l_{inh}^2}{\tau_D^3 \pi^2 v_A^2} \le \frac{6l^2 a^2}{\tau_D^3 \pi^2 v_A^2} \tag{7.5.9}$$

which yields a value of $\nu_{visc} \le 4 \times 10^{12}$ cm^2 s^{-1}, using the parameters above. This is about an order of magnitude below the traditional value of the coronal viscosity, $\nu_{visc}^{cor} = 4 \times 10^{13}$ cm^2 s^{-1}. Nakariakov et al. (1999) used a different scaling law for the coronal viscosity, based on calculations with a numeric visco-resistive MHD code (Ofman et al. 1994),

$$\tau_D = 32.6\tau_A R^{0.22} = 32.6\frac{l}{v_A} \left(\frac{l_{inh} v_A}{\nu_{visc}} \right)^{0.22} , \tag{7.5.10}$$

but this yields the same value of $\nu_{visc} \approx 4 \times 10^{12}$ cm^2 s^{-1} for the viscosity.

The *viscous Reynolds number* R_L based on the Alfvén speed is a measure of viscous convection over a length scale l_{inh} with Alfvén speed v_A,

$$R_L = \frac{l_{inh} v_A}{\nu_{visc}} , \tag{7.5.11}$$

which yields, for the oscillating loop, a value of $R_L \approx 10^4$, which is far below traditionally believed values of the coronal (magnetic) Reynolds number (*Lundquist number*), $R_L^{cor} = 10^{12} - 10^{14}$. The authors suggest that this enhanced value applies to the *shear viscosity*. Thus, the shear viscosity needs to be enhanced by $8-9$ orders of magnitude in the loop compared with classical viscosity to explain the strong damping, if phase mixing is employed as the primary damping mechanism (Nakariakov et al. 1999). An observational test of this model can be made from the expected scaling law for phase mixing, which by inserting the oscillation period $P = 2l/c_k \approx \sqrt{2}l/v_A$ into Eq. (7.5.8) yields (see § 7.5.6),

$$\tau_D(P) = \left[\frac{3 P^2 l_{inh}^2}{\nu_{visc} \pi^2} \right]^{1/3} . \tag{7.5.12}$$

It was pointed out that it would be actually more appropriate to consider the *shear viscosity coefficient* instead of the *kinematic* or *compressional viscosity* to estimate the damping due to phase mixing, considering the strong anisotropy of the coronal plasma (Ofman et al. 1994; Roberts 2002).

7.5.5 Resonant Absorption

Another mechanism that could substantially damp kink-mode oscillations is the process of *resonant absorption*, where the energy of the kink-mode motion can be transferred into (predominantly azimuthal) Alfvén oscillations of the inhomogeneous layers at the loop boundary, where the density, and thus the Alfvén speed, varies drastically. The damping of loop oscillations indicates a mode conversion process (i.e., global kink-mode oscillations transfer energy into localized Alfvén waves, Lee & Roberts 1986). Following the approach of an *initial value problem* (e.g., Sedlacek 1971; Ionson 1978; Rae & Roberts 1982; Lee & Roberts 1986; Hollweg 1987; Hollweg & Yang 1988; Steinolfson & Davila 1993), the damping time τ_D of a kink mode oscillating loop with a thin boundary layer ($l_{skin} \ll r_{loop}$) was calculated due to resonant absorption by Ruderman & Roberts (2002), leading to the damping formula

$$\left(\frac{\tau_D}{P} \right)_{thin} = \frac{2}{\pi} \left(\frac{r_{loop}}{l_{skin}} \right) \frac{(1 + q_n)}{(1 - q_n)} , \tag{7.5.13}$$

where r_{loop} is the loop radius, l_{skin} the *skin depth* or thickness of the loop boundary with varying density, and $q_n = n_e/n_i$ is the density ratio for the external (n_e) to internal loop density (n_i). We see that damping is stronger for larger skin depth ratios, because the extent of the inhomogeneous region where resonant transfer of Alfvén wave energy takes place is relatively large. In addition, the damping is strongest for the largest density contrast ($q_n = 0$), while it is weakest ($\tau_D \mapsto \infty$) for oscillating loops that have the same density as the environment ($n_i \mapsto n_e$, $q_n \mapsto 1$). This damping formula was generalized for elliptical cross sections (Ruderman 2003), and for thick boundaries $0 < l_{skin} \leq 2r_{loop}$ (Van Doorsselaere et al. 2004) using numerical MHD calculations (LEDA code),

$$\left(\frac{\tau_D}{P}\right)_{thick} = q_{TB}\frac{2}{\pi}\left(\frac{r_{loop}}{l_{skin}}\right)\frac{(1+q_n)}{(1-q_n)}, \qquad (7.5.14)$$

where the correction factor q_{TB} depends on the thickness of the boundary layer (l_{skin}/r_{loop}) as well as on the density ratio q_n (e.g., having a value of $q_{TB}(l_{skin}/r_{loop} = 2, q_n = 1/3) = 0.75$ in the fully nonuniform limit $l_{skin}/r_{loop} = 2$). In both studies, the density profile $n(r)$ across the loop cross section is parameterized with a sine function within the skin depth,

$$n(r) = \begin{cases} n_i & \text{for } r < (r_{loop} - l_{skin}) \\ n_i\left[\frac{(1+q_n)}{2} - \frac{(1-q_n)}{2}\sin\frac{\pi}{2}\frac{(2r+l_{skin}-2r_{loop})}{l_{skin}}\right] & (r_{loop} - l_{skin}) < r < r_{loop} \\ n_e & \text{for } r > r_{loop} \end{cases}$$

$$(7.5.15)$$

Oscillating loops observed with TRACE have been found to have typical skin depths of $l_{skin}/r_{loop} = 1.5 \pm 0.2$, which is close to the fully nonuniform limit of $l_{skin}/r_{loop} = 2$. A lower limit of their density ratio was measured at $q_n \leq 0.30 \pm 0.16$ (Aschwanden et al. 2003b). Inserting these mean values into Eq. (7.5.14) yields a predicted ratio of the damping time to the oscillation period of $\tau_D/P \approx 2.0$, which is very close to the average observed value (i.e., $(\tau_D/P)_{obs} = 580/321 = 1.8$, see Table 7.3). So, the mechanism of resonant absorption predicts a damping time of kink-mode oscillations that is close to the observed one, which provides a more natural explanation than the alternative mechanism of phase mixing with an excessively low coronal Reynolds number.

7.5.6 Observational Tests

For MHD fast kink-mode oscillations, there are 11 events observed with TRACE for which the damping time could be determined (Aschwanden et al. 2002a; shown in Fig. 7.22), while for MHD slow-mode (acoustic) oscillations, the damping time has been measured in 54 events from SUMER observations (Wang et al. 2003b). The mean ratio of damping time to oscillation period is $\tau_D/P = 1.8 \pm 0.8$ for the fast kink-mode oscillations (Table 7.3) and $\tau_D/P = 0.84 \pm 0.34$ for the slow-mode (acoustic) oscillations (Table 7.8).

Attempts have been made to determine the scaling law of theoretically predicted damping times and to compare it with the observed damping times for the MHD fast kink-mode oscillation events (Ofman & Aschwanden 2002; Aschwanden et al. 2003b), but the small number statistics of 11 events does not allow us to discriminate between competing damping theories, as the scatter of the data points in Fig. 7.23 illustrates. However, we can determine the average ratios between the predicted and observed damping times, as shown in Fig. 7.23. The results for the radiative cooling time τ_{rad}

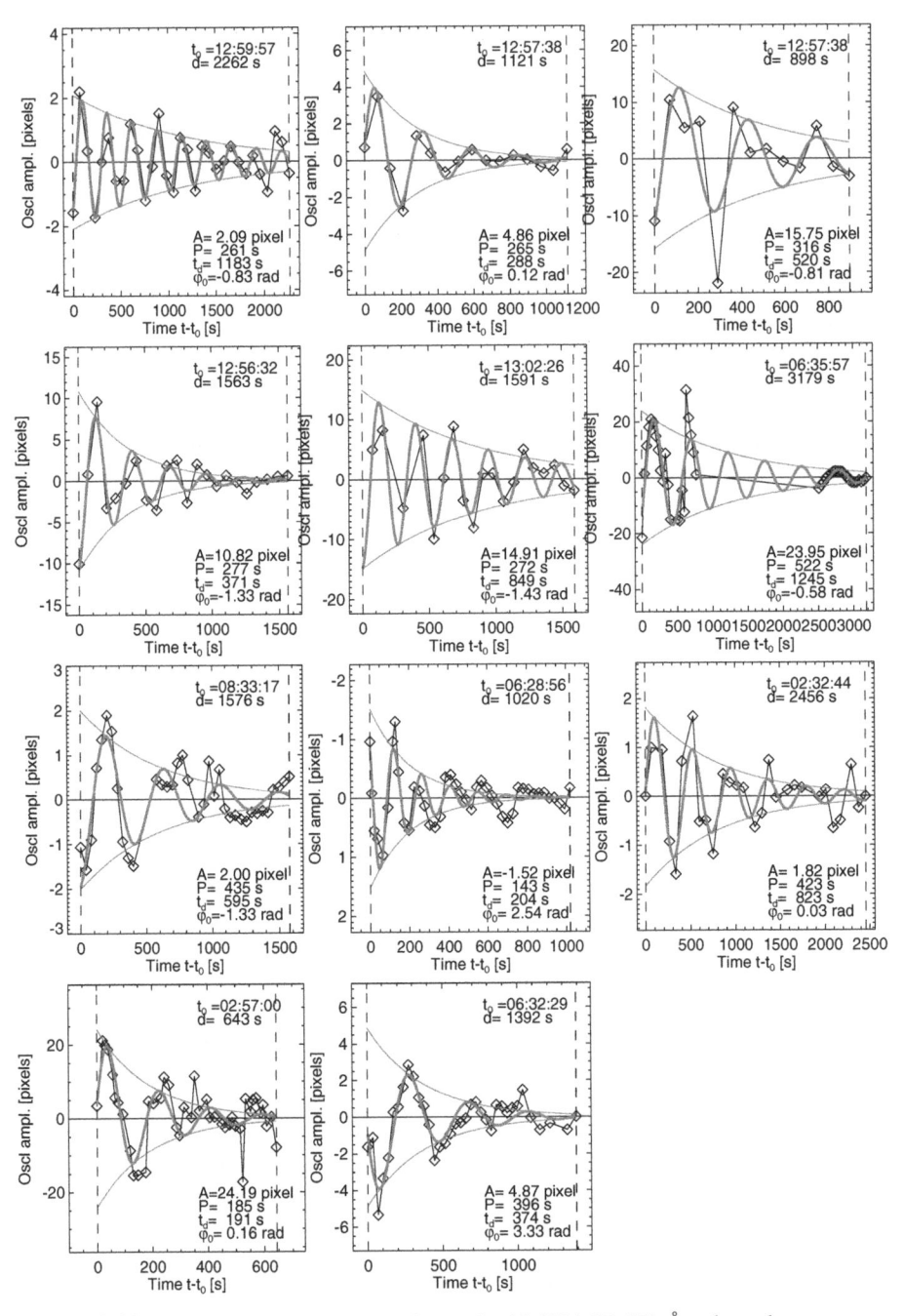

Figure 7.22: Eleven oscillation events observed with TRACE 171 Å, where the transverse MHD kink-mode oscillation amplitude $A(t)$ is fitted with a damped sine function plus a low-order polynomial function. The polynomial trend function is subtracted and only the oscillatory fit (thick curve) to the data points (diamonds) is shown (adapted from Aschwanden et al. 2002a).

Table 7.10: Physical parameters measured for 11 events with damped oscillations observed with TRACE 171 Å (Aschwanden et al. 2003b).

Parameter	Mean and standard deviation
Loop curvature radii R_{curv}	57 ± 21 Mm
Oscillation period P	317 ± 114 s
Damping time τ_D	574 ± 320 s
Observed number of oscillations τ_D/P	1.8 ± 0.8
Predicted minimum of ratio $(\tau_D/P)_{min}$	0.32 ± 0.05
Outer loop radius a	4.5 ± 3.5 Mm
Inner loop radius $a - l$	0.6 ± 0.5 Mm
Mean loop width $w_{loop} = 2a - l$	5.1 ± 3.9 Mm
Loop skin depth l	3.9 ± 3.1 Mm
Relative loop skin depth l/R	1.5 ± 0.2
Loop density n_i	$1.4 \pm 0.7 \ 10^9$ cm^{-3}
External plasma density $n_e(T = 1MK)$	$0.36 \pm 0.18 \ 10^9$ cm^{-3}
Predicted external plasma density $n_e = n_i q_n^{LEDA}$	$0.76 \pm 0.36 \ 10^9$ cm^{-3}
Density ratio $q_n = n_e(T = 1MK)/n_i$	0.30 ± 0.16
Predicted density ratio $q_n^{LEDA} = n_e/n_i$	0.53 ± 0.12
Prediction ratio $n_e/n_e(T = 1MK) = q_n^{LEDA}/q_n$	2.5 ± 2.1

(Eq. 7.5.3), the footpoint leakage decay time τ_{leak} (Eq. 7.5.7), for the phase mixing decay time τ_{PM} (Eq. 7.5.12), and the resonant absorption decay time τ_{RA} (Eq. 7.5.14) are, based on the observational parameters listed in Tables 7.1 and 7.10:

$$\tau_{rad}/\tau_D = 9.78 \pm 7.67 \,,$$

$$\tau_{leak}/\tau_D = 6.16 \pm 2.58 \,,$$

$$\tau_{PM}/\tau_D = 0.79 \pm 0.19 \,,$$

$$\tau_{RA}/\tau_D = 0.37 \pm 0.15 \,. \tag{7.5.16}$$

These statistical results tell us that the radiative cooling time and footpoint leakage times are much longer and thus not relevant for the observed damping. The footpoint leakage time was calculated for a chromospheric scale height of $\lambda = 500$ km, so in principle an effective scale height ($\lambda = 3000$ km) that is 6 times longer could produce enough damping. On the other hand, both the phase mixing and resonant absorption produce damping times comparable or even shorter than the observed ones. For phase mixing, however, extremely low Reynolds numbers would be required, so it is not clear whether the concept of compressional viscosity is correctly applied. The most natural interpretation of the damping mechanism seems to be the concept of resonant absorption, a conclusion that was suggested by Ruderman & Roberts (2002) and Goossens et al. (2002a).

For MHD slow-mode acoustic waves, damping by thermal conduction has been shown to yield damping times that are comparable with the observed ones (i.e., $\tau_D \approx 5 - 30$ min; Ofman & Wang 2002). Note that the observations all refer to hot ($T > 6.3$ MK) flare-like loops observed in Fe XIX with SUMER. Since the thermal conduction

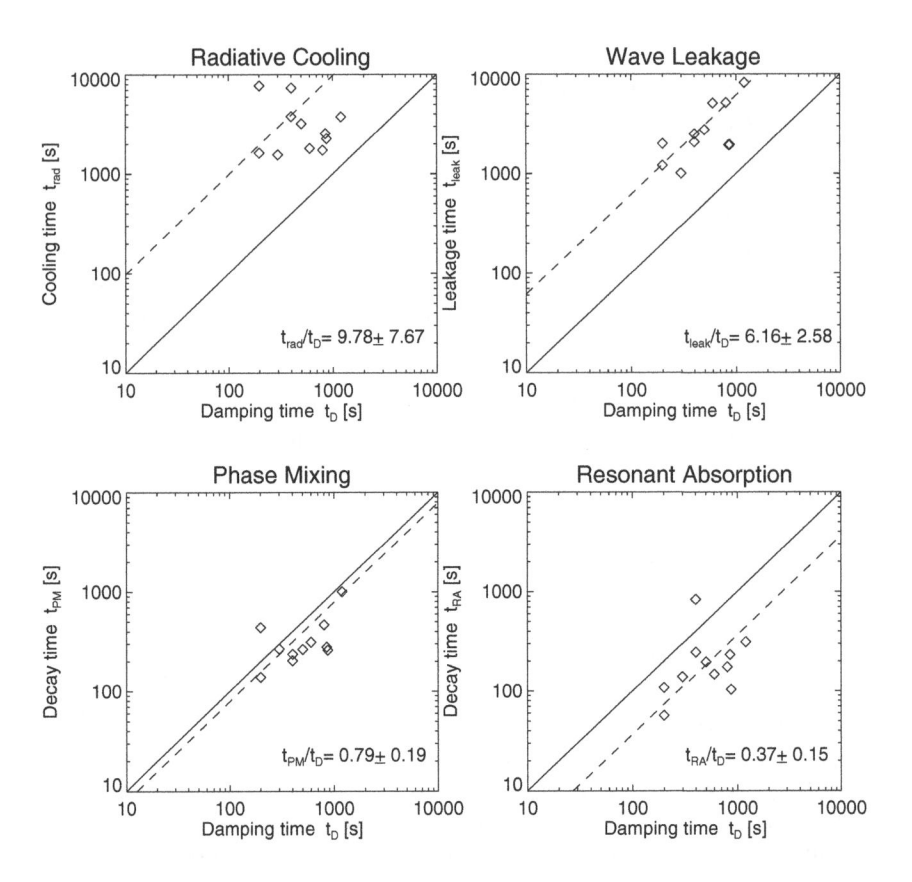

Figure 7.23: Damping times predicted by theory (radiative cooling time τ_{rad}, footpoint leakage damping τ_{leak}, phase mixing decay time τ_{PM}, and resonant absorption τ_{RA}) compared with observed damping times τ_D for 11 events with MHD fast kink-mode oscillations observed with TRACE (see data in Table 7.1 and 7.10). The average ratio is marked with a dashed line with the means and standard deviations (excluding extremal values) indicated in each frame.

times are scaling with $\tau_{cond}(T) \propto T^{-5/2}$, they would be a factor of $6.3^{2.5} = 100$ longer for the cooler ($T \approx 1.0$ MK) EUV loops with detected kink-mode oscillations, and thus would not play a role in the damping of EUV oscillations.

7.6 Summary

Coronal seismology is a prospering new field, stimulated by the first imaging observations of oscillating coronal loops made with TRACE and SOHO/SUMER since about 1999. While the theory of MHD oscillations was developed several decades earlier, only the new imaging observations provide diagnostics on length scales, periods, damping times, and densities that allow a quantitative application

of the theoretical dispersion relations of MHD waves (§ 7.1). The theory of MHD oscillations has been developed for homogeneous media, single interfaces, slender slabs, and cylindrical fluxtubes. There are four basic speeds in fluxtubes: (1) the Alfvén speed $v_A = B_0/\sqrt{4\pi\rho_0}$, (2) the sound speed $c_s = \sqrt{\gamma p_0/\rho_0}$, (3) the cusp or tube speed $c_T = (1/c_s^2 + 1/v_A^2)^{-1/2}$, and (4) the kink or mean Alfvén speed $c_k = [(\rho_0 v_A^2 + \rho_e v_{Ae}^2)/(\rho_0 + \rho_e)]^{1/2}$. For coronal conditions, the dispersion relation reveals a slow-mode branch (with acoustic phase speeds) and a fast-mode branch of solutions (with Alfvén speeds). For the fast-mode branch, a symmetric (sausage) mode and an asymmetric (kink) mode can be distinguished. The fast kink mode (§ 7.2) produces transverse amplitude oscillations of coronal loops which have been detected with TRACE, having periods in the range of P=2−10 min, and can be used to infer the coronal magnetic field strength, thanks to its non-dispersive nature. The fast sausage mode (§ 7.3) is highly dispersive and is subject to a long-wavelength cutoff, so that standing wave oscillations are only possible for thick and high-density (flare and postflare) loops, with periods in the range of P≈1 s to 1 min. Fast sausage-mode oscillations with periods of P≈10 s have recently been imaged for the first time with the Nobeyama Radioheliograph, while there exist numerous earlier reports on non-imaging detections with periods of P≈0.5−5 s. Finally, slow-mode acoustic oscillations (§ 7.4) have been detected in flare-like loops with SUMER, having periods in the range of P≈ 5 − 30 min. All loop oscillations observed in the solar corona have been found to be subject to strong damping, typically with decay times of only 1−2 periods. The relevant damping mechanisms (§ 7.5) are resonant absorption for fast-mode oscillations (or alternatively phase mixing, although requiring an extremely low Reynolds number), and thermal conduction for slow-mode acoustic oscillations. Quantitative modeling of coronal oscillations offer exciting new diagnostics on physical parameters.

Chapter 8

Propagating MHD Waves

The terms *"waves"* and *"oscillations"* are often used interchangably, because the general wave form is often decomposed into Fourier components, each one representing an oscillatory solution, $A(\mathbf{x}, t) = \sum_n A_n \exp(-i[\mathbf{kx} - \omega_n t])$. In this book we use a stricter definition, reserving the term *oscillations* only for *standing waves* with fixed nodes (§ 7), while *propagating waves* have moving nodes (Fig. 8.1). All the MHD oscillation modes we described in chapter 7 have fixed nodes, anchored at both endpoints of a coronal loop, forced by the photospheric line-tying conditions of the magnetic field, analogous to the strings of a violin. In principle, clean harmonic oscillations are only warranted if either the excitation profile along a loop matches the sine function of a harmonic wave solution, or once an initial arbitrary displacement settles into a fundamental harmonic oscillation, after the higher harmonic components are damped out. This time interval can be quite long, for instance it amounts to about 40 oscillation periods for a clarinet, as measured with a high-speed camera. Since coronal loop oscillations have been found to be strongly damped within a few oscillation periods, they probably never have sufficient time to settle into a clean harmonic eigen mode, besides the unavoidable damping due to finite dissipation. We expect a series of short-wavelength disturbances to propagate along the loop, especially when the excitation occurs at one side of a coronal loop on a time scale much shorter than the reflection time over the entire loop length. Hence, there is a gradual transition from harmonic oscillations to propagating waves, depending on the time scale and spatial symmetry of the initial displacement. In this chapter we deal exclusively with propagating waves, a field that experienced a major breakthrough after the recent *SoHO* and *TRACE* observations, including the discoveries of *EIT* (or coronal *Moreton*) waves (Thompson et al. 1998a; Wills−Davey & Thompson 1999), compressible waves in polar plumes (Ofman et al. 1997; DeForest & Gurman 1998; Ofman et al. 1999), wave trains in coronal loops (Berghmans & Clette 1999; Robbrecht et al. 2001; De Moortel et al. 2002a,b,c), as well as with the first detection of propagating wave trains during a solar eclipse (Williams et al. 2001, 2002; Pasachoff et al. 2002). Recent reviews on the subject can be found in Roberts (2000; 2002) and Roberts & Nakariakov (2003).

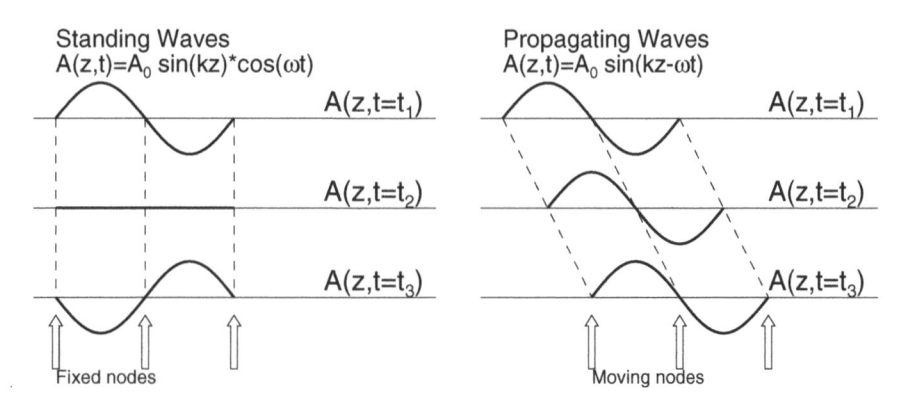

Figure 8.1: Definition of *"standing wave"* or *"oscillation"* (left) and *"propagating wave"* (right). A standing wave has fixed nodes, while a propagating wave has moving nodes as a function of time. Standing waves can also be composed by superposing two oppositely directed propagating waves.

8.1 Propagating MHD Waves in Coronal Loops

8.1.1 Evolutionary Equation for Slow-Mode MHD Waves

We derived the general dispersion relation for magneto-acoustic waves in cylindrical fluxtubes in § 7.1.4, which showed two branches of phase speed solutions ω/k: a fast-mode branch (with Alfvén speeds) and a slow-mode branch (with acoustic speeds), as shown in Fig.7.4 for coronal conditions. In this section we study the propagating waves of the slow mode for the special geometry of coronal loops, which involves gravitational stratification in the vertical direction for fluxtubes curved along closed magnetic field lines, while the case for open magnetic field lines is considered in the next section (§ 8.2). Making some simplifying assumptions, such as neglecting the coupling of the slow magneto-acoustic mode with other wave modes, 2D effects (including wave dispersion), loop curvature, whilst assuming wavelengths much shorter than the gravitational scale height, Nakariakov et al. (2000a) derived the *evolutionary equation*, using the following form of the resistive MHD equations (see § 6.1.5):

$$\frac{\partial \rho}{\partial t} + \frac{\partial}{\partial s}(\rho v) = 0 \,, \tag{8.1.1}$$

$$\rho \left(\frac{\partial v}{\partial t} + v \frac{\partial v}{\partial s} \right) = -\frac{\partial p}{\partial s} - g\rho + \frac{4}{3}\eta_0 \frac{\partial^2 v}{\partial s^2} \,, \tag{8.1.2}$$

$$\frac{1}{(\gamma - 1)} \frac{\partial p}{\partial t} - \frac{1}{(\gamma - 1)} \frac{\gamma p}{\rho} \frac{\partial \rho}{\partial t} = \frac{\partial}{\partial s} \left(\kappa_\parallel \frac{\partial T}{\partial s} \right) \,, \tag{8.1.3}$$

where s is the loop length coordinate, $\rho(s)$ the plasma density, $v(s)$ the longitudinal speed, $p(s)$ the plasma pressure, $T(s)$ the plasma temperature, γ the adiabatic index, $\kappa_\parallel = \kappa T^{5/2}$ the thermal conductivity along the magnetic field, η_0 the compressive

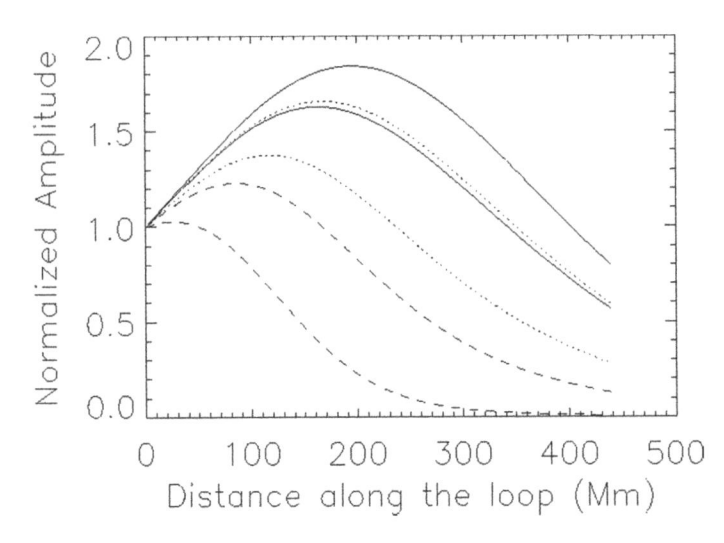

Figure 8.2: Evolution of the amplitude of slow magneto-acoustic waves with the initial amplitude $v(0) = 0.02 \, c_s$ for three wave periods: 900 s (solid curves), 600 s (dotted curves), and
300 s (dashed curves). The upper curve of each kind corresponds to the normalized dissipation
coefficient $\bar\eta = 4 \times 10^{-4}$, and the lower curve to $\bar\eta = 4 \times 10^{-3}$. The amplitude of each wave is
measured in units of the initial amplitude. The loop radius is $r_{curv} = 140$ Mm (Nakariakov et
al. 2000a).

viscosity coefficient, and $g(s)$ the gravitational acceleration projected along the loop
coordinate s for a semi-circular geometry (with curvature radius r_{curv}),

$$g(s) = g_\odot \cos\left(\frac{s}{r_{curv}}\right)\left(1 + \frac{r_{curv}}{R_\odot}\sin\frac{s}{r_{curv}}\right)^{-2} . \qquad (8.1.4)$$

Combining the equations $(8.1.1-3)$ into a wave equation, Nakariakov et al. (2000a)
obtained an evolutionary equation for the density perturbation in the form of a modified
Burgers equation,

$$\frac{\partial v}{\partial s} - \frac{1}{2\lambda_n}v + \frac{\gamma+1}{2c_s}v\frac{\partial v}{\partial\xi} - \frac{R_\odot\rho_0(0)\bar\eta}{2\rho_0(s)}\frac{\partial^2 v}{\partial\xi^2} = 0 , \qquad (8.1.5)$$

where $\xi = s - c_s t$ is the coordinate co-moving with a wave crest with sound speed
c_s, $\lambda_n(s) = c_s^2(\gamma g)^{-1}$ the local density scale height, $R_{gas} = p/\rho T = 2k_B/\mu m_p$ the
gas constant, and $\rho_0(0)$ the equilibrium density at the base of the corona ($s = 0$). The
linearized version of Eq. (8.1.5) can be solved under the assumption of a harmonic
wave, $v(s) \propto \cos(k\xi) = \cos(ks - \omega t)$, propagating with sound speed $\omega/k = c_s$ with
wave number k,

$$v(s) = v(0)\exp\left[\int_0^s \left(\frac{1}{2\lambda_n(x)} - \frac{k^2\bar\eta\rho_0(0)R_\odot}{2\rho_0(x)}\right)dx\right] , \qquad (8.1.6)$$

Table 8.1: Observations of slow-mode (acoustic) waves in coronal structures. N is the number of analyzed events.

Observer	N	Frequency or wavelength	Wave speed v [km/s]	Instrument
DeForest & Gurman (1998)	1	171 Å	$\approx 75 - 150$	SoHO/EIT
Berghmans & Clette (1999)	3	195 Å	$\approx 75 - 200$	SoHO/EIT
De Moortel et al. (2000b)	1	171 Å	$\approx 70 - 165$	TRACE
De Moortel et al. (2002a)	38	171 Å	122 ± 43	TRACE
De Moortel et al. (2002b)	4	195 Å	150 ± 25	TRACE
De Moortel et al. (2002c)	38	171 Å	122 ± 43	TRACE
Robbrecht et al. (2001)	4	171, 195 Å	$\approx 65 - 150$	EIT, TRACE
Berghmans et al. (2001)	1	171, 195 Å	...	EIT, TRACE
Sakurai et al. (2002)	1	5303 Å	≈ 100	Norikura
King et al. (2003)	1	171, 195 Å	...	TRACE
Marsh et al. (2003)	1	171, 368 Å	$\approx 50 - 195$	CDS, TRACE

where the normalized dissipation coefficient $\bar{\eta}$ is defined by

$$\bar{\eta} = \frac{1}{\rho_0(0)c_s R_\odot} \left[\frac{4\eta_0}{3} + \frac{\kappa_\parallel(\gamma - 1)^2}{R_{gas}\gamma} \right] . \qquad (8.1.7)$$

The linearized solution of the evolutionary equation (8.1.5) yields a proportional perturbation in density, pressure, and temperature (according to the continuity equation and ideal gas equation),

$$\frac{\rho}{\rho_0} = \frac{v}{c_s} , \quad \frac{p}{p_0} = \gamma \frac{v}{c_s} , \quad \frac{T}{T_0} = (\gamma - 1)\frac{v}{c_s} . \qquad (8.1.8)$$

The evolution of each normalized quantity (Eq. 8.1.8) as a function of the loop coordinate s is shown in Fig. 8.2. The growth rate of each amplitude (in density, velocity, or pressure) is determined by the balance between the vertical gravitational stratification and dissipation (by thermal conduction and viscosity). Waves of shorter wavelengths (larger wave numbers k) grow slower than long-wavelength waves. Sufficiently short-wavelength perturbations, with $k > 1/\sqrt{\bar{\eta}\lambda_n(0)}$, do not grow at all, but decay with height. So the evolution of upward propagating slow-mode (acoustic) waves, whether they grow or decay, depends on the value of the dissipation coefficient η, thermal conduction coefficient $\kappa_\parallel$, and base density $\rho_0(0)$, as combined in the normalized dissipation coefficient $\bar{\eta}$ (Eq. 8.1.7). Nakariakov et al. (2000a) estimate a lower limit of $\bar{\eta} \approx 4 \times 10^{-4}$, using $\eta_0 = 0.352$ g cm^{-1} s^{-1} according to Braginskii's theory for $n_0 = 5 \times 10^8$ cm^{-3}, $T_e = 1.6$ MK, and neglecting thermal conduction ($\gamma = 1$). Evolutions of slow-mode acoustic waves for $\bar{\eta} = 4 \times 10^{-4}$ and 10^{-3} are shown in Fig. 8.2.

8.1.2 Observations of Acoustic Waves in Coronal Loops

Acoustic waves propagating in coronal loops were probably first noticed in EUV images of *SoHO/EIT* observations, when time sequences of flux profiles $F(s, t)$ along

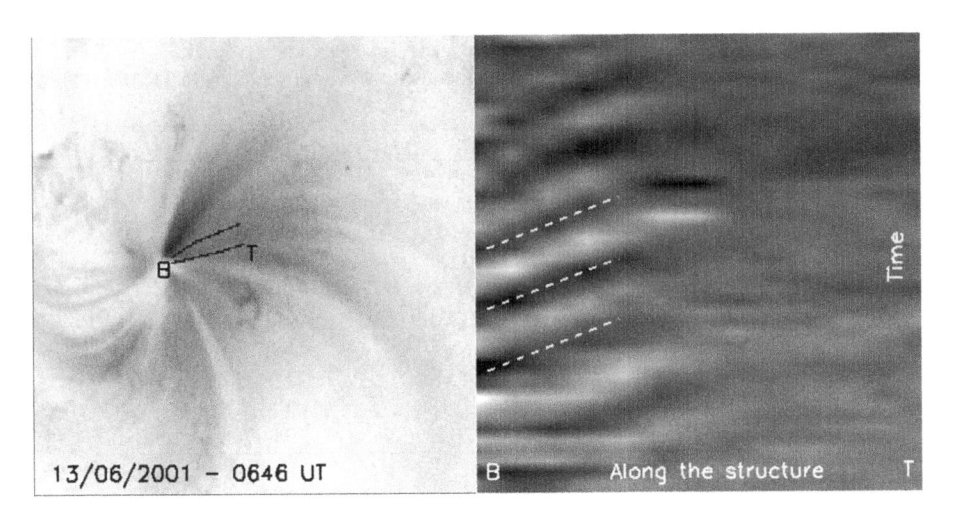

Figure 8.3: TRACE 171 Å observation of a slow-mode (acoustic) wave recorded on 2001 June 13, 06:46 UT. *Left:* the diverging fan-like loop structures emerge near a sunspot, where the acoustic waves are launched and propagate upward. *Right:* a running difference plot is shown for the loop segment marked in the left frame, with time running upward in the plot. Note the diagonal pattern which indicates propagating disturbances (De Moortel et al. 2000a).

loops with transient features were plotted with sufficiently high cadence (e.g., with $\Delta t = 15$ s, Berghmans & Clette 1999). In such space-time diagrams, diagonal patterns were noticed (e.g., Fig. 8.3 right), which exhibited slopes in the range of $v = ds/dt \approx$ 150 km s^{-1} (Berghmans & Clette 1999), corresponding to a speed slightly below the sound speed of $c_s \approx 180$ km s^{-1} at $T \approx 1.5$ MK expected in the used EIT 195 Å temperature band. A compilation of related observations are given in Table 8.1.

A number of propagating waves were also analyzed from TRACE data, starting in active regions and propagating upward into diverging, fan-like bundles of loops that fade out with height (Fig. 8.3, left), while no downward propagating waves were detected (De Moortel et al. 2000b, 2002a,b,c). Typical speeds of $v \approx 122 \pm 43$ km s^{-1} (De Moortel et al. 2000b) were measured from TRACE 171 Å data, where the mean sound is expected to be $c_s \approx 147$ km s^{-1} at $T \approx 1.0$ MK. Multi-wavelength observations with both EIT and TRACE confirm that the diverging fan structures consist of multiple loop threads with different temperatures and corresponding sound speeds (Robbrecht et al. 2001; King et al. 2003). Time periods of $P = 172 \pm 32$ s were found for loops rooted near sunspots, which coincide with the 3-minute p-mode oscillations detected in sunspots (Brynildsen et al. 2000, 2002; Fludra 2001; Maltby et al. 2001), while waves that start further away from sunspots (in active region plages) have periods of $P = 321 \pm 74$ min, which coincide with the global 5-minute p-mode oscillations. This result clearly proves that subphotospheric acoustic p-mode oscillations penetrate through the chromosphere and transition region and excite coronal acoustic waves. The energy flux associated with these propagating waves was estimated to $d\varepsilon_{wave}/dt \approx (3.5 \pm 1.2) \times 10^2$ erg cm^{-2} s^{-1}, far below the requirement for

Table 8.2: Statistical parameters of slow-mode (acoustic) waves observed with TRACE 171 Å in 38 structures (De Moortel et al. 2002a).

Parameter	Average	Range
Length of loop segment L	26.4 ± 9.7 Mm	$10.2-49.4$ Mm
Average footpoint width w	8.1 ± 2.8 Mm	$3.9-14.1$ Mm
Divergence gradient dw/ds	0.28 ± 0.16	$0.07-0.71$
Oscillation period P	282 ± 93 s	$145-525$ s
Propagation speed v_{wave}	122 ± 43 km/s	$70-235$ km/s
Wave amplitude dI/I	0.041 ± 0.015	$0.007-0.146$
Brightness change I_{max}/I_{min}	7.4 ± 5.8	$1-22.7$
Detection length L_{det}	8.9 ± 4.4 Mm	$2.9-23.2$ Mm
Detection ratio L_{det}/L	0.367 ± 0.188	$0.08-0.814$
Energy flux $d\varepsilon_{wave}/dt$	342 ± 126 erg/(cm^2 s)	$194-705$ erg/(cm^2 s)

coronal heating (§ 9.1). The statistical means and ranges of the parameters measured in De Moortel et al. (2002a) are compiled in Table 8.2. The wave trains were found to fade out quickly with height, partially an effect of the decreasing flux amplitude due to the diverging geometry of the loop fans, combined with the damping caused by thermal conduction (De Moortel et al. 2002b; De Moortel & Hood 2003, 2004). The interpretation in terms of slow-mode (acoustic) waves is based on: (1) the observed propagation speed roughly corresponding to the expected sound speed in the used temperature band, and (2) slow-mode (acoustic) waves being compressional waves, producing a modulation of the density and EUV flux, and thus observed as EUV intensity modulation (which is not the case for Alfvén waves).

Slow sound waves were possibly also detected in optical wavelengths (in the green line at 5303 Å) with spectroscopic methods using the Norikura Solar Observatory, with periods of $P \approx 3 - 5$ min and speeds of $v \approx 100$ km s^{-1} (Sakurai et al. 2002), but the confusion in white light seems to be much larger than in narrow-band EUV filters. Similarly, searches for waves with CDS data, which have substantially less spatial resolution than TRACE and EIT data, have only revealed marginal signals of oscillatory wave activity (Ireland et al. 1999; O'Shea et al. 2001; Harrison et al. 2002; Marsh et al. 2003), due to the overwhelming confusion with other spatially unresolved and time-varying loop structures.

8.1.3 Propagating Fast-Mode Waves in Coronal Loops

Fast mode MHD waves have Alfvén phase speeds, which can vary over a considerable range in coronal conditions, between the minimum Alfvén speed value v_A inside of a loop and the maximum speed v_{Ae} outside of the loop (i.e., $v_A \leq v_{ph} = \omega/k \leq v_{Ae}$) (Fig. 7.4). We discussed the standing waves or eigen frequencies of this fast MHD wave mode in § 7.2 (kink mode) and § 7.3 (sausage mode). Now, what about propagating fast MHD waves. We quote Roberts et al. (1984): "Propagating waves, rather than standing modes, will result whenever disturbances are generated impulsively. Such waves may arise in a coronal loop, if the motions have insufficient time to reflect from the far end of

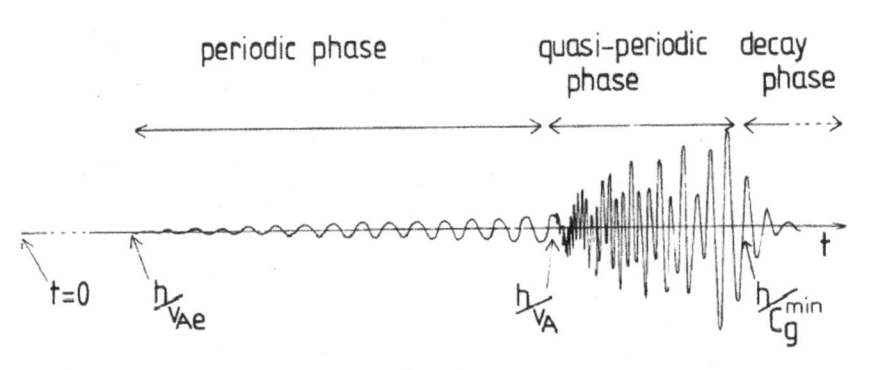

Figure 8.4: The evolution of a signal produced by a propagating fast-mode MHD wave in a coronal loop, which originates at height $h = 0$ and is observed at height $h = z$. The time intervals of the three phases depend on the characteristic velocities v_A, v_{Ae}, and c_g^{min} (Roberts et al. 1984).

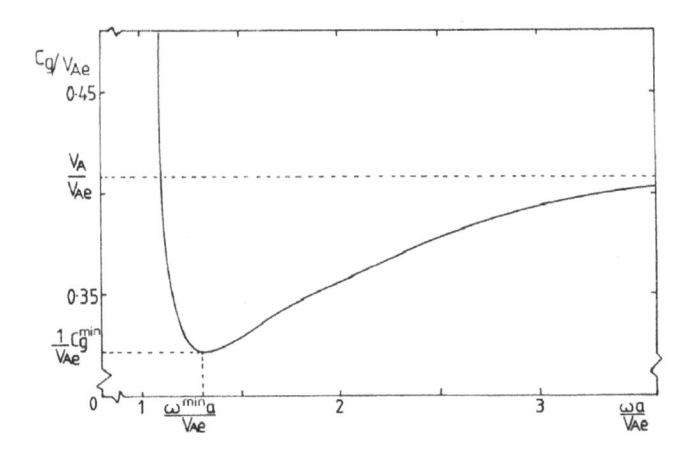

Figure 8.5: The group speed $v_g = \partial\omega(k)/\partial k$, normalized by the external Alfvén velocity v_{Ae}, as a function of the dimensionless wave number frequency $k_e a = \omega a/v_{Ae}$, calculated for coronal conditions $c_s \ll v_A$ and $\rho_0/\rho_e = 6$. Note the occurrence of a minimum in the group speed, c_g^{min} (Roberts et al. 1984).

the loop, or in open field regions. An obvious source of such an impulsive disturbance is the flare (providing either a single or a multiple source of disturbances), but less energetic generators should not be ruled out. If the waves are generated impulsively, then the resulting disturbance may be represented as a Fourier integral over all frequencies ω and wave numbers k. In general, a wave packet results, its overall structure being determined by the dispersive nature of the modes." Roberts et al. (1984) calls this type of wave an *impulsively generated fast wave*. Such propagating fast-mode MHD waves display a bewildering variety of evolutionary scenarios, which have not been explored much in the solar context, but their hydrodynamic analogs have been widely studied in

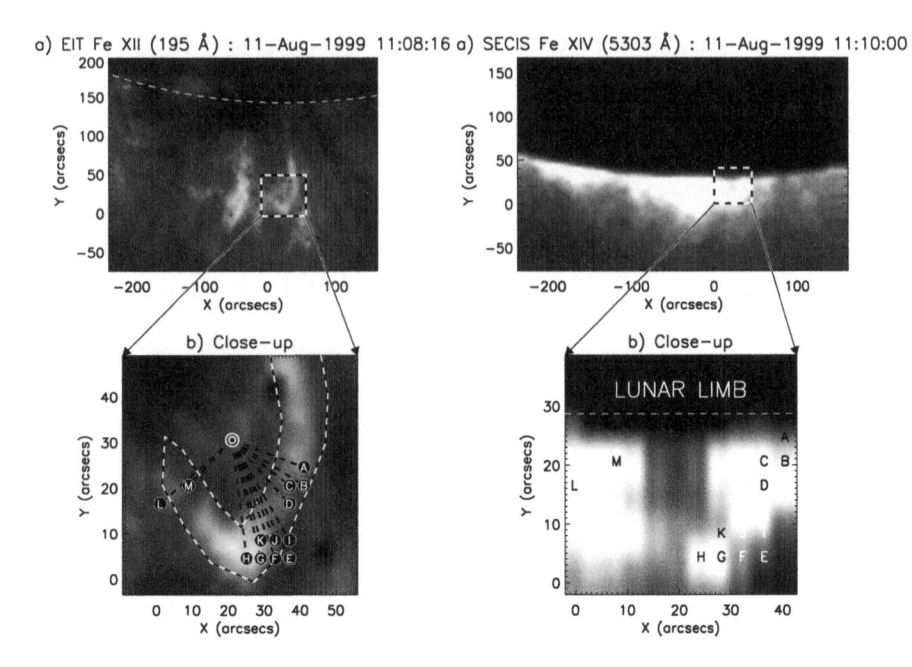

Figure 8.6: SoHO/EIT 195 Å observations (left) and *Solar Eclipse Corona Imaging System (SECIS)* observations of an active region loop during the total solar eclipse on 1999 Aug 11 (Shabla, Bulgaria), which showed propagating waves along the loop with rapid oscillations ($P = 6$ s). The SECIS image is taken in Fe XIV 5303 Å, is averaged over 50 consecutive frames (1.1 s), and is contrast-enhanced. The loop is enlarged and loop positions A−M are marked in 3×3 macropixels with a scale of $4.07''$, while the time profiles at positions A−M are shown in Fig. 8.7 (Williams et al. 2002).

oceanography. Here we summarize just some salient features as described in Roberts et al. (1984).

Let us assume that an impulsive disturbance, in the form of a magnetic field fluctuation $\mathbf{B}(z,t) = \mathbf{B}_0(z) + \mathbf{B}_1(z = z_0, t)$, launches an Alfvén wave near the footpoint of a coronal loop. As we learned in § 7.1.2, a surface wave at the boundary between the overdense loop and the less dense coronal environment will then propagate along the loop (in an upward direction), with a phase speed $v_{ph} = \omega(k)/k$ that depends on the wave number k of the disturbance, which could be a broadband spectrum and excite the whole range of Alfvén velocities $v_A \leq v_{ph} = \omega/k \leq v_{Ae}$. Let us watch the response of the loop plasma at some height $z = h$. The first signal that arrives at a height $z = h$ is that with the fastest phase speed, which is the external Alfvén speed v_{Ae}, having a frequency of $\omega_c = k_c v_{Ae}$, arriving at time $t_1 = h/v_{Ae}$. This is the start time of local periodic oscillations with frequency ω_c. After that, waves with slower phase speeds arrive, down to a minimum speed $v_{ph} = v_A$ after time $t_2 = h/v_A$. This time interval ($t_1 < t < t_2$) is called *periodic phase* (Fig. 8.4) , during which the oscillation amplitude steadily grows. However, there is a Fourier spectrum of wave frequencies ω, but the key for the understanding of the evolution is the group speed, $c_g = \partial\omega(k)/\partial k$,

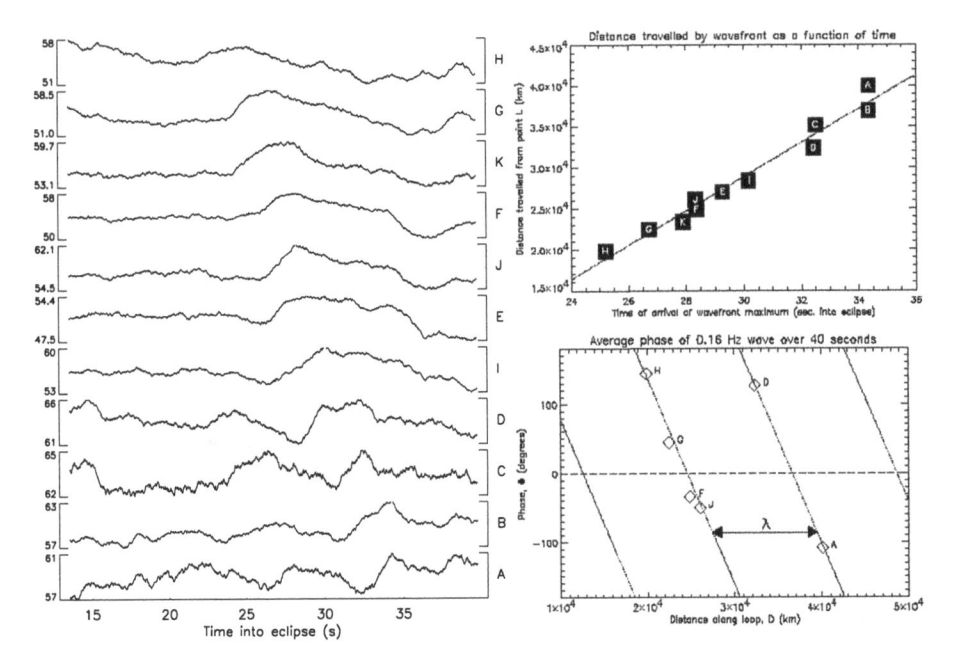

Figure 8.7: Time profiles of the intensity (left) at position A−M of the SECIS image shown in Fig. 8.6. The mean period in each time profile is $P \approx 6$ s, but the peaks shift as a function of position, indicating a propagating wave. The distance traveled by the wave maximum along the positions A−H is shown in the top right diagram, where the slope indicates a velocity of v = 2100 km s^{-1}. The average phase as a function of the distance along the loop is shown in the bottom right diagram, yielding a wavelength of $\lambda = 12$ Mm (Williams et al. 2002).

the observed speed with which the signal of the disturbance is propagating. This group speed $v_g = \partial\omega(k)/\partial k$ has a minimum value c_g^{min} at some wave vector k, as shown in Fig. 8.5, which will arrive at time $t_3 = h/c_g^{min}$ at height $z = h$. The time interval $t_2 < t < t_3$ is called the *quasi-periodic phase* (Fig. 8.4). After time t_3 the amplitude of the disturbance will decline, a phase called the *decay* (or *Airy*) phase (Fig. 8.4). These various phases of an impulsively generated fast wave have actually been observed in oceanography (Pekeris 1948). Numerical simulations of the initial stage confirm this evolutionary scenario (Murawski & Roberts 1993; 1994; Murawski et al. 1998).

The interpretation of solar observations in terms of this evolutionary scenario of fast-mode MHD waves is not trivial. Roberts et al. (1983; 1984) emphasize that the cutoff frequency ω_c and the frequency ω_{min} of the minimum group velocity c_g^{min} are the most relevant time scales to be observed and associate the periods ($P \approx 0.5 - 3.0$ s) observed in radio wavelengths to this mode of (impulsively generated) propagating fast-mode MHD waves. Propagating fast-mode MHD waves imply that a magnetic field disturbance travels at Alfvén speeds. If it modulates gyrosynchrotron emission, the corresponding radio emission should show a frequency-time drift of some ripple in the gyrosynchrotron spectrum, which perhaps has been observed in the form of a

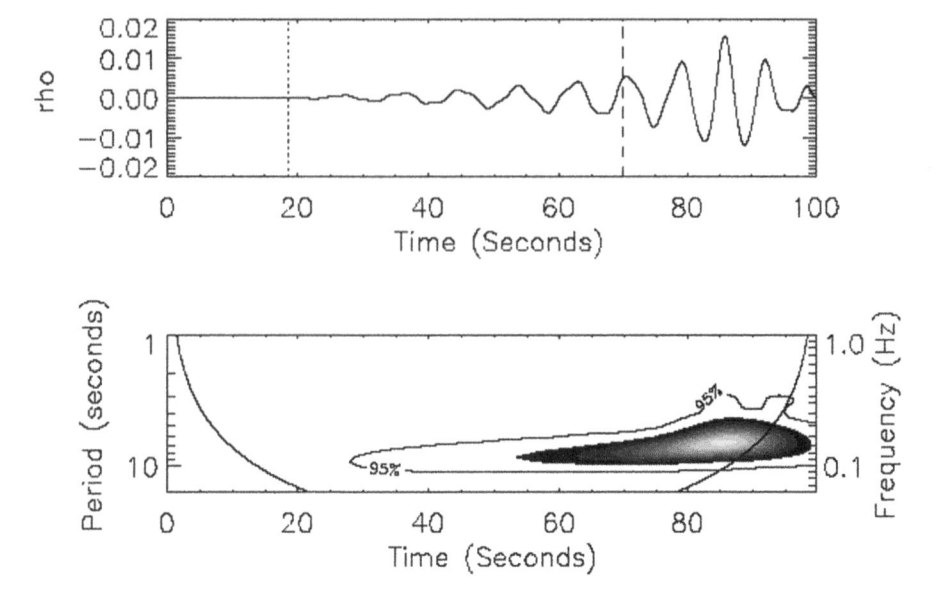

Figure 8.8: Numerical simulation of an impulsively generated fast-mode MHD wave train propagating along a corona loop with a density contrast of $\rho_0/\rho_e = 14$. *Top:* the time series is recorded at location $z = 70w$ (with w the semi-width of the loop). *Bottom:* wavelet transform analysis of the signal, exhibiting a "tadpole" wavelet signature similar to the observations shown in Fig. 8.9 (Nakariakov et al. 2003b).

quasi-periodic fine structure called *fiber bursts* (Rosenberg 1972; Bernold 1980; Slottje 1981). Most of the fast oscillation events, however, have been observed in metric and decimetric frequencies, where plasma emission dominates, but since Alfvén MHD waves are non-compressional (in contrast to the slow-mode acoustic waves), is not clear how they would modulate the plasma emission, which is only a function of the local electron density. Another problem is, even if fast-mode MHD waves modulate plasma emission, that the average density, and thus the total flux, integrated over a loop oscillating in the sausage mode would be conserved, and could not be perceived as an intensity modulation by non-imaging radio instruments, as long as they do not spatially resolve a sausage node (with spatial scale $\lambda = 2\pi/k$).

The first imaging observations that have been interpreted in terms of propagating fast-mode MHD waves (Nakariakov et al. 2003b) are the SECIS eclipse observations of Williams et al. (2001, 2002). During this eclipse, a loop has been observed with propagating wave trains in intensity, with a period of $P \approx 6$ s and a propagation speed of v ≈ 2100 km s^{-1} (Figs. 8.6 and 8.7). The evolution of the propagating fast-mode MHD oscillation has been modeled with a numeric MHD code by Nakariakov et al. (2003b), which confirmed the formation of quasi-periodic wave trains predicted by Roberts et al. (1983, 1984) and Nakariakov & Roberts (1995). The evolution of the loop density as a function of time and oscillation periods, $\rho(t, P)$ is displayed in the

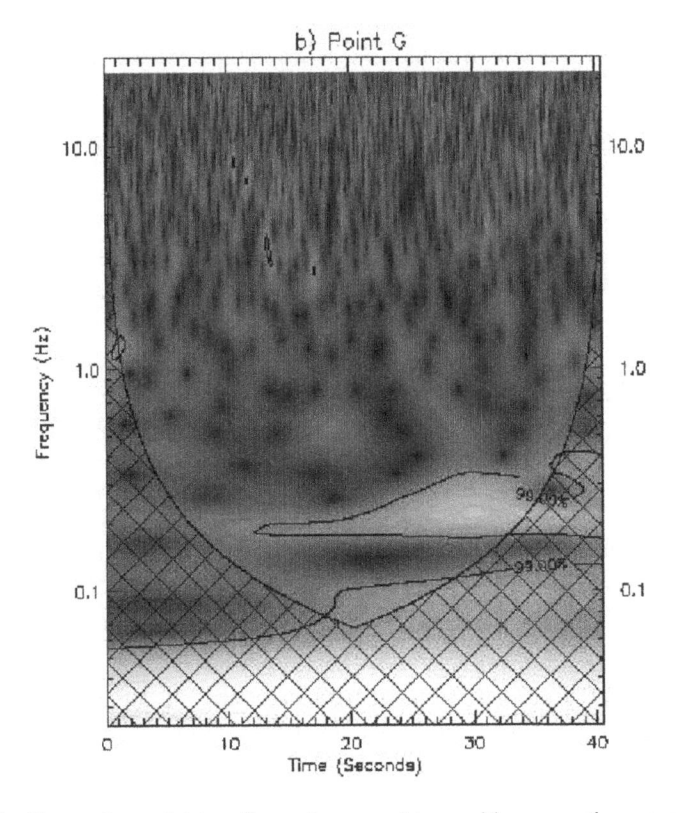

Figure 8.9: Observed wavelet transform of a coronal loop with propagating waves during the eclipse on 1999-Aug-11 observed with SECIS. Note the "tadpole" signature at $P = 1/0.16$ Hz=6 s (Katsiyannis et al. 2003).

form of a wavelet transform in Fig. 8.8, which exhibits at the dominant period $P \approx 6$ s a "tadpole" feature that is also observed by SECIS (Fig. 8.9). The SECIS observations were made with $4''$ pixels ($\approx 8''$ resolution) and averaged over 1.1 s (Katsiyannis et al. 2003). There are no detections of fast-mode MHD waves in coronal loops reported from SoHO/EIT or TRACE, probably because they are rarely operated at their highest possible cadence of seconds. We expect that more detections of fast-mode MHD waves will be accomplished with instruments of comparable spatial resolution and cadence in the future.

8.2   Propagating MHD Waves in the Open Corona

8.2.1   Evolutionary Equation of MHD Waves in Radial Geometry

While closed coronal structures have two boundaries, which control the energy balance and provide fixed nodes for standing waves, open field structures have only a single boundary where waves propagate in one direction without ever being reflected. An-

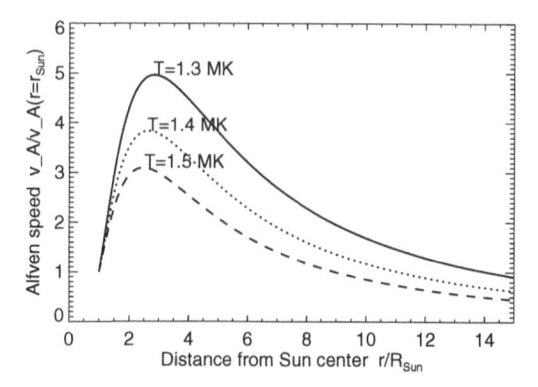

Figure 8.10: Alfvén speed as a function of radial distance from Sun center, computed for
isothermal ($T_e = 1.3, 1.4, 1.5$ MK) open-field structures with radial geometry and in hydrostatic
equilibrium. Note that the Alfvén speed peaks at a few solar radii.

other significant difference is the radial divergence of the open magnetic field (Fig. 1.14),

$$B_0(r) = B_0(R_\odot)\frac{R_\odot^2}{r^2} , \qquad (8.2.1)$$

which can often be neglected in closed field structures. In an isothermal [$T(r) =$
const] plasma in hydrostatic equilibrium, the density follows the same radial dependence as the pressure (i.e., $p(r) = 2n_e(r)k_BT$, Eq. 3.1.9), and thus has the radial
dependence (using Eq. 3.1.15 and $r = R_\odot + h$),

$$\rho(r) = \rho_0(R_\odot)\exp\left[-\frac{R_\odot}{\lambda_p}\left(1 - \frac{R_\odot}{r}\right)\right], \qquad (8.2.2)$$

with λ_p the pressure scale height for a given temperature (Eq. 3.1.16). Note that the radial divergence has no effect on the pressure scale height (see also hydrostatic analogy
of water vessels in Fig. 3.12). Combining Eqs. (8.2.1) and (8.2.2) yields the variation
of the Alfvén speed $v_A(r)$ as a function of the radial distance r from the Sun (shown
in Fig. 8.10),

$$v_A(r) = \frac{B_0(R_\odot)}{[4\pi\rho_0(R_\odot)]^{1/2}}\frac{R_\odot^2}{r^2}\exp\left[\frac{R_\odot}{2\lambda_p}\left(1 - \frac{R_\odot}{r}\right)\right]. \qquad (8.2.3)$$

In this approximation of the open magnetic field with a radial unipolar geometry, the
Alfvén speed reaches a maximum at a distance of a few solar radii, while a semi-
circular dipolar geometry yields a minimum in the lower corona (Fig. 5.10).

To study the propagation of magneto-acoustic waves in an open field structure with
radial geometry, it is useful to transform the ideal MHD equations (§ 6.1.3) into spherical coordinates (r, θ, φ) and to choose the direction $\theta = 0$. For purely radial propagation, the ideal MHD equation in spherical coordinates can then be simplified to two
uncoupled (linearized) wave equations, of which one describes Alfvén waves, characterized by magnetic perturbations B_φ and v_φ (e.g. Nakariakov et al. 2000b; Ofman et

al. 2000b; Ofman & Davila 1998; Roberts & Nakariakov 2003),

$$\frac{\partial^2 v_\varphi}{\partial t^2} - \frac{B_0(r)}{4\pi\rho_0(r)r}\frac{\partial^2}{\partial r^2}[rB_0(r)v_\varphi] = 0 \,, \tag{8.2.4}$$

and the other describes slow-mode (acoustic) waves, characterized with density perturbations ρ and v_r,

$$\frac{\partial^2 \rho}{\partial t^2} - \frac{c_s^2}{r^2}\frac{\partial}{\partial r}\left(r^2\frac{\partial\rho}{\partial r}\right) - g(r)\frac{\partial\rho}{\partial r} = 0 \,. \tag{8.2.5}$$

The right-hand side of these two equations is zero here because all dissipative effects (such as viscosity) are neglected. It is convenient to solve these two wave equations in the *Wentzel−Kramers−Brillouin (WKB)* approximation [i.e., assuming that the wavelength is much smaller than the scale of density variation of the medium ($\lambda \ll \lambda_p$)], as well as using the approximation $\lambda_p \ll R_\odot$.

In the reference frame of an upward moving Alfvén wave with local speed $v_A(r)$, the transformed time variable is,

$$\tau = t - \int \frac{dr}{v_A(r)} \,. \tag{8.2.6}$$

The wave equation for Alfvén waves can then be written in the WKB approximation with the variable $R = r(\lambda/\lambda_p) \ll r$,

$$\frac{dv_\varphi}{dR} - \frac{R_\odot^2}{4\lambda_p R^2}v_\varphi = 0 \,, \tag{8.2.7}$$

which is the linearized evolutionary equation for an Alfvén wave with solution (Nakariakov et al. 2000b),

$$v_\varphi(r) = v_\varphi(R_\odot) \exp\left[\frac{R_\odot}{4\lambda_p}(1 - \frac{R_\odot}{r})\right] \,, \tag{8.2.8}$$

which indicates an Alfvén wave amplitude that is growing with height. This has the implication that Alfvén waves can propagate large distances and deposit energy and momentum several radii away from the Sun. The growth of Alfvén waves with height has also the consequence that nonlinear effects come into play, for instance wave energy transfer of higher harmonics to shorter wavelengths, where dissipation by viscosity matters (Hollweg 1971). When the wave amplitude grows, compressional waves will be driven by Alfvén waves (Ofman & Davila 1997, 1998). Such dissipative effects, which have been neglected in the simplified wave equations [i.e., the right-hand side of Eqs. (8.2.4) and (8.2.5) are set to zero], have been included for weak nonlinearity and viscosity ν_{visc} by Nakariakov et al. (2000b), leading to a more general wave equation that is the spherical scalar form of the *Cohen−Kulsrud−Burgers equation*,

$$\frac{\partial v_\varphi}{\partial R} - \frac{R_\odot^2}{4\lambda_p R^2}v_\varphi - \frac{1}{4v_A(v_A^2 - c_s^2)}\frac{\partial v_\varphi^3}{\partial\tau} - \frac{\nu_{visc}}{2v_A^3}\frac{\partial^2 v_\varphi}{\partial\tau^2} = 0. \tag{8.2.9}$$

An example of a typical evolution of an initially harmonic Alfvén wave during its propagation in an open radial magnetic field is shown in Fig. 8.11, showing three phases

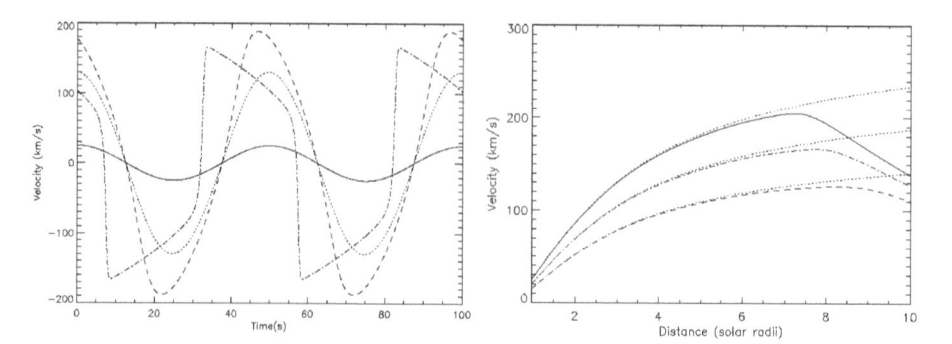

Figure 8.11: *Left:* evolution of weakly nonlinear Alfvén waves propagating in a radial magnetic field away from the Sun. The Alfvén waves have an initial speed of $v_A(R_\odot = 1000 \text{ km s}^{-1})$, an initial wave period of 50 s, an amplitude of $v_\varphi = 25 \text{ km s}^{-1}$, and the corona has an isothermal temperature of $T = 1.4$ MK. The evolution is shown near the Sun (solid line), at $r = 2R_\odot$ (dotted line), at $r = 5R_\odot$ (dashed line), and at $r = 9R_\odot$ dotted-dashed line. *Right:* dependence of the nonlinear spherical Alfvén wave amplitude on the distance from the Sun, for 3 different initial amplitudes: $v_\varphi = 25, 20$, and 15 km s^{-1} (solid, dotted-dashed, dashed) (Nakariakov et al. 2000b).

of nonlinear evolution: (1) linear wave growth, (2) saturation and overturn, and (3) nonlinear dissipation. The theoretically predicted growth rate of Alfvén waves can be tested with observations of the evolution of line broadening as a function of height above coronal holes, assuming that the line broadening is associated with transverse motions caused by Alfvén waves.

8.2.2   Observations of Acoustic Waves in Open Corona

Probably the first detection of propagating MHD waves in (open) coronal structures was made with SoHO/EIT in 1996. Plotting the EUV brightness of polar plumes (Fig. 8.12 top) as a function of time (Fig. 8.12, bottom), using the EIT 171 Å wavelength, propagating features were noticed which had an outward speed of $v \approx 75 - 150$ km s^{-1} and occurred quasi-periodically with periods of $P \approx 10 - 15$ min (DeForest & Gurman 1998). Based on the speed, which is close to the sound speed expected in this temperature band ($T \approx 1.0$ MK, $c_s = 147$ km s^{-1}), and the density modulation inferred from the EUV brightness variation, it was concluded that these wave trains in plumes correspond to propagating slow-mode magneto-acoustic waves, which are compressive waves. The energy flux associated with these wave trains was estimated to $d\varepsilon_{wave}/dt = (1.5 - 4.0) \times 10^5$ erg cm^{-2} s^{-1}, which is comparable to the heating requirement of coronal holes. The evolution of these slow-mode magneto-acoustic waves can be modeled with the theoretical wave equation (8.2.5), derived (with neglect of dissipative effects) for a radially diverging geometry, as appropriate for the these observed wave trains in polar plumes. Ofman et al. (1999) performed a numerical 2D MHD simulation of the evolution of slow-mode magneto-acoustic waves in plumes, found that the waves experience nonlinear dissipation, and concluded that they signif-

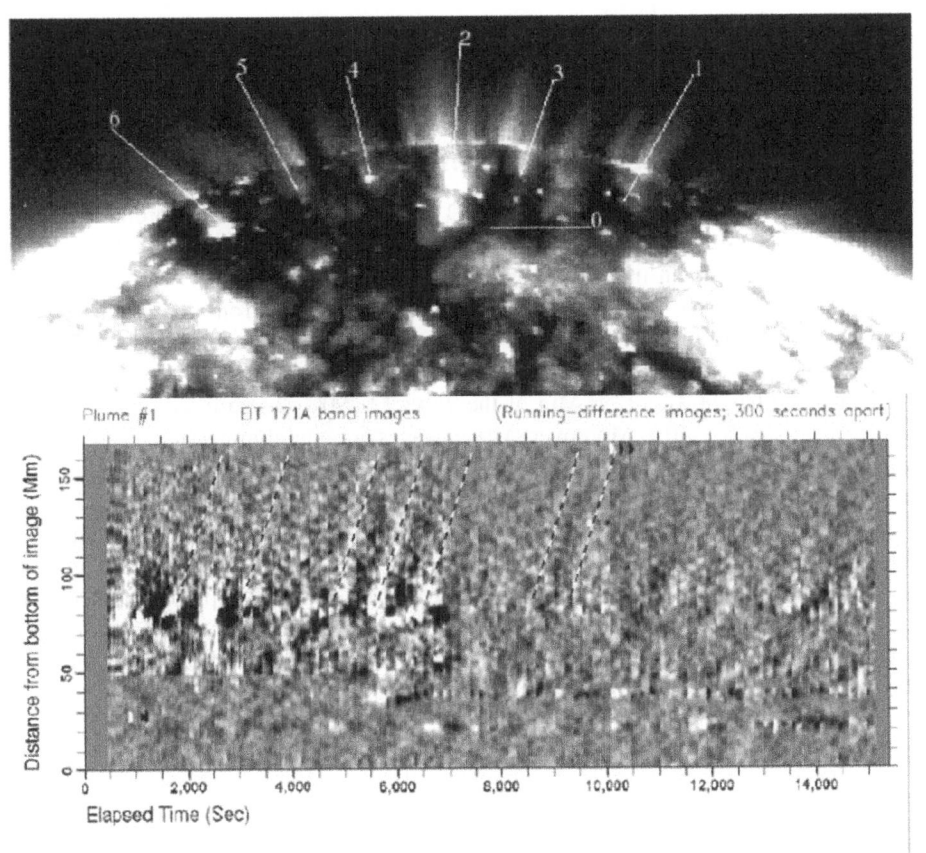

Figure 8.12: *Top:* polar plumes observed over the South Pole of the Sun with SoHO/EIT on 1996 Mar 7 at a wavelength of 171 Å, after subtraction of the radial background model. *Bottom:* running time difference images of plumes #1 and #5, with strips averaged over 360 s. Diagonal features have velocities of v ≈ 100 km s^{-1} (DeForest & Gurman 1998).

icantly contribute to the heating of the lower corona by compressive viscosity. This dissipation mechanism leads to damping of the waves within the first solar radii above the surface (Ofman et al. 2000b). Cuntz & Suess (2001) modeled slow-mode magneto-acoustic waves in plumes with a "basal-spreading" geometry and found that shocks form as a consequence at low coronal heights ($r \lesssim 1.3 R_\odot$), in contrast to models that assume weak nonlinearity.

Further away from the Sun, a search for slow-mode compressional MHD waves was carried out with the SoHO/UVCS white-light channel (Ofman et al. 1997, 2000a). Within a heliocentric distance of $r = 1.9 - 2.45 \ R_\odot$, Fourier power spectra of *polarized brightness* time series revealed significant power at a period of $P \approx 6$ min (Ofman et al. 1997). A wavelet analysis of the same and additional UVCS data confirmed periods in the range of $P \approx 6 - 10$ min, with coherence times of the fluctuations over $\Delta t \approx 30$ min. Banerjee et al. (2001) found long-period oscillations in inter-plume regions with periods of $P \approx 20 - 50$ min up to a height $h \lesssim 20$ Mm above the limb, and interprets them also as slow-mode (acoustic) waves. These observations corroborate the presence of compressional waves high above the limb, which are probably the continuation of the slow-mode magneto-acoustic waves detected in plumes with EIT.

8.2.3 Spectral Observations of Alfvén Waves in the Open Corona

After we have discovered slow-mode MHD waves in the open-field corona (e.g., in plumes, § 8.2.2), the question arises whether there also exist fast-mode MHD waves, which could provide an interesting probe for high-frequency-driven heating and acceleration of the solar wind. So far there is no direct report from imaging observations, probably because of the high time cadence and high-density contrast needed. Vertical Alfvén waves with a speed of $v_A = 1000 - 10,000$ km s^{-1} would cross a vertical scale height $\lambda_n \approx 50$ Mm of the $T \approx 1.0$ MK plasma in coronal holes in $\Delta t = \lambda_n/v_A = 5 - 50$ s. Moreover, Alfvén waves are non-compressional and do not modulate the plasma density, in contrast to slow-mode (acoustic) waves, while fast-mode MHD waves behave somewhere inbetween, but generally modulate the plasma density to a lesser degree than acoustic waves. On the other hand, both compressive magneto-acoustic (slow mode) and incompressive (fast-mode) Alfvén waves perturb the plasma velocity (v_1), which causes positive and negative Doppler shifts that can be detected as line broadening. If the distribution of plasma velocity perturbations is random, it broadens the natural line width in quadrature, so that the broadened line can be fitted by an effective temperature T_{eff},

$$T_{eff} = T_i + \frac{m_i}{2k_B} < \Delta v^2 > , \qquad (8.2.10)$$

where T_i is the temperature of line formation for an ion i, and $< \Delta v^2 >$ is the average line-of-sight component of the unresolved perturbation velocities (e.g., caused by Alfvén waves).

If the line broadening Δv is caused by Alfvén waves, the theory predicts a correlation between the Alfvén velocity disturbance $\Delta v(r) = v_\varphi(r)$ (Eq. 8.2.8) and the mean density $\rho(r) = mn \approx m_i n_i(r)$ (Eq. 8.2.2), which according to the evolutionary equation in radial geometry derived in § 8.2.1 is

$$\Delta v(r) = v_\varphi(r) \propto \rho_0^{-1/4}(r) \propto n_e^{-1/4}(r) . \qquad (8.2.11)$$

Nonthermal broadening of UV and EUV coronal lines have been measured with *Skylab*, where nonthermal velocities of $\Delta v \approx 20$ km s^{-1} were reported in coronal holes and quiet Sun regions (Doschek & Feldman 1977). More recent measurements with SoHO/SUMER (for a review see, e.g., Spadaro 1999) reveal that the nonthermal velocity increases systematically with the altitude above the limb (e.g., from $\Delta v = 24$ km s^{-1} at the limb to $\Delta v = 28$ km s^{-1} at a height of $h = 25$ Mm; Doyle et al. 1998), corresponding to a velocity increase that is consistent with the theoretical prediction of undamped radially propagating Alfvén waves (i.e., $[n_e(h_2)/n_e(h_1)]^{-1/4} \approx [\exp(-h/\lambda_T)]^{-1/4} = \exp(+h/4\lambda_T) \approx \exp(1/8) = 1.13$, $[(\Delta v(h_2)/\Delta v(h_1)] = 28/24 = 1.17$). Erdélyi et al. (1998b) detected a similar Alfvén scaling in the center-to-limb variation of the line broadening in transition region lines. Banerjee et al. (1998)

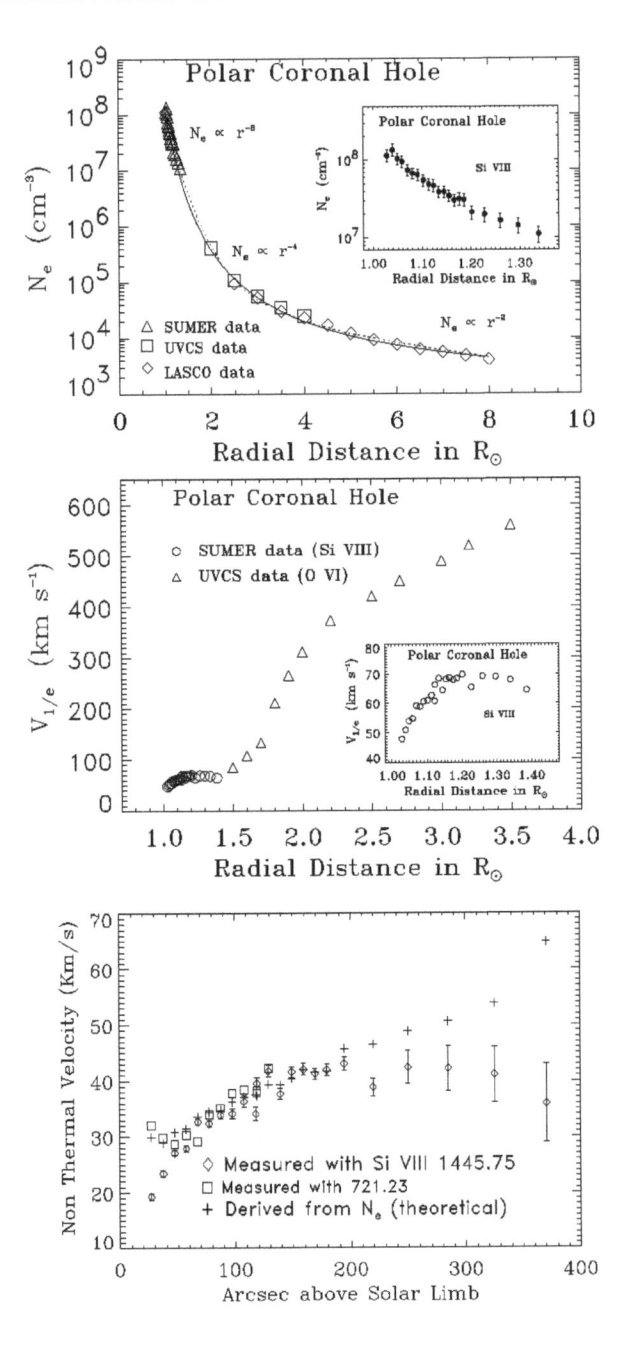

Figure 8.13: *Top:* electron density profile $n_e(h)$ above a coronal hole measured with SoHO/SUMER, UVCS, and LASCO. *Middle:* nonthermal line widths Δv measured with SUMER and UVCS. *Bottom:* comparison of measured nonthermal velocities Δv and predicted line widths from the density model $n_e(h)$ and the relation for Alfvén waves in radially diverging magnetic fields, $\Delta v(h) \propto n_e^{-1/4}(r)$ (Eq. 8.2.11) (Doyle et al. 1999).

confirmed the results from Doyle et al. (1998) over a larger height range, finding an increase of the nonthermal velocity of the Si VIII line from $\Delta v(h_1) = 27$ km s^{-1} at $h_1 = 20$ Mm to $\Delta v(h_2) = 46$ km s^{-1} at $h_2 = 180$ Mm, over which range the density decreased from $n_e(h_1) = 1.1 \times 10^8$ cm^{-3} to $n_e(h_2) = 1.6 \times 10^7$ cm^{-3}; so the observed velocity increase $\Delta v(h_2)/\Delta v(h_1) = 46/27 = 1.70$ agrees well with the theoretical prediction $[n_e(h_2)/n_e(h_1)]^{-1/4} = (0.16/1.1)^{-1/4} = 1.62$. Similar nonthermal velocities were also measured by Chae et al. (1998a) with SoHO/SUMER, by Esser et al. (1999) with SoHO/UVCS (nonthermal velocity widths of $20 - 23$ km s^{-1} at $r = 1.35 - 2.1 R_\odot$), and Doschek et al. (2001) with SoHO/SUMER. Combining the Si VIII with O VI line width measurements, Doyle et al. (1999) found that the Alfvén scaling (Eq. 8.2.11) agrees well only in the height range of $h = 30 - 150$ Mm ($r = 1.04 - 1.2 R_\odot$), suggesting nonlinear evolution of the Alfvén waves at $r \gtrsim 1.2 R_\odot$ (see Fig. 8.13, bottom). Taking all these spectroscopic measurements together, there seems to be strong support for the presence of fast-mode or shear (Alfvén) MHD waves in the open field structures of the solar corona. We will discuss the relevance for coronal heating in § 9.

8.3   Global Waves

So far we have considered MHD waves that propagated inside waveguides, either in coronal loops (§ 8.1) or along vertically open structures with radial divergence (§ 8.2). However, waves have also been discovered that propagate spherically over the entire solar surface, very much like the spherical water waves you produce when you throw a stone in a pond. Obviously, the origin of these spherical waves is very localized, caused by a flare or a *coronal mass ejection (CME)* at the center of the circular waves. These global waves were first discovered in chromospheric Hα emission (called *Moreton waves*) and were recently in coronal EUV images from SoHO/EIT (called *EIT waves*). The big challenge is the physical understanding of the 3D propagation of these global waves in the complex topology of our corona, which is structured by vertical stratification, horizontal inhomogeneities, and magnetic instabilities during CMEs.

8.3.1   Moreton Waves, EIT Waves, and CME Dimming

The discovery of global waves goes back to Moreton & Ramsey (1960), who reported the finding of 7 flare events (out of 4068 flares photographed in Hα during 1959/1960) with disturbances that propagated through the solar atmosphere over distances of the order of 500,000 km at speeds of v ≈ 1000 km s^{-1}. More such reports noted expanding arc features originating in flares and traveling distances of 200,000 km or more with lateral velocities of v $\approx 500 - 2500$ km s^{-1} (Moreton 1961; Athay & Moreton 1961; Moreton 1964; Harvey et al. 1974), or v $\approx 330 - 4200$ km s^{-1} (Smith & Harvey 1971). Reviews on early Hα observations of this type of flare waves can be found in Svestka (1976, § 4.3) and Zirin (1988; § 11). Recent observations of a Moreton wave in Hα and Hβ even revealed a velocity increase from v $= 2500$ km s^{-1} to 4000 km s^{-1} (Zhang 2001). Today it is believed, based on the high propagation speeds which are in

1997 APRIL 7   .   1997 MAY 12

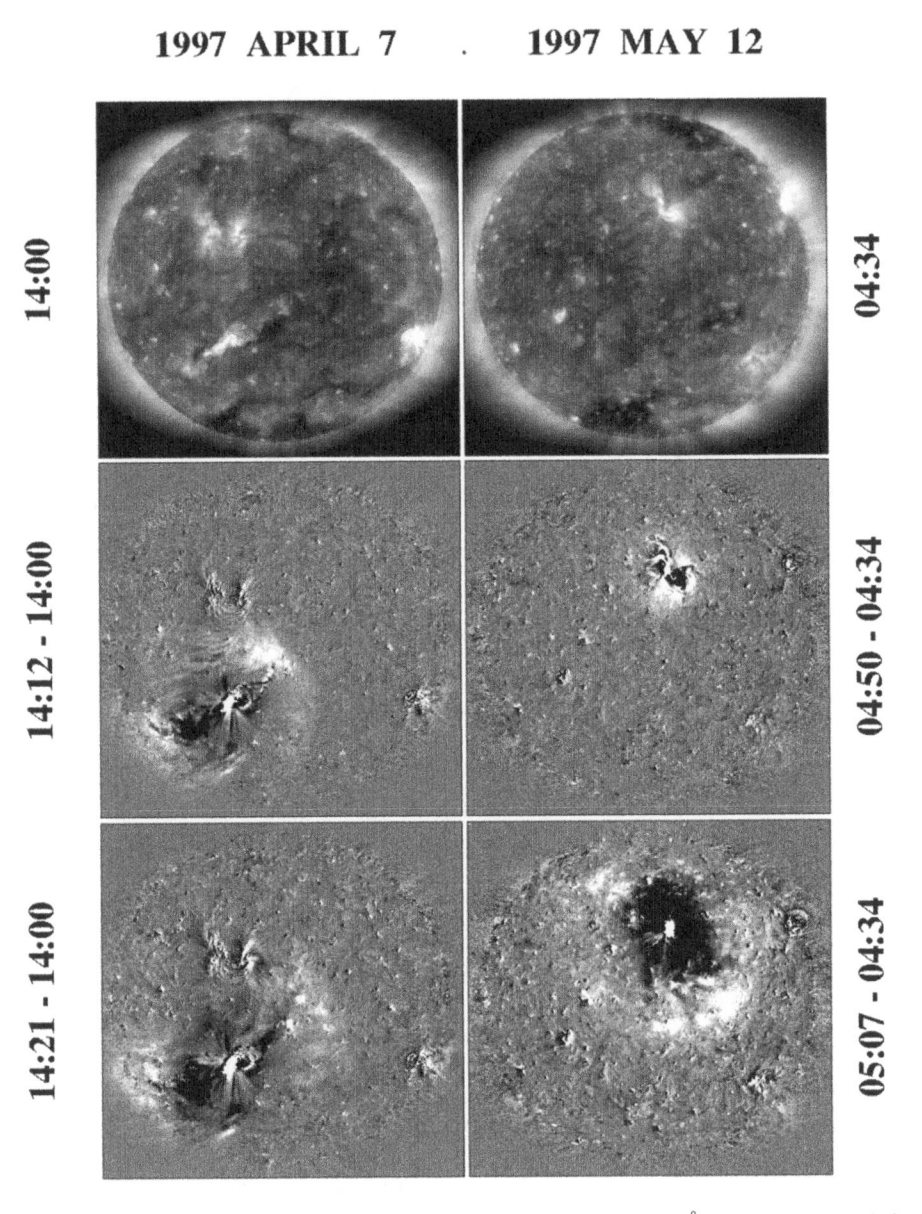

Figure 8.14: Two global wave events observed with SoHO/EIT 195 Å, on 1997 Apr 7 (left) and 1997 May 12 (right). The intensity images (top) were recorded before the eruption, while the difference images (middle and bottom) show differences between the subsequent images, enhancing emission measure increases (white areas) and dimming (black areas) (Wang 2000).

the range of coronal Alfvén speeds, that the phenomenon of Moreton waves represent a tracer of a coronal disturbance, rather than a chromospheric origin (Thompson 2001).

Recent observations by SoHO/EIT (Fig. 8.14) have provided unambiguous evi-

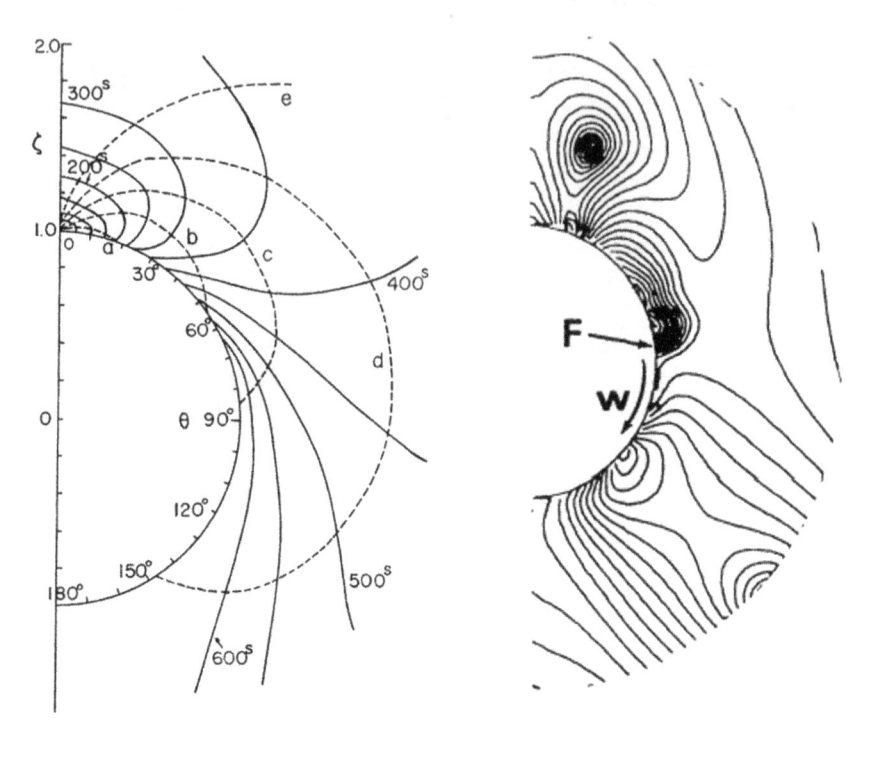

Figure 8.15: *Left:* propagation model of a Moreton wave (Uchida 1968); *Right:* iso-Alfvén speed contours calculated for a coronal portion through which a Moreton wave propagates (Uchida et al. 1973).

dence for global waves, initiated by flares and CMEs. One of the first events was observed during the Earth-directed CME of 1997-May-12, which was characterized as a bright wavefront propagating quasi-radially from the source region, leaving a dimmed region behind, and having a radial speed of v $= 245 \pm 40$ km s^{-1} (Thompson et al. 1998a). More observations of such global waves followed from SoHO/EIT (Thompson et al. 1999; 2000a; Klassen et al. 2000; Biesecker et al. 2002). The catalog of 19 EIT wave events compiled by Klassen et al. (2000) investigated the correlation of radio type II events with EIT waves. Radio type II bursts are believed to trace coronal shock waves and were found to have speeds of $v_{II} \approx 300 - 1200$ km s^{-1}, much faster than the EIT waves which were found to have speeds of $v_{EIT} \approx 170 - 350$ km s^{-1}. Biesecker et al. (2002) investigated correlations between 175 EIT wave events and associated phenomena (CMEs, flares, and radio type II bursts). Wills−Davey & Thompson (1999) observed a global wave with a high spatial resolution using TRACE 195 Å and traced the detailed trajectories of the propagating wave fronts, finding anisotropic deviations from radial propagation and speed variations from v ≈ 200 km s^{-1} to 800 km s^{-1}, clearly illustrating the inhomogeneity of the coronal medium. Two cases of global waves have been analyzed where the wave front of Moreton waves in Hα and

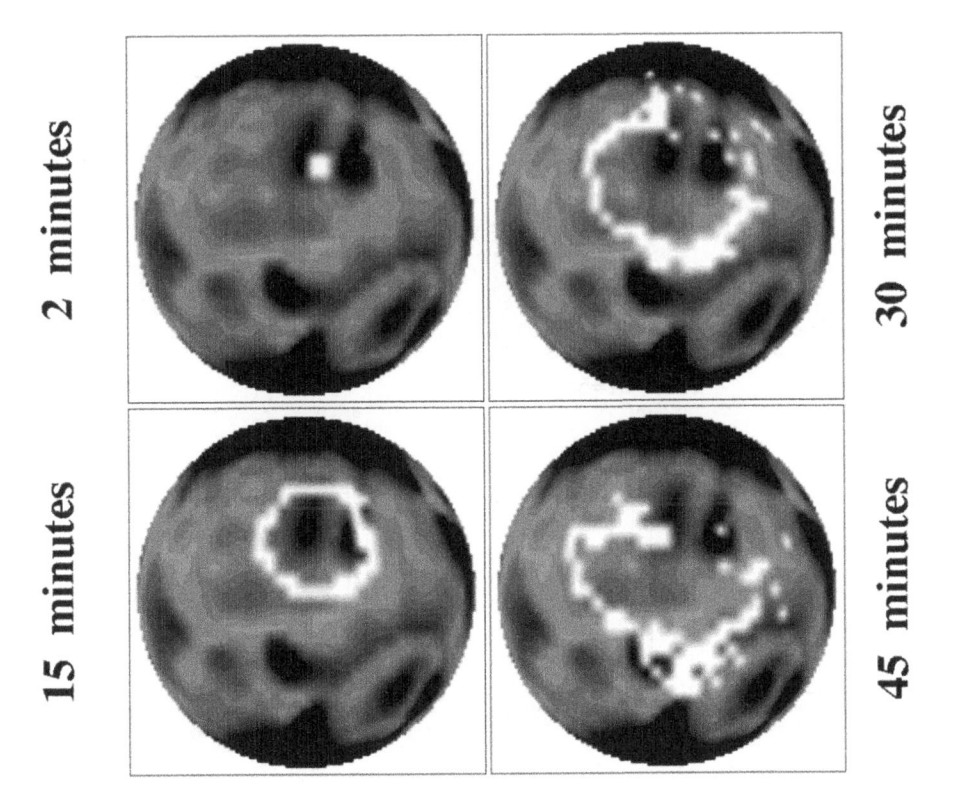

Figure 8.16: Simulation of the EIT wave event of 1997 April 7 (Fig. 8.14) by a ray-tracing method of fast-mode MHD waves. The color range indicates wave speeds v > 500 km s^{-1} (black) and lower speeds (white). Gaps appear in the wave fronts after t > 45 min when waves become reflected back into the chromosphere (Wang 2000).

EIT waves were found to be co-spatial, both experiencing a subsequent deceleration, which was interpreted in terms of a fast-mode shock ("blast wave") scenario (Warmuth et al. 2001), rather than in terms of CME-associated magnetic field adjustment.

An intriguing feature of global waves seen with EIT is the dimming region (e.g., Thompson et al. 2000b), which the wave front leaves behind (see, e.g., Fig. 8.14). If the global wave would be just a compressional wave front, a density enhancement would occur at the front and a rarefaction slightly behind, while the density would be restored in the trail of the wake. The fact that a long-term dimming occurs behind the global waves indicates that material has been permanently removed behind the wave front, probably due to the vertical expulsion of the accompanying CME. This scenario is strongly supported by recent Doppler shift measurements in O V and He I, indicating vertical velocities of v = 100 km s^{-1} in the dimming region that was feeding the CME (Harra & Sterling 2001, 2003).

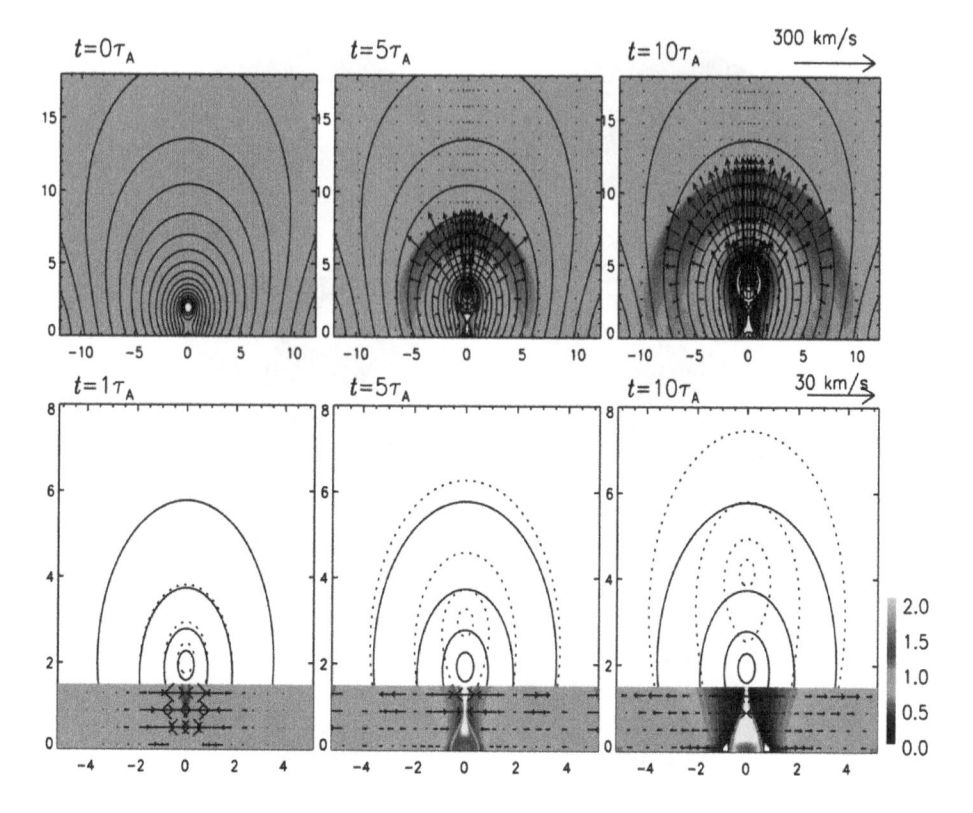

Figure 8.17: MHD simulation of a CME where a piston-driven shock forms at the envelope of the expanding CME, according to the model of Chen & Shibata 2000). The simulated case has $\beta_0 = 0.25$ and $v_{rope} = 100$ km s^{-1}. *Top:* global evolution of the density (greyscale), magnetic field (solid lines), and velocity (arrows). *Lower panel:* local evolution in the lower corona and chromosphere, where the initial magnetic field is shown with solid lines (Chen et al. 2002).

8.3.2   Modeling and Simulations of Global Waves

Global waves in the solar corona were modeled early on in terms of a spherically expanding fast-mode MHD shock wave, from which the shock front is detected as an EIT wave, while the upward propagating shock is manifested in radio type II bursts (Uchida 1974), whereas the Moreton waves seen in Hα represent the chromospheric ground tracks of the dome-shaped coronal shock front (Uchida et al. 1973). Uchida (1974) derived the wave equations for such a spherically propagating fast-mode MHD wave in a radially diverging magnetic field (similar to § 8.2.1) and calculated the wave propagation in the WKB approximation (an example of a calculation of propagating wave fronts is shown in Fig. 8.15 left). Furthermore, observed electron density distributions and magnetograms were used to constrain models of the global wave propagation (Fig. 8.15 right) and the trajectories of the accompanying radio type II bursts and Moreton waves (Uchida et al. 1973; Uchida 1974). The scenario of a flare-produced initial

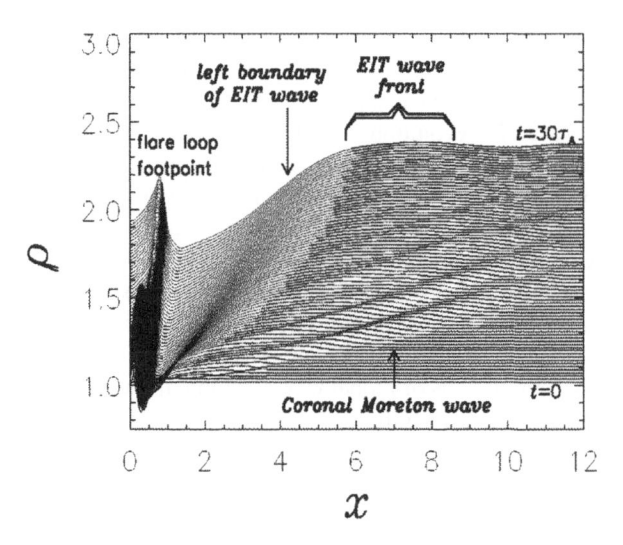

Figure 8.18: Evolution of the horizontal density $\rho(x)$ distribution obtained in the MHD simulation shown in Fig. 8.17, with an increment in time. Note the two wave features: a fast coronal Moreton wave and a slower broader EIT wave front (Chen et al. 2002).

pressure pulse that triggers a fast-mode shock propagating through the corona, the so-called *blast-wave scenario*, was further simulated with a full numerical MHD code by Steinolfson et al. (1978), the ignition of the resulting shock was modeled by Vrsnak & Lulic (2000a,b), and the formation of the expelled blobs in the coronal streamer belt were further modeled with LASCO observations by Wu et al. (2000).

A realistic numeric simulation of the EIT signature of global fast-mode MHD waves was realized by Wang (2000). Fig. 8.16 shows the result for a specific event (1997 April 7), where the global magnetic field was constrained by a photospheric magnetogram and the EUV emission by EIT 195 Å images (Fig. 8.14, left). The fast-mode MHD wave speed is defined by the dispersion relation given in Eq. (7.1.32), which has the quadratic solution (see Eq. 7.1.35 for the special case of $\theta = 0$),

$$v_{ph} = \frac{1}{2}\left[v_A^2 + c_s^2 + \sqrt{(v_A^2 + c_s^2)^2 - 4v_A^2 c_s^2 \cos^2\theta}\right] . \qquad (8.3.1)$$

Wang (2000) used the distribution of phase speeds $v_{ph}[\mathbf{B}(r,\theta,\varphi)]$ constrained by the photospheric magnetic field $\mathbf{B}(r,\theta,\varphi)$ as a lower boundary condition and calculated the propagation of fast-mode wave fronts using a ray-tracing method (Fig. 8.16), which closely ressembles the observations (Fig. 8.14, left). These simulations reproduce the initial horizontal speeds of $v \approx 300$ km s^{-1} of observed EIT waves, which are then found to decelerate to $v = 50 - 200$ km s^{-1} in weak-field regions. The speeds simulated by Wang (2000) are consistent with the observed EIT waves, but are about $2 - 3$ times lower than those simulated by Uchida (1974) for Moreton waves. This discrepancy was reconciled by a numeric MHD simulation which mimics a CME by an initial strong upward-directed external force on a fluxrope, which drives the evacuation of the fluxrope with subsequent magnetic reconnection underneath (Fig. 8.17; Chen et

al. 2002; see model of Chen & Shibata 2000). A piston-driven shock appears strad-
dling over the fluxrope, which moves upward with a super-Alfvén speed of v ≈ 360
km s^{-1}, while near the solar surface, the piston-driven shock degenerates to a finite amplitude MHD fast wave. The evolution of the horizontal density distribution is shown in
Fig. 8.18, where two wave-like features are seen: (1) a piston-driven shock wave with
a speed of v ≈ 400 km s^{-1}, which corresponds to the coronal Moreton wave, and (2)
a second wave with an initial speed of v = 115 km s^{-1}, which becomes increasingly
blurred with time and corresponds to the EIT wave. This simulation explains the fast
Moreton wave in terms of a shock wave that comes from the expanding CME, not from
the flare itself, while the EIT wave front is explained in terms of an adjustment to the
successive opening of CME field lines (Délannée & Aulanier 1999; Délannée 2000).

Other numerical simulations of global waves explore the stability of active regions
under the impact of global waves (Ofman & Thompson 2002) or the global distribution
of the coronal magnetic field (at the height of propagating EIT waves) and the coronal
viscosity (Ballai & Erdélyi 2003), a new discipline that might be called "*global coronal
seismology*".

8.4 Summary

**Propagating MHD waves have moving nodes, in contrast to standing modes with
fixed nodes. Propagating MHD waves result mainly when disturbances are generated impulsively, on time scales faster than the Alfvén or acoustic travel time
across a structure.**

**Propagating slow-mode MHD waves (with acoustic speed) have been recently
detected in coronal loops using TRACE and EIT, usually being launched with
3-minute periods near sunspots, or with 5-minute periods away from sunspots.
These acoustic waves propagate upward from a loop footpoint and are quickly
damped, never being detected in downward direction at the opposite loop side.
Propagating fast-mode MHD waves (with Alfvén speeds) have recently been discovered in a loop in optical (SECIS eclipse) data, as well as in radio images (from
Nobeyama data).**

**Besides coronal loops, slow-mode MHD waves have also been detected in plumes
in open field regions in coronal holes, while fast-mode MHD waves have not yet
been detected in open field structures. However, spectroscopic observations of line
broadening in coronal holes provide strong support for the detection of Alfvén
waves, based on the agreement with the theoretically predicted scaling between
line broadening and density, $\Delta v(h) \propto n_e(h)^{-1/4}$.**

**The largest manifestation of propagating MHD waves in the solar corona are
global waves that spherically propagate after a flare and/or CME over the entire
solar surface. These global waves were discovered earlier in Hα, called Moreton
waves, and recently in EUV, called EIT waves, usually accompanied with a coronal
dimming behind the wave front, suggesting evacuation of coronal plasma by the
CME. The speed of Moreton waves is about three times faster than that of EIT
waves, which still challenges dynamic MHD models of CMEs.**

Chapter 9

Coronal Heating

When Bengt Edlén and Walter Grotrian identified Fe IX and Ca XIV lines in the solar spectrum (Edlén 1943), a coronal temperature of $T \approx 1$ MK was inferred from the formation temperature of these highly ionized atoms, for the first time. A profound consequence of this measurement is the implication that the corona then consists of a fully ionized hydrogen plasma. Comparing this coronal temperature with the photospheric temperature of 6000 K (or down to 4800 K in sunspots), we are confronted with the puzzle of how the 200 times hotter coronal temperature can be maintained, the so-called *coronal heating problem*. Of course, there is also a *chromospheric heating problem* and a *solar wind heating problem* (Hollweg 1985). If only thermal conduction was at work, the temperature in the corona should steadily drop down from the chromospheric value with increasing distance, according to the second law of thermodynamics. Moreover, since we have radiative losses by EUV emission, the corona would just cool off in matter of hours to days, if the plasma temperature could not be maintained continuously by some heating source. In this section we will specify the energy requirement for coronal heating, review a fair number of theoretical models that provide coronal heating mechanisms, and scrutinize them with observational tests if possible. However, all we have available for observational testing are mostly measurements of basic physical parameters, such as density, temperatures, and flow speeds, while theoretical heating models are expressed in parameters that are often not directly measurable in the corona, such as the magnetic field strength, azimuthal field components, nonpotential fields, currents, resistivity, viscosity, turbulence, waves, etc. However, the detection of MHD waves in the corona by TRACE and EIT, the spectroscopic measurements of line widths by SUMER, and the ion temperature anisotropy measurements with UVCS opened up powerful new tools that promise to narrow down the number of viable coronal heating mechanisms in the near future.

Reviews on the coronal heating problem can be found in Withbroe & Noyes (1977), Hollweg (1985), Narain & Ulmschneider (1990, 1996), Ulmschneider et al. (1991), Zirker (1993), Parker (1994), Chiuderi (1996), Mandrini et al. (2000), and Heyvaerts (2001).

9.1 Heating Energy Requirement

We start to analyze the coronal heating problem by inquiring first about the energy requirements. A coronal heating source E_H has to balance at least the two major loss terms of radiative loss E_R and thermal conduction E_C, as we specified in the energy equation (3.6.2) for a hydrostatic corona,

$$E_H(\mathbf{x}) - E_R(\mathbf{x}) - E_C(\mathbf{x}) = 0 , \qquad (9.1.1)$$

where each of the terms represents an energy rate per volume and time unit (erg cm^{-3} s^{-1}), and depends on the spatial location $\mathbf{x}$. Because the corona is very inhomogeneous, the heating requirement varies by several orders of magnitude depending on the location. Because of the highly organized structuring by the magnetic field (due to the low plasma-β parameter in the corona), neighboring structures are fully isolated and can have large gradients in the heating rate requirement, while field-aligned conduction will smooth out temperature differences so that an energy balance is warranted along magnetic field lines. We can therefore specify the heating requirement for each magnetically isolated structure separately (e.g., a loop or an open fluxtube in a coronal hole), and consider only the field-aligned space coordinate s in each energy equation, as we did for the energy equation (3.6.2) of a single loop,

$$E_H(s) - E_R(s) - E_C(s) = 0 . \qquad (9.1.2)$$

Parameterizing the dependence of the heating rate on the space coordinate s with an exponential function (Eq. 3.7.2) (i.e., with a base heating rate E_{H0} and heating scale length s_H), we derived scaling laws for coronal loops in hydrostatic energy balance, which are known as *RTV laws* for the special case of uniform heating without gravity (Eqs. 3.6.14−15), and have been generalized by Serio et al. (1981) for nonuniform heating and gravity (Eqs. 3.6.16−17). It is instructional to express the RTV law as a function of the loop density n_e and loop half length L, which we obtain by inserting the pressure from the ideal gas law, $p_0 = 2 n_e k_B T_{max}$, into Eqs. (3.6.14−15),

$$T_{max} \approx 10^{-3} \, (n_e L)^{1/2} \qquad (9.1.3)$$

$$E_{H0} \approx 2 \times 10^{-17} n_e^{7/4} L^{-1/4} \qquad (9.1.4)$$

This form of the RTV law tells us that the heating rate depends most strongly on the density, $E_{H0} \propto n_e^{7/4}$, and very weakly on the loop length L. Actually, we can retrieve essentially the same scaling law using a much simpler argument, considering only radiative loss, which is essentially proportional to the squared density (Eq. 2.9.1),

$$E_{H0} \approx E_R = n_e^2 \Lambda(T) \approx 10^{-22} n_e^2 \quad (\text{erg cm}^{-3} \, \text{s}^{-1}) \qquad (9.1.5)$$

where the radiative loss function can be approximated by a constant $\Lambda(T) \approx 10^{-22}$ [erg cm^{-3} s^{-1}] in the temperature range of $T \approx 0.5 - 3$ MK that characterizes most parts of the corona. This gives us a very simple guiding rule: the coronal heating rate requirement is essentially determined by the squared density. The rule (Eq. 9.1.5) gives us the following estimates: in coronal holes the base density is typically $n_e \approx 10^8$ cm^{-3} and the heating rate requirement is thus $E_{H0} \approx 10^{-6}$ (erg cm^{-3} s^{-1}).

Table 9.1: Chromospheric and coronal energy losses, in units of (erg cm^{-2} s^{-1}) (Withbroe & Noyes 1977).

Parameter	Coronal hole	Quiet Sun	Active region
Transition layer pressure [dyn cm^{-2}]	7×10^{-2}	2×10^{-1}	2
Coronal temperature [K], at $r \approx 1.1 R_\odot$	10^6	1.5×10^6	2.5×10^6
Coronal energy losses [erg cm^{-2} s^{-1}]			
– Conductive flux F_C	6×10^4	2×10^5	$10^5 - 10^7$
– Radiative flux F_R	10^4	10^5	5×10^6
– Solar wind flux F_W	7×10^5	$\lesssim 5 \times 10^4$	$(< 10^5)$
– Total corona loss $F_C + F_R + F_W$	8×10^5	3×10^5	10^7
Chromospheric radiative losses [erg cm^{-2} s^{-1}]			
– Low chromosphere	2×10^6	2×10^6	$\gtrsim 10^7$
– Middle chromosphere	2×10^6	2×10^6	10^7
– Upper chromosphere	3×10^5	3×10^5	2×10^6
– Total chromospheric loss	4×10^6	4×10^6	2×10^7
Solar wind mass loss [g cm^{-2} s^{-1}]	2×10^{-10}	$\lesssim 2 \times 10^{-11}$	$(< 4 \times 10^{-11})$

Since the heating flux is quickly distributed along a magnetic field line, we can just specify a heating rate per unit area at the coronal base, by integrating the volume heating rate in the vertical direction. For hydrostatic structures, we can integrate the heating rate in the vertical direction simply by multiplying it with the density scale height λ_T, which is proportional to the temperature (Eq. 3.1.16). We denote the heating flux per unit area with the symbol F_{H0} (also called *Poynting flux*),

$$F_{H0} = E_{H0}\lambda_T \approx 5 \times 10^3 \left(\frac{n_e}{10^8 \text{ cm}}\right)^2 \left(\frac{T}{1 \text{ MK}}\right) \quad [\text{erg cm}^{-2} \text{ s}^{-1}] \qquad (9.1.6)$$

Thus for a coronal hole, with $n_e = 10^8$ cm^{-3} and $T = 1.0$ MK, we estimate a required heating flux of $F_{H0} = 5 \times 10^3$ erg cm^{-2} s^{-1}, and in an active region with a typical loop base density of $n_e = 2.0 \times 10^9$ cm^{-3} and $T = 2.5$ MK, we estimate $F_{H0} \approx 5 \times 10^6$ (erg cm^{-2} s^{-1}). Thus the heating rate requirement varies by about 3 orders of magnitude between the two places.

Another conclusion we can immediately draw about the heating function is that the height dependence of the heating has roughly to follow the hydrostatic equilibrium. The heating scale height s_H required in hydrostatic equilibrium is therefore half of the density scale height λ_T, because the radiative loss scales with the squared density, $E_H(h) = E_{H0}\exp{(-h/s_H)} \propto E_R(h) \propto n_e(h)^2 \approx [n_0 \exp{(-h/\lambda_T)}]^2$,

$$s_H \approx \frac{\lambda_T}{2} \approx 23 \left(\frac{T}{1 \text{ MK}}\right) \quad [\text{Mm}] . \qquad (9.1.7)$$

This simple theoretical prediction, assuming that radiative loss is the dominant loss component in the coronal part of loops, is also confirmed by hydrostatic modeling of 40 loops observed with TRACE, where including the effect of thermal conduction yielded only slightly smaller values (i.e., $s_H = 17 \pm 6$ Mm, Aschwanden et al. 2000d).

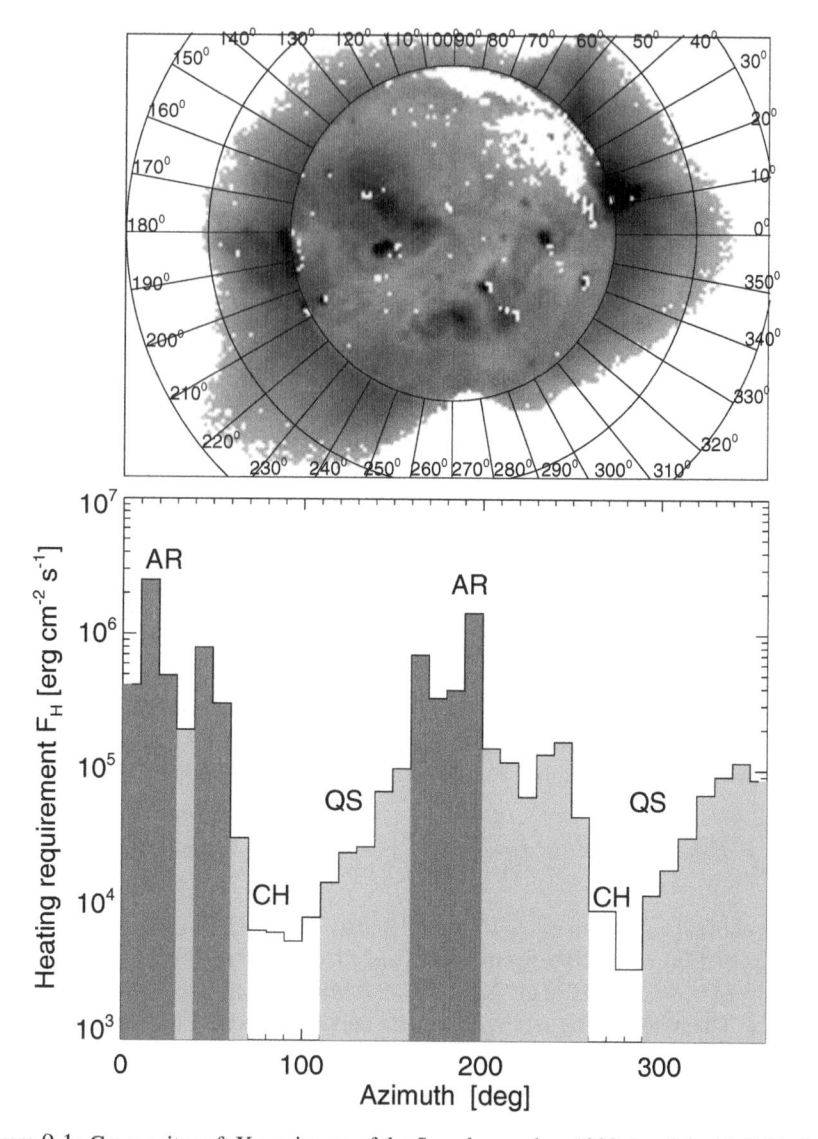

Figure 9.1: Composite soft X-ray image of the Sun observed on 1992 Aug 26 with Yohkoh (*top panel*). The histogram shows the heating rate requirement (*bottom panel*) in the 36 azimuthal sectors around the Sun. The labels indicate the locations of active regions (AR; dark grey), quiet Sun regions (QS; light grey), and coronal holes (CH; white) (Aschwanden 2001b).

The spatial variation of the coronal heating requirement is illustrated in Fig. 9.1, where we deconvolved the mean coronal base density n_{e0} and differential emission measure distribution $dEM(T)/dT$ in 36 different sectors of the corona from Yohkoh SXT two-filter measurements (see § 3.3 and Aschwanden & Acton 2001), and determined the heating requirement F_{H0} for these 36 different sectors, finding $5 \times 10^3 \lesssim F_{H0}$

$< 1 \times 10^4$ (erg cm^{-2} s^{-1}) in coronal holes, $1 \times 10^4 \lesssim F_{H0} < 2 \times 10^5$ (erg cm^{-2} s^{-1}) in quiet Sun regions, and $2 \times 10^5 \lesssim F_{H0} < 2 \times 10^6$ (erg cm^{-2} s^{-1}) in active regions. These measurements agree with the radiative losses found in other observations (e.g., Jordan 1976; Withbroe & Noyes 1977; see Table 9.1).

So, we have a quite specific perception of the heating requirement in the solar corona. The simplest rule is the dependence on the squared electron density, $F_{H0} \propto n_e^2$, which is also proportional to the optically thin emission measure in EUV and soft X-rays, and thus to the observed flux. This sounds trivial, that the heating rate is directly proportional to the observed brightness, if we associate radiation as the major loss, but it would not be true for optically thick radiation, where the observed brightness temperature is lower than the actual electron temperature. A direct consequence of the squared density dependence is that most of the heating is required in the lowest half density scale height. When we ask what the dependence of the heating rate is on the temperature, the RTV law (Eq. 3.6.15) predicts a dependence with the three-and-a-half power, $F_{H0} \propto T^{3.5}$. Thus a soft X-ray-bright loop with a typical temperature of $T = 3$ MK needs about 50 times more heating flux than an EUV-bright loop with $T = 1$ MK.

In Table 9.1 we list the energy losses in the corona and chromosphere for comparison, given separately for coronal holes, quiet Sun regions, and active regions. We see that the radiative losses are fully comparable with the conductive losses (within a factor of 2) in the quiet Sun and active regions. Only in coronal holes, radiative loss is substantially less than the losses by thermal conduction and the solar wind flux, because of the low density. So, we can summarize that the minimum heating requirement at any place on the solar surface is $P_{H0} \gtrsim 3 \times 10^5$ erg cm^{-2} s^{-1}, mostly needed in the lowest half density scale height, and the heating requirement increases up to two orders of magnitude in dense loops in active regions, roughly scaling with the squared density.

9.2 Overview of Coronal Heating Models

In Table 9.2 we categorize theoretical models of coronal heating processes into 5 groups, according to the main underlying or driving physical processes. It became customary to classify coronal heating models into *DC (Direct Current)* and *AC (Alternating Current)* types, which characterize the electromechanic coronal response to the photospheric driver that provides the ultimate energy source for heating. Magnetic disturbances propagate from the photosphere to the corona with the Alfvén speed v_A. If the photospheric driver, say random motion of magnetic field line footpoints, changes the boundary condition on time scales much longer than the Alfvén transit time along a coronal loop, the loop can adjust to the changing boundary condition in a quasi-static way, and thus the coronal currents are almost direct ones, which defines the *DC models*. On the other hand, if the photospheric driver changes faster than a coronal loop can adjust to (e.g., by damping and dissipation of incident Alfvén waves), the coronal loop sees an alternating current, which is the characteristic of *AC models*. For each of the two model groups there are a number of variants of how the currents are dissipated, either by Ohmic dissipation, magnetic reconnection, current cascading, and viscous turbulence in the case of DC models, or by Alfvénic resonance, resonance

Table 9.2: Coronal heating models (adapted from Mandrini et al. 2000).

Physical process	References
1. DC stressing and reconnection models:	
– Stress-induced reconnection	Sturrock & Uchida (1981)
	Parker (1983, 1988); Berger (1991, 1993)
	Galsgaard & Nordlund (1997)
– Stress-induced current cascade	Van Ballegooijen (1986)
	Hendrix et al. (1996)
	Galsgaard & Nordlund (1996)
	Gudiksen & Nordlund (2002)
– Stress-induced turbulence	Heyvaerts & Priest (1992)
	Einaudi et al. (1996a,b)
	Inverarity & Priest (1995a)
	Dmitruk & Gomez (1997)
	Milano et al. (1997, 1999); Aly & Amari (1997)
2. AC wave heating models:	
– Alfvénic resonance	Hollweg (1985, 1991)
– Resonant absorption	Ionson (1978, 1982, 1983), Mok (1987))
	Davila (1987), Poedts et al. (1989)
	Goossens et al. (1992, 1995)
	Steinolfson & Davila (1993)
	Ofman & Davila (1994); Ofman et al. (1994, 1995)
	Erdélyi & Goossens (1994, 1995, 1996)
	Halberstadt & Goedbloed (1995a,b)
	Ruderman et al. (1997)
	Bélien et al. (1999)
– Phase mixing	Heyvaerts & Priest (1983)
	Parker (1991); Poedts et al. (1997)
	De Moortel et al. (1999, 2000a)
– Current layers	Galsgaard & Nordlund (1996)
– MHD turbulence	Inverarity & Priest (1995b)
	Matthaeus et al. (1999)
	Dmitruk et al. (2001, 2002)
– Cyclotron resonance	Hollweg (1986), Hollweg & Johnson (1988)
	Isenberg (1990), Cranmer et al. (1999a)
	Tu & Marsch (1997, 2001a,b)
	Marsch & Tu (1997a,b,2001)
3. Acoustic heating:	Schatzman (1949)
– Acoustic waves	Kuperus, Ionson, & Spicer (1981)
4. Chromospheric reconnection:	Litvinenko (1999a)
	Longcope & Kankelborg (1999)
	Furusawa & Sakai (2000)
	Sakai et al. (2000a,b, 2001a,b)
	Brown et al. (2000)
	Tarbell et al. (1999)
	Ryutova et al. (2001)
	Sturrock (1999)
5. Velocity filtration:	Scudder (1992a,b; 1994)

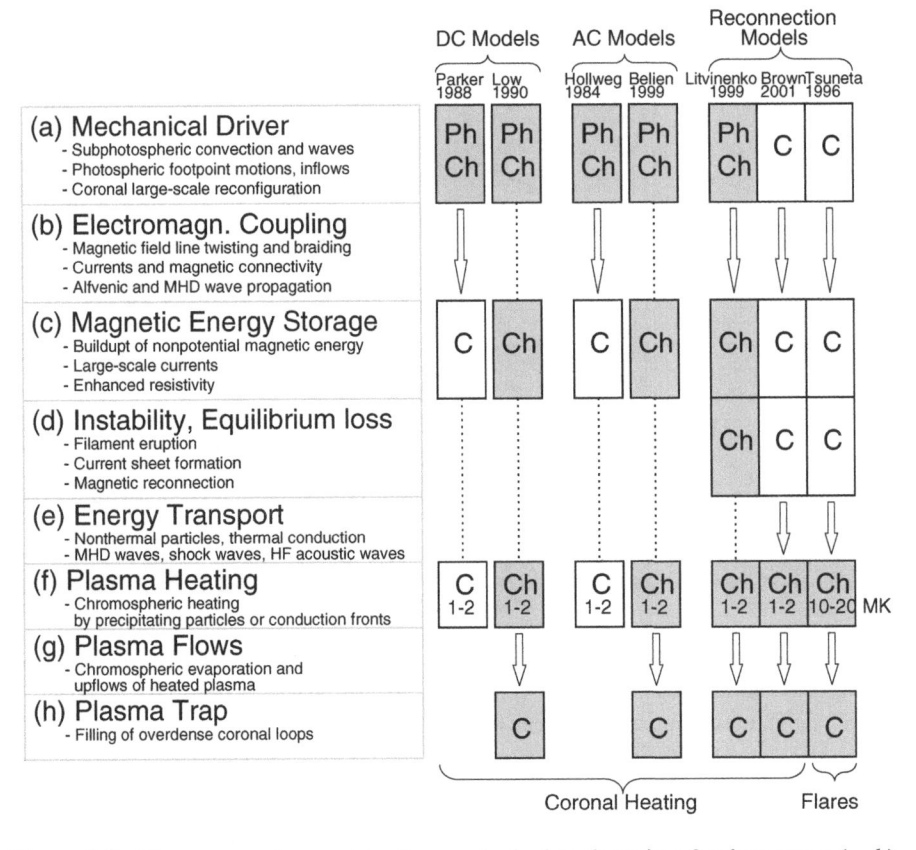

Figure 9.2: The process of coronal heating can be broken down into 8 subprocesses (a−h). Theoretical models include different subsets of these subprocesses, but only models that include the last step and can parameterize the physical parameters of the heated plasma can be compared with observations. The right side of the diagram shows a flow chart for the major heating models (with a typical representative listed at the top). Boxes mark physical steps that are part of the model, arrows mark transport processes between different locations, and dotted lines mark co-spatial locations. The boxes are colored in grey if the physical process takes place in a high-density region (Ph=photosphere, Ch=chromosphere, overdense coronal loops) and appear white for low-density regions (C=coronal background plasma) (Aschwanden 2001b).

absorption, phase mixing, current layer formation, and turbulence in the case of AC models. As an alternative to current dissipation, some heating could also be produced by compressional waves (i.e., by acoustic waves or shocks). Finally, a completely different physical mechanism is that of velocity filtration, which is based on the influence of the gravitational potential field in the corona on a postulated non-Maxwellian chromospheric velocity distribution. We will describe these various coronal heating models in the following sections.

Before we describe the physical processes of individual heating models in more detail, let us first have a look at their compatibility and completeness. In most of the

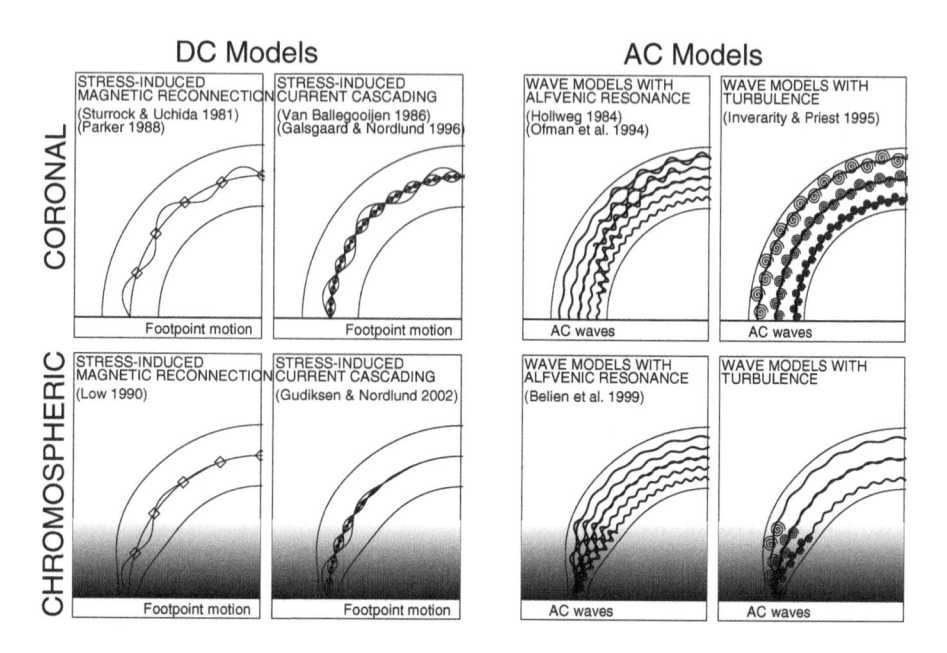

Figure 9.3: Categories of DC (left panels) and AC models (right panels), subdivided into coronal (top row) and chromospheric versions (bottom row). The greytones demarcate high-density regions (chromosphere and transition region) (Aschwanden 2002b).

theoretical models, coronal heating is a multi-stage process, which can be conceptually organized in a scheme with 8 steps, as illustrated in Fig. 9.2: the initial energy comes from a mechanical driver (a), which has an electromagnetic coupling (b) to the location of magnetic energy storage (c). At some point, a magnetic instability and loss of equilibrium (d) occurs, with possible energy transport (e), before plasma heating (f) starts. The resulting overpressure forces plasma flows (g), which become trapped (h) in coronal loops, where they are eventually observed. Various coronal heating models cover only an incomplete subset of these steps, so that these concepts first have to be combined with specific geometric and physically quantified models of coronal structures before they can be applied or fitted to observations. Therefore, observational tests of theoretical heating models are still in their infancy.

An aspect of over-riding importance for modeling coronal heating is the treatment of a realistic chromospheric and transition region boundary. This is visualized in Fig. 9.3 for some standard models. Early versions of coronal heating models usually approximate a coronal loop with a uniform fluxtube (Fig. 9.3, top row), which produces a more or less uniform energy dissipation for stressing of magnetic field lines and has rather large dissipation lengths for Alfvén waves. In other words, these highly idealized models produce an almost uniform heating function that stands in stark contrast to the observations. Recent, more realistic, models include gravity and the density and temperature structure of the chromosphere/transition region at the lower boundary (Fig. 9.3, bottom row), which changes the resulting heating function drastically. Typically, the heating rate is much more concentrated near the footpoints, because of

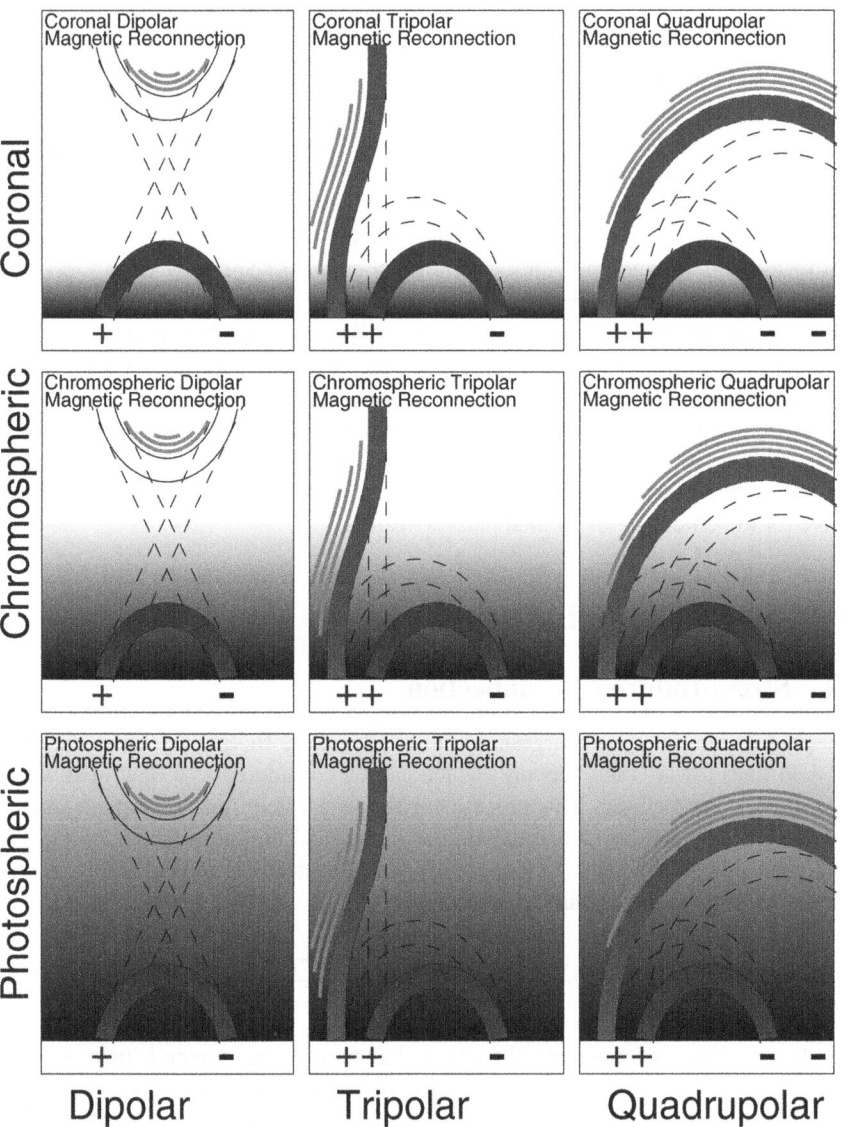

Figure 9.4: Basic categories of magnetic reconnection models, subdivided into coronal (top row), chromospheric (middle row), and photospheric versions (bottom), and according to bipolar (left column), tripolar (middle column), and quadrupolar (right column) configurations. The magnetic polarities are indicated with + and − signs. The greytones show high-density regions (photosphere, chromosphere, transition region). The dashed lines symbolize the preconnection configuration, and the solid lines show the postreconnection configuration, rendered in grey if the new-configured field lines have been filled with heated high-density plasma. The grey curves indicate shock waves or high-frequency acoustic waves (Aschwanden 2002b).

stronger stressing in the canopy-like magnetic field in the transition region, or due to vertical gradients in the density and Alfvén velocity caused by gravitational stratification.

The consideration of the transition region in coronal heating models also plays a crucial role for all models that involve magnetic reconnection. Essentially, the transition region is a dividing line between collisional (chromospheric) and collisionless (coronal) regimes, as illustrated in Fig. 9.4. Magnetic reconnection in collisionless regimes leads, besides plasma heating, to particle acceleration, which in turn, can efficiently contribute to chromospheric plasma heating (e.g., by chromospheric evaporation or thermal conduction fronts, as known for flares). The very same process is also believed to be responsible for heating of the quiet corona to some extent, as the nonthermal signatures of nanoflares in the quiet Sun suggest. However, if the same magnetic reconnection process happens inside the chromosphere, no particles can be accelerated because their collision time is shorter than their escape time out of the chromosphere. So, no secondary heating via accelerated particles is possible for reconnection processes in collisional plasmas. Therefore, the location of the magnetic reconnection region with respect to the transition region (above or below) is extremely decisive for the efficiency of coronal plasma heating.

9.3   DC Heating Models

9.3.1   Stress-Induced Reconnection

Photospheric granular and supergranular flows advect the footpoints of coronal magnetic field lines towards the network, which can be considered as a flow field with a *random walk* characteristic. This process twists coronal field lines by random angles, which can be modeled by helical twisting of cylindrical fluxtubes (see § 5.5.1). The rate of build-up of nonpotential energy (dW/dt) integrated over the volume $V = \pi r^2 l$ of a cylindrical fluxtube is (Sturrock & Uchida 1981),

$$\int \frac{dW}{dt} dV = \frac{\Phi B_0 <v^2> \tau_c}{4\pi} , \qquad (9.3.1)$$

where $\Phi = \pi r^2 B_0$ is the magnetic flux, B_0 is the photospheric magnetic field strength, l the length of the fluxtube, r its radius, $<v>$ the mean photospheric random velocity, and τ_c the correlation time scale of random motion. Sturrock & Uchida (1981) estimate that a correlation time of $\tau_c \approx 10 - 80$ min is needed, whose lower limit is comparable with the lifetimes of granules, to obtain a coronal heating rate of $dW/dt \approx 10^5$ (erg cm^{-2} s^{-1}), assuming small knots of unresolved photospheric fields with $B_{ph} \approx 1200$ G. Furthermore, they predict a scaling law of $p_0 \propto B_c^{6/7} L^{-1}$ based on the RTV law, which compares favorably with the measurements by Golub et al. (1980) with $p_0 \propto B_c^{0.8 \pm 0.2}$.

The idea of topological dissipation between twisted magnetic field lines that become wrapped around each other (Fig. 9.5) has already been considered by Parker (1972). Similar to Sturrock & Uchida (1981), Parker (1983) estimated the build-up

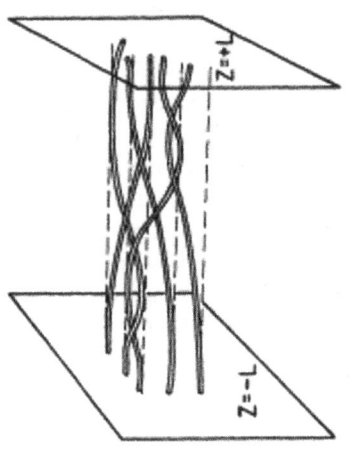
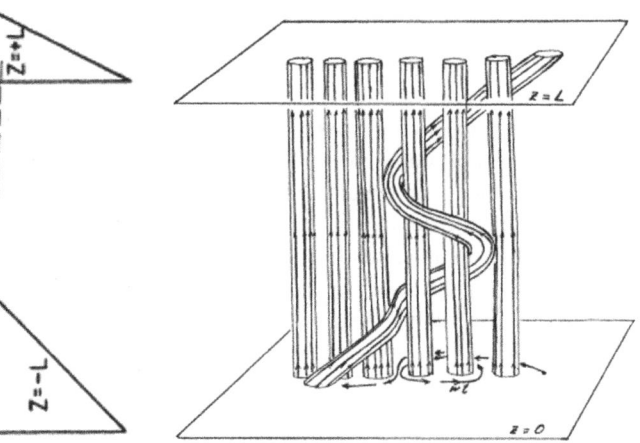

Figure 9.5: Topology of magnetic fluxtubes that are twisted by random walk footpoint motion (left; Parker 1972), leading to a state where fluxtubes are wound among its neighbors (right; Parker 1983).

of the magnetic stress energy $B_0 B_t / 4\pi$ of a field line with longitudinal field B_0 and transverse component $B_t = B_0 \mathrm{v} t / l$,

$$\frac{dW}{dt} = \frac{B_0 B_t}{4\pi} \mathrm{v} = \frac{B_0^2 \mathrm{v}^2 t}{4\pi l} , \qquad (9.3.2)$$

and estimated an energy build-up rate of $dW/dt = 10^7$ (erg cm^{-2} s^{-1}), based on $B_0 = 100$ G, $\mathrm{v} = 0.4$ km s^{-1}, $l = 10^{10}$ cm, and assuming that dissipation is sufficiently slow that magnetic reconnection does not begin to destroy B_t until it has accumulated random motion stress for 1 day. The manifestation of such sporadic dissipation events of tangential discontinuities in the coronal magnetic field in the form of tiny magnetic reconnection events is then thought to be detectable as *nanoflares* in the soft X-ray corona, whenever the twist angle

$$\tan \theta(t) \approx \frac{\mathrm{v} t}{l} \qquad (9.3.3)$$

exceeds some critical angle. Parker (1988) estimates, for a critical angle given by a moderate twist of $B_t = B_z / 4$, corresponding to $\theta = 14°$, that the typical energy of such a nanoflare would be

$$W = \frac{l^2 \Delta L \, B_t^2}{8\pi} \approx 6 \times 10^{24} \quad (\text{erg}) , \qquad (9.3.4)$$

based on $l = \mathrm{v}\tau = 250$ km, $\mathrm{v} = 0.5$ (km s^{-1}), $\tau = 500$ s, $\Delta L = 1000$ km, and $B_t = 25$ G. Thus, the amount of released energy per dissipation event is about nine orders of magnitude smaller than in the largest flares, which defines the term *nanoflare*.

There are several variants of random stressing models. A spatial random walk of footpoints produces random twisting of individual fluxtubes and leads to a *stochastic build-up of nonpotential energy* that grows linearly with time, with episodic random dissipation events (Sturrock & Uchida 1981; Berger 1991). The random walk step size is short compared with the correlation length of the flow pattern in this scenario, so that field lines do not wrap around each other. The resulting frequency distribution of processes with linear energy build-up and random energy releases is an exponential function, which is not consistent with the observed powerlaw distributions of nanoflares (§ 9.8). On the other hand, when the random walk step size is large compared with the correlation length, the field lines become braided (Fig. 9.5 right) and the energy builds up quadratically with time, yielding a frequency distribution that is close to a powerlaw (§ 9.8). In this scenario, energy release does not occur randomly, but is triggered by a critical threshold value (e.g., by a critical twist angle; Parker 1988; Berger 1993), or by a critical number of (end-to-end) twists before a kink instability sets in (Galsgaard & Nordlund 1997). The expected EUV spectrum and DEM of a microflare-heated corona have been estimated by Sturrock et al. (1990) and Raymond (1990).

9.3.2 Stress-Induced Current Cascade

The random footpoint motion would stir up a potential magnetic field, if there is one, and thus nonpotential fields and associated currents would occur in any case. Given the omnipresence of currents, we may ask to what degree *Ohmic dissipation* (also called *Joule dissipation*) of these currents may contribute to coronal heating. The volumetric heating rate E_H requirement for a loop with length l that is heated at both footpoints is (using $F_{H0} = 10^7$ erg cm^{-2} s^{-1}, see Table 9.1),

$$E_{H0} = \frac{2F_{H0}}{l} = 2 \times 10^{-3} \left(\frac{l}{10^{10} \text{ cm}} \right)^{-1} \quad (\text{erg cm}^{-3} \text{ s}^{-1}) . \tag{9.3.5}$$

The required current density j for Joule dissipation, $E_H = j^2/\sigma$, using the classical conductivity of $\sigma = 6 \times 10^{16}$ s^{-1} for a $T = 2$ MK corona is then,

$$j = \sqrt{\sigma E_{H0}} = 1.1 \times 10^7 \left(\frac{l}{10^{10} \text{ cm}} \right)^{-1/2} \quad (\text{esu}) . \tag{9.3.6}$$

On the other hand, the current density $j = (c/4\pi)(\nabla \times \mathbf{B})$ produced by random footpoint motion, say by twisting the field line by an angle $\Delta\varphi$, so that $(\nabla \times \mathbf{B}) \approx 2B_0\Delta\varphi/l)$, is

$$j = \frac{c}{2\pi} \frac{B_0}{l} \Delta\varphi = 48 \times \left(\frac{B_0}{100 \text{ G}} \right) \left(\frac{l}{10^{10} \text{ cm}} \right)^{-1} \Delta\varphi \quad (\text{esu}) \tag{9.3.7}$$

which is about 5 orders of magnitude smaller than required to satisfy the coronal heating requirement (Eq. 9.3.6). Therefore, Joule dissipation is generally inefficient in the corona, unless 10^5 smaller transverse length scales can be produced. This is the main motivation of the model by Van Ballegooijen (1986), who proposes a *current cascade model*, where free magnetic energy is transferred from large to small length scales in the corona as a result of the random motion of photospheric footpoints. From a statistical model of current density fluctuations, Van Ballegooijen (1986) finds that magnetic

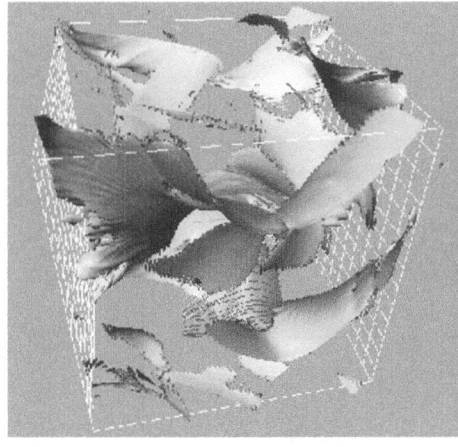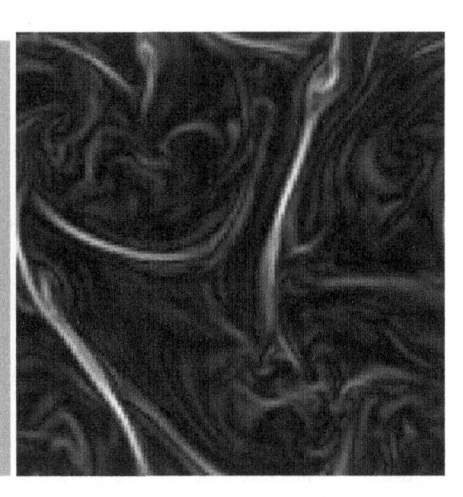

Figure 9.6: Snapshot of 3D MHD simulations of stress-driven current cascading. *Left:* 3D cube with isosurfaces of strong electric current densities. Magnetic field lines are traced from start points just outside the isosurfaces. *Right:* 2D slice of the electric current density in a cross section halfway between the boundaries from the 136^3-cube shown in the left panel (Galsgaard & Nordlund 1996).

energy is fed into the system linearly with time, so that the stored energy (integrated in time) increases quadratically, consistent with Parker (1983), see Eq. (9.3.2). The power spectrum of current-density fluctuations $P(\ln k, t)$ as a function of time t was numerically evaluated by Van Ballegooijen (1986), where he found a *cascading* towards smaller length scales, where energy is transferred from wave number $\ln(kl)$ to wave number $\ln(kl) + 1$ during a braiding time. The r.m.s current density in the stationary state was found to scale with the square root of the magnetic Reynolds number $R_m \approx 10^{10}$,

$$j_{rms} \approx \frac{c}{4\pi} \frac{B_0}{l} R_m^{1/2} , \qquad (9.3.8)$$

and thus approximately provides the enhancement factor of 2×10^5 needed to account for coronal heating. The heating rate was found to scale with

$$E_H \approx 0.19 \frac{B_0^2 D}{l^2} \approx 5 \times 10^{11} \frac{B_0^2}{l^2} , \qquad (9.3.9)$$

where $D \approx 150 - 425$ (km^2 s^{-1}) is the diffusion constant of the photospheric magnetic field. However, using a typical value of $D \approx 250$ (km^2 s^{-1}) and $l = 10^{10}$ cm, the obtained heating rate still falls (by a factor of 40) short of the coronal heating requirement (Eq. 9.3.5), but the efficiency is higher if one corrects for the canopy effects of the magnetic field (Van Ballegooijen 1986).

MHD simulations of this current cascading process were performed by Hendrix et al. (1996), Galsgaard & Nordlund (1996), and Nordlund & Galsgaard (1997). Hendrix et al. (1996) found a weak positive-exponent scaling of the Poynting flux $\mathbf{P} = (c/4\pi)(\mathbf{E} \times \mathbf{B})$ (and Ohmic dissipation rate) with the magnetic Reynolds number

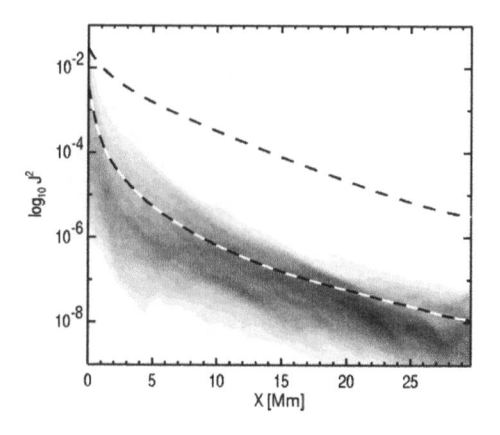

Figure 9.7: The current density $\log [j^2(x)]$ as a function of height x, obtained by the 3D MHD simulation of stress-driven current cascading by Gudiksen & Nordlund (2002). Note that the averaged heating function is strongly concentrated neat the footpoints, similar to the coronal heating scale heights $s_H \approx 10 - 20$ Mm inferred from TRACE observations (Aschwanden et al. 2000d, 2001a).

(Lundquist number), and conclude that line-tied photospheric convection can drive sufficiently large current densities to heat the corona. Also, Galsgaard & Nordlund (1996) found a scaling of the Poynting flux similar to Parker (1983) and Van Ballegooijen (1986) (i.e., $P \propto dW/dt \propto B_0^2 v^2 t/l$), but interpret the velocity v differently. Van Ballegooijen (1986) interpreted the factor $v^2 t$ as a diffusion constant, while Galsgaard & Nordlund (1996) in their experiment interpret v as the amplitude of a shearing step and t as the duration of a shearing episode. A snapshot of their simulations is shown in Fig. 9.6.

   While the early MHD simulations of current cascading by Galsgaard & Nordlund (1996) were restricted to a uniform fluxtube, more recent simulations include the gravitational stratification and a more realistic treatment of the transition region (Gudiksen & Nordlund 2002). An example of their simulation is shown in Fig. 6.10 and described in § 6.2.6. The new effects that arise in these simulations are stronger shear in the canopy-like parts of the transition region magnetic field, as well as much higher electron densities near the transition region, leading to enhanced Joule heating in the transition region and lower corona (Fig. 9.7), which is consistent with the observational finding of footpoint-concentrated heating in TRACE data (Aschwanden et al. 2000d, 2001a).

9.3.3   Stress-Induced Turbulence

Convection and turbulence are important in fluids with high Reynolds numbers. Since the *magnetic Reynolds number* $R_m = l_0 v_0 / \eta_m$ (Eq. 5.1.15) (or *Lundquist number*) is high in the coronal plasma ($R_m \approx 10^8 - 10^{12}$), turbulence may also develop in coronal loops (although there is the question whether turbulence could be suppressed in coronal loops due to the photospheric line-tying). Theoretical models and numer-

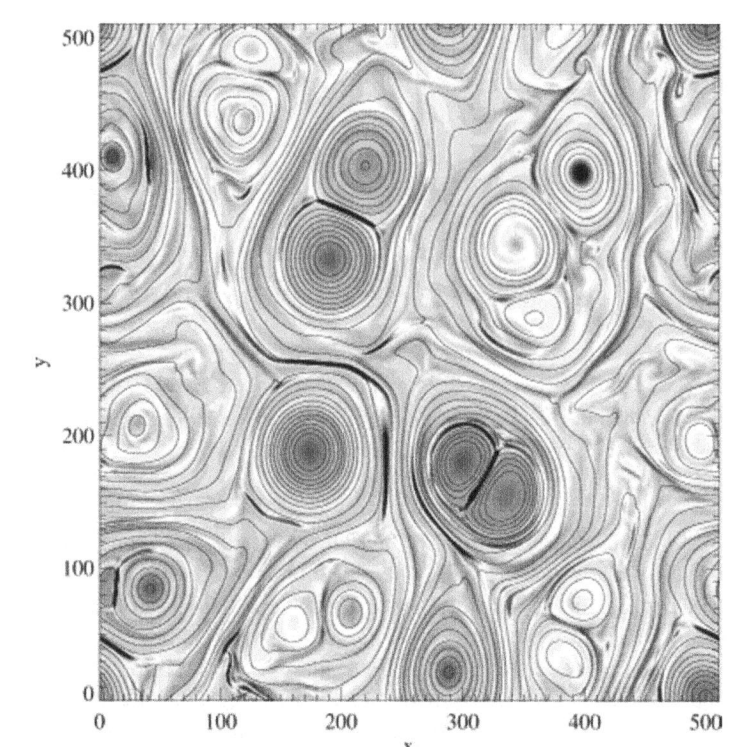

Figure 9.8: Example of a numeric 2D simulation of MHD turbulence with a Reynolds number of $R_m = 2000$, showing the magnetic field lines (contours) and electric current density (greyscale). Reconnection zones have formed between a number of adjacent islands that are coalescing, triggering localized nonsteady reconnection events throughout the simulation box (Matthaeus 2001a).

ical simulations that study MHD turbulence include the *kinematic viscosity* or *shear viscosity* ν_{visc} in the MHD momentum equation (6.1.15), and the *magnetic diffusivity* $\eta_m = c^2/4\pi\sigma$ in the MHD induction equation (5.1.14),

$$\rho\frac{D\mathbf{v}}{Dt} = -\nabla p - \rho\mathbf{g} + (\mathbf{j}\times\mathbf{B}) + \nu_{visc}\rho\left[\nabla^2\mathbf{v} + \frac{1}{3}\nabla(\nabla\cdot\mathbf{v})\right], \qquad (9.3.10)$$

$$\frac{\partial\mathbf{B}}{\partial t} = \nabla\times(\mathbf{v}\times\mathbf{B}) + \eta\nabla^2\mathbf{B}. \qquad (9.3.11)$$

Similar to the models of stress-induced current cascades (§ 9.3.2), random footpoint motion is assumed to pump energy into a system at large scales (into eddies the size of a granulation cell, ≈ 1000 km), which cascade due to turbulent motion into smaller and smaller scales, where the energy can be more efficiently dissipated by friction, which is quantified by the kinematic or shear viscosity coefficient ν_{visc}. Friction and shear are dynamical effects resulting from the nonlinear terms $(v_{1,i}v_{1,j})$, $(v_{1,i}B_{1,j})$, $(B_{1,i}v_{1,j})$, and $(B_{1,i}B_{1,j})$ in Eqs. (9.3.10−11) and are only weakly sensitive to the detailed dynamics of the boundary conditions. Analytical (3D) models of MHD turbulence have

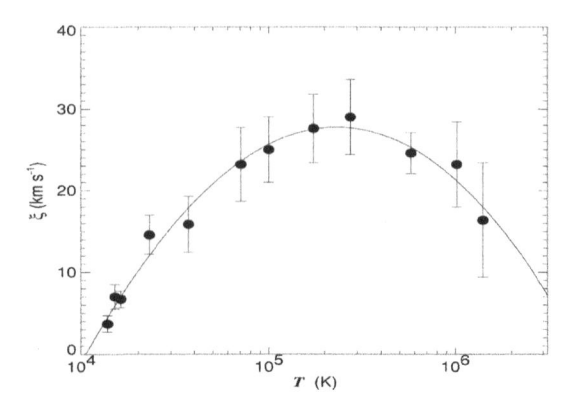

Figure 9.9: Observed nonthermal excess velocities ξ in the transition region and corona of the quiet Sun (adapted from Chae et al. 1998a, 2002b).

been developed by Heyvaerts & Priest (1992), Inverarity et al. (1995), Inverarity & Priest (1995a), and Milano et al. (1997), where the nonlinear viscosity terms are specified as diffusion coefficients. These turbulent diffusion coefficients are free parameters, which are constrained self-consistently by (1) assuming that the random footpoint motion has a turbulence power spectrum (e.g., a Kolmogorov spectrum $P(k) \propto k^{5/3}$); and (2) by matching the observed macroscopic parameters (i.e., velocity of footpoint motion, density, and magnetic field). Heyvearts & Priest (1992) predict turbulent velocities of $v_{turb} \approx 20 - 30$ km s^{-1}, which are consistent with the excess broadening of lines observed with SUMER, which shows a peak of $\xi = 30$ km s^{-1} at a transition region temperature of $T \approx 3 \times 10^5$ K (e.g., Chae et al. 1988a, shown in Fig. 9.9).

Analytical models of turbulent heating are applied to sheared arcades (Inverarity et al. 1995) and twisted fluxtubes (Inverarity & Priest 1995a). Turbulent heating has been numerically simulated in a number of studies, which exhibit a high degree of spatial and temporal intermittency (Einaudi et al. 1996a,b; Dmitruk & Gomez 1997). An example of such a simulation is shown in Fig. 9.8, where it can be seen how larger eddies fragment into smaller ones, forming current sheets and triggering magnetic reconnection during this process. Heating occurs by Ohmic dissipation in the thinnest current sheets. Milano et al. (1999) emphasize that the locations of heating events coincide with quasi-separatrix layers. The formation of such current sheets has also been analytically studied in the context of turbulent heating by Aly & Amari (1997). Numerical simulations reveal intermittent heating events with energies of $E_H = 5 \times 10^{24}$ to 10^{26} erg and a frequency distribution with a powerlaw slope of $\alpha \approx 1.5$, similar to observed nanoflare distributions in EUV (Dmitruk & Gomez 1997; Dmitruk al. 1998).

Most of the theoretical studies on coronal heating by MHD turbulence ignore the spatial distribution at coronal heights. A first attempt to extract the height dependence of MHD turbulent heating has been made by Chae et al. (2002b). Interpreting the observed distribution of nonthermal velocities $\xi(T)$ (Fig. 9.9), which can be characterized by a second-order polynomial in $\log (T)$,

$$\xi(T) = 27.7 - 15.56(\log T - 5.36)^2 \quad (\text{km s}^{-1}) , \tag{9.3.12}$$

as a manifestation of turbulent velocities (Chae et al. 2002b), he equated the heating rate $E_H(T)$ at a given temperature T to the turbulent dissipation rate of the turbulent velocity $\xi(T)$ at the same temperature,

$$E_H(T) = \frac{\varepsilon_0 \rho \, \xi^4(T)}{l_{in} v_A} \, , \tag{9.3.13}$$

where ξ is the nonthermal speed, l_{in} the length scale over which energy injection occurs, and ε_0 a dimensionless factor. Combining a temperature model $T(h)$ of a coronal loop in hydrostatic equilibrium (e.g., Eqs. 3.6.13 or 3.6.18) with $\xi(T)$ from Eq. (9.3.12), then yields the height dependence of the heating rate, $E_H(h) = E_H(T[h])$. With this method, Chae et al. (2002b) found that the heating occurs concentrated strongly near the footpoints of coronal loops, similar to the TRACE findings (Aschwanden et al. 2000d, 2001a).

9.4 AC Heating Models

9.4.1 Alfvénic Resonance

We discussed propagating MHD waves in § 8, where we also reviewed observational evidence for propagating magneto-acoustic and Alfvén waves in the corona, in closed loops as well as in open field regions, and thus in the solar wind. Whenever the time scale of the wave excitation is shorter than the Alfvén (or magneto-acoustic) travel time back and forth a coronal structure, the Alfvén (or magneto-acoustic) waves propagate, rather than locking into a standing wave. For the coronal heating problem (or the solar wind heating problem), we need upward propagating waves that have been generated by a chromospheric or subphotospheric energy source. However, one problem is that magneto-acoustic (i.e., fast-mode MHD) waves tend to be totally internally reflected somewhere in the chromosphere, and thus are evanescent there (e.g., Hollweg 1978). The Alfvén waves do not suffer these difficulties, a substantial fraction can be transmitted through a resonant cavity with a nearly standing Alfvén wave, and in this way can propagate into the corona or solar wind (Hollweg 1984a,b; 1985). In § 8.2.3 we found spectroscopic evidence for line broadening $< \Delta v > \approx 30$ (km s^{-1}) which fulfilled the scaling between line broadening and density expected for Alfvén waves (i.e., $\Delta v(r) \propto n_e^{-1/4}(r)$; Eq. 8.2.11), observed with SUMER and UVCS (e.g., Doyle et al. 1999; Banerjee et al. 1998; Chae et al. 1998b; Esser et al. 1999). Given this evidence for Alfvén waves in coronal holes, we can estimate the energy flux (Poynting flux) that is carried by these Alfvén waves,

$$F_A = \rho < \Delta v^2 > v_A \, , \tag{9.4.1}$$

which is found to be $F_A = 4 \times 10^6$ (erg cm^{-2} s^{-1}), (based on $\rho = \mu m_p n_i$ with $n_i = 10^9$ cm^{-3}, $< \Delta v > = 30$ km s^{-1}, and $v_A = 2000$ km s^{-1}). So the energy flux carried by Alfvén waves is sufficient to heat coronal holes and quiet Sun regions, and even a substantial fraction of active region loops (see Table 9.1), if it can be dissipated over an appropriate height range. However, Alfvén waves are notoriously difficult to

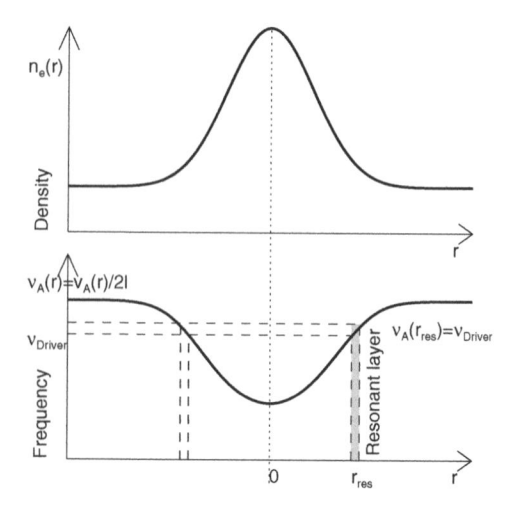

Figure 9.10: The concept of resonant absorption is illustrated for a coronal loop with a density enhancement in the cross-sectional profile, $n_e(r)$ (top). The resulting Alfvén frequency in each layer r is $\nu_A(r) = v_A(r)/2l \propto 1/\sqrt{n_e(r)}$ (bottom). Resonant absorption occurs in layers at locations $r = r_{res}$ where the Alfvén frequency matches the photospheric driver frequency, $\nu_A(r = r_{res}) = \nu_{driver}$.

dissipate in a homogeneous part of the solar corona, because Alfvén waves are shear waves and the coronal shear viscosity is very low (Hollweg 1991). The coronal shear viscosity is only efficient if there are strong cross-field gradients, which is the basic idea of the *resonant absorption* mechanism.

9.4.2   Resonant Absorption

In the low plasma-β corona, the magnetic field varies very little across a loop structure, but the density can vary by orders of magnitude (see derivation based on pressure equilibrium in § 6.2.1). Therefore, the Alfvén velocity is very nonuniform across a loop because of the radial density variation (Fig. 9.10). The phenomenon of *resonant absorption* is based on the idea that absorption of Alfvén waves (by Ohmic or viscous dissipation) is enormously enhanced in narrow layers where the local Alfvén resonance frequency ν_{res} matches the oscillation frequency of the (photospheric) driver ν_{driver} (e.g., the global p-mode oscillation),

$$\nu_{driver} = \nu_{res} = N \frac{v_A(r = r_{res})}{2l} , \qquad (9.4.2)$$

where l is the loop length and $N = 1, 2, 3, ...$ the harmonic number. This is similar to the standing mode of the fast kink-mode oscillation (Eq. 7.2.4), except that we deal here with a single resonant loop layer with uniform Alfvén speed v_A, while fast kink-mode MHD oscillations discussed in § 7.2 are produced by surface waves at a loop surface between internal and external Alfvén speeds v_A and v_{Ae}.

Models for resonant absorption have been developed by Ionson (1978, 1982, 1983), Mok (1987), Davila (1987), Sakurai et al. (1991a,b), Goossens et al. (1992, 1995),

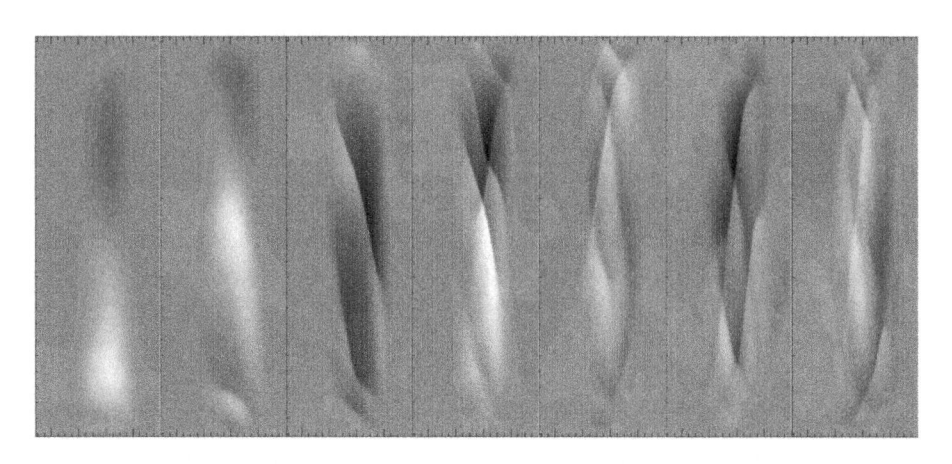

Figure 9.11: An example of an MHD simulation of resonant absorption in a coronal loop. The 7 panels show the evolution of the velocity component v_z of resonant Alfvén waves in the fluxtube, with the two footpoints at the bottom and top. Note that the resonant layers occur at particular locations but drift through the entire loop due to the varying density conditions (Bélien et al. 1999).

Steinolfson & Davila (1993), Ofman & Davila (1994), Ofman et al. (1994, 1995), Erdélyi et al. (1995), Halberstadt & Goedbloed (1995a,b), Ruderman et al. (1997), and Bélien et al. (1999). The driver is assumed to be a broadband spectrum $P(\nu_{driver})$, which covers at least the $P \approx 3 - 5$ min range known from global p-mode oscillations. The efficiency of dissipation in the resonant layer depends of course on the width of the resonant layer (which is reciprocal to the gradient of the Alfvén velocity) and the spectral power found at the resonance frequency $P(\nu_{driver} = \nu_{res})$. For the dissipation mechanism it has been shown that ion viscosity is the dominant contributor (Ionson 1982; Mok 1987), which can heat the resonant layer to a coronal temperature (Ionson 1984). The volumetric heating rate E_H (or dissipation rate) by resonant absorption has been derived as (Hollweg 1984a; Hollweg & Sterling 1984),

$$E_H = 2\pi^{5/2}\rho_{Ph}^{1/2} < \Delta\mathrm{v}_{Ph}^2 > \lambda_{Ch} \, B \, \mathrm{v}_A \, L^{-3} \sum_{N=1}^{\infty} NP(\nu_N) \,, \qquad (9.4.3)$$

where ρ_{Ph} is the photospheric density (where the Alfvén waves are assumed to be generated), $\Delta\mathrm{v}_{Ph}$ the r.m.s velocity associated with the driver in the photosphere, λ_{Ch} the scale height of the Alfvén speed in the chromosphere-transition region, and $P(\nu_N)$ is a weighting factor of the normalized power spectrum for the N−th harmonic resonance.

Numerical simulations have been conducted to derive scaling laws between the dissipation rate and the various loop parameters, and to quantify the relative importance of dissipation by resistivity and shear viscosity, which have been found to be comparable (Poedts et al. 1989; Steinolfson & Davila 1993; Ofman & Davila 1994; Ofman et al. 1994, 1995; Erdélyi & Goossens 1994, 1995, 1996). While early models of resonant absorption were applied to a static resonant layer, more self-consistent models include the density variation due to chromospheric evaporation during the heating, which has the effect that resonant layers drift dynamically throughout the loop (Ofman et al. 1998). The effects of chromospheric coupling were simulated with a 2.5-D MHD code by Bélien et al. (1999) (Fig. 9.11), who finds that about 30% of the incoming Poynting flux is absorbed within the loop structure, and only $\approx 1 - 3\%$ is converted into heat. The resonant Alfvén waves were also found to generate slow magneto-acoustic waves (Bélien et al. 1999; Ballai & Erdélyi 1998a; Erdélyi et al. 2001), which are compressional waves and thus give rise to density variations, which further fragment the resonant layers. The largest uncertainty in determining the heating efficiency by resonant absorption is the unknown power spectrum $P(\nu_{driver})$ of the photospheric driver.

9.4.3 Phase Mixing

When shear Alfvén waves propagate in a structure with a large gradient in Alfvén velocity, $v_A(r)$, as caused by the cross-sectional density profile $n_e(r)$ in a coronal loop (Fig. 9.10, top), the wave oscillations of neighboring field lines suffer friction (from kinematic and shear viscosity) because they have slightly different phase speeds, a process that is called *phase mixing*. This process was analytically studied by Hey-vaerts & Priest (1983), who find dramatically enhanced viscous and Ohmic dissipation. Moreover, they find that MHD instabilities such as the Kelvin–Helmholtz and tearing instability (§ 6.3) occur in the vicinity of velocity nodes. They conclude that phase mixing is a likely process that dissipates shear Alfvén waves in coronal loops and in open structures with strong wave reflectivity. However, including gravitational stratification of the solar corona leads to a reduction of the transverse density gradient with height, so that the oscillation wavelengths become longer and the effect of phase mixing becomes weaker (De Moortel et al. 1999). Including the curvature of the (radially diverging) solar corona shifts the effects of phase mixing to lower altitudes (De Moortel et al. 2000a). Therefore, the effect of phase mixing depends very much on the amplitude of the excited waves, the geometry, and scale heights, being strongest where the steepest density gradients occur.

Phase mixing is an essential ingredient of resonant absorption (Poedts 2002), but it does not need resonances to be effective. The combination of both processes, resonant absorption and phase mixing, has been simulated by Poedts et al. (1997). Phase mixing causes a cascade of energy to small length scales, where the dissipation becomes more efficient. In contrast, Parker (1991) argues that the filamentary structure in coronal holes couples the plane shear Alfvén waves into a coordinated mode, so that phase mixing does not provide efficient heating in the first $1 - 2$ solar radii, but only at a radial distance of $\gtrsim 10$ solar radii, where it can accelerate the solar wind.

9.4.4 Current Layers

In § 9.3.2 we discussed the stress-induced current cascade, leading to current layers with strongly enhanced Ohmic dissipation. The same process can be driven in the slow limit of a DC driver, as well as in the fast limit of an AC driver, compared with the Alfvén travel time through the system. Galsgaard & Nordlund (1996) performed numerical simulations in both regimes. In the AC limit, where the driver was simulated

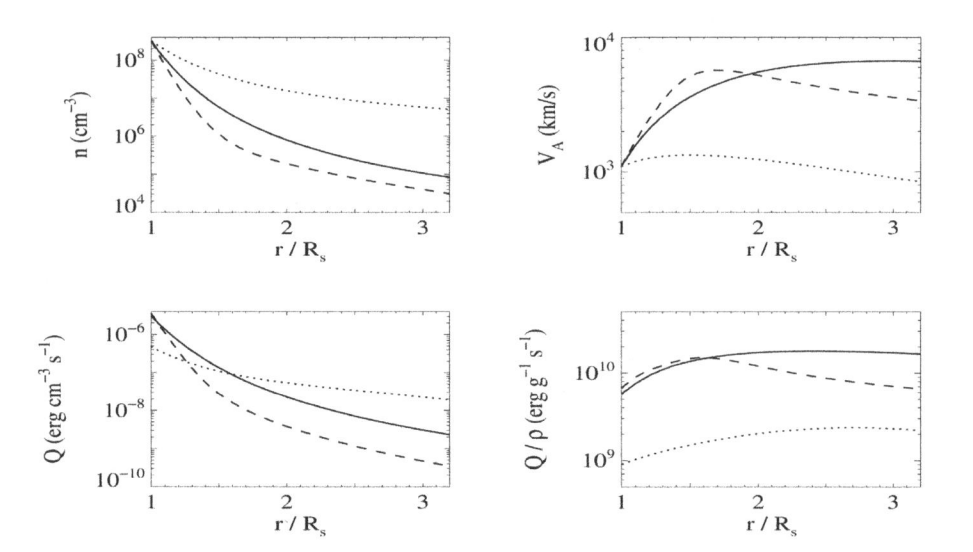

Figure 9.12: Simulations of low-frequency wave-driven MHD turbulence in a radially diverging open magnetic field, including wave reflections. The panels show 3 different density profiles $n_e(r)$ (top left), and the associated Alfvén speed $v_A(r)$ (top right) assumed in the simulations, and the resulting heating per unit volume (bottom left) and per unit mass (bottom right). Note efficient heating in low altitudes $r < 2R_\odot$ due to the presence of reflected waves (Dmitruk et al. 2002).

by a series of random shearing motion steps (instead of a powerlaw spectrum of wave vectors), the total dissipation in the loop was found to become independent of its length.

9.4.5 MHD Turbulence

Also the stress-induced turbulence models (§ 9.3.3) can be operated in the AC limit, where the footpoint motions have a high enough frequency to produce waves rather than a quasi-static evolution as seen in the DC limit. In the model of Inverarity & Priest (1995b), a photospheric driver with a velocity of $v = 1$ km s^{-1} was found to produce sufficient damping by turbulence to satisfy the heating requirement for the quiet corona. In coronal holes, the distinction between the AC and DC limit breaks down, because there is no second boundary that defines an Alfvén reflection time. Matthaeus et al. (1999), however, argue that low-frequency waves (say with periods of $P \approx 100 - 1000$ s) in an inhomogeneous open magnetic structure (with gradients in the Alfvén speed) lead to efficient wave reflections that critically control the formation of MHD turbulent cascades and the associated heating of coronal holes and the solar wind at $r < 2R_\odot$ (Dmitruk et al. 2001, 2002; Oughton et al. 2001) (Fig. 9.12). In the presence of vertical gradients of the Alfvén speed $v_A(h)$, MHD nonlinearities develop that cause upward propagating Alfvén waves to reflect on length scales of $\lambda_A = dh/d\ln v_A$, estimated to be of the order $\lambda_A \approx 15$ Mm, in agreement with heating scale heights measured by TRACE (Matthaeus et al. 2002).

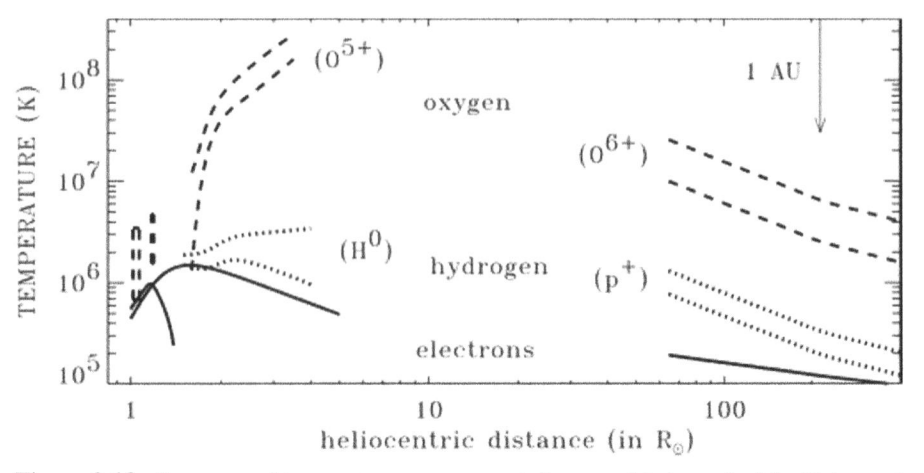

Figure 9.13: Summary of temperature measurements in coronal holes and of the high-speed solar wind: electron temperatures (solid lines), hydrogen (dotted lines), and oxygen O VI,VII (dashed lines) (Cranmer 2001).

9.4.6 Cyclotron Resonance

If we go to the high-frequency AC regime of $\nu = 10^2 - 10^4$ Hz, we come to the ion gyrofrequencies,

$$f_{gi} = \frac{eB}{2\pi m_i c} = 1.52 \times 10^3 \left(\frac{B}{\mu}\right) \qquad [\text{Hz}] , \qquad (9.4.4)$$

which yield for a typical magnetic field strength of $B \approx 1$ G (at a height of $r \approx 1 R_\odot$ in the solar wind) gyrofrequencies of $f_{gi} = 1.5 \times 10^3$ Hz for protons ($\mu = 1$) or $f_{gi} = 1.0 \times 10^2$ Hz for oxygen ions ($\mu = m_i/m_p = 16$). It is therefore conceivable, if a high-frequency driver with such frequencies in the range of $\nu \approx 10^2 - 10^4$ Hz exists in a collisionless plasma (such as in the outer solar wind), that ions can resonate with this driver and obtain high perpendicular velocities $v_\perp$, which corresponds to a higher perpendicular temperature $T_\perp$ of their velocity distribution. In particular the new measurements by SoHO/UVCS reveal that: (1) protons and other ions have high perpendicular temperatures $T_\perp > T_\parallel$, with $T_\perp = 2 \times 10^8$ K at $3 R_\odot$; (2) that positive ions (O VI, Mg X) have a higher temperature than protons by at least their mass ratio (i.e., $(T_{ion}/T_p) \gtrsim (m_i/m_p)$); and (3) that the outflow velocity of these ions is about twice that of the protons (Kohl et al. 1997; 1998; 1999; Li et al. 1998; Cranmer et al. 1999a,b) (Fig. 9.13).

A theoretical model that uses the resonance between left-hand polarized Alfvén waves and the Larmor gyrations of positive ions was developed earlier, but was first applied to the solar corona and acceleration region of the solar wind by Hollweg (1986), Hollweg & Johnson (1988), and Isenberg (1990), and further modeled for the high-speed solar wind by Tu & Marsch (1997, 2001a,b), Marsch & Tu (1997a,b, 2001), and Cranmer et al. (1999a,b). The evidence for perpendicular heating by cyclotron resonance in the solar wind is overwhelming, but the origin of the high-frequency driver is less clear. It could be Alfvén wave generation at the base of the corona, or alternatively a gradual growth or replenishment through the extended corona. The spectroscopic identification of Alfvén waves by the line broadening scaling law $\Delta v(r) \propto n_e^{-1/4}$ (Doyle et al. 1998, 1999; Banerjee et al. 1998) down to altitudes of $h \lesssim 30$ Mm (see Fig. 8.13, bottom) provides evidence for the origin of Alfvén waves in the lower corona or transition region.

9.5 Other Coronal Heating Scenarios

9.5.1 Acoustic Heating

A possible source of energy for coronal heating that was considered early on is acoustic waves from the chromosphere (driven by global p-mode oscillations with $P \approx 3 - 5$ min). However, this option was ruled out when Athay & White (1978, 1979) showed with their analysis of UV spectroscopic data from OSO-8, that the acoustic wave flux does not exceed $\approx 10^4$ erg cm^{-2} s^{-1}, which is about $2 - 3$ orders of magnitude below the coronal heating requirement (see Table 9.1). Acoustic fluctuations with periods shorter than the acoustic cutoff period, $P_{cutoff} = 4\pi c_s/\gamma g \approx 200 - 300$ s, however, heat the chromosphere (Schatzman 1949; Ulmschneider 1971; Kuperus et al. 1981). Upward propagating acoustic waves steepen nonlinearly to shocks in the short photospheric scale height ($\lambda_{Ch} \approx 100$ km) and are dissipated in the chromosphere: short-period waves with $P \approx 40 - 60$ s are dissipated in the lower chromosphere, and long-period waves with $P \approx 300$ s in the upper chromosphere.

On the other hand, the recent observations from TRACE clearly show evidence that acoustic waves propagate upwardly into coronal loops, and that they are of photospheric origin, because they are driven with $P \approx 3$ min periods near sunspots, and with $P \approx 5$ min far away from sunspots (De Moortel et al. 2000b, see § 8.1.2). However, the energy flux of these acoustic waves was estimated to $d\varepsilon_{wave}/dt \approx (3.5 \pm 1.2) \times 10^2$ erg cm^{-2} s^{-1}, far below the requirement for coronal heating.

9.5.2 Chromospheric Reconnection

A number of recent studies deal with *chromospheric reconnection* processes that subsequently contribute to coronal heating, either by generating magneto-acoustic shock waves or upflows of heated plasma.

Litvinenko (1999a) considers a magnetic reconnection scenario that is driven by collisions of photospheric magnetic features with opposite polarities (*magnetic flux cancellation*). He uses a standard VAL-C chromospheric density model, calculates the lateral inflow speed for the Sweet−Parker reconnection model, and finds a maximum inflow speed at a height of $h = 600$ km above the photosphere, which he considers as the most favorable height for chromospheric reconnection. This process may hold for those observed *transition region explosive events* that have been found to be co-spatial with magnetic cancelling features (Dere et al. 1991).

Longcope & Kankelborg (1999) envision a similar model where randomly moving photospheric magnetic flux elements of opposite magnetic polarity collide and cause a chromospheric reconnection, perhaps manifested in the appearance of a *soft X-ray bright point*. Two approaching bipoles reconnect in a quadrupolar reconnection geometry and release a heat flux that is estimated as

$$F_{XBP} \approx 0.1 B_+ B_- v_0 \approx 10^4 \ (\text{erg cm}^{-2} \ \text{s}^{-1}), \qquad (9.5.1)$$

for typical magnetic fields $B_\pm = 1.6$ G and photospheric motion velocities of $v_0 = 2 \times 10^4$ cm s^{-1}, in agreement with the observations of soft X-ray bright points, but about 2 orders of magnitude below the flux required to supply the heating of the quiet Sun.

A number of numerical MHD simulations have been performed by Sakai and co-workers that envision similar scenarios of chromospheric reconnection processes that visualize the microprocesses that could be relevant for heat input to the corona. An example is shown in Fig. 9.14, with a scenario of photospheric magnetic fluxtubes that collide in the network, causing parallel and X-type reconnection events. 3D neutral MHD simulations of this scenario were performed by Furusawa & Sakai (2000), who find that the chromospheric reconnection produces fragmented fluxtubes, strong currents, fast magneto-sonic waves, and upward plasma flows. The resulting shock waves can collide with other fluxtubes and produce surface Alfvén waves (Sakai et al. 2000a). Further simulations reveal the generation of shear Alfvén waves (Sakai et al. 2000b), torsional waves (Sakai et al. 2001a), upflows colliding in the corona (Sakai et al. 2001b), fragmentation of loop threads (Sakai & Furusawa 2002), and loop interactions driven by solitary magnetic kinks (Sakai et al. 2002).

The collision of small-scale magnetic fluxtubes might lead to complex topologies of 3D shock waves, with curved surfaces rather than plane waves. The curved upward propagation trajectories can then lead to a self-focusing of these shocks. The curved shock fronts experience strong gradients in their acceleration, turn to each other after upward propagation, and collide, producing local heating and ejecting plasma jets into the corona (Tarbell et al. 1999, 2000; Ryutova & Tarbell 2000; Ryutova et al. 2001). Another possible by-product of chromospheric reconnections are high-frequency sound waves that could be generated in an oscillatory relaxation phase after reconnection, which could contribute to coronal heating (Sturrock 1999).

9.5.3 Velocity Filtration

An alternative explanation of the coronal heating problem is the view that the solar corona is a result of a non-Maxwellian particle distribution generated in the chromosphere-transition zone, with a nonthermal tail characterized by a *kappa function*, formed by the process of velocity filtration in the gravitational potential (Scudder 1992a,b; 1994). The existence of a non-Maxwellian tail, however, is a somewhat arbitrary assumption that requires an unknown acceleration mechanism in the chromosphere that energizes ions to thermal velocities of $v_{th} \approx 200 - 400$ (km s^{-1}). If we adopt this non-Maxwellian tail population, the kinetic energy and thus the temperature of the non-Maxwellian particles is predicted to increase linearly with height, because the potential energy is proportional to the height difference,

$$T(h) = T_0 + \frac{m_i g_\odot}{k_B (2\kappa_i - 3)} \frac{h}{(1 + h/R_\odot)} \approx T_0 + \frac{m_i g_\odot}{k_B (2\kappa_i - 3)} h, \qquad (9.5.2)$$

(a)

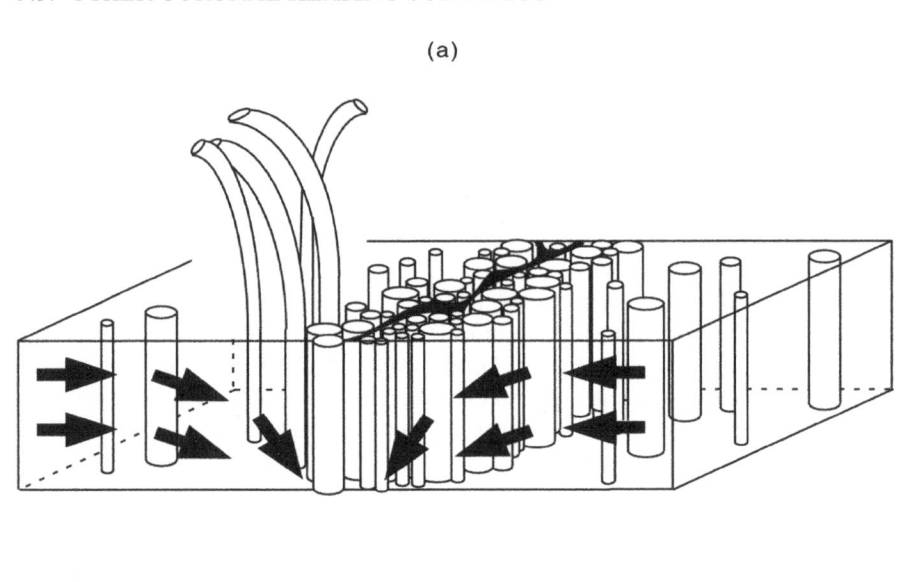

(b) (c)

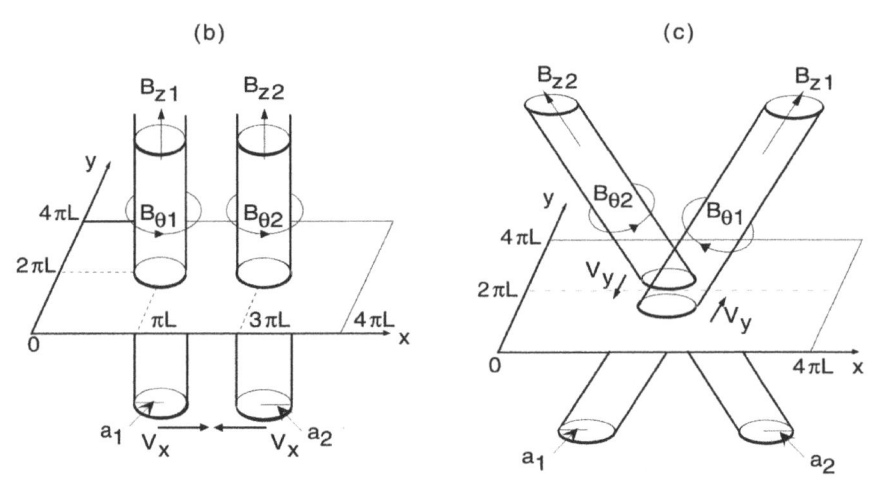

Figure 9.14: Concept of colliding magnetic fluxtubes in the downflows in the network: vertical fluxtubes from the interior of photospheric intra-network cells drift to the network and collide there in the converging flows (top), triggering reconnections between collinear fluxtubes (bottom left) or X-type reconnection between inclined fluxtubes (bottom right) (Furusawa & Sakai 2000)

where T_0 is the temperature at height $h = 0$, κ_i the index of the kappa function of the suprathermal tail for ion species i, and m_i the ion mass. The basic effect is that gravitational stratification filters out in the upper corona cooler nonthermal particles that do not have sufficient energy to overcome the gravitational potential. Since gravity is an undebatable reality, the effect of velocity filtration would definitely occur if all dynamic effects could be eliminated in the corona and if we could wait until everything has settled down into gravitational stratification. There are two strong testable predictions: (1) the linear temperature increase with height, and (2) the density scale heights of various ion species are reciprocal to the ion masses (see Eq. 3.1.16 with $m_i = \mu m_p$),

$$\lambda_p = \frac{2k_b T_i}{m_i g_\odot} , \qquad (9.5.3)$$

(e.g., an iron ion with atomic mass $m_i = 56 m_p$ has a 56 times shorter gravitational scale height than a hydrogen atom).

How are these predictions testable? For EUV emission with TRACE 171 Å, for instance, the emission measure is proportional to the product of the electron density and iron ion density, $EM \propto n_i n_e$ (Eq. 2.9.1). Electrons, which have the same abundance as the hydrogen ions, excite an eight-times ionized iron ion (Fe XI) by collisions, and a 171 Å photon is emitted after the excited electron falls back to its original state. The coronal electrons are distributed in height as the protons (to maintain charge neutrality), so with a pressure scale height of $\lambda_p \approx 50$ Mm for a 1 MK plasma. If the rare iron ions were to be settled into gravitational equilibrium, they would have a 56 times shorter scale height, so $\lambda_p \lesssim 1$ Mm, and the collisions with iron atoms would also be confined to the scale height of the iron ions, which would restrict EUV emission at 171 Å to the bottom of the corona at $\lambda_p \lesssim 1$ Mm. This is of course in contradiction to the observations, which show a scale height of 171 Å emission, that is the same as expected for a hydrogen-helium atmosphere with a scale height of $\lambda_p \approx 50$ Mm (see Fig. 3.2). So, what is missing in the velocity filtration theory? It appears that the iron ions never have the time to settle into their gravitational potential, but are constantly mixed with the hydrogen protons and electrons, either by flows or turbulence. Also, the other prediction of a linear temperature increase with height contradicts the observations, which rather show an isothermal temperature profile in the coronal part of EUV loops (see Fig. 3.23). Perhaps the theory of velocity filtration will never be testable in our highly dynamic corona, because the time scales of dynamic processes (intermittent heating, flows, and radiative cooling) seem to be much shorter than the time scale needed for gravitational settling of the heavy ions (or rare elements) in the corona.

9.6 Observations of Heating Events

Systematic detection and statistical analysis of small-scale phenomena in the solar transition region and corona have only been explored in recent years with soft X-ray and EUV high-resolution imaging, with the motivation to discover direct signatures of coronal heating processes and to evaluate their energy budget. Because the physical nature of these small-scale phenomena is not yet fully understood, they have been given many different names, although they may turn out to belong to identical physical processes, but they exhibit complementary signatures in different wavelength regimes. Among them are: *ephemeral regions, emerging flux events, cancelling magnetic flux events, explosive events, blinkers, soft X-ray bright points, nanoflares, microflares, soft X-ray jets, active region transient brightenings*, etc. The sites of their occurrences (quiet Sun, active region, photosphere, transition region, corona) are listed in Table 9.3. Some recent reviews on these small-scale phenomena can be found in Parnell (2002a,b), Shimizu (2002a,b), and Berghmans (2002).

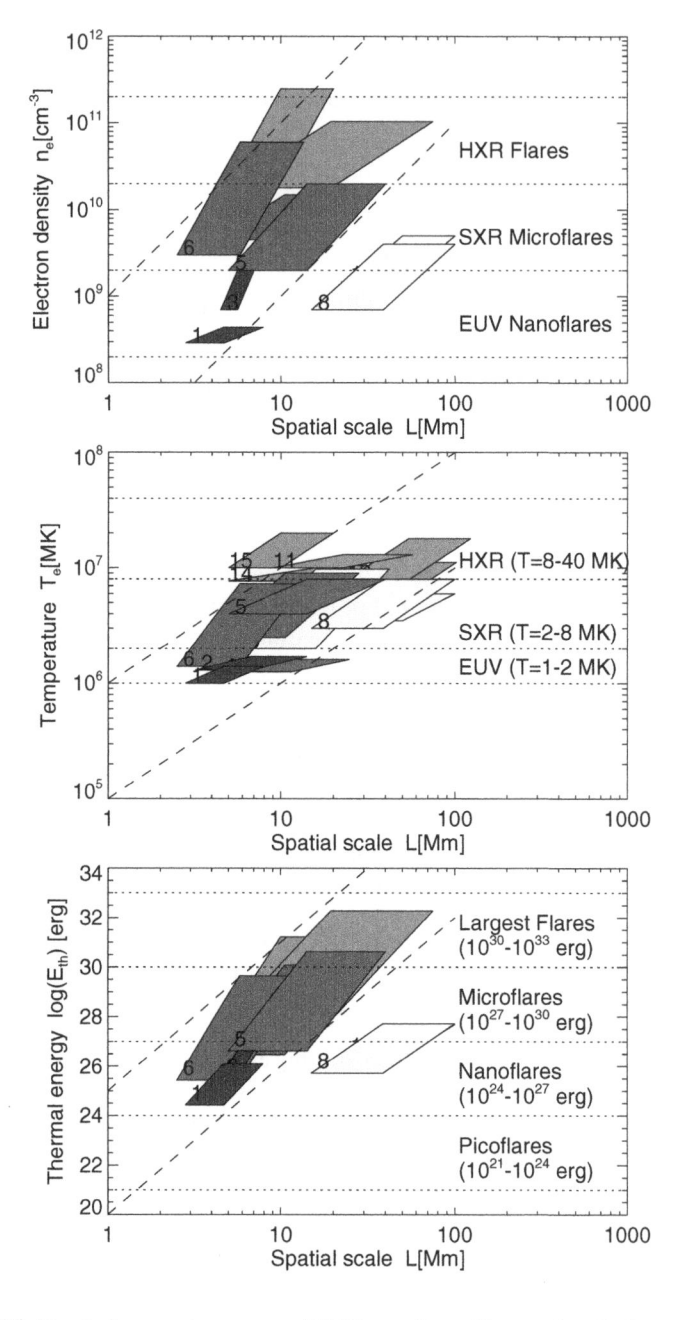

Figure 9.15: Physical parameter ranges of EUV nanoflares (data set $1 - 3$; dark grey), soft X-ray brightenings (data set $4 - 7$; middle grey), SXR jets (data set 8; white), non-flare loops (data set $9 - 10$; white) and flare loops (data set $11 - 15$; light grey), see references and numerical parameter ranges of data sets $1 - 15$ in Table 9.4.

Table 9.3: Small-scale phenomena and their occurrence domain (QS=quiet Sun, AR=active region, Ph=photosphere, TR=transition region, and C=corona) and primary wavelength range (optical, EUV=extreme ultraviolet, SXR=soft X-rays) (adapted from Parnell 2002b).

Phenomenon	Horizontal domain	Vertical domain	Wavelengths
Ephemeral regions	QS	Ph	Optical
Emerging flux events	QS, AR	Ph	Optical
Cancelling magnetic flux events	QS, AR	Ph	Optical
Explosive events	QS	TR	EUV
Blinkers	QS, AR	TR	EUV
Nanoflares, microflares	QS, AR	C	EUV, SXR
X-ray bright points	QS	C	SXR
Soft X-ray jets	QS, AR	C	SXR
Active region transient brightenings	AR	C	SXR

Let us first define some nomenclature, before we enter the zoo of small-scale phenomena. In Table 9.4 we have compiled physical parameters (spatial scale l, electron temperature T_e, and electron density n_e) of reported coronal small-scale phenomena, along with flare events for comparison, and we estimate the upper limits of their total thermal energy,

$$\varepsilon_{th} \leq 3n_e k_B T_e V \; , \tag{9.6.1}$$

based on a volume estimate of $V \leq l^3$ for flares, and $V \leq lw^2$ for linear features (single loops, jets), respectively. The parameter ranges are also displayed in graphical form in Fig. 9.15. We wan subdivide flare-like events into three magnitude groups, according to their total thermal energy content (Fig. 9.15), which can be characterized with the following typical temperature and density ranges:

Large flares: $\varepsilon_{th} = 10^{30} - 10^{33}$ erg, $T_e \approx 8 - 40$ MK, $n_e \approx 0.2 - 2 \times 10^{11}$ cm^{-3}
Microflares: $\varepsilon_{th} = 10^{27} - 10^{30}$ erg, $T_e \approx 2 - 8$ MK, $n_e \approx 0.2 - 2 \times 10^{10}$ cm^{-3}
Nanoflares: $\varepsilon_{th} = 10^{24} - 10^{27}$ erg, $T_e \approx 1 - 2$ MK, $n_e \approx 0.2 - 2 \times 10^{9}$ cm^{-3}

The wavelength regimes in which these heating events are detected are a clear function of the maximum temperature that is reached (SXR, EUV) and whether nonthermal particles are produced (HXR): large flares are detected in all three wavelength regimes (HXR, SXR, EUV), microflares are detected in SXR and EUV, and nanoflares only in EUV wavelengths.

9.6.1 Microflares – Soft X-ray Transient Brightenings

We can define a *microflare* as a miniature version of a large flare, having an energy content with a fraction of $10^{-6} < \varepsilon_{th}/\varepsilon_{flare}^{max} < 10^{-3}$ of the largest flares. In large flares there is always hard X-ray emission produced by nonthermal particles, which may be below the detection threshold of current hard X-ray detectors for small flares or microflares. Microflares with nonthermal energies of $\varepsilon_{nth} \approx 10^{27} - 10^{31}$ erg were

Table 9.4: Physical parameters of coronal small-scale phenomena: spatial scale L, temperature T_e, electron density n_e, and upper limit of total thermal energy $\varepsilon_{th} = 3n_3 k_B T_e V$ (non-flare loops and flares are also included) (adapted from Aschwanden 1999a).

Phenomenon	Number of events N	Spatial Scale L [Mm]	Electron Temperature T_e [MK]	Electron Density $10^8 n_e$ [cm^{-3}]	Thermal Energy $log(\varepsilon_{th})$ [erg]
Nanoflares[1]	281	$2.8 - 7.9$	$1.0 - 1.4$	$2.9 - 4.4$	$24.4 - 26.1$
QS transient bright.[2]	228	$3.2 - 14.1$	$1.3 - 1.7$	...	...
QS heating events[3]	24	$4.5 - 7.9$	$1.2 - 1.5$	$7 - 20$	$25.5 - 26.6$
SXR bright points[4]	23	$5.4 - 24.7$	$1.26 - 1.61$	...	...
AR transient bright.[5]	≈ 200	$5 - 40$	$4 - 8$	$20 - 200$	$26.6 - 30.6$
AR transient bright.[6]	16	$2.5 - 13.5$	$1.4 - 7.3$	$30 - 600$	$25.4 - 29.6$
AR transient bright.[7]	41	$4 - 28$	$2.5 - 9$	$45 - 150$	$26.5 - 30.0$
SXR jets[8]	16	$15 - 100$	$3 - 8$	$7 - 40$	$25.7 - 27.7$
Non-flare SXR loops[9]	47	$6 - 42$	$2.0 - 9.8$	...	...
Non-flare SXR loops[10]	32	$25 - 100$	$3.5 - 6.0$	$15 - 50$	$26.3 - 27.7$
Flares[11]	20	$9 - 57$	$10 - 13$	...	...
Flares[12]	14	$24 - 123$	$7.9 - 17.9$	...	...
Flares[13]	19	$23 - 102$	$5.3 - 11.2$	...	...
Flares[14]	31	$5 - 75$	$7.7 - 10.5$	$180 - 1040$	$27.8 - 32.2$
Flares[15]	44	$5 - 20$	$10 - 20$	$200 - 2500$	$28.0 - 31.2$

References:[1] Aschwanden et al. (2000b), [2] Berghmans et al. (1998), [3] Krucker & Benz (2000), [4] Kankelborg et al. (1997), [5] Shimizu (1997), [6] Shimizu (2002a,b), [7] Berghmans et al. (2001), [8] Shimojo & Shibata (2000), [9] Porter & Klimchuk (1995), [10] Kano & Tsuneta (1995), [11] Reale et al. (1997), [12] Garcia (1998), [13] Metcalf & Fisher (1996), [14] Pallavicini et al. (1977), [15] Aschwanden & Benz (1997).

detected down to 8 keV with the high-sensitivity *CGRO/BATSE Spectroscopy Detectors (SPEC)*, which showed all the characteristics of larger flares [i.e., they exhibit impulsive time profiles and hard (nonthermal) spectra, Lin et al. 2001]. With RHESSI, microflares were recently measured in an even lower energy range, revealing temperatures of $T \approx 6 - 14$ MK from the spectral fits in the $3 - 15$ keV energy range (Benz & Grigis 2002).

Active Region Transient Brightenings (ARTB) is a group of events that seem to coincide with microflares, because they all have flare-like attributes (Fig. 9.16): they appear in active regions, they consist of small loops ($l \approx 5 - 40$ Mm) that become impulsively filled with heated plasma ($T = 4 - 8$ MK) and cool down subsequently, with durations comparable to small flares ($\tau \approx 2 - 7$ min), and have thermal energy contents of $\varepsilon_{nth} \approx 10^{27} - 10^{31}$ erg (Shimizu et al. 1992, 1994; Shimizu 1995, 1997; Shimizu & Tsuneta 1997), comparable to the 8-keV microflares (Lin et al. 2001). About a third of the ARTBs also show flare-like radio and hard X-ray signatures, such as thermal gyroresonance, gyrosynchrotron, and nonthermal microwave emission at $1 - 18$ GHz (Gopalswamy et al. 1994, 1997; White et al. 1995; Gary et al. 1997), as well as hard

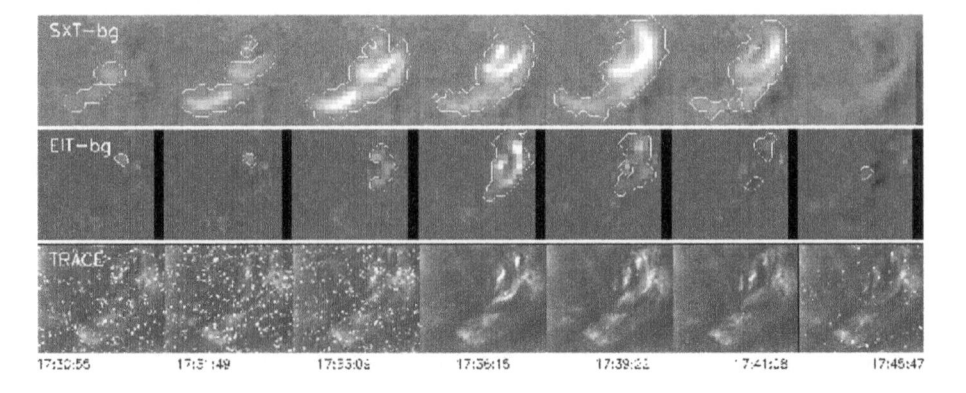

Figure 9.16: Multi-wavelength observations (in soft X-rays with Yohkoh/SXT and in EUV with SoHO/EIT and TRACE) of an *Active Region Transient Brightening (ARTB)*. A slowly varying background has been subtracted in the SXT and EIT data, and the first 3 TRACE images are contaminated by particle radiation spikes. Note that the geometry shows typical small-scale flare loops and the evolution exhibits heating and cooling of soft X-ray-bright plasma over a time scale of 15 min (Berghmans et al. 2001).

X-ray emission, with soft X-ray emission corresponding to GOES class A and B levels (Nitta 1997). Berghmans et al. (2001) also discovered that some microflares launched a slow-mode MHD wave that propagated to the remote footpoint of a pre-existing soft X-ray loop and caused a gradual brightening there, which they called *indirect ARTBs*. Based on all these flare-like characteristics established by multi-wavelength observations, ARTBs can clearly be identified as microflares, the low-energy extension of the classical flare distribution. The implication for coronal heating is that ARTBs cannot heat the corona, if their frequency distribution, which is the low-energy extension of larger flares, has the same slope (see § 9.7).

X-ray Bright Points (XBP) are small-scale phenomena detected in soft X-rays, but they can occur everywhere in the quiet Sun or in coronal hole regions, in contrast to ARTBs which by definition are domiciled in active regions. While it was traditionally believed that flares (and thus microflares) require significant magnetic field concentrations (e.g., $B \approx 10 - 1000$ G), as they can be found in active regions, it was not clear a priori whether microflares can also be produced in weak magnetic fields (e.g., at $B \approx 1 - 10$ G), as is typical for quiet Sun regions and coronal holes. So the question arose whether the ubiquitous *X-ray bright points* found all over the solar surface represent the counterpart to microflares known in active regions (ARTBs), or whether they are a fundamentally different process. X-ray bright points, characterized as small, compact, and bright features, were discovered in soft X-rays and EUV during rocket flights and were analyzed from *Skylab* data (Golub et al. 1974, 1976a,b, 1977; Nolte et al. 1979). Their spatial size and energy released during their lifetime ($\varepsilon_{th} \approx 10^{26} - 10^{28}$ erg) is fairly comparable with ARTBs. Their average lifetime is 8 hours (Golub et al. 1974) to 20 hours (Zhang et al. 2001), which seems to represent the total lifetime of a "mini-active region", rather than the duration of a microflare

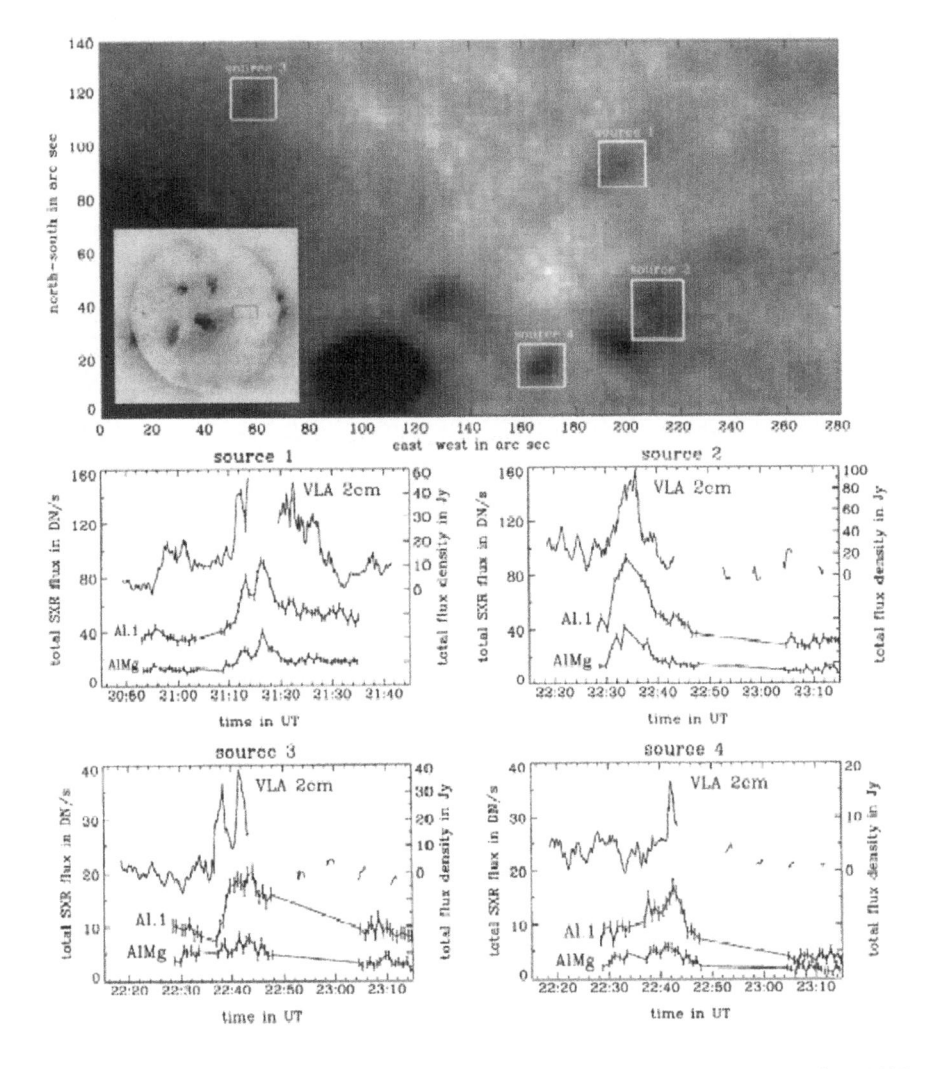

Figure 9.17: Temporal evolution of microflares in the quiet Sun network, observed on 1995-Feb-20 in radio (VLA, 15 GHz) and in soft X-ray (Yohkoh/SXT: Al.1, AlMg filters). The time profiles are extracted from 4 different microflares, with the locations indicated in the field-of-view above. Note the correlated impulsive peaks in radio and soft X-rays, representing the nonthermal and thermal counterparts of flare-like processes. This measurement presents the first clear evidence for flare-like processes in the quiet Sun (Krucker et al. 1997).

episode. Their magnetic structure has been identified to be associated with an *emerging magnetic bipole (ephemeral region)* in 66% or with a *cancelling magnetic bipole* in 33% (Harvey 1996; Shimizu 2002a,b). The occurrence rate of XRPs was initially believed to be in anti-phase to the solar cycle (Golub et al. 1979), but turned out to be uncorrelated to the solar cycle after contrast correction for the scattered light from active regions (Nakakubo & Harra 2000). Their association with magnetic emergence

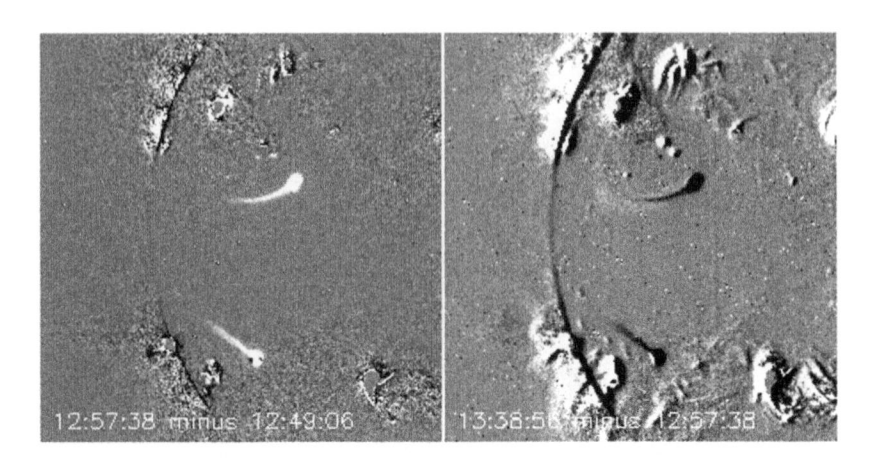

Figure 9.18: Two soft X-ray jets were observed almost simultaneously with Yohkoh/SXT on 1998 Jul 11, 12:57 UT, shown in two difference images. The difference brightness is white in the left frame because the flux increases, while it is dark in the right frame because the flux decreases. They represent outflows from soft X-ray bright points, following the trajectory of large-scale magnetic field lines. They fade out with height due to a combination of field divergence, gravitational stratification, and cooling (courtesy of Sam Freeland and Hugh Hudson).

or cancellation implies that a magnetic reconnection process is the driver, and thus puts the XBP in the category of flare-like energy release processes (models of the underlying reconnection processes will be discussed in more detail in § 10). Nonthermal signatures of XBP were explored with radio observations, which provided evidence for thermal free-free emission at 1.4 GHz (Nitta et al. 1992), thermal emission at 17 GHz (Kundu et al. 1994), and nonthermal (impulsive) gyrosynchrotron emission at 15 GHz (Fig. 9.17; Krucker et al. 1997; Benz & Krucker 1999). Correlations between the radio and soft X-ray locations of XBPs with the photospheric magnetic field revealed that they are preferentially located above the network (Benz et al. 1997), and thus could be triggered by emerging flux (Tang et al. 2000) or by magnetic reconnection of colliding downflows in the network (Fig. 9.14). Detailed correlations of the time profiles between radio, O V, He I, and Fe IX/X lines led to the conclusion that a flare-like process happens, where precipitating electrons first produce (undetectable) hard X-rays and correlated O V and He I responses in the chromosphere, followed by heating of flare plasma and cooling through the coronal Fe IX/X line (Krucker & Benz 1999). Taking together all this evidence for flare-like behavior in XBP events, in the form of nonthermal radio emissions and correctly timed thermal soft X-ray emissions, we can safely conclude that the XBP in the quiet Sun regions represent the counterpart to the ARTBs in active regions, both being governed by flare-like processes. So it appears that these phenomena can be unified by the term *microflares*, which suggests the same physical process, only differing from classical flares by their much smaller energy budget. We will discuss the consistency of their frequency distribution of energies with larger flares in § 9.7, which is crucial for extrapolating whether they can match the coronal heating requirement.

Another small-scale phenomenon observed in soft X-rays are the so-called *X-ray jets* (Shibata et al. 1992a; Strong et al. 1992; Shimojo et al. 1996, 1998, 2001; Shimojo & Shibata 1999, 2000). X-ray jets are found in X-ray bright points and small active regions, exhibiting outflows of heated plasma with velocities of v $= 10 - 1000$ km s^{-1} (Fig. 9.18). Theoretical models envision that X-ray jets are generated in magnetic reconnection regions where newly emerged loops interact with a slanted overlying field region (Yokoyama & Shibata 1995, 1996). So they are dynamic features that can be considered as by-products of microflares. They have similar temperatures to microflares, but have longer dimensions ($l \approx 15 - 100$ Mm) than X-ray bright points, because of their streaming motion along open (or large-scale closed) magnetic field lines. They probably originate in similar densities to those in the soft X-ray microflare location $n_e \approx (0.2 - 2) \times 10^{10}$ cm^{-3}, but exhibit lower average densities [$n_e = (0.7 - 4) \times 10^9$ cm^{-3}; Shimojo & Shibata 2000] because of their streaming dispersion. Consequently, they also have a lower thermal energy content than the microflares from which they originate (note that the event group #8 of SXR jets lies below the microflare and flare population in Fig. 9.15, bottom panel). The fact that their energy cannot exceed that of their "mother" microflare, thus ranks them less important for coronal heating than microflares themselves.

9.6.2 Nanoflares – EUV Transient Brightenings

The diagrams shown in Fig. 9.15 illustrate that EUV small-scale events are found at the bottom of the energy scale, about 9 orders of magnitude below the largest flares, and thus constitute the category of *nanoflares*, with energies of $\varepsilon_{th} \approx 10^{24} - 10^{27}$ erg. They are also found at the bottom of the temperature scale, at $T \approx 1 - 2$ MK, and at the bottom of the density scale, with $n_e \approx (0.2 - 2) \times 10^9$ cm^{-3}. Such tiny EUV brightenings, which we call nanoflares, were first studied with SoHO/EIT data (Krucker & Benz 1998, 2000; Benz & Krucker 1998, 1999, 2002; Berghmans et al. 1998; Berghmans & Clette 1999, Aletti et al. 2000) and with TRACE (Parnell & Jupp 2000; Aschwanden et al. 2000b,c; Aschwanden & Parnell 2002) (e.g., see Fig. 9.19, or with a combination of EIT, TRACE, and SXT together; Berghmans et al. 2001). From the observations it appears that these transient EUV brightenings have all the properties of larger flares: they occur in small-scale loops ($l \approx 3 - 8$ Mm); they exhibit an impulsive rise and decay (on time scales of a few minutes), consistent with plasma heating and subsequent cooling (Fig. 9.20), manifested as a cooling delay $d\tau_{cool}/dT < 0$ in the temperature range $T \approx 1.0 - 1.5$ MK observed in 195 Å and 171 Å. Some nanoflares also exhibit plasma outflows, called EUV jets (Chae et al. 1999). EUV nanoflares do not reach temperatures higher than $T \lesssim 2$ MK, and thus are not visible in soft X-rays (as microflares) or in hard X-rays (as normal flares). Of course, there seems to be a continuous distribution from large flares down to the tiny nanoflares. A fundamental difference is perhaps that large flares occur only in active regions, requiring larger magnetic fields, while microflares and nanoflares can occur everywhere in the quiet Sun or even in coronal holes.

The theoretical interpretation of nanoflares can be made very much as a scaled-down version of large flares. If we employ for nanoflares the standard scenario of classical flares (i.e., coronal magnetic reconnection followed by coronal particle accel-

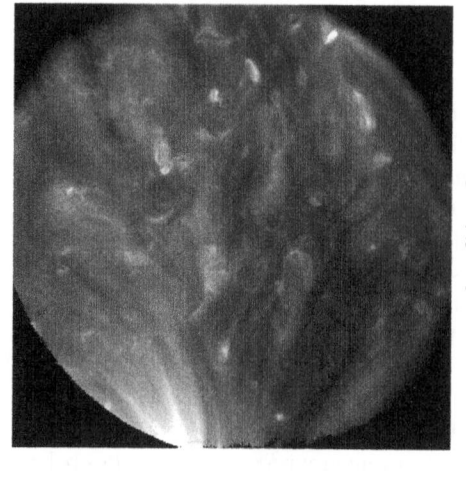

Figure 9.19: TRACE Fe XII (195 Å) image of a quiet Sun region containing nanoflares, observed on 1998-Jun-16, 20:25 UT (left) and contrast-enhanced version to visualize the spatial structure of nanoflares (right). The longer curved features at the bottom belong to coronal loops. The circular boundary is given by the field-of-view (8′) of the TRACE telescope (Parnell & Jupp 2000).

eration and chromospheric evaporation; Brown et al. 2000), nanoflares would require a scaled-down geometry of small flare loops that barely stick out of the transition region. So the location of nanoflares would be confined to the lowest layer of the corona just above the transition region. On the other hand, theoretical models of coronal heating postulate nanoflares throughout the corona (Levine 1974; Parker 1988), which can only be reconciled with the observed tiny nanoflare loops if DC interactions (stress-induced reconnection, current cascades, and turbulence) produce magnetic dissipation events preferentially in the tangled magnetic field in the canopy geometry of the transition region. It is estimated that 95% of the photospheric magnetic flux closes within the transition region (also called *"magnetic carpet"*), while only 5% form large-scale connections over coronal loops (Priest et al. 2002; Schrijver & Title 2002; see also Plates 10 and 11), which would indeed produce a preponderance of magnetic dissipation events in the canopy region suitable for EUV nanoflares. It has been envisioned that such EUV nanoflares result from explosions of sheared magnetic fields in the cores of initially closed bipoles (Moore et al. 1999), which occur preferentially in the network (Falconer et al. 1998).

Numerical simulations of such small-scale heating events are able to reproduce the observed physical parameters and statistical frequency distributions (e.g., Sterling et al. 1991; Walsh et al. 1997; Galtier 1999; Mendoza−Briceno et al. 2002).

9.6.3   Transition Region Transients − Explosive Events, Blinkers

A number of transient small-scale phenomena have also been detected in cooler EUV lines in the temperature range of $T \approx 2 \times 10^4 - 2.5 \times 10^5$ K (e.g., in O V, O IV, He

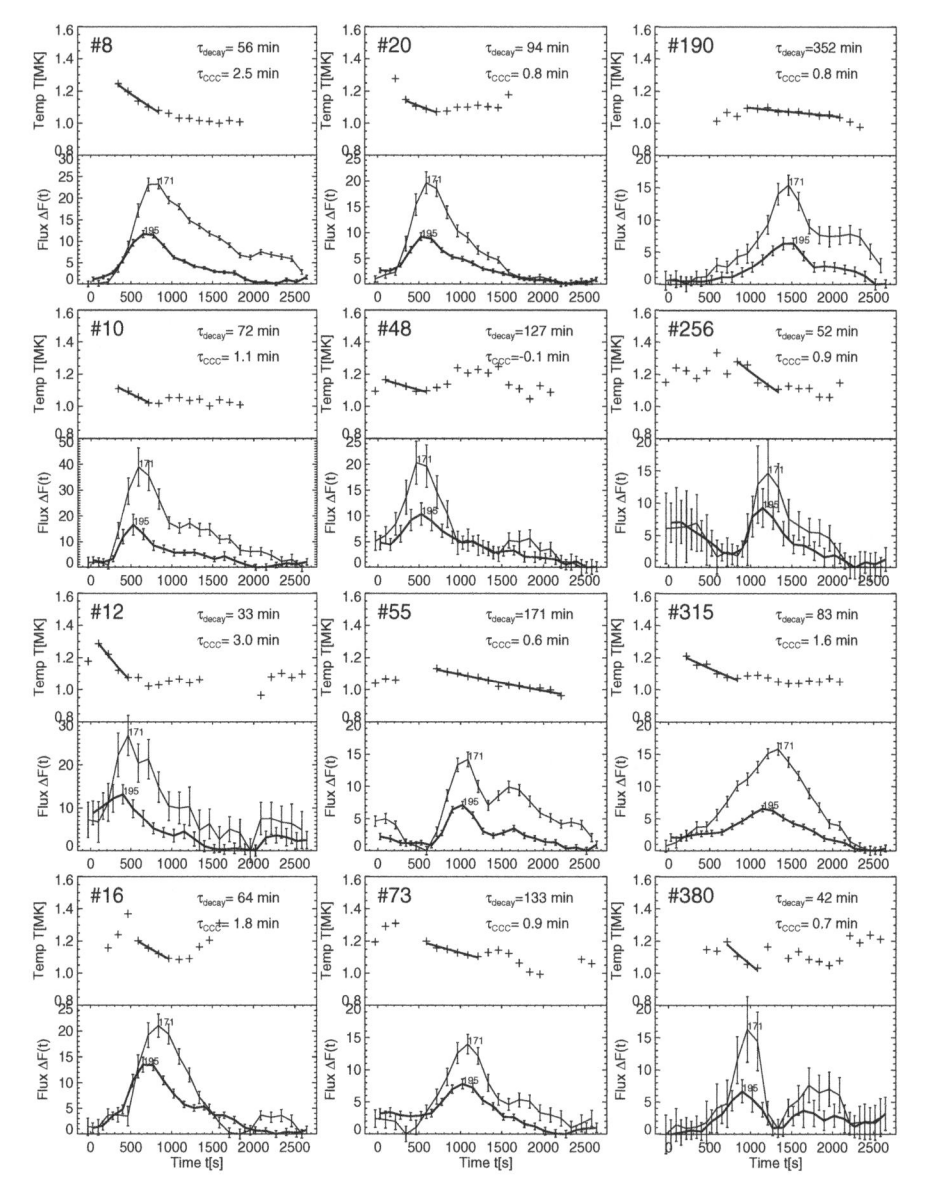

Figure 9.20: Light curves of 16 nanoflares observed with TRACE 171 Å and 195 Å on 1999-Feb-17, 02:16−02:59 UT. Note that all nanoflares peak first in 195 Å ($T \approx 1.5$ MK) and subsequently in 171 Å ($T \approx 1.0$ MK), indicating plamsa cooling. The filter-ratio temperature is indicated in the top of each channel, confirming the temperature decrease after the peak (Aschwanden et al. 2000c).

I, C IV, Si IV, H Lyα, C II, He II, Si III, and N V lines). Based on this temperature range, these phenomena are likely to occur inside the transition region, in particular if their source location is stationary. Alternatively they could be associated with upward ejected cool plasma (e.g., spicules, macrospicules, jets, sprays, surges) or with downward falling plasma from rapidly cooling coronal loops (catastrophic cooling, see § 4.5.4), but these are both dynamic processes that would produce observable source motions.

Jets of cool material (or exploding small loops), that carry typical energies of $\varepsilon_{kin} \approx 2.5 \times 10^{26}$ and show supersonic velocities of v $\approx$ 400 km s^{-1} and turbulent line broadening of Δv $\approx$ 250 km s^{-1}, were observed in quiet Sun regions, a phenomenon that was dubbed *explosive events* (Brueckner & Bartoe 1983; Dere et al. 1989, 1991; Fontenla et al. 1989). Such events have been interpreted as the results of magnetic reconnection in the transition region (Dere 1996), which is also supported by observations of bi-directional jets (Innes et al. 1997; Innes 2001; Perez et al. 1999) that are expected to mark a symmetric X-type reconnection geometry. Some EUV jets have been observed to shoot up high into the corona (Harrison et al. 2001), similar to erupting macrospicules (Parenti et al. 2002).

Longer lived EUV brightenings with modulated time profiles, but otherwise exhibiting similar temperatures and energies as explosive events found in the quiet Sun, were called *blinkers* and were studied with SoHO/CDS (Harrison 1997; Harrison et al. 1999; Bewsher et al. 2002; Parnell et al. 2002; Brkovic et al. 2001). Blinkers are a distinctly different phenomenon from nanoflares, because their brightness or emission measure increase was not found to be caused by a temperature increase (like in nanoflares), but rather due to an (quasi-isothermal) increase in filling factor (Bewsher et al. 2002), more like in compressional waves. They were found both in quiet Sun and active regions (Parnell et al. 2002). There are common features among explosive events, blinkers, and spicules that suggest a unified process, while observational differences are related to their location (network border versus intranetwork, on-disk versus above limb; Moses et al. 1994; Gallagher et al. 1999; Madjarska & Doyle 2002, 2003; Chae et al. 2000c) or flow directions and associated line profile asymmetries (Chae et al. 1998b; Peter & Brkovic 2003).

Regarding the coronal heating problem, all these dynamic phenomena contain relatively cool plasma with transition region temperatures ($T \lesssim 0.5$ MK) and seem not to have a counterpart in coronal temperatures (Teriaca et al. 1999; 2002), so they contribute to cooling of the corona rather than to heating. Also the energy flux of explosive events seems to be insignificant for coronal heating ($F \approx 4 \times 10^4$ erg cm^{-2} s^{-1}; Winebarger et al. 1999, 2002).

9.7 Scaling Laws of Heating Events

The relevance of microflares, nanoflares, and other transient events to the coronal heating problem can only be assessed reliably by quantitative tests of the scaling of their physical parameters and by the statistics of their frequency distributions, which are both related. We will therefore in this section first explore the scaling of various observable parameters: area $A(l)$, volume $V(l)$, density $n_e(l)$, temperature $T_e(l)$, thermal energy $\varepsilon_{th}(l)$, magnetic field strength $B(l)$, with their size (length scale l), which provides a physical understanding of the observed frequency distributions (§ 9.8).

Table 9.5: Geometric scaling between length scale l and fractal area A measured in EUV nanoflares (Aschwanden & Parnell 2002), in the Bastille-Day flare, in numerical avalanche simulations, and in photospheric magnetic flux diffusion.

Phenomenon	Length scaling index $1/D_2$ $l(A) \propto A^{1/D_2}$	Area scaling index D_2 $A(l) \propto l^{D_2}$
TRACE 171 Å[1]	0.67 ± 0.03	1.49 ± 0.06
TRACE 195 Å[1]	0.65 ± 0.02	1.54 ± 0.05
Yohkoh SXT[1]	0.61	1.65
Bastille-Day flare[2]		1.56 ± 0.04
2D Avalanche simulation[3]		1.58 ± 0.04
Magnetic flux diffusion[4]		1.56 ± 0.08
Magnetic flux concentrations[5]		1.54 ± 0.05

[1] Aschwanden & Parnell (2002); [2] Mean from 7 box-counting runs shown in Fig. 9.21; [3] Charbonneau et al. (2001); McIntosh & Charbonneau (2001); [4] Lawrence (1991); Lawrence & Schrijver (1993); [5] Balke et al. (1993).

9.7.1 Geometric Scaling

Because the thermal energy content of a heating event scales with the volume V, the measurement of geometric parameters such as length l and area A need careful treatment to infer the appropriate volume. If the topology of the structure of interest is linear (e.g., filaments, loops), the volume would scale proportionally, $V(l) \propto l^1$; if we deal with a flat structure (i.e., sunspot area, photospheric magnetic flux area), the volume scales quadratically, $V(l) \propto l^2$; if we deal with a 3D plasma structure, (i.e., postflare loop arcade, CME), it could scale with the 3D Euclidian space, $V(l) \propto l^3$, or possibly with a fractal *Haussdorff dimension*, $V(l) \propto l^D$, with $1 \leq D \leq 3$. The fractal scaling is a natural property of many fragmented structures that grow with some self-similar properties, such as fern branches, snow crystals, coastal fjords, and perhaps solar flare arcades. The fractal area over which a nonlinear dissipation process spreads has been explored with simulations of avalanche models (Charbonneau et al. 2001).

In Table 9.5 we compile measurements of geometric scaling laws between the length l and area $A(l)$, projected along the line-of-sight, that have been measured for a set of EUV nanoflares with TRACE and Yohkoh/SXT (Aschwanden & Parnell 2002), along with a numerical simulation of an avalanche model (Charbonneau et al. 2001; McIntosh & Charbonneau 2001) and measurements of the diffusion of magnetic flux elements on the photospheric solar surface (Lawrence 1991; Lawrence & Schrijver 1993; Balke et al. 1993), which give a very similar fractal dimension of $D_2 \approx 1.5$ (to within 10%). For the latter process it was concluded that the measured dimension excludes a Euclidian 2D diffusion process but that it is consistent with percolation theory for diffusion on clusters at a density below the percolation threshold. In Fig. 9.21 we show a fractal area measurement for a larger flare, where the fractal area was determined by a box-counting method with different macro-pixel sizes (from 1 to $64 \times 0.5''$). The constancy of the fractal dimension measured with different pixel sizes indicates that we deal with a truly fractal structure, which can be characterized by a single fractal

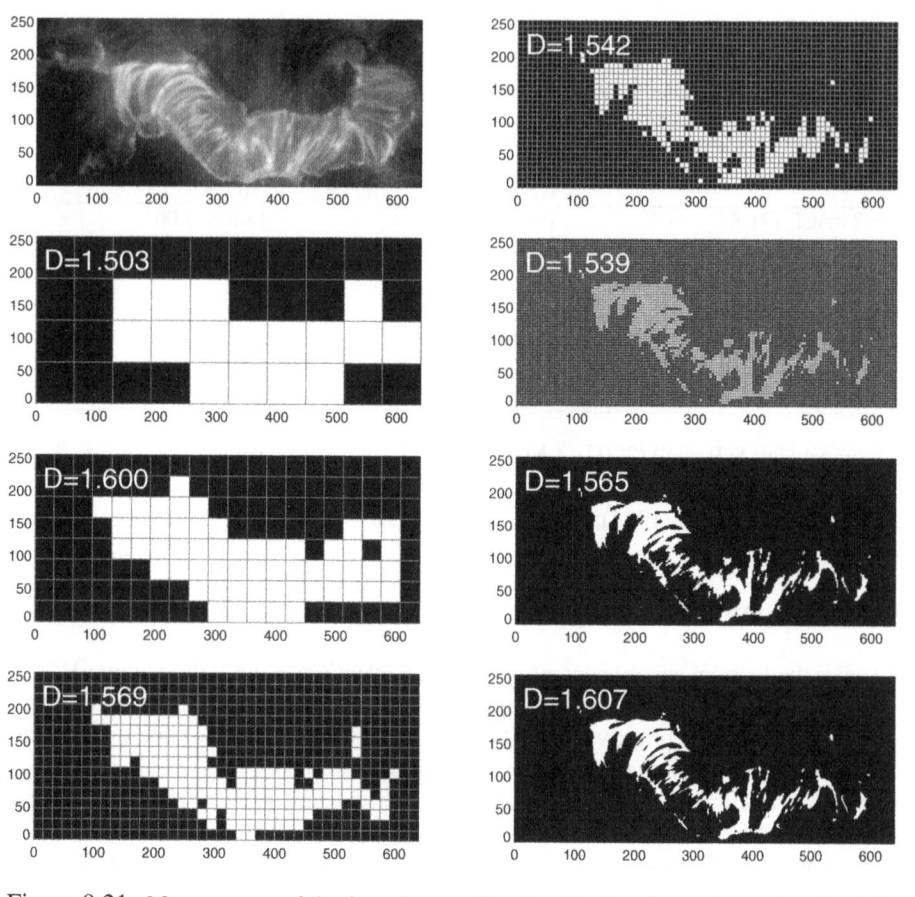

Figure 9.21: Measurement of the fractal area of the Bastille-Day flare, observed by TRACE 171 Å on 2000-Jul-04, 10:59:32 UT. The Haussdorff dimension is evaluated with a box-counting algorithm for pixels above a threshold of 20% of the peak flux value. Note that the Haussdorff dimension is invariant when rebinned with different macro-pixel sizes (64, 32, 16, 8, 4, 2, 1), indicated with a mesh grid. The full resolution image is in the top left frame.

dimension, the so-called *Haussdorff dimension* D_2 (e.g., Mandelbrot 1977; Schroeder 1991; Schuster 1988, p. 54),

$$A(l) \propto l^{D_2} \quad \text{for } l \mapsto 0 . \tag{9.7.1}$$

So, the fractal dimension D_2 can be measured at an arbitrary scale from the logarithmic ratio of the number of 2D-pixels N_A that are covered by the structure area and the number of 1D-pixels N_l that are covered by the maximum structure length,

$$D_2 = \frac{\ln N_A}{\ln N_l} . \tag{9.7.2}$$

The next step, namely to infer the scaling of the volume $V(l)$, is more difficult, because it cannot be measured directly in astrophysical observations, unless we have stereoscopic observations available. In principle, one could use the solar rotation to determine the projected area of a coronal structure from different angles to infer the 3D volume. Alternatively, statistical studies of center-to-limb variations of the projected area could also reveal the 3D volume. For instance, the area of a flat sunspot varies as $A(\theta) \propto \cos(\theta)$ with the inclination angle θ to the solar vertical. We define the fractal dimension of the volume scaling with the symbol D_3,

$$V(l) \propto l^{D_3} \quad \text{for } l \mapsto 0 \,. \tag{9.7.3}$$

An upper limit is the Euclidian scaling, $D_3 = 3$. It is more difficult to establish a lower limit. If we assume that the volume increases at least with the size of the (fractal) area, we would have a lower limit of $D_3 \geq D_2 \approx 1.5$. However, for linear structures with a constant thickness, the dimension could be as low as $D_3 \geq 1$. It is not clear a priori over what range of size scales an object is fractal. Here we are dealing with nanoflares, which can have a size as small as the height of the chromosphere ($L_{min} \gtrsim 3$ Mm), to the largest flares that could have the size of an active region ($L_{max} \lesssim 100$ Mm), so we have a size range of $L_{max}/L_{min} = 100/3 \approx 30$. This implies a range of Euclidian volumes $V_{max}/V_{min} \approx 30^3 \approx 3 \times 10^4$. Upper limits on the filling factor can be estimated from the ratio of the fractal to the Euclidian volume,

$$q_{fill}(l) = \frac{l^{D_3}}{l^3} = l^{D_3-3} \,, \tag{9.7.4}$$

which would amount to $q_{fill} \lesssim 18\%$ for the largest structures and a fractal dimension of $D_3 = 2.5$, but could be as small as $q_{fill} \approx 3\%$ for a fractal dimension of $D_3 = 2.0$. So we could estimate the lower limit of the fractal dimension by measuring the filling factor of flare loops. In the following we adopt an estimate of $D_3 = 2.5 \pm 0.5$.

The fractal scaling of the volume V with the area A is then

$$V(A) \propto l(A)^{D_3} = A^{D_3/D_2} \approx A^{5/3} \,. \tag{9.7.5}$$

So with the fractal dimensions $D_2 \approx 1.5$ and $D_3 \approx 2.5$ we have a scaling of $V(A) \approx A^{5/3}$, which is not too much different from the Euclidian scaling $V(A) = A^{3/2}$. Some nanoflare studies used a pill-box geometry with constant depth, $V(A) \propto A$ (e.g., Benz & Krucker 1998; Parnell & Jupp 2000). In another study, fractal dimensions of $D_2 = 1.5$ and $D_3 = 2.0$ were determined, leading to a scaling of $V(A) = A^{4/3}$ (Aschwanden & Parnell 2002).

9.7.2 Density, Temperature, and Energy Scaling

The RTV scaling laws express an energy balance between heating and (radiative and conductive) losses, using the approximations of constant pressure, no gravity, and uniform heating. While this scaling law breaks down for large cool EUV loops, where the loop length is larger than the pressure scale height, we expect that the approximation should hold for the smaller and hotter flare loops, because the higher temperature produces a larger pressure scale height, so that the assumption of a constant pressure is better fulfilled. In addition, the assumption of uniform heating can be better realized

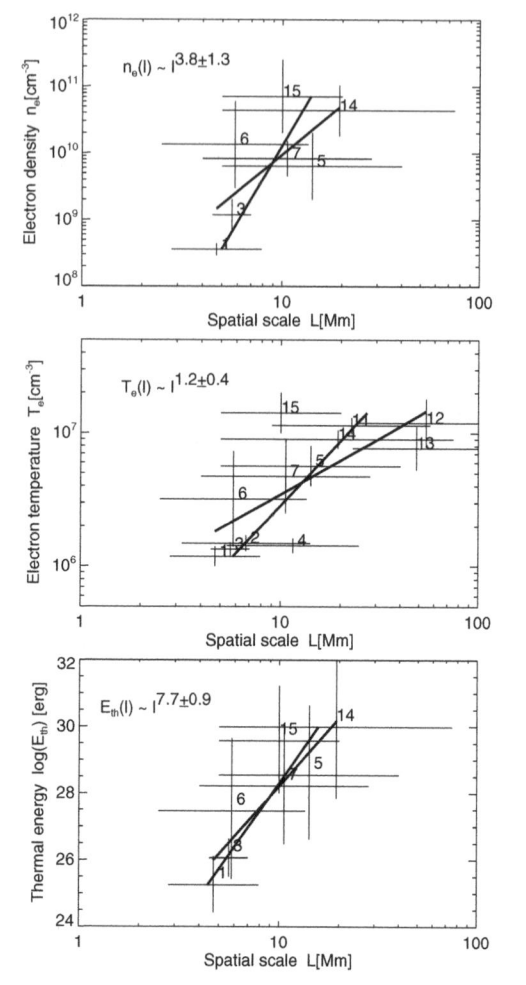

Figure 9.22: Scaling laws between mean electron densities (top), electron temperatures (middle), and thermal energies (bottom) versus mean length scales for the data sets with flare-like events listed in Table 9.4. The regression fit is performed on the logarithmic values and the uncertainties are given by the difference in the fits of $y(x)$ and $x(y)$.

in flare loops than in footpoint-heated active region loops. Another assumption of the RTV laws is the stationary condition for energy balance. This assumption seems to be inappropriate for flares at the first glance. However, the energy balance between heating and cooling terms has to be fulfilled at the temperature peak of the flare, which represents a turning point between dominant heating (i.e., $E_{heat} > E_{loss}$ during the risetime of the flare) and dominant cooling (i.e., $E_{heat} < E_{loss}$ during the decay time of the flare). So, when the flare reaches the temperature maximum at $t = t_{peak}$, the energy balance $E_{heat}(t) = E_{loss}(t)$ actually is an equivalent definition of the turning point $dT(t)/dt = 0$ or temperature maximum $T(t = t_{peak}) = T_{max}$.

So, to explore the scaling of the density $n_e(l)$, temperature $T_e(l)$, and heating rate

$E_H(l)$ as a function of the length scale l for flare-like events, let us use the RTV laws expressed as a function of the density n_e and temperature T_e (Eqs. 9.1.3−4), and relate the volumetric heating rate E_H to the enthalpy change per unit volume, $d\varepsilon_{th}/dV$,

$$T(l) \propto [n_e(l)\, l]^{1/2} \,, \tag{9.7.6}$$

$$E_{H0}(l) \propto n_e^{7/4} l^{-1/4} \,, \tag{9.7.7}$$

$$E_{H0}(l) \propto \frac{d\varepsilon_{th}}{dV} = 3 n_e(l) k_B T_e(l) \,, \tag{9.7.8}$$

Prescribing a powerlaw dependence for all three functions, this system of three coupled equations yields a unique solution for the powerlaw indices,

$$n_e(l) \propto l^3 \,, \tag{9.7.9}$$

$$T_e(l) \propto l^2 \,, \tag{9.7.10}$$

$$\frac{d\varepsilon_{th}(l)}{dV} \propto l^5 \,, \tag{9.7.11}$$

and the total thermal energy integrated over the flare volume is

$$\varepsilon_{th}(l) = 3 n_e(l) k_B T(l) V(l) \propto l^{5+D_3} \approx l^{7.5\pm0.5} \,, \tag{9.7.12}$$

We now perform a linear regression fit through the logarithmic means of the densities $n_e(l)$, temperatures $T_e(l)$, and total thermal energies $\varepsilon_{th}(l)$ from the flare-like data sets listed in Table 9.4 and obtain the following relations (Fig. 9.22; with the uncertainties quantified from the differences of the $y(x)$ and $x(y)$ fits),

$$n_e(l) \propto l^{D_n} \approx \begin{cases} l^3 & \text{RTV law} \\ l^{3.8\pm1.3} & \text{observations} \end{cases} \,, \tag{9.7.13}$$

$$T_e(l) \propto l^{D_T} \approx \begin{cases} l^2 & \text{RTV law} \\ l^{1.2\pm0.4} & \text{observations} \end{cases} \,, \tag{9.7.14}$$

$$\varepsilon_{th}(l) \propto l^{D_n+D_T+D_3} \approx \begin{cases} l^{7.5\pm0.5} & \text{RTV law} \\ l^{7.7\pm0.9} & \text{observations} \end{cases} \,, \tag{9.7.15}$$

So, the agreement between the RTV model and the synthesized data (with nanoflares, microflares, and large flares combined) is reasonable. This result explains the large range of flare energies observed between nanoflares and large flares. Although the spatial scale of flares varies only by a factor of ≈ 20, the energy scaling varies by a factor of $20^{7.5} \approx 5 \times 10^9$, because each physical parameter (n_e, T_e, V) scales with a positive power with the length scale, representing the *"big flare syndrom"*. For nanoflare scaling laws formulated in terms of current sheet parameters or magnetic reconnection models see Litvinenko (1996a), Vekstein & Katsukawa (2000), Craig & Wheatland (2002), and Katsukawa (2003).

There are very few measurements of such scaling laws published in the literature. One scaling law was measured from transient brightenings in soft X-rays observed with Yohkoh/SXT, yielding $\varepsilon_{th}(l) \propto l^{1.6\pm0.1}$ (Ofman et al. 1996). Other samples of EUV nanoflares yielded $\varepsilon_{th}(l) \propto l^{2.11}$ (Aschwanden et al. 2000c) and $\varepsilon_{th}(l) \propto A^{1.634} \approx l^{3.2}$ (Aschwanden & Parnell 2002). We suspect that the scaling laws derived in single data subsets are subject to the temperature bias of a single instrument with limited temperature coverage and to the truncation bias in the limited energy ranges, and thus are not consistent with the synthesized relation derived theoretically and from multiple data sets (Eq. 9.7.15 and Fig. 9.22).

9.7.3 Magnetic Scaling

For both DC and AC coronal heating models the magnetic field is probably the most important quantity determining the amount of deposited energy. A number of studies have therefore been performed to quantify the scaling between the magnetic field and dissipated energy. If we assume a one-to-one conversion of total dissipated magnetic energy $\varepsilon_m \propto B^2 V$ to thermal energy $\varepsilon_{th} = 2n_e k_B T V$, we can predict the following scaling

$$B(l) \propto [n_e(l)T(l)]^{1/2} \propto l^{(D_n+D_T)/2} \approx \begin{cases} l^{2.5} & \text{RTV law} \\ l^{2.5\pm0.7} & \text{observations} \end{cases} , \qquad (9.7.16)$$

or as a function of the area A,

$$B(A) \propto l^{2.5}(A) \propto A^{2.5/D_2} \approx A^{1.7} . \qquad (9.7.17)$$

where the lower limit is $B(A) \propto A^{1.25}$ for Euclidian scaling $D_2 = 2$, and the upper limit is $B(A) \propto A^{2.5}$ for maximum fractality $D_2 = 1$. Observationally, we have to distinguish whether magnetic scaling laws have been derived for entire active regions, for individual coronal loops, or for flare-like events.

Looking at entire active regions, Golub et al. (1980) determined a scaling law for the total thermal energy content ε_{th} in coronal soft X-ray loops with a photospheric magnetic flux $\Phi = BA \approx Bl^2$ (i.e., $\varepsilon_{th} \propto \Phi^{1.5}$). This implies $\varepsilon_{th} \propto nTl^3 \propto (Bl^2)^{3/2} \propto B^{3/2}l^3$, so the dependence on the spatial scale l drops out and the thermal energy per volume, $d\varepsilon_{th}/dV \propto nT \propto B^{1.5}$, is close to a one-to-one conversion of the magnetic energy density, which is expected to scale as $\propto B^2$. Fisher et al. (1998), however, find that the soft X-ray luminosity in active regions correlates best with the total unsigned magnetic flux, $L_{SXR} \approx 1.2 \times 10^{26}$ erg s^{-1} $(|\Phi|_{tot}/10^{22}$ Mx$)^{1.2}$, rather than with B^2. The magnetic flux averaged over active regions has been measured to have a very weak dependence on the area, $\Phi(A) \approx A^{1.1}$ (Harvey 1993; Schrijver & Zwaan 2000), so the magnetic field is almost indpendent of the area of the active region. Van Driel−Gesztelyi et al. (2003) measured the long-term evolution of an active region and found a scaling of $p(B) \approx B^{1.2\pm0.3}$ from BCS measurements, also below the theoretical expectation $p(B) \propto d\varepsilon_{th}/dV \propto B^2$ for a one-to-one conversion of magnetic into thermal energy. We have to be aware that these statistical studies of active regions involve averaging over large areas with variable filling factors, and thus may not reflect the intrinsic scaling between magnetic fields and heating.

Isolating individual active region loops, Porter & Klimchuk (1995) measured with Yohkoh/SXT a scaling law of $p(l) \propto l^{-1.0}$ between the pressure and length of 47 non-flaring soft X-ray loops. Assuming a one-to-one conversion of magnetic to thermal

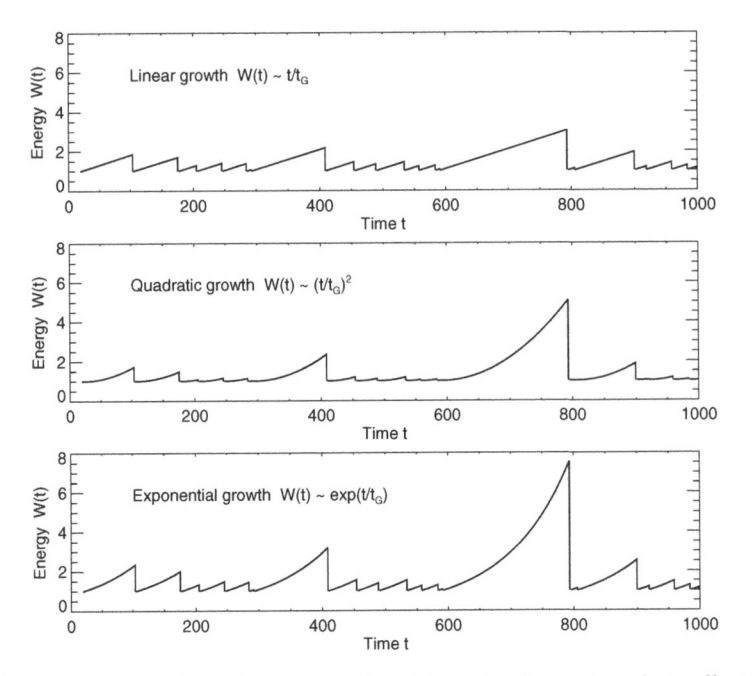

Figure 9.23: Time evolution of energy build-up interrupted by random times: linear (top), quadratic (middle), and exponential growth (bottom). Note that linear growth produces exponential distributions of saturation energies, while nonlinear growth produces powerlaw-like distributions, see Fig. 9.24.

energy, this yields $B(l) \propto [n_e(l)T(l)]^{1/2} \propto p(l)^{1/2} \propto l^{-0.5}$. Mandrini et al. (2000) measured the average magnetic field strengths in some 1000 different coronal loops and found a scaling law of $B(l) \propto l^{-0.9 \pm 0.3}$. Both studies yield relations that are significantly below the theoretical expectation for flare-like events (Eq. 9.7.15). Démoulin et al. (2003) inferred the dependence $E_H(B) \propto B^{\alpha_B}$ from active region loops with Yohkoh SXT and BCS data and corroborate observationally that the index is consistent with the theoretical expectation $a_B = 2$ for a one-to-one conversion of dissipated magnetic energy to heating.

The magnetic scaling for flare loops or flare-like events has not been explored much, for a number of reasons that complicate the analysis: (1) an accurate mapping of flare loop footpoints to the photospheric footpoints is required, which is often ambiguous because of the height difference and canopy-like divergence of the magnetic field; (2) the large-scale currents (and thus the nonpotential magnetic field) has been found to be displaced from hard X-ray flare loop footpoints (Leka et al. 1993); and (3) 95% of the photospheric magnetic field closes inside the transition region, so that the coronal connectivity is relatively sparse (Schrijver & Title 2002). So, there is no quantitative test of the predictions (Eq. 9.7.16) of the the magnetic scaling in flare-like events available yet.

9.8   Statistics of Heating Events

9.8.1   Theory of Frequency Distributions

The statistics of energies in the form of frequency distributions became an important tool for studying nonlinear dissipative events. A *frequency distribution* is a function that describes the occurrence rate of events as a function of their size, usually plotted as a histogram of the logarithmic number $log(N)$ versus the logarithmic size $log(S)$, where the size S could be a length scale l, an area A, a spatial volume V, or a volume-integrated energy E. The two most common functional forms of such frequency distributions are the exponential and the powerlaw function. We will demonstrate that an exponential distribution results from linear or incoherent processes, while a powerlaw distribution results from nonlinear or coherent processes. The latter function has therefore been established as the hallmark of nonlinear dissipative systems. A powerlaw function has no characteristic spatial scale, in contrast to an exponential function, which has an e-folding scale length. The size range over which a powerlaw function applies is called the *inertial range*. We will see that this inertial range extends over more than 8 orders of magnitude in energy for solar flares and nanoflares. Nonlinear dissipative systems, which are constantly driven by some random energy input evolve into a critical state that is maintained as a powerlaw distribution. The fluctuations of the input does not change the powerlaw slope of the dissipated energy events that make up the output, but are just adjusted by a scale-invariant number factor and by a slow shift of the upper cutoff of the distribution. The maintenance of an invariant powerlaw slope is also called *self-organized criticality* and is a property that is inherent to nonlinear dissipative systems. The principle of self-organized criticality has been first applied to solar flare phenomena by Lu & Hamilton (1991).

We can build a simple mathematical model of a nonlinear dissipative system just by two rules: (1) energy is dissipated in random time intervals, and (2) energy builds up with a nonlinear power as a function of time. So, let us consider linear and nonlinear time evolutions (e.g., a quadratic and an exponential function) for the build-up of energy $W(t)$, see Fig. 9.23,

$$W(t) = W_1 \times \begin{cases} (t/\tau_G) & \text{linear} \\ (t/\tau_G)^2 & \text{quadratic} \\ \exp(t/\tau_G) & \text{exponential} \end{cases} , \qquad (9.8.1)$$

where τ_G represents an exponential growth time. If we let each process grow to randomly distributed saturation times $t = t_S$, we will obtain a distribution of saturation energies $W_S = W(t = t_S)$. The distribution of random times t_S obeys Poisson statistics and can be approximated with an exponential distribution with an e-folding time constant t_{Se} (in the tail $t_S \gtrsim t_{Se}$),

$$N(t_S)dt_S = N_0 \exp\left(-\frac{t_S}{t_{Se}}\right)dt_S \qquad (9.8.2)$$

where N_0 is a normalization constant. With these two definitions (Eqs. 9.8.1–2) we can derive the frequency distribution of dissipated energies $N(W_S)$ by substituting the

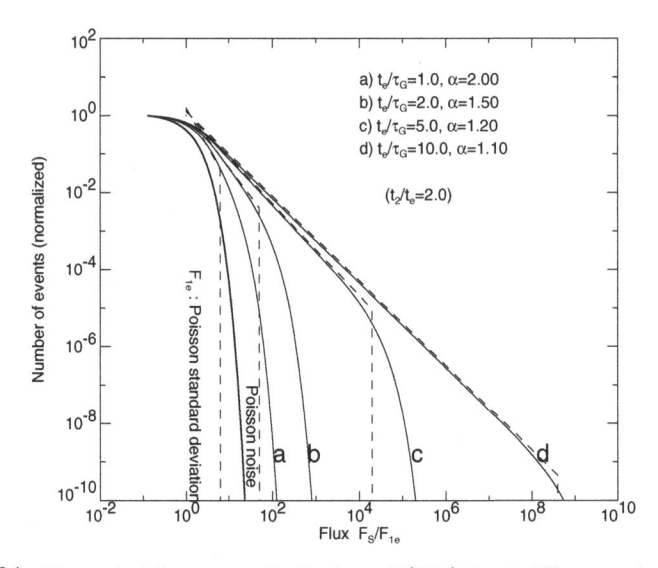

Figure 9.24: Theoretical frequency distributions $N(W_S)$ for 4 different ratios of the mean saturation times t_{Se} to the growth time τ_G: (a) $t_{Se}/\tau_G = 1.0$ (linear case), (b) $t_{Se}/\tau_G = 2.0$, (c) $t_{Se}/\tau_G = 5.0$, (d) $t_{Se}/\tau_G = 10.0$ (nonlinear cases). Note that the frequency distribution evolves from an exponential to a powerlaw distribution the higher the nonlinear saturation time is (Aschwanden et al. 1998a).

saturation times t_S with the energy variable $W_S(t_S)$ in the distribution of saturation times $N(t_S)$ in Eq. (9.8.2),

$$N(W_S)dW_S = N[t_S(W_s)] \left| \frac{dt_S}{dW_S} \right| dW_S . \qquad (9.8.3)$$

So we have to invert the energy evolution time profile $W_S(t_S)$ (Eq. 9.8.1),

$$t_S(W_S) = \tau_G \times \begin{cases} (W_S/W_1) & \text{linear} \\ (W_S/W_1)^{1/2} & \text{quadratic} \\ \ln(W_S/W_1) & \text{exponential} \end{cases} , \qquad (9.8.4)$$

and to calculate the derivatives of the inversions, dt_S/dW_S,

$$\left(\frac{dt_S}{dW_S} \right) = \tau_G \times \begin{cases} (1/W_1) & \text{linear} \\ (1/2W_1)(W_1/W_S)^{1/2} & \text{quadratic} \\ (1/W_S) & \text{exponential} \end{cases} , \qquad (9.8.5)$$

which then can be plugged into Eq. (9.8.3) to yield the frequency distributions of energies:

$$N(W_S) \propto \begin{cases} \exp\left[-(\tau_G/t_{Se})(W_S/W_1)\right] & \text{linear} \\ \exp\left[-(\tau_G/t_{Se})(W_S/W_1)^{1/2}\right] \times W_S^{-1/2} & \text{quadratic} \\ W_S^{-(1+\tau_G/t_{Se})} & \text{exponential} \end{cases} . \qquad (9.8.6)$$

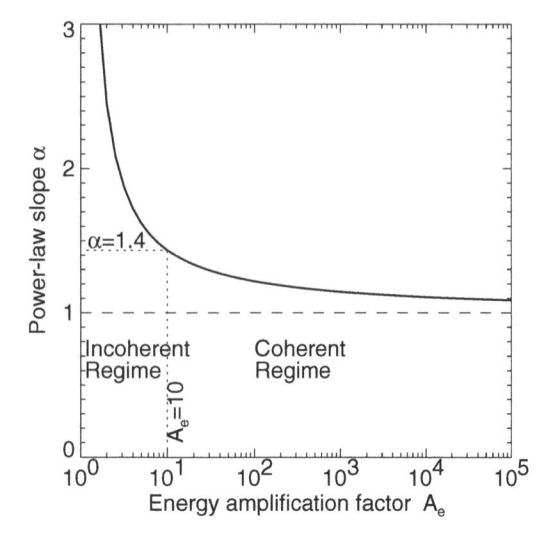

Figure 9.25: Dependence of the powerlaw index α_W of the frequency distributions $N(W_S) \propto W_S^{-\alpha}$ on the ratio of the saturation time t_{Se} to the exponential growth time τ_G (amplification factor). Note that the distribution becomes the flatter the higher the amplification factor (Aschwanden et al. 1998a).

The resulting frequency distribution for linear growth is an exponential distribution, similar to the exponential distribution of saturation times. This is trivial, because the energy W_S is proportional to the saturation time t_S for linear growth. For exponential growth, however, the resulting frequency distribution becomes a powerlaw (Eq. 9.8.6; Fig. 9.24) with an index

$$\alpha_W = \left(1 + \frac{\tau_G}{t_{Se}}\right) , \qquad (9.8.7)$$

(Rosner & Vaiana 1978; Aschwanden et al. 1998a). So the powerlaw slope is determined by the ratio of the exponential growth time τ_G of the nonlinear energy evolution and by the e-folding saturation time t_{Se} of the random distribution of saturation times, with the limit of $\alpha_W \geq 1$ for $t_{Se} \gg \tau_G$ (Fig. 9.25). The linear growth case can also be mimicked by the exponential model for $t_{Se} \ll \tau_G$. We illustrate the relation between the time profiles of the energy evolution $W(t_S)$ and the distribution of saturation energies in Figs. 9.23 to 9.25. The theoretical relation (9.8.7) for the powerlaw slope gives us a diagnostic as to whether the underlying nonlinear dissipative process is incoherent (if $\alpha_W \gg 1$) or coherent with a high amplification factor (if $\alpha_W \gtrsim 1$).

9.8.2   Frequency Distributions of Flare Parameters

We started out to specify the relation between different physical parameters in flares (n_e, T_e, ε_{th}) as a function of the spatial scale l (§ 9.7.2). Therefore, once we have measured the distribution of spatial scales, which might be characterized by a powerlaw distribution,

$$N(l)dl \propto l^{-\alpha_l} dl \qquad 1 \le \alpha_l \le 3 , \qquad (9.8.8)$$

and know the relation of a parameter $y(l)$ as a function of this independent variable l, for example,

$$y(l) \propto l^D \begin{cases} V(l) \propto l^{D_3} & \text{volume} \\ n_e(l) \propto l^{D_n} & \text{density} \\ T_e(l) \propto l^{D_T} & \text{temperature} \\ \varepsilon_{th}(l) \propto l^{D_E} = l^{D_n+D_T+D_3} & \text{thermal energy} \end{cases} . \qquad (9.8.9)$$

we can predict the frequency distributions $N(y)$ of these parameters in the same way as we did in Eqs. (9.8.3−6). We need only to calculate the inversion of $y(l)$ and its derivative, which for powerlaw functions is straightforward,

$$l(y) \propto y^{1/D} , \qquad \frac{dl}{dy} \propto y^{1/D-1} \qquad (9.8.10)$$

and yields the desired frequency distributions,

$$N(y)\, dy = N[l(y)]\frac{dl}{dy}dy = l(y)^{-\alpha_l}\frac{dy}{dl}dy = y^{-[1+(\alpha_l-1)/D]}dy \qquad (9.8.11)$$

So we obtain again powerlaw distributions with the following slopes,

$$N(y)dy \propto y^{-\alpha_y}dy , \qquad \alpha_y = \left(1 + \frac{\alpha_l - 1}{D}\right) . \qquad (9.8.12)$$

Therefore, for the RTV model described in § 9.7.2 we predict the following powerlaw slopes α_y of the frequency distributions $N(y)$ of various physical parameters y,

$$\begin{array}{ll} \underline{\text{RTV model}} & \underline{\text{Volume model}} \\ \alpha_l = 2.5 \pm 0.5 & \alpha_l = 2.5 \pm 0.5 \\ D_3 = 2.5 \pm 0.5 & D_3 = 2.5 \pm 0.5 \\ D_n = 3.0 & D_n = 0.0 \\ D_T = 2.0 & D_T = 0.0 \\ D_E = D_n + D_t + D_3 = 7.5 \pm 0.5 & D_E = 2.5 \pm 0.5 \\ \alpha_V = 1 + \frac{\alpha_l - 1}{D_3} = 1.67 \pm 0.33 & \alpha_V = 1.67 \pm 0.30 \\ \alpha_n = 1 + \frac{\alpha_l - 1}{D_n} = 1.50 \pm 0.17 & \ldots \\ \alpha_T = 1 + \frac{\alpha_l - 1}{D_T} = 1.75 \pm 0.25 & \ldots \\ \alpha_E = 1 + \frac{\alpha_l - 1}{D_E} = 1.21 \pm 0.08 & \alpha_E = 1.67 \pm 0.33 \end{array} \qquad (9.8.13)$$

Thus we predict a powerlaw distribution of energies with a slope of $\alpha_E = 1.21 \pm 0.08$ for the RTV model, where the error bars include only the propagation errors of the fractal dimension $D_3 = 2.5 \pm 0.5$ and the length distribution $\alpha_l = 2.5 \pm 0.5$. Let us define also an alternative model, the so-called *Volume model*, where the thermal energy is directly proportional to the volume, without any dependence on the density and temperature. Such a model may be representative if statistics is done in a subset of the data, say in a narrowband filter with a small temperature range and with a flux threshold (which restricts the range of detected densities). In such a restricted data subset we would predict a slope of $\alpha_E = 1.67 \pm 0.33$. Real data with some temperature range and some moderate flux range, of course, could produce any value between these two cases, $1.21 \leq \alpha_E \leq 1.67$. Powerlaw indices in the range of $1.1 - 1.64$ have also been derived from magnetic braiding and twisting models (Zirker & Cleveland 1993a,b).

9.8.3 Energy Budget of Flare-like Events

The frequency distributions specify the number of events $N(W_i)$ in an energy bin $[W_i, W_{i+1}]$. If we want to know the total energy budget over some range that is bracketed by the minimum W_1 and maximum W_2, we have to integrate the energy powerlaw distribution,

$$W_{tot} = \int_{W_1}^{W_2} N(W)\, W\, dW = \int_{W_1}^{W_2} N_1 W_1 \left(\frac{W}{W_1}\right)^{1-\alpha_W} dW =$$

$$= N_1 W_1^2 \begin{cases} \frac{1}{(2-\alpha_W)} \left[\left(\frac{W_2}{W_1}\right)^{2-\alpha_W} - 1\right] & \text{if } \alpha_2 \neq 2, \\ [\ln (W_2/W_1)] & \text{if } \alpha_2 = 2, \end{cases} \tag{9.8.14}$$

From this expression we see immediately that the integral is dominated by the upper limit W_2 for flat powerlaw indices $\alpha_W < 2$, and by the lower limit W_1 in the case of steep powerlaw indices $\alpha_W > 2$. This implies that nanoflares are important for coronal heating if the frequency distribution of their energy has a slope steeper than 2, a necessary condition that was pointed out by Hudson (1991a).

Let us calculate some practical cases that are relevant for the coronal heating problem. The energy requirements for coronal heating are given in Table 9.1. For the quiet Sun we need a heating rate of $F_{QS} \approx 3 \times 10^5$ erg cm^{-2} s^{-1}, and for active regions we need $F_{AR} \approx 10^7$ erg cm^{-2} s^{-1}. We know that flares occur only in active regions and could have a maximum energy up to $W_{2,AR} \approx 10^{32}$ erg (Fig. 9.27), while the largest microflares occurring in the quiet Sun are the so-called X-ray bright points, which have energies up to $W_{2,QS} \approx 10^{30}$ erg. For the lower energy limit we take the smallest nanoflares that have been observed so far, which have energies of $W_{1,QS} \approx W_{1,AR} \approx 10^{24}$ erg. With these values we obtain with (Eq. 9.8.14) the following rates,

$$N_1 = \frac{W_{tot}}{W_1^2} \frac{1}{(2-\alpha_W)} \frac{1}{\left(\frac{W_2}{W_1}\right)^{2-\alpha_W} - 1} = \begin{cases} 4.3 \times 10^0 & \text{QS : (RTV model)} \\ 1.0 \times 10^3 & \text{QS : (Volume model)} \\ 1.4 \times 10^2 & \text{AR : (RTV model)} \\ 3.5 \times 10^4 & \text{AR : (Volume model)} \end{cases} \tag{9.8.15}$$

We visualize these four distributions in Fig. 9.26. Because the powerlaw slopes are all below the critical value of $\alpha_E = 2$, the total energy is dominated by the upper energy cutoff W_2, so that the lower energy cutoff has almost no effect, since $W_1 \ll W_2$. Consequently, the integrated total energy is also not very sensitive to the exact value of the powerlaw slope. In essence, the occurrence rate at the high energy cutoff $N(W_2)$ is the most decisive parameter determining the energy budget.

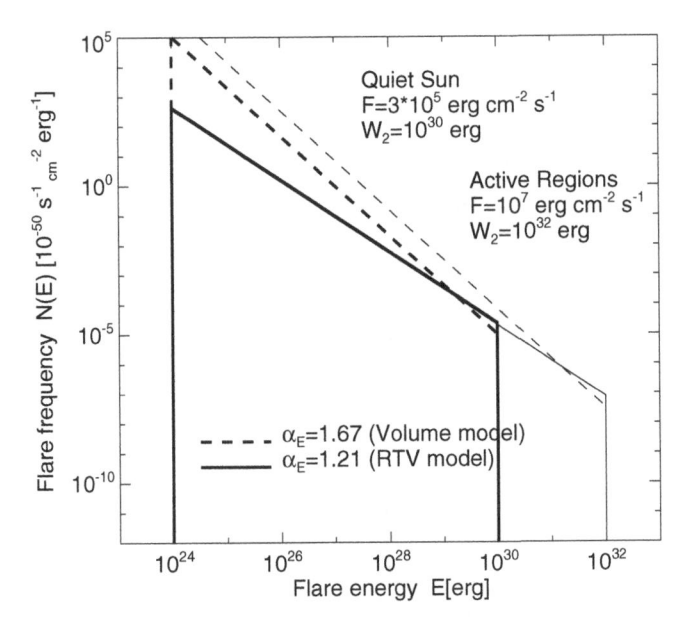

Figure 9.26: Predicted frequency distributions that fulfill the coronal heating requirement, for quiet Sun regions ($F = 3 \times 10^5$ erg cm^{-2} s^{-1}) and active regions ($F = 10^7$ erg cm^{-2} s^{-1}), according to the RTV model (with a powerlaw slope of $\alpha_E = 1.21$ and the volume model ($\alpha_E = 1.67$), see text.

9.8.4 Measurements of Frequency Distributions

A compilation of some recent frequency distributions of nanoflare energies is shown in Fig. 9.27, which all have a powerlaw slope of approximately $\alpha_E \approx 1.55$. In the same figure we also show the energy distribution of the coronal heating requirement for the quiet Sun (grey area in Fig. 9.27), for the same powerlaw slope and the parameters: $F = W_{tot} = 3 \times 10^5$ erg cm^{-2} s^{-1}, $W_1 = 10^{24}$ erg, $W_2 = 10^{30}$ erg, $\alpha_E = 1.55$. We see that the observed nanoflare distribution lies about a factor of 10 below the theoretical occurrence rate, or shifted to the left by about a factor of ≈ 3 in energy. Now we have to be aware that the thermal energy is calculated based on the radiation we detect in EUV and soft X-rays, so it characterizes only the energy equivalent to the radiative losses, while it does not include energy losses due to conduction to the chromosphere or the solar wind flux. The radiative losses in the quiet Sun alone are indeed about a factor of 3 lower than the total coronal energy losses (i.e., $F = 1 \times 10^5$ erg cm^{-2} s^{-1}; Table 9.1). So we can conclude that the detected radiation of the EUV and SXT nanoflares roughly corresponds to a third of the total coronal heating requirement in quiet Sun regions, which covers approximately the radiative losses. Because there are many uncertainties involved in the quantification of observed frequency distributions, this result still needs to be corroborated. If this result holds up, it has the important consequence that we have localized the coronal heating sources in the form of detectable nanoflares in EUV and soft X-rays with a sufficient rate, and thus we do not need to

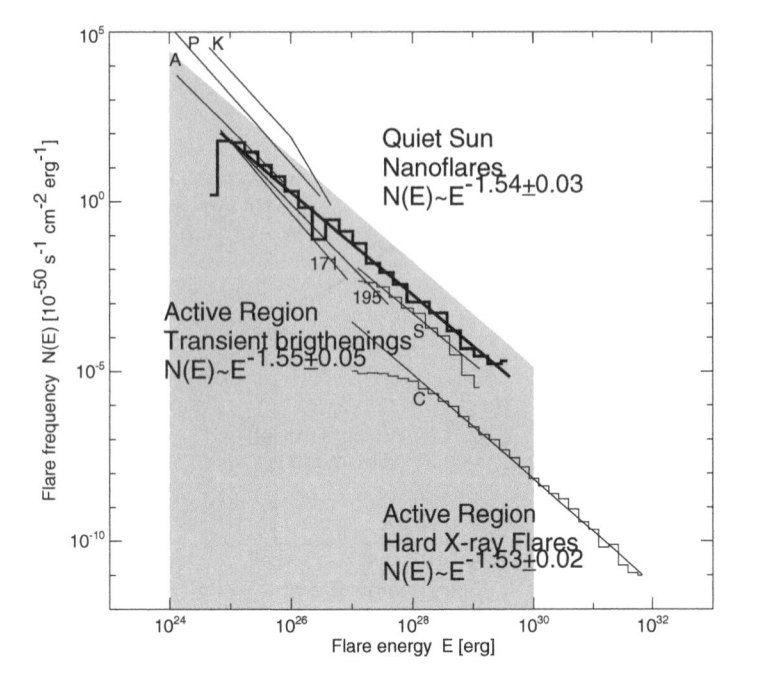

Figure 9.27: Compilation of frequency distributions of thermal energies from nanoflare statistics in the quiet Sun, active region transient brightenings, and hard X-ray flares. The labels indicate the following studies: K=Krucker & Benz (1998), Benz & Krucker (2002); P=Parnell & Jupp (2000) (corrected for an erroneous factor of 100 in the original paper); A=Aschwanden et al. (2000c); S=Shimizu (1995); C=Crosby et al. (1993), and 171, 195=Aschwanden & Parnell (2002). The overall slope of the synthesized nanoflare distribution, $N(E) \propto E^{-1.54\pm0.03}$, is similar to that of transient brightenings and hard X-ray flares. The grey area indicates the coronal heating requirement of $F = 3 \times 10^5$ erg cm^{-2} s^{-1} for quiet Sun regions. Note that the observed distribution of nanoflares falls short of the theoretical requirement by a factor of 10 in occurrence rate or a factor of ≈ 3 in energy.

invoke invisible energy sources such as heating by Alfvén waves to explain the radiation of the heated plasma, at least not in quiet Sun regions. In coronal holes, the total energy losses are much higher due to the solar wind fluxes (Table 9.1), where heating by Alfvén waves is probably required in addition to nanoflare heating.

In Table 9.6 we compile frequency distributions of small-scale phenomena that have been reported from the quiet Sun and calculate their total energy flux F based on the observed energy ranges $[W_1, W_2]$, powerlaw slopes α_E, and rate $N_1 = N(W_1)$. EUV transients, nanoflares and microflares generally are found in the energy range of $W \approx 10^{24} - 10^{26}$ erg and the integrated flux over the entire frequency distribution lies in the range of $F \approx (0.5 \pm 0.2)10^5$ erg^{-1} cm^{-2} s^{-1}, which makes up about one to two-thirds of the total heating requirement of the quiet corona, roughly covering the radiative losses in the quiet Sun corona. This corroborates our finding in Fig. 9.27. A similar flux was also measured for explosive events in C III, Ne IV, and

Table 9.6: Frequency distributions of small-scale phenomena observed in quiet Sun regions.

Phenomenon	Number of events N	Powerlaw slope α_E	Energy range E_1, E_2 10^{24} [erg]	Total flux F [erg cm^{-2} s^{-1}]
EUV transients, EIT, 171+195[1]	233	2.45 ± 0.15	$10 - 300$	0.7×10^5
EUV transients, EIT 195[2]	228	1.35 ± 0.20	$1 - 100$	...
EUV transients, EIT 195[3]	277	1.45 ± 0.20	$10 - 100$	...
Nanoflares, TRACE, 171+195[4]	5131	2.48 ± 0.11	$0.3 - 60$	0.2×10^5
Nanoflares, TRACE+SXT[5]	281	1.53 ± 0.02	$10 - 10^6$	0.5×10^5
Blinkers, CDS, O V[6]	790	1.34 ± 0.08	$0.01 - 0.3$	...
Explosive ev., SUMER C III[7]	3403	2.8 ± 0.1	$0.05 - 2$	0.45×10^5
Explosive ev., SUMER Ne IV[7]	2505	2.8 ± 0.1	$0.6 - 10$	0.16×10^5
Explosive ev., SUMER O VI[7]	5531	3.3 ± 0.4	$0.1 - 2$	0.79×10^5
Explosive ev., SUMER Ne VIII[7]	2907	2.8 ± 0.5	$0.06 - 1$	0.03×10^5

[1] Krucker & Benz (1998); [2] Berghmans et al. (1998); [3] Berghmans & Clette (1999); [4] Parnell & Jupp (2000) [corrected for a factor of 100 in original paper]; [5] Aschwanden et al. (2000b); [6] Brkovic et al. (2001); [7] Winebarger et al. (2002).

O VI (Winebarger et al. 2002), which fits into the picture that explosive events and nanoflares are probably controlled by the same physical process as a magnetic reconnection process in the transition region, which is manifested with comparable amounts of thermal plasma inside the transition region (as detected in the cooler EUV lines in C III, Ne IV, and O VI) as well as in the lower corona (in the hotter EUV lines of Fe IX/X and Fe XII). Other phenomena such as blinkers carry several orders of magnitude less energy ($\varepsilon_{th} \approx 10^{22} - 3 \times 10^{23}$; Brkovic et al. 2001), and thus seem to be energetically less important for coronal heating.

There are some significant variations in the powerlaw slope of the frequency distributions, ranging from as low as $\alpha_E \approx 1.34$ up to $\alpha_E \lesssim 2.6$ (Table 9.6). Our theoretical RTV model predicts a slope in the range of $1.21 \leq \alpha_E \leq 1.67$ (Fig. 9.26), depending on the sampling over a broad or narrow temperature and flux range. There are a number of systematic effects in the data analysis and modeling of the thermal energy that affect the resulting powerlaw slope, such as: (1) event definition and discrimination, (2) sampling completeness, (3) observing cadence and exposure times, (4) pattern recognition algorithm, (5) geometric, density, and thermal energy model, (6) line-of-sight integration, (7) extrapolation in undetected energy ranges, (8) wavelength and filter bias, (9) fitting procedure of frequency distributions, and (10) error estimates of powerlaw slopes. Technical details about these issues are discussed and compared in a number of papers (e.g., Aschwanden & Parnell 2002; Aschwanden & Charbonneau 2002; Benz & Krucker 2002; Parnell 2002a, 2002b; Berghmans 2002). The main lesson is that extrapolation of the powerlaw to unobserved energies that are many orders of magnitude smaller than the observed energy ranges remains questionable. The integrated energy flux over the observed energy range is less susceptible to the powerlaw slope, because the total flux of the sum of all measured events is conserved, regardless how the fine structure is subdivided into discret subevents. Fortunately, since the total energy of the observed nanoflare distributions is commensurable with the radiative losses, there is no need to extrapolate the distribution to unobserved energy ranges, and thus the question whether the powerlaw slope is below or above the critical value of 2 is not decisive for the heating budget. Another lesson is the completeness of temperature coverage, which generally requires coordinated multi-wavelength observations. For instance, a statistical analysis of coronal bright points with EIT 195 Å revealed that bright points cover only about 1.4% of the quiet Sun area, and their radiation accounts for about 5% of the quiet Sun radiation (Zhang et al. 2001), while the multi-wavelength data sets listed in Table 9.5 reproduce almost all of the quiet Sun flux.

9.9 Summary

The coronal heating problem has been narrowed down by substantial progress in theoretical modeling with MHD codes, new high-resolution imaging with the SXT, EIT and TRACE telescopes, and with more sophisticated data analysis using automated pattern recognition codes. The total energy losses in the solar corona range from $F = 3 \times 10^5$ erg cm^{-2} s^{-1} in quiet Sun regions to $F \approx 10^7$ erg cm^{-2} s^{-1} in active regions (§ 9.1). Theoretical models of coronal heating mechanisms include the two main groups of DC (§ 9.3) and AC models (§ 9.4), which involve as a primary energy source chromospheric footpoint motion or upward leaking Alfvén waves, which are dissipated in the corona by magnetic reconnection, current cascades, MHD turbulence, Alfvén resonance, resonant absorption, or phase mixing. There is also strong observational evidence for solar wind heating by cyclotron resonance, while velocity filtration seems not to be consistent with EUV data (§ 9.5). Progress in theoretical models has mainly been made by abandoning homogeneous fluxtubes, but instead including gravitational scale heights and more realistic models of the transition region, and taking advantage of numerical simulations with 3D MHD codes. From the observational side we can now unify many coronal small-scale phenomena with flare-like characteristics, subdivided into microflares (in soft X-rays) and nanoflares (in EUV) solely by their energy content (§ 9.6). Scaling laws of the physical parameters corroborate the unification of nanoflares, microflares, and flares; they provide a physical basis to understand the frequency distributions of their parameters and allow estimation of their energy budget for coronal heating (§ 9.8). Synthesized data sets of microflares and nanoflares in EUV and soft X-rays have established that these impulsive small-scale phenomena match the radiative loss of the average quiet Sun corona, which points to small-scale magnetic reconnection processes in the transition region and lower corona as primary heating sources of the corona.

Chapter 10

Magnetic Reconnection

The solar corona has dynamic boundary conditions: (1) The solar dynamo in the interior of the Sun constantly generates new magnetic flux from the bottom of the convection zone (i.e., the tachocline) which rises by buoyancy and emerges through the photosphere into the corona; (2) the differential rotation as well as convective motion at the solar surface continuously wrap up the coronal field with every rotation; and (3) the connectivity to the interplanetary field has constantly to break up to avoid excessive magnetic stress. These three dynamic boundary conditions are the essential reasons why the coronal magnetic field is constantly stressed and has to adjust by restructuring the large-scale magnetic field by topological changes, called *magnetic reconnection* processes. Of course, such magnetic restructuring processes occur wherever the magnetic stresses build up (e.g., in the canopy-like divergent field in the transition region, in highly tangled coronal regions in active regions, or at coronal hole boundaries). A classical example is a transequatorial coronal hole that sometimes is observed to rotate almost rigidly during several solar rotations, although the underlying photosphere displays the omnipresent differential rotation (in latitude): The shape of the coronal hole can only be preserved quasi-statically, if the photospheric magnetic field constantly disconnects and reconnects at the eastern and western boundaries. Topological changes in the form of magnetic reconnection always liberate free nonpotential energy, which is converted into heating of plasma, acceleration of particles, and kinematic motion of coronal plasma. Magnetic reconnection processes can occur in a slowly changing quasi-steady way, which may contribute to coronal heating (§ 9), but more often happen as sudden violent processes that are manifested as *flares* and *coronal mass ejections* (§ 11-17). These dynamic processes are the most fascinating plasma processes we can observe in the universe, displaying an extreme richness of highly dynamic phenomena observable in all wavelengths.

We concentrate here mainly on the magnetic reconnection processes in the solar corona, including associated processes in the photospheric and chromospheric boundaries, but there is also a rich literature on magnetic reconnection processes in planetary magnetospheres (e.g., Kivelson & Russell 1995, § 9; Treumann & Baumjohann 1997, § 7; Scholer 2003), in laboratory tokomaks and spheromaks (e.g., Bellan 2002), as well as in other astrophysical objects such as comets, planets, stars, accretion disks, etc. Introductions into magnetic reconnection in the solar corona can be found in textbooks (Priest 1982, § 10.1; Sturrock 1994, § 17; Priest & Forbes 2000; Somov 2000, § 16-22; Tajima & Shibata 2002, § 3.3), or in the following recent reviews and encyclopedia articles (Forbes 2001; Schindler & Hornig 2001; Ugai 2001; Hood et al. 2002; Biskamp 2003; Kliem et al. 2003), while specific applications to solar flares can be found in the proceedings of some Yohkoh conferences (Bentley & Mariska 1996; Watanabe et al. 1998; Martens & Cauffman 2002).

10.1 Steady 2D Magnetic Reconnection

Quasi-steady reconnection of magnetic fields enables the coronal plasma to dissipate magnetic energy, a process that has been proposed to yield direct plasma heating of the corona (e.g., Parker 1963a, 1972, 1979, 1983; Sturrock & Uchida 1981; Van Ballegooijen 1986) or to supply direct plasma heating in flares (e.g., Sweet 1958; Parker 1963a; Petschek 1964; Carmichael 1964; Sturrock 1966). This concept represents one of the most fundamental building blocks that has been used in many theoretical models of coronal heating and solar flares, which we outline in the following.

When a new magnetic flux system is pushed towards a pre-existing old magnetic flux system (e.g., as the solar wind runs into the magnetopause at the Earth's bow shock), or as a new emerging flux region pushes through the chromosphere upwards into a pre-existing coronal magnetic field, a new dynamic boundary is formed where the magnetic field can be directed in opposite directions at both sides of the boundary. The magnetic field has then necessarily to drop to zero at the boundary to allow for a continuous change from a positive to a negative magnetic field strength. Thus the balance between the magnetic and thermal pressure (Eq. 6.2.4) across the *neutral boundary layer*,

$$\frac{B_1^2}{8\pi} + p_1 = p_{nl} = \frac{B_2^2}{8\pi} + p_2 \,, \tag{10.1.1}$$

yields a higher thermal pressure (p_{nl}) in the neutral layer (where $B = 0$) than on both sides with finite field strengths B_1 and B_2. In a 1D model we would have an infinite neutral boundary layer. In reality, however, the process of bringing two oppositely directed magnetic flux systems together will always have a finite area of first contact, which limits the extent of the neutral boundary layer and channels outflows to both sides, so that the simplest scenario is a 2D model as shown in Fig. 10.1, where the lateral inflows (driven by external forces) will create outflows along the neutral line in an equilibrium situation. The plasma-β parameter $\beta = p_{th}/(B_1^2/8\pi)$ becomes larger than unity in the central region (because $B_1 \mapsto 0$), so that the plasma can flow across the magnetic field lines, which is called the *diffusion region*, and is channeled into the outflow regions along the neutral boundary. Outside the diffusion region the plasma-β again drops below unity and the magnetic flux is frozen-in. The highly pointed magnetic field lines in the outflow region experience a high curvature force that tries to smooth out the cusps in the outflow region until a balance between the outward-directed magnetic tension force and the inward-directed magnetic pressure force plus thermal pressure is achieved (see § 6.2.2 for the Lorentz force near the X-point). This magnetic field line relaxation process is also called the *sling shot effect*, which is the

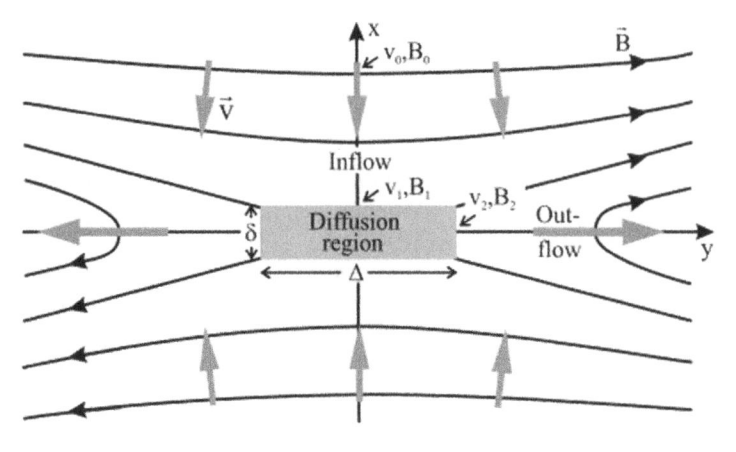

Figure 10.1: Basic 2D model of a magnetic reconnection process, driven by two oppositely directed inflows (in x-direction), which collide in the diffusion region and create oppositely directed outflows (in y-direction). The central zone with a plasma-β parameter of $\beta > 1$ is called the diffusion region (grey box) (Schindler & Hornig 2001).

basic conversion mechanism of magnetic into kinetic energy. The stationary outflows are sandwiched between two *standing slow shocks* (which do not propagate). The end result is a thin diffusion region with width δ and length Δ (Fig. 10.1). The whole process can evolve into a steady-state equilibrium with continuous inflows and outflows, driven by external forces. Since the Lorentz force creates an electric field E_0 in a direction perpendicular to the 2D-plane of the flows (i.e. perpendicular to the image plane of Fig. 10.1), a current j_{nl} in the neutral layer is associated with the electric field E_0 according to Ohm's law (Eq. 5.1.10),

$$E_0 = \frac{1}{c} v_1 B_1 = \frac{1}{c} v_2 B_2 = \frac{j_{nl}}{\sigma} \, , \tag{10.1.2}$$

which is termed the *current sheet* for the diffusion region. The finite resistivity σ requires, strictly speaking, a treatment in the framework of resistive MHD (§ 6.1.5), although the processes outside the diffusion region can be approximated using the ideal MHD equations (§ 6.1.3).

10.1.1   Sweet–Parker Reconnection Model

There exists no full analytical solution for the steady-state situation of the reconnection geometry shown in Fig. 10.1 using the full set of resistive MHD equations (§ 6.1.5), but separate analytical solutions for the external (ideal MHD) region and special solutions for the (resistive MHD) diffusion region are available that can be matched with some simplifications. One such solution is the *Sweet–Parker* model (Sweet 1958; Parker 1963a), where it is assumed that the diffusion region is much longer than it is wide, $\Delta \gg \delta$. For steady, compressible flows ($\nabla \cdot \mathbf{v} \neq 0$), it was found that the outflows roughly have Alfvén speeds,

$$v_2 = v_A = \frac{B_2}{\sqrt{4\pi\rho_2}} , \tag{10.1.3}$$

and that the outflow speed v_2 relates to the inflow speed v_1 reciprocally to the cross sections δ and Δ (according to the continuity equation),

$$\rho_1 v_1 \Delta = \rho_2 v_2 \delta , \tag{10.1.4}$$

and that the *reconnection rate* M_0, defined as the Mach number ratio of the external inflow speed v_0 to the (Alfvén) outflow speed v_A, is (with the approximation $B_1 \approx B_0$, $v_1 \approx v_0$, and $S_1 \approx S_0$),

$$M_0 = \frac{v_0}{v_A} = \frac{1}{\sqrt{S_0}} . \tag{10.1.5}$$

The *Lundquist number S* (or *magnetic Reynolds number*) is defined by

$$S = v_A L/\eta , \tag{10.1.6}$$

analogous to the Reynolds number $R = vL/\eta$ defined for a general fluid velocity v. From Eqs. (10.1.4−6) it follows the relation

$$v_0 = \frac{\eta}{\delta} . \tag{10.1.7}$$

So, for typical coronal conditions (with a large Lundquist number of $S_0 = R_m \approx 10^8 - 10^{12}$) the reconnection rate is typically $M_0 \approx 10^{-4} - 10^{-6}$, which yields inflow speeds in the order of $v_0 \approx v_A M_0 \approx 1000$ km s^{-1} $\times 10^{-5} \approx 0.01$ km s^{-1} (using Eq. 10.1.5) and yields extremely thin current sheets with a thickness of $\delta = \Delta(v_A/v_1) \approx \Delta \times 10^{-5}$ (using Eq. 10.1.4). So, a current sheet with a length of $\Delta \approx 1000$ km would have a thickness of only $\delta \approx 10$ m. In typical flares, energies of $\varepsilon_{tot} \approx 10^{28} - 10^{32}$ erg (Table 9.4, Fig. 9.27) are dissipated over flare durations of $\Delta t \approx 10 - 10^2$ s, which imply much larger dissipation rates than obtained with the Sweet−Parker current sheet,

$$\frac{d\varepsilon_m}{dt} = \frac{B^2}{8\pi}\frac{dV}{dt} \approx \frac{B^2}{8\pi}L^2 v_0 \approx \frac{B^2}{8\pi}\frac{L^2 v_A}{\sqrt{S_0}} \approx 10^{22} \left(\frac{B}{100\ \mathrm{G}}\right)^2 \left(\frac{L}{1\ \mathrm{Mm}}\right)^2 \left(\frac{v_A}{1\ \mathrm{Mm/s}}\right) , \tag{10.1.8}$$

so the Sweet−Parker reconnection rate is much too slow to explain the magnetic dissipation in solar flare events.

10.1.2   Petschek Reconnection Model

A much faster reconnection model was proposed by Petschek (1964), which involved reducing the size of the diffusion region to a very compact area ($\Delta \approx \delta$) that is much shorter than the Sweet−Parker current sheet ($\Delta \gg \delta$) (Fig. 10.2). Summaries of the Petschek model can be found, see example in Priest (1982, p. 351), Jardine (1991), Priest & Forbes (2000, p. 130), Treumann & Baumjohann (1997, p. 148), and Tajima & Shibata (2002, p. 225). Because the length of the current sheet is much shorter, the

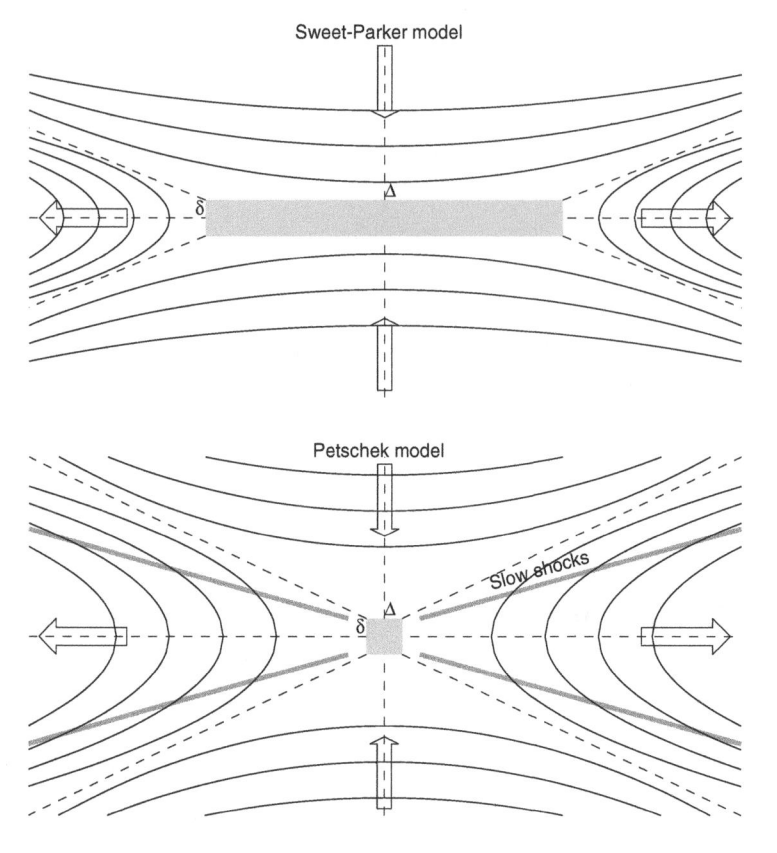

Figure 10.2: Geometry of the Sweet−Parker (top) and Petschek reconnection model (bottom). The geometry of the diffusion region (grey box) is a long thin sheet ($\Delta \gg \delta$) in the Sweet−Parker model, but much more compact ($\Delta \approx \delta$) in the Petschek model. The Petschek model also considers slow-mode MHD shocks in the outflow region.

propagation time through the diffusion region is shorter and the reconnection process becomes faster. However, in a given external area with size L_e comparable with the length Δ_{SP} of the Sweet−Parker current sheet, a much smaller fraction of the plasma flows through the Petschek diffusion region with size Δ_P, where finite resistivity σ exists and field lines reconnect. Most of the inflowing plasma turns around outside the small diffusion region and *slow-mode shocks* arise where the abrupt flow speed changes from v_1 to v_2 in the outflow region (Fig. 10.2, bottom). The shock waves represent an obstacle in the flow and thus are the main sites where inflowing magnetic energy is converted into heat and kinetic energy. Simple energy considerations show that inflowing kinetic energy is split up roughly in equal parts into kinetic and thermal energy in the outflowing plasma (Priest & Forbes 2000). Petschek (1964) estimated the maximum flow speed v_e by assuming a magnetic potential field in the inflow region and found that at large distance L_e the external field $B_0(L_e)$ scales logarithmically with distance L_e,

$$B_0(L_e) = B_0 \left[1 - \frac{4M_0}{\pi} \ln \left(\frac{L_e}{\Delta} \right) \right]. \tag{10.1.9}$$

Petschek (1964) estimated the maximum reconnection rate M_0 at a distance L_e where the internal magnetic field is half of the external value (i.e., $B_0(L_e) = B_0/2$), which yields using Eq. (10.1.9),

$$M_0 = \frac{\pi}{8 \ln (L_e/\Delta)} \approx \frac{\pi}{8 \ln (R_{me})}. \tag{10.1.10}$$

So, the reconnection rate $M_0 = v_0/v_A$ depends only logarithmically on the magnetic Reynolds number $R_{me} = L_e v_{Ae}/\eta$. Therefore, for coronal conditions, where the magnetic Reynolds number is very high (i.e., $R_{me} \approx 10^8 - 10^{12}$), the Petschek reconnection rate is $M_0 \approx 0.01 - 0.02$ according to Eq. (10.1.10), yielding an inflow speed of $v_0 \approx v_A M_0 \approx 10 - 20$ km s^{-1} for typical coronal Alfvén speeds of $v_A \approx 1000$ km s^{-1}. Thus, the Petschek reconnection rate is about three orders of magnitude faster than the Sweet−Parker reconnection rate.

10.1.3   Generalizations of Steady 2D Reconnection Models

The semi-quantitative Petschek model has been generalized in a number of mathematically more rigorous formulations that are summarized in Priest (1982) and Priest & Forbes (2000). The generalizations concern the magnetic field discontinuity in the slow-mode shock regions, the matching of flow velocities between the diffusion region and the external region, and the compressibility of the plasma (Green & Sweet 1967; Petschek & Thorne 1967; Sonnerup 1970; Yeh & Axford 1970; Cowley 1974a, b; Yang & Sonnerup 1976; Roberts & Priest 1975; Priest & Soward 1976; Soward & Priest 1977; Mitchell & Kan 1978). The structure of the diffusion region has been modeled in greater detail (Priest & Cowley 1975; Milne & Priest 1981; Priest 1972; Somov 1992).

A unification of fast, steady, *almost-uniform reconnection* solutions of the MHD equations was accomplished by Priest & Forbes (1986), who derived the following expression for the reconnection rate M_e in the external diffusion region,

$$\left(\frac{M_e}{M_i} \right)^2 \approx \frac{4M_e(1-b)}{\pi} \left[0.834 - \ln \tan \left(\frac{4R_{me}M_e^{1/2}M_i^{1/2}}{\pi} \right)^{-1} \right], \tag{10.1.12}$$

which contains the Sweet−Parker, the Petschek ($b = 0$), and Sonnerup solution ($b = 1$) as special cases. Solutions with $b < 0$ produce *slow-mode compression*, solutions with $b > 1$ produce *slow-mode expansions*, also called the *flux pile-up regime* (Litvinenko 1999b; Litvinenko & Craig 2000; Craig & Watson 2000b), while the intermediate range of $0 < b < 1$ produces hybrid solutions of slow-mode and fast-mode expansions. The unified solutions have been extended by including nonlinearity effects in the inflow, compressibility, energetics, and reverse currents (Jardine & Priest 1988a, b, c, 1989, 1990; Jardine 1991). The *almost-uniform reconnection* solutions refer to the magnetic field in the inflow region (for which Petschek assumed a potential field). As

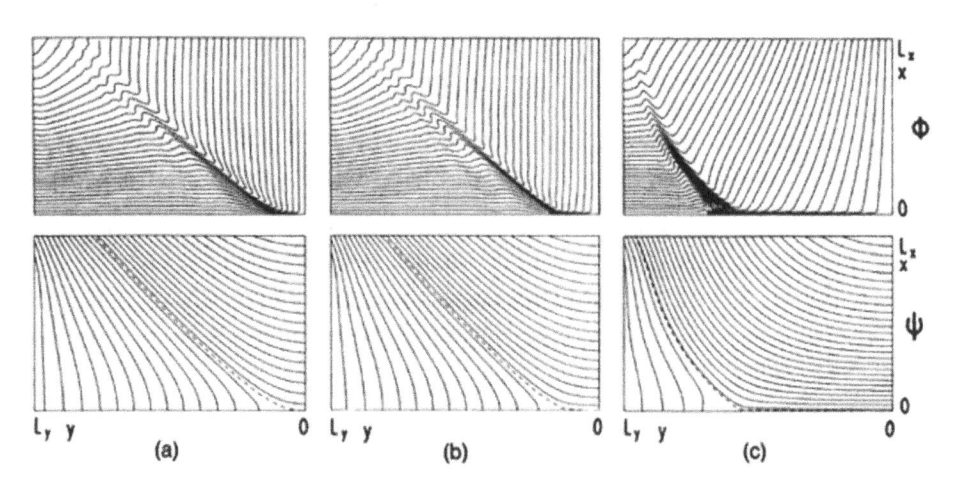

Figure 10.3: Numerical simulations of steady 2D reconnection, showing the flow trajectories (top) and magnetic field lines (bottom) in the top left quadrant of the symmetric configuration shown in Figs. 10.1 and 10.2. The simulations are performed for an external reconnection rate of $M_e = v_e/v_{Ae} = 0.042$ and for three different values of the magnetic Reynolds number: (a) $R_{me} = 1746$; (b) $R_{me} = 3492$; and (c) $R_{me} = 6984$. Note that the current sheet (feature at bottom right of each frame) becomes more elongated with increasing magnetic Reynolds number R_{me}, asymptotically approaching the Sweet−Parker solution (Biskamp 1986).

an extension, *nonuniform reconnection* solutions for highly curved magnetic field lines in the inflow region have also been calculated (Priest & Lee 1990; Strachan & Priest 1994). Solutions with no flows across the separatrices have also been found (Craig & Rickard 1994; Craig & Henton 1994), which confirm the *anti-reconnection theorem*: "Steady MHD reconnection in two dimensions with plasma flow across separatrices is impossible in an inviscid plasma with a highly sub-Alfvén flow and a uniform magnetic diffusivity" (Priest & Forbes 2000). To circumvent this problem, reconnection solutions have been sought by including viscosity in the central diffusion region and separatrix layers (Priest & Forbes 2000). Alternative 2D models also explore asymmetric geometries (Watson & Craig 1998; Ji et al. 2001), cylindrical geometries (Watson & Craig 2002), strongly sheared configurations instead of the conventional *stagnation point flow* topology (Craig & Henton 1995; Craig & McClymont 1997), and solutions for partially ionized plasmas (Ji et al. 2001), applicable to the photosphere. The latter effect is interesting because partially ionized plasmas with ambipolar diffusion can enhance the reconnection rate (Zweibel 1989; Brandenburg & Zweibel 1994).

10.1.4 Numerical Simulations of Steady 2D Reconnection

Numerical simulations of magnetic reconnection have been performed for a variety of boundary conditions by a number of groups (e.g., Ugai & Tsuda 1977; Sato 1979; Biskamp 1986; Scholer 1989; Yan et al. 1992, 1993), and analytical solutions have been compared with the numerical results (Forbes & Priest 1987; Lee & Fu 1986; Jin & Ip 1991). An example is shown in Fig. 10.3, showing the steady-state situation that occurred after formation of the standing shock. Note that numerical simulations generally have a much lower magnetic Reynolds number ($R_m \lesssim 10^5$) than the coronal plasma ($R_m \approx 10^8 - 10^{12}$). The simulations by Biskamp (1986) shown in Fig. 10.3 found a slow reconnection rate as predicted by the Sweet—Parker model, as well as an elongation of the current sheet with higher magnetic Reynolds numbers. This simulation could not reproduce the Petschek model, but a fast reconnection rate could be obtained by appropriate choice of the boundary conditions and using nonuniform (enhanced) resistivity in the diffusion region (Yan et al. 1992), consistent with the Petschek model. It remains puzzling that simulations with uniform resistivity cannot reproduce the analytical solutions of the Petschek type within the limits of a high magnetic Reynolds number R_m (Priest & Forbes 2000).

10.2 Unsteady/Bursty 2D Reconnection

When the diffusion region gets too long (such as in the Sweet—Parker model), it becomes unstable to *secondary tearing* (Furth et al. 1963) and an *impulsive bursty regime* of reconnection ensues (Priest 1986; Lee & Fu 1986; Kliem 1995; Priest & Forbes 2000, § 6-7). Such unsteady reconnection modes are very likely to operate in solar flares, because bursty and intermittent pulses (on time scales of seconds to subseconds) have been observed in hard X-ray and radio signatures of particle acceleration during virtually all flares (§ 13, 15). In the folling we describe a few of those unsteady reconnection modes, such as tearing instability (§ 10.2.1), coalescence instability (§ 10.2.2), and their combined dynamics (i.e., the regime of *bursty reconnection*, § 10.2.3). There are also other unsteady reconnection types, such as X-type collapse (Dungey, 1953; Craig & McClymont 1991, 1993; Craig & Watson 1992a; McClymont & Craig 1996; Priest & Forbes, 2000, p. 205), resistive reconnection in 3D (e.g. Schumacher et al. 2000; Priest & Forbes, 2000, p. 230), or collisionless reconnection (e.g. Drake et al. 1997; Haruki & Sakai, 2001a, b). The latter has not yet been applied to solar flares, but has been discovered in the Earth's magneto-tail (Øieroset et al. 2001).

10.2.1 Tearing-Mode Instability and Magnetic Island Formation

In current sheet formations, resistive instabilities (§ 6.3.8) can occur, where the magnetic field lines can move independently of the plasma due to the non-zero resistivity (opposed to the *frozen-flux theorem* for zero resistivity). In magnetic reconnection regions with high magnetic Reynolds numbers ($R_m = \tau_d/\tau_A$), where the outward diffusion (on a time scale of $\tau_d = l^2/\eta$, with $2l$ the width of the current sheet and $\eta = (\nu\sigma)^{-1}$ the magnetic diffusivity) is much larger than the Alfvén transit time $\tau_A = l/v_A$ (i.e., $\tau_d \gg \tau_A$), three different types of resistive instabilities can occur (§ 6.3.8, Fig. 6.11): *gravitational, rippling,* and *tearing mode* (Furth et al. 1963). Essentially, an Alfvén disturbance can trigger an instability before it can be stabilized by magnetic diffusion, when $\tau_d \gg \tau_A$ (i.e., for large Reynolds numbers $R_m = \tau_d/\tau_A$). The tearing mode, which has a wavelength greater than the width of the sheet ($kl < 1$),

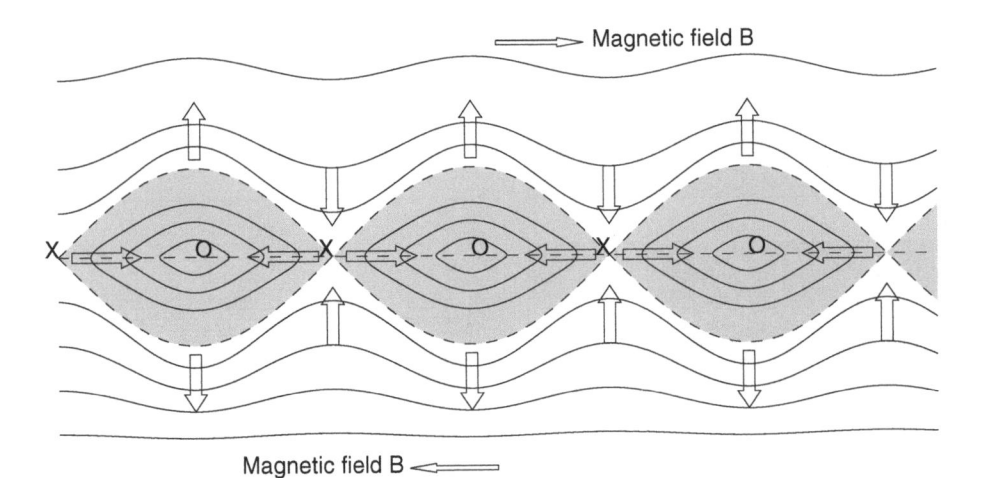

Figure 10.4: Magnetic island formation by tearing-mode instability in the magnetic reconnection region. Magnetically neutral X and O points are formed at the boundary between regions of an oppositely directed magnetic field, with plasma flow in the directions indicated by the arrows (after Furth et al. 1963).

has a growth time τ_G^{tear} of

$$\tau_G^{tear} = [(kl)^2 \tau_d^3 \tau_A^2]^{1/5} \, , \tag{10.2.1}$$

for wave numbers in the approximate range $(\tau_A/\tau_d)^{1/4} < kl < 1$ (e.g., see derivations in Furth et al. 1963; Priest 1982, p. 272; White 1983; and Sturrock 1994, p. 272). Thus, the mode with the longest wavelength has the fastest growth rate,

$$\tau_{G,min}^{tear} = [\tau_d \tau_A]^{1/2} \, . \tag{10.2.2}$$

Tearing mode produces magnetic islands in 2D (see Fig. 10.4) or magnetic fluxropes in 2.5D, respectively. These structures saturate in the nonlinear phase of the tearing mode (if coalescence is not permitted) and their subsequent diffusion at the diffusive time scale τ_d is extremely slow (since $R_m \gg 1$ in the corona). The energy release of tearing-mode instability occurs during the process of island formation. Tearing modes have been applied to solar flares in a number of theoretical studies (e.g. Sturrock 1966; Heyvaerts et al. 1977; Spicer 1977a, b, 1981a; Somov & Verneta, 1989; Kliem 1990), and numerical MHD simulations have been performed (Biskamp & Welter 1989). Kliem (1995) estimated the growth time of tearing mode for coronal conditions ($n_e = 10^{10}$ cm^{-3}, $T = 2.5 \times 10^6$ K, $B = 200$ G, with smallest current sheet half-widths of $l \approx 7 \times 10^3$ cm), which yields $\tau_G^{tear} \approx 0.4$ s. This time scale is comparable with the duration of elementary time structures observed in the form of hard X-ray pulses and radio type III bursts. Because tearing mode has a threshold current density orders of magnitude below the threshold of kinetic current-driven instabilities, it will occur first. Continued shearing and tearing may reduce the width of the current sheet until the threshold of kinetic instability is reached (Kliem 1995).

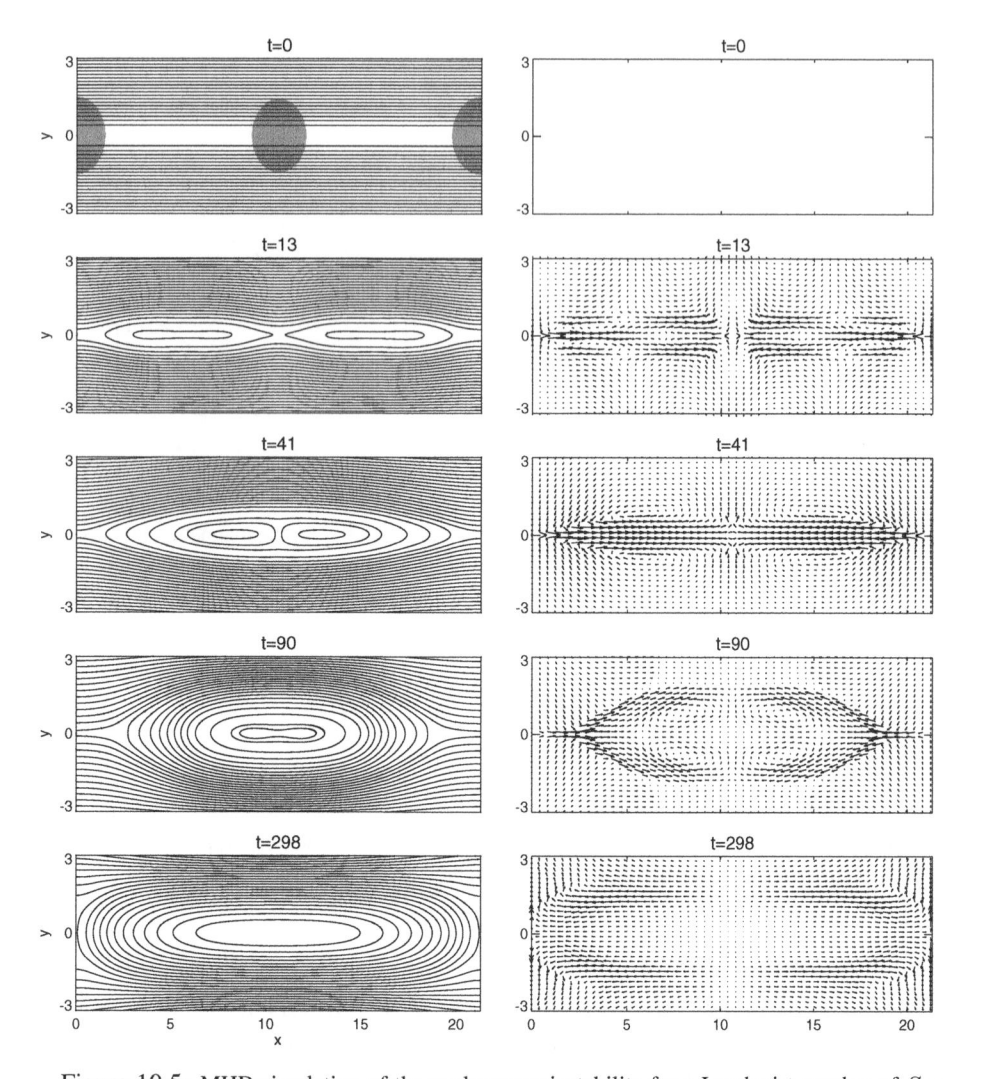

Figure 10.5: MHD simulation of the coalescence instability for a Lundquist number of $S = 1000$ and a plasma-β of 0.1. The magnetic field is shown in left-hand panels, the velocity field in the right-hand panels. Initial resistivity perturbation is shown shaded (Schumacher & Kliem 1997a).

10.2.2 Coalescence Instability

While tearing mode leads to filamentation of the current sheet, the resulting filaments are not stable in a dynamic environment. If two neighboring filaments approach each other and there is still non-zero resistivity, they enter another instability, the *coalescence instability*, which merges the two magnetic islands into a single one (Pritchett & Wu 1979; Longcope & Strauss 1994; Haruki & Sakai 2001a, b). An example of an MHD simulation is shown in Fig. 10.5 (Schumacher & Kliem 1997a). Coalescence instability completes the collapse in sections of the current sheet, initiated by tearing-mode instability, and thus releases the main part of the free energy in the current sheet (Leboef et al. 1982). There is no complete analytical description of coalescence instability, but numerical MHD simulations (Pritchett & Wu, 1979; Biskamp & Welter 1979, 1989; Leboef et al. 1982; Tajima et al. 1982, 1987; Schumacher & Kliem 1997a) show that the evolution consists of two phases: first the pairing of current filaments as an ideal MHD process, and then a resistive phase of pair-wise reconnection between the approaching filaments. The characteristic time scale of the ideal phase is essentially the Alfvén transit time through the distance l_{coal} between the approaching current filaments,

$$\tau_{coal} = \frac{1}{q_{coal}} \frac{l_{coal}}{v_A} , \qquad q_{coal} = \frac{u_{coal}}{v_A} \approx 0.1 - 1 \qquad (10.2.3)$$

where u_{coal} is the velocity of the approaching filaments. For coronal conditions (say $n_e = 10^{10}$ cm^{-3}, $B = 200$ G, $l_{coal} = 1000$ km) we estimate coalescence times of $\tau_{coal} \approx 0.2 - 2.0$ s, which is again the typical time for the observed modulation of hard-X ray pulses and type III electron beams in flares.

10.2.3 Dynamic Current Sheet and Bursty Reconnection

In praxis, the two previously described processes of tearing instability and coalescence instability occur iteratively, leading to a scenario of *dynamic current sheet evolution*, also known as *impulsive bursty reconnection* (Leboef et al. 1982; Priest 1985a; Tajima et al. 1987; Kliem 1988, 1995). A long current sheet is first subject to tearing that creates many filaments, while rapid coalescence clusters and then combines groups of closely spaced filaments, which are once again unstable to secondary tearing, to secondary coalescence, and so forth. MHD simulations reproduce this iterative chain of successive tearing and coalescence events (Malara et al. 1992; Kliem et al. 2000). An example of such a numerical simulation from the study of Kliem et al. (2000) is shown in Fig. 10.8 (magnetic field evolution). Let us review three key studies (Tajima et al. 1987; Karpen et al. 1995; Kliem et al. 2000), where numerical MHD simulations of this process have been applied to solar flares.

Tajima et al. (1987) performed numerical MHD simulations of the nonlinear coalescence instability between current-carrying loops and derived an analytical model of the temporal evolution of electromagnetic fields [see also two comprehensive reviews on this subject by Sakai & Ohsawa (1987) and Sakai & De Jager (1996), and references therein]. This nonlinear system evolves into an oscillatory relaxation dynamics, driven by the interplay of the $\mathbf{j} \times \mathbf{B}$ force and the hydrodynamic pressure response, which was modeled analytically by Sakai & Ohsawa (1987). The oscillatory behavior is very appealing, because it provides a possible explanation for the numerous quasi-periodic time structures observed in radio and hard X-rays during flares. An oscillatory regime of fast reconnection has also been found from other theoretical work on current instabilities in current sheets (Smith 1977) and X-point relaxation (Craig & McClymont 1991, 1993).

Karpen et al. (1995) performed 2.5-dimensional numerical MHD simulations of shear-driven magnetic reconnection in a double arcade with quadrupolar magnetic topology. For strong shear, the initial X-point was found to lengthen upward into a

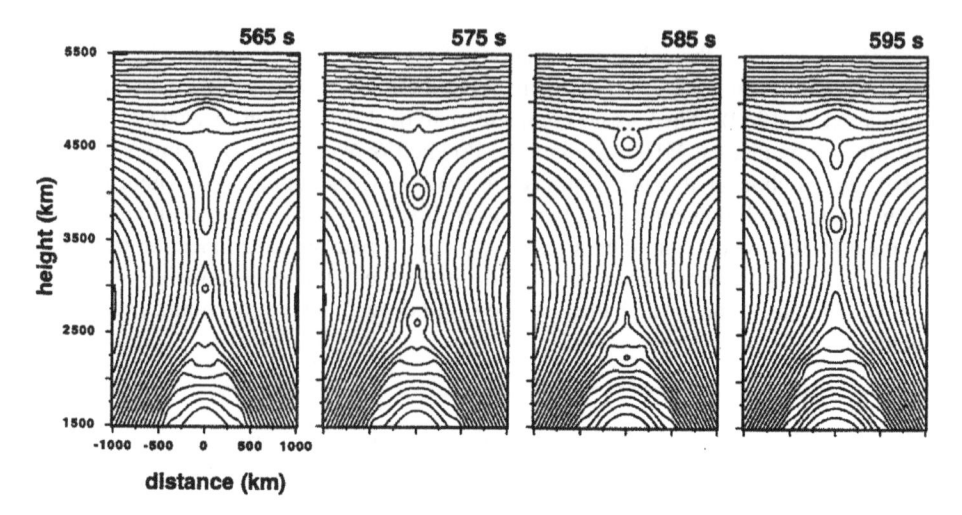

Figure 10.6: Magnetic field lines near the reconnection region at four different times (565, 575, 585, 595 s) during a strong-shear MHD simulation by Karpen et al. (1995). Note the tearing along the vertical current sheet (first frame), which forms two magnetic islands (second frame), which are ejected from the sheet and merge with the flux systems above or below the sheet (third frame), followed by another tearing plus magnetic island formation (fourth frame). (Karpen et al. 1995).

current sheet, that reconnects gradually for a while but then begins to undergo multiple tearing. Several magnetic islands develop in sequence, move towards the ends of the sheet, and disappear through reconnection with the overlying and underlying field (Fig. 10.6). A second study with similar quadrupolar configuration was performed, but with asymmetric shear in dipoles with markedly unequal field strengths (Karpen et al. 1998). Similar intermittency was found in the shear-driven magnetic reconnection process, and the simulations moreover show that each dissipated magnetic island leaves a footprint in the form of fine filaments in the overlying separatrix layer (Fig. 10.7). This dynamic behavior is essentially identical to the pattern of repeated tearing and coalescence first investigated by Leboef et al. (1982) and dubbed *impulsive bursty reconnection* by Priest (1985b). In Fig. 10.7 there are also some other dynamic processes present: (a) a thin region along the slowly rising inner separatrix is compressed; (b) a downflow with $v \approx 30$ km s^{-1}; (c) this is followed by an upflow along the same field lines. Although these simulations by Karpen et al. (1995, 1998) are carried out using parameters corresponding to chromospheric conditions, it demonstrates that magnetic reconnection in sheared flare arcades occurs in a bursty and intermittent mode, and not in a quasi-stationary Sweet−Parker or Petschek mode. The physical origin of this intermittent reconnection dynamics is most essential to understanding the rapidly varying time structures of accelerated particles.

A recent work on *impulsive bursty reconnection* applied to solar flares was carried out by Kliem et al. (2000). Fig. 10.8 shows the evolution of tearing, magnetic island formations, magnetic island coalescence, secondary tearing, and so forth. Tear-

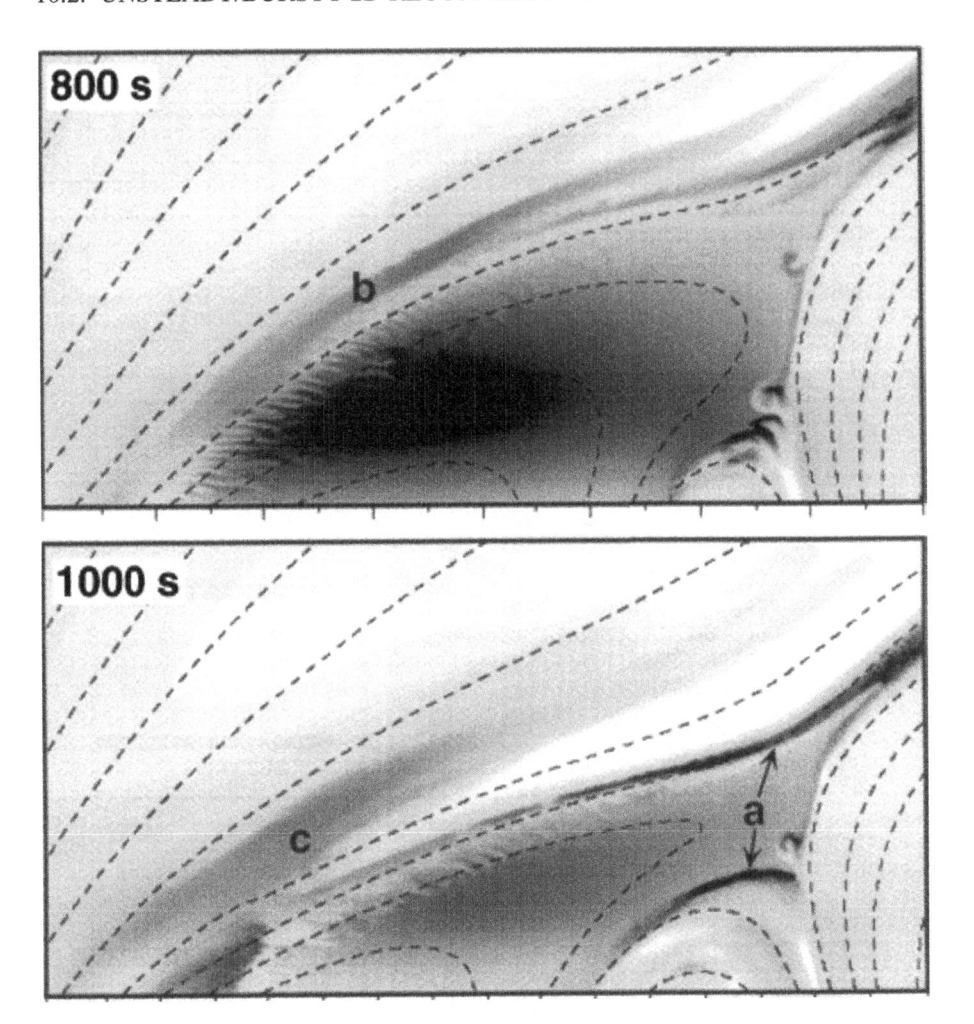

Figure 10.7: Mass density difference ratio (greyscale) and projection of magnetic field lines into the image plane (dashed lines) at 800 s and 1000 s in the vicinity of the reconnection region, during an MHD simulation of a sheared arcade. The location **a** corresponds to a thin compressed region along the slowly rising inner separatrix, **b** to a narrow downflow falling outside of the left outer separatrix, and **c** indicates a broader upflow that follows along the same field lines (Karpen et al. 1998).

ing and coalescence in the bursty magnetic reconnection mode also modulates particle acceleration on time scales that are observed in radio and hard X-rays, and is more consistent with flare observations than steady reconnection scenarios. The iterative processes of tearing and colaescence may repeat down to microscopic scales (of the ion Larmor radius or the ion inertial length), producing a *fractal current sheet* (Shibata & Tanuma 2001). A similar concept is that of MHD turbulent cascading, which leads to similar high fragmentation at the smallest spatial scales, called *turbulent reconnec-*

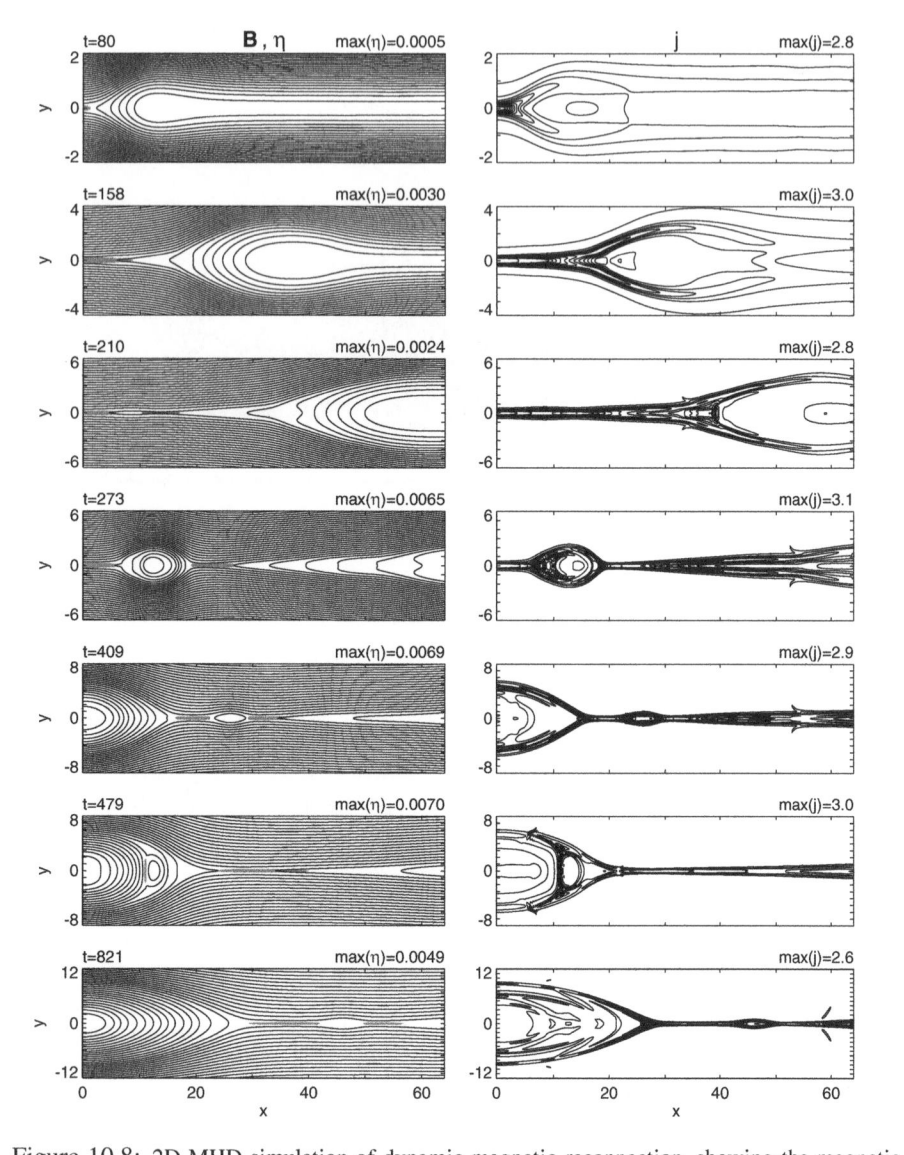

Figure 10.8: 2D MHD simulation of dynamic magnetic reconnection, showing the magnetic field (left panels) and current density (right panels). Regions with anomalous resistivity are shown shaded in the magnetic field plot (at y=0) (Kliem et al. 2000).

tion (e.g., Kim & Diamond 2001; Matthaeus 2001b; see also Fig. 9.8) and applied to flares (Moore et al. 1995; Somov & Oreshina 2000). The two concepts of fractal current sheets and turbulent reconnection could possibly be discriminated observationally from the frequency distribution of time scales, since fractal processes are scale-free and generally produce powerlaw distributions, while turbulent processes are controlled by incoherent random processes that generally produce exponential distributions (§ 9.8.1).

10.3 3D Magnetic Reconnection

In the previous sections we described magnetic reconnection in two dimensions. Such 2D concepts can approximate 3D reality as long as the magnetic field configuration has a translational symmetry in the third dimension: the ignorable coordinate. There are two types of nullpoints in 2D reconnection, X-points and O-points, but there is a much richer variety of magnetic 3D topologies, where 3D volumes with oppositely directed magnetic fields are divided by *2D separatrix surfaces*, intersection lines of two separatrix surfaces form *1D separator lines*, and the intersections of separator lines form *3D nullpoints*. The field of 3D magnetic reconnection is still in a very exploratory phase, encompassing a complex variety of mathematical topologies (e.g., Priest & Forbes 2000, § 8; and Parnell 1996), while only a few of all the possible mathematical topologies have been identified by observations. Here we focus mainly on 3D topologies that seem to be most relevant for solar flares.

10.3.1 3D X-Type Reconnection

There are different theoretical definitions of 3D reconnection, in terms of: (1) changes in magnetic connectivity; (2) the electric field component; (3) plasma flows across the separatrices; or (4) changes in magnetic helicity (e.g., Priest & Forbes 2000, § 8.1). A simple practical discrimination rule between 2D and 3D X-type reconnection can be determined on the basis of whether both prereconnection and postreconnection field lines can be mapped in the same 2D plane or not. We illustrate this in Fig. 10.9 for reconnection between open (having only one footpoint on the solar surface) and closed field lines (with both conjugate footpoints on the solar surface). For 2D reconnection we can have bipolar (open with open), tripolar (open with closed), and quadrupolar (closed with closed) reconnection (Fig. 10.9 top). For 2D reconnection, a necessary condition is that the sequence of magnetic polarities of the reconnecting field lines is alternating in a common 2D plane (e.g., $+/-/+/-$ for a quadrupolar geometry: Fig. 10.9, top right). If the sequence is not alternating (e.g., $+/+/-/-$: Fig. 10.9, bottom right), the reconnection geometry cannot be represented in a 2D plane, because either the prereconnection or the postreconnection field lines would intersect, so they can only be separated topologically in 3D space.

2D reconnection geometries (Fig. 10.9, top) can have arbitrary extensions in the third dimension (along the neutral line), which are called *loop arcades*. 3D reconnection geometries (Fig. 10.9, bottom), in contrast, have more complicated topological constraints in the third dimension, so that they generally involve *interacting loops* rather than *loop arcades*. Sakai & De Jager (1991) classify the geometries of coalescing loops into 1D, 2D, and 3D cases, which correspond to the quadrupolar geometries shown in Fig. 10.9 (right panels). The angle between the magnetic fields at the reconnection points, which is simply $180°$ (anti-parallel fields) in the 2D-case, can take any arbitrary value in 3D reconnection geometries. The reconnection rate depends very much on this angle, being most efficient for anti-parallel fields, but it scales with $\propto [\sin{(\theta/2)}]^{1/2}$ for skewed angles θ in the Petschek model (Soward 1982). Thus, reconnection can still occur for almost parallel magnetic fields, although with reduced efficiency, a phenomenon that is called *component reconnection* in magnetospheric physics. Sakai &

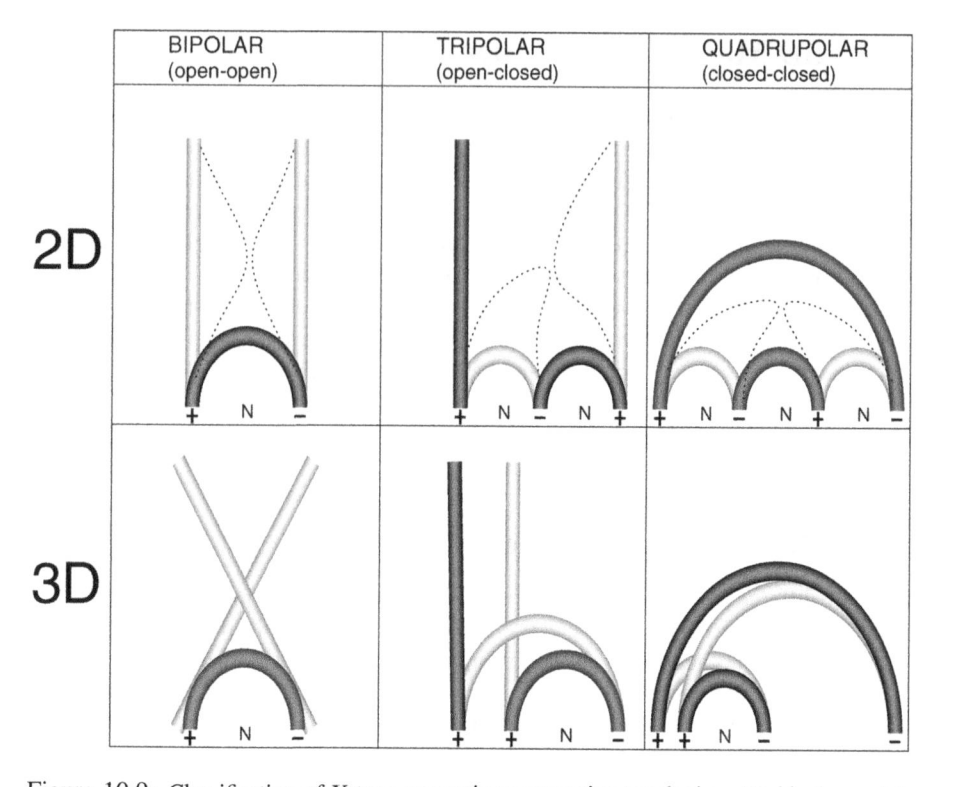

Figure 10.9: Classification of X-type magnetic reconnection topologies: (1) *bipolar* models have reconnection between two open field lines; (2) *tripolar* models have reconnection between an open and a closed field line; and (3) *quadrupolar* models have reconnection between two closed field lines. Prereconnection field lines are rendered in light grey and at the time of reconnection with dotted lines, while the postreconnection field lines, as they occur after relaxation into a near-potential field state, are rendered in dark grey. 2D versions, invariant in the third dimension (forming arcades) are shown in the upper row, while 3D versions are captured in the lower row. Prereconnection field lines (light grey) are located behind each other in the 3D versions, but approach each other in the image plane during reconnection. Note that the number of neutral lines (marked with symbol N, perpendicular to the image plane) is different in the corresponding 2D and 3D cases (Aschwanden 2002b).

Koide (1992) classify magnetic reconnection between two coalescing loops into six types, depending on the parallelity or anti-parallelity of the magnetic field component B_z, the current I_z along the loop, and the azimuthal field component B_θ.

10.3.2   Topology of 3D Nullpoints

Wherever multiple magnetic dipoles occur, each one defines a domain that contains a volume of magnetic field lines with the same connectivity of positive to negative footpoints, as illustrated in Plate 10. Different dipolar domains are separated by separatrix surfaces in 3D space. Intersections of 2D separatrix surfaces form 1D separators, and

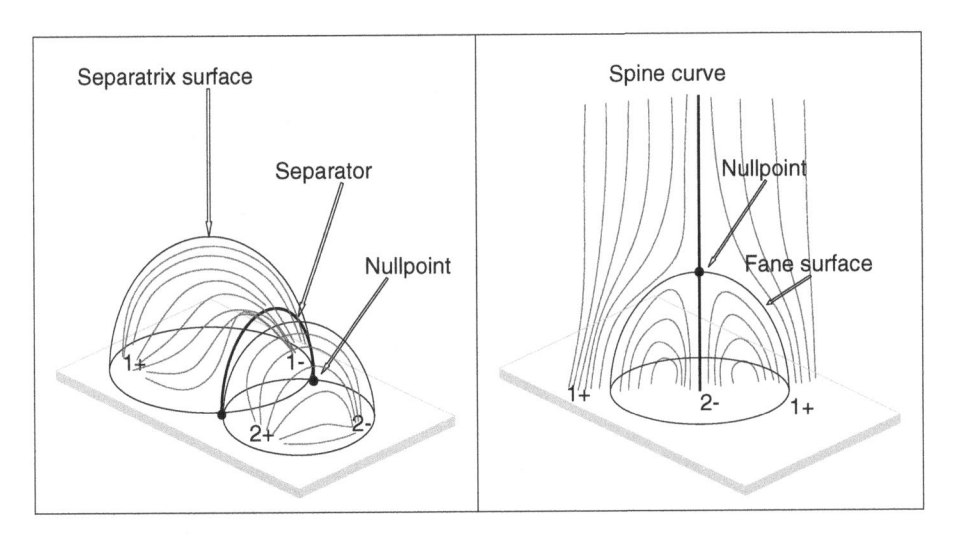

Figure 10.10: Topology of 3D reconnection features for a quadrupolar region (left) and a parasitic region (right). In the quadrupolar region (left), a new emerging dipole $(2+, 2-)$ joins a pre-existing, older dipole region $(1+, 1-)$, which are separated by a separatrix surface. The intersection of the two separatrix surfaces embedding both the old and new dipole region intersects at the separator line, which intersects with the photospheric surface at magnetic nullpoints. In the parasitic region (right) a unipolar flux region $(2-)$ emerges in the center of a pre-existing open field region with polarity $(1+)$. The new regions is shielded from the pre-existing open field by a dome-like *fan surface*, where the symmetry axis is called the *spine*, containing a nullpoint at the intersection with the fan dome.

intersections of 1D separators form 3D nullpoints. The most natural example is the emergence of a secondary bipole in the neighborhood of a pre-existing dipole, which form together a new quadrupolar configuration, as shown in Fig. 10.10 (left). The new dipole region $(2+, 2-)$ pushes the old pre-existing field lines $(1+, 1-)$ aside in the coronal volume and forms a new separatrix as a dividing surface. Strictly speaking, there is also a third domain of magnetic field lines with connections $(1-, 2+)$ inbetween, and perhaps a fourth with connections $(1+, 2-)$ above, depending on the relative strength of the magnetic fluxes.

If the new emerging region is unipolar (which is nothing more than a vertically oriented dipole) and emerges in an open-field region with opposite polarity, it forms a parasitic polarity (Fig. 10.10 right), a configuration also called the *anemone region* in quiet Sun regions, or δ-*spot* in flaring regions. In such a unipolar region, surrounded by opposite magnetic polarity, the new magnetic domain is separated by a dome-like separatrix surface, called the *fan dome*. The symmetry axis of the unipolar region is called the *spine curve*, which intersects the fan dome at a 3D nullpoint and continues above the fan dome (Fig. 10.10, right). A field line that connects two nullpoints is called a *separator*. Nullpoints have been studied in multiple sources (with hexagonal geometry) in the photospheric network with up to 12 cell configurations (Inverarity & Priest 1999).

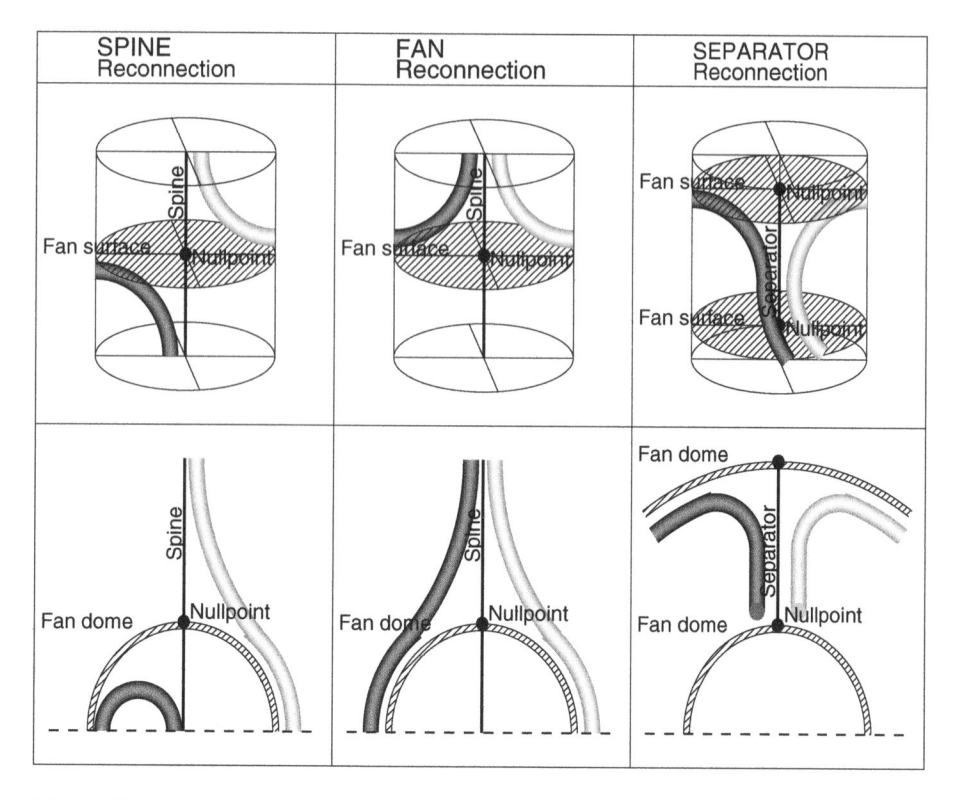

Figure 10.11: Classification of 3D nullpoint reconnection topologies: (1) *spine reconnection* (left), (2) *fan reconnection* (middle), and (3) *separator reconnection* (right), shown for a cylindrical geometry (top row) and for a dome-like fan surface geometry on the solar surface (bottom row). Spine and separator curves are marked with thick lines, fan surfaces with hatched areas, and 3D nullpoints with black dots. The preconnection line is rendered in light grey and the postreconnection field line in dark grey.

10.3.3   3D Spine, Fan, and Separator Reconnection

There is an infinite variety of such 3D topologies, because the complexity increases with the emergence of every new dipole that pushes into the pre-existing maze of coronal magnetic topologies (see Fig. 5.23 for more examples). However, the topological complexity cannot grow to infinity. At some point magnetic stresses will break up highly sheared structures and the neighboring field lines will reconnect to a simpler topology corresponding to a lower energy state. Field lines that are stressed or pushed towards a 3D separatrix layer, a fan surface, a spine, or a separator, will experience a high plasma-β near the zero-magnetic field zone and can slip through the nullpoints and reconnect to new field lines on the opposite side. Three special types of 3D reconnections are illustrated in Fig. 10.11: *spine reconnection*, *fan reconnection*, and *separator reconnection*. In the case of *spine reconnection*, a field line penetrates the fan surface, swirls around the spine, and reconnects at the opposite side of the fan surface and spine curve (Fig. 10.10, left). In the case of *fan reconnection*, the field line merely

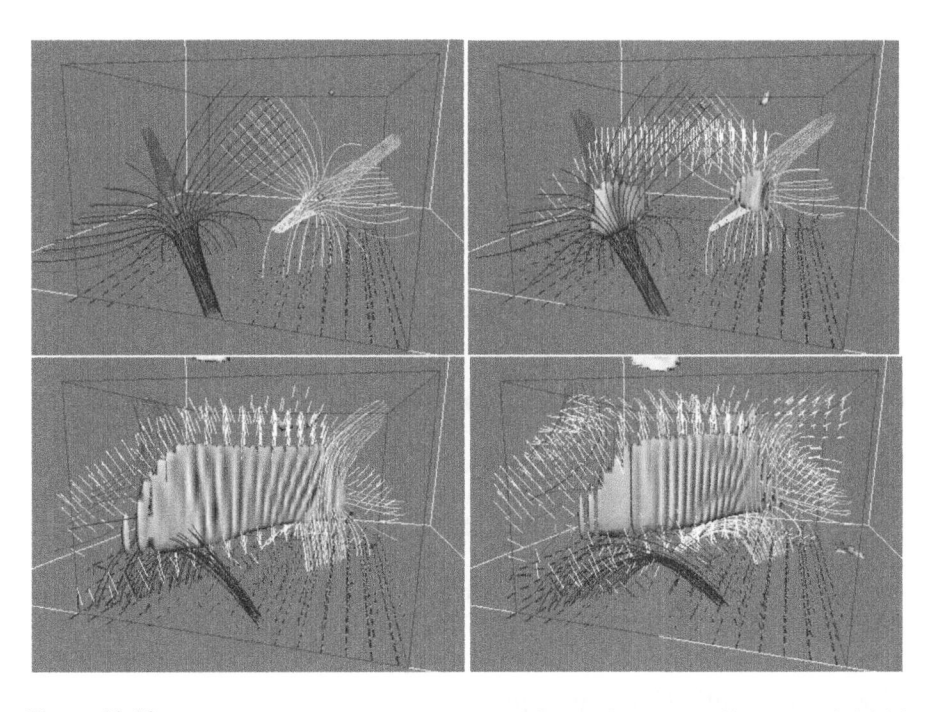

Figure 10.12: This numerical simulation shows the collapse of a separator line. *Top left:* initial magnetic field configuration with a pair of 3D nullpoints. *Top right:* currents start to accumulate in the nullpoints (shown with isosurfaces). *Bottom left:* the separator surface is stretched in the vertical direction (where the thin vectors indicate the driving velocity and the thick white vectors the high-velocity outflow jets driven by the reconnection). *Bottom right:* the process of outflow jets continues until no magnetic energy is left and the current sheet then fades away (Galsgaard et al. 2000).

swirls around the spine, rotates around the fan dome, and reconnects at the other side (Fig. 10.10 middle). The case of *separator direction* is a special case of fan reconnection, where the spine curve is replaced by a separator line (Fig. 10.10 right). Since 3D nullpoints are created in pairs, they are always (at least in the initial stage) connected with a separator field line. We will show related observations in § 10.6, which seem to fit these theoretical reconnection modes.

The theory or reconnection in 3D is presented in greater detail in the textbook of Priest & Forbes (2000, § 8), while shorter reviews and introductions can be found in Priest (1996), Priest & Schrijver (1999), Brown & Priest (1999), Forbes (2000a), Schindler & Hornig (2001), and Hood et al. (2002). Analytical studies quantify the magnetic field in 3D nullpoint topologies (Brown & Priest 2001), the current distributions near 3D nullpoints (Rickard & Titov 1996), and solutions for fan, spine, and separator reconnection (Craig & McClymont 1999; Craig et al. 1999; Craig & Watson 2000a; Ji & Song 2001). The classical Sweet−Parker reconnection rate is found to be the slowest possible in the present 3D reconnection models, but it not clear what type of 3D reconnection yields the fastest reconnection rate suitable for flares. Numerical simulations of 3D reconnection topologies have been performed by Galsgaard et al. (1997b, 2000), which showed that reconnection is not restricted to singular points (nullpoints) but can also occur along separators (Plate 12). The experiments show (Fig. 10.12) that current accumulation is generated by a shear flow across the fan plane of the two nulls and the spine axis through the null points becomes disrupted by the development of the separator current sheet, loses its identity as a singular line, and becomes integrated into the separator surfaces. The experiments of Galsgaard et al. (2000) also suggest that separator reconnection is the most important type of null reconnection among the three possible types (spine, fan, separator).

10.4   Magnetic Reconnection in the Chromosphere

There are a number of small-scale phenomena in the photosphere, chromosphere, and transition region that involve magnetic reconnection processes and may produce signatures in the lower corona such as microflares and nanoflares. An overview of such small-scale phenomena is given in Table 9.3 and their role for coronal heating is discussed in § 9.6. Here we concentrate on the aspect of magnetic reconnection of these small-scale events.

10.4.1   Magnetic Flux Emergence

The magnetic dynamo in the solar interior constantly generates magnetic fluxtubes that emerge through the photosphere and add new magnetic flux systems to the corona, probably moored in deep subsurface structures for the lifetime of an active region (Schrijver & Title 1999). Fig. 10.13 shows the frequency distribution of the emergence rate of magnetic dipoles, which are called *ephemeral regions* (Harvey & Martin 1973; Martin 1988) if they have an area smaller than 2.5 deg^2, or *active regions* if they are larger. Thus, ephemeral regions are essentially "mini-active regions" in quiet Sun areas. The size distribution shown in Fig. 10.13 encompasses four orders of magnitude in magnetic flux ($\Phi \approx 5 \times 10^{18} - 5 \times 10^{22}$ Mx) and eight orders of magnitude variation in the occurrence rate (Hagenaar et al. 2003). A comparison of the magnetic flux emergence rate with the network flux implies an overall mean replacement time of $\approx 8 - 19$ hr in the quiet Sun (Hagenaar et al. 2003).

Emerging flux systems in active regions are aligned with overlying arch filament systems (Strous et al. 1996; Strous & Zwaan 1999), emerging in the form of Ω-loops, U-loops, or "seaserpent-like" shapes (Zwaan 1987; Van Driel–Gesztelyi et al. 2000; Van Driel–Gesztelyi 2002). Newly emerging magnetic dipoles appear at the edges of supergranulation cells (Hagenaar 2001). During emergence, magnetic dipoles grow in size and their rate of divergence is of order v $\approx$ 2.3 km s^{-1}, while magnetic flux increases with a rate of $d\Phi/dt \approx 1.6 \times 10^{15}$ Mx s^{-1} (Hagenaar 2001). The emergence of growing new magnetic flux structures necessarily forces topological changes in the magnetic field of the overlying corona, which may involve magnetic reconnection processes. MHD simulations of such magnetic fluxtubes have been performed in subphotospheric zones (see § 6.2.5 and references therein) and in coronal heights (Shibata et al. 1989b, 1990). When the emerging field has the same orientation as the

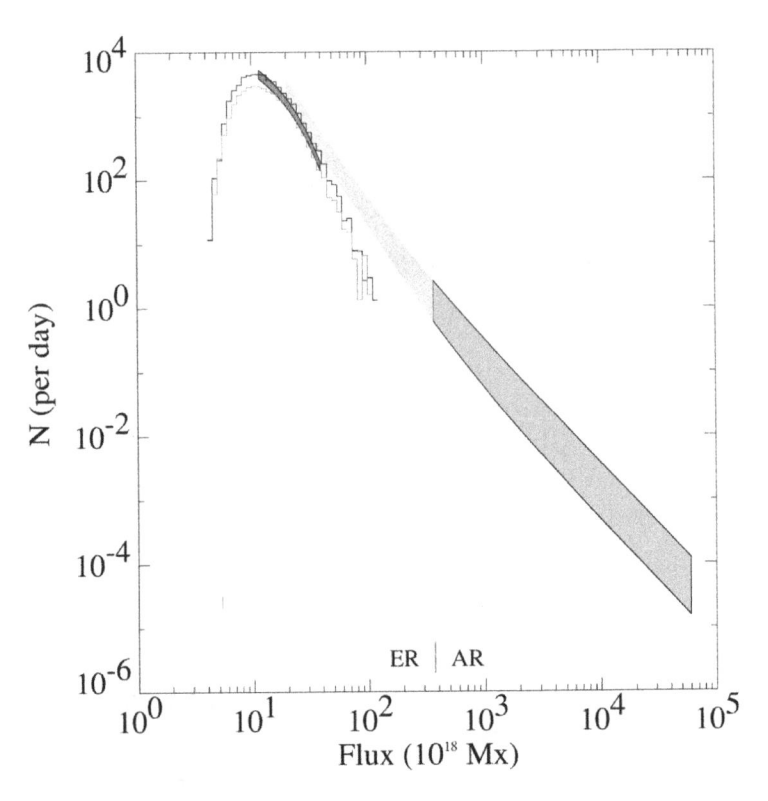

Figure 10.13: Frequency distribution of emerging magnetic bipoles per day, per flux unit of $\Phi[10^{18}$ Mx]. The distribution includes *ephemeral regions* (ER: $\Phi \lesssim 3 \times 10^{20}$ Mx) and active regions (AR: $\Phi \gtrsim 3 \times 10^{20}$ Mx, area $A \gtrsim 2.5$ deg^2). The variation by a factor of 8 is mainly caused by the solar cycle. The histograms include ephemeral regions studied using MDI, with a detection threshold of $\Phi \gtrsim 4 \times 10^{18}$ Mx (Hagenaar et al. 2003).

overlying coronal field, an approximately current-free field forms in the interaction region. When the emerging field is anti-parallel, however, a current sheet forms which could initiate magnetic reconnection (Shibata et al. 1989b), as shown in Fig. 10.14. The increased velocity is however relatively small (in the order of $v_z \approx 1$ km s^{-1}). It is driven by the Parker instability (undular mode of magnetic buoyancy instability) in the model of Shibata et al. (1989a, b). The emergence of a bipole does not necessarily trigger a microflare or flare event, because the emerging flux region could be too small or could have the wrong orientation (Martin et al. 1984). Harvey (1996) found that only 8% of soft X-ray bright points (which are the sites of microflares) were associated with an emerging bipole, while a much larger fraction was associated with magnetic cancellation features (Harvey et al. 1994). Karpen & Boris (1986) concluded that pre-flare brightenings and flares probably do not result directly from the emerging fluxtubes themselves, but from coupled energy release processes, such as current-driven plasma micro-instabilities. Altogether it appears that magnetic flux emergence does not trigger flares directly, but increases the magnetic complexity locally (e.g., by adding a new

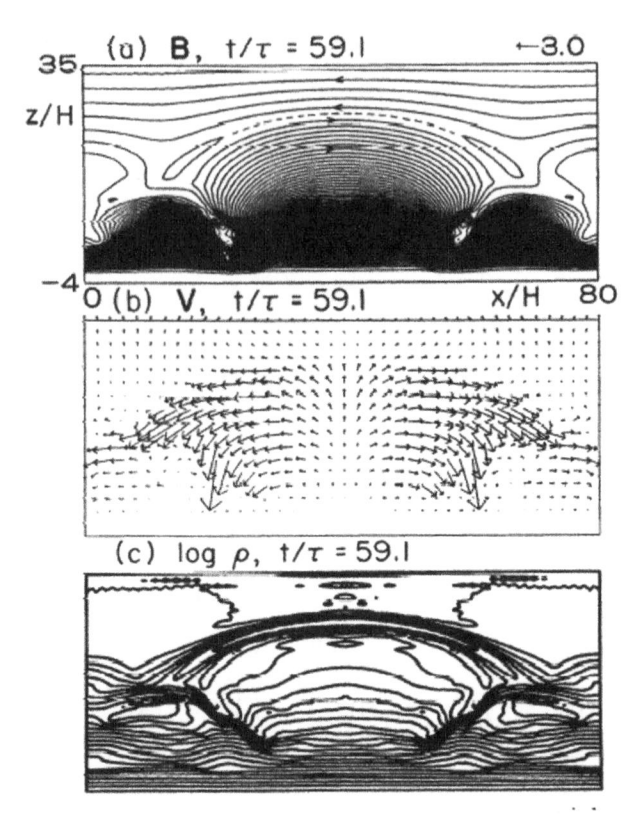

Figure 10.14: Numerical 2D MHD simulation of an emerging dipole into an anti-parallel coronal field: magnetic field (top), velocity field (middle), and density (bottom). A current sheet forms (dashed line in top panel) that enables magnetic reconnection (Shibata et al. 1989b).

separator surface and creating new nullpoints: Fig. 10.10), which after further increases in magnetic complexity ultimately may escalate into a flare-like event. However, flare models that are directly driven by flux emergence have also been proposed (Heyvaerts et al. 1977), see § 10.5. Furthermore, 3D magnetic modeling of quasi-separatrix layers in a flare event and comparison with Hα and soft X-ray images was found to be consistent with the concept of magnetic reconnection driven by the emergence of sheared magnetic field (Schmieder et al. 1997a, b).

10.4.2 Magnetic Flux Cancellation

The reverse process to *magnetic flux emergence* is *magnetic flux cancellation*, which can occur in at least three different manifestations (Martin et al. 1985; Livi et al. 1985; Van Driel−Gesztelyi 2002; Parnell 2002b): (1) by submergence, subduction, or retraction of fluxtubes; (2) by converging flows; or (3) by flux dispersion or diffusion. The first process essentially corresponds to a downward motion of a dipole, which sinks through the photosphere, but does not necessarily involve a magnetic reconnection process. Evidence for such a process was found from the timing of disappearance in different heights of chromospheric and photospheric magnetograms (Harvey et al. 1999). The second process, however, corresponds to a collision between two conjugate magnetic polarity elements and thus involves a magnetic reconnection process (e.g., the model of Litvinenko 1999a). The third one is a fragmentation process driven by surface random flows, which may or may not involve magnetic reconnection between individual fragments.

A theoretical model of magnetic reconnection for converging flows that produce magnetic cancellation was quantified by Litvinenko (1999a). From the equations of continuity, momentum, and Ohm's law, Litvinenko (1999a) derives a reconnection inflow speed v_1 of

$$v_1 = \left[\frac{c^2 B}{4\pi \Lambda \sigma (4\pi m_p n_0)^{1/2}} \right]^{1/2} \left[1 + \frac{B^2}{8\pi k n_0 T} \right]^{1/4} \qquad (10.4.1)$$

where $\Lambda(z) = n_0/(dn_0/dz) \approx 100$ km is the chromospheric density scale height. Using a VAL-C standard atmospheric density model, a maximum of the inflow speed of $v_1 \approx 35$ m s^{-1} is found at a height of $h \approx 600$ km above the photosphere (assuming a magnetic field of $B \approx 30$ G, see Fig. 10.15). This approaching speed corresponds to the observed speeds of some cancelling features. The height of $h \approx 600$ km above the photosphere (in the temperature minimum region) is thus considered as the most favorable location for chromospheric reconnection processes. This process could explain mass upflows into an associated filament during a photospheric cancellation event (Litvinenko & Martin 1999).

A more general model of chromospheric reconnection driven by converging flows has been developed by Chae et al. (2003), including the adiabatic energy equation. This model contains the case of an isothermal current sheet used by Litvinenko (1999a) as a special case (for a specific heat ratio of $\gamma = 1$). Moreover, Chae's model (2003) is formulated only in terms of observable parameters (except for magnetic diffusivity η). The model of Chae et al. (2003) assumes a (vertical) Sweet–Parker current sheet in the chromosphere with length Δ, width δ, inflow speed v_1, and outflow speed v_2, which obey the relations given in Eqs. (10.1.1–7). One observable parameter is the magnetic flux loss $d\Phi/dt$, which is defined as the magnetic flux that is processed through the current sheet,

$$\frac{d\Phi}{dt} = v_1 B_1 L_z = v_2 B_2 L_z , \qquad (10.4.2)$$

where L_z is the length of the current sheet along the ignorable coordinate z in the 2D model shown in Fig. 10.2. The electric field (Eq. 10.1.2) that characterizes the reconnection rate is (using Eqs. 10.4.2 and 10.1.7),

$$E_0 = v_1 B_1 = \frac{1}{L_z} \frac{d\Phi}{dt} = \frac{\eta}{c} \frac{B_1}{\delta} . \qquad (10.4.3)$$

The momentum conservation across the current sheet is

$$\frac{B_1^2}{8\pi} + \frac{1}{2} \rho_1 v_1^2 + p_1 = p_{nl} , \qquad (10.4.4)$$

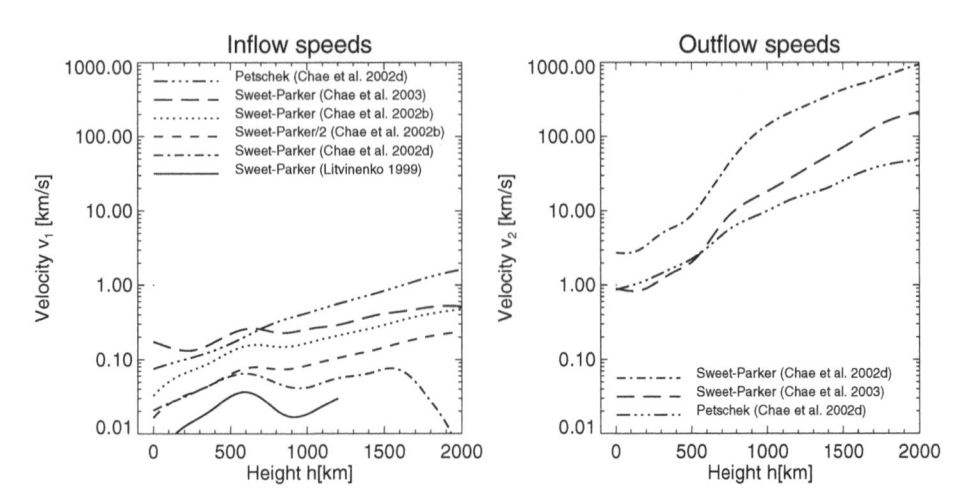

Figure 10.15: Inflow speeds ($v_1(h)$, left) and outflow speeds ($v_2(h)$, right) of chromospheric magnetic reconnection models computed as a function of height h, for various theoretical models. All models use a VAL density model. The model of Litvinenko (1999a) is calculated for a magnetic field of $B = 30$ G. The models of Chae et al. (2002b) match magnetic flux loss rates of $v_1 B_1 = 5.0 \times 10^6$ G cm s^{-1} and 1.2×10^6 G cm s^{-1}. The Sweet–Parker and Petschek models in Chae et al. (2002d) match a flux loss rate of $v_1 B_1 = 1.2 \times 10^6$ G cm s^{-1}. The model of Chae et al. (2003) match a flux loss rate of $v_1 B_1 = 1.5 \times 10^6$ G cm s^{-1}, assume anomalous resistivity $\eta = 50 \eta_{classical}$ and an adiabatic index of $\gamma = 4/3$.

and the momentum conservation along the current sheet is

$$\frac{1}{2}\rho_2 v_2^2 + p_2 = p_{nl} + \frac{B_1 B_2}{8\pi}\frac{\Delta}{\delta} \ , \tag{10.4.5}$$

where p_1, p_2, p_{nl} are the pressures in the inflow, outflow, and neutral layer (or current sheet). Chae et al. (2003) used the adiabatic energy equation (see Eqs. 4.1.14 and 4.1.29),

$$\left(\frac{1}{2}\rho_1 v_1^2 + \frac{\gamma}{\gamma-1}p_1 + \frac{B_1^2}{4\pi}\right) v_1 \Delta = \left(\frac{1}{2}\rho_2 v_2^2 + \frac{\gamma}{\gamma-1}p_2 + \frac{B_2^2}{4\pi}\right) v_2 \delta \ . \tag{10.4.6}$$

Assuming a constant pressure ($p_1 = p_2 = p_{nl}$) and the Sweet–Parker geometry $\delta \ll \Delta$, the following density ratio between the inflow and outflow region is found (by combining Eqs. 10.1.1−7 and 10.4.3−6),

$$f = \frac{\rho_2}{\rho_1} = 1 + \frac{1}{(\gamma - 1 + \gamma\beta_1)} \ , \tag{10.4.7}$$

where the plasma-β parameter is $\beta_1 = 8\pi p_1/B_1^2$. The inflow speed becomes

$$v_1 = \left(\frac{\eta B_1}{\Delta\sqrt{4\pi\rho_1}}\right)^{1/2} f^{1/2} \tag{10.4.8}$$

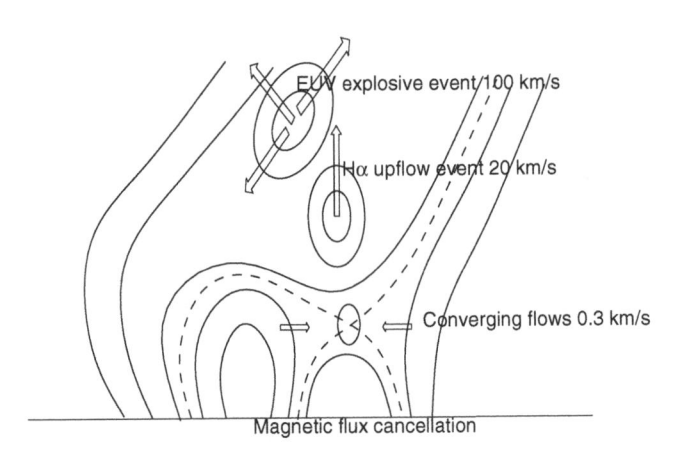

Figure 10.16: Schematic of the chromospheric magnetic reconnection model that is driven by converging flows and produces photospheric magnetic flux cancellation, $H\alpha$ upflow events, and UV explosive events (after Chae 1999).

and the temperature increase due to reconnection is found to be

$$\Delta T = T_0 - T_1 = \frac{1}{2}\frac{\gamma - 1}{\gamma}\frac{\mu m_H}{k_B}v_2^2 , \qquad (10.4.9)$$

with μ the mean molecular weight. Since the magnetic field B_1 in the inflow region is hard to measure, it is easier to obtain the magnetic flux loss rate $d\Phi/dt$ and the cancellation interface length L_z from the observations. Using these observables, the magnetic field B_1 and inflow speed v_1 can be expressed as

$$B_1 = \left[\frac{L\sqrt{4\pi\rho_1}}{\eta}\left(\frac{1}{L_z}\frac{d\Phi}{dt}\right)^2\right]^{1/3} f^{1/3} , \qquad (10.4.10)$$

$$v_1 = \left[\frac{\eta}{L\sqrt{4\pi\rho_1}}\frac{1}{L_z}\frac{d\Phi}{dt}\right]^{1/3} f^{1/3} , \qquad (10.4.11)$$

From observations of cancelling magnetic features Chae et al. (2003) derived a magnetic flux loss rate of $d\Phi/dt = 1.8 \times 10^{18}$ Mx h^{-1} (or 2×10^6 G cm s^{-1} per unit length of the interface), an approaching speed of $v_1 = 0.22$ km s^{-1}, and the length of the interface $L_z = 2.5$ Mm. From this model, magnetic field strengths of $B_1 \approx 8$ G (Eq. 10.4.10) and inflow speeds of $v_1 \approx 0.1$ km s^{-1} are derived for a typical resistivity value of $\eta = 2 \times 10^7$ cm^2 s^{-1}. Since the observations indicate a faster inflow speed, Chae et al. (2003) argued that chromospheric reconnection occurs faster than the Sweet–Parker model predicts, possibly because of anomalous resistivity that is about 20 times higher. In an alternative study it was found that inflow speeds and outflow speeds are more compatible with the reconnection rates predicted by the Petschek model (Chae et al. 2002d). The inflow and outflow speeds calculated in the different studies of Litvinenko (1999a), Chae et al. (2002d, 2003) are shown in Fig. 10.15. Both the inflow and outflow speeds systematically increase with height, as a simple consequence of the increasing Alfvén velocity due to the decreasing density with height. Therefore, the models of Litvinenko (1999a) and Chae et al. (2003) provide a natural explanation for Hα upflows (e.g., Kurokawa & Sano 2000) with moderate speed (v $\approx$ 20 km s^{-1}) in the lower chromosphere ($T \approx 10^4$ K) and for faster flows (v $\approx$ 100 km s^{-1}) detected in explosive EUV events ($T \approx 10^5$ K) in the upper chromosphere. Chae (1999) suggests a scenario as shown in Fig. 10.16.

Very similar chromospheric reconnection scenarios driven by converging flows (Fig. 10.17) have also been envisioned to explain short-lived brightenings observed in Hα, usually found in emerging flux regions. These brightenings are called *Ellerman bombs* (Ellerman 1917; Nindos & Zirin 1997; Qiu et al. 2000; Georgoulis et al. 2002). Ellerman bombs are estimated to have total energies of $\approx 10^{27} - 10^{28}$ erg, produce temperature enhancements of $\Delta T \approx$ 2000 K, have short radiative cooling times of a few seconds, and are believed to contribute to significant heating of the lower chromosphere (Georgoulis et al. 2002). Ellerman bombs provide a sensitive diagnostic of chromospheric reconnection events, and reveal about $\approx$100 events per hour in an emerging flux region, almost randomly distributed, preferentially above neutral lines of supergranulation boundaries where moving magnetic features collide. Such small-scale variability in Hα is also observed simultaneously in C IV (Wang et al. 1999), which establishes a chromospheric temperature for the associated reconnection sites.

Essentially the same scenario of converging (conjugate) magnetic sources that form a nullpoint and trigger reconnection (Fig. 10.18) has also been employed to explain soft X-ray bright point events (Priest et al. 1994, 1996). The model shown in Fig. 10.18 has only a 2D geometry, but more realistic 3D reconnection geometries have also been attempted, involving the separatrix surfaces of quadrupolar regions (Fig. 10.10 left; Van Driel−Gesztelyi et al. 1996), four magnetic domains in a "cloverleaf" arrangement (Fig. 5.23a; Parnell et al. 1994; Parnell 1996), or quasi-separatrix layers in more complex 3D reconnection configurations (Mandrini et al. 1996, 1997; Démoulin et al. 1996, 1997a, b; Bagalá et al. 2000). One line of X-ray bright point models envisions that plasma is heated in the reconnecting current sheets and thus would spread along the separatrix layers (Longcope 1996,1998; Longcope & Kankelborg 1999; Longcope & Noonan 2000; Longcope et al. 2001), which is distinctly different from standard flare models, where nonthermal particles accelerated near a coronal current sheet first precipitate to the chromospheric footpoints and cause chromospheric evaporation to fill the newly reconnected, relaxing loops with soft X-ray emitting plasma. The latter flare model has no difficulty in explaining the density increase in the soft X-ray-emitting flare loops by orders of magnitude, while the separatrix heating model can only produce high densities if the separatrix layers are heated inside the dense chromosphere. Also the fact that 64% of soft X-ray bright points were found to be associated with cancelling magnetic bipoles (Harvey 1996) suggests that the underlying magnetic reconnection process is confined to the chromosphere.

The role of chromospheric magnetic reconnection processes (with flux cancellation) in the context of larger flares is less obvious. There exists some statistics that flares begin at or near opposite polarity features that are cancelling (Livi et al. 1989). Rapid changes of the magnetic fields associated with six X-class flares were detected by Wang et al. (2002a), but the magnetic flux in the leading magnetic polarity was

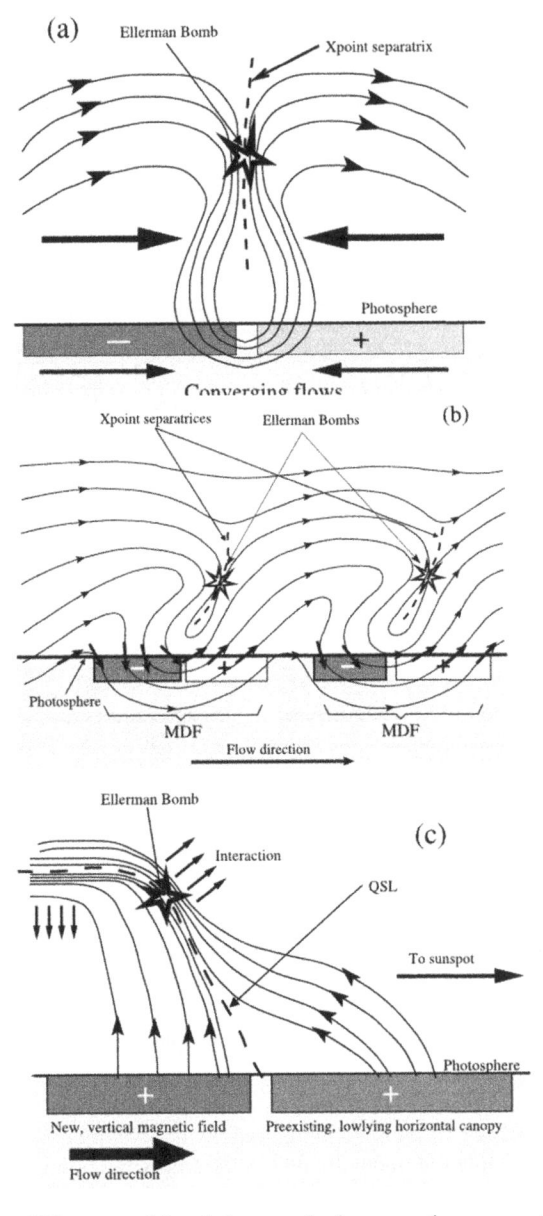

Figure 10.17: Three different models of chromospheric magnetic reconnection at separatrices and quasi-separatrix layers that can explain *Ellerman bombs*: (a) Magnetic reconnection in strong, dipolar magnetic fields sustained by converging horizontal flows; (b) Ellerman bombs triggered by magnetic reconnection above *small-scale moving dipolar magnetic features (MDF)*; and (c) Ellerman bombs triggered by the interaction of two topologically different, unipolar, magnetic configurations (Georgoulis et al. 2002).

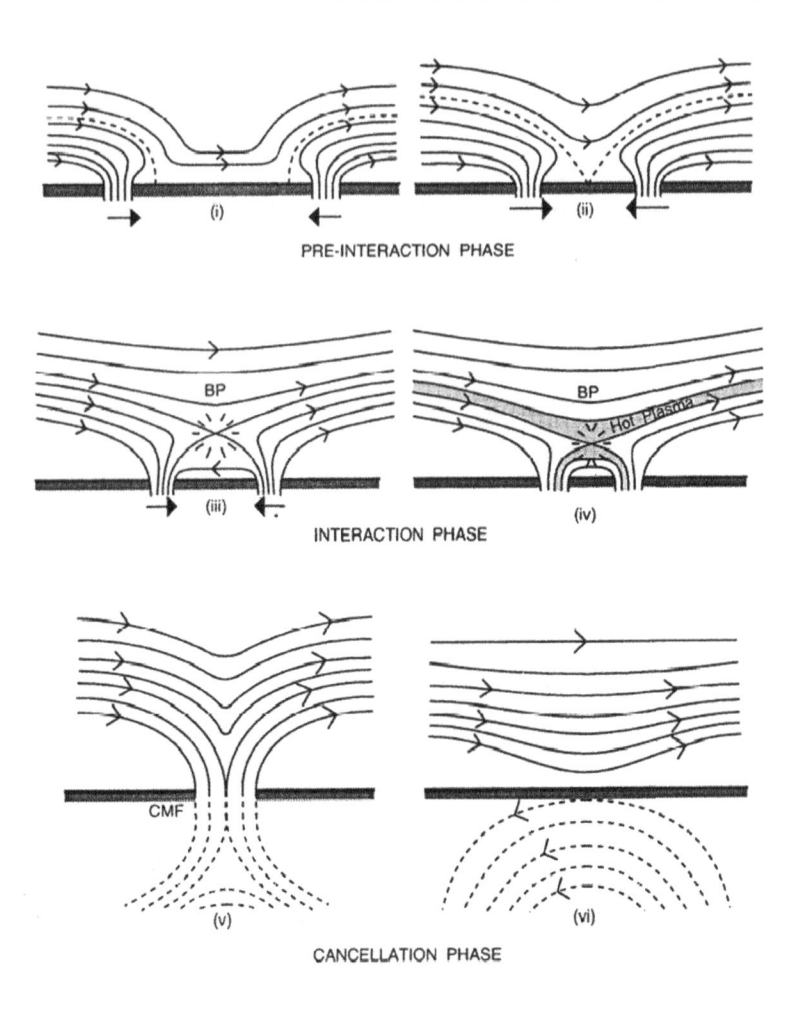

Figure 10.18: Stages of the magnetic flux cancellation process: pre-interaction phase of converging flows (top); interaction phase with X-point reconnection and heating of soft X-ray plasma (middle), and cancellation phase of magnetic flux in photosphere (bottom). Note that the resulting soft X-ray bright point event precedes the photospheric flux cancellation (Priest et al. 1994).

found to increase, while the magnetic flux in the following polarity tended to decrease, by a much smaller amount. So, these events do not support a simple scenario with (symmetric) magnetic flux cancellation. In another (M2.4) flare, a small sunspot was observed rapidly to disappear during the flare, which was interpreted as a possible consequence of a magnetic reconnection process with subsequent submergence (Wang et al. 2002b). In summary, there is a lot of evidence that chromospheric reconnection with magnetic flux cancellation is the most likely mechanism to explain chromospheric variabilities (UV explosive events, Hα upflows, spicules, and soft X-ray brightenings), but this mechanism is probably not the primary driver for larger flares.

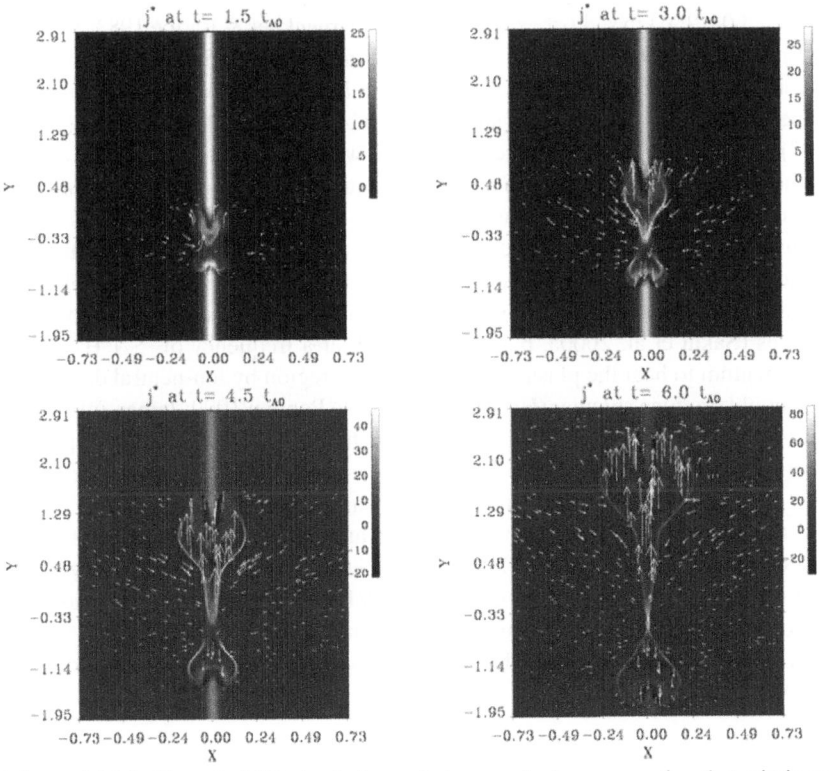

Figure 10.19: Numerical 2D simulations of chromospheric reconnection that mimic explosive events. The greyscale visualize images of the z-component of the current $j_z(x, y)$, while the short streak lines represent the velocity field, at four different time intervals of the simulation (Roussev et al. 2001a).

10.4.3 Chromospheric Reconnection Jets

A theoretical prediction of any X-type reconnection process are two-sided outflows in the cusps of the new-reconnected field lines with Alfvén speeds (Eq. 10.1.3 and Fig. 10.2). If the reconnection occurs inside the chromosphere, there is a steep vertical density gradient and the outflow speed monotonically increases with height. Litvinenko (1999a) estimates at a height of $z \approx 600$ km an inflow speed of $v_1 \approx 40(B/30 \text{ G})^{1/2}$ m s^{-1} and an Alfvén outflow speed of $v_2 \approx 4(B/30)$ km s^{-1}. Since the ion density drops several orders of magnitude in the upper chromosphere (Figs. 1.19 and 4.28), the Alfvén outflow speed $v_A \propto n_i^{-1/2}$ increases at least an order of magnitude and we expect values of $v_2 \approx 10 - 100$ km s^{-1} in the upward direction (Fig. 10.15, right).

Numerical simulations of magnetic reconnection in the chromosphere reproduce intermittent outflows with Alfvén speed (Karpen et al. 1995, 1998; see also Figs. 10.6 and 10.7; Jin et al. 1996; Innes & Toth 1999; Sarro et al. 1999; Roussev et al. 2001a, b, c; 2002; Galsgaard & Roussev 2002; Roussev & Galsgaard 2002; Marik & Erdélyi 2002; see Fig. 10.19) that ressemble the observed cool ($T \approx 2 \times 10^4 - 5 \times 10^5$ K) high-speed outflows (v $\approx 50 - 250$ km s^{-1}) of explosive events (Brueckner & Bartoe 1983; Dere et al. 1991; Erdélyi et al. 1998a, b). Because of the symmetry of X-point reconnection geometries, outflow jets are actually expected to have a *bi-directional structure*, which indeed has been observed in the form of simultaneous blue and red Doppler shifts (Innes et al. 1997). Roussev et al. (2001a) included the upward/downward asymmetry of the chromospheric density stratification in his MHD simulations and predicts up to one order of magnitude less redshift than blueshift (Fig. 10.19). Numerical simulations show that the upward-propagating reconnection jet may collide with the bottom of the U-shaped field lines generating slow-mode MHD waves, and may produce solar spiculae (Takeuchi & Shibata 2001). A related side-effect is the production of upward-traveling Alfvén waves (Sakai et al. 2000a, b; 2001b), which at a frequency of $\lesssim 1$ Hz carry enough momentum to heat the plasma in the transition region by ion-neutral damping, which also could produce *spicules* (Haerendel 1992; De Pontieu 1999; James & Erdélyi 2002), among other possible spicule generation mechanisms driven by chromospheric reconnection (Blake & Sturrock 1985). Further side-effects of chromospheric reconnection processes are the generation of upward-propagating shocks, which have curved fronts due to strong gradients in their acceleration and which may collide (Tarbell et al. 1999, 2000; Ryutova & Tarbell 2000; Ryutova et al. 2001). A recent analysis of the timing of Doppler shifts in C IV, C II, and O VI lines suggests that the majority of observed UV explosive events occur in the presence of the resulting behind-shock downflows and fewer than 10% by the direct collision of shock fronts (Ryutova & Tarbell 2000).

10.5   Flare/CME Models

In this section we discuss the most eminent physical models for flare and CME processes, which all involve magnetic reconnection in some form. What distinguishes the different flare models are mainly the initial magnetic topologies, which are prone to specific instabilities or drivers. This section covers mainly the theoretical aspects of flare models, while supporting observations are compiled in § 10.6. Theoretical reviews on flare/CME models can be found in Brown & Smith (1980), Melrose (1993), Shibata (1998), Priest (2000), Forbes (2000b, 2001), Klimchuk (2001), Low (1999b, 2001b), Priest & Forbes (2002), or in textbook chapters (Svestka 1976, § 6; Priest 1982, § 10; Priest & Forbes 2000, § 11; Tajima & Shibata 2002, § 3.3).

10.5.1   The Standard 2D Flare Model

Although not all flares can be explained by a single model, it is justified to establish a standard model that fits most of the observations and has a well-understood theoretical foundation. The most widely accepted standard model for flares is the 2D magnetic reconnection model that evolved from the concepts of Carmichael (1964), Sturrock (1966), Hirayama (1974), Kopp & Pneuman (1976), called the *CSHKP model* according to the initials of these five authors. This has been further elaborated by Tsuneta (1996a; 1997) and Shibata (1995) based on the modeling of Yohkoh observations.

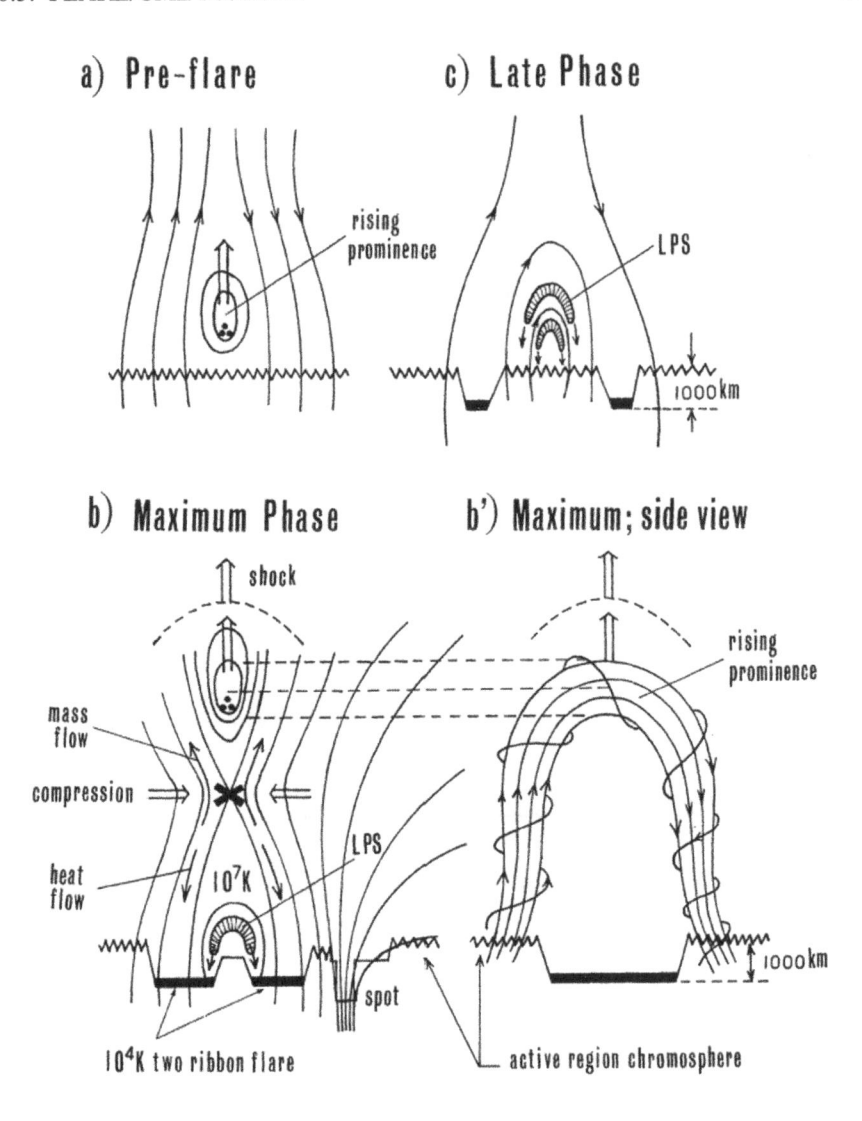

a) Pre-flare

c) Late Phase

b) Maximum Phase

b') Maximum; side view

Figure 10.20: Temporal evolution of a flare according to the model of Hirayama (1974), which starts from a rising prominence (a), triggers X-point reconnection beneath an erupting prominence (b), shown in sideview (b'), and ends with the draining of chromospheric evaporated, hot plasma from the flare loops (c) (Hirayama 1974).

The initial driver of the flare process is a rising prominence above the neutral line in a flare-prone active region (Fig. 10.20a). The rising filament stretches a current sheet above the neutral line, which is prone to Sweet−Parker or Petschek reconnection (Fig. 10.20b). In the model of Sturrock (1966), a helmet streamer configuration was assumed to exist at the beginning of a flare, where the tearing-mode instability (induced by footpoint shearing) near the Y-type reconnection point triggers a flare, ac-

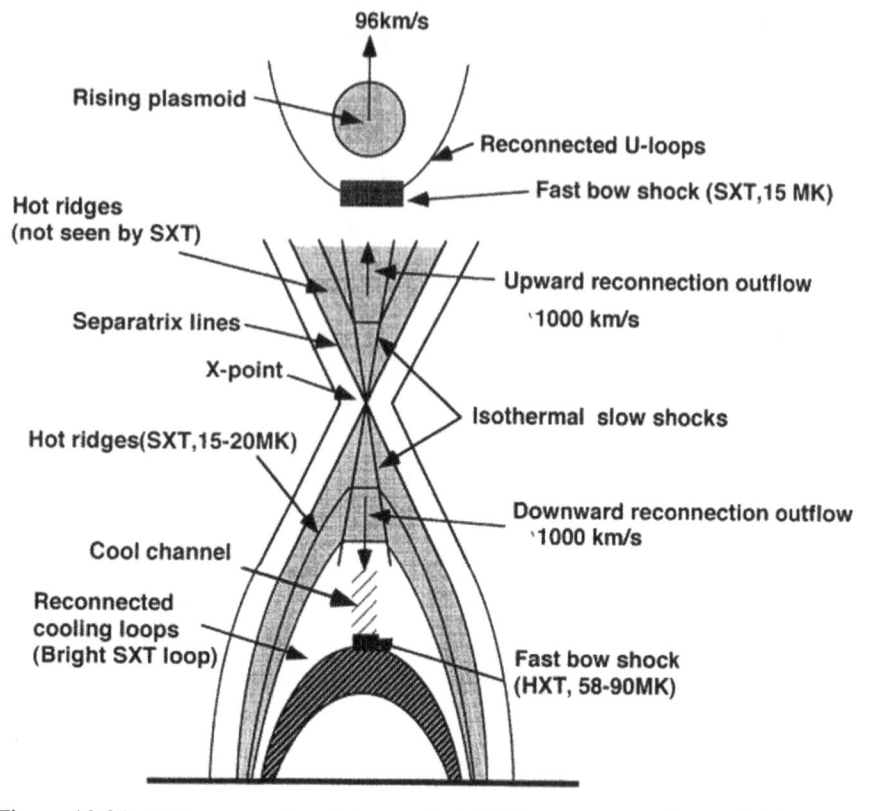

Figure 10.21: Elaborate version of the standard 2D X-type reconnection model that also includes the slow and fast shocks in the outflow region, the upward-ejected plasmoid, and the locations of the soft X-ray bright flare loops (Tsuneta 1997).

celerating particles in a downward direction and producing shock waves and plasmoid ejection in an upward direction. Hirayama (1974) explains the preflare process as a rising prominence above a neutral line (between oppositely directed open magnetic field lines), which carries an electric current parallel to the neutral line and induces a magnetic collapse on both sides of the current sheet after eruption of the prominence. The magnetic collapse is accompanied by lateral inflow of plasma into the opposite sides of the current sheets. The X-type reconnection region is assumed to be the location of major magnetic energy dissipation, which heats the local coronal plasma and accelerates nonthermal particles. These two processes produce thermal conduction fronts and precipitating particles which both heat the chromospheric footpoints of the newly reconnected field lines. As a result of this impulsive heating, chromospheric plasma evaporates (or ablates) and fills the newly reconnected field lines with over-dense heated plasma, which produces soft X-ray-emitting flare loops with temperatures of $T_e \approx 10 - 40$ MK and densities of $n_e \approx 10^{10} - 10^{12}$ cm^{-3}. Once the flare loops cool down by thermal conduction and radiative loss, they also become detectable in EUV ($T_e \approx 1 - 2$ MK) and Hα ($T_e \approx 10^4 - 10^5$ K). Kopp & Pneuman (1976) refined this scenario further and predicted a continuous rise of the Y-type reconnection point, due to the rising prominence. As a consequence, the newly reconnected field lines beneath the X or Y-type reconnection point have an increasingly larger height and wider footpoint separation. Tsuneta (1996a; 1997) and Shibata (1995) elaborated on the temperature structure, upward-ejected plasmoids, slow shocks, and fast shocks in the outflow region of the X-type reconnection geometry (Fig. 10.21). The heated plasma in the reconnection outflow produces hot ridges ($T \approx 15 - 20$ MK) along the separatrices with the slow shocks, sandwiching the denser soft X-ray flare loops that occupy the newly reconnected relaxing field lines, which are filled with chromospheric evaporated plasma. The fast shocks in the reconnection outflows collide with the previously reconnected field lines and may produce hot thermal (as well as nonthermal) hard X-ray sources above the flare looptops. Numerical hydrodynamic simulations of this model reproduce heat conduction fronts and slow-mode shocks (Yokoyama & Shibata 1997) and chromospheric evaporation (Magara et al. 1996; Yokoyama & Shibata 1998; 2001)

This model is essentially a 2D model that describes the evolution in a vertical plane, while evolution along the third dimension (in the direction of the neutral line) can be independently repeated for multiple flare loops (where footpoints extend to a double ribbon) or can be stopped (in the case of a single-loop flare). It is likely that the extension in the third dimension is not continuous (in the form of a giant 2D Sweet–Parker current sheet), but rather highly fragmented into temporary magnetic islands (due to tearing-mode and coalescence instabilities, see § 10.2). Numerical simulations of enhanced resistivity in the current sheet enables the fast reconnection regime (Magara & Shibata 1999) that is required to explain the observed fast (subsecond) time structures. This model fits a lot of the observational features in hard X-rays, soft X-rays, Hα, and radio wavelengths, provides a physical mechanism to explain self-consistently the processes of filament eruption, magnetic reconnection, and coronal mass ejection, but does not specify what drives the initial magnetic system to become unstable. This model fits single-loop and double-ribbon arcade geometries, but is not appropriate for quadrupolar flare loop interactions and 3D nullpoint topologies.

10.5.2 The Emerging Flux Model

The most decisive criterion to judge the relevance of a particular flare model is the driver mechanism that dictates the magnetic evolution, the loss of stability, and subsequent magnetic reconnection process. While the driver is a rising filament/prominence in the Kopp & Pneuman (1976) model, the process of flux emergence has been considered as a driver in the model of Heyvaerts et al. (1977). The model of Heyvaerts et al. (1977) consists of three phases: (1) a preflare heating phase where a new magnetic flux emerges beneath the flare filament and continuously reconnects and heats the current sheet between the old and new flux; (2) the impulsive phase starts when the heated current sheet loses equilibrium at a critical height and turbulent electrical resistivity causes the current sheet rapidly to expand, accelerating particles and triggering chromospheric evaporation; and (3) the main phase where the current sheet reaches a new steady state with marginal reconnection [Fig. 10.22(a) and (b)]. A requirement of this model is the pre-existence of a stable current sheet (with very low resistivity) for

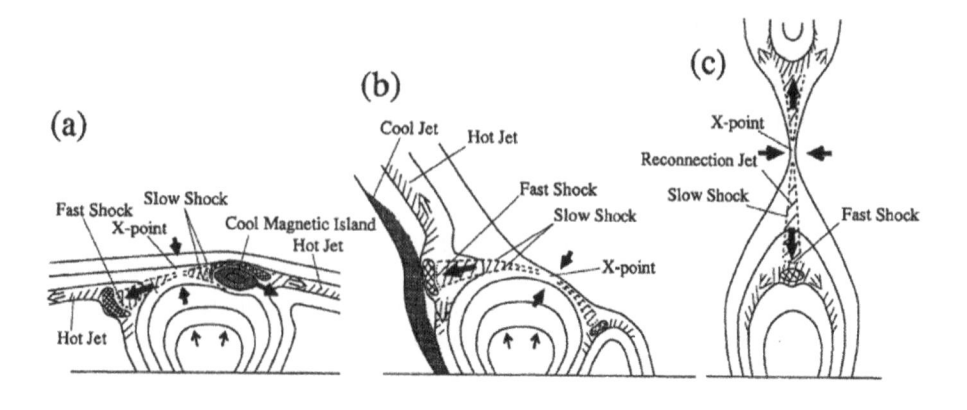

Figure 10.22: X-type reconnection scenarios for three different orientations of the external magnetic field: horizontal (a), oblique (b), and vertical (c). The two versions (a) and (b) represent geometries used in the emerging flux model of Heyvaerts et al. (1977), while version (c) corresponds to the Kopp & Pneuman (1976) model (Yokoyama & Shibata 1996).

periods of the order of a day or more. However, numerical simulations indicate that the current sheets reconnect almost as quickly as they are formed (Forbes & Priest 1984; Shibata et al. 1990). It is therefore believed that the Heyvaerts model can only apply to small flares (Priest & Forbes 2000).

The geometry of the pre-existing magnetic field is assumed to have a horizontal or oblique angle. A consequence of this geometry is the expulsion of two oppositely directed plasma jets during the impulsive phase [for a horizontal orientation see Fig. 10.22(a)], or a single jet in the upward direction [for an oblique orientation see Fig. 10.22(b)]. This model was further elaborated in terms of reconnection outflow characteristics by Shibata et al. (1996c), inspired by the numerous plasma jets that have been observed with Yohkoh/SXT (e.g., Shibata et al. 1992a, 1994a, 1996a, b). The initial driver in Shibata's emerging flux model is the nonlinear evolution of the magnetic buoyancy (Parker) instability (see § 10.4.1 and Fig. 10.14) simulated by Shibata et al. (1989b). This instability was applied to the reconnection between the emerging flux and the overlying coronal field, leading to formation and ejection of magnetic islands or plasmoids (Shibata et al. 1992b). Further numerical hydrodynamic simulations succeeded in modeling coronal X-ray jets and Hα surges (Yokoyama & Shibata 1995, 1996). The locations of the slow-mode and fast-mode shocks of reconnection outflows are indicated in Fig. 10.22. Shibata et al. (1994a, 1996a, b, c) distinguishes between hot and cool jet structures, where the hot jets emerge from the reconnection region, while the cooler jets result from chromospheric evaporation into open field lines. Yokoyama & Shibata (1995, 1996) pointed out that the hot jet ejected from the current sheet region is not the reconnection jet itself, but a secondary jet accelerated by the enhanced gas pressure behind the fast shock, which prevents a direct escape of the primary reconnection jet.

There are other variants of reconnection-driven jet models. The model of Priest et al. (1994) also produces two-sided, soft X-ray jets, but the drivers are converging flows at the footpoint (Fig. 10.18), which is motivated by the observed correlation with magnetic flux cancellation, while the driver in the model of Shibata et al. (1994b, 1996a, b, c) is flux emergence caused by the upward-pushing Parker instability. The model of Karpen et al. (1995, 1998; Figs. 10.6−7) can also produce jets, but is driven by shearing motion at the footpoints, which drives magnetic reconnection in a quadrupolar geometry. Thus, the production of plasma jets is a common characteristic of many flare models, which provides a useful diagnostic of the geometric orientation and spatial location of the involved magnetic reconnection regions (Fig. 10.22).

10.5.3 The Equilibrium Loss Model

The driver in the Kopp−Pneuman flare model (§ 10.5.1) is a rising filament, but the magnetic pre-evolution that leads up to flare instability is not quantified in the various concepts of the CSHKP models. The driver in the Heyvaerts model is emerging flux (§ 10.5.2), but the onset of flare instability is not quantified in terms of a magnetic field evolution. Another criticism of the latter model is that stable current sheets are unlikely to exist over extended periods of time (as numerical simulations demonstrate), which implies that free magnetic energy has to be stored in the form of field-aligned currents (i.e., force-free fields; Forbes 1996). An evolutionary model that starts with a stable (force-free) magnetic field configuration, then applies converging flows as a continuous driver, and demonstrates how the (force-free) evolution passes a critical point where the system becomes unstable and triggers the rise of a filament, has been developed by Forbes & Priest (1995) in 2D. The initial situation of the magnetic field is shown in Fig. 10.23 (b), where the magnetic field is quantified by the 2D equilibrium of a fluxrope at a stationary height, described by the Grad−Shafranov equation. The two footpoints of the field lines that envelop the fluxrope are then driven closer together, while the system evolves through a series of equilibrium solutions. The height h of the fluxrope as a function of the separation half-distance λ is shown in Fig. 10.23(a), which monotonically decreases while the source separation is made smaller from $\lambda = 4 \mapsto 1$. Once the source separation passes the critical point at $\lambda = 1$, the fluxrope enters a loss of equilibrium and jumps in height (from $h = 1$ to $h \approx 5$), while forming a current sheet beneath [Fig. 10.23(d)]. In *ideal* MHD, the rising fluxrope would stop at a higher equilibrium position, because the tension force associated with the current sheet is always strong enough to prevent the fluxrope from escaping (Lin & Forbes 2000). If there is some resistivity, magnetic reconnection is enabled, and even a fairly small reconnection rate is sufficient to allow the fluxrope to escape (Lin & Forbes 2000). Magnetic reconnection in the current sheet releases most ($\approx 95\%$) of the magnetic energy that has been built up from the initial force-free configuration by the converging motion of the footpoints before the loss of equilibrium. This model is formulated fully analytically and yields reasonable amounts of released energies, suitable to explain flares and CMEs. Although this analytical model is restricted to 2D (with a fluxrope that is not anchored at both ends), it demonstrates quantitatively how a loss of magnetic equilibrium leads to a rapid energy release, which probably also takes place in more complex 3D configurations. The question is whether the driver in terms of converging flows is realistic, because typically observed photospheric flows are in the order of ≈ 1 km s^{-1}, which could be too slow or may be randomly oriented. Also, shear flows with

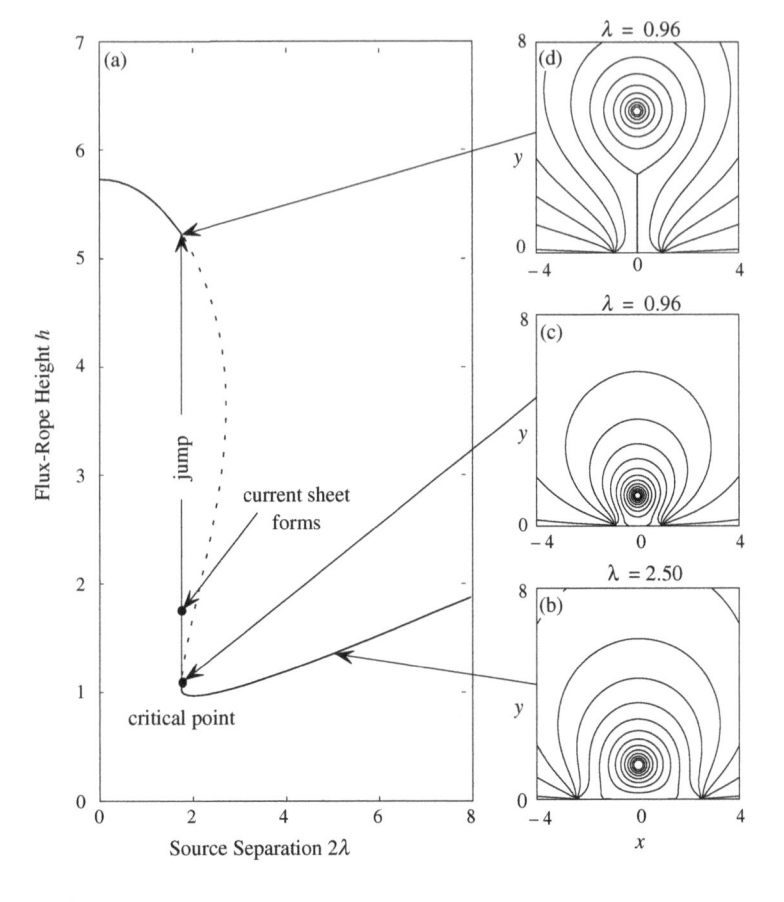

Figure 10.23: Flare dynamics in the model of Forbes & Priest (1995), inferred from the ideal MHD evolution of a 2D arcade containing an unshielded fluxrope (a)–(c). The fluxrope arcade jumps upwards when the two photospheric field sources are pushed too close to one another. (d) The vertical current sheet is subject to magnetic reconnection if enhanced or anomalous resistivity occurs (Forbes & Priest 1995).

subsequent tearing instability have been found to be important drivers of flares, which would require a generalization of Forbes's model to 3D. Numerical 3D simulations of a similar dipolar configuration driven by converging flows have been performed by Birn et al. (2000).

The analytical model of Forbes & Priest (1995) predicts a specific height evolution of the fluxrope $h(t)$, which grows initially as $h(t) \propto t^{5/2}$, or $v(t) \propto t^{3/2}$, and reaches an asymptotic constant speed of order $v_{term} \approx 1500$ km s^{-1}. The solutions of the height $h(t)$, velocity $v(t)$, dissipated energy $dW(t)/dt$, electric field $E_0(t)$, Alfvén speed $v_A(t)$, and reconnection speed $v_R(t)$ are shown in Fig. 10.24, calculated for a model with initial half-separation $\lambda_0 = 50$ Mm, fluxrope length $L = 100$ Mm, fluxrope radius $a_0 = 0.4\lambda_0$, density $n_e = 5 \times 10^{10}$ cm^{-3}, and magnetic field $B_0 = 100$ G.

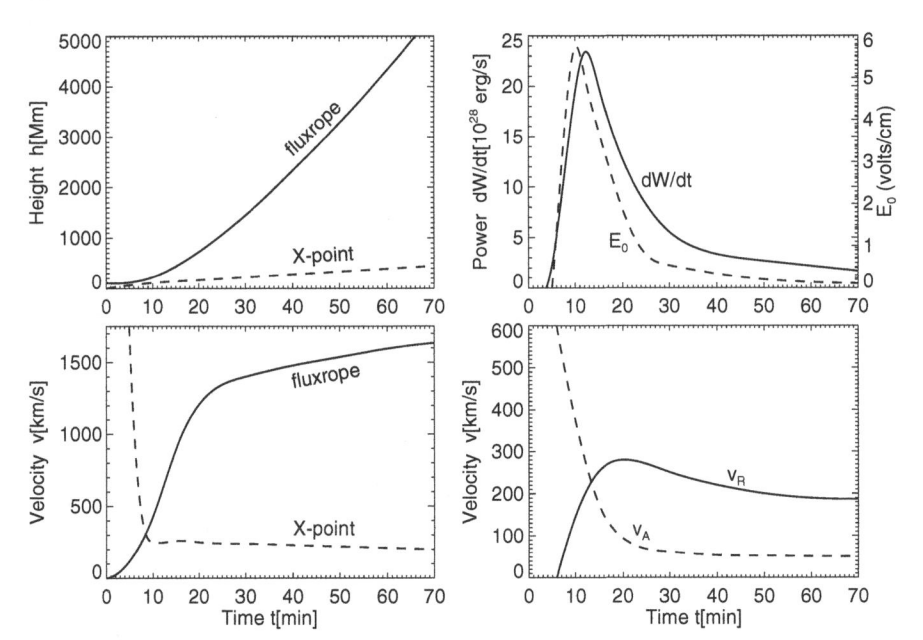

Figure 10.24: Evolution of height $h(t)$ (top left) and velocities $v(t)$ (bottom left) of fluxrope and magnetic reconnection X-point in the equilibrium loss model of Forbes & Priest (1995). The upper limit of the dissipated energy dW/dt in the current sheet and the generated electric field E_0 at the X-point are shown (top right), as well as the reconnection speed $v_R(t)$ and the ambient Alfvén speed (bottom right). See parameters in text (Priest & Forbes 2000).

This calculation represents an upper limit, in the case of an unlimited reconnection rate, so that all magnetic energy goes into the kinetic energy of the upward-accelerated fluxrope. The upward motion of the unstable fluxrope with associated reconnection also predicts a *shrinkage* of flare loops, characterized by the height ratio of the cusp at the beginning of reconnection to the height (of the relaxed dipolar field line) in the postflare phase, which was found to be 20% and 33% in two flares (Forbes & Acton 1996).

10.5.4   2D Quadrupolar Flare Model

Among 2D models, which we classified in Fig. 10.9 into *bipolar* (e.g., Kopp−Pneuman § 10.5.1; Priest−Forbes § 10.5.3), *tripolar* (e.g., Heyvaerts et al. § 10.5.2), and *quadrupolar* ones, we describe here a representative of the latter category (namely, the *quadrupolar photospheric source model*), which was first proposed by Uchida (1980), and later developed further by Uchida et al. (1998a, b) and Hirose et al. (2001). The initial configuration consists of two parallel arcades (as shown in Fig. 10.9, top right), which altogether requires three parallel neutral lines. Formation of such double arcades with current sheets inbetween have been inferred from neighboring active regions (Sakurai & Uchida 1977) and from polar crown filaments with arcades (Uchida et al. 1996). As in the Forbes−Priest (1995) model, the principal driver is a converging flow pat-

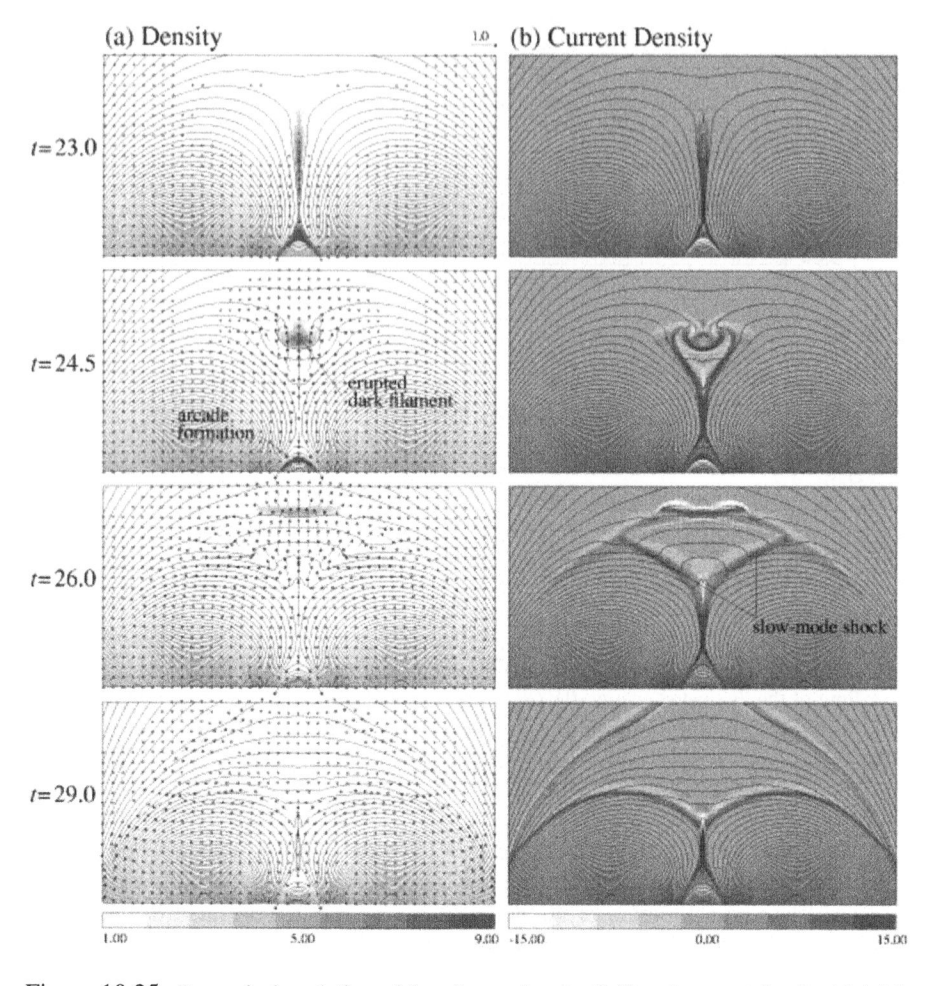

Figure 10.25: Dynamical evolution of the plasma density (left) and current density (right) in the quadrupolar magnetic reconnection model of Uchida (1980), simulated with a 2.5D MHD code. Note that the erupted dark filament transforms into a CME with two slow-mode shocks at both sides (Hirose et al. 2001).

tern that pushes the two arcades together. The X-point above the middle neutral line supports a dark filament. The two flanking arcades that suspend the filament might be partially observable as the "barbs" of the filament channel (Fig. 6.18). While the two arcades push together, the dark filament transforms into a thin vertical current sheet, which at some point becomes unstable due to tearing-mode instability, triggering anomalous resistivity and fast reconnection. The dark filament with helical field lines is accelerated upward in the expanding field structure with a rounded shape and transforms into a CME. The reconnected field lines below the X-point shrink and form the postflare arcade. Numerical simulations of the driven reconnection in quadrupolar arcades or between interacting loops have been performed by Rickard & Priest (1994),

Sakai et al. (1995), Sakai & De Jager (1996), and for the quadrupolar model of Uchida specifically by Hirose et al. (2001), see Fig. 10.25 for an example.

There are several motivations for this 2D quadrupolar model. The Kopp–Pneuman (1976) model cannot explain the magnetic field in the dark filaments seen from the side, because the direction of the magnetic field at the lower side is opposite to what is expected from the polarity of photospheric sources (Leroy et al. 1984). A proposed solution was to introduce a fluxrope (Kuperus & Raadu 1974), which reverses the polarity at the lower dip of the prominence (Fig. 6.13). Therefore, the involvement of a fluxrope-like filament is an essential element in a flare model, which naturally evolves in the Forbes–Priest (1995) and quadrupolar model of Uchida (1980). Further it is argued that the 2D quadrupolar model of Uchida (1980) solves the energy problem to open up the field and to accelerate the filament to escape velocity (Uchida et al. 1996; 1998a, b). In Sturrock's (1966) model, the erupting filament is required to have an energy equal to or greater than that of the flare itself, since the open-field configuration (which is the final state in Sturrock's 1966 model) is conjectured to be the state of maximum energy (Aly 1984; Sturrock 1991). In the numerical simulation of Hirose et al. (2001) it is actually found that the major part of the stored magnetic energy is converted into kinetic energy carried away by the CME (containing the erupted dark filament), while only a minor part is left for heating of the associated arcade flare.

10.5.5 The Magnetic Breakout Model

A further development of the 2D quadrupolar model of Uchida (1980) is the so-called *magnetic breakout* model of Antiochos et al. (1999b) and Aulanier et al. (2000b), which involves the same initial quadrupolar magnetic configuration, but undergoes an asymmetric evolution with the opening up of the magnetic field on one side. The asymmetric evolution is driven by footpoint shearing of one side arcade, where reconnection between the sheared arcade and the neighboring (unsheared) flux systems triggers an eruption. In this magnetic breakout model, reconnection removes the unsheared field above the low-lying, sheared core flux near the neutral line, which then allows the field above the core flux to open up (Antiochos et al. 1999b). Thus, this model addresses the same energy problem as Uchida's model: How very low-lying magnetic field lines can open up (down to the photospheric level) into an open-field configuration during the eruption. Moreover, the eruption is solely driven by free magnetic energy stored in a closed, sheared arcade. It circumvents the Aly–Sturrock energy limit by allowing external, disconnected magnetic flux from a neighboring sheared arcade (which is not accounted for in the "closed-topology" model of Aly and Sturrock) to assist in the opening-up process. Thus, a key point of the magnetic breakout model is the interaction of a multi-flux system (e.g., in a quadrupolar double arcade). It has the same initial configuration as Uchida's model, but is driven by asymmetric shear.

The magnetic topology of the magnetic breakout model has been applied to the Bastille-Day flare by Aulanier et al. (2000b), who found a more complex 3D topology than the 2D quadrupolar model of Antiochos et al. (1999b). Aulanier et al. (2000b) actually identified a magnetic nullpoint in the corona above the flare arcade which was connected with a "spine" field line to a photospheric location where the flare brightens up first. The other side of the coronal nullpoint sits on a dome-like "fan" surface,

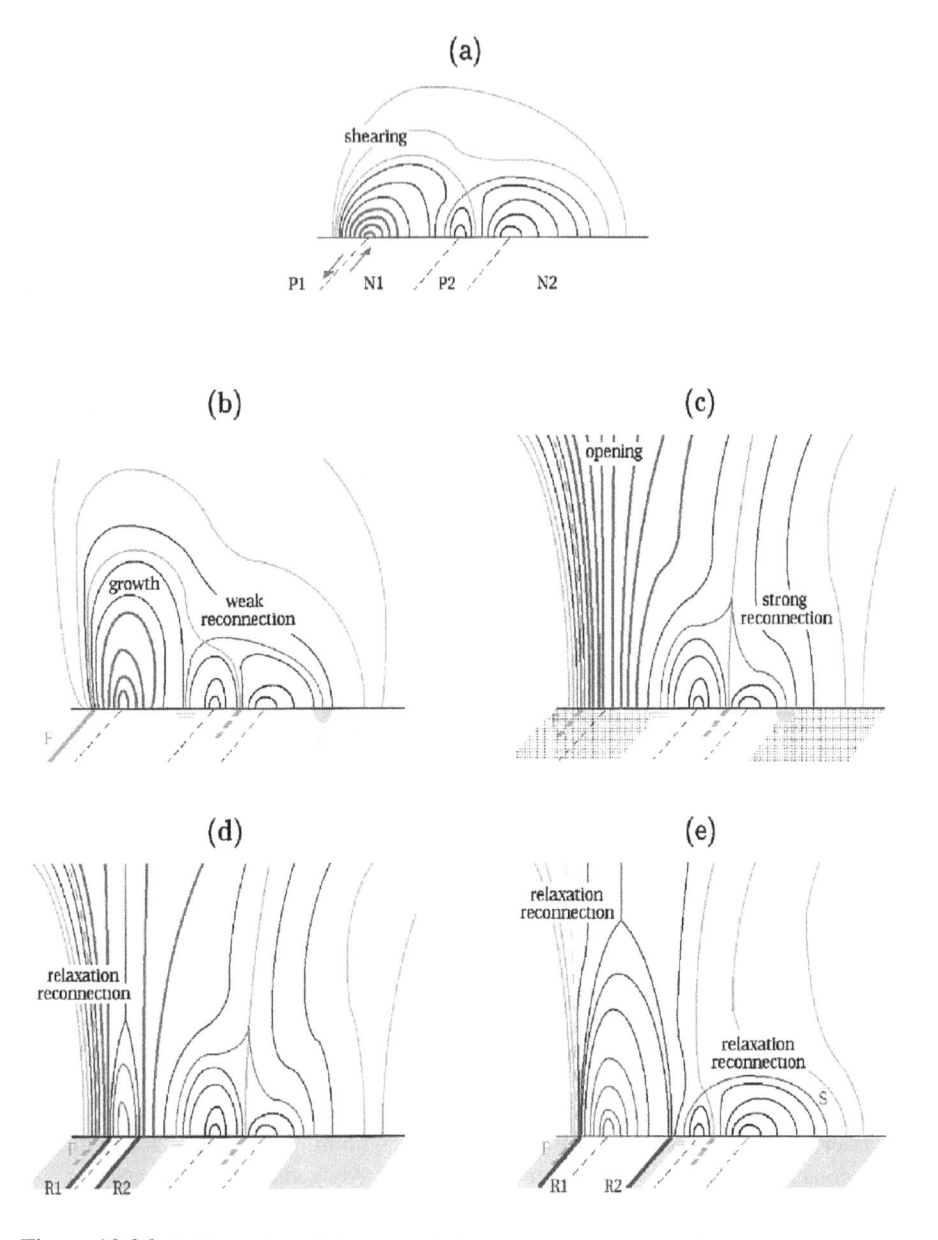

Figure 10.26: 2.5D version of the magnetic breakout model: (a) The initial quadrupolar potential state, with shear applied on both sides of the neutral line P1/N1; (b) the shear triggers some weak reconnection at the coronal nullpoint; (c) fast reconnection at the nullpoint leads to opening up of the field; (d) relaxation reconnection in the opening field lines, forming footpoint ribbons and flare loops; (e) ongoing formation of postflare loops and reconnection at the null. The Kopp–Pneuman (1976) model is a special case in which the magnetic breakout does not occur [eliminating phase (c)] (Aulanier et al. 2000b).

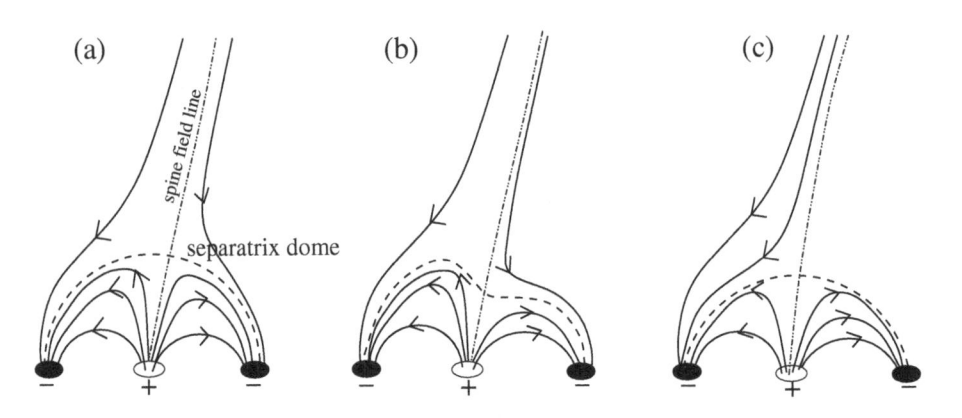

Figure 10.27: Magnetic field topology inferred in the 1993-May-3, 23:05 UT, flare by Fletcher et al. (2001). The sequence shows a 2D representation of the reconnection process via a separator dome (Fletcher et al. 2001)

which encloses the main flare arcade. This topology (shown in Plate 13) can be considered as one of the many possible 3D reconnection scenarios (Fig. 5.23) in which the magnetic breakout model can be realized. Aulanier et al. (2000b) suggest a more general definition: *"A magnetic breakout is the opening of initially low-lying sheared fields, triggered by reconnection at a nullpoint that is located high in the corona and that defines a separatrix enclosing the sheared fields"*. This represents a generalization of the 2.5D version (Fig. 10.26) into 3D reconnection topologies. Obviously, observations are crucial to pinning down the involved magnetic configurations, which are now becoming available increasingly clearly from TRACE postflare loop observations. The reconstruction of the preflare configuration, which is necessary to track down the reconnection process, however, is hampered by the unavailability of high-resolution observations at the much higher flare temperatures ($T \approx 10 - 40$ MK) during the impulsive flare phase. Nevertheless, the 3D reconnection topology could be reconstructed for some cases, clearly showing evidence for 3D reconnection involving a separatrix dome (Fig. 10.27; Fletcher et al. 2001).

10.5.6 3D Quadrupolar Flare Models

Some flares clearly show an interaction between two flare loops (or closed-field systems), which can most simply be interpreted as the outcome of a quadrupolar reconnection process. The magnetic configuration corresponds to a 3D reconnection case (Fig. 10.9, bottom right) that can be represented by a single, common, neutral line for the two interacting flare loops, which is different from the 2D case in Uchida's model, which has three neutral lines (Fig. 10.9, top right). The two observed flare loops represent, of course, the postreconnection situation, but the prereconnection topology can be straightforwardly reconstructed by switching the polarities according to the scheme shown in Fig. 10.28. Thus, magnetic geometry is fully constrained for this type of 3D reconnection and can be reconstructed from the observed postflare loops. A number of

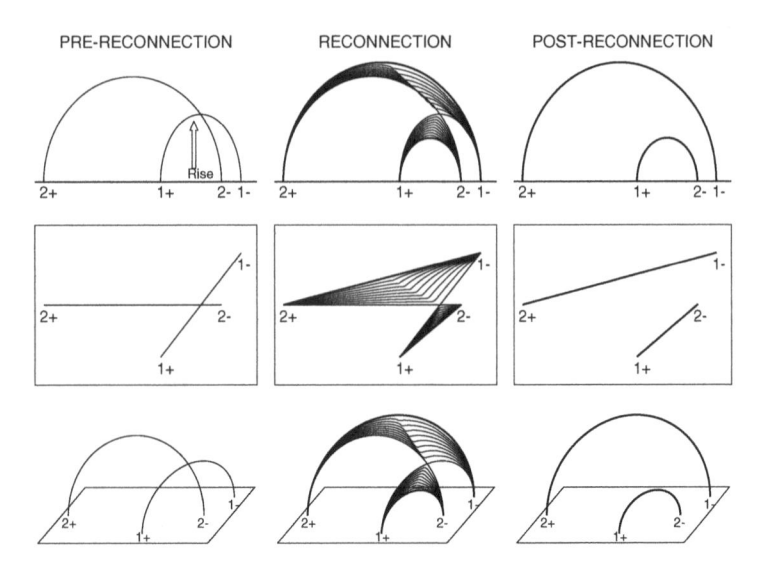

Figure 10.28: The concept of 3D magnetic reconnection in a quadrupolar geometry is visualized for two semi-circular loops (1 and 2), with initial footpoints $(1+, 1-)$ and $(2+, 2-)$. The three stages of prereconnection (left column), main reconnection phase with subsequent relaxation process (middle column), and final postreconnection phase (right column) are depicted for three different views [i.e., side view (top row), top view (middle row), and perspective view (bottom row)]. Note that all loop shapes in the initial and final phase are represented by circular segments, while the intermediate stages of relaxing field lines are rendered by linear interpolation (Aschwanden et al. 1999c).

flares was found to fit quadrupolar geometry (Hanaoka 1996, 1997; Nishio et al. 1997; Aschwanden et al. 1999c).

A theoretical model for this type of 3D quadrupolar reconnection was developed by Melrose (1997) in terms of two interacting current-carrying loops. A fundamental assumption in Melrose's (1997) model is the conservation of the large-scale currents that flow through coronal loops and close below the photosphere. A consequence of this assumption is that magnetic reconnection processes only redistribute the current paths, while the net current flowing into and out of the corona remains fixed. In Melrose's model, magnetic flux $(\Delta\Psi)$ and electric current (ΔI) are transferred during reconnection, where the mutual inductance between interacting flux loops depends on their geometry, relative distance, and preflare current. Flux transfer can be calculated by the mutual inductance between two current-carrying wires, as in electrodynamics. Aschwanden et al. (1999b) found that the transferred free magnetic energy ΔE^I in Melrose's model depends strongest on the self-induction term of the largest involved loop, which can be approximated by

$$\Delta E^I = (0.70 \pm 0.05)\mu_0 C r_2 (I_2 - I_1) I_1 \approx 10^{29.63} \left(\frac{r_2}{10^9 \text{ cm}}\right) \text{ [erg]}, \quad (10.5.1)$$

where r_2 is the curvature radius of the larger loop, I_1 and I_2 are the currents of the two interacting loops (of order $I_2 - I_1 \approx I_1 \approx 0.5 \times 10^{11}$ A), and $C = 1.94$ a capacity constant.

This 3D quadrupolar reconnection model only describes the interaction between two closed loops, which was found to match closely the observed topology of some flares. Obviously the model does not include any open field lines and thus cannot explain the simultaneous rise of a filament and expulsion of a CME, which may occur in a detached magnetic field domain above the interacting loop system. However, a key aspect of this model is that it relates the currents of the prereconnection to the postreconnection field lines in a highly sheared configuration. It also quantifies the efficiency of the reconnection rate as a function of shear angles. Interestingly, most of the relevant observations indicate shear angles between the reconnecting field lines which range from near-parallel to near-perpendicular (Aschwanden et al. 1999b), rather than being anti-parallel as expected in the standard Kopp–Pneuman flare scenario. This observational fact, however, does not violate the Petschek reconnection model, since the reconnection rate can still operate at small angles θ, with efficiency scaling as $\propto [\sin(\theta/2)]^{1/2}$ (Soward 1982).

Similar 3D quadrupolar reconnection models are also described in Somov et al. (1998), Somov (2000, § 16.5.2), and Kusano (2002), where loops that are sheared along the central neutral line of a flare arcade reconnect with the overlying less sheared arcade. Alternatively, 3D quadrupolar reconnection in large flares could also be driven by emerging current loops (Mandrini et al. 1993). Such 3D quadrupolar configurations are particularly suitable to explain double-ribbon flares, but it could not yet be decided observationally whether the primary driver of this type of reconnection is a rising filament (Kopp–Pneuman 1976) or the shear along the neutral line (Sturrock 1966; Somov et al. 1998).

10.5.7 Unification of Flare Models

In Table 10.1 we sort the previously discussed flare/CME models according to the driver mechanisms and dimensionality of magnetic reconnection geometry. There are essentially two locations of drivers: (1) above the flare site (in the form of a rising filament, prominence, or plasmoid); and (2) below the flare site (in the form of photospheric emergence, convergence flows, or shear flows). The three photospheric drivers can essentially be discriminated by their directions: (1) flux emergence corresponds to a flow in the vertical direction (v_z); (2) convergence flows are counter-directed perpendicular to the neutral line ($+v_x, -v_x \perp NL$); and (3) shear flows are counter-directed parallel to the neutral line ($+v_y, -v_y \parallel NL$). The classification in Table 10.1 also shows that 2D models can only be constructed when the driver force is in the 2D plane of the loops (e.g., converging flows in the the x-direction or emergence in the z-direction), while a driver force perpendicular to the 2D loop plane (e.g., shear in the y-direction) requires 3D models. Table 10.1 is by no means a complete list of flare/CME models; in principle there could be for every type of driver at least one 3D model, and moreover multiple models could be conceived for any combination of multiple loops (open or closed, and arcades). The 2D models are probably all idealized approximations, but more accurate future observations might require generalizations of each one to a 3D version. Nevertheless, the classification in Table 10.1 indicates

Table 10.1: Classification of flare/CME Models according to the driver mechanisms and dimensionality of magnetic reconnection geometry.

Driver mechanism:	2D models:	3D models:
Rising filament or prominence $v_z(h \gg h_{Ph})$	**X-type reconnection** (Hirayama 1974) (Kopp & Pneuman 1976)	
Photospheric flux emergence $v_z(h = h_{Ph})$	**Emerging flux model** (Heyvaerts et al. 1977)	**Quadrupolar flux transfer** (Melrose 1995)
Photospheric converging flows $+v_x, -v_x \perp NL$	**Equilibrium loss model** (Forbes & Priest 1995) **Quadrupolar double-arcade** (Uchida 1980)	
Photospheric shear motion $+v_y, -v_y \parallel NL$		**Tearing-mode instability** (Sturrock 1966) **Magnetic breakout model** (Antiochos et al. 1999b) **Sheared loops inside arcade** (Somov et al. 1998)

that at least models with the same driver mechanism could be unified into a 3D model. In future we might even distill a single unified flare/CME model by combining all the important drivers.

A partial unification of flare models has been proposed by Shibata (1998; 1999a), envisioning a *grand unified model* of larger flares (*long-duration events (LDEs)* and *impulsive flares*) and smaller flares (microflares, subflares, soft X-ray jets), produced by fast magnetic reconnection driven by plasmoid ejection as the key process (also called *plasmoid-induced reconnection model*). The unification scheme is visualized in Fig. 10.22, which combines X-type reconnection driven by the emerging flux (Heyvaerts et al. 1977) or by a rising filament (Kopp & Pneuman 1976).

Assuming that the energy release is dominated by magnetic reconnection, "universal scaling laws" were derived for solar and stellar flares (Yokoyama & Shibata 1998; Shibata & Yokoyama 1999b). These scaling laws are based on the assumptions that (1) the flare temperature at the loop apex is balanced by the conduction cooling rate ($Q = d/ds(\kappa_0 T^{5/2} dT/ds) \approx (2/7)\kappa_0 T_A^{7/2}/L^2$) (see Eq. 3.6.3), yielding

$$T_A \approx \left(\frac{2QL^2}{\kappa_0}\right)^{2/7} , \tag{10.5.2}$$

and (2) the heating rate Q is given by the reconnection rate in Petschek's theory (see Eq. 10.1.8),

$$Q = \left(\frac{B^2}{4\pi}\right) \left(\frac{v_{in}}{L}\right) \left(\frac{1}{\sin(\theta)}\right) \approx \left(\frac{B^2}{4\pi}\right) \left(\frac{v_A}{L}\right) , \tag{10.5.3}$$

which yields the following scaling law for the apex temperature T_A (using the definition

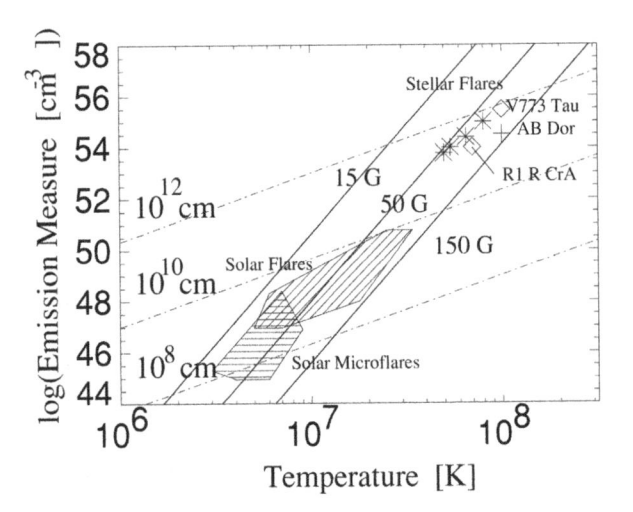

Figure 10.29: The scaling law between the total (volume-integrated) emission measure EM versus electron temperature T_e according to the scaling law of Shibata & Yokoyama (1999). Data are taken from solar active region transient brightenings, flares, and stellar flares (R1 R CrA, AB Dor, 773 Tau) (Shibata & Yokoyama 1999).

of the Alfén speed, $v_A = B/\sqrt{4\pi\mu m_p n_0}$),

$$T_A \approx 10^7 \left(\frac{B}{50\,\text{G}}\right)^{6/7} \left(\frac{n_0}{10^9\,\text{cm}^{-3}}\right)^{-1/7} \left(\frac{L}{10^9\,\text{cm}}\right)^{2/7} \ [\text{K}]\,, \qquad (10.5.4)$$

Further, assuming that (3) the upper limit of the gas pressure is given by the magnetic energy density,

$$p_{th} = 2nk_B T_e = p_m = \left(\frac{B^2}{8\pi}\right)\,, \qquad (10.5.5)$$

and (4) estimating the volume of the heated plasma by a simple cube geometry,

$$EM \approx n^2 L^3 \qquad (10.5.6)$$

a scaling law can be found for the total emission measure EM (by eliminating n and L from Eqs. 10.5.4−6),

$$EM \approx 10^{48} \left(\frac{B}{50\,\text{G}}\right)^{-5} \left(\frac{n_0}{10^9\,\text{cm}^{-3}}\right)^{3/2} \left(\frac{T}{10^7\,\text{K}}\right)^{17/2} \ [\text{cm}^{-3}]\,. \qquad (10.5.7)$$

This scaling law relates the observed total emission measure EM to the flare temperature T, which can be measured in solar and stellar flares. Observations of solar microflares and flares, along with stellar flares, are shown in Fig. 10.29, where the scaling law (10.5.7) (i.e., $EM \propto T^{8.5}$) is plotted for some range of $B = 15 - 150$ G and $n_0 = 10^8 - 10^{12}$ cm^{-3}. Apparently, both solar and stellar flares obey Shibata's scaling law for similar magnetic field strengths $B \approx 50$ G, but differ only in density, which is found in the range of $n_0 \approx 10^8 - 10^{10}$ cm^{-3} for solar microflares and flares, while it is higher ($n_0 \approx 10^{12}$ cm^{-3}) for stellar flares due to the instrumental detection threshold. Thus, the magnetic reconnection scenario is a viable way to explain the scaling of flare plasma temperatures in solar and stellar flares.

10.6 Flare/CME Observations

Overwhelming evidence for magnetic reconnection has been established from Yohkoh observations over the last decade. In this section we touch on those observations of flares and CMEs that provide substantial evidence for various aspects of magnetic reconnection (see compilation in Table 10.2). Reviews on flare and CME observations with particular emphasis on magnetic reconnection processes can be found in Tsuneta (1993a, b; 1994a, b; 1995a, b; 1996b, c, d, e), Tsuneta & Lemen (1993), Shibata (1991; 1994; 1995; 1996; 1999a, b), Shibata et al. (1996a), Kosugi & Somov (1998), McKenzie (2002), Scholer (2003), Kliem et al. (2003), and Van Driel–Gesztelyi (2003).

10.6.1 Evidence for Reconnection Geometry

X-Point Geometry

Let us first discuss observations that constrain the geometry of magnetic reconnection regions. A collection of flare images that are consistent with X-point reconnection in the framework of the Kopp–Pneuman model (§ 10.5.1) are shown in Plate 14, showing a cusp-like structure when imaged during or shortly after reconnection (in soft X-rays) or a relaxed, dipolar-like, postflare arcade when imaged later in the postflare phase (in EUV). A *helmet streamer structure* was observed in *long-duration events (LDEs)*, which have a candlelight shape (Fig. 10.30); this implies an X-type or Y-type reconnection point above the detectable cusp (Tsuneta et al. 1992a; Tsuneta 1996a). Although the appearance of a cusp is not exclusively an indication of a magnetic reconnection process (it could also be cross-field diffusion in a high-β plasma, see Fig. 1.23), other characteristics like the gradual heating during the LDE, the gradual increase of the cusp height, the temperature structure, and loop shrinkage provide additional evidence that the energy is provided by an ongoing reconnection process near the top of the cusp (Tsuneta et al. 1992a; Tsuneta 1996a).

An immediate consequence of the reconnection process is that newly reconnected field lines relax from the initial cusp shape to a more dipolar shape, which has been quantified by the *shrinkage ratio* and was measured to amount to a fraction of 20% and 32% in two flares (Forbes & Acton 1996). By the same token, if particle acceleration is occurring in the highly dynamic electromagnetic fields near the cusp (§ 11), one expects that the particle travel time from the cusp to the footpoints is longer than the loop half-length by about the same shrinkage ratio. Such ratios between the electron time-of-flight time and the (relaxed, soft X-ray bright) flare loop half-lengths were indeed measured [i.e., $L_{TOF}/L_{loop} = 1.43 \pm 0.30$, and $L_{TOF}/L_{loop} = 1.6 \pm 0.6$, respectively (Aschwanden et al. 1999c, d)]. These ratios were found to be invariant in a large range of flare loop heights, $h_{loop} \approx 5 - 50$ Mm. Another manifestation

Table 10.2: Key observations that provide evidence for magnetic reconnection in solar flares and CMEs.

Physical aspect:	Observational signature:	References:
X-point geometry	Cusp in LDE events	Tsuneta et al. (1992a)
X-point altitude	Time-of-flight measurements	Aschwanden et al. (1996c)
	Above-the-looptop HXR sources	Masuda et al. (1994)
X-point rises with time	Increasing footpoint separation	Sakao et al. (1998)
	or double-ribbon separation	Fletcher & Hudson (2001)
X-point symmetry, horizontal	Simultaneous HXR emission	Sakao (1994)
	at conjugate footpoints	
X-point symmetry, vertical	Bi-directional type III bursts	Aschwanden et al. (1995b)
	Coincidence HXR + type III	Aschwanden et al. (1993)
Post-reconnection relaxation	Loop shrinkage ratio	Forbes & Acton (1996)
	cooling loops below hot loops	Svestka et al. (1987)
Quadrupolar geometry	Interacting flare loops	Hanaoka (1996,1997)
		Nishio et al. (1997)
		Aschwanden et al. (1999c)
3D nullpoint geometry	Fan dome and spine morphology	Fletcher et al. (2001)
Reconnection inflows	EUV inward motion	Yokoyama et al. (2001)
Reconnection outflows	Supra-arcade downflows	McKenzie & Hudson (1999)
Slow-mode standing shocks	High-temperature ridges	Tsuneta (1996a)
Fast-mode standing shocks	High density above looptop	Tsuneta (1997)
	Above-the-looptop HXR	Masuda et al. (1994)
Plasmoid ejection	Upward-moving plasmoid	Shibata et al. (1992a)
	Streamer blobs	Sheeley et al. (1997)
Downward conduction	Downward thermal fronts	Rust et al. (1985)
Chromospheric evaporation	Line broadening	Antonucci et al. (1986)
	SXR upflows	Acton et al. (1982)
	SXR blueshifts	Czaykovska et al. (1999)
	$H\alpha$ redshifts	Zarro & Canfield (1989)
	Momentum balance	Wülser et al. (1994)
	HXR/$H\alpha$ ribbons	Hoyng et al. (1981a)

of particle acceleration and temporary trapping in the cusp seems to be the above-the-looptop hard X-ray sources discovered in limb flares (Masuda et al. 1994, 1995, 1996).

A prediction of the Kopp−Pneuman (1976) model is that the X-point progressively rises with time, which implies that newly reconnected field lines should show a progressively larger apex height and an increasing footpoint separation with time. Consequently, lower flare loops are cooling and shrinking, while new hotter loops originate above the cooler loops (Bruzek 1969; Dere & Cook 1979; Svestka et al. 1987; Van Driel−Gesztelyi et al. 1997). Sakao et al. (1998) measured footpoint motion and found that about half of the flares show an increasing footpoint separation with a nonthermal, powerlaw, hard X-ray spectrum, which he interpreted in terms of the Kopp−Pneuman scenario, while the other half exhibited a footpoint separation de-

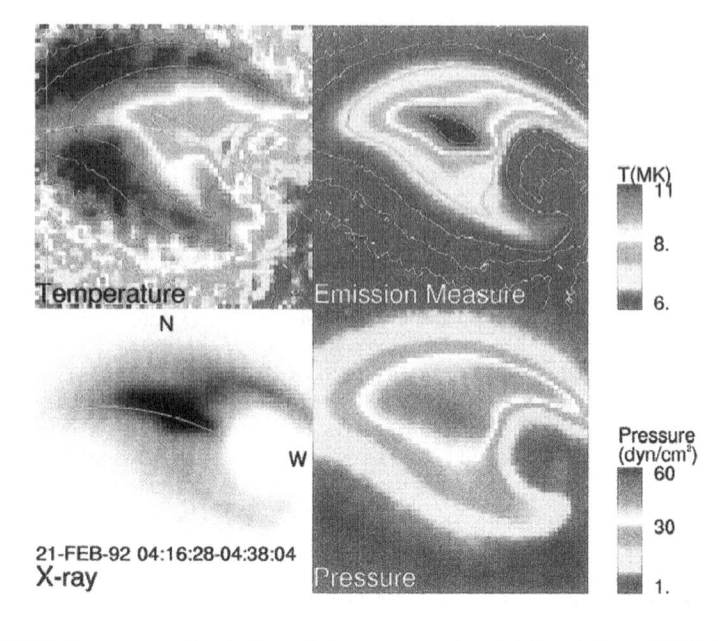

Figure 10.30: Temperature (top left), emission measure (top right), soft X-ray (thick aluminium filter) intensity (bottom left), and pressure maps (bottom right) of the 1992-Febr-21 long-duration flare, which occurred at solar position (N09, E88) near the eastern limb. Note the candlelight shape that is taken as evidence for the cusp geometry in the lower reconnection outflow region (Tsuneta 1996a).

crease as well as a more exponential hard X-ray spectrum, which he interpreted in terms of the emerging flux model (Sakao et al. 1998). During the Bastille-Day flare 2000-Jul-14, a systematic increase of the separation of the flare ribbons was clearly observed in EUV (Fig. 10.33, bottom) as well as in hard X-rays (Fletcher & Hudson 2001). A somewhat more complicated motion was observed in Hα footpoint kernels, but with an overall trend of increasing footpoint separation (Qiu et al. 2002).

The standard Kopp–Pneuman scenario (Fig. 10.20) is symmetric with respect to the vertical trajectory of the rising filament, which also implies symmetric particle path lengths from the acceleration site in the cusp to the two magnetically conjugate chromospheric footpoints. This symmetry is largely confirmed by simultaneity measurements of the hard X-ray evolution at both conjugate footpoints, which generally is coincident within $\lesssim 0.1$ s (Sakao 1994). The Kopp–Pneuman model is also symmetric in the vertical direction (except for gravitational stratification), which is supported by observations of simultaneously accelerated electron beams in the upward direction (type III bursts) and in the downward direction (reverse slope (RS) radio bursts or hard X-ray pulses). Such bi-directional electron beams have been measured to originate at identical plasma frequencies or electron densities (Aschwanden et al. 1995b; Aschwanden & Benz 1997), as well as to exhibit a simultaneous start of $\lesssim 0.1$ s in the upward and downward direction (Aschwanden et al. 1993).

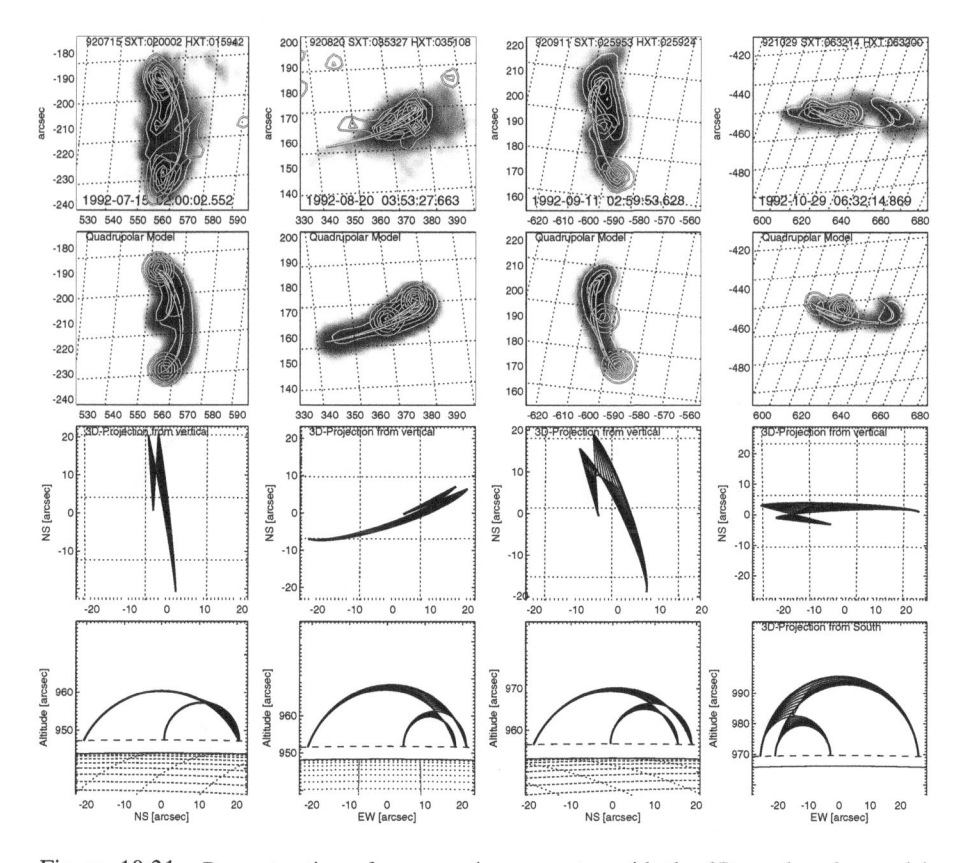

Figure 10.31: Reconstruction of reconnection geometry with the 3D quadrupolar model (Fig. 10.28). *Top row:* soft X-ray image (logarithmic greyscale and thin contours) from Yohkoh/SXT and hard X-ray image (thick contours) from Yohkoh/HXT. The thin circular segments represent the geometric solutions of the preconnection field lines, and the thick circular segments show the corresponding postreconnection field lines, which coincide with the flare loops. *Second row:* simulated SXR and HXR maps constrained by the 3D quadrupolar model, represented by identical greyscales and contour levels just like the original data (in the top row). *Third row:* the geometric solution of the 3D quadrupolar model is rotated so that the vertical z-axis coincides with the line-of-sight. Ten field lines are interpolated between the preconnection and postreconnection state, visualizing the relaxation process of field lines after reconnection. *Bottom row:* the same 3D model is rotated so that either the x-axis (view from west) or the y-axis (view from south) coincides with the line-of-sight. The spacing of the heliographic grid is $1°$ in all frames (corresponding to 12,150 km). (Aschwanden et al. 1999c).

Quadrupolar Geometry

Observations of two interacting flare loops provide the best constraints to reconstruction of magnetic reconnection geometry. Interacting flare loops have been reported in a number of flare observations (e.g., Duijveman et al. 1982; Strong et al. 1984; Benz 1985; Nakajima et al. 1985; Dennis 1985; Machado & Moore 1986; Vrsnak et

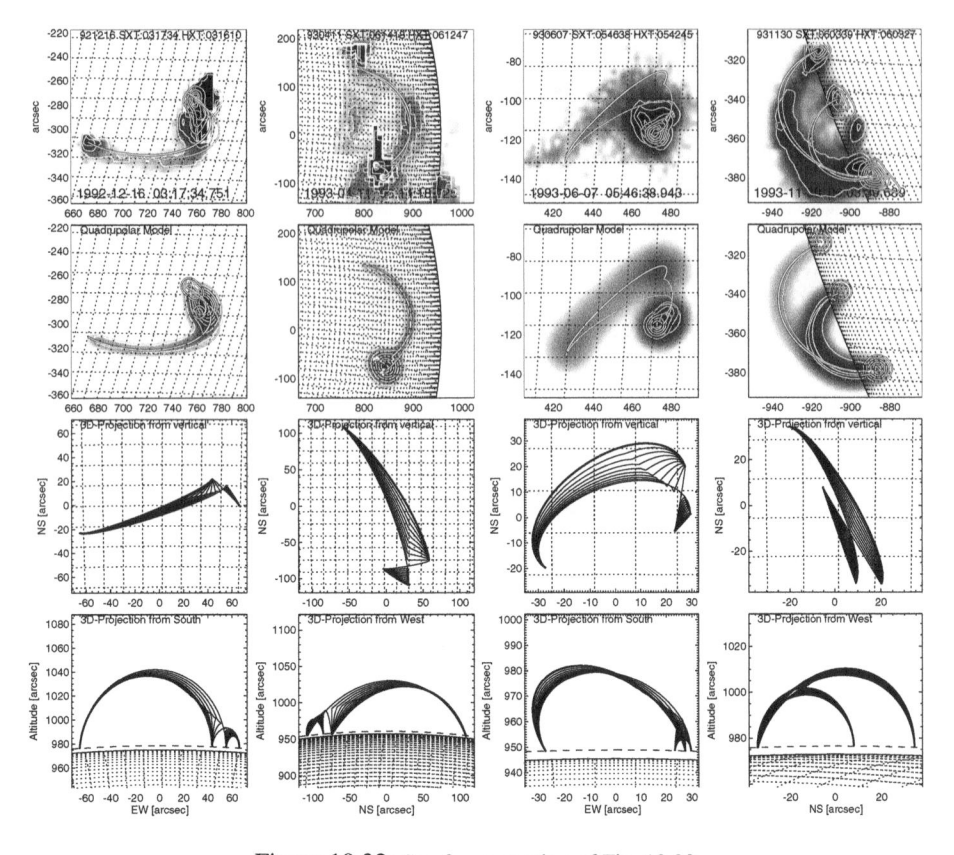

Figure 10.32: See figure caption of Fig. 10.30.

al. 1987a, b; Machado et al. 1988; De Jager et al. 1995; Kundu et al. 1991, 1995; Gopalswamy et al. 1995; Wang et al. 1995), but the 3D geometry of the magnetic reconnection process could not be reconstructed due to insufficient spatial resolution, temperature confusion, or projection effects. The preconnection configuration is generally not visible, because the field lines are not yet illuminated by heated flare plasma. Also the configuration change during reconnection is generally not observable, because it takes some time ($\lesssim 20$ s to 1 min) to enable chromospheric evaporation and to fill the flare loops. So, what one generally observes is the final postreconnection configuration, after the newly reconnected, cusp-shaped field lines relax into a more dipolar shape. Then, interpreting such a quadrupolar configuration change as a simple switch of the connectivities between opposite polarities, one can reconstruct the preconnection geometry with the scheme indicated in Fig. 10.28. A data set of about 10 flares with interacting flare loops has been observed in soft X-rays with Yohkoh/SXT, in radio with Nobeyama, in H-α, and in magnetograms (Hanaoka 1996, 1997; Nishio et al. 1997). From careful studies of the magnetic polarities it was found that the interaction generally happens between a small and a large bipolar flare loop that both share the same neutral line and have one footpoint close together, so that it looks like a *three-legged*

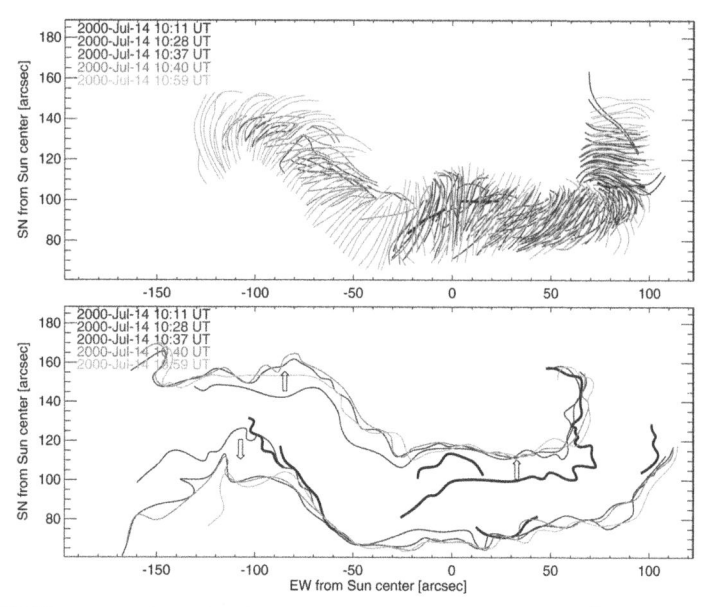

Figure 10.33: *Top:* tracings of individual flare loops from *TRACE* 171 Å images of the Bastille-Day flare 2000-Jul-14. The five sets of loops traced at five different times are marked with different greytones. Note the evolution from highly sheared to less sheared loops. *Bottom:* the position of the two flare ribbons traced from 171 Å images. Note the increasing footpoint separation with time (Aschwanden 2002b).

structure (Hanaoka 1996). However, these "three-legged structures" could be modeled with a quadrupolar configuration (Figs. 10.31 and 10.32; Aschwanden et al. 1999c), which reveals both the prereconnection and postreconnection geometry. From these 3D reconstructions it was found that: (1) the length ratio of interacting flare loops is $L_1 : L_2 \approx 1 : 4$, (2) the angle between the prereconnection field lines is nearly collinear in half the cases (Fig. 10.31) and nearly perpendicular in the other half of the cases (Fig. 10.32), but never anti-parallel, (3) the shear angle between the interacting field lines reduces by $\approx 10° - 50°$ during reconnection, and (4) the smaller loop experiences shrinkage by a factor of $\approx 1.3 \pm 0.4$. These parameters are not predicted by any Petschek-type model, but the magnetic configuration is consistent with a 3D reconnection process in a sheared pair of flare loops (or bundles). Similarly, the interaction between an emerging small-scale loop and a pre-existing larger loop was modeled by Longcope & Silva (1997), who found that the interaction during the flare is consistent with a magnetic flux transfer across the new separatrix surface. In another case, magnetic modeling of the preflare configuration revealed that a highly sheared magnetic fluxtube erupted and triggered reconnection with the overlying large-scale magnetic field (Zhang et al. 2000). On a larger scale, a twisted sigmoid loop was observed to expand and to interact with an overlying loop structure that produced a series of non-thermal radio bursts, which was interpreted as a clear sign of large-scale reconnection (Manoharan et al. 1996).

A generalization of the quadrupolar interaction between two sheared loops is an

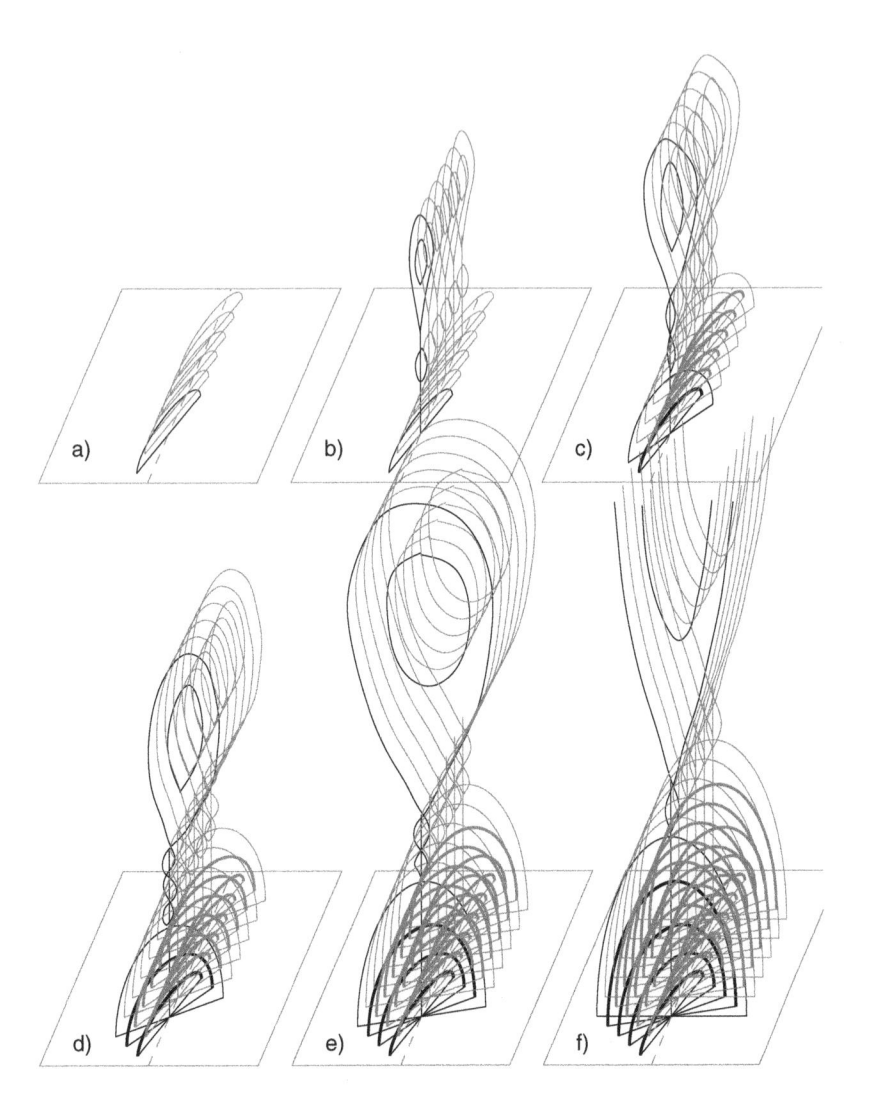

Figure 10.34: Scenario of the dynamic evolution during the Bastille-Day 2000-Jul-14 flare: *(a)* low-lying, highly sheared loops above the neutral line first become unstable; *(b)* after loss of magnetic equilibrium the filament jumps upward and forms a current sheet according to the model by Forbes & Priest (1995). When the current sheet becomes stretched, magnetic islands form and coalescence of islands occurs at locations of enhanced resistivity, initiating particle acceleration and plasma heating; *(c)* the lowest lying loops relax after reconnection and become filled due to chromospheric evaporation (loops with thick linestyle); *(d)* reconnection proceeds upward and involves higher lying, less sheared loops; *(e)* the arcade gradually fills up with filled loops; *(f)* the last reconnecting loops have no shear and are oriented perpendicular to the neutral line. At some point the filament disconnects completely from the flare arcade and escapes into interplanetary space (Aschwanden 2002b).

extension of this configuration along the direction of the neutral line, which characterizes the interaction between two overlying arcades with different shear, as described in Somov et al. (1998), Somov (2000, § 16.2), and Kusano (2002). Such a 3D configuration seems to fit the Bastille-Day flare 2000-Jul-14, whose magnetic structure has been studied by magnetic field extrapolations (Deng et al. 2001; Yan et al. 2001) as well as by tracing EUV postflare loops (Fig. 10.33; Aschwanden & Alexander 2001). The 3D magnetic geometry of this large double-ribbon flare exhibits strongly sheared, low-lying loops above the neutral line, which enter magnetic reconnection first and then progressively affect subsequent layers of less sheared loop arcades with increasing height. The magnetic reconnection process is highly fragmented, because it leaves some 200 different EUV postflare loops in the aftermath. The entire flare seems to consist of a superposition of at least 200 individual magnetic reconnection processes between sheared field lines, corresponding to the bursty reconnection mode we described in § 10.2. The height of individual, local, magnetic reconnection processes tends to increase with height during the flare. The situation is sketched in Fig. 10.34. So, this flare has a lot of aspects in common with the standard Kopp−Pneuman model, but shearing also plays an important role, such that it also has aspects in common with the sheared-arcade model of Sturrock (1966), the magnetic breakout model (Antiochos et al. 1999b; Aulanier et al. 2000b), and the sheared arcade model of Somov et al. (1998). Clearly, the observations constrain the 3D geometry of magnetic reconnection processes in this complex flare to a highly detailed degree and demand a more complex configuration than the most widely used models discussed in § 10.5.

3D Nullpoint Geometry

While most previous flare models with magnetic reconnection were restricted to X-type configurations and current sheet geometries (with 2D nullpoints), recent flare studies also show evidence for 3D nullpoints, saddle points (also called *bald patches*), separator lines, or quasi-separatrix surfaces. One of the first suggestions for the presence of a 3D nullpoint has been inferred from *SoHO/EIT* observations in an active region (Filippov, 1999). However, despite the fact that a saddle-like or hyperbolic magnetic configuration is considered as a necessary condition for magnetic field line reconnection, no heating or flaring was observed in this case. Magnetic field extrapolations of the first Bastille-Day flare (1998-Jul-14) led Aulanier et al. (2000b) to the conclusion that magnetic reconnection occurred along a *spine field line* and the *fan surface* associated with a coronal nullpoint (Plate 13). The existence of a coronal 3D nullpoint was also inferred in the 1999-May-3 flare by Fletcher et al. (2001), based on the emergence of a positive magnetic flux concentration inside a separatrix dome above the surrounding ring of negative magnetic polarity (Fig. 10.27). Relaxation of the stressed magnetic field was interpreted in terms of spine or fan reconnection during this flare. Upflows of heated plasma were also observed along the spine field line, which can be considered a consequence of reconnection-driven heating along the spine field line. These examples offer convincing evidence that the geometry of magnetic reconnection in 3D nullpoints can now be constrained by observations.

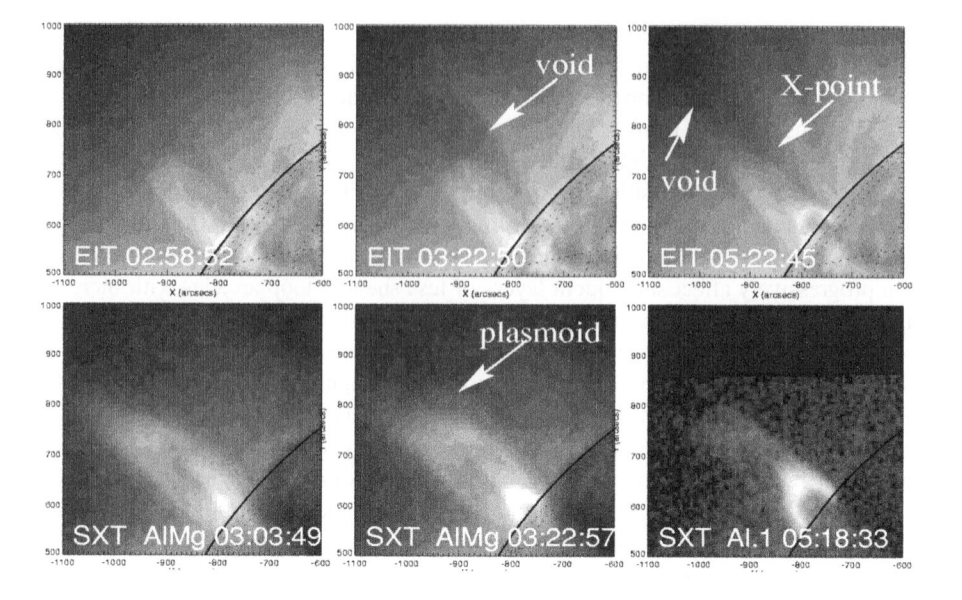

Figure 10.35: Time evolution of the magnetic reconnection process with detected inflows recorded in soft X-rays (Yohkoh/SXT, bottom row) and EUV Fe 195 Å (SoHO/EIT, top row). The field-of-view of each panel is 35×35 Mm (Yokoyama et al. 2001).

10.6.2   Evidence for Reconnection Flows and Jets

All magnetic reconnection models require inflows into the sides of the current sheet and predict near-Alfvénic outflows. In solar flares, however, it turned out to be extremely difficult to detect reconnection inflows, either because of low density, low contrast, slow inflow speed, incomplete temperature coverage, or projection effects (Hara & Ichimoto 1996). Only one study has been published that shows evidence for reconnection inflows in a solar flare (Fig. 10.35−36), where an inflow speed of $v_{in} \lesssim 5$ km s^{-1} was measured and a reconnection rate of $M_A \approx 0.001−0.03$ was estimated (Yokoyama et al. 2001), consistent with the range of other estimates, $M_A \approx 0.001 − 0.01$ (Isobe et al. 2002).

Also, direct reconnection outflows are not readily observable. The first evidence of high-speed downflows above flare loops is believed to have been observed by Yohkoh/ SXT during the 1999-Jan-20 flare, showing dark voids flowing down from the cusps, with speeds of $v_{out} \approx 100−200$ km s^{-1} (McKenzie & Hudson 1999; McKenzie 2000), about an order of magnitude slower than expected from coronal Alfvén speeds. The downward outflows hit a high-temperature ($T \approx 15 − 20$ MK) region, which might be evidence of the fast-mode shock (Tsuneta 1997), sandwiched between the two ridges of the slow-mode shock (Tsuneta 1996a). There is also evidence for thermal wave fronts in the flare loops beneath which are heading downward to the chromosphere with velocities of $v = 800 − 1700$ km s^{-1} (Rust et al. 1985). The subsequent response to downward-moving conduction fronts and precipitating nonthermal particles is the pro-

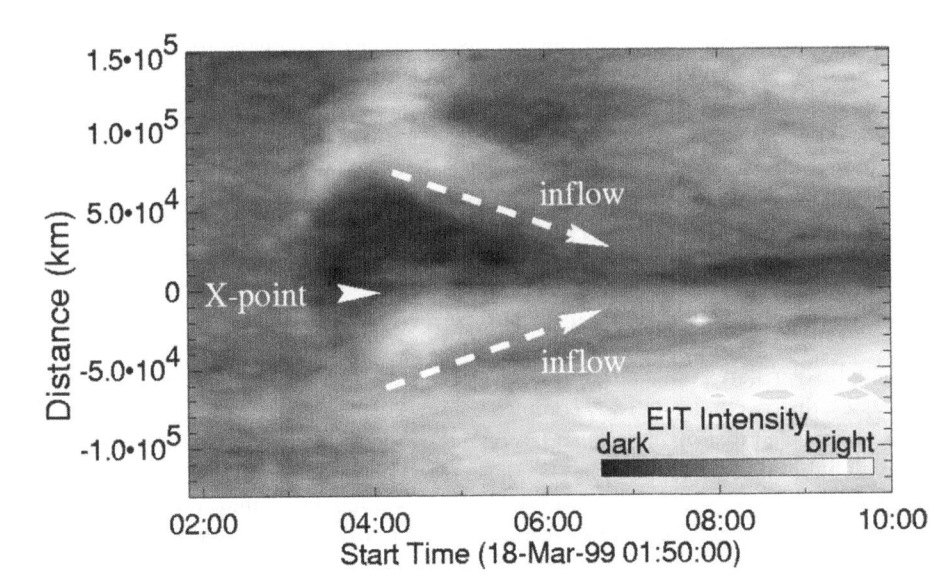

Figure 10.36: Time evolution of a horizontal 1D slice of EUV images (Fig. 10.34), located above the reconnection region, showing a space-time diagram of the inflows (Yokoyama et al. 2001).

cess of chromospheric evaporation, which is observationally witnessed by line broadening (Antonucci et al. 1986, 1996) and blueshifted lines of high-temperature upflows (Acton et al. 1982; Czaykovska et al. 1999) that balance the momentum of the downflows (Zarro & Lemen 1988; Zarro & Canfield 1989; Wülser et al. 1994). Blueshifted Fe XXV lines were also interpreted as direct outflows from Petschek-type reconnection regions (Lin et al. 1996). The precipitating nonthermal electrons are then stopped in the chromosphere, detectable in co-spatial hard X-ray and Hα ribbons (Hoyng et al. 1981a; Duijveman et al. 1982; Sakao 1994; Sakao et al. 1998; Mariska et al. 1996).

Better observable dynamic features of magnetic reconnection are upward-ejected jets and plasmoids, which have been observed abundantly (Shibata et al. 1992a, 1993, 1994a, b, 1995a, 1996a, b, c; Ohyama & Shibata 1996) with typical apparent velocities of $v \approx 200 - 500$ km s^{-1}, possibly connected with outflows (*streamer blobs*) detected with similar speed in the streamer belt out to 30 solar radii (Sheeley et al. 1997; Wang et al. 1998, 1999; Van Aalst et al. 1999).

The radio signatures of magnetic reconnection processes are less certain to interpret, because highly nonlinear emission processes are often involved which are not suitable as regular tracers. Nevertheless, supersonic reconnection outflows are expected to produce shock waves, which possibly could be traced by radio type II bursts. A type II burst structure with no frequency-time drift was observed after one flare, which was interpreted as an outflow termination shock from a reconnection region (Aurass et al. 2002a), but the puzzling fact is why it was only detected an hour after the flare, lasting for 30 minutes. The highly time-variable bursty reconnection is also expected to produce density variations that can modulate the plasma frequency of associated radio emission in a highly dynamic manner, possible manifested as so-called *decimetric pulsation events* (Kliem et al. 2000). However, a more reliable diagnostic of magnetic reconnection processes by means of radio emission requires high-resolution imaging with high cadence aided by detailed 3D modeling, which might become available from future solar-dedicated radio interferometers such as the planned *Frequency-Agile Solar Radiotelescope (FASR)* (White et al. 2003).

10.6.3   Large-Scale Magnetic Restructuring

Although solar flares offer the most convincing evidence for magnetic reconnection processes, they occur in relatively confined volumes inside an active region. In contrast, Yohkoh revealed that there are also large-scale magnetic restructuring processes that have global effects on the corona. For instance, Tsuneta et al. (1992b) observed the formation of a large-scale closed-loop arcade (with a height of almost a solar radius) out of an open-field structure during a time interval of 20 hours, while a polar crown filament disappeared. Tsuneta et al. (1992b) suggest that the eruption of the polar crown filament creates a large-scale current sheet which triggers X-type reconnection and continuously forms closed-field loops beneath in a non-explosive way. So, this is a "non-flaring" magnetic reconnection process, the only difference to flares being the much larger size and lower magnetic field. The timing can extend over extended periods. For instance, Sterling et al. (2001) observed that the onset of a filament rise started more than 6 hours earlier and that the filament disappeared into interplanetary space before the arcade formation became detectable in soft X-rays and EUV.

There are a number of interesting physical aspects that surround large-scale global magnetic reconnection processes and complement our picture of magnetic reconnection. They either occur without flaring, are absent on the smaller scales of flares, or organize flares on a larger scale, such as: (1) remote flare brightenings (Nakajima et al. 1985; Sterling & Moore 2001) ; (2) homologous flares, which are triggered over large global distances (e.g., Zhang et al. 1998); (3) spotless flares, triggered by global restructuring (Sersen 1996); (4) helmet streamer formation (Hiei et al. 1993, 1996); (5) termination and inhibition of upward progressing reconnection at greater heights (Klimchuk 1996); (6) triggering of *radio noise storms* (type I bursts); by coronal restructuring processes (Marque et al. 2001); (7) large-scale dimmings caused by global restructuring processes (Chertok & Grechnev 2003); or (8) helicity conservation from corona to interplanetary space during large-scale reconnection processes (Rust 1996).

10.6.4   Open Issues on Magnetic Reconnection

Although great strides have been made to interpret dynamic flare processes in terms of magnetic reconnection, there are also a number of puzzling observations, or lack of expected observations, which might indicate inappropriate, incomplete, or biased models and physical understanding. A compilation of such critical observational problems and open issues can be found in Hudson & Khan (1996), Tsuneta (1996c), and McKenzie (2002). Here a selection of some open issues:

1. *Initial configuration:* How is the initial vertical current sheet formed in the Kopp—Pneuman model? A long current sheet is unstable in tearing mode, so it is not clear how a stable initial vertical current sheet can be formed in the Kopp—Pneuman model. Also the model of Heyvaerts et al. (1977) requires an initial stable current sheet, but numerical simulations show that reconnection starts as soon as the current sheet is formed.

2. *Anomalous resistivity:* Magnetic reconnection requires enhanced resistivity in the current sheet. What is the source of enhanced resistivity and under what coronal conditions does it occur?

3. *Reconnection inflows:* There is only one published event with a detection of reconnection inflows (Yokoyama et al. 2001). Is this lack of observed inflows an observational problem: low density, low contrast, insufficient temperature contrast, incomplete temperature coverage, low velocity, projection effects?

4. *Reconnection outflows:* There is only one report of supra-arcade downflows that could be interpreted as a direct signature of reconnection outflows (McKenzie & Hudson 1999). Moreover, the inferred outflow speed is about an order of magnitude lower than expected for coronal Alfvén speeds? Are there observational problems for the detection of outflows (confusion from multiple flows, low temperature and density contrast)? Is the magnetic field in the outflow region much lower (≈ 10 G) than typically assumed in flares, in order to explain the Alfvén outflow speed?

5. *Open fields at onset:* Frequent type III bursts and detection of interplanetary particles during the impulsive flare phase require open field regions at the beginning of flares. However, in several flare models (Forbes & Priest 1995; the magnetic breakout model; Antiochos et al. 1999b; quadrupolar models, Uchida 1980, Melrose 1995) there are no open field lines in the beginning. In models with erupting prominences, the initial rising filament is tied down by closed field lines, which need first to be opened up during the flare.

6. *Number problem of energetic particles:* The number of accelerated particles in flares, as inferred from the hard X-ray yield, requires either a high reconnection inflow speed, a large reconnection volume, or many reconnection sites, all requirements that are difficult to reconcile with observations.

7. *Homologous flares:* If magnetic reconnection changes the magnetic topology, how can homologous flares occur that have a matching co-spatial morphology (e.g., identical Hα ribbons?)

10.7 Summary

Theory and numerical simulations of magnetic reconnection processes in the solar corona have been developed for steady 2D reconnection (§ 10.1), bursty 2D reconnection (§ 10.2), and 3D reconnection (§ 10.3). Only steady 2D reconnection models can be formulated analytically (§ 10.1), which provide basic relations for inflow

speed, outflow speed, and reconnection rate, but represent oversimplifications for most (if not all) observed flares. A more realistic approach seems to be bursty 2D reconnection models (§ 10.2), which involve the tearing-mode and coalescence instability and can reproduce the sufficiently fast temporal and small spatial scales required by solar flare observations. The sheared magnetic field configurations and the existence or coronal and chromospheric nullpoints, which are now inferred more commonly in solar flares, require ultimately 3D reconnection models, possibly involving nullpoint coalescence, spine reconnection, fan reconnection, and separator reconnection (§ 10.3). Magnetic reconnection operates in two quite distinct physical parameter domains: in the chromosphere during magnetic flux emergence, magnetic flux cancellation, and so-called explosive events (§ 10.4), and under coronal conditions during microflares, flares, and CMEs (§ 10.5). The best known flare/CME models entail magnetic reconnection processes that are driven by a rising filament/prominence, by flux emergence, by converging flows, or by shear motion along the neutral line (§ 10.1). Flare scenarios with a driver perpendicular to the neutral line (rising prominence, flux emergence, convergence flows) are formulated as 2D reconnection models (Kopp–Pneuman 1976; Heyvaerts et al. 1977; Forbes & Priest 1995; Uchida 1980), while scenarios that involve shear along the neutral line (tearing-mode instability, quadrupolar flux transfer, the magnetic breakout model, sheared arcade interactions) require 3D descriptions (Sturrock 1966; Antiochos et al. 1999b; Somov et al. 1998). Ultimately, most of these partial flare models could be unified in a 3D model that includes all driver mechanisms. Observational evidence for magnetic reconnection in flares includes the 3D geometry, reconnection inflows, outflows, detection of shocks, jets, ejected plasmoids, and secondary effects like particle acceleration, conduction fronts, and chromospheric evaporation processes (§ 10.6). Magnetic reconnection not only operates locally in flares, it also organizes the global corona by large-scale restructuring processes.

Chapter 11

Particle Acceleration

In the previous sections we dealt mostly with MHD descriptions of the coronal plasma, which is a fluid concept to describe a collisional plasma, where collisions among electrons and ions are so frequent that the particle distribution maintains a Maxwellian or Boltzmann distribution that can be characterized by a single temperature locally. In this sense we talk about *thermal particles* and *thermal plasma*, which governs most parts of the corona. Strong electromagnetic fields, as they are generated during magnetic reconnection processes, however, have the ability to accelerate particles out of the thermal distribution to higher energies, which we then call *nonthermal* particles. They can survive in the corona for the duration of a collision time and may even escape along open field lines into interplanetary space thanks to their relativistic speed. The fate of nonthermal particles depends very much on the direction in which they are accelerated, and whether the field line of their trajectory is open or closed. Particles that are accelerated in a downward direction or along closed field lines inevitably slam into the highly collisional chromosphere, where they provide a valuable diagnostic in hard X-rays and gamma-rays. Nonthermal particles that propagate along coronal or interplanetary field lines are part of nonthermal tail distributions that are prone to many kinds of plasma instabilities, which produce electrostatic and electromagnetic radiation in radio wavelengths. Moreover, relativistic particles that propagate in magnetic fields produce gyroresonance and gyrosynchrotron emission, which also provides a valuable diagnostic of energetic particles in solar flares. Thus, nonthermal particles provide a wealth of information that can be quantitatively exploited by means of kinetic plasma theory and radiation mechanisms in gamma-rays, hard X-rays, and radio wavelengths. Nonthermal particles, however, are only produced in significant numbers by highly dynamic processes such as magnetic reconnection in solar flares and shocks in coronal mass ejections (CMEs), which is the main subject of the following sections. We start with the basic physics of particle acceleration in this chapter (§ 11), and then proceed to propagation (§ 12) and to secondary radiation mechanisms produced by nonthermal particles later on, such as hard X-rays (§ 13), gamma-rays (§ 14), and radio emission (§ 15).

11.1 Basic Particle Motion

The basic physics of the dynamics of relativistic particles in electromagnetic fields is covered in textbooks (e.g., Jackson 1962, § 12), and more specific applications to laboratory plasma, the magnetosphere, or the corona can be found in Chen (1974, § 2), Schmidt (1979, § 1), Benz (1993, § 2), Sturrock (1994, § 3-5), Baumjohann & Treumann (1997, § 2), or Boyd & Sanderson (2003, § 2).

11.1.1 Particle Orbits in Magnetic Fields

Since magnetic fields are ubiquitous in the solar corona, charged particles like electrons and ions experience a Lorentz force that makes them gyrate around the guiding magnetic field lines. The basic (relativistic) equation of motion thus includes acceleration by an electric field $\mathbf{E}$ and the Lorentz force exerted by the magnetic field $\mathbf{B}$,

$$\frac{d(m\gamma\mathbf{v})}{dt}(\mathbf{x}, t) = q \left[\mathbf{E}(\mathbf{x}, t) + \frac{1}{c}\mathbf{v} \times \mathbf{B}(\mathbf{x}, t) \right] , \tag{11.1.1}$$

where γ is the relativistic Lorentz factor defined in § 11.1.3. Neglecting the electric field ($E = |\mathbf{E}| = 0$) and assuming a uniform (constant) magnetic field, $\mathbf{B}(\mathbf{x}, t) = B$, we can isolate the gyromotion of the particle, which is a circular orbit with tangential velocity $v_\perp$ at a gyroradius R. The gyroradius R follows directly from Eq. (11.1.1), using the proportionality relation for the velocity change dv in a circular orbit, $dv/v_\perp = v_\perp dt/R$,

$$R = \frac{m\gamma c}{|q|B}v_\perp . \tag{11.1.2}$$

The resulting gyrofrequency is then

$$\Omega_g = 2\pi f_g = \frac{v_\perp}{R} = \frac{qB}{m\gamma c} , \tag{11.1.3}$$

which is independent of the particle velocity. Plugging in the physical constants (see Appendix A and D) we obtain the following simple formulas for the electron and ion gyrofrequencies,

$$\begin{aligned} f_{ge} &= 2.80 \times 10^6 \, B \quad [\text{Hz}] \\ f_{gi} &= 1.52 \times 10^3 \, B/\mu \quad [\text{Hz}] , \end{aligned} \tag{11.1.4}$$

where μ is the molecular weight of the ion mass, $m_i = \mu m_p$. Thus for typical coronal field strengths of $B = 10 - 100$ G the electron gyrofrequency amounts to $f_{ge} = 28 - 280$ MHz, reaching a maximum value in fields of $B \approx 1000 - 2000$ G above sunspots, with $f_{ge} \approx 2.8 - 5.6$ GHz. The proton gyrofrequency is a factor of $m_p/m_e = 1836$ lower, so in the $f_{gi} = 15 - 150$ kHz range for coronal conditions. In the solar wind ($B = 1 - 10$ G), proton gyration times can fall in the range of $\approx 1 - 10$ kHz, and thus produce high-frequency modulations on time scales of $\tau_g \approx 0.1 - 1$ ms.

To evaluate gyroradii, we set the perpendicular velocity $v_\perp$ equal to the thermal velocity, which is $v_{Te} = k_B T_e/m_e$ for electrons and $v_{Ti} = k_B T_i/\mu m_p$ for ions. With this definition, and defining the charge state by $Z = q/e$, the gyroradii of electrons and ions are,

$$
\begin{aligned}
R_e &= v_{Te}/(2\pi f_{ge}) &= 2.21 \times 10^{-2} T_e^{1/2} B^{-1} \quad \text{(cm)} \\
R_i &= v_{Ti}/(2\pi f_{gi}) &= 9.49 \times 10^{-1} T_i^{1/2} \mu^{1/2} Z^{-1} B^{-1} \quad \text{(cm)} ,
\end{aligned}
\tag{11.1.5}
$$

which for typical coronal conditions ($T_e \approx T_i \approx 1.0$ MK, $B = 10 - 100$ G), amounts to $R_e = 0.2 - 2$ cm for electrons, and $R_i = 0.1 - 1$ m for protons, respectively.

Thus, in the absence of any electric field, the motion of particles is controlled by the gyromotion, which is a circular or helical orbit around the guiding magnetic field, with the electrons and ions rotating in opposite directions. Since the Lorentz force is perpendicular to the magnetic field direction and the particle velocity vector, magnetic fields cannot do any work to accelerate particles.

11.1.2 Particle Drifts in Force Fields

In order to accelerate particles, a force is necessary that acts either parallel ($\mathbf{F}_\parallel$) or perpendicular ($\mathbf{F}_\perp$) to the magnetic field. A parallel force accelerates a particle just along the magnetic field, without any interference with the gyromotion. A perpendicular force, however, produces a drift of the charged particle that is perpendicular to both the magnetic field $\mathbf{B}$ and the force direction $\mathbf{F}_\perp$ (see derivations in any of the textbooks quoted at the beginning of this section),

$$
\mathbf{v}_{drift} = \frac{c}{q} \frac{\mathbf{F}_\perp \times \mathbf{B}}{B^2} = \frac{1}{\Omega_g} \left(\frac{\mathbf{F}_\perp}{m} \times \frac{\mathbf{B}}{B} \right) ,
\tag{11.1.6}
$$

and the motion of the particle can be represented by a superposition of the motion along the magnetic field ($\mathbf{v}_\parallel$), the gyromotion ($\mathbf{v}_{gyro}$), and the drift ($\mathbf{v}_{drift}$),

$$
\mathbf{v} = \mathbf{v}_\parallel + \mathbf{v}_{gyro} + \mathbf{v}_{drift} ,
\tag{11.1.7}
$$

The perpendicular force $\mathbf{F}_\perp$ could be an electric force $\mathbf{F}_E$, a polarization drift force $\mathbf{F}_P$, the gravitational force $\mathbf{F}_g$, a magnetic field gradient force $\mathbf{F}_{\nabla B}$, or a curvature force $\mathbf{F}_R$. Inserting the definitions of these forces we obtain with Eq. (11.1.6) the following drift rates,

$$
\begin{aligned}
\mathbf{F}_E &= q\mathbf{E} & \mathbf{v}_{drift} &= (c/B^2)(\mathbf{E} \times \mathbf{B}) \\
\mathbf{F}_P &= m(d\mathbf{E}/dt) & \mathbf{v}_{drift} &= (mc/qB^2)(d\mathbf{E}/dt \times \mathbf{B}) \\
\mathbf{F}_g &= m\mathbf{g} & \mathbf{v}_{drift} &= (mc/qB^2)(\mathbf{g} \times \mathbf{B}) \\
\mathbf{F}_{\nabla B} &= (mv_\perp^2/2B)\nabla\mathbf{B} & \mathbf{v}_{drift} &= (mcv_\perp^2/2qB^3)(\mathbf{B} \times \nabla\mathbf{B}) \\
\mathbf{F}_R &= (mv_\parallel^2/R^2)\mathbf{R} & \mathbf{v}_{drift} &= (mcv_\parallel^2/qR^2B^2)(\mathbf{R} \times \mathbf{B})
\end{aligned}
\tag{11.1.8}
$$

Thus, a force $\mathbf{F}_\parallel$ along the magnetic field will accelerate particles to higher energies, while the perpendicular component $\mathbf{F}_\perp$ will only cause a drift velocity $\mathbf{v}_{drift}$ without energy gain, because the acceleration in the first half of the gyroperiod cancels out with the deceleration during the second-half of the gyroperiod. Since the $\mathbf{E} \times \mathbf{B}$-drift does not depend on the sign of the charge q, both electrons and ions drift into the same direction. The $\mathbf{E} \times \mathbf{B}$-drift thus does not contribute to any net acceleration or energy gain of the particles. However, it can have an important effect in removing particles from the acceleration region, especially for acceleration regions with a small transverse dimension, such as current sheets.

11.1.3   Relativistic Particle Energies

Nonthermal particles are accelerated up to high relativistic speeds in solar flares. It is therefore useful to remember the basic relativistic notations. A reference energy of relativistic particles is the rest mass of the electron,

$$m_e c^2 = 511 \text{ keV} .   \tag{11.1.9}$$

The total energy ε_{total} of a relativistic electron is composed of the rest mass $m_e c^2$ and the kinetic energy ε,

$$\varepsilon_{total} = m_e c^2 + \varepsilon = m_e c^2 \gamma ,   \tag{11.1.10}$$

also specified in terms of the relativistic *Lorentz factor* γ,

$$\gamma = \frac{1}{\sqrt{1 - \beta^2}} = \frac{1}{\sqrt{1 - v/c^2}} ,   \tag{11.1.11}$$

where the relativistic velocity v is often expressed by the dimensionless variable $\beta = v/c$ (not to be confused with the plasma β-parameter). So, the kinetic energy ε is related to the Lorentz factor γ or velocity v by

$$\varepsilon = m_e c^2 (\gamma - 1) = m_e c^2 \left(\frac{1}{\sqrt{1 - (v/c)^2}} - 1 \right) \approx \frac{1}{2} m_e v^2 + ...   \tag{11.1.12}$$

Often we want to know the electron speed v for a given kinetic electron energy ε, which follows from Eqs. (11.1.9−12),

$$v(\varepsilon) = c\sqrt{1 - \frac{1}{\gamma^2}} = c\sqrt{1 - \frac{1}{(1 + \frac{\varepsilon}{m_e c^2})^2}} .   \tag{11.1.13}$$

Let us consider the acceleration by a constant electric field E. The equation of motion (Eq. 11.1.1) can then be integrated using the definition of $\gamma(v)$ (Eq. 11.1.11) yielding,

$$v(t) = c \frac{1}{\sqrt{1 + (mc/qEt)^2}} \approx \frac{qE}{m} t ,   \tag{11.1.14}$$

where the nonrelativistic approximation is given on the right-hand side. A second integration yields the trajectory length of the accelerated particle,

$$x(t) = \frac{mc^2}{qE} \left[\left(1 + \left[\frac{qEt}{mc} \right]^2 \right)^{1/2} - 1 \right] .   \tag{11.1.15}$$

The acceleration time for a particle with Lorentz factor γ follows from Eq. (11.1.14),

$$t_{acc}(\gamma) = \frac{mc}{qE}\sqrt{\gamma^2 - 1}\,. \qquad (11.1.16)$$

A conversion table between relativistic energies ε, speeds β, and Lorentz factors γ relevant for solar flares is given in Appendix E. From this we see that a typical thermal electron in the solar corona (with a temperature of $T_e \approx 1.0$ MK) has a nonrelativistic speed of $\beta = \mathrm{v}/c = 0.018$ and a kinetic energy of $\varepsilon = 0.086$ keV. If such an electron is accelerated to about 4 times the thermal speed, so $\beta \approx 0.07$, which is still nonrelativistic, it can form a bump-in-the-tail in the velocity distribution and produce radio emission at the plasma frequency. Hard X-ray emission is produced by electrons with typical energies of $\varepsilon \approx 20 - 100$ keV, so they have mildly relativistic speeds of $\beta \approx 0.2 - 0.5$. Gyrosynchrotron emission and electron bremsstrahlung in gamma-rays are produced by highly relativistic electrons, with energies of $\varepsilon \approx 0.3 - 10$ MeV and with relativistic speeds of $\beta \gtrsim 0.8$.

In particle acceleration processes, an equipartition of kinetic energies is often assumed for accelerated electrons and protons (i.e., $\varepsilon_p = \varepsilon_e$),

$$\varepsilon_p = m_p c^2 (\gamma_p - 1) = \varepsilon_e = m_e c^2 (\gamma_e - 1)\,, \qquad (11.1.17)$$

which yields a velocity ratio β_p/β_e that is roughly inverse to the square root of the mass ratio (in the case of equipartition: $\varepsilon_p = \varepsilon_e$),

$$\frac{\beta_p}{\beta_e} = \frac{\sqrt{1 - [1 + \frac{\varepsilon_p}{m_p c^2}]^{-2}}}{\sqrt{1 - [1 + \frac{\varepsilon_e}{m_e c^2}]^{-2}}} \approx \sqrt{\frac{\varepsilon_p}{\varepsilon_e} \cdot \frac{m_e}{m_p}} = \frac{1}{43}\sqrt{\frac{\varepsilon_p}{\varepsilon_e}} = \frac{1}{43}\,. \qquad (11.1.18)$$

11.2 Overview of Particle Acceleration Mechanisms

In the context of solar flares, there are essentially three major groups of particle acceleration mechanisms that have been studied over the years: (1) *DC electric field acceleration*, (2) *stochastic acceleration*, and (3) *shock acceleration*. Shock acceleration is sometimes also called *first-order Fermi acceleration*, because the relative momentum gain per shock crossing is linear to the velocity ratio (u/v) (of the shock speed u to the particle speed v). For particles that cross a shock once, this process is also called *shock drift acceleration*, while particles that are scattered multiple times across a shock structure experience *diffusive shock acceleration*. This latter type of *diffusive shock acceleration* is not much different from *stochastic acceleration* (Jones 1994), which is also called *second-order Fermi acceleration*, because the relative momentum gain per collision is proportional to the second power of the velocity ratio (u/v).

A compilation of acceleration mechanisms according to this grouping is given in Table 11.1. The first group envisions acceleration in electric fields, which can be generated in current sheets, during magnetic reconnection processes, or in current-carrying loops. Although acceleration in a static electric field is conceptually the simplest model, an application to solar flares is rather complicated; the current thinking is

Table 11.1: Overview of particle acceleration mechanisms in solar flares.

Acceleration Mechanisms	Electromagnetic fields
DC electric field acceleration:	
− Sub-Dreicer fields, runaway acceleration[1]	$E < E_D$
− Super-Dreicer fields[2]	$E > E_D$
− Current sheet (X-point) collapse[3]	$E = -u_{inflow} \times B$
− Magnetic island (O-point) coalescence[4]	$E_{conv} = -u_{coal} \times B$
− (Filamentary current sheet: X- and O-points)	
− Double layers[5]	$E = -\nabla V$
− Betatron acceleration (magnetic pumping)[6]	$\nabla \times E = -(1/c)(dB/dt)$
Stochastic (or second-order Fermi) acceleration:	
Gyroresonant wave-particle interactions (weak turbulence) with:	
− whistler (R-) and L-waves[7]	$k \parallel B$
− O- and X-waves[8]	$k \perp B$
− Alfvén waves (transit time damping)[9]	$k \parallel B$
− Magneto-acoustic waves[10]	$k \perp B$
− Langmuir waves[11]	$k \parallel B$
− Lower hybrid waves[12]	$k \perp B$
Shock acceleration:	
Shock-drift (or first-order Fermi) acceleration[13]	
− Fast shocks in reconnection outflow[14]	
− Mirror-trap in reconnection outflow[15]	
Diffusive-shock acceleration[16]	

[1]) *Holman (1985), Tsuneta (1985), Benka & Holman (1994);* [2]) *Litvinenko & Somov (1995), Litvinenko (1996b);* [3]) *Tajima & Sakai (1986), Sakai & Ohsawa (1987), Sakai & De Jager (1991);* [4]) *Furth et al. (1963), Pritchett & Wu (1979), Biskamp & Welter (1979), Kliem (1994), Kliem et al. (2000);* [5]) *Block (1978), Volwerk & Kuijpers (1994), Volwerk (1993);* [6]) *Brown & Hoyng (1975), Karpen (1982);* [7]) *Melrose (1974), Miller & Ramaty (1987), Steinacker & Miller (1992), Hamilton & Petrosian (1992);* [8]) *Karimabadi et al. (1987);* [9]) *Lee & Völk (1975), Fisk (1976), Achterberg (1979), Barbosa (1979), Stix (1992), Miller et al. (1990), Hamilton & Petrosian (1992), Steinacker & Miller (1992), Miller et al. (1997);* [10]) *Zhou & Matthaeus (1990); Eichler (1979); Miller & Roberts (1995);* [11]) *Melrose (1980a,b);* [12]) *Papadopoulos (1979), Lampe & Papadopoulos (1977), Benz & Smith (1987), McClements et al. (1990);* [13]) *Fermi (1949), Jokipii (1966), Bai et al. (1983), Ellison & Ramaty (1985);* [14]) *Tsuneta & Naito (1998);* [15]) *Somov & Kosugi (1997);* [16]) *Ramaty (1979), Achterberg & Norman (1980), Decker & Vlahos (1986); Cargill et al. (1988), Lemberge (1995).*

that the relevant electric fields are generated during magnetic reconnection processes, and therefore their nature is highly intermittent in space and time. The second group of stochastic acceleration processes involve random energy gains and losses in a turbulent plasma, where gyroresonant wave-particle interactions provide a net energy gain to resonant particles. The third group of shock acceleration mechanisms involve a particular geometry and inhomogeneous boundary (at a shock front) that is suitable to transfer momentum and energy to intercepting particles.

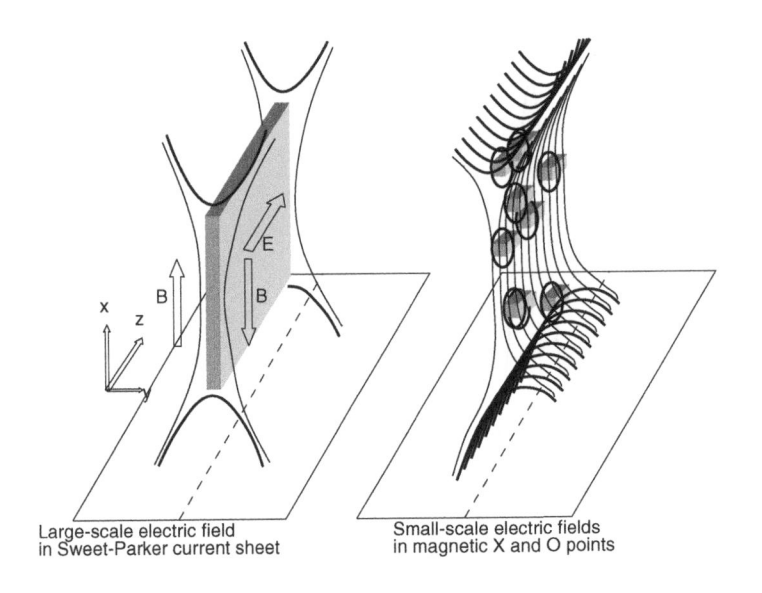

Large-scale electric field
in Sweet-Parker current sheet

Small-scale electric fields
in magnetic X and O points

Figure 11.1: Paradigm shift of current sheet structure and associated particle acceleration regions: *Left:* classical models assume large-scale electric fields based on Sweet−Parker magnetic reconnection, which have a much larger extent in the x and z-direction than their width in the y-direction. *Right:* theory and MHD simulations, however, imply small-scale electric fields in magnetic X-points and coalescing islands with magnetic O-points (Aschwanden 2002b).

Although these three groups seem to distinctly define different physical concepts, the distinction becomes progressively blurred the more realistic theoretical models are used. For instance, large homogeneous current sheets with a well-defined DC electric field are now replaced with more inhomogeneous, fragmented current sheets and magnetic islands in the impulsive bursty reconnection mode (Fig. 11.1), so that the associated electric fields also become highly fragmented and approach the limit of turbulent fields as it is used in the second group of stochastic acceleration. Also some shock acceleration models are highly diffusive in nature and have much in common with stochastic acceleration.

Theoretical reviews on solar particle acceleration mechanisms can be found in Vlahos et al. (1986), Benz (1993), Kirk (1994), Miller et al. (1997), Petrosian (1996, 1999), Miller (2000a,b), Priest & Forbes (2000, § 13), Cargill (2001), Schlickeiser (2003), while there exists a much larger set of observational reviews on particle acceleration signatures in solar flares, to which we will refer to in the following chapters at the relevant places.

11.3 Electric DC-Field Acceleration

Particle acceleration in electric fields applied to solar flares can be categorized according to (1) the electric field strength (weak sub-Dreicer versus strong super-Dreicer

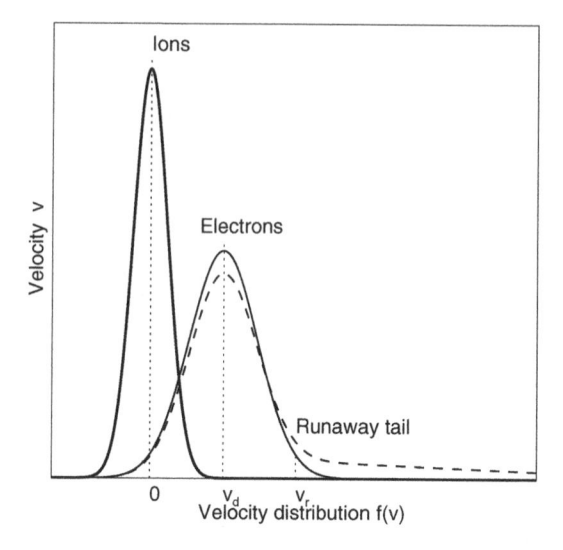

Figure 11.2: Schematic of the parallel velocity distribution of electrons and ions in the presence of an electric field. The bulk of the electron distribution drifts with a velocity v_d, but is not accelerated because the frictional drag force is stronger than the electric field. Above the critical velocity v_r defined by the Dreicer field, the electric force overcomes the frictional force and electrons can be accelerated freely out of the thermal distribution, in the the so-called *runaway acceleration* regime.

fields), (2) the time variability (static versus dynamic electric fields), or (3) the magnetic geometry (current sheets, X-points, O-points).

11.3.1   Sub-Dreicer DC Electric Fields

When an electric DC field is applied to a plasma, electrons and ions become accelerated in opposite directions, but the attraction between opposite electric charges causes an impeding *electric drag force* or *frictional force*, which depends strongly on the ion-collision frequency. However, for large relative velocities the frictional force can become smaller than the accelerating force, and electrons can be accelerated freely out of the thermal distribution, which is called the regime of *runaway acceleration* (Fig. 11.2). The critical runaway velocity v_r is given by the criterion where the frictional force equals the electric force eE, where the frictional force can be expressed by the change of the momentum $(m_e v_r)$ during a *collisional slowing-down time* $t_s^{e,i} = v / < \Delta v_\parallel / \Delta t >$ of an electron test particle in an ambient electric field of ions,

$$eE = \frac{m_e v_r}{t_s^{e,i}(v_r)} \; . \tag{11.3.1}$$

Inserting the expressions for the collisional slowing-down time scale (Dreicer 1959, 1960; for a derivation see, e.g., Benz 1993, § 2 and 9) we obtain the definition of the

Dreicer electric field (e.g., Holman 1985),

$$E_D = \frac{q_i \ln \Lambda}{\lambda_D^2} = 2.33 \times 10^{-8} \left(\frac{n_e}{10^9 \text{ cm}^{-3}} \right) \left(\frac{T_e}{10^7 \text{ K}} \right)^{-1} \left(\frac{\ln \Lambda}{23.2} \right) \quad (\text{statvolt cm}^{-1})$$

$$(11.3.2)$$

where $\ln \Lambda$ is the Coulomb logarithm, $\lambda_D = v_{Te}/\omega_{pe}$ the Debye length, $v_{Te} = (k_B T_e/m_e)^{1/2}$ the thermal speed of the electron, and $\omega_{pe} = (4\pi n_e e^2/m_e)^{1/2}$ the electron plasma frequency. The runaway speed can then be expressed in terms of the Dreicer field E_D (Knoepfel & Spong 1979),

$$v_r = v_{Te} \left(\frac{E_D}{E} \right)^{1/2} . \quad (11.3.3)$$

Runaway acceleration in sub-Dreicer fields has been applied specifically to solar flares by Kuijpers et al. (1981), Heyvaerts (1981), Spicer (1982), Holman (1985; 1995; 1996), Tsuneta (1985), Moghaddam−Taaheri & Goertz (1990), Holman & Benka (1992), Benka & Holman (1992, 1994), Zarro et al. (1995), Zarro & Schwartz (1996a,b), Kucera et al. (1996), and has been reviewed in Norman & Smith (1978), Benz (1987a), Miller et al. (1997), and Holman (2000).

Following the model of Holman (1985), the sub-Dreicer electric field is of order $E \approx 3 \times 10^{-10}$ (statvolt cm^{-1}), extending over spatial scales of $L \approx 30$ Mm, the size of a typical flare loop. Holman (1985) finds that the energy gain scales with

$$(W - W_c) = 7.0 \left(\frac{T}{10^7 \text{ } K} \right)^{1/2} \left(\frac{\nu_e}{10 \text{ } s^{-1}} \right) \left(\frac{L}{10^9 \text{ } cm} \right) \left(\frac{v_r}{v_{Te}} \right)^{-2} \quad (\text{keV}) \quad (11.3.4)$$

yielding electron energies of $W \approx 100$ keV for an electron temperature of $T = 10^7$ K, a collision frequency of $\nu_e \approx 2 \times 10^3$ s^{-1}, a ratio $v_r/v_{Te} \approx 4$ for the critical runaway speed v_r to the thermal speed v_{Te}, a critical energy $W_c = 8$ keV, and a length scale of $L \approx 10^9$ cm. This energy gain is sufficient for the bulk of electrons observed in most hard X-ray flares, supposing that such large-scale electric fields over distances of $L \approx 10$ Mm exist. Higher energies could be achieved by assuming anomalous resistivity, which enhances the value of the effective Dreicer field, and thus the maximum values of sub-Dreicer electric fields. Regarding the transverse extent of current channels, Holman (1985) concludes that a fragmentation of about $\approx 10^4$ current channels is needed, to comply with the maximum magnetic field limit imposed by Ampère's law.

Thus, in principle, the sub-Dreicer DC electric field model can explain the velocity distribution of nonthermal particles as observed in hard X-ray spectra, and excellent spectral fits of the thermal-plus-nonthermal distributions to observed hard X-ray spectra have been obtained with this model (Benka & Holman 1994), as well as for fits to microwave spectra (Benka & Holman 1992). However, there are a number of open issues on the spatial geometry, temporal evolution, stability of the invoked DC electric fields, and the associated magnetic field configuration in the context of suitable flare models, that have not been satisfactorily addressed in previous studies. The major issue is that the model requires a large-scale electric field of the size of flare loops. If one employs a huge Sweet−Parker current sheet (Fig. 11.1, left), a large extent along the current sheet is required ($L \approx 10^9$ cm), which is likely to be unstable to tearing mode and will fragment into many magnetic islands, which coalesce and dissipate in a bursty reconnection mode (Fig. 11.1, right). Alternatively, if one assumes a large-scale DC electric field between the coronal reconnection site and chromospheric footpoints, the energy-dependent timing of electrons accelerated in this DC electric field (§ 12.3) contradicts the observed time-of-flight delays (Aschwanden 1996). A possible way out is to assume that many parallel strands with different electric fields are produced, which would produce an energy-dependent timing of precipitating electrons that is consistent with the measured electron time-of-flight delays in hard X-rays (Aschwanden et al. 1996a), but no model explains how such inhomogeneous large-scale DC electric fields can be generated in the first place. Moreover, any model with static or steady electric DC fields faces the problem that the *electron beam current* require counter-streaming *return currents* (Brown & Melrose 1977; Knight & Sturrock 1977; Hoyng et al. 1978; Spicer & Sudan 1984; Brown & Bingham 1984; LaRosa & Emslie 1989; Van den Oord 1990; Litvinenko & Somov 1991), which can limit the acceleration efficiency severely. To solve the problem of return currents in large-scale electric fields, current closure between adjacent current channels (over distances of a few meters) was proposed, enabled by the cross-field drift of protons at the chromospheric footpoints (Emslie & Hénoux 1995).

11.3.2   Super-Dreicer DC Electric Fields

Since weak DC electric fields (such as in the sub-Dreicer regime) require large distances to accelerate electrons to hard X-ray energies (because the electric potential energy is simply proportional to the distance (i.e., $W \propto L$ in Eq. 11.3.4), strong DC electric fields can produce the same hard X-ray energies over much smaller distances. This is the primary motivation for exploring particle acceleration in the super-Dreicer regime. The compactness of the acceleration region is also consistent with the observed time-of-flight delays (Aschwanden 1996), which show no signs of acceleration delays for high-energy particles. Applications of electron acceleration in super-Dreicer fields ($E \gg E_D$) have been demonstrated by Litvinenko (1996b). He calculated particle orbits in standard current sheet geometries (Fig. 11.3, left), but assumed besides the longitudinal magnetic field $\mathbf{B}_\parallel = (0, 0, B_\parallel)$ and the parallel electric field $\mathbf{E} = (0, 0, E_\parallel)$, also a weaker perpendicular magnetic component, $\mathbf{B}_\perp = (0, B_\perp, 0)$, which serves to scatter electrons accelerated along the E-field direction out of the current sheet before they reach the end of the current sheet (Fig. 11.3, right). The magnetic field inside the current sheet is thus $\mathbf{B} = [(-y/\Delta w_y)B_0, B_\perp, B_\parallel]$. In this configuration, the maximum particle energy W is determined not only by the electric field $E_\parallel$, but also by the ratio of the parallel to the perpendicular magnetic field,

$$W = \frac{B_\parallel}{B_\perp} e \Delta w_y E_\parallel \, , \tag{11.3.5}$$

with Δw_y the width of the current sheet. The acceleration time is

$$\Delta t = \sqrt{\frac{B_\parallel}{B_\perp} \frac{2 \Delta w_y m_e}{e E_\parallel}} \, . \tag{11.3.6}$$

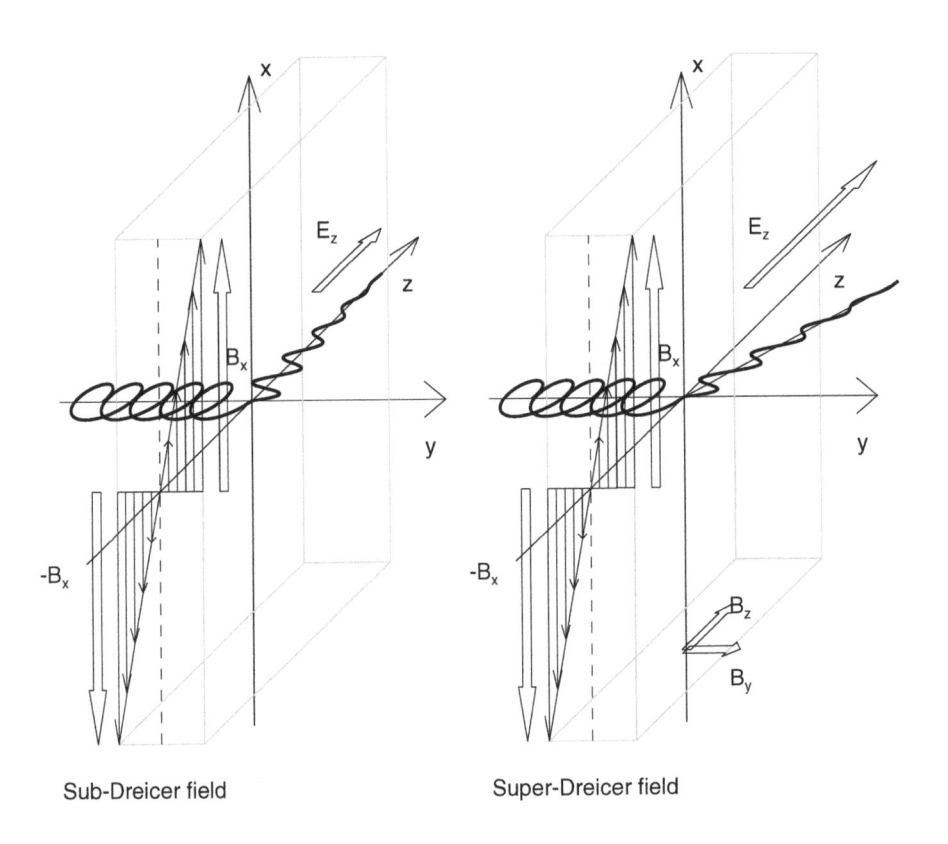

Sub-Dreicer field   Super-Dreicer field

Figure 11.3: Particle orbit in sub-Dreicer electric DC field (left) and super-Dreicer electric DC field (right). In sub-Dreicer fields, the particle is accelerated the full length of the current sheet in the z-direction (see, e.g., model by Holman 1985), while it is scattered out of the current sheet after a short distance in the super-Dreicer field, due to an additional weak perpendicular magnetic field component $(0, B_y = B_\perp, 0)$ to the longitudinal field $(0, 0, B_z = B_\parallel)$ (see, e.g., model by Litvinenko 1996b).

For typical values, $B_\parallel = 100$ G, $B_\perp = 1$ G, $E_\parallel = 10$ V cm^{-1}, $\Delta w = 10^2$ cm, one obtains electron energies of $W \approx 100$ keV. A typical particle orbit for such a configuration is shown in Fig. 11.3 (right). Essentially the particle spirals around the guiding magnetic field during acceleration, until the $E \times B$ drift scatters it out of the thin current sheet. The typical acceleration length in the direction of the electric field is only

$$\Delta l = \Delta w_y \frac{B_\parallel}{B_\perp} \, , \qquad (11.3.7)$$

which is about $\Delta l \approx 10^4$ cm for the parameters above. Thus, the electric field or current sheet does not need a large distance as required in sub-Dreicer field acceleration ($L \approx 10^9$ cm). Therefore, hard X-ray energies can easily be obtained with super-Dreicer fields in relatively small spatial regions (e.g., in fragmented current sheets or in coalescing magnetic islands), as they occur in the *bursty reconnection mode* (Fig. 11.1, right; § 10.2.3).

Depending on the relative magnitude of the magnetic field components across ($B_\perp$) and along ($B_\parallel$) the current sheet, the motion of the accelerated particles can be regular (as shown in Fig. 11.3 right) or chaotic (Litvinenko 1995; Somov 2000, § 18), in particular when the width Δw_y of the current sheet is commensurable with the gyroradius of the particle. Acceleration along the singular line (in the z-direction where $E_\parallel = B_\parallel$) can accelerate electrons up to energies of MeV (Litvinenko 2000b). Further details of the particle acceleration kinematics in current sheets with super-Dreicer fields can be found in Martens (1988), Martens & Young (1990), Jardine & Allen (1996), Litvinenko (1997, 1999c, 2002a,b, 2003a,b), Litvinenko & Somov (1995), and Somov (1996; 2000, § 18).

11.3.3   Acceleration near Magnetic X-Points

For the magnetic configuration in particle acceleration regions, geometries of the Sweet–Parker current sheet type (Fig. 10.2, top) have been adopted in the studies of Litvinenko and Somov. However, since magnetic reconnection occurs at a much faster reconnection rate in Petschek-type (quadrupolar) configurations (Fig. 10.2 bottom), as required to explain the fast time structures observed in radio and hard X-rays during flares, it is of high interest to study particle acceleration near a magnetic X-point.

The behavior of particle orbits near an X-point is distinctly different in the adiabatic regime (when the gyroradius is much smaller than the magnetic field scale length) compared with the non-adiabatic regime. In the adiabatic case the particle performs the regular gyromotion and is accelerated along the electric field that is generally assumed along the singular line (Neukirch 1996). In the non-adiabatic case, the gyromotion interferes with transits between the unmagnetized central region of the X-point and the magnetized outer regions (where the plasma β-parameter is less than unity), and leads to chaotic orbits (Martin 1986; Chen 1992; Hannah et al. 2002). Particle orbits near magnetic X-points have been studied in the adiabatic regime for a static X-point (Vekstein & Browning 1996; Mori et al. 1998; Heerikhuisen et al. 2002), for collapsing X-points (Sakai & Tajima 1986; Sakai & Ohsawa 1987), for quasi-periodic X-point collapses (Tajima et al. 1987; Fletcher & Petkaki 1997), and in the adiabatic regime with chaotic orbits (Moses et al. 1993; Hannah et al. 2002). Examples of numerically simulated particle orbits near a quadrupolar X-point are shown in Fig. 11.4 [i.e., a regular orbit in 2D (Fig. 11.4, top right), a chaotic orbit in 2D (Fig. 11.4, bottom left), and a chaotic orbit in 3D (Fig. 11.4, bottom right), with no electric fields present]. In the presence of an electric field along the singular line, particles that come close to the singular line experience the largest amount of acceleration, because the $\mathbf{E} \times \mathbf{B}$ drift that removes the particles from the accelerating field is minimal near the singular line. This behavior was also verified by trajectory calculations in analytical field solutions near 2D current sheets, where only particles that encounter the current sheet were found to achieve significant acceleration (Vekstein & Browning 1996; Heerikhuisen et al. 2002). The accelerated particles escape along a separatrix line (Hannah et al. 2002). The magnetic moment is not conserved for particles that pass through the central unmagnetized region around the singular X-line (Moses et al. 1993). The energy spectrum of particles

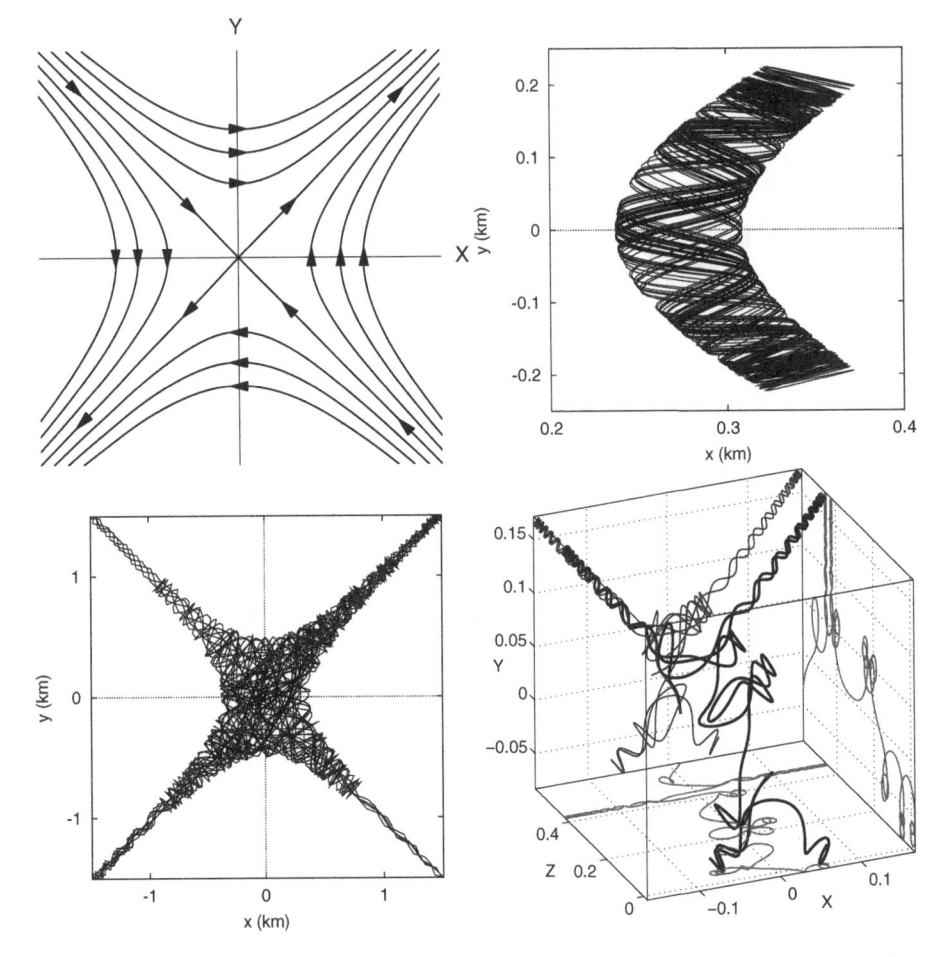

Figure 11.4: Particle orbits near a (quadrupolar) magnetic X-point (top left): a regular orbit in 2D (top right), a chaotic orbit in 2D (bottom left), and a chaotic orbit in 3D (bottom right). In all the cases no electric fields are present (Hannah et al. 2002).

accelerated near an X-point is found to have powerlaw functions from $N(E) \propto E^{-1.3}$ (Fletcher & Petkaki 1997), $N(E) \propto E^{-1.5}$ (Heerikhuisen et al. 2002), $N(E) \propto E^{-1.7}$ (Vekstein & Browning 1996), to $N(E) \propto E^{-2}$ (Mori et al. 1998), a range that is also predicted theoretically (Litvinenko 2003a, see Eqs. 27 and 28 therein). Energy spectra of protons accelerated in an X-point characterized by a generalized solution of the Craig & McClymont (1991) model have been calculated by Hamilton et al. (2003), finding maximum energies up to the γ-ray regime (> 1 MeV). The acceleration of particles in X-point geometries has also been verified experimentally in laboratory plasmas (Brown et al. 2002b).

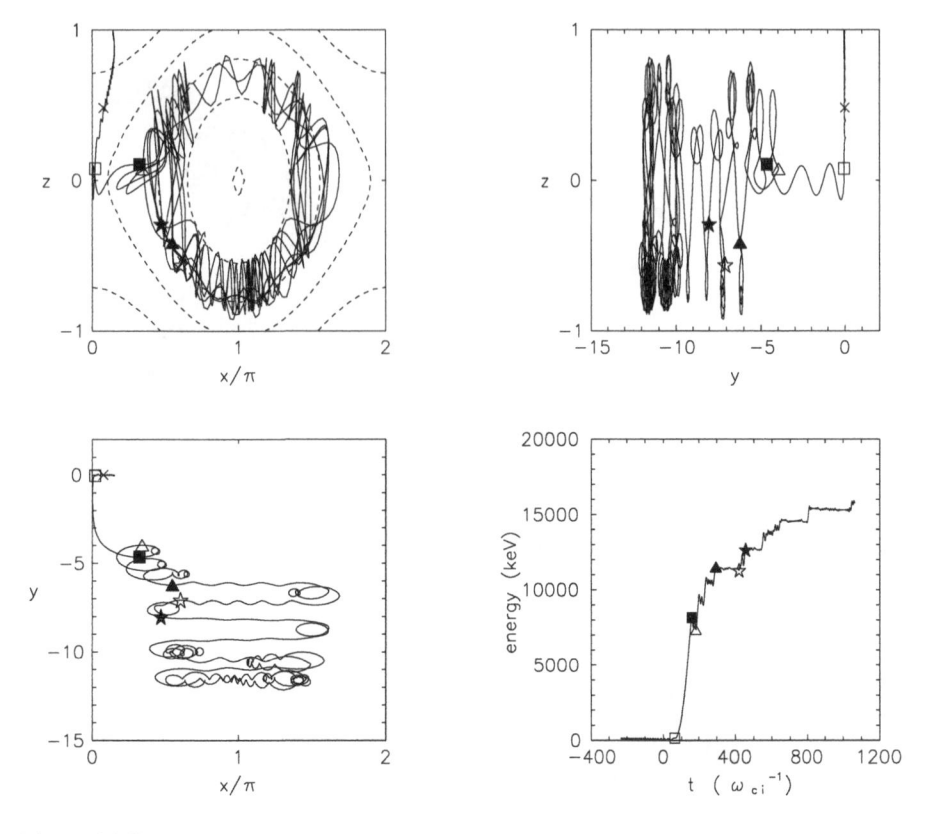

Figure 11.5: Test particle orbit near a magnetic O-point in a magnetic island configuration as shown in Fig. 10.4. The particle is carried into the vicinity of the X line by the $\mathbf{E} \times \mathbf{B}$ drift (equivalent to the reconnection inflow). There it experiences a ∇B drift acceleration with a meander component (top right), which leads to the largest acceleration kick (bottom right). The acceleration continues due to further ∇B drifts in the trapped orbit around the O line. The symbols mark characteristic points of the orbit. (Kliem 1994).

11.3.4 Acceleration near Magnetic O-Points

An equally important configuration of magnetic nullpoints are magnetic islands, which naturally form as a consequence of the tearing-mode instability (Furth et al. 1963) (see Fig. 10.4). The presence of magnetic islands adds an interesting feature to the dynamics of accelerated particles, namely that it allows for temporary trapping inside the magnetic islands, which prolonges the acceleration time and thus requires less demanding DC electric fields to achieve the same final kinetic energy, compared with electric DC fields in X-point geometries.

The particle motion for this configuration of two approaching magnetic islands has been numerically computed in a *Fadeev equilibrium* (Kliem 1993; 1994; Kliem et al. 1996; Kliem & Schumacher 1997). Three types of net motion in directions across the magnetic fields are found: meander orbits at magnetic X and O-lines, the magnetic

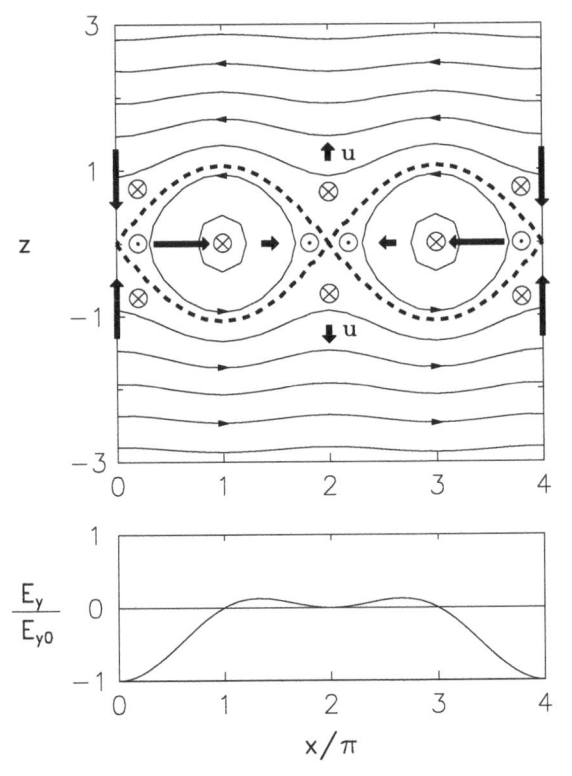

Figure 11.6: Electric field components (arrow heads ⊙ and tails ⊗), magnetic field $\mathbf{B}(x, y)$ (solid lines), separatrices (dashed lines), and mass flows (thick arrows, $\mathbf{u}$) in a pair of approaching magnetic islands (Kliem 1994).

gradient and curvature drift, and the $\mathbf{E} \times \mathbf{B}$ drift. The drift motion in inhomogeneous magnetic fields near the X-lines can be characterized by a superposition of the gradient-B drift $\mathbf{F}_{\nabla B}$ and curvature drift $\mathbf{F}_R$ (Eq. 11.1.8),

$$\mathbf{v}_\nabla = \mathbf{v}_{\nabla B} + \mathbf{v}_R = \frac{mc}{2q}(\mathrm{v}_\perp^2 + 2\mathrm{v}_\parallel^2)\frac{\mathbf{B} \times \nabla \mathbf{B}}{B^3} \qquad (11.3.8)$$

which is exactly parallel to $\pm \mathbf{E}$ in the 2D field case. The drift and meander orbit of a test particle is shown in Fig. 11.5, obtained from numeric calculations by Kliem (1994). It shows a particle that enters the separatrix of a magnetic island and becomes trapped inside. The particle experiences the largest acceleration kick during the first part of the orbit after entering the magnetic island near the X-point, and then *meanders* around the magnetic O-point in a stable orbit, slowly drifting along the y-axis, dragged by the ∇B and $\mathbf{E} \times \mathbf{B}$ force. The stability of meander orbits essentially traps particles sufficiently long so that they can experience the full energy gain provided by the strength and extent of the convective electric field near the O-lines. On the other hand, only particles with a sufficiently high threshold energy (probably requiring a preaccelerated seed population) enter a meander-like orbit within one characteristic coalescence time, but those gain the highest kinetic energies. Also, only particles that come close to an X-point experience the largest acceleration kick (e.g., Fig. 11.5).

In the bursty reconnection mode (§ 10.2.3), the formation of magnetic islands (by the tearing-mode instability) and coalescence of neighboring magnetic islands occur in rapid succession (Pritchett & Wu 1979; Biskamp & Welter 1979; Longcope & Strauss 1994), limiting the acceleration time of a particle trapped in a magnetic island to the lifetime of a magnetic island (also called *coalescence time*). The coalescence dynamics produces additional electric field components that contribute to particle acceleration. The basic situation is shown in Fig. 11.6 for a pair of approaching magnetic islands. The plasma motion of the approaching islands also forces an outflow at the intervening X-point (marked with vector **u** in Fig. 11.6. This plasma flow produces an extra **u** × **B** Lorentz force, which is called the *convective electric field*,

$$\mathbf{E}_{conv} = -\frac{1}{c}\mathbf{u} \times \mathbf{B} \qquad (11.3.9)$$

and for fast reconnection ($u \approx 0.01 - 0.1 v_A$) is typically orders of magnitude larger than the Dreicer field. The directions of the electric field components at various locations inside and outside the magnetic islands are indicated in Fig. 11.6 for a pair of approaching islands. The maximum energy particles can obtain in coalescing magnetic islands is essentially given by the convective electric field, which is

$$E \approx u_{inflow} B_0/c \approx 0.03 v_{A0} B_0/c \approx 3 \times 10^3 \qquad (\text{V m}^{-1}) \qquad (11.3.10)$$

for a main magnetic field of $B_0 = 200$ G and a density of $n_0 = 10^{10}$ cm^{-3}. The half width of a current sheet l_{CS} was estimated using the minimum critical current j_{cr} for the onset of anomalous resistivity,

$$l_{CS} \approx \frac{cB}{4\pi j_{cr}} \approx 7 \times 10^3 \text{ cm} , \qquad (11.3.11)$$

and the typical acceleration length was numerically found to be of the order of the magnetic island (or coalescence) length, $\approx L_{CI}$ (Kliem 1994). The maximum energy a particle can obtain in such a static, convective electric DC field is then $W_{max} \approx 200$ keV, which is significantly higher than the particles would gain by the parallel field ($W_{max} \approx eL_z E_\| \lesssim 20$ keV). Acceleration to higher energies would require multiple magnetic islands or dynamic reconnection.

In summary, while (sub-Dreicer) parallel electric fields could be responsible for bulk (or pre-)acceleration, the convective electric fields provide a more powerful acceleration from the superthermal seed population to higher energies, for a subset of particles that pass close to an X-line and become trapped inside magnetic islands around O-lines. We should also not forget that the tearing instability that drives the entire process of magnetic island coalescence requires anomalous resistivity (e.g., provided by turbulence from current-driven instabilities at X-lines). Recent work describes the topology of reconnection in multiple magnetic islands in terms of a fractal structure, because the iterative processes of tearing, current sheet thinning, and Sweet–Parker sheet formation are thought to repeat iteratively by cascading down to microscopic scales, either to the ion Larmor radius or the ion inertial length (Shibata & Tanuma, 2001).

11.3.5 Acceleration in Time-Varying Electromagnetic Fields

Since a flare is governed by the highly dynamic evolution of magnetic reconnection processes, every model based on static electric fields must be considered as unrealistic. Electric DC fields that accelerate particles are likely to be generated by a number of dynamic processes (e.g., driven by tearing and coalescence of current sheets, by collapses and relaxational oscillations of X-points, or by the electromagnetic fields induced by magneto-acoustic waves or solitons). We describe a few of the physical mechanisms that may faciliate particle acceleration by time-varying electromagnetic fields during flares.

Betatron Acceleration

In a laboratory betatron, an oscillating magnetic field induces an AC electric field that accelerates electrons in a synchronized way during each Larmor orbit (according to Maxwell's induction equation, see Eqs. 5.1.3−4), while the electrons spiral to larger gyroradii and gain energy with each orbit. By analogy, it was proposed that a time-varying magnetic field in a solar flare loop (e.g., caused by MHD oscillations), could pump energy into gyrating electrons (Brown & Hoyng 1975). Since the adiabatic moment $\mu = m_e \mathrm{v}_\perp^2 / 2B$ is conserved in a collisionless plasma, a time-varying longitudinal magnetic field $B(t)$ modulates the perpendicular kinetic energy proportionally,

$$\frac{\mathrm{v}_\perp^2(t)}{\mathrm{v}_{\perp,0}^2} = \frac{B(t)}{B_0} \ . \tag{11.3.12}$$

So, a particle will gain perpendicular energy when it moves from a lower to a higher magnetic field location, or when the local magnetic field $B(t)$ increases. If a flare loop oscillates in a fast sausage mode (§ 7.3), the longitudinal magnetic field would indeed be modulated, because the magnetic flux Φ over the cross section $A(t)$ is conserved,

$$\Phi = A_0 B_0 = A(t) B(t) \ . \tag{11.3.13}$$

In a collisionless plasma, a mirroring trapped electron would just gain and lose perpendicular energy in synchronization with the MHD period, without net energy gain. If the plasma is slightly collisional or subjected to wave turbulence, however, the perpendicular momentum can be irreversibly transferred to parallel momentum during each cycle, so that the electron experiences a net gain with each cycle. This process is called *magnetic pumping* and could in principle accelerate trapped particles to higher energies. The application to solar flares was mainly motivated by the observation of quasi-periodic time structures seen in hard X-rays, which moreover showed a correlated modulation of the slope in the energy spectrum, which can be theoretically explained in terms of a betatron model (Brown & Hoyng 1975; Brown & McClymont 1976). This theoretically expected correlation, however, was found to hold only during the decay phase of the flare, based on the analysis of a larger set of flare events (Karpen 1982). Moreover, the variation of the hardness of the hard X-ray spectrum can also be explained by a modulation of the relative ratio of directly precipitating to trapped-precipitating electrons during an injection cycle, without betatron acceleration.

Field-Aligned Electric Potential Drops

Acceleration in electric potential drops that are aligned with the magnetic field is a common acceleration mechanism in astrophysics, believed to operate in auroral arcs of the Earth's magnetosphere (e.g., Mozer et al. 1977; Block 1978), in pulsar magneto-spheres (e.g., Goldreich & Julian 1969), and also occasionally thought to occur in solar flares (e.g., Colgate 1978; Haerendel 1994; Volwerk 1993; Volwerk & Kuijpers 1994; Tsuneta 1995c). Electric potential drops can be caused in many ways; by ambipolar diffusion of electrons and ions, by the difference of the mirroring heights of electrons and ions (which produces the so-called *V-events* in the aurora), by shear flows (which convert magnetic stress energy into kinetic energy), by wave solitons, by double layers, by shock waves, or return-current electric fields (Karlický 1993). Electric potentials $E_\parallel$ that are aligned with the magnetic field can accelerate particles directly without being impeded by the magnetic field, while perpendicular potentials $E_\perp$ would merely cause an $\mathbf{E}_\perp \times \mathbf{B}$ drift (§ 11.1.2).

While the mechanism of acceleration by field-aligned potential drops is widely accepted in the magnetospheric community thanks to *in situ* spacecraft measurements, the application to solar flares is less clear. There is the concept of *electrostatic double layers*, which are small-scale high-intensity electric fields in a current-carrying plasma or at the boundary of two plasmas with different characteristics (e.g., temperature, density, or chemical composition; Volwerk 1993). A solar flare cannot be modeled by a single large double layer, because its generation would require enormously high current densities, such that the drift velocity of the current-carrying particles is well above the thermal velocity (Volwerk 1993). Therefore, concepts with many small double layers or highly filamented current channels have to be envisioned. Holman (1985) and Emslie & Hénoux (1995) proposed a flare system with many oppositely directed current channels, but there is no physical mechanism known that generates them in the first place. Moreover, strong double layers are highly unstable and are estimated to have lifetimes of the order of milliseconds only (Volwerk & Kuijpers 1994).

A more specific scenario for particle acceleration by field-aligned potential drops in solar flares has been proposed by Haerendel (1994). Coronal magnetic field reconfig-urations during a flare are thought to generate Alfvén waves which propagate towards the chromosphere and set up a parallel electric potential drop when they are reflected and dissipated in the chromosphere, generating supercritical field-aligned currents in the chromosphere that can accelerate particles to high energies. The maximum energy that can be obtained in such a scenario is estimated to be

$$\Phi_\parallel = 5 \text{ MV} \times \left(\frac{B_{Chr}}{1000 \text{ G}} \right) \left(\frac{\Delta B_{\perp,Cor}}{100 \text{ G}} \right)^2 , \qquad (11.3.14)$$

which provides gamma-ray emitting electrons with energies of up to ≈ 60 MV, as observed. These megaVolt field-aligned potential drops, however, are very short-lived (essentially the dissipation time of the reflected Alfvén waves at the top of the chromosphere), so acceleration is highly transient and the launch of many coronal Alfvén waves are required. The advantages of the model are: (1) that it produces instantaneous acceleration to gamma-ray energies, without requiring a preacceleration mechanism to a threshold energy (as it is needed in some other acceleration models); and (2) that the proximity of the acceleration site to the chromosphere ameliorates the electron number problem far better than any other model with a coronal acceleration site. However, the proximity of the acceleration site to the chromosphere seems to contradict the observed electron time-of-flight delays, which require an acceleration site high up in the corona above the soft X-ray flare loops (Aschwanden et al. 1996a,b,c).

There are a number of flare models based on dissipation of electric currents in flare loops, sometimes conceived as an LRC electric circuit (Alfvén & Carlqvist 1967; Carlqvist 1969; Spicer 1977b, 1981; Colgate 1978; Zaitsev et al. 1998, 2000). By analogy to oscillations in an LRC circuit, oscillating magnetic fields $[B_z(t), B_\varphi(t)]$ and associated electric currents $[j_z(t), j_\varphi(t)]$ are thought to modulate field-aligned electric fields that produce quasi-periodic bunches of accelerated electrons (Zaitsev et al. 1998, 2000). However, such closed-loop flare models ignore magnetic reconnection processes and electron time-of-flight measurements, which both suggest an acceleration site *above* the flare loops.

A field-aligned potential drop scenario that is more in line with the magnetic reconnection scenario suggested by the Yohkoh/SXT and HXT observations and the electron time-of-flight measurements measured with CGRO/BATSE, is the model of Tsuneta (1995c). The fast reconnection outflows from the cusp in a downward direction are heading towards the underlying flare loops that have already been filled by the chromospheric evaporation process. These downflows collide with the loop and stream at the outer surface towards the chromosphere, generating strong shear flows at the interface with the soft X-ray emitting flare loops. In this process they generate small-scale time-varying shear flows (vortices), which create oppositely directed field-aligned current channels and associated voltage drops of the order ≈ 100 keV. These field-aligned potential drops can then accelerate electrons (in the runaway regime) to hard X-ray energies. Since acceleration is localized near the top of the soft X-ray flare loops it is consistent with the electron time-of-flight measurements.

Coalescence and X-Point Collapse

The dynamics of accelerating electric fields has been studied for the case of two coalescing current-carrying loops during an *explosive magnetic reconnection process* in great detail by Tajima & Sakai (1986), Sakai & Tajima (1986), Tajima et al. (1987), Sakai & Ohsawa (1987), and Sakai & DeJager (1991). For the initial configuration of the current sheet they assume a 2D geometry with width w_y and length l_x (as illustrated in Figs. 11.1 and 11.3). The reconnection process is driven by the lateral inflow v_y, producing a reconnection outflow v_x with the local Alfvén velocity, $v_x = v_A \approx B_x/\sqrt{n_e}$, which are related to the geometric dimensions of the current sheet by the mass conservation law (in an incompressible plasma),

$$l_x v_y = w_y v_x . \tag{11.3.15}$$

If the lateral magnetic influx, $v_y B_x$, is constant, the reconnection rate would be just the Sweet−Parker reconnection rate in the case of long current sheets, $l_x \gg w_y$, or the Petschek reconnection rate, in the case of X-type short current sheets ($l_x \approx w_y$). The essential feature of nonsteady reconnection processes is the dynamics of the driver. In the treatment of Sakai & Ohsawa (1987) it is assumed that the lateral magnetic influx

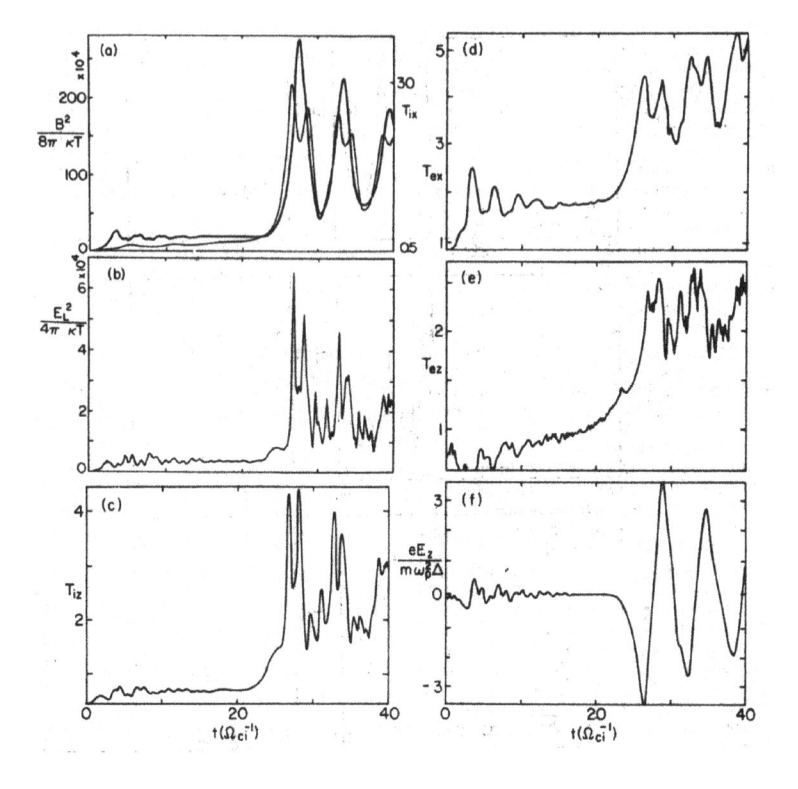

Figure 11.7: The dynamic evolution during a magnetic current loop collapse. (a) The effective potential describing the magnetic collapse. The schematic pattern of the nonlinear oscillations after the explosive phase for (b) the magnetic energy $B^2 \approx B_y^2$, (c) the electric field $E_L \approx E_x^2$, and (d) the ion temperature T_{iz}. The oscillation period is of the order of the explosion time t_0 (Tajima et al. 1987).

increases *explosively* (defined by a time-dependence $\propto (t-t_0)^p$ with a negative power p, so that a singularity, speak *explosion*, occurs at time $t = t_0$),

$$v_y(t) \propto -\frac{y}{(t-t_0)} , \tag{11.3.16}$$

dynamically driving the dimensions of the current sheet as

$$w_y(t) \propto w_{y0}\eta(t_0 - t) , \tag{11.3.17}$$

$$l_x(t) \propto l_{x0}\eta(t_0 - t)^2 . \tag{11.3.18}$$

Once the time t approaches t_0, the length of the current sheet $l_x(t)$ decreases faster than the width $w_y(t)$, so that the Sweet–Parker current sheet ($l_x \gg w_y$) makes a transition to a Petschek type ($l_x \approx w_y$). The associated change in magnetic flux, $\Delta\Phi(t)$, then becomes explosive,

$$\Delta\Phi(t) \propto \Delta\Phi_0(t_0 - t)^{-4/3} \tag{11.3.19}$$

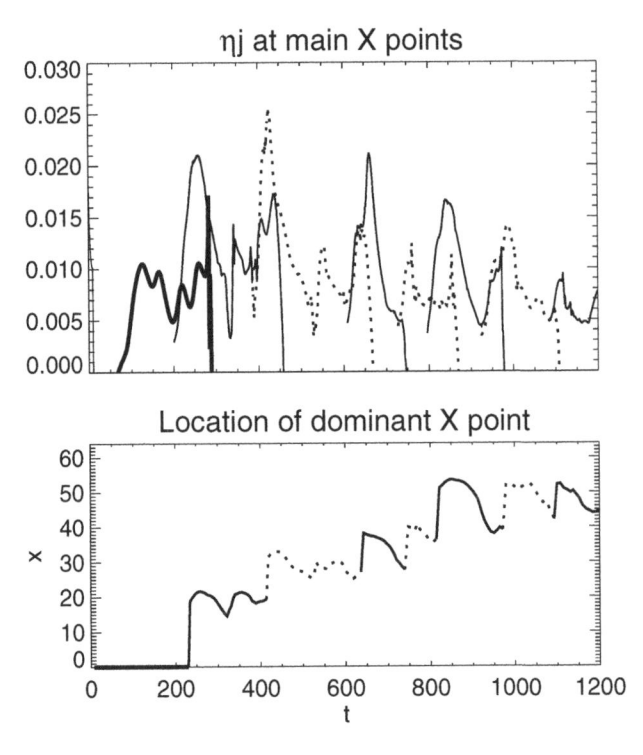

Figure 11.8: The evolution of the electric field in the main X-points is shown in a numerical simulation of the bursty reconnection mode (see also Fig. 10.8) (Kliem et al. 2000).

which evolves into a very rapid reconnection rate, independent of the plasma resistivity η. The associated electric and magnetic field components show the following nonlinear time dependencies,

$$E_y(t) \propto -\frac{y}{(t_0 - t)^2} \,, \tag{11.3.20}$$

$$B_x(t) \propto \frac{x}{(t_0 - t)^{4/3}} \,, \tag{11.3.21}$$

$$E_z(t) \propto c_1 \frac{x^2}{(t_0 - t)^{7/3}} + c_2 \frac{1}{(t_0 - t)^{5/3}} \,, \tag{11.3.22}$$

producing complicated particle orbits that require numerical computations. Sakai & Ohsawa (1987) calculate the evolution of this dynamic reconnection process with analytical approximations, as well as by performing numerical simulations, using the standard MHD equations for a two-fluid (electrons and ions) plasma. They found a variety of dynamical evolutions: (1) explosive collapse, (2) nonlinear oscillations between magnetic collapses and recoveries, and (3) double-peak structures during nonlinear oscillations (Fig. 11.7), depending on the particular value of the plasma-β parameter $\beta = c_s^2/v_A^2$. Oscillatory evolutions result from the counteracting forces of the $\mathbf{j} \times \mathbf{B}$-term, which drives the magnetic collapse, and the pressure gradient term ∇p inside the current sheet (Sakai & Ohsawa 1987; Litvinenko 2003b). These theoretical results are supported by two particular flare observations, which showed indeed quasi-periodic sequences of double-peak structures (i.e., 1980-Jun-7, and 1982-Nov-26; Sakai & Ohsawa 1987). Since the high-resolution observations of *TRACE* show a high degree of spatially-fragmented loop arcades, we think that most of the (generally aperiodic) fast hard X-ray time structures result from spatially-separated magnetic collapses, rather than from a quasi-periodic oscillation of a single collapse region. Also, multiple, spatially separated magnetic collapses ease the number problem of accelerated particles considerably, compared with the operation of a single collapse region which has to be driven by a fast inflow to constantly replenish particles. Nevertheless, the work of Sakai & Ohsawa (1987) yields valuable quantitative physical insights into the nonlinear evolution of the 3D electric and magnetic field components that accelerate electrons and ions in a collapsing current sheet.

More advanced work that includes multiple magnetic collapses in a magnetic reconnection region has been performed by Karpen et al. (1998) and Kliem et al. (2000), dubbed the *"bursty reconnection regime"* (§ 10.2.3). Numerical simulations show highly time-varying currents and electric field components in the X-point regions (Fig. 11.8), which produce quasi-periodic pulses of accelerated particles, probably detected during *broadband decimetric pulsations* in radio wavelengths (Kliem et al. 2000). Particle acceleration is expected to be more efficient in a scenario with multiple X-point reconnection processes, because the particles stay trapped longer in the meander-like orbits around the O-points (Ambrosiano et al. 1988; Kliem 1994; Kliem et al. 1996), and thus are exposed longer to the accelerating fields, opposed to single X-point configurations, where they can escape quickly. Kliem et al. (2000) scaled the physical parameters of the numerical MHD simulation to solar conditions to estimate the acceleration time scales of radio-emitting electrons. The time intervals between subsequent peaks in the reconnection rate were found to be $t_R \approx 200\tau_A \approx 13\delta_x/v_A$ using an average distance of $\delta_x \approx 15\, l_{CS}$ between neighboring X-points, where l_{CS} is the current sheet half width. Translating this time scale $t_R \approx 13\delta_x/v_A$ to solar flare conditions ($n_0 = 10^{10}$ cm^{-3}, $T_0 = (2.5 - 9) \times 10^6$ K, $B_0 = 70$ G), and estimating δ_x from the mean free path length λ_{mfp} (which implies $t_R \approx n_0^{-1/2}T_0^2 B_0^{-1}$), they found typical time intervals of $t_R = 0.4 - 4$ s between subsequent radio bursts. This quantitative example demonstrates that tearing and coalescence in the *bursty magnetic reconnection mode* can modulate particle acceleration on time scales that are observed in radio and hard X-rays.

11.4   Stochastic Acceleration

While the source of energy is an external DC electric field in the previously described particle acceleration mechanisms, we turn now to AC fields of waves as an external energy source, which can transfer energy to particles, in particular for wave frequencies that are in resonance with the gyrofrequencies of the particles. Since there is generally a broadband spectrum of waves present in real nature, some waves will have constructive interference and others destructive interference with the gyromotion of the particles, so that the energy transfer between waves and particles is a stochastic

process. Some parts of the velocity distribution of particles, however, experience a
net gain of energy transfer, at the expense of resonant wave energies, which we call
stochastic acceleration. The governing theory of *wave-particle interactions* in a magnetized plasma can describe both (1) the coherent growth of waves by absorption of
free energy from unstable particle distributions, and (2) the *stochastic acceleration* process for particles that are resonant with parts of the wave spectrum. This theory of
wave-particle interactions and *quasi-linear diffusion* is covered most rigorously in the
textbook of Melrose (1980a,b), and in more concise form in Benz (1993), Sturrock
(1994), or Somov (2000).

11.4.1 Gyroresonant Wave-Particle Interactions

We outline the theory of wave-particle interactions to some degree so that the reader
becomes aware of what basic physics is involved, while the full derivation is referred
to the literature, which is rather extensive because of the rich variety of possible wave
modes in a magnetized plasma and its wave dispersion relations.

The general starting point in the theory of wave-particle interactions is a coupled
pair of rate equations, which describe the changes of the wave photon spectrum $N(\mathbf{k}, t)$
due to interactions with particles, and vice versa the changes in the particle distribution
function $f(\mathbf{p}, t)$ due to interactions with waves,

$$\frac{\partial N(\mathbf{k})}{\partial t} + \mathbf{v}_g(\mathbf{k}) \frac{\partial N(\mathbf{k})}{\partial \mathbf{r}} = \Gamma(\mathbf{k}, f[\mathbf{p}]) N(\mathbf{k}) - \Gamma_{coll}(\mathbf{k}) N(\mathbf{k}) , \qquad (11.4.1)$$

$$\frac{\partial f(\mathbf{p})}{\partial t} + \mathbf{v}(\mathbf{p}) \frac{\partial f(\mathbf{p})}{\partial \mathbf{r}} = \frac{\partial}{\partial p_j} \left[\hat{D}_{ij}(N[\mathbf{k}]) \frac{\partial f(\mathbf{p})}{\partial p_i} \right] + \left(\frac{\partial f(\mathbf{p})}{\partial t} \right)_{Source} + \left(\frac{\partial f(\mathbf{p})}{\partial t} \right)_{Loss} ,$$
$$(11.4.2)$$

where $N(\mathbf{k}) = W(\mathbf{k})/\hbar\omega$ represents the occupation number of photons in the wave
energy range $W(\mathbf{k})$ in **k**-space, $\mathbf{v}_g = \partial\omega(\mathbf{k})/\partial\mathbf{k}$ is the group velocity of emitted
waves, $\Gamma(\mathbf{k}, f[\mathbf{p}])$ the wave amplification growth rate, $\Gamma_{coll}(\mathbf{k})$ the wave damping rate
due to collisions, $f(\mathbf{p})$ the particle density distribution in momentum **p** space, $\mathbf{v}(\mathbf{p}) = \mathbf{p}/m$ is the particle velocity, $\hat{D}_{ij}(N[\mathbf{k}])$ the diffusion tensor, and the last two terms in
Eq. (11.4.2) represent source and loss terms of particles. The coupling between these
two rate equations resides in the wave growth rate $\Gamma(\mathbf{k}, f[\mathbf{p}])$ which depends on the
particle distribution function $f(\mathbf{p})$, and vice versa in the diffusion tensor $\hat{D}_{ij}(N[\mathbf{k}])$
that depends on the wave spectrum $N(\mathbf{k})$.

In principle, the evolution of the particle distribution function $f(\mathbf{p}, t)$ and wave
photon spectrum $N(\mathbf{k}, t)$ can only be described self-consistently by a simultaneous
solution of the coupled rate equations (11.4.1−2). The solution for the wave photon
spectrum provides information on the coherent growth of waves that are driven by
particular shapes of unstable particle distribution functions and are detectable in radio wavelengths (e.g., *plasma emission* that is generated by *electron beam* distribution
functions, or *electron-cyclotron maser emission* that is driven by *losscone* distribution
functions, see § 15). In the context of stochastic particle acceleration, the evolution
of the wave spectrum is generally neglected (Eq. 11.4.1), and an isolated homogeneous volume is considered without particle sources or losses, so that the change in the

particle distribution function is only controlled by the quasi-linear diffusion term (in Eq. 11.4.2), which characterizes the diffusion of wave-resonant particles in momentum space,

$$\frac{\partial f(\mathbf{p})}{\partial t} = \frac{\partial}{\partial p_j} \left[\hat{D}_{ij}(N[\mathbf{k}]) \frac{\partial f(\mathbf{p})}{\partial p_i} \right] , \tag{11.4.3}$$

where the diffusion tensor includes transitions of the photon distribution $N(\mathbf{k})$ from momentum $(\hbar k_i)$ to $(\hbar k_j)$, integrated over the entire wave vector space $d^3\mathbf{k}$,

$$D_{ij}(\mathbf{p}) = \int \frac{d\mathbf{k}^3}{(2\pi)^3} w^\sigma(\mathbf{p},\mathbf{k}) N^\sigma(\mathbf{k})(\hbar k_i)(\hbar k_j) , \tag{11.4.4}$$

where $w^\sigma(\mathbf{p},\mathbf{k})$ denotes the transition probability for a specific magneto-ionic mode σ,

$$w^\sigma(\mathbf{p},\mathbf{k},s) = \frac{1}{\hbar} \left[A_s^{(\sigma)}(\mathbf{p},\mathbf{k}) \, \delta\left(\omega - \frac{s\Omega}{\gamma} - k_\| v_\| \right) \right] , \tag{11.4.5}$$

which can be expressed in terms of an anisotropy factor $A_s^{(\sigma)}$ and a resonance condition,

$$\omega - \frac{s\Omega}{\gamma} - k_\| v_\| = 0 . \tag{11.4.6}$$

The so-called *Doppler resonance condition* is the most decisive term for resonant particle acceleration. It essentially expresses the resonance of the wave vector with the gyromotion of the particle, which simply is $\omega = s\Omega$ for the harmonics s in the nonrelativistic limit. The additional corrections are for the relativistic gyrofrequency Ω/γ, and for Doppler motion of the propagating particles, which shifts the gyrofrequency to $s\Omega/\gamma + k_\| v_\|$. The cases with positive harmonic number $(s > 0)$ are called *normal Doppler resonance*, the cases with negative harmonic number $(s < 0)$ *anomalous Doppler resonance*, and the degenerate case with $s = 0$ is referred to as *Landau damping* or *Cerenkov resonance*.

For the calculation of the anisotropy factor $A_s^{(\sigma)}$ in Eq. (11.4.5), which is a measure of the "coupling strength" in the (gyroresonant) wave-particle interaction, we have to consider the gyromagnetic emission that is radiated by particles due to their spiralling motion in a magnetic field. The situation is sketched in Fig. 11.9, where a gyrating particle is shown, that rotates with the gyrofrequency Ω around the guiding magnetic field B_0 and interacts with an electromagnetic wave vector $\mathbf{k}$, which has AC components E_1 and B_1 rotating with the wave frequency ω around the wave vector direction $\mathbf{k}$. The spiral motion of the gyrating particle can then be expressed by the spatial vector $\mathbf{r}_g(t)$,

$$\mathbf{r}_g(t) = \mathbf{r}_0 + \left(R\sin[\phi_0 + \Omega t], \pm R\cos[\phi_0 + \Omega t], v_\| t \right) , \tag{11.4.7}$$

with gyroradius R, gyrofrequency Ω, and the sign $\pm$ of the electric charge. A spiralling charge causes a current density,

$$\mathbf{j}(\mathbf{r},t) = q\,\mathbf{v}(t)\,\delta(\mathbf{r} - \mathbf{r}_q[t]) . \tag{11.4.8}$$

The relative orientation of the wave vector $\mathbf{k}$ is characterized by the angle θ to the direction of the magnetic field B_0 and the phase angle $\psi(t)$ (in azimuthal direction

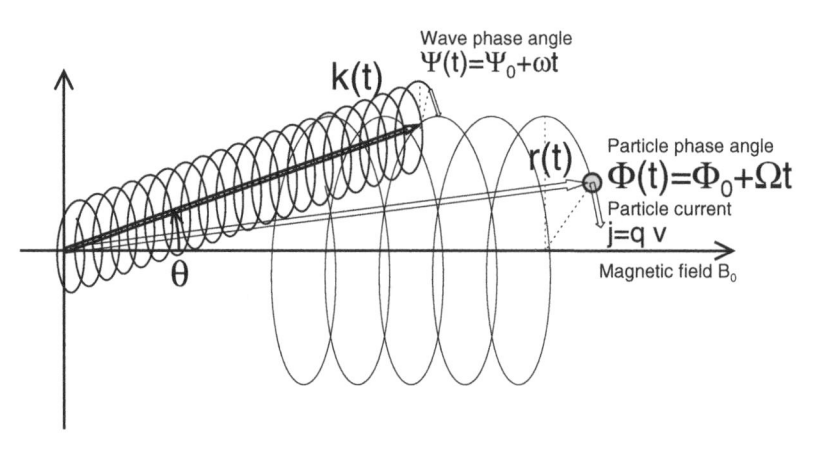

Figure 11.9: The wave-particle interaction between a wave with a propagation vector $\mathbf{k}(t)$ and a wave frequency ω, which has an angle θ to the magnetic field direction B_0, with a particle gyrating at position $\mathbf{r}(t)$ with gyrofrequency Ω around the guiding field B_0. The (azimuthal) phase angle of the wave vector is $\Psi(t)$ and of the particle is $\Phi(t)$.

around the wave propagation vector $\mathbf{k}$),

$$\mathbf{k}(t) = k(\sin\theta\cos\psi(t), \sin\theta\sin\psi(t), \cos\theta) . \tag{11.4.9}$$

The resulting current density $\mathbf{j}(\mathbf{r}, t)$ can then be expressed by the Fourier transform $\mathbf{j}(\mathbf{k}, \omega)$,

$$\mathbf{j}(\mathbf{k}, \omega) = q \int dt \int d^3r \, \exp[-i(\mathbf{k}\cdot\mathbf{r} - \omega t)] \, \mathbf{v}(t) \, \delta^3(\mathbf{r} - \mathbf{r}_q[t])$$

$$= q \int dt \, \mathbf{r}_q(t) \exp[-i(\mathbf{k}\cdot\mathbf{r}_q - \omega t)] . \tag{11.4.10}$$

This current density $\mathbf{j}(\mathbf{k}, \omega)$ can be evaluated by first expanding into Bessel functions (Melrose 1980a), yielding the expression

$$\mathbf{j}(\mathbf{k}, \omega) = 2\pi q \exp(-i\mathbf{k}\cdot\mathbf{r}_0) \sum_{s=-\infty}^{\infty} \exp[-is(\phi_0 \pm \psi)]\mathbf{V}(s, \mathbf{p}, \mathbf{k}) \, \delta\left(\omega - \frac{s\Omega}{\gamma} - k_{\parallel}v_{\parallel}\right) ,$$

$$\tag{11.4.11}$$

where the velocity quantity $\mathbf{V}(s, \mathbf{p}, \mathbf{k})$ for random phase approximation (by choosing $\mathbf{r}_0 = 0, \phi_0 = 0$, and $\psi = 0$) is expressed by the Bessel function $J_s(z)$ and its derivative $J'_s(z)$, with the Bessel argument $z = k_{\perp}R$,

$$\mathbf{V}(s, \mathbf{p}, \mathbf{k}) = \left[v_{\perp}\frac{s}{z}J_s(z), \pm(-i)v_{\perp}J'_s(z), v_{\parallel}J_s(z) \right] . \tag{11.4.12}$$

From the current density $\mathbf{j}(\mathbf{k}, \omega)$ we find an implicit solution of the electric field by inserting the conductivity tensor σ_{ij} (which can be computed from the dispersion relation for cold plasma),

$$j_i^{ind}(\mathbf{k}, \omega) = \sigma_{ij}(\mathbf{k}, \omega)\mathbf{E}_j(\mathbf{k}, \omega) . \tag{11.4.13}$$

From this electric field the radiated power $U^\sigma(\mathbf{k})$ in k-space (or emissivity) can be derived using the power theorem of the Fourier transform,

$$U^\sigma(\mathbf{k}) = \int dt \int d^3\mathbf{r} \, [\mathbf{j}^{ext}(\mathbf{r}, t)\mathbf{E}(\mathbf{r}, t)] \,. \tag{11.4.14}$$

On the other hand, the radiated power in the quantum-mechanical description is

$$P(\mathbf{k}) = \lim_{\tau \mapsto \infty} \frac{1}{\tau} U^\sigma(\mathbf{k}) = (\hbar\omega) \, w^\sigma(\mathbf{k}, \mathbf{p}) = \omega \, A_s^{(\sigma)}(\mathbf{p}, \mathbf{k}) \, \delta\left(\omega - \frac{s\Omega}{\gamma} - k_\parallel v_\parallel\right) \,. \tag{11.4.15}$$

The steps of Eqs. (11.4.7−15) outline the formal derivation of the anisotropy factor $A_s^{(\sigma)}(\mathbf{k}, \mathbf{p})$ specified in Eq. (11.4.5). A more detailed derivation can be found in Melrose (1980a, p. 98−112; 1980b, p. 256−280), which leads to the following expression for the anisotropy factor (used with the same parameterization in Melrose & Dulk 1982; Aschwanden 1990a),

$$A_s^{(\sigma)}(p, k) = \frac{4\pi^2 e^2 c^2 \beta_\perp^2}{\omega n_\sigma \frac{d(\omega n_\sigma)}{d\omega}(1 + T_\sigma^2)} \left[\frac{K_\sigma \sin\theta + (\cos\theta - n_\sigma \beta_\parallel)T_\sigma}{n_\sigma \beta_\perp \sin\theta} J_s(z) + J_s'(z)\right]^2 \,, \tag{11.4.16}$$

where the argument of the Bessel function is $z = (\omega/\Omega_e)n_\sigma \beta_\perp \sin\theta$. The waves are described by their refractive index n_σ, the longitudinal part of their polarization is K_σ, and the axial ratio of their polarization ellipse is T_σ. The dispersion relation for electromagnetic waves in a cold plasma can be parameterized by (Melrose 1980b, p. 256−263),

$$k(\omega) = \frac{\omega}{c} n_\sigma(\omega, \theta) \,, \tag{11.4.17}$$

$$n_\sigma^2 = 1 - \frac{XT_\sigma}{T_\sigma - Y \cos\theta} \,, \tag{11.4.18}$$

$$K_\sigma = \frac{XY \sin\theta}{1 - X} \frac{T_\sigma}{T_\sigma - Y \cos\theta} \,, \tag{11.4.19}$$

$$T_\sigma = -\sigma(x^2 + 1)^{1/2} - x \,, \tag{11.4.20}$$

$$x = \frac{Y \sin^2\theta}{2(1 - X)\cos\theta} \,, \tag{11.4.21}$$

$$X = \frac{\omega_p^2}{\omega^2} \,, \tag{11.4.22}$$

$$Y = \frac{\Omega_e}{\omega} \,. \tag{11.4.23}$$

The equations 11.4.16−23 fully define the anisotropy factor $A_s^{(\sigma)}$ for a specific magneto-ionic mode ($\sigma = 1$ for the ordinary mode [O-mode] and $\sigma = -1$ for the extraordinary mode [X-mode]), harmonic number s of the gyrofrequency, wave vector $\mathbf{k} = (k, \theta)$ and particle momentum $\mathbf{p} = mc(\beta_\parallel, \beta_\perp)$, which is required for the calculation of the transition probability specified in Eq. (11.4.5).

The next step in the calculation of the quasi-linear diffusion coefficient (Eq. 11.4.4) is the integration over the entire wave vector space $d^3\mathbf{k}$,

$$d^3\mathbf{k} = (2\pi k^2 \sin\theta) \, dk \, d\theta = \frac{2\pi\omega^2 n_\sigma^2 \sin\theta}{c^3} \frac{d(\omega n_\sigma)}{d\omega} \, d\omega \, d\theta \, , \qquad (11.4.24)$$

which involves the resonance condition (Eq. 11.4.6). This resonance condition specifies a particular subset of wave vectors $\mathbf{k} = (k, \theta)$ for every particle momentum $\mathbf{p} = mc(\beta_\parallel, \beta_\perp)$. The mathematical nature of the resonance condition can be derived by inserting $k_\parallel = k(\omega) \cos\theta$ and the dispersion relation $k(\omega)$ (Eqs. 11.4.17–23), the relativistic Lorentz factor $\gamma = 1/\sqrt{1 - \beta^2}$, and the parallel velocity $\mathrm{v}_\parallel = \mathrm{v} \cos\alpha$. The simplest parameterization is to choose the wave vector parameters (ω, θ) as independent parameters, which leads to ellipse solutions for the resonant particles, called *resonance ellipses* (Melrose & Dulk 1982). This means that the subset of resonant particles that fulfil the *Doppler resonance condition* (Eq. 11.4.6) have the following velocity components that can be characterized by an ellipse in velocity space,

$$\beta_\parallel = \mathrm{v}_0 - V\sqrt{1 - e^2} \cos\Psi \, , \qquad (11.4.25)$$

$$\beta_\perp = V \sin\Psi \, , \qquad (11.4.26)$$

which has the center at velocity v_0 (on the axis in the direction of the magnetic field), the eccentricity e, and the semi-major axis V (Melrose & Dulk 1982; Aschwanden 1990a),

$$\mathrm{v}_0 = \frac{e^2}{n_\sigma \cos\theta} \, . \qquad (11.4.27)$$

$$e = \left[1 + \left(\frac{sY}{n_\sigma \cos\theta} \right)^2 \right]^{-1/2} \, , \qquad (11.4.28)$$

$$V = \left[1 - \left(\frac{sY}{n_\sigma \cos\theta} \right)^2 \right]^{1/2} \, . \qquad (11.4.29)$$

Examples of resonance ellipses are shown in Fig. 11.10, for subsets of particles that are resonant with wave vectors $\mathbf{k} = (\omega, \theta)$ near the gyrofrequency $\omega/\Omega_e = 1.01 - 1.04$ and emission angles $\theta = 68° - 77°$. The mathematical solutions of resonance ellipses implies that quasi-linear diffusion coefficients (Eq. 11.4.4) have to be calculated by integrating the anisotropy factors $A_s^{(\sigma)}$ along these resonance ellipses, summing over all wave vectors $\mathbf{k} = (\omega, \theta)$.

In order to perform the calculation of the diffusion coefficients we have also to express the integral equation (Eq. 11.4.4) in spherical coordinates. Following the semi-classical treatment of Melrose (1980a, p. 151), the emission (or absorption) of a photon with the quantum-mechanical energy,

$$\epsilon = \hbar\omega \, , \qquad (11.4.30)$$

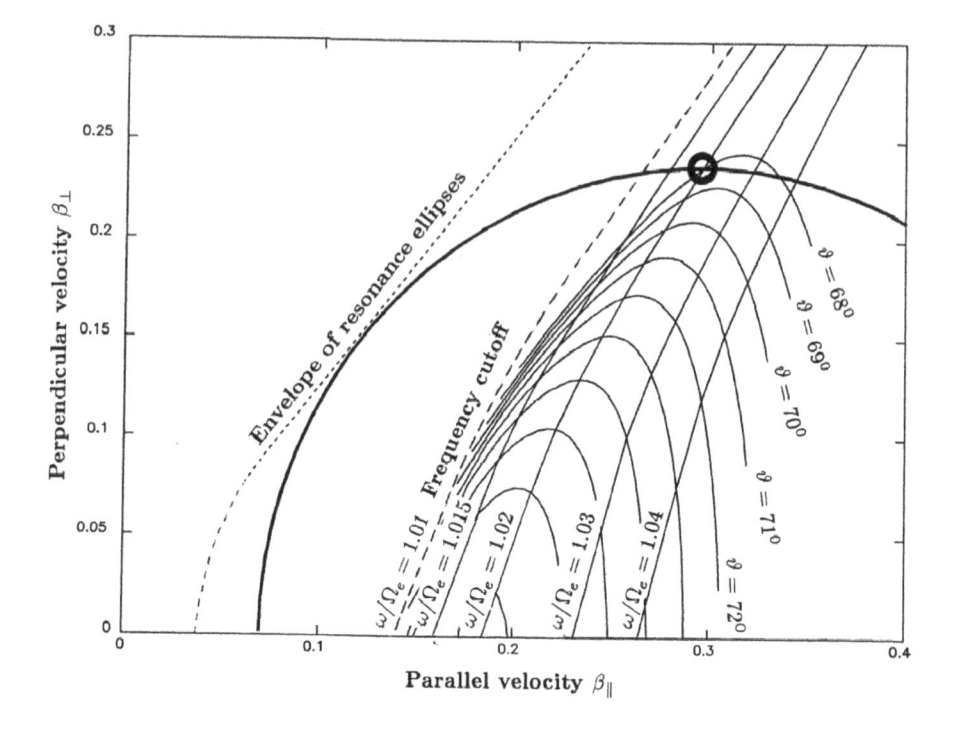

Figure 11.10: A set of resonance ellipses in velocity space $(\beta_\parallel, \beta_\perp)$ for particles that fulfill the Doppler gyroresonance condition with electromagnetic waves near the gyrofrequency. The subset of electromagnetic waves is specified by the frequency $\omega/\Omega_e = 1.01 - 1.04$ and emission angles $\theta = 68° - 77°$ to the guiding magnetic field. Particles can only resonate with the fundamental $(s = 1)$ magneto-ionic ordinary and extraordinary mode above the frequency cutoff $\omega/\Omega_e \gtrsim 1.01$, which excludes thermal or suprathermal particles with subrelativistic speeds of $\beta_\parallel \lesssim 0.15$. Thus, stochastic acceleration by gyroresonant wave particle interactions requires a minimum energy threshold of $\varepsilon \gtrsim 10$ keV for these magneto-ionic modes (Aschwanden 1990a).

causes a momentum change in the resonant particle, which is most usefully split into a parallel and perpendicular component to the magnetic field, $(\Delta p_\parallel, \Delta p_\perp)$, because the perpendicular component entails the momentum of the gyromotion,

$$\Delta p_\parallel = \hbar \Delta k_\parallel = \hbar k_\parallel \,, \tag{11.4.31}$$

$$\Delta p_\perp = \hbar \Delta k_\perp = \hbar \left(\frac{s\Omega_e}{\gamma v_\perp} \right) \,. \tag{11.4.32}$$

Because of the gyromotion it is most useful to employ spherical polar coordinates (p, α, Φ), where p represents the particle momentum and α the pitch angle, while the azimuthal phase Φ can be neglected in the random-phase approximation. The changes in momentum (Eqs. 11.4.31−32) for gyromagnetic emission in spherical coordinates are then

$$(\Delta p) = \left(\frac{\hbar \omega}{v} \right) , \qquad (11.4.33)$$

$$(p\Delta \alpha) = \left(\frac{\hbar \omega}{v} \right) \left(\frac{\cos \alpha - k_\| v/\omega}{\sin \alpha} \right) , \qquad (11.4.34)$$

and the four components of the diffusion tensor (Eq. 11.4.4) can be expressed as

$$\begin{pmatrix} D_{\alpha\alpha} \\ D_{\alpha p} = D_{p\alpha} \\ D_{pp} \end{pmatrix} = \int \frac{d\mathbf{k}^3}{(2\pi)^3} \, w^\sigma(\mathbf{p}, \mathbf{k}) \, N^\sigma(\mathbf{k}) \begin{pmatrix} (p\Delta\alpha)^2 \\ (p\Delta\alpha)(\Delta p) \\ (\Delta p)^2 \end{pmatrix} , \qquad (11.4.35)$$

and the diffusion equation (11.4.3) reads in spherical coordinates

$$\frac{\partial f(p, \alpha)}{\partial t} = \frac{1}{\sin \alpha} \frac{\partial}{\partial \alpha} \left[\frac{\sin \alpha}{p^2} \left(D_{\alpha\alpha} + p D_{\alpha p} \frac{\partial}{\partial p} \right) f(p, \alpha) \right]$$

$$+ \frac{1}{p^2} \frac{\partial}{\partial p} \left[p \left(D_{p\alpha} \frac{\partial}{\partial \alpha} + p D_{pp} \frac{\partial}{\partial p} \right) f(p, \alpha) \right] \qquad (11.4.36)$$

Now we have the analytical tool to calculate stochastic acceleration of an arbitrary particle distribution function $f(\alpha, \beta)$ for a specific magneto-ionic wave mode (σ, s), which can include electromagnetic waves (O, X-waves), Langmuir waves, or their low-frequency counterparts (L, R-waves, and whistler waves). An overview of the most common wave modes in the cold plasma approximation is given in Fig. 11.11, and a summary of the corresponding dispersion relations of electrostatic and electromagnetic waves is given in Table 11.2. For typical coronal conditions, the order of the collision frequency ω_{coll}, plasma frequencies ω_p, and gyrofrequencies Ω for electrons (e) and ions (i) is: $\omega_{coll} < \Omega_i < \omega_p^i < \Omega_e < \omega_p^e$, as shown in Fig. 11.11. This diagram illustrates which waves can resonate with the gyrofrequencies of particles. Derivations of the wave dispersion relations (Eqs. 11.4.17–23) can be found in many textbooks on plasma physics (e.g., Chen 1974; Schmidt 1979; Priest 1982; Benz 1993; Sturrock 1994; Baumjohann & Treumann 1997; Boyd & Sanderson 2003).

11.4.2 Stochastic Acceleration of Electrons

A glance at the dispersion relations in Fig. 11.11 immediately shows which waves are important for stochastic acceleration of electrons and ions: essentially, Alfvén waves, magneto-sonic waves, and ion-sound waves can resonate with the gyrofrequency of ions, while whistler waves, Langmuir waves, and electromagnetic waves can resonate with the (fundamental or harmonic) gyrofrequency of electrons. Stochastic acceleration of electrons is reviewed in Miller et al. (1997, § 3.1.1).

Whistler waves, which have frequencies in the range of $\Omega_H \ll \omega \lesssim \Omega_e$, can fulfill the Doppler resonance condition (Eq. 11.4.6) with electrons, and thus can accelerate electrons up to relativistic energies (Melrose 1974; Miller & Ramaty 1987). For the (unknown) wave spectrum it is generally assumed that wave turbulence is isotropic and that the spectral energy density is a powerlaw,

$$W(k) = \frac{(q-1)}{k_0} \left(\frac{k_0}{k} \right)^q W_{tot} , \quad W_{tot} = \int_{k_0}^\infty W(k) dk , \qquad (11.4.37)$$

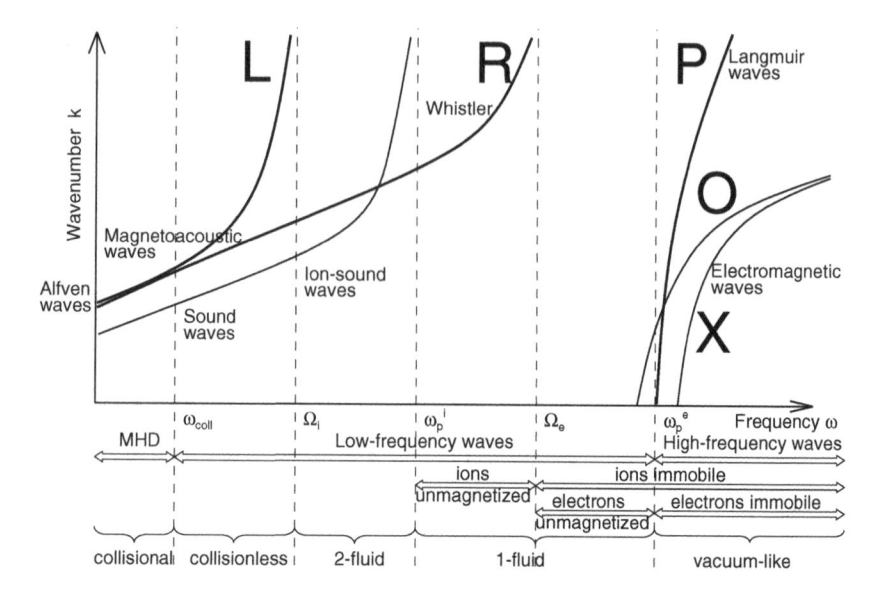

Figure 11.11: Overview of wave modes in cold plasma playing a role in stochastic wave-resonant particle acceleration. Ions can resonate with low-frequency waves, while electrons resonate mostly with high-frequency waves. Indicated are the frequency regimes separated by the collision frequency ω_{coll}, the ion gyrofrequency Ω_i, the ion plasma frequency ω_p^i, the electron gyrofrequency Ω_e, and the electron plasma frequency ω_p^e. L and R are left and right-hand circular polarized low-frequency electromagnetic waves, and O and X are the ordinary and extraordinary mode of the high-frequency electromagnetic waves.

where the lower wave vector threshold for whistler waves is $k_0^W = k_{th} = \Omega_p/c\beta_A$ (Hamilton & Petrosian 1992). For the choice of the powerlaw index, either *Kolmogorov spectra* (with slope $-5/3 = -1.67$) or *Kraichnan spectra* are used (with slope $-3/2 = -1.50$),

$$W(k) \propto \begin{cases} \left(\dfrac{k}{k_0}\right)^{-5/3} , & \text{(Kolmogorov)} \\[2ex] \left(\dfrac{k}{k_0}\right)^{-3/2} . & \text{(Kraichnan)} \end{cases} \qquad (11.4.38)$$

In the Kolmogorov treatment, the spectral energy transfer time at a particular wavelength is the turbulence eddy turnover time $\lambda/\delta v$, where δv is the velocity fluctuation of the wave, while the transfer time is longer by a factor of $v_A/\delta v$ in the Kraichnan treatment (Zhou & Matthaeus 1990; Miller et al. 1996). Stochastic acceleration of test particles in turbulent MHD plasmas have been demonstrated by Ambrosiano et al. (1988) and Kobak & Ostrowski (2000).

   Numerical simulations of stochastic acceleration of electrons by whistler waves, using the turbulence wave spectra specified in Eqs. (11.4.37−38) and computing the quasi-linear diffusion coefficients (Eqs. 11.4.36), have been performed by Hamilton & Petrosian (1992), Miller et al. (1996), Park et al. (1997), and Pryadko & Petrosian

Table 11.2: Summary of dispersion relations of elementary waves for parallel and perpendicular directions in cold plasma (Chen 1974).

Wave direction	Dispersion relation	wave mode
Electron waves (electrostatic)		
$\mathbf{B}_0 = 0$ or $\mathbf{k} \parallel \mathbf{B}_0$	$\omega^2 = \omega_p^2 + \frac{3}{2}k^2 v_{th}^2$	Langmuir waves (P)
		(Plasma oscillations)
$\mathbf{k} \perp \mathbf{B}_0$	$\omega^2 = \omega_p^2 + \omega_c^2$	Upper hybrid waves
Ion waves (electrostatic)		
$\mathbf{B}_0 = 0$ or $\mathbf{k} \parallel \mathbf{B}_0$	$\omega^2 = k^2 v_s^2$	Ion sound waves
	$\omega^2 = k^2 \left(\frac{\gamma_e k_B T_e + \gamma_i k_B T_i}{M} \right)$	(Acoustic waves)
$\mathbf{k} \perp \mathbf{B}_0$	$\omega^2 = \Omega_p^2 + k^2 v_s^2$	Electrostatic ion
		cyclotron waves
	$\omega^2 = \omega_l^2 = \Omega_c \omega_c$	Lower hybrid waves
Electron waves (electromagnetic)		
$\mathbf{B}_0 = 0$	$\omega^2 = \omega_p^2 + k^2 c^2$	Light waves
$\mathbf{k} \perp \mathbf{B}_0, \mathbf{E}_1 \parallel \mathbf{B}_0$	$\frac{c^2 k^2}{\omega^2} = 1 - \frac{\omega_p^2}{\omega^2}$	Ordinary waves (O)
$\mathbf{k} \perp \mathbf{B}_0, \mathbf{E}_1 \perp \mathbf{B}_0$	$\frac{c^2 k^2}{\omega^2} = 1 - \frac{\omega_p^2}{\omega^2} \frac{\omega^2 - \omega_p^2}{\omega^2 - \omega_h^2}$	Extraordinary waves (X)
$\mathbf{k} \parallel \mathbf{B}_0$	$\frac{c^2 k^2}{\omega^2} = 1 - \frac{\omega_p^2/\omega^2}{1 - \omega_c^2/\omega^2}$	Right hand waves (R)
		(whistler mode)
	$\frac{c^2 k^2}{\omega^2} = 1 - \frac{\omega_p^2/\omega^2}{1 + \omega_c^2/\omega^2}$	Left hand waves (L)
Ion waves (electromagnetic)		
$\mathbf{B}_0 = 0$	none	
$\mathbf{k} \parallel \mathbf{B}_0$	$\omega^2 = k^2 v_A^2$	Alfvén waves
$\mathbf{k} \perp \mathbf{B}_0$	$\frac{\omega^2}{k^2} = c^2 \left(\frac{v_s^2 + v_A^2}{c^2 + v_A^2} \right)$	Magneto-sonic waves

(1997), yielding nonthermal tails in the electron spectrum above energies of ≈ 10 keV, as observed in hard X-ray flares. Hamilton & Petrosian (1992) find that fits to observed hard X-ray spectra require loop lengths of $L \approx 100$ Mm, magnetic fields of $B \approx 100$ G, and densities of $n > 3.6 \times 10^{10}$ cm^{-3} for this acceleration mechanism, and that the acceleration times are less than 1 s for these parameters. To make sure that the pitch angle scattering rate by whistler waves is significantly larger than the collision rate in these high-density flare loops, a prerequisite for stochastic acceleration, a collision term has additionally to be included in the diffusion equation (Eq. 11.4.36), which then corresponds to a Fokker–Planck-type equation (Hamilton et al. 1990; Hamilton & Petrosian 1992; Park & Petrosian 1995, 1996). Further simulations by Steinacker & Miller (1992) demonstrated that acceleration times of the order of seconds could be reproduced if the whistler turbulence energy density was about 10% of the magnetic field energy density, and that acceleration to gamma-ray energies could occur by also including the lower frequency waves. Further studies on ion acceleration and abundance enhancements were pursued by Steinacker et al. (1993) and Miller & Viñas (1993).

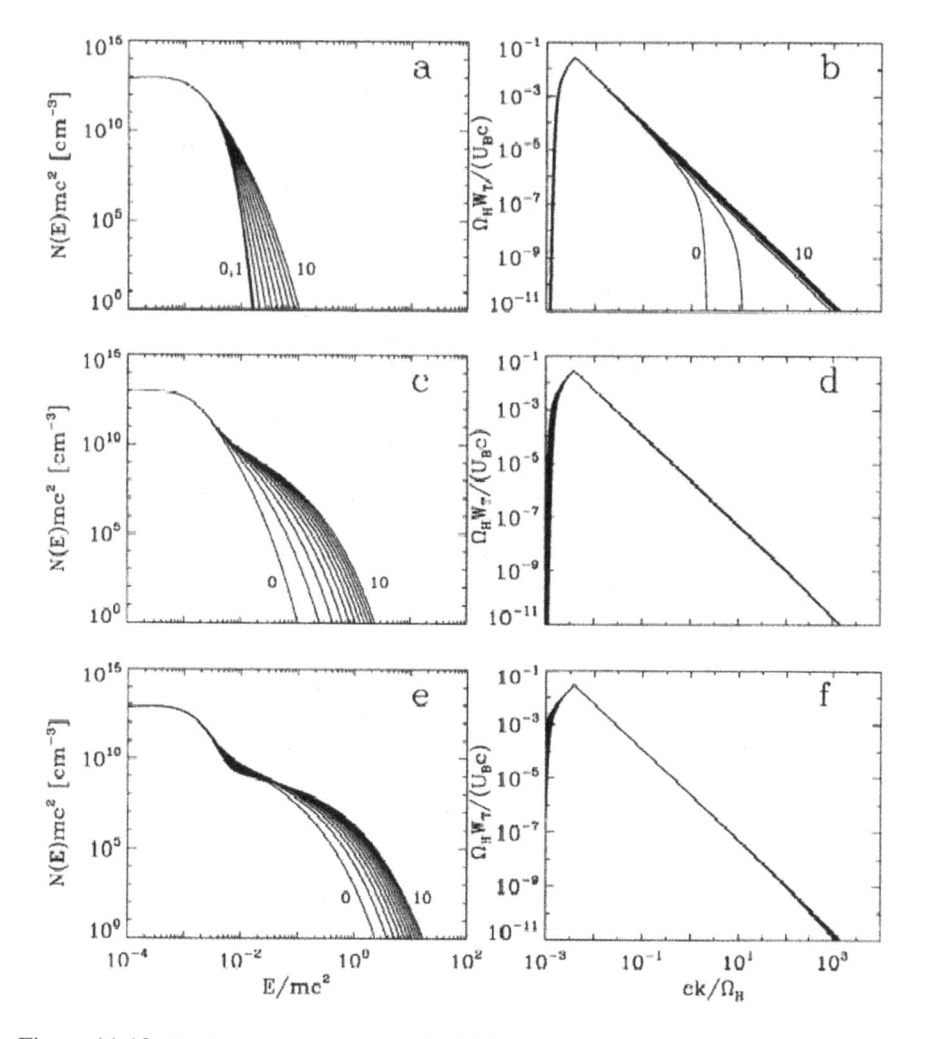

Figure 11.12: Electron energy spectrum $N(E)$ (left) and wave spectral densities W_T (right) resulting from cascading and transit time damping of fast-mode waves. The waves were injected at a wavelength of $\approx 10^7$ cm, at a rate of about 19 ergs cm^{-3} s^{-1}, and over a time of 0.6 s. The ambient electron density was 10^{10} cm^{-3}. (a) and (b) Evolution from $t = 4 \times 10^5 T_H$ to $5 \times 10^5 T_H$. N and W_T are shown at times $t_n = (4 \times 10^5 + 10^4 n)T_H$, for $n = 0 - 10$. (c) and (d) Evolution form $t = 10^6 T_H$ to $3 \times 10^6 T_H$. N and W_T are shown at times $t_n = (10^6 + 2 \times 10^5 n)T_H$, for $n = 0 - 10$. Here $T_H = \Omega_H^{-1} \approx 2.1 \times 10^{-7}$ s and $U_B = B_0^2/8\pi$ is the ambient magnetic field energy density (Miller et al. 1996).

A remaining question is how electrons are first accelerated from their thermal distribution ($T \approx 1 - 10$ MK, i.e. $E \approx 0.04 - 0.4$ keV) to mildly relativistic energies. Because gyroresonant stochastic acceleration seems not to be efficient for such small energies, the *Landau* or *Cerenkov resonance* ($s = 0$) was considered (i.e., $\omega = k_\parallel v_\parallel$ in

Eq. 11.4.6), using the compressive magnetic field component of (magneto-sonic) fast-mode waves. The resonance condition can be written as $v_\parallel = v_A/\eta$, where $\eta = k_\parallel/k$, which shows that magneto-sonic waves (which have similar speeds as Alfvén waves, i.e., $v_A \approx 1000$ km/s), can resonate with thermal electron speeds. This process is the magnetic equivalent of Landau damping and is called *transit-time damping* (Lee & Völk, 1975; Fisk 1976; Achterberg 1979; Stix 1992), because the transit time of a particle across a wavelength is equal to the period of the wave. The only drawback of this mechanism is that electrons are only accelerated in the parallel direction, so that a highly beamed distribution would result ($v_\parallel \gg v_\perp \approx v_{th}$), and thus some additional (unknown) pitch angle scattering is required (Miller 1997) to transfer momentum to the perpendicular component ($v_\perp \approx v_\parallel$). Miller et al. (1996) have conducted numerical simulations of transit-time damping, starting from an MHD-turbulent cascade wave spectrum, and could demonstrate electron acceleration out of the thermal distribution up to relativistic energies during subsecond time intervals (see example in Fig. 11.12).

While electron transit-time damping represents a stochastic acceleration process that works for weak turbulence, the strong-turbulence case with large amplitudes of MHD waves ($\delta B/B \approx 1$) corresponds to the classic *Fermi mechanism* (Fermi, 1949) of collisions between electrons and magnetic scattering centers, which was also applied to solar flares (Ramaty 1979; LaRosa & Moore 1993).

Stochastic acceleration of electrons by high-frequency waves ($\omega \geq \Omega_e$, i.e., plasma waves, ordinary, and extraordinary electromagnetic waves, see Fig. 11.11), have also been examined for waves propagating obliquely to the magnetic field ($k \perp B_0$), but only a small fractions of ambient electrons ($< 10^{-3}$) were energized, which could be sufficient for electron beams detected as radio type III bursts rather than the precipitating electrons detected in hard X-rays (Sprangle & Vlahos 1983; Karimabadi et al. 1987).

11.4.3 Stochastic Acceleration of Ions

For low-frequency waves, when the wave frequency is much smaller than the ion gyrofrequency, with $\omega \ll \Omega_H$, the Doppler resonance condition (Eq. 11.4.6) simplifies to $\Omega_H \approx k_\parallel v_\parallel$. Since the dispersion relation for Alfvén waves is $\omega^2 = k^2 v_A^2$ (Table 11.2), the two conditions imply $v_\parallel \gg v_A$, which is a threshold for ion speeds. Thus, ions need to have a threshold energy of $\varepsilon_i \gtrsim (1/2)m_p v_A^2 \approx 20$ keV for typical coronal Alfvén speeds, $v_A \approx 2000$ km s^{-1}, before they can be accelerated by gyroresonant, stochastic acceleration. Above this threshold, however, Alfvén waves can accelerate protons to γ-ray energies of GeV per nucleon on time scales of $\approx 1 - 10$ s (Barbosa 1979; Miller et al. 1990; Steinacker & Miller 1992).

Similar to electrons, there is an injection problem in the sense that a suitable acceleration mechanism needs to be found to accelerate the ions from their thermal energy ($\lesssim 1$ keV) to the Alfvénic acceleration threshold (≈ 20 keV). It was proposed that nonlinear Landau damping of Alfvén waves can lead to rapid proton heating and energization above the Alfvénic acceleration threshold (Lee & Völk 1973; Miller 1991; Miller & Ramaty 1992; Smith & Brecht 1993). Alternatively, higher frequency waves of the Alfvénic branch type (i.e., magneto-acoustic waves with $k \perp B$) through gyroresonant interaction were also shown to be able to accelerate protons directly from thermal ener-

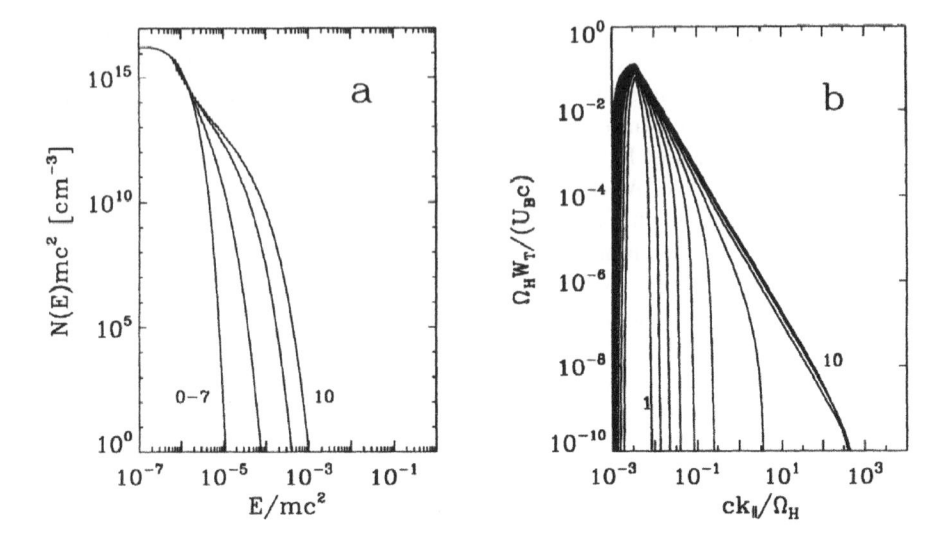

Figure 11.13: Proton energy spectrum $N(E)$ (left) and wave spectral density W_T (right) resulting from cascading and cyclotron damping of Alfvén waves. The waves were injected at a wavelength of $\approx 10^7$ cm, at a rate of about 100 ergs cm^{-3} s^{-1}, and over a time period of 2 s. The ambient proton density was 10^{10} cm^{-3}. (a) N at times $t_n = n(5 \times 10^4 T_H)$, for $n = 0 - 10$. The leftmost curve is the spectrum for $n \leq 7$, and the remaining curves, from left to right, are the spectra for $n = 8, 9$ and 10. (b) Spectral densities at the same times (Miller & Roberts 1995).

gies (Eichler 1979; Zhou & Matthaeus 1990; Miller & Roberts 1995). An example of numerical simulations of proton acceleration employing an *MHD-turbulent cascade* to generate the initial wave spectrum with a quasi-linear code is shown in Fig. 11.13. Acceleration of protons to energies > 30 MeV on time scales of ≈ 1 s is found, consistent with gamma-ray observations. However, the detailed shape of the spectrum depends on the balance between the acceleration rate and escape rate, which requires a detailed trap-plus-precipitation model.

An attractive feature of gyroresonant stochastic acceleration is its ability to explain the enhanced ion abundances, which is not easy to accomplish with DC electric field acceleration. In the scenario of *turbulent MHD cascades*, where long-wavelength Alfvén waves cascade to shorter wavelengths, gyroresonant interactions are first enabled for the lowest gyrofrequencies of the heaviest ions such as iron (see mass dependence of ion gyrofrequency $\Omega_i = \Omega_H/\mu$ in Eq. 11.1.4), and proceed then to higher gyrofrequencies of the lighter ions (Si, Mg, Ne, O, N, C, ^{4}He, see atomic masses in Table 1.2). Such ion enhancements have been reproduced with quasi-linear codes that simultaneously solve all ion diffusion equations, and have been found to be consistent with the observed ion enhancements during impulsive and gradual flares (Table 11.3; Miller & Reames 1996). Some problems still remain with reproducing the observed ^{3}He/^{4}He abundance.

Table 11.3: Enhanced elemental abundances during flares (Miller & Reames 1996).

Element ratio	Impulsive flares	Gradual flares (corona)
^{3}He/^{4}He	≈ 1 ($\times 2000$ increase)	≈ 0.0005
^{4}He/O	≈ 46	≈ 55
C/O	≈ 0.436	≈ 0.471
N/O	≈ 0.153	≈ 0.128
Ne/O	≈ 0.414 ($\times 2.8$ increase)	≈ 0.151
Mg/O	≈ 0.413 ($\times 2.0$ increase)	≈ 0.203
Si/O	≈ 0.405 ($\times 2.6$ increase)	≈ 0.155
Fe/O	≈ 1.234 ($\times 8.0$ increase)	≈ 0.155
H/He	≈ 10	≈ 100

11.4.4 Acceleration by Electrostatic Wave-Particle Interactions

The previously discussed gyroresonant wave-particle interactions involve all *electro-magnetic waves*, which are defined as solutions of Maxwell's equations with both an oscillating electric field $E_1(t)$ and magnetic field $B_1(t)$ component, and include O, X, L, and R *(whistler) waves* for electrons, and *Alfvén waves* and *magneto-sonic waves* for ions (Table 11.2). Wave solutions with no oscillating magnetic field, $B_1(t) = 0$, are called *electrostatic waves*, which include *Langmuir waves* (plasma oscillations) and *upper-hybrid waves* for electrons, and *ion-sound (acoustic) waves*, *electrostatic ion-cyclotron waves*, and *lower-hybrid waves* for ions (Table 11.2). In the following we discuss stochastic acceleration of particles by such *electrostatic waves*.

Langmuir waves ($\mathbf{k} \parallel \mathbf{B}_0$) have been found to be very efficient for electron beams (Benz 1977; Melrose 1980b, § 8), because they have phase speeds in the range of $v_{Te} \ll \omega/k \lesssim c$ and thus can fulfil the *Cerenkov resonance condition* $\omega - \mathbf{k} \cdot \mathbf{v} = 0$ for nonthermal electrons. However, a major problem is the identification of a plausible source of primary Langmuir waves for this process. Since electron beams produce secondary Langmuir waves by themselves, another source of primary Langmuir waves is needed for acceleration. Nevertheless, the stochastic acceleration of electrons with velocities that are close to the phase speed of electrostatic (or electromagnetic) waves (i.e., $v \approx v_{ph} = \omega/k$), can efficiently resonate with the strong DC electric fields of the waves and in this way produce electron beam distributions that radiate electrostatic and electromagnetic emission, which is called *direct radiation* (Tajima et al. 1990; Wentzel 1991; Güdel & Wentzel 1993).

Takakura (1988) proposed that a strong field-aligned electric field is induced during the anomalous decay of force-free magnetic fields in the presence of *ion-sound waves* ($\mathbf{k} \parallel \mathbf{B}_0$), which could accelerate electrons to hard X-ray energies.

Other electrostatic waves are *lower-hybrid waves* (with $\mathbf{k} \perp \mathbf{B}_0$), which were found to be easily excited by two-stream instabilities between electron and ions (e.g., when ions drift across the magnetic field with a drift rate of a few ion thermal velocities v_{Ti}; Papadopoulos 1979; Tanaka & Papadopoulos 1983). The excited lower-hybrid waves can then stochastically accelerate electrons, but only to low energies of $\lesssim 6v_{Te}$. In addition, the efficiency is limited to $\lesssim 10^{-3}$, so it might be a viable acceleration mechanism for radio type II and III bursts (Lampe & Papadopoulos 1977; Vlahos et al. 1982; Benz & Smith 1987; Meister 1995). More powerful acceleration of electrons by lower-hybrid waves can be achieved with unstable ion ring distributions (McClements et al. 1990, 1993), which could be initiated by a quasi-perpendicular shock (Goodrich 1985) or by collisionless ion motions in a current sheet (Chen et al. 1990), but the shock requirement is considered as a strong restriction (Miller et al. 1997). In the Earth's magnetosphere, *lower-hybrid waves* were found to accelerate preferentially H^+ ions, while *hydrogen cyclotron waves* were found to prefer O^+ ions, thought to be responsible for acceleration of auroral ion beams (Varvoglis & Papadopoulos 1985).

11.4.5 General Evaluation of Stochastic Acceleration Models

In summary, the concept of stochastic acceleration accomplishes a number of advantages over the concept of (large-scale) DC electric field acceleration: (1) the accelerating fields occur on a microscopic scale and completely average out over a macroscopic volume, so that no return current problems occur, which require a strong filamentation of the acceleration region in DC electric field models; (2) arbitrary large energies (up to gamma-ray producing energies) can be obtained for stochastically accelerated particles, while sub-Dreicer electric fields cannot produce gamma-ray energies; and (3) gyroresonant particle interactions can naturally explain enhancements of heavy ions (Table 11.3), for which no explanation exists in DC electric field models. On the negative side, the major criticism of the stochastic acceleration model is mainly concerned with the existence of sufficiently strong wave turbulence sources and the efficiency of the turbulent MHD cascading, which are *ad hoc* assumptions that cannot easily be measured, and thus are arbitrarily tuned to reproduce the observed spectra.

11.5 Shock Acceleration

Shocks occur at discontinuous boundaries, such as in the spherical front of an expanding supernova remnant, in the bow shock of planetary magnetospheres that are circumvented by the solar wind, or at the front of coronal mass ejections (CMEs) that propagate into interplanetary space. Particle acceleration in such shock structures has been theoretically and observationally investigated and became a well-established astrophysical acceleration mechanism for high-energy particles. For the case of solar flares, evidence for particle acceleration in shock structures was mainly established from the observations of radio type II bursts, for which the shock speed could be measured from the plasma frequency and imaging observations. However, type II bursts only trace shock waves that propagate outward into interplanetary space. What is new about shock acceleration in the context of solar flares is that the conspicuous evidence for magnetic reconnection processes also implies the existence of standing fast-mode and slow-mode shocks in the reconnection outflow regions, which are believed now to play an important role for particle acceleration during solar flares.

Basic introductions into astrophysical shock waves can be found in textbooks (e.g., Priest 1982, § 5; Benz 1993, § 10; Kirk 1994; Burgess 1995, § 5; Baumjohann & Treumann 1997, § 8; Priest & Forbes 2000, § 13.3), Boyd & Sanderson (2003; § 10.5).

Shocks are waves with nonlinear amplitudes that propagate faster than the sound (or magneto-sonic) speed of the ambient medium. Shocks have been classified either by (1) the change in magnetic field direction from the upstream to the downstream region (*slow-mode, intermediate-mode, fast-mode*), (2) the particle velocity distribution (*collisional* or *collisionless shocks*), (3) the ion acceleration (*subcritical* or *supercritical shocks*), or (4) the driving agent (*blast wave* or *piston-driven wave*). Regarding particle acceleration in shock waves, the classical theories deal with single encounters of particles with shock waves (*first-order Fermi acceleration*, § 11.5.1 and § 11.5.2) or with multiple encounters (*diffusive shock acceleration*, § 11.5.3). In the context of solar flares, we discuss shock acceleration in *reconnection outflows* and *chromospheric evaporation fronts* (§ 11.5.4), as well as in shock waves that propagate in CMEs and interplanetary space (§ 11.5.5).

11.5.1 Fermi Acceleration

Fermi (1949) explained the acceleration of cosmic-ray particles by reflections on moving magnetized clouds. This idea is generally valid for charged particles that encounter a moving boundary with higher magnetic fields, because magnetic mirroring then produces a reflection for adiabatic particle motion. For the energy spectrum of cosmic-ray particles, Fermi (1949, 1954) derived a powerlaw function, based on statistical arguments that a head-on collision of a particle with a randomly moving magnetic field is more likely than an overtaking one, so that the average particle will be accelerated. He derived the spectrum from the law of conservation of momentum. If $\mathbf{c}_{sh}$ is the velocity of the shock structure (i.e., a magnetized cloud with high magnetic field that acts as magnetic mirror in Fermi's original model), then the change in particle energy $\Delta\varepsilon$ for one collision of a relativistic particle is (see e.g., Lang 1980, p. 474),

$$\Delta\varepsilon = -2\varepsilon\frac{\mathbf{c}_{sh}\cdot\mathbf{v}_{\parallel}}{c^2}\,, \qquad (11.5.1)$$

where ε is the particle energy and $\mathbf{v}_{\parallel}$ is the parallel velocity of the particle. Since the probability of a head-on collision is proportional to $v + c_{sh}$, and that for an overtaking one is proportional to $v - c_{sh}$, the average energy gain, $< \Delta\varepsilon >$, per collision is

$$< \Delta\varepsilon > \approx \frac{v + c_{sh}}{2v}\Delta\varepsilon - \frac{v - c_{sh}}{2v}\Delta\varepsilon \approx 2\frac{c_{sh}^2}{c^2}\varepsilon\,. \qquad (11.5.2)$$

Thus, if τ_{coll} is the mean time between multiple collisions with shock structures (or magnetized clouds), then the average rate of energy gain is

$$\frac{d\varepsilon}{dt} \approx \frac{2c_{sh}^2}{\tau_{coll}\,c^2}\varepsilon, \qquad (11.5.3)$$

which leads to an exponential growth in mean energy as a function of the acceleration time t_A,

$$\varepsilon(t_A) = \varepsilon_0 \exp\left(\frac{t_A}{\tau_G}\right), \qquad (11.5.4)$$

with an e-folding growth time τ_G of

$$\tau_G = \tau_{coll} \frac{c^2}{2\, c_{sh}^2} \,. \tag{11.5.5}$$

Assuming that such encounters of particles with cosmic clouds occur at random, is it natural to assume a random distribution for the total acceleration time t_A of a particle, which can be characterized by an exponential function for rare events (i.e., low probability events in the tail of a Gaussian normal distribution), see also Eq. (9.8.2),

$$N(t_A)dt_A = N_0 \exp\left(-\frac{t_A}{t_{Ae}}\right), \tag{11.5.6}$$

where t_{Ae} is the e-folding time constant of random acceleration times. Given these two mathematical properties, (1) an exponential growth in energy $\varepsilon(t_A)$ (Eq. 11.5.4), and (2) a random distribution $N(t_A)$ of (acceleration) time scales (Eq. 11.5.6), it follows immediately that the distribution $N(\varepsilon)$ of energies is a powerlaw function, according to the theory of frequency distributions described in § 9.8.1,

$$N(\varepsilon)d\varepsilon \propto \varepsilon^{-\delta}d\varepsilon\,, \qquad \delta = \left(1 + \frac{\tau_G}{t_{Ae}}\right)\,. \tag{11.5.7}$$

For cosmic ray spectra, powerlaw slopes of $\delta \approx 2$ are observed (e.g., see Lang 1980, p. 471−472), which correspond to the hardest spectra observed in solar flares. The model of first-order Fermi acceleration was also applied to the hydromagnetic bow shock in the Earth's magnetosphere, explaining the >30 keV electrons at a few Earth's radii beyond the bow shock (Fan et al. 1964; Jokipii 1966; Anderson et al. 1979). Fermi acceleration has been applied to solar flare loops that trap high-energetic protons (Bai et al. 1983), as well as to reconnection outflows in solar flares (Somov & Kosugi 1997; Tsuneta & Naito 1998), as described in more details below (§ 11.5.4).

11.5.2   Shock-Drift (or First-Order Fermi) Acceleration

Let us understand Fermi's acceleration mechanism in terms of particle kinematics. Most shocks in the solar corona have a sufficiently low density so that they are essentially collisionless during the passage of a particle, and thus adiabatic particle orbit theory can be applied. The normal component of the magnetic field is continuous across the shock front ($B_{d,norm} = B_{u,norm}$), while the tangential component varies, most strongly for *fast shocks* ($B_{d,tang} \gg B_{u,tang}$). Therefore, the total magnetic field strength increases across the shock front, $B_d \gg B_u$, and the particle gains perpendicular velocity ($v_{d,\perp} \gg v_{u,\perp}$) due to the conservation of the magnetic moment,

$$\mu = \frac{\frac{1}{2}mv_{u,\perp}^2}{B_u} = \frac{\frac{1}{2}mv_{d,\perp}^2}{B_d} = const\,, \tag{11.5.8}$$

which explains the definition of *slow shocks* and *fast shocks* (Fig. 11.14): In a fast shock, the magnetic field lines are closer together when bent away from the shock normal, which increases the magnetic field strength (due to conservation of the magnetic

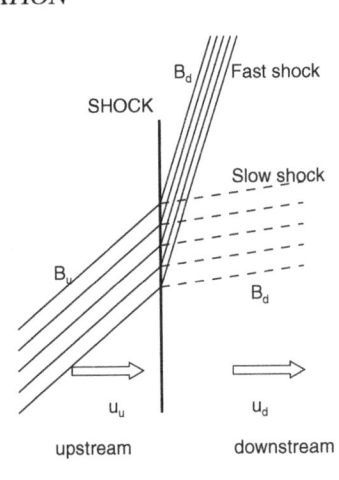

Figure 11.14: Definition of fast and slow shocks (left) and adiabatic orbit of a particle during shock passage (right). The magnetic field lines are closer together for a fast shock, indicating that the field strength increases when the field is bent away from the shock normal.

flux $\Phi = AB$), and thus increases the perpendicular particle velocity (due to conservation of particle moment, Eq. 11.5.8). In a slow shock, the downstream magnetic field is weaker.

The relations between the physical parameters (density, velocities, pressure, and magnetic field) on both sides of the shock front can be derived from the conservation of mass, momentum, energy, and magnetic flux, which are called *jump* or *Rankine−Hugoniot* relations, derived in many MHD textbooks (e.g., Priest 1982, § 5; Baumjohann & Treumann 1997, § 8; Boyd & Sanderson 2003, § 5.5).

A charged particle that crosses a shock front (Fig. 11.15, left) experiences a drift by an electric field $\mathbf{E}$ that is produced by the Lorentz force $\mathbf{c}_{sh} \times \mathbf{B}$ of the shock motion with velocity $\mathbf{c}_{sh}$ in a magnetic field $\mathbf{B}$,

$$\mathbf{E} = \frac{1}{c}\mathbf{c}_{sh} \times \mathbf{B} , \qquad (11.5.9)$$

and can be transformed away in a coordinate system that is co-moving with the drift speed perpendicular to the shock normal, so that the electric field $\mathbf{E}$ due to the fluid motion with velocity $\mathbf{c}_{sh}$ vanishes. This moving coordinate system is called *de Hoffmann−Teller (HT) frame* (Fig. 11.15, right). In the HT reference frame, the incident particle has a transformed velocity $\mathbf{u}' = \mathbf{u} - \mathbf{u}_{HT}$ that is parallel to the upstream magnetic field $\mathbf{B}_u$ by definition, so that $\mathbf{u}' \times \mathbf{B}_u = 0$ with no resulting electric field. To fulfil this condition, the transformation velocity $\mathbf{v}_{HT}$ of the HT reference system therefore has to be,

$$\mathbf{v}_{HT} = \frac{\mathbf{n} \times (\mathbf{u}_u \times \mathbf{B}_u)}{\mathbf{n} \cdot \mathbf{B}_u} = u_u \tan\theta_{Bn} , \qquad (11.5.10)$$

where θ_{Bn} is the angle between the upstream magnetic field direction $\mathbf{B}_u$ and the shock normal direction $\mathbf{n}$.

The acceleration of a particle in a shock interaction is easiest described in the HT coordinate system. Let us consider a particle that collides with a shock front and is

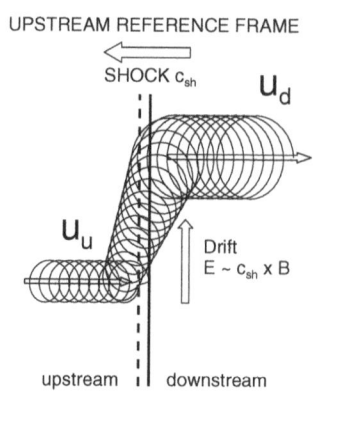
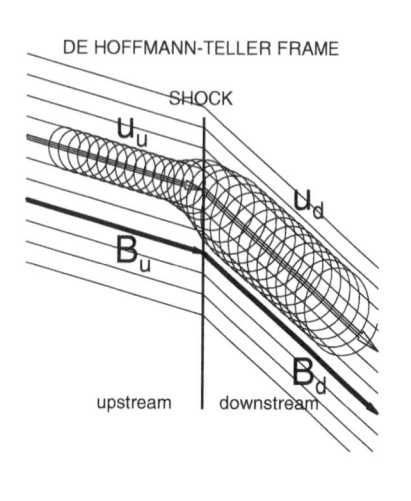

Figure 11.15: The passage of a particle through a shock front is shown in the *upstream reference frame* (at rest to the observer), where a charged particle experiences a drift along the shock front due to the electric field $\mathbf{E} = -(\mathbf{c}_{sh}/c) \times \mathbf{B}$ induced by the motion of the shock plasma with velocity c_{sh} (left). This electric field is transformed away in the *De Hoffmann–Teller frame* (co-moving with the shock and drifting with $\mathbf{v}_{HT}$ parallel to the shock front, where the incident particle motion is by definition parallel to the incident magnetic field, $\mathbf{u}_u \parallel \mathbf{B}_u$ (right).

reflected, which are the particles that gain more energy than the overpassing ones in the Fermi model. In the HT reference frame, the incident particle has a velocity $\mathbf{u}_\parallel$ that is aligned with the magnetic field $\mathbf{B}_u$, and the reflected particle has a collinear velocity $\mathbf{v}_\parallel$, which would be equal for an adiabatic reflection mechanism, or lower if the magnetic moment is not conserved. We may quantify the relative ratio with the (positive) constant α (Boyd & Sanderson 2003, § 10.5.3),

$$\mathbf{v}_\parallel = -\alpha \mathbf{u}_\parallel \ . \tag{11.5.11}$$

The corresponding velocities in the upstream rest frame have the additional component of the HT transformation velocity $\mathbf{v}_{HT}$,

$$\mathbf{u} = \mathbf{u}_\parallel + \mathbf{v}_{HT} \ , \tag{11.5.12}$$

$$\mathbf{v} = \mathbf{v}_\parallel + \mathbf{v}_{HT} \ , \tag{11.5.13}$$

The difference between incident and reflected (squared) velocities is then (using Eqs. 11.5.11−13)

$$\begin{aligned}
u^2 - v^2 &= (\mathbf{u}_\parallel + \mathbf{v}_{HT})^2 - (\mathbf{v}_\parallel + \mathbf{v}_{HT})^2 \\
&= u_\parallel^2 - v_\parallel^2 + 2v_{HT}(\mathbf{u}_\parallel - \mathbf{v}_\parallel) \\
&= u_\parallel^2 - v_\parallel^2 + 2(\mathbf{u} - \mathbf{u}_\parallel)\mathbf{u}_\parallel(1 - \alpha) \ .
\end{aligned} \tag{11.5.14}$$

The ratio of the reflected to the incident kinetic energy is thus in the rest frame of the shock,

$$\frac{v^2}{u^2} = 1 + (1+\alpha)^2 \frac{u_\parallel}{u^2} - 2(1+\alpha)\frac{\mathbf{u}_\parallel \cdot \mathbf{u}}{u^2} \ , \tag{11.5.15}$$

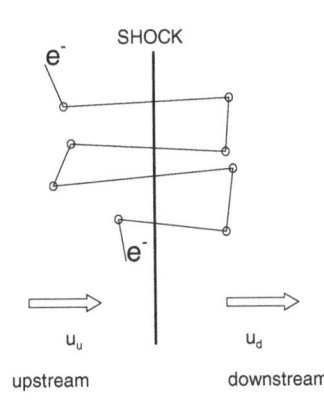

Figure 11.16: Particle orbit during diffusive shock acceleration. The electron crosses the shock front multiple times, scattered by wave turbulence, entering it from the upstream direction.

which can be expressed by the angles θ_{Bn}, θ_{un}, and θ_{vn} between $\mathbf{B}$, $\mathbf{u}$, $\mathbf{v}$ and the shock normal $\mathbf{n}$, which obey $u_\| \cos\theta_{Bn} = u\cos\theta_{vn}$ and $\mathbf{u}_\| \cdot \mathbf{u} = u_\| u \cos\theta_{Bn}$, yielding

$$\frac{v^2}{u^2} = 1 + (1+\alpha)^2 \frac{\cos^2\theta_{vn}}{\cos^2\theta_{Bn}} - 2(1+\alpha)\frac{\cos\theta_{Bv}\cos\theta_{vn}}{\cos\theta_{Bn}} . \qquad (11.5.16)$$

This expression demonstrates, that for quasi-perpendicular shocks (i.e., $\theta_{Bn} \lesssim 90°$ or $\cos\theta_{Bn} \gtrsim 0$) the second term becomes dominant and leads to large increases of the kinetic energy of the reflected particles ($v^2 \gg u^2$). This confirms that slow-mode shocks or quasi-parallel shocks yield little amount of acceleration, while fast-mode shocks or quasi-perpendicular shocks lead to efficient acceleration.

   If electrons pass the shock front only in a single encounter, the energy gain is limited to the downstream/upstream ratio B_1/B_2 of the magnetic field strengths, which is typically a factor of 4. Higher energies can be achieved if the magnetic field has a trapping region upstream of the shock, so that particles are mirrored multiple times at the shock front and gain energy each time. When the shock propagation becomes near-perpendicular ($1° - 2°$) to the upstream magnetic field (fast shock), acceleration becomes most efficient (Wu 1984; Krauss−Varban & Wu 1989), but on the other hand, this small angle restriction limits the high acceleration efficiency only to $\approx 1\%$ of the electrons. Given these restrictions, shock-drift acceleration was mainly applied to the Earth's bow shock (Jokipii 1966; Krauss−Varban & Burgess, 1991) and to radio type II bursts (Holman & Pesses 1983; Melrose & Dulk 1987; Benz & Thejappa 1988). If particles are preaccelerated and injected into a shock with an energy of 100 keV, however, electron and protons can be accelerated up to γ-ray energies of 100 MeV by first-order Fermi acceleration (Ellison & Ramaty 1985).

11.5.3 Diffusive Shock Acceleration

The basic problem of first-order Fermi acceleration, the limited energy gain during a single shock encounter, can be overcome in inhomogeneous plasmas, where particles are scattered many times back and forth across the shock front, so that they experience a cumulative acceleration effect during multiple shock encounters (Fig. 11.16). The original concept of Fermi is based on particles that encounter a cloud of scattering centers moving in random directions, where colliding particles gain more energy than they lose on average, a statistical effect. A similar approach with a spatial random distribution of multiple shock fronts was also formulated by Anastasiadis & Vlahos (1991, 1994). In solar flares, wave turbulence or low-frequency waves are used as a means to randomize anisotropic acceleration, which can cause particle scattering in pitch angle and energy. The mechanism of *diffusive shock acceleration* invokes efficient particle scattering in the regime of *strong wave turbulence*, while *stochastic acceleration* based on gyroresonant wave-particle interactions (described in § 11.4) works in the regime of *weak wave turbulence*.

Mathematically, diffusive shock acceleration has much in common with stochastic acceleration by gyroresonant wave-particle interactions (Jones 1994). Particles encounter multiple traversals of shock fronts and pick up each time an increment of momentum that is proportional to its momentum. Therefore, after N shock crossings, a particle will have a momentum of

$$p(N) = p_0 \, \Pi_{i=1}^N \left(1 + <\delta p/p>_i\right) , \qquad (11.5.17)$$

where the term of the momentum increase $<\delta p/p>$ is averaged over the particle flux. Each cycle includes a back and forth-crossing of a shock front. During each cycle there is a small probability ϵ_i of escaping from the shock. So the probability $P(N)$ that a particle undergoes N cycles is,

$$P(N) = \Pi_{i=1}^N (1 - \epsilon_i) . \qquad (11.5.18)$$

Taking the ratio of the logarithms of the two equations (Eqs. 11.5.17−18) we obtain

$$\frac{\ln P(N)}{\ln[p(N)/p_0]} = \frac{\sum_{i=1}^N (1 - \epsilon_i)}{\sum_{i=1}^N (1 + <\delta p/p>_i)} \approx \frac{-\sum_{i=1}^N \epsilon_i}{\sum_{i=1}^N <\delta p/p>_i} = \Gamma(N) , \quad (11.5.19)$$

which leads to a powerlaw function for the integral spectrum, which is the probability that a particle has a momentum $\geq p$,

$$P(p) = \left(\frac{p}{p_0}\right)^{-\Gamma(N)} , \qquad (11.5.20)$$

where the powerlaw exponent $\Gamma(N)$ could depend on the number of crossings N. Defining the time interval τ_i for a shock-crossing cycle, the incremental momentum gain is,

$$\frac{\sum_{i=1}^N <\delta p/p>_i}{\sum_{i=1}^N \tau_i} = \frac{1}{p}\frac{dp}{dt} = \alpha(N) , \qquad (11.5.21)$$

while the probability of escape per cycle is

$$\frac{\sum_{i=1}^N \epsilon_i}{\sum_{i=1}^N \tau_i} = \frac{1}{T(N)} , \qquad (11.5.22),$$

which yields the powerlaw exponent

$$\Gamma(N) = \frac{1}{\alpha(N)T(N)} \qquad (11.5.23)$$

and thus a powerlaw spectrum for the particle momentum,

$$f(p) \propto \left(\frac{p}{p_0}\right)^{-1+\Gamma(N)} . \qquad (11.5.24)$$

The spectrum is only a true powerlaw function when $\Gamma(N)$ or the product $\alpha(N)T(N)$ is independent of the number of crossings N. It has been shown that for a particle bound to a flowing plasma by scattering (Parker 1963b; Gleeson & Axford 1967) or by electromagnetic forces (Jones 1990), that the probability of escaping from a shock depends only on the compression ratio u_d/u_u of the shock (Jones 1994),

$$\Gamma(N) = \frac{3}{u_d/u_u - 1} , \qquad (11.5.25)$$

and thus is independent of the number of crossings. In strong shocks as they occur in astrophysical objects, the compression ratio is bounded by $u_d/u_u \leq 4$, so that the powerlaw exponent is $\Gamma \geq 1$, but can be close to $\Gamma \gtrsim 1$ as often observed. The fact that the particle spectrum is a powerlaw spectrum (Eq. 11.5.24) shows us that the process of *diffusive shock acceleration* is very similar to *stochastic acceleration* or *second-order Fermi acceleration*, where we also derived a powerlaw function for the energy spectrum (Eq. 11.5.7).

Note that the original Fermi acceleration mechanism is based on a statistical argument of many random collisions with magnetic clouds, which leads to a relative momentum gain proportional to the second power of the (small) velocity ratio $< \delta p/p > \propto (c_{sh}/v)^2$ (Eq. 11.5.2), and is therefore called a *second-order Fermi* type, while *shock-drift acceleration* yields only a momentum gain that is proportional to the velocity ratio, $< \delta p/p > \propto (c_{sh}/v)$ (Eq. 11.5.11), and is therefore called a *first-order Fermi* type.

The cursory derivation above summarizes only the statistical argument that leads to the spectrum of accelerated particles, but ignores the detailed equations of the conservation of particle momentum and energy. The general process of diffusive shock acceleration has to be described by the evolution of the particle distribution function $f(\mathbf{x}, \mathbf{v}, t)$ in terms of the *diffusion convection equation* (e.g. Parker, 1965; Priest & Forbes, 2000, § 13.3),

$$\frac{\partial f}{\partial t} + \mathbf{u} \cdot \nabla f = \nabla \cdot (\kappa \nabla f) + \frac{1}{3}(\nabla \cdot \mathbf{u}) \, p \frac{\partial f}{\partial p} + \frac{1}{p_i^2} \frac{\partial}{\partial p_i} \left(p_j^2 D_{ij} \frac{\partial f}{\partial p_j}\right) + I - L , \quad (11.5.26)$$

where the terms describe the time dependence of the particle distribution f, spatial advection (∇f), spatial diffusion (κ), adiabatic expansion or compression (at the shock front), momentum diffusion (D_{ij}), particle injection (I), and escape or loss (L).

Some first energy spectra of ions accelerated in diffusive shocks were calculated by Ramaty (1979) which could reproduce observed gamma-ray spectra and interplanetary particle spectra, and was found to be a viable mechanism to accelerate ions up

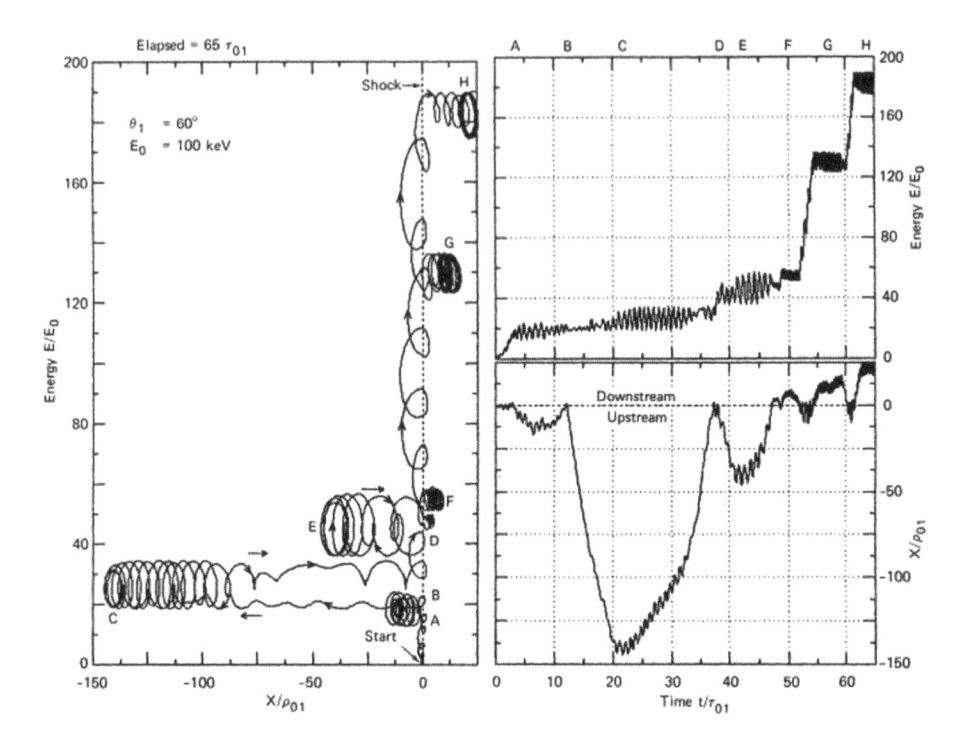

Figure 11.17: Particle orbit of a particle that undergoes *diffusive shock acceleration* in a quasi-perpendicular shock with $\theta = 60°$. *Left:* energy versus X-orbit in shock frame, where X is the distance from the shock, with $X < 0$ upstream and $X > 0$ downstream. The elapsed time of orbit is 270 gyroperiods. *Right:* time history or energy gain and X-orbit (Decker & Vlahos 1986).

to $\approx 100\,\text{MeV}$ within $\lesssim 1$ s. Wave-particle interactions in diffusive shocks (e.g., scattering by whistlers and Alfvén waves), was included by Achterberg & Norman (1980), Decker & Vlahos (1986), and Decker (1988). An example of the history of a particle acceleration in a quasi-perpendicular shock calculated by Decker & Vlahos (1986) is shown in Fig. 11.17. However, an important problem that was identified was the preacceleration to $\varepsilon \approx 20$ keV for electrons, requiring another mechanism for *first-phase acceleration*, while protons and ions can directly be accelerated out of the Maxwellian distribution. Diffusive-shock acceleration has been simulated numerically, mostly for cosmic rays (e.g., Kang & Jones 1995; Baring et al. 1994). For solar flares, the simulations confirmed that ions can promptly be accelerated to gamma-ray energies (Cargill et al. 1988), with an electron/proton ratio of $\approx 1\% - 10\%$ at 1 GeV (Levinson 1994), and that shock waves can form electron streams capable of radio type II emission (Lemberge 1995).

11.5.4 Shock Acceleration in Coronal Flare Sites

An early application of shock acceleration to solar flares has been proposed by Bai et al. (1983). They used the concept of the original Fermi acceleration model to explain the gamma-ray emission, which was found to be slightly delayed with respect to the hard X-ray emission, a phase during a solar flare that was called *second-step accelera-tion*. The basic idea was that two shock fronts would propagate upward from the two flare loop footpoints, trapping energetic protons in between, which then gain energy at every reflection between the converging mirrors. The average time interval τ_{coll} be-tween two successive collisions with shock fronts is the proton travel time between the two mirrors, $\tau_{coll} = d/v_{p,\parallel}$. Bai et al. (1983) argued that the interpretation in terms of Fermi acceleration is supported by the facts that: (1) the protons gain more energy than electrons at the same speed, (2) that the gamma-ray emission from energetic protons (with the higher energy) is delayed to the hard X-ray emission from electrons, and (3) that interplanetary protons are associated with those events with a gamma-ray delay. These characteristics are all consistent with a Fermi acceleration model, but the ob-served gamma-ray delay Δt of a few seconds requires extremely small flare loop sizes $d \lesssim c_{sh}\Delta t$, and it is also not clear how the trapped high-energy protons can escape into interplanetary space.

During the era of Yohkoh observations when magnetic reconnection models be-came popular, first-order Fermi acceleration has been applied to the slow-mode and fast-mode shocks in the reconnection outflows (Somov & Kosugi 1997; Tsuneta & Naito 1998). Essentially, it is assumed that particles are trapped in the reconnection outflow between the coronal reconnection region and the standing fast shock wave above the soft X-ray flare loop (Fig. 11.18). Since the newly reconnected field lines are relaxing, the cusp region represents a collapsing trap, where particles are mirrored between the standing shock beneath and the converging cusp region above. During every reflection at the standing shock, the trapped particles are accelerated by the first-order Fermi acceleration mechanism at every bounce time, and thus represents a very efficient acceleration mechanism. Applying Fermi's relation (Eq. 11.5.5) we obtain an estimate of the e-folding energization growth time τ_G, using the collisional deflection time $\tau_{coll} \approx \tau_{defl}(E) \propto E^{3/2}/n_e$ and a magnetic mirror speed of $u = v_A/\cos\theta$ with $v_A \approx 1000$ km s^{-1} as the Alfvén speed, where θ is the angle of the magnetic mirror to the magnetic field,

$$\tau_G = \tau_{coll}\frac{c^2}{2c_{sh}^2} \approx \left(\frac{\varepsilon}{100 \text{ keV}}\right)^{3/2}\left(\frac{n_E}{10^{11} \text{ cm}^{-3}}\right)^{-1}\left(\frac{c^2\cos^2\theta}{2v_A^2}\right) \quad (s) . \quad (11.5.27)$$

We see that sufficiently short acceleration times can only be achieved in fast-mode shocks, where $\cos\theta \ll 1$. To accelerate an electron from a suprathermal energy of $\varepsilon_1 \approx 3$ keV to a typical hard X-ray emitting energy of $\varepsilon = 100$ keV, an acceleration time of $t = t_G \ln(\varepsilon/\varepsilon_1) \approx \tau_G \ln(100/3) = 3.5\tau_G$ is needed (according to Eq. 11.5.4). The observed durations of hard X-ray pulses set an upper limit of $t \lesssim 1$ s on the acceleration time, or $t_G \lesssim 0.3$ s on the growth time, requiring a fast shock angle of $\theta \lesssim 87°$. This scenario, in terms of first-order Fermi acceleration, implies that the fast-mode standing shocks above the soft X-ray flare loops, which provide a highly perpendicular shock structure, are the most efficient locations for particle acceleration (Fig. 11.19), while the

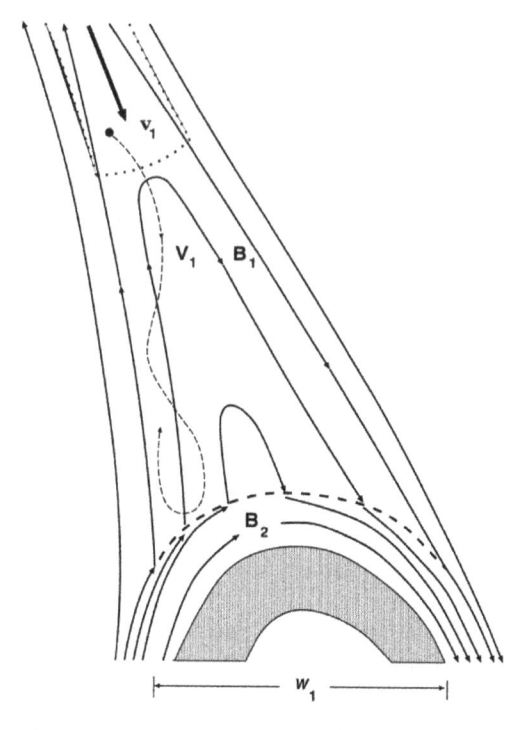

Figure 11.18: Magnetic trap between the *High-Temperature Turbulent-Current Sheet (HTTCS)* and the shock front in the downward-directed outflow of a reconnection region. Accelerated particles move with velocity v_1 along the field lines, are reflected at the fast shock above the soft X-ray flare loop (grey), and experience first-order Fermi acceleration (Somov & Kosugi 1997).

slow-mode shocks that "sandwich" the reconnection outflow may serve as a mechanism for preacceleration.

The model of first-order Fermi acceleration is viable for $10 - 100$ keV electrons if the following four observational requirements are met (Tsuneta & Naito, 1998): (1) The net acceleration rate has to overcome the (relatively high) collisional loss rate,

$$\frac{d\varepsilon}{dt} = \left(\frac{d\varepsilon}{dt}\right)_{acc} - \left(\frac{d\varepsilon}{dt}\right)_{coll} > 0 , \qquad (11.5.28)$$

(2) the energy gain has to be sufficiently high to explain the ≈ 50 keV hard X-ray emission of Masuda's above-the-looptop sources, (3) the acceleration time has to be sufficiently fast ($\lesssim 1$ s) to explain the impulsive hard X-ray bursts, and (4) the number of accelerated electrons has to meet the hard X-ray inferred electron injection rates of $\approx 10^{34} - 10^{35}$ electrons s^{-1}. The acceleration rate in the fast shock was estimated to

$$\left(\frac{d\varepsilon}{dt}\right)_{acc} = \frac{\Delta\varepsilon}{\Delta t} = \frac{2}{3}\varepsilon\frac{u}{l\cos\theta} , \qquad (11.5.29)$$

where $u \approx v_A \approx 1000$ km s^{-1} (Tsuneta 1996a) is the speed of the outflow from the reconnection region, $l \approx 500$ km is the estimated diffusion length, and θ the angle be-

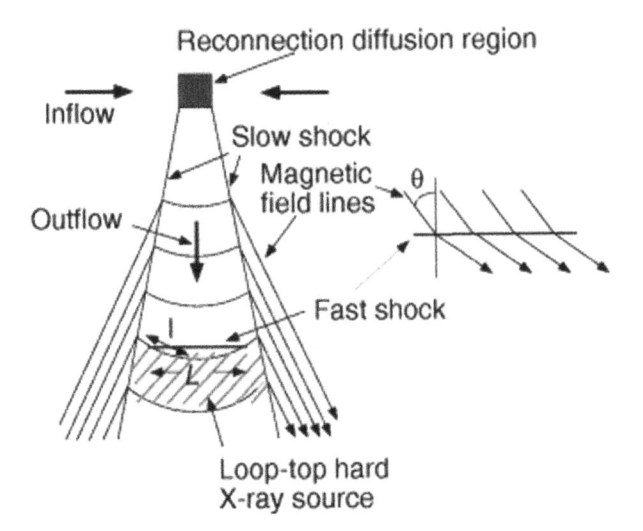

Figure 11.19: Magnetic field line configuration of the reconnection region. An Alfvénic down-ward outflow is sandwiched by the two steady slow shocks. A fast shock with length L forms between the slow shocks. Magnetic disturbances both upstream and downstream of the fast shock scatter the electrons, which become accelerated. The total length of the diffusion region along the field lines is l (Tsuneta & Naito 1998).

tween the fast-shock normal and the magnetic field line crossing the shock (Fig. 11.19). The net acceleration rate (Eq. 11.5.28), after subtraction of the collisional losses,

$$\left(\frac{d\varepsilon}{dt}\right)_{coll} \approx 47 \frac{n_{10}}{\sqrt{\varepsilon_{keV}}} , \tag{11.5.30}$$

is shown in Fig. 11.20, for shock angles of $\theta = 0°, 60°$, and $85°$. The diagram shows that the net acceleration exceeds the collisional loss at energies of $\varepsilon \gtrsim 4$ keV for shock angles of $\theta = 85°$. So, if a bulk energization mechanism exists that preaccelerates electrons out of the thermal distribution ($T \approx 10 - 20$ MK, $\varepsilon \approx 0.5 - 1$ keV) to $\varepsilon_0 \gtrsim 4$ keV, first-order Fermi acceleration will accelerate them up to ≈ 1 MeV energies at the fast shock. Tsuneta & Naito (1998) estimate that the preacceleration could be provided by the slow shocks. A diffusion length of $l \approx 500$ km is needed at a fast shock angle of $\theta \approx 85°$ to satisfy the maximum energies ($\varepsilon \lesssim 1$ MeV), for acceleration time scales of $t_{acc} \approx 0.3 - 0.6$ s and a number of accelerated electrons of $N_{acc} = n_e u L^2 \approx 5 \times 10^{35}$ electrons s^{-1}. They point out that this scenario considerably ameliorates the injection problem of earlier first and second-order Fermi acceleration scenarios, where relatively high initial energies ($\varepsilon_0 \approx 20 - 100$ keV) were required (Bai et al. 1983; Ramaty 1979).

Shock acceleration in the context of magnetic reconnection regions in solar flares has also been studied for electrons and ions separately (Sakai & Ohsawa 1987; Ohsawa & Sakai 1987, 1988a,b). Besides the fast-mode shocks, reconnection outflows also produce strong wave turbulence (e.g., of whistler waves), which can accelerate particles by the first-order Fermi mechanism (LaRosa & Moore 1993; LaRosa et al. 1994). The magnetic reconnection outflows propagate in the form of thermal conduction fronts

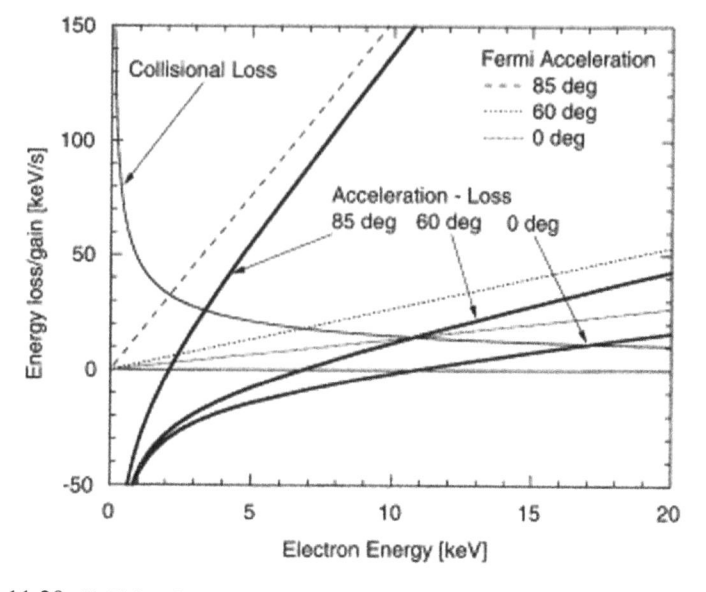

Figure 11.20: Collisional energy loss rate and Fermi acceleration rate for three different shock angles. The net energy gain rate (thick lines) is the energy gain rate (dotted and dashed lines) minus the loss rate (thin line) (Tsuneta & Naito 1998).

down to the chromosphere, which generate ion-acoustic turbulence, ambipolar electric fields (McKean et al. 1990a,b), and return currents (Karlický 1993), which also can contribute to suitable conditions for first-order and diffusive Fermi acceleration.

11.5.5   Shock Acceleration in CMEs and Type II Bursts

Since shock waves are ubiquitous in the flare process, particle acceleration by shocks may occur in a variety of associated transient phenomena, such as in reconnection out-flows, flare-initiated *EIT* and Moreton waves, coronal mass ejections, filament erup-tions, radio type II bursts, interplanetary shocks, etc. Some reviews on coronal shock acceleration can be found in Nelson & Dulk (1985, § 13), Benz (1993, § 10.2), Mann (1995), or Priest & Forbes (2000, § 13.4.2).

The first evidence of shock acceleration in the solar corona came from *radio type II bursts*, which were identified as shock structures from imaging observations of their outward motion with speeds of $c_{sh} \approx 200 - 2000$ (km s^{-1}). The frequency-time (dynamic) spectrum of a type II burst generally shows a central ridge that drifts with the shock speed (called *backbone*) and bifurcating fast-drift structures (called *herring-bones*) that indicate electron beams that propagate in the forward and backward direc-tion of the shock front. Thus, both the accelerator (shock wave) as well as the accel-erated particles (by plasma emission of electron beams) are identified. The backbone usually shows a split-band structure with a near-harmonic frequency ratio, indicative of plasma emission at the fundamental and harmonic frequency, probably produced by electrostatic waves from electrons with a shifted ring distribution resulting from a first-

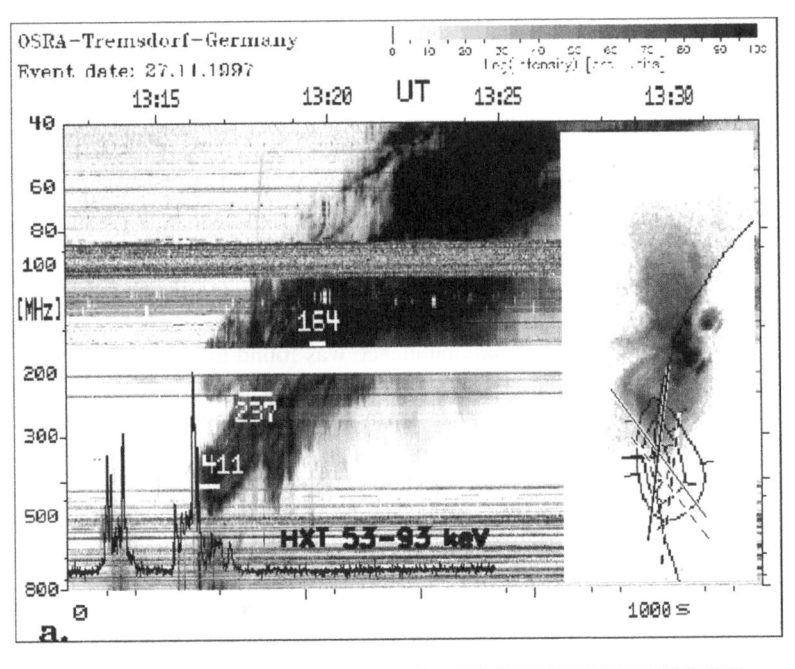

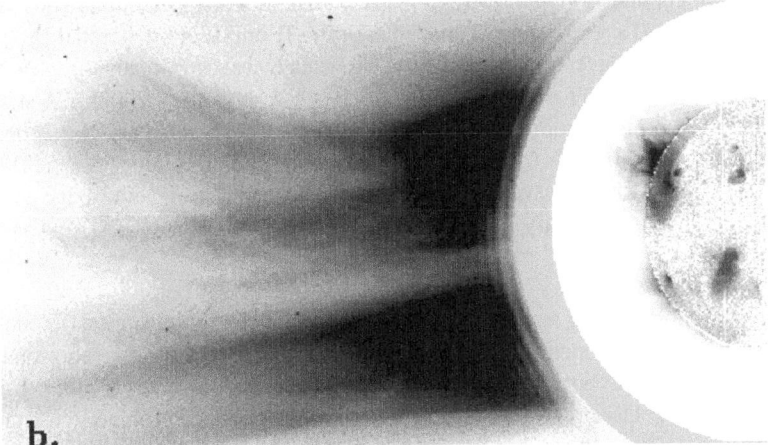

Figure 11.21: (a) Dynamic spectrogram (Tremsdorf Radio spectrograph) and hard X-ray lightcurve (Yohkoh HXT) during the 1997-Nov-27 type II event. Insert to (a): Yohkoh SXT negative image of AR 8113 (Al.1 filter) at 12:40:36 UT and harmonic type II source during the time intervals indicated by horizontal bars at the appropriate frequencies in the spectrogram: contours of equal brightness at 411 MHz (10, 50, 90% of maximum brightness in the image) and source centroid positions and half widths of the harmonic type II sources at 237 (dashed) and 164 MHz (solid). (b) Yohkoh SXT image of (a), superposed on SoHO/LASCO-C1 and C2 images at 1328 and 1330 UT (negative image). Note that the shock waves that produce radio type II emission are imaged in the lower corona very close to the flare site (top right), while the CME bowshock is about 5 solar radii away (bottom left), and thus cannot be responsible for the type II emission (Klein et al. 1999b).

order Fermi acceleration upstream (Benz & Thejappa 1988). From the bandwidth of the backbone, the density compression and Mach number of the shock can be computed (Mann et al. 1995). The occasional absence of a backbone structure was explained by the lack of reflected particles in almost perpendicular shocks (Holman & Pesses 1983). The propagation of the shock front could be monitored through interplanetary space, and the onset of geomagnetic storms on Earth was found to coincide with the arrival time of shock fronts (e.g., Malitson et al. 1973; Maxwell & Rinehart 1974). Meanwhile, it was identified that the driver of interplanetary shocks are *coronal mass ejections (CME)* (Hundhausen 1972). However, the detailed interplay between CMEs and shock fronts spawned an ongoing debate, because the exact timing, association, point of origin, geometry, and elemental abundance was found to be inconsistent in many cases (Cane et al. 1981, 1986, 1991, 2000, 2002; Cane 1984, 1988; Cane & Reames 1988a,b, 1990; MacDowall et al. 1987; Reames 1990a; Kundu et al. 1990; Reames et al. 1991a,b, 1999; Klein & Trottet 1994; Klein et al. 1999a). Cane (1988) estimated that the percentage of shocks detected at 1 AU that can be tracked back to originate in solar flares is less than 50%. The origin of type II bursts is often believed to be tied to the bow shock of the CME, but detailed imaging observations suggest that some originate directly in the reconnection sites of flares (Fig. 11.21), well below the CME bowshock (Klein et al. 1999b; Pick 1999; Klein & Trottet 2001; Klein et al. 2001). Stationary type II burst emission has been interpreted as a radio signature of the standing shock in reconnection outflow regions (Aurass et al. 2002b). Comparisons of CME speeds and type II speeds suggest that many type II bursts are generated by (piston-driven) flare blast waves rather than by CME shock waves (Gopalswamy & Kundu 1991), or by coronal EIT waves (Mann et al. 1999; Aurass et al. 2002b). A statistical study of CME images from LASCO and interplanetary type II bursts recorded with the WIND spacecraft leads to a distinction between three originators of type II bursts: flare-related blast shock waves (30%), shocks driven by the leading edge of CMEs (30%), and shocks driven by internal parts or flanks of the CMEs (30%) (Classen & Aurass 2002).

Theoretical modeling was performed by numerical simulations of fast-mode MHD shock waves in interplanetary space that attempted to match the observed propagation speed of type II radio bursts (Dryer & Maxwell 1979; Smith & Dryer 1990; Karlický & Odstrčil 1994; Dryer et al. 2001). *In situ* measurements of proton pitch angle distributions near shock fronts that propagate in interplanetary space (e.g., at *co-rotating interaction regions*, the interface between the slow and fast solar wind) have been modeled with numerical simulations and theoretical models and were found to be consistent with first-order Fermi acceleration in the electric field of the shock front (Pesses 1979; Mann et al. 2002). The intensity time profiles of protons accelerated in CME-driven shocks, however, have quite different characteristics eastward and westward of the shock location, as mapped with multi-spacecraft observations (Reames et al. 1996). It was argued that shock-drift and diffusive shock acceleration are inefficient in slow CMEs (Kundu et al. 1989b). From the analysis of *solar energetic particle (SEP)* events, Kahler & Reames (2003) conclude that fast CMEs (> 900 km s^{-1}) that propagate in slow solar-wind regions produce stronger shocks than in fast solar-wind regions, because they produce a higher compression ratio.

11.6 Summary

Particle acceleration in solar flares is mostly explored by theoretical models, because neither macroscopic nor microscopic electric fields are directly measurable by remote-sensing methods. The motion of particles can be described in terms of acceleration by parallel electric fields, drift velocities caused by perpendicular forces (i.e., $E \times B$-drifts), and gyromotion caused by the Lorentz force of the magnetic field (§ 11.1). Theoretical models of particle acceleration in solar flares can be broken down into three groups (§ 11.2): (1) DC electric field acceleration, (2) stochastic or second-order Fermi acceleration, and (3) shock acceleration. In the models of the first group, there is a paradigm shift from large-scale DC electric fields (of the size of flare loops) to small-scale electric fields (of the size of magnetic islands produced by the tearing mode instability). The acceleration and trajectories of particles is studied more realistically in the inhomogeneous and time-varying electromagnetic fields around magnetic X-points and O-points of magnetic reconnection sites, rather than in static, homogeneous, large-scale Parker-type current sheets (§ 11.3). The second group of models entails stochastic acceleration by gyroresonant wave-particle interactions, which can be driven by a variety of electrostatic and electromagnetic waves, supposed that wave turbulence is present at a sufficiently enhanced level and that the MHD turbulence cascading process is at work (§ 11.4). The third group of acceleration models includes a rich variety of shock acceleration models, which is extensively explored in magnetospheric physics and could cross-fertilize solar flare models. Two major groups of models are studied in the context of solar flares (i.e., first-order Fermi acceleration or shock-drift acceleration, § 11.5.2, and diffusive shock acceleration, § 11.5.3). New aspects are that shock acceleration is now applied to the outflow regions of coronal magnetic reconnection sites, where first-order Fermi acceleration at the standing fast shock is a leading candidate (§ 11.5.4). Traditionally, evidence for shock acceleration in solar flares came mainly from radio type II bursts. New trends in this area are the distinction of different acceleration sites that produce type II emission: flare blast waves, the leading edge of CMEs (bow-shock), and shocks in internal and lateral parts of CMEs (§ 11.5.5). In summary we can say that (1) all three basic acceleration mechanisms seem to play a role to a variable degree in some parts of solar flares and CMEs, (2) the distinction between the three basic models become more blurred in more realistic models, and (3) the relative importance and efficiency of various acceleration models can only be assessed by including a realistic description of the electromagnetic fields, kinetic particle distributions, and MHD evolution of magnetic reconnection regions pertinent to solar flares.

Chapter 12

Particle Kinematics

Particle acceleration, a universal theme in high-energy astrophysics, remains a black box in all astronomical observations, even in solar flares, from our nearest astrophysical laboratory. We are just now starting to understand the complicated electromagnetic field dynamics in magnetic reconnection regions, which probably provide the most prolific sources of accelerated particles, but we still have no reliable method to map out the relevant electric and magnetic fields. We do not know the field strengths nor the exact locations of the accelerating fields. In laboratory accelerators, electromagnetic field strengths, the geometry of the fields, and particle trajectories are usually known with high precision by design. Laboratory measurements concentrate mainly on the kinematic reconstruction of collisional products of accelerated particle beams that hit a target, which sometimes leads to discoveries of new particles, whose identities are constrained by the kinematics of their trajectories, momentum, energy, and parity conservation laws. In solar flares, we observe analogous collision experiments, where electrons and ions are accelerated in a coronal, collisionless magnetic reconnection region, which then propagate in the low-β plasma along magnetic field lines to chromospheric footpoints, corresponding to the targets in laboratory accelerators. Progress has been made over recent years to measure the small time-of-flight differences of energized electrons between the coronal acceleration site and the chromospheric target. Because the time-of-flight is a direct function of the flight distance and the velocity (or kinetic energy) of the electrons, we have a powerful new diagnostic on the so far unlocalized particle accelerators. Because the tiny time-of-flight differences are only in the order of $\approx 10 - 100$ ms, high-precision timing measurements have to be conducted which generally require a photon statistics of $\gtrsim 10^4$ photons per second. This requirement calls for large-area detectors with high sensitivity, which were available in the *Burst and Transient Source Experiment (BATSE)* on the *Compton Gamma-Ray Observatory (CGRO)*, which operated between 1991 and 2000. However, the photon count rate in large flares is sufficiently high that also smaller detectors, such as *HXRBS/SMM* or *RHESSI*, can accomplish kinematic measurements. In the following we lay out the fundamentals of particle kinematics applied to solar flares, including the propagation processes during acceleration, free-streaming, injection, trapping, and precipitation. A review on this novel topic can be found in Aschwanden (2002b).

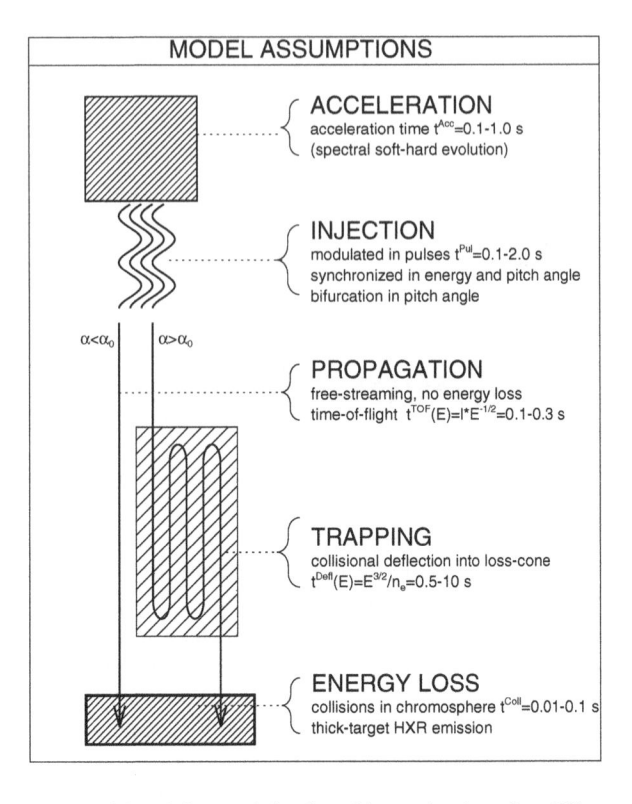

Figure 12.1: Conceptual breakdown of the flare kinematics into five different physical processes, described in this section (Aschwanden 1998a).

12.1   Overview on Particle Kinematics

In solar flares, the kinematics of nonthermal particles can generally be characterized by at least five different physical processes, as sketched in Fig. 12.1: (1) acceleration, (2) injection, (3) free-streaming propagation, (4) trapping, and (5) precipitation and energy loss, which we will quantify in turn in the following subsections. The acceleration phase is defined by the time interval in which a thermal particle gains velocity and kinetic energy up to a maximum value. At the end of the acceleration phase, we define an injection mechanism that allows a particle to leave the acceleration region and to enter a magnetic field line where it freely propagates. The third and fourth processes could happen in parallel, some particles propagate freely (free-streaming) on a field line, while others become trapped and bounce back and forth in a magnetic mirror region, depending on the initial pitch angle at injection. The fifth step entails the phase of energy loss, which inevitably occurs when the particles precipitate towards the chromosphere, where high density thermalizes all particles very rapidly, causing the bremsstrahlung that is detectable in hard X-rays (§ 13) and gamma-rays (§ 14). Trapped particles can be detected by thin-target bremsstrahlung in hard X-rays (§ 13) or gyrosynchrotron emission (§ 15) in radio wavelengths.

In principle one can define for each kinematic process a time scale that strongly depends on the velocity and thus on the kinetic energy ε of the particle, so that the total elapsed time is composed of the sum of each time interval,

$$t(\varepsilon) = t_{acc}(\varepsilon) + t_{inj}(\varepsilon) + \left\{ \begin{array}{l} t_{prop}(\varepsilon) \\ t_{trap}(\varepsilon) \end{array} \right\} + t_{loss}, \qquad \begin{array}{l} \text{for } \alpha < \alpha_0 \\ \text{for } \alpha \geq \alpha_0 \end{array}, \qquad (12.1.1)$$

where particles bifurcate into free-streaming propagation for small pitch angles ($\alpha < \alpha_0$) or trapping for large pitch angles ($\alpha \geq \alpha_0$). In practice, the acceleration time may not be directly measurable if acceleration occurs continuously, while the injection mechanism may synchronize the release of bunches of particles, as suggested by the common appearance of the subsecond pulses detected in hard X-rays. Also the energy loss time may not be measurable in the case of thick-target collisions, because the stopping distance in the chromosphere is much shorter than the propagation distance from the coronal acceleration site down to the chromosphere. In this most likely scenario, the time difference between a synchronized pulsed injection at the "exit" of a coronal magnetic reconnection site and the detection of hard X-rays by chromospheric thick-target bremsstrahlung is only dominated by the propagation time interval, either by free-streaming or trapped particle motion,

$$t(\varepsilon) \approx \left\{ \begin{array}{ll} t_{prop}(\varepsilon) & \text{for } \alpha < \alpha_0 \\ t_{trap}(\varepsilon) & \text{for } \alpha \geq \alpha_0 \end{array} \right. . \qquad (12.1.2)$$

The results of recent time delay analysis indicates that these processes dominate the timing of hard X-ray-producing electrons [namely, electron time-of-flight differences of free-streaming electrons $t_{prop}(\varepsilon)$, or trapping times of electrons $t_{trap}(\varepsilon)$]. The bimodality in the timing can easily be verified by the energy dependence, because time-of-flight differences have an energy dependence of $t_{prop} \propto l/v \propto l\,\varepsilon^{-1/2}$ (in the nonrelativistic limit), while trapping times scale with the collisional deflection time, $t_{trap}(\varepsilon) \propto \varepsilon^{3/2}/n_e$,

$$t(\varepsilon) \propto \left\{ \begin{array}{ll} \varepsilon^{-1/2} & \text{for } \alpha < \alpha_0 \\ \varepsilon^{3/2} & \text{for } \alpha \geq \alpha_0 \end{array} \right. . \qquad (12.1.3)$$

Although these two processes seem to be dominant based on the observational analysis, we do not want to bias ourselves to consider this as the only possibility, but will also discuss the timing of other processes (acceleration, injection, and energy loss), which could dominate in specific flare models. For instance, in large-scale DC electric fields, particles experience acceleration all the way on their trajectory and may never reach a free-streaming orbit phase. Thus, large-scale DC electric fields produce a distinctly different timing than small-scale acceleration regions of any kind. We will see that the timing of particle kinematics also depends strongly on the geometry and topology of the magnetic field, which is different for each flare model, and thus can be used as a powerful method to discriminate competing flare models. We emphasize that such kinematic tests of particle acceleration models are relatively novel, while traditional approaches attempt to extract information on acceleration mechanisms by spectral modeling.

12.2 Kinematics of Free-Streaming Particles

12.2.1 Definition of Time-of-Flight Distance

We start with the simplest kinematic case, which is the propagation of free-streaming particles in a collisionless plasma. We consider first just the simplest case of constant velocity. Two particles that are injected with different velocities v_1 and v_2 at the "exit" of a coronal acceleration site onto a magnetic field line experience a velocity dispersion that translates into a time-of-flight difference Δt_{prop} over a common distance l_{TOF},

$$\Delta t_{prop} = \frac{l_{TOF}}{v_1} - \frac{l_{TOF}}{v_2} = \frac{l_{TOF}}{c}\left(\sqrt{1 - \frac{1}{\gamma_1^2}} - \sqrt{1 - \frac{1}{\gamma_2^2}}\right), \qquad (12.2.1)$$

where γ_1 and γ_2 are the relativistic Lorentz factors as defined in Eqs. (11.1.11) and (11.1.13). Obviously, the faster particles (with higher kinetic energy) arrive first, preceding the slower ones. Thus, if propagation delays dominate the total energy-dependent timing (Eq. 12.1.1), hard X-ray pulses from electron bremsstrahlung should peak first at the highest energies. Conversion of the kinetic energy of electrons, ε_i, into photon energies detected in hard X-rays, $\epsilon_{X,i} = h\nu_X$, can be calculated from the bremsstrahlung cross section, which we will treat in § 13. This conversion factor depends on the energy, the spectral slope of the electron spectrum, and the assumed bremsstrahlung cross section, and amounts typically to $q_\varepsilon = \epsilon_{X,i}/\varepsilon \approx 0.5$, since electrons can only produce hard X-ray photons of lower energy. As Fig. 12.2 illustrates, this conversion factor is needed to evaluate the correct kinetic energies and velocities of the particles from a measured hard X-ray delay Δt_X.

The time-of-flight distance l_{TOF} evaluated using Eq. (12.2.1) measures the path length of an electron spiraling along the magnetic field trajectory with some pitch angle (Fig. 12.2, middle). In order to obtain the length l_{mag} of the magnetic field line, we need to correct for the parallel velocity component, $v_\parallel$,

$$l_{mag} = l_{TOF}\, q_\alpha, \qquad q_\alpha = \frac{v_\parallel}{v} = \cos\alpha \le \frac{2}{\pi} = 0.64. \qquad (12.2.2)$$

The numerical value of $q_\alpha \le 0.64$ is obtained by averaging the pitch angle over a loop bounce time in the limit of large mirror ratios (see derivation in § 12.5).

An additional length correction needs to be applied if the magnetic field line is helically twisted (Fig. 12.2, bottom). If we straighten the flare loop to a cylindrical geometry with radius r, the projected length l_{loop} of the cylinder is related to the length of the helical field line l_{mag} by

$$l_{loop} = l_{mag}\, q_h, \qquad q_h = \cos\theta = \sqrt{1 - \left(\frac{2\pi n r}{l_{mag}}\right)^2} \approx 0.85, \qquad (12.2.3)$$

with n being the number of complete twists by 2π radians. The helicity of coronal loops cannot exceed a few radians before they become magnetically unstable. For instance, a critical twist of $\approx 2.5\pi$ is predicted for erupting prominences (Vrsnak et al. 1991). The kink instability is predicted for twists in excess of $\approx 4.8\pi$ (Mikić et al. 1990). Thus

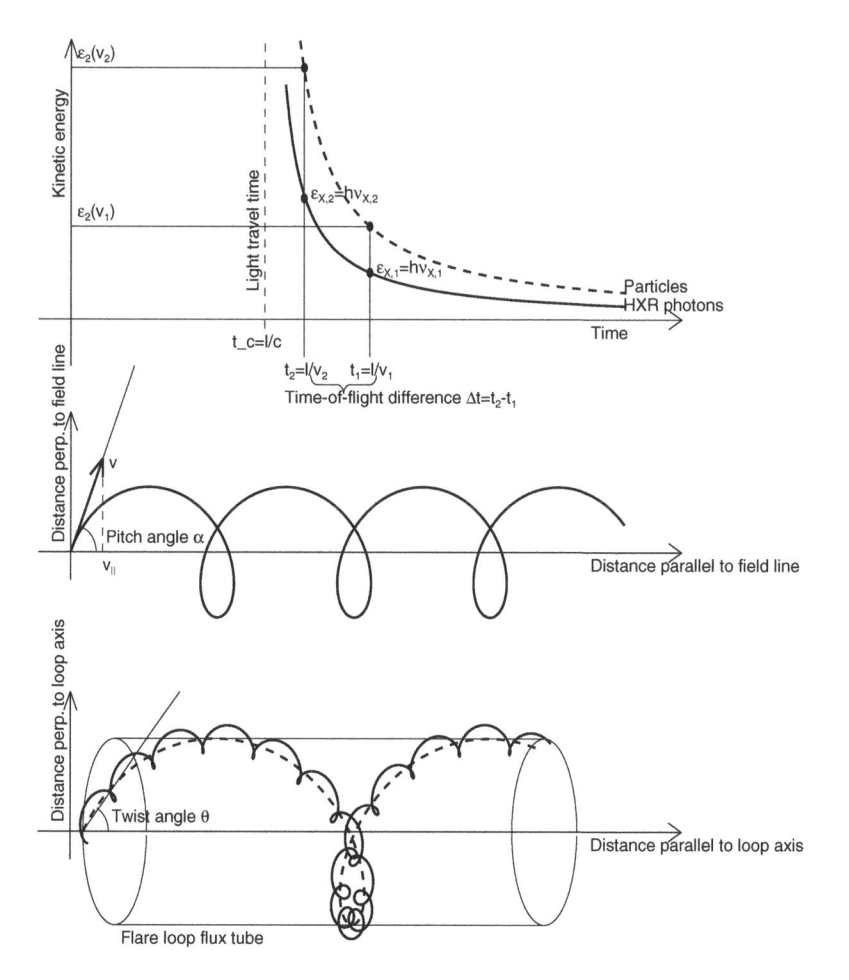

Figure 12.2: Geometrical effects that determine the timing in free-streaming particle propagation: *(top)* velocity dispersion between particles with different kinetic energies ε_i, which translate into a time difference between the detected hard X-ray photon energies $\varepsilon_{X,i}$; *(middle)* pitch angle α of particle, which defines the parallel velocity $v_\parallel = v \cos \alpha$; *(bottom)* twist angle θ of magnetic field line, which defines the projected trajectory length $l_{loop} = l_{mag} \cos \theta$.

we might use an estimated twist of 2π (i.e., $n = 1$ for the number of twists) and a loop aspect ratio of $2r/l_{loop} \approx 0.2$ (observed for the Masuda flare), which yields a length correction factor of $q_h \approx 0.85$ in Eq. (12.2.3). Combining the two factors we have

$$l_{loop} = q_h \, l_{mag} = q_\alpha q_H \, l_{TOF} \approx 0.54 \, l_{TOF} \,, \tag{12.2.4}$$

so the length of the particle trajectory projected onto the loop axis is about half the effective time-of-flight distance.

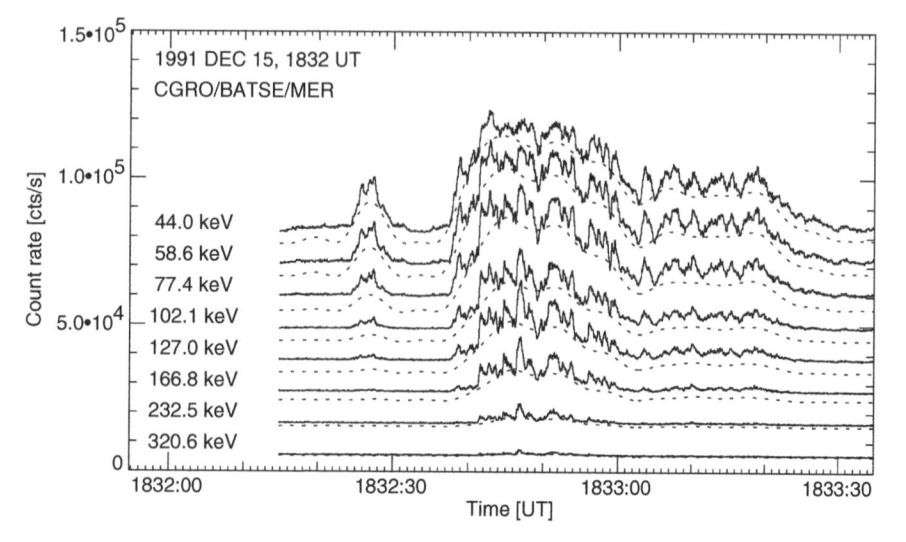

Figure 12.3: Hard X-ray observations of the 1991-Dec-15, 18:32 UT flare, recorded with *BATSE/CGRO* in the *Medium Energy Resolution (MER)* mode that contains 16 energy channels. Channels #3 − 10 are shown, with the low-energy edges indicated on the left. The channels are incrementally shifted and a lower envelope is indicated, computed from a Fourier filter with a cutoff at a time scale of 4 s. The fine structure (with the envelope subtracted) is cross-correlated in the delay measurements shown in Fig. 12.4 (Aschwanden 1996).

12.2.2   Time-of-Flight Measurements

The previously described pure *time-of-flight (TOF) model*, where propagation of free-streaming electrons dominates the energy-dependent timing in Eq. (12.1.1), with all other terms negligible, seems to explain the relative time delays of fast time structures seen in hard X-ray time profiles satisfactorily. The smoothly varying component in hard X-ray time profiles usually exhibits a different energy-dependent timing that is attributed to trapped particles (§ 12.5). An example is shown in Fig. 12.3, where the smoothly varying background is subtracted (dashed curves in Fig. 12.3), and the remaining fast time structures are cross-correlated in different energies, leading to the delay curve shown in Fig. 12.4. Taking the conversion factor of photon to electron energies into account, $q_\varepsilon = \epsilon_{X,i}/\varepsilon \approx 0.5$, and fitting the TOF model (Eq. 12.2.1), a time-of-flight distance of $l_{TOF} = 29.0 \pm 1.8$ Mm is found, with a reduced χ^2 or 0.59 (Aschwanden 1996).

As Fig. 12.4 shows, the relative time delays between adjacent energy channels are generally very small [e.g., $\Delta t = t(\varepsilon_X = 46 \text{ keV}) - t(\varepsilon_X = 60 \text{ keV}) = 19 \pm 2$ ms between the lowest two channels]. How is it possible to measure reliably such small time delays beyond the temporal resolution or time binning of the instrument, which is 64 ms for the *BATSE/MER* data used here? The trick is simply to use a cross-correlation technique with interpolation at the maximum of the *cross-correlation coefficient (CCC)*, which yields sub-binning accuracy if the photon statistics are sufficiently

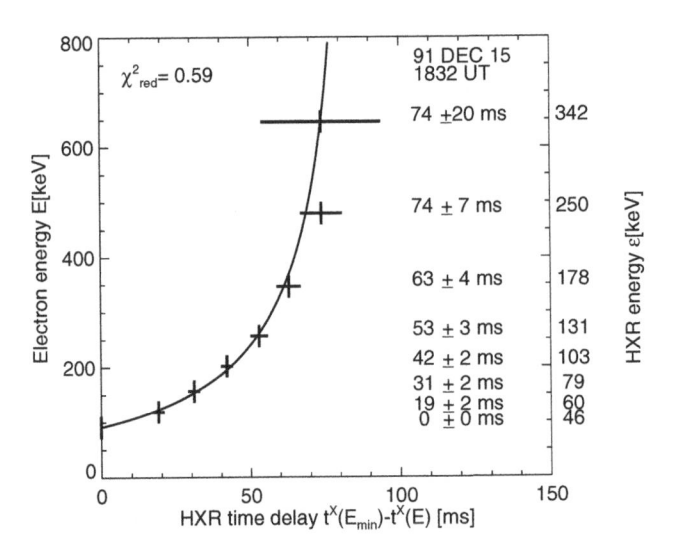

Figure 12.4: Hard X-ray time delay measurements between channel #3 and #4 − 10 from the data shown in Fig. 12.3. The delays and statistical uncertainties are indicated on the right. The hard X-ray energies ϵ_X on the right axis represent the medians of the count spectra in each channel, while the electron energies ε on the left axis are computed using the conversion factor $q_\varepsilon = \epsilon_{X,i}/\varepsilon \approx 0.5$. The solid line represents a fit of the 1-parameter model given in Eq. (12.2.1), yielding a TOF distance of $l = 29.0 \pm 1.8$ Mm, with $\chi^2_{red} = 0.59$ (Aschwanden 1996).

high. The cross-correlation coefficient between two flux profiles is defined by

$$CCC(\tau_j) = \frac{\sum_{i=i_1}^{i_2} F_1(t_i) F_2(t_{i-j})}{\sqrt{\sum_{i=i_1}^{i_2} F_1^2(t_i) \sum_{i=i_1}^{i_2} F_2^2(t_{i-j})}} , \tag{12.2.5}$$

where the index i only runs over the overlapping time interval in both energy channels to avoid a bias from aliasing effects. Empirically we can measure the accuracy of the delay measurement by repeating the cross-correlation by adding random noise (Aschwanden & Schwartz 1995). If the correlated time profiles contain pulses with Gaussian-like shape, characterized by a Gaussian width σ, we can assert the uncertainty of the mean just as for a normal distribution of N events, which is $m_t = \pm\sigma/\sqrt{N}$. For the delay measurement $\tau = t_1 - t_2$ we can then use the law of error propagation,

$$m_\tau = \sqrt{\left(\frac{d\tau}{dt_1}\right)^2 m_{t1}^2 + \left(\frac{d\tau}{dt_2}\right)^2 m_{t2}^2} = \sigma \sqrt{\frac{1}{N_1} + \frac{1}{N_2}} . \tag{12.2.6}$$

Thus, for the data shown in Fig. 12.3 we roughly have count rates of order $N \approx 4 \times 10^4$ cts s^{-1} in the lowest channels and pulses with Gaussian widths of $\sigma \approx 1000$ ms, so we estimate an uncertainty of $m_\tau \approx 1000\sqrt{2/4 \times 10^4} \approx 7$ ms for a single pulse. Averaging over 12 pulses increases the accuracy by another factor of $\sqrt{12} = 3.5$ and yields the obtained accuracy of 2 ms, as it was evaluated by repeating with added

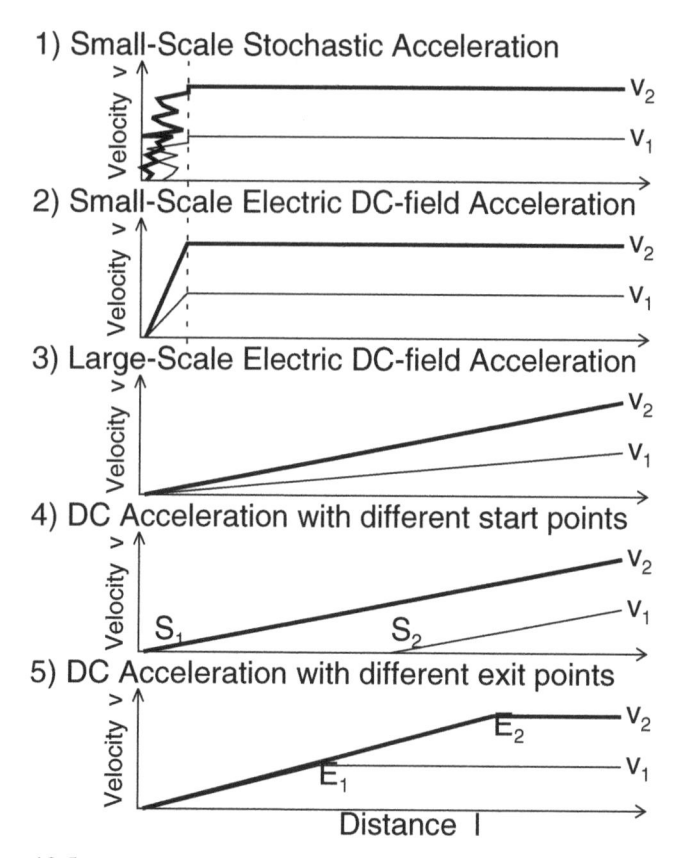

Figure 12.5: Five different models for the timing of electron acceleration and propagation. The velocity changes of a low-energy (v_1) and a high-energy electron (v_2) are shown along a 1D path from the beginning of acceleration (left side) to the thick-target site (right side). Models 1 and 2 characterize small-scale acceleration processes, while models $3-5$ depict scenarios with large-scale acceleration. Models 4 and 5 illustrate different start (S_1, S_2) and exit positions (E_1, E_2) for the accelerated electrons (Aschwanden 1996).

random noise. These measurements were accomplished with the *BATSE* detectors on *CGRO*, which had the largest collecting area (2000 cm^2). Other hard X-ray detectors have smaller areas [e.g., *HXRBS/SMM* with 71 cm^2 and *RHESSI* with an effective area of 38 cm^2 × 9 detectors × (0.5^2) (bi-grid transmission) = 85 cm^{-2}], so count rates are a factor of ≈ 25 smaller and the timing accuracy is reduced by a factor of ≈ 5.

12.3   Kinematics of Particle Acceleration

Let us now refine our kinematic models by including finite acceleration times and study how this affects the timing of hard X-ray emission, or the arrival times of electrons at the thick-target site. Basically one can distinguish between two opposite scenarios: small-scale and large-scale acceleration processes.

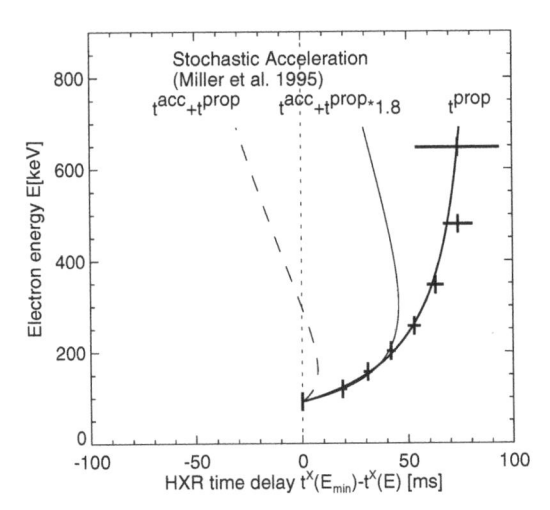

Figure 12.6: Fit of the stochastic acceleration model of Miller et al. (1996) to the same data as in Fig. 12.4. Adding the predicted acceleration time to the propagation time (thick line) yields negative delays (dashed line). Adjusting the propagation time by a factor of 1.8 (thin line) yields an acceptable fit in the $100 - 200$ keV range, but not at higher energies (Aschwanden 1996).

In small-scale acceleration processes, one can generally neglect acceleration time scales compared with propagation time scales, $t_{acc} \ll t_{prop}$, as long as the acceleration path is small compared with the free-flight propagation path. Such situations are depicted in model 1 and 2 (Fig. 12.5). Hard X-ray timing can then be described with the time-of-flight model (§ 12.2) that was found to be fully consistent with the data for the flare shown in Fig. 12.3.

12.3.1 Stochastic Acceleration

We consider a stochastic acceleration process as it can occur in coronal regions with enhanced wave turbulence (or similarly in shock fronts). Even when the spatial scale of the acceleration region is small compared with the TOF propagation distance (model 1 in Fig. 12.5), this does not necessarily imply that the acceleration time is also much smaller than the propagation time. In the case of diffusive stochastic acceleration the particles can be bounced around in a turbulent region significantly longer than the travel time through this region. For instance, LaRosa et al. (1995) estimated the bulk energization time of electrons in a reconnection-driven MHD-turbulent cascade to $t_{acc}(\varepsilon=20 \text{ keV}) \approx 300$ ms, which is comparable with the propagation time inferred in our flare (Fig. 12.4), $t_{prop}(\varepsilon=20 \text{ keV}) = l/v = 29$ Mm $/(0.27 \text{ c}) = 360$ ms. More specifically, Miller et al. (1996) estimated acceleration times of 70 ms to energize electrons from ≈ 5 to 50 keV, or about 180 ms to 511 keV. They specified an energy dependence of

$$t_{acc}(\varepsilon) = \left[\left(\frac{\varepsilon}{m_e c^2} \right)^{1/6} - 0.48 \right] \cdot 350 \text{ ms} \quad (\varepsilon > 5 \text{ keV}), \qquad (12.3.1)$$

to energize electrons by gyroresonant interactions with fast-mode waves in an MHD-turbulent cascade. We fit this model to the 1991-Dec-15 flare and show the expected hard X-ray timing in Fig. 12.6. First we add the acceleration time to the same propagation time inferred in Fig. 12.4 (based on a TOF distance of $l = 29$ Mm). The expected hard X-ray delay (dashed curve in Fig. 12.6) becomes negative above 200 keV, meaning that the high-energy electrons arrive later than the low-energy electrons at the chromosphere due to the longer acceleration time. If we perform a fit of the combined expression $t_X = t_{acc} + t_{prop}$ (using Eqs. 12.2.1 and 12.3.1) we find that the data can be reasonably fitted in the $100 - 200$ keV range with a 1.8 times larger TOF distance (to compensate for the acceleration time), but the hard X-ray delay decreases above 200 keV significantly below measured values. Thus, the energy-dependent scaling of the acceleration time specified in the stochastic acceleration model of Miller et al. (1996) cannot fit the observed delays over the entire energy range of $80 - 800$ keV for this specific flare.

This example illustrates that the observed hard X-ray delays require that the higher energy electrons arrive earlier than the low-energy electrons, which is a natural outcome for time-of-flight dispersion, but is an opposite trend to most acceleration models, where it takes statistically longer to accelerate to higher energies. This is a strong indication that acceleration times might not be dominant for the observed hard X-ray timing.

12.3.2   Electric DC Field Acceleration

Acceleration mechanisms employing DC electric fields have been studied by various researchers (e.g., Holman 1985; Tsuneta 1985; or Emslie & Hénoux 1995), but there is no detailed comparison of the predicted timing with observations. Models $2 - 5$ in Fig. 12.5 depict various scenarios where acceleration and propagation time scales have different weighting, depending on the spatial location and extent of the DC field. Because the free-flight path of electrons is complementary to the acceleration path length in unidirectional DC fields in a more direct fashion than in the case of stochastic acceleration, the resulting hard X-ray timing provides a crucial test between different models.

The simplest case is given in model 2 (Fig. 12.5), where the spatial extent of the DC field is small compared with the TOF distance, and necessarily also implies that the acceleration time is small compared with the free-flight time ($t_{acc} \ll t_{prop}$), and thus can be neglected. Hard X-ray timing can then be adequately described using Eq. (12.2.1), which fits the data satisfactorily for the 1991-Dec-15 flare (Fig. 12.4).

Another simple approach is to assume that an electric field extends over the entire flare loop and that electrons are accelerated from one end of the loop to the other. Assuming a constant electric field, the electrons would end up with a monoenergetic spectrum and coincident timing and thus cannot explain the observed delays (Fig. 12.4). A first variant is to assume that a flare loop consists of a number of current channels with different electric fields. In this scenario electrons are accelerated in separate channels with different electric fields E, obeying the force equation (Eq. 11.1.1),

$$F = m_e \, a = \frac{d}{dt}(\gamma m_e v) = eE = \frac{\varepsilon}{l} \, , \qquad (12.3.2)$$

where ε represents the kinetic energy of an electron gained by the electric field E over a distance l. Acceleration a can then be expressed as a function of the kinetic electron energy ε or Lorentz factor γ (using Eq. 11.1.12) by

$$a(\gamma) = \frac{\varepsilon}{m_e\, l} = \frac{c^2}{l}(\gamma - 1)\,. \tag{12.3.3}$$

Acceleration time t_{acc} as a function of energy can then be derived by integrating the force equation (Eq. 12.3.2) and inserting Eq. (11.1.13),

$$t_{acc}(\gamma) = \frac{c}{a(\gamma)}\sqrt{\gamma^2 - 1}\,. \tag{12.3.4}$$

Inserting the acceleration $a(\gamma)$ from Eq. (12.3.3) into Eq. (12.3.4) then yields

$$t_{acc}(\gamma) = \frac{l}{c}\sqrt{\frac{\gamma + 1}{\gamma - 1}}\,. \tag{12.3.5}$$

In model 3 (Fig. 12.5), hard X-ray timing is entirely determined by this energy dependence on the acceleration process, corresponding to the approximation $t_X \approx t_{acc}$ in the general timing equation (Eq. 12.1.1). We fit this model to the observed hard X-ray timing of the 1991-Dec-15 flare and show the results in Fig. 12.7 (left panel). Interestingly, this model shows a very similar energy dependence to the TOF propagation model (shown in Fig. 12.4) and thus fits the data equally well. The inferred acceleration path length is a factor of 0.44 shorter than the path length in the TOF propagation model, because the average electron speed is about half the final speed applied in the propagation model (being exactly half in the nonrelativistic limit). Thus, the two models cannot be distinguished from the timing alone, but the inferred distance scale is a factor of ≈ 2 different. However, there is no model that explains the existence of many current channels with very different large-scale electric fields in the first place, which have to be switched on and off on subsecond time scales to explain the observed hard X-ray pulses.

In model 4 and 5 (Fig. 12.5) we investigate two further variants of electric DC field acceleration, where the accelerated electrons are allowed to enter into (model 4) or exit from (model 5) an electric field channel at different locations. In both models the resulting electron energy is proportional to the acceleration path length, assuming a constant (mean) electric field E in all current channels. Model 4 is a natural situation in the sense that all electrons in a current channel experience acceleration once the electric field is turned on. In this scenario, the acceleration path length of each electron is defined by the distance between its start position and the loop footpoint. Defining the acceleration a by the maximum electron energy $\varepsilon_{max} = m_e c^2 (\gamma_{max} - 1)$ obtained by the electric field E over the loop length l (or an electron energy ε gained over a proportionally smaller distance x_{acc}),

$$a = \frac{\varepsilon_{max}}{m_e\, l} = \frac{\varepsilon}{m_e\, x_{acc}}\,, \tag{12.3.6}$$

we find the following timing for electrons in model 4:

$$t_{acc}(\gamma) = \frac{c}{a}\sqrt{\gamma^2 - 1} = \frac{l}{c}\frac{\sqrt{\gamma^2 - 1}}{(\gamma_{max} - 1)}\,. \tag{12.3.7}$$

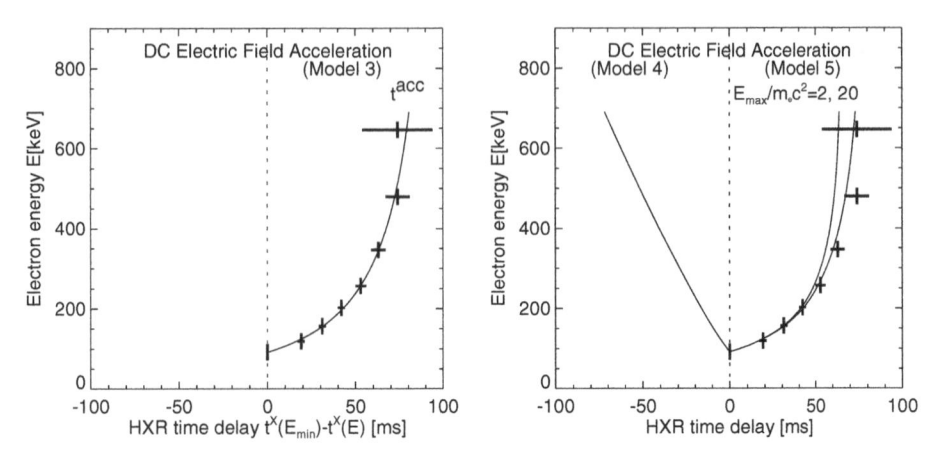

Figure 12.7: Fit of the DC electric field model 3 (left) and model 4 and 5 (right) for the same data as shown in Fig. 12.4. With respect to the TOF distance $l^{prop} = 29$ Mm fitted in Fig. 12.4, the best fit of model 3 (left) yields an acceleration distance of $l = l^{prop} \cdot 0.44$, model 4 is shown for $l = l^{prop}$, and model 5 yields $l = l^{prop} \cdot 1.1$ (for $E_{max}/m_e c^2 = 2$) and $l = l^{prop} \cdot 1.0$ (for $E_{max}/m_e c^2 = 20$), (Aschwanden 1996).

The resulting timing in model 4 is shown in Fig. 12.7 (right panel) for the same TOF distance $l = 29$ Mm used in Fig. 12.4 and for $\gamma_{max} - 1 = 2$ (or $\varepsilon_{max} \approx 1$ MeV). High-energy electrons arrive later at the thick-target site than low-energy electrons for every parameter combination and, thus, cannot fit the data. Therefore, model 4 can clearly be rejected for all flares where high-energy electrons arrive first.

In model 5 the electrons are allowed to exit a current channel with an accelerating electric field at an arbitrary location. Because electron spectra always have negative slopes, this means that more electrons leave the current channel after a short distance than after longer distances. This model may mimic a realistic situation when the current channels are relatively thin (for the most extreme aspect ratios see Emslie & Hénoux 1995), so that electrons exit a current channel by cross-field drifts. The length of the acceleration path can then be determined from the final electron energy ε using Eq. (12.3.6),

$$x_{acc}(\gamma) = l\frac{\varepsilon}{\varepsilon_{max}} = l\frac{(\gamma - 1)}{(\gamma_{max} - 1)} \ . \tag{12.3.8}$$

The resulting timing of electrons arriving at the thick-target site is then composed of the sum of the acceleration time t_{acc} over the acceleration path length x_{acc} and the free-flight propagation time t_{prop} over the remaining path length $l - x_{acc}$,

$$t_X(\gamma) = t_{acc}(\gamma) + t_{prop}(\gamma) = \frac{c}{a(\gamma)}\sqrt{\gamma^2 - 1} + \frac{l - x^{acc}(\gamma)}{v(\gamma)}$$

$$= \frac{l}{c}\left[\frac{\sqrt{\gamma^2 - 1}}{(\gamma_{max} - 1)} + \frac{1 - \frac{(\gamma - 1)}{(\gamma_{max} - 1)}}{\sqrt{1 - \gamma^{-2}}}\right] \ . \tag{12.3.9}$$

The fit of the timing in model 5 is also shown in Fig. 12.7 (right panel) for two different parameter combinations ($\varepsilon_{max} = 1$ MeV and 10 MeV). The essential result is that model 5 fits the data the better the smaller the acceleration time is relative to the propagation time, a situation that asymptotically approaches model 2 for high electric field strengths. Consequently, the best fit is consistent with a small-scale acceleration region, as in model 2.

Acceleration by field-aligned DC electric fields only increases the parallel velocity $v_\parallel$ of a particle, while the perpendicular velocity given by gyromotion, approximately equal to thermal speed, $v_\perp \approx v_{Te}$, remains unaffected. Thus the pitch angle of an accelerated particle would change as a function of the energy gained $\varepsilon = m_e c^2(\gamma - 1)$ as,

$$\alpha_{acc}(\gamma) = \arcsin\left(\frac{v_\perp}{v_{acc}(\gamma)}\right). \tag{12.3.10}$$

For every electron that is accelerated out of a thermal distribution to relativistic energies, this would produce very small pitch angles in the order of a few degrees, such that all accelerated electrons would immediately precipitate and no trapping would be possible, in contradiction to the ubiquitous trapping observed in virtually all flares.

In summary, we find that all large-scale DC electric field acceleration models have some problems in accomodating the observed energy-dependent, hard X-ray time delays. The only model that yields consistent timing (model 3) requires the existence of many current channels with different electric fields, which need to be switched on and off on subsecond time scales in synchronization. Such a requirement is difficult to realize, since currents cannot be switched off faster than with Alfvén speed (Melrose 1992). The problems go away for small-scale electric fields, whether there are super-Dreicer DC fields in reconnection regions as envisioned by Litvinenko (1996b) or convective electric fields in coalescing magnetic islands (Kliem 1994), as long as acceleration times are substantially smaller than propagation times to the chromosphere, $t_{acc} \ll t_{prop}$. In this case, however, our time delay measurements are no longer sensitive to any particular acceleration time dependence $t_{acc}(\varepsilon)$, other than being able to place upper limits.

12.4 Kinematics of Particle Injection

12.4.1 Scenarios of Synchronized Injection

In the previous sections we essentially concluded that observed energy-dependent hard X-ray delays (where the lower energies of pulses are delayed with respect to the higher energies) are not sensitive to acceleration models, because the models generally predict the opposite timing (i.e., longer acceleration times for high-energy particles than for low-energy particles). On the other hand, acceleration times are estimated to be comparable with or longer than propagation times [e.g., for stochastic acceleration (Eq. 12.3.1) or for sub-Dreicer electric fields in large current sheets]. The insensitivity of the observed energy-dependent timing to these acceleration time scales can thus only be reconciled by an intermediate injection mechanism that synchronizes the injection of accelerated particles at the "exit" of the acceleration region (Fig. 12.1).

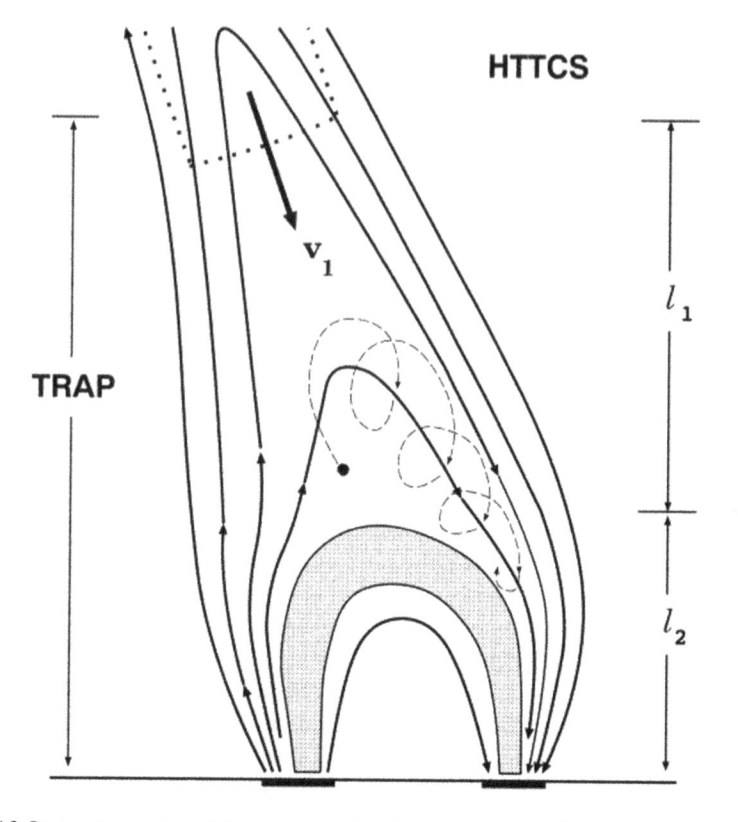

Figure 12.8: Accelerated particles are trapped in the cusp region while they undergo first-order Fermi acceleration. Particles leave the acceleration-plus-trap region once they become injected into the magnetic field lines that connect to the footpoints. The injection mechanism could be related to magnetic island dissipations in the reconnection region, fast shock waves, or pitch angle evolution (Somov & Kosugi, 1997).

In other words, the acceleration region could be a black box that accelerates particles continuously or with an arbitrary time cadence, while an injection mechanism opens a gateway in a pulsed fashion onto magnetic field lines that connect to the chromospheric footpoints of flare loops. The start time of particle propagation is then decoupled from the acceleration time.

The physical nature of the postulated particle injection mechanism is little explored and may be different for various acceleration and flare models. For parallel DC electric field acceleration, the particle leaves the acceleration path directly and continues propagation (e.g., model 5 in the previous section) without any intervening delay, so there is no injection mechanism present. Also, for perpendicular DC electric field acceleration (Fig. 11.3, right panel; Litvinenko 1996b), there is no time delay between the acceleration inside the current sheet and subsequent propagation once the particle exits sideward from the current sheet. In contrast, for DC field acceleration in magnetic islands (Figs. 11.5 and 11.6; Kliem 1994), for stochastic acceleration (e.g., Miller et

al. 1996), for shock acceleration (Fig. 11.19; Tsuneta & Naito 1998), and in particular in collapsing traps (Figs. 11.18 and 12.8; Somov & Kosugi, 1997), particles are
temporarily trapped during acceleration and do not escape onto a magnetic field line
(leading to the chromospheric footpoints) before an injection mechanism deflects the
particle out of the acceleration trap. This injection mechanism could be the dynamics of
a macroscopic structure (e.g., the coalescence of two magnetic islands into one), which
changes the locations of the separatrix surfaces that divide trapped from free-streaming
electrons. Alternatively, particles can also leave the acceleration region by microscopic
changes, such as by changing the pitch angles or by transit near an X-point. Generally,
when particles are accelerated in a direction parallel to the magnetic field, their pitch
angles become smaller and the probability of being mirrored in converging magnetic
bottles becomes smaller. So every parallel acceleration mechanism also controls the
pitch angle evolution of particles in such a way that they automatically escape after a
finite time from the magnetic trap. However, the observation of subsecond pulses in
hard X-rays and radio bursts that appear to be strictly synchronized at all (nonthermal)
energies (Fig. 12.3) suggests a macroscopic mechanism rather than a microscopic kinetic effect, because microscopic effects are statistical and not synchronized between
particles of different energies.

12.4.2 Model of Particle Injection During Magnetic Reconnection

Let us derive a quantitative model (Aschwanden 2004) that explains particle injection
in solar flares in the context of a Petschek-type magnetic reconnection scenario. In
the standard flare reconnection model (§ 10.5.1; Carmichael 1964; Sturrock 1966; Hirayama 1974; Kopp & Pneuman 1976; Tsuneta 1996a; Tsuneta et al. 1997; Shibata
1995), an X-type reconnection occurs at a coronal height h_X and the newly reconnected magnetic field lines relax into a force-free configuration which later becomes
(after chromospheric evaporation) the soft-X-ray-bright flare loop. The height ratio of
the reconnection point height h_X to the flare loop height h_L,

$$q_h = \left(\frac{h_X}{h_L}\right) \approx 1.5 \,, \tag{12.4.1}$$

has been determined by electron time-of-flight measurements, using pulse background
subtraction methods ($h_X/h_L = 1.7 \pm 0.4$; $l_{TOF}/l_{half-loop} = 1.4 \pm 0.3$; Aschwanden
et al. 1996c), pulse deconvolution methods ($l_{TOF}/l_{half-loop} = 1.6 \pm 0.6$; Aschwanden
et al. 1999d), or loop shrinkage measurements ($l_{cusp}/l_{half-loop} = 1.25, 1.5$; Forbes &
Acton 1996).

   The reconnection outflow has initially an Alfvén speed $v_A(h = h_X)$ and carries the
magnetic flux of the newly reconnected field lines. The newly reconnected magnetic
field lines experience a strong Lorentz force $\mathbf{j} \times \mathbf{B}$ in the initial cusp shape, which gradually reduces when the field line relaxes into a force-free shape, because the curvature
force is reciprocal to the curvature radius r_{loop} if the magnetic field only has a parallel
component (Eqs. 6.2.17−19),

$$\mathbf{j} \times \mathbf{B} = -\nabla\left(\frac{B^2}{8\pi}\right) + \frac{1}{4\pi}(\mathbf{B} \cdot \nabla)\mathbf{B} = \frac{B^2}{4\pi}\frac{1}{r_{loop}}\mathbf{e}_n \,. \tag{12.4.2}$$

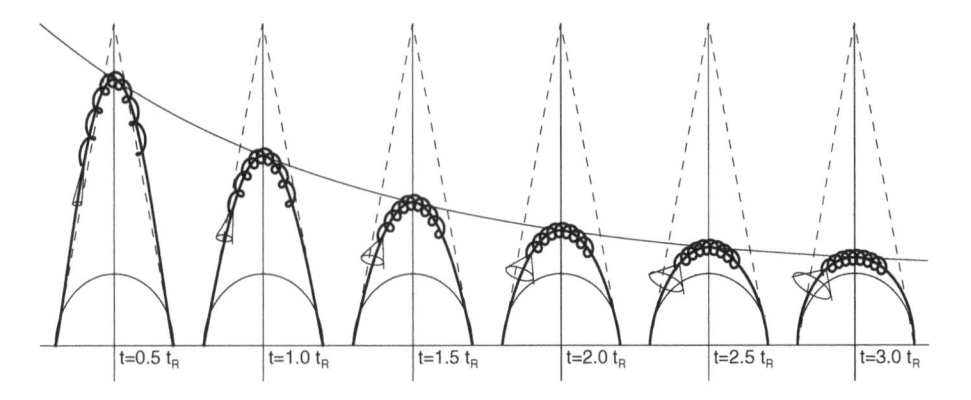

Figure 12.9: Scenario of field line relaxation after an X-point reconnection: The apex height of the field line relaxes exponentially into a force-free state from the initial cusp shape to the final quasi-circular geometry. The losscone angle of the trapped particles gradually opens up and releases more particles from the trap (Aschwanden 2004).

Thus, in our dynamic model we assume that the motion of the postreconnection field line occurs initially with Alfvén speed v_A, when the curvature radius is minimal in the cusp, and then decreases asymptotically to zero at the end of the relaxation phase, when the force-free field line reaches the maximum curvature radius in a quasi-circular flare loop geometry (Fig. 12.9).

We approximate the height dependence of the apex of the relaxing field line with an exponential function (Fig. 12.9) with an e-folding relaxation time scale t_R,

$$h(t) = h_L + (h_X - h_L) \exp\left(-\frac{t}{t_R}\right), \qquad (12.4.3)$$

which matches the initial condition $h(t = 0) = h_X$ (demarcating the height of the X-point) and final asymptotic limit $h(t = \infty) = h_L$ (the height of the relaxed flare loop). The speed of the height change is simply the derivative of Eq. (12.4.3),

$$v(t) = \frac{dh(t)}{dt} = -\frac{(h_X - h_L)}{t_R} \exp\left(-\frac{t}{t_R}\right) = -v_X \exp\left(-\frac{t}{t_R}\right), \qquad (12.4.4)$$

which defines the relaxation time scale t_R,

$$t_R = \frac{h_X - h_L}{v_X}, \qquad (12.4.5)$$

so that Eq. (12.4.4) fulfills the initial condition $v(t = 0) = -v_X$ and the final asymptotic limit $v(t = \infty) = 0$. Reconnection theories predict that the outflows have Alfvén speed, so we can set the initial value of the relaxing field line approximately equal to the reconnection outflow speed, assuming frozen-in flux conditions,

$$v_X \approx v_A = 2.18 \times 10^{11} \frac{B_{ext}}{\sqrt{n_X}}, \qquad (12.4.6)$$

where B_{ext} denotes the external magnetic field strength on both sides of the outflow region and n_X denotes the internal electron density in the X-point.

The magnetic field in the X-point is zero by definition and gradually increases with increasing distance from the X-point. Correspondingly, the losscone angle α is zero at the X-point and then gradually increases with increasing distance from the X-point or as a function of time for the progressively relaxing field line. A convenient parameterization for this evolution of the losscone angle $\alpha(h[t])$ is an exponential function, with an e-folding time scale t_I that we call *injection time*,

$$\alpha(t) = \frac{\pi}{2}\left[1 - \exp\left(-\frac{t}{t_I}\right)\right] . \qquad (12.4.7)$$

The pitch angle is initially zero, $\alpha(t = 0) = 0$ and then opens up to a full half-cone, $\alpha(t \mapsto \infty) = \pi/2$, for relaxation times $t \gg t_I$. We define a mirror ratio between the magnetic field $B(h)$ at an arbitrary height h in the cusp and B_L at the flare looptop, which relates to the losscone angle $\alpha(h[t])$ by,

$$R(h[t]) = \frac{B_L}{B(h[t])} = \frac{1}{\sin^2 \alpha(h[t])} . \qquad (12.4.8)$$

With this parameterization, the evolution of the magnetic field strength at the apex of the relaxing field line is (using Eqs. 12.4.7–8),

$$B(t) = B_L \sin^2 \alpha(t) = B_L \sin^2\left(\frac{\pi}{2}\left[1 - \exp\left(-\frac{t}{t_I}\right)\right]\right), \qquad (12.4.9)$$

or, inserting the transformation $h(t)$ from Eq. (12.4.3), we obtain the height dependence of the magnetic field

$$B(h) = B_L \sin^2\left(\frac{\pi}{2}\left[1 - \left(\frac{h - h_L}{h_X - h_L}\right)^{t_R/t_I}\right]\right) . \qquad (12.4.10)$$

The magnetic field strength in the X-point is zero by definition, $B_X = B(h = h_X) = 0$, and then increases in the reconnection outflow region to the ambient value, which we denote by $B_L = B(h = h_L)$ at the height of the flare loop (at the position after final relaxation). So the magnetic field monotonically increases from the X-point towards the flare looptop.

For the spatial variation of the magnetic field near the X-point (Eq. 12.4.10) it is actually more instructive to express the injection time t_I in terms of a spatial scale. We define a *magnetic length scale* L_B where the magnetic field increases from zero at the center of the X-point towards half of the value at infinity, $B(h = h_X - L_B) = B_L/2$ (Fig. 12.10, bottom panel). This can be calculated straightforwardly from Eq. (12.4.10) and we find the relation,

$$t_I = t_R \frac{\ln(1 - q_B)}{\ln(1/2)} \approx t_R \tan\left(\frac{\pi}{2} q_B\right) \approx t_R \frac{\pi}{2} q_B , \qquad (12.4.11)$$

where q_B expresses the fraction of the magnetic length scale L_B to the cusp distance $(h_X - h_L)$,

$$q_B = \frac{L_B}{(h_X - h_L)} . \qquad (12.4.12)$$

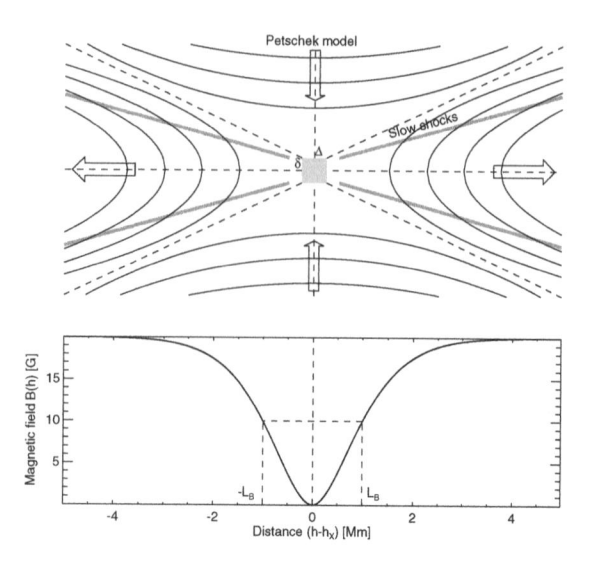

Figure 12.10: Definition of the magnetic length scale L_B at the X-point in the context of a Petschek-type X-point geometry (top panel), with height in the horizontal direction. The magnetic field $B(h - h_X)$ as a function of height near the X-point is shown (bottom panel) for the model parameters used in Fig. 12.11. The magnetic length scale L_B is defined by the range where the magnetic field increases to half the asymptotic value far away from the X-point (Aschwanden 2004).

The relaxation of the field line has the effect that the narrow losscone at the reconnection point opens up and allows particles with gradually larger pitch angles to precipitate. In other words, the particles that are accelerated near the X-point see initially a very small losscone angle and are thus fully trapped, while the gradual opening forced by the relaxation of the field line untraps them as a function of relaxation time. If we assume that the accelerated particles have all been accelerated near the X-point and have an isotropic pitch angle distribution, their initial distribution is

$$N(\alpha) \, d\alpha \propto 2\pi \sin \alpha \, d\alpha \ . \tag{12.4.13}$$

The precipitation rate $dN_P(t)/dt$ is then proportional to the time derivative, where the losscone angle has the time dependence given in Eq. (12.4.7),

$$\frac{dN_P(t)}{dt} = \frac{dN(\alpha)}{d\alpha} \frac{d\alpha(t)}{dt} \propto \cos\left[\alpha(t)\right] \frac{d\alpha(t)}{dt} = \cos\left(\frac{\pi}{2}\left[1 - \exp\left(-\frac{t}{t_I}\right)\right]\right) \exp\left(-\frac{t}{t_I}\right). \tag{12.4.14}$$

This implies that the opening speed of the losscone angle is fastest at the beginning and slows down with time.

To account for a finite acceleration time, we assume that the number of particles accelerated to an arbitrary energy increases with some power a of the trapping time t, so the fraction of accelerated particles increases as,

$$N_A(t) \propto t^a \ . \tag{12.4.15}$$

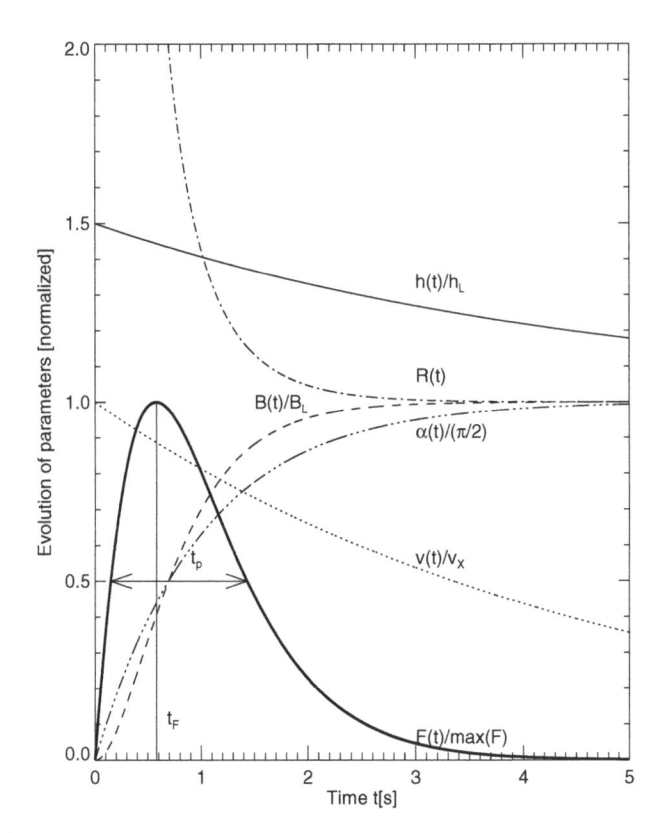

Figure 12.11: The dynamic evolution of parameters is shown for the field-line relaxation model (§ 12.4.2). The parameters are: height of X-point $h_X = 30$ Mm, height of relaxed flare loop $h_L = 20$ Mm, density $n_X = 10^9$ cm^{-3}, magnetic field at apex of flare loop $B_L = 20$ G, and magnetic scale length in the reconnection region $L_B/(h_X - h_L) = 0.1$. The curves show the evolution of height $h(t)/h_L$, relaxation velocity $v(t)/v_X$ with $v_X = 2070$ km s^{-1}, magnetic field at the apex of the relaxing field line $B(h)/B_L$, losscone angle $\alpha(t)/(\pi/2)$, magnetic mirror ratio $R(t)$, and precipitation flux $F(t)/max(F)$. The relaxation time is $t_R = 4.84$ s, the injection time scale $t_I = 1.00$ s, and the FWHM pulse duration is $t_w = 1.28$ s (Aschwanden 2004).

The flux of precipitating accelerated particles then has the time dependence

$$F(t) = N_A(t)\frac{dN_P(t)}{dt} \propto t^a \, cos\left(\frac{\pi}{2}\left[1 - \exp\left(-\frac{t}{t_I}\right)\right]\right) \exp\left(-\frac{t}{t_I}\right). \quad (12.4.16)$$

For a given acceleration mechanism, parameter a is a constant, and thus the pulse profiles are scale-invariant if scaled by normalized time (t/t_I). Therefore, the FWHM of the flux profile, t_w, is also scale-invariant [i.e., the ratio (t_w/t_I)]. We calculate this scale-invariant ratio $q_a = (t_w/t_I)$ numerically for different acceleration power indices

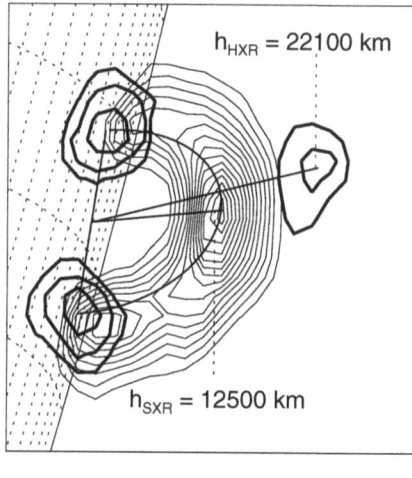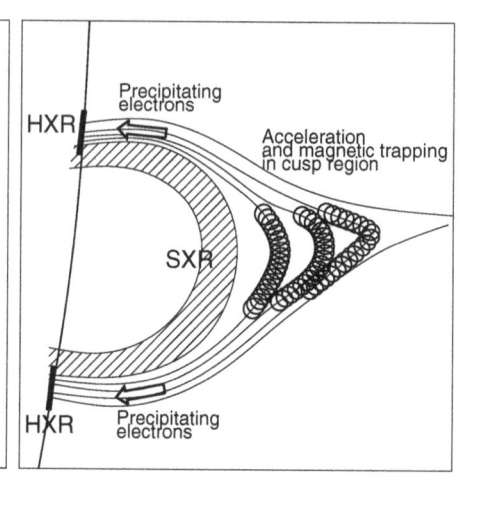

Figure 12.12: Temporary trapping occurs in the acceleration region in the cusp region below the reconnection point in our model (right panel), which can explain the coronal hard X-ray emission observed during the Masuda flare (left panel; Masuda et al. 1994). The observations in the left panel show a *Yohkoh/HXT* 23 − 33 keV image (thick contours) and Be119 *SXT* image (thin contours) of the 92-Jan-13, 17:28 UT flare (Aschwanden 1998b).

$a = 0.0, ..., 2.0,$

$$q_a = \frac{t_w}{t_I} = \begin{cases} 0.493 & \text{for } a = 0.0 \\ 1.014 & \text{for } a = 0.5 \\ 1.284 & \text{for } a = 1.0 \\ 1.503 & \text{for } a = 1.5 \\ 1.697 & \text{for } a = 2.0 \end{cases} \qquad (12.4.17)$$

We show the time evolution of the various parameters in Fig. 12.11: height $h(t)/h_L$, relaxation velocity $v(t)/v_X$, magnetic field at apex of the relaxing field line $B(h)/B_L$, losscone angle $\alpha(t)/(\pi/2)$, magnetic mirror ratio $R(t)$, and precipitation flux $F(t)$.

We can now express the FWHM pulse duration t_w as just a function of the independent parameters t_R and q_B, using Eqs. (12.4.11−12) and (12.4.17),

$$t_w = t_R \, q_a \left(\frac{\ln(1 - q_B)}{\ln(1/2)} \right) \approx q_a \frac{\pi}{2} \frac{L_B}{v_X} \approx \frac{2L_B}{v_A} . \qquad (12.4.18)$$

The right-hand approximation expresses most succinctly the relation of the pulse duration to the underlying physical parameters: the pulse duration is proportional to the Alfvén transit time (with velocity $v_X \approx v_A$) through the magnetic length scale L_B of the X-point region. The fastest time structures are produced by the smallest X-point regions (and presumably for the smallest flare loops). For instance, an X-point region with a spatial extent of $L_B = 500$ km, an Alfvén velocity of $v_A = 2000$ km s^{-1}, yields pulses with time scales of typically $t_w \approx 0.5 \pm 0.25$ s, depending on the acceleration power index a. If the Alfvén speed is known, the pulse duration can be used to estimate the magnetic length scale L_B of a Petschek-type X-point (Fig. 12.10), which entails the

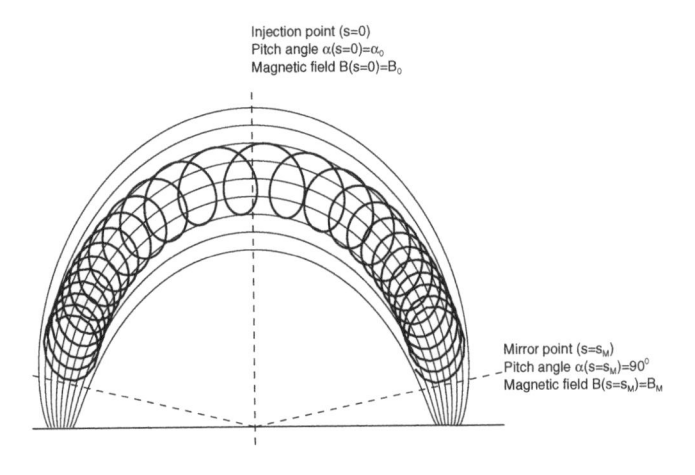

Injection point (s=0)
Pitch angle α(s=0)=α₀
Magnetic field B(s=0)=B₀

Mirror point (s=s_M)
Pitch angle α(s=s_M)=90°
Magnetic field B(s=s_M)=B_M

Figure 12.13: The magnetic mirror ratio $R = B_M/B_0$ is defined by the ratio of the magnetic field strengths between the mirror point, $B(s = s_M) = B_M$, and the injection point, $B(s = 0) = B_0$, which define the losscone angle α_0. Particles injected with a larger pitch angle $\alpha > \alpha_0$ are mirrored, while particles with smaller pitch angles precipitate through the losscone.

diffusion region of a magnetic reconnection process. Thus, this model provides a direct diagnostic of magnetic reconnection geometry. This model also provides a natural explanation for the existence of above-the-looptop hard X-ray emission (Fig. 12.12), as discovered by Masuda et al. (1994) and reviewed in Fletcher (1999). The model also demonstrates that the observed pulse durations are controlled by the injection time rather than by the acceleration time scale.

12.5 Kinematics of Particle Trapping

12.5.1 Magnetic Mirroring

After a magnetic reconnection process in a flare, the closed field lines always relax into a more force-free state, which corresponds to dipole-like field geometries with the strongest magnetic fields at the footpoints and a weaker magnetic field in the coronal segments inbetween (Fig. 9.4). This implies that each closed magnetic field line naturally forms a *magnetic trap* (Fig. 12.13), where particles mirror back and forth as long as the trapped plasma is collisionless and adiabatic particle motion is ensured. At the collisionless limit, particle motion is adiabatic and the magnetic moment μ is conserved along the loop coordinate s,

$$\mu = \frac{\frac{1}{2}m_e v_\perp^2(s)}{B(s)} = \frac{\frac{1}{2}m_e v^2 \sin^2 \alpha(s)}{B(s)} = const . \tag{12.5.1}$$

The pitch angle $\alpha(s)$ changes as a function of the magnetic field $B(s)$ along the field line, while the velocity v is constant for adiabatic motion.

To understand the basic kinematics of magnetic traps we assume a dipole-like magnetic field, which at its lowest order can be approximated by a parabolic equation (e.g., Trottet et al. 1979),

$$B(s) = B_0 \left[1 + (R - 1) \frac{s^2}{s_M^2} \right] , \qquad (12.5.2)$$

where $B_0 = B(s = 0)$ represents the (minimum) magnetic field strength at the looptop at $s = 0$, and s_M is the length from the looptop to the mirror point. The location of the mirror point $s = s_M$ can be defined at the interface between the (coronal) collisionless and (chromospheric) collisional regime, which usually coincides with the zone of largest magnetic divergence (i.e., the canopy structure in the transition region: Fig. 4.25). The *magnetic mirror ratio R* is defined by the ratio of the magnetic field strengths at the mirror point and looptop,

$$R = \frac{B(s = s_M)}{B_0} = \frac{1}{\sin^2(\alpha_0)} . \qquad (12.5.3)$$

The losscone angle α_0 is defined by the critical pitch angle α_0 of a particle at the looptop which determines whether a particle is mirrored and trapped (if it has a larger pitch angle, $\alpha > \alpha_0$) or whether it escapes through the losscone and becomes untrapped (if it has a smaller pitch angle, $\alpha < \alpha_0$). This losscone angle simply follows from comparing the magnetic moment (Eq. 12.5.1) at the looptop, $\alpha(s = 0) = \alpha_0$, with that of the mirror point, $\alpha(s = s_M) = \pi/2$,

$$\alpha_0 = \arcsin \sqrt{\frac{1}{R}} . \qquad (12.5.4)$$

We will see later on (§ 13) that the mirror ratio R in flare loops can be determined from measurements of the losscone angle α_0, using the ratios between directly precipitating and trapped electrons from the corresponding hard X-ray fluxes.

Another important quantity is the *bounce time* of a mirroring particle back and forth a mirror trap, which, for our parabolic field (Eq. 12.5.2) and for a loop half-length L is

$$t_B = 4 \int_0^L \frac{ds}{v_\parallel(s)} = 4 \int_0^L \frac{ds}{\sqrt{v^2 - v_\perp^2}} = \frac{2\pi L}{v \sqrt{1 - \frac{1}{R}}} . \qquad (12.5.5)$$

The ratio of the travel time $t_0 = 4L/v$ along the magnetic field line to the bounce time t_B is,

$$q_\alpha(R) = \frac{t_0}{t_B} = \frac{2}{\pi} \left(1 - \frac{1}{R} \right)^{1/2} , \qquad (12.5.6)$$

which approaches the value $q_\alpha = 0.64$ for large mirror ratios $R \gg 1$. This ratio q_α also yields the relevant *pitch angle correction factor* to convert an electron-time-of flight distance to the length of a magnetic field line [i.e., $q_\alpha = l_{mag}/l_{TOF} = t_0/t_B$ (Eq. 12.2.2)].

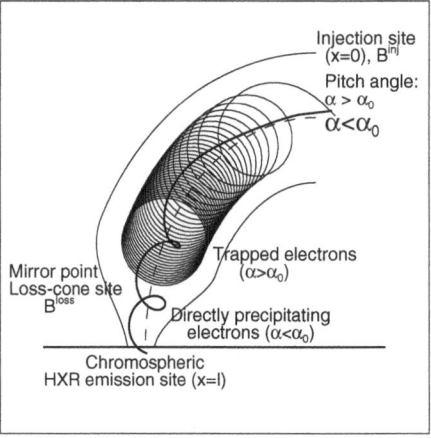 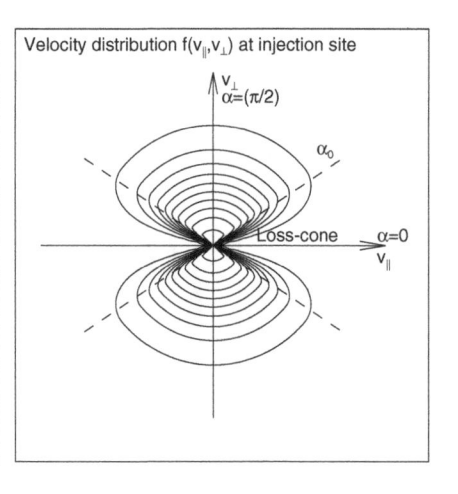

Figure 12.14: The velocity distribution of a trapped particle distribution forms a losscone with a critical pitch angle α_0 (right frame). Electrons with small pitch angles ($\alpha < \alpha_0$) precipitate directly to the chromosphere, while electrons with large pitch angles ($\alpha > \alpha_0$) are intermediately trapped and precipitate after they become scattered in the losscone (Aschwanden 1998a).

12.5.2 Bifurcation of Trapping and Precipitation

The criterion for the bifurcation of particle trajectories, whether they propagate free-streaming directly to the footpoints or become trapped (Fig. 12.1, 12.14), is controlled by the initial pitch angle α_0 at the injection site. Those particles that have a pitch angle larger than the critical losscone angle ($\alpha > \alpha_0$) become trapped, until they are pitch angle-scattered into the losscone ($\alpha < \alpha_0$) after an energy-dependent trapping time $t_{trap}(\varepsilon)$. They subsequently escape from the trap and precipitate (in the so-called *trap-plus-precipitation* model). In the simplest model without energy loss in the trap, for a δ-like injection, the number of electrons $N_{trap}^{\delta}(\varepsilon, t)$ (with kinetic energy ε) in the trap decreases exponentially, with an e-folding time constant that corresponds to the trapping time $t_{trap}(\varepsilon)$ (Melrose and Brown 1976), i.e.

$$N_{trap}^{\delta}(\varepsilon, t) = N_{trap}^{\delta}(\varepsilon, 0) \, \exp[-t/t_{trap}(\varepsilon)] . \qquad (12.5.7)$$

The precipitation rate, $n_{trap}^{\delta}(\varepsilon, t)$, is defined by the time derivative of $N_{trap}^{\delta}(\varepsilon, t)$, i.e.

$$n_{trap}^{\delta}(\varepsilon, t) = -\frac{dN_{trap}^{\delta}(\varepsilon, t)}{dt} = n_{trap}^{\delta}(\varepsilon, 0) \, \exp[-t/t_{trap}(\varepsilon)] , \qquad (12.5.8)$$

with $n_{trap}^{\delta}(\varepsilon, 0) = N_{trap}^{\delta}(\varepsilon, 0)/t_{trap}(\varepsilon)$. Consequently, it is also exponentially decreasing. For a general injection function $f(\varepsilon, t)$, the precipitation rate from the trap, $n_{trap}(\varepsilon, t)$, can be described by convolution with the trapping time $t_{trap}(\varepsilon)$,

$$n_{trap}(\varepsilon, t) = \frac{1}{t_{trap}(\varepsilon)} \int_0^t f(\varepsilon, t') \, \exp[-\frac{(t - t')}{t_{trap}(\varepsilon)}] \, dt' . \qquad (12.5.9)$$

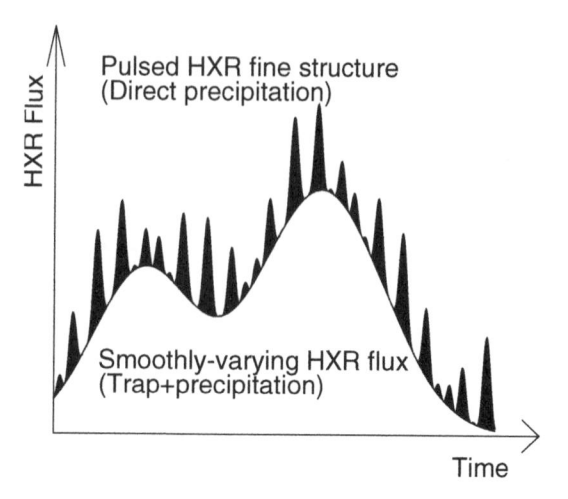

Figure 12.15: Relation between electron propagation and observed hard X-ray time structures: Electrons with small pitch angles (injected near the flare looptop) precipitate directly and produce rapidly varying hard X-ray pulses with time-of-flight delays, while electrons with large pitch angles become trapped and produce a smoothly varying hard X-ray flux when they eventually precipitate, with a timing that corresponds to trapping time scales (Aschwanden 1998b).

If the acceleration mechanism produces a wide range of pitch angles, there is always a mixture of small and large pitch angles, which then produces a mixture of directly precipitating and trap-precipitating particles. This generally seems to be the case based on the data analysis of observed hard X-ray delays (§ 13). This dichotomy is also present in the morphological structure of hard X-ray time profiles, where fast pulses indicate the fraction of directly precipitating electrons that preserve the pulsed time profile of the injection mechanism, while the smoothly varying lower envelope represents a measure of the trap-precipitating electrons that smear out the time profiles of the injection mechanism (Fig. 12.15). Therefore, the total electron precipitation rate $n(\varepsilon, t)$ is composed of a combination of the two components (Fig. 12.16): (1) a fraction q_{prec} of electrons that precipitate directly, and (2) the complementary fraction $(1 - q_{prec})$ that precipitate after some temporary trapping,

$$n(\varepsilon, t) = q_{prec} f(\varepsilon, t) + (1 - q_{prec}) n_{trap}(\varepsilon, t)$$

$$= q_{prec} f(\varepsilon, t) + \frac{(1 - q_{prec})}{t_{trap}(\varepsilon)} \int_0^t f(\varepsilon, t') \exp[-\frac{(t - t')}{t_{trap}(\varepsilon)}] \, dt' \, . \qquad (12.5.10)$$

The dichotomy of the direct-precipitating and trap-precipitating electrons can be deconvolved from observed hard X-ray time profiles [e.g., by forward-fitting of the convolution function (Eq. 12.5.10)], (Aschwanden 1998a).

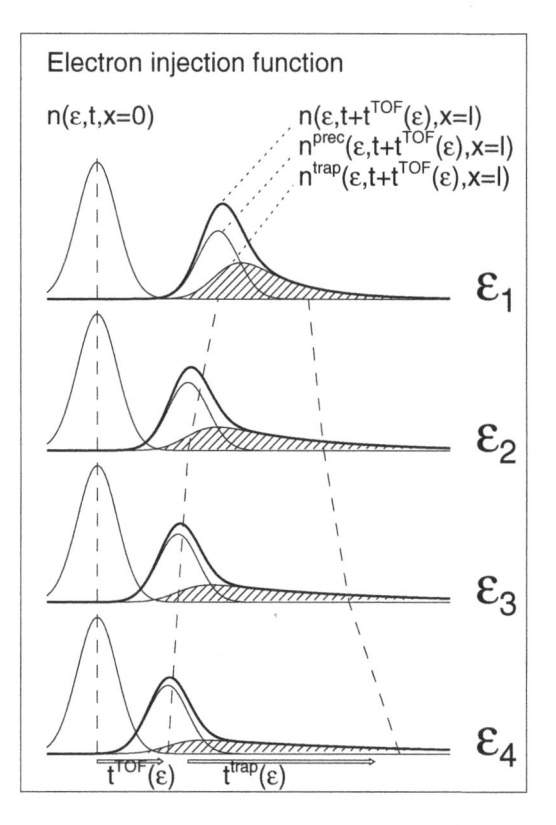

Figure 12.16: Temporal relation of the electron injection function $n(\varepsilon, t, x = 0)$ at the injection site and at the hard X-ray emission site $n(\varepsilon, t+t_{TOF}(\varepsilon), x = l)$, delayed by an energy-dependent electron time-of-flight interval $t_{TOF}(\varepsilon)$, schematically shown for four different energies $\varepsilon_1 < \varepsilon_2 < \varepsilon_3 < \varepsilon_4$. The injection function at the hard X-ray emission site (thick curve) is broken down into a directly precipitating component (thin curve) and the trap-precipitating component (hatched curve). Note that the (e-folding) trapping time $t_{trap}(\varepsilon)$ increases with energy ε, while the time-of-flight delay $t_{TOF}(\varepsilon)$ decreases (Aschwanden 1998a).

12.5.3 Trapping Times

Theoretical treatments of trapping models can be found for solar flares (e.g., Benz 1993; § 8) or for magnetospheric applications (e.g., Baumjohann & Treumann 1997; § 3). Let us briefly review the theoretical development of solar flare models. Trapping times of electrons were estimated from various pitch angle scattering mechanisms into the losscone, for example by *Coulomb collisional deflection* (Benz & Gold 1971), by quasi-linear diffusion (via resonant wave-particle interactions) induced by electrostatic (hydrodynamic) waves (Wentzel 1961; Berney & Benz 1978), whistler waves (Kennel & Petschek 1966; Wentzel 1976; Berney & Benz 1978; Kawamura et al. 1981; Chernov 1989), lower-hybrid waves (Benz 1980), electron-cyclotron maser (Aschwanden & Benz 1988a, b; Aschwanden et al. 1990), or plasma turbulence (e.g., review by Ramaty

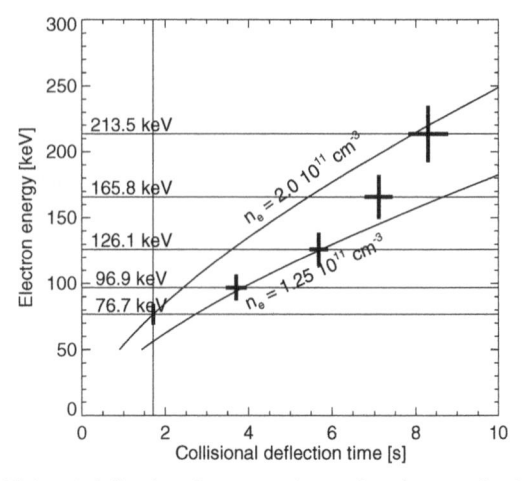

Figure 12.17: Collisional deflection times are shown for electron densities in the range of $n_e = (1.25 - 2.0) \times 10^{11}$ cm^{-3}, roughly fitting the time delays of the lower envelopes of the hard X-ray time profiles observed during the Masuda flare on 1992-Jan-13, 17:29 UT, and thus are interpreted as trapping time differences (Aschwanden et al. 1996c).

& Mandzhavidze 1994; Petrosian 1996).

Coulomb collisional deflection time is considered to be an upper limit for trapping times, $\tau_{trap} \lesssim \tau_{defl}$ (also called *weak-diffusion limit*). The trapping time (in the weak-diffusion limit) is given by the electron collisional deflection time $t_{defl}(\varepsilon)$ (Trubnikov 1965; Spitzer 1967; Schmidt 1979; Benz 1993)

$$t_{trap}(\varepsilon) \lesssim t_{defl}(\varepsilon) = 0.95 \cdot 10^8 \Big(\frac{\varepsilon_{keV}^{3/2}}{n_e}\Big) \Big(\frac{20}{\ln \Lambda}\Big) , \tag{12.5.11}$$

where $\ln \Lambda$ is the *Coulomb logarithm*,

$$\ln \Lambda = \ln[8.0 \cdot 10^6 \ (T_e \ n_e^{-1/2})] , \qquad T_e > 4.2 \cdot 10^5 \ K . \tag{12.5.12}$$

The observed time delays between the slowly varying components of hard X-ray time profiles (§ 13) generally fit the energy-dependence of collisional deflection times (Eq. 12.5.11) for reasonable plasma densities (i.e., $n_e \approx 10^{11}$ cm^{-3}: Fig. 12.17), which are found to be consistent with densities measured independently from the emission measures of soft X-ray flare loops (Aschwanden et al. 1997). Trapping in solar flare loops thus seems to be controlled by collisions in the weak-diffusion limit.

The lower limits of trapping times are controlled by the diffusion rate into the loss-cone (with angle α_c), which can occur as fast as every bounce time t_b (the so-called *strong-diffusion limit*), (i.e., Kennel 1969),

$$\tau_{trap} \geq \frac{t_b}{2\alpha_c^2} . \tag{12.5.13}$$

The temporal dynamics of trap-plus-precipitation models, which includes the temporal and spectral evolution of trapped and escaping particle distributions and related hard

X-ray fluxes, has been analytically described in a number of papers (Melrose & Brown 1976; Bai 1982a; MacKinnon et al. 1983; Zweibel & Haber 1983; Craig et al. 1985; Vilmer et al. 1986; Ryan 1986; MacKinnon 1986; 1988; Alexander 1990; McClements 1990a, b; 1992; McClements & Baynes 1991; Ryan & Lee 1991; Hamilton et al. 1990; Lu & Petrosian 1988; 1990; Hamilton & Petrosian 1990), as well as simulated numerically (MacKinnon & Craig 1991). Particle dynamics in traps have been modeled to study asymmetries of hard X-ray sources (Melrose & White 1979), or trajectories of radio type N bursts (Hillaris et al. 1988). Besides purely magnetic traps, models with additional electrostatic potentials have also been proposed for solar flares (Spicer & Emslie 1988), similar to models of *auroral kilometric radiation (AKR)* (e.g. Louarn et al. 1990).

12.6 Kinematics of Particle Precipitation

In the following first we consider particle precipitation first in symmetric traps (with a symmetric magnetic field; § 12.6.1), for the sake of simplicity, as well as for asymmetric traps (with an asymmetric field; § 12.6.2), which occur more frequently in solar flares.

12.6.1 Symmetric Traps

The fraction of directly precipitating electrons q_{prec} can be self-consistently related to the critical losscone angle α_0. For the sake of simplicity, we first consider the case of an isotropic pitch angle distribution at the injection site and symmetric loop geometries,

$$q_{prec}(\alpha_0) = \frac{\int_0^{\alpha_0} \sin \alpha \, d\alpha}{\int_0^{\pi/2} \sin \alpha \, d\alpha} = (1 - \cos \alpha_0) \,. \qquad (12.6.1)$$

This case corresponds to a double-sided losscone distribution (Fig. 12.14 right panel), as it occurs in symmetric flare loops. With a deconvolution method we can measure q_{prec} directly. The corresponding losscone angle α_0 is then (for isotropic pitch angle distributions and symmetric loops),

$$\alpha_0(q_{prec}) = \arccos(1 - q_{prec}) \,, \qquad (12.6.2)$$

leading to the mirror ratio $R(\alpha_0)$ defined in Eq. (12.5.3).

The inference of the magnetic mirror ratio $R = B_M/B_0$ [i.e., the ratio of the magnetic field ($B(s = s_M) = B_M$) at the mirror point to that at the injection point at the looptop ($B(h = h_0) = B_0$)], together with the (projected) time-of-flight distance l_{TOF} between the injection site and chromospheric energy loss site (which is presumably close to the losscone site), yields a measure of the magnetic scale height λ_B. Defining the scale height by an exponential model

$$B_0 = B_M \exp\left[-\frac{(h_0 - h_M)}{\lambda_B}\right] \approx B_M \exp\left[-\frac{l_{TOF}}{\lambda_B}\right] \,, \qquad (12.6.3)$$

and using the definition of the mirror ratio $R = B_M/B_0$, we obtain

$$\lambda_B = \frac{l_{TOF}}{\ln(R)} .$$ (12.6.4)

The magnetic field can be approximated by a dipole field, parametrized with dipole depth h_D and photospheric field B_{ph},

$$B(h) = B_{ph} \left(1 + \frac{h}{h_d} \right)^{-3} ,$$ (12.6.5)

to which the magnetic mirror ratio can be related by $R = B_0/B_M \approx B(h)/B_{ph}$. The resulting magnetic scale height λ_B is then

$$\lambda_B(h) = -\frac{B}{\nabla B} = \frac{(h_D + h)}{3} .$$ (12.6.6)

Typical dipole depths are of order $h_D \approx 100$ Mm, leading to magnetic scale heights of $\lambda_B \approx 60$ Mm, magnetic mirror ratios of $R \approx 1.6$, losscone angles of $\alpha_0 \approx 50°$, and precipitation ratios of $q_{prec} \approx 0.4$, for symmetric traps (Aschwanden et al. 1997).

12.6.2 Asymmetric Traps

The timing analysis of *CGRO* data, based on the total hard X-ray flux without spatial information, provides a global electron-trapping time scale. The spatial structure, however, can often be described by two magnetically conjugate footpoint sources, which often have asymmetric hard X-ray fluxes according to the *Yohkoh/HXT* images. These unequal double-footpoint sources indicate electron precipitation sites in a flare loop with asymmetric magnetic field geometry. We need therefore to develop an asymmetric trap model to relate trapping time information from *CGRO* data to the asymmetric double-footpoint sources seen in *Yohkoh/HXT* data.

In order to mimic an asymmetric trap model we rotate the reference system of a symmetric dipole-like magnetic field by an angle ψ, as shown with three examples in Fig. 12.18 (top): the symmetric case with $\psi = 0$ (left), a weakly asymmetric case where the dipole coil is rotated by $\psi = 30°$ (middle), and strongly asymmetric case where the coil is rotated by $\psi = 60°$. (A spherically symmetric sunspot with a "unipolar" vertical field would correspond to the extreme case of $\psi = 90°$.) The acceleration or injection site into the trap is assumed to be midway (with a magnetic field B_A) between the two footpoints (with magnetic fields B_1 and B_2). These three magnetic field values B_A, B_1, B_2 determine which fraction of electrons are trapped or precipitate to the two footpoints. Because the magnetic moment is conserved, $\mu = \frac{1}{2} m_e v_\perp^2 / B \propto \sin\alpha(s)^2 / B(s) = const$, the critical pitch angles that separate precipitating from trapped particles at the two footpoints are defined by the *magnetic mirror ratios*

$$R_1 = \frac{B_1}{B_A} = \frac{1}{\sin(\alpha_1)^2} ,$$ (12.6.7)

$$R_2 = \frac{B_2}{B_A} = \frac{1}{\sin(\alpha_2)^2} .$$ (12.6.8)

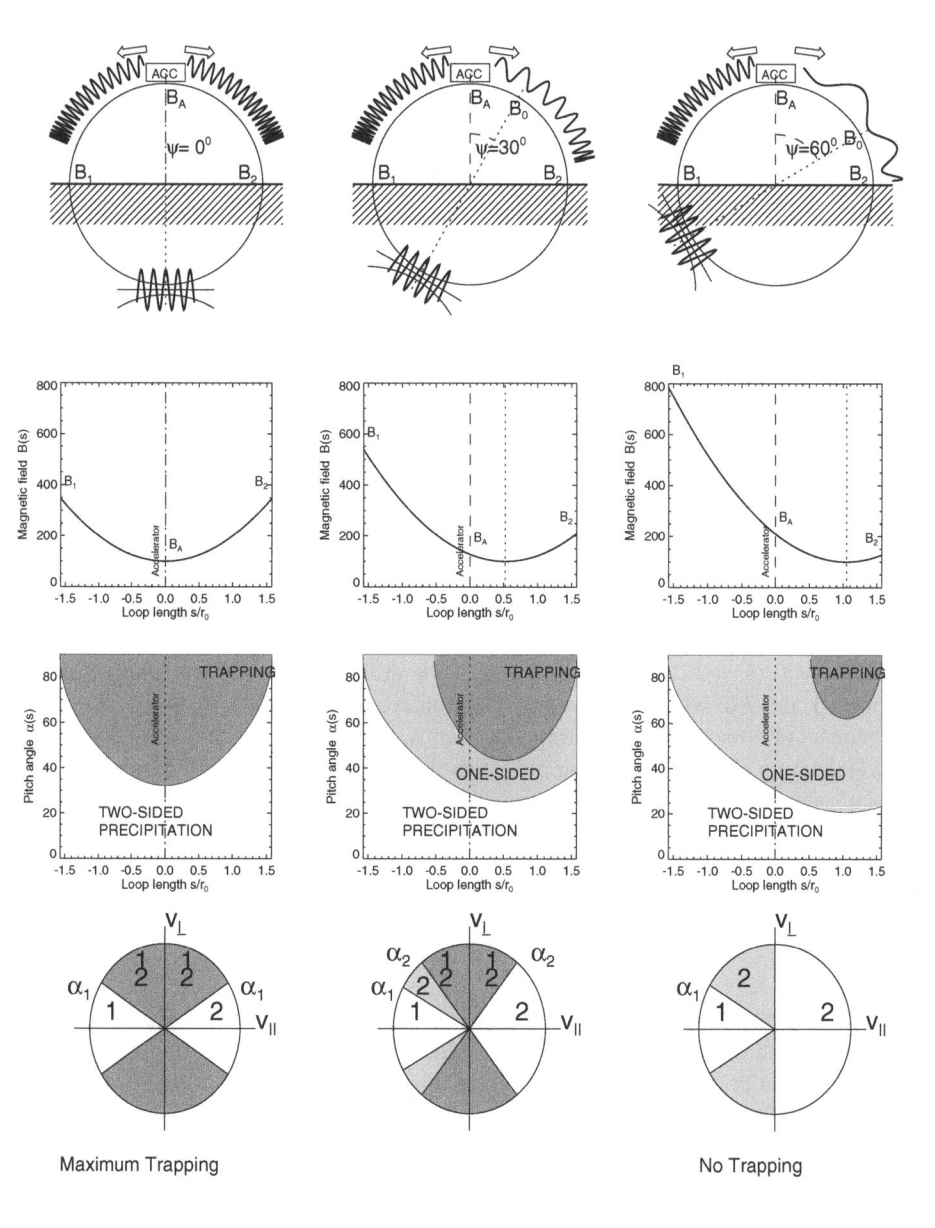

Figure 12.18: Three model scenarios for a symmetric (left column), a slightly asymmetric (middle column), and a strongly asymmetric magnetic trap (right column). The figure shows the spatial configuration of a buried dipole and the resulting pitch angle motion of trapped and/or precipitating electrons (top row), the magnetic field $B(s)$ parametrized as a function of the loop coordinate s (second row), the pitch angle variation $\alpha(s)$ as a function of the loop coordinate and three pitch angle regimes (third row), and the corresponding pitch angle regimes in velocity space ($v_{\parallel}, v_{\perp}$) (bottom row). The numbers 1 and 2 correspond to the left and right losscone site, with the stronger magnetic field located at footpoint 1 (Aschwanden et al. 1999d).

For positive rotation angles $\psi > 0$ the magnetic field at footpoint B_1 is stronger ($B_1 > B_2, R_1 > R_2$), and the critical angle, also called the *losscone angle*, is smaller ($\alpha_1 < \alpha_2$) than at the opposite footpoint 2. The asymmetric magnetic field $B(s)$ along a flare loop is visualized in Fig. 12.18 (second row) using a quadratic model. Note that in the case of strong asymmetry ($\psi = 60°$), trapping is not possible because $B_2 < B_A$, corresponding to a mirror ratio $R_2 < 1$.

Pitch angle variation $\alpha(s)$ along the loop according to conservation of the magnetic moment is shown in Fig. 12.18 (third row). Generally, three regimes can be distinguished in the $\alpha - s$ plane: (1) a direct precipitation regime for initial pitch angles $0 < \alpha(s = 0) < \alpha_1$, (2) a secondary precipitation regime after a single mirror bounce for initial pitch angles $\alpha_1 < \alpha(s = 0) < \alpha_2$, and (3) a trapping regime for initial pitch angles $\alpha_2 < \alpha(s = 0) < \pi/2$. These three regimes are clearly discernible in the case of weak asymmetry ($\psi = 30°$, Fig. 12.18, middle column). For the symmetric case, the secondary precipitation regime collapses to zero because $\alpha_1 = \alpha_2$ (Fig. 12.18 left column). For the strongly asymmetric case ($\psi = 60°$) no trapping is possible because there is no solution for $\alpha_2 < \pi/2$ with $R_2 < 1$.

Once we have a quantitative description of the pitch angle ranges that contribute to trapping and precipitation at both footpoints, we can calculate the relative fractions of precipitating electrons at the two footpoints and obtain quantitative expressions for the hard X-ray flux asymmetry A. Let us visualize the pitch angle regimes in velocity space $(v_\parallel, v_\perp)$ (Fig. 12.18 bottom) and label the different regimes with the footpoint numbers 1 and 2 to which the electrons precipitate, either directly, after a single mirror bounce, or after intermediate trapping. Let us now determine the relative fractions of precipitating electrons by integration over the corresponding pitch angle ranges in velocity space. Here and in the following we assume an isotropic pitch angle distribution at the acceleration/injection site [$f(\alpha) = const$]. The fraction q_{DP1} of directly precipitating electrons at footpoint 1, which has the smaller losscone ($\alpha_1 \leq \alpha_2$), is

$$q_{P1} = q_{DP1} = \frac{\int_{\pi - \alpha_1}^{\pi} f(\alpha) \sin(\alpha)\, d\alpha}{\int_0^\pi f(\alpha) \sin(\alpha)\, d\alpha} = \frac{(1 - \cos\alpha_1)}{2}. \tag{12.6.9}$$

The fraction of directly precipitating electrons at footpoint 2, which has the larger losscone angle (α_2), includes not only those electrons that precipitate without bouncing (q_{DP2}) but also those that bounce once at the mirror site 1 (q_{MP2}) and then precipitate at footpoint 2 [i.e., with initial pitch angles of $(\pi - \alpha_2) < \alpha(s = 0) < (\pi - \alpha_1)$],

$$q_{P2} = q_{DP2} + q_{MP2} = \frac{\int_0^{\alpha_2} f(\alpha) \sin(\alpha)\, d\alpha + \int_{\pi - \alpha_2}^{\pi - \alpha_1} f(\alpha) \sin(\alpha)\, d\alpha}{\int_0^\pi f(\alpha) \sin(\alpha)\, d\alpha}$$

$$= \frac{(1 + \cos\alpha_1 - 2\cos\alpha_2)}{2}. \tag{12.6.10}$$

From spatially unresolved data (e.g., from *CGRO*), only the combined fraction q_{prec} of directly precipitating electrons at both footpoints can be measured, equivalent to the sum of both footpoint components q_{P1} and q_{P2},

$$q_{prec} = q_{P1} + q_{P2} = 1 - \cos\alpha_2. \tag{12.6.11}$$

The total fraction of trapped electrons is determined by the pitch angle range of the larger losscone [i.e., $\alpha_2 < \alpha(s = 0) < (\pi - \alpha_2)$],

$$q_T = \frac{\int_{\alpha_2}^{\pi - \alpha_2} f(\alpha) \sin(\alpha) \, d\alpha}{\int_0^{\pi} f(\alpha) \sin(\alpha) \, d\alpha} = \cos \alpha_2 = (1 - q_{prec}) . \tag{12.6.12}$$

Trapped electrons are randomly scattered, but their pitch angle increases statistically with time, until they diffuse into a losscone. Because collisional deflection is an accumulative process of many small-angle scattering deflections that can add up to a small or large net value after each loop transit, there is some probability that trapped electrons can escape on either losscone side. Escape probability through the side of the larger losscone is higher than from the opposite side with the smaller losscone. To first order, we estimate that the relative escape probabilities from the trap towards the two losscone sides is proportional to the probabilities for direct precipitation at the two losscone sides (for particles with pitch angles $\alpha \leq \alpha_2$),

$$q_{T1} = q_T \left(\frac{q_{P1}}{q_{prec}} \right) , \tag{12.6.13}$$

$$q_{T2} = q_T \left(\frac{q_{P2}}{q_{prec}} \right) . \tag{12.6.14}$$

The proportionality also implies symmetric escape probabilities for symmetric losscones. We emphasize that this proportionality assumption $q_{T1}/q_{T2} \approx q_{P1}/q_{P2}$ represents an approximation that is mathematically simple and seems to agree well with (unpublished) numerical simulations of the asymmetric escape probabilities using the Fokker–Planck equation. This proportionality assumption is strictly true for the *strong-diffusion limit* (where scattering time into the losscone is shorter than the bounce time, see Eq. 12.5.13), but it also seems to hold approximately for the *weak-diffusion limit*.

To estimate the relative hard X-ray fluxes at both footpoints, we have to sum the precipitating and trapped contributions at both sides. We denote the combined fractions at both footpoints by q_1 and q_2,

$$q_1 = q_{P1} + q_{T1} = q_{P1}(1 + \frac{q_T}{q_{prec}}) = q_{P1}(\frac{1}{q_{prec}}) , \tag{12.6.15}$$

$$q_2 = q_{P2} + q_{T2} = q_{P2}(1 + \frac{q_T}{q_{prec}}) = q_{P2}(\frac{1}{q_{prec}}) . \tag{12.6.16}$$

Neglecting differences in the spectral slope (e.g., arising from asymmetric accelerators or asymmetric coronal energy loss), the hard X-ray flux at a given energy ε is proportional to the number of (nonthermal) electrons with energies $\varepsilon \gtrsim \epsilon_X = h\nu_X$. Assuming a similar spectral slope of the electron injection spectrum towards the two opposite directions 1 and 2, the hard X-ray fluences $F_1 = \int f_1(t)dt$ and $F_2 = \int f_2(t)dt$ observed at the two footpoints are then expected to be proportional to the precipitating electron fluxes q_1 and q_2. This constitutes a relation between the observed hard X-ray flux asymmetry A and the losscone angle α_2,

$$A = \frac{F_2}{(F_1 + F_2)} = \frac{q_2}{(q_1 + q_2)} = q_2 = (1 - q_1) . \tag{12.6.17}$$

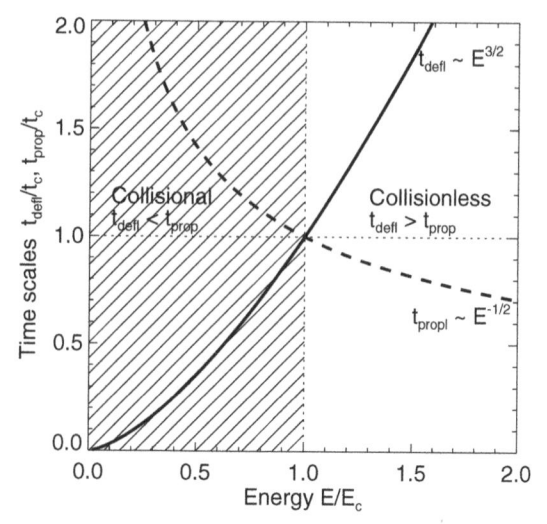

Figure 12.19: Energy dependence of the collisional deflection time (thick curve) and the propagation time (dashed curve). The critical energy E_c where they intersect separates the collisional from the collisionless regime.

With these relations we have a simple method of determining the two losscone angles α_1 and α_2 from the observables q_{prec} and A,

$$\alpha_1 = \arccos\left[1 - 2(1-A)q_{prec}\right] , \qquad (12.6.18)$$

$$\alpha_2 = \arccos\left[1 - q_{prec}\right] , \qquad (12.6.19)$$

as well as the corresponding magnetic mirror ratios R_1 and R_2 with Eqs. (12.6.7−8), or the ratios of the magnetic fields B_1/B_A and B_2/B_A, respectively. Most observed double-footpoint flares indeed show asymmetric hard X-ray fluxes (Aschwanden et al. 1999d). The asymmetry of magnetic field configurations in flare loops produces a higher precipitating electron flux from the side with higher hard X-ray fluxes, but increases the trapping efficiency and related emission from the other side, such as gyrosynchrotron emission from trapped highly relativistic electrons. This asymmetric radio emission has been modeled by the so-called *cornupia model* (Li et al. 1997), and the complementarity of asymmetric hard X-ray and radio emission has been verified in several observations (e.g. Wang et al. 1995).

12.6.3   Collisional Limit

Hard X-ray emission $\gtrsim 25$ keV is observed at loop footpoints in most flares, which indicates that the propagation path is essentially collisionless during the propagation time. For nonthermal electrons with energies $\lesssim 25$ keV, energy loss in the corona is generally not negligible. Setting the propagation time equal to the collisional deflection time,

$$t_{prop}(\varepsilon) = \frac{L}{v_{\parallel}(\varepsilon)} \leq t_{defl}(\varepsilon) = 2 \times 10^8 \left(\frac{\varepsilon_{keV}^{3/2}}{n_e}\right)\left(\frac{20}{\ln\Lambda}\right) \quad \text{(s)} \qquad (12.6.20)$$

yields a lower limit for the particle energy required to cross a propagation path of length L across a density n_e. This critical energy ε_c is,

$$\varepsilon_c \geq 20 \sqrt{\left(\frac{L}{10^9 \text{ cm}}\right)\left(\frac{n_e}{10^{11} \text{ cm}^{-3}}\right)\left(\frac{0.7}{\cos\alpha}\right)} \quad \text{(keV)} \quad (12.6.21)$$

The dependence of the two time scales is shown in Fig. 12.19, where the intersection defines the critical energy ε_c. Hard X-ray images from *Yohkoh/HXT* usually show emission along the full length of the flare loops in the lowest energy channel ($Lo = 14 - 23$ keV), because electrons lose their energy before they reach the footpoints at these low energies, $\varepsilon < \varepsilon_c$ (Eq. 12.6.21).

Energy loss is also significant for trapped particles. So far we have treated particle kinematics mainly in the collisionless regime, where adiabatic particle motion is applied. This is justified in low-density regions in the corona, where particle propagation times are much shorter than energy loss times. For particle trapping, however, the collisional time scale for pitch angle scattering into the losscone, τ_{defl},

$$\tau_{defl} = \frac{\text{v}^2}{<\Delta\text{v}_\perp^2>} \quad (12.6.22)$$

is defined as half the energy loss time τ_{loss} (Trubnikov 1965; Spitzer 1967; Schmidt 1979),

$$\tau_{loss} = \frac{\varepsilon^2}{<\Delta\varepsilon^2>} \approx 2\tau_{defl} \, . \quad (12.6.23)$$

Thus, trapped particles lose a significant fraction of their energy in the trap before they precipitate, a second-order effect that is ignored in the estimate of trapping times and trap densities in § 12.6.2.

12.7 Summary

Particle kinematics, the quantitative analysis of particle trajectories, has been systematically explored in solar flares by performing high-precision energy-dependent time delay measurements with the large-area detectors of the Compton Gamma-Ray Observatory. There are essentially five different kinematic processes that play a role in the timing of nonthermal particles energized during flares: (1) acceleration, (2) injection, (3) free-streaming propagation, (4) magnetic trapping, and (5) precipitation and energy loss. The time structures of hard X-ray and radio emission from nonthermal particles indicate that the observed energy-dependent timing is dominated either by free-streaming propagation (obeying the expected electron time-of-flight dispersion) or by magnetic trapping in the weak-diffusion limit (where the trapping times are controlled by collisional pitch angle scattering). There is no evidence that any of the observed energy-dependent hard X-ray timing corresponds to a theoretical particle acceleration process. The most plausible explanation for this fact is that particles are ejected from the acceleration region by an energy-independent synchronization process (e.g., by the relaxation

of newly reconnected magnetic field lines in the reconnection outflow). In such
a relaxation process, the magnetic mirror ratio opens up the losscone during the
Alfvén transit of the diffusion region and injects particles during subsecond pulses
into free-streaming orbits away from the reconnection region. Particles with small
pitch angles precipitate directly and preserve the pulse structure of the injection
process, while particles with large pitch angles become intermediately trapped
and precipitate latest after a collision deflection time. For particles with kinetic
energies of $\lesssim 20$ keV, the propagation time to the flare loop footpoints may exceed the collisional deflection time, and thus they are stopped in the coronal part
of the flare loop. Particle kinematics is a novel key tool for the interpretation of
nonthermal hard X-ray, γ-ray, and radio emission.

Chapter 13

Hard X-Rays

Photons with energies in hard X-ray wavelengths ($\epsilon_x \approx 10 - 300$ keV) are produced by particles in a collisional plasma, mostly by collisions between relativistic electrons and thermal ions. Since we cannot place *in situ* particle detectors into the solar corona, we have to obtain information on energetic particles from solar flares by remote-sensing of hard X-ray photons and radio photons. Moreover, since the Earth's atmosphere absorbs hard X-rays, solar hard X-rays can only be detected by space-borne detectors. Major contributions to hard X-ray observations have been made over the last 25 years from hard X-ray spectrometers on board *SMM* and *CGRO*, and from the hard X-ray imagers *SMM/HXIS*, *Hinotori*, *Yohkoh/HXT*, and *RHESSI*. Solar flare hard X-ray emission provides us with a key diagnostic on particle acceleration and propagation processes. However, since hard X-ray emission is produced most prolifically when nonthermal electrons precipitate from collisionless coronal sites towards the highly collisional chromosphere, most information comes from mapping the electron precipitation sites, their energy-dependent timing, their energy spectra, and their temporal correlation with flare signatures in other wavelengths. From these pieces of information, we have to work backward to reconstruct the magnetic topology in flare regions, to localize the acceleration sites with respect to magnetic reconnection diffusion regions, the particle trajectories of free-streaming and trapped particles, and ultimately to attempt a diagnostic on the physical processes of energization and acceleration in the first place. A major breakthrough during the last decade was the discovery of coronal above-the-looptop sources and electron time-of-flight measurements, which both pinpoint consistently the acceleration and injection of particles in reconnection regions. At the time of writing we are witnessing pioneering results from *RHESSI*, which provides the first hard X-ray high-resolution imaging spectroscopy, the first high-resolution γ-ray line spectroscopy, and the first imaging at energies above 100 keV.

Reviews on flare-related hard X-ray emission can be found in Emslie & Rust (1980), Dennis (1985, 1988), De Jager (1986), Dennis et al. (1987), Vilmer (1987), Dennis & Schwartz (1989), Bai & Sturrock (1989), Brown (1991), Culhane & Jordan (1991), Hudson & Ryan (1995), Aschwanden (1999b, 2000b, 2002b), Lin (2000), and Vilmer & MacKinnon (2003).

Table 13.1: Compilation of hard X-ray and gamma-ray detectors and imagers used for solar flare observations.

Spacecraft	Instrument or Detector	Energy range	Time resolution	Detector area	Imaging resolution	Operation period
OSO-5	CsI(Na)[1]	20−250 keV	1.8 (0.2) s	71 cm^2	−	1969−75
OSO-7	NaI(Ti)[2]	10−300 keV	10, (2.5) s	9.57 cm^2	−	1971−74
(Balloon)	NaI(Ti)[3]	> 30 keV	0.1 s	60 cm^2	−	1974
(Balloon)	Ge[4]	13−300 keV	0.008 s	300 cm^2	−	1981
Hinotori	SXT[5]	10−40 keV	7 s	113 cm^2	20″	1981−82
	HXM[5]	17−340 keV	0.125 s	57 cm^2	−	1981−82
	SGR[5]	0.21−6.7 MeV	2 s	62 cm^2	−	1981−82
ISEE-3	NaI(Tl)[6]	12−1250 keV	0.125 s	22 cm^2	−	1978−...
SMM	HXIS[7]	3.5−30 keV	0.5−7 s	1.44 cm^2	8″, 32″	1980−89
	HXRBS[8]	20−260 keV	0.128 s	71 cm^2	−	1980−89
	GRS[9]	0.3−9 MeV	16, (2) s	200 cm^2	−	1980−89
CGRO	BATSE[10]	20−300 keV	0.064 s	2025 cm^2	−	1991−00
	OSSE[11]	0.05−10 MeV	2, (0.016) s	2620 cm^2	−	1991−00
Yohkoh	HXT[12]	14−93 keV	0.5 s	70 cm^2	5″	1991−01
	WBS[13]	X, γ	2, (0.25) s	12 cm^2	−	1991−01
RHESSI	HPGe[14]	3 keV−20 MeV	2 s	90 cm^2	2.3″	2002−...

Sources: [1]*) Frost (1969), Frost et al. (1971);* [2]*) Datlowe et al. (1974); Harrington et al. (1972);* [3]*) Hurley & Duprat (1977);* [4]*) Lin et al. (1981);* [5]*) Makishima (1982), Takakura et al. (1983a), Enome (1983), Tsuneta (1984);* [6]*) Anderson et al. (1978);* [7]*) Van Beek et al. (1980);* [8]*) Orwig et al. (1980);* [9]*) Forrest et al. (1980);* [10]*) Fishman et al. (1989);* [11]*) Kurfess et al. (1998);* [12]*) Kosugi et al. (1991);* [13]*) Yoshimori et al. (1991);* [14]*) Lin et al. (1998).*

13.1 Hard X-ray Instruments

In Table 13.1 we provide a compilation of hard X-ray detectors and imagers that have made major contributions to the study of solar flares. Most of them are space-based instruments, attached to solar-dedicated or all-sky astrophysical missions, and a few instruments have also been flown on balloon flights. In the following we describe in more detail the four instruments that collected most of the flare observations.

13.1.1 SMM − HXRBS, GRS, HXIS

The *Solar Maximum Mission (SMM)* was the first solar flare-dedicated mission, lasting a full decade, from 1980-Feb-4 to 1989-Dec-2. The scientific highlights of the SMM mission are reviewed in the monograph of Strong et al. (1999). The instrument suite contained three hard X-ray instruments, the *Hard X-Ray Burst Spectrometer (HXRBS)* (Orwig et al. 1980), the *Gamma-Ray Spectrometer (GRS)* (Forrest et al. 1980), and the *Hard X-Ray Imaging Spectrometer (HXIS)* (Van Beek et al. 1980). SMM was in an orbit with an initial altitude of 524 km and an inclination of $28.6°$.

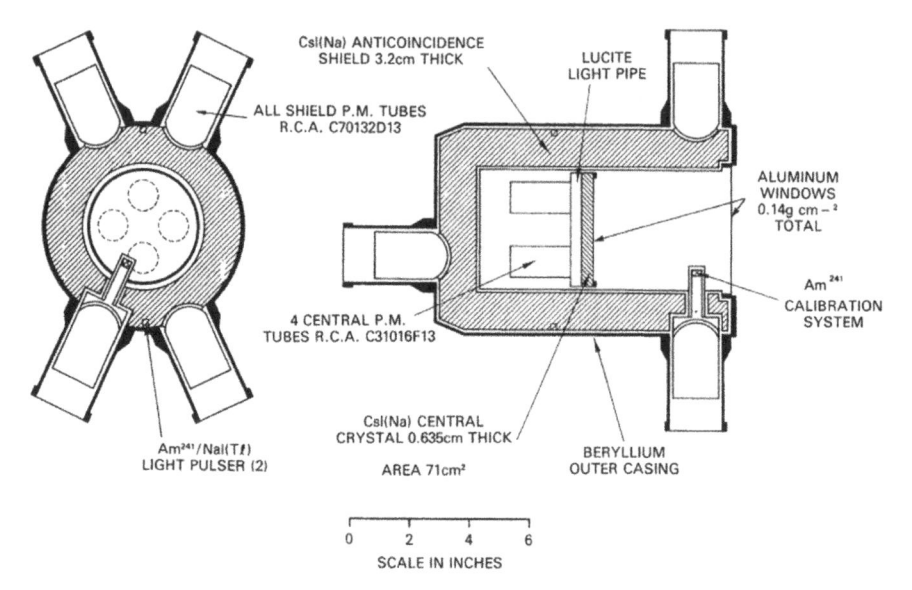

Figure 13.1: Cross-sectional views of the *Hard X-Ray Burst Spectrometer (HXRBS)* on board the *Solar Maximum Mission (SMM)* spacecraft (Orwig et al. 1980).

HXRBS provided high-time resolution (0.128 s) histories of hard X-ray time profiles in 16 channels in the energy range of $\approx 20 - 255$ keV. The *HXRBS* instrument consisted of a central CsI(Na) detector surrounded by a CsI active collimator element (Fig. 13.1). The CsI crystal was viewed by four photomultipliers, operated in anticoincidence. The duty cycle of the instrument was about 50% and *HXRBS* recorded over 12,000 solar flare events (see catalog by Dennis et al. 1991).

GRS observed at higher energies, in the range of $0.3-9$ MeV, and recorded some 270 γ-ray flares at > 300 keV. *HXIS* was the first instrument to image hard X-ray flares, with a resolution of $32''$, and in a fine FOV with $8''$-pixels. Because the *HXIS* energy range was $3.5 - 30$ keV, the images are dominated by thermal emission.

13.1.2 Yohkoh – HXT

Shortly after the SMM decade, the first hard X-ray imager at energies > 30 keV was launched on board the *Yohkoh* spacecraft, called the *Hard X-Ray Telescope (HXT)* (Kosugi et al. 1991), while the context images in soft X-rays were recorded with the *Soft X-Ray Telescope (SXT)*. The *Yohkoh* mission was a solar flare-dedicated mission, and both of the instruments switched to higher time and spatial resolution in flare mode. *HXT* is a Fourier synthesis imager with 64 collimator detectors, each one with a bi-grid collimator in front (Fig. 13.2), providing measurements of the sine and cosine of 32 independent spatial Fourier components of the source. Images can be reconstructed

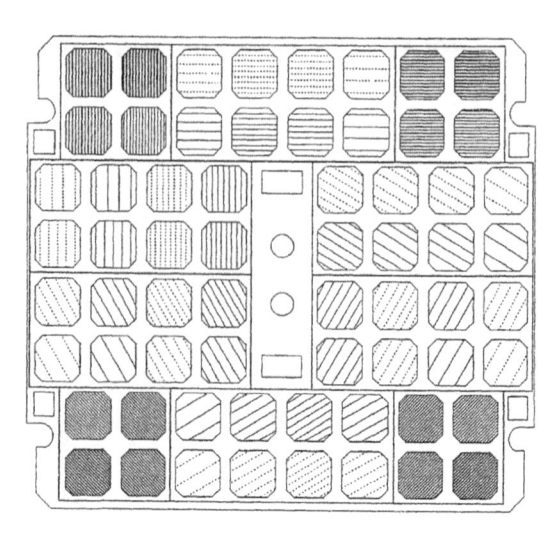

Figure 13.2: Arrangement of the 64 subcollimators of *Yohkoh/HXT* which record 64 (1D) Fourier components. The cosine and sine Fourier element pairs are shown with solid and dashed linestyles (Kosugi et al. 1991).

from these measurements using algorithms such as the *Maximum Entropy Method (MEM)* and *CLEAN*. *HXT* was able to produce images in four energy bands, $15 - 24$ keV, $24 - 35$ keV, $35 - 57$ keV, and $57 - 100$ keV, with an angular resolution of $\approx 5'' - 8''$ and a time resolution of 0.5 s. The lifetime of *Yohkoh* extended over a full solar cycle, and *HXT* recorded a total of 3,112 flare events during the period from 1991-Oct-1 to 2001-Dec-14, documented in the *Yohkoh HXT/SXT Flare Catalogue* (Sato et al. 2003).

13.1.3   CGRO – BATSE

The *Burst and Transient Source Experiment (BATSE)* on the *Compton Gamma-Ray Observatory* is the most sensitive all-sky hard X-ray and γ-ray detector system ever flown. It consists of eight *large-area detectors (LADs)*, with an area of 2025 cm^2 each, placed at the eight corners of the spacecraft (see Fig. 1.3). In addition, *BATSE* is also equipped with *spectroscopy detectors (SDs)*, which consist of NaI(Tl) scintillators with a front area of 127 cm^2, which cover a broad energy range of 15 keV-110 MeV and have 7.2% energy resolution. *BATSE* was operated in different energy and time binning modes, triggering automatically to a higher rate after a burst trigger. Weak solar flare events were recorded with a low time resolution of 1.024 s in four energy channels ($25 - 50, 50 - 100, 200 - 300, >300$ keV), while burst trigger events were recorded with 16 ms and 64 ms time resolution in 16 energy channels. *CGRO* operated during the period from 1991-Apr-5 to 2000-Jun-4, and *BATSE* recorded a total of 8021 burst triggers with the following identification of events: 2704 astrophysical gamma-ray bursts, 1192 solar flares, 1717 magnetospheric events, 78 terrestrial gamma flashes, 2003 transient sources, and 185 soft gamma repeaters.

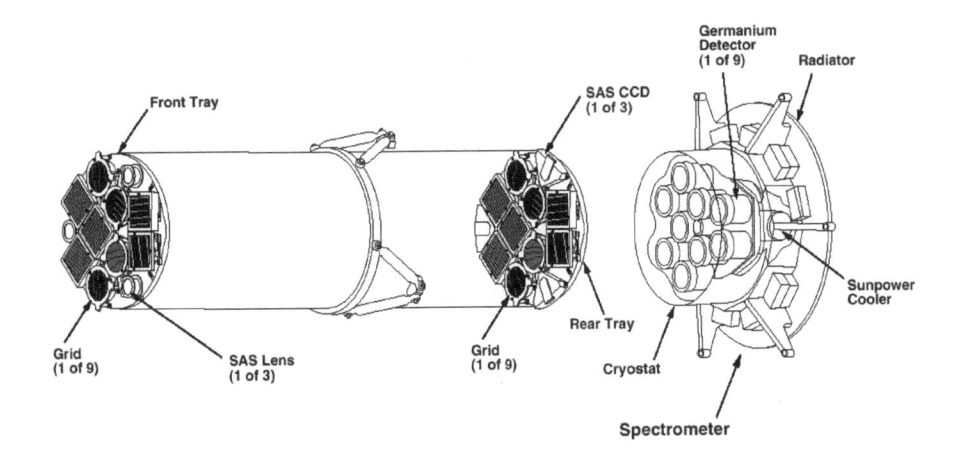

Figure 13.3: Layout of the RHESSI telescope that was mounted on a rotating spacecraft. The telescope (left) contains a set of nine front grids and nine identical rear grids which together modulate the incoming hard X-ray photons. The mounting of the nine grids (left) is also shown: The grid pitch (slit and slat) increases by a factor of $\sqrt{3}$ from grid 1 to 9, so that each one modulates a particular angular Fourier period. The modulated throughput is detected by nine cooled germanium detectors, one behind each of the rear grids (right), (Hurford et al. 2002).

13.1.4 RHESSI

The *Reuven Ramaty High Energy Solar Spectroscopic Imager (RHESSI)*, launched on 2002-Feb-5, is a Fourier imager of the class of *rotation-modulated collimators (RMCs)*, which spins with a period of ≈ 4 s. *RHESSI* uses nine collimators, each one made up of a pair of widely separated grids. Each grid is a planar array of equally spaced, X-ray-opaque slats separated by transparent slits (Fig. 13.3). The slits of each pair of grids are parallel to each other and their pitches are identical, so that transmission through the grid pair depends on the direction of the incident X-rays. Different Fourier components are measured at different rotation angles and with grids of different pitches. For RHESSI, the grid pitches range from $p = 34$ μm to 2.75 mm in steps of $\sqrt{3}$. This provides angular resolutions spaced logarithmically from 2.3″ to 180″. The images are reconstructed with algorithms like *Clean, Maximum Entropy Method (MEM), Maximum Entropy Method Visibilities (MEMVis), Pixon*, and *Forward-Fitting*. RHESSI is the first telescope to provide hard X-ray imaging at such high angular resolution.

Another prime capability of *RHESSI* is its high energy resolution, thanks to the cooled germanium detectors (i.e., $\lesssim 1$ keV FWHM at 3 keV, increasing to ≈ 5 keV at 5 MeV). This allows the many γ-ray lines with typical FWHMs of $2 - 100$ keV in the $1 - 10$ MeV range to be resolved for the first time. Instrumental descriptions of the RHESSI instrument can be found in Lin et al. (1998, 2002), the imaging concept is described in Hurford et al. (2002), and the spectrometer in Smith et al. (2002).

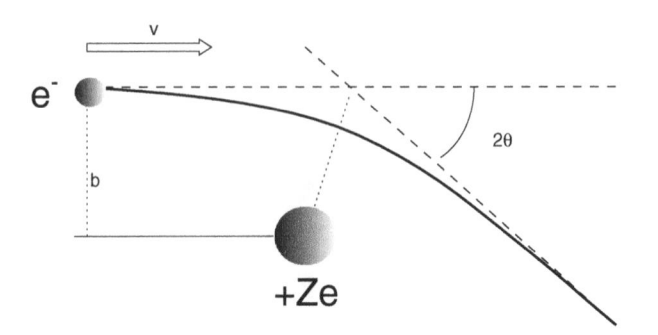

Figure 13.4: Elastic scattering of an electron (e^-) off a positively charged ion $(+Ze)$. The electron moves with velocity v on a path with impact parameter b and is deflected by an angle of 2ϑ, with $\tan \vartheta/2 = Ze^2/(mv^2b)$, according to the Rutherford formula. Electromagnetic radiation (bremsstrahlung) is emitted as a consequence of the acceleration of the particle during the swing-by around the ion.

13.2 Bremsstrahlung

The most important radiation mechanism that produces a continuum of emission in hard X-ray wavelengths is *bremsstrahlung*, which results from the emission of photons when electrons are elastically scattered in the electric Coulomb field of ambient ions (Fig. 13.4). We distinguish three different situations: (1) *Thermal bremsstrahlung* results when the colliding electrons have the same temperature as the ambient plasma (§2.3); (2) *thick-target bremsstrahlung* occurs when the incident electrons have first been accelerated to a much higher (nonthermal) energy (in a collisionless plasma) and then become collisionally stopped when they hit a thermal plasma; and (3) *thin-target bremsstrahlung* occurs when electrons are continuously accelerated in a collisional plasma and the X-ray spectrum is nearly unchanged from the acceleration or injection spectrum.

13.2.1 Bremsstrahlung Cross Sections

Elastic scattering of a single electron $(-e)$ off an ion with charge $(+Ze)$ is quantified by the *differential scattering cross section* $d\sigma_s/d\Omega$ using the *Rutherford formula* (Eq. 2.3.5). The derivation of this *differential scattering cross section* for bremsstrahlung in Coulomb collisions can be found in standard textbooks on classical electrodynamics (e.g., Jackson 1962; §15.2). The *radiation cross section* $Q_r(v, \nu)$, which specifies how much energy is radiated in bremsstrahlung photons at frequency $\nu, ..., \nu + d\nu$ by an incident electron with velocity v is calculated by integration over all scattering angles ϑ or impact parameters b within the possible range of $b_{min} \leq b \leq b_{max}$ (Eq. 2.3.7; Fig. 2.5). In the classical derivation of bremsstrahlung, the upper limit is estimated by the maximum momentum transfer $b_{max} = 2p = 2m_ev$, and the lower limit is set by the collision time. The integral of this radiation cross section is also called the *Coulomb integral* $\ln(\Lambda)$ (or *Gaunt factor* $g(\nu, T)$, see definition in Eq. 2.3.8). The classical derivation is nonrelativistic, and conservation of momentum and photon energy are not considered; it is thus only valid for thermal bremsstrahlung (§2.3).

For nonthermal bremsstrahlung, however, conservation of energy and momentum must be considered, which is (for weakly relativistic electrons),

$$\varepsilon = \varepsilon' + h\nu \,, \tag{13.2.1}$$

$$b^2 = (\mathbf{p} - \mathbf{p'} - \mathbf{k})^2 \approx (\mathbf{p} - \mathbf{p'})^2 \,, \tag{13.2.2}$$

where $\varepsilon = p^2/2m_e$ and $\varepsilon' = p'^2/2m_e$ are the kinetic energies of the electrons before and after the collision, $h\nu$ and $\mathbf{k} = h\nu/c$ are the energy and momentum of the photon, and b is the momentum transferred to the scattering center. The ratio of the maximum to the minimum momentum transfer is given by the kinematic limits $(p+p')$ and $(p-p')$,

$$\frac{b_{max}}{b_{min}} = \frac{p+p'}{p-p'} = \frac{\sqrt{\varepsilon}+\sqrt{\varepsilon'}}{\sqrt{\varepsilon}-\sqrt{\varepsilon'}} = \frac{(\sqrt{\varepsilon}+\sqrt{\varepsilon'})^2}{\varepsilon - \varepsilon'} = \frac{(\sqrt{\varepsilon}+\sqrt{\varepsilon - h\nu})^2}{h\nu} \,, \tag{13.2.3}$$

which leads to the nonrelativistic (or weakly relativistic) *Bethe–Heitler cross section* (Jackson 1962; Eq. 15.29)

$$Q_r(\mathrm{v},\nu) = \frac{16}{3}\frac{Z^2e^2}{c}\left(\frac{z^2e^2}{mc^2}\right)^2\frac{1}{\beta^2}\ln\left[\lambda'\frac{(\sqrt{\varepsilon}+\sqrt{\varepsilon-h\nu})^2}{h\nu}\right] \,. \tag{13.2.4}$$

The numerical factor $\lambda' = 1$ corresponds to the quantum-mechanical result in the *Born approximation*, first calculated by Bethe & Heitler in 1934. Inserting the fine structure constant $\alpha_{FS} = e^2/\hbar c = 1/137$, the classical electron radius $r_e = e^2/m_ec^2 = 2.8 \cdot 10^{-13}$ cm, the kinetic energy of an electron $\varepsilon = m_ec^2(\gamma - 1) \approx (1/2)m_ec^2\beta^2$, considering only collisions between electrons ($z = 1, m = m_e$) and protons ($Z = 1$), and using the Born approximation ($\lambda' = 1$), the cross section reads,

$$Q_r(\mathrm{v},\nu) = \frac{16}{3}\hbar\,\alpha_{FS}\,r_e^2\frac{\frac{1}{2}m_ec^2}{\varepsilon}\ln\left[\frac{(\sqrt{\varepsilon}+\sqrt{\varepsilon-h\nu})^2}{h\nu}\right] \quad (\mathrm{cm}^2\ \mathrm{erg}\ \mathrm{Hz}^{-1}) \,. \tag{13.2.5}$$

The *radiation cross section* $Q_r(\mathrm{v},\nu)$ (per frequency unit ω) can be converted into a *photon cross section* $\sigma(\varepsilon,\epsilon_x)$, per photon energy $\epsilon_x = h\nu = \hbar\omega$, which is defined by (e.g., Lang 1980, p. 43),

$$\sigma(\varepsilon,\epsilon_x) = \frac{Q_r(\mathrm{v},\nu)}{\hbar^2\omega} = \frac{1}{\hbar}\frac{Q_r(\mathrm{v},\nu)}{\epsilon_x} \tag{13.2.6}$$

which yields the commonly used form of the Bethe–Heitler cross section (e.g., Brown 1971; Hudson et al. 1978)

$$\sigma(\varepsilon,\epsilon_x) \approx \frac{\sigma_0}{\epsilon_x\varepsilon}\ln\left[\sqrt{\frac{\varepsilon}{\epsilon_x}}+\sqrt{\frac{\varepsilon}{\epsilon_x}-1}\right] \quad (\mathrm{cm}^2\ \mathrm{keV}^{-1}) \tag{13.2.7}$$

$$\sigma_0 = \frac{8}{3}\alpha_{FS}\,r_e^2\,m_ec^2 = 0.78 \times 10^{-24} \quad (\mathrm{cm}^2\ \mathrm{keV}) \,. \tag{13.2.8}$$

A simpler form can sometimes be used which neglects the logarithmic term; this is the *Kramers cross section*,

$$\sigma(\varepsilon,\epsilon_x) \approx \frac{\sigma_0}{\epsilon_x\varepsilon} \,. \tag{13.2.9}$$

The *Bethe–Heitler* and *Kramers* cross sections are only applicable to nonrelativistic or mildly relativistic electrons. For higher energies, a fully relativistic cross section has been derived (Elwert 1939; Koch & Motz 1959), which was applied in the form of a multiplicative *Elwert factor* by Holt & Cline (1968). An expansion of the fully relativistic cross section up to sixth order of the momentum $p = \sqrt{\varepsilon/2m_e}$ is given in Haug (1997). Comparisons show that the relative error using the nonrelativistic cross section can exceed 10% even for mildly relativistic energies of $\varepsilon_1 > 30$ keV (Haug 1997).

13.2.2 Thick-Target Bremsstrahlung

Total X-ray emission from an emitting volume V is proportional to the number of collisions between electrons and ions (mainly protons) [i.e., to the product of their densities integrated over the volume, $\int n_p n_e(\varepsilon) dV$]. The bremsstrahlung cross section $\sigma(\varepsilon, \epsilon_x)$ has the unit of a target area per photon energy (cm^2 keV^{-1}), see Eq. (13.2.7), so the product of the cross section $\sigma(\varepsilon, \epsilon_x)$ with the velocity v(ε) of the incident electrons corresponds to a target volume per time unit. To obtain the total number of emitted photons (of a given photon energy $\epsilon_x = h\nu$) we also have to integrate over all contributions from electrons with energies higher than the photon energy (i.e., $\varepsilon \geq \epsilon_x$),

$$\frac{dN_{phot}}{dt\, d\epsilon_x} = \int_{\epsilon_x}^{\infty} \sigma(\varepsilon, \epsilon_x)\, \mathrm{v}(\varepsilon) \left(\int n_p n_e(\varepsilon) dV \right) d\varepsilon \qquad (\mathrm{s}^{-1}\, \mathrm{keV}^{-1}) , \quad (13.2.10)$$

which has the unit of photons per time and energy (s^{-1} keV^{-1}). Assuming a uniform target density over the volume, so that n_p is constant and defining $n_0 = \int n_p dV$, the mean photon count rate at Earth distance, $r = 1$ AU, has to be scaled by a factor of $1/4\pi r^2$, yielding an observed hard X-ray intensity $I(\epsilon_x)$ of

$$I(\epsilon_x) = \frac{dN_{phot}}{4\pi r^2\, dt\, d\epsilon_x} = \frac{1}{4\pi r^2} n_0 \int_{\epsilon_x}^{\infty} \sigma(\varepsilon, \epsilon_x)\, \mathrm{v}(\varepsilon) n_e(\varepsilon) d\varepsilon , \qquad (13.2.11)$$

which has the units of photons per detector area, time, and energy. If we plug the nonrelativistic electron velocity v$(\varepsilon) = \sqrt{2\varepsilon/m_e}$ and the Bethe–Heitler cross section $\sigma(\varepsilon, \epsilon_x)$ (Eq. 13.2.7) into the integral of the hard X-ray spectrum $I(\epsilon_x)$ (Eq. 13.2.11), we obtain

$$I(\epsilon_x) = \frac{1}{4\pi r^2} n_0 \int_{\epsilon_x}^{\infty} \frac{\sigma_0}{\epsilon_x \varepsilon} \ln\left[\sqrt{\frac{\varepsilon}{\epsilon_x}} + \sqrt{\frac{\varepsilon}{\epsilon_x} - 1} \right] \left(\frac{2\varepsilon}{m_e} \right)^{1/2} n_e(\varepsilon) d\varepsilon . \quad (13.2.12)$$

This equation implicitly defines the *instantaneous nonthermal electron spectrum* $n_e(\varepsilon)$ that is present in the hard X-ray emitting source. For thick-target emission, however, the *electron injection spectrum* $f_e(\varepsilon)$ that is externally injected into the hard X-ray emitting source is different from $n_e(\varepsilon)$. The transformation from the injection spectrum $f_e(\varepsilon)$ to the source spectrum $n_e(\varepsilon)$ is defined by the energy loss process. If energy losses are purely Coulomb-collisional, we have the following energy loss function,

$$\frac{d\varepsilon}{dt} = -\frac{Kn(\varepsilon)\mathrm{v}(\varepsilon)}{\varepsilon} , \qquad K = 4\pi e^4 \Lambda . \qquad (13.2.13)$$

The total number of photons emitted from an electron with initial energy ε_0 at photon energy ϵ_x during braking, as long as $\varepsilon > \epsilon_x$, is

$$\nu(\epsilon_x, \varepsilon_0) = \int_{t(\varepsilon=\varepsilon_0)}^{t(\varepsilon=\epsilon_x)} \sigma(\epsilon_x, \varepsilon) n(\varepsilon) \mathrm{v}(\varepsilon) dt , \qquad (13.2.14)$$

which can be written as an energy integral by substituting $n(\varepsilon)\mathrm{v}(\varepsilon)dt = -d\varepsilon\,(\varepsilon/K)$ from Eq.(13.2.13),

$$\nu(\epsilon_x, \varepsilon_0) = \int_{\epsilon_x}^{\varepsilon_0} \sigma(\epsilon_x, \varepsilon) \frac{\varepsilon}{K} d\varepsilon . \qquad (13.2.15)$$

Since the electrons are decelerated until they are at rest in the target, the ambient plasma density is not relevant for the emitted photon flux $\nu(\epsilon_x, \varepsilon_0)$ (Eq. 13.2.15), which is thus independent of n_p. The total photon emission rate from the region $I(\epsilon_x)$ can then be expressed in terms of the *electron injection spectrum* $f_e(\varepsilon)$ incident into the X-ray-emitting region per second (with Eq. 13.2.15),

$$I(\epsilon_x) = \frac{dN_{phot}}{4\pi r^2\,dt\,d\epsilon_x} = \frac{1}{4\pi r^2} \int_{\epsilon_x}^{\infty} \nu(\epsilon_x, \varepsilon_0)\,f_e(\varepsilon_0) d\varepsilon_0 \qquad (13.2.16)$$

$$= \frac{1}{4\pi r^2} \int_{\epsilon_x}^{\infty} f_e(\varepsilon_0) \left[\int_{\epsilon_x}^{\varepsilon_0} \sigma(\epsilon_x, \varepsilon) \frac{\varepsilon}{K} d\varepsilon \right] d\varepsilon_0 . \qquad (13.2.17)$$

This equation implicitly defines the *injection spectrum* $f_e(\varepsilon)$ by the observed photon spectrum $I(\epsilon_x)$.

The implicit equations for the *instantaneous source spectrum* $n_e(\varepsilon)$ (Eq. 13.2.12) and the *electron injection spectrum* $f_e(\varepsilon)$ (Eq. 13.2.17) can be transformed into *Abelian integral equations* (the analytical solution was calculated by Brown 1971). He assumed a powerlaw function for the observed hard X-ray spectrum $I(\epsilon_x)$ (i.e., left-hand side of Eqs. 13.2.12 and 13.2.17),

$$I(\epsilon_x) = I_1 \frac{(\gamma - 1)}{\epsilon_1} \left(\frac{\epsilon_x}{\epsilon_1} \right)^{-\gamma} \qquad \text{(photons cm}^{-2}\,\text{s}^{-1}\,\text{keV}^{-1}) , \qquad (13.2.18)$$

where ϵ_1 is a reference energy, above which the integrated photon flux is I_1 (photons cm^{-2} s^{-1} keV^{-1}), and γ is the powerlaw slope (not to be confused with the Lorentz factor). The parameters ϵ_1 and γ of the hard X-ray spectrum can be time-dependent. The total number of photons above a lower cutoff energy ϵ_1 is the integral of Eq. (13.2.18),

$$I(\epsilon_x \geq \epsilon_1) = \int_{\epsilon_1}^{\infty} I(\epsilon_x) d\epsilon_x = I_1 \qquad \text{(photons cm}^{-2}\,\text{s}^{-1}) . \qquad (13.2.19)$$

Brown (1971) solved the inversion of Eqs. (13.2.12) and (13.2.17) and found the following *instantaneous nonthermal electron spectrum* $n_e(\varepsilon)$ present in the X-ray-emitting region, with the associated *electron injection spectrum* $f_e(\varepsilon)$,

$$n_e(\varepsilon) = 3.61 \times 10^{41} \gamma(\gamma-1)^3\,B\left(\gamma - \frac{1}{2}, \frac{3}{2}\right) \frac{I_1 \sqrt{\varepsilon}}{n_0 \epsilon_1} \left(\frac{\varepsilon}{\epsilon_1} \right)^{-\gamma} \quad \text{(electrons keV}^{-1}) ,$$

$$(13.2.20)$$

$$f_e(\varepsilon) = 2.68 \times 10^{33} \gamma^2 (\gamma-1)^3 \, B\left(\gamma - \frac{1}{2}, \frac{3}{2}\right) \frac{I_1}{\epsilon_1^2} \left(\frac{\varepsilon}{\epsilon_1}\right)^{-(\gamma+1)} \text{(electrons keV}^{-1}\text{ s}^{-1}),$$

(13.2.21)

with n_0 (cm^{-3}) the mean electron or proton density in the emitting volume, ϵ_1 [keV] the lower cutoff energy in the spectrum, I_1 (photons cm^{-2} s^{-1} keV^{-1}) the total X-ray photon flux at energies $\epsilon \gtrsim \epsilon_1$, and $B(p, q)$ is the *Beta function*,

$$B(p, q) = \int_0^1 u^{p-1}(1 - u)^{q-1} du \,,$$

(13.2.22)

which is calculated in Hudson et al. (1978) for a relevant range of spectral slopes γ and is combined in the auxiliary function $b(\gamma)$,

$$b(\gamma) = \gamma^2 (\gamma - 1)^2 \, B\left(\gamma - \frac{1}{2}, \frac{3}{2}\right) \approx 0.27 \, \gamma^3 \,.$$

(13.2.23)

So the powerlaw slope of the electron injection spectrum ($\delta = \gamma + 1$) is steeper than that (γ) of the photon spectrum in the thick-target model. With this notation we can write the electron injection spectrum as

$$f_e(\varepsilon) = 2.68 \times 10^{33} \, (\gamma - 1) b(\gamma) \frac{I_1}{\epsilon_1^2} \left(\frac{\varepsilon}{\epsilon_1}\right)^{-(\gamma+1)} \text{(electrons keV}^{-1}\text{ s}^{-1})\,.$$

(13.2.24)

The total number of electrons above a cutoff energy ε_c is then

$$F(\varepsilon \geq \varepsilon_c) = \int_{\varepsilon_c}^{\infty} f_e(\varepsilon) d\varepsilon = 2.68 \times 10^{33} \, b(\gamma) \frac{(\gamma - 1)}{\gamma} \frac{I_1}{\epsilon_1} \left(\frac{\varepsilon_c}{\epsilon_1}\right)^{-\gamma} \text{(electrons s}^{-1})\,.$$

(13.2.25)

The power in nonthermal electrons above some cutoff energy ε_c is

$$P(\varepsilon \geq \varepsilon_c) = \int_{\varepsilon_c}^{\infty} f_e(\varepsilon) \, \varepsilon \, d\varepsilon = 2.68 \times 10^{33} \, b(\gamma) I_1 \left(\frac{\varepsilon_c}{\epsilon_1}\right)^{-(\gamma-1)} \text{(keV s}^{-1})\,.$$

(13.2.26)

or a factor of (keV/erg)$=1.6 \times 10^{-9}$ smaller in cgs units,

$$P(\varepsilon \geq \varepsilon_c) = \int_{\varepsilon_c}^{\infty} f_e(\varepsilon) \, \varepsilon \, d\varepsilon = 4.3 \times 10^{24} \, b(\gamma) I_1 \left(\frac{\varepsilon_c}{\epsilon_1}\right)^{-(\gamma-1)} \text{(erg s}^{-1})\,.$$

(13.2.27)

Solar flares have typical photon count rates in the range of $I_1 = 10^1 - 10^5$ (photons s^{-1} cm^{-2}) at energies of $\varepsilon \geq 20$ keV and slopes of $\gamma \approx 3$. Thus, for $\varepsilon_c = \epsilon_1 = 20$ keV, and using $b(\gamma) \approx 0.27\gamma^3 \approx 7$ (Eq. 13.2.23), we estimate using Eq. (13.2.27) a nonthermal power of $P(\varepsilon \geq 20$ keV$) \approx 3 \times 10^{25} - 3 \times 10^{30}$ erg s^{-1}. Integrating this power over typical flare durations of $\tau_{flare} \approx 10^2$ s yields a range of $W = P(\varepsilon \geq 20$ keV$) \times \tau_{flare} \approx 3 \times 10^{27} - 3 \times 10^{32}$ [erg] for flare energies. A frequency distribution of total nonthermal flare energies in electrons (> 25 keV) which covers this range has been determined in Crosby et al. (1993), see Fig. 9.27.

In some work, the photon spectrum (Eq. 13.2.18) is specified with the variable A, which gives the photon flux at 1 keV and relates to I_1 in Eq. (13.2.18) by,

$$A = I_1(\gamma - 1)\epsilon_1^{(\gamma - 1)} , \qquad (13.2.28)$$

The observed hard X-ray photon spectrum $I(\epsilon_x)$ observed at Earth (Eq. 13.2.18), the thick-target electron injection spectrum $f_e(\epsilon)$ (Eq. 13.2.24), and the total power in nonthermal electrons above some cutoff energy ε_c [i.e., $P(\varepsilon \geq \varepsilon_c)$, Eq. 13.2.27], are then

$$I(\epsilon_x) = A\,\epsilon_x^{-\gamma} \qquad \text{(photons cm}^{-2}\,\text{s}^{-1}\,\text{keV}^{-1}) , \qquad (13.2.29)$$

$$f_e(\varepsilon) = 2.68 \times 10^{33}\, b(\gamma) A\varepsilon^{-(\gamma+1)} \qquad \text{(electrons keV}^{-1}\,\text{s}^{-1}) , \qquad (13.2.30)$$

$$P(\varepsilon \geq \varepsilon_c) = 4.3 \times 10^{24} \frac{b(\gamma)}{(\gamma - 1)} A\,(\varepsilon_c)^{-(\gamma-1)} \qquad \text{(erg s}^{-1}) . \qquad (13.2.31)$$

13.2.3 Thin-Target Bremsstrahlung

For the thin-target case, the electrons are continuously accelerated and the hard X-ray spectrum is nearly identical to the acceleration or injection spectrum. The *thin-target electron injection spectrum* $f_e(\varepsilon)$ is given in Lin & Hudson (1976) and Hudson et al. (1978),

$$f_e(\varepsilon) = 1.05 \times 10^{42}\, C(\gamma) A \frac{1}{n_0} \varepsilon^{-(\gamma-1/2)} \qquad \text{(electrons keV}^{-1}) , \qquad (13.2.32)$$

and the function $C(\gamma)$ is defined in terms of the *Beta function*,

$$C(\gamma) = \frac{(\gamma - 1)}{B(\gamma - 1, \frac{1}{2})} \approx (\gamma - 1.5)^{1.2} . \qquad (13.2.33)$$

We see that a given photon spectrum (e.g., with a slope of $\gamma = 3$), implies a flatter electron injection spectrum in the thin-target case ($\delta = \gamma - \frac{1}{2} = 2.5$) than in the thick-target case ($\delta = \gamma + 1 = 4$). Hard X-ray sources observed in chromospheric heights are generally interpreted in terms of thick-target bremsstrahlung. In occulted flares, however, when the hard X-ray sources at the supposed chromospheric flare loop footpoints are blocked, extended sources of hard X-ray and γ-ray emission are sometimes observed over a considerable coronal height range, which are likely to be produced by thin-target bremsstrahlung (e.g., Datlowe & Lin 1973; Barat et al. 1994; Trottet et al. 1996). The fact that coronal hard X-ray sources are generally much weaker than chromospheric footpoint sources implies that thick-target bremsstrahlung is generally dominant (in non-occulted flares).

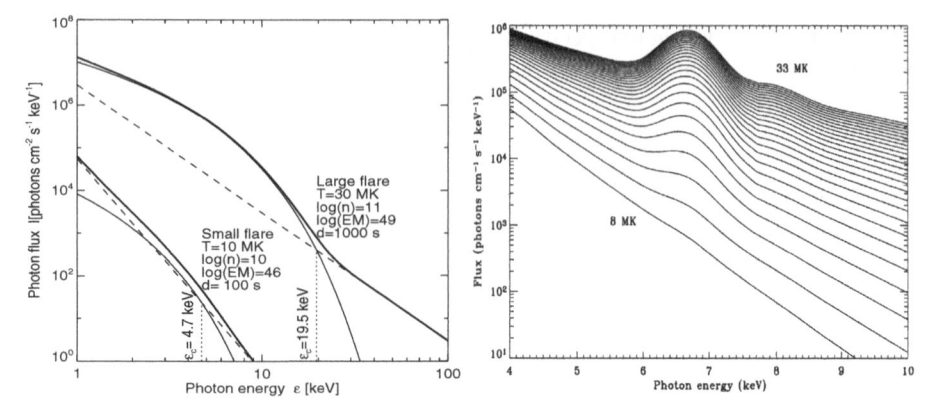

Figure 13.5: *(Left)* Theoretical hard X-ray spectrum consisting of a thermal and a nonthermal (powerlaw) component with equal energy content above the cutoff energy ε_c. The parameters are chosen for a large flare with $T_e = 30$ MK, $n_e = 10^{11}$ cm^{-3}, $EM_V = 10^{49}$ cm^{-3}, $\gamma = 3$, and duration $\tau_{flare} = 1000$ s; and for a small flare with $T_e = 10$ MK, $n_e = 10^{10}$ cm^{-3}, $EM_V = 10^{46}$ cm^{-3}, $\gamma = 5$, and duration $\tau_{flare} = 100$ s. *(Right)* Synthetic spectra between 4 and 10 keV show line features at 6.7 keV (Fe) and at 8.2 keV (Fe/Ni) that become progressively stronger for increasing flare temperatures. The spectra have been calculated with the CHIANTI code, with coronal abundances of Fe and Ni, and smoothed with a Gaussian filter with a width of FWHM=0.8 keV. Spectra are given in 1 MK intervals from 8 to 33 MK. Fluxes are those at the mean solar distance and for a flare with volume emission measure $\int_V n_e^2 dV = 10^{49}$ cm^{-3} (Phillips 2004).

13.3   Hard X-ray Spectra

13.3.1   Thermal-Nonthermal Spectra

Soft X-ray measurements show that the flare plasma has typical electron temperatures in the range of $T_e \approx 10 - 30$ MK (see compilation in Table 9.4; e.g., Pallavicini et al. 1977; Metcalf & Fisher 1996; Reale et al. 1997; Garcia 1998; Sterling et al. 1997; Nitta & Yaji 1997). This temperature range corresponds to electron energies of $\varepsilon = k_B T_e \approx 0.9-2.6$ keV (Appendix E). In the thick-target bremsstrahlung model (§13.2.2), nonthermal populations of electrons and ions accelerated in the corona precipitate to the chromosphere, heat up the plasma at flare loop footpoints, and cause an overpressure that drives upflows of heated plasma into the flare loops seen in soft X-rays. In this so-called *chromospheric evaporation* process we expect that the energy of the precipitating nonthermal electrons has to exceed the thermal energy of the heated plasma that is produced as a consequence. Let us compute such a combined thermal-plus-nonthermal hard X-ray spectrum in order to understand the energy ranges in which the two components dominate: the cutoff energy that separates them and the overall spectral shape.

    In §2.3 we defined the thermal flux spectrum $F_{th}(\epsilon)$ (Eq. 2.3.13), which yields the photon number by dividing by the photon energy ϵ [i.e., the photon spectrum is $I_{th}(\epsilon) = F_{th}(\epsilon)/\epsilon$]. Assuming a uniform temperature throughout the flare volume and using the units $EM_{49} = EM_V/10^{49}$ cm^{-3}, $T_7 = T_e/10^7$ K, $k_bT_7 = 0.86$ keV, and photon energy ϵ in keV, we obtain from Eq. (2.3.13) the following thermal photon spectrum,

$$I_{th}(\epsilon) = 2.6 \times 10^7 \left(\frac{EM_{49}}{T_7^{1/2}}\right) \frac{1}{\epsilon_{keV}} \exp\left(-\frac{\epsilon_{keV}}{0.86\,T_7}\right) \quad (\text{photons cm}^{-2}\,\text{s}^{-1}\,\text{keV}^{-1}).$$

(13.3.1)

The total energy in thermal electrons is (with Eq. 9.6.1),

$$W_{th} = 3n_e k_B T_e V = 4.1 \times 10^{29} \left(\frac{EM_{49}T_7}{n_{11}}\right) \quad (\text{erg}), \quad (13.3.2)$$

where the *volume emission measure* is defined as $EM_V = n_e^2 V$ (assuming a filling factor of unity).

On the other hand we can quantify the total energy in nonthermal electrons above some cutoff energy ε_c: multiplying the power (Eq. 13.2.31) by the flare duration (τ_{flare}) gives,

$$W_{nth}(\epsilon \geq \varepsilon_c) = P(\epsilon \geq \varepsilon_c) \times \tau_{flare} = 1.1 \times 10^{26} \frac{\gamma^3}{(\gamma - 1)} A\,\varepsilon_c^{-(\gamma-1)}\,\tau_2 \quad (\text{erg})$$

(13.3.3)

where we used the approximation $b(\gamma) \approx 0.27\gamma^3$ and denote $\tau_2 = \tau_{flare}/10^2$ s.

If we assume energy equivalence between thermal and nonthermal energies (i.e., $W_{th} = W_{nth}(\epsilon \geq \varepsilon_c)$ using Eqs. 13.3.2–3), we find a condition for the photon flux constant A, which then yields the nonthermal photon spectrum (Eq. 13.2.29),

$$I_{nth}(\epsilon) = 3.5 \times 10^5 \frac{EM_{49}T_7}{n_{11}\tau_2} \frac{(\gamma - 1)}{\gamma^3} \frac{1}{\varepsilon_c} \left(\frac{\epsilon}{\varepsilon_c}\right)^{-\gamma}. \quad (13.3.4)$$

The cutoff energy ε_c of the nonthermal spectrum can be defined by the intersection of the thermal with the nonthermal photon spectrum [i.e., $I^{th}(\epsilon = \varepsilon_c) = I^{nth}(\epsilon = \varepsilon_c)$], which yields the following expression,

$$\varepsilon_c = 0.86\,T_7\,\ln\left(74\frac{\gamma^3}{(\gamma - 1)}\frac{n_{11}\tau_2}{T_7^{3/2}}\right). \quad (13.3.5)$$

In Fig. 13.5 we plot such a theoretical thermal-plus-nonthermal photon spectrum based on the energy equivalence between both components. For a large flare ($T_7 = 3, n_{11} = 1, EM_{49} = 1, \gamma = 3$, and $\tau_2 = 10$) we find a nonthermal cutoff energy of $\varepsilon_c = 19.5$ keV, and for a small flare ($T_7 = 1, n_{11} = 0.1, EM_{49} = 0.001, \gamma = 5$, and $\tau_2 = 1$) we find a nonthermal cutoff energy of $\varepsilon_c = 4.7$ keV. Thus, in such a model where nonthermal energy is fully converted into thermal energy, we expect nonthermal cutoff energies in the range of $\varepsilon_c \approx 5 - 20$ keV. At lower energies $\epsilon < \varepsilon_c$ we expect the thermal (near-exponential) spectrum to dominate, while the nonthermal (powerlaw-like) spectrum dominates at higher energies $\epsilon > \varepsilon_c$.

Spectral fitting to observed hard X-ray spectra have indeed confirmed the existence of the generic thermal-plus-nonthermal model described above. Early hard X-ray detectors, such as NaI(Tl) and CsI(Na) on ISEE-3 or HXRBS/SMM, did not have

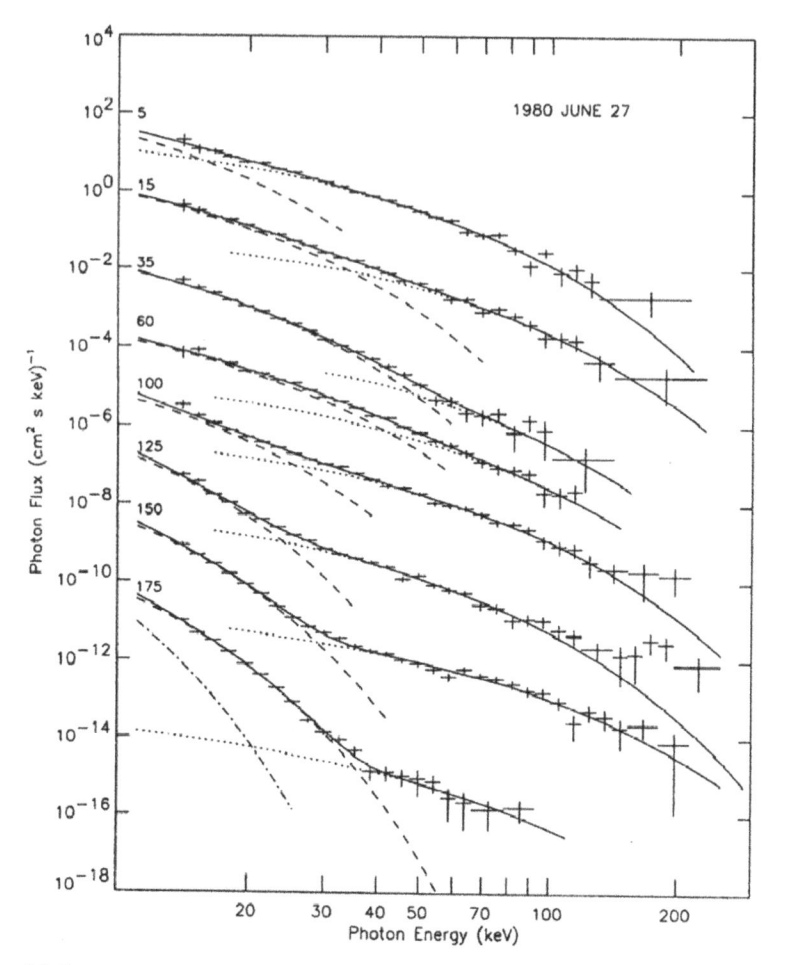

Figure 13.6: Observed hard X-ray spectra from the 1980-Jun-27 flare, labeled with the time in s after 16:14:41.87 UT. The flux of each spectrum is offset by a factor of 10^{-2}. The dashed lines show fits of the thermal spectrum and the dotted lines fits of a thick-target bremsstrahlung spectrum produced by DC electric field acceleration in the runaway regime (Benka & Holman 1994).

sufficient spectral resolution to resolve the steep thermal spectrum at any energy, but high-resolution spectra with cooled germanium detectors from balloon flights (Lin et al. 1981) and RHESSI clearly reveal the detailed spectral shape of the thermal-plus-nonthermal spectrum as calculated in Fig. 13.5. The hard X-ray spectrum of the 1980-Jun-27 flare was observed with high spectral resolution in the $13 - 300$ keV range and revealed the presence of a *"superhot temperature component"* with a maximum $T_e \approx 34$ MK and an emission measure of $EM = 2.9 \times 10^{49}$ cm^{-3} (Lin et al. 1981; Lin & Schwartz 1987). The same flare was also fitted with a DC electric field acceleration model for the nonthermal component (Fig. 13.6; Benka & Holman 1994; Kucera et al. 1996). For such "superhot" temperatures of $T_e \gtrsim 30$ MK, the thermal compo-

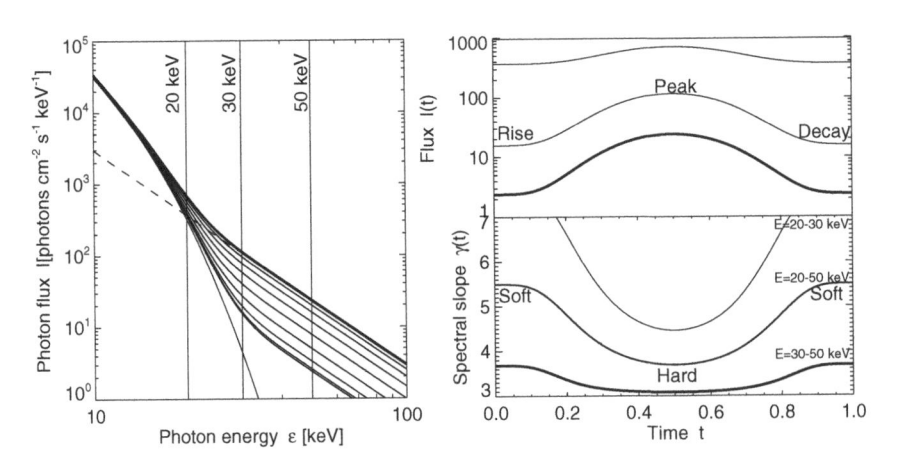

Figure 13.7: Simulation of spectral evolution during a flare: The thermal-nonthermal spectrum is identical to the large flare calculated in Fig. 13.5, except that the nonthermal component varies in time between 10% and 100% of the total thermal energy. The resulting spectra are shown (left), and the temporal evolution of the fluxes at 20, 30, and 50 keV are shown (top right). The evolution of the spectral powerlaw slopes $\gamma(t)$ in the three energy ranges $20 - 30$ keV, $20 - 50$ keV, and $30 - 50$ keV is shown (right). Note the soft-hard-soft evolution due to the changing ratio of the thermal to nonthermal spectrum. Note, however, that the soft-hard-soft evolution at higher energies ($\gtrsim 30$ keV) cannot be explained with this model and probably is an intrinsic property of the acceleration mechanism.

nent could even be fitted by using instruments with poorer spectral resolution, such as with *Hinotori*, *HXRBS/SMM*, and *Yohkoh* (Nitta et al. 1989, 1990; Nitta & Yaji 1997). Combined fitting of the thermal and nonthermal component (e.g., using recent RHESSI data), then allows comparison of the energy budget of both components, which were found to be comparable: for example $W_{th} \approx W_{nth} \approx 2.6 \times 10^{31}$ erg in the 2002-Jul-23 flare (Holman et al. 2003), or $W_{nth}(\epsilon \geq 10$ keV$)=2.6 \times 10^{30}$ erg in the 2002-Feb-26 flare at 10:26 UT (Saint–Hilaire & Benz 2002; Dennis et al. 2003).

13.3.2 Soft-Hard-Soft Spectral Evolution

The hard X-ray spectra of flares often initially show a steep spectral slope (soft), which flattens at the peak of the flare (hard), and then becomes steeper again (soft) in the decay phase of the flare. This evolutionary pattern has been called *soft-hard-soft* evolution. In other words, the value of the spectral powerlaw slope is anti-correlated with the hard X-ray flux. Such observations have been reported by Parks & Winckler (1969), Benz (1977), Brown & Loran (1985), Dennis (1985), Fletcher & Hudson (2002), and Hudson & Farnik (2002). A soft-hard-soft evolution was observed for every flare subpeak in the 1980-Jun-27 flare, suggesting that two different electron populations dominate at the flare peaks and valleys (Lin & Johns 1993). There are a number of physical effects that can explain spectral changes in the hard X-ray spectrum: (1) Particle trapping favors the presence of higher energy particles that have a longer collisional time, thus

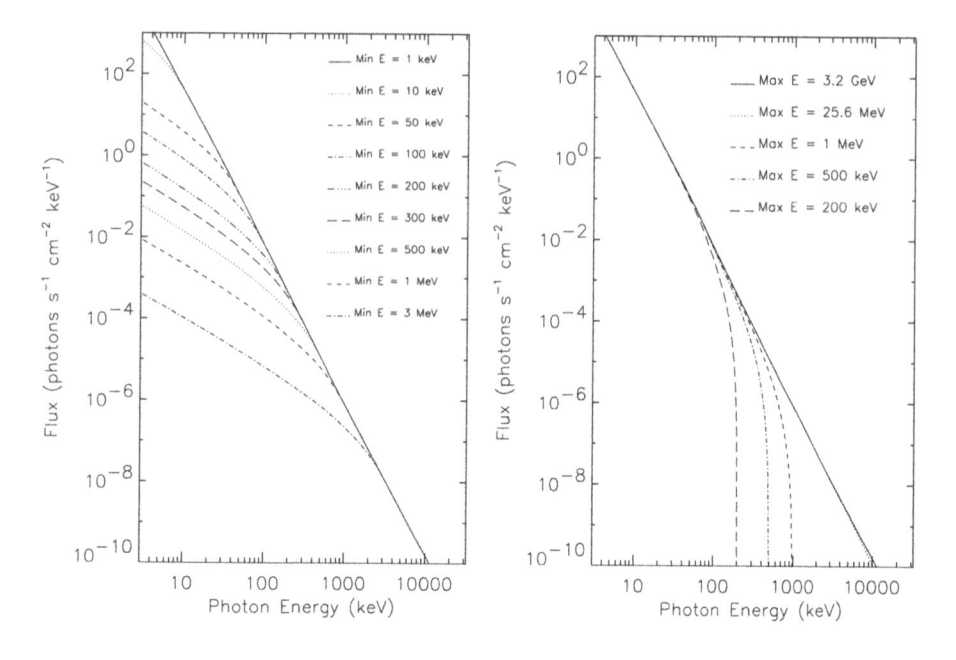

Figure 13.8: Thick-target bremsstrahlung spectra for a powerlaw slope $\delta = 5$ of the electron injection spectrum and low-energy cutoffs between $\varepsilon_{min} = 1$ keV and 3 MeV (left) and for high-energy cutoffs between $\varepsilon_{max} = 200$ keV and 3.2 GeV (right), (Holman 2003).

producing a hardening of the spectrum. (2) A hardening or flattening of the spectrum could also be produced by increasing the efficiency of the acceleration mechanism. (3) A variation of the relative weighting of thermal and nonthermal contributions, as it is perceived with detectors with poor energy resolution (illustrated in Fig. 13.7). The thermal spectrum has a much steeper slope than the nonthermal powerlaw in the 20-30 keV range. If the relative contribution of the nonthermal component decreases, the relative contribution from the thermal spectrum increases, and thus the spectrum is softening. In the example in Fig. 13.7 the spectral slope changes from $\gamma = 3.7$ to $\gamma = 5.5$ in the $20 - 50$ keV range after the flare, when the 50 keV flux drops down to 10%.

Although the *soft-hard-soft* spectral evolution is very common, it does not apply to all flares. In some large flares there are also *soft-hard-harder* patterns observed, in particular those where solar proton events are also detected, possibly related to extended acceleration phases, more efficient trapping, or higher proton escape probabilities in larger flare loops that get involved in the late flare phase (Kiplinger 1995).

13.3.3   Low-Energy and High-Energy Cutoffs

The classical inversion of thick-target bremsstrahlung spectra by Brown (1971) applies to powerlaw spectra without boundaries. For an observed hard X-ray photon spectrum with a powerlaw slope of γ, the inverted electron injection spectrum has a slope of

$\delta = \gamma + 1$ in the thick-target case (Eq. 13.2.24) and $\delta = \gamma - \frac{1}{2}$ for the thin-target case
(Eq. 13.2.32). The presence of low ε_{min} or high-energy cutoffs ε_{max} in the electron injection spectrum, however, can modify the slope of the photon spectrum considerably,
as shown in Fig. 13.8. The photon spectra in Fig. 13.8 were calculated by Holman
(2003) using the thick-target model (Eq. 13.2.17) and the relativistic cross section of
Haug (1997) with the Elwert (1939) correction factor. For an unbounded powerlaw
with a slope of $\delta = 5$ in the electron spectrum we expect a photon spectrum with a
powerlaw slope of $\gamma = \delta - 1 = 4$. The relativistic cross section, however, yields a
slope of $\delta = 3.8$ (solid line in Fig. 13.8, left). Introducing low-energy cutoffs in the
electron spectrum ($\varepsilon_{min} = 1, 10, 500$ keV, ... , 3 MeV in Fig. 13.8, left) progressively
flattens the photon spectrum, yielding a slope of $\gamma = 1.15$ for the highest cutoff at
$\varepsilon_{min} = 3$ MeV. Exceptionally flat hard X-ray spectra have been observed with slopes
as low as $\gamma = 1.98$ at $\epsilon_x = 33$ keV (Farnik et al. 1997). Fitting BATSE energy spectra,
low-energy cutoffs in the range of $\varepsilon_{min} \approx 45 - 97$ keV were found, with an average
of $\varepsilon_{min} \approx 60$ keV (Gan et al. 2002).

The existence and particular value of the low-energy cutoff ε_{min} is decisive for the
estimate of total injected power $P(\varepsilon \geq \varepsilon_{min})$ (Eq. 13.2.27), because the integral is
dominated by contributions at the lower boundary for powerlaw slopes of $\gamma \geq 1$. The
existence of nonthermal spectra down to energy ranges of $\approx 8 - 15$ keV in microflares
observed with *RHESSI* (Krucker et al. 2002) suggests that the value of the low-energy
cutoff depends very much on flare size. A physical model for the low-energy cutoff en-
ergy can be derived from the collisional limit $\tau_{prop} = l/v_{\parallel}(\varepsilon) \leq \tau_{defl}(\varepsilon)$ (Eq. 12.6.20)
of the maximum propagation path l of energized electrons between the acceleration
site and the thick-target site, yielding a critical energy $\varepsilon_c \approx \varepsilon_{min}$ (Eq. 12.6.21). The
concept of a low-energy cutoff has been criticized, because there is a gradual tran-
sition from a "warm-target energy loss" at $\varepsilon \gtrsim k_B T_e$ to "cold-target energy loss" at
$\varepsilon \lesssim 5k_B T_e$, which modifies the estimate of total injected power P (contained in precipitating electrons) (Emslie 2003).

Introducing a high-energy cutoff ($\varepsilon_{max} = 0.2, 0.5, 1$ MeV, ..., 3.2 GeV in Fig. 13.8
right) leads to a steepening of the photon spectrum near the upper cutoff, causing deviations from an ideal powerlaw function approximately an order of magnitude below
the cutoff (i.e., $0.1\varepsilon_{max} \lesssim \epsilon \lesssim \varepsilon_{max}$: Holman 2003).

13.3.4 Spectral Inversion

The forward-calculation of photon spectra $I(\varepsilon_x)$ from assumed electron spectra $f_e(\varepsilon)$
requires various assumptions on the shape of the spectral function, the cutoffs, and the
choice of a model (thermal, thick-target, thin-target). It is therefore desirable to develop
model-independent inversion methods that yield unbiased electron flux spectra $f_e(\varepsilon)$
solely based on the observed photon spectra $I(\varepsilon_x)$ (Brown et al. 2003). Analytical
studies on the inversion problem tell us that: (1) The inversion of a powerlaw photon
spectrum yields a powerlaw electron spectrum, though with a different slope (Brown
1971); (2) a powerlaw photon spectrum can also be produced by multi-thermal spectra
(Brown 1974); (3) "bumps" in photon spectra could indicate "upturns" (with positive
slope) in electron spectra (Brown et al. 1991); (4) nonuniform ionization in the chromospheric thick-target regions modifies spectral inversion at higher energies (Brown et

al. 1998b); and (5) there are analytical limits to physically acceptable solutions by the constraints of higher derivatives (Brown & Emslie 1988). The key problem of spectral inversion, of course, lies in the uniqueness and stability of numerical inversion techniques in the presence of data noise. Because the photon spectrum is an integral of the electron spectrum convolved with the bremsstrahlung cross section, an inversion technique essentially has to reliably determine the derivative in energy, and this amplifies data noise. Spectra with low-energy resolution (i.e., $d\epsilon/\epsilon \approx 0.1 - 0.3$ in SMM/HXRBS) are not suitable for inversion and are restricted to double (or "broken") powerlaw fits in the nonthermal range (Dulk et al. 1992), but high-resolution spectra with $d\epsilon/\epsilon \approx 0.01$ obtained from germanium-cooled detectors, such as were obtained during a balloon flight (Lin et al. 1981; Lin & Schwartz 1987) and available from RHESSI (Lin et al. 1998, 2001, 2002), should facilitate sufficiently accurate inversions to reveal deviations from single- and double-powerlaw functions in the nonthermal energy range ($\epsilon \gtrsim 20$ keV). Spectral features in the form of "knees" have been explained in terms of an acceleration high-energy limit, solar albedo backscattering (Alexander & Brown 2002; Zhang & Huang 2003), or transport effects through a region of nonuniform ionization (Conway et al. 2003; Kontar et al. 2002, 2003). Numerical inversion techniques have been applied to such spectra using regularization (or Bayesian) approaches (Johns & Lin 1992), being criticized as an ill-posed inverse problem (Craig & Brown 1986; Thompson et al. 1992), or Tikhonov's regularization techniques (Piana 1994; Piana et al. 1995, 2003; Massone et al. 2003).

13.4 Hard X-ray Time Structures

13.4.1 Pulse Observations

While hard X-ray time profiles obtained from earlier observations often did not show fast time structures in flare light curves due to photon noise limitations, more sensitive instruments such as *BATSE/CGRO* clearly revealed pulsed time structures on subsecond time scales in virtually all flares, at nonthermal energies $\gtrsim 20$ keV (sometimes down to $\gtrsim 8$ keV: Lin et al. 2001). Four examples of flare time profiles are shown in Fig. 13.9 (left frame), where the fast time structures are enhanced with a high-pass filter by subtracting a lower envelope (Fig. 13.9, right frame). The fastest time structures discovered with the *Hard X-Ray Burst Spectrometer (HXRBS)* on the *Solar Maximum Mission (SMM)* spacecraft have been found down to rise times and decay times of ≈ 20 ms and with pulse widths of $\gtrsim 45$ ms (Kiplinger et al. 1983b). Such fast time structures were not very common in *SMM* data, because only about 10% of all flares detected with *HXRBS* had a sufficiently high count rate to detect significant variations on a subsecond time scale (Dennis 1985). With the much more sensitive *BATSE/CGRO* detectors, which had a sensitive area of 2025 cm^2 each, subsecond time structures were detected virtually in all flares recorded in flare mode (activated by a burst trigger). A systematic study of 640 *BATSE* flare events with an automated pulse detection algorithm (Aschwanden et al. 1995c) revealed a total of 5430 individual pulses, the distribution of whose pulse widths is shown in Fig. 13.10. This distribution shows that hard X-ray pulses with durations of $\tau_x \approx 0.3 - 1.0$ s are most typical. As the examples in

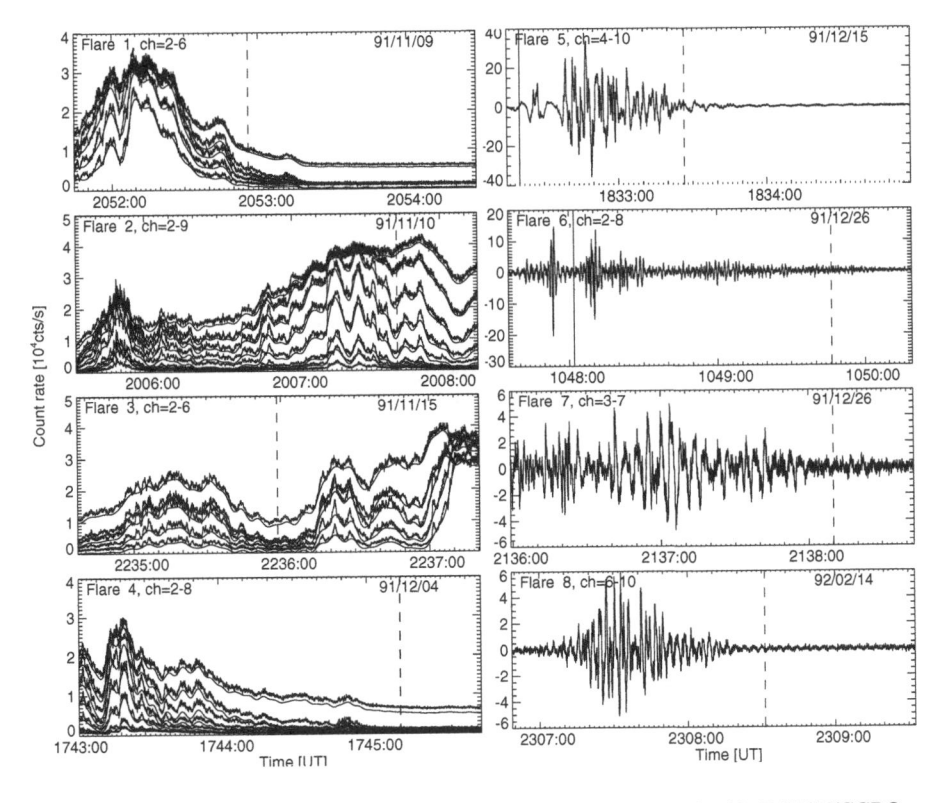

Figure 13.9: *(Left)* Hard X-ray time profiles of four flares observed with *BATSE/CGRO* at $\gtrsim 20$ keV. The time profiles were recorded in the 16-channel *Medium Energy Resolution (MER)* mode with a time resolution of 64 ms. The frames have a duration of 163.84 s. The lower envelope to the rapid fluctuations was constructed with an FFT low-pass filter with cutoffs of $\tau_{filter} = 1.5 - 3.6$ s. *(Right)* The high-pass filtered time profiles (with the lower envelope subtracted) are shown, revealing numerous subsecond pulses (Aschwanden et al. 1996b).

Fig. 13.9 (left panel) show, the *modulation depth* of the pulses varies typically in the range of $dF/F \approx 10\% - 50\%$.

13.4.2 Distribution of Pulse Durations

The distribution of observed pulse durations can be characterized by an exponential function in the range of $\tau_x \approx 0.3 - 1.0$ s, with an average value of $< \tau_x > \approx 0.5$ s. The distributions are almost identical at different energies (Fig. 13.10; Aschwanden et al. 1995c),

$$N(\tau_x) \propto \begin{cases} \exp\left(-\tau_x/0.41 \text{ s}\right) & \text{for } \epsilon > 25 \text{ keV} \\ \exp\left(-\tau_x/0.44 \text{ s}\right) & \text{for } \epsilon > 50 \text{ keV} \end{cases} \qquad (13.4.1)$$

What is the physical origin of these hard X-ray pulses? In most flare models it is assumed that magnetic reconnection plays a key role in primary energy release (§10.5). Because large current sheets are not stable, but prone to tearing-mode instability, we

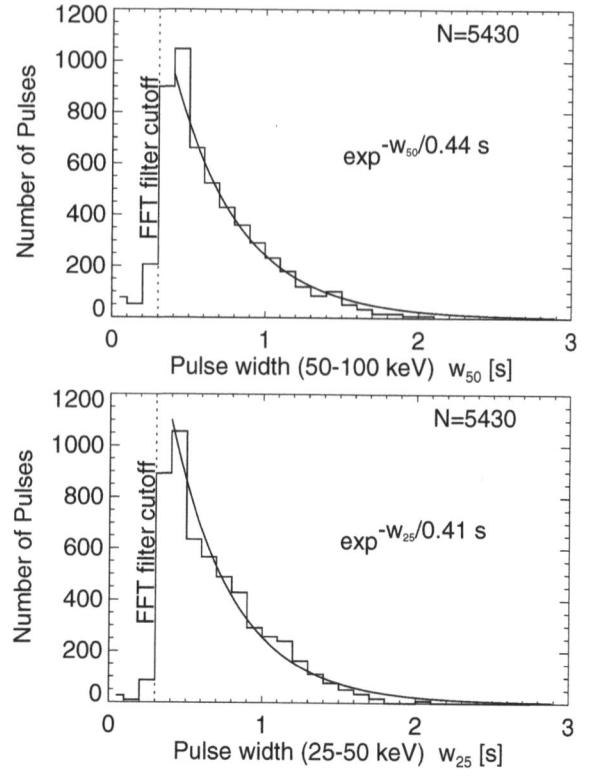

Figure 13.10: Observed distribution of hard X-ray pulse durations detected in the $25 - 50$ keV and $50 - 100$ keV channels from *BATSE/CGRO* during 647 solar flares. The cutoff of detected pulse widths at $\lesssim 0.3$ s is caused by the Fourier (FFT) filter used for structure detection (Aschwanden et al. 1995c).

expect that magnetic reconnection occurs in the *impulsive bursty regime* (§10.2) rather than in a steady (Parker-type or Petschek-type) mode (§10.1). In the *impulsive bursty regime*, many localized reconnection events take place in a flare region, each one separated in space and time from the other, but activated by a common trigger that produces a clustering in time. Each localized reconnection event energizes a bunch of particles which either precipitate directly to the chromosphere or after some intermediate trapping. The time signature of energized particles is consequently a temporal cluster of pulses, which are detectable in hard X-rays by bremsstrahlung and in radio by gyrosynchrotron emission and beam-driven plasma emission. In §12.4.2 we developed a model of the particle injection mechanism during localized magnetic reconnection events of the Petschek type. We argued that the relaxation of newly reconnected magnetic field lines opens up the magnetic trap of energized particles in the cusp region of a Petschek-type X-point and synchronizes the injection of these energized particles into free-streaming magnetic field lines. From this model we derived the pulse duration and found that it scales roughly with the Alfvén transit time through the diffusion region of the X-point (Eq. 12.4.18). Using this model to predict the pulse durations, $\tau_x \approx \tau_w \approx q_a(\pi/2)L_B/v_x$ (Eq. 12.4.18), the scaling ratio for X-point heights, $q_h = (h_x/h_L) \approx 1.5$ (Eq. 12.4.1), and the definition of the Alfvén velocity v_A (Eq. 12.4.6), we expect a scaling of

$$\tau_w = 0.50 \left(\frac{h_L}{10 \text{ Mm}}\right) \left(\frac{B_{ext}}{30 \text{ G}}\right)^{-1} \left(\frac{n_e}{10^9 \text{ cm}^{-3}}\right)^{1/2} \left(\frac{q_h - 1}{0.5}\right) \left(\frac{q_a}{1.3}\right) \left(\frac{q_B}{0.1}\right) \text{ (s)}.$$

$$(13.4.2)$$

This model predicts an approximate correlation of pulse width τ_w with loop height h_L [i.e., $\tau_w \approx 0.5(h_L/10 \text{ Mm})$ s]. It reproduces the observed average pulse duration of $< \tau_x > \approx 0.5$ s for an average magnetic field strength $B_{ext} \approx 30$ G, electron density $n_e \approx 10^9$ cm^{-3} (above flare loops), cusp height ratio $q_h = h_X/h_L \approx 1.5$, and X-point length scale $q_B = L_B/(h_X - h_L) \approx 0.1$.

The next test is whether our model can also reproduce the observed exponential distribution (Eq. 13.4.1). We performed a Monte Carlo simulation by assuming a normal distribution for each of the five parameters (x) in Eq. (13.4.2) and varied the Gaussian widths σ_x. For a Gaussian width ratio of $\sigma_x/ < x >= 0.5$ we could reproduce the same distribution as the observed one (Fig. 13.10) with the same exponential fit $\exp(-\tau_w/0.4$ s) above a cutoff of $\tau_{filter} > 0.3$ s (Aschwanden 2004). So, making the minimal assumption that each parameter has the same relative scatter $\sigma_x/ < x > \approx 0.4$, we find that the following distribution of parameters is most consistent with the observations: $h_L = 10 \pm 5$ Mm, $B_{ext} = 30 \pm 15$ Mm, $n_e = (1.0 \pm 0.5) \times 10^9$ cm^{-3}, $q_h = 1.5 \pm 0.25$, and $q_B = 0.10 \pm 0.05$. The latter ratio translates in absolute units into a range of $L_B = q_B * h_L * (q_h - 1) = 500 \pm 400$ km for the magnetic length scale of the X-point (i.e., the extent of the diffusion region).

13.4.3 Scaling of Pulse Duration with Loop Size

The majority of flares show double, hard X-ray sources, which are usually interpreted as the *conjugate footpoints* of flare loops. This interpretation is strengthened when they are located in areas with opposite magnetic polarity and the connecting flare loop is visible in soft X-rays. Assuming a semi-circular loop, the footpoint separation $d = 2r$ thus provides a simple way of estimating the flare loop radius r. For a set of 46 flares simultaneously observed with *Yohkoh/HXT*, *SXT*, and *BATSE/CGRO*, flare loop radii were determined in the range of $r \approx 2.5 - 25$ Mm (see range of x-axis on left panel of Fig. 13.11).

For the same flares, the time structures were analyzed using a wavelet technique (Aschwanden et al. 1998c). Wavelet analysis represents a multi-resolution method that yields dynamic decomposition of the power at different time scales $\Delta t = \nu^{-1}$ in the form of a *scalogram* or *dynamic power spectrum* $P(\nu, t)$, in contrast to a Fourier spectrum $P(\nu)$ that averages over the entire analyzed time interval. The *wavelet scalogram* can also be converted into a distribution of time scales $N(\nu)$ with proper normalization. This is a suitable tool to characterize the most dominant time scales τ_x of pulses in hard X-ray time profiles. This wavelet study revealed pulse time scales of $\tau_x \approx 0.1 - 0.7$ (see range of y-axis on left panel of Fig. 13.11). Moreover, a spatio-temporal correlation was found between these pulse time scales τ_x and the flare loop curvature radii

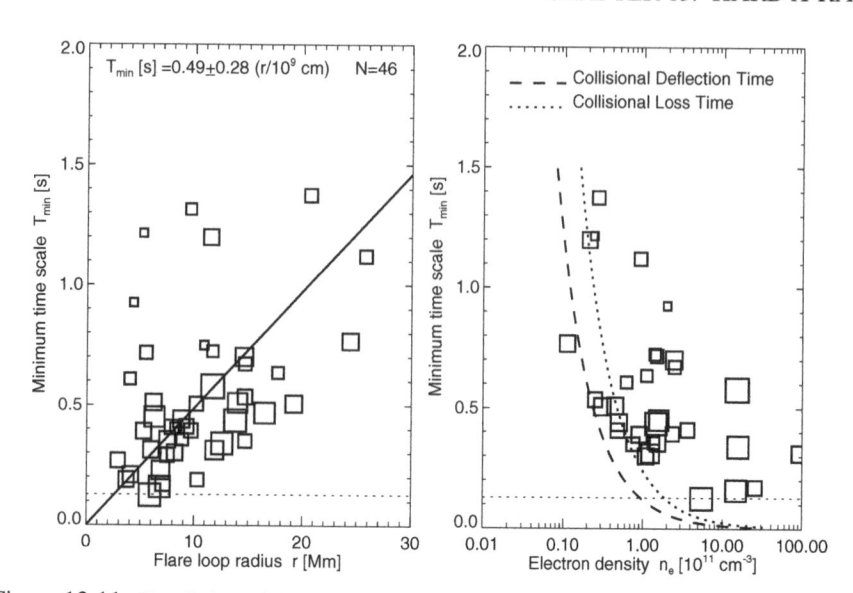

Figure 13.11: Correlation of the minimum time scale T_{min} with the flare loop radius r (left) and trap electron density n_e (right) for a set of 46 flares simultaneously observed with *BATSE/CGRO* and *Yohkoh*. The symbol size of the data points is proportional to the logarithm of the count rate. The mean ratio T_{min}/r is indicated (solid line in left panel), and collisional time scales for 25 keV electrons are shown (dashed and dotted line in right panel) (Aschwanden et al. 1998c).

r_{curv}, which can be expressed by the relation (Aschwanden et al. 1998c),

$$\tau_x = (0.5 \pm 0.3) \left(\frac{r_{curv}}{10^9 \text{ cm}} \right) s .\tag{13.4.3}$$

The correlation is shown in the left panel of Fig. 13.11.

On the other hand, we have derived a theoretical model where the duration of particle injection from a Petschek-type reconnection region also predicts a spatio-temporal relation between the Alfvén transit time $\tau_A = 2L_B/v_A$ through the diffusion region with a geometric size L_B (Eq. 12.4.18). Since the geometric ratio of the cusp height h_X was found to be proportional to the flare loop height $h_L = r_{curv}$, having an approximate ratio of $q_h = (h_X/h_L) \approx 1.5$ (Eq. 12.4.1), we can conclude that the reconnection geometry is *scale-invariant*, which can explain the observed spatio-temporal correlation (Eq. 13.4.3). Matching the theoretical relation (Eq. 12.4.18) with the observed relation (Eq. 13.4.3), we find the average flare parameters given in Eq. (13.4.2). A most interesting parameter is the length scale of the reconnection diffusion region, $L_B = 500 \pm 400$ km which is based on fitting the observed pulse duration distribution (Eq. 13.4.1) shown in Fig. 13.10.

The pulse time scales analyzed in the same study also obey another constraint: they are always larger than the collisional deflection time scales in the flare loops, based on the soft X-ray electron density measurements (Fig. 13.11, right). This fact is consistent with the interpretation that the density in the cusp region of the relaxing field line is always lower ($n_e^{cusp} \approx 10^9$ cm^{-3}) than in the soft X-ray-emitting flare loops which

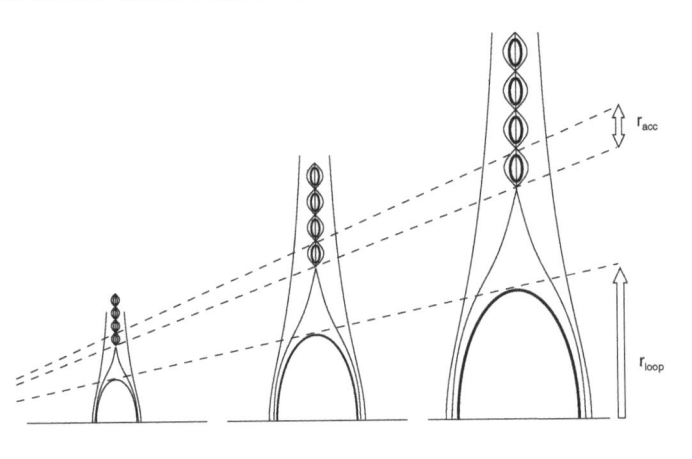

Figure 13.12: Proportionality between the spatial scale of the acceleration region r_{acc} and the flare loop curvature radius r_{curv}, as implied by the spatio-temporal relation between the hard X-ray pulse time scale τ_x and loop size r_{curv}, is visualized in the context of a scale-invariant reconnection geometry with magnetic islands in the current sheet region. The scale-invariant geometry is consistent with tearing mode. For larger loops at the base of the current sheet, the width l_{CS} is also larger. The length scale of the islands, which corresponds to the tearing mode with the largest growth rate and scales with l_{CS}, is then also larger (Aschwanden 2002b).

have been filled with evaporated plasma, $n_e^{SXR} \approx 10^{10} - 10^{12}$ cm^{-3} (Fig. 13.11, right), so the potential trapping time in the cusp is always longer than the trapping time in the soft X-ray flare loop.

In Fig. 13.12 we apply the observed spatio-temporal relation to an *impulsive, bursty reconnection* scenario by Kliem (1995). The tearing-mode instability produces a string of magnetic islands which undergo coalescence. We can interpret the pulse duration τ_x of an elementary particle injection as the coalescence time τ_{coal} (Eq. 10.2.3) of a magnetic island, which roughly represents the lifespan from formation and dissipation of a magnetic island, which determines the approximate duration τ_x of an elementary particle acceleration episode. Thus, setting $\tau_{coal} \approx \tau_x$ and inserting the observed scaling of Eq. (13.4.3) $\tau_x \approx (r_{curv}/10^9$ cm s$^{-1})$ into Eq. (10.2.3), we find the following scaling for the size of a coalescing magnetic island,

$$l_{coal} = \tau_{coal} q_{coal} v_A \approx 0.2 \, r_{curv} \, q_{coal} \left(\frac{B}{100 \text{ G}} \right) . \qquad (13.4.4)$$

This is illustrated in Fig. 13.12, as applied to a string of Petschek-type X-points that result in the *impulsive, bursty reconnection regime* due to tearing-mode instability. We can interpret the pulse duration τ_x of an elementary particle injection as the coalescence time τ_{coal} (Eq. 10.2.3) of a magnetic island. This roughly represents the lifespan from formation to dissipation of a magnetic island, and this in turn determines the approximate duration τ_x of an elementary particle acceleration episode. Thus, setting $\tau_{coal} \approx \tau_x$ and inserting the observed scaling of Eq. (13.4.3), $\tau_x \approx (r_{curv}/10^9$ cm s$^{-1})$ into Eq. (10.2.3), we find the following scaling for the size of a coalescing magnetic island,

$$l_{coal} = \tau_{coal} q_{coal} \mathrm{v}_A \approx 0.2 \, r_{curv} \, q_{coal} \left(\frac{B}{100 \, \mathrm{G}}\right) \left(\frac{n_e}{10^{10} \, \mathrm{cm}^{-3}}\right)^{-1/2} . \quad (13.4.5)$$

This yields a range of $l_{coal} \approx 60 - 6000$ km for the sizes of magnetic islands, based on a range of loop sizes $r_{curv} \approx 3 - 30$ Mm and $q_{coal} = 0.1 - 1$, for $B = 100$ G and $n_e = 10^{10}$ cm^{-3}, using the same parameters estimated in Kliem (1995).

With the same spatio-temporal scaling law, we can estimate the rate R_{acc} of accelerated particles per magnetic island, defined by the product of the volume $V \approx l_{coal}^3$, the electron density n_e, and the acceleration efficiency q_{acc}, which yields using Eq. (13.4.4),

$$R_{acc} = \frac{q_{acc} l_{coal}^3 n_e}{\tau_{coal}} = 8 \times 10^{34} \, q_{acc} q_{coal}{}^3 \left(\frac{r_{curv}}{10^9 \, \mathrm{cm}}\right)^2 \left(\frac{B}{100 \, \mathrm{G}}\right)^3 \left(\frac{n_e}{10^{10} \, \mathrm{cm}^{-3}}\right)^{-1/2} (\mathrm{s}^{-1}).$$

$$(13.4.6)$$

This relation indicates that large flares require fast coalescence speeds $q_{coal} \lesssim 1$ (or $u_{coal} \lesssim \mathrm{v}_A$) and high magnetic fields $B \gtrsim 100$ G, which both scale with the third power to the acceleration rate, while the acceleration efficiency q_{acc} only scales linearly. Also, the number of magnetic islands would increase the rate of accelerated particles only with linear scaling. Interestingly, a lower density yields a higher acceleration rate than a higher density, because a lower density increases the Alfvén velocity v_A or coalescence speed $u_{coal} = q_{coal} \mathrm{v}_A$.

13.5   Hard X-Ray Time Delays

13.5.1   Time-of-Flight Delays

Although the energized nonthermal electrons propagate with relativistic speed, it takes them a finite time to travel from a coronal acceleration site to the chromospheric precipitation (and hard X-ray emission) site. For a typical height difference of $h_x \approx 15$ Mm, say, it takes an electron at half the speed of light (i.e., $\beta \approx 0.5$ at $\varepsilon \approx 80$ keV), about 100 ms to reach the chromosphere. Compared with an electron that has half the speed (i.e., $\beta \approx 0.25$ at $\varepsilon \approx 17$ keV), which needs twice the travel time, we expect a time difference of $\Delta t \approx 100$ ms. Such energy-dependent time delays due to velocity dispersion, also called *time-of-flight (TOF) delays*, have been measured from *BATSE/CGRO* data in virtually all flares with pulse structures (Aschwanden et al. 1995c, 1996a, b, c; Aschwanden & Schwartz 1995, 1996). The kinematics of free-streaming particles and the basic measurement technique of TOF delays between pulses with different electron velocities has been described in §12.2. Let us now consider the relation between electrons and hard X-ray photons, which are needed to infer TOF distances from energy-dependent hard X-ray timing.

We approximate the nonthermal hard X-ray spectrum with a power law, $I(\epsilon_x) \propto \epsilon_x^{-\gamma}$ (Eq. 13.2.29). In the thick-target model (§13.2; Brown 1971), the electron injection spectrum also obeys a powerlaw form $f_e(\varepsilon) \propto \varepsilon^{-\delta}$ with a slope of $\delta = \gamma + 1$ (Eq. 13.2.30). Let us now consider a time-dependent electron injection spectrum,

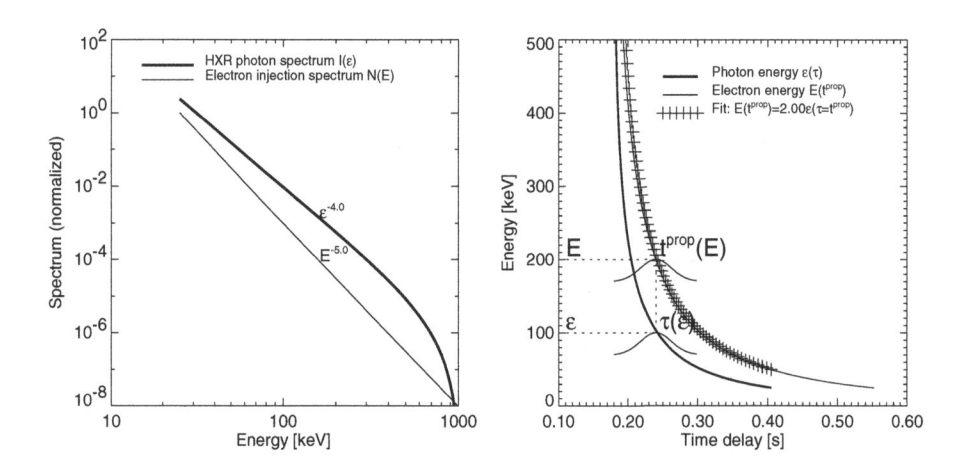

Figure 13.13: A time-dependent hard X-ray photon spectrum $I(\epsilon_x, t)$, computed from a time-dependent electron injection spectrum $N(\varepsilon, t)$, represented in the spectral (left) and temporal domain (right). *(Left)* The electron injection spectrum $N(\varepsilon, t) \propto \varepsilon^{-5}$ (with an upper cutoff energy of $\varepsilon_0 = 1$ MeV) and the numerically computed hard X-ray photon spectrum $I(\epsilon_x, t = 0) \propto \epsilon_x^{-4}$. *(Right):* The peak time $\tau(\epsilon_x)$ of the hard X-ray photon spectrum $I(\epsilon_x, t)$ is marked with the thick curve. The propagation delay $t^{prop}(\varepsilon)$ of electrons is shown with the thin curve, which has a similar functional dependence to the hard X-ray delay $\tau(\epsilon_x)$, and can be fitted by multiplying the photon energies $\epsilon_x(\tau)$ with a factor of $q_\varepsilon \approx 2.0$ (crosses). The Gaussians shown at $\epsilon_x = 100$ keV and $\varepsilon = 200$ keV symbolize coincident peak times of photons and electrons, but the width of the Gaussian pulses is reduced by a factor of 100 for clarity (Aschwanden & Schwartz 1996).

which we characterize with a power law in energy and with a Gaussian pulse shape in time (with a Gaussian width τ_G), that is,

$$f_e(\varepsilon, t, x = 0) = f_0 \varepsilon^{-\delta} \exp\left[-\frac{(t - t_0)^2}{2\tau_G^2} \right] . \qquad (13.5.1)$$

After the electrons propagate over a distance l to the thick-target site, they have a velocity dispersion corresponding to the time-of-flight $t^{TOF} = l/\mathrm{v}(\varepsilon)$, and the instantaneous electron injection spectrum at $x = l$ is,

$$f_e(\varepsilon, t, x = l) = f_0 \varepsilon^{-\delta} \exp\left[-\frac{(t - t_0 - l/\mathrm{v}[\varepsilon])^2}{2\tau_G^2} \right] . \qquad (13.5.2)$$

From this instantaneous electron injection spectrum $f_e(\varepsilon, t, l)$ at the thick-target site, the hard X-ray photon spectrum $I(\epsilon_x, t, l)$ can be calculated by a convolution with the Bethe−Heitler bremsstrahlung cross section (Eq. 13.2.12),

$$I(\epsilon_x, t, l) = I_0 \frac{1}{\epsilon_x} \int_{\epsilon_x}^{\epsilon_0} f_e(\varepsilon, t, l) \left(\int_{\epsilon_x}^{\varepsilon} \ln\frac{1 + \sqrt{1 - \epsilon_x/\varepsilon'}}{1 - \sqrt{1 - \epsilon_x/\varepsilon'}} d\varepsilon' \right) d\varepsilon , \qquad (13.5.3)$$

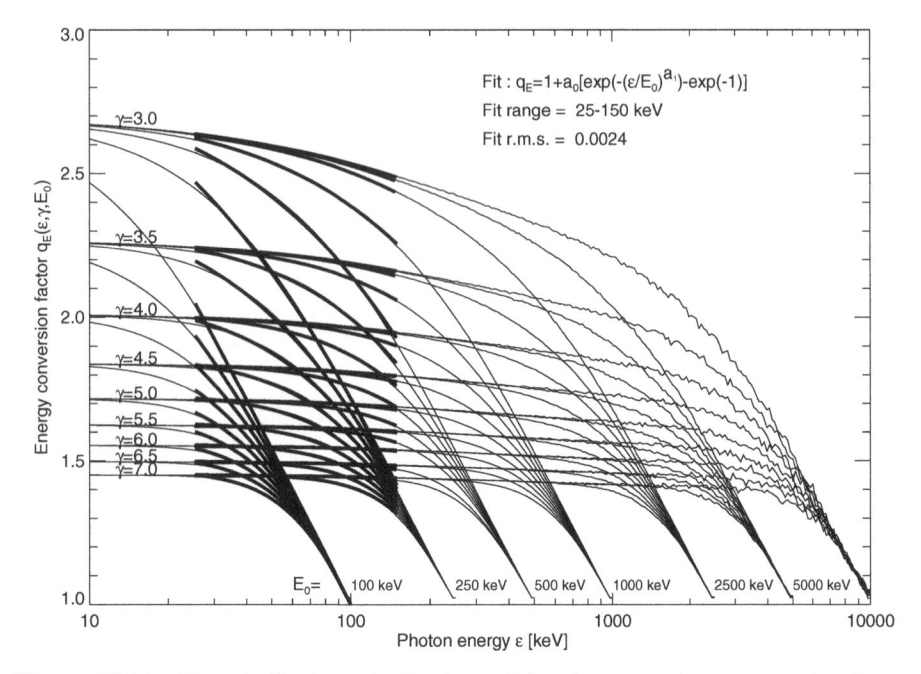

Figure 13.14: Numerically determined values of the *electron-to-photon conversion factor* $q_\varepsilon(\epsilon, \gamma, \epsilon_0)$ (thin lines) as a function of the photon energy ϵ_x for different spectral indices $\gamma = 3.0, 3.5, ..., 7.0$ and high-energy cutoffs $\epsilon_0 = 0.1, 0.25, 0.5, 1.0, 2.5, 5.0, 10.0$ MeV. Analytical approximations $q_\varepsilon(\epsilon, \gamma, \epsilon_0)$ described in Aschwanden & Schwartz (1996) are optimized for the energy range of $\epsilon = 25 - 150$ keV (thick lines).

with ϵ_0 being the high-energy cutoff of the electron injection spectrum. Next, we convolve the hard X-ray photon spectrum at each time t with the instrumental response functions $R_i(\epsilon)$ of the energy channels i to obtain the count rate profiles $C_i(\epsilon, t)$,

$$C_i(\epsilon_x, t) = I(\epsilon_x, t, x = l) \otimes R_i(\epsilon_x) . \qquad (13.5.4)$$

To extract the time-of-flight distance l_{TOF} from observed hard X-ray counts, we need either an inversion technique or a forward-fitting method.

In data analysis, we measure the relative time delays between different energy channels i and j by cross-correlation of count rate time profiles, seeking the maximum of the cross-correlation coefficient $CCC(\tau_{ij})$,

$$CCC(\tau_{ij}) = C_i(\epsilon_i, t) \otimes C_j(\epsilon_j, t + \tau_{ij}) , \qquad (13.5.5)$$

which yields the energy-dependent time delays $\tau_{ij} = \tau(\epsilon_{x,i}, \epsilon_{x,j})$. In Fig. 13.13 we show the peak time of the Gaussian time profile $f_e(\varepsilon, t, x = l)$ as a function of the electron energy, $\tau_{prop}(\varepsilon)$, as well as the peak times of the photon spectrum $I(\epsilon, t, x = l)$ as a function of the photon energy, $\tau(\epsilon_x)$, which essentially contains the energy-dependent time delay that is measured with the cross-correlation function (Eq. 13.5.5). A convenient concept for our analysis is to specify a quantitative relation between the

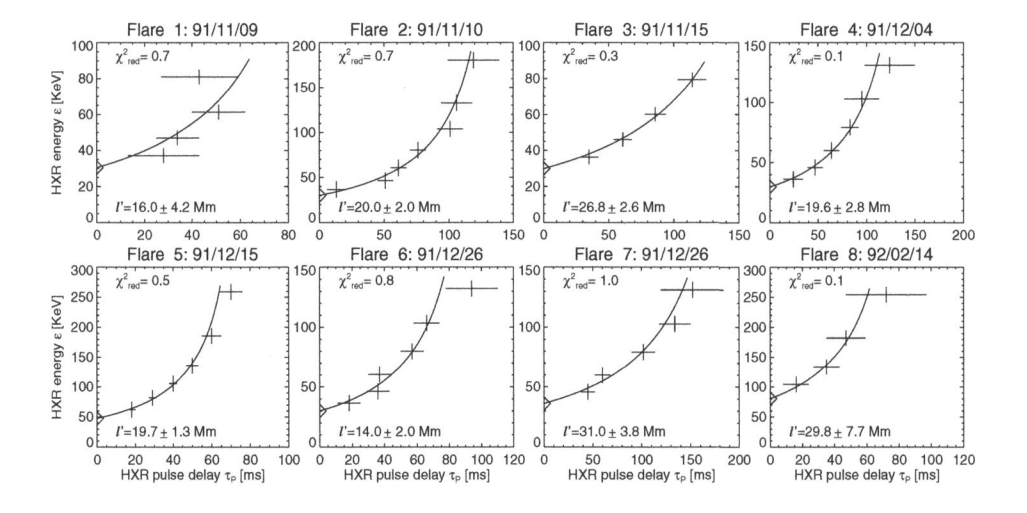

Figure 13.15: Energy-dependent time delays $\tau_{ij} = t(\epsilon_i) - t(\epsilon_j)$ of the filtered hard X-ray pulses $F^P(t)$ of eight flares (some shown in Fig. 13.9), measured from the cross-correlation between the energy channels ϵ_i and ϵ_j (with $i = j_{min}$) during selected time segments. The horizontal bars represent the uncertainties of the delay measurement caused by Poisson noise. The curve represents the best fit of the TOF model $\tau_{ij} = (l/c)(1/\beta_i - 1/\beta_j)$. The projected TOF distance $l_{TOF} = l \times 0.54$ and the χ^2_{red} of the best fit are indicated in each panel (Aschwanden et al. 1996b).

electron energy ε and photon energy ϵ_x. Of course, such a relation depends on the photon energy ϵ_x as well as on spectral parameters, such as the powerlaw slope γ and high-energy cutoff ε_0 of the electron spectrum. Thus, we define such a *electron-to-photon energy conversion factor* q_ε,

$$q_\varepsilon(\epsilon_x, \gamma, \epsilon_0) = \frac{\varepsilon(t = t_{peak})}{\epsilon_x(t = t_{peak})} \qquad (13.5.6)$$

The example shown in Fig. 13.13, for which an electron injection spectrum with a slope of $\gamma = 4$ and a cutoff of $\varepsilon_0 = 1$ MeV was used, shows a ratio of $q_\varepsilon \approx 2.0$ at an electron energy of $\varepsilon = 200$ keV. To quantify this ratio q_ε for general applications, this ratio has been numerically computed in a large parameter space (i.e., $\varepsilon = 10 - 10,000$ keV, $\epsilon_0 = 100 - 10,000$ keV, and $\delta = 3.0 - 7.0$). The results are shown in Fig. 13.14 and are quantified by analytical approximation formulae in Aschwanden & Schwartz (1996).

The measurement of a time-of-flight distance can now be carried out using the following steps: (1) First we measure a time delay τ_{ij} between two hard X-ray energy channels by interpolating the maximum of the cross-correlation function (13.5.5) between the two time profiles; (2) the "mean" photon energies $\epsilon_{x,i}$ and $\epsilon_{x,j}$ of the two hard X-ray channels are then determined from the mean or median of the contribution

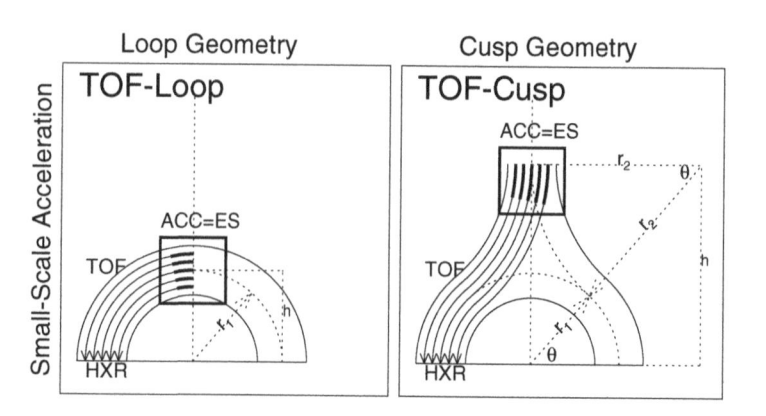

Figure 13.16: Geometric models of electron propagation paths with injection at the looptop of semi-circular flare loops (left) and injection from a symmetric cusp above the looptop (right). The radius of the flare loop is r_1 and the curvature radius of the cusp field line is r_2, with both segments tangentially matching at a loop azimuth angle θ (Aschwanden et al. 1996a).

functions $C_i(\epsilon_x, t)$ and $C_j(\epsilon_x, t)$, based on the convolution of channel response functions with the photon spectrum (Eq. 13.5.4); (3) the local powerlaw slope of the photon spectrum between these two is $\gamma = -\ln(I_2/I_1)/\ln(\epsilon_{x,2}/\epsilon_{x,1})$; (4) estimating some high-energy cutoff energy ϵ_0 from the photon spectrum, we can then use the *photon-to-electron energy conversion* factors (calculated in Fig. 13.14) to obtain the electron energies $\varepsilon_i = \epsilon_{x,i} \times q_\varepsilon(\epsilon_{x,i}, \gamma, \epsilon_0)$ and $\varepsilon_j = \epsilon_{x,j} \times q_\varepsilon(\epsilon_{x,j}, \gamma, \epsilon_0)$ that correspond to the peak times of the electron injection pulses (Fig. 13.13); (5) from the electron energies ε_i and ε_j we obtain the velocities using the relativistic relations $\beta_i = \mathrm{v}(\varepsilon_i)/c$ and $\beta_j = \mathrm{v}(\varepsilon_i)/c$ from (Eq. 11.1.13); (6) finally, we can apply the velocity dispersion formula to evaluate the time-of-flight distance l_{TOF},

$$l_{TOF} = c\tau_{ij} \left(\frac{1}{\beta_i} - \frac{1}{\beta_j} \right)^{-1} , \qquad (13.5.8)$$

which then also needs to be corrected for the electron pitch angle q_α (Eq. 12.2.2) and helical twist q_h of the magnetic field line to obtain the *projected loop distance* $l_{loop} = q_\alpha q_h l_{TOF}$.

Examples of such time-of-flight measurements are shown in Fig. 13.15, where we obtained projected TOF distances in the range of $l' \approx 15 - 30$ Mm from energy-dependent hard X-ray time delays in the energy range of $\epsilon_x \approx 25 - 250$ keV. Note that HXR time delays are always measured with respect to the lowest energy (Fig. 13.15), which provides the most accurate reference time due to it having the best photon statistics.

13.5.2   Scaling of TOF Distance with Loop Size

We have developed an accurate method to measure the electron propagation distance from a coronal acceleration site to the chromospheric thick-target site in the previ-

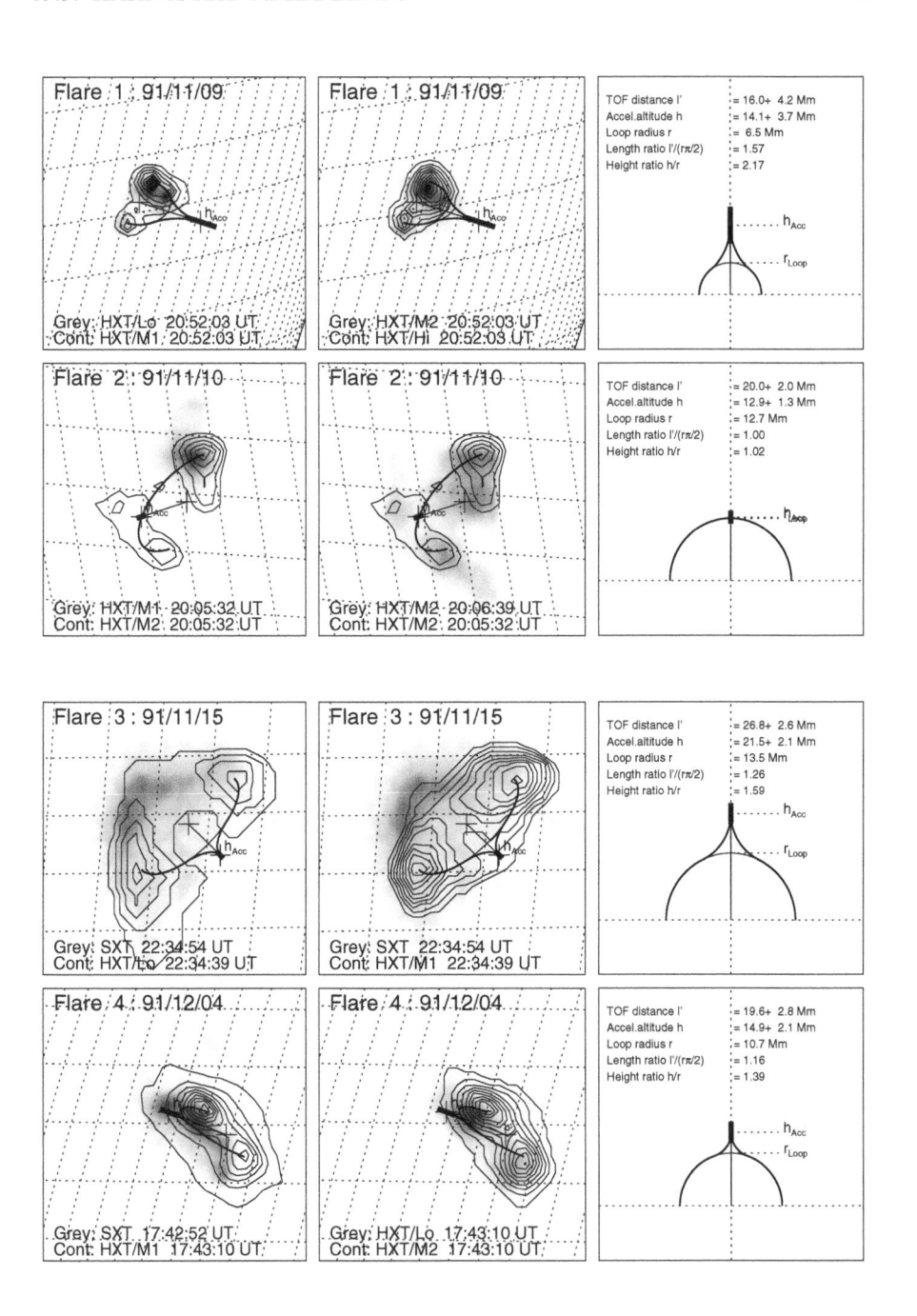

Figure 13.17: Overlays of co-registered *HXT* and *SXT* images (left and middle columns). The geometry of magnetic cusp field lines is modeled with circular segments and matches the projected time-of-flight distance l_{TOF}. The vertical projection is shown in the right panels and uncertainties are marked with thick bars (Aschwanden et al. 1996b).

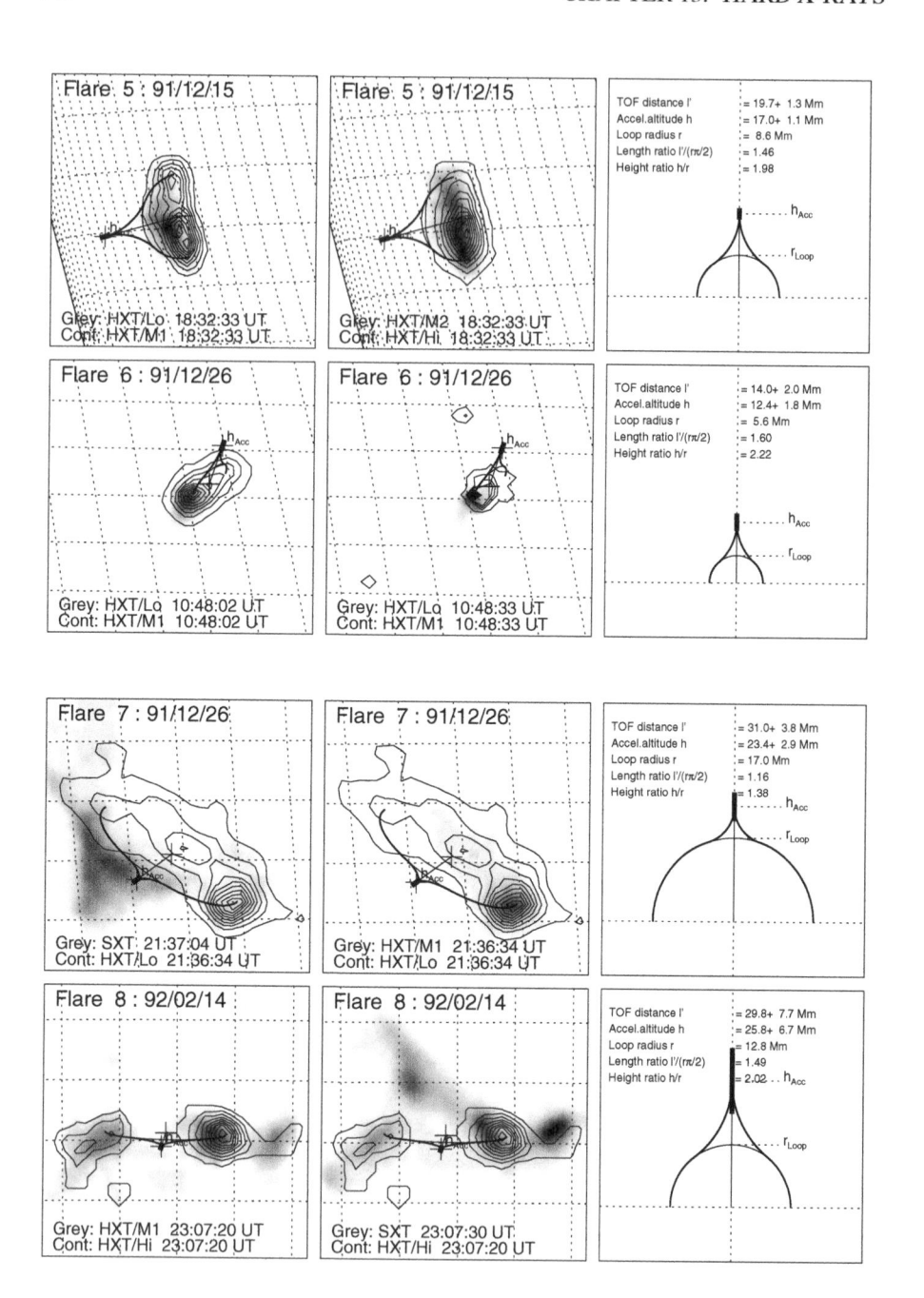

Figure 13.18: Overlays of co-registered *HXT* and *SXT* images (left and middle columns). The geometry of the magnetic cusp field lines is modeled with circular segments and matches the projected time-of-flight distance l_{TOF}. Representation similar to Fig. 13.17 (Aschwanden et al. 1996b).

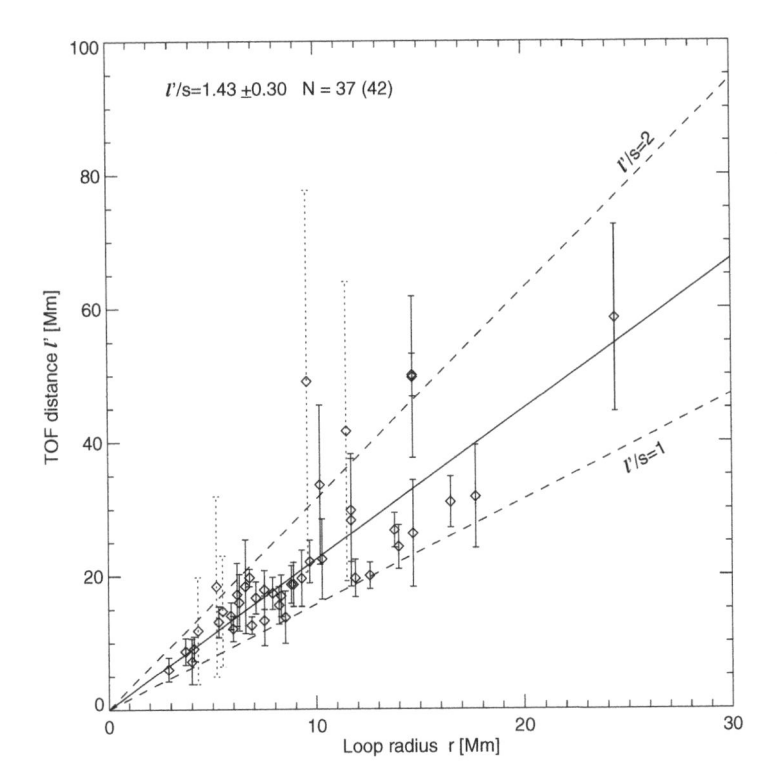

Figure 13.19: Scaling law of TOF distance l_{TOF} versus loop radius r in 42 flare events. The average ratio of the TOF distance l_{TOF} and loop half-length s is indicated with a solid line; the dashed lines indicate a loop half-length ($l_{TOF}/L = 1$) or a full loop length ($l_{TOF}/L = 2$) (Aschwanden et al. 1996c).

ous section (assuming synchronized injection, no dependence on acceleration times, and negligible energy loss times). It is of immediate interest to compare these electron propagation distances with flare loop sizes. Two geometric models are shown in Fig. 13.16: Injection at the looptop of a semi-circular flare loop (Fig. 13.16, left), and injection in the cusp above a flare loop (Fig. 13.16, right). A cusp-like field line that connects the footpoints with the acceleration site can simply be composed of two circular segments with opposite curvature, joined together at a common tangent (see dotted curves in Fig. 13.16, right). The length of such a field line, which corresponds to the projected time-of-flight distance l_{TOF}, is then

$$l_{TOF} = \frac{r}{2} \left(1 + \frac{h^2}{r^2} \right) \arctan \left(\frac{2hr}{h^2 - r^2} \right) . \tag{13.5.9}$$

In the limit $h \mapsto r$, this geometric model also includes the case of injection at the looptop of a semi-circular loop without a cusp (Fig. 13.16, left).

Some examples of such geometric reconstructions of acceleration heights with the TOF method are shown in Figs. 13.17 and 13.18; they are based on double-footpoint sources observed with *Yohkoh/HXT*, soft X-ray images of the flare loops observed with *Yohkoh/SXT*, and hard X-ray time delay measurements carried out with *BATSE/CGRO* data in the energy range $\approx 20 - 120$ keV. In Figs. 13.17 and 13.18 we show the co-aligned maps (left and middle) and the vertical geometry of the TOF distance (right). Interestingly, the propagation distances always turn out to be longer than the loop half-length, and thus require a cusp model if one assumes symmetric precipitation to both footpoints.

The statistics of 42 suitable events, for which both the TOF distance l_{TOF} and the flare loop radius r could be determined, reveals a remarkable result, shown in Fig. 13.19: There is a strong correlation between the TOF distance l_{TOF} and the flare loop half-length $L = r \times (\pi/2)$, revealing a linear relation with a ratio of

$$l_{TOF}/L = 1.43 \pm 0.30 . \tag{13.5.10}$$

This correlation clearly demonstrates a physical relation between the independently measured spatial and temporal parameters and, moreover, implies a scaling law between the location of the acceleration site (which is supposedly close to the reconnection point) and the relaxed flare loop size after reconnection. This corroborates the scale-free property of reconnection geometries (Fig. 13.16, right), where the length ratio of a newly reconnected (cusp-like) field line to the relaxed (force-free) field line is essentially invariant for different sizes of the system. One could argue that the accuracy of this result may be hampered by the separation method of oppositely signed energy-dependent time delays $dt/d\epsilon$ in the analyzed time profiles. In order to test the robustness of the scaling law, a full kinematic model of directly precipitating and trap-plus-precipitating electrons was developed (Aschwanden 1998a) and time-of-flight analysis was repeated with a deconvolution method of the two electron components. A very similar result was found using this different method (Aschwanden et al. 1998b, 1999a),

$$l_{TOF}/L = 1.6 \pm 0.6 . \tag{13.5.11}$$

This scaling law can be applied to various reconnection geometries. The fact that the scaling law is closer to $l_{TOF}/L \approx 1.5$, which roughly corresponds to the length of the cusp-shaped field lines, rather than to $l_{TOF}/L \approx 1.0$, which is the value for injection at the top of a flare loop, indicates that the particle acceleration site is located near the reconnection point, rather than near the soft X-ray-bright flare loop. The location of the acceleration region significantly above SXR-bright flare loops is to be expected, because the densities in soft X-ray-bright flare loops are much higher ($n_e \approx 10^{10}-10^{12}$ cm^{-3}, Aschwanden et al. 1998a; $n_e \approx (0.2 - 2.5) \times 10^{11}$ cm^{-3}, Aschwanden et al. 1997) than in the cusp above, $n_e \approx (0.6-10) \times 10^9$ cm^{-3}, Aschwanden et al. 1997). Had the acceleration site be located inside the dense, soft X-ray-bright flare loops, as assumed in earlier simple flare models (e.g., see Fig. 1 in Dennis & Schwartz 1989), the collisional energy loss would be so high that electrons could not be accelerated to high energies or could not propagate to the footpoints, where the brightest hard X-ray emission is generally seen from the thick-target bremsstrahlung of precipitating electrons.

The ratio between time-of-flight distance and relaxed postflare loop half-length is also consistent with analysis of field line shrinkage in two bipolar flares with "candle-flame" morphology, where a shrinkage of 20% and 32% was found (Forbes & Acton

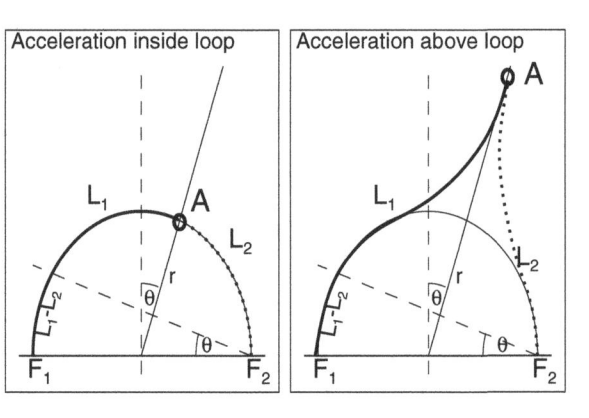

Figure 13.20: Geometry of electron path trajectories (with lengths L_1, L_2) from a common acceleration site (A) to two conjugate footpoints (F_1, F_2). The inclination angle θ of an asymmetric location A is related to the path difference by $(L_2 - L_1) = 2r\theta$. The two cases illustrate the geometry for an acceleration source inside a semi-circular flare loop (left) and for an acceleration site above the flare loop (right) (Aschwanden 1998b).

1996). These values correspond in our definition to a ratio of $l_{TOF}/L = 1.25$ and 1.5, respectively, a range that is expected in dipole field models with a vertical (Syrovatskii-type) current sheet. Other models constrained by the (quite narrow) opening angle of reconnection outflows yield ratios of $l_{TOF}/L \approx 2 - 4$ (Tsuneta 1996a).

13.5.3 Conjugate Footpoint Delays

While electron time-of-flight measurements mainly provide 1D constraints to the altitude of the acceleration site, relative time delays between conjugate footpoint hard X-ray emission yield additional information on the horizontal offset or asymmetry of the acceleration site with respect to the vertical symmetry axis of the flare loop (Fig. 13.20). Hard X-ray light curves from magnetically conjugate footpoints are shown in Fig. 13.21. The detailed correlation and near-simultaneity of conjugate pulses suggests a common injection mechanism that connects to both footpoints, which can be naturally explained with magnetic reconnection geometries (e.g., Fig. 13.16, right).

Sakao (1994) measured the simultaneity of hard X-ray emission from two conjugate footpoints by cross-correlating the hard X-ray fluxes from both footpoints (F_1, F_2). Cross-correlation time delays were found to be near-simultaneous in six cases out of seven investigated flares, with uncertainties of $\pm(0.1 - 0.5)$ s. Using the geometry indicated in Fig. 13.20 we obtain the following relation between the asymmetry angle θ and the travel times $t_1 = L_1/v$ and $t_2 = L_2/v$ for particles propagating with speed v from a common acceleration site A to two conjugate footpoints F_1 and F_2:

$$\theta = \frac{v}{2r_{curv}} \frac{(t_1 - t_2)}{q_\alpha q_H} \qquad (13.5.12)$$

where r_{curv} is the semi-circular loop curvature radius, $q_\alpha \approx 0.64$ is a correction factor for the average pitch angle α, and $q_H \approx 0.85$ is a correction factor for helical

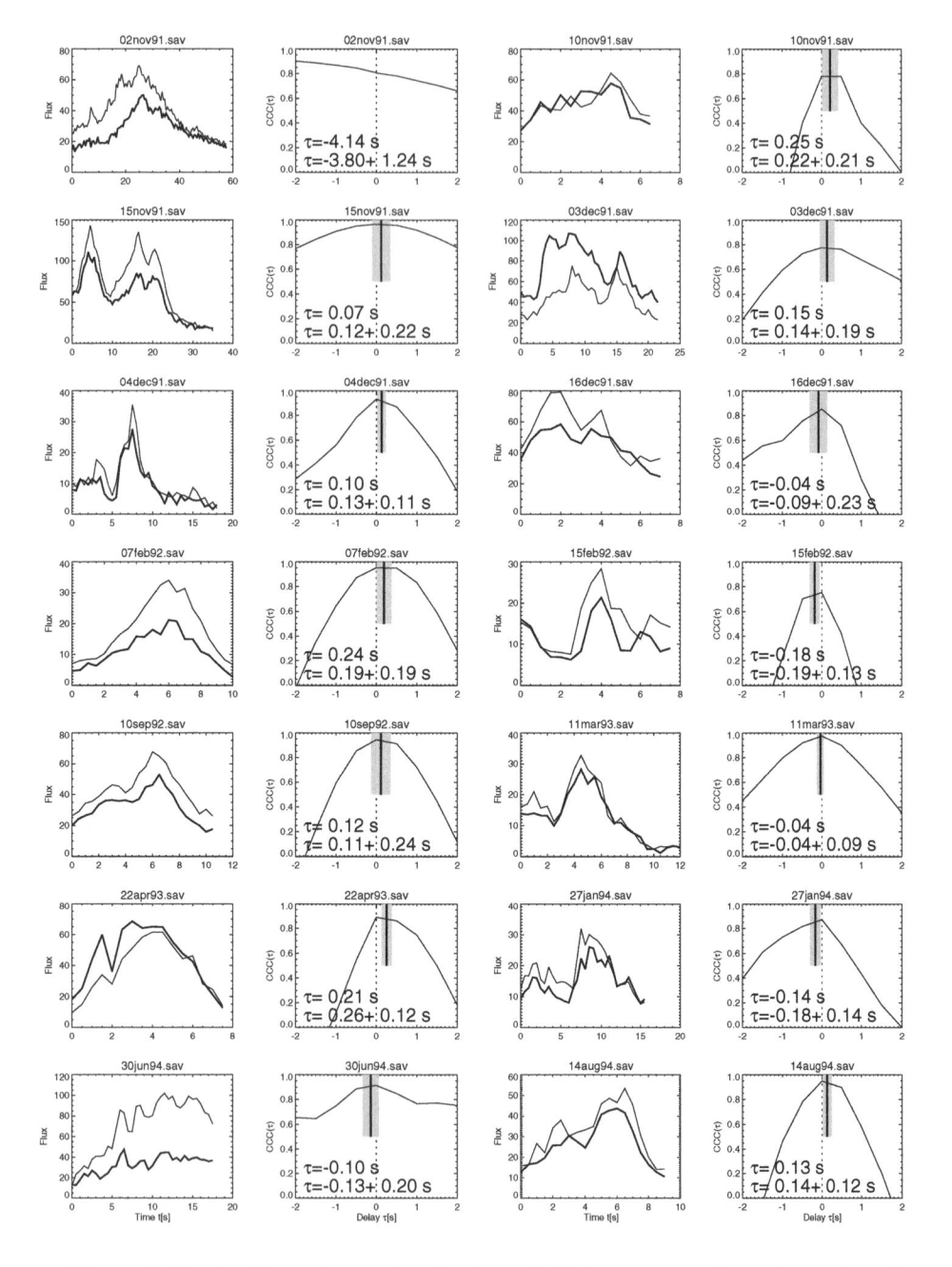

Figure 13.21: Hard X-ray time profiles $F_1(t)$ and $F_2(t)$ from magnetically conjugate footpoint sources, extracted from *Yohkoh/HXT* images for 14 flares (first and third column; courtesy of Taro Sakao). The two conjugate time profiles are cross-correlated. The cross-correlation coefficient $CCC(\tau)$, relative time delay τ, and uncertainty σ_τ (assuming a dynamic range of 1:10 in *HXT* images) is shown (second and fourth column) (Aschwanden 2000).

twist of the magnetic field line along the trajectory (Aschwanden et al. 1996a). The path difference $(L_1 - L_2)$ does not depend on the height of the acceleration source or on whether acceleration takes place inside or outside of the loop. Based on the upper limits of the travel time differences $(t_2 - t_1)$ measured by Sakao (1994) and the observed loop curvature radii r_{curv} of the corresponding flares, one can calculate (using Eq. 13.5.12) upper limits for the asymmetry angles θ as a function of the speed v. Since the measured footpoint delays are of the order of the loop-crossing times for electrons, no high symmetry is required for the acceleration site with respect to flare loop footpoints. Slower agents, however, such as protons or thermal conduction fronts, would require a very high symmetry to explain the near-simultaneity of the hard X-ray footpoint pulses.

13.5.4 Trapping Time Delays

In §12.5 we described the kinematics of trapped particles in a flare loop with magnetic mirroring. In the weak-diffusion limit, the trapping times correspond roughly to the collisional deflection times (Eq. 12.5.11), and thus have a timing, $\tau_{defl}(\varepsilon) \propto \varepsilon^{3/2}$, with an opposite trend as a function of energy, compared with time-of-flight delays, $\tau_{prop}(\varepsilon) \propto \varepsilon^{-1/2}$. Thanks to this opposite trend, which translates into an opposite sign in the derivative $d\tau_{delay}/d\varepsilon$, the two components of directly precipitating and trap-precipitating electron populations can be cleanly separated in hard X-ray time profiles, either with a lower envelope subtraction method (Fig. 12.16; Aschwanden et al. 1997) or with a deconvolution method (Aschwanden 1998a). For the deconvolution method, one has to invert the convolution of the two-population injection function (Eq. 12.5.10) with the bremsstrahlung cross section (Eq. 13.2.17).

Data analysis of some 100 flares simultaneously observed with *CGRO* and *Yohkoh* are highly consistent with this model, where the observed hard X-ray flux represents a convolution of directly precipitating and trapped-plus-precipitating components. The two components can be separated by subtracting a lower envelope to the fast (subsecond) time structures (Fig. 12.16). In Fig. 13.22 we show a selection of hard X-ray, energy-dependent time delay measurements of the slowly varying, hard X-ray flux from 20 flares (Aschwanden et al. 1997). The delays have been fitted with the expression for the collisional deflection time (Eq. 12.5.11), which seems to be consistent for all analyzed data at energies $25 \lesssim \epsilon \lesssim 200$ keV.

The fits of collisional deflection times to trapping delays also also yield an electron density measurement n_e^{trap} in the trap region, typically in the range of $n_e^{trap} \approx 3 \times 10^{10}$ cm^{-3} $- 3 \times 10^{11}$ cm^{-3} (Fig. 13.23, top). Comparison of simultaneous electron density measurements in the soft X-ray-bright flare loops with *Yohkoh/SXT* yields similar densities (Fig. 13.23, middle). This result indicates that trap density is on average comparable with flare loops that have been filled with chromospheric upflows. It does not necessarily mean that the trap regions have to be co-spatial with filled, soft X-ray-bright flare loops. The trapping loops could be collisionless and only the mirror points could have a density comparable with the chromospheric upflows, producing efficient collisional deflection near the mirror points, consistent with our fits. The fact that the trapping times are consistent with collisional deflection times is a strong argument that trapping is controlled in the weak-diffusion limit. For the strong-diffusion limit, where

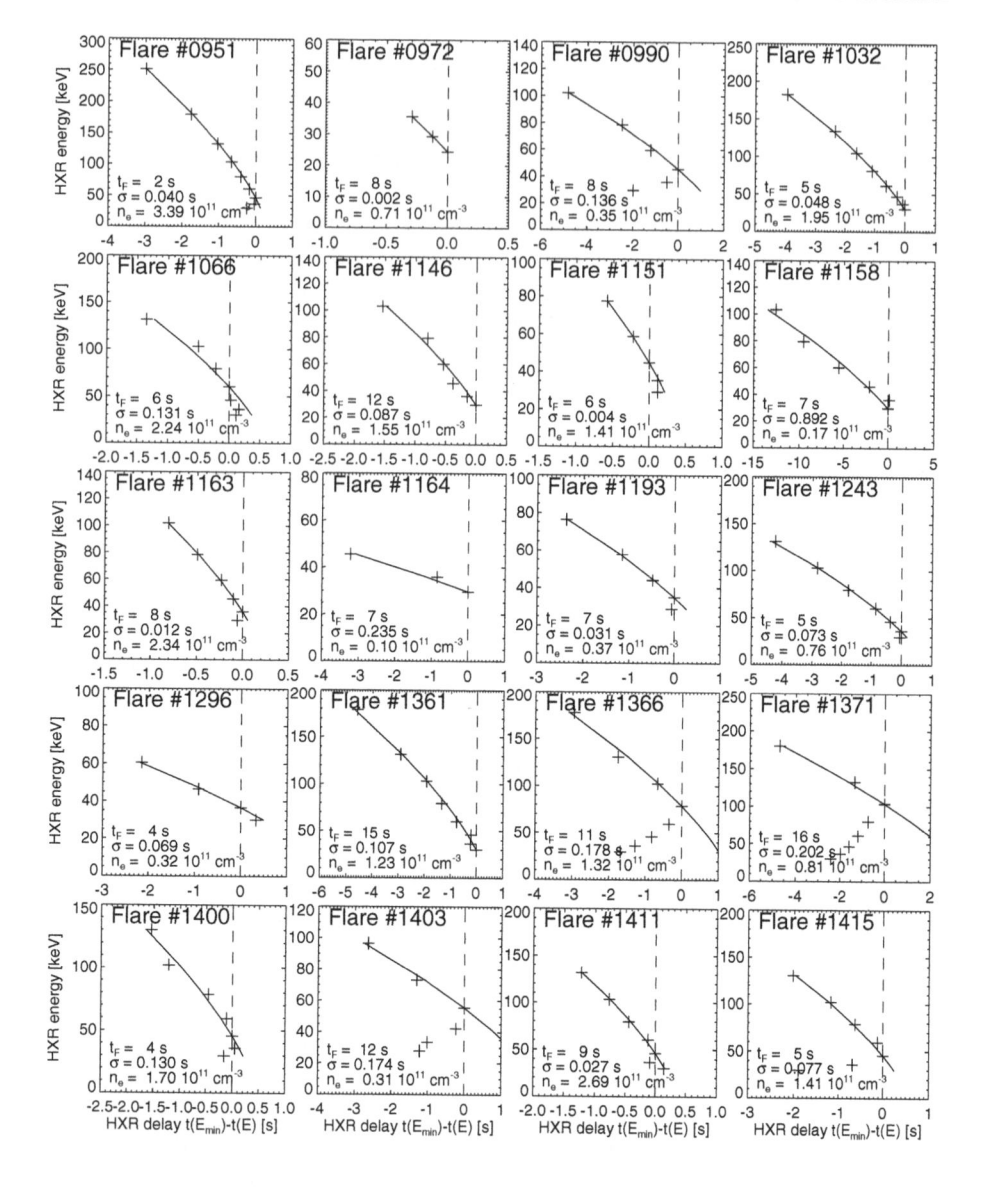

Figure 13.22: Energy-dependent time delay measurements of the *smooth hard X-ray compo-nent*, after filtering out the *pulsed hard X-ray component* with a Fourier filter time scale τ_{filter} (third line from bottom in each panel) from all 44 flare events. The flares are identified (in chronological order) with the *BATSE/CGRO* burst trigger number (top of each panel). Each cross indicates a time delay measurement of one of the 16 *MER* energy channels with respect to a low-energy reference channel (indicated with a vertical dashed line at zero delay). The best model fit is drawn with a solid line, quantified by the trap electron density n_e (lowest line in each panel) and the mean standard deviation σ (second lowest line in each panel) (Aschwanden et al. 1997).

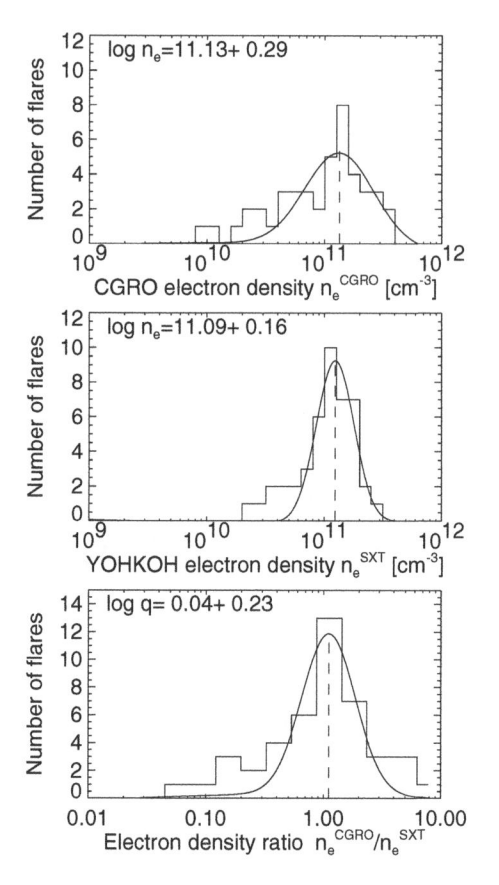

Figure 13.23: Distributions of the *CGRO*-inferred trap densities n_e^{CGRO} (top), the *Yohkoh*-inferred electron densities n_e^{SXT} (middle), and their ratios $q_{n_e} = n_e^{CGRO}/n_e^{SXT}$, fitted with Gaussians (Aschwanden et al. 1997).

pitch angle scattering is controlled by wave turbulence, trapping times would be expected to be significantly shorter.

13.5.5 Thermal-Nonthermal Delays (Neupert Effect)

In the thick-target bremsstrahlung model, hard X-ray emission reaches a peak when most of the nonthermal electrons hit the chromosphere and cause impulsive heating during their energy loss. As a consequence, the heated chromospheric plasma evaporates into the corona due to the local overpressure, which is detectable in the form of soft X-ray emission from the heated plasma that fills the coronal parts of the flare loops. The peak of soft X-ray emission, therefore, lags the peak of hard X-ray emission, due to this causal relationship. In fact, if all nonthermal energy was converted into heating, the energy in soft X-rays should actually mimic the time integral of the energy depo-

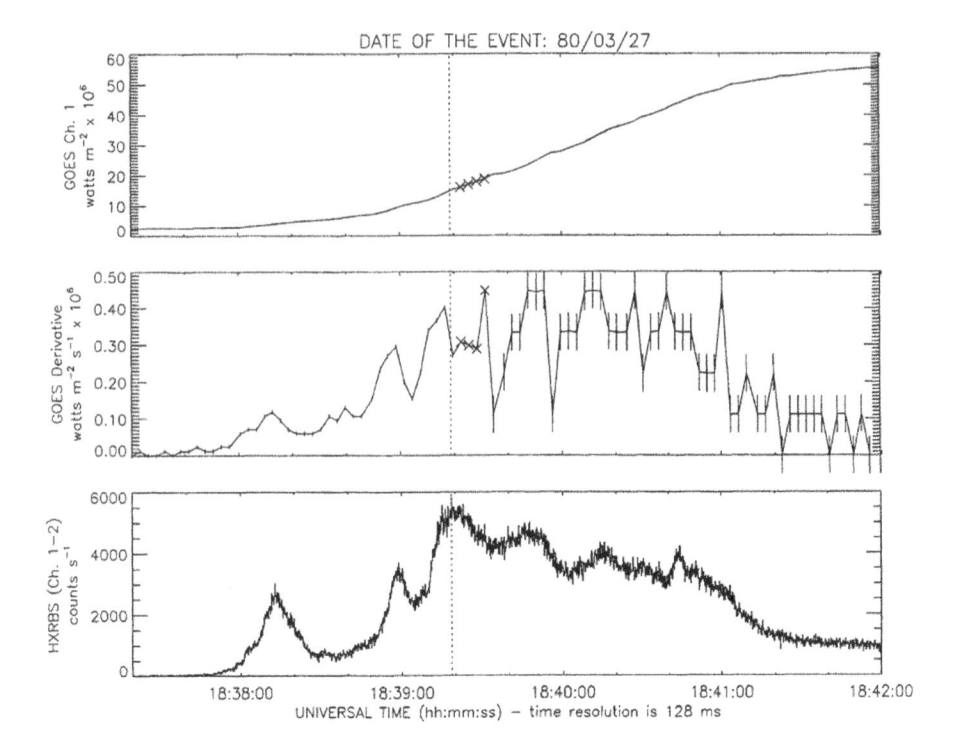

Figure 13.24: The Neupert effect is illustrated using the following three time profiles. *(Top)* GOES 1-8 Å soft X-ray time profile plotted with a time resolution of 3 s for a flare on 1980-Mar-27. *(Middle)* Time derivative of GOES soft X-ray flux shown in the top plot. *(Bottom) HXRBS/SMM* hard X-ray time profile in the $26 - 51$ keV energy channels. Note that each peak in the time derivative of the soft X-rays (middle) closely matches the peaks in the hard X-ray light curve (bottom) (Dennis & Zarro 1993).

sition rate seen in hard X-rays. In practice, however, there also occurs conductive and radiative energy loss during the heating and evaporation process, but the "time integral effect" should still be present if cooling times are longer than impulsive heating time scales. The relation between the evolution of the soft X-ray flux $F_{SXR}(t)$ and hard X-ray flux $F_{HXR}(t)$ can be modeled with an empirical cooling time τ_{cool},

$$F_{SXR}(t) = q_F \int F_{HXR}(t) \ \exp^{-t/\tau_{cool}} \ dt \ . \qquad (13.5.13)$$

This effect was first pointed out by Werner Neupert (1968), who noticed that time-integrated microwave fluxes closely match the rising portions of soft X-ray emission curves, which was later dubbed the *Neupert effect* by Hudson (1991b). In a statistical study (68 events) it was found that 80% of the flares display the expected Neupert effect, showing $\lesssim 20$ s time difference between hard X-ray peaks and the peaks of soft X-ray time derivatives (see example in Fig. 13.24; Dennis & Zarro 1993). Deviations from a pure integral effect can always be explained by cooling processes, heating

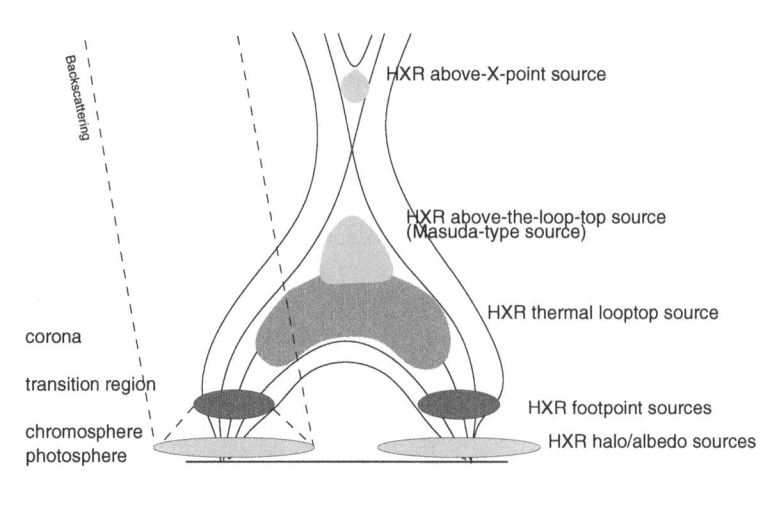

Figure 13.25: Schematic of physically different hard X-ray sources: thermal hard X-rays at the flare looptop, nonthermal hard X-ray sources at flare loop footpoints (above the flare loop and above the X-point), and hard X-ray halo or albedo sources from backscattering at the photosphere.

without detectable hard X-ray emission, or multi-loop confusion. Observations of the Neupert effect are documented in many multi-wavelength studies of flares (e.g., Silva et al. 1996; Tomczak 1999; Kundu et al. 2001; Gallagher et al. 2002; Joshi et al. 2003), *long duration events (LDEs)* (Hudson & McKenzie 2001), microflares (e.g., Krucker & Benz 2000; Benz & Grigis 2002), or in numerical simulations (Li et al. 1997). High-temperature plasma ($T_e \geq 16.5$ MK) was found to be more likely to exhibit the Neupert effect than low-temperature plasma (McTiernan et al. 1999). The integral effect also yields a prediction for the slope of the frequency distributions of hard X-ray fluences (time-integrated fluxes) and soft X-ray peak fluxes, which is not yet properly understood from observational statistics (Lee et al. 1993c, 1995; Veronig et al. 2002a, b; Veronig 2003), due to unknown scaling laws between energy deposition rates, flare temperatures, densities, and energy cutoffs (see §9.8.2). The Neupert effect has also been observed in stellar flares, based on EUV, optical, and radio observations (Hawley et al. 1995, 2003; Güdel et al. 1996, 2002).

13.6 Hard X-ray Spatial Structures

Flare observations have revealed at least five types of hard X-ray sources (Fig. 13.25) which are produced by distinctly different physical mechanisms: (1) the strongest sources are flare loop footpoint sources due to thick-target bremsstrahlung in or near the chromosphere (§13.6.1); (2) thermal hard X-ray sources in the upper part of flare loops (§2.3 and §16); (3) Masuda-type above-the-looptop sources due to temporary trapping during acceleration (§13.6.3); (4) above-the-X-point hard X-ray sources in or near soft X-ray flare ejecta (Sui & Holman 2003); and (5) hard X-ray halo or albedo sources due to backscattering at the photosphere, reported for the first time from recent RHESSI observations (Schmahl & Hurford 2002).

13.6.1 Footpoint and Loop Sources

Flare loop footpoint sources are generally the most prominent, nonthermal hard X-ray sources observed during flares, at typical energies of $\epsilon_x \gtrsim 20$ keV. Due to the magnetic conjugacy of flare loop footpoints, they are generally observed as pairs of point-like or ribbon-like structures on the opposite sides of the main neutral line. The first imaging observations of hard X-ray footpoint sources were accomplished with SMM/HXIS (e.g., Hoyng et al. 1981a, b) and modeled thereafter (e.g., MacKinnon et al. 1985). Sakao (1994) provided the first statistics of flare hard X-ray sources and found that double sources are most frequently observed (43%), while single sources (28%) or multiple ≥ 3 sources (28%) are less frequently observed. A representative set of hard X-ray flare maps obtained with *Yohkoh/HXT* at energies $\gtrsim 20$ keV is shown in the form of contour plots in Fig. 13.26 and in the form of flux profiles across the flare loop baseline in Fig. 13.27. The flux profile of the footpoint sources can be fitted by Gaussians, which have a full width comparable to the *Yohkoh/HXT* spatial resolution of $\approx 5'' - 8''$, so many sources may be unresolved. Hard X-ray footpoint sources coincide with the endpoints of soft X-ray flare loops, so they have to be located in the chromosphere or slightly above in the transition region. This is expected from the thick-target bremsstrahlung model, if acceleration of precipitating electrons takes place at (collisionless) coronal heights. The height of these hard X-ray footpoint sources depends on the electron energy as well as on the chromospheric density $n_p(h)$ as a function of altitude h, which determines the stopping depth of precipitating electrons. A simplified derivation of the height of hard X-rays sources is given in Brown et al. (2002a).

Making the simplifying assumptions of (1) full target ionization , (2) 1D Coulomb collisional transport, neglecting pitch angle changes (pitch angle $\alpha = 0$ or $\mu = \cos \alpha = 1$), (3) no mirroring of particles, and (4) powerlaw function for electron injection flux energy spectrum, the kinetic energy $\varepsilon(\varepsilon_0, N)$ of a decelerated electron in the thick-target region varies as a function of the column depth $N(z) = \int_z^{z_{max}} n(z')dz'$ (Brown 1972),

$$\varepsilon(\varepsilon_0, N) = (\varepsilon_0^2 - 2KN)^{1/2} , \qquad (13.6.1)$$

where $K = 2\pi e^4 \Lambda$ is a constant and Λ the Coulomb logarithm (see Spitzer 1967). This (Eq. 13.6.1) defines a stopping depth N_s for a given initial energy ε_0,

$$N_s(\varepsilon_0) = \frac{\varepsilon_0^2}{2K} . \qquad (13.6.2)$$

The continuity condition then yields (using Eq. 13.6.1)

$$f(\varepsilon, N) = f_0(\varepsilon_0)\frac{d\varepsilon_0}{d\varepsilon} = f_0([\varepsilon^2 + 2KN]^{1/2})\frac{\varepsilon}{(\varepsilon^2 + 2KN)^{1/2}} \qquad (13.6.3)$$

for the local electron spectrum $f(\varepsilon, N)$ as a function of N. Note that we neglect here the fact that Λ may vary somewhat in lower regions where the ionization falls (e.g., Emslie 1978). The photon flux per unit energy ϵ_x and per unit source height range dz

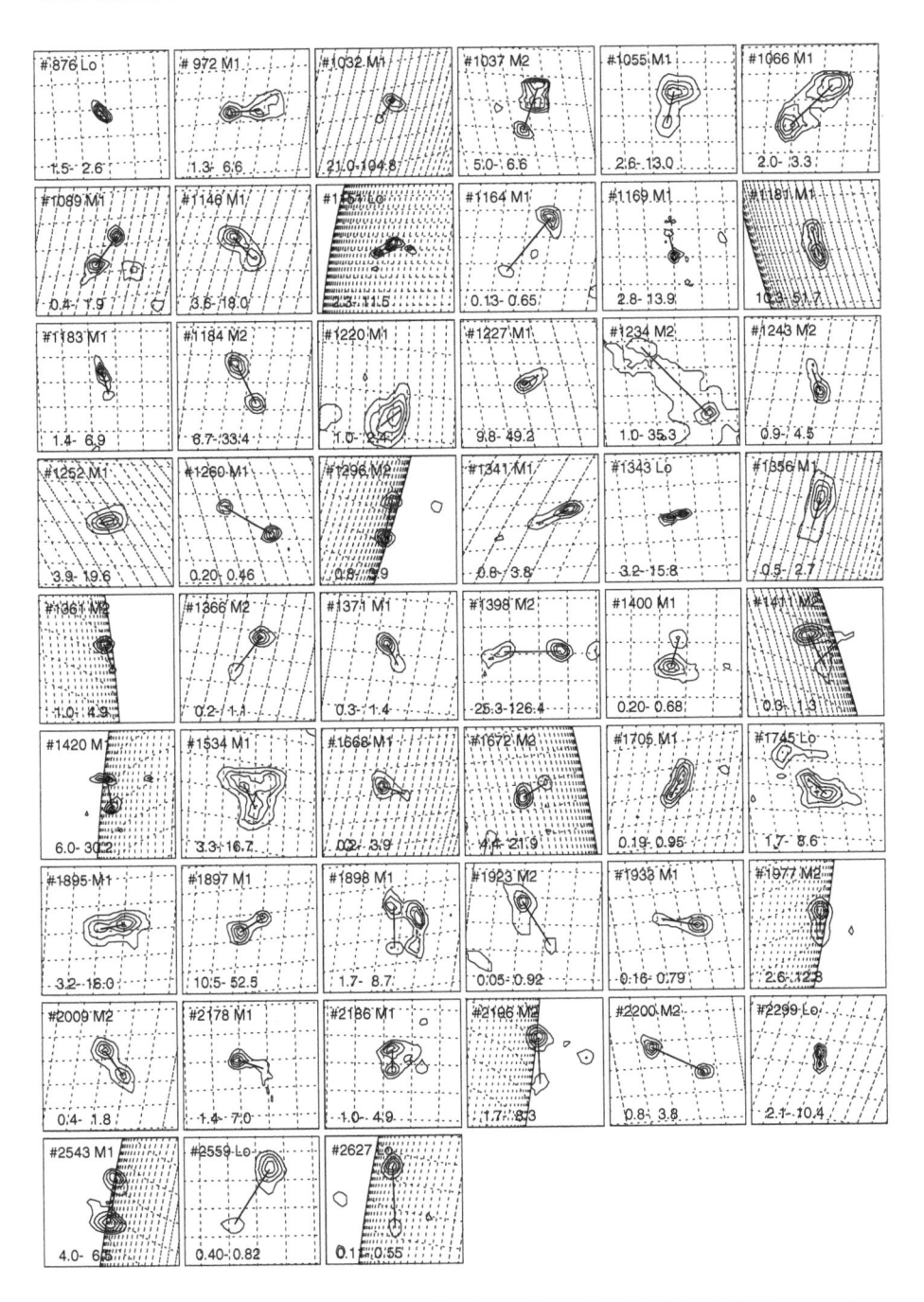

Figure 13.26: *Yohkoh/HXT* maps of 54 solar flares, shown for the highest *HXT* energy channels (Lo=14 − 23 keV, M1=23 − 33 keV, M2=33 − 53 keV) where the footpoint sources are detectable. All frames have the same spatial scale of 32 *HXT* pixels in x and y-axis (i.e. 78.5″ or 55 Mm) (Aschwanden et al. 1999d).

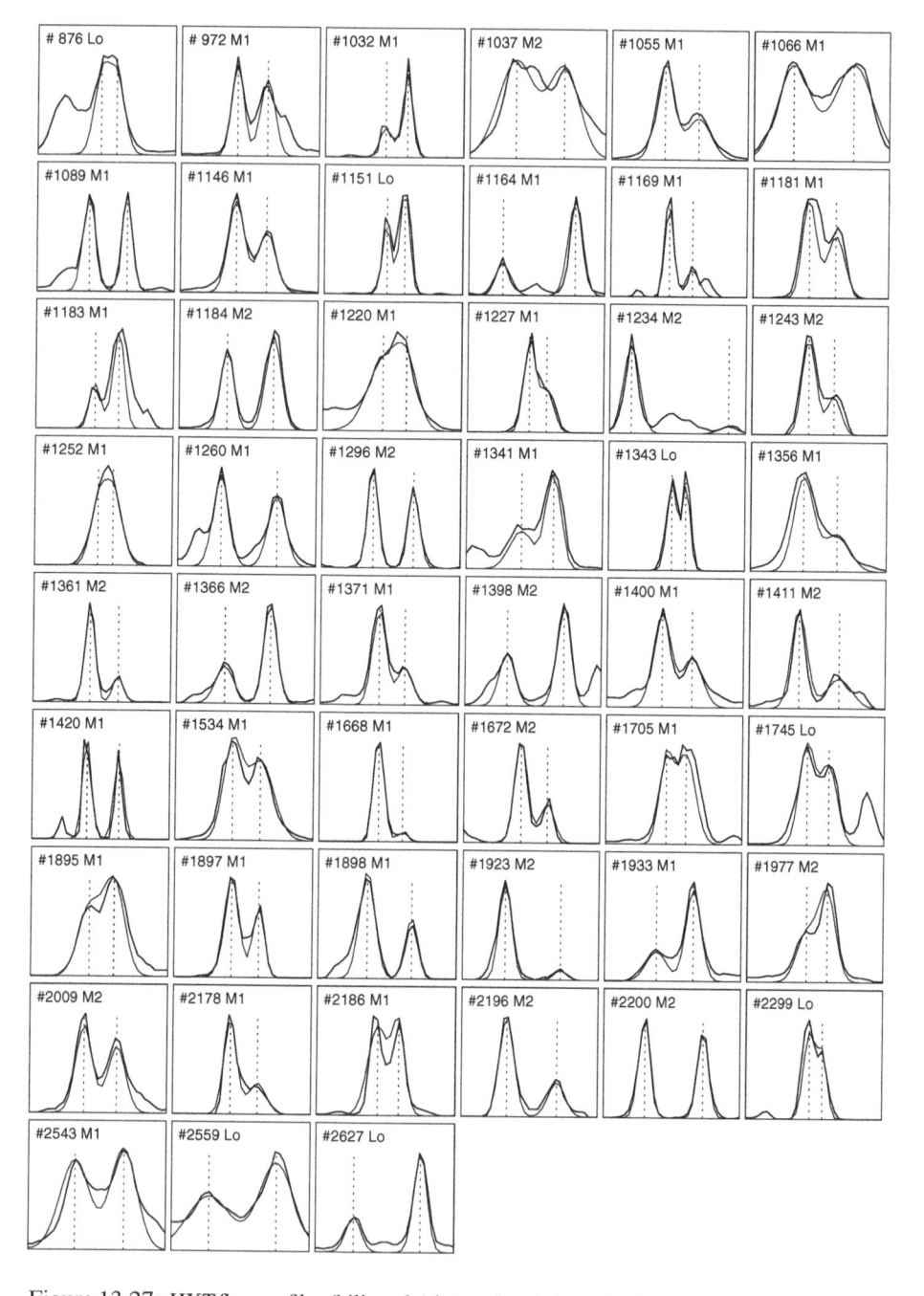

Figure 13.27: *HXT* flux profiles (bilinearly) interpolated along the footpoint baseline computed from the Gaussian fits as shown in Fig. 13.26. The observed flux is shown with a thick solid line, while the fit of the two-component Gaussian model is shown with a thin solid line, and the center positions of the Gaussians are marked with dashed lines (Aschwanden et al. 1999d).

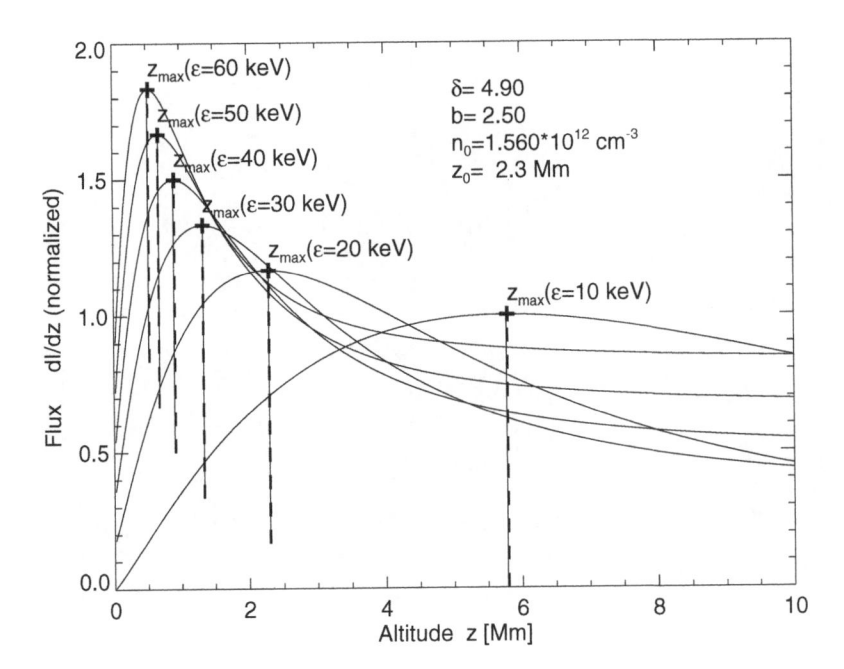

Figure 13.28: Analytical functions of the energy flux per height range, $dI/dz(\varepsilon, z)$ (see Eq. 13.6.14), for six different energies, $\varepsilon = 10, 20, ..., 60$ keV, as a function of altitude z (top). The functions are normalized to unity and shifted by 0.1 per energy value to make them visible. The locations $z_{max}(\varepsilon)$ of these functions are fitted to the observed centroids of the altitudes of footpoint hard X-ray sources in the 2002-Feb-22 flare (Aschwanden et al. 2002b).

for a beam cross section $\sigma(\epsilon_x, \varepsilon)$ and beam area A, at Earth distance $r = 1$ AU is (see Eq. 13.2.16),

$$\frac{dI}{dz}(\epsilon_x, z) = \frac{A}{4\pi r^2} n(z) \int_{\epsilon_x}^{\infty} f(\varepsilon, N)\sigma(\epsilon_x, \varepsilon)d\varepsilon \quad (\text{cm}^{-2} \text{ s}^{-1} \text{ erg}^{-1} \text{ cm}^{-1}).$$

$$(13.6.3)$$

Using the simplified *Kramers* bremsstrahlung cross section (Eq. 13.2.9) and approximating the injection spectrum $f_0(\varepsilon_0)$ with a powerlaw function (Eq. 13.2.18), the photon flux (13.6.2) becomes

$$\frac{dI}{dz}(\epsilon_x, z) = \frac{A\sigma_0}{4\pi r^2} \frac{1}{\epsilon_x} n(z) \int_{\epsilon_x}^{\infty} \frac{f_0((\varepsilon^2 + 2KN)^{1/2})}{(\varepsilon^2 + 2KN)^{1/2}} d\varepsilon. \quad (13.6.4)$$

Using Eqs. (13.6.1) and (13.6.3) we obtain for a powerlaw injection function $f_0(\varepsilon)$,

$$\frac{dI_\varepsilon}{dz} = (\delta - 1)\frac{AF_1}{\varepsilon_1^{\delta+1}} \frac{\sigma_0}{4\pi r^2} \frac{1}{\epsilon_x} n(z) \int_{\epsilon_x}^{\infty} \frac{d\varepsilon}{(\varepsilon^2 + 2KN)^{(\delta+1)/2}}. \quad (13.6.5)$$

The product $AF_1 = \mathcal{F}_1$ is the total flux of electrons s^{-1} at $\varepsilon_0 \geq \varepsilon_1$ over the area A.

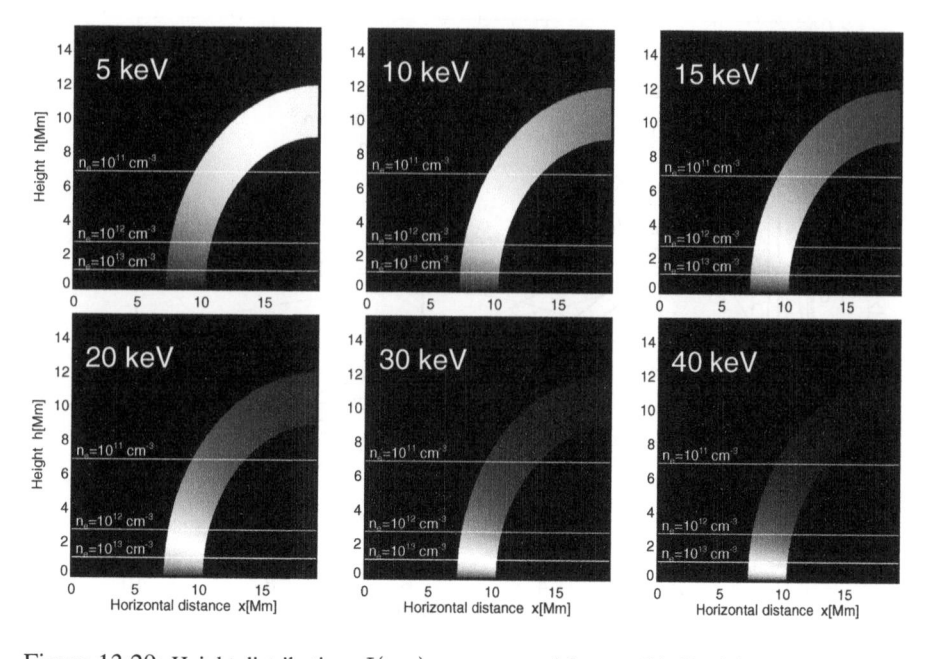

Figure 13.29: Height distributions $I(z, \varepsilon)$ are computed for $\varepsilon = 5$ keV, 10 keV, ... 40 keV and are represented by brightness maps of the flare loops in different energy bands. The vertical axis represents the height scale h, for a flare loop with a loop radius of 12.4 Mm. The corresponding chromospheric densities n_e are indicated on the left side. Note that the brightness distributions of all energies $\varepsilon \gtrsim 20$ keV are fairly concentrated inside the chromosphere (at $h \lesssim 2.0$ Mm), while the $\varepsilon \lesssim 15$ keV emission is spread along the entire flare loop (Aschwanden et al. 2002b).

Introducing $u(\epsilon_x, z) = \epsilon_x^2/2KN(z)$, one obtains

$$\frac{dI}{dz}(\epsilon_x, z) = (\delta - 1)\frac{\mathcal{F}_1}{\varepsilon_1}\frac{\sigma_0}{8\pi r^2}\frac{1}{\epsilon_x}n(z)\left(\frac{2KN}{\varepsilon_1^2}\right)^{-\delta/2}B\left(\frac{1}{1+u}, \frac{\delta}{2}, \frac{1}{2}\right), \quad (13.6.6)$$

where B is the *Incomplete Beta function*,

$$B\left(\frac{1}{1+u}, \frac{\delta}{2}, \frac{1}{2}\right) \equiv \int_0^{1/(1+u)} x^{\delta/2-1}(1 - x)^{-1/2}dx . \quad (13.6.7)$$

Note that the Incomplete Beta function B (Eq. 13.6.7) depends on ϵ_x only in the combination $u = \epsilon_x^2/2KN$. On the other hand, $n(z)[2KN/\varepsilon_1^2]^{-\delta/2}$ depends only on z for a given δ, and is independent of the photon energy ϵ_x.

For a practical application we specify a chromospheric density model $n(z)$, which can be locally approximated using a powerlaw function,

$$n(z) = n_0 \left(\frac{z}{z_0}\right)^{-b} . \quad (13.6.8)$$

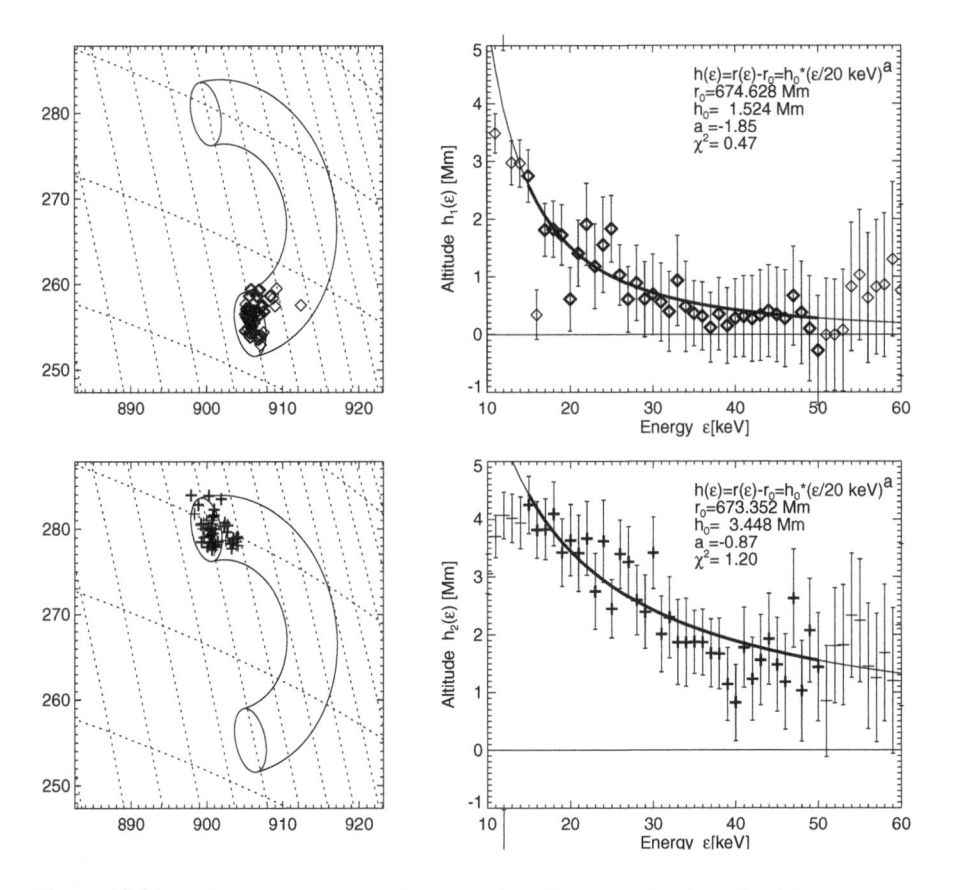

Figure 13.30: Height measurements of the centroids of the Gaussian fits to hard X-ray sources. The spatial location of source centroids is shown (left) and altitudes as functions of energy (right). Measurements at the southern footpoint are shown (top) and at the northern footpoint (bottom). The error bars of the height measurements were estimated from Poisson statistics. The curves indicate powerlaw fits, marked by thick lines in the energy range of the fit ($15 - 50$ keV) (Aschwanden et al. 2002b).

The *column depth* $N(z)$ is then the integrated density,

$$N(z) = \int_z^\infty n(z)dz = N_0 \left(\frac{z}{z_0}\right)^{1-b} , \qquad N_0 = \frac{n_0 z_0}{b - 1} . \qquad (13.6.9)$$

Inserting the chromospheric density model (Eq. 13.6.9) into the hard X-ray intensity versus height function dI/dz (Eq. 13.6.6) yields the following height dependence,

$$\frac{dI}{dz}(\tilde{\epsilon}_x, \tilde{z}) = I_0 \, \tilde{\epsilon}_x^{-1} \tilde{z}^{-b+(b-1)\delta/2} \, B\left(\frac{1}{1 + \tilde{\epsilon}_x^2 \tilde{z}^{b-1}}, \frac{\delta}{2}, \frac{1}{2}\right) , \qquad (13.6.10)$$

using dimensionless variables

$$\tilde{\epsilon}_x = (\epsilon_x/\epsilon_{x,0}) , \qquad (13.6.11)$$

$$\tilde{z} = (z/z_0) \, , \tag{13.6.12}$$

with the reference energy $\epsilon_{x,0}$ and constant I_0,

$$\epsilon_{x,0} = \sqrt{2KN_0} \, , \tag{13.6.13}$$

$$I_0 = (\delta - 1) \frac{\mathcal{F}_1}{E_1} \frac{\sigma_0}{8\pi r^2} \frac{1}{\sqrt{2KN_0}} n_0 \left(\frac{2KN_0}{\varepsilon_1^2} \right)^{-\delta/2} . \tag{13.6.14}$$

The height distribution $dI/dz(\tilde{\epsilon}_x, \tilde{z})$ is shown for different photon energies in Fig. 13.28, where the peaks of the height distributions have been fitted to the heights of the corresponding hard X-ray sources observed with RHESSI in the 2002-Feb-22, 11:06 UT flare (Aschwanden et al. 2002a). Height distributions are also visualized in terms of a circular flare loop geometry in Fig. 13.29. This example, which is probably typical for many flares, shows that $\approx$ 5 keV emission occurs near the looptop, while $\gtrsim$ 15 keV emission is localized close to the footpoints, in chromospheric densities of $n_e \approx 10^{12} - 10^{13}$ cm^{-3}. Thus, the centroid height of hard X-ray footpoint sources is located at progressively lower altitudes with higher energies. Altitude measurements of the hard X-ray sources carried out using RHESSI observations for the 2002-Feb-22 flare are shown in Fig. 13.30. These height measurements also constrain a chromospheric density model $n_e(z)$ that is shown in Fig. 4.28 for this observation. Altitude measurements of hard X-ray sources have been reported earlier by Matsushita et al. (1992), with the same trend of lower heights with higher energies.

Conjugate hard X-ray footpoint sources are generally not equally bright (Figs. 13.26 and 13.27) and may have different spectra (e.g., Emslie et al. 2003 for a recent RHESSI observation). This implies an asymmetry of the magnetic field in the intervening flare loop, and thus asymmetric mirror ratios and asymmetric trapping (§12.6.2). This asymmetry introduces a complementary behavior between particle precipitation and trapping at each footpoint: The side with the weaker magnetic field (lower magnetic mirror ratio) exhibits a higher precipitation rate and a smaller fraction of trapped particles compared with the other footpoint and vice versa. Since hard X-rays are dominantly produced by precipitating electrons, while microwave emission is predominantly produced by gyrosynchrotron emission from trapped high-energy electrons, relative footpoint strengths in hard X-rays are thus complementary to the radio brightness (e.g. observed by Kundu et al. 1995; Wang et al. 1995). Electron precipitation sites, however, do not coincide with locations of high vertical current density, which implies that electrons are not accelerated by large-scale current systems (Li et al. 1997).

13.6.2 Footpoint Ribbons

The classical Kopp–Pneuman model (Figs. 10.20 and 10.21) is a 2D model that makes no prediction about the spatial extent along the neutral axis. However, Hα ribbons on both sides of the neutral line were observed long ago, sometimes with considerable length, but the matching counterparts of hard X-ray double ribbons were not detected until the end of the Yohkoh era. So, it was not clear whether nonthermal hard X-ray emission is more localized in its occurrence or whether there is an instrumental sensitivity problem to detect extended hard X-ray ribbons. At the end of the Yohkoh mission,

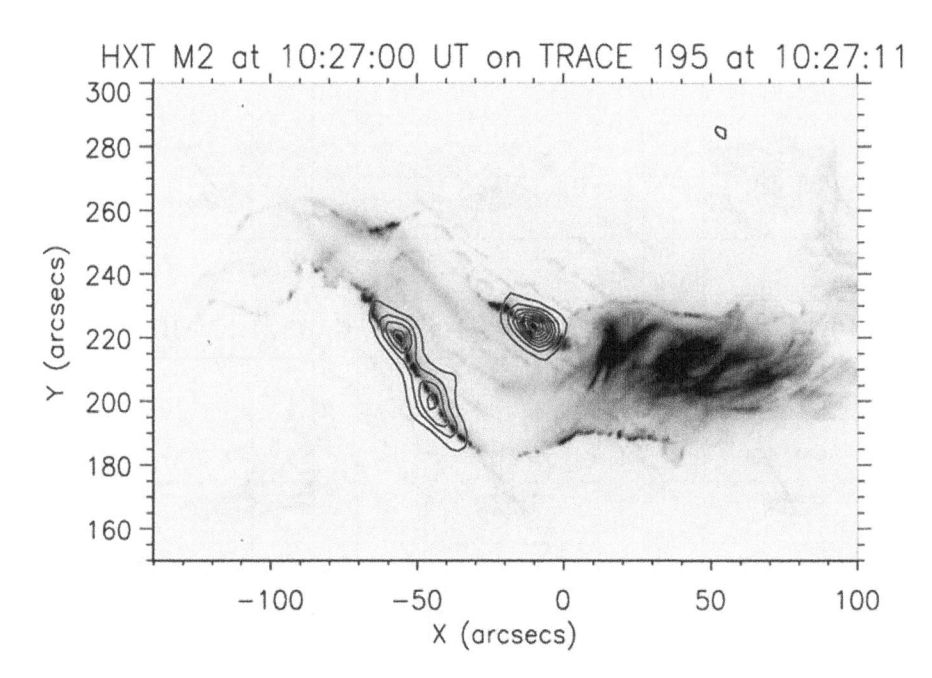

Jul-14 flare is overlaid on EUV images (*TRACE* 195 Å). Note the detailed co-spatiality with the flare ribbons observed in EUV (Fletcher & Hudson 2001).

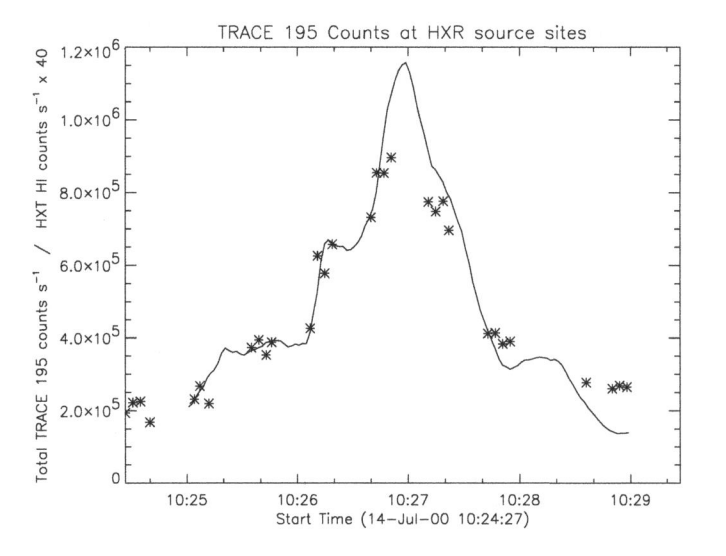

Figure 13.32: The *Yohkoh/HXT* light curve (solid lines) is scaled to the *TRACE* 195 Å source counts/s measured at the ribbon locations of the *HXT* sources (see Fig. 13.31). The HXT energy range of the HI channel is $\gtrsim 50$ keV. Note that the hard X-ray light curve represents the total flux (Fletcher & Hudson 2001).

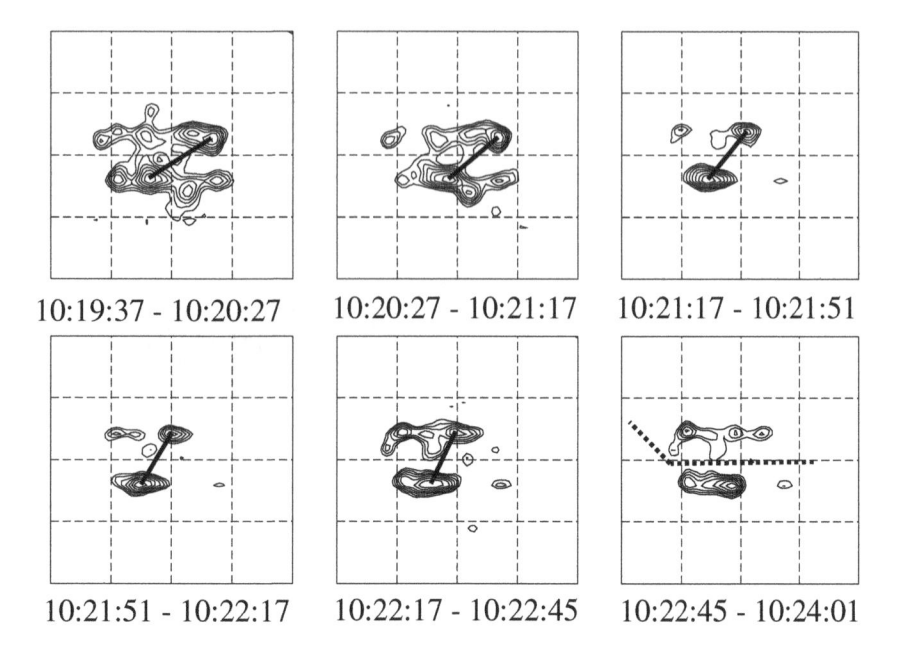

10:19:37 - 10:20:27   10:20:27 - 10:21:17   10:21:17 - 10:21:51

10:21:51 - 10:22:17   10:22:17 - 10:22:45   10:22:45 - 10:24:01

Figure 13.33: Sequence of *Yohkoh/HXT/Hi* (53-93 keV) images observed during the beginning phase of the 2000-Jul-14 flare, during 10:19-10:24 UT. The field of view of each panel is 157″ × 157″. The contours levels are 9%, 13%, 18%, 25%, 35%, 50%, and 71% of the peak intensity for each panel. A thick line connects the most intense kernels in the two-ribbon structure. In the final panel, the magnetic neutral line is indicated with a thick dotted line according to SOHO/MDI observations (Masuda et al. 2001).

ribbon-like hard X-ray sources were reported for the first time during the Bastille-Day flare (Masuda et al. 2001; Fletcher & Hudson 2001), which were aligned with the EUV and Hα ribbons that extended over a length of some 200 Mm (Figs. 10.33, 13.31). *Yohkoh/HXT* and SXT maps clearly outline the flare arcade, with low-energy hard X-ray emission (*HXT:Lo*=14−23 keV) concentrated at the looptop, while all high-energy emission (*HXT*: M2, Hi) is concentrated near the flare ribbon footpoints. The hard X-ray ribbons, however, appear not to cover the full length of the EUV or Hα ribbons or flare arcade, but rather partial segments at intermittent times (Figs. 13.31 and 13.33), progressing from west to east during the course of this flare. Nevertheless, these hard X-ray ribbons were more extended than ever detected before.

A detailed study on these hard X-ray ribbons has been performed by Masuda et al. (2001). Fig. 13.33 shows the evolution of hard X-ray sources during a first spike of the flare, which occurs in the western half of the flare arcade. Connecting the strongest kernel in the northern ribbon to the strongest kernel in the southern ribbon, deemed to be conjugate footpoints, one clearly sees an evolution from a highly sheared hard X-ray footpoint pair (10:19:37 UT) to a less and less sheared footpoint pair (10:22:17 UT). This is the same evolution as outlined in Fig. 10.34, where the general pattern evolves from initially highly sheared, low-lying arcade loops to less sheared, higher loops. This pattern, which is not predicted in the 2D standard model of Kopp−Pneuman, may indicate some important physics of the trigger mechanism. The highly sheared arcade part is probably most unstable to tearing mode and triggers magnetic island formation and subsequent coalescence first, as simulated by Karpen et al. (1998) and Kliem et al. (2000), see § 10.2. A recent study on the spatio-temporal evolution of conjugate footpoints can be found in Asai et al. (2003).

Another study on this flare concentrated on the relation between hard X-ray ribbons and EUV ribbons (Fletcher & Hudson, 2001). While hard X-ray ribbons mark the chromospheric sites of direct bombardment of nonthermal particles, EUV ribbons can also be produced by conductive heating. So they do not need necessarily to be identical in the two wavelengths, and differences may tell us something about the different exciter roles. For instance, Czaykowska et al. (1999, 2001) demonstrated that EUV ribbons were heated dominantly by heat conduction (or possibly by protons) in one flare, without any detectable > 20 keV hard X-ray emission, a puzzling exception to the majority of flares. However, evidence for preflare EUV footpoint brightening preceding hard X-ray emission was also reported by Warren & Warshall (2001). In the 2000-Jul-14 Bastille-Day flare, nevertheless, Fletcher & Hudson (2001) find an extremely good spatial (Fig. 13.31) and temporal correlation (Fig. 13.32) between hard X-ray and EUV fluxes, suggesting that both signatures are excited by the same precipitating electrons. If the EUV ribbons were produced by thermal conduction, one would expect the peak of the EUV flux to be delayed with respect to the peak of the hard X-ray flux, according to the Neupert effect, but this is not the case here (see Fig. 13.32). The "smooth" hard X-ray images (Fig. 13.31) overlaid onto the "sharp" EUV ribbons leave the impression that the hard X-ray images are compromised by *HXT's* limited spatial resolution and insufficient uv-coverage. In particular, fine and long flare ribbons cannot properly be imaged with sparse uv-coverage. If EUV and hard X-ray emission are produced by the same precipitating electrons, we expect to see the same sharp flare ribbons in both wavelengths. Some parts of EUV ribbons, however, seem to be activated in the preflare phase without hard X-ray signatures (Warren & Warshall 2001).

A dynamical effect that is predicted in almost every reconnection model is that the footpoints or footpoint ribbons should separate as a function of time when reconnection progresses to larger altitudes, because higher X-points connect to the chromosphere at larger footpoint separation (see Fig. 10.21 and 10.34). Tracking footpoint separation during flares, Sakao (1999) found that it increases sometimes, as expected in the Kopp−Pneuman reconnection scenario due to the rise of the reconnection point, but sometimes it decreases. Interestingly, Sakao (1999) found that the sign of footpoint separation correlates with spectral evolution. Based on this correlation, he concluded that flares with separating footpoints have hard X-ray spectra with a superhot thermal component as expected in the Kopp−Pneuman model, while flares with approaching footpoints have no superhot thermal component as expected in the emerging flux model of Heyvaerts et al. (1977). The separation of footpoint ribbons is also clearly seen during the Bastille-Day flare (Fig. 10.33), although some effects are not understood, such as the bifurcations of some ribbon segments or the asymmetries in the associated positive and negative magnetic flux changes (Fletcher & Hudson 2002). Also, the electric field $E = v_\parallel B_n$ inferred from the ribbon motion is not found to correlate with the

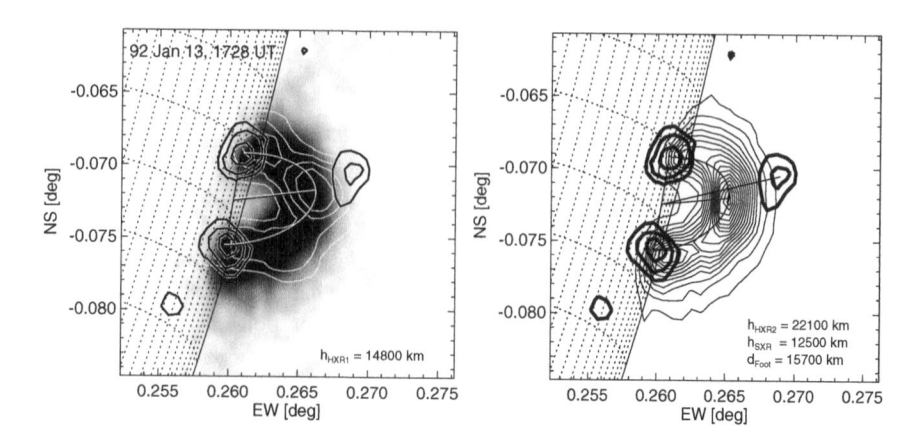

Figure 13.34: The *Masuda flare* of 1992-Jan-13: Co-registered overlays of Yohkoh/SXT (1728:07 UT, Be 119 filter; greyscale in left frame and thin contours in right frame) and Yohkoh/HXT (17:28:04−17:28:40 UT, Lo 14 − 23 keV, white contours in left frame; and M1 23 − 33 keV, black contours in both frames) (Aschwanden et al. 1996a)

hard X-ray flux in a simple way (Qiu et al. 2002). This complicates the application of electric field acceleration models (§11.3).

13.6.3   Above-the-Looptop (Masuda) Sources

The discovery of hard X-ray sources located above the loop of soft X-ray-bright flare loops (see Figs. 13.25 and 13.34) by Masuda et al. (1994, 1995) represented a major breakthrough in the localization of particle acceleration sources near magnetic reconnection sites (see Fletcher 1999 for a review). Previously it was not clear whether hard X-ray-emitting electrons are accelerated inside or outside prominent soft X-ray flare loops. From reconnection models one would expect that magnetic energy is converted into heating and particle acceleration near the X-points, which have to be located above soft X-ray flare loops (that are generated by chromospheric evaporation after the newly reconnected field lines relax into a force-free state, Fig. 10.21). However, since plasma densities ($n_e \approx 10^8 - 10^9$ cm^{-3}) above flare loops are relatively low, no hard X-ray emission was expected in this collisionless plasma above flare loops. Therefore, Masuda's discovery of hard X-ray emission at coronal locations *above* soft X-ray-bright flare loops changed our minds. Masuda et al. (1994) discovered an *above-the-looptop* hard X-ray source at energies of $\gtrsim 20 - 50$ keV with *Yohkoh/HXT* in about 10 flares (four examples are shown in Plate 15), besides the well-known (usually double) chromospheric footpoint sources. Initially, it was not clear how electrons can emit collisional bremsstrahlung in such low plasma densities above flare loops. An interpretation in terms of thermal hard X-ray emission was ruled out based on the required temperatures $T \approx 200$ MK, for which there was no evidence from other soft X-ray instruments (*Yohkoh/SXT* and *BCS*). Also, the time variability of looptop hard X-ray emission was too rapid to be consistent with thermal cooling times and thus required a nonthermal interpretation (Hudson & Ryan, 1995). Therefore, a plausible explanation

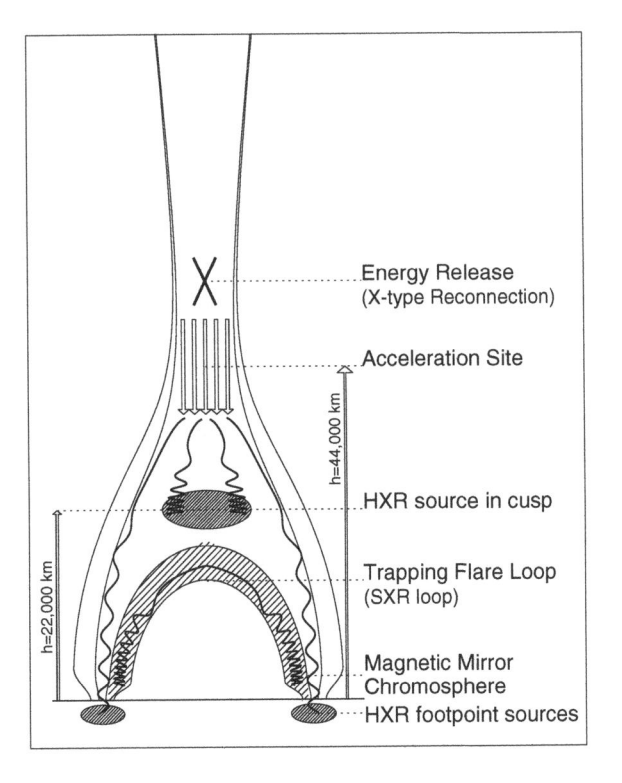

Figure 13.35: Diagram of a possible flare scenario for the Masuda flare (Fig. 13.34), constrained by the height ($h = 22$ Mm) of the *above-the-looptop* hard X-ray source and the height ($h = 44$ Mm) of the acceleration source inferred from electron time-of-flight measurements (Aschwanden et al. 1996a).

is collisional bremsstrahlung from trapped electrons, which are directly fed in from the accelerator in the cusp region beneath the reconnection point (Fig. 13.35). The location of the coronal hard X-ray source was measured to be about $10''$ (7250 km) above the soft X-ray loop, and in slightly higher altitudes in images taken in higher ($\gtrsim 50$ keV) energy bands (Masuda et al. 1994). This location is fully consistent with the cusp geometry in bipolar reconnection models (Fig. 10.21), and thus the coronal hard X-ray emission has to be emitted relatively close to the acceleration region associated with the reconnection point, rather than from a secondary trap somewhere else in the corona. The height of electron acceleration sources inferred from time-of-flight measurements yields a location above the soft X-ray flare loops (§13.5.2, Fig. 13.19; Aschwanden et al. 1996c), while the Masuda hard X-ray sources lay typically beneath these inferred acceleration sites (Plate 15), as expected in reconnection-driven injection (§12.4) or outflow-shock models (Tsuneta et al. 1997; Somov et al. 1999).

The spectrum of the coronal hard X-ray source in the 1992-Jan-13 flare was analyzed in detail by Alexander & Metcalf (1997). To overcome the problem of analyzing a weak source that is suppressed by the much (ten times) brighter footpoint sources nearby, they employed the photometrically more accurate pixon method to reconstruct images from *HXT* data. Their conclusions were that the coronal hard X-ray source: (a) has an impulsive temporal profile (similar to the footpoint sources), (b) has a non-thermal spectrum, (c) has a very hard spectrum or a low-energy cutoff in the electron injection spectrum, and (d) the looptop and footpoint hard X-ray sources are produced by two distinct particle populations. All this information indicates that coronal hard X-ray emission is produced directly by nonthermal electrons in or near the acceleration region. Consequently, these observations are one of the most direct witnesses of the acceleration process itself. Apparently, electrons are not simply accelerated in large-scale electric fields, because the required background density for free-streaming electrons to produce $\gtrsim 30$ keV hard X-ray emission would be $n_e \approx 3 \times 10^{11}$ cm^{-3} (Fletcher, 1995; Fletcher & Martens 1998) to $n_e \approx 10^{12}$ cm^{-3} (Wheatland & Melrose 1995). This is about $2 - 3$ orders of magnitude higher than what is inferred at the locations of Masuda's hard X-ray sources. A more likely acceleration mechanism is stochastic acceleration, which allows the electrons to stay sufficiently long in a local trap to produce detectable amounts of nonthermal bremsstrahlung in the low-density coronal plasma. Specific models envision particle trapping by wave scattering in the MHD turbulent regions above flare loops (Petrosian & Donaghy 1999, 2000; Jakimiec et al. 1998; Jakimiec 1999).

13.6.4   Occulted Flares

Obviously, coronal hard X-ray sources are most easily observed for limb flares, because the separation between footpoint, looptop, and above-the-looptop sources is least confused for such a view. This is the reason why Masuda's sample of 10 flares includes only limb flares. However, there are also examples found on the solar disk (e.g., the 1991-Nov-19, 09:29 UT, flare shown in Plate 15, top right). A comprehensive analysis of 18 limb flares exhibited 15 events with detectable impulsive looptop emission, so that it was concluded that the detection of above-the-looptop hard X-ray emission is merely a sensitivity problem and is probably a common feature of all flares (Petrosian et al. 2002).

An even better opportunity to map the coronal part of hard X-ray emission concerns flares with occulted footpoints (Masuda 2000), but they are rare. The hard X-ray spectra of occulted flares were found to be softer and the nonthermal broadening of soft X-ray lines was found to be smaller (Mariska et al. 1996), clearly indicating different physical conditions above the looptop than at the footpoints, but the thermal properties are indistinguishable from non-occulted flares (Mariska & McTiernan 1999). The hard X-ray time profiles of occulted flares were also occasionally found to exhibit spiky emission, perhaps a signature of turbulent flare kernels (Tomczak 2001). Earlier stereoscopic multi-spacecraft measurements of occulted flares yielded maximum heights of $h \lesssim 30$ Mm at which impulsive hard X-ray emission was identified (Kane et al. 1979, 1982; Hudson 1978; Kane, 1983). A record height of $\gtrsim 200$ Mm was determined for an occulted flare (1984-Feb-16, 09:12 UT) at hard X-ray energies of $\gtrsim 5$ keV, by combined stereoscopic measurements with the *ICE* and ISEE-3 spacecraft (Kane et al. 1992). Typical altitudes of acceleration regions are found in the range $h_{acc} \approx 5 - 35$ Mm, based on electron time-of-flight measurements, which is fully consistent with the

Table 13.2: Observed frequency distributions of peak fluxes F_P in solar flares (Aschwanden et al. 1998a).

Power-law slope α	Elements	Observations, data set	Literature reference
1.9	Flares	OSO V, 1969−71	Dennis (1985)
1.8	Flares	OSO VII, 20 keV, 1971−72	Datlowe et al. (1974)
2.0	Flares	Balloon, 20 keV, 1980	Lin et al. (1984)
1.8	Flares	HXRBS/SMM, > 30 keV, 1980−85	Dennis (1985)
1.66−1.75	Flares	HXRBS/SMM, > 30 keV, 1980−89	Schwartz et al. (1992)
1.67−1.73	Flares	HXRBS/SMM, > 30 keV, 1980−89	Crosby et al. (1993)
1.70−1.86	Flares	HXRBS/SMM, > 30 keV, 1980−84	Bai (1993)
1.74	Flares	HXRBS/SMM, > 30 keV, 1980−89	Kucera et al. (1997)
1.75	Flares	ISEE3/ICE, > 26 keV, 1978−86	Lee et al. (1993c)
1.86−2.00	Flares	ISEE3/ICE, > 26 keV, 1978−86	Bromund et al. (1995)
1.61	Flares	BATSE/CGRO,> 25 keV, 1991	Schwartz et al. (1992)
1.60−1.74	Flares	BATSE/CGRO,> 25 keV, 1991−92	Biesecker (1994)
1.61−1.66	Pulses	BATSE/CGRO,> 25 keV, 1991−94	Aschwanden et al. (1995c)
1.69−1.82	Pulses	BATSE/CGRO,> 50 keV, 1991−94	Aschwanden et al. (1995c)
1.56±0.43	Pulses	BATSE/CGRO,> 25 keV, 1991−94	Aschwanden et al. (1998a)
1.46±0.34	Pulses	BATSE/CGRO,> 50 keV, 1991−94	Aschwanden et al. (1998a)
1.6	Microflares	BATSE/CGRO,> 8 keV, 1991	Lin et al. (2001)

height limit of $h \lesssim 30$ Mm in earlier stereoscopic measurements. The most extreme case with a stereoscopic height of $h \approx 200$ Mm (Kane et al. 1992) could be associated with thin-target emission from trapped electrons in a large-scale flare loop.

13.7 Hard X-Ray Statistics

13.7.1 Flare Statistics of Nonthermal Energies

We discussed the statistics of small-scale heating events in §9.8. They generally show powerlaw frequency distributions for their thermal energy content ε_{th}, with typical powerlaw slopes in the range of $\alpha_\varepsilon \approx 1.5 - 2.5$ (Table 9.6). Energy content is simply estimated by integrating thermal energy over volume, $\varepsilon_{th} = 3n_e k_B T_e V$. We also derived two theoretical models that predict powerlaw slopes between $\alpha_\varepsilon = 1.21$ and 1.67. For solar flare events, we expect that there is roughly a one-to-one conversion between nonthermal and thermal energy, based on the chromospheric evaporation model and the Neupert effect (§13.5.5). It is therefore instructive to study the frequency distributions of nonthermal flare energies, which can be obtained by deriving the electron injection spectrum from the observed hard X-ray photon spectrum, and by integrating over the nonthermal energy range (above some cutoff) and over the flare duration (Eq. 13.2.31). A compilation for the frequency distributions of flare peak fluxes F_P, the most commonly measured flare parameter, is given in Table 13.2. The frequency distribution of nonthermal flare energies ε_{nth}, is expected to be somewhat different, because spectral

Table 13.3: Frequency distributions of flare parameters obtained from hard X-ray data (Crosby et al. 1993).

Parameter	Symbol	Powerlaw slope α	Scaling law
Peak count rate	P	1.73 ± 0.01	–
Total duration	D	2.17 ± 0.05	–
Peak hard X-ray flux at 25 keV	$I(25\,\mathrm{keV})$	1.62 ± 0.02	$I \propto P^{1.01}$
Peak hard X-ray flux > 25 keV	$F_P(25\,\mathrm{keV})$	1.59 ± 0.01	$F_P \propto P^{1.01}$
Peak energy flux in electrons > 25 keV	$F(25\,\mathrm{keV})$	1.67 ± 0.04	$F \propto P^{0.94}$
Total energy in electrons > 25 keV	$W(25\,\mathrm{keV})$	1.53 ± 0.02	$W \propto P^{1.25}$

slope and flare duration are not independent of flare peak flux F_P. The powerlaw slopes of the frequency distributions of different flare parameters and scaling laws determined from hard X-ray data are shown in Table 13.3, obtained from $\approx 10^4$ flares recorded by *HXRBS* during the SMM era (Crosby et al. 1993). We see that the frequency distribution of flare energies is somewhat flatter ($\alpha_\varepsilon \approx 1.53$, Fig. 9.27) than that of the hard X-ray peak count rates ($\alpha_P = 1.73$), so a similar correction of $\Delta\alpha \approx -0.2$ can be applied to the results given in Table 13.2 to estimate the distribution of flare energies. What is remarkable is that all distributions are flatter than a powerlaw slope of 2, so total energy is always dominated by the largest flares (Hudson 1991a). This implies that there is only negligible energy in microflares and nanoflares compared with energies in larger events, if all physical parameters obey the same scaling laws for smaller events.

An interesting quantity to compare would be the frequency distribution of thermal flare energies. Curiously, no such distribution is available in the literature, probably because of the difficulty in measuring flare volume for a large number of events, which could be fractal (Fig. 9.21 and §9.7.1) and requires careful volume modeling. Once both the thermal and nonthermal frequency distributions are known, we could easily verify whether there is one-to-one conversion from nonthermal to thermal energies, as expected in the chromospheric evaporation model and from the Neupert effect (Veronig et al. 2002b).

13.7.2 Flare Statistics During Solar Cycles

The (monthly averaged) solar flare rate varies about a factor of 20 between solar maximum and minimum, similar to the monthly sunspot number (Fig. 13.36). Since the sunspot number (defined by an empirical formula that weights the sunspot area and number of sunspot groups) is a measure of the total magnetic flux present on the solar surface at a given time (DeToma et al. 2000), the basic correlation with the solar flare rate implies, of course, that the number of produced flares is associated with magnetic flux, and thus flare energy is ultimately of magnetic origin. A closer look at Fig. 13.36 reveals that the flare rate also fluctuates much more (with a modulation depth of $\lesssim 90\%$) than the sunspot number (with a modulation depth of $\lesssim 20\%$). This property implies a highly nonlinear response of flare rate as a function of available magnetic flux. The total soft X-ray luminosity was found to scale with the magnetic

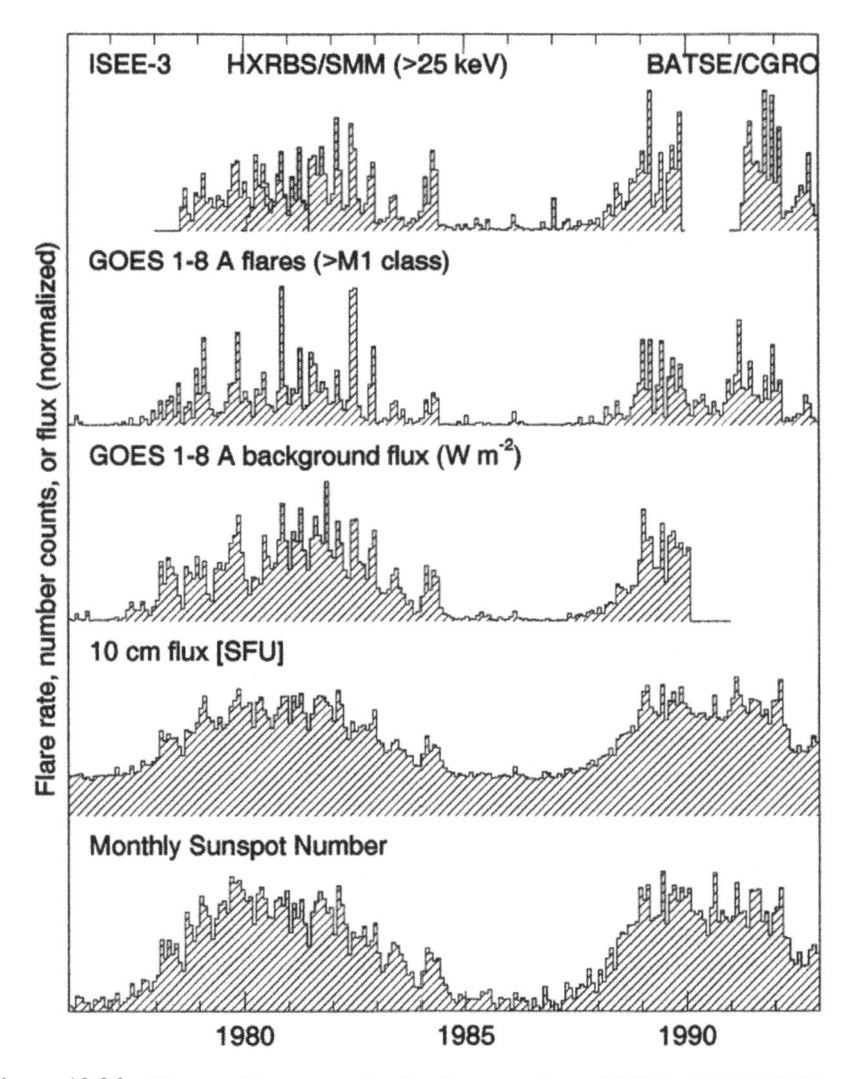

Figure 13.36: The monthly averaged solar flare rate (top: ISEE-3, HXRBS/SMM, and BATSE/CGRO) compared with the soft X-ray flare rate (second panel: GOES $>$ $M1$ class flares), soft X-ray background flux (third panel: GOES $1 - 8$ Å flux), radio 10-cm flux (fourth panel), and monthly sunspot number (Aschwanden 1994).

field as $I_{SXR} \approx < |B_{\parallel}| >^2$ (Fig. 1.13; Benevolenskaja et al. 2002). Since the soft X-ray peak flux is proportional to the time-integrated hard X-ray flux according to the Neupert effect (§13.5.5), we can understand that the flare rate counted in soft X-rays or hard X-rays is also somehow related to the available magnetic flux on the solar surface. The detailed relation, however, is certainly more complex than a simple proportionality. Flare rate must be related in some form to magnetic complexity (sheared magnetic fields and nonpotential magnetic energy), rather than to total magnetic flux *per se*.

13.8 Summary

Hard X-ray emission is produced by energized electrons via collisional bremsstrahlung, most prominently in the form of thick-target bremsstrahlung when precipitating electrons hit the chromosphere. Thin-target bremsstrahlung may be observable in the corona for footpoint-occulted flares. Thermal bremsstrahlung dominates only at energies of $\lesssim 15\,\mathrm{keV}$. Hard X-ray spectra can generally be fitted with a thermal spectrum at low energies and with a single- or double-powerlaw nonthermal spectrum at higher energies. A lot of physical insights can be obtained from time structures and energy-dependent time delays. Virtually all flares exhibit fast (subsecond) pulses in hard X-rays, which scale proportionally with flare loop size and are most likely spatio-temporal signatures of bursty magnetic reconnection events. The energy-dependent timing of these fast subsecond pulses exhibit electron time-of-flight delays from the propagation between the coronal acceleration site and the chromospheric thick-target site. The inferred acceleration site is located about 50% higher than the soft X-ray flare loop height, most likely near X-points of magnetic reconnection sites. The more gradually varying hard X-ray emission exhibits an energy-dependent time delay with opposite sign, which corresponds to the timing of the collisional deflection of trapped electrons. In many flares, the time evolution of soft X-rays roughly follows the integral of the hard X-ray flux profile, which is called the "Neupert effect". Spatial structures of hard X-ray sources include: (1) footpoint sources produced by thick-target bremsstrahlung; (2) thermal hard X-rays from flare looptops; (3) "above-the-looptop" (Masuda-type) sources that result from nonthermal bremsstrahlung from electrons that are either trapped in the acceleration region or interact with reconnection shocks; (4) hard X-ray sources associated with upward soft X-ray ejecta; and (5) hard X-ray halo or albedo sources due to backscattering at the photosphere. In spatially extended flares, the footpoint sources assume ribbon-like morphology if mapped with sufficient sensitivity. The monthly hard X-ray flare rate varies about a factor of 20 during the solar cycle, similar to magnetic flux variations implied by the monthly sunspot number, as expected from the magnetic origin of flare energies.

Chapter 14

Gamma-Rays

Electromagnetic radiation in gamma-rays (also denoted as γ-rays) occupies the high-energy end of the wavelength spectrum, which conveys us information from MeV to GeV particles generated in large solar flares. In solar flares, and similarly in other astrophysical sites, gamma-ray emission is mostly produced by interactions of high-energy particles with an ambient plasma: by Couloumb collisions between electrons and ions (bremsstrahlung), by collisions between accelerated ions and thermal ions (producing nuclear de-excitation lines), or by collisions between protons (which can produce neutrons, pions, and positrons). The latter processes are detectable by the neutron capture line, by pion-decay radiation, and positron annihilation radiation. Gamma-ray emission provides us information on both electrons and ions, because both continuum and line spectra are present, in contrast to the hard X-ray spectrum, which is exclusively produced by electron bremsstrahlung without any significant radiative signatures from ions. Thus, gamma-rays provide two lines of diagnostic during solar flares: (1) properties of the acceleration mechanisms (maximum energy, electron/ion ratios, electron/ion acceleration efficiencies, directivity, pitch angle distributions), and (2) properties of the ambient plasma (chromospheric ion abundances, coronal trapping times).

Prime gamma-ray observations have been obtained from the *Gamma-Ray Spectrometer (GRS)* on *SMM*, the *Hinotori* spacecraft, the *Wide-Band Spectrometer (WBS)* on the *Yohkoh* spacecraft, the *Burst and Transient Source Experiment (BATSE)* and *Oriented Scintillation Spectrometer Experiment (OSSE)* on *CGRO*, and most recently from the *Reuven Ramaty High Energy Spectroscopic Imager (RHESSI)* spacecraft. Some of the gamma-ray lines in solar flares are, for the first time, resolved with the cooled germanium detectors of *RHESSI*.

Review articles on gamma-ray emission in solar flares can be found in Lingenfelter & Ramaty (1967), Ramaty (1986, 1996), Ramaty & Murphy (1987), Hudson (1989), Hudson & Ryan (1995), Chupp (1995, 1996), Ramaty & Lingenfelter (1996), Share et al. (1997), Trottet & Vilmer (1997), Vestrand & Miller (1999), Share & Murphy (2000), Ramaty & Mandzhavidze (1994, 2000b, 2001), Aschwanden (2002b), and Vilmer & MacKinnon (2003). For proceedings on solar high-energy particle workshops see Zank & Gaisser (1992), Ryan & Vestrand (1994), Ramaty et al. (1996), and Ramaty & Mandzhavidze (2000a).

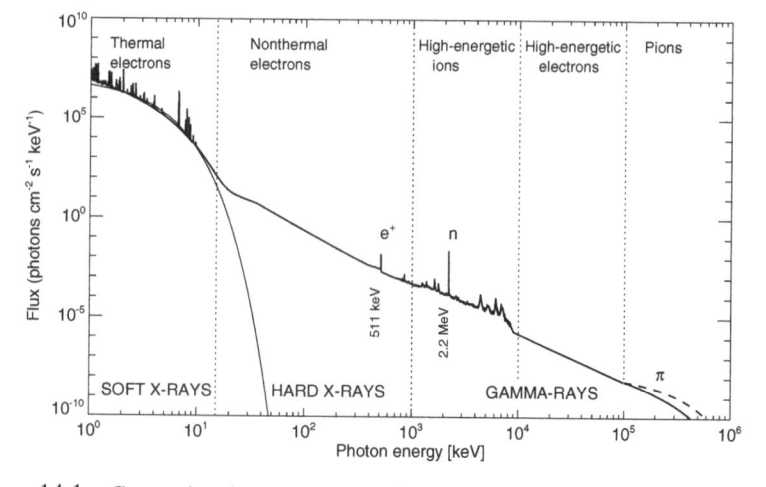

Figure 14.1: Composite photon spectrum of a large flare, extending from soft X-rays (1−10 keV), hard X-rays (10 keV−1 MeV), to gamma-rays (1 MeV−100 GeV). The energy spectrum is dominated by different processes: by thermal electrons (in soft X-rays), bremsstrahlung from nonthermal electrons (in hard X-rays), nuclear de-excitation lines (in ≈ 0.5 − 8 MeV gamma-rays), by bremsstrahlung from high-energetic electrons (in ≈ 10 − 100 MeV gamma-rays), and by pion-decay (in ≳ 100 MeV gamma-rays). Note also the prominent electron-positron annihilation line (at 511 keV) and the neutron capture line (at 2.2 MeV).

14.1   Overview on Gamma-Ray Emission Mechanisms

An overview on the most relevant gamma-ray emission processes in solar flares is given in Table 14.1, taken from Ramaty & Mandzhavidze (1994). There are at least six distinctly different physical mechanisms that produce photons in gamma-ray wavelengths: electron bremsstrahlung continuum emission, nuclear de-excitation line emission, neutron capture line emission, positron annihilation line emission, pion-decay radiation, and neutron production processes. A complete solar flare spectrum is shown in Fig. 14.1, extending from soft X-rays over hard X-rays to gamma-rays, (although only few flares exhibit all the features present in Fig. 14.1). We see that line emission occurs only in soft X-rays (by atomic transitions) and in gamma-rays (by nuclear transitions). The gamma-ray spectrum (Fig. 14.1) contains a background spectrum produced by electron bremsstrahlung that can be dominated by gamma-ray lines and pion-decay emission at particular gamma-ray energies during large flares. An expanded spectrum with fits of nuclear de-excitation lines from ^{56}Fe, ^{24}Mg, ^{20}Ne, ^{28}Si, ^{12}C, ^{16}O, ^{15}N, the electron-positronium line at 511 keV, and the neutron capture line at 2.223 MeV is shown in Fig. 14.2, using a gamma-ray spectrum observed by *OSSE/CGRO* (Murphy et al. 1997). Let us now give a brief description of the various gamma-ray emission processes, in the order listed in Table 14.1 (which is also the order of the following sections in this chapter).

Electron bremsstrahlung continuum: The bremsstrahlung of electrons consists of two types: (1) electrons directly accelerated in the flare and (2) secondary electrons and

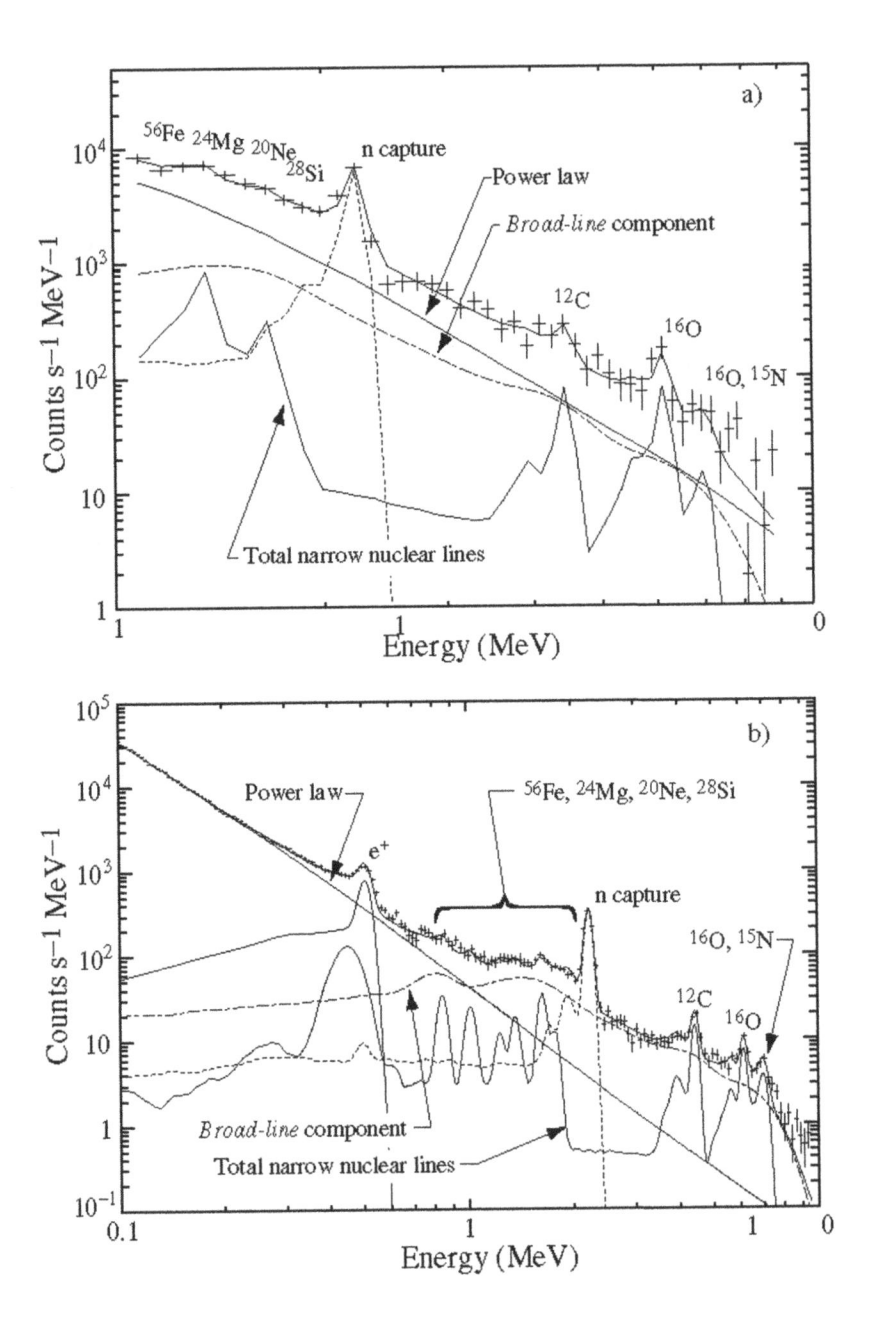

Figure 14.2: Count spectra and best fits obtained using the photon models described in Murphy et al. (1997). (a) An 8-s *CGRO/OSSE* spectrum accumulated soon after the peak of emission in the first orbit by detector 2 while pointing 45° off the Sun. The spectrum has been rebinned into larger energy intervals to improve the statistics for fitting. (b) A 2-minute spectrum accumulated late in the first orbit by detector 1 while pointing at the Sun (Murphy et al. 1997).

Table 14.1: High-energy photon and neutron production mechanisms (from Ramaty & Mandzhavidze 1994).

Emissions	Processes	Observed photons or neutrons	Primary ion or electron energy range
Continuum	Primary electron bremsstrahlung	20 keV − 1 MeV > 10 MeV	20 keV − 1 GeV
Nuclear de-excitation lines	Accelerated ion interactions, e.g., ^{4}He$(\alpha,n)^7$Be* ^{4}He$(\alpha,p)^7$Li* ^{20}Ne$(p,p')^{20}$Ne* ^{12}C$(p,p')^{12}$C* ^{16}O$(p,p')^{16}$O*	Lines at e.g., 0.429 MeV 0.478 MeV 1.634 MeV 4.439 MeV 6.129 MeV	1 − 100 MeV/nucl
Neutron capture line	Neutron production by accelerated ions followed by ^{1}H$(n,\gamma)^2$H	Line at 2.223 MeV	10 − 100 MeV/nucl
Positron annihilation radiation	β^+ Emitter or π^+ production by accelerated ions, e.g., ^{12}C$(p,pn)^{11}$C $\mapsto$ ^{11}B$+e^+ + \nu$ p+p $\mapsto$ $\pi^+...\pi^+ + \mapsto \mu^+ \mapsto e^+$ followed by $e^+ + e^- \mapsto 2\gamma$ $e^+ + e^- \mapsto$ Ps$+h\nu$ or $e^+ + ^1$H $\mapsto$ Ps$+p$ Ps $\mapsto 2\gamma, 3\gamma$	Line at 0.511 MeV Orthopositronium Continuum < 511 keV	10 − 100 MeV/nucl
Pion decay radiation	π^0 and π^+ production by accelerated particles, e.g., p+p $\mapsto \pi^0, \pi^{\pm}...$ followed by $\pi^0 \mapsto 2\gamma$, $\pi^{\pm} \mapsto \mu^{\pm} \mapsto e^{\pm}$ $e^+ \mapsto \gamma_{brems}, \gamma_{ann}$ in flight $e^- \mapsto \gamma_{brems}$	10 MeV − 3 GeV	0.2−5 GeV
Neutrons	Accelerated particle interactions, e.g., ^{4}He$(p,pn)^3$He $p + p \mapsto \pi + n + ...$ ^{22}Ne$(\alpha, n)^{25}$Mg	neutrons in space (10−500 MeV) neutron induced atmospheric cascades (0.1−10 GeV) Neutron decay protons in space (20−200 MeV)	10 MeV − 1 GeV 0.1−10 GeV 20−400 MeV

positrons that arise from high-energy reactions (involving pions decay and muon production). The latter often dominates above 10 MeV. This continuum emission, which can be detectable up to 1 GeV, is produced by collisional bremsstrahlung from relativistic (20 keV − 1 GeV) electrons that lose their energy by collisions with chromospheric protons and ions. The spectral and temporal consistency of this continuum radiation between hard X-rays and gamma-rays suggests that the primary relativistic electrons and ions originate from the same coronal acceleration source during flares.

Nuclear de-excitation lines: Most of the gamma-ray line emission in the 0.5−8 MeV energy range is produced by protons and ions that have been accelerated in the corona and precipitate (like the electrons) to the chromosphere, where they collide with other ions and produce nuclear de-excitation lines (e.g., ^{56}Fe at 0.847 MeV, ^{24}Mg at 1.369 MeV, ^{20}Ne at 1.634 MeV, ^{28}Si at 1.779 MeV, ^{12}C at 4.439 MeV, or ^{16}O at 6.129 MeV). Narrow lines result from the bombardment of chromospheric nuclei by accelerated protons or α (He) particles, while broad lines result from the inverse reaction in which accelerated carbon (C) and heavier nuclei collide with ambient hydrogen (H) and helium (α). The broadening of the de-excitation lines is a consequence of the Doppler shift in the frame of the observer.

Neutron capture line: The very narrow 2.2 MeV line emission is not a prompt process, it is emitted only after neutrons become thermalized in the photosphere and become captured by protons to produce deuterium, which has a binding energy of 2.223 MeV, so that a delayed photon with an energy of $\epsilon = h\nu = 2.223$ MeV is emitted.

Positron annihilation: Positrons are produced in solar flares by the decay of radioactive nuclei and charged pions. Annihilation of these positrons with free electrons produces the 511 keV line emission and a continuum below 511 keV.

Pion decay radiation: Above 10 MeV there is, besides the electron bremsstrahlung continuum, also significant pion-decay radiation detected (Fig. 14.1). Charged and neutral pions (π^+, π^- mesons) are produced by collisions between protons and ions with energies $\gtrsim 300$ MeV/nucleon in the chromosphere. They subsequently decay into muons (μ^+, μ^-). The secondary electrons and positrons produce bremsstrahlung, while almost all neutral pions (π^0) decay electromagnetically into two γ-rays, each with a (Doppler-broadened) energy of 67 MeV (i.e., half of the pion rest mass of $m_\pi^0 c^2 = 135$ MeV).

Neutrons: Neutrons are produced by interactions between accelerated ions and protons or α-particles, mostly in chromospheric precipitation sites. Neutrons that escape the chromospheric flare site can propagate directly to Earth, unimpeded by the heliospheric magnetic field (because of their neutral charge) producing atmospheric showers (cascades) that in intense flares are detectable on ground by neutron monitors.

14.2 Electron Bremsstrahlung Continuum

The continuum spectrum is dominated by electron bremsstrahlung below 0.5 MeV in hard X-rays and in the energy range of 10−50 MeV in gamma-rays (Fig. 14.1). We have discussed the theory of the collisional electron bremsstrahlung emission mechanism already in §13 for the hard X-ray range, which also applies to the gamma-ray range. A difference is that we deal with highly relativistic electrons (with $\beta = v/c \gtrsim 0.9$ for energies $\varepsilon \gtrsim 1$ MeV), and thus have to apply the fully relativistic cross sections (Koch & Motz 1959; Haug 1975, 1997) to calculate gamma-ray spectra. Of course, gamma-rays are not always detected during flares, partially because the electron spectrum falls off so steeply with energy that the sensitivity of gamma-ray detectors is insufficient for small flares, or because there is indeed a high-energy cutoff in the acceleration efficiency in smaller flares. The two possibilities can often not be distinguished from gamma-ray spectra alone.

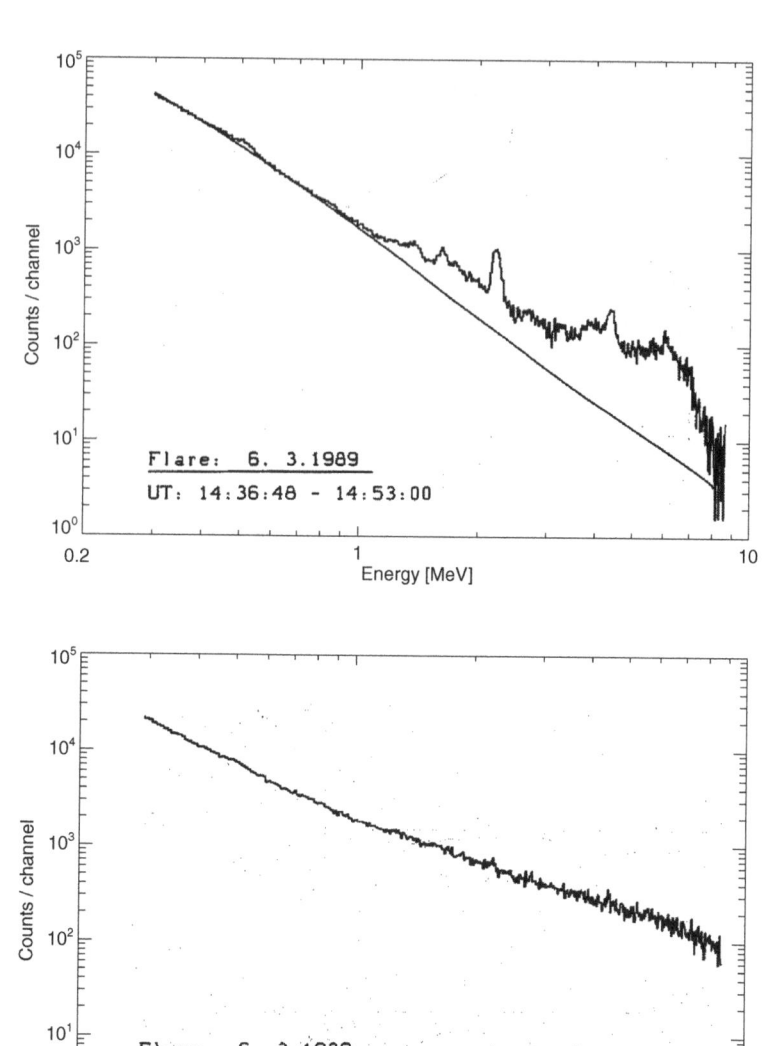

Figure 14.3: Comparison of a gamma-ray line-dominated phase (end phase of flare, 14:36:48−14:53:00 UT; top panel) and an electron-dominated phase (peak phase of flare, 13:57:12−13:58:50 UT; bottom panel) of the 1989-Mar-06 flare. The drop at $\gtrsim 7$ MeV in the end phase (top panel) is due to the absence of strong nuclear lines above this energy. During the flare peak (bottom panel), the electron-dominated spectrum flattens above $\gtrsim 1$ MeV and there is only a very weak indication of the 2.223 MeV neutron capture line and the 4.439 MeV carbon line, compared with the later flare phase (top panel), when gamma-ray lines dominate the $2 - 8$ MeV range (Rieger & Marschhäuser 1990).

14.2.1 Electron-Dominated Flares

In some flares, the electron bremsstrahlung continuum dominates even in the $0.5 - 8$ MeV range that includes the gamma-ray lines. The gamma-ray spectra of such flares show a featureless continuum spectrum (e.g., Fig. 14.3, bottom panel) and are called *electron-rich* events (Rieger & Marschhäuser 1990). Such events exhibit the electron bremsstrahlung spectrum in its least contaminated form. Two such typical impulsive events were observed with *CGRO* on 1991-Jun-30 and 1991-Jul-2, both being not exceptionally large flares (GOES M-class), but exhibiting a very hard spectrum ($\epsilon^{-1.5,...,-2.2}$) above ≈ 0.8 MeV (which is flatter than at lower energies), with only small contributions from nuclear lines (Marschhäuser et al. 1994; Dingus et al. 1994; Ryan 1994). A comprehensive catalog of gamma-ray spectra from 185 flare events observed with GRS on SMM is published in Vestrand et al. (1999). The spectral hardening above 0.8 MeV can vary from peak to peak within a flare (e.g., from 0.8 MeV to 0.5 MeV in just 20 s during the 1990-Jun-11 flare; Trottet et al. 1998). This represents a major challenge for acceleration mechanisms and has not yet been explained. No correlation between the spectral slope in the gamma-ray range and heliospheric position has been found, implying that the degree of anisotropy of the radiating electrons is low (Rieger et al. 1998). The upper energy cutoff provides a true lower limit of the maximum energy of accelerated electrons, which is 50 MeV in the 1991-Jun-30 flare and 10 MeV in the 1991-Jul-2 flare (Dingus et al. 1994). Electron-dominated events have a relatively short duration of a few seconds to a few tens of seconds (Pelaez et al. 1992; Yoshimori et al. 1992; Marschhäuser et al. 1994; Rieger 1994). In *interplanetary particle events*, electron-rich events are generally believed to originate in coronal flare sites along with *impulsive hard X-ray events*, in contrast to proton-rich events that originate in CMEs and are associated with *gradual hard X-ray events* (Reames 1992).

14.2.2 Maximum Acceleration Energy

There is no known high-energy cutoff of the electron bremsstrahlung spectrum; the highest energies of observed bremsstrahlung are around several 100 MeV (Forrest et al. 1985; Akimov et al. 1991, 1994a,b,c, 1996; Dingus et al. 1994; Trottet 1994a; Kurt et al. 1996; Rank et al. 2001; see also Fig. 2 in the review by Ramaty & Mandzhavidze 1994). Gamma-rays were reported up to energies above 1 GeV with the *Energetic Gamma-ray Experiment Telescope (EGRET)* on *CGRO* during the 1991-Jun-11 flare (Kanbach et al. 1992). The spectrum of the flare could be fitted with a composite of a proton generated pion neutral spectrum and a primary electron bremsstrahlung component (Kanbach et al. 1992).

Do these highest observed energies constrain or rule out any acceleration mechanism? For DC electric field acceleration in sub-Dreicer fields, the maximum velocity to which electrons can be accelerated is limited by the value of the Dreicer field, which depends on the density and temperature of the plasma, $E_D \approx 2 \times 10^{-10} n/T$ (statvolts cm^{-1}), see Eq. (11.3.2). Holman (1996) argues that electron energies up to $10 - 100$ MeV can be attained for high densities of $n_e \approx 10^{12}$ cm^{-3} and low temperatures $T \approx 2$ MK (yielding a Dreicer field of $E_D = 1 \times 10^{-4}$ statvolt cm^{-1}, i.e., 3 V m^{-1}), if electrons are continuously accelerated over a current channel with a length of $L = 10 - 100$

Mm. The requirement for such large-scale DC electric fields, however, conflicts with the observed time-of-flight delays of hard X-ray pulses and the short inductive switch on/off time scales required for the observed subsecond hard X-ray pulses, as discussed in the kinematic Section 12.3.2.

Alternatively, Litvinenko (1996b) envisions super-Dreicer electric fields, which can generate arbitrarily high maximum electron energies within much smaller spatial scales, and thus can be consistent with the time-of-flight delays of the observed hard X-ray pulses. The maximum energy for relativistic electrons obtained over an acceleration time t is approximately,

$$\varepsilon(t) = eEl(t) \approx eEct \,. \tag{14.2.1}$$

Litvinenko (2003) identifies plausible physical conditions with super-Dreicer fields of order $E \gtrsim 100$ V m^{-1} in reconnecting current sheets that lead to electron acceleration with gamma-ray energies of a few 10 MeV in electron-rich flares or to the generation of protons with energies up to several GeV in large gradual flares. Also stochastic acceleration can generate 10 MeV electrons and 1 GeV protons (Miller et al. 1997), if a sufficiently high wave turbulence level is assumed (which could not be observationally constrained so far). Thus, the maximum observed gamma-ray energies imply severe constraints only for the acceleration mechanism of sub-Dreicer electric DC fields.

14.2.3 Long-Term Trapping Sources

In §12.5 we described the kinematics of particle trapping. For the case of Coulomb collisional scattering, the energy dependence of the trapping time is proportional to $\varepsilon^{3/2}$, so we expect much longer trapping times for high-relativistic electrons that emit gamma-ray bremsstrahlung than for low-relativistic electrons in hard X-rays. The collisional deflection time sets an upper limit on the trapping time (Eq. 12.5.11), which is for electrons

$$t_{trap}(\varepsilon) \lesssim t_{defl}(\varepsilon) = 0.95 \left(\frac{\varepsilon}{100 \text{ keV}}\right)^{3/2} \left(\frac{10^{11} \text{ cm}^{-3}}{n_e}\right) \left(\frac{20}{\ln \Lambda}\right) \quad (s) \quad (14.2.2)$$

and a factor of ≈ 60 longer for ions. A major observational result of the energy-dependent time delays of the smoothly varying hard X-ray emission is the agreement of the observed delays with the energy dependence of the collisional deflection time (according to Eq. 14.2.2), for reasonable flare densities of $n_e \approx 10^{11}$ cm^{-3} (see §13.5.4). This suggests that trapping of flare electrons is controlled by pitch angle scattering in the *weak-diffusion limit* (by Coulomb collisions) rather than in the *strong-diffusion limit* (by wave turbulence; e.g., see models of turbulent trapping by Ryan & Lee 1991). Thus, trapping times for $\varepsilon = 20 - 200$ keV hard X-ray-producing electrons amount only to $\tau_{trap} \approx 0.2 - 2.0$ s in typical flare loops (see also measurements in Fig. 13.22). If we extrapolate this trapping time to particles with higher energies [e.g., to gyrosynchrotron-producing highly relativistic electrons ($\varepsilon \approx 0.3 - 1.0$ MeV) detected at microwaves], we expect trapping times of $\tau_{trap} \approx 3 - 10$ s, and for gamma-ray-producing electrons ($\varepsilon \approx 1 - 100$ MeV) we expect $\tau_{trap} = 10 - 10^4$ s (Fig. 14.4). In large-scale traps in the upper corona, where the density drops down to $n_e \approx 10^9$

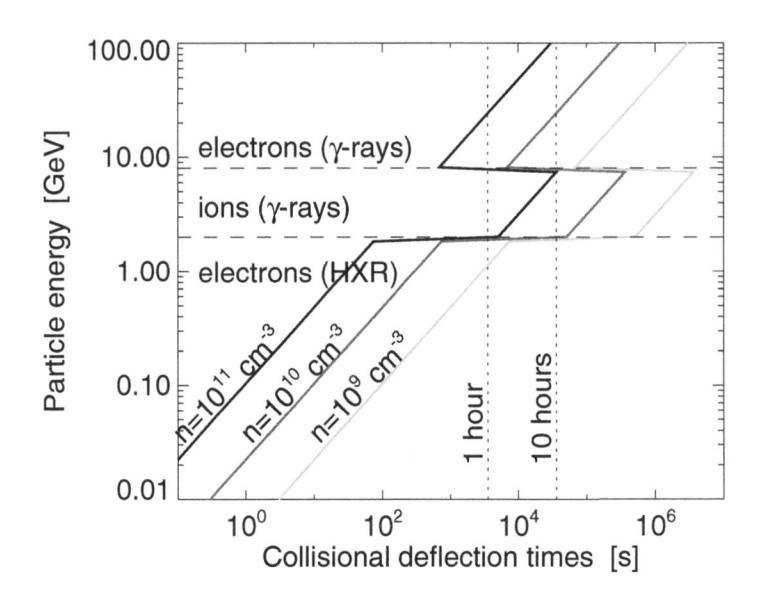

Figure 14.4: Collisional deflection times for electrons and ions as a function of energy, which represent the upper limits on the trapping times. Note that trapping times of $1 - 10$ hours are feasible for gamma-ray-producing ions in the $\varepsilon = 2 - 8$ GeV energy range and for gamma-ray-producing electrons in the $\varepsilon > 10$ GeV energy range.

cm^{-3}, gamma-ray-producing electrons with $\varepsilon = 1 - 100$ MeV could be trapped up to $\tau_{trap} \approx 1 - 10^3$ hours, if there is no other pitch angle scattering mechanism present. Since the ion collision times are about a factor of 60 longer than for electrons, the nuclear gamma-ray lines can easily be produced over time intervals of many hours due to ion trapping in coronal flare loops.

For such long trapping times, however, there may be some containment problems due to magnetic field fluctuations ($\Delta B/B$) from Alfvén waves, *magnetic gradient drift*, and *curvature drift* (for the latter two effects see, e.g., Sturrock 1994, p. 53),

$$\mathbf{v}_D = \frac{\gamma mc}{qB} \left(\frac{\mathrm{v}_\parallel^2}{B} (\mathbf{B} \times \mathbf{k}) + \frac{1}{2} \frac{\mathrm{v}_\perp^2}{B^2} (\mathbf{B} \times \nabla \mathbf{B}) \right) , \qquad (14.2.3)$$

where $\mathbf{k}$ is the curvature vector. This confinement problem can be cured, like in *tokamaks*, by a compensating electric field that produces a counteracting $\mathbf{E} \times \mathbf{B}$ drift, or by a twisted field that satisfies the force-free equilibrium equation $\mathbf{B} \times (\nabla \times \mathbf{B}) = 0$, or

$$\nabla \times \mathbf{B} = \lambda \mathbf{B} . \qquad (14.2.4)$$

Particle orbits in such force-free fields have been simulated by Lau et al. (1993), and it was found that long-term containment of energetic protons in a coronal loop is possible if the magnetic field lines have enough twist (i.e., $\approx 2\pi$ between the mirror points of a bounce orbit). It was also calculated that the amount of matter encoutered by a 1-GeV proton (the grammage) is sufficiently low during several hours trapping time so that the protons still have sufficient kinetic energy to produce pions.

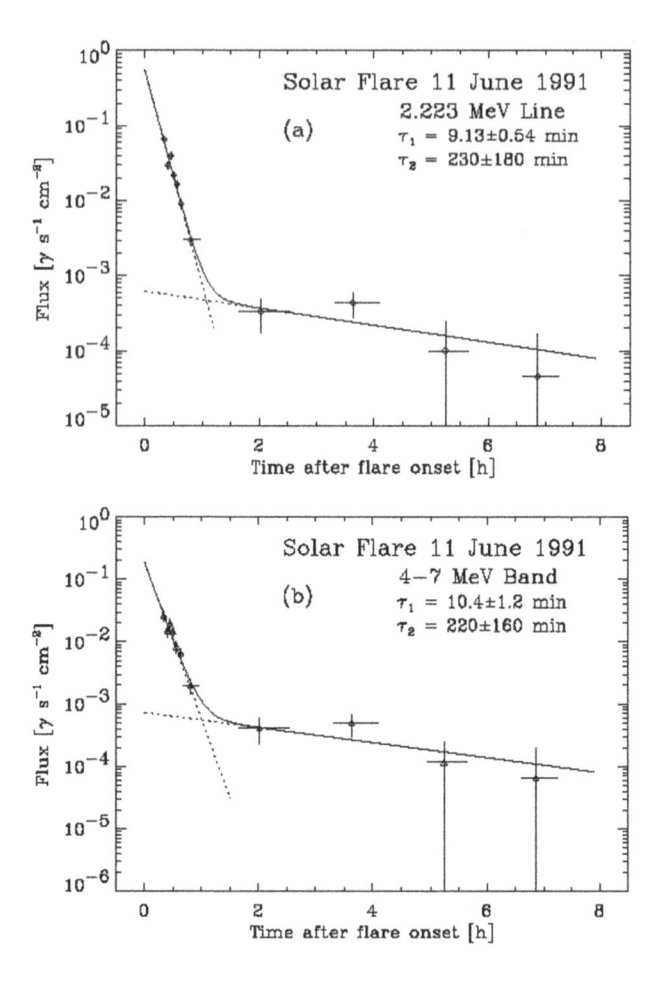

Figure 14.5: Extended γ-ray emission as measured by *COMPTEL* for the 1991-Jun-11 flare in the 2.223 MeV neutron capture line (top) and the $4-7$ MeV nuclear line flux (bottom). The data have been corrected for primary and secondary bremsstrahlung. A two-fold exponential decay has been fitted. The origin of the time axis is 01:56 UT, and the flare onset reported by GOES. Only data of the extended phase (after 02:13 UT) are shown (Rank et al. 2001).

Observations of long-duration gamma-ray flares, with extended >25 MeV emission over several hours, much longer than the impulsive phase seen in hard X-rays (typically a few minutes), were interpreted as evidence for prolonged acceleration of high-energy particles. Bai (1982b) suggested that this requires a distinct second-phase acceleration process, in addition to the impulsive phase. However, an interpretation in terms of trapping in large-scale (low-density) loops is also plausible considering the long collisional time scales (Ramaty & Mandzhavidze 1996). In the 1991-Jun-11 flare, 50 MeV-2 GeV γ-ray emission was observed by *CGRO/EGRET* and *CGRO/COMPTEL* for more than 8 hours (Kanbach et al. 1993), and 2.223 MeV neutron line emission

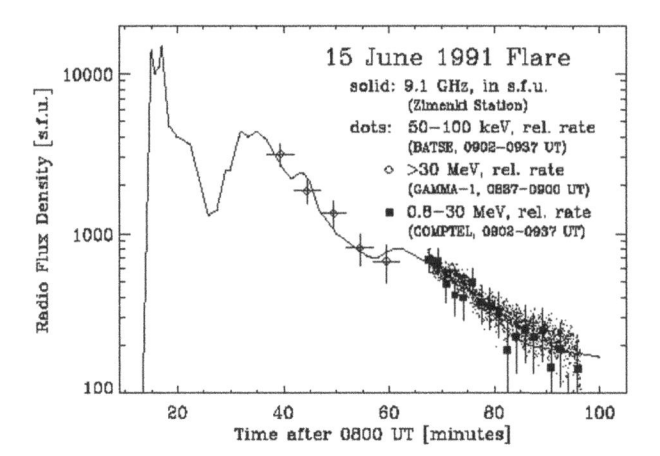

Figure 14.6: The beginning of the extended emission during the 1991-Jun-15 flare shows that the gamma-ray flux is proportional to the microwave flux at 9.1 GHz (solid curve, in solar flux units, s.f.u.). The diamonds represent > 30 MeV emission from GAMMA-1 (Akimov et al. 1991; Kocharov et al. 1994) and the black squares represent $0.8 - 30$ MeV COMPTEL data (Rank et al. 2001).

(Fig. 14.5) over 5 hours (Rank et al. 1996, 2001). Since gamma-ray emission in this energy range is dominated by pion decay, either a continuous source of proton acceleration or a long-term trapping of accelerated protons is needed (Ryan 2000). Ramaty & Mandzhavidze (1994) calculate that the following conditions are required for an interpretation in terms of long-term trapping: (1) A low level of plasma turbulence and a relatively high mirror ratio ($W_A < 2 \times 10^{-8}$ ergs cm^{-3} for $B_p/B_c = 50$) to prevent the fast precipitation of the particles through the losscones; (2) a matter density in the coronal part of the loop $n_c < 5 \times 10^{10}$ cm^{-3} to prevent Coulomb and nuclear losses. Arguments against continuous or second-step acceleration over 8 hours can be construed based on the lack of hard X-ray emission from lower energy particles and the smooth exponential decay of the 8-hour gamma-ray emission which is a natural characteristic of a trap population. The extended gamma-ray emission of the three flares of 1991 June 9, 11, and 15 has been reanalyzed in the most comprehensive study by Rank et al. (2001), who found different spectral slopes, e/p ratios, and pion emission during subsequent flare phases. Based on these spectral changes they subdivided the flare evolution into three phases (impulsive, intermediate, and extended phase) and concluded that the extended phase cannot be explained by long-term trapping from a single acceleration phase alone, but rather requires additional injections hours after the flare onset. Akimov et al. (1996) made a similar argument in favor of prolonged acceleration (rather than long-term trapping) based on the variability of associated radio emission (Fig. 14.6). The proportionality of radio and gamma-ray emission (e.g., Chupp et al. 1993; Akimov et al. 1996; Kaufmann et al. 2000) suggests a common population of highly relativistic electrons in an optically thin region producing gyrosynchrotron and bremsstrahlung emission.

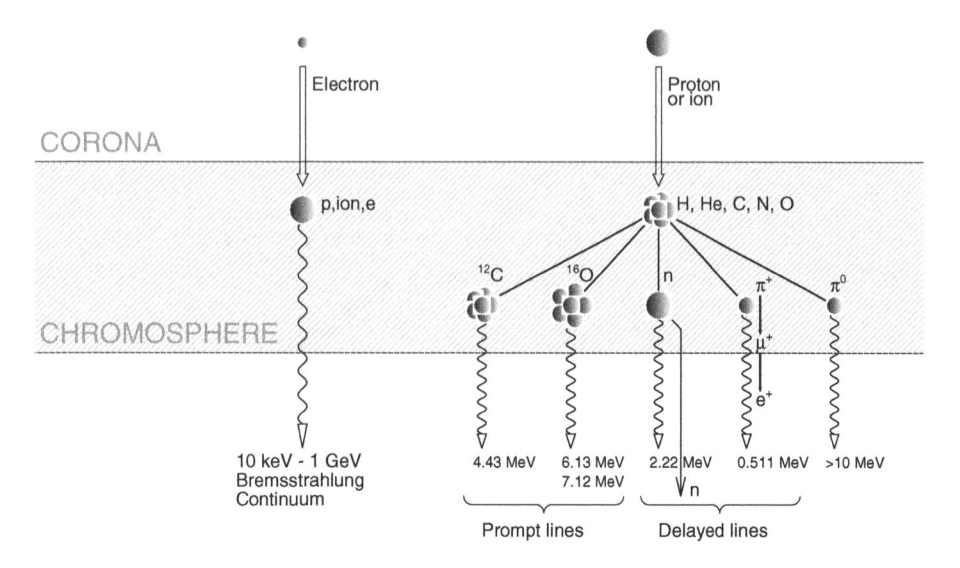

Figure 14.7: Schematic overview of hard X-ray, gamma-ray, and neutron production mechanisms. See also overview of processes in Table 14.1 (adapted from Rieger 1989).

14.3   Nuclear De-excitation Lines

14.3.1   Gamma-Ray Line Spectroscopy

In contrast to the continuum spectrum of electron bremsstrahlung, narrow gamma-ray lines in the $0.5 - 8.0$ MeV energy range result from the interaction of accelerated protons and α-particles with ambient He and heavier nuclei. The high-energy protons and α-particles are supposedly accelerated in coronal magnetic reconnection sites, concomitantly with the electrons, then precipitate to the chromosphere, and produce gamma-rays while interacting with hydrogen, helium, and heavier nuclei of the ambient chromosphere, which is fully ionized in the upper (transition) region, but only partially ionized in the deeper layers. The strongest lines result from the de-excitation of ^{12}C nuclei at 4.439 MeV and ^{16}O at 6.129 MeV. The required proton (kinetic) energies are $5 - 30$ MeV. Because the lifetimes of the excited nuclear states are $\approx 10^{-12}$ s or shorter, called *prompt lines* (Fig. 14.7). Concise reviews on gamma-ray line spectroscopy can be found, for example, in Lingenfelter (1994), Ramaty & Mandzhavidze (1994, 1996, 2000a, 2000b, 2001), or Share & Murphy (2000, 2004).

Nuclear de-excitation lines are calculated from nuclear cross sections, measured in laboratory accelerator experiments. Narrow lines result from the bombardment of ambient nuclei by accelerated protons and α-particles, while broad lines result from the inverse reactions in which accelerated C and heavier nuclei collide with ambient H and He. The broadening of the de-excitation lines is the consequence of the Doppler shifting of the essentially monochromatic radiation produced in the rest frame of the excited nuclei. In the case of the narrow lines, the broadening is due to the recoil velocity of the excited nuclei, which is small. The widths of the broad lines are much larger because

Table 14.2: Prompt nuclear de-excitation lines (compiled from Kozlovsky et al. 2002) and line widths (Smith et al. 2000, 2003).

Energy (MeV)	Reaction	Transition	Mean lifetime (s)	Line width (keV)
0.339	$^{56}Fe(\alpha,n)^{59}Ni^*$	$^{59}Ni^*0.339 \mapsto$ g.s.	9.8×10^{-11}	4
$0.429^{a,b}$	$^{4}He(\alpha,n)^{7}Be^*$	$^{7}Be^*0.429 \mapsto$ g.s.	1.9×10^{-13}	5
$0.478^{a,b}$	$^{4}He(\alpha,p)^{7}Li^*$	$^{7}Li^*0.478 \mapsto$ g.s.	1.1×10^{-13}	10
0.451	$^{24}Mg(p,x)^{23}Mg^*$	$^{23}Mg^*0.451 \mapsto$ g.s.	1.6×10^{-12}	30
0.847	$^{56}Fe(p,p')^{56}Fe^*$ $^{56}Fe(\alpha,\alpha')^{56}Fe^*$	$^{56}Fe^*0.847 \mapsto$ g.s.	8.9×10^{-12}	$5 (1.2+2.9)^c$
0.931	$^{56}Fe(p,x)^{55}Fe^*$	$^{55}Fe^*0.931 \mapsto$ g.s.	1.2×10^{-11}	5
1.369	$^{24}Mg(p,p')^{24}Mg^*$ $^{25}Mg(p,pn)^{24}Mg^*$ $^{26}Mg(p,p2n)^{24}Mg^*$ $^{24}Mg(\alpha,\alpha')^{24}Mg^*$ $^{28}Si(p,x)^{24}Mg^*$	$^{24}Mg^*1.369 \mapsto$ g.s.	2.0×10^{-12}	$16 (21+8)^c$
1.634	$^{20}Ne(p,p')^{20}Ne^*$ $^{20}Ne(\alpha,\alpha')^{20}Ne^*$ $^{24}Mg(p,x)^{20}Ne^*$ $^{24}Mg(\alpha,x)^{20}Ne^*$ $^{28}Si(p,x)^{20}Ne^*$	$^{20}Ne^*1.634 \mapsto$ g.s.	1.1×10^{-12}	$20 (17.6+4.3)^c$
1.779	$^{28}Si(p,p')^{28}Si^*$ $^{28}Si(\alpha,\alpha')^{28}Si^*$ $^{32}S(p,x)^{28}Si^*$	$^{28}Si^*1.779 \mapsto$ g.s.	6.9×10^{-13}	$20 (16.7+4.5)^c$
2.614	$^{20}Ne(p,p')^{20}Ne^*$ $^{20}Ne(\alpha,\alpha')^{20}Ne^*$ $^{24}Mg(p,x)^{20}Ne^*$ $^{28}Si(p,x)^{20}Ne^*$	$^{20}Ne^*4.248 \mapsto ^{20}Ne*1.634$	9.2×10^{-14}	60
4.439	$^{12}C(p,p')^{12}C^*$ $^{12}C(\alpha,\alpha')^{12}C^*$ $^{14}N(p,x)^{12}C^*$ $^{14}N(\alpha,x)^{12}C^*$ $^{16}O(p,x)^{12}C^*$ $^{16}O(\alpha,x)^{12}C^*$	$^{12}C^*4.439 \mapsto$ g.s.	6.1×10^{-14}	$145 (92+42)^c$
6.129	$^{16}O(p,p')^{16}O^*$ $^{16}O(\alpha,\alpha')^{16}O^*$ $^{20}Ne(p,x)^{16}O^*$	$^{16}O^*6.129 \mapsto$ g.s.	2.7×10^{-11}	$145 (122+68)^c$

^{a} Narrow lines for a downward beam or a fan beam (Smith et al. 2000).
^{b} Broad line (when 0.429 and 0.478 MeV combined) for an isotropic distribution (Smith et al. 2000).
^{c} Measured by RHESSI for the 2002-Jul-23 flare (Smith et al. 2003).

the excited nuclei continue to move rapidly after their excitation. Additional narrow lines are produced by accelerated ^{3}He, whose abundance can be enhanced during impulsive flares. Comprehensive treatments of the nuclear de-excitation gamma-ray line emission can be found in Ramaty et al. (1975) and Kozlovsky et al. (2002), including proton and α-particle reactions with He, C, N, O, Ne, Mg, Al, Si, S, Ca, and Fe. In Table 14.2 we list the most prominent de-excitation processes with their energies, nuclear transitions, lifetimes, and line widths. Nuclear cross sections of these gamma-ray lines as a function of energy are given in Kozlovsky et al. (2002). An overview of the gamma-ray line spectrum in the energy range of $0.5 - 8$ GeV is shown in Fig. 14.2, observed with *CGRO/OSSE*, where most of the lines are not resolved, except the broadest

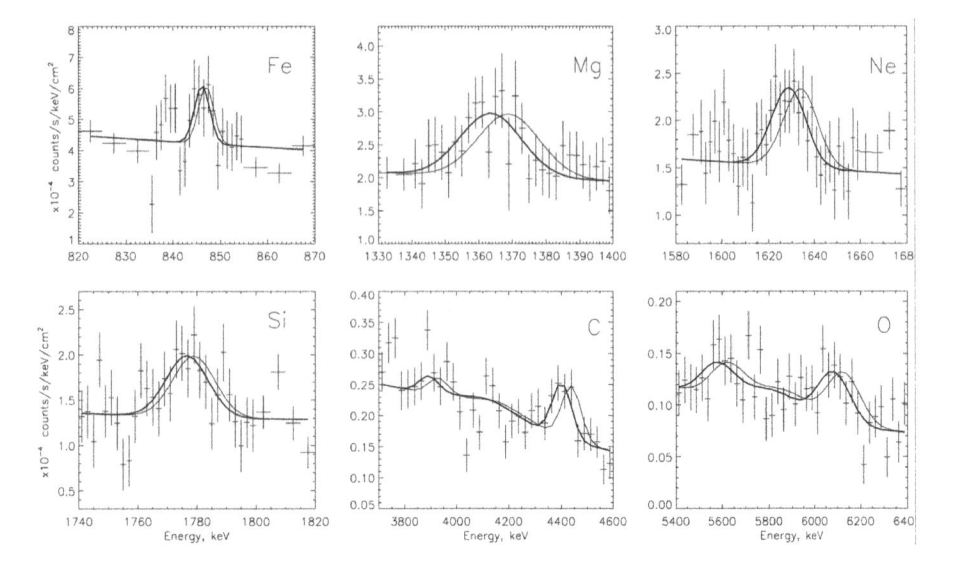

Figure 14.8: RHESSI background-subtracted count spectra from 00:27:20 to 00:43:20 UT on 2002-Jul-23. Each panel is labeled with the element primarily responsible for the line shown. The carbon and oxygen lines also show the secondary peak from the escape of a 511 keV positron-annihilation photon, which also contains information on the line shape. The thick curve shown in each panel is the Gaussian fit plus underlying bremsstrahlung continuum and broad lines, convolved with the instrument response. The thinner line is the same fit forced to zero redshift for comparison. The error bars are $\pm 1\sigma$ from Poisson statistics (Smith et al. 2003).

ones with line widths of $\approx 150\,\text{keV}$ (^{12}C, ^{16}O). Most of the prominent gamma-ray lines have been spectrally resolved for the first time in a solar flare with the cryogenically cooled germanium detectors of *RHESSI*, which have an energy resolution of $\approx 0.2\%$ FWHM from $1-6\,\text{MeV}$ (Smith et al. 2002). Spectrally resolved gamma-ray lines have been published in Smith et al. (2003) for the 2002-Jul-23 flare (a GOES X4.8 class) and are shown in Fig. 14.8. The line fits show a Doppler shift (typically 1% redshift) due to the nuclear recoils from the ion interaction and the emission of gamma-rays. The elemental abundances of the accelerated ions as well as of the ambient gas can be calculated from the gamma-ray line fluences. This provides unique information on enrichments of coronal over chromospheric abundances (the FIP effect, see §2.10).

14.3.2   Ion-Electron Ratios

The relative ratio of accelerated electrons to protons is not well known. Generally there is a significant correlation (but with broad scattering) between the fluences of hard X-ray producing electrons ($> 50\,\text{keV}$) and gamma-ray lines ($4-8\,\text{MeV}$), which suggests a common acceleration mechanism for $\approx 100\,\text{keV}$ electrons and $\approx 10\,\text{MeV}$ protons (e.g., Vestrand 1988; Cliver et al. 1994; Rieger et al. 1998). There are so-called *electron-rich events* that are so intense in electron bremsstrahlung that they ob-

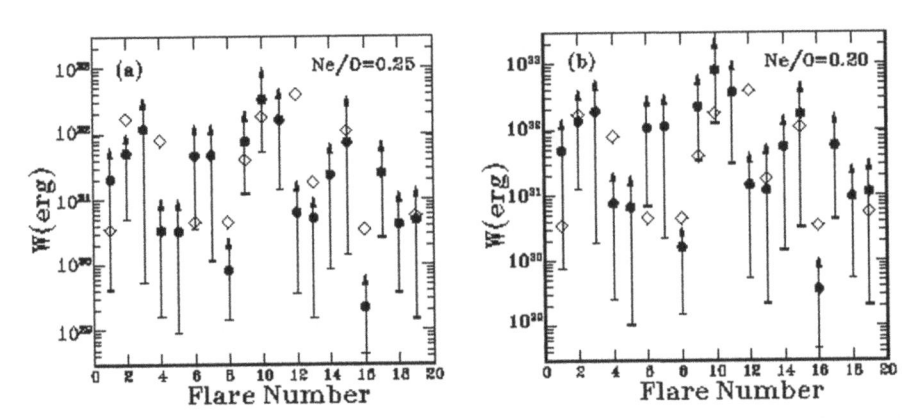

Figure 14.9: Calculated energy contents in ions and electrons. The solid circles and the lower limits indicated by horizontal bars are, respectively, the ion energy contents assuming low-energy cutoffs at 1 MeV/nucleon and at the maximum ϵ_c allowed by the observed Ne/O line fluence ratios. The open diamonds are electron energy contents assuming a cutoff at 20 keV (Ramaty & Mandzhavidze 2000a).

scur gamma-ray line emission (Rieger & Marschhäuser 1990). On the other hand, there are arguments that protons dominate the energy budget (Simnett 1986a). So, what is the relative proton-to-electron acceleration ratio? The bulk of the γ-ray line emission is believed to be produced by ions with energies of $5 - 100$ MeV/nucleon that contain only a small fraction of the energy in > 20 keV electrons (e.g. Lin 2000). However, systematic studies of γ-ray lines observed with *SMM* (Share & Murphy 1995) show that the 1.634 MeV ^{20}Ne line is unexpectedly enhanced. Because the cross section for ^{20}Ne has an unusually low energy threshold (≈ 2.5 MeV), this effect may be due to large fluxes of low-energy ions with a total energy content perhaps comparable to that in accelerated electrons (Ramaty et al. 1995; Emslie et al. 1997). On the other hand, warm thick-target effects could also mimic an enhanced ^{20}Ne line strength (MacKinnon & Toner 2003). To estimate the relative proton-to-electron ratio, or the relative energy content of their spectra, an extrapolation of the spectra has to be made in the $\epsilon < 1.6$ MeV range. Earlier spectral modeling with Bessel functions yielded a small proton-to-electron energy content, $W_i \ll W_e$ (Ramaty 1986). However, when more gamma-ray data constraining the relative Ne/O abundance became available, the spectral fits in the gamma-ray range favored an unbroken powerlaw, which led to a much larger ion energy content, comparable to equipartition, $W_i \approx W_e$ (Ramaty et al. 1995). The total energy contained in ions and electrons was calculated for 19 flares, based on a single-powerlaw spectrum with the slope constrained by the Ne/O ratio (Fig. 14.9). For a range of Ne/O=0.2 − 0.25, the energy in ions seems to be larger than the energy in ≥ 20 keV electrons in about 5 out of the 19 flares. It should also be recalled that the energy content in nonthermal electrons strongly depends on the assumed lower cutoff energy, which is traditionally chosen at ≈ 20 keV, but could be below ≈ 10 keV in small flares (Fig. 13.5). A lower cutoff energy can dramatically increase the energy content in nonthermal electrons. For discussions of e/p-ratios in solar flares see, for example, reviews by Vestrand & Miller (1999) and Vilmer & MacKinnon (2003).

14.3.3 Ion-Electron Timing

If gamma-ray emission is detected during flares, both electrons and ions are involved, so the question arises of what are the differences in the acceleration properties for both species, regarding the acceleration times, efficiency, and maximum obtained energies. In well-observed flares, the data clearly show that both species of particles, electrons, and ions, are accelerated to relativistic velocities relatively promptly (within a few seconds), and nearly simultaneously (Forrest & Chupp 1983). This is illustrated in Fig. 14.10 for the 1980-Jun-7 flare, which consists of a sequence of seven hard X-ray pulses, each one (except for the first one) followed by a gamma-ray pulse detected in the energy range of $4.2 - 6.4$ MeV, with only a slight delay of about $\approx 1 - 2$ seconds. Forrest & Chupp (1983) found that the peaks in hard X-rays and gamma-rays during the two intense flares of 1980-Jun-7 and 1980-Jun-21 coincided within ± 2.2 and ± 0.8 s. Also in the 1992-Feb-8 flare, individual peaks between 40 keV and 40 MeV coincided within ± 1 s (Kane et al. 1986). It is generally argued that this observed delay of $\lesssim 2$ s represents an upper limit for the acceleration time scale of protons and ions (that are responsible for the $4.2 - 6.4$ MeV gamma-ray line emission). In Section 12.3 we found that only acceleration times that are significantly smaller than the electron propagation times (from the coronal acceleration site to the chromospheric hard X-ray emission site), $t_{acc}(\varepsilon) \ll t_{prop}(\varepsilon)$, can satisfy the observed energy-dependent hard X-ray delays. This necessarily also implies that the spatial extent l_{acc} of the acceleration region is significantly smaller than the propagation path length L (i.e., $l_{acc} \ll L$). Based on this argument we can assume that both electrons and ions are accelerated in the same coronal acceleration region, regardless of the type of acceleration mechanism, such as small-scale electric fields, stochastic acceleration, or shocks.

For co-spatial acceleration, say in the cusp near a coronal reconnection point, accelerated electrons and protons have to propagate the same distance L to the flare loop footpoints, where electrons produce thick-target bremsstrahlung in hard X-rays, while the protons and α-particles produce nuclear de-excitation lines with ambient chromospheric ions in gamma-rays. What is the relative timing between hard X-rays and gamma-rays for this simplest scenario? For hard X-rays (e.g., for the $\epsilon_x = 35 - 114$ keV channel as shown in Fig. 14.10), we know that the hard X-ray photons with energy ϵ_x are produced by relativistic electrons with kinetic energies of $\varepsilon = \epsilon_x \, q_\epsilon \approx (35 - 140$ keV$) \times 1.124 \approx 40 - 160$ keV (from the photon-to-electron energy conversion factor for bremsstrahlung, defined in Eq. 13.5.6). For gamma-ray emission, however, which is dominated by the nuclear de-excitation lines of ^{12}C and ^{16}O ions in the $\varepsilon = 4.2 - 6.4$ MeV energy range (e.g., Hudson et al. 1980), hit by precipitating protons (see Table 14.2), the kinetic energy of the accelerated protons can have a large possible range. Because the standard composition of coronal plasma is made of 90% protons and 10% helium we consider only protons here. As a first approximation, let us assume equipartition for the kinetic energy of the accelerated electrons and protons (i.e., $\varepsilon_p = \varepsilon_e$),

$$\varepsilon_p = m_p c^2 (\gamma_p - 1) = \varepsilon_e = m_e c^2 (\gamma_e - 1) \,, \qquad (14.3.1)$$

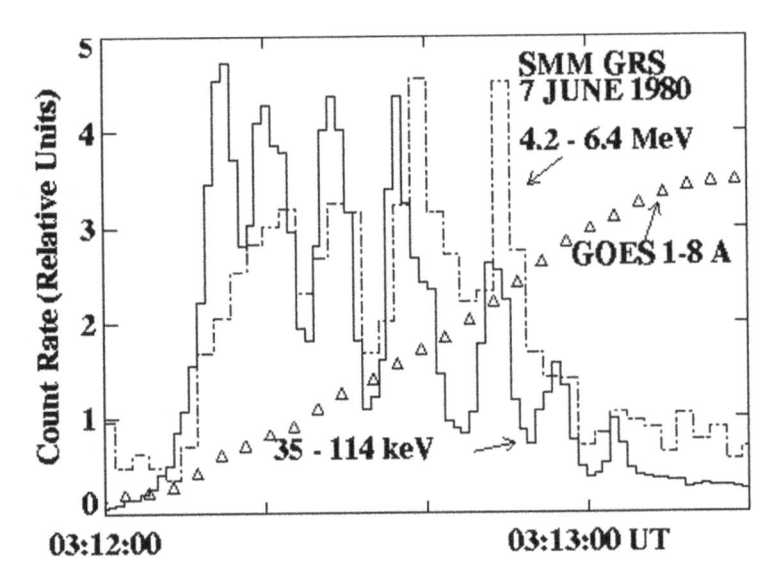

Figure 14.10: Time profiles for a flare showing near-coincidence impulsive peaks in $35 - 114$ keV hard X-rays (from energetic electrons) and $4.2 - 6.4$ MeV gamma-rays (from energetic ions). The time binning is 1 s. Note that the gamma-ray pulses are delayed by only $\approx 1 - 2$ s each (Forrest & Chupp, 1983).

which yields the following velocity ratio β_p/β_e for protons and electrons,

$$\frac{\beta_p}{\beta_e} = \frac{\sqrt{1 - [\frac{\varepsilon_p}{m_p c^2} + 1]^{-2}}}{\sqrt{1 - [\frac{\varepsilon_e}{m_e c^2} + 1]^{-2}}} \approx \sqrt{\frac{\varepsilon_p}{\varepsilon_e} \cdot \frac{m_e}{m_p}} = \frac{1}{43}\sqrt{\frac{\varepsilon_p}{\varepsilon_e}} , \qquad (14.3.2)$$

where the right-hand approximation applies to the nonrelativistic (or mildly relativistic) case. For equipartition, the velocity ratio is $\beta_p/\beta_e \approx \sqrt{m_e/m_p} = \sqrt{1/1836} \approx 1/43$. Therefore, the time-of-flight difference τ_{ep} between electrons and protons over a distance l' is

$$\tau_{ep} = \frac{l'}{c}(\frac{1}{\beta_e} - \frac{1}{\beta_p}) = \frac{l'}{\beta_e c}\left(1 - \sqrt{\frac{\varepsilon_e}{\varepsilon_p} \cdot \frac{m_p}{m_e}}\right) , \qquad (14.3.3)$$

which, in the case of (kinetic energy) equipartition, would reduce to

$$|\tau_{ep}^{equip}| \approx 42\frac{l'}{\beta_e c} \, [s] . \qquad (14.3.4)$$

Given the example of the flare shown in Fig. 14.10, where we obviously do not have equipartition, we have electrons with kinetic energies of $\varepsilon_e \gtrsim 35$ keV (Lorentz factor $\gamma_e = 1 + \varepsilon_e/m_e c^2 = 1 + 35/511 = 1.0685$ and a speed $\beta_e = \sqrt{1 - 1/\gamma_e^2} = 0.35$) and protons with $\varepsilon_p \gtrsim 4.2$ MeV ($\gamma_p = 1 + \varepsilon_p/m_p c^2 = 1 + 4.2/938 = 1.0045$ and $\beta_p = \sqrt{1 - 1/\gamma_e^2} = 0.094$). We do not know the flare loop size for the 1980-Jun-7 flare (Fig. 14.10), but if we substitute typical flare loop radii observed with *Yohkoh* ($r = 3 - 25$ Mm, Fig. 13.19) and the canonical scaling law for electron time-of-flight distances, $l' \approx 1.5r \times (\pi/2)$ (Eqs. 13.5.10 and 13.5.11), we obtain propagation distances of $l' \approx 7 - 60$ Mm. Thus we expect for the smallest loops with $l' = 7$ Mm an electron-proton delay of $\tau_{ep} = (l'/c)(1/\beta_e - 1/\beta_p) \approx -0.2$ s, and for the largest loops with $l' = 60$ Mm a delay of $\tau_{ep} \approx -1.6$ s. Therefore, we can predict that we generally expect gamma-ray delays of $|\tau_{ep}| \approx 0.2 - 2$ s for a typical range of flare sizes, just based on the proton time-of-flight delay from the coronal acceleration site to the chromosphere. So it is likely that the gamma-ray delay is a proton propagation delay rather than an acceleration delay. If shorter gamma-ray delays are observed, this could be explained by faster protons than the equipartition principle predicts, and the measured delay can then be used to put an upper limit on the maximum energies of the accelerated protons (with Eq. 14.3.3). Similar values for the gamma-ray delays were also obtained by other transport models (e.g., Ryan & Lee 1991; Vilmer et al. 1982), which support our conclusions here. More stringent time delays could be predicted by estimating the primary proton or ion energies involved in a given nuclear de-excitation line from the nuclear cross sections.

14.3.4   Directivity and Pitch Angle Distributions

The bremsstrahlung radiation from relativistic particles is highly directional, limited to within an angle of $\vartheta \approx 1/\gamma$. Thus, we do not expect any measurable directivity for hard X-rays ($\gamma \gtrsim 1$), while it should be clearly detectable in gamma-rays at energies of $\epsilon \gtrsim 0.5$ MeV ($\gamma \gtrsim 2$). For gamma-rays above 10 MeV ($\gamma > 20$), which are dominantly produced by electron bremsstrahlung, the relativistic beaming is very strong. The directivity of bremsstrahlung radiation in solar flares has been discussed in various studies (Elwert & Haug 1971; Brown 1972; Petrosian 1973, 1985), including the geometric effects of curvature and convergence of magnetic field lines and pitch angle scattering due to Coulomb collisions (Leach & Petrosian 1981, 1983).

Particles with small pitch angles are expected to enter the chromosphere nearly vertically, and thus would have the highest emission in the downward direction, which does not coincide with any line-of-sight direction, neither at the disk center nor at the limb. However, if particles have large pitch angles (ring, pan-cake, or losscone distributions), which are naturally produced near the magnetic mirrors at the footpoints of flare loops, we expect that the strongest emission coincides with the line-of-sight direction for limb flares, while it is weakest at the disk center. Therefore, a larger number of gamma-ray flares is expected to be detected near the limb than near disk center. This was indeed observed for flare events before 1987, but not confirmed for later events. Vestrand et al. (1987) performed statistics on 150 gamma-ray flares detected at > 300 keV with the *SMM Gamma-Ray Spectrometer (GRS)* and found an excess of gamma-ray events detected near the limb, as well as a slightly harder spectrum for limb flares. This observational result is consistent with losscone distributions, but inconsistent with beam distributions. Center-to-limb distributions of hard X-ray and gamma-ray flares are shown for three different energies (>30 keV, >300 keV, and >10 MeV) in Fig. 14.11, where the directivity effect is clearly most prominent for the highest energies. The percentage of flares at heliocentric distances of $> 64°$ is

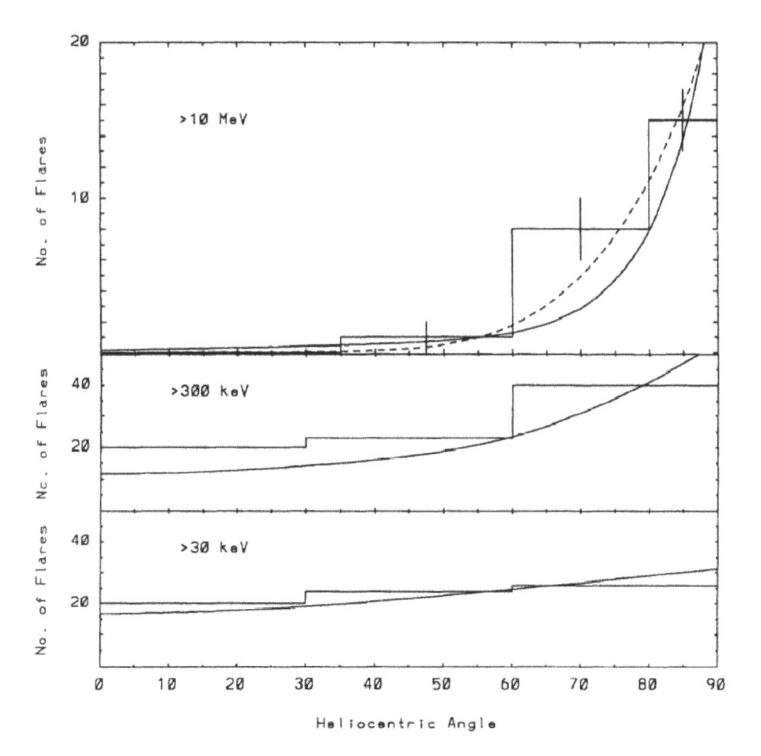

Figure 14.11: The number distribution of flares as a function of the heliocentric angle for three different minimum energies, subdivided into $3 - 4$ bins. The data of GRS flares with energies ≥ 300 keV are taken from Vestrand et al. (1987). The solid curves represent model distributions with different pitch angles and magnetic mirror ratios (McTiernan & Petrosian 1991).

expected to be $\approx 30\%$. Vestrand et al. (1987) observed $\approx 80\%$ of > 10 MeV flares during solar Cycle 21 and $\approx 55\%$ during Cycle 22 near the limb. Models by Petrosian (1985) and Dermer & Ramaty (1986) reproduce these directivity effects with pancake distributions.

While the center-to-limb effect confirms the directivity only on a statistical basis, it could also be directly measured by stereoscopic observations, which however did not confirm evidence for strong directivity, probably due to the lower energies used in the measurements, or possibly also due to the unknown orientation of the magnetic field as well as Compton backscattering.

Another method to determine the directivity is the modeling of gamma-ray line profiles, which should show different amounts of redshifts and blueshifts depending on the pitch angle distribution of the impinging ions. For instance, the different spectral line profiles of the 0.429 and 0.478 MeV gamma-ray lines from ^{7}Be and ^{7}Li (produced by accelerated α-particles colliding with ambient α particles in the chromosphere) have been calculated for different pitch angle distributions by Murphy et al. (1988). Fig. 14.12 shows the different line shapes for four different pitch angle distributions:

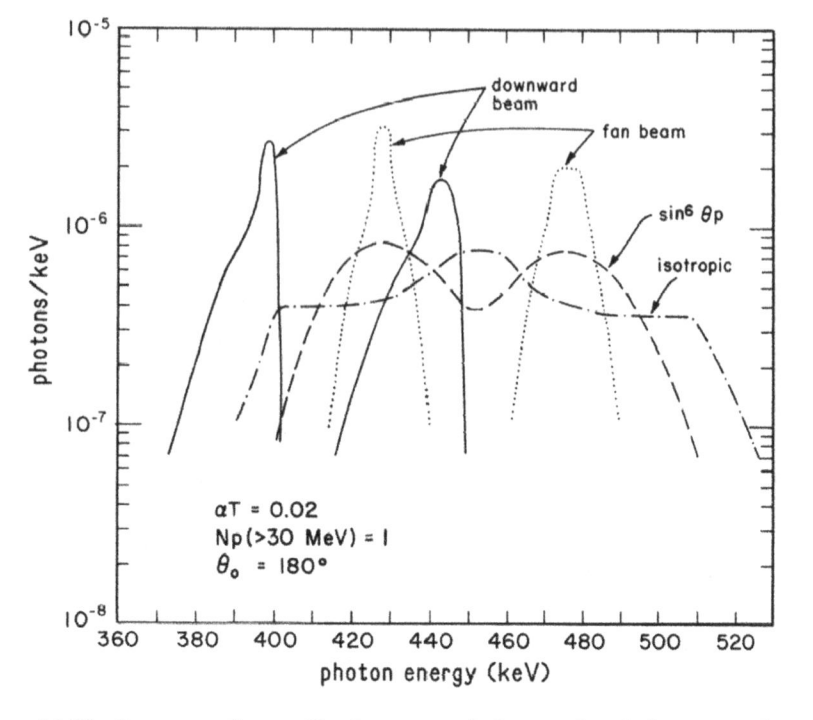

Figure 14.12: Gamma-ray line profiles from $\alpha - \alpha$ fusion reactions in flares occurring at the center of the solar disk for four different angular distributions of accelerated α-particles (Murphy et al. 1988).

for isotropic, downward beam, losscones (also called *fan-beam* or *pan-cake*), and for a $\sin^6$-distribution. Clearly, the line widths and centroids are quite different for these four distributions and thus provide a sensitive diagnostic for the anisotropy of accelerated α-particles, although the density gradient in a stratified atmosphere adds complications. Recent fits to the $\alpha - \alpha$ lines in the $0.3 - 0.7$ MeV range have been performed with this method for 19 flares by Share & Murphy (1997), and it was found that a downward beam distribution could be ruled out with high confidence in 4 flares, while losscone distributions and isotropic distributions provide acceptable fits to the data. The interpretation of this result depends on magnetic field models at the loop footpoints. For a high magnetic mirror ratio one would expect that the precipitating α-particles have large perpendicular velocities (i.e., losscone distributions), consistent with the observed anisotropy. Similar studies with fits on ^{7}Be, ^{7}Li (Share et al. 2003) and ^{12}C, ^{16}C, and ^{20}Ne lines (Share et al. 2002) corroborated the same result that a downward beam distribution can be ruled out and that other distributions (isotropic or downward-isotropic) yield acceptable fits. The directivity results rule out field-aligned DC electric field acceleration and favor stochastic acceleration processes. Brief reviews on the directivity results can be found in Rieger (1989), Ramaty & Mandzhavidze (1994), and Vestrand & Miller (1999).

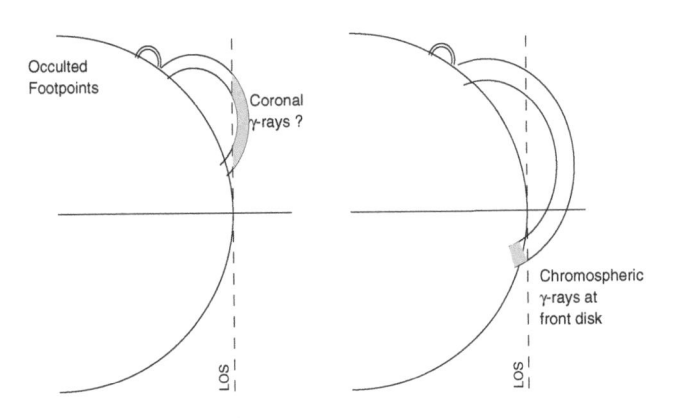

Figure 14.13: Two scenarios to explain γ-ray emission from behind-the-limb flares. *Left:* because the footpoints of the primary flare loop are occulted behind the limb, γ-ray emission was postulated to originate from a non-occulted coronal part of a connected large-scale trapping loop. *Right:* although the primary flare site is behind the limb, some large-scale trapping flare loop could connect to the front side, where γ-ray emission is produced in the chromospheric footpoint.

14.3.5 Gamma-Ray Line Emission in Occulted Flares

Because gamma-rays in the $2 - 7$ MeV range are generally dominated by nuclear de-excitation lines which require chromospheric ion densities, some observations of occulted (behind-the-limb) flare sites raised the problem of how the gamma-ray emission could be explained for such flares. The issue is whether gamma-rays are produced in an extended coronal trap region connected to the flare site (Fig. 14.13, left), or whether a large-scale loop was connecting the behind-the-limb flare site with a remote footpoint on the front side of the solar disk, where it could produce the usual chromospheric gamma-ray emission (Fig. 14.13, right). A special class of gradual hard X-ray emission originating in an occulted flare ($20°$ behind the limb) was already reported earlier (Hudson 1978), before gamma-ray observations were available, interpreted as a pure coronal (large-scale) trapping structure that is capable of trapping electrons and protons for extended periods of time.

Gamma-ray emission from an occulted flare was observed on 1989-Sep-29, 10:47 UT, and was associated with NOAA region 5698 at position W105°, so about 15° behind the limb (Vestrand & Forrest 1993, 1994). The gamma-ray emission showed electron bremsstrahlung continuum, a positron annihilation line, prompt nuclear emission in the $1 - 8$ MeV range, and a neutron capture line at 2.22 MeV that was surprisingly strong. Since the bulk of the prompt gamma-ray emission requires densities of $n_H > 10^{14}$ cm^{-3} to efficiently capture neutrons within their 900 s lifetime, it was concluded that the observations require a spatially extended loop that connects from behind-the-limb to the front side, covering a distance of $\approx 30°$ heliographic degrees (i.e., ≈ 370 Mm), as illustrated in Fig. 14.13 (right).

Intense gamma-ray emission of prompt lines in the $1 - 2$ and $4 - 7$ MeV energy range was also reported from PHEBUS on GRANAT during the 1991-Jun-1, 14:46 UT, GOES class X12.0 event, associated with NOAA region 6659, located 7° behind the east limb (Barat et al. 1994). This corresponds to occultation heights ranging from 3000 to 7000 km above the photosphere. Although prompt gamma-ray line emission requires densities of $n_H > 10^{12}$ cm^{-3}, which places the gamma-ray line emission region at a height of < 1500 km, it was concluded that the gamma-ray emission comes from coronal heights > 3000 km, because no 2.2 MeV neutron capture line was observed, that could reveal a possible front-side footpoint connected with the behind-the-limb flare site. Modeling of the gamma-ray line emission with a thin-target model with densities of $n_e \approx 1 - 5 \times 10^{11}$ cm^{-3} could reproduce the observed gamma-ray line fluxes (Trottet et al. 1996), although a very hard spectrum for the accelerated particles is required (Murphy et al. 1999).

An electron-dominated event was observed during the occulted 1991-Jun-30, 02:55 UT flare, with significant emission in the $10 - 100$ MeV range from a height of $\gtrsim 10^4$ km (between 2° and 12° behind the east limb), but no gamma-ray line emission in the $2 - 7$ MeV range was detected (Vilmer et al. 1999). This flare was also stereoscopically observed by *Ulysses* at energies >28 keV, revealing that the spatial structure was so extended that 0.07% of the total emission >28 keV (i.e., 10% of the occulted emission) originated in on-the-disk sources (Trottet et al. 2003), as indicated in Fig. 14.13 (right).

14.3.6 Abundances of Accelerated Ions

The ratio of certain gamma-ray line fluxes can be used to probe abundance ratios. Particularly interesting are abundance ratios of low-FIP to high-FIP elements (Fig. 2.15), which provide us information about what the enhancement factor is between coronal and chromospheric abundances during flares, for a given element. In the following we describe some combinations of gamma-ray lines that have been used to determine abundance ratios in accelerated particles.

The ratio of α-particles to protons (α/p) can be determined from the fluence ratio of the 0.339 MeV and 0.847 MeV lines. According to the processes listed in Table 14.2 we see that both lines result from de-excitations in ambient ^{56}Fe, so the line ratio is independent of the ambient abundances. However, it depends on α/p, because the 0.847 MeV line is produced by both protons and α particles (i.e., ^{56}Fe(p,p')^{56}Fe* and ^{56}Fe(α, α')^{56}Fe*), while the 0.339 MeV line is only produced by α-particles (i.e., ^{56}Fe(α,n)^{59}Ni*), which is thus called a *pure α-particle line*. Studies of this line find ratios of $\alpha/p \gtrsim 0.5$ or He/H $\gtrsim 0.1$ (Share & Murphy 1998).

The strong, narrow gamma-ray lines at 6.13 MeV (^{16}O), 4.44 MeV (^{12}C), 1.78 MeV (^{28}Si), 1.63 MeV (^{20}Ne), 1.37 MeV (^{24}Mg), and 0.847 MeV (^{56}Fe) result from de-excitations in nuclei of the relatively abundant constituents of the solar atmosphere. This has allowed the determination of the abundances of these elements (Murphy et al. 1996; Ramaty et al. 1995, 1996), showing that the low-FIP elements Mg, Si, and Fe are enriched relative to C and O in the gamma-ray producing region, a result that implies that the FIP bias known in the corona (Fig. 2.15) already sets in at lower altitudes, in the chromosphere.

A surprising result was that the Ne/O≈0.25 ratio was found to be higher (Ramaty et al. 1996) than the commonly accepted photospheric value of Ne/O≈ 0.15 (Meyer

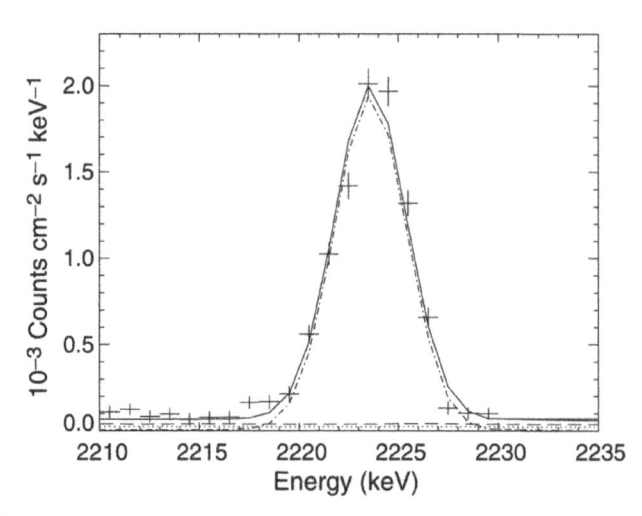

Figure 14.14: The 2.223 MeV neutron capture line count spectrum observed during the 2003-Jul-23, 00:27:20-00:43:20 UT, flare with RHESSI. The line profile fit includes the nuclear de-excitation line component (dotted), the electron bremsstrahlung powerlaw (dashed), and a Gaussian for the line (dash-dotted). The sum of all components is rendered with a solid curve. Note the high-resolution binning of 1 keV, providing a spectral resolution of $\approx 0.05\%$ (Murphy et al. 2003).

1996), implying either a large ion/electron ratio (which makes the ion energies dominant), a FIP effect issue (Shemi 1991; Share & Murphy 1995; Share et al. 1996), or an effect of a warm thick-target model (MacKinnon & Toner 2003).

The largest enhancement factors during gamma-ray flares have been measured for ^{3}He/^{4}He ratios. While solar wind values are typically ^{3}He/^{4}He$\approx (4.5 \pm 0.3) \times 10^{-4}$, flares with gamma-ray lines [^{16}O(^{3}He,p)^{18}F$^* \mapsto 0.937, 1.04, 1.08$ MeV] show that the ratio could be consistent with ^{3}He/^{4}He ≈ 0.1 in essentially all flares, in some cases up to ^{3}He/^{4}He ≈ 1 (Mandzhavidze et al. 1999). This huge enhancement by a factor of $\approx 10^3$, which is not observed in gradual *solar energetic particle (SEP)* events near the Earth (Reames et al. 1994; Reames 1995a), calls for a specific ^{3}He-production process (e.g., stochastic acceleration through gyroresonant wave-particle interactions). For discussions of abundance measurements in gamma-rays see, for example, Meyer (1996), Mandzhavidze & Ramaty (2000a), Ramaty & Mandzhavidze (2000b), and Reames (2000).

14.4 Neutron Capture Line

A prominent, very narrow gamma-ray line appears at 2.223 MeV which results from neutron capture. Neutrons are produced from all accelerated ions (protons and heavier nuclei such as C, N, O, Fe, etc.). The dominant neutron production process at higher energies in solar flares comes from the breakup of He nuclei, both in the accelerated particles and in the ambient medium. Like the gamma-ray lines, neutrons are also most likely produced in the high-density chromosphere at the footpoints of flare loops, where precipitating ions enter the collisional regime. The produced neutrons can propagate upwards and escape the Sun, or they can propagate downwards into the photosphere, where they become thermalized by elastic collisions with hydrogen. Subsequently, a significant fraction is captured by protons [i.e., $^1H(n,p)^2H$], to produce deuterium (^2H) and monoenergetic photons at 2.223 MeV, the binding energy of the deuteron. The never resolved 2.223 MeV neutron capture line is narrow (a few eV) because it is broadened by only the relatively low photospheric temperature of ≈ 6000 K. Because the production site (photosphere) of the 2.223 MeV line is much lower than that of nuclear de-excitation lines (chromosphere), a significant limb darkening of the 2.223 MeV results for the much longer column depth near the limb. Thus, while the 2.223 MeV line is stronger than the nuclear de-excitation lines for disk flares, the role is reversed for limb flares.

Because the 2.223 MeV line is very narrow (a few eV), it was never resolved. The line width observed with high spectral-resolution instruments, such as with HEAO-3 (Prince et al. 1982) or with the cooled germanium detectors of RHESSI (Fig. 14.14; Murphy et al. 2003) is of instrumental nature. Since the neutrons have first to slow down before they can be captured by ambient hydrogen, the line occurs delayed by $\approx 50 - 300$ s. The delay is mostly affected by the energy distribution of the interacting particles and the photospheric ^3He abundance. A competing process in the photosphere is capture by ^3He [i.e., $^3He(n,p)^3H$]. Neutron capture by ^3He does not produce radiation but reduces the delay of the 2.223 capture line, and thus can be used to infer the $^3He/H$ abundance ratio. Making some assumptions on the chromospheric density, the angular (pitch angle) distribution of the interacting particles, the magnetic convergence in the chromosphere, and fitting the time history of the 2.223 neutron capture line (e.g., Fig. 14.5 top; Rank et al. 2001; Murphy et al. 1994), values in the range of $^3He/H$ $\approx (4 \pm 3) \times 10^{-5}$ have been inferred (Ramaty & Kozlovsky 1974; Chupp et al. 1981; Prince et al. 1983; Hua & Lingenfelter 1987; Trottet et al. 1993; Murphy et al. 1997; Rank et al. 2001).

The spatial source of the 2.223 MeV emission is expected to coincide with the flare loop footpoints where also nonthermal hard X-rays and other gamma-ray lines are emitted, if one assumes that electrons and ions are concomitantly accelerated and precipitate along the same magnetic field lines. The first imaging observations of the 2.223 MeV line, which were made with RHESSI during the 2002-Jul-23, 00:30 UT, flare (GOES class X4.8), however, revealed a displacement of $\approx 20'' \pm 6''$ (≈ 15 Mm) between the centroid of the 2.223 MeV emission and the $50 - 100$ keV, $0.3 - 0.6$ MeV, and $0.7 - 1.4$ MeV sources (Fig. 14.15; Hurford et al. 2003). This puzzling observation suggests that electrons and ions are either not jointly accelerated or propagate in different directions. A possible explanation was proposed in terms of stochastic acceleration driven by MHD-turbulence cascading ($\S11.4.2-3$), which favors ion acceleration in larger loops and electron acceleration in shorter loops (Emslie et al. 2004).

Theoretical studies and reviews on the neutron capture line can be found in Lingenfelter et al. (1965), Lingenfelter & Ramaty (1967), Ramaty et al. (1975), Ramaty (1986), Ramaty & Mandzhavidze (2001).

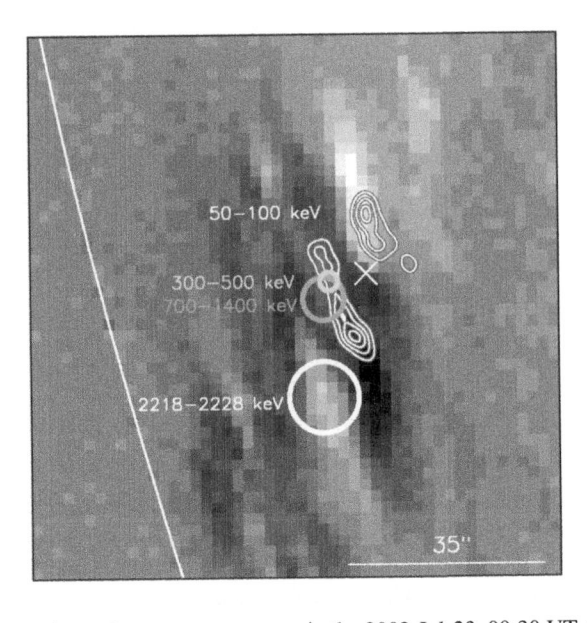

Figure 14.15: Locations of gamma-ray sources in the 2002-Jul-23, 00:30 UT, flare, overlaid on an *SoHO/MDI* magnetogram. The hard X-ray sources (50−100 keV) seem to outline a double-ribbon structure as typical for flare arcades. Note the displacement of the 2.223 MeV neutron capture line, which is displaced by $\approx 20'' \pm 6''$ to the south (Hurford et al. 2003).

14.5 Positron Annihilation Line

Positrons (e^+) are the antiparticles to electrons (e^-), having the same mass but opposite electric charge (historically also called β^+ particles). Wolfgang Pauli showed in 1930 that the radioactive beta-decay could only be explained if the proton (p) and neutron (n) have *weak interactions in nuclei*,

$$n \mapsto p + e^- + \overline{\nu_e} \,, \tag{14.5.1}$$

$$p \mapsto n + e^+ + \nu_e \,, \tag{14.5.2}$$

where ν_e is the electron neutrino and $\overline{\nu_e}$ the anti-electron neutrino. The beta-decay process occurs also for radioactive nuclei (Z, A) with charge Z and mass number A,

$$(Z - 1, A) \mapsto (Z, A) + e^- + \overline{\nu_e} \,, \tag{14.5.3}$$

$$(Z + 1, A) \mapsto (Z, A) + e^+ + \nu_e \,. \tag{14.5.4}$$

Abundant ions accelerated in flares are ^{12}C, ^{14}N, and ^{16}O. The accelerated ions then collide mostly with ambient thermal protons and helium (α) particles and produce radioactive ions such as ^{11}C, ^{13}N, ^{15}O, which then produce positrons (e^+) by beta-decay processes according to Eq. (14.5.4). A list of such principal beta-decay processes in the solar chromosphere that occur during gamma-ray flares is given in Table 14.3, taken from Lingenfelter & Ramaty (1967), also reproduced in Lang (1980). Nuclear cross

Table 14.3: Principal positron emitters for abundant nuclei in gamma-ray flares (Lingenfelter & Ramaty 1967).

Production mode	β^+ emitter and decay mode	Maximum β^+ energy (MeV)	Half life time (min)	Threshold energy (MeV)	Production cross section (mb)
$C^{12}(p,pn)C^{11}$	$C^{11} \mapsto B^{11} + \beta^+ + \nu$	0.97	20.5	20.2	50
$N^{14}(p,2p2n)C^{11}$				13.1	30
$N^{14}(p,\alpha)C^{11}$				2.9	–
$O^{16}(p,3p3n)C^{11}$				28.6	10
$N^{14}(p,pn)N^{13}$	$N^{13} \mapsto C^{13} + \beta^+ + \nu$	1.19	9.96	11.3	10
$O^{16}(p,2p2n)N^{13}$				5.54	8
$N^{14}(p,n)O^{14}$	$O^{14} \mapsto N^{14} + \beta^+ + \nu$	1.86	1.18	6.4	–
$O^{16}(p,pn)O^{15}$	$O^{15} \mapsto N^{15} + \beta^+ + \nu$	1.73	2.07	16.54	50

sections for the production of radioactive proton emitters resulting from protons and α-particles and the ambient medium are treated in detail in Kozlovsky et al. (1987).

An alternative nuclear interaction that leads to the production of positrons are π-meson production processes, of which a typical reactions is

$$p + p \mapsto p + n + \pi^+ , \tag{14.5.5}$$

where p are protons, n neutrons, and π^+ is a positive π-meson. After π-mesons are produced, μ-mesons, electrons, positrons, photons (γ), and neutrinos can be produced by the decay reactions

$$\pi^\pm \mapsto \mu^\pm + \nu_\mu/\overline{\nu_\mu} , \tag{14.5.6}$$

$$\pi^0 \mapsto \gamma + \gamma , \tag{14.5.7}$$

$$\mu^\pm \mapsto e^\pm + \nu_e/\overline{\nu_e} + \nu_\mu/\overline{\nu_\mu} . \tag{14.5.8}$$

Positively charged pions are created in solar flares with accelerated protons with energies ≥ 200 MeV, producing positrons that contribute to the 511 keV emission line (Murphy et al. 1987).

The positrons (e^+) that are produced in the chromosphere by the decay processes described above, slow down by Coulomb interactions and either directly annihilate with electrons,

$$e^+ + e^- \mapsto 2\gamma , \tag{14.5.9}$$

or form *positronium (Ps)* by attachment to a free electron (e^-),

$$e^+ + e^- \mapsto Ps + h\nu , \tag{14.5.10}$$

or by charge exchange with a hydrogen atom ^1H (also called *charge exchange in flight*),

$$e^+ + {}^1H \mapsto Ps + p . \tag{14.5.11}$$

Positronium is formed in the singlet or triplet spin state. Both the direct annihilation process (Eq. 14.5.3) and annihilation from the singlet state contribute to the 511 keV line. When annihilation takes place from the triplet state, three photons are emitted,

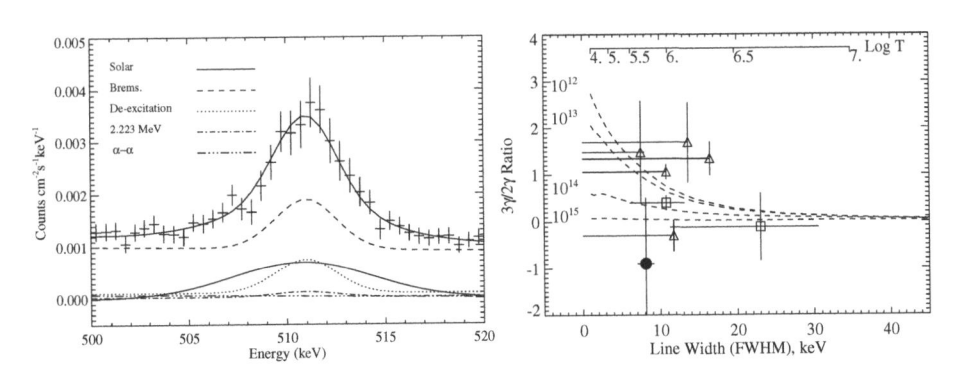

Figure 14.16: *Left:* fit of the 511 keV annihilation line profile observed with RHESSI during the 2002-Jul-23, 00:27:20−00:43:20 UT, flare, including the total background-corrected count spectrum and the best fit (thin solid curve). *Right:* the $3\gamma/2\gamma$ ratio versus the 511 keV line width and temperature for a fully ionized medium is shown for SMM flares (triangles) and the RHESSI observed flare of 2002-Jul-23 (black dot). The (dashed) curves show model calculations for different densities (Share et al. 2003).

producing a continuum. The number of photons observed in this continuum divided by the number of photons in the line is known as the $3\gamma/2\gamma$ ratio. The line width and time profile of the 511 keV line depends on the temperature, density, and ambient composition of the ambient medium, which determine the slow-down time of the positrons and the formation process of the positronium (Crannell et al. 1976). The FWHM width of the 511 keV line is

$$FWHM = 1.1 \text{ keV} \sqrt{\frac{T}{10^4 \text{ K}}}, \qquad (14.5.12)$$

if the line width is entirely determined by thermal broadening during the positronium formation in a hot ionized medium (Crannell et al. 1976).

The 511 keV line has been for the first time fully resolved during a solar flare with the germanium-cooled detectors of RHESSI, namely during the 2002-Jul-23 flare, which was a prolific emitter of the 511 keV annihilation line (Share et al. 2003). The line profile could be fitted with two models in vastly different environments: (1) formation of positronium by charge exchange in flight with hydrogen at a photospheric/chromospheric temperature of ≈ 6000 K, which yields a line width of $\approx 7.5 \pm 0.5$ keV, or (2) positronium formation in a hotter thermal plasma with transition region temperatures of $T \approx (4 - 7) \times 10^5$ K, which reproduces the observed (best-fit) line width of 8.1 ± 1.1 keV (according to Eq. 14.5.12). The line profile fits and the $3\gamma/2\gamma$ ratio of the RHESSI measurement by Share et al. (2003) are shown in Fig. 14.16. As Fig. 14.16 illustrates, the $3\gamma/2\gamma$ ratio -0.8 ± 0.5 of the RHESSI measurement requires high densities ($\gtrsim 10^{14}$ cm^{-3}) where the positronium slows down, which is difficult to reconcile with the lower densities $\lesssim 10^{12}$ cm^{-3} usually measured at transition region temperatures. Adopting the second interpretation thus raises the question why annihilation does not take place in the lower chromosphere (at densities of $\gtrsim 10^{14}$ cm^{-3}) where the positrons are traditionally expected to be produced and where there is sufficient material to slow them down (Share et al. 2003)?

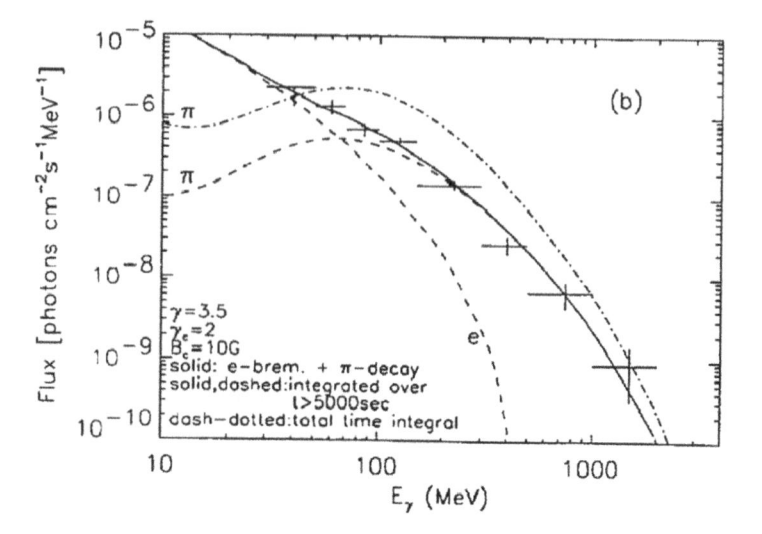

Figure 14.17: A gamma-ray spectrum observed with *EGRET/CGRO* during the 1991-Jun-11, 02:04 UT flare, accumulated during 03:26−06:00 UT (Kanbach et al. 1993). The spectrum is fitted with a combination of primary electron bremsstrahlung and pion-decay radiation. Note that pion decay is dominant at energies $\gtrsim 40$ MeV (Mandzhavidze & Ramaty 1992a).

Reviews and discussions of the 511 keV annihilation line in solar flares can be found in Lingenfelter & Ramaty (1967), Ramaty et al. (1975), Ramaty (1986), Ramaty & Mandzhavidze (1994).

14.6 Pion Decay Radiation

The rest mass of π-mesons is 139.57 MeV for charged pions ($\pi^{\pm}$) and 134.96 MeV for neutral pions (π^{0}). Therefore, if protons or heavier ions are accelerated to sufficiently high energies of $\gtrsim 260$ MeV/nucleon, the kinetic energy can create pions. A typical pion production process is given in Eq. (14.5.5), where accelerated protons colliding with ambient nuclei are capable of producing both charged and neutral π-mesons. The $\pi^{\pm}$-mesons decay succesively into muons $\mu^{\pm}$ (Eq. 14.5.6) and then into an electron or positron ($e^{\pm}$) and neutrinos (Eq. 14.5.8). The secondary electrons (positrons) in turn radiate by bremsstrahlung in collisions with ambient protons and ions. Neutral pions (π^{0}), on the other hand, decay electromagnetically into two gamma-rays (Eq. 14.5.7), each with the half pion energy (67 MeV), with some Doppler broadening. Other pion decay modes are rare ($< 1\%$). Pion-decay radiation usually dominates the flare gamma-ray spectrum above energies of $\gtrsim 67$ MeV (Fig. 14.1). The pion-decay radiation requires either a very hard proton spectrum or an independent population of high-energy particles. Therefore, spectral fits of theoretical pion-decay spectra provide constraints for the spectrum of accelerated protons (required for pion production). Theoretical models of pion production have been calculated for specific acceleration processes, such as for stochastic or shock acceleration (Murphy et al. 1987), whilst taking the time evolution of trapping (by magnetic mirroring) and precipitation into account (Mandzhavidze & Ramaty 1992a).

The first detection of pion radiation in solar flares was accomplished during the 1982-Jun-03 flare (Forrest et al. 1985, 1986), which showed evidence for proton and ion acceleration up to 1 GeV and high-energy neutrons $\gtrsim 500$ MeV (Chupp et al. 1985, 1987). The most prominent event for modeling of pion-decay radiation was the 1991-Jun-11, 02:04 UT, flare (Fig. 14.17), which was observed by *EGRET/CGRO* in the energy range of ≈ 10 MeV-2 GeV (Kanbach et al. 1993). The time profile and the gamma-ray spectrum have been modeled self-consistently with a long-term trap model of accelerated particles (Ryan & Lee 1991; Ryan 2000; Rank et al. 2001), where pion production is controlled by precipitating protons and α-particles, which have an energy-dependent trapping time and are required to meet the necessary threshold energy for pion production. The best fit was found for a hard proton injection spectrum with a powerlaw slope of 3.5 (Fig. 14.17). More fine-tuned models distinguish between at least three different injection phases during the 8-hour-long event, which reconciles the two models of continuous (or episodic) acceleration and long-term trapping (Mandzhavidze et al. 1996; Akimov et al. 1996; Rank et al. 2001; see also §14.2.3).

The theory of pion-decay emission in solar flares is treated in detail in Murphy et al. (1987), Mandzhavidze & Ramaty (1992a), and Heristchi & Boyer (1994). Observations of solar flare pion-decay radiation are described in Mandzhavidze & Ramaty (1992b, 1993), Ramaty & Mandzhavidze (1994), Hudson & Ryan (1995), and Mandzhavidze et al. (1996).

14.7 Summary

The energy spectrum of flares in gamma-ray wavelengths (0.5 MeV-1 GeV) is more structured than in hard X-ray wavelengths ($20 - 500$ keV), because it exhibits both continuum emission as well as line emission. There are at least six different physical processes that contribute to gamma-ray emission: (1) electron bremsstrahlung continuum emission (§14.2); (2) nuclear de-excitation line emission (§14.3); (3) neutron capture line emission at 2.223 MeV (§14.4); (4) positron annihilation line emission at 511 keV (§14.5); (5) pion-decay radiation at $\gtrsim 50$ MeV (§14.6); and (6) neutron production. The ratio of continuum to line emission varies from flare to flare and gamma-ray lines can be completely overwhelmed in electron-rich flares or flare phases. When gamma-ray lines are present, they provide a diagnostic of the elemental abundances, densities, and temperatures of the ambient plasma in the chromosphere, as well as of the directivity and pitch angle distribution of the precipitating protons and ions that have been accelerated in coronal flare sites, presumably in magnetic reconnection regions. Critical issues that have been addressed in studies of gamma-ray data are the maximum energies of coronal acceleration mechanisms, the ion/electron ratios (because selective acceleration of ions indicate gyroresonant interactions), the ion/electron timing (to distinguish between simultaneous or second-step acceleration), differences in ion/electron transport (e.g., neutron sources were recently found to be displaced

from electron sources), and the first ionization potential (FIP) effect of chromospheric abundances (indicating enhanced abundances of certain ions that could
be preferentially accelerated by gyroresonant interactions). Although detailed
modeling of gamma-ray line profiles provides significant constraints on elemental abundances and physical properties of the ambient chromospheric plasma, as
well as on the energy and pitch angle distribution of accelerated particles, little information or constraints could be retrieved about the time scales and geometry of
the acceleration mechanisms, using gamma-ray data. Nevertheless, the high spectral and imaging resolution of the recently launched RHESSI spacecraft faciliates
promising new data for a deeper understanding of ion acceleration in solar flares.

Chapter 15

Radio Emission

"War is the father of all things" (Heraclitus, 535-475 BC). So it was during World War II that solar radio astronomy was born, when English radar stations received strongly directed noise signals, first suspected from enemy transmitters, but finally identified to be of solar origin. Grote Reber, a pioneer in galactic radio astronomy, first reported solar radio radiation in 1944. After this discovery, solar radio astronomy was enthusiastically undertaken by construction of radio antennas and spectrometers, in Europe, the U.S.A., Japan, Russia, and Australia. Radio spectrometers expanded their frequency range from metric to decimetric, microwave, and millimeter wavelengths. These radio spectrometers are non-imaging instruments, but provide very useful dynamic spectra, many being still in operation today. A further breakthrough came with the construction of radio interferometers, which allow for Fourier-type imaging. The most prominent radio interferometers used for solar observations over extended periods of time are the Dutch instrument in *Westerbork*, the *Culgoora radioheliograph* in Australia, the *Very Large Array (VLA)* in New Mexico, the *Owens Valley Radio Observatory (OVRO)* in California, the *Nançay radioheliograph* in France, and the *Nobeyama radioheliograph* in Japan. Solar observations in radio wavelengths represent the second-best explored wavelength regime (besides optical), because it could be done with ground-based instruments. A future solar-dedicated, frequency-agile radio array is being planned, the so-called *Frequency-Agile Solar Radiotelescope (FASR)*, which will provide imaging with about 100 antennas over a very broad frequency range from 30 MHz to 30 GHz (see web-page *http://www.ovsa.njit.edu/fasr/* for instrumental descriptions and updates).

Radio observations range from quiet Sun emission to flares, CMEs, and interplanetary particles. Radio emission is produced by thermal particle distributions (free-free emission), by mildly relativistic particles (gyrosynchrotron emission), as well as by unstable anisotropic particle distributions, such as electron beams (producing plasma emission) or losscones (producing electron-cyclotron maser emission). Since solar radio astronomy has been developed over an epoch of sixty years, there are a number of textbooks available, focussing either more on observations (Kundu 1965; Zheleznyakov 1970; Krueger 1979; McLean & Labrum 1985) or more on the theoretical underpinning in plasma physics (Benz 1993; Melrose 1980a,b). Reviews on

Table 15.1: Radio emission mechanisms during solar flares, (gyrofrequences are given in units of angular frequencies, $\omega = 2\pi\nu$) (Aschwanden 2002b).

Emission mechanism	Frequency	Source/Exciter
(1) Incoherent radio emission:		
(1a) Free-free emission (bremsstrahlung)	$\nu \gtrsim 1$ GHz	Thermal plasma
− Microwave postbursts		Thermal plasma
(1b) Gyroemission	$\omega = s\Omega_e$	
Gyroresonance emission	($s = 1, 2, 3, 4$)	Thermal electrons
Gyrosynchrotron emission	($s \approx 10 - 100$)	Mildly relativistic electrons
− Type IV moving		Trapped electrons
− Microwave type IV		Trapped electrons
(2) Coherent radio emission:		
(2a) Plasma emission	$\nu_{pe} = 9000\sqrt{n_e}$	Electron beams
− Type I storms		Langmuir turbulence
− Type II bursts		Beams from shocks
− Type III bursts		Upward propagating beams
− Reverse slope (RS) bursts		Downward propagating beams
− Type J bursts		Beams along closed loops
− Type U bursts		Beams along closed loops
− Type IV continuum		Trapped electrons
− Type V		Slow electron beams
(2b) Electron-cyclotron maser:	$\omega = s\Omega_e/\gamma + k_\parallel v_\parallel$	Losscones
- Decimetric ms spike bursts		Losscones

flare-related radio emission can also be found in Wild & Smerd (1972), Marsh & Hurford (1982), Kundu & Vlahos (1982), Wu (1985), Kundu (1985), Dulk (1985), Simnett (1986b; 1995), Goldman & Smith (1986), Aschwanden (1987a; 2002b), Benz (1987b; 1993), Trottet (1994b), Aschwanden & Treumann (1997), Bastian et al. (1998), and Fleishman & Mel'nikov (1998). Short encyclopedic articles have been written by Bastian (2000), Benz (2000), Melrose (2000), and Bougeret (2000).

15.1   Overview on Radio Emission Mechanisms

Radio emission can generally be classified into *coherent emission* and *incoherent emission mechanisms*. Incoherent emission results from continuum processes, such as thermal particle distributions that produce through Coulomb collisions *free-free emission* in microwave and mm wavelengths, or thermal/mildly relativistic electron distributions that produce through their random-phase gyromotion *gyroresonance/gyrosynchrotron* emission. Coherent emission, in contrast, occurs by kinetic instabilities from unstable particle distributions. When a particle distribution function $f(v_\parallel, v_\perp)$ evolves with a positive slope ($\partial f/\partial v_\parallel > 0$) in a parallel direction to the magnetic field, it is called a *beam distribution*, and the bump in parallel direction drives a so-called *bump-in-tail in-*

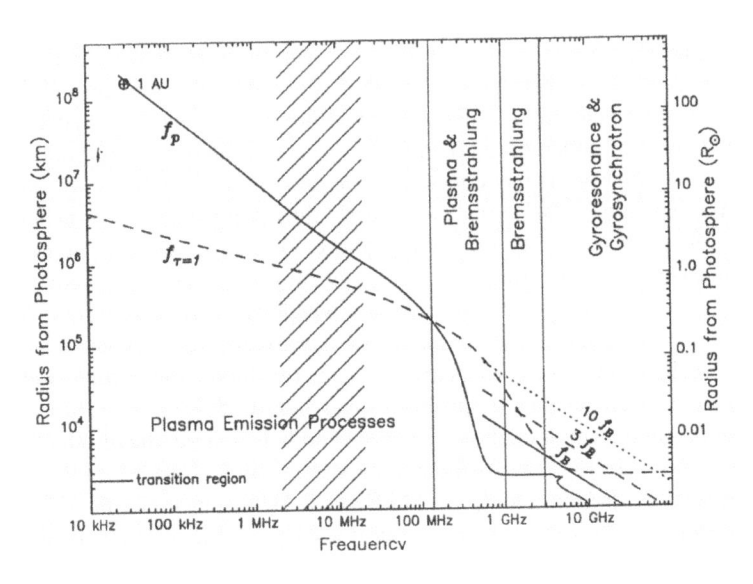

Figure 15.1: Characteristic frequencies for radio emission in the solar corona and heliosphere. The plasma frequency (solid line) and the frequency where the free-free opacity is unity (dashed line) are shown as distance from the photosphere versus frequency. The regimes of dominant plasma emission, bremsstrahlung emission, and gyroresonance or gyrosynchrotron emission are marked as a function of frequency (Gary & Hurford 1989).

stability that produces *plasma emission*. When a particle distribution function evolves with a positive slope ($\partial f / \partial v_\perp > 0$) in a perpendicular direction, for example *losscone*, *ring*, or *Dory−Guest−Harris distributions*, it is prone to *electron-cyclotron maser emission*. The term *"maser"* (*microwave amplified stimulated emission radiation*) refers to an induced emission stimulated by an inverted particle distribution function, analogous to the optical *laser* emission that is pumped by inverted quantum-mechanical level populations. A quick overview of these radio emission mechanisms relevant to solar flares is given in Table 15.1, which also lists the characteristic frequencies and sources or physical exciter mechanisms. We will structure this chapter according to the same groups of emission mechanisms.

The classification into these four major groups of physical emission mechanisms tells us right away in what environment the corresponding radio emission or bursts occur. Incoherent emission mechanisms require thermal (or isotropic suprathermal) distributions, which are present in the quiet Sun and in active regions. Thermal free-free emission tends to dominate over other mechanisms at low frequencies, at high densities, or in cool regions. Thermal free-free emission is the dominant mechanism in quiet Sun regions, coronal holes, and in prominences, and sometimes can be relevant in active regions. Gyroemission, on the other hand, is most efficient in high magnetic fields, and thus is expected to dominate radio emission above sunspots. Gyrosynchrotron emission requires mildly relativistic particles, which are naturally produced during flares. For coherent emission mechanisms, the most natural way to produce anisotropic particle distributions is either by velocity dispersion, which produces electron beams and thus plasma emission, or by mirroring in a magnetic trap geometry, which produces losscone distributions and thus is prone to losscone instabilities (such as *electron-cyclotron emission*). Such magnetic mirror geometries are expected in flare loops. Thus, most of coherent radio emission occurs during flares, intermittently and bursty, driven by *bursty magnetic reconnection processes* (§ 10.2) and the associated flare plasma dynamics. The term *"radio burst"* thus dominated the nomenclature of observed radio emission right from the beginning. Of course, there is also non-flaring or quiet, quasi-steady radio emission, such as gyroresonance emission above sunspots (§ 5.7.2) or free-free emission in active regions, which we discussed earlier (§ 5.7.1 and § 2.3). Thus we focus in this chapter mainly on flare-related radio emission.

A diagram that shows the coronal and heliospheric altitude range where radio emission occurs as a function of the frequency is given in Fig. 15.1 (Gary & Hurford 1989). Using a canonical density model of the corona and heliosphere, the plasma frequency run $f_p(h)$ is shown in Fig. 15.1 (solid curve), starting at $f_p \lesssim 1$ GHz in the transition region and steadily dropping to $f_p \approx 30$ kHz at 1 AU distance. Using a density and temperature model, the run of the optically thick layer with $\tau_{ff} = 1$ is also shown in Fig. 15.1 (dashed curve), which illustrates that the corona is optically thin down to frequencies of a few 100 MHz, The gyrofrequency extends up to $f_B \lesssim 7$ Ghz, the third harmonic up to $3f_B \lesssim 20$ GHz, and gyrosynchrotron with harmonics $s \leq 10$ extends up to $10f_B \lesssim 70$ GHz. In active regions, we expect that plasma emission dominates at frequencies of $f \lesssim 1$ GHz, bremsstrahlung dominates in the range of $f \approx 1 - 3$ GHz, and gyroemission at $f \gtrsim 3$ GHz. Typical radio fluxes at different frequencies are shown in the solar irradiance spectrum of Fig. 1.25, for the quiet Sun and flare-related bursts. Most of the radio spectra drop off with a high negative power towards higher frequencies.

15.2 Free-Free Emission (Bremsstrahlung)

15.2.1 Theory of Bremsstrahlung in Microwaves

Bremsstrahlung is produced when individual electrons are deflected in the Coulomb field of ambient ions due to the acceleration they experience by the Coulomb force. Since both the test particle (electron) as well as the field particle (ion) are free particles in a (partially or fully) ionized plasma, this emission is also called *free-free emission*. When the test particle is part of the same thermal distribution, we talk about *thermal bremsstrahlung*, which we treated in § 2.3. In contrast, if the test particle has a much higher energy, and thus is part of a nonthermal distribution, we are talking about *nonthermal bremsstrahlung*. Moreover we distinguish in hard X-rays between *thick-target* and *thin-target bremsstrahlung*, depending on whether the particle impinges from a un-collisional plasma onto a collisional target plasma (§ 13.2.2), or if the test particle is continuously accelerated in a thin-target plasma (§ 13.2.3). Bremsstrahlung produced by nonthermal particles is observable in microwave and hard X-ray wavelengths (§ 13), while bremsstrahlung produced by thermal particles is detectable in soft X-rays (e.g., see Fig. 2.6) and in microwaves, usually dominant at millimeter wavelengths.

We derived the *free-free absorption coefficient* $\alpha_\nu(z)$ for thermal electrons in § 2.3, which depends on the ambient ion density $n_i(z)$ and electron temperature $T_e(z)$, at position z along a given observer's line-of-sight, and radio frequency ν (Eq. 2.3.16),

$$\alpha_{ff}(z, \nu) \approx 9.786 \times 10^{-3} \frac{n_e(z) \sum_i Z_i^2 n_i(z)}{\nu^2 T^{3/2}(z)} \ln \Lambda , \qquad (15.2.1)$$

where the Coulomb integral $\ln \Lambda(z)$ is for coronal temperatures approximately ($T \gtrsim 3 \times 10^5$ K),

$$\ln \Lambda(z) = \ln \left[4.7 \times 10^{10} \left(\frac{T_e(z)}{\nu} \right) \right] = 24.5 + \ln T_e(z) - \ln \nu \qquad (\text{cm}^{-1}) .$$
$$(15.2.2)$$

The corona is commonly assumed to be fully ionized, so that $n_i(z) = n_e(z)$, the so-called *coronal approximation*. From these two expressions (Eq. 15.2.1–2) we straight-forwardly obtain the *free-free opacity* $\tau_{ff}(z, \nu)$ as a function of position z by integrating over the column depth range $z' = [-\infty, z]$,

$$\tau_{ff}(z, \nu) = \int_{-\infty}^{z} \alpha_{ff}[T_e(z'), n_e(z'), \nu] \, dz' . \qquad (15.2.3)$$

Note that the temperature $T_e(z)$ and density $n_e(z)$ along the line-of-sight are usually very inhomogeneous and cannot directly be measured. In principle, one could measure the *differential emission measure (DEM) distribution* from many EUV and soft X-ray wavelengths to constrain it. In practice, simple models with a single temperature or a two-component plasma have been used. A more realistic approach is to use a sta-tistical multi-temperature and multi-density model that matches the DEM distribution of co-spatial EUV and soft X-ray measurements (see Fig. 3.17 and § 3.5.3). Once the free-free opacity (Eq. 15.2.3) is calculated, one can then determine the *radio bright-ness temperature* $T_B(\nu)$ at the observer's frequency ν with a further integration of the opacity along the line-of-sight,

$$T_B(\nu) = \int_{-\infty}^{0} T_e(z) \exp^{-\tau_{ff}(z, \nu)} \alpha_{ff}(z, \nu) \, dz . \qquad (15.2.4)$$

The observed quantity is the *flux density* $I(\nu)$, which can be calculated from the *bright-ness temperature* $T_B(\nu)$ with the *Rayleigh–Jeans approximation* at radio wavelengths (Eq. 2.2.5, with $S_\nu = B_\nu$), integrated over the solid angle Ω_S of the radio source (Eq. 2.2.9),

$$I(\nu) = \frac{2\nu^2 k_B}{c^2} \int T_B d\Omega_S . \qquad (15.2.5)$$

At low frequencies where the radio source becomes optically thick, the *brightness tem-perature* T_B equals the *electron temperature* T_e in a single-temperature plasma, and depends on the temperature and density. The optically thick radio spectrum is there constant, $T_B(\nu) = const$, and the radio flux density increases with $I(\nu) \propto \nu^2$. At high frequencies, where the plasma becomes optically thin, the brightness temperature drops with $T_B(\nu) \propto \nu^{-2}$ (because of its proportionality to the free-free absorption co-efficient, see Eq. 15.2.1), and the radio flux density becomes constant, $I(\nu) = const$.

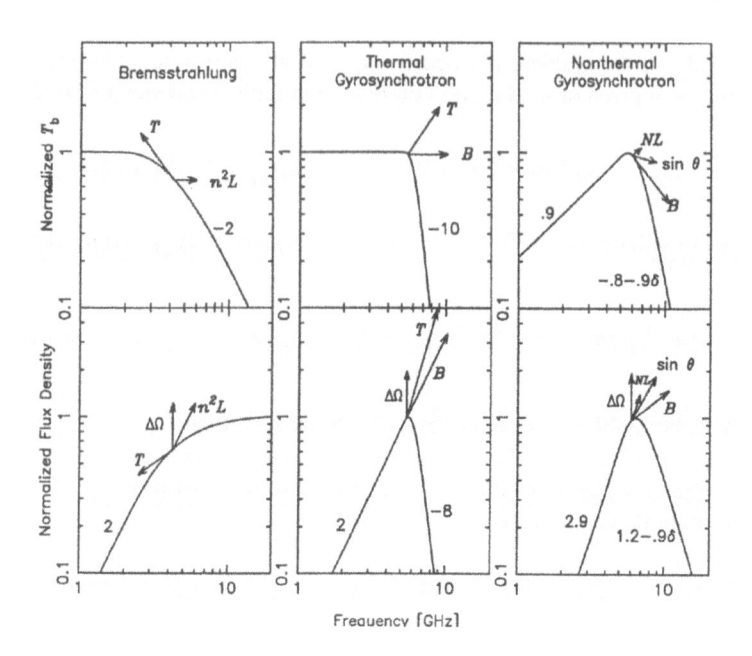

Figure 15.2: Universal frequency spectra of a homogeneous source in radio brightness temperature (top row) and radio flux density (bottom row), for three different emission mechanisms: free-free emission (left), thermal gyrosynchrotron (middle), and nonthermal gyrosynchrotron emission. The arrows represent the magnitude and directions of shift for an increase of parameters by a factor of two (Gary & Hurford 1989).

This typical spectrum of the radio brightness temperature $T_B(\nu)$ and flux intensity $I(\nu)$ is summarized in Fig. 15.2 (left frame).

An additional complication is introduced by the magnetic field B, which changes the refractive index $n_\nu(B)$ and thus the phase speed of electromagnetic waves in a magnetized plasma, leading to different brightness temperatures for the two magneto-ionic (ordinary and extraordinary) modes, which is quantified by the degree of polarization V. The theoretical expressions are given in § 5.7.1 and can be used to infer the magnetic field strength B from the observed degree of polarization in free-free emission.

15.2.2 Radio Observations of Free-Free Emission

Free-free emission in the radio wavelength is expected to be ubiquitous in the solar corona, wherever the coronal plasma is sufficiently dense (since the absorption coefficient scales with the squared density, see Eq. 15.2.1). However, because the dynamic range of radio imaging is generally limited (typically a few times 100:1), active regions dominate over the quiet Sun emission, and the presence of flare plasma outshines everything else. So the free-free emission of active regions can only be mapped in the absence of flares, and free-free emission from the quiet Sun only in the absence of active regions.

Free-free emission from solar flares has been imaged in microwave frequencies. An

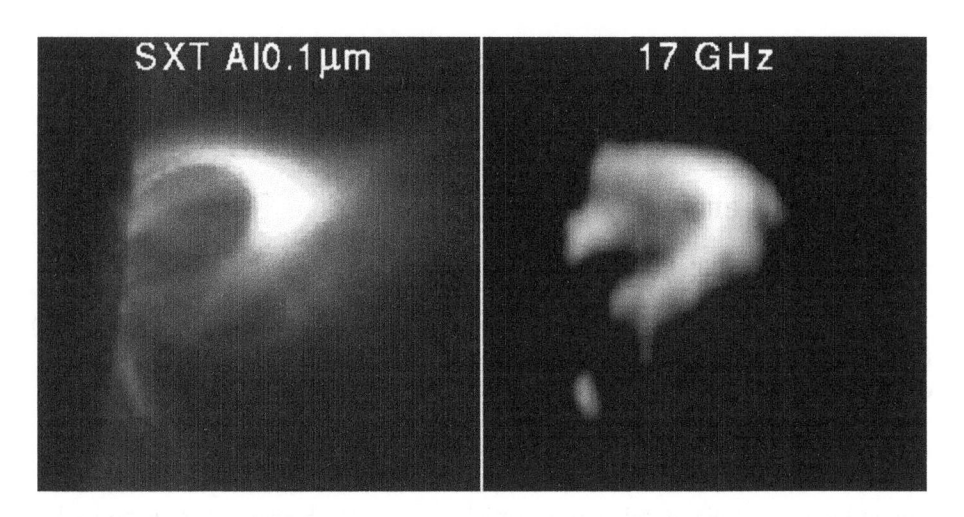

Figure 15.3: Observations of the long-duration flare event of 1993-Mar-16 with the *Nobeyama Radioheliograph* at 17 GHz and *Yohkoh/SXT*. Note the common loop-like structure, although there are little differences due to the different instrumental temperature responses (Hanaoka 1994).

example is shown in Fig. 15.3, obtained with the *Nobeyama radioheliograph* at 17 GHz during the *long-duration event (LDE)* of the 1993-Mar-16 flare (Hanaoka 1994). The close similarity with a near-simultaneous soft X-ray image obtained with *Yohkoh/SXT* clearly illustrates that the same thermal plasma produces free-free emission in both wavelengths. Of course, the match is not perfect because the response functions of the two instruments have a different temperature dependence: the SXT response steeply increases with temperature (Fig. 3.25), while the temperature weighting at a given radio frequency roughly follows with $\propto n_e^2 T^{-3/2}$. Thus, the radio flux is more sensitive to cooler plasma (at equal densities), which might explain why the flare loop footpoints are more pronounced in the radio image than in the soft X-ray image. This example illustrates that radio images are quite complementary in temperature coverage to soft X-ray and EUV images. Modeling of free-free emission using combined radio and other multi-wavelength data thus enables a quantification of the inhomogeneous plasma in flares, active regions, or the quiet Sun. Most impulsive flares show an increase of radio flux in millimeter wavelengths during the postflare phase, which is produced by optically thin free-free emission from thermal and nonthermal electrons residing in the hot flare plasma evaporated from the chromosphere.

There is abundant literature on radio observations of free-free emission, addressing various aspects such as flare temperature studies with radio and soft X-ray images (e.g., Gary et al. 1996; Silva et al. 1996, 1997a,b), motions of thermal flare plasmas (e.g., Bastian & Gary 1992), temperature studies of active regions and loops (e.g., Kundu & Velusamy 1980; Lang et al. 1983, 1987; Gary et al. 1990; Gopalswamy et al. 1991), magnetic field measurements from radio polarization (see § 5.7.1 and Table 5.1 with references therein), dominance of free-free emission in millimeter wavelengths (White & Kundu 1992), position-dependent identification of free-free emission versus gyroemission (e.g., Gary & Hurford 1987; 1994), chromospheric density models (e.g., Bastian et al. 1993a, 1996), and density structure of prominences or filaments (e.g., Kundu et al. 1986; Lang & Willson 1989; Bastian et al. 1993b).

15.3 Incoherent Gyroemission

15.3.1 Theory of Gyrosynchrotron Emission

The fact that the plasma-β parameter is less than unity in most coronal regions (Fig. 1.22) forces electrons to gyrate around the magnetic field lines. The circular gyromotion is equal to a radial acceleration of a charged particle, and thus, instead of free-free emission and absorption, we have *gyroresonance* (or *cyclotron*) emission and absorption in the case of nonrelativistic particles (with a Lorentz factor $\gamma \gtrsim 1$), *gyrosynchrotron emission* in the case of mildly relativistic particles ($1 \gtrsim \gamma \gtrsim 3$), and *synchrotron emission* in the case of highly relativistic particles ($\gamma \gg 1$). Because there is no back-reaction of the gyroemission on the particle distribution, the radiation output is proportional (linear) to the number of particles, and thus is called an *incoherent* (random-like) emission mechanism. In contrast, resonant (in particular gyroresonant) wave-particle interactions can lead to the coherent (nonlinear) growth of resonant waves in unstable particle distributions, which are called *coherent* emission mechanisms (§ 15.5).

We derived the gyroresonance absorption coefficient $\tau(\nu, s, \theta)$ for a thermal (Maxwellian) particle distribution in § 5.7.2, and calculated the corresponding radio spectra as a function of the emission angle θ in Fig. 5.26 for different harmonics ($s = 2, 3, 4$). The gyroabsorption coefficients $\tau(\nu, s, \theta)$, the brightness temperatures T_B, and peak frequencies ν_{peak} of the spectrum are summarized in the form of approximative expressions in the review article of Dulk (1985), for the cases of gyroresonance emission ($s \approx 1, 2, 3, 4$) from thermal electrons, gyrosynchrotron emission from thermal and powerlaw electrons ($s \approx 10 - 100$), and synchrotron emission from powerlaw electrons. The gyrosynchrotron emissivity is not simply proportional to the number of energetic electrons, but is also highly sensitive to the pitch angle distribution and the magnetic field. For a powerlaw electron spectrum (with slope δ and electron density n_e) the gyrosynchrotron emissivity η varies as a function of the angle θ to the magnetic field B (or electron gyrofrequency f_{ge}) approximately as (Dulk, 1985),

$$\eta(\nu, \theta, \delta) \approx 3.3 \times 10^{-24} 10^{-0.52\delta}\, B\, n_e\, (\sin\theta)^{-0.43+0.65\delta} \left(\frac{\nu}{f_{ge}}\right)^{1.22-0.90\delta},$$

$$(15.3.1)$$

and the brightness temperature is $T_B = T_{eff}(1 - e^{-\tau})$, with

$$T_{eff}(\nu, \theta, \delta) \approx 2.2 \times 10^9\, 10^{-0.31\delta}\, (\sin\theta)^{-0.36-0.06\delta} \left(\frac{\nu}{f_{ge}}\right)^{0.50+0.58\delta}, \quad (15.3.2)$$

where both approximations have an accuracy of $\approx 20\%$. So, for a typical powerlaw spectrum with a slope of $\delta = 4$ the spectral and angular dependence is $\eta(\nu, \theta) \propto B\, n_e \sin\theta^{2.2}\nu^{-2.4}$ and $T_{eff}(\nu, \theta) \propto \sin\theta^{-0.6}\nu^{0.84}$. A graphical summary of the universal spectral shapes are shown in Fig. 15.2 (middle and right frames). Note that a

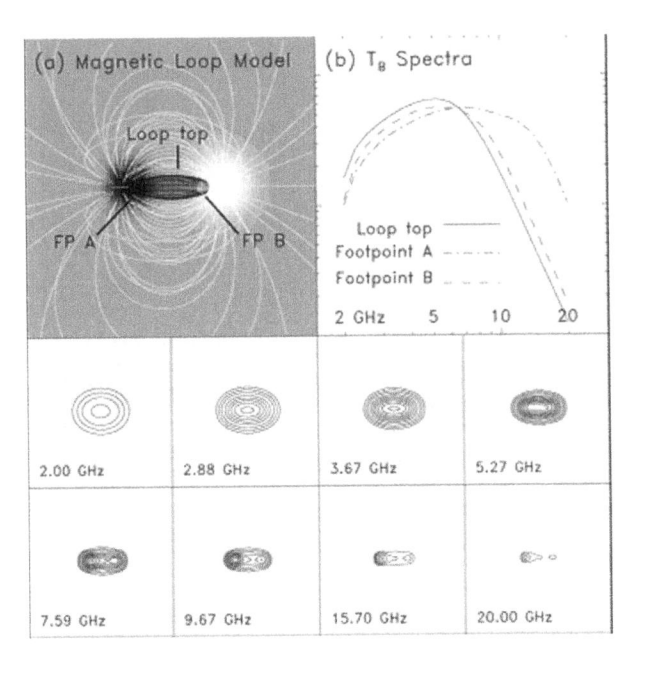

Figure 15.4: A theoretical model of gyrosynchrotron emission from a single asymmetric magnetic loop. *Top left:* the magnetic loop model. *Top right:* brightness temperature spectra at the two footpoints and the looptop. *Bottom panels:* the brightness distribution in total intensity at 8 different microwave frequencies between 2 GHz and 20 GHz (Bastian 2000).

spectral slope of $T_B(\nu) \propto \nu^{0.9}$ is indicated in the optically thick part of the nonthermal gyrosynchrotron spectrum in Fig. 15.2 (top right), which applies for an electron spectrum with $\delta \approx 4.7$ according to Eq. (15.3.2). The main difference of gyrosynchrotron flux spectra to free-free spectra is the pronounced peak (at ≈ 5 GHz in Fig. 15.2) and their steeper fall-off at higher frequencies.

At the low-frequency end of the gyrosynchrotron spectrum, a sharp cutoff can be produced by (1) self-absorption, (2) free-free absorption, or by (3) *Razin−Tsytovitch suppression* (e.g., Ginzburg & Syrovatskii 1964; Melrose 1980b, p. 100). The theory of gyrosynchrotron emission is extensively covered in a number of textbooks (Kundu 1965; Zheleznyakov 1970; Krueger 1979; McLean & Labrum 1985; Benz 1993; Melrose 1980a,b).

15.3.2 Radio Observations of Gyrosynchrotron Emission

Gyrosynchrotron emission is typically produced by nonthermal electrons with energies of $\varepsilon \approx 100$ keV − 10 MeV, and thus requires particle acceleration mechanisms like those for hard X-ray and gamma-ray producing electrons in flares. Gyrosynchrotron emission is commonly observed as a broadband microwave spectrum in a typical frequency range of $\nu \approx 2 - 20$ GHz. Below $\lesssim 1$ GHz it is self-absorbed and masked by free-free absorption from the overlying plasma. The spectrum of gyrosynchrotron

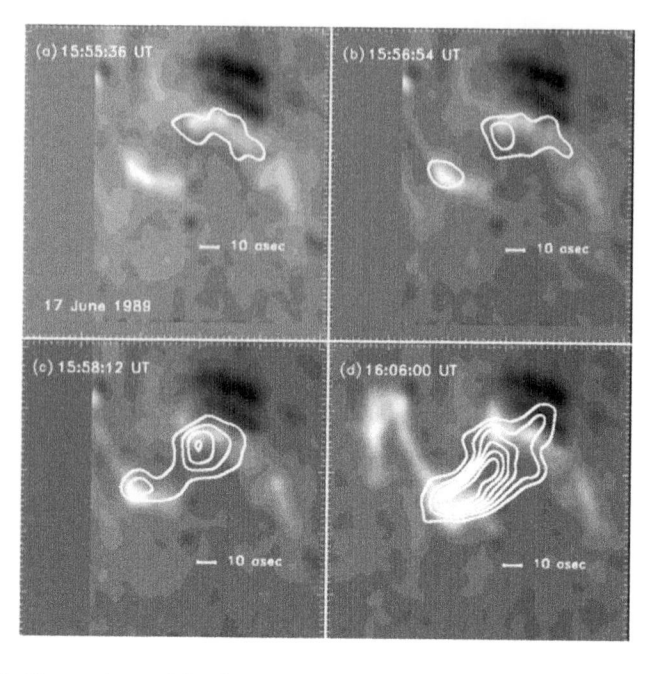

Figure 15.5: Observations of the time evolution of gyrosynchrotron emission from a double-ribbon flare arcade, during the 1989-Jun-17 flare, 15:55 UT. The contours represent the 4.9 GHz ($\lambda = 6.1$ cm) microwave emission observed with the *Very Large Array (VLA)* and the underlying greyscale shows the Hα emission (Bastian 2000).

emission peaks typically around $\nu_{peak} \approx 5 - 10$ GHz ($\lambda \approx 3 - 6$ cm), being optically thick at lower frequencies and optically thin at higher frequencies. Examples of gyrosynchrotron spectra are computed in Fig. 15.4 (top right), based on an asymmetric dipolar magnetic field model (Fig. 15.4, top left). The corresponding radio maps in the $2 - 20$ GHz range (Fig. 15.4, bottom) show that the gyrosynchrotron emission dominates at the top of the flare loop at low frequencies ($\nu = 2$ GHz), and at the footpoints for high frequencies ($\nu = 20$ GHz). This shift in the centroid position of the gyrosynchrotron source results from optical depth effects. The source maximum lies between the footpoints at low frequencies because (1) the gyrosynchrotron is optically thick there, and (2) because the magnetic field is weaker there. Higher harmonics contribute to the radiation at a fixed frequency and hence, the brightness temperature is higher than at the footpoints. At higher frequencies, though, points between the footpoints become optically thin, and they are consequently less bright than the footpoints.

An example of a microwave observation with gyrosynchrotron emission is shown in Fig. 15.5, obtained with the *VLA* at $\nu = 4.9$ GHz. The underlying Hα maps outline the double ribbons at the footpoints of the flare arcade. The evolutionary set of 4 radio maps (Fig. 15.5) shows initially gyrosynchrotron emission at the footpoints, which then bridges the double ribbons and traces out the flare loop arcade. This evolution could be interpreted in the way that directly precipitating electrons dominate gyrosynchrotron emission in the beginning, while trapped electrons dominate in the postflare phase. A

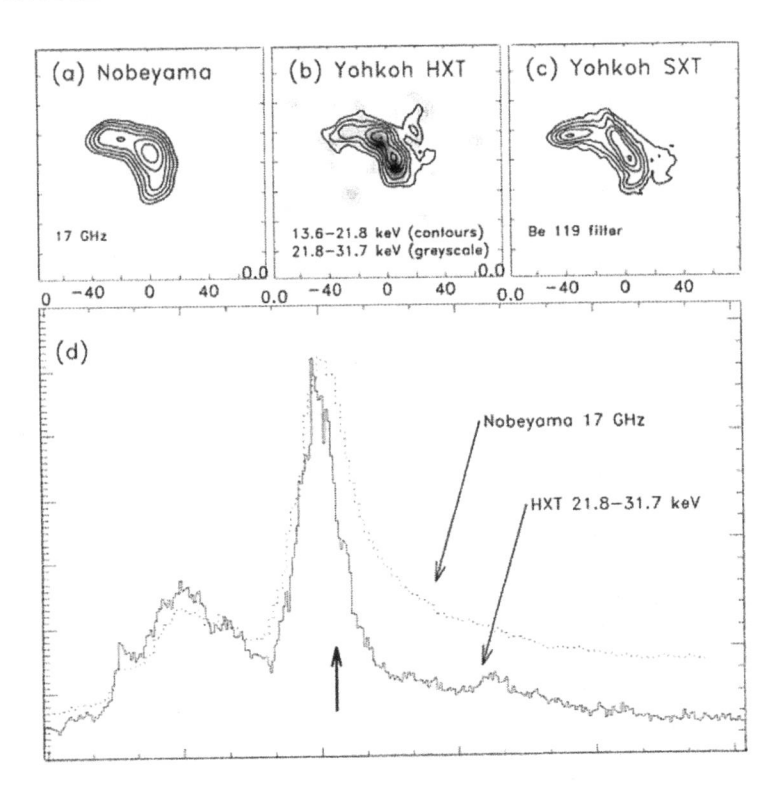

Figure 15.6: Observations of the 1992-Oct-27 flare in microwaves at 17 GHz (*Nobeyama radioheliograph*, top left), in hard X-rays at 13.6 − 21.8 keV and 21.8 − 31.7 keV (*Yohkoh/HXT*, top middle), and in soft X-rays (*Yohkoh/SXT*, top right). Note the close similarity of hard X-ray and microwave time profiles (bottom panel: 01:44:30-01:47:00 UT, with 10-min tickmarks), both being produced by the same population of nonthermal electrons (Bastian 2000).

physical understanding of this evolution requires dynamic modeling of the evolution of injected and trapped energetic particle distributions (see § 12 on particle kinematics) and the resulting optical depth effects.

Since both hard X-ray bremsstrahlung und gyrosynchrotron emission are produced by high-energetic electrons, there is often a close co-evolution between the two emissions. An example of a hard X-ray light curve (*HXT* 21.8 − 31.7 keV) and a microwave time profile (*Nobeyama radioheliograph* 17 GHz) is shown in Fig. 15.6, observed during the 1992-Oct-27, 01:44 UT, flare. There is a detailed correlation between the two time profiles, but the peak of the microwave emission seems to lag behind the peak of the hard X-ray emission by a few seconds, which has some important ramifications for the underlying particle dynamics. As we discussed in § 13, hard X-ray bremsstrahlung in flare time profiles consists of (1) intermittent pulses that are driven by directly precipitating electrons and (2) exponential tails that are driven by trap-plus-precipitating electrons with typical delays of a few seconds, making up most of the lower envelope or smoothly varying component of the hard X-rays. The trapped electrons generally

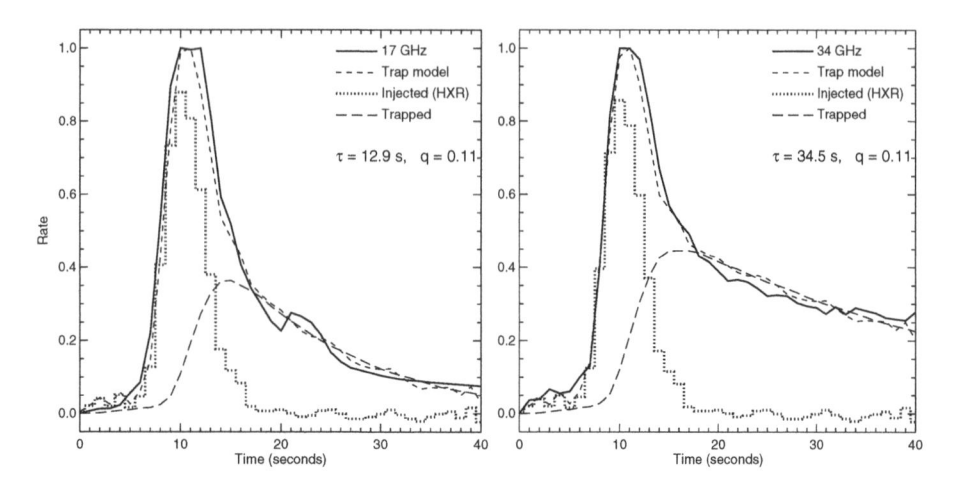

Figure 15.7: Comparison of the radio time profiles for the event on 1998-Jun-13 for 17 GHz (left) and 34 GHz (right) with a trap model (short-dashed line) derived using the *Yohkoh/HXT* 53 − 93 keV hard X-ray time profile (dotted histogram) as an injection function. The radio time profile is modeled as the sum of a component identical to the hard X-ray time profile (the injection function) and a trapped component (long-dashed line) derived by integrating over the injection function convolved with an exponential decay term (time constant τ). All time profiles are normalized to a peak of unity, and the parameter q specifies the relative contributions of the injected and trapped components (Kundu et al. 2001).

produce less hard X-ray bremsstrahlung than the precipitating electrons, because of the lower density in the trap. Gyrosynchrotron emission, however, is emitted from all energetic electrons, regardless whether they precipitate or are trapped. Therefore, there is a bias that gyrosynchrotron emission is dominated by trapped electrons, because they accumulate and spend a longer time in the trap than on the precipitation path, in opposition to the hard X-ray bremsstrahlung yield, which is dominated by the precipitating electrons that hit the dense chromosphere. This explains the general delay of the microwave peak with respect to the hard X-ray peak in flare time profiles. The difference amounts only to a few seconds, based on the typical trapping times that have been measured from hard X-rays (§ 12.5.3). The two time profiles shown in Fig. 15.6 also reveal a longer exponential decay time for microwaves than for hard X-rays, which again is consistent with the larger relative fraction of trapped electrons that contribute to the observed total emission. When the trap is asymmetric (§ 12.6.2), the situation is somewhat more complicated, but a statistical rule is that the asymmetric gyrosynchrotron emission is complementary to the hard X-ray footpoint fluxes, because the footpoint with the stronger magnetic field has a higher mirror ratio and is more efficient for trapping (which increases the gyrosynchrotron emission and reduces the hard X-ray emission).

A flare with a simple spiky time structure that has been modeled in terms of a trap-plus-precipitation model is shown in Fig. 15.7, observed with *Yohkoh/HXT* and the *Nobeyama* radio telescope at 17 and 34 GHz (Kundu et al. 2001). This simple time

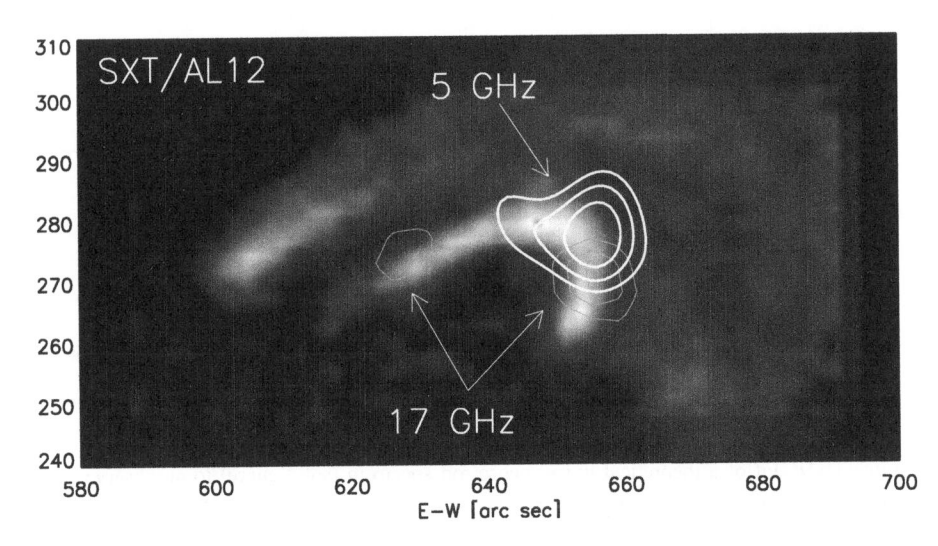

Figure 15.8: Microwave data during the 1993-Jun-3 flare, showing the radio intensity peaks (contours) on top of a soft X-ray image from a filtered *Yohkoh SXT*/Al12 at 23:39 UT. Contours are 80% to 99% of the maximum intensities: 1.8×10^7 K at 5 GHz and 1.2×10^5 K at 17 GHz, respectively. The 5 GHz (looptop) source is produced by gyrosynchrotron emission, while the 17 GHz (footpoint) sources could be a combination of gyrosynchrotron and free-free emission (Lee & Gary 2000).

profile clearly shows the two components of an impulsive peak and a gradual decay, which can be naturally modeled with the simple trap model outlined in § 12.5.2. The fitting of a trap model reveals a trapping time of $\tau = 12.9$ s at 17 GHz and $\tau = 34.5$ s at 34 GHz. These time scales are compatible with collisional deflection time scales t_{trap} of $E \approx 100 - 250$ keV electrons in densities of $n_e \approx 10^{10}$ cm^{-3} (Eq. 12.5.11), and thus may explain the slowly decaying tail of the emission in terms of trapping of gyrosynchrotron emitting electrons. The difference in trapping times between the two frequencies of 17 and 34 GHz can be understood qualitatively: The higher frequency emission results from higher energy electrons, which have a smaller collision frequency than those responsible for the 17 GHz emission. In the same study, the relative ratio of the radio flux produced by direct-precipitating and trap-precipitating electrons was also determined, yielding a ratio of $q = 0.11$. The gyrosynchrotron emissivity also depends on the observing angle, the pitch angle distribution, and the magnetic field, and the radio flux ratio is thus additionally weighted by the gyrosynchrotron emissivity function $\eta(\theta, \delta)$. Generally there are different observing angles in the trapping region near the looptop (θ_T) and in the precipitation sites near the footpoints (θ_P). Thus the time evolution of gyrosynchrotron emission $S(t)$ could be described as (Kundu et al. 2001),

$$S(t) = q \int d\varepsilon \, \eta(\varepsilon, \theta_I) f_I(\varepsilon, t) \frac{L}{2\mathrm{v}} + (1 - q) \int d\varepsilon \, \eta(\varepsilon, \theta_T) n_{trap}[\varepsilon, t_{trap}(\varepsilon), t] \,,$$

$$(15.3.3)$$

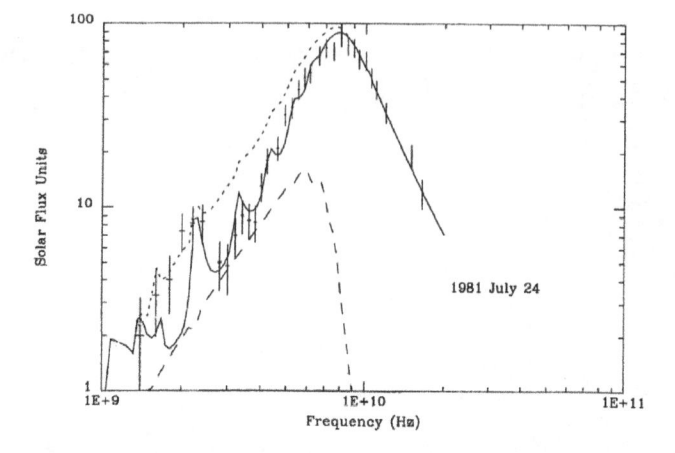

Figure 15.9: Fit of a theoretical gyrosynchrotron spectrum (solid curve) to an observed microwave spectrum (crosses) during the 1981-Jul-24 flare (Stähli et al. 1989). The derived parameters are: $T = 76$ MK, $n_e = 6 \times 10^9$ cm^{-3}, $\delta = 4.5$, $B = 340$ G, $A = 5.0 \times 10^{17}$ cm^{-2}, and $\theta = 57°$. Note the spectral substructures due to low gyroharmonics at the optically thick side of the spectrum (Benka & Holman 1992).

where $\eta(\varepsilon, \theta_I)$ represents the gyroemissivity of the injected electron distribution $f_I(\varepsilon, t)$ (which can be modeled using the hard X-ray time profile as a proxi, see Fig. 15.7), $\eta(\varepsilon, \theta_T)$ is the gyroemissivity of the trapped electron distribution $n_{trap}[\varepsilon, t_{trap}(\varepsilon), t]$, and q is the fraction of directly precipitating electrons. The time evolution of the trapped electron distribution can be described by a convolution with an e-folding (energy-dependent) trapping time $t_{trap}(\varepsilon)$,

$$n_{trap}(\varepsilon, t) = \frac{1}{t_{trap}(\varepsilon)} \int_0^t f_I(t', \alpha > \alpha_0) \, \exp[-\frac{(t - t')}{t_{trap}(\varepsilon)}] \, dt' \, . \qquad (15.3.4)$$

as we used it to quantify the evolution of trapped hard X-ray-emitting electrons (Eq. 12.5.9). Another flare that has been modeled with such a trap-plus-precipitation model is the 1993-Jun-3, 23:22 UT, flare (Fig. 15.8), where looptop emission from trapped electrons was observed at $\nu = 5$ GHz and footpoint emission near the precipitation sites at $\nu = 17$ GHz (Lee & Gary 2000).

There is a large body of related radio observations, but very few studies exist with quantitative modeling of gyrosynchrotron emission. Modeling efforts consider 3D models of gyrosynchrotron-emitting flare loops (e.g., Kucera et al. 1993), the complementarity of gyrosynchrotron emission and hard X-ray emission in asymmetric (dipolar) flare loops (Wang et al. 1995), or in quadrupolar flare loop configurations (Hanaoka 1997; Kundu et al. 2001; Lee et al. 2003), the pitch angle distribution function of injected electrons (Lee & Gary 2000; Lee et al. 2000; 2002), and combined modeling of gyrosynchrotron and hard X-ray emission (Silva et al. 1996, 2000; Wang et al. 1994, 1995, 1996; Benka & Holman 1992). An example of a detailed spectral model fit of gyrosynchrotron emission from a thermal plus nonthermal particle distribution is shown in Fig. 15.9. However, since source inhomogeneities are not expected to pro-

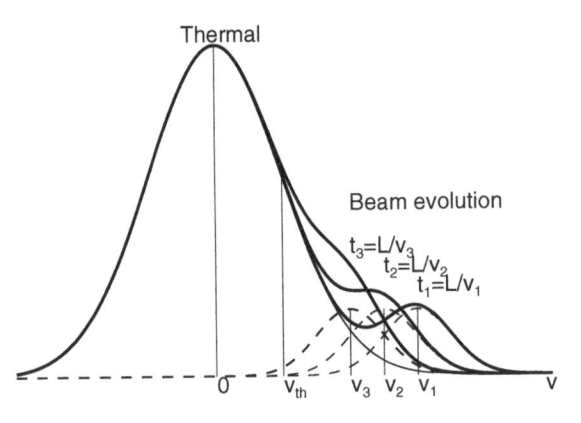

Figure 15.10: The evolution of a beam in the tail of a thermal distribution is shown, starting with the arrival of the fastest electrons at time $t_1 = L/v_1$, producing a positive slope $\partial f/\partial v > 0$ and which is unstable. At later times, slower electrons arrive at $t_2 = L/v_2$ and $t_3 = L/v_3$, but the slowest ones do not produce a positive slope and are stable (adapted from Lin et al. 1981b).

duce narrowband spectral structures for incoherent gyrosynchrotron emission, it is not clear whether the observed spectral peak at $\nu \gtrsim 2$ GHz could be due to an instrumental artifact (Bastian et al. 1998).

Historically, radio bursts produced by gyrosynchrotron emission have been referred to as *type IV bursts*, *microwave type IV bursts*, or *flare continuum*. A rarer subgroup was called *moving type IV bursts*, based on the outward motion of the source, believed to be an erupting plasmoid with trapped electrons that emit either plasma or gyrosynchrotron emission. Earlier observations based on this nomenclature are summarized in Kundu (1965), Zheleznyakov (1970), Krueger (1979), McLean & Labrum (1985), and Dulk (1985).

15.4 Plasma Emission

15.4.1 Electron Beams

Since particle acceleration in solar flares occurs in a quasi-collisionless plasma, supposedly near magnetic reconnection sites in the corona, the energized (nonthermal) electrons can propagate along the magnetic field lines, either in an upward direction into interplanetary space (in case of open field lines) or to remote footpoints (in the case of closed field lines), or in a downward direction towards the chromosphere. Electron propagation in a collisionless plasma obeys adiabatic motion and thus the velocity dispersion [or the time-of-flight difference $\Delta t = \Delta s(1/v_1 - 1/v_2)$] allows the higher energy electrons to race ahead of the lower energy electrons, which creates a so-called *bump* or *beam* in the forward direction of the particle distribution function (Fig. 15.10). Such transient beams are unstable to the *bump-in-tail instability* in the 1D velocity distribution function, $\partial f/\partial v_\parallel > 0$. *Landau resonance* with the unstable electron beam generates *Langmuir waves*, which are believed to undergo nonlinear wave-wave inter-

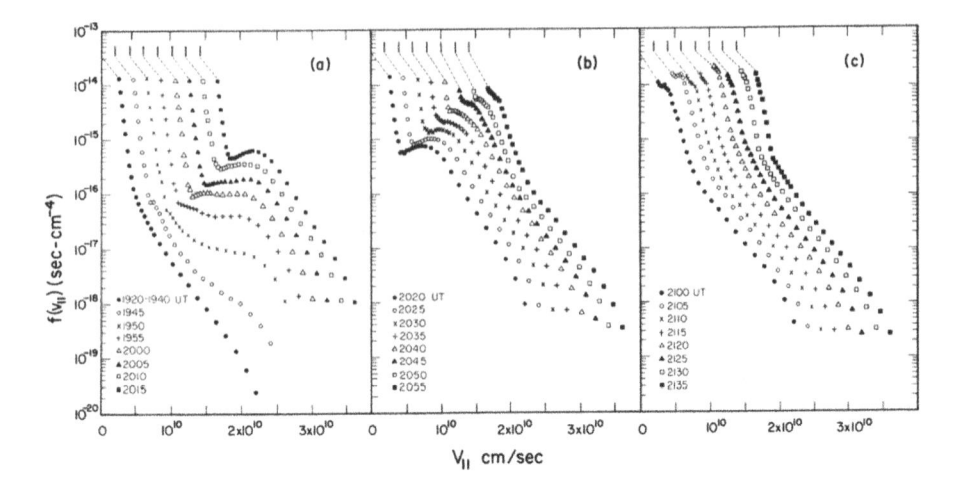

Figure 15.11: *In situ* measurements of the electron beam distribution in interplanetary space by ISEE-3 on 1979-Feb-17. The sequence of 24 distribution functions is taken every 64 s and is displayed incrementally shifted to the right by amounts of $\Delta v = 2 \times 10^9$ cm s^{-1}. Note that a positive slope occurs mainly between 20:00 and 20:30 UT (Lin et al. 1981b).

actions that produce electromagnetic emissions at the local *electron plasma frequency* (f_{pe}) and its second harmonic ($2f_{pe}$), generally called *type III radio bursts*.

Although beams are the primary drivers of the manyfold type III-like radio emissions observed in the solar corona, the kinetic evolution of electron and ion beam velocity distributions could only be studied by interplanetary *in situ* measurements, such as with the *ISEE-3*, *Ulysses*, or *WIND* spacecraft, or with numerical simulations of time-dependent particle distributions $f(\mathbf{x}, \mathbf{v}, t)$ in terms of the *Fokker–Planck equation*. If a (Gaussian) electron beam distribution $f_b(v_\parallel) \propto \exp[-(v_\parallel - v_b)^2/2\Delta v^2)]$ is superimposed on a thermal distribution function, $f_{th}(v) \propto \exp[-(v^2/2v_{th}^2)]$, a positive gradient $\partial f/\partial v > 0$ occurs only if the beam is sufficiently strong and has a minimum beam speed of $v_b \gtrsim 3v_{th}$ (Fig. 15.10). A bump is first produced at a distance L from the source at the time of arrival of the fastest electrons, while slower electrons arrive successively later after a time-of-flight $t = L/v$, so that the bump in the tail proceeds to lower velocities (Fig. 15.10). Because the positive slope becomes progressively flatter towards lower velocities (due to the superposition on the systematically steeper thermal distribution at lower velocities), the instability criterion $\partial f/\partial v > 0$ is not satisfied anymore at lower velocities $v \lesssim 3v_{th}$, and thus the bump-in-tail instability is quenched. Such an evolution of the electron velocity distribution has indeed been measured in interplanetary space (e.g., with ISEE-3 during the 1979-Feb-17 event: Fig. 15.11; Lin et al. 1986). For coronal and interplanetary type III bursts it is measured that the speed of typical beams is mildly relativistic, $v_b \approx 0.1 - 0.3$, and the beam electron number is only a fraction of $n_b/n_{th} \approx 10^{-9} - 10^{-6}$ of the ambient thermal electrons. Since such electron beam distributions with a positive slope are highly unstable towards Landau resonance, they generate Langmuir waves by wave-particle interactions almost as fast

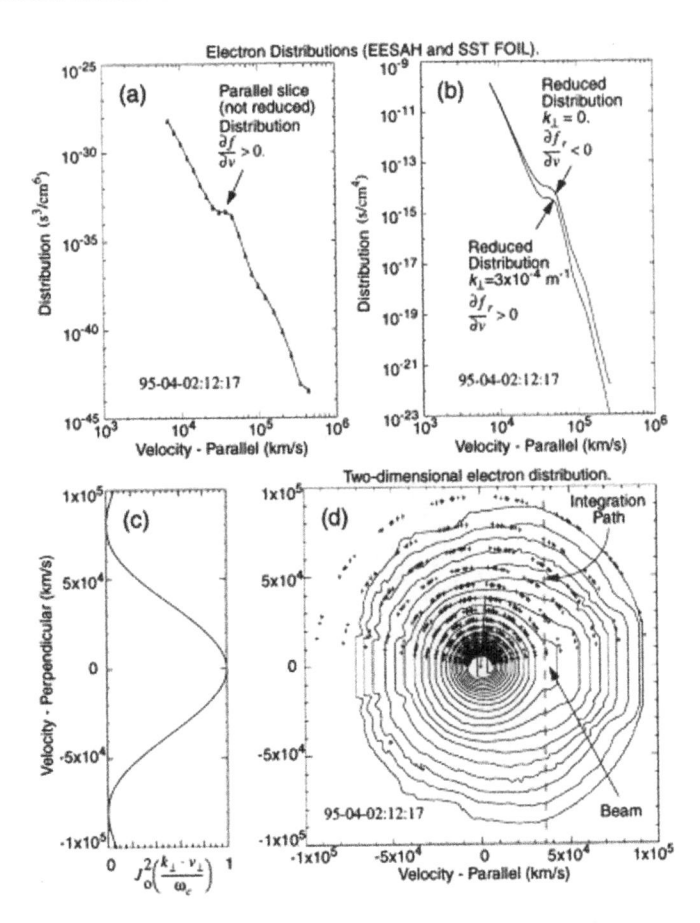

Figure 15.12: Electron velocity distribution functions that produce solar type III bursts measured with the *WIND* spacecraft on 1995 April 2, 12:17 UT. (a) parallel slice $f(v_\parallel, v_\perp = 0)$ (not reduced) with small positive slope $\partial f(v)/\partial v > 0$; (b) reduced distributions at $k_\perp = 0$ and $k_\perp = 3 \times 10^{-4}$ m^{-1}; (c) perpendicular distribution $f(v_\perp)$; and (d) 2D velocity distribution $f(v_\parallel, v_\perp)$. Note the asymmetry of the distribution with a plateaued beam component in the forward direction of $v_\parallel$ (Ergun et al. 1998).

as they form, so unstable beams are rarely observed directly, while the relaxed plateaus are more frequently seen (Figs. 15.11 and 15.12), which are in a state of *marginal stability*. The formation of beams and their apparent stability over large propagation distances (since they are observed at distances of 1 AU) therefore could not initially be understood (*Sturrock's dilemma*, Sturrock 1964), but later work includes a number of effects (induced scatter of ions, plasma inhomogeneities, large-scale fluctuations due to ion-acoustic waves, resonant back-reaction of strong turbulence waves, stochastic growth theory, return currents, etc.) to generalize the linear theory based on quasilinear diffusion (Eq. 11.4.3) with weak turbulence (e.g., Grognard 1985).

Observations of interplanetary type III bursts allowed us to test the numerical sim-

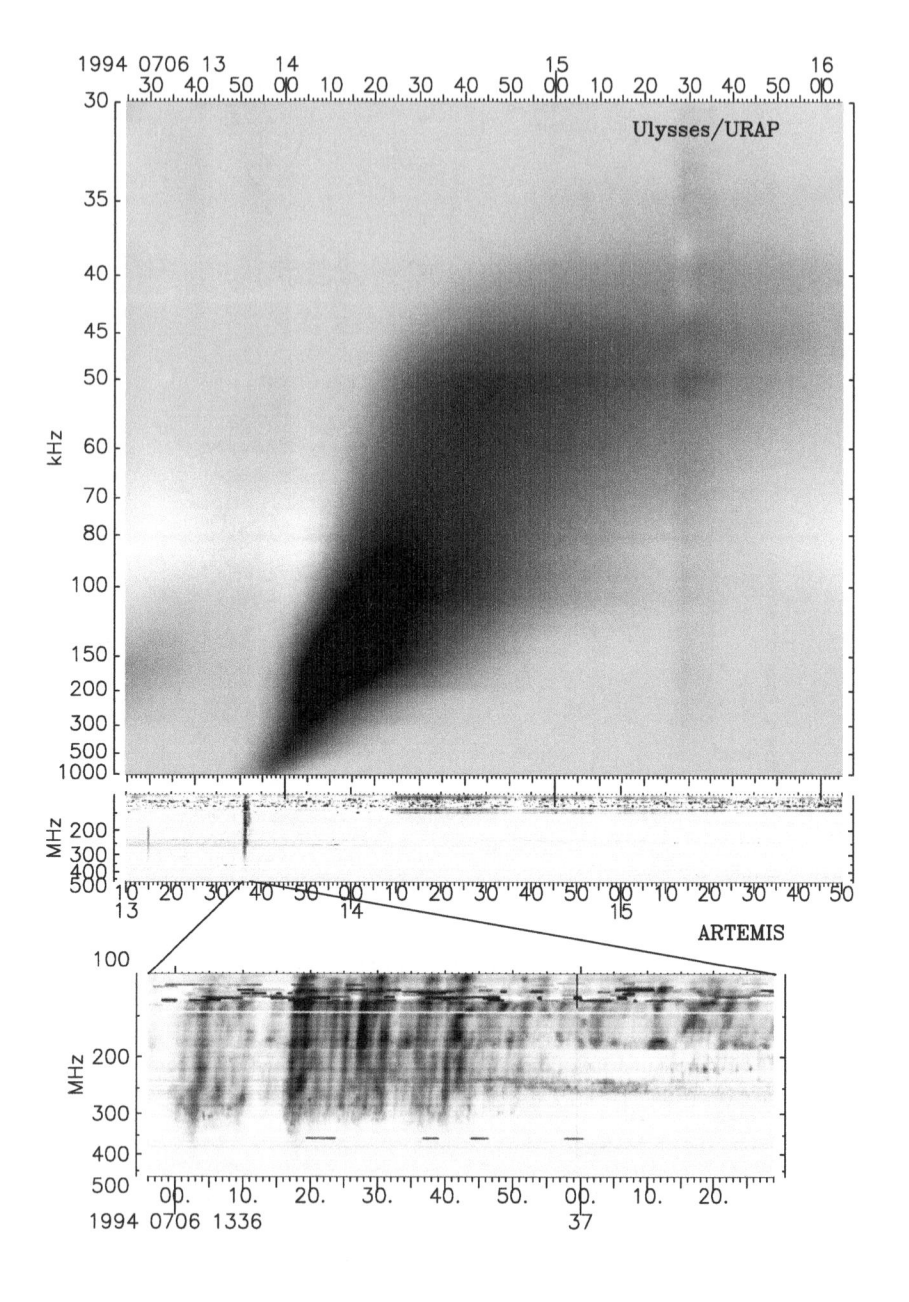

Figure 15.13: *Artemis* and *Ulysses* observations of a group of solar metric type III bursts and an interplanetary type III radio burst. The latter is represented with frequency decreasing, as is common for solar type III bursts. The scale is chosen in such a way that its prolongation to higher frequencies gives the starting point of the interplanetary burst in the corona. It is seen that this point coincides with the low-frequency starting point of the solar type III group. The lowest panel shows the high time resolution dynamic spectrum of the type III group identifying it as a large group consisting of many individual type III bursts (Poquérusse & McIntosh 1995).

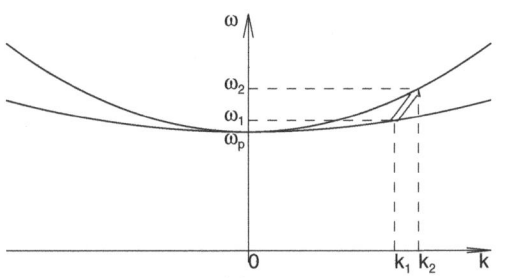

Figure 15.14: The dispersion relations $\omega(k)$ for two Langmuir waves associated with particles of slightly different thermal velocities v_{th} are shown. An ion sound wave (S) with a frequency $\omega_S = \omega_2 - \omega_1$ and wave vector $k_s = k_2 - k_1$ can couple with the first Langmuir wave L (ω_1, k_1) to produce a second Langmuir wave L' (ω_2, k_2), and in this way transfer energy via the process $L + S \mapsto L'$.

ulations by measurements of *in situ* particle distribution functions, but also revealed inhomogeneous clumps (with sizes $\approx$ 400 km), intermittent Langmuir waves, and the association of ion-acoustic waves with Langmuir bursts (Lin et al. 1981b, 1986; Gurnett & Anderson 1977). As the electron beam travels upward in the corona and away from the Sun into interplanetary space, it encounters a decreasing plasma density, and thus the radio frequency of the observed type III bursts drifts to lower frequencies with time (Fig. 15.13), so a general characteristic is the negative *frequency-time drift rate* $\partial \nu / \partial t < 0$.

Detailed studies on the formation, evolution, and propagation of electron beams can be found in Ryutov & Sagdeev (1970), Takakura (1979), Grognard (1985), Spicer & Sudan (1984), Goldman (1989), LaRosa (1988), Hillaris et al. (1990a,b), McTiernan & Petrosian (1990), Van den Oord (1990), Ledenev (1994, 1998), Robinson (1996), Syniavskii & Zharkova (1994), Zharkova & Syniavskii (1997), in textbooks (Kaplan & Tsytovitch 1973; Melrose 1980b, § 10−11; McLean & Labrum 1985, § 8, 11; Benz 1993, § 5−7; Sturrock 1994, § 8−9) and reviews (Muschietti 1990; Melrose 1990; Pick & Van den Oord 1990).

15.4.2 Langmuir Waves

Plasma emission is a multi-stage process, which includes, e.g., (1) formation of (unstable) beam distributions by velocity dispersion, (2) generation of Langmuir turbulence, and (3) its nonlinear evolution and conversion into escaping (electromagnetic) radiation (*plasma emission*). We described the dispersion relation for electrostatic and electromagnetic waves in § 11.4.1, summarized also in Fig. 11.11 and Table 11.2. The dispersion relation for *Langmuir waves* is

$$\omega^2(k) = \omega_p^2 + \frac{3}{2}k^2 v_{th}^2 = \omega_p^2(1 + 3k^2\lambda_D^2)\,, \qquad (15.4.1)$$

where we used the definition of the *Debey length*, $\lambda_D = v_{th}/\omega_p$. Electrons with velocities v_{th} can undergo (gyroresonant) wave-particle interactions with waves $\omega(k)$ when they fulfill the *Doppler resonance condition* $\omega - s\Omega/\gamma - k_{\parallel}v_{\parallel} = 0$ (Eq. 11.4.6), and the *wave growth rate* $\Gamma[\mathbf{k}, f(\mathbf{v})]$ (or *absorption rate* if $\Gamma < 0$) can be calculated for a given particle velocity distribution function $f(\mathbf{v})$ with the quasi-linear equations (Eqs. 11.4.1−2). For any resistive instability involving *Langmuir waves* and nonrelativistic electrons, the growth rate $\Gamma(\mathbf{k})$ or absorption coefficient is given by (e.g., Benz 1993),

$$\Gamma(\mathbf{k}) = \left(\frac{\pi}{2}\right) \frac{\omega_p^2}{k^2} \frac{\omega}{n_0} \left(\frac{\partial f_0(\mathbf{v}_\parallel)}{\partial \mathbf{v}_\parallel}\right)_{\omega/k} \approx \left(\frac{\pi}{2}\right) \frac{n_b}{n_0} \left(\frac{\mathbf{v}_{ph}}{\Delta \mathbf{v}}\right)^2 \omega , \qquad (15.4.2)$$

where $\omega_p = 2\pi f_{pe}$ is the plasma frequency, n_0 the ambient electron density, n_b the beam electron density, and $\mathbf{v}_{ph} = \omega/k$ the phase speed. So, a positive slope in the velocity distribution generates Langmuir waves, which can interact with other waves by a number of processes. For every wave-wave interaction, the energy and momentum equations have to be fulfilled, that is, the matching conditions of frequencies and wave vectors,

$$\omega_1 + \omega_2 = \omega_3 , \qquad (15.4.3)$$

$$\mathbf{k}_1 + \mathbf{k}_2 = \mathbf{k}_3 . \qquad (15.4.4)$$

For instance, a primary Langmuir wave $(\omega_1, \mathbf{k}_1)$ can couple with an ion acoustic wave $(\omega_2, \mathbf{k}_2)$ to generate a secondary Langmuir wave $(\omega_3, \mathbf{k}_3)$ (Fig. 15.14). Because the phase speed of ion acoustic (sound) waves is much smaller than for Langmuir waves, the primary and secondary Langmuir waves have similar frequencies and wave vectors. For plasma emission, we need ultimately a conversion into an escaping (electromagnetic) wave. Among Langmuir waves (L), ion acoustic waves (S), and (electromagnetic) transverse waves (T), the following three-wave interactions have been considered for plasma emission (e.g., Melrose 1987),

$$L + S \mapsto L' \qquad (15.4.5)$$

$$L + S \mapsto T \qquad (15.4.6)$$

$$T + S \mapsto L \qquad (15.4.7)$$

$$T + S \mapsto T \qquad (15.4.8)$$

$$L + L' \mapsto T \qquad (15.4.9)$$

where the first process is important to generate Langmuir turbulence, while the second and third are of interest for fundamental plasma emission, the forth for scattering of transverse waves, and the last for second harmonic plasma emission. Harmonic plasma emission has been explained in terms of *two-stream instabilities*. However, even if Langmuir waves are converted into electromagnetic waves, they could be absorbed in the solar corona by free-free absorption and do not reach the observer. Only when the density scale height along the line-of-sight is sufficiently small so that the free-free opacity does not exceed unity, plasma radiation from electron beams can be observed.

The theory of plasma emission for solar type III bursts was first pioneered by Ginzburg & Zheleznyakov (1958), Zheleznyakov & Zaitsev (1970a,b), and is extensively described in the textbooks of Zheleznyakov (1970), Krueger (1979), Melrose (1980b), McLean & Labrum (1985), Benz (1993) and reviews by Takakura (1967), Smith (1974), Goldman & Smith (1986), Dulk (1985), Melrose (1987), and Robinson (1997).

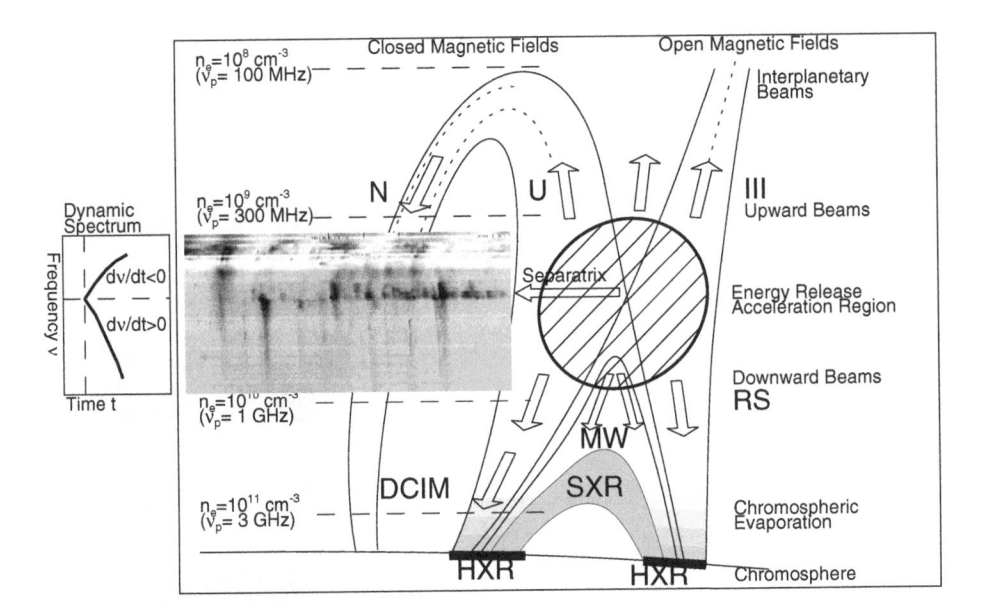

Figure 15.15: Radio burst types in the framework of the standard flare scenario: The acceleration region is located in the reconnection region above the soft X-ray-bright flare loop, accelerating electron beams in the upward direction (type III, U, N bursts) and in the downward direction
(type RS, DCIM bursts). Downward moving electron beams precipitate to the chromosphere
(producing hard X-ray emission and driving chromospheric evaporation), or remain transiently
trapped, producing microwave (MW) emission. Soft X-ray loops become subsequently filled up,
with increasing footpoint separation as the X-point rises. The insert shows a dynamic radio spectrum (*ETH* Zurich) of the 92-Sept-06, 1154 UT, flare, showing a separatrix between type III and
type RS bursts at ≈ 600 MHz, probably associated with the acceleration region (Aschwanden
1998b).

15.4.3 Observations of Plasma Emission

Metric Type III, J, U, and RS Bursts

From the previously described theory we expect that propagating electron beams generate plasma emission, which is a coherent emission mechanism, reaching much higher
radio brightness temperatures T_B than incoherent emission mechanisms (such as free-
free emission, § 15.2, or gyrosynchrotron emission, § 15.3). Thus, the theory makes a
number of specific predictions that allows us to distinguish beam-driven plasma emission from other radio emission: (1) the propagating electron beams have mildly relativistic speeds which implies specific frequency-time drift rates; (2) the plasma frequency range is a strict function of the ambient electron density (yielding densities
of $n_e \approx 10^8 - 10^{10}$ cm^{-3} for plasma emission in the $\nu_p = 100 - 1000$ MHz range);
(3) plasma emission produces higher brightness temperatures than incoherent emission,
and (4) since the generation of relativistic electrons requires an acceleration mechanism
as is available in flares and CME shocks, we expect a high correlation with the occur-

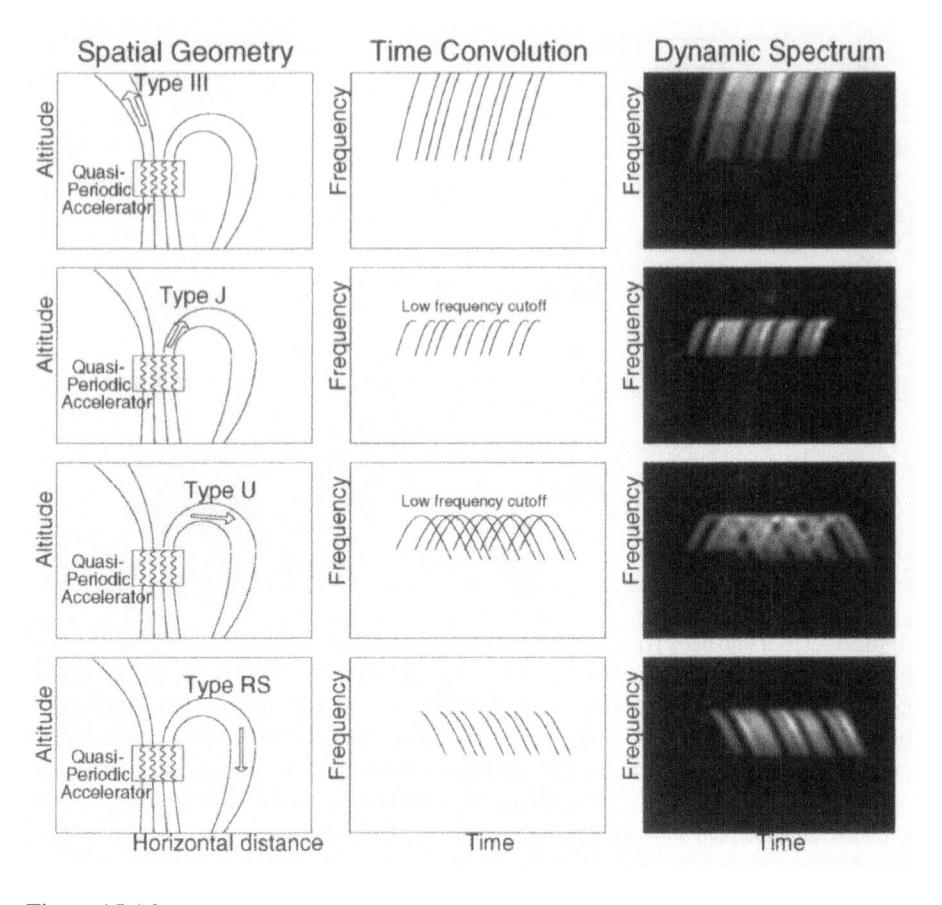

Figure 15.16: Spatial geometry and schematic dynamic spectrum of quasi-periodically injected electron beams, for the case of injection into open (top row) or closed (second to forth row) magnetic field lines. Note the typical low-frequency cutoff and curvature in the drift rate for electron beams propagating along closed field lines, for the case of J, U, or RS bursts. The flux and quasi-periodicity is randomly modulated to simulate a realistic representation of a dynamic spectrum (right) (Aschwanden et al. 1994a).

rence of other beam-driven emissions, such as with hard X-ray pulses from electron beams that precipitate to the chromosphere. Let us consider some of these theoretical predictions in some more detail and relate them to the relevant observed radio signatures (Fig. 15.15), which historically have been classified as *type III*, *type J*, *type U*, or *type RS* bursts (Wild et al. 1963).

Since the majority of solar radio observations have been conducted with radio spectrometers, which record the dynamic spectra of the radio flux $S(\nu, t)$ as a function of radio frequency ν and time t, an important quantity to characterize the observed radio bursts is the frequency-time drift rate $d\nu/dt$. For plasma emission, the observed radio frequency ν is close to the fundamental plasma frequency or its harmonic (i.e.,

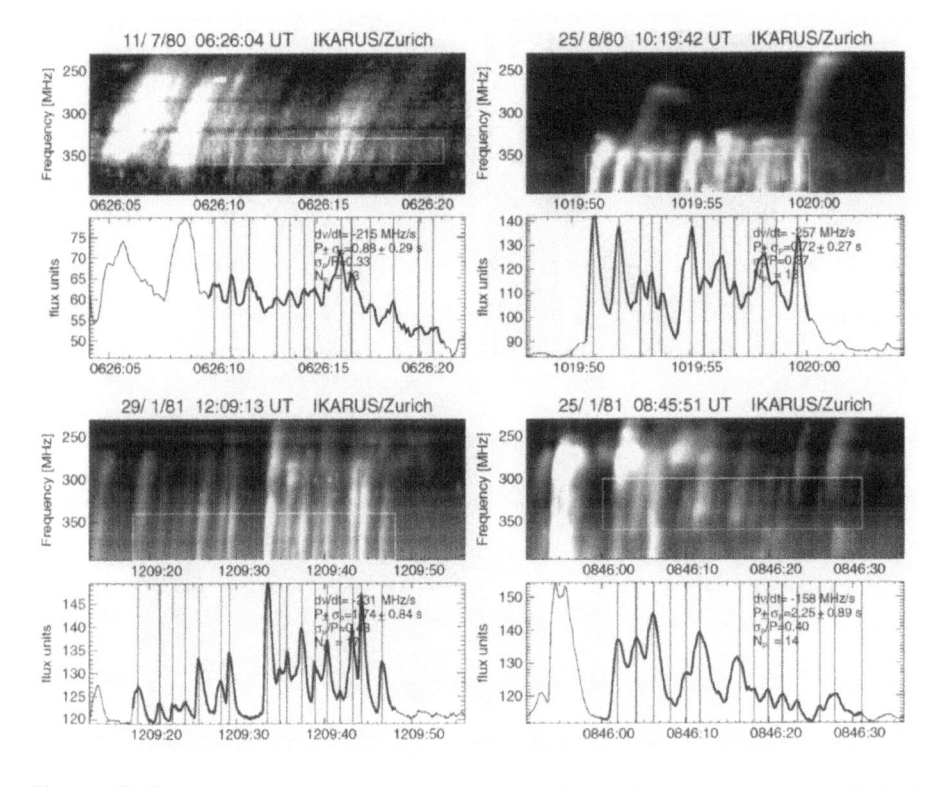

Figure 15.17: Dynamic spectra (panels with black background) and time series (profiles) of four events with sequences of type III or J bursts, recorded by *ETH* Zurich. The times of the type III peak fluxes are marked with thin vertical lines. Statistics on the intervening time intervals have been performed and show often quasi-periodic rather than random time sequences (Aschwanden et al. 1994a).

$\nu \approx s\nu_p$), which depends only on the ambient electron density (i.e., $\nu_p(n_e) \propto n_e^{1/2}$, Eq. 5.7.9). Since electron beams propagate along magnetic field lines, the density changes as a function of the position $n_e(s)$, which could be modeled by a barometric density model as function of height $n_e(h) \propto exp(-h/\lambda_T)$ in the simplest case, where λ_T represents the exponential density scale height (§ 3). The positional coordinate s depends on the time t according to the beam velocity v_B (i.e., $s(t) = \int v(s)dt \approx v_B dt$). Combining these three assumptions, we find the following simple expression for the frequency-drift rate,

$$\left(\frac{\partial \nu}{\partial t}\right) = \left(\frac{\partial \nu(n_e)}{\partial n_e}\right) \left(\frac{\partial n_e(h)}{\partial h}\right) \left(\frac{\partial h(s)}{\partial s}\right) \left(\frac{\partial s(t)}{\partial t}\right) \approx -\frac{\nu \, v_B \cos(\theta)}{2\lambda_T} .$$
(15.4.10)

Thus the frequency drift rate is proportional to the frequency ν, the beam speed v_B, and reciprocal to the density scale height λ_T. Note that the sign of the frequency drift rate is negative for upward propagation (towards lower densities), which is called

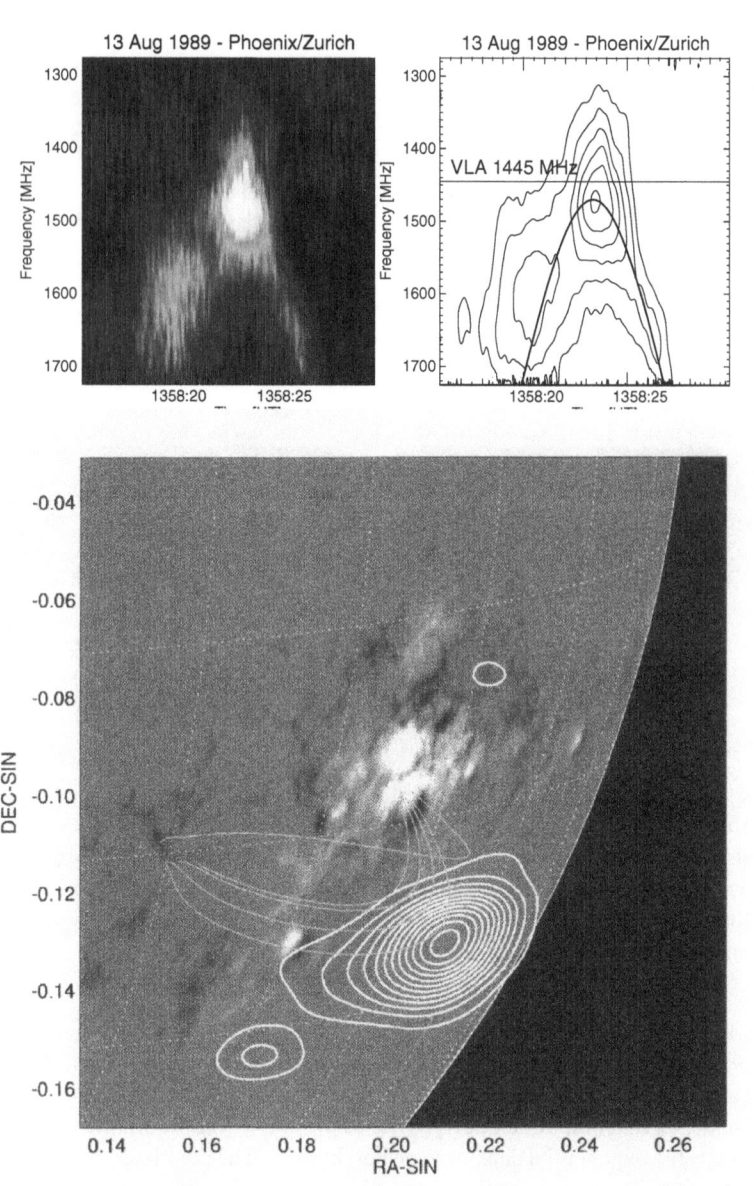

Figure 15.18: *Top:* radio observations of a type U-burst on 1989-Aug-13 with the *ETH* Zurich spectrometer. The ascending branch is broader than the descending branch. The loop transit time is about 6 s at 1.7 GHz for this U-burst. The same type U-burst has been imaged with the *VLA* at a frequency of 1446 MHz. *Bottom:* photospheric magnetic field in AR 5629 from KPNO/NSO on 1989-Aug-13, 1500 UT, about 1 hour after the type U-bursts. The contours show an overlaid *VLA* map recorded at a frequency of 1446 MHz integrated over the same time interval. The thin lines show a subset of extrapolated magnetic (potential field) field lines that intersect with the type U-burst emission, which seems to be at the top of the field lines. The apices of the field lines range from 75,000 to 142,000 km, with magnetic fields of 25 G to 8 G (Aschwanden et al. 1992b).

normal-drifting and defines the classical *type III bursts*. If a beam is propagating downward (towards higher density), the sign is positive and the radio signature is called a *reverse-drifting (RS) burst*. Of course, some electron beams may propagate along a closed magnetic field line and thus show first a negative drift rate and later a positive drift rate after they crossed the apex, which appears shaped as an inverted letter U, and thus are named *U-bursts*. A partial *U-burst* which does not show a fully developed downward branch is called a *J-burst*. So the historical nomenclature is based on the morphological appearence in the dynamic spectra, but the physical relevance is actually their diagnostic of whether the underlying electron beams propagate along open or closed magnetic field lines. A tutorial for the translation of magnetic structures (that guide electron beams) into the observed radio burst nomenclature is given in Fig. 15.16. It is also interesting to note that type U bursts often occur as sequences with almost identical turnover frequencies (e.g., Fig. 15.17, top right), which suggests repetitive injection of electron beams into the same closed-loop system. An example of a type U burst observation with simultaneous radio spectrometer and interferometer coverage is shown in Fig. 15.18, where the turnover frequency ($\nu \approx 1.4$ GHz) of the type U-burst seen in the dynamic spectrum (Fig. 15.18, top) turned out to be identical to the used imaging frequency, and thus allowed to mapping of the spatial position of the electron beam at the apex position of the closed magnetic field structure (Fig. 15.18, bottom).

Another consequence of the electron beam interpretation of radio type III bursts is their detailed one-to-one correlation to other beam signatures, such as hard X-ray pulses. An example is given in Fig. 15.19, where five events with apparent bi-directional beams are shown, along with the detailed evolution of the simultaneous hard X-ray time profiles. The radio dynamic spectra clearly show a starting point at some frequency between 600 MHz and 1000 MHz from where a normal-drifting type III burst as well as a reverse-drifting (RS) burst start, coinciding with simultaneous hard X-ray pulses. Detailed cross-correlation analysis revealed that the start of the radio bursts is always delayed to the hard X-ray pulses by $\Delta t = 270 \pm 150$ ms, which was modeled in terms of different kinetic energies ($\varepsilon = 5.7$ keV for the radio-emitting electrons and $\varepsilon \geq 25$ keV for the hard X-ray emitting electrons) and a finite growth time for Langmuir waves ($\Delta t < 150$ ms) (Aschwanden et al. 1993). Thus, radio type III bursts and hard X-ray pulses can be interpreted in terms of a simultaneous injection of accelerated electrons in an upward and downward direction. The fraction of electrons escaping in an upward direction into interplanetary space is usually found to be much smaller ($\approx 10^{-2} - 10^{-3}$; Lin 1974) than the downward propagating electrons detected in hard X-rays. It is not yet clear whether this asymmetry is caused by the acceleration process or by the dominance of closed magnetic field topology.

Other Metric and Decimetric Bursts

As Table 15.1 shows, there are a number of other radio burst types that have been interpreted in terms of plasma emission, such as *type I, type II, type IV continuum,* and *type V* bursts, as well as a variety of *decimetric bursts (DCIM)*. Most of the radio burst classification is based on their morphology as they appear in dynamic spectra, while the physical interpretation in terms of emission mechanisms is mostly guided

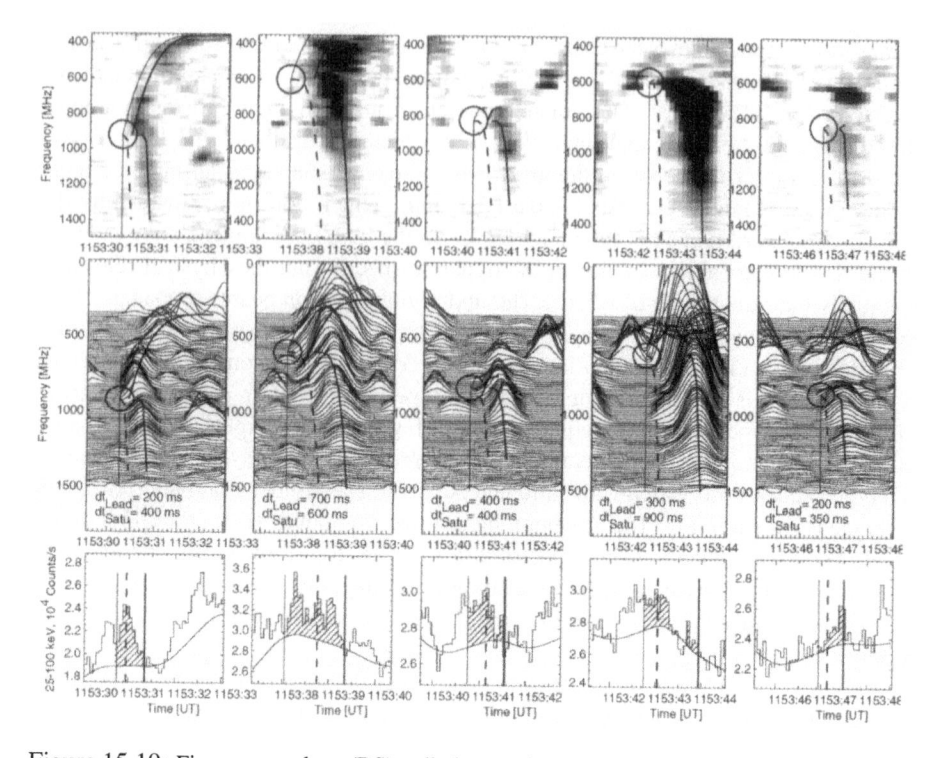

Figure 15.19: Five reverse slope (RS) radio bursts observed with *ETH* Zurich (top and middle panels) and 25–100 keV hard X-ray count rate observed with *BATSE/CGRO* (bottom panels). All time intervals have a length of 3.0 s, the time resolution of the radio data is 100 ms and for hard X-rays 64 ms. The start time of the radio bursts is marked with a thin line, the leading edge with a dashed curve, and the peak time with a thick solid curve. The envelope-subtracted hard X-ray flux during the radio start and peak are hatched, consisting of a single of multiple pulses (Aschwanden et al. 1993).

by the frequency range, brightness temperature, polarization, and frequency-drift rates. Concise summaries on the observed properties and physical interpretations of radio burst types are given in the two reviews by Dulk (1985) and Bastian et al. (1998). Moreover, extensive accounts on solar radio bursts can be found in the textbooks by Kundu (1965), Krueger (1979), McLean & Labrum (1985), and Benz (1993). Catalogs of decimetric burst types can be found in Bernold (1980), Slottje (1981), and Güdel & Benz (1988).

An overview of metric radio burst types is given in Fig. 15.20. The most rapid response to a particle acceleration phase are type III bursts, which generally appear as long as hard X-ray emission is produced, typically during a few minutes in the impulsive flare phase. Type III bursts are most easily recognizable in dynamic spectra from their fast frequency-drift rate, corresponding to the speed of mildly relativistic electron beams. Another established class are the *type II* bursts, which have a much lower frequency-drift rate corresponding to speeds of v ≈ 200 − 2000 km s^{-1}, and

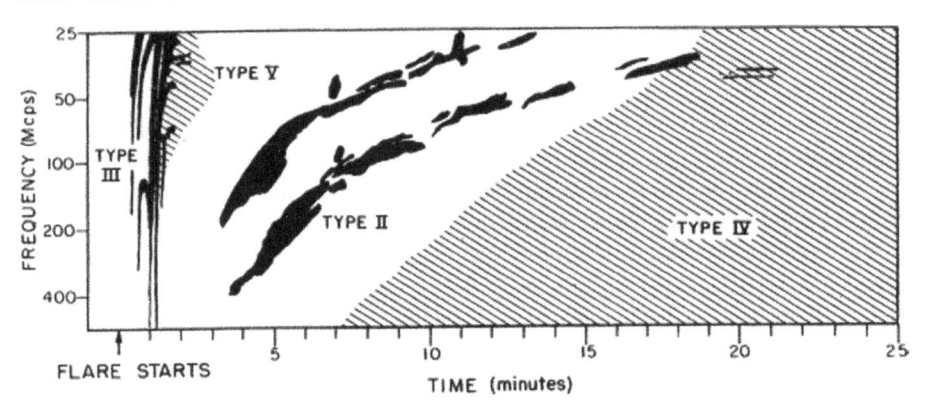

Figure 15.20: Overview of metric radio burst types during a fully developed eruptive flare, including type II, III, IV, and V emission (Cliver 2001).

thus have been associated with exciters from MHD shocks. The *backbone* structure of type II bursts often shows harmonic bands (Fig. 15.20), corresponding to the fundamental (s=1) and harmonic (s=2) plasma frequency (Plate 16). Sometimes fast-drifting type III-like bursts emanate from the central backbone, called a *herringbone* structure, which indicate escaping electron beams that have been accelerated in the shock front. Another class are *type IV bursts* or *flare continuum*, which appear as a stationary, broadband, long-lasting emission after the start of type II bursts (Fig. 15.20), which is generally interpreted in terms of plasma radiation from energetic electrons trapped in large postflare loops. A further class are *type V bursts*, which last a few minutes after type III bursts (Fig. 15.20), which seem to be a close by-product of the preceding type III bursts, but are probably produced by slower electrons (see beam evolution in Fig. 15.10).

Besides the classical metric burst types (I, II, III, IV, V), there is a variety of decimetric burst types, which have been subdivided into type III-like and type IV-like groups, indicating that they are also produced by plasma emission from either propagating electron beams (type III-like) or from electrons trapped in (post)flare loops (type IV-like). The major difference between decimetric and metric bursts is that they are produced in plasmas with higher densities ($n_e \approx 10^9 - 10^{11}$ cm^{-3} for decimetric frequencies of $\nu \approx 0.3 - 3$ GHz), and thus originate in magnetic reconnection regions and postflare loops, while metric bursts originate in the higher corona above. Decimetric bursts also show a lot of interesting time structures, such as (1) parallel drifting bands (so-called *zebra bursts* and *fiber bursts*) which indicate multiple gyroharmonics of propagating Alfvén waves, or (2) pulsating broadband structures (called *decimetric pulsations*), which could be attributed to modulations by MHD oscillations or nonlinear relaxational oscillations of wave-particle interactions. So, waves and oscillations can modulate the emission from trapped electrons, regardless of whether they emit incoherently (e.g., gyrosynchrotron emission) or coherently (e.g., beam-driven plasma emission or losscone-driven gyroemission). As a general rule, coherent emission can easily be distinguished from incoherent emission from the much higher (nonthermal) radio brightness temperature ($T_B \approx 10^8 - 10^{12}$ K). As further criterion, the degree of circular polarization can be used. For instance, plasma emission is expected in the magneto-ionic *ordinary mode*, while electron-cyclotron maser emission is generally dominated by the *extraordinary mode*.

Most of the radio burst studies have been performed with non-imaging instruments, and thus are subject to major uncertainties regarding the identification of the radiation emission mechanism and the physical interpretation of involved waves and particles. Limited progress has been made with radio imaging instruments, mainly because the observed radio source sizes are subject to substantial wave scattering in coronal inhomogeneities (Bastian 1994), and because the current radio imaging instruments have too few frequencies and some instruments are not solar-dedicated and thus are rarely able to catch a flare. The latter two obstacles could be overcome with the planned solar-dedicated *Frequency-Agile Solar Radiotelescope (FASR)* (White et al. 2003).

15.5   Losscone Emission

Coherent radio emission is driven by kinetic instabilities of unstable particle distribution functions, which can be grouped into two major categories: *beams* (with positive slopes in a parallel direction to the magnetic field, $\partial f/\partial v_\parallel > 0$) and *losscones* (with positive slopes in a perpendicular direction, $\partial f/\partial v_\perp > 0$). Losscone distributions develop in magnetic mirror regions in the weak scattering regime. Here we focus on losscone-driven emission, while beam-driven emission is treated in § 15.4.

15.5.1   Electron Cyclotron Maser Emission

A losscone distribution (Fig. 15.21) naturally originates in a magnetic trap, such as in a flare loop with magnetic mirrors above the footpoints due to the diverging magnetic field with height. We discussed the kinematics of particle trapping in such magnetic mirror configurations in § 12.5.1 in the context of hard X-ray producing electrons. Now we consider the consequences for radio emission. The basic mechanism is described in the framework of *gyroresonant wave-particle interactions* (§ 11.4.1), which quantifies the wave growth and absorption rate $\Gamma[\mathbf{k}, f(\mathbf{v})]$ in an unstable particle velocity distribution function $f(\mathbf{v})$ with the so-called *quasi-linear diffusion* equation system (Eqs. 11.4.1−2). For a given wave $(\mathbf{k}, \omega)$ there is a subset of particle velocities $(v_\parallel, v_\perp)$ that fulfill the *Doppler resonance condition* $\omega - s\Omega/\gamma - k_\parallel v_\parallel = 0$ (Eq. 11.4.6), and thus can contribute to wave growth or absorption. The solution space for each wave $(\mathbf{k}, \omega)$ is an ellipse in $(v_\parallel, v_\perp)$ space, a so-called *resonance ellipse*. Therefore, the key for coherent wave growth is that the slope in the velocity distribution, either $\partial f/\partial v_\parallel > 0$ or $\partial f/\partial v_\perp > 0$, has be to dominantly positive to produce a positive wave growth rate, which in the theoretical framework described in § 11.4.1 is defined by,

$$\Gamma_s^\sigma[\mathbf{k}, f(\mathbf{p})] = \int d\mathbf{p}^3\, A_s^\sigma(\mathbf{p}, \mathbf{k})\, \delta(\omega - s\Omega/\gamma - k_\parallel v_\parallel)\, \mathbf{k}\frac{df(\mathbf{p})}{d\mathbf{p}}\,, \qquad (15.5.1)$$

where $\mathbf{p} = \gamma m\mathbf{v}$ is the particle momentum. This is naturally fulfilled in losscones, where the tangent resonance ellipses to the losscone edge mostly face a positive value $\partial f/\partial v_\perp > 0$ (see examples of resonance ellipses in Fig. 11.10). The velocity space can therefore be subdivided into a domain of dominant growth (called *undamped regime* in

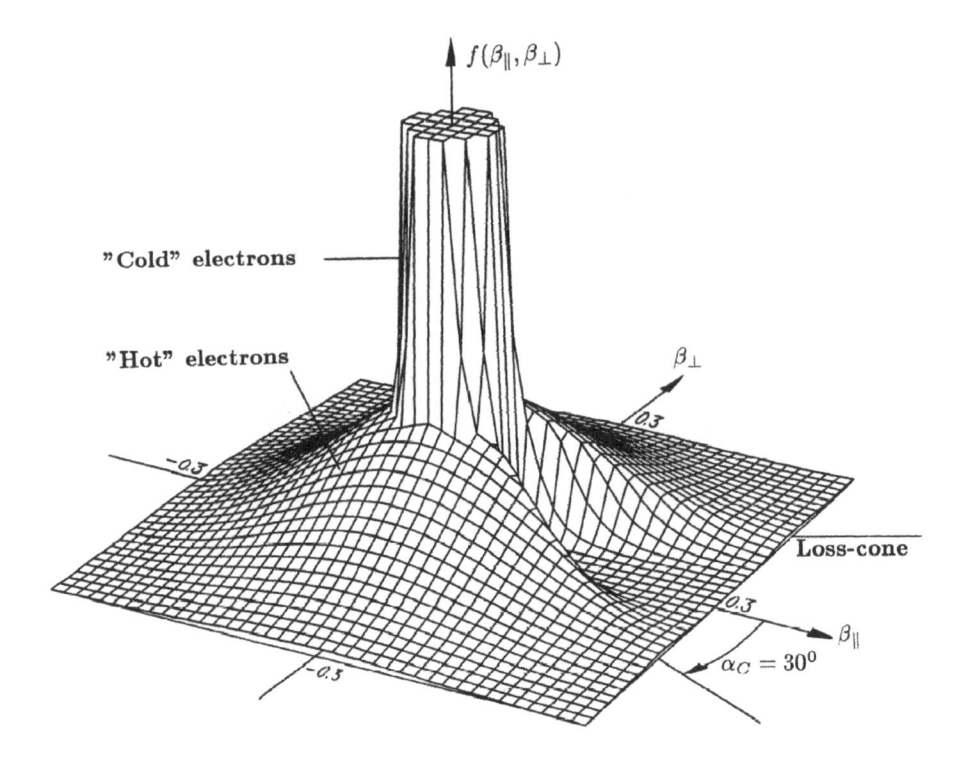

Figure 15.21: A particle velocity distribution $f(\beta_\parallel, \beta_\perp)$ is shown that contains a *"cold electron"* or thermal distribution, given by an isotropic Maxwellian distribution of temperature $T_c = 10^6$ K, and a *"hot electron"* distribution that is shaped by a losscone, characterized by a Maxwellian distribution of $T_h = 10^8$ K outside the critical pitch angle of $\alpha_c = 30°$, shaped by a $\sin^6(\alpha/\alpha_c \cdot \pi/2)$ pitch angle distribution inside the losscone angle α_c. The losscone is one-sided, as it occurs after a magnetic mirror reflection (Aschwanden 1990a).

Fig. 15.22), while the remaining regime is dominated by absorption (denoted as *wave damping* in Fig. 15.22).

The maximum maser growth rates $\Gamma_{\max}$ have been calculated as a function of the ratio of the plasma frequency ω_p to the gyrofrequency Ω_e, which characterizes physical parameter regimes with different electron densities n_e and magnetic fields B. For the particular losscone distribution shown in Fig. 15.21, it was found that the X-mode ($s = 1$) dominates at $\omega_p/\Omega_e \lesssim 0.3$, the Z-mode ($s = 1$) competes around $\omega_p/\Omega_e \approx 0.3$, the O-mode ($s = 1$) at $0.3 \approx \omega_p/\Omega_e \lesssim 1.0$, and harmonic modes ($s = 2$) at higher values of $1.0 \lesssim \omega_p/\Omega_e \gtrsim 1.4$ (Melrose et al. 1984). The growth rate is fastest ($\tau_{growth} = 1/\Gamma_{max} \approx 10^{-5}$ s) for $\omega_p/\Omega_e \ll 1$, but drops to $\tau_{growth} \approx 10^{-2}$ s for $\omega_p/\Omega_e \approx 1$. In solar flare conditions, a ratio of $\omega_p/\Omega_e \approx 1$ requires about $n_e \approx 10^{10}$ cm^{-3} ($\omega_p/2\pi \approx 1$ GHz) and $B = 357$ G ($\Omega_e/2\pi \approx 1$ GHz). Once an unstable losscone exists, gyroresonant waves grow exponentially at the expense of the free kinetic energy of the resonant particles at the losscone edge by quasi-linear diffu-

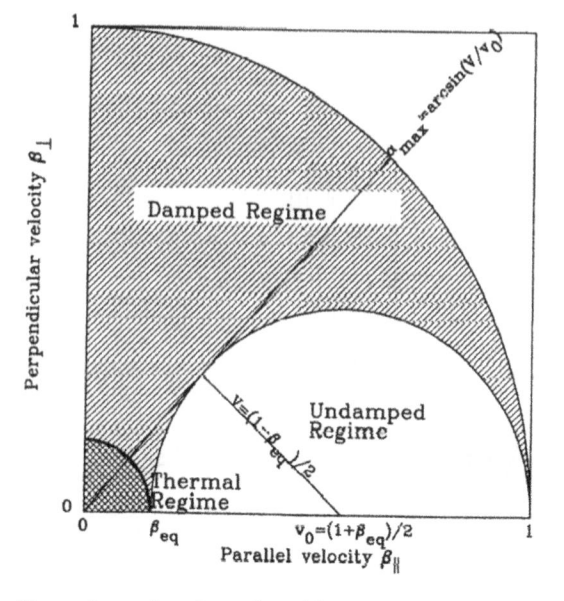

Figure 15.22: The undamped regime of positive growth rates for electron-cyclotron maser emission in velocity space. Approximating the resonance ellipses by circles, the undamped regime is confined by a circle with radius $V = (1 - \beta_{eq})/2$. This is the envelope of all undamped resonance circles. The geometrical sketch shows which part of the thermal tail can not be affected by quasi-linear maser diffusion, mainly low-energetic particles with higher pitch angles (Aschwanden 1990a).

sion. While the resonant particles lose perpendicular momentum, they drift inside the losscone and fill it up until the distribution inside the losscone forms a plateau, at which point the maser saturates and becomes quenched. Given these fast maser growth rates, the question arises whether such initial unstable conditions with empty losscones can be built up sufficiently fast. The build-up time of a losscone depends on the advection time of replenishing particles. For mildly relativistic electrons, say $v/c \approx 0.1$, and a maser saturation time of $\tau_{sat} \approx 10\tau_{growth} \approx 10^{-4} - 10^{-1}$ s, relatively small source sizes of $L \approx 3 - 3000$ km are required to yield advection times shorter than maser saturation times. Thus, unstable losscones that produce electron-cyclotron maser emission are expected to have rather small spatial scales and short lifetimes, but produce high radio brightness temperatures due to the coherent wave growth. So the expected observational characteristics are short (millisecond), narrowband, and intense spikes. The location of such maser sources is expected above the footpoints of flare loops where losscones are formed (Fig. 15.23).

More recent work involve other wave modes than the originally proposed electromagnetic X- and O-modes, such as the electrostatic upper-hybrid waves (see Table 11.2), which provide a natural explanation of the harmonic ratios occasionally observed in decimetric millisecond spike events (e.g., Willes et al. 1996; Fleishman & Yastrebov 1994a,b; Fleishman & Arzner 2000; Fleishman & Mel'nikov 1998).

There is an extensive literature on the theory of losscone-driven radio emission ap-

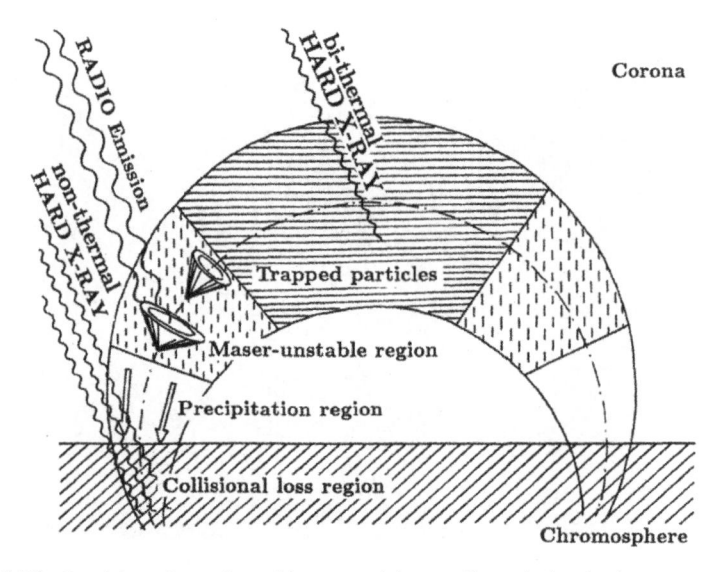

Figure 15.23: Spatial configuration of losscone-driven radio emission in the context of a flare loop with a magnetic trap (Aschwanden et al. 1990).

plied to solar flare plasmas. The reader is referred to Berney & Benz (1978), Benz (1980), Melrose & Dulk (1982), Hewitt et al. (1982, 1983), Sharma et al. (1982), Hewitt & Melrose (1983, 1985), MacKinnon et al. (1983), Sharma & Vlahos, (1984), Melrose et al. (1984), Pritchett (1984, 1986), Wu (1985), Winglee (1985a,b), Winglee & Dulk (1986a,b,c), Robinson (1986, 1988, 1989, 1991a,b), Vlahos & Sprangle (1987), Vlahos (1987), Aschwanden & Benz (1988a,b), Aschwanden (1990a,b), Smith & Benz (1991), Charikov & Fleishman (1991), Fleishman & Charikov (1991), Benz (1993, § 8), Charikov et al. (1993), Willes & Robinson (1996), Conway & MacKinnon, (1998), Fleishman & Mel'nikov (1998), Fleishman & Arzner (2000), Conway & Willes (2000).

Time-dependent models of losscone-driven radio emission require *trap-plus-precipitation models*, which have been developed by a number or authors: Kennel & Petschek (1966), Benz & Gold (1971), Melrose & Brown (1976), Alexander (1990), McClements (1990a, 1992), Lee et al. (2000). The particle dynamics of trapped electrons is described in Melrose & White (1979, 1981), White et al. (1983), Craig et al. (1985), Vilmer et al. (1986), McClements (1990b), Hamilton & Petrosian (1990), Hamilton et al. (1990).

15.5.2   Decimetric Observations

Decimetric radio emission with spiky fine structure that is very narrowbanded ($\Delta\nu/\nu \approx 1\%$), fast ($\Delta t \lesssim 100$ ms), and has an opposite circular polarization (X-mode) to type III emission (O-mode), has been interpreted in terms of losscone-driven coherent emission (e.g., *electron-cyclotron maser emission*: Holman et al. 1980; Melrose & Dulk 1982). This burst type shows the narrowest frequency bandwidths, which translates into small spatial source sizes L, regardless of whether the underlying emission is produced near

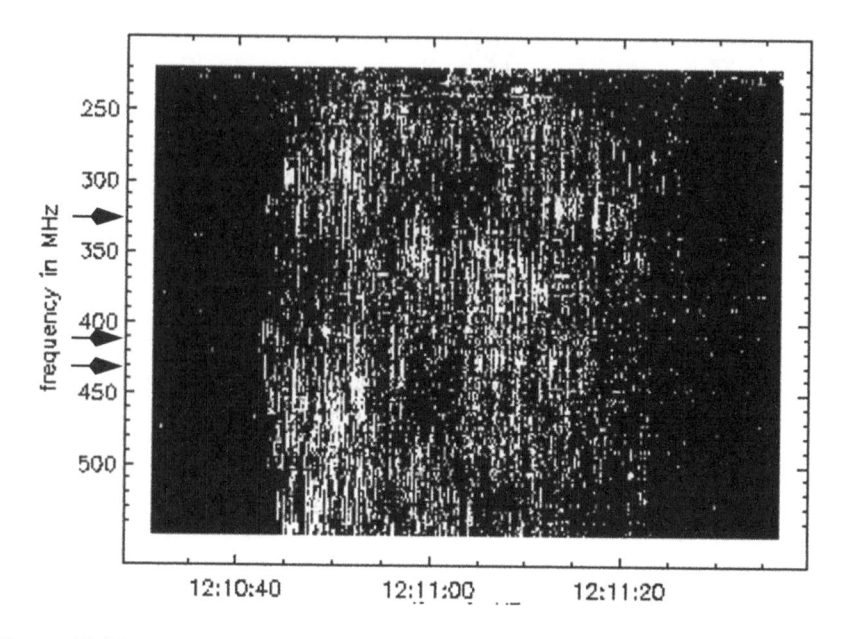

Figure 15.24: Radio dynamic spectrum $F(t, \nu)$ of a cluster of *decimetric millisecond spikes*, recorded with the Zurich radio spectrometer PHOENIX-2 in the frequency range of $\nu = 800 - 1000$ MHz during the 1980-Apr-06 flare, during a time interval of 24 s. The frequency bandwidth of individual spikes amounts to $d\nu \approx 10$ MHz and the time resolution is $\Delta t = 0.1$ s. Most of the spikes are probably not resolved in time (Benz et al. 2002).

the plasma frequency f_{pe} or the gyrofrequency f_{ge},

$$L \le \frac{\Delta \nu}{\nu} \left\{ \begin{array}{ll} \times \lambda_B & \text{if } \nu \approx f_{ge} \\ \times 2\lambda_T & \text{if } \nu \approx f_{pe} \end{array} \right. . \tag{15.5.2}$$

Therefore, for typical magnetic or density scale heights of $\lambda_B \approx \lambda_T \approx 10^5$ km, relative bandwidths of $\Delta \nu / \nu \approx 1\%$ translate into source sizes of $L \approx 200$ km (Benz 1986). Using this estimate of the source size and the measured fluxes, one obtains (with Eq. 15.2.5) very high brightness temperatures of $T_B \approx 10^{13}$ K, which is six orders of magnitude higher than flare plasma temperatures ($T_e \approx 10^7$ K), and thus the emission mechanism is clearly nonthermal or coherent emission. Since moreover type III bursts have much broader (drifting) bandwidths and an opposite circular polarization, a losscone-driven emission mechanism with magneto-ionic X-mode is an obvious interpretation.

An example of a radio dynamic spectrum of *decimetric millisecond spikes* is shown in Fig. 15.24. It shows not only a highly fragmented cluster of individual spikes, but also some drifting chain structures, which have been found to have harmonic frequency ratios for harmonics $s = 2, ..., 6$ (Güdel 1990), which seem to be related to multiple gyroharmonics. This is not consistent with the conventional model of semi-relativistic maser emission (Melrose & Dulk 1982), but could be interpreted in terms of Z-mode emission and upper-hybrid waves, which produce higher harmonics (Winglee & Dulk

1986b). Another interesting observation revealed that the decay time scale of decimetric millisecond spikes scales reciprocally with the frequency (i.e., $\tau_{decay} \propto \nu^{-1}$), which in the case of plasma emission $\nu \propto s f_p$ yields a dependence of $\tau_{decay} \propto n_e^{-1/2}$, which is a similar trend to collisional processes (Güdel & Benz 1990). Decimetric millisecond spikes have been detected exclusively during ($\gtrsim 20$ keV) hard X-ray emission of flares, but only in $\approx 2\%$ of the flares (Güdel et al. 1991), so they are co-produced with accelerated electrons. Although the temporal modulation of the spike radio flux mimics closely the hard X-ray pulses, the radio spike emission is typically delayed by $2 - 5$ s (Aschwanden & Güdel 1992), which could be related to a trapping time, implying that the high spatial fragmentation occurs at the precipitation site rather than in the primary acceleration or injection site. Unfortunately there are no clear-cut imaging observations available from decimetric millisecond spike sources that confirm whether they originate at the expected losscone sites above flareloop footpoints. Instead, some observations locate sources of narrowband millisecond spikes in large coronal heights (e.g., Krucker et al. 1995; Benz et al. 2002). Imaging observations with high spatial resolution, cadence, and frequency-agile coverage are needed to provide a deeper physical understanding of losscone-driven radio emission.

15.6 Summary

Radio emission in the solar corona is produced by thermal, nonthermal, up to high-relativistic electrons, and thus provides a lot of useful diagnostics complementary to EUV, soft X-rays, hard X-rays, and gamma-rays. Thermal or Maxwellian distribution functions produce in radio wavelengths either free-free emission (bremsstrahlung) for low magnetic field strengths and gyroresonance emission in locations of high magnetic field strengths, such as above sunspots, which are both called incoherent emission mechanisms. While EUV and soft X-ray emission occurs in the optically thin regime, the emissivity adds up linearly along the line-of-sight. Free-free emission in radio is somewhat more complicated, because the optical thickness depends on the frequency, which allows direct measurement of the electron temperature in optically thick coronal layers in metric and decimetric frequencies up to $\nu \lesssim 1$ GHz. Above ≈ 2 GHz, free-free emission becomes optically thin in the corona, but gyroresonance emission at harmonics of $s \approx 2, 3, 4$ dominates in strong-field regions. In flares, high-relativistic electrons are produced that emit gyrosynchrotron emission, which allows for detailed modeling of precipitating and trapped electron populations in time profiles recorded at different microwave frequencies.

Unstable non-Maxwellian particle velocity distributions, which have a positive gradient in parallel (beams) or perpendicular (losscones) direction to the magnetic field, drive gyroresonant wave-particle interactions that produce coherent wave growth, detectable in the form of coherent radio emission. Two natural processes that provide these conditions are dispersive electron propagation (producing beams) and magnetic trapping (producing losscones). The wave-particle interactions produce growth of Langmuir waves, upper-hybrid waves, and electron-cyclotron maser emission, leading to a variety of radio burst types (type I, II,

III, IV, V, DCIM), which have been mainly explored from (non-imaging) dynamic spectra, while imaging observations have been rarely obtained. Although there is much theoretical understanding of the underlying wave-particle interactions, spatio-temporal modeling of imaging observations is still in its infancy. A solar-dedicated, frequency-agile imager with many frequencies (FASR) is in planning stage and might provide more comprehensive observations.

Chapter 16

Flare Plasma Dynamics

A solar flare is a catastrophic event that is triggered by an instability of the underlying magnetic field configuration (§ 10) and evolves then into a more stable state by changing and reconnecting the magnetic topology. This change in magnetic topology provides free magnetic energy that is released in the form of currents that spawn *primary plasma heating* and *particle acceleration* (§ 11). In most flare models these primary processes take place in the corona, in the immediate environment of magnetic reconnection points and associated magnetic separator lines or separatrix surfaces. In a second step, the accelerated particles and thermal conduction fronts propagate to the chromosphere where they heat up the chromospheric plasma, which is a *secondary heating process*, driven by the energy loss of the precipitating particles or by thermal conduction of the impinging ion-acoustic waves. This chromospheric heating process triggers a third step, an upflow of heated chromospheric plasma, which is called *chromospheric evaporation* (or physically more correctly, *chromospheric ablation*). This third step fills up what appears as prominent flare loops in soft X-ray wavelengths. In principle, also the primary coronal heating process could be detected in soft X-rays, but is usually outshone by the much brighter upflows of chromospheric plasma, due to its higher density and emission measure. The second step of chromospheric heating is most prominently observed in gamma-rays (§ 13) and hard X-rays (§ 14), and sometimes even in UV and white light. Thus we have to keep in mind that most of the soft X-ray observations document only the third step in this chain reaction, and we still have very insufficient diagnostic about the first step of how the flare is initiated. Once the flare passes its peak in soft X-ray emission, plasma cooling processes start to dominate over heating. When the plasma cools down from the initial $10 - 30$ MK temperatures at the peak of the flare down to $1 - 3$ MK, the postflare loop system becomes prominently detectable in EUV, showing the beautiful fractal structures of postflare arcades seen in high-resolution TRACE movies (Plate 17). Once the temperature drops below 0.5 MK, instabilities occur in the postflare loops that cause a rapid break-up and precipitation of the cooling plasma, visible in UV and Hα. These five steps may occur in successive order for a simple single-loop flare, but usually occur parallel and time-overlapping in multi-loop flares, and thus are hard to disentangle. In this section we focus on the various heating and cooling processes of the flare plasma.

Table 16.1: Primary plasma heating processes in solar flares.

Heating process	References
Resistive or Joule heating	Spicer (1981a,b), Holman (1985)
– Anomalous resistivity heating	Coppi & Friedland (1971), Duijveman et al. (1981)
– Ion-acoustic waves (AC)	Rosner et al. (1978b)
– Electron ion-cyclotron waves (IC)	Shapiro & Knight (1978), Hinata (1980)
Shock heating	Petschek (1964), Tsuneta (1997)
– Slow-shock heating	Cargill & Priest (1983), Hick & Priest (1989)
Electron beam heating	
– Coulomb collisional loss	Fletcher (1995, 1996), Fletcher & Martens (1998)
Proton beam heating	
– Kinetic Alfvén waves, MHD turbulence	Voitenko (1995, 1996), Voitenko & Goossens (1999)
Inductive current heating	Melrose (1995; 1997)

16.1   Coronal Flare Plasma Heating

We discussed heating mechanisms of coronal plasmas in § 9, of which some could play a role during flares. However, the chief difference between flare plasma heating and (quiet Sun) coronal heating is the *impulsiveness*. Therefore, simple application of *steady-state* or *quasi-steady* magnetic reconnection processes, such as Sweet−Parker or Petschek-type (§ 10.1), cannot explain the flare dynamics. What is called for in flares are *unsteady* and *bursty magnetic reconnection modes*, such as the *tearing-mode instability* and *coalescence instability* (§ 10.2). We discussed various flare/CME models in § 10.5, which provide a framework for physical modeling of *primary plasma heating processes*, as they occur locally or in the immediate neighborhood of reconnection sites. In Table 16.1 we compile a list of *primary plasma heating mechanisms*, along with some representative theoretical studies or numerical simulations. Observational evidence for heated plasma regions at the beginning of flares has indeed been found near the expected reconnection regions above flare loop arcades (Warren & Reeves 2001).

16.1.1   Resistive or Joule Heating

Flares are thought to occur by release of nonpotential magnetic energy and their associated currents $\mathbf{j} = (1/4\pi)\nabla \times \mathbf{B}$. If DC electric fields $\mathbf{E}$ arise, which are necessary ingredients in DC electric field and runaway acceleration models (§ 11.3), they carry an associated macroscopic current $\mathbf{j} = \sigma\mathbf{E}$ that can be dissipated by *Joule heating* of the thermal plasma. The *resistivity* $\eta = c^2/(4\pi\sigma)$ or *electrical conductivity* $\sigma = n_e e^2 \tau_{ce}/m_e \approx 6.96 \times 10^7 \ln(\Lambda)^{-1} Z^{-1} T_e^{3/2}$, which is given by the *electron collision time* τ_{ce} or *electron collision frequency* $f_{ce} = 1/\tau_{ce} = 3.64\, n_e \ln \Lambda T_e^{-3/2}$ determines the Joule heating rate. Since the energy dissipated by a current density $\mathbf{j}$ is $\mathbf{j} \cdot \mathbf{E}$, a *Joule heating time scale* $\tau_J = n_e k_B T_e/(\mathbf{j} \cdot \mathbf{E})$ can be derived from the definitions of classical resistivity η and the electron thermal velocity v_{Te}, which is in terms of the *electric Dreicer field* $\mathbf{E}_D$ (§ 11.3.2) (Holman 1985),

$$\tau_J = \frac{n_e k_B T_e}{\mathbf{j} \cdot \mathbf{E}} = \left(\frac{E_D}{E}\right)^2 \tau_{ce} = \left(\frac{v_c}{v_{Te}}\right)^4 \tau_{ce}, \qquad (16.1.1)$$

where v_c is the critical velocity for run-away electrons, so it is in the order of the *electron collision time scale* τ_{ce} for thermal electrons. Significant plasma heating can only be obtained when the *Joule heating time* τ_J is shorter than the radiative or conductive cooling time. From this constraint, Holman (1985) finds that the resistivity in a current sheet must be much greater than the classical resistivity in (hot) quiet corona conditions ($n_e \approx 10^9$ cm^{-3}, $T_e = 10^7$ K). For flares, a plasma density of either $n_e \gtrsim 10^{11}$ cm^{-3} is required, or *anomalous resistivity* in the case of lower densities. But anomalous resistivity driven by a parallel current $\mathbf{j}_{\parallel}$ first requires heating of the electron plasma to a temperature that is at least an order of magnitude higher than the ion temperature. Thus, most of the MHD flare simulations assume anomalous resistivity as an initial condition. Joule heating in solar flares has been studied by Spicer (1981a,b), Duijveman et al. (1981), Holman (1985), Tsuneta (1985), and Holman et al. (1989).

16.1.2 Anomalous Resistivity

While the classical resistivity is determined by frictional forces in Coulomb collisions between electrons and ions of the same temperature, *anomalous resistivity* occurs in plasmas in a turbulent state (Coppi & Friedland 1971). In a turbulent plasma, the electrons that carry the current will also interact with the electric field of waves, which change the resistivity and other transport coefficients, depending on the type of growing waves. For waves that affect the resistivity, *electrostatic ion-cyclotron waves (IC)* and *ion-acoustic waves (AC)* have been considered (Rosner et al. 1978b; Shapiro & Knight 1978; Hinata 1980; Duijveman et al. 1981). It was found that ion-acoustic heating can produce an exponentiating electric field and in this way can generate electron temperatures that are much higher than the ion temperature, $T_e/T_i \gg 1$, the maximum limit being restricted by the saturation level of ion-acoustic or ion-cyclotron wave growth. There are very few analytical theories about anomalous resistivity with solar flare applications, one being applied to a Petschek-type reconnection scenario (Kulsrud 2001; Uzdensky 2003). Anomalous resistivity is commonly assumed in numerical MHD simulations because it is one of the easiest ways to reproduce Petschek-type fast reconnection, while uniform resistivity would lead to the much slower Sweet−Parker type reconnection (Yokoyama & Shibata 1994). Particular MHD simulations investigated the scaling of the reconnection rate with anomalous resistivity (Yokoyama & Shibata 1994), the scaling of the Joule heating rate with anomalous resistivity (Roussev et al. 2002), the influence of heat conduction on the magnetic reconnection rate (e.g., Chen et al. 1999a; Miyagoshi & Yokoyama 2003), the dynamics of X-points in current sheets (e.g., Schumacher & Kliem 1996b, 1997b), or the coalescence of magnetic islands (e.g., Schumacher & Kliem 1997).

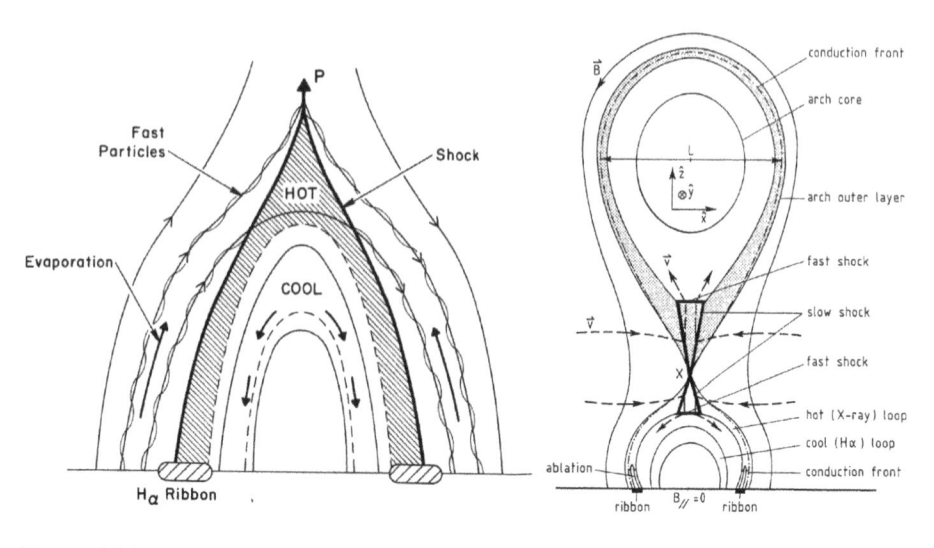

Figure 16.1: *Left:* slow-shock heating in the downward-directed reconnection outflows. The reconnection point P is rising vertically, trailing behind two slow MHD shocks (thick lines). Behind the shock there is a series of loops with heated plasma (shaded), while the cooled plasma in the inner loops falls down as seen in Hα (Cargill & Priest 1983). *Right:* slow-shock heating in upward-directed reconnection outflows. The heated plasma connected with the slow MHD shocks forms a bubble-like conduction front of a postflare arch (Hick & Priest 1989).

16.1.3   Shock Heating

In § 10.5.1 we described the standard 2D X-type reconnection model, where a *fast-shock* front develops in the reconnection outflow and standing *slow shocks* at the V-shaped, lateral boundaries (ridges) of the reconnection outflow region (Fig. 10.21; Tsuneta 1997). The critical importance of slow shocks in converting magnetic energy to plasma kinetic and thermal energies was pointed out theoretically by Petschek (1964), Cargill & Priest (1983), and Hick & Priest (1989). The heating rate at a slow-shock interface can be computed from the jump conditions (i.e., the conservation of the normal magnetic field, normal and tangential momentum, energy, and tangential electric field). The heating rate was found to be sufficient to explain the continued heating of postflare loops (Fig. 16.1 left; Cargill & Priest 1983) as well as for postflare arches (Fig. 16.1 right; Hick & Priest 1989). Numerical MHD simulations by Yokoyama & Shibata (1997, 1998) showed that adiabatic slow MHD shocks that emanate from the neutral point become dissociated into heat conduction fronts and isothermal slow MHD shocks, due to heat conduction effects. Further confirmation of slow-shock heating was obtained by soft X-ray observations of high-temperature ridges with temperatures of $T \approx 10$ MK above and below the X-point during and after flares (Tsuneta 1996a, 1997; Tsuneta et al. 1997).

16.1.4 Electron Beam Heating

Energized (nonthermal) electrons lose their kinetic energy in the Coulomb field of ambient (thermal) ions and electrons when they enter a collisional plasma. This energy loss not only stops the fast electrons, but also heats the ambient target plasma. This heating process is observed most prominently in hard X-rays when nonthermal electrons precipitate to the chromosphere and produce free-free bremsstrahlung during the multi-encounter energy loss (see thick-target bremsstrahlung process in § 13.2.2). However, although most of the coronal regions are collisionless to a good approximation, some rare Coulomb collisions and thus energy loss and plasma heating already takes place in the coronal regions that the electrons cross during their journey to the chromosphere. The coronal energy loss is particularly enhanced in high-density regions or in trapping regions, where the trapping time can be much longer than the time-of-flight crossing time. Evidence for coronal energy loss was mainly established from the observations of *above-the-looptop* hard X-ray sources (Masuda et al. 1994, 1995). Modeling of such coronal hard X-ray emission was performed by Fletcher (1995, 1996), Fletcher & Martens (1998), and Brown et al. (1983, 2002a), from which the associated coronal energy loss and coronal plasma heating rate can be inferred. Alexander & Metcalf (1997) attempted to isolate the hard X-ray spectrum of the (Masuda-type) above-the-looptop from the looptop sources and found a nonthermal spectrum for the Masuda-type source, which indicates that collisions are too infrequent there to thermalize the plasma. The looptop sources are generally found to have a thermal spectrum (with $T \approx 40$ MK for the Masuda flare), but it is difficult to separate the plasma that is directly heated by electron beams in the corona from the upflowing plasma that was heated in a secondary phase in the chromosphere.

16.1.5 Proton Beam Heating

Protons and electrons are thought to be accelerated concomitantly in magnetic reconnection regions, so that proton beams as well as electron beams emerge. Proton beams that propagate in a downward direction towards the chromosphere are expected to excite strong *kinetic Alfvén waves (KAW)* (Voitenko 1995, 1996; Voitenko & Goossens 1999; Voitenko et al. 2003). The proton beam-driven instability of *kinetic Alfvén waves (KAW)* can be saturated by the velocity-space (quasi-linear) diffusion of the beam protons. This saturation amplitude of KAWs, however, is too low to explain the observed nonthermal velocities in flares, but an inverse MHD-turbulent cascade formed by three-wave interactions is envisioned to spread the turbulence spectrum into the low wave number domain with enhanced amplitudes. The resulting turbulence of KAWs contains enough energy to produce the typical nonthermal velocities of v $\approx 200 - 400$ km s^{-1} observed in flares (Voitenko & Goossens 1999), and thus could contribute to impulsive primary plasma heating between the reconnection outflow regions and the flare loop footpoints. Diagnostic on flare protons, however, is difficult to come by, but estimates of the proton flux have been attempted by using the ^{20}Ne 1.634 MeV line (Emslie et al. 1997).

16.1.6   Inductive Current Heating

Heating of plasma in coronal flare loops can also be accomplished by *current induction* according to Maxwell's laws. Typical coronal loops carry currents of the order $I \approx 10^{12}$ A and changes in the current path $\partial I/\partial t$ and the associated inductance $\partial L/\partial t$ are faciliated by reconnecting magnetic field lines. According to the simple circuit equation,

$$\frac{d}{dt}(LI) + RI + \frac{Q}{C} = EMF \,, \qquad (16.1.2)$$

a coronal loop with electric resistivity R and voltage V then dissipates the current with a power of $P = IV = RI^2$, according to Ohm's law, which can be converted into plasma heating. Early flare models based on current induction were pioneered by Alfvén & Carlquist (1967). The change in inductance $\partial L/\partial t$ can occur in at least three ways in a flare environment: either like current-carrying loops move apart; the current path shortens; or unlike currents move closer together (e.g., Melrose 1995). A model of magnetic flux or electric current transfer in quadrupolar flare loop pairs was derived in Melrose (1997) and applied to solar flares with interacting flare loops by Hardy et al. (1998) and Aschwanden et al. (1999c; see also § 10.5.6). An interesting prediction of this model is that plasma heating occurs along the separator field lines and separatrix surfaces (Longcope & Silva 1997). Further aspects of induction between current-carrying loops, such as rising loops, loop oscillations, and stabilizing effects are studied in Khodachenko et al. (2003). Observational verification of these processes are difficult, because simultaneous measurements of current changes and temperature changes in flare loops have not been managed yet.

16.2   Chromospheric Flare Plasma Heating

Let us proceed now to secondary flare heating processes, which occur when accelerated particles or thermal conduction fronts during their downward propagation hit the transition region and chromosphere. Because of the large density gradient at this interface, this secondary heating process produces far more heated flare plasma than any of the primary heating mechanisms that operate in the corona, as we described in § 16.1. There are two competing agents for chromospheric heating, nonthermal particles versus thermal conduction fronts, which both seem to be important in flares. We review these two agents separately in the following two sections.

16.2.1   Electron and Proton Precipitation–Driven Heating

There are also two competing scenarios for beam-driven chromospheric heating during flares, the *electron beam hypothesis* and the *proton beam hypothesis* (Brown et al. 1990).

The salient feature of the electron beam model is a stream of fast electrons ($\varepsilon \gtrsim 20$ keV), carrying $\lesssim 10^{36}$ electrons s^{-1} in large flares, a kinetic power of $P(\varepsilon > 20$ keV) $\approx 10^{29}$ erg s^{-1}, an associated current of $I \approx 10^{26}$ statamps over an area of $A \lesssim 10^{18}$ cm^2, during a time interval of $\Delta t \approx 10^2 - 10^3$ s (Hoyng et al. 1976; 1981b), as derived

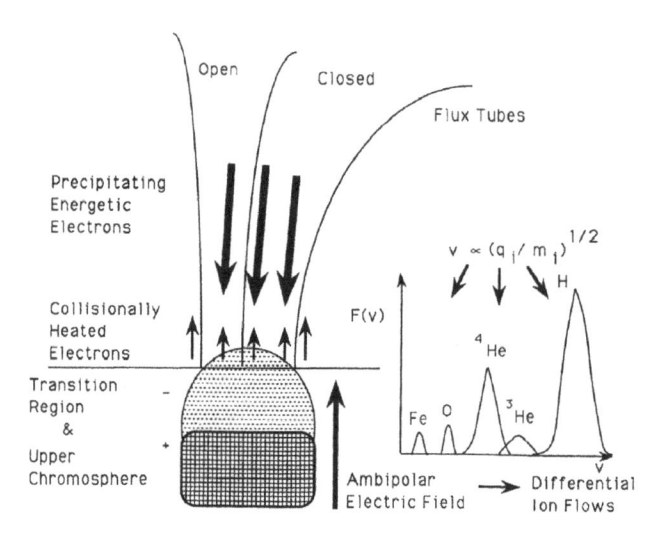

Figure 16.2: Schematic diagram of the physical processes during the heating of chromospheric flare plasma: An ambipolar electric field develops when the collisionally heated electrons propagate outward faster than the ions. The ambipolar field drags the ions outward in an effort to maintain quasi-neutrality. The velocity of an ion is dependent on its charge-to-mass ratio, with the heaviest ions in general moving more slowly through the potential drop than the lighter ions (Winglee 1989).

from the thick-target bremsstrahlung model (§ 13.2.2). The collisional interaction of a beam of charged particles with a hydrogen target of arbitrary ionization level has been quantified in a number of theoretical studies (e.g., Emslie 1978, 1983, 2003; Emslie et al. 1981). The detailed amount of energy deposition as a function of the chromospheric height of course depends on the atmospheric density model (Emslie 1981; Emslie et al. 1981; Brown et al. 2002a; Aschwanden et al. 2002b) and magnetic field model (Emslie et al. 1992).

Proton beam models are less constrained because we do not observe any direct radiative signature in hard X-rays, such as bremsstrahlung in the case of electrons. Nevertheless, proton beam models have the advantage (over electron beam models) that a lower particle flux and beam current is needed, and thus a number of proton beam models have been postulated (Emslie & Brown 1985; Emslie et al. 1996, 1997; Brown et al. 1990), which can be subdivided into low-energy ($\approx$ 1 MeV) and high-energy ($\gtrsim$ 40 MeV) proton beam models. Problems with high-energy protons are that they produce heating and hard X-rays much too deep in the atmosphere, lack the impulsive (subsecond) time scales, and predict too high gamma-ray fluxes (Brown et al. 1990), while low-energy protons cannot explain the hard X-ray emission. Thus, electron beam models are necessary to explain the observed hard X-rays, while an unknown number of low-energy protons are likely to be concomitant, both contributing to the heating of chromospheric plasma during flares.

The detailed kinetic evolution of beam-driven chromospheric heating requires nu-

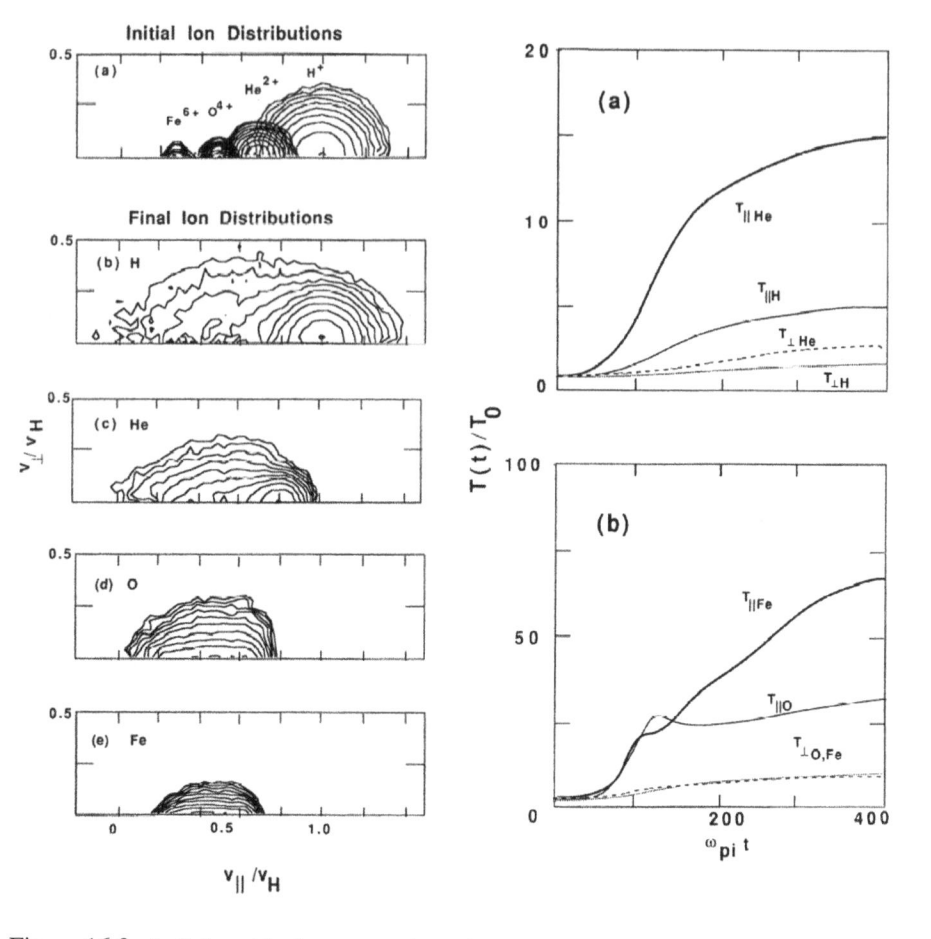

Figure 16.3: *Left:* logarithmic contour plots of initial and final velocity distributions of H^+, He^{2+}, O^{4+}, and Fe^{6+} ions during chromospheric heating by precipitating particles. The evolution of the particle velocity distributions is simulated with a particle-in-cell code that mimics the wave-particle interactions in the microscopic electromagnetic fields between electrons and different ion species. *Right:* time histories of the parallel and perpendicular temperatures normalized to their initial temperatures (Winglee 1989).

merical simulations with particle codes that self-consistently treat the properties of the current system and associated electric fields during the propagation and energy transport of the energetic electrons and ions from the corona to the chromosphere. Such numerical simulations have been performed [e.g., by Winglee (1989) and Winglee et al. (1991a, b)], from which we show the conceptual setup in Fig. 16.2 and some results of time histories in Fig. 16.3. In this simulation, a preflare density and temperature of $n_e = 10^{10}$ cm^{-3} and $T_e = 10^5$ K is assumed in the transition region, and $n_e = 10^{11}$ cm^{-3} and $T_e = 10^4$ K in the upper chromosphere, with heating initiated by precipitating energetic ($\gtrsim 10$ keV) electrons. The primary (precipitating) electrons heat secondary electrons. The energy of these heated secondary electrons goes in part into collisional heating of the ambient ions, and in part into upward propagation due to the overpressure. As the secondary electrons propagate upward, a return current drags the ions with them through an ambipolar field, in order to maintain charge neutrality. The ambipolar electric field accelerates lighter ions to higher speeds than the heavier ions, so that different ion species form separated beams in the velocity distribution (see Fig. 16.2, right, and Fig. 16.3, left), which become the subject of streaming instabilities. Some of the ions become decelerated while others become accelerated up to speeds comparable to the initial speed of the light ions. This mechanism reproduces some of the observed abundance enhancements (such as He^3; Reames et al. 1985) and yields bulk speeds in the upflowing plasma that are comparable with observed soft X-ray Doppler shifts (Winglee 1989; Winglee et al. 1991a,b).

16.2.2 Heat Conduction-Driven Heating

The magnetic reconnection process in a coronal X-point is thought to accelerate particles as well as to heat up the local plasma, to temperatures of $T \approx 10^7$ K. The overpressure in the heated plasma will cause an expansion, with thermal conduction fronts that have a steep temperature gradient at the leading edges. The leading edge is expected to propagate with the ion-sound speed c_s (Smith 1977; Smith & Lilliequist 1979), leading to an anomalous heat flux of $F_{an} = (3/2)nk_BTc_s$. The consequences of anomalous flux limitations have motivated a *dissipative thermal flare model* (Brown et al. 1979; Smith & Lilliequist 1979; see also the review by Machado 1991), where the impulsively heated coronal plasma is confined by the relatively slowly moving conduction fronts. A substantial fraction of the observed soft X-rays (10 eV − 1 keV) are then produced by thermal bremsstrahlung of the bottled-up electrons. A number of observational tests have been performed for this model, where a proportionality between the hard X-ray rise time and the (microwave) flare size was found, which was interpreted in terms of a constant source expansion speed (e.g., the ion-sound speed c_s in the conduction-front model; Batchelor 1989). Another argument in favor of the conduction-front model was brought forward for a flare that showed all signatures of chromospheric evaporation upflows but a lack of ($\gtrsim 15$ keV) hard X-ray emission (Czaykowska et al. 2001).

Recent numerical MHD simulations of magnetic reconnection processes include heat conduction and reproduce the evolution and propagation of conduction fronts in detail (Yokoyama & Shibata 1997, 1998, 2001; Chen et al. 1999a). The first numerical simulation of magnetic reconnection including heat conduction showed the propagation of both the conduction front and isothermal slow shocks (Yokoyama & Shibata 1997), which was originally predicted by analytical work (Forbes et al. 1989). Yokoyama & Shibata (1998) further succeeded to include chromospheric evaporation in the numerical simulation of reconnection coupled with heat conduction. The 2.5D simulation of Chen et al. (1999a) also includes field-aligned heat conduction and developments of slow shocks as well as a heat conduction front were found, where the conduction front heats the plasma along the field lines. The 2D simulations of Yokoyama & Shibata (2001) include the effects of anisotropic heat conduction and chromospheric evaporation (Fig. 16.4). They find that the energy transported by heat conduction causes an increase in temperature and pressure of the chromospheric plasma, according

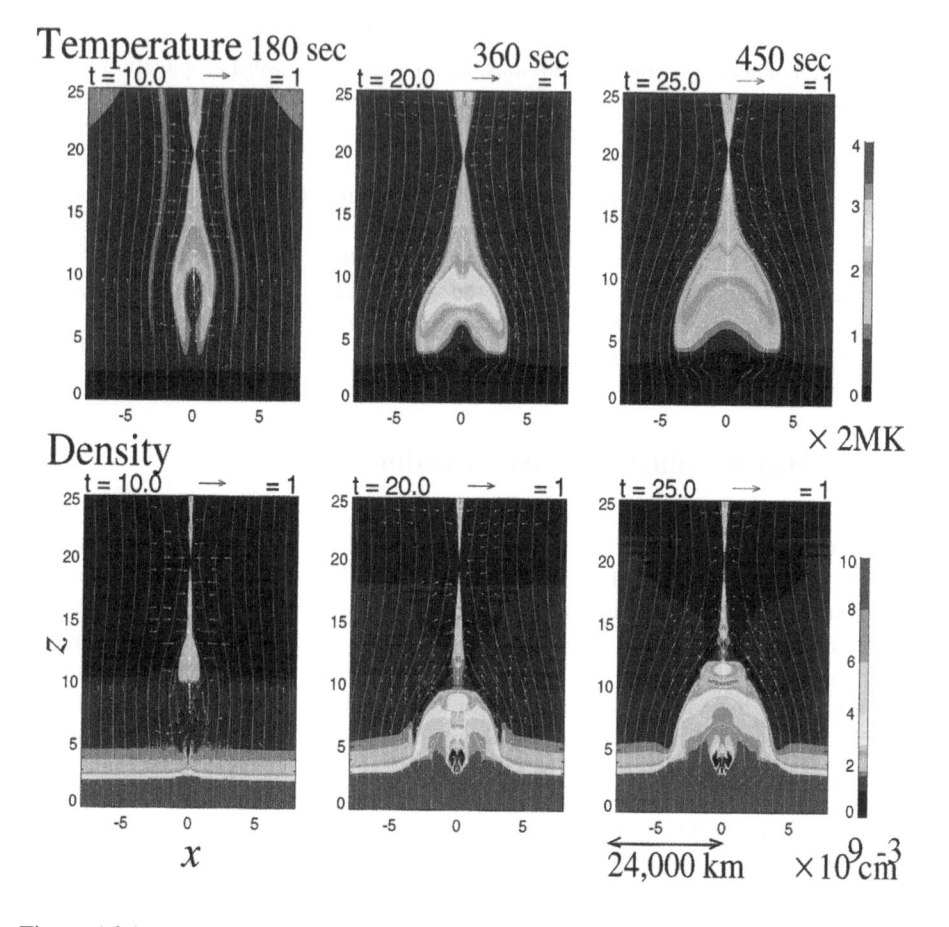

Figure 16.4: 2D numerical MHD simulation of a solar flare with chromospheric evaporation and anisotropic heat conduction in the framework of a 2D magnetic reconnecting geometry. The temporal evolution of the plasma temperature (top row) and density (bottom row) is shown. The temperature and density scale is shown in the bars on the right side. The simulation illustrates the propagation of thermal conduction fronts and the upflows of chromospheric plasma in response (Yokoyama & Shibata 2001).

to the scaling

$$T_{top} = \left(\frac{B^3 L}{2 \pi \kappa_0 \sqrt{4 \pi \rho}} \right)^{2/7} , \qquad (16.2.1)$$

where T_{top} is the temperature of the flare looptop, B the coronal magnetic field strength, ρ the coronal mass density, and κ_0 the heat conduction coefficient, respectively. Thus, these MHD simulations confirm that chromospheric evaporation can also be produced by heat conduction fronts, which is important for flares without detectable hard X-rays. In flares with detectable hard X-rays, both drivers (i.e., particle precipitation and heat conduction) may compete in spawning chromospheric evaporation.

16.2.3   Hα Emission

From the hydrogen lines (Fig. 2.8) we see that the lowest atomic levels in visible light are the Balmer series (with transitions between $n = 2$ and $n \geq 3$), with the Hα line ($n = 2 \mapsto n = 3$) at 6563 Å having the lowest energy levels. Therefore, the solar chromosphere can most easily be observed with ground-based instruments in the Hα line, showing chromospheric fine structure in the quiet Sun, filaments, active region loops, surges, prominences, and flare ribbons with high contrast. The high opacity of this line makes phenomena easily observable above the limb. Observations are made in the center of the Hα line, which appears in absorption in the quiet Sun and in active region plages, but in emission during major flares. The center of the Hα line is often saturated during flares, so that the kernels of flares are better observed in the line wings, say at +2.5 Å. Historically, flare research has thus been pioneered from ground mostly in the Hα line [e.g., see textbooks of Svestka (1976) and Zirin (1988)].

Quantitative analysis of Hα data, however, is complicated because the contributions to Hα line emission span from the photosphere in the line wing up to the upper chromosphere in the line core. Furthermore the Hα line involves high-excitation states ($n \geq 2$), and heating processes produce both emission and absorption features. The evolution of the Hα line profile at the footpoint of flare loops in response to the precipitation of nonthermal electrons, heat conduction, and chromospheric evaporation has been studied in detail by Ricchiazzi & Canfield (1983), Canfield et al. (1984), Ichimoto & Kurokawa (1984), and Canfield & Gunkler (1985). These studies provide us a diagnostic on the effects of electron precipitation and heat conduction based on the intensity, width, wing, and central reversal of the Hα line profile (e.g., see evolution of Hα profile during a flare in Fig. 16.5). Canfield & Gunkler (1985) concluded in one flare that chromospheric evaporation is controlled by heat conduction rather than by thick-target nonthermal electron heating. In other flares it was demonstrated that the Hα energy flux is consistent with an energy flux deposited by thick-target electrons over an area of $A \approx 2 \times 10^{17}$ cm^2 = (4500 km)^2 (Wülser et al. 1992). Chromospheric evaporation produces an upflow of heated plasma (e.g., as seen as blueshift in Ca XIX) and simultaneously downflows of chromospheric gas (e.g., as seen as redshift in Hα) to balance the momentum, also called *"chromospheric condensation"* (Fisher 1989). This momentum balance between Ca XIX and Hα emission was verified in a number of solar flares (Zarro et al. 1988b; Canfield et al. 1990; Wülser et al. 1992, 1994).

The precipitation of high-energy particles (mostly protons) also produces a linear polarization of the Hα line, which is called *impact polarization*. Successful measurements of the Hα linear polarization therefore would provide support in favor of precipitating protons as drivers of chromospheric evaporation, rather than electrons. The observed linear polarization in Hα is weak, in the order of a few percent, and is difficult to measure, but detections were claimed by Hénoux et al. (1990), Metcalf et al. (1992, 1994), Vogt & Hénoux (1999), Emslie et al. (2000), Vogt et al. (2002), and Hanaoka (2003).

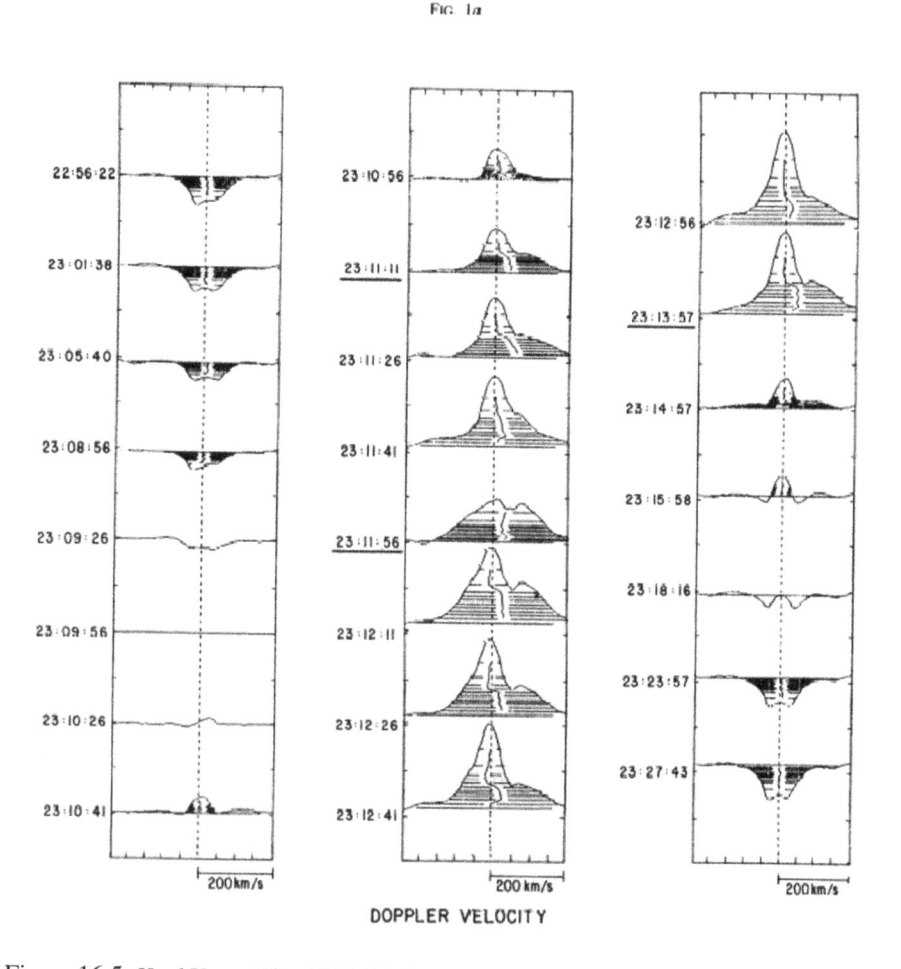

Figure 16.5: Hard X-ray (28 – 101 keV) time profile (top) and Hα line profile evolution during the 1980-Jun-23, 23:10 UT, flare. Note that the Hα line shows first absorption in the preflare phase, but strong emission during the hard X-ray phase. The Doppler shift of the line center is indicated with a bisector line at different flux levels (Canfield et al. 1990).

16.2.4 White-Light Emission

On very rare occasions, flares can even be observed in white light, which must originate deep in the chromosphere. The excitation mechanism of white-light flare emission is still unknown, but there is some consensus that the origin of the optical continuum of white-light flares should be associated with accelerated particles that penetrate deep into the dense chromosphere, based on the good temporal correlation between white light, hard X-rays, and microwave emission (Hudson 1972; Rust & Hegwer 1975; Neidig 1989; Hudson et al. 1992; Neidig & Kane 1993; Rieger et al. 1996; Ding et al. 1999). Hudson (1972) finds that the $\gtrsim 5$ keV electrons in major flares have enough energy to create long-lived excess ionization in the heated chromosphere high above the photosphere to enhance the free-free and free-bound continuum as occasionally seen in white light. *Yohkoh* observations, (the first white-light observations of solar flares from space), actually show white-light emission more frequently, down to the weakest flares (Hudson et al. 1992; Sylwester & Sylwester 2000; Matthews et al. 2002; Metcalf et al. 2003). White-light flares have been characterized into two types. The more common type I events show strong and broadened hydrogen Balmer lines and a Balmer and Paschen jump and demonstrate a good correlation with signatures of nonthermal electrons (in hard X-rays and microwaves). Type I white-light flares thus have been interpreted in terms of flare energy deposition in the chromosphere, where the white-light continuum is produced predominantly by hydrogen recombination, with the energy transport to lower levels in the chromosphere accomplished by photo-ionization (Fig. 16.6), also called *radiative backwarming* (Hudson 1972; Metcalf et al. 1990a,b, 2003; Ding & Fang 1996). Type II white-light flares do not show this strong chromospheric effect and are suspected to be produced by an energy release in the photosphere or temperature minimum region (Ding et al. 1999).

16.2.5 UV Emission

Ultraviolet (UV) covers the wavelength range shorter than blue visible light (i.e., $\lambda \approx 100 - 3000$ Å), where the shorter wavelength range of $\lambda \approx 100 - 300$ Å is also called *extreme ultraviolet (EUV)* (Fig. 1.25). The UV range includes strong lines with formation temperatures typical for the transition region [e.g., O V (629 Å, $T \approx 0.25$ MK), O IV (1404 Å, $T \approx 0.17$ MK)], which have been studied with *SMM/UVSP* during flares (e.g., see the reviews by Cheng 1999, Dennis 1988). Other instruments in the UV range were designed to observe active regions (*SoHO/SUMER*, Wilhelm et al. 1995), or the solar wind (*SoHO/UVCS*, Kohl et al. 1995). Flare-related UV emission was also imaged with *TRACE* (e.g., Warren & Warshall 2001).

A key observation for the understanding of UV line emission during flares is the detailed coincidence of hard X-ray pulses with UV time profiles, such as in O V, which shows a cross-correlation delay of as little as $\lesssim 0.1 - 0.3$ s (Fig. 16.7), as reported by Woodgate et al. (1983), Orwig & Woodgate (1986), and Cheng et al. (1988). Despite this detailed coincidence, the causal relation between the two emissions is difficult to understand. In the framework of the thick-target model (§ 13.2.2), hard X-ray electrons lose their energy in the transition region and upper chromosphere at heights of $h \approx 2000 - 5000$ km (Fig. 13.28), while the UV continuum emission is expected to

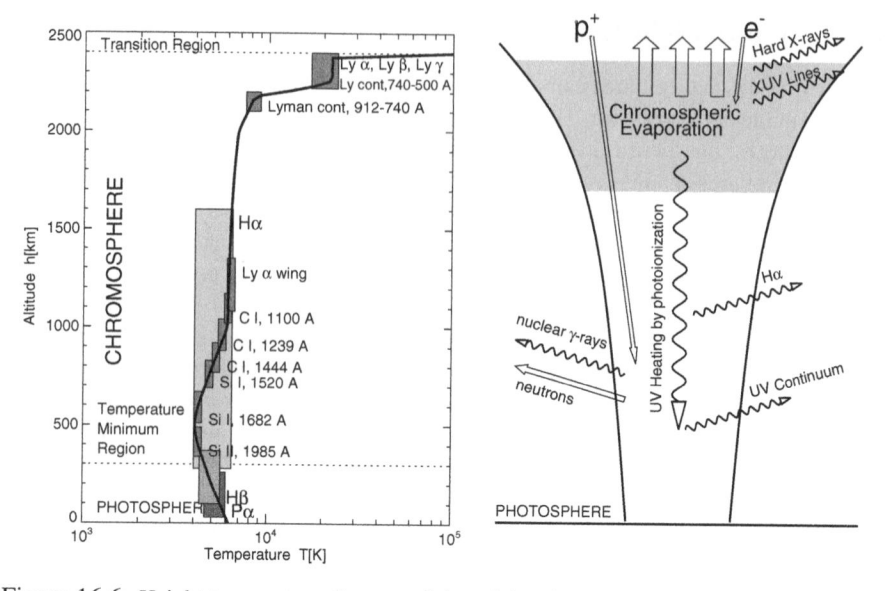

Figure 16.6: Height-temperature diagram of the origin of Hα, Lyman line emission (Lα, Lβ, Lγ), Lyman continuum, UV lines ($T \gtrsim 20,000$ K), and UV continuum (Si I, Si II, C I) are shown (left), along with a temperature model (thick line) according to the model of Vernazza et al. (1973). The corresponding heights of the various emissions (hard X-rays, XUV lines, nuclear γ-rays, neutrons, Hα, and UV continuum) in a flare loop footpoint are sketched (right). Chromospheric evaporation is caused by heating of precipitating electrons, nuclear γ-rays by precipitating protons, and UV continuum emission by photo-ionization from collisional bremsstrahlung in the upper chromosphere, also called *backwarming* (adapted from Vernazza et al. 1973; Brown & Smith 1980; Dennis 1988).

originate in the temperature minimum region at lower heights of $h \approx 500 - 800$ km (Figs. 16.6 and 1.19). So the nonthermal hard X-ray electrons do not penetrate deep enough to cause direct heating and collisional excitation of UV lines in the temperature minimum region. Neither does thermal conduction work, because it takes too long and is ineffective at these depths. So, an alternative explanation was suggested by Machado & Mauas (1987). Since the $\lambda = 1350 - 1680$ Å UV continuum radiation of the quiet Sun as well as in flare conditions originates in the temperature minimum region and is primarily due to Si I, after electron capture by Si II, they propose that UV line emission from the transition region (mainly the C IV resonance line at 1549 Å) increases the amount of Si II in the temperature minimum region by UV photo-ionization (Fig. 16.6), which is also called *radiative backwarming* (e.g., Metcalf et al. 2003). Thus the increase in UV emission during flares can more conveniently be explained by photo-ionization of the hard X-ray producing electrons, because (1) it requires about four orders of magnitude less energy input than by collisional excitation from the same hard X-ray electrons, and (2) because photo-ionization occurs more rapidly. Evidence for this model of photo-ionization of neutral silicon atoms (Si I) near the temperature minimum region by enhanced UV emission has been verified by quan-

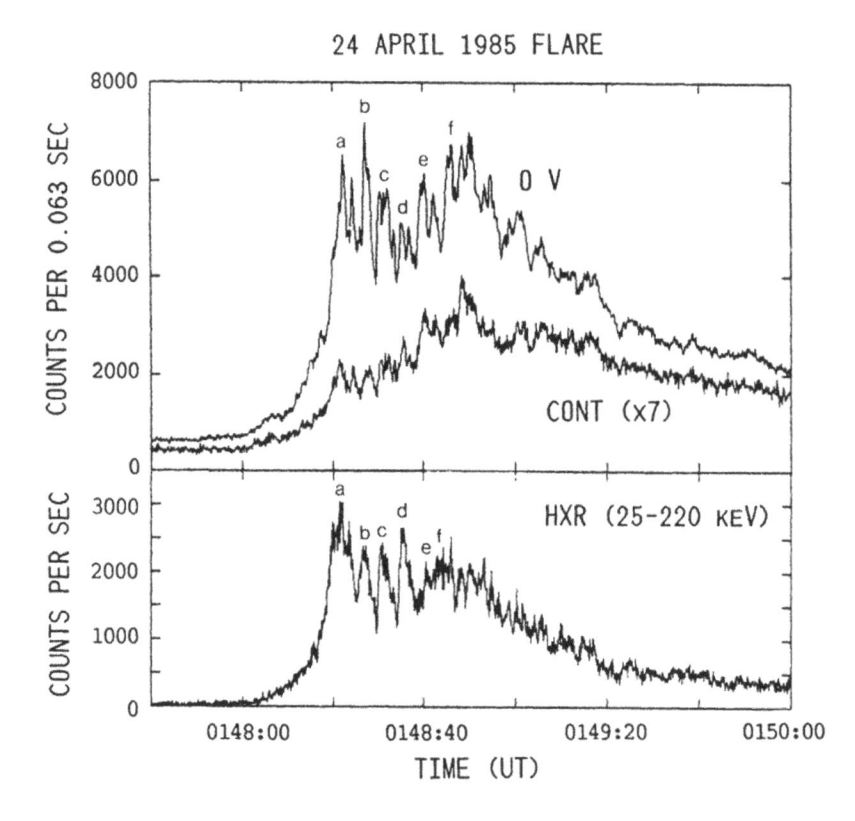

Figure 16.7: Time profiles of (25 − 200 keV) hard X-rays (bottom), O V line, and UV continuum (top) during the 1985-Apr-24, 01:48 UT, flare. Note the detailed one-to-one correspondence of six impulsive peaks (labeled with a through f), (Cheng et al. 1988; Dennis 1988).

titative correlations between C II (1335 Å) or C IV (1584 Å) fluxes and Si I (1520 Å) fluxes in two flares (Doyle & Phillips 1992).

16.3 Chromospheric Evaporation

Although the term *"evaporation"* designates a change of state in classical physics (i.e., from liquid to a gaseous state), the same term became popular in solar flare physics, although it corresponds actually not to a change of state, and should rather be called *"ablation"*. According to the theoretical model of the *"chromospheric evaporation"* process, the primary flare energy is conveyed to the chromosphere in the form of particle precipitation or heat conduction. It heats up the chromospheric material at a sufficiently rapid rate that it reaches coronal and flare temperatures ($T \approx 5 - 35$ MK), and driven by the overpressure, subsequently expands upward into the coronal flare loops, where it emits soft X-ray emission. Various reviews related to the topics of chromospheric evaporation have been spawned by observations from the *Skylab*, *Hinotori*, and *SMM*

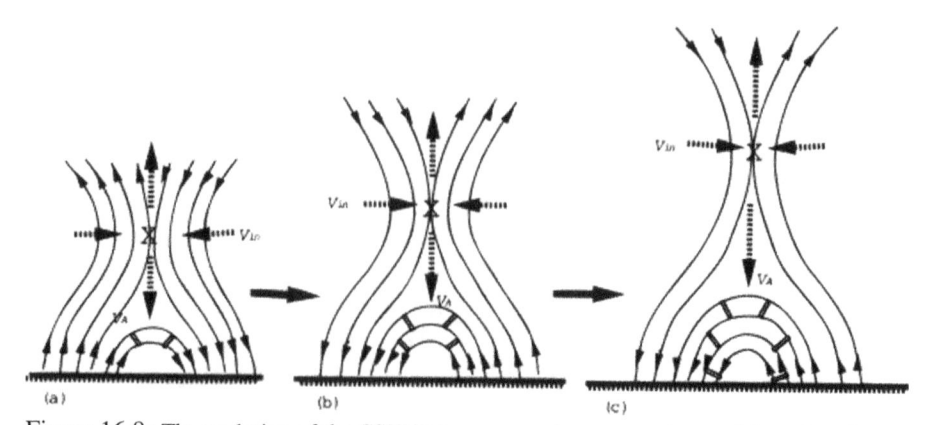

Figure 16.8: The evolution of the CSHKP-type magnetic reconnection model predicts that the magnetic reconnection point rises in altitude with time, and that chromospheric evaporation is initiated sequentially in overlying loops with increasing footpoint separation (Hori et al. 1997).

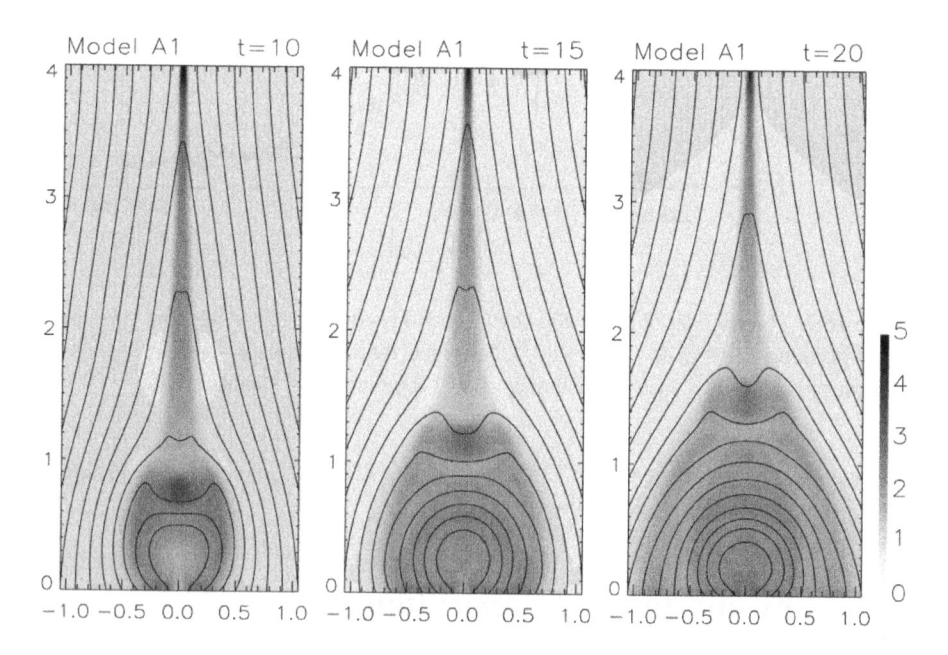

Figure 16.9: Temperature (greyscale) and magnetic field evolution (contours) in a dynamic 2.5-dimensional numerical hydrodynamic simulation of magnetic reconnection. Note that the hot flare loops show an increasing footpoint separation with time (Chen et al. 1999b).

missions (Sturrock 1973, 1980; Canfield et al. 1980; Moore et al. 1980, Doschek et al. 1986; Doschek 1990, 1991; Wu et al. 1986; Canfield et al. 1986; Canfield 1986a,b; Tanaka 1987; Watanabe 1987; Antonucci 1989; Antonucci et al. 1999; Emslie 1989; Zarro 1992; Bornmann 1999). Theoretical treatments can be found in, for example, Brown & Emslie (1989).

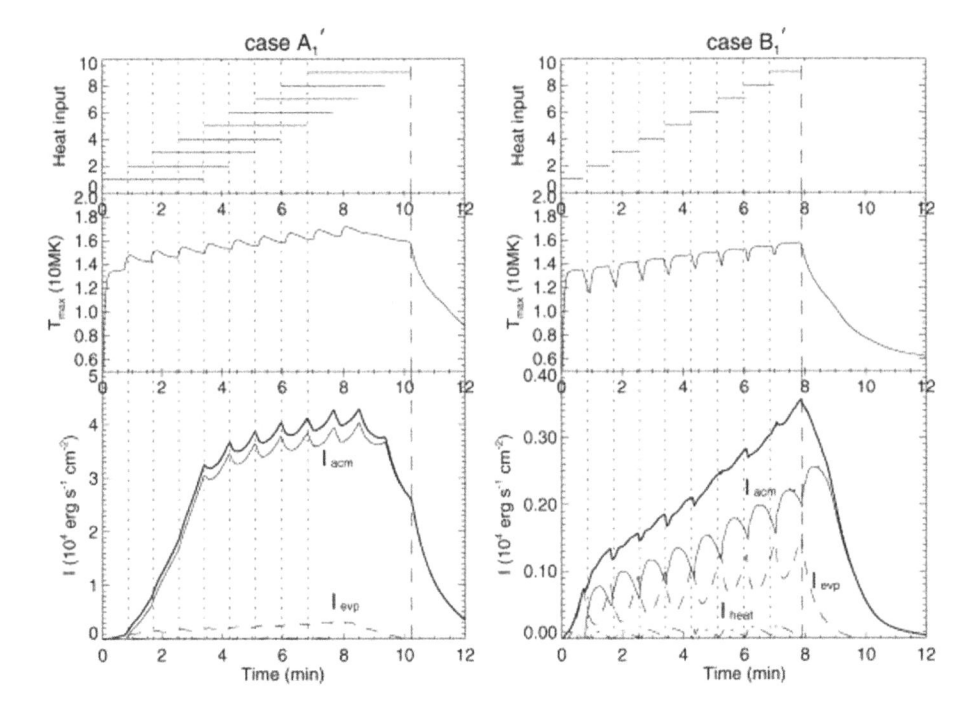

Figure 16.10: Temperature and intensity (or emission measure) evolution in the scenario of sequentially reconnecting flare loops as shown in Figs. 16.8 and 16.9. The heating functions of sequentially heated flare loops are overlapping (case A) or exactly sequential (case B). The intensity is calculated for the Ca XIX line, with the directly heated (dash-dotted curve), evaporated (dashed), accumulated (thin solid curve), and sum of the components (thick solid line) (Hori et al. 1998).

16.3.1 Hydrodynamic Simulations of Chromospheric Evaporation

The process of impulsive heating of chromospheric plasma has been simulated with hydrodynamic (HD) codes, which demonstrated the response of the chromosphere in the form of rapid pressure build-up that drives upflows of heated plasmas into coronal loops. Most of the early numerical HD simulations used a 1D hydrodynamic code, which solved the time-dependent hydrodynamic equations of particle, momentum, and energy conservation (Eqs. 4.1.24−26) for a given initial condition of rapid heating, by incorporating the effects of plasma heating, radiative cooling, thermal conduction (in the energy equation), and viscous damping (in the momentum equation, see Eqs. 6.1.12−14). The basic evolution, that impulsive heating would not immediately be dissipated in the chromosphere by radiative loss, but rather would heat up the local plasma and drive significant mass motions due to the high overpressure (i.e., upflows into the coronal parts of the connected loops) was already recognized earlier, and has been called *chromospheric evaporation* since (Neupert 1968; Antiochos & Sturrock 1978; Acton et al. 1982; Doschek et al. 1986). The heating function $E_H(s, t)$ is usually split into two components, containing a steady low heating rate to maintain a steady state corresponding to the background corona, and an impulsive high heating rate to mimic either the collisional energy loss by precipitating particles or a thermal conduction front originating from the energy release at the looptop. It was found that the chromospheric response produces a *gentle upflow* for heating rates below $E_H \lesssim 10^{10}$ (erg cm^{-2} s^{-1}), because the plasma is heated only to temperatures $T \lesssim 0.2$ MK, which is below the peak of the radiative loss function (Fig. 2.14) and thus the increasing radiative loss rate with temperature limits the temperature and pressure increase. Above this critical heating rate, *explosive upflows* are generated with speeds exceeding the sound speed c_s, with a supersonic limit of v$/c_s \lesssim 2.35$ (Fisher et al. 1984; 1985b). Parametric studies show that flare loop densities up to $n_e \lesssim 5 \times 10^{11}$ cm^{-3} can be achieved (Tsiklauri et al. 2004), which is about two orders of magnitude higher than the ambient corona in active regions. Numerical HD simulations of the chromospheric evaporation process have been conducted for two different drivers (or heating scenarios): for nonthermal particle precipitation (e.g., Somov et al. 1981; Bloomberg et al. 1977; MacNeice et al. 1984; Nagai & Emslie 1984; Fisher et al. 1985a,b,c; Mariska & Poland 1985) and for heat conduction from the looptop (e.g., Nagai 1980; Somov et al. 1982; Cheng et al. 1984; Pallavicini & Peres 1983; MacNeice 1986; Fisher 1986; Mariska et al. 1989; Gan et al. 1991; Falchi & Mauas 2002).

The various simulations differ in their numerical techniques and attempt to achieve sufficient temporal and spatial resolution in the critical transition region with adaptive grids. More advanced models of the chromospheric structure include radiative loss of optically thick emission, which demands radiative transfer calculations (McClymont & Canfield 1983b; Fisher et al. 1985a,b,c; Fisher 1986). A concise summary of design issues and problems of numerical HD simulations can be found in Antonucci et al. (1999).

Some numerical simulations include a two-fluid plasma, so that partial ionization of hydrogen in the chromosphere can be reproduced. MacNeice et al. (1984) uses two temperature equations, one for the electrons and the other for the neutral atoms and positive ions of hydrogen. Pure proton beams (with energies of ≈ 1 MeV), however, seem not to reproduce the observed velocity differential emission distribution better than electron beam models (Emslie et al. 1998).

One issue is whether the *Spitzer–Härm formula* (Eq. 3.6.3) for thermal conduction is valid in all parts of a flare loop. Strictly speaking, the Spitzer–Härm solution is only accurate in a strongly collisional regime where the ratio of the mean free path length of a thermal electron to the thermal scale height (the *Knudsen parameter*) is small ($\lesssim 2\%$). This issue was investigated by Ljepojevic & MacNeice (1989), who found significant departures from the Spitzer–Härm solution in the upper transition region and lower corona of a preflare loop, but not in the lower transition region where the thermal scale height is much shorter.

The newer generation of hydrodynamic codes are 2D and 2.5D, which also allows inclusion of the dynamics of the reconnection region (Figs. 16.8 and 16.10), such as the altitude rise of the reconnection point and footpoint separation of the flare loops (Yokoyama & Shibata 1998, 2001; Chen et al. 1999b), and the co-existence of soft X-ray emitting loops produced in sequential order with altitudes increasing with time (Hori et al. 1997, 1998). Ultimate versions may attempt to incorporate self-consistently

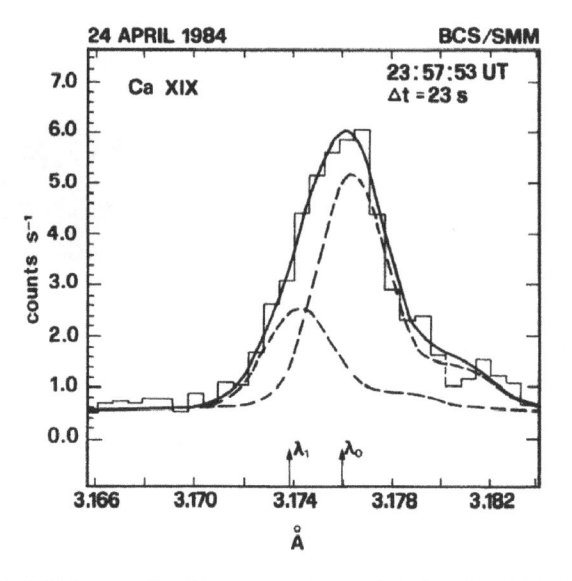

Figure 16.11: Ca XIX line profile (histogram) observed during the 1984-Apr-24, 23:52 UT, flare, fitted with the sum (solid curve) of a stationary and a blueshifted component (dashed lines) corresponding to a line-of-sight velocity of v = 210 km s^{-1} (Antonucci 1989).

the 3D-MHD, ionization balance, atomic physics, radiation transport, particle kinetics, and nuclear physics.

16.3.2 Line Observations of Chromospheric Evaporation

Once a time-dependent hydrodynamic simulation is calculated, in terms of the density $n_e(s,t)$, velocity v(s,t), and temperature evolution $T_e(s,t)$, the next step is to convolve these physical parameters with the instrumental response function of a given filter passband, in order to mimic the observables. For instance, the time evolution of soft X-ray and EUV line intensities has been simulated for O V, 1371 Å (MacNeice 1986; Mariska & Poland 1985), Ca XIX 3.177 Å (Nagai & Emslie 1984; Li et al. 1989; Mariska 1995), O VIII, Mg XI, Ne IV, Si XIII, S XV, Ca XIX, Fe XXV (Pallavicini & Peres 1983), Fe XXV, Ca XIX, Fe XXI (Cheng et al. 1984), or *Yohkoh/SXT* filters (Peres & Reale 1993a,b; Bornmann & Lemen 1994). These are mostly soft X-ray lines in the 1 − 22 Å wavelength range, which are formed due to transitions in the higher ionization stages of the most abundant ions, emitted in plasmas at temperatures of $T \gtrsim 10$ MK. Quantitative analysis of the soft X-ray line profiles (i.e., measurements of the line flux, Doppler shifts, and line broadening) provides crucial information on flow velocities, turbulent velocities, electron temperatures ($T \approx 2 - 40$ MK), differential emission measure distributions, ionization states, and in some cases electron densities ($n_e \approx 10^{10} - 3 \times 10^{12}$ cm^{-3}), filling factors ($q_{fill} \approx 3 \times 10^{-4} - 1$), and elemental abundances. The analysis of Ca XIX, Fe XXV, Fe XXVI, and Mg XI spectra is described in a number of studies (Antonucci et al. 1982, 1984b, 1985, 1990a; Tanaka et al. 1982; Antonucci & Dennis 1983; Canfield et al. 1983; Zarro et al. 1988a,b; Fludra

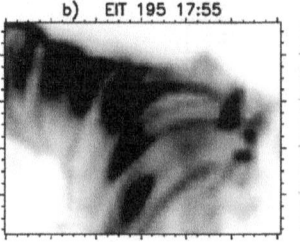
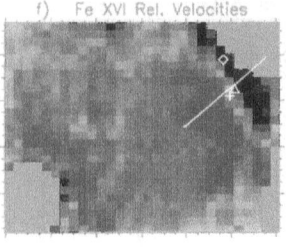
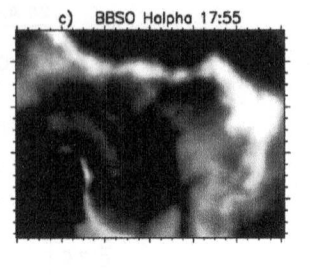

Figure 16.12: Flare of 1998-Apr-29, 16:37 UT, observed in EUV (SoHO/EIT 195 Å, left), in Doppler shift (SoHO/CDS Fe XVI; middle), and in Hα (BBSO; right). Note that the magnetic configuration corresponds to a loop arcade, curved around the sunspot in the south-east of the image. Blueshifts are observed at the outer and inner ribbons of the curved arcade (black color in middle image), while redshifts are found in the interior of the arcade (Czaykowska et al. 1999).

et al. 1989) and reviews on the subject are given in Antonucci (1986, 1989), Canfield (1986a,b), Doschek et al. (1986), Doschek (1990), and Antonucci et al. (1999).

A puzzle that existed for a long time is that the hydrodynamic simulations produced strong blueshifts and redshifts in soft X-ray and EUV lines due to the upflows of evaporated material from the flare loop footpoints, while the observations revealed only relatively small blueshifted components (e.g., Cheng et al. 1984). An example of a prominent blueshift is shown in Fig. 16.11, corresponding to an upflow speed of $v = 210$ km s^{-1}, measured in the Ca XIX line (Antonucci 1989). For Fe XXV, an even higher line-of-sight velocity of $v = 480$ km s^{-1} was measured in the same flare. Such velocity measurements in different lines have been generalized with the concept of a continuous function, the so-called *velocity differential emission measure (VDEM)* distribution $dEM(v)/dv$, which in flares covers a range of $v \approx 100 - 1000$ km s^{-2} (Newton et al. 1995). The blueshift, which marks an upflow, is usually only seen in the rise of the impulsive flare phase (for typically $\Delta t \approx 10 - 20$ s), though it can occur repetitively in multi-loop flares. There are also *"gentle"* upflows that have been observed in the late flare phase (Schmieder et al. 1987, 1990), which are thought to be driven by downward heat conduction in postflare loops (Forbes & Malherbe 1986a,b; Forbes et al. 1989). Observational evidence for conduction-driven evaporation was also inferred from a linear correlation between the upward enthalpy flux and the downward thermal conductive flux (Zarro & Lemen 1988) and from the lack of hard X-ray emission (Czaykowska et al. 2001). The reason why large blueshifts never have been observed in flares might be a result of the confusion of separating upflows and downflows in space and time. If a number of overlying or neighboring loops experience time-shifted heating phases (Figs. 16.8 and 16.10), most of the blue and redshifts cancel out temporally and spatially. In a large two-ribbon flare observed with CDS, EIT, and MDI (Fig. 16.12), dominant blueshifts could only be spatially resolved and isolated at the outer edges of the flare ribbons, which advance outwards like a bush-fire front and map back to the latest (rising) magnetic reconnection points in the corona (Czaykowska et al. 1999).

Besides the blueshift feature, some lines also show a so-called *"nonthermal line broadening"* in excess of the natural line width and instrumental broadening (Antonucci et al. 1999). This term implies an interpretation in terms of nonthermal processes, such as turbulent motion or unresolved flows. However, the interpretation is ambiguous, because a superposition of many multiple thermal plasmas can also lead to an excessive line broadening. This line broadening is expressed in terms of a nonthermal velocity v_{nt},

$$v_{nt} = \sqrt{\frac{2k_B(T_D - T_e)}{m_i}} \, , \qquad (16.3.1)$$

where T_D is the Doppler temperature, T_e the electron temperature obtained from the line ratio diagnostics, and m_i is the mass of the ion considered. Typical values of line broadening are $v_{nt} = 100 - 200$ km s^{-1}, as observed in Ca XIX and Fe XXV lines (Ding et al. 1996; Harra–Murnion et al. 1997). One argument that the line broadening is a nonthermal effect is given by the fact that there is no significant difference found in the excess widths in Ca XIX ($T \approx 14-20$ MK) and Fe XXV lines ($T \approx 16-26$), which suggests identical ion and electron temperatures, so it cannot be explained by multithermal effects (Antonucci & Dodero 1986; Saba & Strong 1991). Models of Alfvén wave turbulence have been applied to soft X-ray lines by Alexander & MacKinnon (1993), Alexander & Matthews (1994), and MacKinnon (1991). In particular, Alfvénic outflows and *high-temperature turbulent current sheet* models (Somov 1992; Antonucci & Somov 1992; Antonucci et al. 1994) produce flow velocities that are compatible with the observed Doppler shifts in soft X-ray lines, but a spatial localization could not be established yet.

16.3.3 Imaging Observations of Chromospheric Evaporation

Early imaging observations of flare loops in soft X-rays, which trace the chromospheric evaporation process (also called *coronal explosions*) were accomplished with *SMM/ HXIS* at energies ≥ 3.5 keV with a time resolution of $\Delta t \approx 0.5 - 7$ s (e.g., Hoyng et al. 1981a; Duijveman et al. 1982; De Jager et al. 1984; De Jager & Boelee 1984; De Jager 1985; or in the reviews of De Jager 1986 and Antonucci et al. 1999). The heated plasma typically expands with a velocity of $v \approx 100 - 400$ km s^{-1} up into the loop, so it fills a loop with a length of $L = 10 - 20$ Mm in about $\Delta t = L/v \gtrsim 25$ s. The cooling of the hot flare loops, with a typical temperature of $T \approx 15$ MK, has been observed to occur with an e-folding time scale of $t_{cool} \approx 45$ s (De Jager 1985). In most of the flares there are multiple loops involved which sequentially become filled and cool down, which could not spatially be separated properly with the angular resolution $(8'', 32'')$ of *SMM/HXIS*.

The heliographic position of flares is expected to show a systematic center-to-limb variation of the Doppler blueshift, if the chromospheric upflows are detected in the vertical part near the flare loop footpoints. Such a center-to-limb effect $v = v_0 \cos{(l - l_0)}$ was indeed observed in Ca XIX, Fe XXV, and S XV, with a mean radial velocity of $v_0 \approx 60 - 80$ km s^{-1} in the middle of the flare rise phase (Mariska et al. 1993; Mariska 1994). Occulted flares were found to have a smaller nonthermal line broadening (Mariska et al. 1996) and a slightly lower temperature (Mariska & McTiernan 1999) than non-occulted flares, so the chromospheric upflows near the footpoints exhibit more

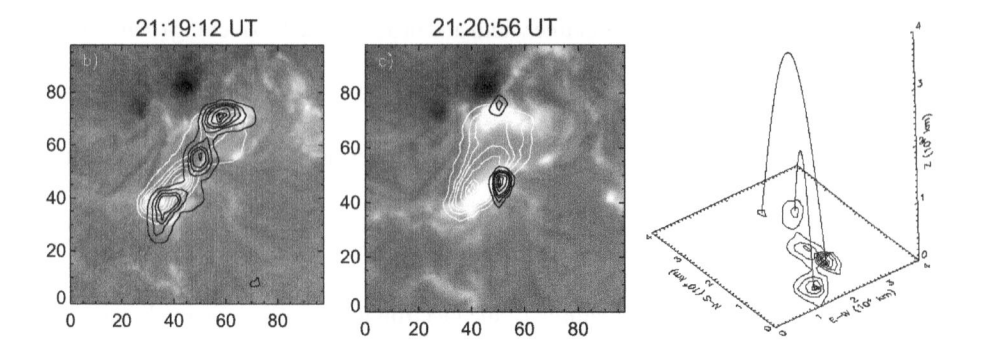

Figure 16.13: Evolution of chromospheric evaporation from a SE−NW oriented loop structure (left) to a N−S oriented loop structure (middle), possibly corresponding to the 3D configuration shown (right). The greyscale represents Hα images (BBSO), the black contours the hard X-ray footpoints (HXT), and the white contours the filled soft X-ray flare loops (SXT) (Silva et al. 1997b).

turbulent motion (or a larger spread of unresolved flow speeds and directions) and they cool down when reaching higher altitudes.

The context of hard X-ray and soft X-ray emission in the flare loop footpoints was studied in more detail with *Yohkoh/HXT*, *SXT*, and *BCS*. Maximum blueshifts in Ca XIX, which correspond to maximum upflow speeds in the chromospheric evaporation scenario, were noted two minutes prior to the hard X-ray peak (Silva et al. 1997b), which implies that the least confusion for upflows occurs at the beginning of the flare, while unresolved upflows and downflows almost cancel out during the flare peak. In the 1994-Jun-30, 21:19 UT flare, the process of chromospheric evaporation was observed to proceed from a lower lying flare loop to an overarching higher flare loop (Fig. 16.13; Silva et al. 1997b), similarly to the scenario of interacting flare loops in a quadrupolar 3D-configuration (Fig. 10.28, § 10.5.6). In another event (2000-Mar-16) an asymmetric behavior was observed in the heating of chromospheric footpoints, leading to asymmetric Hα emission in more or less symmetric soft X-ray and hard X-ray flare loops (Qiu et al. 2001). The asymmetry in the Hα footpoints was thus ascribed to asymmetric heating rates rather than to a magnetic mirror asymmetry. The chromospheric heating in flares actually produces two signatures, one is the upflowing heated plasma which shows an integral time profile of the hard X-rays (Neupert effect), and the other are *impulsive soft X-ray brightenings*, which is a direct radiation enhancement produced by the energy deposition in the ambient chromosphere around the footpoints (Hudson et al. 1994; Tomczak 1999; Mrozek & Tomczak 2002). Most of the studies on the *Neupert effect* (Neupert 1968) are restricted to soft X-ray and hard X-ray time profiles integrated over the entire flare location (Dennis & Zarro 1993; Li et al. 1993; Plunckett & Simnett 1994; Lee et al. 1995; McTiernan et al. 1999; Veronig et al. 2002a,b; Veronig 2003) and find significant deviations from the expected *empirical Neupert effect* (§ 13.5), which possibly could be resolved by discriminating the detailed spatial structure of multiple flare loops (*spatial Neupert effect*). The variation of plasma flow speeds as a function of height in flare loop footpoints has been for the first time probed

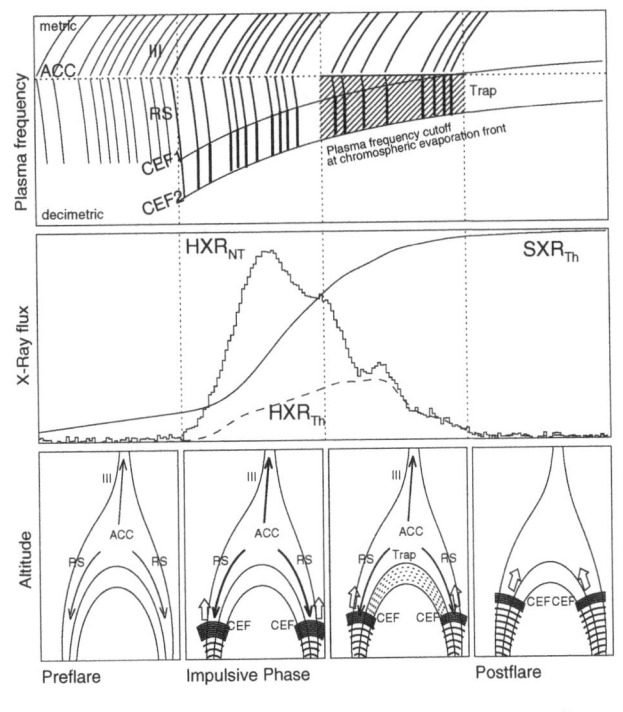

Figure 16.14: The effect of the chromospheric evaporation process on radio emission near the plasma frequency: the upward propagation of the chromospheric evaporation front (bottom) corresponds to a slow drift of the plasma frequency cutoff (top). Radio emission produced by downward propagating electron beams see a slowly drifting frequency cutoff (top). The time profile of the concomitant hard X-ray and soft X-ray emission is shown (middle) (Aschwanden & Benz 1995).

with CDS over a comprehensive temperature range of $log(T) = 4.3 - 6.9$, where strong downflows (v $\lesssim +40$ km s^{-1}) at chromospheric levels, strong upflows (v $\lesssim -100$ km s^{-1}) at transition region levels and (v $\lesssim 160$ km s^{-1}) at coronal levels were measured (Teriaca et al. 2003), as expected in theoretical chromospheric evaporation models.

16.3.4 Radio Emission and Chromospheric Evaporation

The chromospheric evaporation process also has its manifestation at radio wavelengths, as illustrated in Fig. 16.14. The basic effect is that the local disturbance of the electron density and temperature, introduced by the upflowing chromospheric plasma, is detectable from radio bursts emitted at the local plasma frequency. Plasma emission produced by electron beams has been observed in the lower corona up to a frequency of 8.4 GHz (Benz et al. 1992). The detection of plasma emission at such high frequencies requires overdense fluxtubes, so that plasma emission can escape in a direction perpendicular to the fluxtube axis, where the density scale height is much shorter than in a homogeneous corona, and thus, free-free absorption is substantially reduced. In the

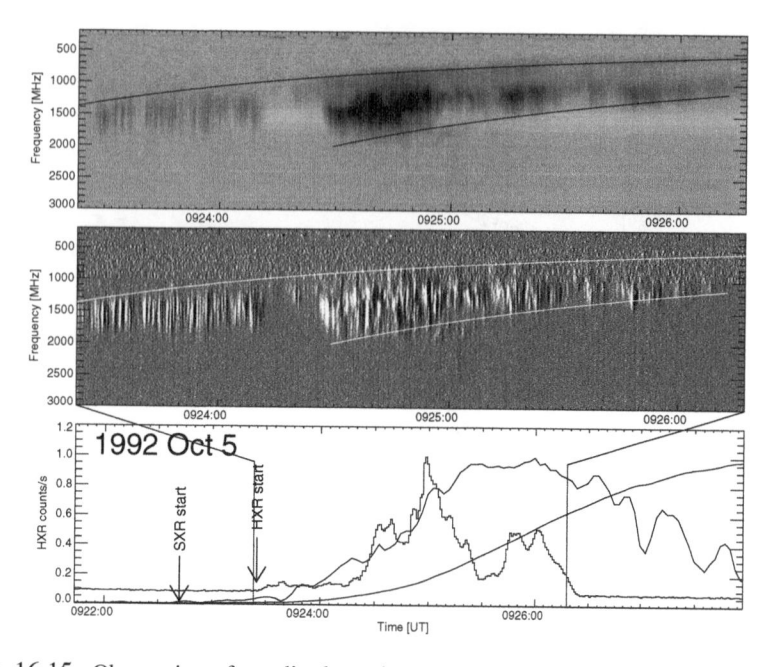

Figure 16.15: Observation of a radio dynamic spectrum during the 1992-Oct-05, 0924 UT, flare with the *Zurich* radio spectrometer *Phoenix*, shown in total flux (black in top panel), and as a gradient-filtered difference image (middle), with the hard X-ray ≥ 25 keV time profile observed with *CGRO* and 0.5−4 Å soft X-ray time profiles observed with *GOES* (bottom panel). Note the slowly drifting high-frequency cutoff (from 2000 to 1200 MHz) in the radio dynamic spectra, which indicates a plasma-frequency cutoff most likely resulting from the chromospheric evaporation front (Aschwanden & Benz 1995).

event of chromospheric evaporation, we expect that the upflowing plasma surrounds overdense fluxtubes and seals off escape routes for plasma emission, because the additional plasma material, if sufficiently dense, makes the escape routes optically thick due to free-free absorption. Since the evaporating plasma propagates upwards with a bulk speed of ≈ 300 km s^{-1}, it is expected to produce a slowly drifting high-frequency cutoff for plasma emission. This high-frequency cutoff is thought to apply to any kind of plasma emission originating in "evaporating" flare loops (e.g., to type III bursts excited by precipitating electron beams). The drift rate of this high-frequency cutoff for plasma emission (Eq. 15.4.10) is estimated to be

$$\left(\frac{d\nu}{dt}\right)_{CE} = \frac{\partial\nu}{\partial h}\frac{\partial h}{\partial t} \approx -\frac{\nu}{2\lambda}v_{CE}\cos\theta \;, \qquad (16.3.2)$$

where ν is the observed frequency (assumed to be at the fundamental plasma frequency), λ the local density scale height, h the altitude, v_{CE} the velocity of the chromospheric evaporation front, and θ the propagation angle with respect to the vertical. Typical physical parameters of the chromospheric evaporating plasma are listed in Table 16.2 (from Antonucci et al. 1984b). Based on these electron densities we calculate

Table 16.2: Typical parameters of flare plasma during chromospheric evaporation (Antonucci et al. 1984b).

Time phase	Velocity v_{CE} (km s^{-1})	Temperature T_e (MK)	Density n_e (10^{11} cm^{-3})	Plasma frequency ν_p (GHz)
Flare start			0.56−0.96	2.1−2.8
Flare end			2.0−3.5	4.0−5.3
Average	270±90	16.5±2.4	0.8−1.5	2.5−3.5

that the plasma frequency of the upflowing plasma varies in the range of $\nu_p = 2.5-3.5$ GHz, and can be as low as 2.1 GHz at the start of the flare. For an upward moving ($\theta = 0$) evaporation front, a velocity of $v_{CE} \approx 270$ km s^{-1}, a scale height range of $\lambda = 5 - 66$ Mm (inferred from decimetric type III bursts; Aschwanden & Benz 1986), and a frequency of $\nu = 2$ GHz, we estimate a drift rate of $(\partial\nu/\partial t)_{CE} = -8$ to -108 MHz s^{-1} for the high-frequency cutoff. From this considerations we expect slowly drifting high-frequency cutoffs at decimetric frequencies for various radio bursts that are related to plasma emission, preferentially at the start of the impulsive phase of flares. An example of a radio observation is shown in Fig. 16.15, which displays a sequence of decimetric pulsations or fast-drift bursts that have a slowly drifting high-frequency cutoff which moves from 2000 MHz at the beginning of the impulsive flare phase towards 1200 MHz at the end. This corresponds to a density change from $n_e = 5.0 \times 10^{10}$ cm^{-3} to 1.8×10^{10} cm^{-3} in the upward moving evaporation front, in the case of fundamental plasma emission. A detailed study on the plasma cutoff frequency and related free-free opacity in the context of a specific density model for a chromospheric evaporation model is given in Aschwanden & Benz (1995). Alternative studies on the influence of chromospheric evaporation on radio bursts can be found in Karlický et al. (2001, 2002a,b) or Karlický & Farnik (2003).

16.4 Postflare Loop Cooling

16.4.1 Cooling Delays of Flux Peaks

The temporal evolution of a flare can be subdivided into a heating and a cooling phase, separated by the temperature peak time t_p. In the heating phase ($t < t_p$), the heating rate exceeds the absorption rate due to thermal conduction and radiative loss, and thus, the temperature of the heated (evaporating) plasma rises. During the cooling phase ($t > t_p$), the heating rate drops below the sum of the conduction and radiative loss rate, and consequently, the flare plasma temperature drops. A consequence of this evolution is that the flux time profiles measured in different temperature filters exhibit a peak delay that progressively increases with decreasing temperature of the filters. This progressive cooling delay can be tracked in different wavelengths, from hard X-rays, soft X-rays, to EUV. An example is shown for the 2000-Jul-14 flare in Fig. 16.16 (top), which contains the co-registered time profiles from *Yohkoh/HXT*, *Yohkoh/SXT*,

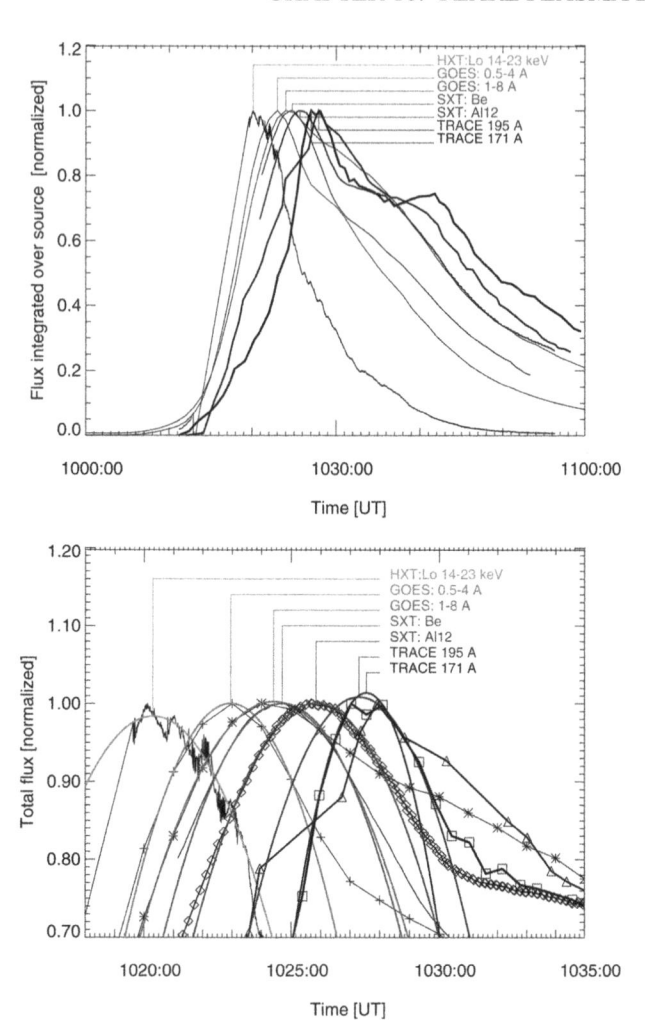

Figure 16.16: *Top:* Co-registered light curves of the Bastille-Day flare 2000-Jul-14, 10:10 UT, flare, from Yohkoh/HXT, Yohkoh/SXT, GOES, and TRACE, normalized to unity. *Bottom:* enlarged portion of the flare peak times, fitted by parabolic curves to determine the mean peak times for each wavelength. Only data points in the top 20% of the peak fluxes are considered for the parabolic fits. The relative time delays of the peaks are listed in Table 16.3. Note the systematic delay with decreasing temperature (Aschwanden & Alexander 2001).

GOES, and *TRACE*, normalized to unity. All light curves were integrated over the entire field-of-view of each instrument, which generally encompass most of the flaring region.

The co-registered time profiles show that emission in all wavelengths peak during 10:20−10:28 UT (Fig. 16.16, top). The time profiles in hard X-rays, soft X-rays, and EUV are all similar, except for an increasingly longer decay towards longer wavelengths or lower temperatures. To evaluate the centroid peak time of each light curve,

Table 16.3: Peak times of total flux in different instruments and wavelengths, and time delays relative to the hard X-ray HXT/Lo peak for the 2000-Jul-14 flare (Aschwanden & Alexander 2001).

Instrument Wavelength	Peak time t_i (MK)	Time delay Δt_i	Peak Response temperature T_i (MK)
HXT Lo 14−23 keV	10:20:18 UT	0 s	28.7
GOES 0.5−4 Å	10:22:57 UT	159 s	20.1
GOES 1.0−8 Å	10:24:24 UT	246 s	16.2
SXT Be	10:24:42 UT	264 s	14.4
SXT Al12	10:25:50 UT	332 s	11.2
TRACE 195 Å	10:27:18 UT	420 s	1.5
TRACE 171 Å	10:27:33 UT	435 s	1.0

a parabolic curve is fitted to the peak at each wavelength (Fig. 16.16, bottom), using the data points in the top 20% near the peak flux. One finds that there is a well-defined progression of peak times between 10:20 and 10:28 UT, with the higher temperatures peaking first, starting with thermal hard X-rays, followed by soft X-rays and EUV, in order of decreasing temperature (see Table 16.3). So the flare peaks in different wavelengths are dispersed over about 7 minutes, with an increasing time delay for filters with lower temperatures, as expected for a cooling flare plasma.

16.4.2 Differential Emission Measure Evolution

If we assume that all loops are heated approximately to the same peak temperature and cool down with a similar time evolution, we can determine the cooling function $T(t)$ of a single loop just from the statistical average of all simultaneously brightening loops, according to the superposition principle. In essence, if we add up the flux profiles $F_n(T[t - t_n])$ of many loops (indexed with subscript n) with arbitrary relative phases of their maximum heating time t_n, then the superposition $\Sigma F_n(T[t - t_n])$ will exhibit (in first order) the same temperature-dependent time delays $\Delta t(T)$ as a single loop. The flux time profiles shown in Fig. 16.16 (top) consist of such a superposition of a multitude of flare loops.

In the next step we determine the temperatures T_i of the peak response of each filter i. The instrumental response functions $R_i(T)$ are shown in Fig. 16.17 (middle). Then we model a differential emission measure distribution $dEM(T)/dT$ (Fig. 16.17, top) that we convolve with the instrumental response functions $R_i(T)$ to obtain the contribution functions $dF_i(T)/dT$ (Fig. 16.17 bottom) for each instrument i, and then the instrument fluxes F_i by integrating over the temperature T,

$$F_i = \int \frac{dF_i(T)}{dT} dT = \int R_i(T) \otimes \frac{dEM(T)}{dT} dT \,. \qquad (16.4.1)$$

where the response function is defined by $R_i(T) = dF_i(T)/dEM(T)$. Thus, the average emission measure distribution $dEM(T)/dT$ (Fig. 16.17, top or Fig. 16.18, top

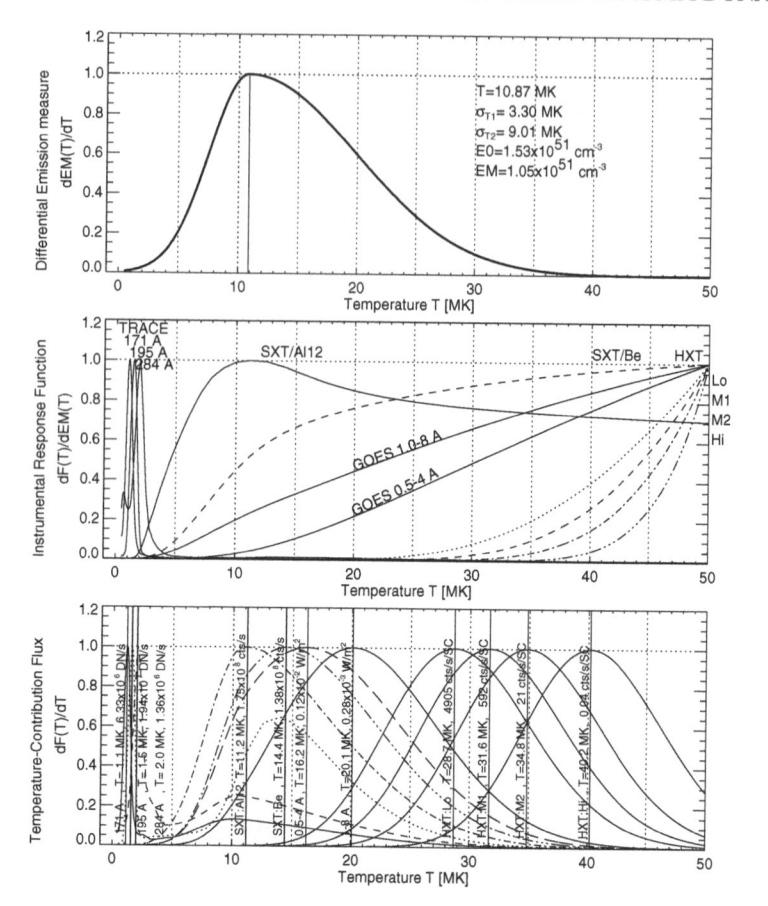

Figure 16.17: *Top:* best-fit model of differential emission measure distribution during the Bastille-Day flare (2000-Jul-14), constrained by the fluxes observed with TRACE, GOES, and YOHKOH. *Middle:* instrumental response functions, normalized to unity in the temperature range of $T = 0.1 - 50$ MK. *Bottom:* temperature contribution functions of the fluxes $dF_i(T)/dT$ detected by each instrument filter, normalized to unity. The absolute flux values and peak temperatures are labeled for each instrument filter (Aschwanden & Alexander 2001).

right) is constrained by the instrument fluxes F_i, as indicated in Fig. 16.17 (bottom). The peak response temperature T_i of the contribution functions are also indicated in Fig. 16.17 (bottom) and are listed in Table 16.3. From the observations we measure a cooling delay $\Delta t_i(T_i)$ (Table 16.3). Inverting this cooling delay, we obtain a cooling function $T_i(\Delta t_i)$ in Fig. 16.18 (top left). Since this cooling function is monotonic, we can also substitute the cooling function $T_i(\Delta t)$ into the differential emission measure distribution $dEM(T_i)/dT$ to obtain the evolution function of the differential emission measure, $dEM(T_i)/dT = dEM(T_i[\Delta t_i])/dT = dEM(\Delta t_i)/dT$, which is shown in Fig. 16.18 (bottom left). Using the standard definition of the volumetric emission measure, $EM_V = n_e^2 dV$, we can then also infer the evolution of the mean electron density in the flare volume, $n_e(t)$, shown in Fig. 16.18 (bottom right).

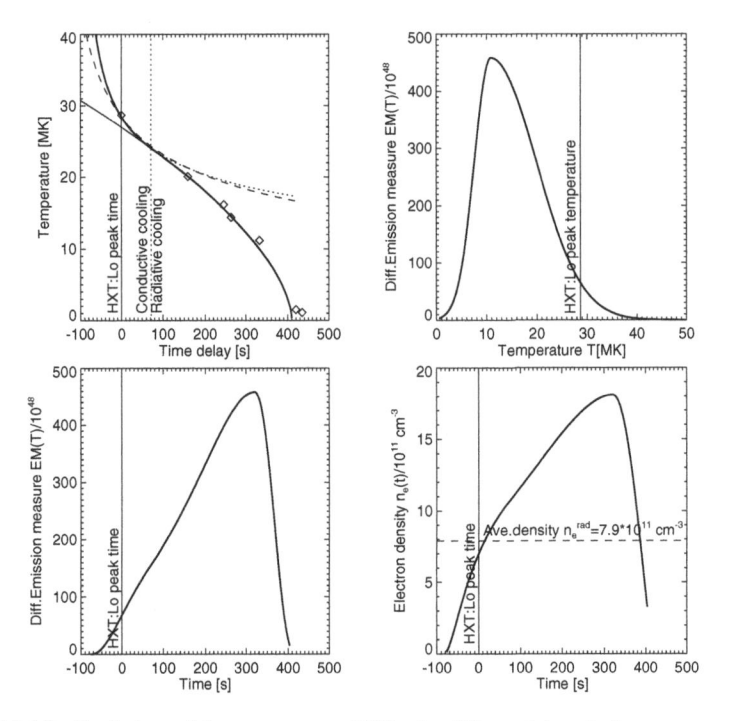

Figure 16.18: Evolution of the temperature $T(t)$, the differential emission measure $EM(t)$, and electron density $n_e(t)$ of an average flare loop during the peak of the Bastille-Day flare on 2000-Jul-14, 10:10 UT. Theoretical models of the cooling function $T(t)$ include conductive cooling models: Antiochos & Sturrock 1978, dotted line; Eq. (16.4.6); Culhane et al. 1994, dashed line; Eq. (16.4.3); and the radiative cooling model, solid curves; Eqs. (16.4.8) and (16.4.13), fitting the observed peak time delays $t_i(T)$ (diamonds) (Aschwanden & Alexander 2001).

16.4.3   Conductive Cooling

Let us understand the physical nature of this observed cooling function $T(t)$. Because the temperature T_1 and temperature gradients $\partial T(t = t_1)/\partial s$ of a loop heated at time $t = t_1$ are initially highest, it is expected that conductive cooling dominates initially. Neglecting all other terms in the hydrodynamic energy equation, the change in internal energy $de(t)/dt$ is then mainly balanced by the thermal conduction loss rate [using Eqs. (3.6.3), (4.1.13), and (4.1.23)],

$$mn\frac{de(t)}{dt} = \frac{d}{dt}[3n_e(t)k_B T(t)] = -\frac{d}{ds}\left(\kappa T^{5/2}\frac{dT(t)}{ds}\right) \approx -\frac{2}{7}\kappa\frac{T(t)^{7/2}}{L^2}\,, \quad (16.4.2)$$

with κ the Spitzer thermal conductivity and L the loop half length. This differential equation (Eq. 16.4.2) can be directly integrated, if we apply the mean-value theorem to the time dependence of the density $n_e(t)$, by replacing it by the mean value $< n_e >$

outside of the integral. The resulting temperature evolution is then

$$T(t) = T_1 \left(1 + \frac{(t - t_1)}{\tau_{cond}}\right)^{-2/5} , \qquad (16.4.3)$$

with a "conduction time scale" τ_{cond} defined by

$$\tau_{cond} = \frac{21}{5} \frac{<n_e> k_B L^2}{\kappa T_1^{5/2}} . \qquad (16.4.4)$$

This solution was used in Culhane et al. (1994) to fit the cooling of a flare plasma from $T_1 = 22$ MK down to 12 MK. Antiochos & Sturrock (1978) derived a more general cooling function by including the flow velocities v in the hydrodynamic energy equation (using Eq. 4.1.23)

$$-\frac{5}{2} \frac{p}{n_e} \left(\frac{\partial n_e}{\partial t} + v \frac{dn_e}{ds}\right) = \frac{d}{ds}(\kappa \frac{dT}{ds}) . \qquad (16.4.5)$$

Solving this hydrodynamic equation they find a slightly different solution for the cooling curve:

$$T(t) = T_1 \left(1 + \frac{(t - t_1)}{\tau_{cond}}\right)^{-2/7} . \qquad (16.4.6)$$

We fit both solutions from Culhane et al. (1994) and Antiochos (1980) to our measured cooling curve $T(t)$, in the initial time interval at the highest temperature (Fig. 16.18, top left), but do not find agreement for the later time intervals (when the temperature drops below $T \approx 25$ MK), and thus conclude that *the later cooling phase of this flare is not dominated by conductive cooling.*

16.4.4   Radiative Cooling

Alternatively, we consider dominant radiative cooling. Equating the internal energy loss $de(t)/dt$ to the radiative cooling rate,

$$\frac{d}{dt}[3n_e(t)k_B T(t)] = -n_e(t)^2 \Lambda(T[t]) \approx -n_e(t)^2 \Lambda_0 T(t)^{-2/3} , \qquad (16.4.7)$$

according to the piece-wise powerlaw approximation (Fig. 2.14) of the radiative loss function $\Lambda(T)$ by Rosner et al. (1978a), and applying again the mean-value theorem for the time-dependence of the density $n_e(t)$, we can integrate Eq. (16.4.7) analytically and find the solution

$$T(t) = T_1 \left[1 - \frac{(t - t_1)}{\tau_{rad}}\right]^{3/5} \qquad t_1 < t < \tau_{rad} , \qquad (16.4.8)$$

with a "radiative cooling time" τ_{rad} defined by

$$\tau_{rad} = \frac{9k_B T_1^{5/3}}{5 <n_e> \Lambda_0} . \qquad (16.4.9)$$

We fit this radiative cooling function to the data in Fig. 16.18 and find a surprisingly good fit, except for the first data point. The best-fit parameters are $T_1 = 27$ MK and $\tau_{rad} = 420$ s, yielding a mean density of

$$< n_e >_{rad} = \frac{9 k_B T_1^{5/3}}{5 \Lambda_0 \tau_{rad}} , \tag{16.4.10}$$

with a value of $< n_e >= 7.9 \times 10^{11}$ cm^{-3}. Therefore, we conclude *that the later phase of this flare is dominated by radiative cooling.*

For an analytical model we can synthesize the conductive cooling phase ($T_{cond}(t)$, Eq. 16.4.6) in the initial phase of loop heating with the dominant radiative cooling ($T_{rad}(t)$, Eq. 16.4.8) in the later phase. We join the two analytical solutions smoothly together at time t_2, where we require a steady function, $T_{cond}(t = t_2) = T_{rad}(t = t_2)$ and a smooth derivative, $dT_{cond}(t = t_2)/dt = dT_{rad}(t = t_2)/dt$. These two boundary conditions yield with Eqs. (16.4.6) and (16.4.8) the transition time t_2,

$$(t_2 - t_1) = \frac{10 \tau_{rad} - 21 \tau'_{cond}}{31} , \tag{16.4.11}$$

and initial temperature T_2 of the conductive phase,

$$T_2 = T_1 \frac{(1 - t_2/\tau_{rad})^{3/5}}{(1 + t_2/\tau'_{cond})^{-2/7}} . \tag{16.4.12}$$

The synthesized cooling model $T(t)$ reads then (for a reference time t_1),

$$T(t) = \begin{cases} T_2 \left(1 + \frac{(t-t_1)}{\tau'_{cond}} \right)^{-2/7} & \text{for } t_1 < t < t_2, \\[2mm] T_1 \left(1 - \frac{(t-t_1)}{\tau_{rad}} \right)^{3/5} & \text{for } t_2 < t < \tau_{rad}. \end{cases} \tag{16.4.13}$$

We find an initial temperature of $T_2 = 28.4$ MK (at time t_1) and a transition time of $t_2 - t_1 = 71$ s, with a total cooling time of $\tau_{rad} = 420$ s (i.e., 7 minutes).

For comparison, flare plasma cooling was modeled by Culhane et al. (1994) from Yohkoh/HXT and BCS data, and the data were found to be consistent with conductive cooling in the temperature range from 23 MK down to 11 MK, over a time period of 180 s. Because the electron densities are comparable in the two flares within a factor of $\lesssim 2$, the two obtained cooling functions are consistent in the first phase with dominant conductive cooling, while the second phase with dominant radiative cooling could not be reconstructed in the study by Culhane et al. (1994) due to the temperature restriction of BCS. Our case contradicts theoretical models that assume that conductive cooling mostly dominates over radiative cooling (Antiochos & Sturrock 1978), but strongly supports models with dominant radiative cooling (e.g., Antiochos 1980), and models with initial conductive cooling followed by radiative cooling (e.g., Cargill et al. 1995). Our observationally obtained plasma cooling function $T(t)$ has good similarity with theoretically obtained cooling functions, which typically reach a maximum of $T \approx 30$ MK within 200 s, and then cool down to $T = 1$ MK in the subsequent next 800 s (e.g., Fisher & Hawley 1990). A new technique to quantify the flare heating and cooling function has been explored by modeling the time evolution of the EUV brightness (e.g., with 171 Å TRACE data) in particular locations of the flare loop footpoint ribbons (Antiochos et al. 2000b). Cooling times have been measured from $T = 10^7$ K in soft X-rays all the way to $T = 10^4$ K in Hα (Kamio et al. 2003). The cooling of postflare loops is now more realistically modeled with composites that contain superpositions of hundreds of individual flare loops (Reeves & Warren 2002). Moreover, the process of cooling in postflare loops is now perceived as part of a dynamic process, where localized cooling of coronal plasma by the thermal instability triggers magnetic reconnection through the enhanced instability, rather than being a mere passive cooling process (Kliem et al. 2002).

16.5 Summary

The flare plasma dynamics and associated thermal evolution of the flare plasma consists of a number of sequential processes: plasma heating in coronal reconnection sites (§ 16.1), chromospheric flare plasma heating (§ 16.2), chromospheric evaporation in the form of upflowing heated plasma (§ 16.3), and cooling of postflare loops (§ 16.4). The initial heating of the coronal plasma requires anomalous resistivity, because Joule heating with classical resistivity is unable to explain the observed densities and temperatures in flare plasmas (§ 16.1.1 and 16.1.2). Other forms of coronal flare plasma heating, such as by slow shocks, electron beams, proton beams, or inductive currents are difficult to constrain with currently available observables. The second stage of chromospheric heating is more thoroughly explored, based on the theory of the thick-target model, with numeric hydrodynamic simulations, and with particle-in-cell simulations (§ 16.2.1 and 16.2.3). Important diagnostics on chromospheric heating are also available from Hα (§ 16.2.3), white light (§ 16.2.4), and UV emission (§ 16.2.5), but quantitative modeling is very sparsely available. The third stage of chromospheric evaporation has been extensively explored with hydrodynamic simulations (§ 16.3.2), in particular to explain the observed Doppler shifts in soft X-ray lines (§ 16.3.2), while application of spatial models to imaging data is quite sparse (§ 16.3.3). Also certain types of slow-drifting radio bursts seem to contain information on the motion of chromospheric evaporation fronts (§ 16.3.4). The forth stage of postflare loop cooling is now understood to be dominated by thermal conduction initially, and by radiative cooling later on (§ 16.4). However, spatio-temporal temperature modeling of flare plasmas is still in its infancy.

Chapter 17

Coronal Mass Ejections (CMEs)

Every main sequence star is losing mass, caused by dynamic phenomena in its atmosphere that accelerate plasma or particles beyond the escape speed. Inspecting the Sun, our nearest star, we observe two forms of mass loss: the steady *solar wind outflow* and the sporadic ejection of large plasma structures, termed *coronal mass ejections (CMEs)*. The solar wind outflow amounts to $\approx 2 \times 10^{-10}$ (g cm^{-2} s^{-1}) in coronal holes, and to $\lesssim 4 \times 10^{-11}$ (g cm^{-2} s^{-1}) in active regions. The phenomenon of a CME occurs with a frequency of few events per day, carrying a mass in the range of $m_{CME} \approx 10^{14} - 10^{16}$ g, which corresponds to an average mass loss rate of $m_{CME}/(\Delta t \cdot 4\pi R_{\odot}^2) \approx 2 \times 10^{-14} - 2 \times 10^{-12}$ (g cm^{-2} s^{-1}), which is $\lesssim 1\%$ of the solar wind mass loss in coronal holes, or $\lesssim 10\%$ of the solar wind mass in active regions. The transverse size of CMEs can cover from a fraction up to more than a solar radius, and the ejection speed is in the range of $v_{CME} \approx 10^2 - 2 \times 10^3$ (km s^{-1}). Ambiguities from line-of-sight projection effects make it difficult to infer the geometric shape of CMEs. Possible interpretations include fluxropes, semi-shells, or bubbles. There is a general consensus that a CME is associated with a release of magnetic energy in the solar corona, but its relation to the flare phenomenon is controversial. Even big flares (at least GOES M-class) have no associated CMEs in 40% of the cases (Andrews 2003). A long-standing debate focused on the question of whether a CME is a by-product of the flare process or vice versa. This question has been settled in the view that both CMEs and flares are quite distinctly different plasma processes, but related to each other by a common magnetic instability that is controlled on a larger global scale. A CME is a dynamically evolving plasma structure, propagating outward from the Sun into interplanetary space, carrying a frozen-in magnetic flux and expanding in size. If a CME structure travels from a sub-solar point radially towards the Earth, it is called a *halo-CME* or an *Earth-directed* event. CME-accelerated energetic particles reach the Earth most likely when a CME is launched in the western solar hemisphere, since they propagate along the curved *Parker spiral* interplanetary magnetic field. Related geomagnetic storms in the Earth's magnetosphere can cause disruptions of global communication and navigation networks, or failures of satellites and commercial power systems, and are thus of practical importance.

Reviews on CMEs can be found in MacQueen (1980), Howard et al. (1985), Kahler

(1987, 1992), Low (1994, 1996, 2001a), Hundhausen (1999), Forbes (2000c), Klimchuk (2001), Cargill (2001), or in the monographs and proceedings of Crooker et al. (1997), Low (1999a), and Daglis (2001).

17.1 Theoretical Concepts of CMEs

The physical concept of various theoretical CME models can perhaps be best understood in terms of mechanical analogues, as shown in Fig. 17.1. We summarize the essential concepts of five major CME models, following the theoretical review by Klimchuk (2001), see also Low (1999b, 2001a,b).

17.1.1 Thermal Blast Model

Early models proposed that the driving force of a CME is caused by a greatly enhanced thermal pressure, produced by a flare, which cannot be contained by the magnetic field and thus pushes the CME outward into the heliosphere (Dryer 1982; Wu 1982). An analogue to the thermal blast model is the overpressure generated by a bomb explosion (Fig. 17.1, top panel). So, the flare was initially thought to be the primary trigger of a CME. In the meantime, however, many CMEs have been recorded without a preceding flare, or the timing was found such that the CME launch occurred first, and flare-related emission later (e.g., Harrison 1986). Thus, today we think that the thermal blast model cannot be correct in many CME events (Gosling 1993), although the relative timing is sometimes very close (e.g., Dryer 1996; Délannée et al. 2000; Zhang et al. 2001b). A recent MHD simulation that employed hot plasma injection as a driver mechanism of a fluxrope eruption found that this model could not reproduce the interplanetary magnetic cloud data over the range of $0.4 - 5$ AU (Krall et al. 2000).

17.1.2 Dynamo Model

The class of dynamo-driven CME models implies a rapid generation of magnetic flux by real-time stressing of the magnetic field. A mechanical analogue is the stressing of a spring by an external force (Fig. 17.1, second panel). In the solar application, the driver of magnetic stressing is accomplished by an external force (e.g., by rapid displacements of the footpoints of a coronal magnetic field system). A theoretical study demonstrated that shearing of a coronal loop arcade always leads to an inflation of the entire magnetic field (Klimchuk 1990), and thus a sufficiently fast driver is expected to produce a CME-like expulsion. In recent simulations (Chen 1989, 1997a, 2000; Krall et al. 2000) such a driver mechanism is called *flux injection*, which corresponds to one of the three scenarios: (1) pre-existing coronal field lines become twisted, (2) new ring-shaped field lines rise upward in the corona while becoming detached from the photosphere, or (3) new arch-shaped field lines emerge into the corona while staying anchored at their photospheric footpoints. The problem with the first scenario is that the required footpoint motion has to be at least two orders of magnitude faster than the observed one (e.g., Krall et al. 2000). Also the second scenario is unlikely because the amount of entrained mass has never been observed and no obvious forces exist

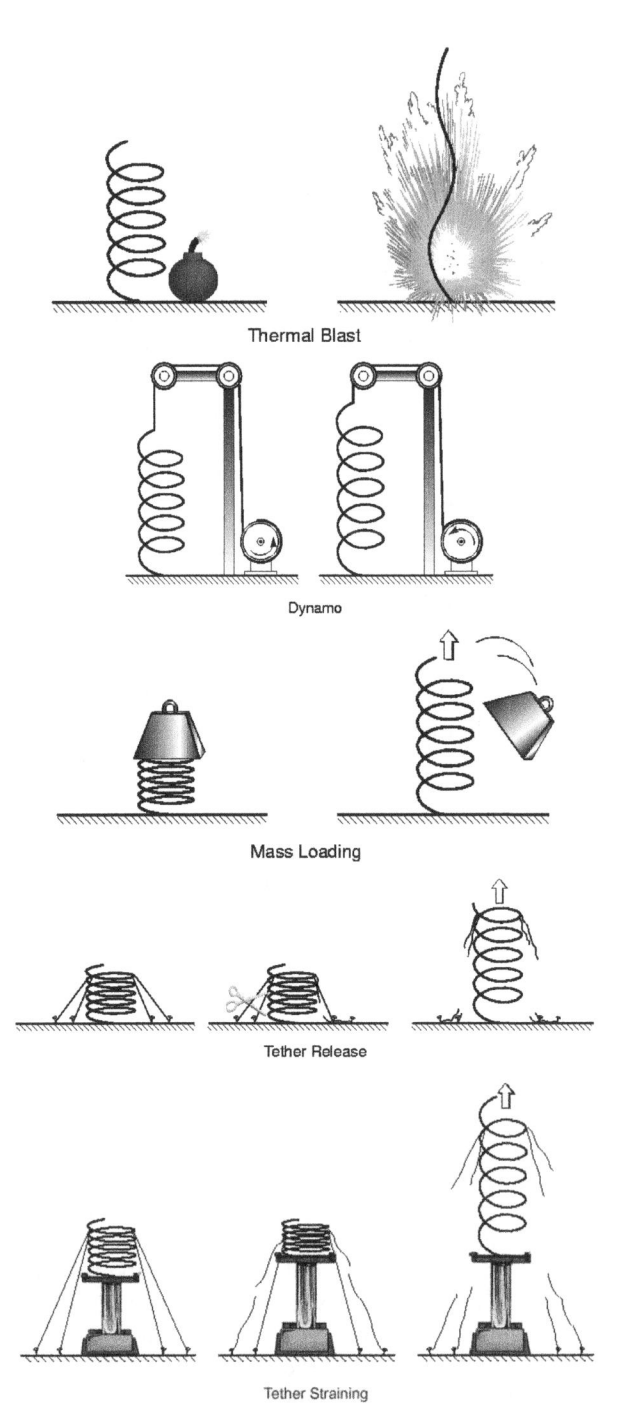

Figure 17.1: Physical (mechanical) analogues of five different *coronal mass ejection (CME)* models: (a) thermal blast model, (b) dynamo model, (c) mass loading model, (d) tether cutting model, and (e) tether straining model (Klimchuk 2001).

to lift the mass. The third possibility with emerging flux is more likely, but there
are issues whether the required increase of vertical flux through the photosphere is
consistent with observations. Blackman & Brandenburg (2003) suggest that the launch
of CMEs balances the conservation of magnetic helicity during the solar cycle, by
simultaneously liberating small-scale twist and large-scale writhe of opposite sign.

17.1.3   Mass Loading Model

The next three CME models are also called *storage and release models*, which entail a
slow build-up of magnetic stress before eruption begins. One of them is loading with
mass (see the analogue of a spring that is compressed by a heavy weight and explosively uncoils when the weight is shifted to one side, Fig. 17.1, middle panel). The
mass loading process during the pre-eruption phase of a CME can be manifested in
the form of a growing quiescent or eruptive filament (§ 6.4), for instance. Theoretical
studies compare the total magnetic energy in pre-eruption and posteruption equilibrium
configurations in order to demonstrate the plausible transition from a higher to a lower
energy state (Low & Smith 1993; Chou & Charbonneau 1996; Wolfson & Dlamini
1997; Wolfson & Saran 1998; Guo & Wu 1998; Low 1999a). There are two forms
of mass loading: (1) by prominences, which are extremely dense, contained in a compact volume, and of chromospheric temperature; and (2) by a relatively higher electron density distributed over a large volume, which is unstable to the Rayleigh–Taylor
or Kruskal–Schwarzschild instability, if it overlays a volume of lower density. The
first concept that prominences play a fundamental role in the launch of CMEs (Low
1996, 1999a) is supported by the observations with coincident starts of prominences
and CMEs. A crucial criterion is the mass of the prominence (Low et al. 2003, Zhang
& Low 2004). The second concept of unstable mass loading over a larger volume
is supported by observations of CMEs from helmet streamers that contain lower density cavities (Hundhausen 1988, 1999), but there are also numerous counter-examples
without any signs of internal low-density regions.

17.1.4   Tether Release Model

As we discussed in § 6.2.2, magnetically dominated configurations like coronal loops
generally involve a balance between the upward-directed force of magnetic pressure,
$-\nabla(p + B^2/8\pi)$, and the downward-directed force of magnetic tension, $(1/4\pi)(\mathbf{B} \cdot \nabla)\mathbf{B}$. The field lines that provide the tension are sometimes called *tethers*, analogous
to the ground-anchored ropes that hold down a buoyant balloon. In our mechanical
analogy, the tether ropes hold down a compressed spring (Fig. 17.1, forth panel). Once
the tethers are released one after the other, the tension on the remaining thethers increases, until the strain becomes eventually so large that the remaining thethers start to
break and the spring uncoils in a catastrophic explosion. This process has been dubbed
tether release, while the earlier term *tether cutting* refers more to the explosive end
phase. A 2D model (with translational symmetry) has been developed which demonstrates how a tether release may work in the solar corona (Forbes & Isenberg 1991;
Isenberg et al. 1993; Lin et al. 1998a; Van Tend & Kuperus 1978; Van Ballegooijen & Martens 1989). We described the loss-of-equilibrium model of Forbes & Priest

(1995) in § 10.5.3, which is a transition through a sequence of equilibria, driven by converging footpoint motion, until a loss of equilibrium occurs and the X-point jumps discontinuously upward into a new equilibrium position. In a non-ideal MHD situation, where enhanced resistivity is present in the X-point, an eruption with a launch of a CME would result, after break-off of the "tethers" during the loss-of-equilibrium phase (Forbes 1991; Lin & Forbes 2000; Mikić & Linker 1999; Amari et al. 2000).

17.1.5 Tether Straining Model

The *tether straining model* is similar to the *tether release model*, except that the strain on the tethers gradually increases by some external force until they brake. In the tether release model, the force on the tethers is constant, but is distributed to fewer and fewer tethers with time until they brake. One physical model of the tether-straining class is the *magnetic breakout model* of Antiochos (1998) and Antiochos et al. (1999b), described in § 10.5.5 (Fig. 10.26). The magnetic breakout model is a quadrupolar structure with two adjacent arcades, having overarching magnetic field lines over the whole system that represent the tethers. One loop arcade is continuously sheared and builds up magnetic stress until magnetic reconnection starts in the overlying X-point between the two loop arcades. The magnetic reconnection process then opens up the magnetic field in an upward direction (i.e., the "break-up" phase), and allows the CME to escape into interplanetary space. There are variants of this magnetic breakout model. A similar breakout effect can also be achieved in a bipolar magnetic field with the mass loading model (Low & Zhang 2002; Zhang et al. 2002). While the original model of Antiochos et al. (1999b) is 2D and symmetric, the version of Aulanier et al. (2000a) involves 3D nullpoints with a separatrix dome beneath the 3D nullpoint and a spine field line above (Plate 13), which can be an open field line (Fig. 10.26) and then marks the escape route of the CME. Another thether-straining model is the equilibrium-loss model of Forbes & Priest (1995), described in § 10.5.3. The straining driver is given by the converging footpoint motion and magnetic reconnection is initialized underneath the erupting structure, while it occurs above the erupting structure in the magnetic breakout model. Other examples of tether-straining models are the sheared arcade models of Mikić & Linker (1994b), Linker & Mikić (1995), Choe & Lee (1996), and Amari et al. (1996), and the fluxrope models of Wu et al. (1995; 2000).

17.2 Numerical MHD Simulations of CMEs

There are two kinds of theoretical simulations on CMEs: (1) analytical time-dependent MHD models, which provide insights into the physical mechanisms, but cannot reproduce the detailed morphology of the observations; and (2) numerical time-dependent MHD simulations, which should be able to reproduce the observations if sufficiently accurate initial conditions and boundary conditions are known. Reviews on the theoretical modeling of CMEs can be found in Low (2001b), and a review on numerical MHD modeling of CMEs in Wu et al. (2001).

17.2.1   Analytical Models of CMEs

The general framework of the ideal and resistive MHD equations is given in § 6.1.3 and § 6.1.5. The simplest description of a CME in fully developed motion was modeled analytically with the one-fluid MHD equations including gravity, but avoiding the complicated energy equation, but instead using the polytropic assumption with an index of $\gamma < 5/3$. The MHD equations for a polytropic index of $\gamma = 4/3$ yield a family of self-similar solutions in 2D and 3D space (Gibson & Low 1998; Low 2001b). In this model, the mass expulsion in the gravitational potential well yields an almost constant speed or mildly accelerating CME, after the hydromagnetic system becomes gravitationally unstable (Low et al. 1982; Low 1984b). Kinematic models with raising filaments that increase the magnetic pressure under a helmet streamer and drive the outward motion have been presented by Pneuman (1980) and Van Tend (1979). Photospheric flows as drivers of CMEs have been considered by Biskamp & Welter (1989). The energetics and causes of CMEs in terms of fluxrope geometries have been studied by Forbes & Isenberg (1991) and Chen (1997a). Analytical solutions of the time-dependent MHD equations that describe the expulsion of a CME have been calculated by Gibson & Low (1998). Low (1984) pointed out that the launch of a CME is a two-step process, consisting of (1) an initial phase where the closed coronal magnetic field is opened up to eject the trapped (prominence) material, which can be an ideal MHD process, and (2) a second phase involving magnetic reconnection of the open field lines beneath the erupted structure, which is a dissipative or resistive MHD process. A further refinement along the same basic evolution is the *magnetic breakout model* of Antiochos (1998) and Antiochos et al. (1999b), although it has not been modeled analytically.

17.2.2   Numerical MHD Simulations of CMEs

A more general approach is to solve the MHD equations with a numerical code, starting from an initial condition and propagating in time, with a least two dimensions in space. There are three generations of numerical MHD simulations of CMEs, based on (1) thermal blast models, (2) helmet streamer configurations, and (3) magnetic fluxrope configurations.

The first generation of numerical MHD models of CMEs assumed the initial corona to be static and potential (i.e., current free) or force-free (i.e., current-aligned) fields, where a pressure pulse was introduced to mimic a flare energy release (Nakagawa et al. 1978, 1981; Dryer et al. 1979; Steinolfson et al. 1978; Wu et al. 1978, 1982). The deficiency of this model is that neither the initial state nor the driver (thermal blast model, § 17.1.1) is realistic and consequently the model cannot reproduce observed morphological features of CMEs (Dryer 1994; Wu et al. 2001).

The second generation of numerical MHD models of CMEs (Steinolfson & Hundhausen 1988; Steinolfson 1992; Mikić et al. 1988; Guo et al. 1992; Wang et al. 1995b; Wu et al. 2000; see example in Fig. 17.2) assume a coronal helmet streamer to be the magnetic configuration of the initial state, where a CME originates from the disruption of global-scale streamers (Illing & Hundhausen 1986; Dere et al. 1997b; Subramanian et al. 1999; Plunkett et al. 2000). This generation of MHD simulations succeded to reproduce a loop-like CME (Steinolfson & Hundhausen 1988) and to reproduce the

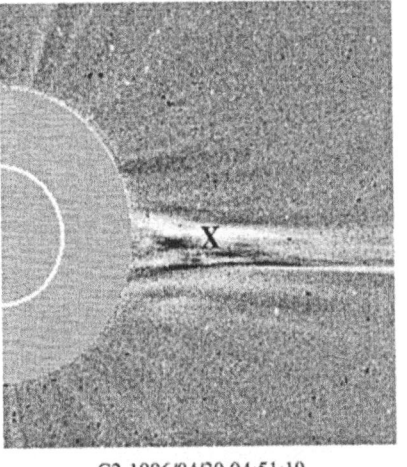

C2-1996/04/30 04:51:19

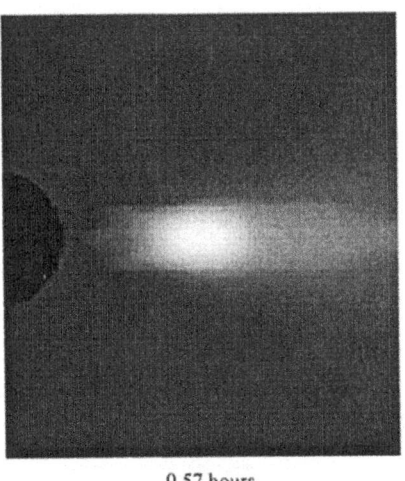

9.57 hours

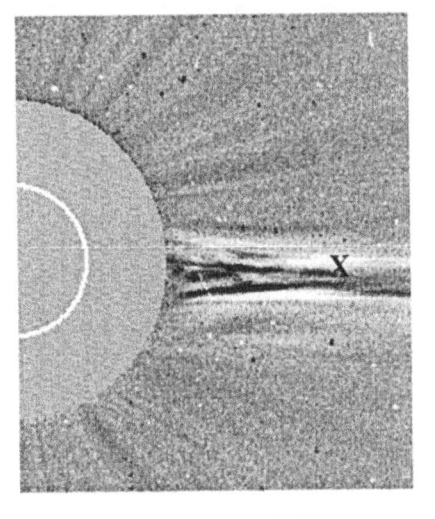

C2-1996/04/30 09:37:59

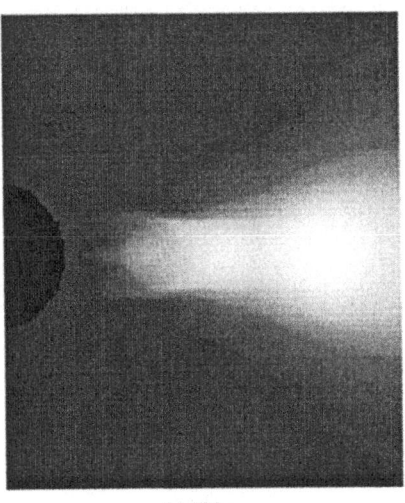

14.17 hours

Figure 17.2: Numerical MHD simulation of a CME with the helmet streamer model is shown in the right panels. A comparison with the observed running difference images of an outward plasma blob, observed with SoHO/LASCO on 1996-Apr-30, 04:51 UT and 09:37 UT, is shown in the left panels. Note that the centroid of the plasma blob (marked with a cross in the left panels) coincides with the centroid in the MHD simulations (Wu et al. 2000).

observed three-part structure: (1) a bright front or leading edge (the pre-eruption helmet structure), surrounding (2) a dark cavity, which contains (3) a bright core, identified as a helical prominence (Hundhausen 1988, 1999; Guo & Wu 1998). MHD simulations of prominences and CMEs demonstrated that the magnetic buoyancy force drives

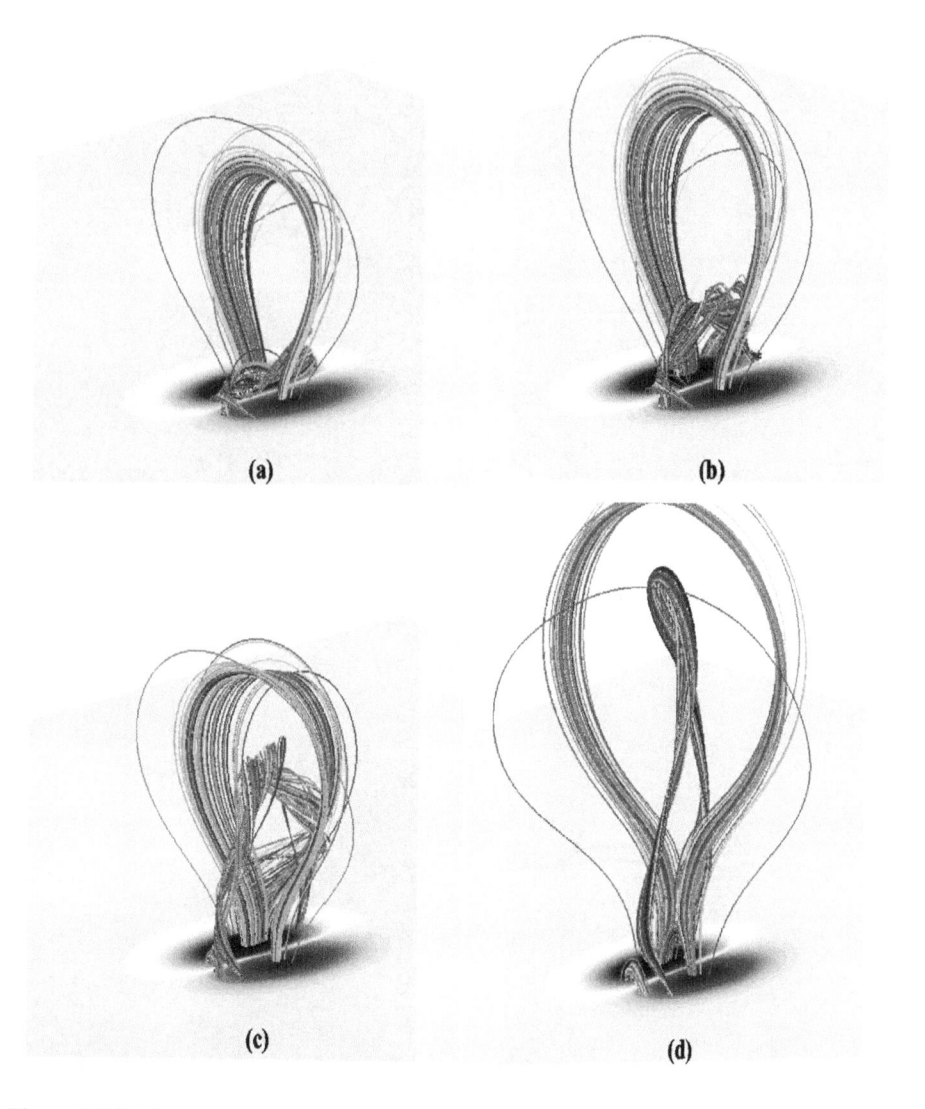

Figure 17.3: Numerical MHD simulation of the evolution of a CME, driven by turbulent diffusion. The four panels correspond to the times (a) t=850, (b) t=950, (c) t=1050, and (d) t=1150, where viscous relaxation is started at t=850, triggering a global disruption involving opening, reconnection through the overlying arcade and below, and the formation of a current sheet, associated with a high dissipation of magnetic energy and a strong increase of kinetic energy (Amari et al. 2003).

the outward motion (Yeh & Wu 1991; Wu & Guo 1997a). Another driver mechanism that can lead to a CME eruption is emerging flux (Chen & Shibata 2000). A number of studies used the shearing of magnetic footpoints to increase the energy of a helmet streamer, which forces a partial opening-up of the coronal magnetic field to launch a

CME (Linker & Mikić 1995; Mikić & Linker 1994; Mikić et al. 1988). However, there is a debate whether a magnetic fluxrope is present prior to eruption, which needs to be included in such helmet streamer models.

A third generation of numerical MHD models of CMEs implements the feature of magnetic helical fluxropes (Chen 1997b; Chen et al. 2000; Low & Smith 1993; Low 1994; Guo & Wu 1998; Wu & Guo 1997b; Wu et al. 1995, 1997b,c, 1999; Krall et al. 2000; see example in Fig. 17.3). One model assumes a magnetic fluxrope with footpoints anchored below the photosphere, where an eruption is driven by increasing the poloidal flux (i.e., magnetic flux injection or dynamo model), which can reproduce the dynamics of observed morphological features near the Sun and in magnetic cloud data in interplanetary space (Chen 1989, 1996, 2000; Krall et al. 2000). Another model simulates the evolution of the 3D magnetic field in a current sheet that undergoes magnetic reconnection above a sheared arcade, leading to topological changes with intertwined open fluxtubes (Birn et al. 2000, 2003), similar to the helical fluxropes observed in CMEs. The kink instability leads then to the eruption of sigmoidal (twisted) fluxrope (Fan & Gibson 2003, 2004; Török & Kliem 2003; Török et al. 2004; Kliem et al. 2004). In another model, the combination of photospheric shearing and opposite-polarity emergence is used to produce erupting twisted magnetic fluxropes (Amari et al. 2000; 2003a,b; see Fig. 17.3), similar to the S-shaped (sigmoid) structures observed in soft X-rays. Some MHD simulations focus on the acceleration mechanism of erupting fluxropes, which can be controlled by enhanced magnetic reconnection rates (Cheng et al. 2003).

17.3 Pre-CME Conditions

The cause of a CME is the key for their physical understanding and should be detectable in pre-CME conditions. Once we have a deeper understanding which pre-CME conditions lead to the magnetic instability that drives a CME eruption, we obtain not only a diagnostic but also a predictive tool for the occurrence and evolution of CMEs. Furthermore we can then justify the assumed drivers that have been used in the numerical MHD simulations described in § 17.2. Thus we concentrate in this section on observational signatures of possible CME drivers during pre-CME conditions.

17.3.1 Photospheric Shear Motion

CMEs originate in active regions, which generally exhibit a roughly bipolar field. In order to provide conditions for eruptive phenomena such as flares and CMEs, free magnetic energy needs to be stored in the form of a stressed and sheared field, which is a prerequisite for several CME models (e.g., the dynamo model § 17.1.2 or the tether-straining model § 17.1.5). The stress of the magnetic field can be observationally determined, after removing the 180° ambiguity, by calculating the shear angle between potential field and transverse field vectors from a vector magnetogram (Fig. 17.4), which contains the information of the full 3D magnetic field vectors at the photospheric boundary. This method has been applied to flaring and flare-quiet regions but no discriminating differences were found (Leka & Barnes 2003a,b). In a slight variation

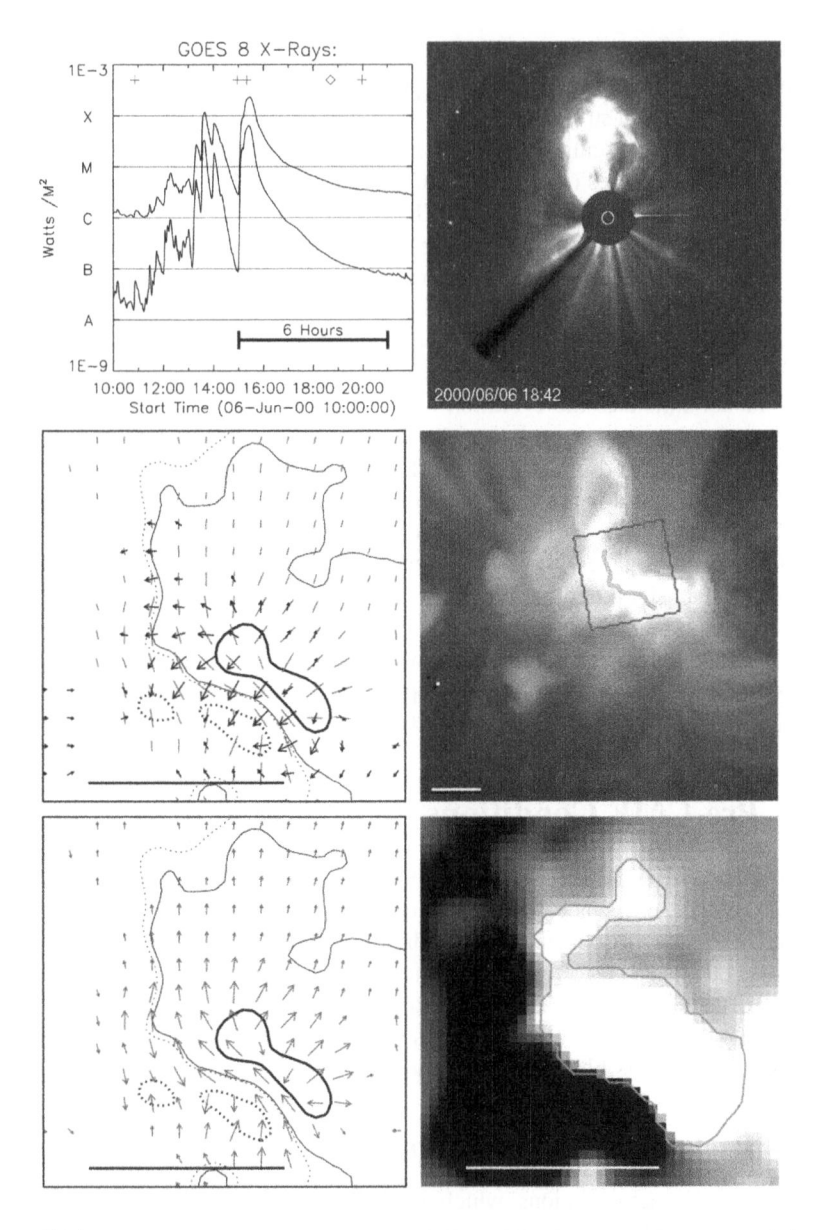

Figure 17.4: GOES soft X-ray light curves (top left) and SoHO/LASCO image (top right) of a halo CME on 2000-Jun-6, 18:42 UT, originating in AR 9026, at heliographic position 21N/14E. Vector magnetograms of AR 9026 are shown in the left panels and soft X-ray *Yohkoh/SXT* images in the right panels in the middle and bottom rows. The contours of the magnetogram are at line-of-sight field strengths of 25 G and 500 G and the observed transverse field strength and direction are marked by dashes (middle left) and arrows (bottom left). The potential transverse field computed from the observed line-of-sight field is shown by the arrows in the middle left panel. Note that the magnetic field is highly sheared near the neutral line, with the transverse field almost parallel to the neutral line. The highly sheared segment of the neutral line is overlaid on the *Yohkoh/SXT* image middle right. The spatial scale is indicated with a bar with 50 Mm length (Falconer et al. 2002).

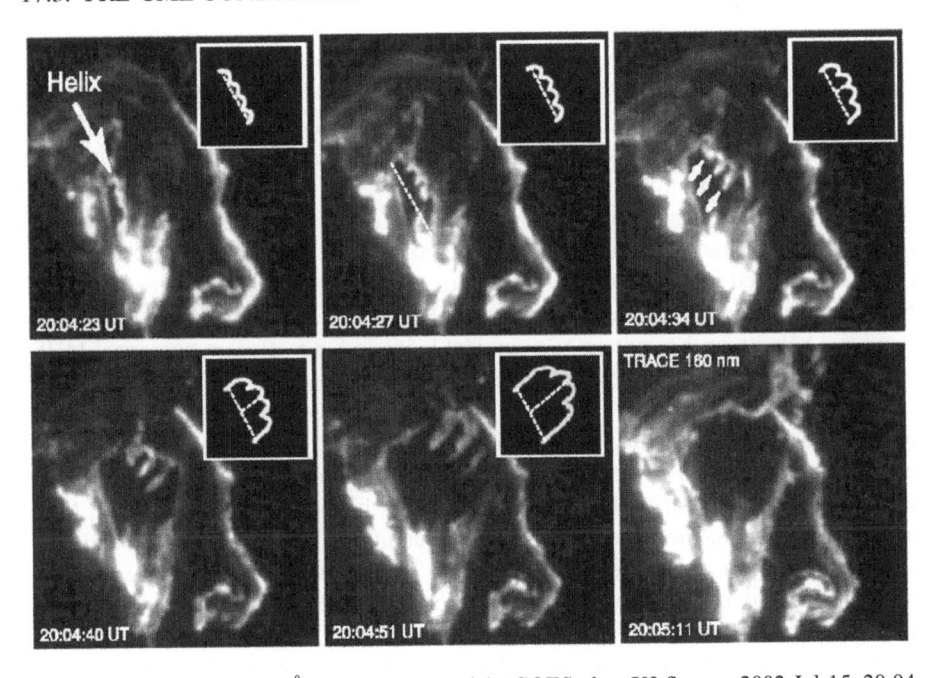

Figure 17.5: TRACE 1600 Å images in C IV of the GOES-class X3 flare on 2002-Jul-15, 20:04 UT. The inserts illustrate the geometry of the helical structure, exhibiting 3 − 4 turns. Note that the helical structure expands, rises, and unwinds during the eruption (Gary & Moore 2004).

of this method, the length l_{SS} of the highly sheared segment of the neutral line was evaluated and a correlation was found with the electric current I_N flowing from one polarity to the other, which is a measure of the nonpotentiality of the active region (i.e., $l_{SS} \propto I_N$; Fig. 17.4; Falconer et al. 2001). In a sample of 17 vector magnetograms it was found that this criterion, applied to segments of the neutral line with strong transverse field (> 150 G), yields a viable proxy for the prediction of the CME productivity of an active region (Falconer et al. 2002).

Evidence for a highly sheared magnetic configuration was found to lead to a filament eruption and flare without the presence of a helmet streamer configuration (Cheng & Pallavicini 1984). Theoretical models explain the eruption of a prominence from a sheared magnetic arcade configuration by the formation of helical field lines with subsequent flux cancellation above the neutral line (§ 6.4.1, Fig. 6.15; Van Ballegooijen & Martens 1989; Roumeliotis et al. 1994). The difference of electric conductivities outside and inside the filaments constitutes a magnetic expulsion force (Litvinenko & Somov 2001). Also a change in field-aligned currents can destabilize a filament (Nenovski et al. 2001).

17.3.2 Kink Instability of Twisted Structures

Shearing and stressing of magnetic field lines above the neutral line leads to helical (S-shaped in projection), so-called *sigmoid* structures. Once the helical twist exceeds some critical angle, the structure becomes susceptible to the kink instability, which produces a disruption of the magnetic field leading to the expulsion of a filament or CME. The sigmoidal shape is regarded as an observational signature of azimuthal currents in twisted coronal structures (i.e., loops, arcades, or filaments). The helicity of twisted loops has been found to have a hemispheric preference: forward (reverse) S shapes dominate in the southern (northern) hemisphere (Rust & Kumar 1996). The sense of the sigmoidal shape (forward-S or backward-S) and the handedness of the magnetic twist (left-handed or right-handed; i.e., positive or negative α in force-free fields, see Eq. 5.3.6) have been found to be correlated (Pevtsov et al. 1997). We discussed the magnetic helicity in the context of sigmoidal loops in § 5.5. Recent numerical MHD simulations of the kink instability applied to twisted loops have been performed (e.g., by Fan & Gibson 2003, 2004; Kliem et al. 2004; Török & Kliem 2003, 2004; Török et al. 2004), finding a critical twist number of $2.5\pi \lesssim \Phi_{twist} \lesssim 3.5\pi$ above which no equilibrium exists, consistent with the analytical (force-free) solution $\Phi_{twist} \lesssim 2.49\pi$ of Gold & Hoyle (1960). They also investigated which loop parameters (e.g., twist angle, resistivity, magnetic field gradient with height) lead to quasi-static (stable) non-eruptive expansion, rather than to an eruption. Some twisted filaments have been observed to expand, but failed to erupt (e.g., observed with TRACE on 2002-May-27, 18 UT; Rust 2003; Török & Kliem 2004).

There is now mounting observational evidence that the kink instability indeed plays a prime role for many eruptive filaments, flares, and CMEs (e.g., Canfield et al. 1999; Rust 2001b; Yurchyshyn 2002). Canfield et al. (1999) established statistically that active regions are significantly more likely to be eruptive if they are either sigmoidal or large. A most conspicuous case of a helical fluxtube with multiple turns associated with the double (X3-class) flare event and double CME on 2002-Jul-15 in AR 10030 (Fig. 17.5) has been described by Gary & Moore (2004) and Lui et al. (2003). The erupting helical structure exhibited up to $3 - 4$ turns (Fig. 17.5; Gary & Moore 2004), and thus is clearly far in the unstable regime of the kink instability. The eruption of the multi-turn helix, however, occurred after the peak of the gyrosynchrotron emission, which is interpreted to be a postreconnection erupting feature below the reconnection region, as one would expect in the *magnetic breakout model*.

17.4 Geometry of CMEs

The geometry of a CME and its dynamic change as a function of time provide the primary input for parameterizing a physical 3D-model. Geometric concepts of CMEs range from semi-spherical shells to helical fluxropes and the observations are often sufficiently ambiguous so that these two opposite concepts cannot easily be discriminated in the data. While CMEs propagating close to the plane of the sky have a relatively simple projected shape, other CMEs propagating in a direction towards the observer have much more complex shapes, the so-called *halo CMEs*. The true 3D configuration is still unclear due to the difficulties of the optically thin coronal plasma and the highly dynamic nature of CMEs. Coronagraphs measure mainly photospheric photons scattered by free electrons in the coronal plasma (*Thomson scattering*), yielding the integrated density along the line-of-sight, providing us only with a white-light image against the

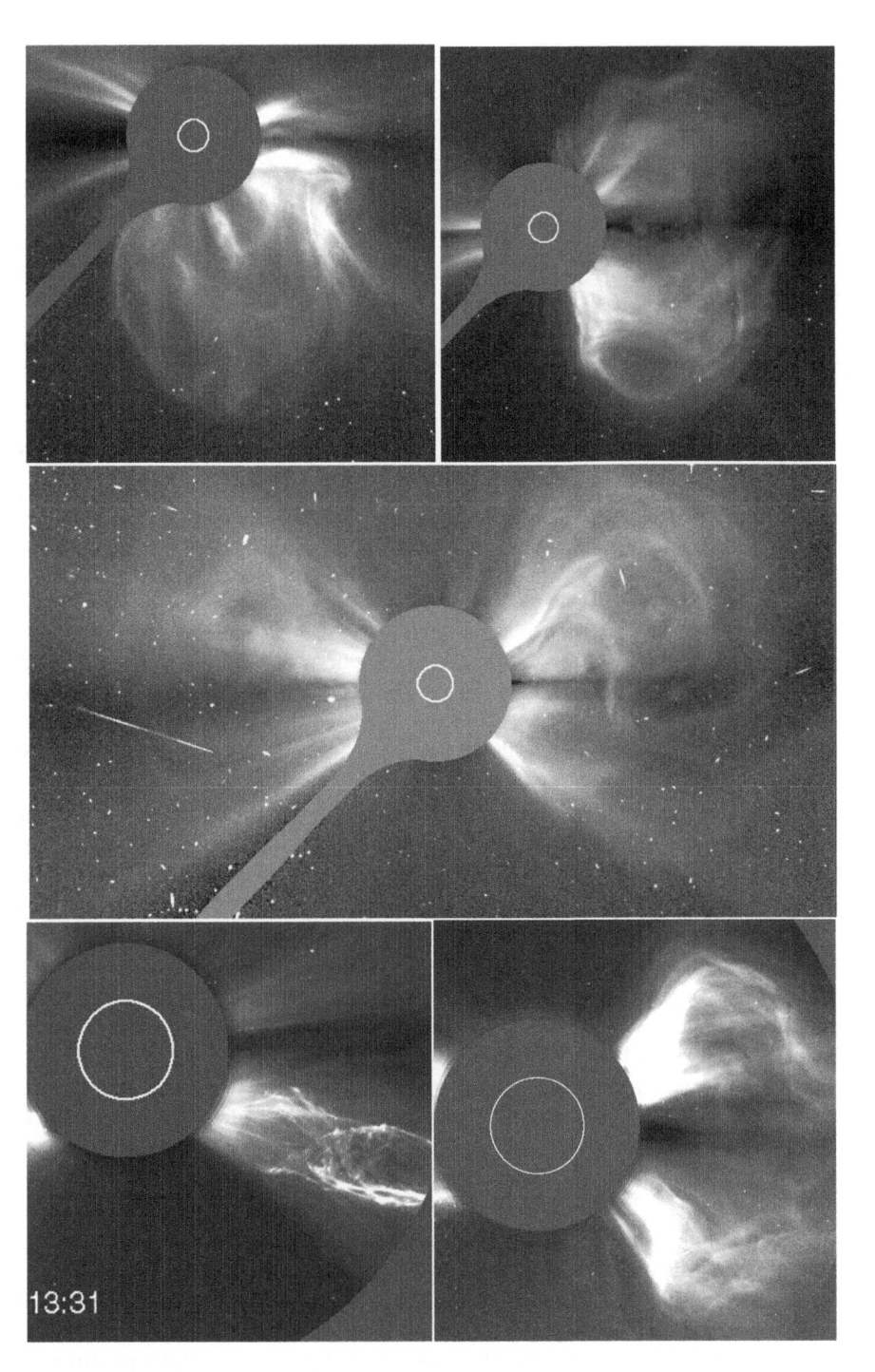

Figure 17.6: LASCO C3 image of a CME on 1998-Mar-29 (top left), a CME of 1998-Apr-20 (top right), a halo CME of 1998-May-6 (middle), an erupting prominence of 1998-Jun-02, 13:31 UT (bottom left), and a large CME of 1997-Nov-06, 12:36 UT (bottom right).

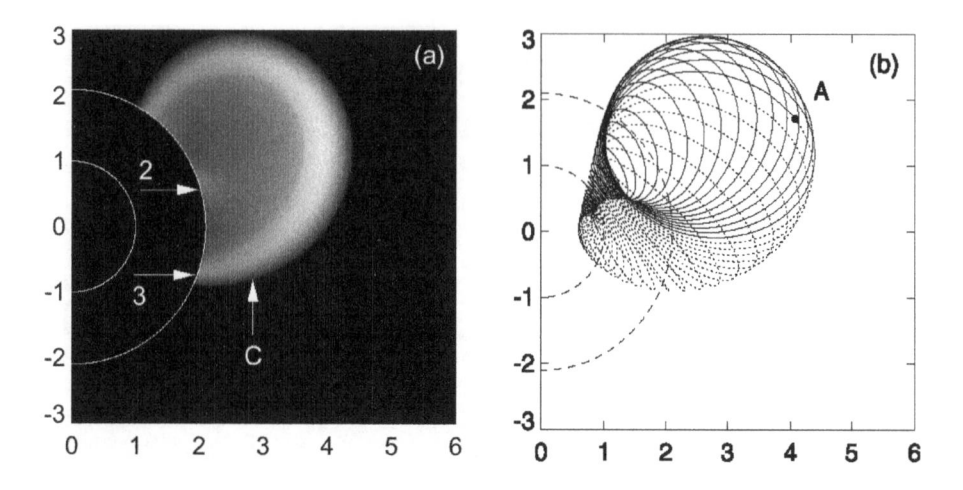

Figure 17.7: Synthetic coronagraph image (a) based on the 3D geometry of a helical fluxrope model (b). Note that the bright leading edge is produced by projection effects (Chen et al. 2000).

plane of the sky which is not trivial to deconvolve. Geometric inversions are only possible by using strong a priori constraints (e.g., spherical symmetry), while forward-modeling requires very flexible dynamic geometric models. A promising new method to derive the 3D geometry of CMEs has just been developed (at the time when this book went to print) by inversion of the polarization from white-ligth images (Moran & Davila 2004).

The first geometric characterizations of CMEs with large statistics were obtained from *SMM Coronagraph/Polarimeter (C/P)* observations, which included some 1300 CME events in 1980 and during 1984−1989. There is a large range of angular widths, with an average of 47°, launched at an average latitude of 35° (Hundhausen 1993). A typical characteristic of many CMEs is the three-part structure, consisting of (1) a bright leading edge, (2) a dark void, and (3) a bright core (Illing & Hundhausen 1985). It was suggested that CMEs have a loop-like geometry in a 2D plane, based on close associations of CMEs with eruptive prominences and disappearing filaments (Trottet & MacQueen 1980). Alternatively, 3D geometries were suggested, such as lightbulb bubbles, arcades of loops, or curved and twisted fluxtubes, particularly from *SoHO/LASCO* observations (Fig. 17.6) that became available after 1995 (e.g., Crifo et al. 1983; Schwenn 1986; Webb 1988; MacQueen 1993; Howard et al. 1997; St.Cyr et al. 2000; Vourlidas et al. 2000; Plunkett et al. 2000; Zhao et al. 2002; Gopalswamy et al. 2003; Cremades & Bothmer 2004).

Geometric modeling of CMEs is still in its infancy. Based on the concept of magnetic fluxropes, which consist of helical field lines wound around a curved cylinder (or a segment of a torus), the evolution of a CME is conceived as a steady expansion of this fluxrope into interplanetary space, with the legs connected to the footpoints on the Sun (Chen 1997a). Simulating the Thomson scattering on such a fluxrope structure, a synthetic coronagraph image was then produced (Fig. 17.7) which approximately reproduces the expanding bright leading edge feature of an observed CME (Chen et

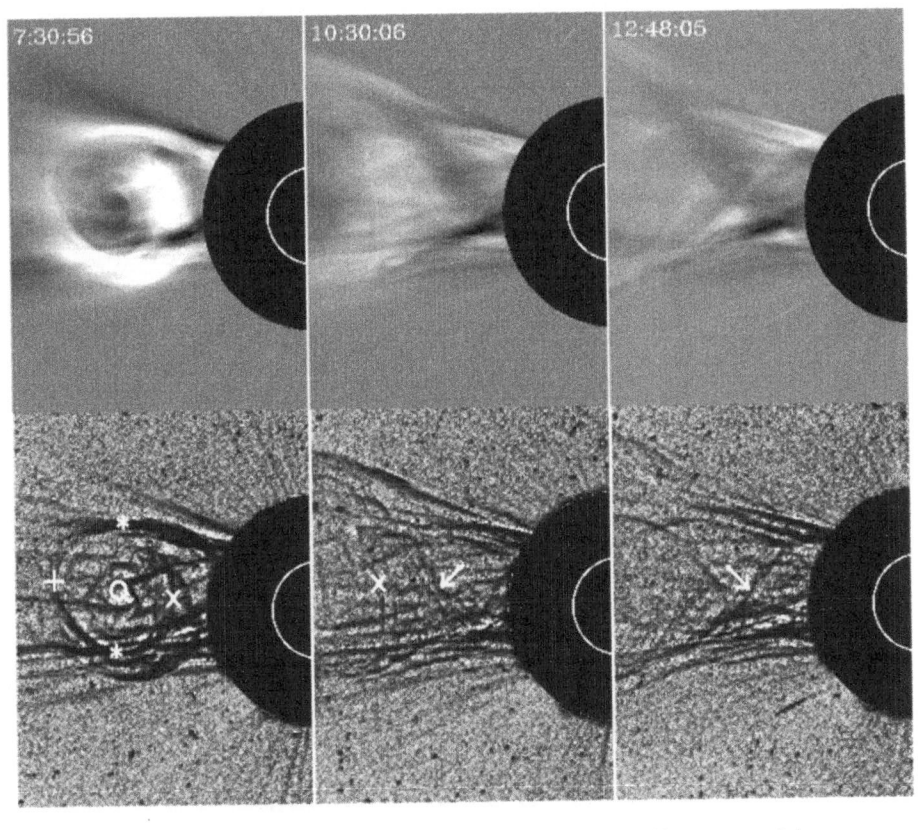

Figure 17.8: LASCO C2 images of the CME of 1997-Apr-30, processed by average-differencing (top row) and edge-enhancing (bottom row). The leading edge is marked +, the trailing edge X, the sides *, and the centroid O. Helical lines (marked with arrows) are seen below the rim that possibly trace the magnetic field (Wood et al. 1999).

al. 2000). Edge-enhancing techniques, however, reveal the detailed fine structure of CMEs, which appear to be composed of numerous helical strands (Fig. 17.8, Wood et al. 1999; Fig. 17.9, Dere et al. 1999). Thus, realistic 3D models of CMEs need to disentangle these multiple helical strands, which can be aided by correlation-tracking of time sequences of edge-enhanced images. The kinematic and morphological properties of the CME observed on 1997-Apr-30 and 1997-Feb-23 seem to confirm the concept of erupting fluxrope models (Wood et al. 1999). Comparisons of MHD modeling and observed CME geometries can be found in, for example, Gibson & Low (1998), Andrews et al. (1999), Wu et al. (2000, 2001), and Tokman & Bellam (2002).

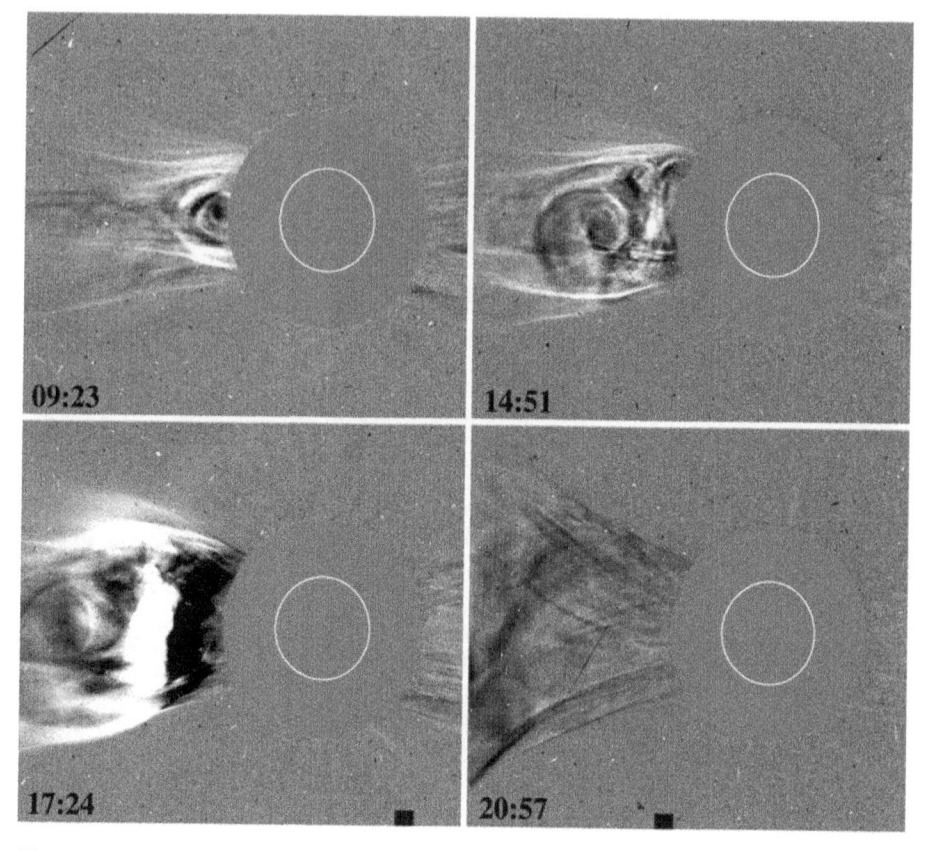

Figure 17.9: LASCO C2 running-difference images of the CME of 1997-Oct-19. The inner circle indicates the solar disk. Note the helical structures in the expanding CME (Dere et al. 1999).

17.5   Density and Temperature of CMEs

17.5.1   Density Measurements of CMEs

The density of CMEs of course is very inhomogeneous and varies by orders of magnitude as a function of the distance from the Sun, or as a function of time. Radial expansion is associated with $n(r) \propto r^{-2}$ for a steady constant expansion speed, or, $n(t) \propto (vt)^{-2}$ for a time-dependent homologous expansion. Masses of CMEs lay in the range of $m_{CME} \approx 10^{14} - 10^{16}$ g. The density is very inhomogeneously distributed, with the highest density in the compressed plasma at the leading edge, and a comparable mass in the bright core structure inside the cavity (Low 1996). Knowledge of the density and temperature allows estimation of the thermal pressure p_{th} in CME structures, and together with estimates of the magnetic field, one obtains the plasma-β parameter $\beta = p_m/p_{th}$, which provides an important diagnostic of whether the morphology of a CME structure is controlled by the magnetic field or by free radial expansion. Since the radiative loss rate is proportional to the squared density, the

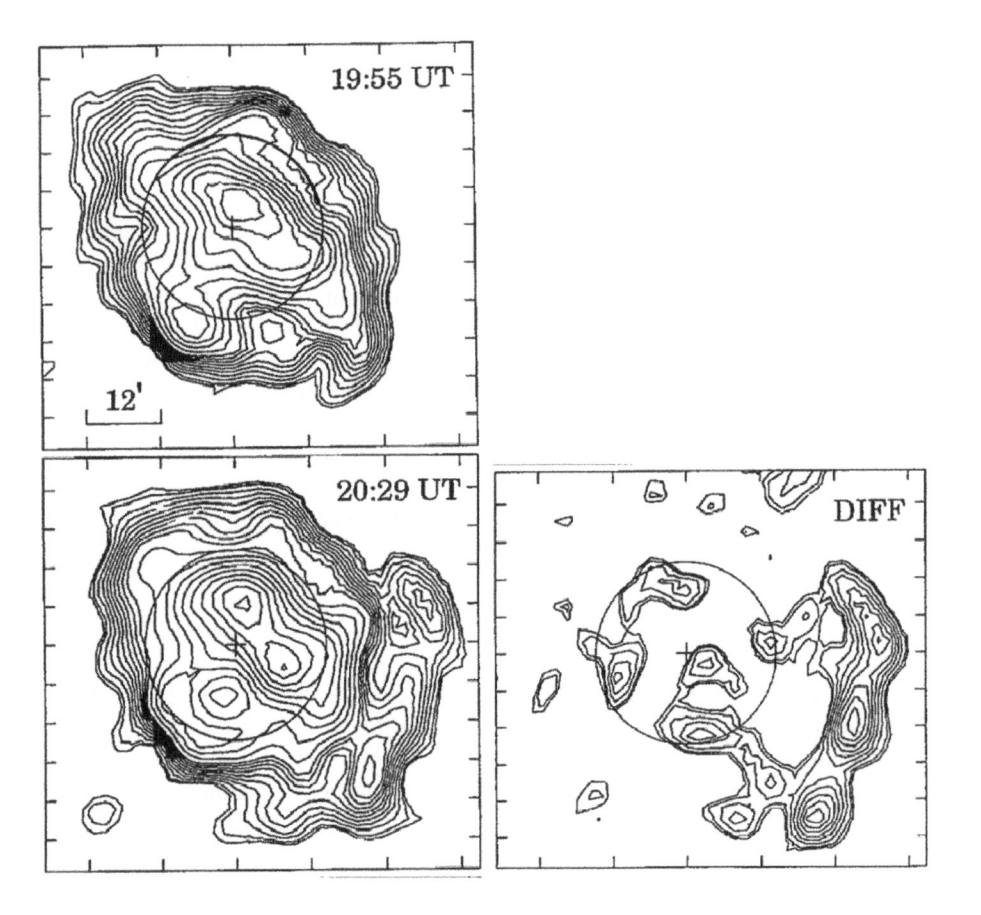

Figure 17.10: Two subsequent radio brightness images and a difference image recorded with the *Clark Lake Radioheliograph* at 73.8 MHz on 1986-Feb-16, before a CME at 19:55 UT, and during the CME at 20:29 UT. The contours range from 15,000 to 300,000 K. The spatial resolution is 4.7 and 5.5 arcmin in E−W and N−S direction, respectively. The radio emission is interpreted as thermal free-free emission from both the quiet Sun corona and from the CME. Note that the difference image subtracts out the quiet Sun component and exhibits the CME leading edge (Gopalswamy & Kundu 1992).

knowledge of the density yields also crucial information on the existence of energy dissipation and heating mechanisms in CMEs. CME masses were mostly estimated from white-light coronagraphs (e.g., Hildner 1977; Poland et al. 1981; Howard et al. 1984). In the following we report on mass estimates of CMEs from two other wavelengths, one in radio (Gopalswamy & Kundu 1992) and one in UV (Ciaravella et al. 2001).

While coronagraphs detect photospheric light that is Thomson-scattered by the CME electrons, radio telescopes can detect thermal bremsstrahlung from CMEs directly. If the brightness temperature of the quiet Sun background (before the CME) is much lower than the brightness temperature of the CME, the background density n_e

can be neglected and the CME density n_{CME} is obtained straightforwardly from the observed brightness temperature T_B

$$T_B = 0.2T_e^{-1/2}f^{-2}\int_0^\infty (n_e + n_{CME})^2 ds \approx 0.2T_e^{-1/2}f^{-2}n_{CME}^2 L \qquad (17.5.1)$$

with L the linear dimension of the CME along the line-of-sight. A radio brightness image before and during a CME is shown in Fig. 17.10 (left panels), and the difference image is shown in Fig. 17.10 (right panel). Depending on the line-of-sight depth L, Gopalswamy & Kundu (1992) estimated a CME mass of $m_{CME} = (2.7 - 4.2) \times 10^{15}$ g. With a temperature of $T = 1.0 \times 10^6$ K and an observed brightness temperature of $T_B = 9,7 \times 10^4$ K at $f = 73.8$ MHz, they estimated a mean electron density of $n_{CME} = 0.5 \times 10^6$ cm^{-3} in the frontal leading edge.

There are three common methods to estimate the density of CMEs from UV spectra: (1) Density-sensitive (and temperature-sensitive) line ratios, (2) emission measure method of optically thin plasma, and (3) ratio of collisional excitation rate coefficient $C_{ex}(T)n_e n_i$ to the radiative scattering rate $(< I_\lambda \sigma_\lambda > n_i)$,

$$q = \frac{C_{ex}(T)n_e}{< I_\lambda \sigma_\lambda >} \qquad (17.5.2)$$

where C_{ex} is the excitation rate coefficient, n_e the electron density, n_i the ion density which cancels out in the ratio, I_λ the illuminating flux, and σ_λ the scattering cross section. Using these methods with *SoHO/UVCS* data, a temperature in the range of $T_{CME} \approx 10^{4.5} - 10^{5.5}$ (Ciaravella et al. 2000) and densities in the range of $n_{CME} \approx (1-3) \times 10^7$ cm^{-3} were determined from C III/O VI and N V/O VI ratios at a distance of $R = 1.7R_\odot$ (Ciaravella et al. 2001).

17.5.2 Temperature Range of CMEs

Most CME observations are made in white light (e.g., with *SMM/CP* or *SoHO/LASCO*), which provides no temperature information. Many CMEs are also seen in *EUV*, so they must have substantial mass within the temperature range of $T_{CME} \approx 0.5 - 2.0$ MK. Recent observations with *SoHO/UVCS* allow us to narrow down the temperature range (e.g., $T_{CME} \approx 10^{4.5} - 10^{5.5}$ from C III, Si III, N V, O VI, and S V line ratios). Coronal mass loss, however, has also been observed at higher temperatures in soft X-ray wavelengths with *Yohkoh/SXT*, which indicates temperatures of $T_{CME} \gtrsim 2$ MK (Hudson & Webb 1997; Hudson 1999). Ciaravella et al. (2000) observe CME plasma at the same time in the intermediate temperature range of $T_{CME} \approx 30,000 - 300,000$ K with *SoHO/UVCS* and with *SoHO/EIT* 195 Å, which has a peak response around $T \approx 1.5$ MK, but argue that the emission seen by EIT 195 Å must result from the sensitivity to cooler temperatures at $T < 0.3$ MK, because cooling plasma would not recombine sufficiently fast to form C III or Si III. On the other hand, prominence material was found to be hot ($T \gtrsim 1.5$ MK) based on similar UVCS diagnostic in Ciaravella et al. (2003) contrary to the assumptions in most CME models, where the core is taken as cold plasma. Thus, more refined work on the temperature diagnostic of CMEs is needed.

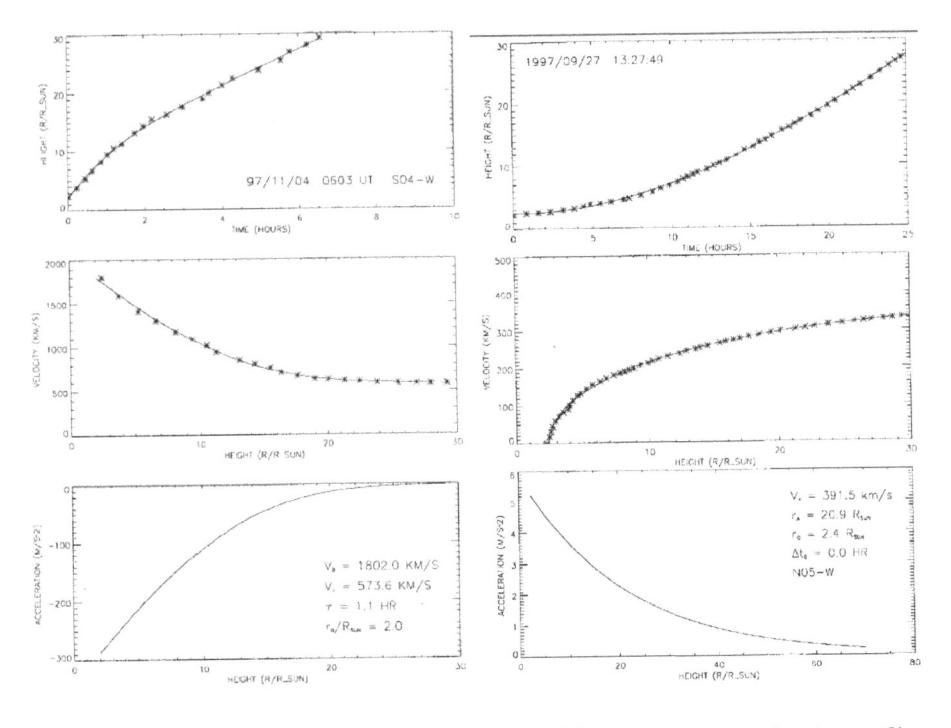

Figure 17.11: Height-time $h(t)$ plot (top), velocity $v(r)$ (middle), and acceleration profiles $a(r)$ (bottom), as a function of distance $r/R_\odot$, are shown for representants of two different CME classes: a gradual CME with initially negative acceleration (right), and an impulsive CME with initially positive acceleration (left) (Sheeley et al. 1999).

17.6 Velocities and Acceleration of CMEs

The height, velocity, and acceleration of a well-defined CME feature, such as the bright leading edge, are observables that can be measured as a function of time relatively easy, in particular for limb events. The time phases of acceleration reveal the height range where accelerating forces operate, and thus might provide crucial insights into the drivers of CMEs.

Based on the observed characteristics of CME velocity $v(t)$ and acceleration profiles $a(t)$ observed with *SoHO/LASCO* over the distance range of $r = 2 - 30 \ R_\odot$ it was proposed that there exist two distinct classes of CMEs (Sheeley et al. 1999): (1) *gradual CMEs*, apparently formed when prominences and their cavities rise up from below coronal streamers, typically attaining slow speeds ($v \approx 400 - 600$ km s^{-1}) with clear signs of gradual acceleration ($a = 3 - 40$ m s^{-2}) at distances $R < 30R_\odot$; and (2) *impulsive CMEs*, often associated with flares and Moreton waves on the visible disk, with speeds in excess of $v \gtrsim 750 - 1000$ km s^{-1}, observed to have a constant velocity or decelerating at distances $R \gtrsim 2R_\odot$ when first seen in coronagraphs (Sheeley et al. 1999). An example of each class is given in Fig. 17.11: a gradual CME shows initially positive acceleration (Fig. 17.11, right), while an impulsive CME shows initial

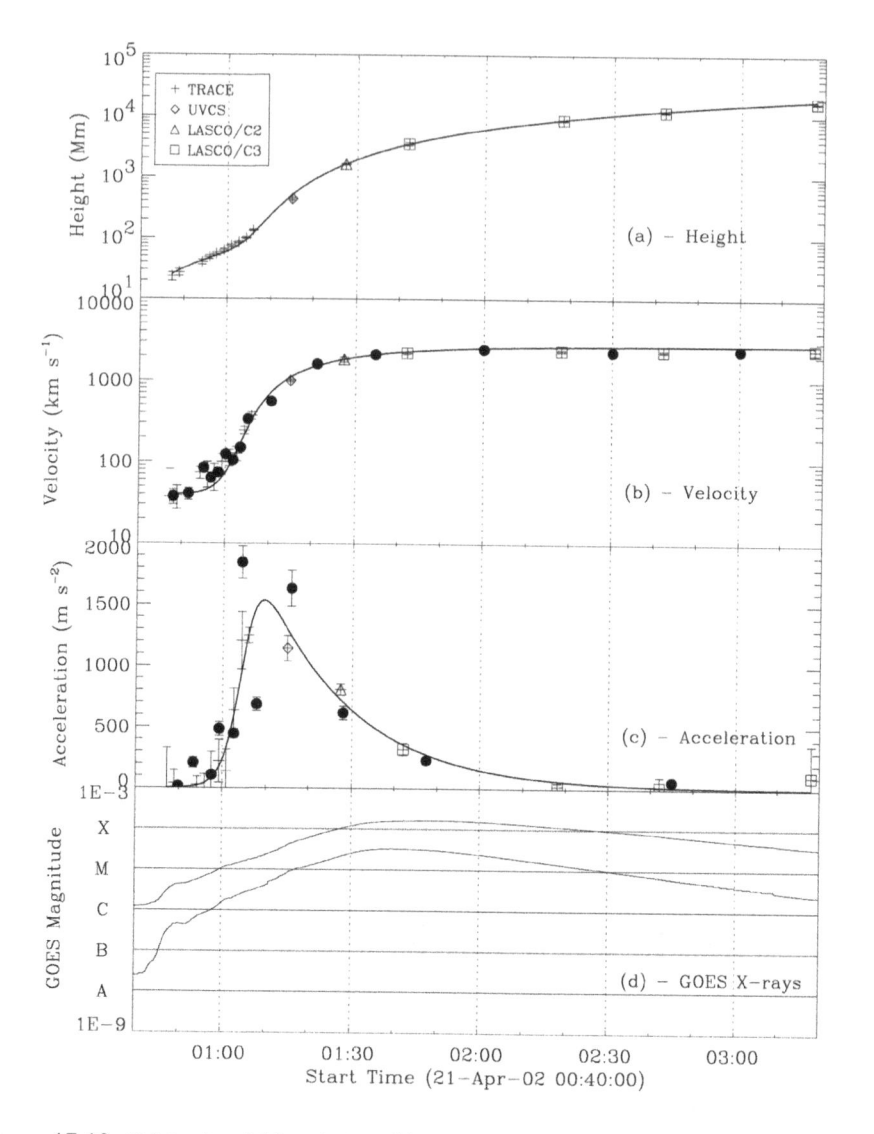

Figure 17.12: Height-time $h(t)$, velocity $v(t)$, and acceleration $a(t)$ profiles as well as GOES-10 soft X-ray flux profiles for a CME observed with TRACE, UVCS, and LASCO during the 2002-Apr-21, X1.5 GOES-class flare, shown during the interval of 00:47−03:20 UT. The solid lines are the best fits using Eqs. 17.6.2−4 (Gallagher et al. 2003).

negative acceleration (Fig. 17.11, left). Sheeley et al. (1999) also found that (Earth-directed) halo versions of the two classes appear as smooth halos (for gradual CMEs) or more ragged structures (for impulsive CMEs).

The observations in Fig. 17.11 suggest that the acceleration profile $a(t)$ can be approximated by either an exponentially increasing or decreasing function, (e.g., Sheeley et al. 1999),

$$a(t) = a_0 \exp\left[-(t - t_0)/t_a\right] . \qquad (17.6.1)$$

The velocity profile $v(t)$ follows then from integrating the acceleration profile $a(t)$,

$$v(t) = v_0 + \int_{t_0}^{t} a(t)\, dt , \qquad (17.6.2)$$

and the height-time profile $h(t)$ from double integration of the acceleration profile $a(t)$,

$$h(t) = h_0 + v_0(t - t_0) + \int_{t_0}^{t} \int_{t_0}^{t} a(t)\, dt\, dt . \qquad (17.6.3)$$

The acceleration profile of CMEs cannot be observed at low heights ($R \lesssim 2R_\odot$) with coronagraphs. However, coordinated measurements in EUV can fill in this gap. A coordinated observation of a CME with TRACE, UVCS, and LASCO revealed both the initial acceleration as well as the later deceleration phase (Fig. 17.12). So the acceleration profile shown in Fig. 17.12 could be fitted with a combined function of exponentially increasing and decreasing acceleration (Gallagher et al. 2003),

$$a(t) = \left[\frac{1}{a_r \exp\left(t/\tau_r\right)} + \frac{1}{a_d \exp\left(t/\tau_d\right)}\right]^{-1} . \qquad (17.6.4)$$

The CME event shown in Fig. 17.12 reaches a final speed of $v \approx 2500$ km s^{-1}, which is among the fastest 1% CME speeds observed with LASCO. The start of the acceleration at 00:47 UT coincides with the start of hard X-ray emission at energies ≥ 25 keV, while the maximum of acceleration at 01:09 UT coincides with the peak of the ≥ 25 keV hard X-ray emission, which suggests a close causal connection between the energy release and CME driving force.

A remarkable observation of a CME event of 1998-Apr-23 05:29 UT has been reported in soft X-rays (Alexander et al. 2002). A variable acceleration model fitted to the data yields a peak acceleration of $a \approx 4865$ m s^{-2} within the first $0.4R_\odot$, which is comparable with the largest reported CME accelerations. Also unusual is the detection in soft X-ray wavelengths, and it is not clear whether the accelerated soft X-ray plasma represents a fluxrope, shock front, or corresponds to other identifiable parts of a CME seen in white light.

A quantitative model for the acceleration of CMEs was developed by Chen & Krall (2003), based on a 3D magnetic fluxrope model (Fig. 17.13, left). The accelerating force $\mathbf{F}$ can be integrated over a toroidal section of the fluxrope from the MHD momentum equation (Eq. 6.1.17),

$$\mathbf{F} = -\nabla p - \rho \mathbf{g} + \mathbf{j} \times \mathbf{B} . \qquad (17.6.5)$$

This model predicts a universal scaling law where maximum acceleration is attained shortly after the expanding loop passes the height of the minimum curvature radius,

$$Z_* = \frac{S_f}{2} , \qquad (17.6.5)$$

where S_f is the footpoint separation distance of the magnetic fluxrope. An example of the application of this model to an observed CME (1997-Feb-23) is shown in Fig. 17.13 (right).

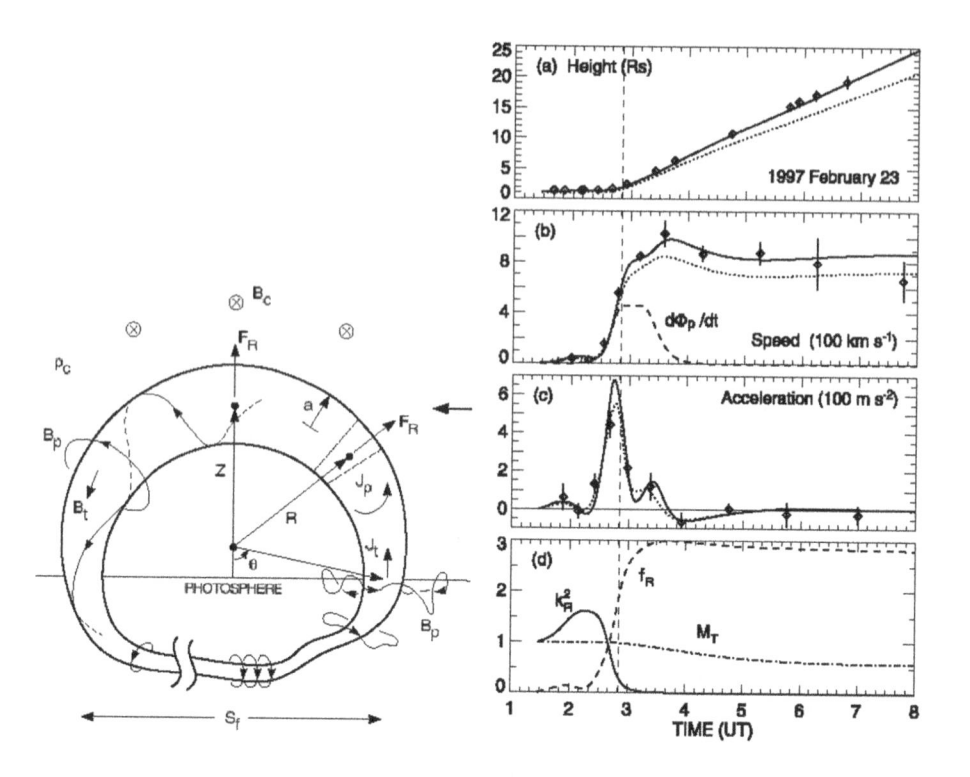

Figure 17.13: Height-time $h(t)$, velocity $v(t)$, and acceleration $a(t)$ profiles fitted from a magnetic fluxrope model to LASCO observations of the 1997-Feb-23 CME event. The solid curve corresponds to the leading edge and the dotted curve to the centroid of the expanding fluxrope. The vertical line indicates the time when the CME reached $2R_\odot$ from the Sun center. The leading edge attains a maximum speed of $v = 1000$ km s^{-1}. A schematic of the magnetic fluxrope model is shown on the left-hand side, indicating the footpoint separation distance S_f and the height Z of the fluxrope centroid (Chen & Krall 2003).

17.7 Energetics of CMEs

A key question is how the required (magnetic) energy storage is achieved and how it is released to produce a CME. The energetic problem has been pointed out by the Aly–Sturrock conjecture (Aly 1984; Sturrock 1991), which implies that a closed force-free magnetic field has less energy than the equivalent fully open field (with an identical photospheric boundary condition). This conjecture severely constrains the occurrence of a CME in a force-free corona if the magnetic field is the primary driver of the eruption. There are three groups of CME models that satisfy this constraint: (1) the pre-CME magneto-static corona is not force-free and cross-field currents are present (Wolfson & Dlamini 1997); (2) the CME involves magnetic flux from several flux systems so that most of the involved field is not opened, such as in the *magnetic breakout model* (Antiochos 1998, Antiochos et al. 1999b); or (3) the CME includes a detached fluxrope (Low 1996). The magnetic energy of a CME can be estimated to some ex-

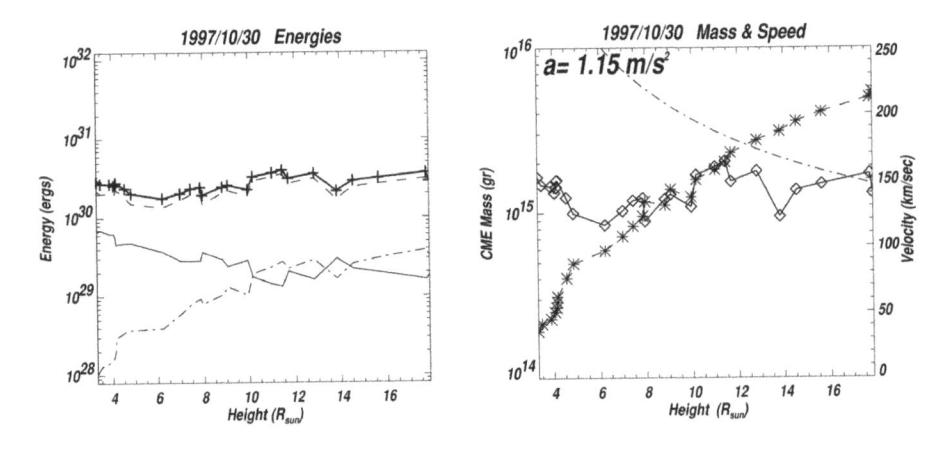

Figure 17.14: LASCO measurements of the evolution of the potential energy (dashed line), kinetic energy (dash-dotted line), magnetic energy (solid lines), and total energy (solid line with crosses) of the CME of 1997-Apr-30 (left frame). The evolution of the CME mass (solid line with diamonds), center-of-mass speed (dashed line with asterisks), and the derived acceleration (dash-dotted line) are shown in the right frame (Vourlidas et al. 2000).

tent from integrating the extrapolated 3D magnetic field over the volume of the CME, but these extrapolations are problematic since the force-free extrapolations are given inputs of the observed photospheric field unlikely to be potential or force-free (Metcalf et al. 1995).

One approach to obtain a better understanding of the dynamical evolution of physical parameters in an erupting CME is the study of the energy budget. A recent study (Vourlidas et al. 2000) indicates that some of the accelerating fluxrope CMEs have conservation of their total energy (i.e., the sum of magnetic, kinetic, and gravitational potential energy is constant; see example in Fig. 17.14, left). In this study, white-light intensities I_{obs}, velocities v_{CME}, and angular widths of CMEs were measured from LASCO observations at distances of $R = 2.5 - 30\ R_\odot$. The mass of a CME is estimated from the ratio of the excess observed brightness I_{obs} (from difference images) to the brightness $I_e(\vartheta)$ of a single electron at angle ϑ from the plane of the sky, which is computed from the Thomson scattering function (Billings 1966). Assuming a standard abundance of fully-ionized hydrogen with 10% helium, the CME mass is

$$m_{CME} = \frac{I_{obs}}{I_e(\vartheta)}\ \mu\ m_p \approx \frac{I_{obs}}{I_e(\vartheta)}\ 2 \times 10^{-24}\ (\text{g})\ . \qquad (17.7.1)$$

The potential energy ε_{grav} of the fluxrope is defined by the amount of energy required to lift its mass from the solar surface, that is,

$$\varepsilon_{grav}(R) = \sum_{fluxrope} \int_{R_\odot}^{R} \frac{GM_\odot m_i}{r_i^2}\ dr_i\ , \qquad (17.7.2)$$

where m_i and r_i are the mass and distance from the Sun center, respectively, for each pixel in the observed difference image. The kinetic energy ε_{kin} is integrated over the

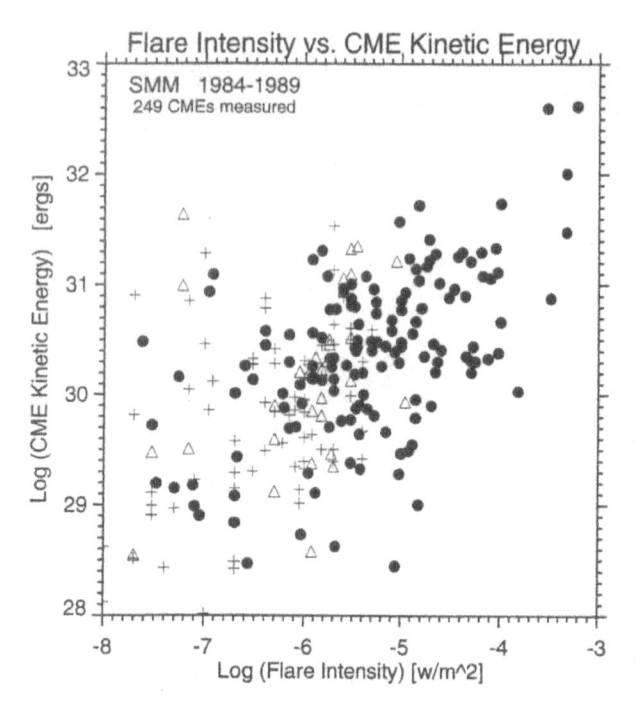

Figure 17.15: A logarithmic scatterplot of kinetic energies of CMEs and the peak intensities of associated X-ray flares seen in the GOES integrated soft X-ray flux. The sample includes 249 CME events observed with *SMM C/P* (Hundhausen 1997).

CME area,

$$\varepsilon_{kin}(R) = \frac{1}{2} \sum_{fluxrope} m_i v_{CME}^2 .$$ (17.7.3)

The magnetic energy is assumed to vary during propagation according to the conservation of magnetic flux, $B(R) \times A(R) = const$, where A is the area of the fluxrope. Expressing the volume with $V = Al$, where l is the length of the fluxrope, the magnetic energy ε_{mag} can be estimated as,

$$\varepsilon_{mag}(R) = \frac{1}{8\pi} \int_{fluxrope} B^2(R) dV \approx \frac{1}{8\pi} \frac{l}{A} (B \times A)^2 ,$$ (17.7.4)

where an average value of the magnetic flux is $< B \times A > = 1.3 \pm 1.1 \times 10^{21}$ G cm^2, obtained from several magnetic clouds observed with the *Wind* spacecraft during 1995–1998 at Earth distance (Lepping et al. 1990). With this method Vourlidas et al. (2000) analyzed the energy budget of 11 CMEs and found that the kinetic energy is smaller than the potential energy for relatively small CMEs, but larger for relatively fast CMEs ($v_{CME} \geq 600$ km s^{-1}). The magnetic energy advected by the fluxrope is converted into kinetic and potential energy for relatively slow CMEs, so that the total

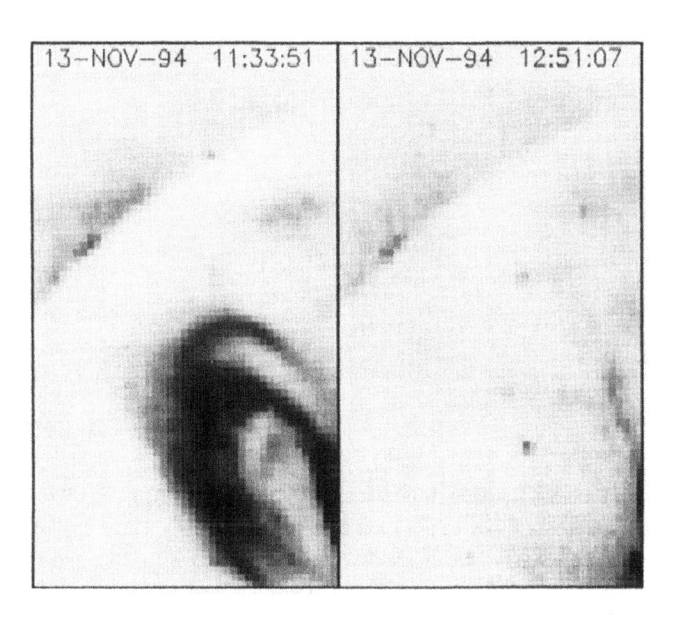

Figure 17.16: Yohkoh/SXT images of a long-duration flare at beginning (left) and after (right) the impulsive flare phase, located in the northeast of the solar disk. Note that most of the flare loops disappear almost completely from the field-of-view as a result of their outward motion (Hudson et al. 1996).

energy is constant, $\varepsilon_{tot} \approx \varepsilon_{grav} + \varepsilon_{kin} + \varepsilon_{mag}$. Thus, the slow CMEs are magnetically driven. For relatively fast CMEs ($v_{CME} \geq 600$ km s^{-1}) the magnetic energy is significantly below the potential and kinetic energy. Typical total energies of CMEs are $\varepsilon_{tot} \approx 10^{29} - 10^{32}$ erg, which is comparable with the range of flare energies estimated from nonthermal electrons (Fig. 9.27). The kinetic energies of CMEs from a larger sample of 294 events is shown in Fig. 17.15, demonstrating an approximate correlation with the total soft X-ray flux and a similar energy range as nonthermal flare energies (Hundhausen 1997). Moore (1988) estimated the energy of CMEs from the nonpotential magnetic energy stored in twisted fluxropes and found similar values (i.e., $\Delta\varepsilon_{twist} \approx 10^{30} - 10^{32}$ erg).

17.8 Coronal Dimming

A powerful diagnostic of the early phase of CMEs is the so-called *coronal dimming*, which is often detectable as a relative deficit of coronal mass or emission measure compared with pre-CME conditions, interpreted as a vacuum-like rarefaction after the launch or "evacuation" of a CME. The effect of coronal dimming is most dramatically seen on the solar disk, but is also detectable above the solar limb in some cases. We discussed the effect of CME dimming previously in the context of global waves (§ 8.3), which originate at CME launch sites and propagate more or less spherically over the solar surface, displaying a density compression at the wavefront and a rar-

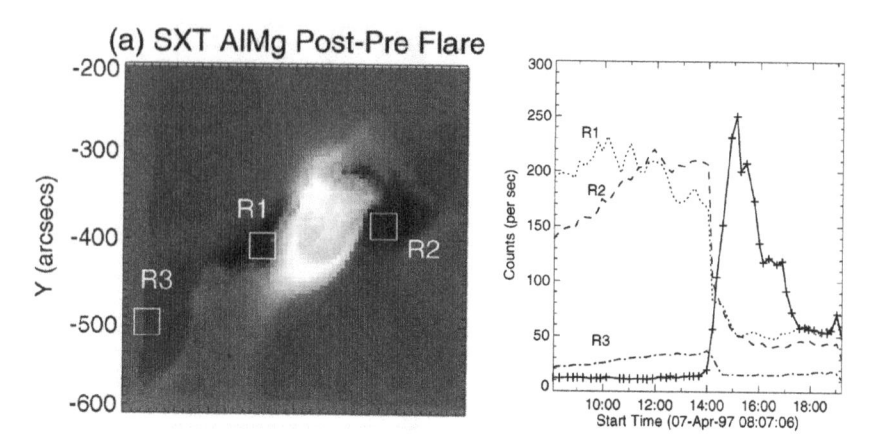

Figure 17.17: *Yohkoh/SXT* Al/Mg difference image of the 1997-Apr-07, 13:50 UT, flare, with the difference taken before (13:28:42 UT) and after (17:40:40 UT). Three spatial locations with notable dimming in soft X-rays are marked with R1, R2, and R2 (left), and the corresponding light curves are shown from *SoHO/EIT* 195 Å (right panel) and compared with the (anti-correlated) flare light curve (thick solid line in right panel) (Zarro et al. 1999).

efaction behind the wavefront. Such waves have been called *Moreton waves* and *EIT waves* (§ 8.3.1) and have been simulated based on theoretical models (§ 8.3.2). Here we concentrate on observations that relate the effect of coronal dimming more specifically to the occurrence of CMEs.

Coronal dimming occurs after a CME launch and were first described as *abrupt depletions of the inner corona* using the HAO K-coronameter data (Hansen et al. 1974), or as a *transient coronal hole* (Rust 1983), using *Skylab* data (Rust & Hildner 1978).

Let us review coronal dimming observed in soft X-ray wavelengths. The disappearance of soft X-ray-bright loops in the long-duration flare of 1994-Nov-13, 11:30 UT, has been witnessed (Fig. 17.16) by Hudson et al. (1996). The disappearance and associated dimming were interpreted as a consequence of outward motion rather than as a cooling process, based on the fact that the radiative cooling time was estimated to be much longer than the disappearance (dimming) time. This event is considered as an example for the counterpart of a CME seen in soft X-rays, with an estimated mass loss of $> 4 \times 10^{14}$ g and a temperature of $T \gtrsim 2.8$ MK (Hudson et al. 1996). A dimming was also observed just prior to a "halo" CME on 1997-Apr-07, using *Yohkoh/SXT* (Fig. 17.17; Sterling & Hudson 1997; Zarro et al. 1999). Here the strongest dimming occurred symmetrically at both sides of the flare volume, close to the ends of a pre-flare S-shaped sigmoid (Sterling & Hudson 1997). The resulting dimmings in these regions persisted for more than 3 days following the flare. At the same time, a dramatic dimming was also noticed in EUV, using the *SoHO/EIT* 195 Å Fe XII images (Zarro et al. 1999). The locations of reduced EUV intensity are co-spatial and simultaneous with those of soft X-ray dimming features. The EIT light curves show a drop down to $\approx 25\%$ of the preflare flux, and are clearly anti-correlated with the flare flux (Fig. 17.17, right). The cause of EUV and SXR coronal dimmings were interpreted

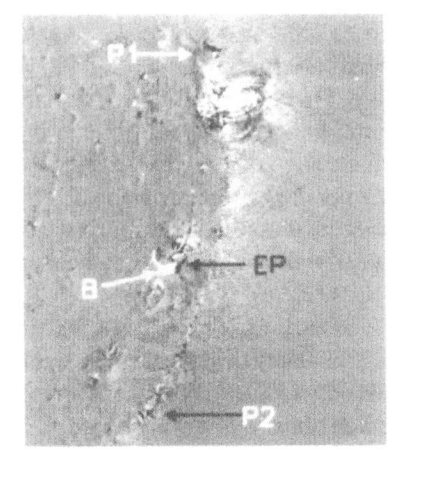
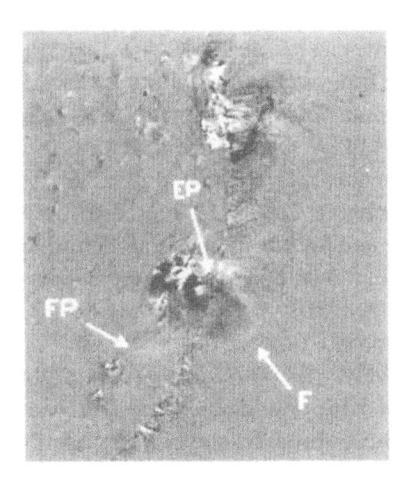

20:20 UT   20:32 UT

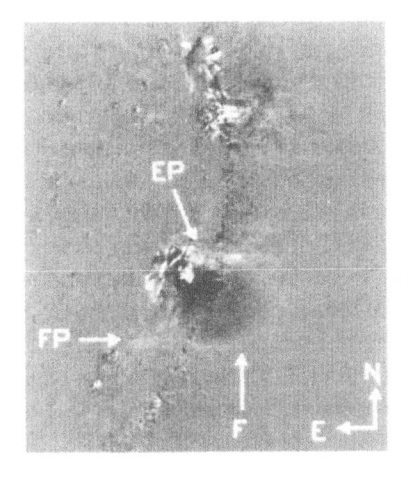
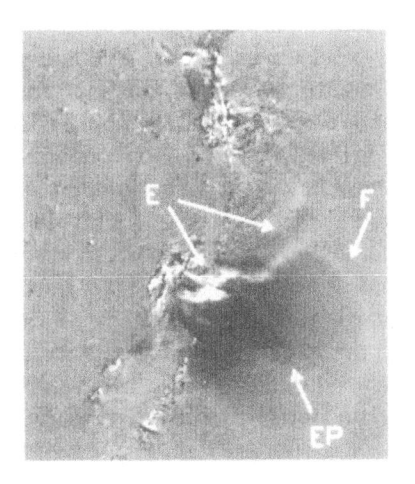

20:41 UT   20:53 UT

Figure 17.18: EIT running difference images of the initiation of a CME. EP=eruptive promi-
nence, B=flare-like brightening, F=outer front of CME, FP=footpoint of one side of CME, and
E=secondary set of ejecta. Note the location of strongest dimming at the center of the expanding
CME bubble (Dere et al. 1997b).

within the framework of a fluxrope eruption, partially controlled by the CME.

The dimming at the launch time of a CME is most conspicuously observed in EUV,
generally associated with a spherically expanding wave over the global solar surface
(§ 8.3; e.g., Thompson et al. 1999). Probably the clearest record of the vertical structure
of coronal dimming regions during the launch of a CME can be seen in the EIT 195 Å
observations of 4 time frames during the CME of 1996-Dec-23, 20:20 UT (Fig. 17.18;
Dere et al. 1997b). The onset of the dimming appears to be coincident with the ini-

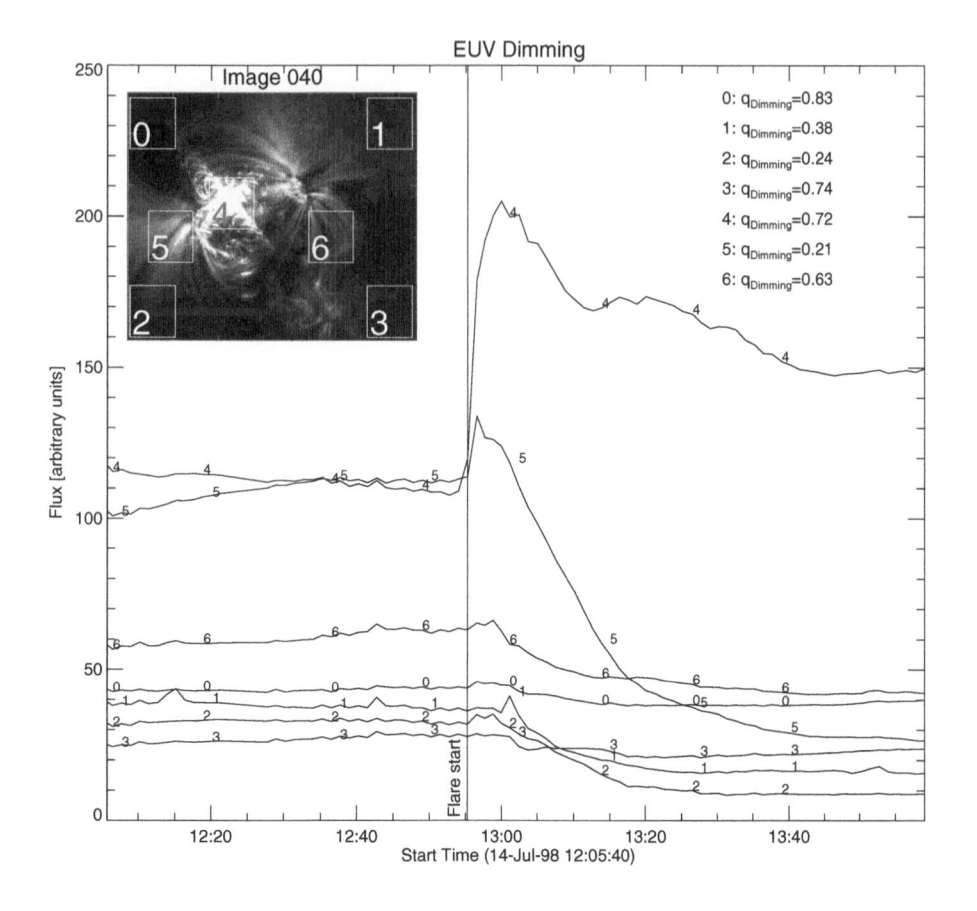

Figure 17.19: EUV dimming measured during the 1998-Jul-14, 12:55 UT, Bastille-Day flare with TRACE 171 Å. Note that the strongest dimming is aligned with the dipole axis of the active region (positions 1, 2, and 5). The numbered light curves correspond to the EUV fluxes integrated over the numbered boxes in the inserted map. The center location of the flare coincides with the center of the diffraction pattern (Aschwanden et al. 1999b).

tiation of global EIT waves, usually continues for hours thereafter, and can exhibit a quite asymmetric and skewed distribution regarding the center of origin (Thompson et al. 1998a), or even forming channels of irregular shapes (Chertok & Grechnev 2003). An analysis of 7 fast (> 600 km s^{-1}) CMEs corroborated that the coronal dimmings generally map out the apparent footprints of the CMEs observed in white light (Thompson et al. 2000b). The coronal dimming after a CME launch was found to coincide in EUV and Hα (Thompson et al. 2000a; Jiang et al. 2003). The Hα dimming is thought to be associated with the material evacuated near the feet of the erupted fluxrope (Jiang et al. 2003). This dipolar symmetry of EUV dimming has also been observed during the 1998-Jul-14, 12:55 UT, Bastille-Day flare, where the strongest dimming (down to a level of $21\% - 38\%$ of the preflare flux) occurred near the leading and following polarity of the dipolar active region, while the dimming was much less pronounced (at a level of 63%–83% of the preflare flux) in orthogonal directions (Fig. 17.19; Aschwanden et al. 1999b; Chertok & Grechnev 2004).

A multi-wavelength study with a broad temperature coverage between 20,000 K and 2 MK using CDS data showed that the dimming after a CME is strongest for plasma with a temperature of ≈ 1.0 MK, and thus the evacuated material comes from coronal heights rather than from transition region heights (Harrison & Lyons 2000). Spectroscopic evidence that coronal dimming at CME onsets represent indeed material outflows (rather than temperature changes) has been proven by measurements of Doppler shifts (e.g., with a velocity of v ≈ 30 km s^{-1} in coronal Fe XVI and Mg IX lines co-spatial with dimming regions; Harra & Sterling 2001). Another line of evidence that dimming corresponds to mass loss (rather than temperature changes) comes from mass loss estimates, which have been found to agree between white-light emission (observed with LASCO) and EUV dimming (observed with CDS), in the range of $m_{CME} \approx 5 \times 10^{13} - 4 \times 10^{15}$ g (Harrison et al. 2003). Also the comparison of the ejected material ($\approx 6 \times 10^{15}$ g) of an eruptive prominence observed in microwaves was found to be comparable with the coronal dimming ($\approx 1.7 \times 10^{15}$ g) estimated from soft X-ray data (Gopalswamy & Hanaoka 1998).

17.9 Interplanetary CME Propagation

Most of the coronal phenomena described in this book occur at a distance of $1R_\odot < r \lesssim 2R_\odot$ from the Sun center. The propagation of CMEs has been observed in white light by using coronagraphs (e.g., with *SoHO/LASCO*, in the range of $2R_\odot \lesssim r \lesssim 30R_\odot$). Many space-based observations of CME-related phenomena are performed from satellites in an Earth orbit, at a distance of 1 AU (i.e., $r \approx 200 \ R_\odot$). The physics of interplanetary and heliospheric phenomena (which is beyond the scope of this book) entails a plethora of plasma physics processes equally as rich as coronal phenomena, and are described in a number of textbooks and monographs (Russell et al. 1990; Schwenn & Marsch 1991a,b; Kivelson & Russell 1995; Crooker et al. 1997; Song et al. 2001; Balogh et al. 2001; Carlowicz & Lopez 2002). In the following section we sketch a short overview of physical concepts that connect coronal to interplanetary CME phenomena. A subset of these phenomena that play a role in solar-terrestrial connectivity are also referred to as *space weather phenomena* (Song et al. 2001), of which the geoeffective ones (e.g., *solar storms*; Carlowicz & Lopez 2002), are of utmost interest for the inhabitants on Earth.

17.9.1 Interplanetary Magnetic Field (IMF)

The *heliospheric 3D magnetic field* is defined by the flow of the solar wind. The field in the regions between the planets near the ecliptic plane is more specifically called the *interplanetary magnetic field*. The basic geometry of the interplanetary magnetic field has the form of an Archimedian spiral, as inferred by Parker (1963b) from the four assumptions: (1) the solar wind moves radially away from the Sun at a constant speed; (2) the Sun rotates with a constant period (i.e., with a synodic period of 27.27

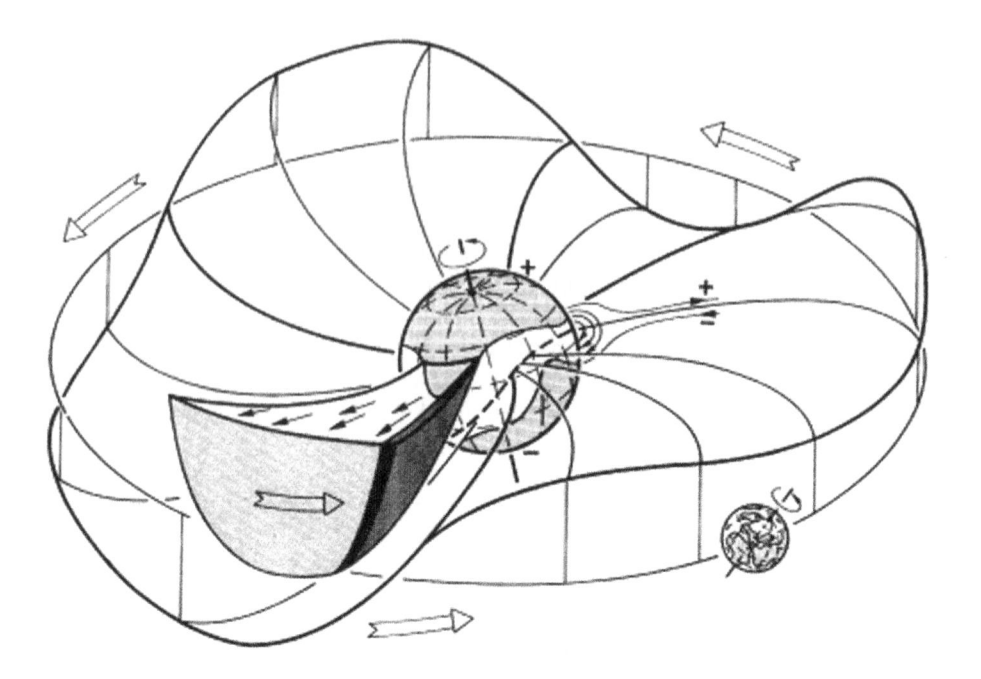

Figure 17.20: The interplanetary magnetic field has a spiral-like radial field and the boundary layer between the two opposite magnetic polarities in the northern and southern hemispheres is warped like a *"ballerina skirt"*. This concept was originally suggested by Hannes Alfvén in 1977.

days at the prime meridian defined by Carrington); (3) the solar wind is azimuthally symmetric with respect to the solar rotation axis; and (4) the interplanetary magnetic field is frozen-in the solar wind and anchored at the Sun. The solar wind stretches the global, otherwise radial field into spiral field lines with an azimuthal field component. The resulting Archimedian spirals leave the Sun near-vertically to the surface and cross the Earth orbit at an angle of $\approx 45°$. Measurements of the magnetic field direction at Earth orbit reveal a *two-sector pattern* during the period of declining solar activity and a *four-sector pattern* during the solar minimum, with oppositely directed magnetic field vectors in each sector. From this ecliptic cut, a warped heliospheric current sheet can be inferred that has the shape of a *"ballerina skirt"* (Fig. 17.20). The solar axis is tilted by $7.5°$ to the ecliptic plane, and the principal dipole magnetic moment of the global field can be tilted by as much as $\approx 20° - 25°$ at activity minimum, and thus the warped sector zone extends by at least the same angle in northerly and southerly direction of the ecliptic plane. A longitudinal cut of the solar magnetic field near the Sun is shown in Fig. 1.14, based on a model by Banaszkiewicz et al. (1998).

The strength of the interplanetary magnetic field, of course, depends on the solar cycle (§ 1.3), varying between $B \approx 6$ nT and 9 nT ($\approx 10^{-5}$ G) at a distance of 1 AU. The interplanetary magnetic field can be heavily disturbed by CME-related shocks and

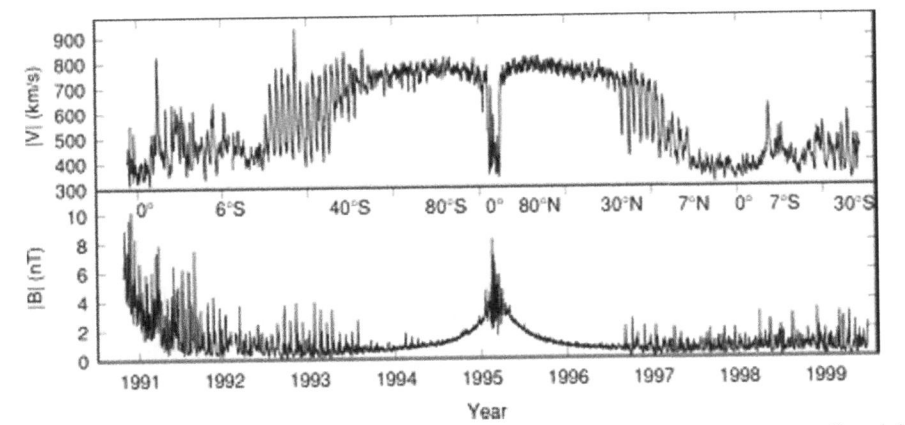

Figure 17.21: The solar wind speed as a function of heliographic latitude (upper panel) and the magnitude of the magnetic field as a function of time (lower panel) measured by *Ulysses* from launch through its first solar orbit. Note the detection of the slow solar wind (v ≈ 400 km s^{-1}) in low latitudes (≲ 20°) and of the fast solar wind (v ≈ 800 km s^{-1}) in high latitudes (≳ 20°) (Balogh 2001).

propagating CMEs. The magnetic field is near-radial near the Sun and falls off with $B(R) \approx B_r(R) \propto R^{-2}$ there, while it becomes more azimuthal at a few AU and falls off with $B(R) \approx B_\varphi(R) \propto R^{-1}$ at larger heliocentric distances according to the model of Parker. Reviews on the interplanetary magnetic field can be found in, for example, Kivelson & Russell (1995; § 4), Burlaga (2001), Ness (2001), Russell (2001), and Schwenn & Marsch (1991a).

17.9.2 Solar Wind

Parker's (1958) theoretical model of the solar wind (§ 4.10; Fig. 4.33) predicts that the coronal plasma outflow expands into a supersonic solar wind, which was confirmed by measurements of *in situ* spacecraft, such as with *Mariner II* in 1962, or with *Ulysses* more recently (Fig. 17.21). The transition into a supersonic wind occurs at $r \approx 5\ R_\odot$. However, the model of Parker (1958) does not address the energy equation and cannot explain the slow and fast solar wind components. The energy balance equation yields a different solution for open field regions, where the fast solar wind originates, and for the corona over closed field regions, where the slow solar wind originates (see shaded area in Fig. 1.14). In magnetically closed regions, downward heat conduction is the dominant energy loss mechanism. In open field regions, energy is taken out with the solar wind in the forms of work done against gravity and kinetic energy of the flow (Table 9.1). Of course, the exact solution of the energy balance equation also depends on the coronal heating function, which is not known. However, to obtain a fast solar wind of v ≈ 800 km s^{-1}, energy needs to be deposited far out in the corona (e.g., by dissipation of Alfvén waves; see § 9.4). Furthermore, the energy deposition is also different for electrons and ions, as measured by the higher ion temperature in the solar wind, compared with the electron temperature (Fig. 9.13). The solar wind flow speed is usually much larger than the local sound speed or Alfvén speed, typically having a Mach number of ≈ 10, which implies that the dynamic pressure is much higher than both the magnetic pressure and the thermal pressure. The magnetic field is frozen-in in the solar wind flow due to the high conductivity.

CMEs represent transient activities that disturb the solar wind. The average CME speed is slightly below the solar wind speed in the corona. The CME plasma is entrained in the interplanetary magnetic field lines and is transported into the solar wind. Various signatures of CMEs in the solar wind include: (1) transient interplanetary shock waves, (2) He abundance enhancements, (3) unusual ionization states (e.g., He^+), (4) brief density enhancements and long-duration density decreases, (5) proton and electron temperature depressions, (6) bi-directional field-aligned flows of halo electrons and low-energy protons, and (7) magnetic field variations associated with *magnetic clouds* or *fluxropes*. The chemical abundance and charge state compositions have been found to be systematically different in CMEs and in the background solar wind.

Reviews on the solar wind and CME disturbances of the solar wind can be found in, for example, Holzer (1989), Schwenn & Marsch (1991a,b), Gosling (1994, 1996), Goldstein et al. (1995), Hundhausen (1995), Winterhalter et al. (1996), Schwadron et al. (1997), Burgess (1997), Neugebauer et al. (2001), Leer (2001), Schwenn (2001), Marsch (2001), Webb (2001), Habbal & Woo (2001), Balogh (2001), Russell (2001), Balogh et al. (2001), Bochsler (2001), Matthaeus (2001b), Esser (2001), Kunow (2001), Cranmer (2002a,b), Ofman (2003), Erdoes (2003).

17.9.3   Interplanetary Shocks

CMEs have typical propagation speeds of $v \approx 300 - 400$ km s^{-1}, but fast CMEs have been measured at speeds in excess of $v = 2000$ km s^{-1}. Since the fast solar wind has a typical speed of $v \approx 800$ km s^{-1}, fast CMEs are super-Alfvénic. Thus, such fast CMEs can drive transient interplanetary shocks. Numerical simulations with HD or MHD codes (for instance see Fig. 17.22), have been able to reproduce the observed speeds and pressure profiles of shocks and CME events out to large distances from the Sun. In such simulations, a pressure pulse is initiated in the lower corona. As the front of a fast CME overtakes the slower solar wind, a strong gradient develops and pressure waves steepen into a forward shock propagating into the ambient wind ahead, and occasionally a reverse shock propagates back through the CME towards the Sun (Riley et al. 1977; 1999). Numerical simulations of CMEs propagating from the corona through the heliosphere can be found in, for example, Mikić & Linker (1994); Linker & Mikić (1995); Linker et al. (2001); Odstrčil et al. (1996, 2002), Toth & Odstrčil (1996), Odstrčil & Pizzo (1999a,b), and Riley et al. (2003).

There are a number of complications that can occur, such as the fact that a faster CME can catch up with a slower CME and interact. Such interactions form compound streams in the inner heliosphere. These systems continually evolve further and merge with other CMEs and shocks as they move outward. In the outer heliosphere, beyond 10-15 AU, such structures form *merged interaction regions*, which become so extensive that they encircle the Sun like a distant belt. Such regions block and modulate galactic cosmic rays (i.e., the flux of high-energy particles that continuously streams into the

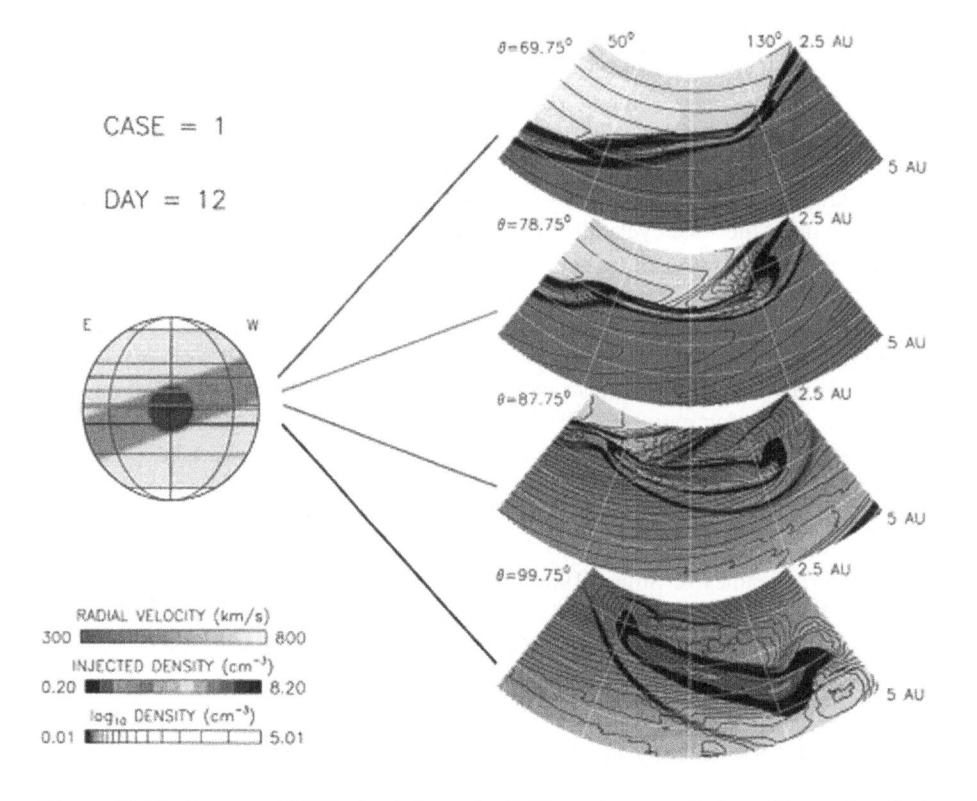

Figure 17.22: Numerical MHD simulations of a CME moving through the ambient solar wind. The CME is injected in the center of the heliospheric current sheet streamer belt (left), which is tilted to the solar axis. The propagating CME is shown at slices in heliolongitude and at a distance of 2.5-5 AU from the Sun 12 days after launch. The slices are 4 different heliolatitudes and show how the CME's shape, pressure and speed vary depending on the ambient solar wind conditions (Courtesy Victor Pizzo, NOAA/SEC).

heliosphere). Finally, a forward interplanetary shock wave that passes the Earth's magnetosphere may cause a sudden commencement of a *magnetic storm* or *substorm* at the Earth and change the electrical and magnetic connection of the interplanetary magnetic field with the Earth's magnetic field. Reviews on interplanetary shocks, CMEs, and related phenomena can be found in, for example, Schwenn & Marsch (1991a,b), Kivelson & Russell (1995), Burlaga (1995), Colburn & Sonett (1996), Crooker et al. (1997), Balogh & Riley (1997), Whang et al. (1998), Balogh et al. (2001), Song et al. (2001), Lepping (2001), and literature referenced therein.

17.9.4 Solar Energetic Particles (SEP)

Solar energetic particle (SEP) events refer to accelerated particles detected in the heliosphere. Some originate in solar flares, while others are accelerated in transient interplanetary shocks, driven by fast CMEs. The acceleration mechanisms are basically the same types we discussed in § 11.5 on shock acceleration. Solar energetic particle events are classified into two types, *gradual* and *impulsive* SEP events, depending on their energy versus time profile. *Gradual SEP events* occur with a rate of $\approx$ 10/year during the maximum of the solar cycle, each one can last several days, and they are likely to be accelerated directly in interplanetary shocks rather than by flares in the corona. *Impulsive SEP events* occur more frequently, with a rate of $\approx$ 100/year during the maximum of the solar cycle, they last only a few hours, and they are much weaker than gradual SEP events. Since they originate along magnetic field lines connected to coronal flare sites, their acceleration could be governed by the same magnetic reconnection process that governs the associated flare. Because their $^3He/^4He$ ratio is much higher than in the normal solar wind, they are also called ^3He-rich events. Interplanetary particles can also be accelerated in the electric fields that are generated at *co-rotating interaction regions (CIR)* between high-speed and low-speed streams. To some extent, the location where acceleration of interplanetary particles takes place can be determined from the velocity dispersion (i.e., time-of-flight effects), $t_{prop} = L/v$, of particles arriving at Earth.

Literature on solar energetic particle events can be found in, for example, Kahler et al. (1984), Reames & Stone (1986), Reames et al. (1988, 1991a,b, 1992, 1994; 1996, 1997, 2001a,b), Reames (1990b, 1995a,b, 1999, 2001a,b, 2002), Gosling (1993), Kahler (1992, 1994, 2001), Tylka (2001), Reames & Tylka (2002), and references therein.

17.9.5 Interplanetary Radio Bursts

There are two sources of energetic particles in interplanetary space, either flare-related magnetic reconnection sites in the solar corona that are connected to interplanetary space via open field lines, or shock acceleration sites associated with super-Alfvénic CME fronts that propagate through interplanetary space. Since the plasma in interplanetary space is collisionless, suprathermal and high-energy particles can propagate unimpeded through interplanetary space and form particle beams (e.g., electron beams or ion beams). The beam free energy is converted into Langmuir waves, and some Langmuir wave energy is converted to radio waves at the fundamental or harmonic local plasma frequency (§ 15.1, 15.4). Thus, beam-driven type III-like radio bursts are common in interplanetary space (Fig. 15.13), and occasionally type IV-like radio bursts also occur (i.e., synchrotron emission caused by energetic electrons confined in a magnetic trap created behind an interplanetary shock wave). The spatial size of interplanetary radio bursts can be very large, since the extent of the radio source grows with distance from the Sun. However, interplanetary type III emission is not produced continuously along the propagation path of electron beams, but rather seems to occur in localized, unresolved regions of the interplanetary medium. Interplanetary type II-like bursts, also called *shock-associated (SA)* events, also occur and are believed to be produced by collisionless shock waves associated with passing CMEs. Thus, interplanetary radio bursts provide a rich diagnostic on the acceleration and propagation of energetic particles and shock waves. However, only radio bursts with plasma frequencies $\gtrsim$ 20 MHz (above the Earth's ionospheric cutoff frequency) can be observed with ground-based radio telescopes, which extends only out to about $1 - 2$ solar radii, while all interplanetary radio bursts further out have lower plasma frequencies and require space-based radio detectors.

Reviews on interplanetary radio bursts can be found in, for example, Lin (1974), Simnett (1986b), Dulk (1990), Schwenn & Marsch (1991), Robinson (1997), Aschwanden & Treumann (1997), Cairns et al. (2000), Reiner (2001), and Bougeret (2001).

17.10 Summary

Coronal Mass Ejections (CMEs) are a new eruptive phenomenon distinct from flares although they are related by the common solar magnetic fields that produce them. Both eruptions involve releases of magnetic energy in comparable magnitudes. CMEs and flares represent complementary phenomena, both being produced as by-products of a common magnetic instability that is controlled on a larger scale in the solar corona. Theoretical models include five categories: (1) thermal blast models, (2) dynamo models, (3) mass loading models, (4) tether release models, and (5) tether straining models (§ 17.1). Numerical MHD simulations of CMEs are currently produced by combinations of a fine-scale grid that entails the corona and a connected large-scale grid that encompasses propagation into interplanetary space, which can reproduce CME speeds, densities, and the coarse geometry (§ 17.2). The trigger that initiates the origin of a CME seems to be related to previous photospheric shear motion and subsequent kink instability of twisted structures (§ 17.3). The geometry of CMEs is quite complex, exhibiting a variety of topological shapes from spherical semi-shells to helical fluxropes (§ 17.4), and the density and temperature structure of CMEs is currently investigated with multi-wavelength imagers (§ 17.5). The height-time, velocity, and acceleration profiles of CMEs seems to establish two different CME classes: gradual CMEs associated with propagating interplanetary shocks, and impulsive CMEs caused by coronal flares (§ 17.6). The total energy of CMEs (i.e., the sum of magnetic, kinetic, and gravitational energy), seems to be conserved in some events, and the total energy of CMEs is comparable to the energy range estimated from flare signatures (§ 17.7). A closely associated phenomenon to CMEs is coronal dimming, which is interpreted in terms of an evacuation of coronal mass during the launch of a CME (§ 17.8). The propagation of CMEs in interplanetary space (which is beyond the scope of this book), provides diagnostics on the heliospheric magnetic field, the solar wind, interplanetary shocks, solar energetic particle (SEP) events, and interplanetary radio bursts (§ 17.9).

Problems

Chapter 1: Introduction

Problem 1.1: Plot the wavelength ranges of all solar instruments mentioned in this Section onto a diagram with a wavelength axis similar to Fig. 1.25. Convert energies (keV) to wavelengths (Å) with Eq. (1.10.4). Do the instruments from all previous solar missions cover the entire wavelength range? Are there still some windows left for new discoveries?

Problem 1.2: Compile a metrics of coronal phenomena (enumerated in Section 1.2) versus wavelengths. Which wavelength domain is richest in displaying coronal phenomena and why?

Problem 1.3: In Fig.1.13, a correlation between the soft X-ray flux I_{SXR} and the longitudinal magnetic field $B_\parallel$ is shown, yielding a powerlaw slope specified in Eq. (1.3.1). Assume that the average magnetic field strength varies with heliographic longitude l as $B_\parallel = B\cos(l)$. If the magnetic field varies over the range of $B = 10 - 1000$ G in active regions during a solar cycle, by what factor is the soft X-ray flux expected to vary?

Problem 1.4: Determine the magnetic field strength B_{foot} (at the photosphere) and the dipole depth h_D (defined by Eq. 1.4.2) that fit closest the empirical model of Dulk & McLean (1978) (given in Eq. 1.4.1) at their lower limit of $1.02\,R_\odot$. Overplot the two models to verify the best fit at the lower limit and characterize how the two models diverge with height.

Problem 1.5: Can you estimate the filling factors of loop strands as a function of a fractal dimension defined in a 2D geometry? (Hint: Use definition of Eq. 9.7.2). What is the fractal dimension D_2 if a filling factor of $q_{fill} = 0.1$ is measured in a quadratic image area with a length of $N_l = 100$ pixels. How does the filling factor q_{fill} change if the same area is imaged with an instrument with 10 times better spatial resolution and the structure has the same fractal dimension D_2?

Problem 1.6: Derive analytically or graphically for what mean temperature the Baumbach−Allen density model (Eq. 1.6.1) is closest to hydrostatic equilibrium near the solar surface? Hint: Use the pressure scale height Eq. (3.1.16) to obtain the temperature.

Problem 1.7: In the differential emission measure distributions shown in Fig. 1.21, at what temperature is most of the coronal plasma found in the Quiet Sun and in active regions? What maximum factor is the electron density of the 1.5 MK plasma higher in active regions than in Quiet Sun regions, assuming the same column depth for the regions shown in Fig. 1.21?

Problem 1.8: If you assume the electron densities n_e and electron temperatures T_e given in Table 1.1, what is the range of magnetic field strengths B that reproduces the spread of the plasma-β parameter $\beta(h)$ shown in Fig.1.22 (from Gary 2001) in photospheric ($h = 250$ km), coronal ($h = 10$ Mm), and heliospheric ($h = 2$ $R_\odot$) altitudes? (Use Eq. 1.8.1).

Problem 1.9: Which chemical elements should have been detected in the solar corona at the time when the sensitivity was $\gtrsim 10^{-4}$ of hydrogen abundance? What elements do you predict to be discovered in the corona when the sensitivity improves to a level of $\gtrsim 10^{-8}$ of the hydrogen abundance? Use Fig.1.24 and Table 1.2.

Problem 1.10: Using the conversion formula given in Eq. (1.10.6), calculate at what wavelengths λ (Å) the following phenomena with their characteristic temperatures should be brightest: (a) sunspot umbra with $T = 4500$ K, (b) photosphere with $T = 6000$ K, (c) transition region with $T = 20,000$ K, (d) coronal loop with $T = 1.0$ MK, and (e) flare loop with $T = 15$ MK. Identify the wavelength domains.

Chapter 2: Thermal Radiation

Problem 2.1: Compute the free-free absorption coefficient α_{ff} (Eq. 2.3.16), the free-free opacity τ_{ff} (Eq. 2.3.18), and the radio brightness temperature T_B (using Eqs. 15.2.3 and 15.2.4) of the coronal plasma with an average temperature of $T_e = 10^6$ K, an average density of $n_e = 10^9$ cm^{-3}, over a scale height of $\lambda_T = 5 \times 10^9$ cm, for radio frequencies of $\nu = 100$ MHz, 1 GHz, and 10 GHz. At what radio frequency does the corona become optically thick?

Problem 2.2: Verify the conversion of Planck's law as a function of wavelength λ (Eq. 2.2.2) to a function of frequency ν (Eq. 2.2.3). Hint: use $\nu = c/\lambda$ and $d\nu = [d\nu(\lambda)/d\lambda]d\lambda$.

Problem 2.3: Derive the wavelength $\lambda_{max}(T)$ where the Planck function peaks for a given temperature T, as given in Eq. (2.2.6). Hint: Substitute the variable $x = \lambda(k_B T/hc)$ in Eq. (2.2.3) and set the derivative $\partial B_\lambda(x)/\partial x = 0$ to zero. The value of x that corresponds to λ_{max} can be calculated numerically or graphically.

Problem 2.4: The hard X-ray spectrum of the flare shown in Fig.2.6 consists of a thermal < 35 keV and a nonthermal (powerlaw) spectral component. Compute the total thermal energy, assuming a typical flare density of $n_e \approx n_i = 10^{11}$ cm^{-3} (use Eq. 13.3.2). Is this an average-sized or a large flare? (Compare with frequency distribution of flare energies in Fig. 9.27).

Problem 2.5: Calculate the Lyman, Balmer, and Paschen series with the Rydberg formula (Eq. 2.4.3) for helium. In what wavelength regimes (SXR, EUV, UV, optical, IR) are the helium series to be found?

Problem 2.6: Calculate the relative abundances of neutral and ionized calcium (with ionization potentials of 6.1 and 11.9 eV, and statistical weights $g_0 = 1$ and $g_1 = 2$, respectively) at photospheric temperatures ($T_e = 6000$ K). Which of the two lines is stronger in the chromosphere, Ca I or Ca II?

Problem 2.7: Estimate the photon energies involved in the emission processes shown in Fig. 2.9 and identify the wavelength ranges for free-bound, free-free, and ionization processes for hydrogen.

Problem 2.8: If you have the *Interactive Data Language (IDL)* software available and have installed the *SolarSoftWare (SSW)*, load the *CHIANTI* package and reproduce the ionization equilibrium for Fe levels shown in Fig. 2.10. Instructions are given in the tutorials provided on one of the *CHIANTI* web pages, e.g., *http://wwwsolar.nrl.navy.mil/chianti.html.*

Problem 2.9: Compare the radiative losses (integrated over a vertical pressure scale height) at a coronal temperature of $T_e = 1.0$ MK and coronal density of $n_e \approx 10^8$ cm^{-3} from the radiative loss rate given in Fig. 2.14 (or Eqs. 2.9.1 and 2.9.2) with the Stefan−Boltzmann law (Eq. 2.2.7). Quantify the difference between line emission and blackbody radiation.

Problem 2.10: Enter all detected coronal elements with known FIP energy (provided in Table 1.2) into a FIP diagram as shown in Fig. 2.15. What sensitivity in abundance measurements is needed in future to verify the coronal enhancement of the new low-FIP elements?

Chapter 3: Hydrostatics

Problem 3.1: Calculate with Eqs. (3.1.2), (3.1.6), and (3.1.16) the pressure scale heights λ_p of various stars, such as for (1) a giant, e.g., αLyr (Vega), spectral class A0 V, with $\log(M_*/M_\odot) \approx +0.51$, $\log(R_*/R_\odot) \approx +0.40$, (2) a bright giant, e.g., βGem (Pollux), spectral class K0 III, with $\log(M_*/M_\odot) \approx +0.6$, $\log(R_*/R_\odot) \approx +1.2$, (3) a white dwarf, e.g., αLeo (Regulus), spectral class B7 V, with $\log(M_*/M_\odot) \approx +0.7$, $\log(R_*/R_\odot) \approx +0.5$, and (4) the Earth, with an atmospheric pressure of $p_0 = 1.02 \times 10^6$ dyne cm^{-2} and a mean density of $\rho_0 = 1.29 \times 10^{-3}$ g cm^{-3}.

Problem 3.2: Give an analytical expression for the flux $F(h)$ observed in a hydrostatic fluxtube as function of the height h above the limb (1) for a narrowband filter with $\Delta T/T \ll 1$, and (2) for a broadband filter with $\Delta T/T \approx 1$, assuming a rectangular response function $R(T)$ and taking the hydrostatic weighting bias into account.

Problem 3.3: Predict the center-limb brightening (Eq. 3.3.10 and Fig.3.9) for various spatial resolutions ($\Delta h = 1''$, $10''$, $100''$) at different temperatures (T=1, 4 MK). Hint: Approximate the peak of the column depth function $z_{eq}(h, T_0)$ shown in Fig. 3.9 with a triangle and read off the values at the disk center and limb from the graphs. Determine the limb brightness by approximating the triangular peak with a rectangle of the given spatial resolution. If you have a *Yohkoh/SXT* or *SoHO/EIT* image at hand, test your prediction of the center-limb brightening.

Problem 3.4: What are the correction factors for the vertical hydrostatic scale height λ_p, which has to be applied to the observed scale height of the following projected semi-circular loops: (1) a loop at disk center with footpoints in North-South direction, when the loop top is displaced from the baseline by a quarter of the footpoint separation; (2) a loop at the west limb with footpoints in North-South direction and loop top displaced from the baseline by half the footpoint separation; and (3) a loop at the west limb with coinciding footpoints when the looptop is located above the limb at a height that corresponds to half of the projected loop size?

Problem 3.5: Derive a relation of the intensity contrast between the brightest loop and the coronal background in the DEM shown in Fig. 3.17 (bottom right). Hint: Assume that the background emission consists of contributions from N loops along a line-of-sight and that fluctuations in the DEM obey Poisson statistics. How does the requirement for instrumental flux contrast vary with the number of resolved loops? What is the flux contrast for N=10, 100, or 1000 loops along a line-of-sight?

Problem 3.6: Derive the RTV scaling law for the temperature (Eq. 3.6.14) from the inverse temperature profile, s(T) (Eq. 3.6.12). Hint: Neglect the heating term, approximate the radiative loss function with the single-powerlaw function $\Lambda(T) \approx 10^{-18.8}T^{-1/2}$, and assume a constant pressure $p_0 = 2n_e k_B T$.

Problem 3.7: Calculate the temperature drop $T(s = L/2)$ in the upper half of a hydrostatic loop with a loop-top temperature of $T_{max} = 1.0$ MK and a heating scale height of $s_H = L$ (use analytical approximation given in Eq. 3.6.18). Compare the result with the value obtained from the graphical curves shown in Fig. 3.22.

Problem 3.8: Use the instrumental response functions graphically shown in Fig. 3.25 and estimate the relative brightness of loops with temperatures of $T = 1, 3, 5, 10$ MK (and equal density) in *TRACE* 171 Å and *Yohkoh/SXT* Al 1265 Å.

Problem 3.9: If the contrast of a TRACE image is 1:10, is the loop top of a hydrostatic loop detectable for the following loop parameters: $T_{max} = 1$ MK, $L = 100$ Mm? How does it compare for a hot loop with $T_{max} = 5$ MK with the same length for a Yohkoh image, assuming the same contrast? Hint: Apply the exponential density model $n_e(s)$ (Eq. 3.2.1-2) with the proper scale height λ_p (Eq. 3.1.16). Does it explain that *TRACE* sees only the lower parts of loops while *Yohkoh/SXT* sees complete loops?

Problem 3.10: If the stellar DEM of ζ Boo A is approximated by two populations of loops, cool loops with $T = 3$ MK and hot loops with $T = 10$ MK, with equal lengths $L = 100$ Mm and heating scale heights of $s_H = 20$ Mm, what is the ratio of hot to cool loops, if all are assumed to be in hydrostatic equilibrium? Hint: Obtain the corresponding DEM components from Fig. 3.30 and use Serio's scaling law (Eq. 3.6.16) to determine the base pressures p_0. The base density n_0 can then be estimated with the ideal gas law (Eq. 3.1.9), and the corresponding emission measures are $EM \propto n_0^2$.

Chapter 4: Hydrodynamics

Problem 4.1: Derive the hydrodynamic equations (4.1.21-24) from the general notation (4.1.2-4.1.4), using the 3D operator for the total derivative (4.1.1) and inserting the definitions for entropy (Eq. 4.1.8) and internal energy (Eq. 4.1.12). Verify all the steps from Eq. (4.1.4) to Eq. (4.1.23).

Problem 4.2: Derive with the continuity equation (Eq. 4.2.16) and a hydrostatic density profile $n_e(h)$ an approximate scaling law between the loop height and looptop temperature in siphon flow loops that develop shocks at the looptop.

Problem 4.3: Use criterion (Eq. 4.3.11) to estimate the density range $n_e(L)$ of thermally stable loops with typical coronal temperatures of $T_0 = 1.0$ MK as a function of their loop length L.

Problem 4.4: Estimate the upflow speed v_0 at the footpoint of Brekke's loop shown in Fig. 4.8, which shows at location A a blueshift of $v(h_A) = 60$ km s^{-1}. Hint: assume a constant cross section, a density profile $n_e(h)$ according to the hydrostatic scale height temperature of the O V line, use the height scale given in arcseconds in Fig. 4.8, and apply the continuity equation.

Problem 4.5: Simulate the time profile $I_{Filter}(t)$ of a cooling loop with an initial temperature of $T = 2.5$ MK and a constant density of $n_e = 10^9$ cm^{-3}, assuming purely radiative cooling (Eq. 4.5.3), an iron FIP enhancement of $\alpha_{FIP} = 4$, and instrumental response functions of the three TRACE passbands approximated by Gaussian shapes with peak temperatures of $T_{filter} = 1.0, 1.5, 2.0$ and a Gaussian width of $\Delta T_{filter} \approx 0.2$ MK. What are the delays when the emission peaks in the different filters? Does the simulated duration in each filter match the observed ones shown in Fig. 4.12 (Loop #2)?

Problem 4.6: What is the expected relation between the flux detection threshold with a signal-to-noise ratio of $\Delta I/I$ and the super-hydrostatic scale height factor $q_\lambda = \lambda/\lambda_T$ shown in Fig. 4.19, required for detection of the looptop? (Hint: Choose a disk center position, where the background column depth is approximately a super-hydrostatic scale height). What is the contrast $\Delta I/I$ at the footpoint and looptop, if a loop near disk center has a density ratio of $n_e^{loop}/n_e^{back} = 2$ to the background at the base, a width to scale height ratio of $w/\lambda_T = 0.1$, a height of $h = \lambda_T$, for both a hydrostatic and a super-hydrostatic ($\lambda = 4\lambda_T$) scale height?

Problem 4.7: Estimate the time scale of an acoustic wave traveling along a typical coronal loop (with half length of $L = 30,000$ km) and back? Compare it with the average time interval between subsequent shock formations or surge flows shown in the simulation of Robb & Cally (1992) in Fig. 4.22.

Problem 4.8: Calculate in first order the correction to the differential emission measure distribution $dEM(T)/dT$ for a loop with the canopy quantified by Chae et al. (1998) as given in Eq. (4.8.6), and shown in Fig. 4.27, compared with a straight loop without canopy. (Hint: Consider only the influence of the cross-sectional area).

Problem 4.9: Assume a hydrostatic corona with an average temperature of $T = 1.0$ MK and quantify upper density limits in coronal holes to explain a factor $\gtrsim 10$ difference of the radio brightness temperature between coronal holes and quiet Sun regions, as observed with Clark Lake at $\nu = 74$ MHz (see Fig. 4.32). Use the Rayleigh-Jeans approximation and consult problem (2.1).

Problem 4.10: How does the density $n(r)$ and velocity $v(r)$ drop off at large distances $r \gg r_c$ in Parker's solar wind solution? Does the resulting pressure $p(r)$ vanish at large distances in the interstellar medium?

Chapter 5: Magnetic Fields

Problem 5.1: Convert the Maxwell's equations (Eqs. 5.1.1-4), Ampère's law (Eq. 5.1.7), and Ohm's law (Eq. 5.1.10) from Gaussian units (cgs) into metric (SI or mks) units. Compare them with Appendix C or standard textbooks written in SI units. How do the Maxwell's equations differ for vacuum and matter?

Problem 5.2: Calculate analytically the loop width variation $w(\theta)$ as a function of the angle θ along the loop relative to the footpoint location θ_0 for a dipole field (see Fig. 5.2 and use Eq. 5.2.22). Can you reproduce the loop width expansion factors q_w given in Fig. 5.14 (bottom) ?

Problem 5.3: Calculate analytically the loop expansion factor between footpoint and looptop for a force-free loop in the framework of a sheared arcade (using Eq. 5.3.19) for $k = l = 1$. Can you reproduce the loop width expansion factors q_w given in Fig. 5.14 (bottom) ?

Problem 5.4: Calculate how the plasma β-parameter varies as a function of height in the hydrostatic dipole model described in §5.4.2. Plot a minimum-maximum range of $\beta(h)$ as a function of height similar to Fig. 5.10.

Problem 5.5: Reproduce the general outline of an S-shaped field line in Fig. 5.20 by plotting semi-circular, helically twisted fluxtubes with constant diameters, with proper projection of the 3D coordinates. Can an S-shaped field line be used to discriminate a potential magnetic field model (e.g., Fig. 5.20) from a linear force-free magnetic field model (e.g., Fig. 5.19)?

Problem 5.6: What is the length of loops that produce the average force-free parameter $|\alpha| \approx 2 \times 10^{-8}$ m^{-1} shown in the observation of Pevtsov et al. (1997) (Fig. 5.21), if we assume one full twist at the loop axis? At what radius r (or ratio r/l) does the number of twists double (in the thin fluxtube approximation)?

Problem 5.7: Search for more observations of coronal nullpoints in literature and data archives similar to the example shown in Fig. 5.25. (e.g., see TRACE and Yohkoh data archive, publications, or web-pages).

Problem 5.8: Derive the lowest-order expressions for the refractive index $n_{\nu,\sigma}$ of the two magneto-ionic modes, with and without magnetic fields (for longitudinal waves, $\psi = 0$), in the cold-plasma approximation (Eq. 5.7.7).

Problem 5.9: Plot an angular emissivity diagram for gyroresonance emission (using Eq. 5.7.21) for different harmonics ($s = 2, 3, 4$). At what angles is the strongest gyroresonance emission observed ?

Problem 5.10: Assuming a chromospheric pressure scale height $\lambda_{p,Ch}$ corresponding to a temperature of $T \approx 10^4$ K, estimate and the canopy-like expansion factor $A(h_{Ch})/A(h = 0)$ over the chromospheric height of $h_{Ch} \approx 2$ Mm. Hint: Use the pressure balance approximation (Eq. 5.8.2) and assume magnetic flux conservation.

Chapter 6: Magneto-Hydrodynamics (MHD)

Problem 6.1: Verify analytically the different steps in the transformation of the energy equation (6.1.23) from the microscopic Boltzmann equation into the macroscopic MHD energy equation (6.1.32).

Problem 6.2: Show that the set of ideal MHD equations (6.1.16-6.1.22) is a special case of the resistive MHD equations (6.1.33-6.1.39).

Problem 6.3: Calculate the part the Lorentz force due to the field line curvature (Eq. 6.2.18) in a dipole field (§5.2.2). By what fraction does the magnetic field vary across the loop width for a typical loop with a width of $w = 1$ Mm and a curvature radius of $r_{curv} = 50$ Mm in order to balance the curvature force in a dipolar loop?

Problem 6.4: Plot geometric projections of twisted fluxtubes (using the parameterization 6.2.22) and try to match up observed loop shapes shown in Fig. 6.4. Do you succeed in measuring the number of turns in some observed loops?

Problem 6.5: If the twisted loop has a length of $L = 100$ Mm a width of $w = 10''$, and is twisted by one turn, what is the required expansion speed to explain the rotational Doppler shift $v_{rot} \lesssim 30$ km s^{-1} as observed with SoHO/SUMER in Fig. 6.5 ?

Problem 6.6: Calculate the vertical speed of an emerging, stiffly twisted fluxtube to explain the observed positional rotation of the data points F10 and P10 shown in Fig. 6.9. Assume that the twisted fluxtube has emerged completely after 9 hours, starting from the bottom of the convection zone ($\approx 0.3R_\odot$). What is the number of twisted turns needed to explain the observations shown in Fig. 6.9, assuming a semi-circular loop width to length ratio of 1:10.

Problem 6.7: Discuss which of the instabilities given in §6.3 are most likely to occur in quiet Sun coronal loops, flare loops, and prominences.

Problem 6.8: Calculate the minimum and maximum pressure in prominences, using the extremal values of electron densities n_e and electron temperatures T_e listed in Table 6.2, and compare them with the observed pressure values in Table 6.2.

Problem 6.9: Calculate the maximum ratio of the mass of a prominence to the total coronal mass (Eqs. 6.4.12-6.4.13), using the maximum density and geometrical parameters given in Table 6.2.

Problem 6.10: Use criterion (4.3.11) to estimate the density range $n_e(L)$ of thermally stable filaments with typical temperatures of $T_0 = 10^4$ K as a function of their length L. Are all filaments thermally stable, according to the observed densities given in Table 6.2?

Chapter 7: MHD Oscillations

Problem 7.1: Plot the polar diagram of the phase speeds of magneto-acoustic waves (Fig. 7.1) for typical coronal conditions, $c_S \approx v_A/10$.

Problem 7.2: Calculate numerically the phase speed $v_{ph} = \omega(k)/k$ of the fast kink mode and sausage mode as shown in Fig.7.4 (as a function of ka) from the dispersion relation specified in Eqs. (7.1.50-51). Choose for density and speed ratios round numbers close to those shown in Fig. 7.4. What are the corresponding tube speeds expressed as ratios of the sound speeds?

Problem 7.3: Determine the kink mode period P_{kink} (Eq. 7.2.4) for a loop with a constant magnetic field of $B = 20$ G and a length of $L = 100$ Mm, an internal density of $n_0 = 10^9$ cm^{-3}, as a function of the external density n_e in the entire possible range ($0 \leq n_e/n_0 \leq 1$). How much is the variation?

Problem 7.4: What are the uncertainties in the magnetic field strength B in the 10 measurements given in Table 7.1, if the density ratio n_e/n_0 is unknown? Assume an additional 10% error in the period P.

Problem 7.5: What fraction of the kink-mode oscillation periods P reported in Table 7.4 could be possibly triggered by global oscillations, either by 5-min p-mode oscillations or 3-min oscillations near sunspots, allowing for a 10% mismatch in resonant periods?

Problem 7.6: Assume an exponential density model $n_e(h)$ for the 1 MK corona with a base density of $n_e = 10^9$ cm^{-3}, a dipole magnetic field $B(h)$ (Eq. 5.4.5 and Fig.5.10) with a base value of $B_0 = 100$ G and a dipole depth of $h_D = 75$ Mm, and calculate the cutoff periods P_{min} for the MHD fast sausage mode (Eq. 7.3.9) for a loop diameter of $a = 1$ Mm. What are the longest sausage periods expected at altitudes of $h = 2$ Mm and $h = 200$ Mm?

Problem 7.7: Estimate the modulation depth dI/I of the soft X-ray flux that is observed in SUMER loops in the case of MHD slow-mode oscillations for the Doppler velocities v reported in Table 7.8, assuming a sound speed corresponding to the Fe XIX line temperature of $T = 6.3$ MK (Use Eq. 7.4.9).

Problem 7.8: Estimate the chromospheric density scale height from the RHESSI measurements shown in Fig. 4.28 and predict the damping times due to footpoint wave leakage (Eq. 7.5.7). Could they explain the observed damping times (Table 7.1) in fast kink-mode oscillations ?

Problem 7.9: What range of theoretical to observed damping time ratios t_{RA}/t_d do you predict for resonant absorption for fully-nonuniform loops ($r_{loop}/l_{skin} = 2$) if the density ratio $q_n = n_e/n_0$ is not known. (Use Eq. 7.5.14).

Problem 7.10: How do radio dynamic spectra of loops with kink oscillations (TRACE observations) and slow-mode oscillations (SUMER observations) loops look like? Assume that the density modulation is reflected in the plasma frequency modulation. What frequency bandwidth of the modulated radio emission do you expect?

Chapter 8: Propagating MHD Waves

Problem 8.1: Verify the form of the MHD equations given in Eqs. (8.1.1−3) from the general set of MHD equations (§6.1.2−5) and viscosity term (Eq. 6.1.14). What definitions and assumptions are used?

Problem 8.2: Predict the range of sounds speed in each of the TRACE 171 and 195 Å passbands, based on the FWHM of their response functions given in Fig. 3.25 (i.e., in the sensitivity range at $> 50\%$ of the peak response, $R(T) > 0.5 \times max[R(T)]$), and compare them with the measured ranges given in Table 8.1.

Problem 8.3: Calculate the time intervals $t_1 = h/v_{Ae}$, $t_2 = h/v_A$, and $t_3 = h/c_g^{min}$ in the evolution of an impulsively generated fast-mode MHD wave as shown in Fig. 8.4. How long lasts the periodic phase and the quasi-periodic phase? Use a loop height of $h = 100$ Mm, a magnetic field of $B = 20$ G, an external electron density of $n_e^{ext} = 10^9$ cm^{-3}, a density ratio of $n_0^{int}/n_e^{ext} = 6$, and a minimum group speed of $c_g^{min} = v_{Ae}/3$ as obtained in Fig. 8.5.

Problem 8.4: Measure the wave speed in the SECIS data shown in Fig. 8.7, using the time information of Fig. 8.7 and positional information given in Fig. 8.6. Can you reproduce the reported speed of v $= 2100$ km s^{-1} as given in the caption of Fig. 8.7?

Problem 8.5: Fit a low-order polynomial function to the density profile $n_e(h)$ shown in Fig. 8.13 (top panel) and to the line width $\Delta v(h)$ shown in Fig. 8.13 (middle panel), and verify the theoretical prediction (8.2.11).

Problem 8.6: Estimate the propagation speeds of the simulated EIT wave shown in the 4 panels of Fig. 8.16. Does the wave accelerate of decelerate?

Chapter 9: Coronal Heating

Problem 9.1: Estimate the variation of the coronal heating requirements in quiet Sun areas and active regions during the solar cycle, assuming that all active regions disappear during the solar minimum and the entire coronal magnetic field opens up like in coronal holes. Hint: Use Table 9.1 and calculate the factors for the changes in conductive flux F_C, radiative flux F_R, solar wind flux F_{SW}, and total flux (in erg cm^{-2} s^{-1}), with reference to the coronal hole values.

Problem 9.2: Calculate the Ohmic dissipation rate due to random footpoint motion. How much smaller is it than the coronal heating rate requirement in the quiet Sun. Hint: Use Eqs. (9.3.5-7) and Table 9.1.

Problem 9.3: Derive the viscosity force (from Eq. 9.3.10) for a turbulent eddy in the corona with a rotation speed of $v_0 = 1$ km s^{-1}, the size of a granulation cell ($l = 1000$ km), and a coronal density of $n_e = 10^9$ cm^{-3}. How does the viscosity force scale with rotation velocity v, size l and density n_e?

Problem 9.4: Perform a Monte-Carlo simulation of flare energies, using a powerlaw distribution $N(l) \propto l^{-D_3}$ of size scales l (Eq. 9.8.8), with fractal dimension $D_3 = 2.5$, and using and the scaling laws given in Eqs. (9.7.9−15). Can you reproduce the frequency distribution of flare energies $N(E_{th}) \propto E_{th}^{-1.21}$ predicted in Eq. (9.8.13)?

Problem 9.5: Use the same Monte-Carlo code of Problem 9.4 and introduce random scatter in the scaling laws. How does the frequency distribution of flare energies change? Can it explain the observed distributions listed in Table 9.6 ?

Problem 9.6: Using the relation Eq. (9.8.12), how can you explain the steepest observed frequency distributions with a powerlaw slope of $\alpha_E = 3.3$ found in Table 9.6 (Winebarger et al. 2002) ?

Problem 9.7: Plot an exponential frequency distribution $N(E) \propto exp(-E/E_0)$ (see Fig. 9.24) and determine the local powerlaw slopes as a function of energy. How much does the powerlaw slope change from $E_1 = E_0/2$ to $E_2 = E_0 \times 2$?

Problem 9.8: When you increase the number of events in a frequency distribution of energies, $N(E) = N_1(E/E_1)^{-\alpha}$, e.g., when you sample over a longer duration of time, how does the upper cutoff energy E_2 change as a function of the number of events N_1 detected (at the lower boundary E_1). What dynamic range (E_2/E_1) have frequency distributions with $N_1 = 10, 100, 1000, 10^4, 10^5$ events and a powerlaw slope of $\alpha = 2$?

Problem 9.9: What are the mean (e-folding) amplification factors (t_{Se}/τ_G) of the observed frequency distributions listed in Table 9.6 (Use Eq. 9.8.7)? What observational bias affects these values most?

Problem 9.10: Estimate the temperature filter bias on the powerlaw slope in the frequency distribution of nanoflare energies. Apply it to a 171 Å EUV filter with a peak response at $T_1 = 1.0$ MK and a maximum sensitivity range of $\Delta T = \pm 0.5$ MK. Assume that the corrected powerlaw slope is $\alpha_E = 1.8$. Estimate the change in the powerlaw slope $\alpha_e = (\log[N(E_2)] - \log[N(E_1)])(\log[E_2] - \log[E_1])$, when then thermal flare energy $E = 3n_e k_B T V$ is underestimated because the EUV filter measures a temperature at $T_1 + \Delta T$ instead of a true temperature T_2, for the temperature range from $T_1 = 1.0$ MK to $T_2 = 2.0$ MK.

Chapter 10: Magnetic Reconnection

Problem 10.1: Calculate the magnetic energy that is dissipated during a flare with a duration of 100 s for a Sweet–Parker current sheet (Eq. 10.1.8), assuming typical coronal parameters, $B = 100$ G, $L = 1$ Mm, and $v_A = 1$ Mm s^{-1}. Is it sufficient to explain a nanoflare $(E = 10^{24}$ erg) or a microflare $(E = 10^{27}$ erg)? How many current sheets are needed for one of the largest flares $(E = 10^{32}$ erg)?

Problem 10.2: Repeat the energy budget calculations for a nanoflare, a microflare, and the largest flares with the *bursty reconnection* mode (Use Eqs. 10.1.8 and 10.2.3). How many magnetic islands would be needed for each event and what coronal length scale would they require?

Problem 10.3: Add to a pre-existing dipole region additional small emerging dipole regions (Fig. 10.10 left). How does the number of separatrix surfaces, separator lines, and nullpoints grow with the number of added emerging dipoles? Distinguish between non-overlapping emerging regions and overlapping ones. How many separatrix surfaces, separator lines, and photospheric nullpoints are created after the emergence of 10 isolated (non-overlapping) and 20 pair-wise overlapping dipoles?

Problem 10.4: Copy Fig. 10.10 (right) and indicate the direction of the magnetic field in each domain. Does it agree with the magnetic field directions shown in Fig. 10.27? Indicate the path of a field line during spine reconnection. From where to where does the footpoint of a reconnecting spine field line move?

Problem 10.5: Approximate the frequency distribution of magnetic fluxes $\Phi = AB$ of Hagenaar et al. (2003) shown in Fig. 10.13 with a powerlaw distribution, e.g., $N(\Phi) \propto \Phi^{-2}$. Assume the scaling law $B(A) \propto A^{1.7}$ (Eq. 9.7.17) between the magnetic field strength B and the area A of a magnetic element. What frequency distribution do you infer for the areas, $N(A)$? Based on the lower limit of active regions given in Fig. 10.13, $\Phi_0 = 3 \times 10^{20}$ Mx and $A_0 = 2.5$ deg^2, how much larger are the area A_{max} and magnetic field B_{max} at the upper end of the distribution (at $\Phi_{max} = 5 \times 10^{22}$ Mx), and how much smaller are the area A_{min}

and magnetic field B_{min} of the smallest ephemeral region at the lower end of the distribution (at $\Phi_{min} = 5 \times 10^{18}$ Mx)?

Problem 10.6: Verify the expressions for chromospheric reconnection inflow speeds v_1 (Eq. 10.4.1) in the model of Litvinenko (1999a), using the MHD equations of continuity and momentum, and Ohm's law, in lowest order applied to a current sheet with length 2Λ and width $2a$.

Problem 10.7: Discuss how the driver speed of reconnection processes (v_x, v_y, or v_z in Table 10.1) relates to the reconnection inflow or outflow speed in the various flare models listed in Table 10.1.

Problem 10.8: Verify the universal scaling laws between emission measure and flare temperature in the reconnection model of Shibata & Yokoyama (1999), expressed in Eq. 10.5.7. Derive the scaling law for the total thermal flare energy E_{th} from the same model. What is the thermal flare energy for a typical solar flare with $B = 50$ G, $n_0 = 10^{11}$ cm^{-3}, and $T = 3 \times 10^7$ K?

Problem 10.9: Label all the observational signatures listed in Table 10.2 (which provide evidence for magnetic reconnection) in a flare model cartoon like Fig.10.21. Discriminate to which of the six flare models discussed in §10.5.1−6 the observations apply.

Problem 10.10: Measure the reduction of shear angles between the prereconnection and postreconnection field lines in the 8 flares reconstructed in Figs.10.31−32 (use third row of Figure panels). Can you confirm the range mentioned in the text?

Chapter 11: Particle Acceleration

Problem 11.1: Calculate the gyration radii of different heavy ions (given in Table 11.3) for typical coronal conditions ($T_i = 10^6$ K, $B = 10$ G) and compare them with the dimensions of Parker-type current sheets (§10.1.1) and magnetic islands caused by the tearing mode (§10.2.1). Predict whether these ions have regular or chaotic orbits near magnetic X- and O-points (§10.3.3−4)?

Problem 11.2: Estimate the proton drift speed for the gravitational force given in Eq.(11.1.8) for a coronal magnetic field of $B = 100$ G. What values need the other parameters ($E, dE/dt, \nabla B, R_{curv}$) to produce a comparable particle drift speed? Are the values depending on the coronal magnetic field? How long is the drift time until a proton becomes scattered out of a current sheet with $\nabla B = B/l$?

Problem 11.3: Predict the electron-to-proton ratio for an observed electron spectrum of $N(\varepsilon) \propto E^{-\delta}$ (for $\delta = 3$), if equipartition of energy (Eq. 11.1.18) applies to a particular acceleration mechanism. Hint: Assume that electrons and protons have the same velocity to produce hard X-rays by $p − e$ bremsstrahlung.

Problem 11.4: How do the Dreicer electric field (Eq. 11.3.2) and the critical threshold velocity v_r (Eq. 11.3.3) for runaway acceleration change for a typical range of flare parameters ($T_e = 10, ...30$ MK, $n_e = 10^9, ..., 10^{12}$ cm^{-3})?

Problem 11.5: Write down the equations of motion in a 3D cartesian coordinate system for the super-Dreicer DC electric field configuration envisioned in the model of Litvinenko (1996). See description in §11.3.2 and Fig. 11.3.

Problem 11.6: Derive analytically the parameters v_0, V, e of the resonance ellipse (Eqs. 11.4.25−29) from the general Doppler resonance condition (Eq. 11.4.6). Hint: Start with inserting the Lorentz-factor $\gamma = 1/\sqrt{1 - \beta^2}$ and splitting the velocity components into parallel and perpendicular components, $\beta_\parallel = v_\parallel/c$ and $\beta_\perp = v_\perp/c$.

Problem 11.7: Plot the dispersion relations $\omega(k)$ of the wave types given in Table 11.2 for typical coronal parameters and verify the diagram shown in Fig. 11.11.

Problem 11.8: Estimate the maximum energy gain of a 10 keV electron for shock-drift acceleration (using Eq. 11.5.16), for angles of $\theta_{Bn} = 80° - 89°$, $\theta_{vn} = 45°$, and $\alpha = 1$. What angles between the magnetic field and the shock normal is most favorable for acceleration?

Problem 11.9: Verify the diagram of energy gain by first-order Fermi acceleration shown in Fig. 11.20, using the model of Tsuneta & Naito (1998) specified by Eqs. (11.5.28−30).

Problem 11.10: Measure the shock speed from the frequency-time drift rate of the type II burst in Fig. 11.21, assuming plasma emission at the fundamental and harmonic (split band feature) and a hydrostatic density model with a scale height of $\lambda_T \approx 46,000$ km. Hint: Use Eq. (15.4.10).

Chapter 12: Particle Kinematics

Problem 12.1: Generalize the time-of-flight difference formula (12.2.1) for particles with velocities β_1, β_2 and arbitrary pitch angles α_1, α_2, traveling on different field lines with twist angles θ_1, θ_2, using the definitions of Eqs. (12.2.2−4).

Problem 12.2: Estimate the time accuracy of the HXR time delay measurements Δt_{ij} between different energy channels $i = 3$ and $j = 4, ..., 10$ based on the error propagation law (Eq. 12.2.6) and the count rate spectrum shown in Fig. 12.3, which can be approximated by $N(\varepsilon) \approx 4 \times 10^4 (\varepsilon/44 \text{ keV})^{-2}$. How well do the estimated uncertainties agree with the values specified in Fig. 12.4 for an average time scale of $\tau \approx 0.2$ s?

Problem 12.3: What electric field strength (in [statvolt cm^{-1}]) follows from the DC acceleration model (with Eq. 12.3.2) in the fit to the data shown in Fig. 12.7 (left panel)? Is this a sub-Dreicer or super-Dreicer field, for typical flare densities and temperatures?

Problem 12.4: Derive the nonrelativistic approximations for acceleration time for the simplest electric DC-field acceleration model (Eq. 12.3.4) and show how the relativistic expression (Eq. 12.3.5) matches the nonrelativistic limit.

Problem 12.5: Express the hard X-ray pulse duration t_w resulting from the field line relaxation after a reconnection process in dimensionless units for magnetic scale heights $(L_B/10$ Mm$)$, magnetic fields $(B/100$ G$)$, and electron densities $(n_e/10^8$ cm$^{-3})$, using the approximation given in Eq. (12.4.18). What is the time scale if all dimensionless parameters are unity?

Problem 12.6: Modify the pulse shape model (Eq. 12.4.16) for anisotropic acceleration: (a) for a ring or Dory$-$Guest$-$Harris distribution, e.g., $N(\alpha) \propto sin^N(\alpha)$, and (b) for a beam distribution, e.g., $N(\alpha) \propto cos^N(\alpha)$.

Problem 12.7: Derive the expression for the bounce time t_B (Eq. 12.5.5) in a parabolic field (Eq. 12.5.2) and plot it as a function of the magnetic mirror ratio R in the range of $R = 1, .., 10$. For what range of mirror ratios $[R_{min}, \infty]$ does the correction factor q_α for the conversion of electron time-of-flight distances to magnetic field line lengths change by less than 10% ?

Problem 12.8: Convolve a Gaussian injection profile $f(\varepsilon, t) = f(\varepsilon) \exp[-(t - t_0)^2/2\sigma_t^2]$ (with a peak at $t_0 = 5.0$ s and a Gaussian width $\sigma_t = 1.0$ s), with a varying trapping time $t_{trap} = 1, 2, ..., 10$ s according to the function given in Eq. (12.5.9). How does the delay of the peak in the convolved time profile change as a function of the trapping time? This exercise can be done analytically, but is easier to do numerically.

Problem 12.9: Determine the losscone angles for asymmetric hard X-ray fluxes at the flare loop footpoints for flux ratios of $A = 0.1, 0.5, 0.9$ and precipitation fraction of $q_{prec} = 0.5$ (use Eqs. 12.6.17$-$19).

Problem 12.10: Calculate the critical energies ε_c (Eq. 12.6.21) for flare loop lengths $L = 5, 50$ Mm and flare loop densities of $n_e = 10^9, 10^{10}, 10^{11}$ cm^{-3}. In which cases are electron energies of $\varepsilon = 25$ keV thermal?

Chapter 13: Hard X-Rays

Problem 13.1: Enumerate the similarities between Fourier-type hard X-ray imagers (SMM/HXIS, Yohkoh/HXT, RHESSI) and radio interferometers, e.g., the Very Large Array (VLA). Compare the concept of a rotation-modulated collimator (RMC) hard X-ray imager with that of Earth-rotation aperture synthesis in radio.

Problem 13.2: Express the hard X-ray spectrum $I(\epsilon_x)$ with the the simple Kramers bremsstrahlung cross section (Eq. 13.2.9), rather than using the Bethe-Heitler cross-section (Eq. 13.2.7) that is used in Eq. (13.2.12). Can you then predict the powerlaw slope γ of the hard X-ray spectrum for a particle injection spectrum with powerlaw slope δ?

Problem 13.3: Calculate the low-energy cutoff $\varepsilon(h)$ of a nonthermal bremsstrahlung spectrum as a function of the height h_A of the acceleration source with height h_X where most X-rays are emitted, assuming that the cutoff is given by the criterion of equal travel and collisional deflection times (Eq. 12.6.21), using a simple hydrostatic density model, e.g., $n_e(h) = n_0 \exp(-h/\lambda_T)$, with $n_0 = 10^9$ cm^{-3} and $\lambda_T = 46$ Mm. For an acceleration source in height $h_A = 100$ Mm, what are the cutoff energies at heights of $h_X = 2$, 20, and 50 Mm? How do the cutoff energies change if the electrons propagate in a high-density fluxtube with a base density of $n_0 = 10^{11}$ cm^{-3}?

Problem 13.4: Carry out a Monte-Carlo simulation of the distribution of hard X-ray pulse durations τ_w (Eq. 13.4.2). Let each parameter be normally distributed with the same gaussian width $\sigma_x/ < x >$ with respect to the mean value $< x >$. Conduct three simulations for $\sigma_x/ < x >= 0.2$, 0.4, and 0.6. How does the e-folding time constant of the resulting pulse width distribution change? Can you reproduce the observed distributions shown in Fig. 13.10?

Problem 13.5: Measure the relative hard X-ray delays for case 5 in Fig. 13.15, between 50 and 300 keV, and use the TOF formula (Eq. 13.5.8), using electron-to-photon energy conversion factors q_ε (Eq. 13.5.6) from Fig. 13.14 for a spectral slope of $\gamma = 3$ and the largest high-energy cutoff, in order to estimate the TOF distance. How close does it agree with the value given in Fig. 13.15 for case 5?

Problem 13.6: What is the time-of-flight distance ratio l_{TOF}/r if the altitude of the acceleration source is a factor $h/r = 1.5$ higher than the flare loop radius r? Hint: use Eq. (13.5.9).

Problem 13.7: Calculate the asymmetry angles of acceleration sources for relative time delays of $\Delta t = 100$ ms between conjugate footpoint sources, for electron energies of $\varepsilon = 25, 50, 100$ keV and a loop radius of $r = 15$ Mm (using Eq. 13.5.12). Are such asymmetries larger or smaller than observed in the Masuda flare (Fig. 13.34)?

Problem 13.8: Calculate what relative trapping time delays due to collisional deflections (Eq. 12.5.11) are expected for trap densities of $n_e \approx 10^{10} - 10^{11}$ cm^{-3} (Fig. 13.23) between 50 keV and 100 keV electrons. Compare with the observed delay in flare # 1361 shown in Fig. 13.22, assuming typical electron-to-photon energy conversion factors of $q_\varepsilon \approx 2$.

Problem 13.9: Estimate from the footpoint asymmetries observed in Flare # 1032 (3rd example shown in Fig. 13.27) the loss cone angles α_1 and α_2 for trapping fractions $q_{prec} = 0.1, 0.5, 0.9$ (using Eqs. 12.6.17–19). Hint: Measure the flux ratios at both footpoints in Fig. 13.27, assuming that the y-axis is a linear flux scale.

Problem 13.10: Estimate the flux contrast F_{top}/F_{max} between the hard X-ray emission between from the looptop (assumed to be at position $z = 10$ Mm) and that from the brightest source (at $z = z_{max}$) from the hard X-ray profiles dI/dz for electron energies of $\varepsilon = 10, 20, ..., 60$ keV (using Fig. 13.28). At what energies can a hard X-ray imager with 20% flux contrast separate the two identically bright footpoint sources?

Chapter 14: Gamma Rays

Problem 14.1: Identify in each of the gamma-ray emission processes listed in Table 14.1 the projectile particles, the target particles, where they originate (corona, chromosphere), and where they interact (corona, chromosphere).

Problem 14.2: Identify the gamma-ray lines in the spectrum shown in Fig. 14.3 (top) based on the energies of nuclear de-excitation lines listed in Tables 14.1 and 14.2 and compare the result with the labeled spectrum in Fig. 14.2. Which lines are not detectable in the spectrum observed in Fig. 14.3?

Problem 14.3: Compare the exponential decay times in the time profiles shown in Figs. 14.5 and 14.6 with the collisional trapping times given in Eq. (14.2.2). Assuming that coronal traps have typical densities of $n_e = n_i \approx 10^9 - 10^{10}$ cm^{-3}, what kinetic energies do you infer for the trapped ions that produce the gamma-ray lines after precipitation to the chromosphere?

Problem 14.4: How can we tell from the time profile shown in Fig. 14.6 when extended continuous acceleration occurs, and when trapping without acceleration is at work?

Problem 14.5: Compare the observed or fitted line widths in the spectra shown in Fig. 14.8 (also given in Table 14.2, footnote c) and compare with the theoretical line widths for the nuclear de-excitation lines given in Table 14.2. Does RHESSI resolve these gamma-ray lines? Note that the instrumental resolution is about 3.0 keV at 847 keV, 4.1 keV at 1779 keV, and ≈ 10 keV at 6129 keV (Smith et al. 2003). What lines are significantly broadened by the instrumental resolution?

Problem 14.6: What time delay is expected between hard X-rays produced by 25 keV electrons and gamma-ray lines produced by 5 MeV ions in a flare loop with a half length of $L = 100$ Mm (see Section 14.3.3)?

Problem 14.7: What are the projectile particles and target particles that produce the 2.223 MeV neutron capture line?

Problem 14.8: Measure the width of the positron annihilation line at 511 keV in the line profile shown in Fig. 14.16 (left) and determine the temperature of the target medium with Eq. (14.5.12)?

Problem 14.9: At what energy does pion decay radiation dominate over electron brems-strahlung, based on the spectrum shown in Fig. 14.17?

Problem 14.10: Make a list of all solar instruments and spacecraft that carried gamma-ray detectors, based on the information given in §14.

Chapter 15: Radio Emission

Problem 15.1: Use a coronal/heliospheric electron density model $n_e(R)$, e.g., the Baumbach-Allen model given in Eq. (1.6.1), the ten-fold Baumbach-Allen model for active regions, or the heliospheric model by Erickson (1964) (Eq. 1.6.2), calculate the plasma frequency as function of the distance R from the Sun, and try to reproduce the curve $f_p(R)$ shown in Fig. 15.1. Which model agrees best at 0.01 and 100 solar radii?

Problem 15.2: At what frequency does each radio spectrum shown in Fig. 15.2 become optically thin, say where $\tau \lesssim 1$?

Problem 15.3: The Yohkoh soft X-ray image in Fig. 15.3 (left) shows a flare loop with an electron temperature of $T_e \approx 11$ MK and a peak electron density of $n_e \approx 2 \times 10^{12}$ cm^{-3}. The field-of-view of the image is about 170,000 km. Calculate the free-free absorption coefficient α_{ff} (Eq. 15.2.1), optical depth τ_{ff} (Eq. 15.2.3), and radio brightness temperature T_B for the radio frequency of the image shown in Fig. 15.3 (right). Hint: Assume a constant absorption coefficient along the line-of-sight and estimate the diameter of the flare loops assuming that both image panels in Fig. 15.3 have the same field-of-view.

Problem 15.4: Gyrosynchrotron spectra of an asymmetric flare loop are shown in Fig. 15.4 for the looptop and the two footpoints. At what frequencies are the footpoints brighter than the looptop? Are the footpoints separable in the simulated images in Fig. 15.4 for these frequencies?

Problem 15.5: Estimate the electron density in the radio-emitting trap from the exponential decay time of the 17 GHz light curve shown in Fig. 15.6 and the theoretical expression for collisional deflection times given in Eq. (14.2.2). Assume a typical energy of $\varepsilon_R \approx 300$ keV for the electrons that are responsible for gyrosynchrotron emission. (The tickmarks of the time axis in Fig. 15.6 mark intervals of 10 s.)

Problem 15.6: Determine the kinetic energy of the electron beam from the velocity spectra shown in Figs. 15.11, using the relativistic formulae given in §11.1.3.

Problem 15.7: Calculate the propagation speed of the type III-producing electrons at the leading edge shown in Fig. 15.13, radiating at a fundamental plasma frequency of $\nu = 125$ kHz, using the heliospheric electron density model $n_e(R)$ of Erickson (1964) (Eq. 1.6.2). How many solar radii away from the Sun is this radio emission expected?

Problem 15.8: What is the plasma density at the top of the coronal loop in which type U burst-producing electron beams propagate during the 25/08/80 event shown in Fig. 15.17 (top right panel)?

Problem 15.9: Explain in words the difference between the process of electron-cyclotron maser emission and stochastic acceleration (compare the maser growth rate in Eq. (15.5.1) and the wave-particle interaction rate in Eq. (11.4.5).

Problem 15.10: How does the radio bandwidth of decimetric millisecond spikes depend on the local magnetic field B and plasma temperature T (use Eq. 15.5.2)?

Chapter 16: Flare Plasma Dynamics

Problem 16.1: Derive an expression for the critical velocity v_c for run-away electrons in terms of the electron collision time τ_{ce}, the electron thermal velocity v_{Te}, and the electric field E. Hint: Use the definition of the Joule heating time scale given in Eq. (16.1.1) and the definition of plasma parameters in Appendix D.

Problem 16.2: Measure the ratios of parallel and perpendicular temperature changes in the evolutionary curves shown in Fig. 16.3 (right panels; between the start and end time of the time axis). What relative changes do you predict for the widths of the parallel and perpendicular velocity distributions of the ions with respect to hydrogen, assuming that all ions initially had the same temperature? Compare them with the velocity distributions shown in Fig. 16.3 (left panels). For which ions do your predictions agree with the simulations?

Problem 16.3: Derive the scaling law for flare looptop temperatures, Eq. (16.2.1), using similar arguments as for the unification of flare models discussed in §10.5.7.

Problem 16.4: In what altitudes do you expect Hα emission, white-light emission, and UV emission in solar flares, based on the discussion in §16.2.3−5?

Problem 16.5: Measure the blueshift in the Ca XIX line profile shown in Fig. 16.11 and verify the derived Doppler velocity given in the Figure caption. Name potential error sources when determining the Doppler shift.

Problem 16.6: Predict the frequency-time drift rates $d\nu/dt$ (Eq. 16.3.2) of decimetric radio emission that occurs in a chromospheric evaporation front, based on the mean parameters measured in Table 16.2, and assuming a mean temperature of $T_e = 1.5$ MK for the background corona. Apply it to the observation in Fig. 16.15, assuming a propagation angle of $\theta \approx 0°$.

Problem 16.7: Derive the expressions for conductive cooling time and radiative cooling time in terms of changes in the thermal energy. Compare them with the expressions given in Eqs. (16.4.4) and (16.4.9). Give your expressions in dimensionless units for typical flare parameters, i.e., electron densities of $n_e = 10^{11}$ cm^{-3}, and flare plasma temperatures of $T_e = 10^7$ K ($T_e = 10^6$ K) for conductive (radiative) cooling, and flare loop half lengths of $L = 10$ Mm.

Problem 16.8: For what flare temperature are the conductive and radiative cooling times (derived in Problem 16.7) equal, using typical flare parameters of $n_e = 10^{11}$ cm^{-3} and $L = 10$ Mm?

Problem 16.9: What is the relative importance of conductive and radiative cooling at the beginning, peak, and decay phase of a flare?

Problem 16.10: The Bastille Day flare (2000-Jul-14) involved the heating and subsequent cooling of a series of over 200 loops (see EUV loop tracings in Fig. 10.33). The time delays of the peak fluxes in cooler and cooler wavebands are shown in Fig. 16.16 and specified in Table 16.3. If each loop cools on time scales indicated by the Table, what is the minimum number of successive flare loops that is required to provide the observed 30 minutes of continuous emission in soft X-rays at temperatures of $T_e \geq 2$ MK? What is the average number of loops seen simultaneously during the flare?

Chapter 17: Coronal Mass Ejections

Problem 17.1: Identify for each of the five theoretical CME models described in §17.1 (represented with mechanical analogs in Fig. 17.1) which hydrodynamic or MHD instability (Table 6.1, Fig. 6.11) or flare model (§10.5 and Table 10.1) represents the physical driver.

Problem 17.2: Estimate the number of helical windings in the interior filament shown in the numerical simulations by Amari et al. (2003) (see Fig. 17.3). Compare with the instability criteria for the kink instability (§6.3.9) and assess whether the simulated filaments fulfil the instability criteria.

Problem 17.3: How do the structures of an erupting filament seen in UV (shown in Fig. 17.5) relate to the theoretical MHD model shown in Fig. 17.3. Identify which magnetic field line shown in Fig. 17.3 corresponds most closely to the erupting filament shown in the TRACE image (Fig. 17.5).

Problem 17.4: Measure the distance of the CME front in the pictures shown in Fig. 17.6 and estimate how much earlier the CMEs were launched on the solar surface, using the CME speeds published in the NRL/CU/GSFC online CME catalog. The speeds (in the plane of sky) are 680, 1863, 792, 751, and 1556 km s^{-1} for the 5 CMEs listed in Fig. 17.6 Which picture was taken closest to the launch? Note that the solar disk is indicated with a white circle and assume that a CME moves at a constant speed since onset.

Problem 17.5: How far is the center of the helical fluxrope located away from Sun center and what is the radius of the helical fluxrope (in units of solar radii) in the three cases shown in Fig. 17.7, 17.8 (left panel), and 17.9 (at 14:51 UT)?

Problem 17.6: Calculate how the average density $n(t)$ of an expanding CME changes as function of time, assuming radial expansion $n(r) \propto r^{-2}$ and average speed $v_{CME} = r/t$. How many hours after launch does a CME have an average density of $n_e = 10^7, 10^5, 10^3$ cm^{-3}, if it propagates with an average speed of $v_{CME} \approx 1000$ km s^{-1} and has an initial coronal density of $n_0 = 10^9$ cm^{-3}? What are the corresponding distances from Sun center?

Problem 17.7: Assuming a stratified atmosphere (with a temperature of $T_e \approx 1.0$ MK and an average base density of $n_0 = 10^9$ cm^{-3}), what mass fraction of an active region needs to be accelerated to explain a CME-mass of $m_{CME} = 10^{14} - 10^{16}$ g, assuming that the mass is ejected vertically above an area with a longitudinal (l) and latitudinal (b) extent of 45° each, bordering the solar equator?

Problem 17.8: Estimate the terminal velocity of fast CMEs, assuming that their total energies are $\varepsilon_{tot} \approx 10^{29} - 10^{32}$ erg and their masses are $m_{CME} \approx 10^{14} - 10^{16}$ g. Assume that the most energetic CME has the highest mass, as well as that the least energetic CME has the lowest mass.

Problem 17.9: Estimate the maximum density rarefaction from the EUV dimming (assuming an EUV intensity of $I_{EUV} \propto EM \propto \int n_e^2 dz$) shown in the time profiles in Figs. 17.17 and 17.19. If the CME has a spherical expansion in the heliosphere, what linear size increase does this density rarefaction correspond to?

Problem 17.10: The Ulysses spacecraft has a high-latitude solar orbit, starting in the aphelion near Jupiter on February 8, 1992, approaching the Sun at 80° southern latitude on September 13, 1994, and traveling though high northern latitudes during June to September 1995. Can you explain the variation of the solar wind speed that Ulysses observed according to the time profile shown in Fig. 17.21 (top).

Solutions

Chapter 1: Introduction

Solution 1.1: Gamma-rays are covered by SMM/GRS, CGRO/OSSE, and RHESSI ($\approx$ 100 keV $-$ 160 MeV $\approx 10^{-4} - 10^{-1}$ Å), hard X-rays by CGRO/BATSE, SMM/HXRBS, Yohkoh/HXT, and RHESSI ($\approx$10 keV $-$ 100 keV $\approx 0.1 - 1$ Å), soft X-rays by SMM/HXIS, BCS, FCS, Yohkoh/SXT ($\approx$0.5 keV $-$ 10 keV $\approx 1 - 25$ Å), EUV by NRL rockets, SoHO/EIT, CDS, TRACE ($\approx 150 - 800$ Å), UV by SMM/UVSP, SoHO/SUMER, TRACE ($\approx 500 - 1600$ Å), white-light by SMM/CP, SoHO/LASCO, TRACE ($\approx$4400-6500 Å), infrared by SMM/ACRIM, and radio by VLA, Culgoora, Nançay, OVRO, Nobeyama, RATAN-600 ($\approx$1 cm $-$ 3 m $\approx$100 MHz$-$30 GHz). Gaps in wavelength ranges that are not covered by these solar missions and instruments can be found in UV ($25 - 150$ Å), in optical ($\approx 1600 - 4400, 6500 - 8000$ Å), in infrared (0.8 μm$-$1 mm), in high-frequency radio ($\approx$1 mm$-$1 cm $\approx 30 - 300$ GHz), and in low-frequency radio ($\lesssim$ 100 MHz, $\gtrsim$ 3 m; decametric, hectometric, and kilometric wavelengths).

Solution 1.2: According to the compilation in Table 1, the wavelength ranges in soft X-rays (SXR), EUV, and radio display the richest variety of coronal phenomena, because of their sensitivity to coronal temperatures ($\approx 1 - 2$ MK).

Table 1: Coronal phenomena versus observability in wavelength regimes

Coronal phenomena	γ-rays	HXR	SXR	EUV,UV	Hα	WL	radio
Active regions			X	X	X	X	X
Quiet Sun			X	X	X	X	X
Coronal holes				X		X	X
Helmet streamers			X	X		X	X
Filaments, prominences			X	X	X	X	X
Coronal, postflare loops			X	X	X		X
Sigmoid loops			X	X			X
Soft X-ray jets			X				X
Nanoflares				X			
Microflares		X	X				X
Flares	X	X	X	X	X	X	X
Coronal mass ejections				X		X	X

Solution 1.3: The longitudinal magnetic field is statistically lower by a factor of $< B_{\|} > / < B > = < \cos\theta > = \int_0^{\pi/2} \cos\theta \, d\theta / \int_0^{\pi/2} d\theta = 2/\pi \approx 0.64$. The expected change of the soft X-ray flux during a solar cycle is a factor of $|1000 \times 0.64|^{1.7} / |10 \times 0.64|^{2.1} \approx 1200$.

Solution 1.4: Using the height variable $h = R - R_\odot$ [Mm], the dipole model is $B_D(h) = B_{foot}(1 + h/h_D)^{-3}$ and the powerlaw model of Dulk & McLean (1978) is $B_P(h) = 0.5(h/696)^{-1.5}$, with a value of $B_P(h = h_1) = 177$ G at the lower limit of $h_1 = 14$ Mm. The best fit at the lower limit can be found by requiring identical values, $B_D(h = h_1) = B_P(h = h_1)$, and identical first derivatives $dB_D/dh(h = h_1) = dB_P/dh(h = h_1) = -15.8$ G/Mm, which yields two equations for B_{foot} and h_D. They can be analytically solved, yielding $h_D = h_1 \approx 14$ Mm and $B_{foot} \approx 1400$ G. The dipole field yields progressively lower magnetic fields with height than the powerlaw model, about a factor of 2 at a height of $h = 60$ Mm.

Solution 1.5: The Haussdorff dimension D_2 is defined as $D_2 = \log(N_A)/\log(N_l)$ in 2D, and the filling factor can be defined as $q_{fill} = N_A/N_l^2$, because N_A is the number of pixels covered by a fractal structure in an Euclidian area of N_l^2 pixels. The filling factor is unity for Euclidian filling, and $q_{fill} = 1 \mapsto N_A = N_l^2$ leads to the Euclidian dimension $D_2 = \log(N_l^2)/\log(N_l) = 2$. Thus, the filling factor can be expressed as a function of the fractal dimension by $q_{fill} = N_l^{(D_2-2)}$, or vice versa, $D_2 = 2 + \log(q_{fill})/\log(N_l)$. So, for $N_l = 10^2$ and $q_{fill} = 0.1$ the fractal dimension is $D_2 = 1.5$, and the filling factor for 10 times better resolution ($N_l = 10^3$) and the same fractal dimension ($D_2 = 1.5$) is $q_{fill} = 0.03$.

Solution 1.6: Substituting the radial distance $r = R/R_\odot$ in the Baumbach-Allen formula (Eq. 1.6.1) yields the base density $n_0(r = 1) = 10^8(2.99 + 1.55 + 0.036) = 4.6 \times 10^8$ cm^{-3} and the density gradient $dn_e/dr(r = 1) = 10^8(-16 * 2.99 - 6 * 1.55 - 1.5 * 0.036) = -5.7 \times 10^9$ cm^{-3}. This yields for an exponential (barometric) density model with $dn_e/dr(r = 1) = n_0/(\lambda/R_\odot)$ a scale height of $\lambda/R_\odot = 0.08$ or $\lambda = 55.7$ Mm, which corresponds to a scale height temperature of $T = 55.7/47 = 1.2$ MK (with Eq. 3.1.16).

Solution 1.7: The peaks of the DEM distribution functions give the temperature of the most abundant plasma in each region, i.e., $T = 10^{6.18} \approx 1.5$ MK for QR 91 and QR 93; $T = 10^{6.2} \approx 1.6$ MK for AR 93; and $T = 10^{6.62} \approx 4.2$ MK for AR 91 and AR 93. The values at $\log(T = 1.5$ MK$)=6.18$ are $\log(dEM/dT) = 21.5$ for AR 93 and $\log(dEM/dT) = 20.75$ for QR 93, yielding a factor $10^{21.5-20.75} = 10^{0.75} = 5.6$ difference in the emission measure, which corresponds to a density factor of $\sqrt{5.6} = 2.4$ in electron density, if the same column depth is assumed (according to Eq. 1.7.1).

Solution 1.8: The plasma-β ranges shown in Fig. 1.22 are: $\beta \approx 0.1 - 100$ at a photospheric height of $h = 0.25$ Mm, $\beta \approx 0.0005 - 0.02$ at a coronal height of $h = 10$ Mm, and $\beta \approx 0.06 - 100$ at a heliospheric height of $h = 1400$ Mm. Inserting the densities n_e and temperatures T_e from Table 1.1, this yields the following ranges

for the magnetic field: $B = 10 * \sqrt{0.07 * 0.5 * 2 \times 10^8 * 0.005/[0.1 - 100]} = $
$187 - 5900\,$G in the photosphere, $B = 10 * \sqrt{0.07 * 1.0 * 1 * 2/[0.0005 - 0.02]} = $
$26 - 167\,$G in the corona, and $B = 10 * \sqrt{0.07 * 1.0 * 0.01 * 1/[0.06 - 100]} = $
$0.03 - 1.0\,$G in the heliosphere.

Solution 1.9: Atoms that have a chemical abundance $A/A_H \gtrsim 10^{-4}$ have abundance
values of $a = 12.0 + {}^{10}\log(A/A_H) \gtrsim 8.0$, which are according to Table 1.2:
He (a=10.93), C (a=8.52), O (8.83), and Ne (a=8.08). In the sensitivity range of
$10^{-8} \lesssim A/A_H \lesssim 10^{-4}$ we expect elements with abundance values of $4 < a < 8$
to be discovered. This includes the following atoms that have been detected in
the photosphere but not in the corona (see Table 1.2): F (a=4.56), K (a=5.12), Ti
(a=5.02), V (a=4.00), Cr (a=5.67), and Mn (a=5.39).

Solution 1.10: With the conversion formulae Eqs.(1.10.6) and (1.10.1) we obtain
(a) $\lambda = 3.2\ \mu$m (infrared) for sunspot umbrae, (b) $\lambda = 2.4\ \mu$m (infrared) for the
photosphere, (b) $\lambda = 7200$ Å (visible light) for the transition region, (c) $\lambda = 144$
Å (EUV) for coronal loops, and (d) $\lambda \approx 10$ Å ≈ 1 keV (soft X-rays) for flare
loops.

Chapter 2: Thermal Radiation

Solution 2.1: The free-free absorption coefficient for $T_e = 10^6$ K, $n_e = 10^9$ cm^{-3},
$\ln \Lambda \approx 20$, and $\nu = [10^8, 10^9, 10^{10}]$ Hz is $\alpha_{ff} \approx 10^{-2}\nu^{-2}n_e^2 T_e^{-3/2} \ln \Lambda \approx$
$[2 \times 10^{-8}, 2 \times 10^{-10}, 2 \times 10^{-12}]$ cm^{-1}. The free-free opacity over a scale height
of $\lambda_T = 5 \times 10^9$ cm is $\tau_{ff} \approx \nu_{GHz}^{-2} n_9^2 T_6^{-3/2}(\ln \Lambda/20)(\lambda_T/5 \times 10^9$ cm$) =$
$[100, 1, 0.01]$. The free-free opacity $\tau_{ff}(z)$ changes along the path z linearly
for a constant absorption coefficient α_{ff}, according to Eq. (15.2.3): $\tau_{ff}(z) =$
$\int_0^z \alpha_{ff}(z')dz' \approx \alpha\,z$, which yields the following integral for the radio brightness temperature (Eq. 15.2.4): $T_B = T_e \int_0^{\lambda_T} \exp^{(-\alpha_{ff}\,z)} \alpha_{ff} dz = T_e[1 -$
$\exp(-\tau_{ff})]$, yielding values of $T_B/T_e = [1.00, 0.63, 0.01]$ for the frequencies
$\nu = [10^8, 10^9, 10^{10}]$ Hz. Thus, the radio brightness temperatures are $T_B =$
$[10^6, 6.3 \times 10^5, 10^4]$ K for these frequencies. The corona becomes optically
thick at radio frequencies of $\nu \lesssim 1$ GHz.

Solution 2.2: Substituting the variable $\nu = c/\lambda$ and inserting $\nu^3 = c^3/\lambda^3$ and the
derivative $d\nu = [d\nu(\lambda)/d\lambda]d\lambda = -(c/\lambda^2)d\lambda$ into the expression $B_\nu(T)d\nu$
(Eq. 2.2.2) yields the desired expression as function of wavelength, $B_\lambda(T)d\lambda$
(Eq. 2.2.3).

Solution 2.3: Substituting the variable $x = \lambda(k_B T/hc)$, i.e., $\lambda = x(hc/k_B T)$, into
the Planck function (Eq. 2.2.3) yields the expression $B_\lambda(x) \propto x^{-5}[\exp(1/x) -$
$1]^{-1}$. The maximum is found by setting the derivative to zero, $\partial B_\lambda(x)/\partial x = 0$,
which yields the equation $-1 + x^{-1}e^{1/x}/5(e^{1/x} - 1) = 0$. For $x \ll 1$ we
have approximately $e^{1/x} \approx (e^{1/x} - 1)$ and the equation simplifies to $-1 +$
$1/5x \approx 0$, which has the solution $x \approx 0.2$. The exact numerical solution is
$x = 0.2013$. From this we can find the wavelength at the peak of the Planck

function, $\lambda_{max} = x(hc/k_BT) = 0.2898/T[K]$ (Eq. 2.2.6), by inserting the constants h, c, k_B given in Appendix A.

Solution 2.4: The total thermal energy is (using Eq. 13.3.2) $W_{th} = 3n_ek_BT_eV = 4.1{\times}10^{29}(EM_{49}T_7/n_{11}) \approx 2{\times}10^{29}$ erg, inserting the emission measure $EM = n_en_iV = 1.8 \times 10^{48}$ cm^{-3} and the flare plasma temperature $T_e = 29.9 \times 10^6$ K from Fig. 2.6. The frequency distribution shown in Fig. (9.27) shows a range of nonthermal flare energies $W_{nth} \approx 10^{27} - 10^{32}$ erg, so the 1980-June-27 flare shown in Fig. 2.6 is a medium-sized flare.

Solution 2.5: Combining the two formulae (2.4.2) and (2.4.3) yields $\lambda = [R_HZ^2 (1/n_1^2 - 1/n_2^2)]^{-1}$, or specifically for helium (Z=2), by inserting the Rydberg constant $R_H = 109,740 \times 10^{-8}$ Å^{-1}, $\lambda_{He} = 228$ Å$\times[1/n_1^2 - 1/n_2^2]^{-1}$ Å. This yields for the He Lyman series ($n_1 = 1, n_2 = 2, 3, 4, ..., \infty$) the wavelengths $\lambda_{He,Ly} = 304, 256, 243, ..., 228$ Å, which are in the EUV domain. For the He Balmer series ($n_1 = 2, n_2 = 3, 4, 5, ..., \infty$), the wavelengths $\lambda_{He,Ba} = 1641, 1216, 1085, ..., 911$ Å are in the UV domain. For the He Paschen series ($n_1 = 3, n_2 = 4, 5, 6..., \infty$), the wavelengths $\lambda_{He,Pa} = 4689, 3205, 2735, ...,$ 2049 Å are in the optical light domain. Note that the exact atomic wavelengths differ from the Rydberg formula by a few Å due to the larger mass of the He nucleus.

Solution 2.6: The population density of excited states at temperature T_e is given by the Saha equation (Eq. 2.7.5). Inserting the values $\epsilon_0 = 6.1$ eV, $\epsilon_1 = 11.9$ eV, $g_0 = 1, g_1 = 2, T_e = 6000$ K into Eq. (2.7.5), with the constants k_B, h, m_e and conversion factor of 1 eV into erg given in Appendix A, leads to a population ratio of $N_1/N_0 \approx 0.6$ for a photospheric density of $n_e = 10^{17}$ cm^{-3}, and $N_1/N_0 \approx 6 \times 10^5$ for a density of $n_e = 10^{11}$ cm^{-3} at the top of the chromosphere. So, the line intensity (Eq. 2.8.1) is expected to be much stronger for (ionized) Ca II at chromospheric densities of $n_e \lesssim 10^{16}$ cm^{-3} than for (neutral) Ca I.

Solution 2.7: The ionization energy for hydrogen is $\epsilon = 13.6$ eV, which corresponds to a wavelength of λ=(12.4 keV/13.6 eV)=912 Å, according to Eq. (1.10.4). So, ionization requires photons with UV wavelengths, while free-bound transitions can be accomplished with free electrons of larger energies or smaller wavelengths ($\lambda < 912$ Å) in UV. Free-free emission and absorption mostly involves photons with larger energies or shorter wavelengths, in EUV and soft X-rays.

Solution 2.8: Consult tutorial and examples given on the CHIANTI website, which will guide you to reproduce the ionization equilibrium curves for Fe V to Fe XIX as shown in Fig. 2.10.

Solution 2.9: The radiative flux per unit area and time is the volumetric radiative loss rate E_R multiplied with the scale height λ_T, i.e., $F_R = E_R\lambda_T \approx n_e^2\Lambda(T)\lambda_T \approx 10^{-17.73}n_e^2T_e^{-2/3} 4.7{\times}10^9(T_e/10^6$ MK$) \approx 10^4$ erg cm^{-2} s^{-1} at $T_e = 1.0{\times}10^6$ K (using Eqs. 2.9.1, 2.9.2, and 3.1.16). The blackbody brightness would be much higher, according to the Stefan$-$Boltzmann law (Eq. 2.2.7), $B(T) \approx \sigma T^4/\pi \approx$

2×10^{19} erg cm^{-2} s^{-1}, but does not apply because the tenuous coronal plasma is not optically thick (as required for blackbody radiation), and thus the brightness is drastically reduced by the small opacity $\tau \ll 1$.

Solution 2.10: Elements for which the coronal abundance has been measured and the first-ionization potential (FIP) is given in Table 1.2, but do not appear in the FIP diagram of Fig. 2.15, are: P (FIP=10.5 eV, a=5.5), Cl (FIP=13.0 eV, a=5.8), and Ni (FIP 7.6 eV, a=6.25). Thus, measurements with a sensitivity in abundances down to a fraction $10^{5.5...6.25}/10^{12} \approx 3 \times 10^{-7}...2 \times 10^{-6}$ of hydrogen are needed to evaluate the FIP effect of these additional elements.

Chapter 3: Hydrostatics

Solution 3.1: Inserting the gravity for the Sun ($g_\odot$) and for stars (g_*) using Eq. (3.1.2) into the expression for the pressure scale height (Eq. 3.1.16) yields $\lambda_p^*(T_e) = 47{,}000$ km $\times (\mu_*/\mu_\odot)^{-1}(R_*/R_\odot)^2(M_*/M_\odot)^{-1}(T_e/1 \text{ MK})$. Assuming $\mu^* = \mu_\odot$ and using the logarithmic mass and radial ratios interpolated for the given spectral classes (e.g., Zombeck 1990, p.65 and p.73), we find (1) $\lambda_p^* = 92$ Mm $(T_e/1 \text{ MK})$ for the giant star Vega, (2) $\lambda_p^* = 3000$ Mm $(T_e/1 \text{ MK})$ for the bright giant Pollux, (3) $\lambda_p^* = 94$ Mm $(T_e/1 \text{ MK})$ for the white dwarf Regulus, and (4) $\lambda_p^E = p_0/(\rho_0 g_e) \approx 8$ km for the Earth.

Solution 3.2: Using Eq. (3.2.4) and a flat response function $R(T) = R_0$ in the temperature range $[T_1, T_2]$, where $T_1 = T - \Delta T/2$ and $T_2 = T + \Delta T/2$, and denoting the DEM at the base of the fluxtube with $DEM_0 = dEM(T, h = 0)/dT$, we have (1) $F(h) = DEM_0 R_0 \Delta T_0 \exp[-2(h/\lambda_0)(T_0/T)]$ for a narrowband filter, and (2) $F(h) = DEM_0 R_0 \int_{T_1}^{T_2} \exp[-2(h/\lambda_0)(T_0/T)]dT$ for a broadband filter, where $T_0 = 1.0$ MK and $\lambda_0 = 47{,}000$ km.

Solution 3.3: Denoting the triangle that approximates the peak of the column depth function $z_{eq}(h = 0)$ with a peak height z_0 and base width $h_0/R_\odot$ we are reading off the peak values $z_0 = 0.35R_\odot$ for $T = 1$ MK, and $z_0 = 0.90R_\odot$ for T=4 MK, and base widths of $h_0 = 0.05R_\odot \approx 50''$ for $T = 1$ MK, and $h_0 = 0.2R_\odot \approx 200''$ for $T = 4$ MK. The column depths at disk center are $z_c = 0.033$ for $T = 1$ MK, and $z_c = 0.13$ for $T = 4$ MK. The ratio of the column depths at the center and limb, which are proportional to the center-limb brightening, are then $I_{limb}/I_{center} \approx (z_0/z_c)[1 - r/(2h_0)]$, which for the spatial resolutions $r = 1'', 10'', 100''$ yield the values $I_{limb}/I_{center} \approx 10.5, 9.5, 0.0$ for T=1 MK, and $I_{limb}/I_{center} \approx 6.8, 6.6, 5.1$ for T=4 MK, respectively.

Solution 3.4: The inclination angles to the vertical are $\vartheta = 30°, 60°$, and $60°$ for the three loop positions. Since we are seeing the horizontally projected scale height at disk center, the correction for the vertical scale height is $\lambda_p^{vertical} = \lambda_p^{horizontal}/\tan(\vartheta)$ or a factor of $\sqrt{3} = 1.73$ for loop (1). At the limb we are seeing the vertical projection of the scale height (check Figs. 3.12 and 3.16), regardless of the azimuthal orientation, and thus the correction factor is unity for both loops (2) and (3).

Solution 3.5: Assuming Poisson statistics, the fluctuations above a background of N loops is proportional to $\sqrt{N}$, so the flux contrast is $F_{loop}/F_{backgr} = c(N) \approx \sqrt{N}/N = 1/\sqrt{N}$, yielding a contrast of $c(N)=32\%$, 10%, or 3.2% for $N = 10, 100, 1000$ loops.

Solution 3.6: Inserting the radiative loss rate $E_R(T) = n_e^2 \Lambda(T)$, the radiative loss function $\Lambda(T) = \Lambda_0 T^{-1/2}$, and the pressure $p_0 = 2n_e k_B T$ into the auxiliary RTV function (3.6.11) and integrating yields $f_R(T) = (\kappa p_0^2 \Lambda_0/2k_B^2)T$. Inserting this function $f_R(T)$ into the inverse temperature profile $s(T)$ (Eq. 3.6.12) and neglecting the heating term, i.e., assuming $f_H(T) \ll f_R(T)$ yields with $L = s_{max} = s(T_{max})$ the RTV scaling law $T_{max} = const(p_0 L)^{1/3}$ with the constant $const = (3/k_B)^{1/3}(\Lambda_0/2\kappa)^{1/6} \approx 1.8 \times 10^3$, which is close to the original result of RTV (Eq. 3.6.14).

Solution 3.7: Inserting $s = L/2$ into Eq. (3.6.18) yields $T(L/2) = T_{max}[1 - 2^{-a}]^b$, and inserting $s_H/L = 1$ into Eqs. (3.6.22-23) with the coefficients from Table 3.1 yields $a = a_0 + a_1 = 2.356$ and $b = b_0 + b_1 = 0.311$, amounting to a temperature drop of $T(s = L/2) = T_{max} \times 0.93$, so a temperature decrease by 7%.

Solution 3.8: The relative TRACE 171 Å flux of loops is $F(T)/F(T = 1\,\mathrm{MK})=[1.00, 0.009, 0.003, 0.001]$, and the relative Yohkoh/SXT Al 1265 Å flux is $F(T)/F(T = 5\,\mathrm{MK})=[0.006, 0.536, 1.000, 0.822]$. So, the brightest structures in TRACE 171 Å are the 1 MK loops, and in Yohkoh/SXT the 5 MK loops, respectively.

Solution 3.9: The flux contrast between looptop [at a height $h_T = L/(\pi/2)$] and footpoint is for an exponential density model: $c(T) = EM(h_T, T)/EM(0, T) = \exp(-2h_T/\lambda_p T)$. The contrast for a loop with a length of $L = 100$ Mm is $c(T = 1\,\mathrm{MK})=0.066$, and $c(T = 5\,\mathrm{MK})=0.58$, respectively. So, if the detection threshold is $c > 0.1$, TRACE will not see the top of the cold 1 MK loop, while Yohkoh will see the top of a hot 5 MK loop.

Solution 3.10: Serio's scaling law (Eq. 3.6.16) yields the base pressure $p_0 = L^{-1}[T^3 (1400 \exp[-0.08(L/s_H) - 0.04(L/\lambda_p)])]^{-3}$, with the numerical values $p_0 = 3.6$ dyne cm^{-2} for the cool ($T = 3.0$ MK), and $p_0 = 124$ dyne cm^{-2} for the hot loops ($T = 10$ MK). The resulting base density is then $n_0 = p_0/(2k_B T) = 4.3 \times 10^9$ cm^{-3}, and 4.5×10^{10} cm^{-3}, respectively. The DEM distribution in Fig. 3.30 yields $dEM(T)/dlog(T) \approx 130 \times 10^{50}$ cm^{-3} for 3 MK loops, and $\approx 25 \times 10^{50}$ cm^{-3} for 10 MK loops, respectively. Assuming $dEM(T)/dlog(T) \propto N_{loop} n_e^2$, we find the following number ratio of cold to hot loops: $N_{cold}/N_{hot} = (130/25) \times (n_{0,hot}/n_{0,cold})^2 = 560$.

Chapter 4: Hydrodynamics

Solution 4.1: All analytical steps from Eq. (4.1.1) through Eq. (4.1.24) are described in detail in Section 4.1.

Solution 4.2: Inserting the density model $n(h) = n_0 \exp(-h/\lambda_T)$ with $\lambda_T = \lambda_0 \times$ (T/1 MK) and $\lambda_0 = 46,000$ km, into the momentum equation (4.2.16), assuming a sonic point $v(s = L) = c_s$ at the looptop $h = L/(\pi/2)$, we obtain the scaling law $h(T) = \lambda_0 \ln(c_s/v_0)$(T/1 MK).

Solution 4.3: The radiative loss function $E_R = n_e^2 \Lambda(T) = n_e^2 \Lambda_0 T^\alpha = \chi \rho^2 T^\alpha$ can be approximated with a constant $\Lambda_0 = 10^{-21.94} = 1.15 \times 10^{-22}$ erg s^{-1} cm^{-3} and $\alpha = 0$ in the temperature range of $10^{5.75} < T < 10^{6.3}$ (see Fig. 2.14 and Rosner et al. 1978). The criterion for radiative instability, $L < L_{max} = (\kappa T_0^{7/2}/n_e^2 \Lambda_0)^{1/2}$ (Eq. 4.3.11) can then be expressed as a criterion for a critical density, $n_e(L) > n_{e,min} = L^{-1}(\kappa T_0^{7/2}/\Lambda_0)^{1/2} = 2.8 \times 10^8$ (100 Mm/L) cm^{-3} for a given loop half length L.

Solution 4.4: Using the density model $n_e(h) = n_0 \exp(-h/\lambda_T)$, the continuity equation $n_e(h)v(h) = n_0 v_0$, the scale height $\lambda_T = 46,000 \times$ (T/1 MK) km = 11,500 km for the formation temperature ($T = 10^{5.4} = 0.25$ MK) of the O V line (Fig. 2.12), and the height $h \approx 60'' \approx 44,000$ km and velocity $v(h) = 60$ km s^{-1} for the blob A shown in Fig. 4.8, we obtain a base velocity of $v_0 = v \exp(-h/\lambda_T) \approx 1.3$ km s^{-1}. Thus, blob A experiences a strong acceleration over this height range.

Solution 4.5: The filter response function $R_i(T)$, which is a function of time t for monotonic cooling, can be approximated with a Gaussian function, $R_i[T(t)] = R_0 \exp(-[T(t)-T_i]^2/2\Delta T_i^2)$. Assuming an exponential cooling process, $T(t) = T_0 \exp(-t/\tau_{cool})$, with a radiative cooling time of $\tau_{cool} \approx 3k_B T_0/n_e \alpha_{FIP} \Lambda_0$ = 37.6 min (see Eqs. 5.4.1-4), we obtain the following delays between the peak responses in the three filters ($T_{171} = 1.0$ MK, $T_{195} = 1.5$ MK, $T_{284} = 2.0$ MK): $t_{195} - t_{284} = 10.8$ min, and $t_{171} - t_{195} = 15.2$ min, respectively. These delays are close to the observed ones of loop #2 in Fig. 4.12.

Solution 4.6: The signal-to-noise ratio is proportional of the emission measure ratio of the loop to the background, i.e., $\Delta I/I = EM_{loop}/EM_{back} = [n_{loop}^2(h)w]/[n_{back}^2 \lambda_T]$ near disk center. For an exponential density model with super-hydrostatic scale height $\lambda = q_\lambda \lambda_T$, i.e., $n_{loop}(h) = n_0 \exp(-h/q_\lambda \lambda_T)$, we obtain for a loop with $(n_e^{loop}/n_e^{back}) = 2$, $(w/\lambda_T) = 0.1$, and $h = \lambda_T$, a contrast of $\Delta I/I = 50\%$ at the loop base $h = 0$, $\Delta I/I = 24\%$ at the looptop $h = \lambda_T$ for a super-hydrostatic scale height ratio $q_\lambda = 4$, and $\Delta I/I = 5\%$ at the looptop for a hydrostatic scale height ($q_\lambda = 1$). Thus, the looptops are easier to detect for super-hydrostatic loops.

Solution 4.7: The acoustic travel time along a loop and back is $\Delta t = 4L/c_s \approx 12$ min, for $L = 30,000$ km (Fig. 4.22) and a sound speed of $c_s = 166$ km s^{-1} (for $T = 1$ MK). The average time interval between two successive surges in the simulation shown in Fig. 4.22 is $\Delta t_{surge} \approx 1000s \approx 17$ min, and thus they could be triggered by reflected (acoustic) disturbances.

Solution 4.8: Insert the area correction factor defined by Chae et al. (1988), i.e., $A(t[s]) = A(T_h)[1 + (\Gamma^2 - 1)(T[s]/T_h)^p]^{1/2}/\Gamma$ (Eq. 4.8.6), into the DEM expression $dEM(T)/dT = A(s[T])n_e^2(s[T])ds(T)/dT$ in Eq. (3.10.3).

Solution 4.9: The free-free opacity for $\nu = 0.074$ GHz at $T = 1.0$ MK is $\tau_{ff} \approx 182$ for $n_e = 10^9$ cm^{-3}, and $\tau_{ff} \approx 1.8$ for $n_e = 10^8$ cm^{-3}, so the corona is optically thick for densities $n_e \gtrsim 10^8$ cm^{-3} at this frequency, and thus the radio brightness temperature is equal to the plasma temperature, $T_B \approx T_e$, according to the Rayleigh-Jeans approximation (Eq. 2.2.5). Since the electron temperature in coronal holes, $T_e^{hole} \gtrsim 0.8$ MK (see Fig. 4.30), and in quiet Sun regions, $T_e^{quiet} \lesssim 3.0$ MK (Fig. 3.10), would only allow for a maximum contrast of $T_e^{quiet}/T_e^{hole} \approx 3/0.8 \approx 4$, the radio emission in coronal holes has to be optically thin for an observed contrast of ≥ 10, and thus the density in coronal holes has to be $n_e < 10^8$ cm^{-3}.

Solution 4.10: At large distances where $v \gg v_c$, the velocity behaves like $v(r) \approx (\ln r)^{1/2}$, for which the first and third terms are dominant in Eq. (4.10.6) and the density falls of like $n(r) \approx r^{-2}(\ln r)^{-1/2}$, so that the pressure vanishes at infinity, as required in the interstellar medium.

Chapter 5: Magnetic Fields

Solution 5.1: For conversion of the Maxwell's equations see Appendix C for rationalized mks-units. The Ampére law in mks-units is $\mathbf{j} = (1/\mu)\nabla \times \mathbf{B}$, and Ohm's law in mks-units is $\mathbf{j} = \sigma(\mathbf{E} + \mathbf{v} \times \mathbf{B})$, see, e.g., Eqs. (2.15) and (2.9) in Priest (1982). The Maxwell's equations in matter contain the material constants μ and ϵ, while these values are unity in vacuum in cgs-units.

Solution 5.2: From the dipole field line parameterization Eq. (5.2.22) we define the distance to a footpoint with $r_0 = r_1 \sin^2 \theta_0$. The dipole is buried at a dipole depth $h_D = r_0 \sin \theta_0$, which yields the angle $\theta_0 = \arcsin[(h_D/r_1)^{1/3}]$. Defining two dipole field lines with height r_1 and r_2 above the dipole center, a loop width can be approximately defined by the difference in distance from the dipole center, i.e., $w(\theta) = (r_2 - r_1) \sin^2 \theta$. The expansion factor referenced to the loop width at the footpoint is then $\Delta w(\theta) = (r_2 - r_1)(\sin \theta/\sin \theta_0)^2$. A more accurate expansion factor can be derived from transformation into cartesian x- and y-coordinates.

Solution 5.3: Choosing the parameters $k = l = 1$, the parameterization of two force-free field lines according to Eq. (5.3.19) is $z_1(x) = \log[\sin(x)] + z_{01}$ and $z_2(x) = \log[\sin(x)] + z_{02}$. The width of the fluxtube bound by these two field lines at the looptop is $\Delta w_{top} = z_2(x = \pi/2) - z_1(x = \pi/2) = (z_{02} - z_{01})$. The footpoint locations are constrained by the requirement $z_1(x_1) = 0$ and $z_2(x_2) = 0$, which yields $x_1 = \arcsin(\exp[-z_{10}])$ and $x_2 = \arcsin(\exp[-z_{20}])$, so the footpoint separation width (with $y(x) = \alpha x + y_0$) is $\Delta w_{foot} = \sqrt{(x_2 - x_1)^2 + (y_2 - y_1)^2}$ $= \sqrt{1 + \alpha^2} \times (x_2 - x_1)$, and the loop expansion factor is $q_w = \Delta w_{top}/\Delta w_{foot}$.

Solution 5.4: The plasma-β parameter (Eq. 1.8.1) as a function of height h, using the hydrostatic density model (Eq. 5.4.4) and magnetic dipole model (Eq. 5.4.5) is $\beta(h) = 16\pi\xi n_0 k_B T_e B_0^{-2} \exp(-h/\lambda_T)(1 + h/h_D)^6$. A lower limit $\beta_{min}(h)$ can be derived with $n_{0,min}$ and $B_{0,max}$, and an upper limit $\beta_{max}(h)$ with $n_{0,max}$ and $B_{0,min}$, respectively.

Solution 5.5: Parameterize a helical field line first in a straight cylinder in cylindrical coordinates, parameterized by its length l, radius r, twist angle θ, and azimuthal angle φ_0 of the starting point, i.e., $\varphi(z) = \varphi_0 + (z/r)\tan(\theta)$. Then transform into a curved cylinder system, and calculate the projections onto a plane. S-shaped field lines can be reproduced with helically twisted (force-free) field lines (Fig. 5.19; $\alpha \neq 0$) as well as with a potential field model (Fig. 5.20; $\alpha = 0$), and thus cannot be used as discriminator between the two models.

Solution 5.6: Using the definition of the α-parameter for a uniformly twisted fluxtube (Eq. 5.5.9) we find $l = 4\pi N_{twist}/\alpha \approx 600$ Mm for $r \ll l$. With the same equation (5.5.9) we find $r/l = \sqrt{(4\pi N_{twist}/\alpha l - 1)}/2\pi N_{twist} = 1/4\pi \approx 0.08$ for $N_{twist} = 2$.

Solution 5.7: Examples can be found in: *Observation of a 3D magnetic nullpoint in the solar corona* by Filippov (1999, Solar Physics **185**, 297); *Evidence for the flare trigger site and 3D reconnection in multiwavelength observations of a flare* by Fletcher et al. (2001; ApJ **554**, 451); or on TRACE webpages. The TRACE *Picture-of-the-Day (POD)* website *http://trace.lmsal.com/POD/* can be searched with the keyword *"X-point"*, which displays X-point configurations in AR 10767 (2005-May-24), AR 10561, AR 9373, AR 9149+9147, AR 9611, and active regions on 2005-Mar-29, and 2000-May-20 11:12 UT.

Solution 5.8: Without magnetic fields ($\nu_B = 0$) the gyrofrequency variables disappear in Eqs. (5.7.2-6) ($Y = 0, Y_T = 0, Y_L = 0$) and the refractive index is in lowest order $n_{\nu,\sigma}^2 = 1 - X + X^2 \approx 1 - X = 1 - (\nu_p/\nu)^2$. With a magnetic field, but longitudinal waves (**k** $\parallel$ **B**, $\psi = 0$), we have $Y_T = 0, Y_L = Y$ and obtain in lowest order ($X \ll 1$) the approximation $n_{\nu,\sigma}^2 \approx 1 - X/(1 + \sigma Y)$.

Solution 5.9: The angular dependence of the gyro-opacity is shown in Fig. 5.26 for $s = 2, 3, 4$. Strongest gyroemission is observed at angles $\theta \approx 90°$ for X-mode, and at $\theta \approx 70°$ for O-mode.

Solution 5.10: The pressure balance (Eq. 5.8.2) leads to the relation $B_{int}^2(h) \propto (p_E - p_0)\exp(-h/\lambda_{T,Ch})$, and assuming magnetic flux conservation $B(h) \propto 1/A(h)$ yields the scaling $A(h) \propto \exp(+h/2\lambda_{T,Ch})$. The chromospheric height is about 4 scale heights, $h_{Ch} \approx 4\lambda_{T,Ch}$, which yields an expansion factor of $A(h_{Ch})/A(h = 0) \approx e^2 \approx 7.4$.

Chapter 6: Magneto-Hydrodynamics (MHD)

Solution 6.1: All the steps of the analytical derivation are described in detail in §6.1.4. The last step from Eq. (6.1.31) to Eq. (6.1.32) can be gathered from §4.1.

Solution 6.2: The equations for particle conservation (Eq. 6.1.33 = 6.1.16), momentum conservation (Eq. 6.1.34 = 6.1.17), Ampère's law (Eq. 6.1.38 = 6.1.19), and magnetic divergence (Eq. 6.1.37 = 6.1.21) are identical in the resistive and ideal MHD equations. In the case of ideal MHD it is assumed that the plasma is a perfect electric conductor. Setting $\sigma \mapsto \infty$ or $1/\sigma \mapsto 0$, transforms the electric field from Eq. (6.1.39) to (6.1.22) and the induction equation Eq. (6.1.36) to $\partial \mathbf{B}/\partial t = \nabla \times (\mathbf{v} \times \mathbf{B})$, where the electric field (Eq. 6.2.22) can be inserted and leads to Eq. (6.1.20). After inserting $1/\sigma \mapsto 0$, the energy equation (6.1.35) becomes identical with Eq. (4.1.18), which corresponds to the form of Eq. (4.1.4). Inserting the entropy S from Eq. (4.1.18) and neglecting heating and cooling ($E_H = 0$, $E_R = 0, \nabla F_C = 0$) leads then to the adiabatic energy equation (6.1.18).

Solution 6.3: The Lorentz force due to the loop curvature is $(\mathbf{j} \times \mathbf{B})_{curv} = (B^2/4\pi)$ $/r_{curv}$ (Eq. 6.2.19). The curvature radius of a dipolar field line is $r_{curv} \approx (h + h_D)/2$, where h is the height of the loop and h_D the dipole depth (see Fig. 5.2). The curvature force is balanced by the radial magnetic pressure difference, $(\mathbf{j} \times \mathbf{B}) = -\nabla(B^2/4\pi) = -[(B+\Delta B)^2 - B^2]/4\pi \Delta r \approx -2B\Delta B/4\pi w$. Thus the radial change of the magnetic field across the loop width is $\Delta B/B \approx w/2r_{curv} = 1\%$ for $w = 1$ Mm and $r_{curv} = 50$ Mm.

Solution 6.4: Parameterize a helical field line in cylindrical coordinates, transform it into curved cylinder, and calculate 2D projections as shown in Fig. 6.3. Most of the loops shown in Fig. 6.4 have a number of twists of $N_{twist} \approx 1$.

Solution 6.5: The twist angle is $\theta \approx \tan \theta = 2\pi(w/2)N_{twist}/L \approx 0.22$ rad (or $\approx 13°$). The required expansion speed to explain rotational Doppler shifts of $v_{rot} \approx 30$ km s^{-1} is then (with Eq. 6.2.25) $v_{exp} \approx v_{rot}/\tan \theta \approx 130$ km s^{-1}.

Solution 6.6: Assuming that the observations show first the top of the loop and after 9 hours the footpoints of a twisted semi-circular loop, we obtain a vertical emergence speed of $v = \Delta h/\Delta t = 0.3R_\odot/9 \times 3600 = 6.5$ km s^{-1}. The footpoint separation according to the diagram in Fig. 6.9 top right is $d_{foot} = \sqrt{(x_2 - x_1)^2 + (y_2 - y_1)^2} \approx \sqrt{(63 - 50)^2 + (16 - 0)^2} \approx 20$ Mm. The length of the semi-circular loop is then $l = d_{foot}(\pi/2) \approx 30$ Mm. The angular change of observed loop direction, which changes from $|\Delta x/\Delta y|(t = 0) = \tan \theta_1 = 6/20$ ($\theta = 17°$) to $|\Delta x/\Delta y|(t = 9hr) = \tan \theta_1 = 15/20$ ($\theta = 37°$) amounts to $\Delta\theta \approx 20°$. The number of turns of the twisted loop is thus $N_{twist} = l \tan(\Delta\theta)/(2\pi r) \approx 1.2$ for a loop with a radius of $r = w/2 = l/20 = 1.5$ Mm according to an aspect ratio of $w/l = 0.1$.

Solution 6.7: The Rayleigh-Taylor instability can occur in reconnection outflows in flares, or at interfaces below prominences. The Kruskal-Schwarzschild instability can occur at the surface of (erupting) loops, filaments, and prominences that are not line-tied. The Kelvin-Helmholtz instability can occur in supersonic CME fronts. The ballooning instability can occur in large-scale flare loops. The convective thermal instability occurs in the convection zone and generates the granulation and super-granulation pattern in the photosphere. The radiatively-driven thermal instability could occur in low-lying compact hot loops. The heating scale-height instability can occur in footpoint-heated loops and filaments. Resistive instabilities occur in thin current sheets in flares. The kink instability is thought to occur during processes associated with sigmoid loops leading to filament eruptions and CMEs. The sausage instability can occur in fat and overdense flare loops.

Solution 6.8: The pressure in prominences, $p = 2\xi n_e k_B T_e$, includes neutrals with a degree of ionization $0.1 < \xi < 1$. Thus, an upper limit for the minimum pressure is $p_{min} \leq 2n_{min}k_B T_{min} = 0.002$ dyne cm^{-2}, and the maximum pressure is $p_{max} \leq 2n_{max}k_B T_{max} = 1.24$ dyne cm^{-2}, which covers the observed range of $p = 0.03 - 0.38$ dyne cm^{-2}.

Solution 6.9: The maximum mass of a prominence is $M_{prom} = n_e m_p q_{fill}(l \times h \times w) \lesssim 9 \times 10^{16}$ g, using the maximum values of each parameter range given in Table 6.2. This amounts to $M_{prom}/M_{cor} = 9 \times 10^{16}/3 \times 10^{17} \lesssim 30\%$ of the coronal mass. (Note that we neglected the neutrals in this estimate, which would increase the estimated prominence mass.)

Solution 6.10: The criterion (4.3.11) yields (see also problem 4.3 and Fig. 2.14) a density limit of $n_e(L) > 10^5 \ (10^{10}$ cm/ L) cm^{-3} at a temperature of $T = 10^4$ K, which is fulfilled for all observed filaments according to Table 6.2.

Chapter 7: MHD Oscillations

Solution 7.1: The dispersion relation (7.1.32) is a quadratic equation of v_{ph}^2, hence the explicit solution is $v_{ph}^2 = [(v_A^2 + c_s)^2 \pm \sqrt{(v_A^2 + c_s)^2 - 4c_s^2 v_A^2 \cos^2 \theta}]/2$. Inserting $c_s = 0.1v_A$ yields $v_{ph}^2(\theta) = v_A^2[1.01 \pm \sqrt{1.01^2 - 0.04 \cos^2 \theta}]/2$, from which the polar diagram can be plotted with $(v_{ph}/v_A) \cos(\theta)$ on the x-axis vs. $(v_{ph}/v_A) \sin \theta$ on the y-axis, with different $\pm$signs (from the two square roots) in each of the four quadrants.

Solution 7.2: Use a mathematical software that provides routines for the Bessel functions I_n and K_n and a root finder algorithm, e.g., IDL. From Fig. 7.4 we choose similar speed ratios, e.g., $v_{Ae} = 2v_A$ (which implies $\rho_0/\rho_e = 4$), $v_A = 2c_0$, and $c_0 = 2c_e$. These ratios imply the following tube speeds (defined with Eq. 7.1.27): $c_{T0} = c_0 v_A/\sqrt{c_0^2 + v_A^2} = (2/\sqrt{5})c_0 \approx 0.89c_0$, and $c_{Te} = c_e v_{Ae}/\sqrt{c_e^2 + v_{Ae}^2} = (8/\sqrt{65})c_e \approx 0.99c_e$, respectively. The expressions m_0 and m_e (Eq. 7.1.51) can then expressed in speed ratios in units of v_A and v_{Ae} and inserted into Eq. (7.1.50). For an array of values $k_z a = 0.0, 0.1, ..., 4.0$, solutions of the dispersion relation (7.1.50) can then be found for ω/k_z using a numerical root finder algorithm, to yield the functions shown in Fig. 7.4. Be aware of the wave number cutoff for $k_z a < k_c$ (see §7.3.1).

Solution 7.3: The Alfvén speed is $v_A = 1224$ km s^{-1} for $B = 20$ G, $n_e = n_i = 10^9$ cm^{-3}, and $\mu = 1.27$ (Eq. 3.1.7). The resulting kink-mode period is $P_{kink} = 142$ s for a mean density ratio of $(n_e/n_0) = 0.5$, and varies only $\lesssim \pm 20\%$ for extremal values $0 \le n_e/n_0 \le 1$.

Solution 7.4: Applying the error propagation law to $B = (1/P)\sqrt{8\pi\rho_0(1+q_e)}$ (Eq. 7.2.5), with $q_e = \rho_e/\rho_0$, we find the derivatives $dB/dP = -B/P$ and $dB/dq_e = B/[2(1+q_e)]$, and with $(m_P/P) = 10\%$, $q_e = 0.5$, and $m_{q_e} = 0.5$ we find a propagation error of $(m_B/B) = \sqrt{(dB/dP)^2 m_p^2 + (dB/dq_e)^2 m_{q_e}^2} \approx 20\%$.

Solution 7.5: About 25% of the periods are found in the range of 5-min p-mode oscillations ($P = 300 \pm 30$ s), about 25% in the range of 3-min sunspot oscillations ($P = 180 \pm 18$ s), and about 50% are found outside these ranges. So, up to half of cases could be resonant with global oscillations.

Solution 7.6: The resulting height dependence of the sausage mode period limit is $P_{saus}(h) = (2.62a/v_{A0}) \exp(-h/2\lambda_T)(1+h/h_D)^3$, with $v_{A0} = 1210 B_{20} n_9^{1/2}$ km s^{-1}. Thus, the maximum sausage period are $P_{saus} \lesssim 0.5$ s at a height of $h = 2$ Mm, and are $P_{saus} \lesssim 2.5$ s at a height of $h = 200$ Mm, respectively.

Solution 7.7: The sound speed is $c_s = 147\sqrt{6.3} = 369$ km s^{-1} and the phase speed is $v_1 = 43 \pm 25$ km s^{-1}. Using Eq. 7.4.9, the soft X-ray intensity variation is then $\Delta I/I = \Delta EM/EM = [(\rho_0+\rho_1)^2 - \rho_0^2)]/\rho_0^2 \approx 2\rho_1/\rho_0 = 2v_1/c_s = 23 \pm 13\%$.

Solution 7.8: The RHESSI measurements in Fig. 4.28 show a density drop from $n_e(h = 1000$ km$) \approx 10^{13}$ cm s^{-1} to $n_e(h = 3000$ km$) \approx 10^{12}$ cm s^{-1}, hence the density scale height is approximately $\lambda = (3000 - 1000$ km$)/\ln(10^{13}/10^{12}) \approx 870$ km. Plugging the 11 values for loop lengths L and periods P from Table 7.1 into Eq. (7.5.7), we obtain a mean damping period of $\tau_D = LP/4\pi^2\lambda = 2200 \pm 1800$ s, which is about a factor of 4.3 longer than the observed periods of $\tau_D^{obs} = 509 \pm 341$ s listed in Table 7.1, and thus this model cannot account for the observed damping.

Solution 7.9: The ratio of damping time to period is expected to be $(\tau_D/P)_{thick} \approx 0.75(2/\pi)(1 + q_n)/(1 - q_n) \gtrsim 1.0$, which can be arbitrarily larger when the densities become comparable (for $q_n \lesssim 1$).

Solution 7.10: Kink-mode oscillations as seen with TRACE do not significantly modulate the electron density, hence there would be no frequency modulation detectable in radio. The plasma frequency modulation due to density variation of slow-mode oscillations would be $\Delta\nu_p/\nu_p = \sqrt{n_e + \Delta n_e(t)}/\sqrt{n_e} \approx (1/2)\Delta n_e/n_e \approx (1/2)v_1/c_s \approx 6 \pm 3\%$, using Eq. (7.4.9), and $v_1/c_s \approx 12 \pm 6\%$ (see Problem 7.7).

Chapter 8: Propagating MHD Waves

Solution 8.1: Inserting the 1D total derivative (Eq. 8.1.1), i.e., $D/Dt = \partial/\partial t + v(\partial/\partial s)$, into the continuity equation (Eq. 6.1.16), where the divergence in 1D is $\nabla v = \partial v/\partial s$, and making use of the product rule $d(\rho v)/ds = \rho(\partial v/\partial s) + v(\partial \rho/\partial s)$, leads to the continuity equation in form of Eq. (8.1.1). - Using the momentum equation (6.1.15), neglecting the magnetic field ($\mathbf{B} = 0$), but including the 1D viscosity term $F_{visc} = \nu_{visc}\rho(4/3)[\partial^2 v/\partial s^2]$ (Eq. 6.1.14), with the viscosity constant defined as $\eta_0 = \nu_{visc}\rho$, leads to the momentum equation in form of Eq. (8.1.2). - Using the energy equation in form of Eq. (4.1.4), but neglecting the heating and radiative loss term, inserting the 1D total derivative, using the adiabatic equation $(\partial p/\partial s) = (\gamma p/\rho)\partial \rho/\partial s$; Eq. (4.2.2), and defining the parallel thermal conductivity as $\kappa_\parallel = \kappa T^{5/2}$ in Eq. (3.6.3), we obtain the energy equation in form of Eq. (8.1.3).

Solution 8.2: If we define the sensitivity range for each filter by the temperature range where the response function exceeds 50% of the peak response, i.e., $R(T) > 0.5 \times max[R(T)]$, we obtain a temperature range of $T = 0.69 - 1.28$ MK for the TRACE 171 Å filter, and $T = 1.08 - 1.79$ MK for the TRACE 195 Å filter, as it can be measured from the graphic response function $R(T)$ (Fig. 3.25) or from calling the IDL procedure *TRACE_T_RESP.PRO*. The corresponding ranges of sound speeds are (see Appendix D and use $\mu = 1.27$, Eq. 3.1.7): $c_s \approx 122 - 166$ km s^{-1} for TRACE 171 Å, and $c_s \approx 152 - 197$ km s^{-1} for TRACE 195 Å, which cover the observed ranges of slow-mode waves reported in Table 8.1.

Solution 8.3: Using the definition of the Alfvén speed as given in Appendix D (with $\mu = 1.27$; Eq. 3.1.7), we obtain $v_A = 500$ km s^{-1}, $v_{Ae} = 1223$ km s^{-1}, $c_g^{min} = 407$ km s^{-1}, $t_1 = 82$ s, $t_2 = 200$ s, and $t_3 = 245$ s, which yield a duration of $t_2 - t_1 = h/v_A - h/v_{Ae} = 118$ s for the periodic phase, and $t_3 - t_2 = h/c_g^{min} - h/v_A = 45$ s for the quasi-periodic phase.

Solution 8.4: From the time profiles shown in Fig. 8.7 we find that a wave train passes the looptop position H at $t_1 \approx 25$ s, and loop position A at $t_2 \approx 35$ s. In Fig. 8.6 (bottom right) we find that pixel A is separated from pixel H by $\Delta y = 5$ pixels in vertical direction, and $\Delta x = 4$ pixels in horizontal direction. If we approximate the curved loop with a diagonal line, we have a distance of $\Delta s = \sqrt{\Delta x^2 + \Delta y^2} = 6.4$ pixels, or using a semi-circular approximation, we estimate a distance of $\Delta s = \Delta y(\pi/2) = 7.9$ pixels. Taking the pixel size of $4.07''$ into account, we estimate a speed of $v = \Delta s/\Delta t = 1900$ km s^{-1}, or $v = 2300$ km s^{-1}, respectively, which bracket the reported value of $v = 2100$ km s^{-1}.

Solution 8.5: Measuring the densities in Fig. 8.13 (top panel) in the range of $r = R/R_\odot = 1 - 2$ we find $n_e(r = 1.0) \approx 10^8$ cm^{-1}, $n_e(r = 1.5) \approx 4 \times 10^6$ cm^{-1}, $n_e(r = 2.0) \approx 4 \times 10^5$ cm^{-1}, which fits a radial density profile of $n_e(r) \propto r^{-8}$. Measuring the line widths in Fig. 8.13 (middle panel) in the same range we find $\Delta v(r = 1.0) \approx 50$ km s^{-1}, $\Delta v(r = 1.5) \approx 100$ km s^{-1}, $\Delta v(r = 2.0) \approx 280$

km s^{-1}, which fits a radial function of $\Delta v(r) \propto r^{2.5}$, which is close to the theoretical prediction $n_e^{-1/4}(r) \propto r^{[-8\times(-1/4)]} = r^{2.0}$.

Solution 8.6: Measuring the diameters of the circular EIT wave pattern in Fig. 8.16 we find $d(t = 2$ min$)\approx 3$ mm, $d(t = 15$ min$)\approx 15$ mm, $d(t = 30$ min$)\approx 25$ mm, $d(t = 45$ min$)\approx 30$ mm, which have to be scaled by the solar diameter $d_\odot = 49$ mm. Thus, the average speed of the EIT wave is $v_{EIT} = R_\odot(d/d_\odot)/t \approx 360, 240, 200, 160$ km s^{-1}, so the EIT wave is decelerating.

Chapter 9: Coronal Heating

Solution 9.1: Assuming that the coronal losses do not change significantly in coronal holes (CH) during the solar cycle, and that quiet Sun areas (QS) and active regions (AR) become similar to coronal holes during the solar minimum, we estimate the following changes from the solar cycle maximum to the minimum, according to Table 9.1: The conductive flux F_C will decrease by a factor of 0.3 in QS and 0.006-0.6 in AR; the radiative flux F_R will decrease by a factor of 0.1 in QS and 0.002 in AR; the solar wind flux F_{SW} will increase by a factor of ≥ 14 QS and ≥ 7 in AR; the total corona losses (and thus the total heating requirement) $F_C + F_R + F_{SW}$ will increase by a factor of 2.7 in QS and decrease by 0.08 in AR.

Solution 9.2: The Ohmic dissipation rate is (from Eqs. 9.3-9.7): $F_H^{Ohm} = E_H^{Ohm} \times (l/2) = j^2 l/2\sigma = 2 \times 10^{-4}(l/10^{10}$ cm$)^{-1}\Delta\varphi^2$, which is about a factor of $F_H^{Ohm}/F_H^{QS} = 2 \times 10^{-4}/3 \times 10^5 \approx 7 \times 10^{-10}$ smaller than the heating rate requirement for the Quiet Sun.

Solution 9.3: The viscosity force for a 1D flow component, e.g., $v_x(x) \neq 0$, is $F_{visc} = \nu_{visc}(4/3)\rho(\partial^2 v_x(x)/\partial x^2)$. For a circular eddy motion with $v(x) = v_0 \sin(kx)$ and $k = (2\pi/l)$, the divergence is zero (∇v), and we obtain $F_{visc} = \nu_{visc}\rho v_0(2\pi/l)^2 = 3 \times 10^{-11}(v/1$ km/s$) (l/1$ Mm$)^{-2}(n_e/10^9$ cm$^{-3})$ [g cm^{-2} s^{-2}].

Solution 9.4: Use a random generator to generate values x_i, $i = 1, ..., N$ (say for $N = 10,000$ values) with a uniform distribution $N(x) = const$ in the range of $0 \leq x_i < 1$. In order to generate a distribution of length scales that obeys a powerlaw function $N(l) \propto l^{-D_3}$, we seek the function $l(x)$ by requiring $dN(l)dl = N(x[l])(dx/dl)dl = l^{-D_3}$. Since $N(x[l]) = const$, we find $x(l) = l^{-D_3+1}$, and the inverse function is $l(x) = x^{1/(1-D_3)}$. Thus, we generate the corresponding length scales with the function $l_i = x_i^{1/(1-D_3)}$, and the corresponding flare energies with $E_i = l_i^{7.5}$, according to the scaling given in Eq. (9.7.11). The distributions $N(l)$ of the values l_i, and $N(E)$ of E_i, can then be sampled in logarithmic histograms, from which the powerlaw slopes can be obtained with a linear regression fit to $\log[N(l)]$ vs. $\log(l)$, and $\log[N(E)]$ vs. $\log(E)$, yielding the power law slopes $\alpha_l = 2.5$ and $\alpha_E = 1.2$, as predicted in Eq. (9.8.13).

Solution 9.5: Introducing random scatter in the scaling laws, e.g., $E_i = l_i^{7.5+r_i}$, where r_i is a random variable from a normal distribution with a mean of zero, does not change the powerlaw slope of the frequency distribution in energy, $N(E) \propto E^{-\alpha_E}$, so it cannot explain other observed slopes as reported in Table 9.6. Thus, other powerlaw slopes require different scaling laws.

Solution 9.6: Eq. (9.8.12) with the observed value $\alpha_E = 3.3$ implies $\alpha_L = 1 + 2.3 * D = 1 + 2.3 * (D_n + D_T + D_3)$. Absolute lower limits are given in the case of uncorrelated parameters ($D_n = 0, D_T = 0$) and highly fractal structures, i.e., curvi-linear structures ($D_3 = 1.0$), leading to $\alpha_L > 3.3$. For typical flare scaling ($D = D_n + D_T + D_3 = 7.5$ the length scale index would be $\alpha_L = 1 + 2.3 \times 7.5 = 18.25$, indicating an extremely steep cutoff. In both cases this implies that the distribution of spatial scales falls off steeper than observed in flares $\alpha = 2.5$, probably caused by a truncation bias from the limited temperature range of a single wavelength filter.

Solution 9.7: Parameterizing in terms of a normalized energy $\epsilon = E/E_0$, the exponential distribution is $N_1(\epsilon) = n_1 \exp(-\epsilon)$ and the powerlaw distribution is $N_2(\epsilon) = n_2\epsilon^{-\alpha}$. Equality in the local value, $N_1(\epsilon) = N_2(\epsilon)$, and local derivative $dN_1(\epsilon)/d\epsilon = dN_2(\epsilon)/d\epsilon$, yields $n_2 = n_1 \exp(-\epsilon)\epsilon^\alpha$ and the solution $\alpha(\epsilon) = \epsilon$. The the powerlaw slope changes from $\alpha(E = E_0/2) = 1/2$ to $\alpha(E = 2E_0) = 2$.

Solution 9.8: If we define the upper cutoff energy E_2 where the probability for detection is at least one single event, we have the requirement $N(E_2) = N_1(E_2/E_1)^{-\alpha} = 1$, which yields the dynamic range $(E_2/E_1) = N_1^{1/\alpha}$. Thus for $N_1 = 10, 10^2, 10^3, 10^4, 10^5$ and $\alpha = 2$ we expect the dynamic ranges $(E_2/E_1) = \sqrt{(N_1)} \approx 3, 10, 30, 100, 300$.

Solution 9.9: Using the definition of the mean (e-folding) amplification factor from (Eq. 9.8.7), $(t_{Se}/\tau_G) = (\alpha_E - 1)^{-1}$, we obtain from Table 9.6 the values $\approx 0.7, 2.9, 2.2, 0.7, 1.9, 2.9, 0.6, 0.6, 0.4, 0.5$. Incomplete sampling at either the lower or upper energy cutoff changes the powerlaw slope, and thus the inferred amplification factor.

Solution 9.10: With a narrowband filter, the true flare energy $E_2 = 3n_e k_B T_2$ is underestimated as $E_2' = 3n_e k_B(T_1 + \Delta T)$. This affects the powerlaw slope $\alpha_E = \log(N_2/N_1)/\log(E_2/E_1) \approx \log(N_2/N_1)/\log(T_2/T_1)$, which is measured as $\alpha_E' \approx \log(N_2/N_1)/\log(T_2/[T_1 + \Delta T])$. So, the biased powerlaw slope is $\alpha_E' = \alpha_E \log(T_2/T_1)/\log(T_2/[T_1 + \Delta T])$, which for $\alpha_E = 2.0, T_1 = 1.0$ MK, $T_2 = 2.0$ MK, $\Delta T = 0.5$ MK yields $\alpha_E' = 1.8 \times \log(2)/\log(1.5) = 3.1$. Thus, event detection with a single narrowband filters tend to yield a steeper powerlaw distribution.

Chapter 10: Magnetic Reconnection

Solution 10.1: Using Eq. (10.1.8), the energy dissipated over $\Delta t = 100$ s in a typical current sheet is $\Delta \varepsilon_m \approx 10^{22} \Delta t = 10^{24}$ erg, which is the equivalent of a nanoflare, but insufficient for a microflare. For a large flare with $\Delta E = 10^{32}$ erg, a total of $N_{CS} = \Delta E / \Delta \varepsilon_m \approx 10^8$ current sheets would be needed.

Solution 10.2: For the bursty reconnection mode we can replace the reconnection inflow speed $v_0 = v_A / \sqrt{S_0}$ in Eq. (10.1.8) with the coalescence speed $u_{coal} = v_A \times q_{coal}$, leading to a dissipation rate of $d\varepsilon_M / dt = 10^{32}$ $(B/100 \text{ G})^2$ $(L/1 \text{ Mm})^2$ $(v_A/1 \text{ Mm/s})$ $(q_{coal}/0.3)$. This requires the dissipation of 1 magnetic island for nanoflares, microflares, or large flares each, with length scales of $L \approx 0.1, 3, 1000$ km for the dissipated magnetic islands.

Solution 10.3: From Fig. 10.10 (left) it can be gathered that each emerging dipole creates a separatrix surface at the interface to the overlying corona, but only overlapping dipole areas create a separator line, which has two chromospheric footpoints. Thus after emergence of 10 non-overlapping and 20 overlapping dipoles we have 30 separatrix surfaces, 10 separator lines, and 20 photospheric nullpoints.

Solution 10.4: The magnetic topology in Figs. 10.10 (right) and 10.27 is identical, but the sequence of magnetic poles is opposite ($+, -, +$ versus $-, +, -$), and thus all magnetic fields point into opposite direction. In Fig. 10.27, a spine field connected with the right ($-$) footpoint polarity flips through the central spine field line and ends up connected with the left ($-$) footpoint polarity.

Solution 10.5: The resulting scaling of the magnetic flux with area is $\Phi(A) = AB(A) \propto A^{(1+\alpha)} = A^{2.7}$ with $\alpha = 1.7$. The frequency distribution of areas is then $N(A)dA = N(\Phi[A]) \, |d\Phi(A)/dA| \, dA = \Phi^{-2}(A)A^\alpha dA = A^{-\alpha-2}dA = A^{-3.7}dA$. The magnetic field for the smallest active regions is $B_0 = \Phi_0/A_0 = 81$ G. The largest active regions have then $A_{max} = A_0(\Phi_{max}/\Phi_0)^{(1/2.7)} = A_0 \times 6.65 \approx 17 \text{ deg}^2$ and $B_{max} = B_0(A_{max}/A_0)^{1.7} = B_0 \times 25 \approx 2000$ G. The smallest ephemeral regions have then $A_{min} = A_0(\Phi_{min}/\Phi_0)^{(1/2.7)} = A_0 \times 0.22 \approx 0.6 \text{ deg}^2$ and $B_{min} = B_0(A_{min}/A_0)^{1.7} = B_0 \times 0.076 \approx 6$ G.

Solution 10.6: Following Litvinenko (1999a), we denote the parameters sideways of a vertical current sheet with n_0, v_0, T_0, and the parameters above the current sheet with n, v, T, assuming $T = T_0$. The current sheet has a vertical height of 2Λ and a width of $2a$. The continuity equation is then $n_0 v_0 \Lambda = nva$, the momentum equation in horizontal direction is $n_0 kT_0 + B^2/8\pi = nkT$, and the momentum equation in vertical direction is $(1/2)m_p nv^2 + n_0 kT_0 = nkT$. Ohm's law $\mathbf{j} = \sigma(1/c)(\mathbf{v_0} \times \mathbf{B}) \approx (\sigma/c)v_0 B$, and the definition of the current density $\mathbf{j} = (1/4\pi)(\nabla \times \mathbf{B})$, yield then $v_0 = c/4\pi\sigma a$. Considering electromagnetic units, $j_{emu} = j_{esu}/c$ (see Appendix C), we have $v_0 = c^2/4\pi\sigma a$, yielding the current sheet width $a = c^2/4\pi\sigma v_0$. Comparing the two momentum equations yields $v = B/\sqrt{(4\pi m_p n)} = \sqrt{(n_0/n)}B/\sqrt{(4\pi m_p n_0)}$. The momentum equation in horizontal direction yields $(n/n_0) = (1 + B^2/8\pi n_0 kT)$. Inserting the expressions a, v, and (n/n_0) into the continuity equation $v_0 = v(n/n_0)(a/\Lambda)$ yields then Eq. (10.4.1), where the variable v_0 (used in Litvinenko 1999a) is denoted as v_1 in Eq. (10.4.1).

Solution 10.7: In the CSHKP (standard) 2D model, the erupting prominence with vertical speed v_z relates to the reconnection outflow speed v_A in upward direction as an upper limit, i.e., $v_z \gtrsim v_A$, but the eruption may be driven by a sideward inflow pressure with speed v_1. In the emerging flux model the current sheet is horizontal or oblique, so the upward motion v_z of the emerging field lines act as driver of the lateral inflow speed v_1, i.e., $v_z \approx v_1$. In most of the other models, i.e., in the equilibrium loss model, in the quadrupolar model, and in shear-driven 3D reconnection models, the shear motion v_x is not aligned with the directions of reconnection inflows or outflows, and thus not directly related. The shearing motion stores non-potential energy, which triggers reconnection processes spontaneously in a nonlinear way.

Solution 10.8: The flare density scales according to Eq. (10.5.5) as $n = B^2/16\pi kT = 3.6 \times 10^{10}(B/50\text{ G})^2(T/10^7\text{ K})^{-1}$ [cm^{-3}], and thus $n^2 = 1.3 \times 10^{21}(B/50$ G$)^4(T/10^7\text{ K})^{-2}$ [cm^{-3}]. The length scale scales according to Eq. (10.5.4) as $L = 10^9(B/50\,\text{G})^{-3}(T/10^7\text{ K})^{7/2}(n_0/10^9\text{ cm}^{-3})^{1/2}$ cm, and $L^3 = 10^{27}(B/50$ G$)^{-9}(T/10^7\text{ K})^{21/2}(n_0/10^9\text{ cm}^{-3})^{3/2}$ cm. The emission measure follows then from inserting these terms for n^2 and L^3, i.e., $EM = n^2L^3 \approx 10^{48}(B/50$ G$)^{-5}(T/10^7\text{ K})^{17/2}(n_0/10^9\text{ cm}^{-3})^{3/2}$ cm^{-3}, which validates Eq. (10.5.7). The thermal energy $E_{th} = 3nk_BTV \approx 3k_BTnL^3$ follows the same way by inserting the terms for n and L^3, yielding $E_{th} \approx 10^{22}(B/50\text{ G})^{-7}(T/10^7\text{ K})^{21/2}(n_0/10^9$ cm$^{-3})^{3/2}$ erg. For a typical solar flare we obtain $E_{th} \approx 10^{30}$ erg.

Solution 10.9: All observations listed in Table 10.2 apply to the 2D standard (CSHKP) model described in §10.5.1, except the observations of quadrupolar geometries and 3D nullpoint geometries. The phenomenon of post-reconnection relaxation applies also to most of the other flare models.

Solution 10.10: The angle between the prereconnection and postreconnection field lines can be measured from the footpoint positions shown in vertical projection in the 3rd rows of Fig. 10.31 ($\approx 10° - 20°$) and Fig. 10.32 ($\approx 40° - 50°$). The range is quoted as $\approx 10° - 50°$ in the text on p.457.

Chapter 11: Particle Acceleration

Solution 11.1: The heavy ions given in Table 11.3 are [He, C, N, O, Ne, Mg, Si, Fe], which have the atomic number or electric charge number $Z = q/e = [2, 6, 7, 8, 10, 12, 14, 26]$, and the ion/proton mass ratio $\mu_i = m_i/m_p = [4, 12, 14, 16, 20, 24, 28, 56]$. Generally, $\mu_i \simeq 2Z$, because the number of protons is equal to the number of neutrons. However, some elements have different isotopes with a larger number of neutrons, such as iron, with 4 additional neutrons in the average. The resulting ion gyro radii are $R_i = 0.95\sqrt{(T_i\mu_i)}/Z B \approx [95, 55, 51, 48, 42, 39, 36, 27]$ cm for $B = 10$ G. Parker-type current sheets are estimated to have a thickness of $\delta \approx 10$ m (§10.1.1), while tearing mode islands are estimated to have smallest widths of $l \approx 70$ m (§10.2.1). In the central zone of Parker-type current sheets, where the magnetic field drops to $B \lesssim 1$ G, ion gyroradii can exceed the width of the current sheet, i.e., $R_i \gtrsim \delta \approx 10$ m, and thus these ions can have chaotic orbits.

Solution 11.2: The drift speed due to solar gravitation is $v_{drift} = m_p c g_\odot / eB = 0.0286$ cm s^{-1} for $B = 100$ G. The equivalent value for electric field change is $dE/dt = g_\odot = 2.74 \times 10^4$ cm s^{-2}. The equivalent value for the electric field is $E = m_p g_\odot / e = 9.5 \times 10^{-11}$ statvolt cm^{-1}. The equivalent magnetic field gradient is $\nabla B = 2 g_\odot / v_\perp^2$. The equivalent curvature radius is $R_{curv} = v_\parallel^2 / g_\odot$. They all do not depend on the magnetic field. The proton drift time out of a current sheet is $t_{drift} = l/v_{drift} = 2 e B l^2 / m_p c v_\perp^2$.

Solution 11.3: At equal speeds ($\beta_p = \beta_e$) the (kinetic) energy ratio is $\varepsilon_p = 1836 \times \varepsilon_e$, and thus the electron-to-proton ratio is $N_e/N_p = N_0 (\varepsilon/\varepsilon_1)^{-\delta} / N_0 (1836\varepsilon/\varepsilon_1)^{-\delta} = 1836^{-\delta} \approx 1.6 \times 10^{-10}$ for $\delta = 3$.

Solution 11.4: The Dreicer electric field changes as $E_D \propto n_e/T_e$, neglecting the small changes in the Coulomb logarithm, so the Dreicer field varies proportionally to the electron density (by a factor of 10^3) and reciprocally to the flare temperature (by a factor of 3). The threshold velocity (Eq. 11.3.3) changes as $v_r \propto v_{Te} \sqrt{E_D/E} \propto \sqrt{T_e} \sqrt{n_e/T_e} \propto \sqrt{n_e}$, so is independent of the flare temperature and varies proportionally to the square root of the density.

Solution 11.5: Following the definitions of Litvinenko (1996) we have the magnetic field components inside the current sheet, $B_x = (-y/\Delta w_y) B_0$, $B_y = \xi_\perp B_0$, and $B_z = \xi_\parallel B_0$, or $\mathbf{B} = [-y/\Delta w_y, \xi_\perp, \xi_\parallel] B_0$, and the electric field $\mathbf{E} = (0, 0, E_0)$. The nonrelativistic equation of motion is $m d\mathbf{v}/dt = e[\mathbf{E} + (1/c)\mathbf{v} \times \mathbf{B}]$, and using the time scale $\Omega_0^{-1} = mc/eB_0$ and length scale Δw_y, the components can be written in a dimensionless 3D cartesian coordinate system as: $d^2x/dt^2 = \xi_\parallel (dy/dt) - \xi_\perp (dz/dt)$, $d^2y/dt^2 = -\xi_\parallel (dx/dt) - y(dz/dt)$, $d^2z/dt^2 = \epsilon + \xi_\perp (dx/dt) + y(dy/dt)$, where the dimensionless electric field is defined by $\epsilon = mc^2 E_0 / \Delta w_y e B_0^2$.

Solution 11.6: Defining $\gamma = 1/\sqrt{1-\beta^2}$, $\beta^2 = \beta_\parallel^2 + \beta_\perp^2$, $\beta_\parallel = v_\parallel/c$, $\beta_\perp = v_\perp/c$, $a = k_\parallel c$, $b = s\Omega_e$, and inserting into the Doppler resonance condition (Eq. 11.4.6) yields $\omega - b\sqrt{1 - \beta_\parallel^2 - \beta_\perp^2} - \beta_\parallel a = 0$, which is a quadratic equation and can be transformed into the form of an ellipse equation $[(v_\parallel - v_0)/A]^2 + (v_\perp/V)^2 = 1$ with eccentricity $e = \sqrt{1 - (A/V)^2}$ (Eq. 11.4.28), ellipse center at v_0 (Eq. 11.4.27), major semi-axis V (Eq. 11.4.29), and minor semi-axis $A = V\sqrt{1 - e^2}$.

Solution 11.7: Choose appropriate frequency ratios $\Omega_i \ll \omega_p^i \ll \Omega_e \ll \omega_p^e$ as indicated in Fig. 11.11 and plot the dispersion relations for the O, X, R, L, P modes, ion sound, and Alfvén waves according to the dispersion relations given in Table 11.2. The dispersion relations for the P mode, ion sound, and Alfvén waves can explicitly be expressed as $k(\omega)$, while the others can be found with numerical methods. The sketch in Fig. 11.11 shows approximately $\log(k)$ on the y-axis versus $\log(\omega)$ on the x-axis.

Solution 11.8: For $\theta_{Bn} \lesssim 90°$ the maximum energy gain is according to Eq. (11.5.16) $v^2/u^2 \approx [(1 + \alpha) \cos\theta_{vn}/\cos\theta_{Bn}]^2 \approx 70 - 7000$. So, a 10 keV particle can gain energies of ≈ 700 keV-70 MeV. Acceleration is most efficient for quasi-perpendicular shocks ($\theta_{Bn} \lesssim 90°$).

Solution 11.9: For an energy range of $\varepsilon = 0.1 - 20$ keV, fast shock speed $u = 1000$ km s^{-1}, diffusion length scale $l = 500$ km, and shock angles $\theta = 0°, 60°, 85°$ plot the collisional loss (Eq. 11.5.30), energy gain by acceleration (Eq. 11.5.29), and net gain (Eq. 11.5.28), yielding the diagram shown in Fig. 11.20.

Solution 11.10: According to Eq. (15.4.10) the exciter speed is $v \geq (\partial\nu/\partial t)(2\lambda_T/\nu)$. The drift rate of the type II burst is $(\partial\nu/\partial t) \approx 100$ MHz min^{-1} at $\nu = 500$ MHz or $(\partial\nu/\partial t) \approx 10$ MHz min^{-1} at $\nu = 50$ MHz, yielding a lower limit of $v \gtrsim 300$ km s^{-1} for the shock speed.

Chapter 12: Particle Kinematics

Solution 12.1: From Eqs. (12.2.2-4) follows $l_{TOF} = l_{loop}/(\cos\alpha\cos\theta)$, which inserted into Eq. (12.2.1) yields the generalized TOF delay $\Delta t_{prop} = (l_{loop}/c)\times [1/(\beta_1\cos\alpha_1\cos\theta_1) - 1/(\beta_2\cos\alpha_2\cos\theta_2)]$.

Solution 12.2: The count spectrum for the energies given in Fig. 12.3, $\varepsilon = 44, 59, 77, 102, 127, 167, 233, 321$ keV (channel 3-10), is $N(\varepsilon) = 40,000; 23,000; 13,000; 7400; 4800, 2800, 1400; 750$ cts s^{-1}, and the time uncertainties in the cross-correlation between channels N_3 and $N_i, i = 4, 10$ are for $\tau = 0.2$ s, according to Eq. (12.2.6), $m_\tau = \tau\sqrt{1/N_3 + 1/N_i} = 1.4, 1.7, 2.0, 2.5, 3.1, 3.9, 5.4,$ and 7.4 ms, which agrees approximately with the correlation measurements in Fig. 12.4: $m_\tau = 2, 2, 2, 3, 4, 7, 20$ ms. The slight differences occur because the assumed time scale $\tau = 0.2$ s is only an approximate average of the time scales in the fine structure of the time profiles that dominate the cross-correlation.

Solution 12.3: The electric field is, according to Eq. (12.3.2), $E = \varepsilon_{max}/(el) = 0.0017$ statvolt cm^{-1}, given the maximum electron kinetic energy of $\varepsilon = 650$ keV and flare loop half length $l = l^{prop} \times 0.44$ with $l^{prop} = 29,000$ km. The Dreicer electric field in typical flare conditions ($n_e \approx 10^{11}$ cm^{-3}, $T_e \approx 10^7$ K) is $E_D \approx 2 \times 10^{-6}$ statvolt cm^{-1} (Eq. 11.3.2), and thus the best-fit electric field is a super-Dreicer field, since $E > E_D$.

Solution 12.4: The acceleration is $a = \varepsilon/(m_e l)$ according to Eq. (12.3.4). Integration of Newton's force equation yields $v = at$ and $l = (1/2)at_{acc}^2$, so the nonrelativistic acceleration time is $t_{acc} = l\sqrt{2m_e/\varepsilon}$. The relativistic expression (Eq. 12.3.5) is in the nonrelativistic limit, using $\varepsilon = m_e c^2(\gamma-1)$ and $(\gamma+1) \approx 2$, $t_{acc} = (l/c)\sqrt{(\gamma + 1)/(\gamma - 1)} \approx (l/c)\sqrt{2/(\varepsilon/m_e c^2)} = l\sqrt{2m_e/\varepsilon}$.

Solution 12.5: Inserting the Alfvén velocity (Eq. 12.4.6) into Eq. (12.4.18) yields $t_w \approx 1.0 \times (L_B/10 \text{ Mm}) (B/100 \text{ G})^{-1} (n_e/10^8 \text{ cm}^{-3})^{1/2}$ [s]. The typical time scale is $t_w \approx 1.0$ s.

Solution 12.6: The pulse shape is $F(t) \propto t^{\alpha}(\sin\alpha)^{N-1}\cos\alpha\,(d\alpha/dt)$ for a Dory–Guest–Harris distribution (a), and $F(t) \propto t^{\alpha}(\cos\alpha)^{N-1}\sin\alpha\,(d\alpha/dt)$ for a beam distribution, with $\alpha(t)$ defined in Eq. (12.4.7).

Solution 12.7: Using the constancy of the magnetic moment, $v_{\perp}^2/B(s) = v^2/B_m$ (Eq. 12.5.1), the definition of the mirror ratio, $R = B_m/B_0$ (Eq. 12.5.3), and a parabolic field, $B(s) = B_0[1 + (R-1)s^2/L^2]$ (Eq. 12.5.2), yields the relation $v_{\perp}^2(s) = v^2(1/R)[1 + (R-1)s^2/L^2]$, which inserted into the definition of the bounce time, $t_B = 4\int_0^L ds/\sqrt{v^2 - v_{\perp}^2(s)}$ leads to the integral $t_B = 4L/v\sqrt{1-1/R}\int_0^L (L^2+s^2)^{-1/2}ds$, which has the solution $t_B = 4L/v\sqrt{1-1/R}\,\arcsin(s/L)|_0^L = 2\pi L/v\sqrt{1-1/R}$ (Eq. 12.5.5). The correction factors are $q_{\alpha}(R=\infty) = 2/\pi$ and $q_{\alpha}(R=R_{min}) = (2/\pi) \times 0.9$, requiring $R_{min} = 1/(1-0.9^2) \approx 5.26$.

Solution 12.8: An IDL version of a numerical program that solves this problem is:

```
nt =2000
t0 =5.0
sig_t =1.0
dt =0.01
t =dt*(findgen(nt)+1)
for j=0,9 do begin
   t_trap =1.+j
   f_inj =exp(-(t-t0)^2/(2.*sig_t^2))
   plot,t,f_inj
   f =fltarr(nt)
   for i=0,nt-1 do begin
       f(i)=(1./t_trap)*total(f_inj(0:i)*exp(-(t(i)-t(0:i))/t_trap)*dt)
   endfor
   oplot,t,f,thick=2
   !noeras=2
   f0=max(f_inj,im0) & fm=max(f,im)
   print,'Trapping time = ',t_trap,' Delay=',t(im)-t(im0)
endfor
```

The resulting delays are $\Delta t = 0.69, 1.01, 1.21, 1.35, 1.46, 1.55, 1.62, 1.68, 1.74, 1.79$ s for trapping times $t_{trap} = 1, 2, ..., 10$ s.

Solution 12.9: The losscone angle on one side is $\alpha_1 = \arccos[1 - 2(1 - A)q_{prec}] = 84°, 60°, 26°$ for $A = 0.1, 0.5, 0.9$, and at the other side $\alpha_2 = \arccos[1 - q_{prec}] = 60°$ for every value A.

Solution 12.10: The critical energy for collisions are $\varepsilon_c = 2.0, 6.3$, and 20 keV for loop lengths of $L = 10$ Mm and densities of $n_e = 10^9, 10^{10}, 10^{11}$ cm^{-3}. The corresponding values are $\varepsilon_c = 4.5, 14$, and 45 keV for loop lengths of $L = 50$ Mm. Thus electrons with energies of $\varepsilon = 25$ keV are thermal ($\varepsilon < \varepsilon_c$) for the longest ($L = 50$ Mm) and densest ($n_e = 10^{11}$ cm^{-3}) loop with $\varepsilon_c = 45$ keV.

Chapter 13: Hard X-Rays

Solution 13.1: Both the Hard X-ray Fourier imagers and the radio interferometers have no direct image capabilities (such as a CCD camera), but reconstruct images by measuring the Fourier components. The image quality depends on the number of independent Fourier components (the number of grids in hard X-rays, or pairs of single antennas in radio interferometers). Both the rotation-modulated collimators (RMC) in hard X-rays and Earth-rotation aperture synthesis in radio use the rotation of the whole instrument to measure the number of Fourier components.

Solution 13.2: Inserting Kramer's cross-section (Eq. 13.2.9) into Eq. (13.2.11) yields $I(\epsilon_x) = n_0\sigma_0/(4\pi r^2\epsilon_x)\sqrt{2/m_e}\int_{\epsilon_x}^\infty \varepsilon^{-1/2}n_e(\varepsilon)d\varepsilon$. For a particle injection spectrum $n_e(\varepsilon) \propto \varepsilon^{-\delta}$ the integral yields then a hard X-ray photon spectrum $I(\epsilon_x) \propto \epsilon_x^{-\gamma}$ with the relation $\gamma = \delta - 1/2$.

Solution 13.3: Using an average density of $<n_e> = n_0 \exp\left[-(h_A + h_X)/(2\lambda_T)\right]$ during propagation between the heights $h = h_A$ and $h = h_X$, with a propagation distance $L = h_A - h_X$, we find with Eq. (12.6.21) a cutoff energy of $\varepsilon_c(h_A, h_X) \approx 20\left[(h_A - h_X)/10 \text{ Mm}\right]^{1/2} (\exp\left[-(h_A + h_X)/(2\lambda_T)\right])^{1/2}[n_0/10^{11} \text{ cm}^{-3}]^{1/2}$, which yields cutoff energies $\varepsilon_c = 3.6, 2.9, 2.0$ keV for $n_0 = 10^9$ cm^{-3}, and $\varepsilon_c = 36, 29, 20$ keV for $n_0 = 10^{11}$ cm^{-3}.

Solution 13.4: An IDL version of a Monte-Carlo simulation of this problem is:

```
n =2^14
for i=0,2 do begin
    fwhm =0.2*(1+i)
    sig =fwhm/2.35
    h = 10.*(1.+sig*randomn(1111,n))
    b = 30.*(1.+sig*randomn(2222,n))
    nel =1.e9*(1.+sig*randomn(3333,n))
    qh = 1.5*(1.+sig*randomn(4444,n))
    qa = 1.3*(1.+sig*randomn(5555,n))
    qb = 0.1*(1.+sig*randomn(6666,n))
    tw =0.5*(h/10.)*(b/30.)^(-1)*(nel/1.e9)^(0.5)*((qh-1.)/0.5)*(qa/1.3)*(qb/0.1)
    dt =0.1
```

```
h =histogram(tw,min=0,max=3.0,binsize=dt)
hn =float(h)/max(h)
t =dt*(findgen(30)+0.5)
plot_io,t,hn,psym=10,thick=i+1,yrange=[1.e-3,1.1]
ind =where((t ge 0.4) and (hn gt 0))
t_fit =t(ind)
h_fit =alog(hn(ind))
c =linfit(t_fit,h_fit)
h0 =exp(c(0))
te =-1./c(1)
oplot,t_fit,h0*exp(-t_fit/te),thick=i+1
print,'fwhm=',fwhm,' exp.time const=',te,' s'
!noeras=1
endfor
```

The exponential decay times of the distributions of pulse durations, obtained from exponential fits to the histograms, are $t_e = 0.11, 0.29, 0.51$ s for $\sigma_x/x = 0.2, 0.4$, and 0.6. Thus, the observed distributions with a decay time of $t_e \approx 0.4$ s can be reproduced with a scatter of $\sigma_x/x \approx 0.5$.

Solution 13.5: From Fig. 13.15 we read off for flare 5 (91/12/15) a relative time delay of $\tau_{ij} \approx 64$ ms between the energies $\epsilon_x = 50$ keV and 300 keV. In Fig. 13.14 we read off energy conversion factors of $q_\epsilon = 2.6$ and 2.5 for a powerlaw slope of $\gamma = 3$. Thus the corresponding electron energies are $\varepsilon = \epsilon_x q_\epsilon = 130$ keV and 750 keV (Eq. 13.5.6). The corresponding Lorentz factors are $\gamma = \varepsilon/m_e c^2 + 1 = 1.25$ and 2.74 (Eq. 11.1.10), and the relativistic speeds are $\beta = \sqrt{1 - 1/\gamma^2} = 0.604$ and 0.914 (Eq. 11.1.11). The resulting time-of-flight propagation distance is then $t_{TOF} = c\tau_{ij}(1/\beta_i - 1/\beta_j)^{-1} = 34$ Mm (Eq. 13.5.8), and the projected loop half length is $l \approx l_{TOF} \times 0.54 = 18$ Mm (Eq. 12.2.4), which is close to the value $l' = 19.7 \pm 1.3$ Mm given in Fig. 13.15. The difference comes from the arbitrarily assumed value of the spectral slope.

Solution 13.6: Using Eq. (13.5.9), the time-of-flight distance ratio is $l_{TOF}/r = (1/2)(1 + q^2) \arctan[2q/(q^2 - 1)] = 1.91$ for a height ratio of $q = h/r = 1.5$.

Solution 13.7: For electron energies of $\varepsilon = 25, 50, 100$ keV, the Lorentz factors (Eq. 11.1.10) are $\gamma = \varepsilon/m_e c^2 + 1 = 1.049, 1.098, 1.196$, the relativistic speeds (Eq. 11.1.11) are $\beta = \sqrt{1 - 1/\gamma^2} = 0.302, 0.413, 0.548$, and the asymmetry angles (Eq. 13.5.12) are $\theta = \beta c \Delta t / (2r \times 0.54) = 24°, 33°$, and $44°$. These asymmetry angles are larger than observed in the Masuda flare, which is $\theta \lesssim 20°$ (Fig. 13.34).

Solution 13.8: The trapping time delay is $\Delta t_{trap} = 0.95 \times 10^8 \left(\epsilon_2^{3/2} - \epsilon_1^{3/2} \right) / n_e = 6.14$ s and 0.61 s between electron energies of $\epsilon_2 = 100$ keV and $\epsilon_1 = 50$ keV and electron densities of $n_e = 10^{10}$ and 10^{11} cm^{-3}. For an energy conversion factor of $q_\varepsilon = 2$, this corresponds to hard X-ray energies of $\epsilon_2 = 50$ keV and $\epsilon_1 = 25$ keV, for which a delay time of $\Delta t_{obs} \approx 1.0$ is observed in Flare #1361 shown in Fig. 13.22.

Solution 13.9: Measuring the relative hard X-ray fluxes of the left and right footpoints in flare # 1032 (third example shown in Fig. 13.27) we obtain the asymmetry ratio $A = F_2 / (F_1 + F_2) = 14$ mm/(5 mm +14 mm)=0.74. The resulting losscone angles are then (with Eqs. 12.6.18-19): $\alpha_1 = \arccos \left[1 - 2(1 - A) q_{prec} \right] = 19°$, $43°, 58°$ for $q_{prec} = 0.1, 0.5, 0.9$, and $\alpha_2 = \arccos \left[1 - q_{prec} \right] = 26°, 60°, 84°$.

Solution 13.10: From Fig. 13.28 we measure peak fluxes of $F_{max} = 1.04, 1.17, 1.33, 1.49, 1.68, 1.84$ and looptop fluxes of $F_{top} = 0.85, 0.45, 0.43, 0.55, 0.69, 0.85$, yielding flux contrast ratios of $F_{top}/F_{max} = 0.82, 0.38, 0.32, 0.37, 0.41, 0.46$, at energies of $\varepsilon = 10, 20, ..., 60$ keV. Thus, an instrument with a flux contrast of $> 20\%$ (i.e., $F_{top}/F_{max} \leq 0.8$), can only separate footpoint sources for energies of $\varepsilon \gtrsim 10$ keV.

Chapter 14: Gamma Rays

Solutions 14.1:

Gamma-Ray emission:	Projectile particles:	Target particles:
electron bremsstrahlung	Corona: e^-	Chrom.: ions
nuclear de-excitation lines	Corona: p, ions	Chrom.: ions (^{56}Fe, ^{24}Mg, ^{20}Ne, ^{28}Si, ^{12}C, ^{16}O)
– narrow lines	Corona: p, He	Chrom.: ions
– broad lines	Corona: C, ions	Chrom.: H, He
neutron capture line	Corona: p	Chrom.: p
pion decay radiation	Corona: p	Chrom.: p,ions
positron annihilation line	Chrom.: e^+	Chrom.: e^-

Solutions 14.2: Gamma-ray lines that are detectable in the spectrum shown in Fig. 14.3 (top frame) are: 0.5 MeV (e$^+$ positron annihilation line), 1.4 MeV (^{24}Mg), 1.6 MeV (^{20}Ne), 2.2 MeV (n capture line), 4.4 MeV (^{12}C), and 6.1 MeV (^{16}O). Gamma-ray lines listed in Table 14.2 that are not detectable in the spectrum are: 0.3 MeV (^{56}Fe), 0.4 MeV (^{4}He, ^{24}Mg), 0.8 MeV (^{56}Fe), 0.9 MeV (^{56}Fe), 1.8 MeV (^{28}Si), and 2.6 MeV (^{20}Ne).

Solutions 14.3: The trapping times for ions [including an additional factor of ≈ 60 in Eq. (14.2.2)] is $t_{trap}^{ions} \approx 60 \times 0.95 (\varepsilon_{ion}/100 \text{ keV})^{3/2} (10^{11} \text{ cm}^{-3}/n_i)$, or expressed explicitly for the energy, $\varepsilon_{ion} \approx 100 \text{ keV } [(t_{trap}^{min}/0.95) (n_i/10^{11} \text{ cm}^{-3})]^{2/3} \text{ keV}$, yielding $\varepsilon_{ion} \approx 20 - 100 \text{ keV}$ for densities of $n_i \approx 10^9 - 10^{10}$ cm^{-3} and a trapping time of $\tau_{trap} \approx 10$ min (as observed in the 1991-Jun-11 and 1991-Jun-15 flare), or $\varepsilon_{ion} \approx 200 - 800 \text{ keV}$ for $\tau_{trap} \approx 230$ min (as observed in the long-duration component of the 1991-Jun-11 flare shown in Fig. 14.5).

Solutions 14.4: Whenever the time profile shows pure exponential decays, it is likely that those time phases are governed by trapping, while every impulsive rise is indicative of a particle injection from a new acceleration phase. So, acceleration seems to be intermittent in the 1991-Jun-15 flare, producing pronounced injections at $\approx$08:15, 08:30, and 09:00 UT, while the intervening time intervals seem to be governed by precipitation out of a trap.

Solutions 14.5: The fitted line widths (the values are given in parentheses in the rightmost column of Table 14.2) are all resolved with RHESSI. The effect of the instrumental resolution has been removed in the measured values, but is negligible, except for the ^{56}Fe line.

Solutions 14.6: The relativistic parameters of an electron with a kinetic energy of $\varepsilon_e = 25 \text{ keV}$ are $\gamma = \varepsilon_e/m_e c^2 + 1 = 1.05$ and $\beta = \sqrt{(1 - 1/\gamma^2)} = 0.30$. A proton with a kinetic energy of $\varepsilon_p = 5 \text{ MeV}$ has $\gamma = \varepsilon_p/m_p c^2 + 1 = 1.005$ and $\beta = \sqrt{(1 - 1/\gamma^2)} = 0.10$. The time-of-flight delay between these particles over a distance of $L = 100 \text{ Mm}$ amount to (Eq. 14.3.3) $\tau_{ep} = (L/c)(1/\beta_e - 1/\beta_p) = -2.2$ s.

Solutions 14.7: In the first step, ions accelerated in the corona act as projectile particles and precipitate into the chromosphere, where they collide with other ions (which are the target) and produce neutrons, primarily from the breakup of He nuclei. In a second step, the neutrons are captured by protons and produce deuterium (^{2}H) and monoenergetic photons at 2.223 MeV, i.e., ^{1}H + n $\mapsto$ ^{2}H + $\gamma_{2.2 \text{ MeV}}$.

Solutions 14.8: The FWHM of the 511 keV line shown in Fig. 14.16 is approximately FWHM≈ 4 keV. Eq. (14.5.12) yields $T = 10^4 \text{ (FWHM}/1.1 \text{ keV})^2) \approx 1.3 \times 10^5$ K, which is a transition region temperature.

Solutions 14.9: Pion decay radiation dominates over electron bremsstrahlung at $\varepsilon \gtrsim 60$ MeV for a flare subinterval (of 5000 s), or at $\varepsilon \gtrsim 40$ MeV if integrated over the entire flare.

Solutions 14.10: A list of gamma-ray detectors includes: GRS (onboard SMM); WBS (onboard Yohkoh); BATSE, OSSE, EGRET, COMPTEL (onboard CGRO); RHESSI; GAMMA-1; PHEBUS (onboard GRANAT), HEAO-3, as well as SPI (onboard INTEGRAL) and CORONAS-F, which are not mentioned in the text.

Chapter 15: Radio Emission)

Solution 15.1: Parameterize the logarithmic height $(h/R_\odot)$ in the range of $[0.01, 100]$. Defining the dimensionless distance from Sun center, $r = R/R_\odot = 1 + h/R_\odot$, the three models read as: (1) $n_{BA}(r) = 10^8[2.99r^{-16} + 1.55r^{-6} + 0.036r^{-1.5}]$; (2) $n_{AR}(r) = 10 \times n_{BA}(r)$; and (3) $n_{HP}(r) = 7.2 \times 10^5 r^{-2}$. The plasmafrequency f_p as function of the distance $r - 1 = h/R_\odot$ is then $f_p(r) \approx 9000\sqrt{n_e(r)}$. Model (3) yields $f_p(h/R_\odot = 100) = 76$ kHz and agrees best with Fig. 15.1 in the low-frequency part. Model (2) yields $f_p(h/R_\odot = 0.01) = 570$ MHz and agrees best with Fig. 15.1 in the high-frequency part.

Solution 15.2: The spectra shown in Fig. 15.2 become optically thin $(\tau \lesssim 1)$ at the high frequency side where $T_B \lesssim T_B^{max} = T_e$ (upper panels), which occurs for bremsstrahlung at $\nu \gtrsim 2$ GHz, for thermal gyrosynchrotron emission at $\nu \gtrsim 5$ GHz, and for nonthermal gyrosynchrotron emission at $\nu \gtrsim 6$ GHz.

Solution 15.3: The average loop width in Fig. 15.3 is $w \approx 30,000$ km. For the radio frequency of $\nu = 17$ GHz, the free-free absorption coefficient is then $\alpha_{ff} \approx 0.01 n_e^2 \nu^{-2} T_e^{-3/2} \ln \Lambda \approx 5 \times 10^{-8}$ cm^{-1}, with a Coulomb logarithm of $\ln \Lambda \approx 17$. The free-free opacity is then $\tau_{ff} \approx \alpha_{ff} w \approx 150 \gg 1$, so the radio emission is optically thick, and thus the radio brightness temperature is expected to be the same as the electron temperature, i.e., $T_B = T_e \approx 11$ MK.

Solution 15.4: The radio brightness temperature of the footpoints dominates over looptop emission for frequencies of $\nu \gtrsim 7$ GHz, according to the spectra shown in Fig. 15.4 (top right). The simulated radio maps (Fig. 15.4 bottom row) show indeed two separate footpoint sources for $\nu = 7.59, ..., 20.00$ GHz, while the footpoint sources cannot be separated from the looptop sources for the lower frequencies $\nu = 2.00, ..., 5.27$ GHz.

Solution 15.5: The e-folding decay time of the 17 GHz time profile, measured at the steepest decay part, is about 1.5 tickmarks, so $\tau_e \approx 15$ s. Using an energy of $\varepsilon_R \approx 300$ keV for the radio-emitting electrons, we obtain with Eq. (14.2.2) the trap density $n_e \approx 10^{11}(0.95/\tau_e)(\varepsilon_R/100 \text{ keV })^{3/2} \approx 3 \times 10^{10}$ cm^{-3}. This indicates a high-density flare loop.

Solution 15.6: An electron beam with a positive slope is most conspicuously seen in the 8th curve in the left panel (shifted by $\Delta v_\parallel = 8 \times 2 \times 10^9$ cm s^{-1}, and in the first curve in the middle panel. The steepest slope in those two curves give a parallel velocity of $v_\parallel \approx 60,000 - 75,000$ km s^{-1}, or a relativistic speed of $\beta = v_\parallel/c = 0.20 - 0.25$, yielding a Lorentz factor of $\gamma = 1/\sqrt{1 - \beta^2} = 1.021 - 1.033$, or a kinetic energy of $\varepsilon = m_e c^2(\gamma - 1) \approx 11 - 17$ keV.

Solution 15.7: The leading edge in Fig. 15.3 shows a frequency-time drift rate of $\partial\nu/\partial t = (150 - 100)$ kHz/60 s = 0.8 kHz s^{-1}. The plasma frequency of $\nu_p = 125$ kHz corresponds to an electron density of $n_e \approx 2 \times 10^2$ cm^{-3}. According to the heliospheric density model by Erickson (1964) we find a distance of $R = R_\odot(n_e/7.2 \times 10^5 \text{ cm}^{-3})^{-1/2} \approx 60 R_\odot$. The local density scale height λ_R can be evaluated from the relation $n_e(R) = n_0 \exp\left[-(R-R_\odot)/\lambda_R\right] = n_0(R/R_\odot)^{-2}$, yielding $\lambda_R = (R-R_\odot)/2\ln(R/R_\odot) \approx 7.2\,R_\odot \approx 5 \times 10^6$ km. Using Eq. (15.4.10), the electron beam velocity can then be evaluated from $v_B \approx (\partial v/\partial t) 2\lambda_R/v \approx 64,000$ km s^{-1}, or $\beta = v/c \approx 0.21$.

Solution 15.8: The turnover frequency of the U-bursts in flare 1980-08-25 (Fig. 15.17 top right) is $\nu \approx 350$ MHz. For fundamental plasma emission this corresponds to an electron density of $n_e = (\nu_p/9000)^2 \approx 1.5 \times 10^9$ cm^{-3}.

Solution 15.9: Both wave-particle interactions are governed by the Doppler resonance condition $\delta(\omega - s\Omega/\gamma - k_\parallel v_\parallel)$, but positive wave growth occurs from free energy provided by the anisotropic (losscone) particle distribution in the case of electron cyclotron maser emission, while the energy transfer occurs in opposite direction for stochastic acceleration, because the particles gain energy from gyroresonant waves.

Solution 15.10: With Eq. (15.5.2) we find $\Delta\nu/\nu \approx L/\lambda_B = L(-\partial B/\partial l)/B$ in the case of gyroemission, or $\Delta\nu/\nu \approx L/2\lambda_T = L/2\lambda_1(T/1 \text{ MK})$ in the case of plasma emission, so the radio bandwidth is reciprocal the the magnetic field B or temperature T.

Chapter 16: Flare Plasma Dynamics

Solution 16.1: Eq. (16.1.1) yields $v_c = v_{Te}(n_e k_B T_e/\tau_{ce} \mathbf{j} \mathbf{E})^{1/4}$. Inserting Ohm's law, $\mathbf{j} = \sigma \mathbf{E}$, the electrical conductivity, $\sigma = n_e e^2 \tau_{ce}/m_e$ (Appendix D), and the electron thermal velocity, $v_{Te} = (k_B T_e/m_e)^{1/2}$ (Appendix D), yields $v_c = v_{Te}(m_e v_{Te}/e E \tau_{ce})^{1/2}$, i.e.. the critical velocity is given by the thermal speed, if the momentum ($m_e v_{Te}$) at the thermal speed equals the induced electric momentum ($eE\tau_{ce}$). Compare also with Eq. (4) in Holman (1985).

Solution 16.2: From Fig. 16.3 (right) we measure the following temperature changes:
$T_{\parallel H}(t)/T_{\parallel H}(t_0) \approx 5$, $T_{\perp H}(t)/T_{\perp H}(t_0) \approx 1.5$,
$T_{\parallel He}(t)/T_{\parallel He}(t_0) \approx 15$, $T_{\perp He}(t)/T_{\perp He}(t_0) \approx 2.5$,
$T_{\parallel O}(t)/T_{\parallel O}(t_0) \approx 15$, $T_{\perp O}(t)/T_{\perp O}(t_0) \approx 5$,
$T_{\parallel Fe}(t)/T_{\parallel Fe}(t_0) \approx 30$, $T_{\perp Fe}(t)/T_{\perp Fe}(t_0) \approx 5$.
If all ions had initially the same temperature T_i, we expect that the thermal ion velocities scale as $v_{Ti} = \sqrt{k_B T_i/\mu_i m_p}$, where the mean molecular weights are $\mu_H = 1$, $\mu_{He} = 4$, $\mu_O = 16$, and $\mu_{Fe} = 56$. Thus the relative changes in the widths of the parallel and perpendicular velocity distributions of the ions with respect to the initial hydrogen velocity are:
$v_{\parallel H}/v_{\parallel H} = \sqrt{5/(5 \times 1)} = 1.0$, $v_{\perp H}(t)/v_{\perp H} = \sqrt{1.5/(1.5 \times 1)} = 1.0$,
$v_{\parallel He}/v_{\parallel H} = \sqrt{15/(5 \times 4)} = 0.87$, $v_{\perp He}(t)/v_{\perp H} = \sqrt{2.5/(1.5 \times 4)} = 0.65$,
$v_{\parallel O}/v_{\parallel H} = \sqrt{15/(5 \times 16)} = 0.43$, $v_{\perp O}(t)/v_{\perp H} = \sqrt{5/(1.5 \times 16)} = 0.46$,
$v_{\parallel Fe}/v_{\parallel H} = \sqrt{30/(5 \times 56)} = 0.33$, $v_{\perp Fe}(t)/v_{\perp H} = \sqrt{5/(1.5 \times 56)} = 0.24$.

The expected changes in the widths of the velocity distributions agree approximately with the diagrams in Fig. 16.3 (left).

Solution 16.3: Inserting the heating rate $Q \approx (B^2 \mathrm{v}_A / 4\pi L)$ (Eq. 10.5.3) and Alfvén velocity $\mathrm{v}_A = B/\sqrt{4\pi\rho}$ into the apex temperature $T_A \approx (2QL^2/\kappa_0)^{2/7}$ (Eq. 10.5.2) yields the scaling law $T_{top} = (B^3 L/2\pi\kappa_0\sqrt{4\pi\rho})^{2/7}$ given in Eq. (16.2.1).

Solution 16.4: Hα emission originates deep in the chromosphere, ranging from the photosphere in the line wing to the upper chromosphere in the line core, i.e., in altitudes of $h \approx 300 - 1600$ km (see Fig. 16.6). White light originates deep in the chromosphere by hydrogen recombination, while the energy transport to lower altitudes in the chromosphere is accomplished by photo-ionization (radiative back-warming). Similarly, UV continuum emission results from radiative backwarming deep in the chromosphere, in the temperature minimum region ($h \approx 500 - 800$ km) based on good correlations between C II or C IV and Si I fluxes (see also Fig. 16.6).

Solution 16.5: The arrows in Fig. 16.11 mark the rest wavelength of the Ca IX line at $\lambda_0 = 3.1760$ Å, and the blueshifted line at $\lambda_1 = 3.1738$ Å. Thus the Doppler velocity is $\mathrm{v} = c(\Delta\lambda/\lambda) = c(\lambda_1 - \lambda_0)/\lambda_0 = -210$ km s^{-1}, as quoted in the Figure caption of Fig. 16.11. Errors can be due to the accuracy of the line fit and due to the value of the rest wavelengths, which are hard to measure in the laboratory at these wavelengths.

Solution 16.6: Based on the average velocity of $\mathrm{v} \approx 270$ km s^{-1} and plasma frequency $\nu \approx 3.0$ GHz (Table 16.2), a coronal scale height of $\lambda_T \approx 47,300 \times 1.5$ km and a propagation angle of $\theta \approx 0°$ we expect a mean drift rate of $\partial\nu/\partial t = -\nu\mathrm{v}\cos\theta/\lambda_T \approx 17$ MHz s^{-1}, according to Eq. (16.3.2). Applied to Fig. 16.15, we obtain for $\nu = 2$ GHz, $\Delta\nu = (\nu_2 - \nu_1) = 2000 - 1000 = 1000$ MHz, $\Delta t \approx 100$ s, and $\theta \approx 0$ a drift of $\partial\nu/\partial t \approx 11$ MHz s^{-1}, which approximately agrees with the observed value of $\Delta\nu/\Delta t \approx 10$ MHz s^{-1}.

Solution 16.7: Defining the conductive cooling time by $d\varepsilon_{th}/dt = \varepsilon_{th}/\tau_{cond} = -2\kappa T^{7/2}/7L^2$ (Eq. 16.4.2) and inserting $\varepsilon_{th} = 3n_e k_B T_e$ we obtain $\tau_{cond} = (21/2)n_e k_B L^2/\kappa T^{5/2}$, or $\tau_{cond} \approx 500 \times (n_e/10^{11} \text{ cm}^{-3})(L/10 \text{ Mm})^2(T_e/10$ MK$)^{-5/2}$ s. Defining the radiative cooling time by $d\varepsilon_{th}/dt = \varepsilon_{th}/\tau_{rad} = -n_e^2\Lambda_0 T_e^{-2/3}$ we obtain $\tau_{rad} \approx 22 \times (T/1 \text{ MK})^{5/3}(n_e/10^{11} \text{ cm}^{-3})^{-1}$ s. Note that the expressions in Eqs. (16.4.4) and (16.4.9) have different numerical factors because the change in enthalpy is considered rather than the change in thermal energy.

Solution 16.8: For $n_e = 10^{11}$ cm^{-3} and $L = 10$ Mm the conductive cooling time (derived in Problem 16.7) is $\tau_{cond} = 500 \ (T_e/10 \text{ MK})^{-5/2} = 157,500 \ (T_e/1$ MK$)^{-5/2}$ s, and the radiative cooling time is $\tau_{rad} = 22 \ (T_e/1 \text{ MK})^{5/3}$. The two time scales are equal at $T_e = 1$ MK $\times(157,500/22)^{6/25} = 8.4$ MK.

Solution 16.9: In the beginning and peak of the flare, when the flare temperature is high ($T_e \gtrsim 10$ MK), the conductive cooling is faster than radiative cooling, and thus conductive cooling dominates. In the late decay phase of the flare, when the flare loops cool down to EUV temperatures ($T_e \lesssim 2$ MK), radiative cooling dominates.

Solution 16.10 Based on a cooling time of $\tau_{cool} \approx 6$ min to cool down to $T_e \lesssim 2$ MK (Table 16.3), we expect a minimum number of subsequent flare loops $N_{min} > \Delta t_{flare}/\tau_{cool} \approx 5$. Thus, we see simultaneously $N_{sim} \approx N_{total}/N_{min} \approx (200/5) \approx 40$ flare loops.

Chapter 17: Coronal Mass Ejections

Solution 17.1: (1) The thermal blast model is driven by plasma heating (in a flare region), producing an overpressure that cannot be confined magnetically. (2) The dynamo model is driven by magnetic stressing, caused for instance by flux emergence, such as in the *emerging flux model* of Heyvaerts et al. (1977). (3) The mass loading model is driven by an unstable filament, caused either by mass drainage or unstable loading, subject to the Rayleigh-Taylor or Kruskal-Schwarzschild instability. (4) The tether release model can be enacted by the *loss-of-equilibrium model* of Forbes & Priest (1995), driven by converging flows towards the underlying neutral line. (5) The tether straining model is a more complex version with gradual increase of external strain, such as in the *magnetic breakout model* of Antiochos et al. (1999b).

Solution 17.2: The inner filament seems to be line-tied during the simulation, so the number of windings is conserved during the expansion. From Fig. 17.3 (panels b-d) the number of windings can be estimated to $N_{twist} \approx 2.5$, corresponding to a twist angle of $\varphi_{twist} \approx 2.5 \times 2\pi = 5\pi$, which is larger than all critical angles for kink instability quoted in §6.3.9: $\varphi_{twist} \gtrsim 3.3\pi$ (for force-free fields); $\varphi_{twist} \gtrsim 4.8\pi$ (Mikić et al. 1990).

Solution 17.3: The central helical filament has ≈ 4 windings in Fig. 17.5 and corresponds to the central filament with ≈ 2.5 windings in Fig. 17.3. This helical filament expands and becomes larger in Fig. 17.5, similar to the simulation in Fig. 17.3. The other features seen in the C IV 1600 Å images (Fig. 17.5) are probably of chromospheric origin and do not correspond to any of the coronal magnetic field lines shown in the simulation in Fig. 17.3.

Solution 17.4: The distances d of the CME front from Sun center can be measured in units of solar radii (i.e., half the diameter of the white circle that marks the solar disk in Fig. 17.6), yielding: (a) 18 $R_\odot$, (b) 20 $R_\odot$, (c) 20 $R_\odot$, (d) > 7.3 $R_\odot$, and (e) 6 $R_\odot$. Using the quoted CME speeds we obtain for the times $t = d/v_{CME}$: (a) 5.1 h, (b) 2.1 h, (c) 4.9 h, (d) 1.9 h, (e) <0.7 h. Picture (e) was taken closest to the launch.

Solution 17.5: The distances d and helical curvature radii r are approximately: (1) Fig. 17.7: $d \approx 3 R_\odot$, $r \approx 1.5 R_\odot$; (2) Fig. 17.8: $d \approx 4 R_\odot$, $r \approx 1.3 R_\odot$; (3) Fig. 17.9: $d \approx 3.7 R_\odot$, $r \approx 1.5 R_\odot$.

Solution 17.6: The distances are $r(t) = R_\odot \sqrt{n_0/n(t)} = 10R_\odot$, $100R_\odot$, and $1000R_\odot \approx 5$ AU (Jupiter orbit). The propagation times are $t = r(t)/v_{CME} = 2, 20$, and 200 h (≈ 8 days).

Solution 17.7: The area of the active region is $A_{AR} = R_\odot^2 \times \int_0^{\pi/4} dl \int_0^{\pi/4} \cos b \, db = 0.55R_\odot^2$, a fraction of 1/22 of the total solar surface. The mass of the active region is $m_{AR} = n_0 m_p A_{AR} \lambda_p \approx 2 \times 10^{16}$ g. So the fraction to be accelerated for a CME is $m_{CME}/m_{AR} \approx 0.5\%$ and 50%.

Solution 17.8: Assuming that their total energy is conserved, the total energy is the sum of gravitational potential, kinetic, and magnetic energy. While slow CMEs are magnetically driven, the magnetic energy can be neglected for fast CMEs. A CME reaches its terminal velocity far away from the Sun when the gravitational potential can be neglected, so that the kinetic energy essentially is equal to the total energy. The terminal velocity is thus $v_{CME} = \sqrt{2\varepsilon_{tot}/m_{CME}} \approx 450 - 1400$ km s^{-1}.

Solution 17.9: The maximum intensity decreases (EUV dimming) are approximately $I(t)/I_0 \approx 0.25$ in Fig. 17.17 and $I(t)/I_0 \approx 0.21$ in Fig. 17.19, corresponding to a density decrease of $n_e(t)/n_0 = \sqrt{I(t)/I_0} = 0.5$ and 0.46. The expansion factors of the CMEs are then $l(t)/l_0 \propto [V(t)/V_0]^{1/3} \propto [n_e(t)/n_0]^{-1/3} \approx 1.3$.

Solution 17.10: Ulysses is near the ecliptic ($\lesssim 10°$) from 1991 to June 1992 and thus sees the equatorial slow solar wind (v $\approx 400 - 500$ km s^{-1}). From June 1992 to June 1993 Ulysses moves from latitude $\approx 10°$ to $\approx 40°$ and sees both slow and fast solar wind speeds, modulated by the solar rotation rate. From June 1993 until December 1995, Ulysses moves through the high southern latitudes ($\gtrsim 40°$) and sees only the fast solar wind component (v ≈ 800 km s^{-1}), reaching perihelion in January 1995, where it sees slow solar wind for a short while, followed by the passage through high northern latitudes during February 1995 until July 1996 (fast solar wind), and moves towards lower latitudes ($\lesssim 30°$) afterwards, gradually picking up more of the slow solar wind component.

– Quod Erat Demonstrandum –

Appendices

Appendix A: Physical Constants

Physical quantity	Symbol	Value	cgs units
Speed of light in vacuum	c	$= 2.9979 \times 10^{10}$	cm s^{-1}
Elementary charge	e	$= 4.8023 \times 10^{-10}$	statcoulomb
Electron mass	m_e	$= 9.1094 \times 10^{-28}$	g
Proton mass	m_p	$= 1.6726 \times 10^{-24}$	g
Proton/electron mass ratio	m_p/m_e	$= 1.8361 \times 10^3$	
Gravitational constant	G	$= 6.6720 \times 10^{-8}$	dyne cm^2 g^{-2}
Boltzmann constant	k_B	$= 1.3807 \times 10^{-16}$	erg K^{-1}
Planck constant	h	$= 6.6261 \times 10^{-27}$	erg s
Rydberg constant	$R_H = me^4/4\pi\hbar^3 c$	$= 1.0974 \times 10^5$	cm^{-1}
Bohr radius	$a_0 = \hbar^2/m_e e^2$	$= 5.2918 \times 10^{-9}$	cm
Electron radius	$r_e = e^2/m_e c^2$	$= 2.8179 \times 10^{-13}$	cm
Stefan−Boltzmann constant	$\sigma = 2\pi^5 k_B^4/(15c^2 h^3)$	$= 5.6774 \times 10^{-5}$	erg cm^{-2} s^{-1} K^{-4}
1 electron Volt	ε_{eV}	$= 1.6022 \times 10^{-12}$	erg
	T_{eV}	$= 1.1604 \times 10^4$	K
	λ_{eV}	$= 1.2398 \times 10^{-4}$	cm
	ν_{eV}	$= 2.4180 \times 10^{14}$	Hz
1 Ångstrøm	(Å)	$= 10^{-8}$	cm
1 Jansky	(Jy)	$= 10^{-23}$	erg s^{-1} cm^{-2} Hz^{-1}
1 Solar flux unit	(SFU)	$= 10^{-19}$	erg s^{-1} cm^{-2} Hz^{-1}
1 Astronomical unit	(AU)	$= 1.50 \times 10^{13}$	cm
Solar radius	$R_\odot$	$= 6.96 \times 10^{10}$	cm
Solar mass	$M_\odot$	$= 1.99 \times 10^{33}$	g
Solar gravitation	$g_\odot = GM_\odot/R_\odot^2$	$= 2.74 \times 10^4$	cm s^{-2}
Solar escape speed	v_∞	$= 6.18 \times 10^7$	cm s^{-1}
Solar age	$t_\odot$	$= 4.60 \times 10^9$	years
Solar radiant power	$L_\odot$	$= 3.90 \times 10^{33}$	erg s^{-1}
Solar radiant flux density	$F_\odot$	$= 6.41 \times 10^{10}$	erg cm^{-2} s^{-1}
Solar constant (flux at 1 AU)	$f_\odot$	$= 1.39 \times 10^6$	erg cm^{-2}
Solar solid angle (at 1 AU)	$\Omega_\odot = \pi R_\odot^2/AU^2$	$= 6.76 \times 10^{-5}$	ster
Photospheric temperature	T_{phot}	$= 5762$	K

Appendix B: Conversion of Physical Units

Physical quantity	Gaussian units [cgs]	Rationalized metric units [mks]
Length	1 cm	$= 10^{-2}$ m
Mass	1 g	$= 10^{-3}$ kg
Time	1 s	$= 1$ s
Force	1 dyne	$= 10^{-5}$ N (Newton)
Energy	1 erg	$= 10^{-7}$ J (Joule)
Power	1 erg s^{-1}	$= 10^{-7}$ W (Watt)
Charge	1 statcoulomb	$= \frac{1}{3} \cdot 10^{-9}$ C (Coulomb)
Electric field	1 statvolt cm^{-1}	$= 3 \cdot 10^4$ V m^{-1}
Current	1 statampere	$= \frac{1}{3} \cdot 10^{-9}$ A (Ampère)
Current density	1 statampere cm^{-2}	$= \frac{1}{3} \cdot 10^{-5}$ A m^{-2}
Electrical conductivity	1 s^{-1}	$= \frac{1}{9} \cdot 10^{-9}$ Siemens m^{-1}
Magnetic induction	1 G (Gauss)	$= 10^{-4}$ T (Tesla)
Magnetic field	1 Oersted	$= \frac{1}{4\pi} \cdot 10^3$ A m^{-1}

Appendix C: Maxwell's Equations in Different Physical Unit Systems

Gaussian units [cgs] (current in *emu* units))	Gaussian units [cgs] (current in *esu* units))	Rationalized metric units [mks]
$\nabla \cdot \mathbf{E} = 4\pi \rho_E$	$\nabla \cdot \mathbf{E} = 4\pi \rho_E$	$\nabla \cdot \mathbf{D} = \rho_E$
$\nabla \cdot \mathbf{B} = 0$	$\nabla \cdot \mathbf{B} = 0$	$\nabla \cdot \mathbf{B} = 0$
$\nabla \times \mathbf{E} = -\frac{1}{c}\frac{d\mathbf{B}}{dt}$	$\nabla \times \mathbf{E} = -\frac{1}{c}\frac{d\mathbf{B}}{dt}$	$\nabla \times \mathbf{E} = -\frac{d\mathbf{B}}{dt}$
$\nabla \times \mathbf{H} = \frac{1}{c}\frac{d\mathbf{D}}{dt} + 4\pi\mathbf{j}$	$\nabla \times \mathbf{H} = \frac{1}{c}\frac{d\mathbf{D}}{dt} + \left(\frac{4\pi}{c}\right)\mathbf{j}$	$\nabla \times \mathbf{H} = \frac{d\mathbf{D}}{dt} + \mathbf{j}$

– emu units: The current $\mathbf{j}$ is measured in *electromagnetic units*.
– esu units: The current $\mathbf{j}$ is measured in *electrostatic units*, the ratio of the values of $\mathbf{j}$ is $\mathbf{j}(esu)/\mathbf{j}(emu) = c \approx 3 \times 10^{10}$ cm s^{-1}.
– *Rationalized mks units* are SI units with the factor 4π removed from the equations.
– The electric displacement is $\mathbf{D} = \varepsilon\mathbf{E}$. In a plasma, $\varepsilon \approx 1$ in cgs units.
– The magnetic field is $\mathbf{H} = \mathbf{B}/\mu$. In a plasma, $\mu \approx 1$ in cgs units.

Appendix D: Plasma Parameters

Physical quantity	Definition	Numerical formula (cgs units)
Thermal pressure	$p_{th} = 2n_e k_B T_e$	$= 2.76 \times 10^{-16} \, n_e \, T$ (dyne cm^{-2})
Magnetic pressure	$p_m = B^2/(8\pi)$	$= 3.98 \times 10^{-2} \, B^2$ (dyne cm^{-2})
Plasma-β parameter	$\beta = (p_{th}/p_m)$	$= 6.94 \times 10^{-15} \, n_e T_e B^{-2}$
Thermal scale height	$\lambda_T = 2k_B T_e/(\mu_C m_p g_\odot)$	$= 4.73 \times 10^3 \, T_e$ (cm)
Electron thermal velocity	$v_{Te} = (k_B T_e/m_e)^{1/2}$	$= 3.89 \times 10^5 \, T_e^{1/2}$ (cm s^{-1})
Ion thermal velocity	$v_{Ti} = (k_B T_i/\mu m_p)^{1/2}$	$= 9.09 \times 10^3 \, (T_i/\mu)^{1/2}$ (cm s^{-1})
Ion mass density	$\rho = n_i m_i = n_i \mu m_p$	$= 1.67 \times 10^{-24} \, \mu \, n_i$ (g cm^{-3})
Sound speed	$c_S = (\gamma p_{th}/\rho)^{1/2}$	$= 1.66 \times 10^4 \, (T/\mu)^{1/2}$ (cm s^{-1})
Alfvén speed	$v_A = B/(4\pi \mu m_p n_i)^{1/2}$	$= 2.18 \times 10^{11} \, B \, (\mu n_i)^{-1/2}$ (cm s^{-1})
Electron plasma frequency	$f_{pe} = (n_e e^2/\pi m_e)^{1/2}$	$= 8.98 \times 10^3 \, n_e^{1/2}$ (Hz)
Ion plasma frequency	$f_{pi} = (n_i Z^2 e^2/\pi \mu m_p)^{1/2}$	$= 2.09 \times 10^2 \, Z(n_i/\mu)^{1/2}$ (Hz)
Electron gyrofrequency	$f_{ge} = eB/(2\pi m_e c)$	$= 2.80 \times 10^6 \, B$ (Hz)
Ion gyrofrequency	$f_{gi} = ZeB/(2\pi \mu m_p c)$	$= 1.52 \times 10^3 \, B/\mu$ (Hz)
Electron collision frequency	f_{ce}	$= 3.64 \times 10^0 n_e \ln \Lambda T_e^{-3/2}$ (Hz)
Ion collision frequency	f_{ci}	$= 5.98 \times 10^{-2} n_i \ln \Lambda Z^2 T_i^{-3/2}$ (Hz)
Electron collision time	$\tau_{ce} = 1/f_{ce}$	$= 2.75 \times 10^{-1} T_e^{3/2}/(n_e \ln \Lambda)$ (s)
Ion collision time	$\tau_{ci} = 1/f_{ci}$	$= 1.67 \times 10^1 T_i^{3/2}/(n_i \ln \Lambda Z^2)$ (s)
Electron gyroradius	$R_e = v_{Te}/(2\pi f_{ge})$	$= 2.21 \times 10^{-2} T_e^{1/2} B^{-1}$ (cm)
Ion gyroradius	$R_i = v_{Ti}/(2\pi f_{gi})$	$= 9.49 \times 10^{-1} T_i^{1/2} \mu^{1/2} Z^{-1} B^{-1}$ (cm)
Debye length	$\lambda_D = (k_B T_e/4\pi n_e e^2)^{1/2}$	$= 6.90 \times 10^0 T^{1/2} n_e^{-1/2}$ (cm)
Dreicer field	$E_D = Ze \ln \Lambda/\lambda_D^2$	$= 1.01 \times 10^{-11} Z \ln \Lambda n_e T_e^{-1}$ (statvolt cm^{-1})
Electrical conductivity	$\sigma = n_e e^2 \tau_{ce}/m_e$	$= 6.96 \times 10^7 \ln(\Lambda)^{-1} Z^{-1} T_e^{3/2}$ (Hz)
Magnetic diffusivity	$\eta = c^2/(4\pi\sigma)$	$= 1.03 \times 10^{12} \ln(\Lambda) Z \, T_e^{-3/2}$ (cm^2 s^{-1})
Magnetic Reynolds number	$R_m = lv/\eta$	$= 9.73 \times 10^{-13} \, l \, v \, T^{3/2} \ln \Lambda^{-1}$
Thermal Spitzer conductivity coeff.	κ	$= 9.2 \times 10^{-7}$ (erg s^{-1} cm^{-1} K$^{-7/2}$)
Thermal conductivity	$\kappa_\parallel = \kappa T^{5/2}$	$= 9.2 \times 10^{-7} T^{5/2}$ (erg s^{-1} cm^{-1} K^{-1})
Radiative loss rate	$\Lambda_0 (T \approx 1 \, \text{MK})$	$= 1.2 \times 10^{-22}$ (erg s^{-1} cm^3)
Coronal viscosity	ν_{visc}	$= 4.0 \times 10^{13}$ (cm^2 s^{-1})

– cgs units: length l (cm), mass m (g), time t (s), Temperature T (K), magnetic field B (G), densities n_i, n_e (cm^{-3}).

– Adiabatic index: $\gamma = c_p/c_v = (N+2)/N = 5/3 = 1.67$.

– Ion/proton mass ratio $\mu = m_i/m_p$: $\mu(H) = 1$, $\mu(He) = 4$, $\mu(Fe) = 56$.

– Mean molecular weight in corona (H:He=10:1): $\mu_C = (10*1 + 1*4)/11 = 1.27$

– Coronal approximation (full ionization): $n_i = n_e$.

– Coulomb logarithm: $\ln \Lambda = 23 - \ln(n_e^{1/2} T_e^{-3/2}) \approx 20$ for $T_e \lesssim 10$ eV.

– Charge state: proton $\mapsto Z = 1$, Fe IX $\mapsto Z = 8$.

Appendix E: Conversion Table of Electron Temperatures into Relativistic Parameters

Temp. T_e (MK)	Energy ε (keV)	Velocity $\beta = v/c$	Lorentz factor γ	Temperature T_e (MK)	Energy ε (keV)	Velocity $\beta = v/c$	Lorentz factor γ
0.1	0.009	0.006	1.000	100.0	8.618	0.181	1.017
0.2	0.017	0.008	1.000	200.0	17.235	0.253	1.034
0.3	0.026	0.010	1.000	300.0	25.853	0.307	1.051
0.4	0.034	0.012	1.000	400.0	34.470	0.350	1.067
0.5	0.043	0.013	1.000	500.0	43.088	0.387	1.084
0.6	0.052	0.014	1.000	600.0	51.705	0.419	1.101
0.7	0.060	0.015	1.000	700.0	60.323	0.447	1.118
0.8	0.069	0.016	1.000	800.0	68.940	0.473	1.135
0.9	0.078	0.017	1.000	900.0	77.558	0.496	1.152
1.0	0.086	0.018	1.000	1000.0	86.175	0.517	1.169
1.0	0.086	0.018	1.000	1000.0	86.175	0.517	1.169
2.0	0.172	0.026	1.000	2000.0	172.350	0.664	1.337
3.0	0.259	0.032	1.001	3000.0	258.526	0.748	1.506
4.0	0.345	0.037	1.001	4000.0	344.701	0.802	1.675
5.0	0.431	0.041	1.001	5000.0	430.876	0.840	1.843
6.0	0.517	0.045	1.001	6000.0	517.052	0.868	2.012
7.0	0.603	0.049	1.001	7000.0	603.227	0.889	2.180
8.0	0.689	0.052	1.001	8000.0	689.402	0.905	2.349
9.0	0.776	0.055	1.002	9000.0	775.577	0.918	2.518
10.0	0.862	0.058	1.002	10,000.0	861.753	0.928	2.686
20.0	1.724	0.082	1.003	20,000.0	1723.505	0.974	4.373
30.0	2.585	0.100	1.005	30,000.0	2585.258	0.986	6.059
40.0	3.447	0.116	1.007	40,000.0	3447.010	0.992	7.746
50.0	4.309	0.129	1.008	50,000.0	4308.763	0.994	9.432
60.0	5.171	0.141	1.010	60,000.0	5170.516	0.996	11.118
70.0	6.032	0.152	1.012	70,000.0	6032.268	0.997	12.805
80.0	6.894	0.163	1.013	80,000.0	6894.020	0.998	14.491
90.0	7.756	0.172	1.015	90,000.0	7755.773	0.998	16.178
100.0	8.618	0.181	1.017	100,000.0	8617.525	0.998	17.864

−The thermal energy of an electron is $\varepsilon_{th} = k_B T_e$ and set $\varepsilon_{th} = \varepsilon_{kin}$.
−The kinetic energy of a nonrelativistic electron is $\varepsilon_{kin} = \frac{1}{2} m_e v^2$.
−The kinetic energy of a relativistic electron is $\varepsilon_{kin} = m_e c^2 (\gamma - 1)$.
−The relativistic Lorentz factor is $\gamma = 1/\sqrt{(1 - \beta^2)}$.
−The relativistic velocity is $\beta = v/c$.
−The electron rest mass is $m_e c^2 = 511$ keV.

Appendix F: EUV Spectral Lines

The following list contains important transition region and coronal EUV lines (e.g.,
observed with SoHO/CDS, see Harrison et al. 1995, *Solar Physics* **162**: 233−290),
sorted according to increasing formation temperature.

Temperature T(MK)	log[T(K)]	Wavelengths λ (Å)	Line	Instrument
0.02	4.3	584.33, 537.03	He I	CDS
0.03	4.4	735.90, 743.70	Ne I	CDS
0.08	4.4	303.78	He II	CDS, EIT
0.08	4.9	702.98, 599.59	O III	CDS
0.08	4.9	489.50	Ne III	CDS
0.10	5.0	765.14	N IV	CDS
0.20	5.3	554.52	O IV	CDS
0.25	5.4	172.17, 629.73	O V	CDS
0.30	5.5	482.10	Ne V	CDS
0.40	5.6	399.83, 401.14	Ne VI	CDS
0.40	5.6	562.83, 558.59	Ne VI	CDS
0.40	5.6	399.20, 400.68	Mg VI	CDS
0.40	5.6	168.18, 186.60	Fe VIII	CDS
0.50	5.7	465.22	Ne VII	CDS
0.63	5.8	770.40, 780.30	Ne VIII	CDS
0.63	5.8	277.04, 278.40	Mg VII	CDS
0.63	5.8	272.60, 275.37	Si VII	CDS
0.63	5.8	557.76	Ca X	CDS
0.80	5.9	313.73, 317.01	Mg VIII	CDS
0.80	5.9	316.22, 319.83	Si VIII	CDS
1.00	6.0	171.07	Fe IX	CDS, EIT, TRACE
1.00	6.0	217.10	Fe IX	CDS
1.00	6.0	368.06, 705.80	Mg IX	CDS
1.00	6.0	296.12, 345.12	Si IX	CDS
1.00	6.0	261.06, 271.99	Si X	CDS
1.10	6.1	624.94	Mg X	CDS
1.10	6.1	341.94	Si IX	CDS
1.30	6.1	347.40, 356.01	Si X	CDS
1.30	6.1	174.53, 177.24	Fe X	CDS
1.30	6.1	358.62	Fe XI	CDS
1.30	6.1	180.40, 188.22	Fe XI	CDS
1.30	6.1	192.81	Fe XI	CDS
1.40	6.1	568.12, 550.03	Al XI	CDS
1.60	6.2	195.12, 193.51	Fe XII	CDS, EIT, TRACE
1.60	6.2	364.47, 346.85	Fe XII	CDS
1.60	6.2	203.79, 213.77	Fe XIII	CDS
1.60	6.2	320.80, 359.64	Fe XIII	CDS
1.60	6.2	348.18	Fe XIII	CDS
1.80	6.3	520.66	Si XII	CDS
2.00	6.3	211.32, 220.08	Fe XIV	CDS
2.00	6.3	334.17, 353.83	Fe XIV	CDS
2.00	6.3	284.16	Fe XV	CDS, EIT, TRACE
2.00	6.3	327.02	Fe XV	CDS
2.50	6.4	200.80	Fe XVI	CDS
2.50	6.4	335.40, 360.76	Fe XVI	CDS

Appendix G: Vector Identities

$\mathbf{a}, \mathbf{b}, \mathbf{c}, \mathbf{d}$ are vectors, $\psi(\mathbf{x})$ and $\phi(\mathbf{x})$ are scalar functions (e.g., a potential function), and $\nabla = (d/dx, d/dy, d/dz)$ the vectors of derivatives.

$$\begin{aligned}
\mathbf{a} \cdot \mathbf{b} &= \mathbf{b} \cdot \mathbf{a} \\
\mathbf{a} \times \mathbf{b} &= -\mathbf{b} \times \mathbf{a} \\
\mathbf{a} \cdot (\mathbf{b} \times \mathbf{c}) &= \mathbf{b} \cdot (\mathbf{c} \times \mathbf{a}) = \mathbf{c} \cdot (\mathbf{a} \times \mathbf{b}) \\
\mathbf{a} \times (\mathbf{b} \times \mathbf{c}) &= (\mathbf{a} \cdot \mathbf{c})\mathbf{b} - (\mathbf{a} \cdot \mathbf{b})\mathbf{c} \\
(\mathbf{a} \times \mathbf{b}) \cdot (\mathbf{c} \times \mathbf{d}) &= (\mathbf{a} \cdot \mathbf{c})(\mathbf{b} \cdot \mathbf{d}) - (\mathbf{a} \cdot \mathbf{d})(\mathbf{b} \cdot \mathbf{c}) \\
(\mathbf{a} \times \mathbf{b}) \times (\mathbf{c} \times \mathbf{d}) &= [(\mathbf{a} \times \mathbf{b}) \cdot \mathbf{d}]\mathbf{c} - [(\mathbf{a} \times \mathbf{b}) \cdot \mathbf{c}]\mathbf{d} \\
\nabla \cdot (\nabla \times \mathbf{a}) &= 0 \\
\nabla \times (\nabla \times \mathbf{a}) &= \nabla(\nabla \cdot \mathbf{a}) - \nabla^2 \mathbf{a} \\
\nabla(\mathbf{a} \cdot \mathbf{b}) &= (\mathbf{a} \cdot \nabla)\mathbf{b} + (\mathbf{b} \cdot \nabla)\mathbf{a} + \mathbf{a} \times (\nabla \times \mathbf{b}) + \mathbf{b} \times (\nabla \times \mathbf{a}) \\
\nabla(\mathbf{b} \cdot \mathbf{b})/2 &= (\mathbf{b} \cdot \nabla)\mathbf{b} + \mathbf{b} \times (\nabla \times \mathbf{b}) \\
\nabla \cdot (\mathbf{a} \times \mathbf{b}) &= \mathbf{b} \cdot (\nabla \times \mathbf{a}) - \mathbf{a}(\nabla \times \mathbf{b}) \\
\nabla \times (\mathbf{a} \times \mathbf{b}) &= \mathbf{a}(\nabla \cdot \mathbf{b}) - \mathbf{b}(\nabla \cdot \mathbf{a}) + (\mathbf{b} \cdot \nabla)\mathbf{a} - (\mathbf{a} \cdot \nabla)\mathbf{b} \\
\mathbf{a} \times (\nabla \times \mathbf{b}) &= (\nabla \mathbf{b}) \cdot \mathbf{a} - \mathbf{a} \cdot (\nabla \mathbf{b}) \\
\nabla(\psi\phi) &= \psi\nabla\phi + \phi\nabla\psi \\
\nabla \times \nabla\psi &= 0 \\
\nabla \cdot (\psi\mathbf{a}) &= \mathbf{a} \cdot \nabla\psi + \psi\nabla \cdot \mathbf{a} \\
\nabla \times (\psi\mathbf{a}) &= \nabla\psi \times \mathbf{a} + \psi\nabla \times \mathbf{a} \\
\nabla^2\psi &= \nabla \cdot (\nabla\psi) \\
\nabla^2\mathbf{a} &= \nabla(\nabla \cdot \mathbf{a}) - \nabla \times (\nabla \times \mathbf{a})
\end{aligned}$$

$\mathbf{x} = (x, y, z)$ is the coordinate of a point, with distance $r = |\mathbf{x}|$ from the origin of the coordinate system $(0, 0, 0)$, and $\mathbf{n} = \mathbf{x}/r$ is the radial unit vector.

$$\begin{aligned}
\nabla \cdot \mathbf{x} &= 3 \\
\nabla \times \mathbf{x} &= 0 \\
\nabla \cdot \mathbf{n} &= \tfrac{2}{r} \\
\nabla \times \mathbf{n} &= 0 \\
(\mathbf{a} \cdot \nabla)\mathbf{n} &= \tfrac{1}{r}[\mathbf{a} - \mathbf{n}(\mathbf{a} \cdot \mathbf{n})] = \tfrac{\mathbf{a}_\perp}{r}
\end{aligned}$$

$\mathbf{T}$ is a second-order tensor, with $(\mathbf{e}_1, \mathbf{e}_2, \mathbf{e}_3)$ the orthonormal unit vectors in the cartesian coordinate system,

$$\mathbf{T} = \sum_{i,j} T_{ij}\mathbf{e}_i\mathbf{e}_j$$

$$(\nabla \cdot \mathbf{T})_i = \sum_j (dT_{ji}/dx_j)$$

$$\nabla \cdot (\psi\mathbf{T}) = \nabla\psi \cdot \mathbf{T} + \psi\nabla \cdot \mathbf{T}$$

Appendix H: Integral Identities

$\phi(\mathbf{x})$ and $\psi(x)$ are scalar functions, $\mathbf{A}(\mathbf{x})$ is a vector function, V the 3D volume with volume element dx^3, S is the 2D surface that encloses the volume, with an area element da, and $\mathbf{n}$ is the outward pointing normal vector on surface element da.

Divergence theorem:

$$\int_V \nabla \cdot \mathbf{A} \, dx^3 = \int_S \mathbf{A} \cdot \mathbf{n} \, da$$

$$\int_V \nabla \psi \, dx^3 = \int_S \psi \mathbf{n} \, da$$

$$\int_V \nabla \times \mathbf{A} \, dx^3 = \int_S \mathbf{n} \times \mathbf{A} \, da$$

Green's first identity:

$$\int_V (\phi \nabla^2 \psi + \nabla \phi \cdot \nabla \psi) \, dx^3 = \int_S \phi \mathbf{n} \cdot \nabla \psi \, da$$

Green's theorem:

$$\int_V (\phi \nabla^2 \psi - \psi \nabla^2 \phi) \, dx^3 = \int_S (\phi \nabla \psi - \psi \nabla \phi) \cdot \mathbf{n} \, da$$

S is an open surface and C is the contour line at the boundary, with line element $d\mathbf{l}$. The normal $\mathbf{n}$ to S is defined by the right-hand side rule in relation to the sense of the line integral around C.

Stokes' theorem:

$$\int_S (\nabla \times \mathbf{A}) \cdot \mathbf{n} \, da = \int_C \mathbf{A} \cdot d\mathbf{l}$$

$$\int_S \mathbf{n} \times \nabla \psi \, da = \int_C \psi \, d\mathbf{l}$$

Appendix I: Components of Vector Operators

$(\mathbf{e}_1, \mathbf{e}_2, \mathbf{e}_3)$ are the orthogonal unit vectors and (A_1, A_2, A_3) the corresponding components of the vector $\mathbf{A}$. $\psi(\mathbf{x})$ is a scalar function (e.g., a potential field). $\nabla\psi$ is the gradient vector, $\nabla \cdot \mathbf{a}$ is the divergence, $\nabla \times \mathbf{a}$ the curl, and $\nabla^2\psi$ or $\Delta\psi$ the Laplacian operator.

Cartesian Coordinate System (x, y, z)

$$\nabla\psi = \mathbf{e}_1\frac{\partial\psi}{\partial x} + \mathbf{e}_2\frac{\partial\psi}{\partial y} + \mathbf{e}_3\frac{\partial\psi}{\partial z}$$

$$\nabla \cdot \mathbf{A} = \frac{\partial A_x}{\partial x} + \frac{\partial A_y}{\partial y} + \frac{\partial A_z}{\partial z}$$

$$\nabla \times \mathbf{A} = \mathbf{e}_1\left(\frac{\partial A_z}{\partial y} - \frac{\partial A_y}{\partial z}\right) + \mathbf{e}_2\left(\frac{\partial A_x}{\partial z} - \frac{\partial A_z}{\partial x}\right) + \mathbf{e}_3\left(\frac{\partial A_y}{\partial x} - \frac{\partial A_x}{\partial y}\right)$$

$$\nabla^2\psi = \frac{\partial^2\psi}{\partial x^2} + \frac{\partial^2\psi}{\partial y^2} + \frac{\partial^2\psi}{\partial z^2}$$

Cylindrical coordinates (r, φ, z)

$$\nabla\psi = \mathbf{e}_1\frac{\partial\psi}{\partial r} + \mathbf{e}_2\frac{1}{r}\frac{\partial\psi}{\partial\varphi} + \mathbf{e}_3\frac{\partial\psi}{\partial z}$$

$$\nabla \cdot \mathbf{A} = \frac{1}{r}\frac{\partial}{\partial r}(rA_r) + \frac{1}{r}\frac{\partial A_\varphi}{\partial\varphi} + \frac{\partial A_z}{\partial z}$$

$$\nabla \times \mathbf{A} = \mathbf{e}_1\left(\frac{1}{r}\frac{\partial A_z}{\partial\varphi} - \frac{\partial A_\varphi}{\partial z}\right) + \mathbf{e}_2\left(\frac{\partial A_r}{\partial z} - \frac{\partial A_z}{\partial r}\right) + \mathbf{e}_3\frac{1}{r}\left(\frac{\partial}{\partial r}(rA_\varphi) - \frac{\partial A_r}{\partial\varphi}\right)$$

$$\nabla^2\psi = \frac{1}{r}\frac{\partial}{\partial r}\left(r\frac{\partial\psi}{\partial r}\right) + \frac{1}{r^2}\frac{\partial^2\psi}{\partial\varphi^2} + \frac{\partial^2\psi}{\partial z^2}$$

Spherical coordinates (r, θ, φ)

$$\nabla\psi = \mathbf{e}_1\frac{\partial\psi}{\partial r} + \mathbf{e}_2\frac{1}{r}\frac{\partial\psi}{\partial\theta} + \mathbf{e}_3\frac{1}{r\sin\theta}\frac{\partial\psi}{\partial\varphi}$$

$$\nabla \cdot \mathbf{A} = \frac{1}{r^2}\frac{\partial}{\partial r}(r^2 A_r) + \frac{1}{r\sin\theta}\frac{\partial}{\partial\theta}(\sin\theta A_\theta) + \frac{1}{r\sin\theta}\frac{\partial A_\varphi}{\partial\varphi}$$

$$\nabla \times \mathbf{A} = \mathbf{e}_1\frac{1}{r\sin\theta}\left[\frac{\partial}{\partial\theta}(\sin\theta A_\varphi) - \frac{\partial A_\theta}{\partial\varphi}\right] + \mathbf{e}_2\left[\frac{1}{r\sin\theta}\frac{\partial A_r}{\partial\varphi} - \frac{1}{r}\frac{\partial}{\partial r}(rA_\varphi)\right]$$

$$+ \mathbf{e}_3\frac{1}{r}\left[\frac{\partial}{\partial r}(rA_\theta) - \frac{\partial A_r}{\partial\theta}\right]$$

$$\nabla^2\psi = \frac{1}{r^2}\frac{\partial}{\partial r}\left(r^2\frac{\partial\psi}{\partial r}\right) + \frac{1}{r^2\sin\theta}\frac{\partial}{\partial\theta}\left(\sin\theta\frac{\partial\psi}{\partial\theta}\right) + \frac{1}{r^2\sin^2\theta}\frac{\partial^2\psi}{\partial\varphi^2}$$

where

$$\frac{1}{r^2}\frac{\partial}{\partial r}\left(r^2\frac{\partial\psi}{\partial r}\right) = \frac{1}{r}\frac{\partial^2}{\partial r^2}(r\psi)$$

Notation

Physical units symbols (see also Appendix B)

A	Ampère, unit for electric current (SI)
Å	Ångstrøm = 10^{-8} cm (Appendix A)
AU	astronomical unit (Appendix A)
C	Coulomb, unit for electric charge (SI)
cm	centimeter, unit for length (cgs)
dyne	unit for force (cgs)
erg	unit for energy (cgs)
eV	electron Volt; keV, MeV, GeV
g	gram, unit for mass (cgs); kg (SI)
G	Gauss, unit for magnetic field (cgs); kG
J	Joule, unit for energy (SI)
Jy	Jansky (Appendix A)
Hz	Hertz = s^{-1}, unit for frequency (SI); kHz, MHz, GHz
K	Kelvin, unit for temperature (cgs, SI); MK
m	meter, unit for length (SI); μm, mm, cm, dm, km, Mm
N	Newton, unit for force (SI)
rad	radian, unit angle π
s	second, unit for time (cgs, SI)
ster	sterad, unit for solid angle (ster=rad^2)
SFU	solar flux unit (Appendix A)
T	Tesla, unit for magnetic field (SI)
V	Volt, unit for electric potential (SI)
W	Watt, unit for power (SI); kW, MW

Latin Symbols

A	magnetic vector potential function
A	area (cm^2)
A	oscillation amplitude (cm)
A_X	elemental abundance of element X, (e.g., A_H = hydrogen abundance)
A_{mn}	Einstein coefficient for spontaneous emission (§2.5)
a	acceleration (cm s^{-2})

a_0	Bohr atom radius (Appendix A)
B	magnetic field vector, magnetic induction (Appendix C)
B	magnetic field strength (G)
$B(x, y)$	Beta function
B_ν, B_λ	brightness function (radiation transfer, §2.2)
B_{mn}	Einstein coefficient for spontaneous emission (§2.5)
b	impact parameter (Rutherford formula, §2.3)
b	heliographic latitude (deg)
b_0	heliographic latitude of Sun center (deg)
C	contribution function (§2.8)
C	electric capacity (cm)
C_{mn}	collision coefficient (§2.7)
c	speed of light (Appendix A)
c_k	kink-mode speed (§7.2.1)
c_p	specific heat at constant pressure
c_s	sound speed (Appendix D)
c_T	MHD tube speed (Eq. 7.1.27)
c_v	specific heat at constant volume
D	electric displacement (Appendix C)
D	fractal dimension
D	diffusion constant ($cm^2\ s^{-1}$)
D_{ij}	diffusion tensor
d	distance (cm)
d	duration (s)
E	electric field vector
E	electric field strength (statvolt cm^{-1})
E_C	conductive loss rate (erg $cm^{-3}\ s^{-1}$)
E_D	Dreicer field (Appendix D)
E_H	heating rate (erg $cm^{-3}\ s^{-1}$)
E_R	radiative loss rate (erg $cm^{-3}\ s^{-1}$)
EM	emission measure $EM = n^2 z$ (cm^{-5})
EM_V	volumetric emission measure $EM_V = n^2 V$ (cm^{-3})
e	unit vector
e	elementary electric charge (Appendix A)
e	internal energy (Eq. 4.1.13)
F	photon flux (erg $s^{-1}\ cm^{-2}\ keV^{-1}$)
F_C	conductive flux (erg $cm^{-2}\ s^{-1}$)
$\mathbf{F}_{grav}$	gravitational force (dyne)
F_H	heating flux or Poynting flux (Eq. 9.1.6)
$\mathbf{F}_L$	Lorentz force (dyne)
$\mathbf{F}_{visc}$	viscosity force (Eq. 6.1.14)
$F_\odot$	solar radiant flux density (Appendix A)
f	frequency (Hz)
f	function
$f(\mathbf{v})$	particle distribution function
$f(p)$	particle momentum distribution function

f_{ce}	electron collision frequency (Appendix D)
f_{ci}	ion collision frequency (Appendix D)
f_{ge}	electron gyrofrequency, also ν_B, (Appendix D)
f_{gi}	ion gyrofrequency (Appendix D)
f_{pe}	electron plasma frequency, also ν_p, (Appendix D)
f_{pi}	ion plasma frequency (Appendix D)
$f_\odot$	solar constant (flux at 1 AU), (Appendix A)
G	gravitational constant (Appendix A)
$G(\mathbf{r}, \mathbf{r}')$	Green's function
$\mathbf{g}$	gravitational acceleration (cm s^{-2})
g	Gaunt factor (§2.3)
g_{eff}	effective gravity (§3.4)
g_n	atomic statistical weigth of atomic level ϵ_n (§2.5)
$g_\odot$	solar gravitation (Appendix A)
$\mathbf{H}$	magnetic field (Appendix C)
H	helicity (Eq. 5.5.11)
h	height above solar surface (cm)
h	heating rate per volume
h	Planck constant (Appendix A)
$\hbar$	Planck constant, $\hbar = h/2\pi$
h_D	dipole depth below solar surface (cm)
I	current (statampere)
$I_n(r)$	modified Bessel function
I_ν	intensity of radiation (erg s^{-1} cm^{-2} Hz^{-1} ster^{-1}) (radiation transfer, §2.1)
$\mathbf{j}$	current density vector
j	current density (statamp)
$j_{0,s}$	zeros of Bessel function (Eq. 7.3.1)
K	constant in stopping depth of thick-target bremsstrahlung (Eq. 13.6.1)
$\mathbf{k}$	wave vector (cm^{-1})
k_B	Boltzmann constant (Appendix A)
k_c	cutoff wave number (cm^{-1})
L	loop half length (cm)
L_B	magnetic scale height (cm)
L_{ij}	electric inductance (s^2 cm^{-1})
$L_\odot$	solar luminosity or radiant power (Appendix A)
l	length (cm)
l	heliographic latitude (deg)
l_0	heliographic latitude of Sun center (deg)
l_{TOF}	time-of-flight distance
M	Mach number
$M_\odot$	solar mass (Appendix A)
m	mass (g)
m_e	electron mass (Appendix A)
m_i	ion mass (e.g., m_H = hydrogen mass)
m_p	proton mass (Appendix A)
N	number of degrees of freedom

$N(\mathbf{k})$	photon number spectrum
$N_s(\varepsilon_0)$	stopping depth for electrons with energy ε_0
N_n	population number of atomic energy level ϵ_n (§2.7)
$N(X^{+m})$	number of atoms X in ionization state (+m)
$N(S)$	frequency or size distribution (log-N versus log-S)
$\mathbf{n}$	normal vector
n	power index
n	total density in plasma ($n = n_e + n_i$)
n	principal quantum-mechanical level
n_e	electron number density (cm^{-3})
n_H	hydrogen density
n_i	ion number density (cm^{-3})
n_n	atomic quantum number of energy level n (Rydberg formula, Eq. 2.4.3)
n_p	proton number density (cm^{-3})
n_ν	refractive index
P	power (erg s^{-1})
P	period (s)
P	probability
P	position angle of solar disk (rad, deg) (§3.4.4)
P_{ij}	pressure tensor (Eq. 6.1.11)
P_{mn}	transition probability from atomic level n to m
p	pressure (dyne cm^{-2})
p_{grav}	gravitational pressure (dyne cm^{-2})
p_{th}	thermal pressure (dyne cm^{-2}), (Appendix D)
p_m	magnetic pressure (dyne cm^{-2}), (Appendix D)
Q	heating rate
Q	electric charge
Q_r	radiation cross section (cm^2 erg Hz^{-1}), (§2.3)
$Q(T)$	temperature filter ratio
q	ratio
q	electric charge
q_{fill}	filling factor (Eq. 4.5.8)
q_p	pressure ratio
q_x	dimensionless ratio or correction factor
q_λ	scale height ratio (Eq. 3.6.24)
R	radial distance from Sun center (cm)
R	rate (s^{-1})
R	reflectivity coefficient
R	magnetic mirror ratio
R_e	electron gyroradius (Appendix D)
R_i	ion gyroradius (Appendix D)
R_m	magnetic Reynoldsnumber (Appendix D)
$R(T)$	instrumental response function (§3.8)
$R_\odot$	solar radius (cm), (Appendix A)
R_H	Rydberg constant (Appendix A)
r	radius or distance from center (cm)
r_e	electron radius (Appendix A)

r_{curv}	curvature radius of flare loop (i.e., height of semi-circular loop)
r_{loop}	cross-sectional radius of loop
S	surface (specifying a surface integral)
S	entropy per unit mass (Eq. 4.1.8)
S_ν	source function (radiation transfer, §2.1)
S_0	Lundquist (magnetic Reynolds number) (Eq. 10.1.8)
s	path distance along curve (cm)
s	harmonic number
s_H	heating scale length (cm)
T	temperature (K)
T_B	radio brightness temperature (K)
T_e	electron temperature (K)
T_i	ion temperature (K)
T_{max}	maximum temperature (K)
T_{phot}	photospheric temperature (Appendix A)
t	time (s)
t_{peak}	peak time of a time profile (s)
$t_\odot$	solar age (Appendix A)
U_ν	energy density (erg cm^{-3})
u	velocity (cm s^{-1})
u	Fourier component (in uv-space)
V	volume (cm^3)
V	circular polarization (Eq. 5.7.14)
v	Fourier component (in uv-space)
$\mathbf{v}, \mathrm{v}$	velocity (cm s^{-1})
v_A	Alfvén speed (Appendix D)
v_{gr}	group speed (cm s^{-1})
v_{ph}	phase speed (cm s^{-1})
v_{Te}	electron thermal velocity (Appendix D)
v_{Ti}	ion thermal velocity (Appendix D)
v_∞	solar escape speed (Appendix A)
W	total radiated power per frequency interval (Eq. 2.3.3)
W	nonpotential magnetic energy (Eq. 9.3.1, 9.8.1)
w	width (cm)
$w^\sigma(\mathbf{p}, \mathbf{k}, s)$	transition probability (Eq. 11.4.5)
X^{+m}	atom X in ionization state (+m)
X	cartesian coordinate in loop plane system
x	cartesian coordinate (in east–west direction)
Y	cartesian coordinate in loop plane system
y	cartesian coordinate (in south–north direction)
Z	Atomic charge number, charge state
Z	cartesian coordinate in loop plane system
z	cartesian coordinate (along line-of-sight)
z	normalized loop coordinate $z = (s - s_0)/(L - s_0)$ (Eq. 3.7.4)
z_{eq}	equivalent width (along line-of-sight)

Greek Symbols

α	particle pitch angle (between velocity and magnetic field)
α	azimuth angle (rad, deg)
α	powerlaw index
α	force-free parameter (Eq. 5.3.18)
α_{Fe}	iron abundance ratio (Eq. 4.5.6)
α_{FIP}	first ionization potential factor (Eq. 2.9.1)
α_ν	absorption coefficient (radiation transfer, §2.1)
$\alpha(\mathbf{r})$	Clebsch variable
β	relativistic velocity $\beta = v/c$
β	plasma-β parameter (Appendix D)
$\beta(\mathbf{r})$	Clebsch variable
Γ	loop cross section expansion factor
γ	relativistic Lorentz factor $\gamma = 1/\sqrt{1 - \beta^2}$
γ	adiabatic index $\gamma = c_p/c_v$
γ	powerlaw index of photon spectrum
Δ	Laplace operator
Δ	current sheet length (cm)
δ	powerlaw index of electron spectrum
δ	current sheet width (cm)
ϵ	photon energy $\epsilon = h\nu$ (keV)
ϵ_x	hard X-ray photon energy $\epsilon = h\nu_x$
ϵ_ν	emission coefficient (erg s^{-1} cm^{-3} Hz^{-1} rad^{-2}), (Eq. 2.3.11)
ε	kinetic energy of particle $\varepsilon = mc^2(\gamma - 1)$, also ε_{kin}, (erg)
ε_{enth}	enthalpy (Eq. 4.1.14), (erg)
ε_{grav}	gravitational energy (erg)
ε_I	ionization energy (eV)
ε_{kin}	kinetic energy of particle, also ε, (erg)
ε_m	magnetic energy $\varepsilon_m = B^2/8\pi$, (erg)
ε_{th}	thermal energy $\varepsilon_{th} = k_B T$, (erg)
η	magnetic diffusivity (Appendix D)
θ	inclination angle (to vertical) in spherical coordinate (rad), (deg)
θ	shear angle (rad), (deg)
ϑ	scattering angle (rad), (deg)
κ	thermal (Spitzer) conductivity coefficient (Appendix D)
$\kappa_\parallel$	thermal conductivity $\kappa_\parallel = \kappa T^{5/2}$, (Appendix D)
$\Lambda(T)$	Coulomb logarithm (Appendix D)
Λ_0	radiative loss rate (Appendix D)
λ	wavelength (cm)
λ_D	Debye length (Appendix D)
λ_p	pressure scale height (cm)
λ_n	density scale height (cm)
λ_T	thermal (or temperature) scale height (cm), (Appendix D)
μ	magnetic moment (Eq. 11.5.8)
μ	molecular weight $\mu = m_i/m_p$
μ_C	molecular weight of corona (Appendix D)

ν	frequency (s^{-1}=Hz)
ν	growth rate (s^{-1}=Hz), (Eq. 4.3.17)
ν_B	electron gyrofrequency, also f_{ge}, (Appendix D)
ν_p	electron plasma frequency, also f_{pe} (Appendix D)
ν_{visc}	coronal viscosity (Appendix D)
ξ	ionization correction of pressure (Eq. 1.8.1)
$\xi(T)$	nonthermal velocity (Eq. 9.3.12)
ρ	mass density, $\rho = n\, m$ (Appendix D)
ρ_E	electric charge density (Appendix C)
σ	electrical conductivity (Appendix D)
σ	Stefan–Boltzmann radiation constant (Appendix A)
σ_s	scattering cross section (Eq. 2.3.4)
σ_g	Gaussian width
σ_T	Gaussian temperature width (K)
τ	time scale or time delay (s)
τ_A	Alfvénic transit time (Appendix D)
τ_{ce}	electron collision time (Appendix D)
τ_{ci}	ion collision time (Appendix D)
τ_{coal}	coalescence time scale (dissipation of magnetic islands), (s)
τ_{cool}	cooling time (s)
τ_{cond}	conductive cooling time (s), (Eq. 4.3.10)
τ_{damp}, τ_D	damping time scale (loop oscillations), (s)
τ_{defl}	collisional deflection time (s)
τ_{ff}	free-free opacity
τ_{filter}	Fourier filter time scale (s)
τ_g	Gaussian width of time pulse
τ_{loss}	loss time scale (s)
τ_{prop}	propagation time (s)
τ_{rad}	radiative cooling time (s), (Eq. 4.3.8)
τ_{tear}	tearing mode time scale (s)
τ_w	width of time pulses (s)
τ_x	time scale of hard X-ray pulses (s)
τ_ν	optical depth (radiation transfer, §2.1)
Φ	magnetic flux (Mx = G cm^2)
$\phi(\mathbf{r})$	potential field (scalar) function (§5)
φ	azimuthal angle in cylindrical or spherical coordinates
χ	radiative loss constant (Eq. 4.3.3)
χ_n	excitation energy of atomic level ϵ_n
$\Psi(\mathbf{r})$	potential field (scalar) function (Appendix I)
Ψ_g	gravitational potential
ψ	angle between line-of-sight and magnetic field line
Ω	gyrofrequency, also f_{ge}, (Hz)
Ω_S	solid angle of source S
$\Omega_\odot$	solid angle of Sun at a distance of 1 AU
Ω_{mn}	collision strength (§2.7)
ω	circular frequency $\omega = 2\pi\nu$

Acronyms

1D, 2D, 3D	one, two, three-dimensional
AC	alternating current
ACRIM	Active Cavity Radiometer Irradiance Monitor (on SMM)
AR	active region
BATSE	Burst and Transient Source Experiment (on CGRO)
BBSO	Big Bear Solar Observatory (in California)
BC	before Christ (before year 0000)
BCS	Bent Crystal Spectrometer (on SMM)
BCS	Bragg Crystal Spectrometer (on Yohkoh)
CCD	Charge Coupled Device (camera)
CDS	Coronal Diagnostic Spectrometer (on SoHO)
CELIAS	Charge, ELement and Isotope Analysis (on SoHO)
CME	coronal mass ejection
COMPTEL	COMPton TELescope (on CGRO)
COMSTOC	COronal Magnetic Structures Observing Campaign
COSTEP	COmprehensive SupraThermal and Energetic Particle analyser (on SoHO)
CP	Coronagraph/Polarimeter (on SMM)
CGRO	Compton Gamma Ray Observatory (spacecraft)
DC	direct current
DEM	differential emission measure (distribution)
EGRET	Energetic Gamma Ray Experiment Telescope (on CGRO)
EIT	Extreme-ultraviolet Imaging Telescope (on SoHO)
ERNE	Energetic and Relativistic Nuclei and Electron experiment (on SoHO)
ESA	European Space Administration
EUV	extreme ultraviolet
EUVI	Extreme-UltraViolet Imager (on SECCHI/STEREO)
FAL	Fontenla−Avrett−Loeser (atmospheric model)
FASR	Frequency-Agile Solar Radiotelescope
FCS	Flat Crystal Spectrometer (on SMM)
FIP	first ionization potential
FITS	Flexible Image Transport System (data file format)
FWHM	full width half maximum
FOV	field-of-view
GOES	Geostationary Orbiting Earth Satellite (spacecraft)

GOLF	Global Oscillations at Low Frequency (on SoHO)
GSFC	Goddard Space Flight Center
GRS	Gamma-Ray Spectrometer (on SMM)
HPGe	Hyper-pure Germanium detector (on RHESSI)
HD	hydrodynamics
HS	hydrostatics
HXIS	Hard X-ray Imaging Spectrometer (on SMM)
HXRBS	Hard X-Ray Burst Spectrometer (on SMM)
HXT	Hard X-ray Telescope (on Yohkoh)
IDL	Interactive Data Language (software used by most solar physicists)
ISAS	Institute of Space and Astronautical Science (Japan)
KPNO	Kitt Peak National Observatory (in Arizona)
KSC	Kagoshima Space Center (in Japan)
LAD	Large Area Detectors (of BATSE instrument on CGRO)
LASCO	Large Angle Solar COronagraph (on SoHO)
MDI	Michelson Doppler Imager (on SoHO)
MHD	Magneto-Hydrodynamics
MSSTA	Multi-Spectral Solar Telescope Array (sounding rocket)
NASA	National Aeronautics and Space Administration
NRL	Naval Research Laboratory (in Washington DC)
NSO	National Solar Observatory (in USA)
OSO	Orbiting Solar Observatory (spacecraft)
OSSE	Oriented Scintillation Spectrometer Experiment (on CGRO)
OVRO	Owens Valley Radio Observatory (radiointerferometer in California)
RHESSI	Reuven Ramaty High Energy Solar Spectroscopic Imager (spacecraft)
RTV	Rosner−Tucker−Vaiana (coronal loop model)
SECCHI	Sun Earth Connection Coronal and Heliospheric Investigation (on STEREO)
SEP	solar energetic particle events
SERTS	Solar EUV Research Telescope and Spectrograph (sounding rocket)
SMM	Solar Maximum Mission (spacecraft)
SoHO	Solar and Heliospheric Observatory (spacecraft)
SOI	Solar Oscillations Investigations (instrument on MDI, SoHO)
SMEX	SMall EXplorer mission (NASA mission category)
SSW	Solar SoftWare (software package in IDL)
STEREO	Solar TErrestrial RElations Observatory (spacecraft)
SUMER	Solar Ultraviolet Measurements of Emitted Radiation (on SoHO)
SXT	Soft X-ray Telescope (on Yohkoh)
SWAN	Solar Wind ANisotropies (on SoHO)
TOF	Time-of-flight (difference)
TRACE	Transition Region And Coronal Explorer (spacecraft)
UV	ultraviolet
UVCS	UltraViolet Coronagraph Spectrometer (on SoHO)
UVSP	UltraViolet Spectrometer/Polarimeter (on SMM)
VIRGO	Variability of solar IRradiance and Gravity Oscillations (on SoHO)
VAL	Vernazza−Avrett−Loeser (atmospheric model)
VLA	Very Large Array (radiointerferometer in Socorro, New Mexico)
WBS	Wideband Spectrometer (on Yohkoh)
XUV	extreme ultraviolet

References

Proceedings List

Proc-1958-Lehnert: (1958) *Electromagnetic Phenomena in Cosmic Physics*, International Astronomical Union (IAU) Symposium 6, (ed. B. Lehnert), New York: Cambridge University Press.

Proc-1958-UN: (1958), United Nations, Vol. 31, Proc. 2nd Internat. Conf. on *Peaceful Uses of Atomic Energy*, United Nations, Geneva.

Proc-1964-Hess: (1964) National Aeronautics and Space Administration Special Publication 50, *The Physics of Solar Flares*, Proc. AAS-NASA Symposium, held at Goddard Space Flight Center, Greenbelt, Maryland, 1963 Oct 28-30, (ed. W.N. Hess), NASA Science and Technical Information Division, Washington DC.

Proc-1967-Shen: (1967) *High Energy Nuclear Reactions in Astrophysics*, Symposium, (ed. Shen, B.S.P.), University of Pennsylvania, Benjamin Inc., New York.

Proc-1968-Kiepenheuer: (1968), *Structure and Development of Solar Active Regions*, International Astronomical Union Symposium 35, held in Budapest, Hungary, 4-8 Sept 1967, (ed. K.O. Kiepenheuer), Reidel, Dordrecht, The Netherlands.

Proc-1969-DeJager: (1969), *Solar Flares and Space Research*, Proc. Symposium, 11th Plenary Meeting of the Committee on Space Research, held in Tokyo, Japan, May 9-11, 1968, (eds. C. de Jager and Z. Svestka), North−Holland Publication, Amsterdam.

Proc-1971-Labuhn: (1971) *New Techniques in Space Astronomy*, IAU Coll. 41, held in Munich, Aug. 10-14, 1970, (ed. F. Labuhn and R. Lust), Reidel, Dordrecht.

Proc-1971-Macris: (1971), *Physics of the Solar Corona*, Proc. NATO Advanced Study Institute, held at Cavouri−Vouliagmeni, Athens, Sept 6-17, 1970, (ed. C.J. Macris), Astrophysics and Space Science Library Vol. 27, Reidel, Dordrecht, The Netherlands.

Proc-1973-Ramaty: (1973), *High Energy Phenomena on the Sun*, NASA Special Publication 342, (eds. R. Ramaty & R.G. Stone), U.S. Govt. Printing Office, NASA/GSFC, Greenbelt, Maryland.

Proc-1974-Newkirk: (1974), *Coronal Disturbances*, Proc. International Astronomical Union Symposium 57, held in Surfers Paradise, Queensland Australia, 1973 Sept 7-11, (ed. G.A. Newkirk), Reidel, Dordrecht, The Netherlands.

Proc-1974-Leningrad: (1974), Proc. *6th Leningrad International Seminar on Particle Acceleration and Particle Acceleration and Nuclear Reactions in the Cosmos*, USSR Academy of Sciences, A.F.Yoffe Physico−Technical Institute, Leningrad.

Proc-1974-Righini: (1974) Osservazioni e Memorie Osservatorio di Arcetri, Vol. 104, *Skylab Solar Workshop*, (ed. G. Righini), Osservatorio di Arcetri, Firenze,

Proc-1976-Bumba: (1976), *Basic Mechanisms of Solar Activity*, Proc. International Astronomical Union Symposium IAU 71, held in Prague, Czechoslovakia, 25-29 August 1975, (eds. Bumba, V. and Kleczek, J.), Reidel, Dordrecht.

Proc-1977-Shea: (1977) *Study of Travelling Interplanetary Phenomena*, (eds. Shea, M.A., Smart, D.F., and Wu, S.T.), Reidel Publishing Company, Dordrecht, The Netherlands.

Proc-1979-Akasofu: (1979), *Dynamics of the Magnetosphere*, (ed. Akasofu, S.I.), Astrophysics and Space Science Library, Vol. 78, Kluwer Academic Publishers, Norwell, Massachussetts.

Proc-1979-Arons: (1979), American Institute of Physics 56, *Particle Acceleration Mechanisms in Astrophysics*, Proc. Workshop, La Jolla, California, 1979 Jan 3-5, (eds. J. Arons, C. Max, & C. McKee), AIP Press, New York.

Proc-1979-ICRC: (1979), Proc. *16th Internat.Cosmic Ray Conference*, Vol. 2.

Proc-1979-Jensen: (1979), *Physics of Solar Prominences*, International Astronomical Union Symposium 44, held in Oslo, Norway, Aug 14-18, 1978, (eds. Jensen, E., Maltby, P., and Orrall, F.Q.), Institute of Theoretical Astrophysics, Blindern, Oslo.

Proc-1980-Kundu: (1980) *Solar Burst Observations at Centimeter Wavelengths*, Proc. IAU Symp. No. 86, College Park, Maryland, August 7-10, 1979, (eds. M.R. Kundu and T.E. Gergely), Reidel, Dordrecht, The Netherlands.

Proc-1980-Sturrock: (1980), *Solar Flares*, A monograph from SKYLAB Solar Workshop II, (ed. P.A. Sturrock), Colorado Associated University Press, Boulder, Colorado.

Proc-1981-Jordan: (1981), *The Sun as a Star*, Monograph series on nonthermal phenomena in stellar atmosphere, (ed. S.D. Jordan), NASA Washington, DC, NASA Special Publication NASA CP-450.

Proc-1981-Orrall: (1981), *Solar Active Regions*, A monograph from SKYLAB Solar Workshop III, (ed. Orrall, F.Q.), Colorado Associated University Press, Boulder, Colorado.

Proc-1981-Priest: (1981), *Solar Flare Magnetohydrodynamics*, (ed. Priest, E.R.), The Fluid Mechanics of Astrophysics and Geophysics. Volume 1, 574p., Gordon and Breach Science Publishers, New York.

Proc-1982-ESA200: (1984) *The Hydromagnetics of the Sun*, Proc. 4th European Meeting on Solar Physics, Noordwijkerhout, Netherlands, 1-3 Oct. 1984, (eds. Guyenne, T.D. and Hunt, J.J.), ESA, ESTEC Noordwijk, The Netherlands, Special Publication ESA SP-200.

Proc-1982-Lingenfelter: (1982) *Gamma-Ray Transients and Related Astrophysical Phenomena*, American Institute of Physics Conference Proceedings 77, Proc. Workshop, La Jolla, CA, August 5-8, 1981, (eds. R. Lingenfelter, H.S. Hudson, and D.M. Worral), AIP Press, New York.

Proc-1982-Tanaka: (1982) *Hinotori Symposium on Solar Flares*, held in Tokyo, Japan, 1982, (ed. Y. Tanaka), ISAS, Tokyo, Japan.

Proc-1983-ICRC18, (1983), International Cosmic Ray Conference, 18th, Bangalore, India, August 22-September 3, 1983, Conference Papers. Volume 4, Bombay, Tata Institute of Fundamental Research.

Proc-1984-Hagyard: (1984), *Measurements of Solar Vector Magnetic Fields*, (ed. M.J. Hagyard), NASA, Washington DC, Conference Publication NASA CP-2374.

Proc-1985-Buti: (1985), *Advances in Space Plasma Physics*, (ed. B. Buti), World Scientific, Singapore.

Proc-1985-DeJager: (1985), *Solar Physics and Interplanetary Travelling Phenomena*, Proc. of Workshop, held in Kunming, China, 21-25 Nov 1983, (eds. C. de Jager.C. & Chen Biao), Science Press, Beijing, China.

Proc-1985-ESA: (1985), *ESA Future Missions in Solar, Heliospheric and Space Plasma Physics*, ESA, Paris, France.

Proc-1985-ICRC19: (1985), 19th International Cosmic Ray Conference, La Jolla, California, NASA CP-2376, Vol. 4.

Proc-1985-Kundu: (1985), *Unstable Current Systems and Plasma Instabilities in Astrophysics*, Proc. International Astronomical Union Symposium IAU 107, held at University of Maryland, College Park, August 8-11, 1983, (eds. Kundu, M.R. and Holman, G.D.), Reidel Publisher, Dordrecht, The Netherlands.

Proc-1985-Priest: (1985), *Solar System Magnetic Fields*, Summer School on Solar System Plasmas, Imperial College of Science and Technology, London, England, September 1984, Lectures, (ed. Priest, E.R.), Reidel, Dordrecht.

Proc-1985-Schmidt: (1985) Max Planch Institute Volume 212, *Theoretical Problems in High Resolution Solar Physics*, (ed. U.H. Schmidt), Max Planck−Institute, Garching/Munich.

Proc-1985-Tsurutani: (1985), *Collisionless Shocks in the Heliosphere: Reviews of Current Research*, (eds. Tsurutani, B.T. and Stone, R.), American Geophysical Union, Washington, DC, AGU Geophysical Monograph Vol. 35.

Proc-1986-ESA251: (1986) European Space Agency Special Publication Vol. 251, *Plasma Astrophysics*, Proc. Joint Varenna−Abastumani School and Workshop, held in Sukhumi, USSR, 19-28 May 1986, ESA-SP 251, ESA, ESTEC, Nordwjik, The Netherlands.

Proc-1986-Kundu: (1986) *Energetic Phenomena on the Sun*, Proc. Solar Maximum Mission Flare Workshops, held at NASA/GSFC, Greenbelt, Maryland, 1983 Jan 24-28, 1983 June 9-14, and 1984 Febr 13-17, (eds. Kundu, M.R., Woodgate, B. and Schmahl, E.J.), NASA, Washington DC, Conference Publication NASA CP-2439.

Proc-1986-Mihalas: (1986), *Radiation Hydrodynamics in Stars and Compact Objects*, (eds. Mihalas, D. and Winkler, K.H.).

Proc-1986-Neidig: (1986), *The Lower Atmosphere in Solar Flares*, Proc. NSO/SMM Summer Symposium, 20-24 August 1985, (ed. Neidig, D.F.), NSO, Sacramento Peak, Sunspot, New Mexico.

Proc-1986-Poland: (1986), *Coronal and Prominence Plasmas*, Proc. Workshops held at NASA GSFC, Greenbelt, Maryland, 1985 April 9-11, 1986 April 8-10, (ed. A.I. Poland), NASA, Washington DC, Conference Publication NASA CP-2442.

Proc-1986-Sturrock: (1986), *The Physics of the Sun II*, (eds. P.A. Sturrock, T.E. Holzer, D. Mihalas, and R.K. Ulrich), Reidel, Dordrecht.

Proc-1986-Swings: (1986), *Highlights of Astronomy*, Vol. 7, (ed. J.P. Swings), Kluwer Academic Publishers, Dordrecht, The Netherlands.

Proc-1987-Athay: (1987) *Theoretical Problems in High Resolution Solar Physics II*, Proc. Workshop, held in Boulder, Colorado, 1986 Sept 15-17, (ed. G. Athay & D.S. Spicer), NASA, Washington DC, Conference Publication NASA CP-2483.

Proc-1987-Dennis: (1987) *Rapid Fluctuations in Solar Flares*, Proc. Workshop, Lanham, Maryland, Sept 30-Oct 4, 1985, (eds. Dennis, B.R., Orwig, L.E. and Kiplinger, A.L.), NASA, Washington DC, Conference Publication NASA CP-2449.

Proc-1988-Ballester: (1988), *Dynamics and Structure of Solar Prominences*, (eds. Ballester, J.L. & Priest, E.R.), Secr. Publ. Interc. Cient., Universitat de les Illes Balears, Mallorca, Spain.

Proc-1988-ESA285: (1988) European Space Agency Special Publication Vol. 285, *Reconnection in Space Plasma*, Proc. Internat. School and Workshop on Reconnection in Space Plasma, ESA Special Publication Vol. 285/II, ESA, ESTEC Noordwijk, The Netherlands.

Proc-1988-Pizzo: (1988) Proc. 6th International Solar Wind Conference, TN-306, (eds. Pizzo, V.J., Holzer, T.E., and Sime, G.D.), National Center for Atmsopheric Research, Boulder, Colorado.

Proc-1989-Johnson: (1989), Proc. CGRO Science Workshop, (ed. W.N. Johnson), NASA Document 4-366-4-373, GSFC, Greenbelt, Maryland.

Proc-1989-Priest: (1989), *Dynamics and Structure of Quiescent Solar Prominences*, Proc. Workshop, Palma de Mallorca, Spain, Nov. 1987, (ed. Priest, E.R.), Astrophysics and Space Science Library Vol. 150, Kluwer Academic Publishers, Dordrecht.

Proc-1989-Waite: (1989), Proc. 1988 Yosemite Conf. on Outstanding Problems in *Solar System Plasma Physics: Theory and Instrumentation*, (eds. J.H. Waite, J.L. Birch, and R.L. Moore), American Geophysical Union Monograph Vol. 54.

Proc-1990-Ruzdjak: (1990), *Dynamics of Quiescent Prominences*, Proc. 117th Colloquium of IAU, held at Hvar, Yugoslavia, Sept. 25-29, 1989, (eds. V. Ruzdjak & E. Tandberg−Hanssen), Springer, Berlin,

Proc-1990-Russell: (1990), *Physics of Magnetic Flux Ropes*, Geophysics Monograph Series Vol.58, (eds. Russell, C.T., Priest, E.R. and Lee, L.C.), AGU Washington DC.

Proc-1990-Winglee: (1990), *Solar Flares: Observations and Theory*, Proc. MAX'91 Workshop #3, Estes Park, Colorado, June 3-7, 1990, (eds. Winglee, R.M. and Kiplinger, A.L.), University of Colorado, Boulder, Colorado. Lecture Notes of Physics, Vol. 363.

Proc-1991-ICRC22: (1991), Proc of 22nd Internat. Cosmic Ray Conference, Dublin, Aug 1991.

Proc-1991-Culhane: (1991), *The Physics of Solar Flares*, Proc. Royal Society Discussion Meeting, held on 13/14 March 1991, (eds. J.L. Culhane & C. Jordan), The Royal Society, London.

Proc-1991-Priest: (1991), *Advances in Solar System Magnetohydrodynamics*, (eds. E.R. Priest and A.W. Hood), Cambridge University Press, Cambridge.

Proc-1991-Schmieder: (1991), *Dynamics of Solar Flares*, Proc. Flares 22 Workshop, held in Chantilly, France, 1990 Oct 16-19, (ed. B. Schmieder & E.R. Priest), Department of Astronomie Solaire of Observatoire the Paris (DASOP), Paris, France.

Proc-1991-Uchida: (1991), *Flare Physics in Solar Activity Maximum 22*, Proc. Internat. SOLAR-A Science Meeting Held at Tokyo, Japan, 23-26 October 1990, (eds. Y. Uchida, R.C. Canfield, T. Watanabe, & E. Hiei), Springer, Berlin, Lecture Notes in Physics, Vol. 387.

Proc-1991-Ulmschneider: (1991), *Mechanisms of Chromospheric and Coronal Heating*, Proc. Internat. Conf., held in Heidelberg, Germany, 5-8 June 1990, (eds. P. Ulmschneider, E.R. Priest, & R. Rosner), Springer, Berlin.

Proc-1991-Winglee: (1991), *MAX'91/SMM Solar Flares: Observations and Theory*, Proc. MAX 91 Workshop #3, held in Estes Park, Colorado, 1990 June 3-7, (eds. R.M. Winglee & A.L. Kiplinger), University of Colorado, Boulder, Colorado.

Proc-1992-EGRET: (1992), Technical Report N94-19462 04-89, EGRET Mission and Data Analysis 5 p, Max-Planck-Inst. fuer Physik und Astrophysik, Munich.

Proc-1992-ESA348: (1992) European Space Agency Special Publication Vol. 348, *Coronal Streamers, Coronal Loops, and Coronal and Solar Wind Composition*, Proc. 1st SoHO workshop, held in Annapolis, Maryland, 1992 Aug 25-28, ESA, ESTEC Noordwijk, The Netherlands.

Proc-1992-Shrader: (1992) *The Compton Observatory Science Workshop*, Proc. Workshop held in Annapolis, Maryland, Sept 23-25, 1991, (eds. Shrader, C.R., Gehrels, N., and Dennis, B.R.), NASA Conference Publication 3137, NASA, Washington DC.

Proc-1992-Svestka: (1992), *Eruptive Solar Flares*, Proc. of IAU Colloquium 133, held at Iguazu, Argentina, 1991 Aug 2-6 Aug, (eds. Z. Svestka, B.V. Jackson, & M.E. Machado), Lecture Notes in Physics, Vol. 399, Springer, Berlin.

Proc-1992-Zank: (1992), *Particle Acceleration in Cosmic Plasmas*, Proc. Conf., held in Newark, Delaware, 1991 Dec 4-6, (eds. G.P. Zank and T.K. Gaisser), American Institute of Physics Conference Proceedings Vol. 264, AIP Press, New York.

Proc-1993-ESA351: (1993) European Space Agency Special Publication Vol. 351, *Plasma Physics and Controlled Nuclear Fusion*, Proc. Conf. held in Tokyo, Japan, 17-20 Nov 1922, ESA, ESTEC Noordwijk, The Netherlands.

Proc-1993-Linsky: (1993), *Physics of Solar and Stellar Coronae*, G.S.Vaiana Memorial Symposium, held 22-26 June, 1992, in Palermo, Italy, (ed. J.L. Linsky and S. Serio), Astrophysics and Space Science Library Vol. 183, Kluwer Academic Publishers, Dordrecht, The Netherlands.

Proc-1993-Zirin: (1993) Astronomical Society of the Pacific Conference Series Vol. 46, *The Magnetic and Velocity Fields of Solar Active Regions*, (eds. H. Zirin, G. Ai, & H. Wang), ASP, San Francisco.

Proc-1994-Balasubramaniam: (1994) Astronomical Society of the Pacific Conference Series Vol. 68, *Solar Active Region Evolution: Comparing Models with Observations*, Proc. 14th Internat. Summer Workshop, National Solar Observatory, Sacramento Peak, Sunspot, New Mexico, USA, 1993 Aug 30-Sept 3, (eds. K.S. Balasubramaniam & G.W. Simon), ASP, San Francisco.

Proc-1994-Belvedere: (1994), *7th European Meeting on Solar Physics: Advances in Solar Physics*, held in Catania, Italy, 1993 May 11-15, (eds. Belvedere, G., Rodono, M., Schmieder, B., and Simnett, G.M.), Springer–Verlag, Berlin.

Proc-1994-Benz: (1994), *Plasma Astrophysics*, Saas–Fee Advanced Course 24, Lecture Notes 1994, Swiss Society for Astrophysics and Astronomy, (eds. A.O. Benz & T.J.-L. Courvoisier), Springer, Berlin.

Proc-1994-Enome: (1994) Nobeyama Radio Observatory Report No. 360, *New Look at the Sun with Emphasis on Advanced Observations of Coronal Dynamics and Flares*, Proc. Kofu Symposium, held in Kofu, 1993 Sept 6-10, (eds. S. Enome and T. Hirayama), NOAJ, Minamisaku, Nagano 384-13, Japan.

Proc-1994-ESA373: (1994) European Space Agency Special Publication Vol. 373, *Solar Dynamic Phenomena and Solar Wind Consequences*, Proc. 3rd SOHO Workshop, Estes Park, Colorado, USA, 26-29 September 1994, (eds. Hunt, J.J. and Domingo, V.), European Space Agency Special Publication ESA SP-373, ESA Paris.

Proc-1994-Fichtel: (1994), *Second Compton Symposium*, College Park, Maryland, (eds. C.E. Fichtel, N. Gehrels, and J.P. Norris), American Institute of Physics (AIP), New York, Vol. 304.

Proc-1994-Pap: (1994), *The Sun as a Variable Star: Solar and Stellar Irradiance Variations*, Proc. International Astronomical Union Symposium IAU 143, held in Boulder, Colorado, June 20-25, 1993, (eds. Pap, J.M., Froehlich, C., Hudson, H.S., & Solanki, S.K.), Solar Physics 152, Special Issue, Kluwer Academic Publishers, Dordrecht.

Proc-1994-Rutten: (1994), *Solar Surface Magnetism*, NATO Advanced Science Institutes (ASI), Series C: Mathematical and Physical Sciences, Proc. NATO Advanced Research Workshop, Vol. 431, held at Soesterberg, The Netherlands, Nov 1-5, 1993, (eds. Rutten, R.J. and Schrijver, C.J.), Kluwer Academic Publishers, Dordrecht.

Proc-1994-Ryan: (1994), *High-Energy Solar Phenomena – A New Era of Spacecraft Measurements*, Proc. Conference, held in Waterville Valley, New Hampshire, 1993 March, (eds. J.M. Ryan and W.T. Vestrand), American Institute of Physics, New York, Conference Proceedings AIP CP-294.

Proc-1995-Benz: (1995), *Coronal Magnetic Energy Releases*, Proc. CESRA Workshop, held in Caputh/Potsdam, Germany, 1994 May 16-20, (eds. A.O. Benz & A. Krüger), Springer, Berlin, Lecture Notes in Physics, Vol. 444.

Proc-1995-Ichimaru: (1995), *Elementary Processes in Dense Plasmas*, (eds. Ichimaru, S. & Ogata, S.), Addison Wesley Publishers.

Proc-1995-Wang: (1995) *3rd China–Japan Seminar on Solar Physics*, Proc. Workshop, Workshop, Dunhuang, China, Sept 1994, (eds. Wang, J.X., Ai, G.X., Sakurai, T., and Hirayama, T.), Internat. Academic Publishers, Beijing, China.

Proc-1995-Watanabe: (1995), Proc. *2nd SOLTIP Symp.*, (eds. Watanabe, Ta.), STEP GBRSC News (SOLTIP).

Proc-1996-Bentley: (1996) Astronomical Society of the Pacific Conference Series Vol. 111, *Magnetic Reconnection in the Solar Atmosphere*, Proc. Yohkoh Conference, held in Bath, England, 1996 March 20-22, (eds. R. Bentley and J. Mariska), ASP, San Francisco.

Proc-1996-Makino: (1996) *X-ray Imaging and Spectroscopy of Cosmic Hot Plasmas*, Internat. Symp. on X-ray Astronomy, (ed. F. Makino), Universal Academy Press, Tokyo.

Proc-1996-Ramaty: (1996) American Institute of Physics Conference Proceedings 374, *High Energy Solar Physics*, Proc. Conference, held in Greenbelt, Maryland, 1995 Aug 16-18, (eds. R. Ramaty, N. Mandzhavidze, & X.-M. Hua), AIP, Woodbury, New York.

Proc-1996-Tsinganos: (1996) NATO ASI Series C-481, *Solar and Astrophysical Magnetohydrodynamic Flows*, (ed. K.C. Tsinganos), Kluwer Academic Publishers, Dordrecht.

Proc-1996-Uchida: (1996), *Magnetohydrodynamic Phenomena in the Solar Atmosphere: Prototypes of Stellar Magnetic Activity*, Proc. International Astronomical Union Symposium 153, held in Makuhari, Tokyo, 1995 May 22-27, (eds. Y. Uchida, T. Kosugi, & H.S. Hudson), Kluwer Academic Publishers, Dordrecht, The Netherlands.

Proc-1996-Williams: (1996), *The Analysis of Emission Lines*, Proc. of the Space Telescope Science Inst. Symp., Baltimore/MD, May 16-18, 1994, (eds. R. Williams and M. Livio), ESO.

Proc-1996-Winterhalter: (1996), *Solar Wind Eight*, Internat. Solar Wind Conference, held in Dana Point, California, June 1995, (eds. D. Winterhalter, J.T. Gosling, S.R. Habbal, W.S. Kurth, and M. Neugebauer), AIP Press, New York, American Institute of Physics Conference Proceedings AIP CP-382.

Proc-1997-Crooker: (1997), American Geophysical Union monograph Vol. 99, *Coronal Mass Ejections: Causes and Consequences*, Proc. Chapman Conference, held in Bozeman, Montana, (eds. N. Crooker, J. Joselyn, & J. Feynman), AGU, Washington DC.

Proc-1997-Dermer: (1997), Proc. 4th Compton Symposium, (eds. C.D. Dermer, M.S. Strickman, and J.D. Kurfess), American Institute of Physics (AIP), New York.

Proc-1997-ESA404: (1997), European Space Agency Special Publication Vol. 404, *The Corona and Solar Wind near Minimum Activity*, Proc. 5th SOHO Workshop, held at Institute of Theoretical Astrophysics, University of Oslo, Norway, 17-20 June, 1997, (ed. Wilson, A.), ESA, ESTEC Noordwijk, The Netherlands.

Proc-1997-Jokipii: (1997), *Cosmic Winds and the Heliosphere*, (eds. J.R. Jokipii, C.P. Sonnett, and M.S. Gianpapa), University of Arizona Press.

Proc-1997-Mouradian: (1997), *Theoretical and Observational Problems Related to Solar Eclipses*, (eds. Z. Mouradian, & Stavinschi, M.), Kluwer Academic Publishers, Dordrecht, The Netherlands.

Proc-1997-Simnett: (1997), *Solar and Heliospheric Plasma Physics*, Proc. CESRA Workshop, held in Halkidiki, Greece, 1996 May 13-18, (eds. G.M. Simnett, C.E. Alissandrakis, & L. Vlahos), Springer, Berlin, Lecture Notes in Physics, Vol. 489.

Proc-1997-Trottet: (1997), *Coronal Physics from Radio and Space Observations*, Proc. CESRA Workshop, held in Nouan−le−Fuzelier, France, 1996 June 3-7, (ed. G. Trottet), Springer, Berlin, Lecture Notes in Physics, Vol. 483.

Proc-1998-Alissandrakis: (1998), *Three-Dimensional Structure of Solar Active Regions*, Proc. 2nd Advances in Solar Physics Euroconference, held in Preveza, Greece, 1997 Oct 7-11, (eds. C. Alissandrakis and B. Schmieder), Astronomical Society of the Pacific Conference Series Vol. 155, ASP, San Francisco.

Proc-1998-Dermer: (1998), *Proc. 4th Compton Symposium*, Proc. 4th Compton Symposium, American Institute of Physics Conference Proc. 410, (ed. C.D. Dermer, M.S. Strickman, and J.D. Kurfess), AIP, New York.

Proc-1998-Donahue: (1998), *Cool Stars, Stellar Systems and the Sun*, Proc. 10th Cambridge Workshop, (eds. Donahue, R.A. and Bookbinder, J.A.), Astronomical Society of the Pacific Conference Series Vol. 154, ASP, San Francisco.

Proc-1998-ESA421: (1998) European Space Agency Special Publication Vol. 421, *Solar Jets and Coronal Plumes*, Proc. Internat. Meeting, held in Guadeloupe, France, 1998 Febr 23-26, (ed. Guyenne, T.D.), ESA, ESTEC Noordwijk, The Netherlands.

Proc-1998-Watanabe: (1998), *Observational Plasma Astrophysics: Five Years of Yohkoh and Beyond*, Proc. Yohkoh Conference, held in Yoyogi, Tokyo, Japan, 1996 Nov 6-8, (eds. Watanabe, T., Kosugi, T., & Sterling, A.C.), Astrophysics and Space Science Library Vol. 229, Kluwer Academic Publishers, Dordrecht, The Netherlands.

Proc-1998-Webb: (1998) *New Perspectives on Solar Prominences*, Proc. IAU Colloquium 167, (eds. D. Webb, B. Schmieder, and D. Rust), Astronomical Society of the Pacific Conference Series Vol. 150, ASP, San Francisco.

Proc-1999-Bastian: (1999) Nobeyama Radio Observatory Report No. 479, *Solar Physics with Radio Observations*, Proc. Nobeyama Symposium, held in Kiyosato, Japan, 1998 Oct 27-30, (eds. T. Bastian, N. Gopalswamy, & K. Shibasaki), NOAJ, Tokyo, Japan.

Proc-1999-Brown: (1999) *Magnetic Helicity in Space and Laboratory Plasmas*, Proc. Coll. IAU 179, AGU Geophysics Monograph Series 111, (eds. Brown, M.R., Canfield, R.C. and Pevtsov, A.A.), AGU, Washington DC.

Proc-1999-ESA446: (1999) European Space Agency Special Publication Vol. 446, *Plasma Dynamics in the Solar Transition Region and Corona*, Proc. 8th SoHO WOrkshop, held in Paris, France, 22-25 June 1999, (eds. B. Kaldeich−Schurmann & J.C. Vial), ESA, ESTEC Noordwijk, The Netherlands.

Proc-1999-ESA448: (1999) European Space Agency Special Publication Vol. 448, *Magnetic Fields and Solar Processes*, Proc. 9th European Meeting on Solar Physics, held in Florence, Italy, 1999 Sept 12-18, (ed. A. Wilson), ESA, ESTEC Noordwijk, The Netherlands.

Proc-1999-Habbal: Solar Wind Nine, Proceedings of the Ninth International Solar Wind Conference, Nantucket, MA, October 1998. (Ed. Shaddia Rifai Habbal, Ruth Esser, Joseph V. Hollweg, and Philip A. Isenberg). AIP Conference Proceedings, Vol. 471.

Proc-1999-Ostrowski: *Plasma Turbulence and Energetic Particles in Astrophysics*, Proc. International Conference, Cracow (Poland), 5-10 September 1999, (eds. M. Ostrowski, R. Schlickeiser), Obserwatorium Astronomiczne, Uniwersytet Jagiellonski, Krakow.

Proc-1999-Rimmele: (1999), *High Resolutions Solar Physics: Theory, Observations, and Techniques*, Proc. Summer Workshop, National Solar Observatory, Sacramento Peak, Sunspot, New Mexico, USA, 1998, (eds. T.R. Rimmele, K.S. Balasubramaniam, & R.R. Radick), Astronomical Society of the Pacific Conference Series Vol. 183, ASP, San Francisco.

Proc-2000-Martens: (2000) International Astronomical Union Symposium 195: *Highly Energetic Physical Processes and Mechanisms for Emission from Astrophysical Plasmas*, Proc. Conference, held at Montana State University, Bozeman, Montana, 1999 July 6-10, (eds. P.C.H. Martens & S. Tsuneta), ASP, San Francisco.

Proc-2000-Ramaty: (2000), *High Energy Solar Physics − Anticipating HESSI*, Proc. HESSI Conference, held in College Park, Maryland, 1999 Oct 18-20, (eds. R. Ramaty & N. Mandhzhavidze), Astronomical Society of the Pacific (ASP) Conference Series Vol. 206, ASP, San Francisco.

Proc-2000-Rozelot: (2000), *Transport and Energy Conversion in the Heliosphere*, (eds. J.P. Rozelot, L. Klein, & J.C. Vial), Springer, Berlin, Lecture Notes in Physics, Vol. 553.

Proc-2001-Ballester: (2001), *MHD Waves in Astrophysical Plasmas*, Proc. INTAS Workshop, helt at Universitat de les Illes Balears, Palma de Mallorca, 2001 May 9-11, (eds. J.L.

Ballester & B. Roberts), Universitat de les Illes Balears.

Proc-2001-Balogh: (2001) *The Heliosphere Near Solar Minimum — The Ulysses perspective*, (eds. Balogh, A., Marsden, R.G. and Smith, E.J.), Springer—Praxis Books in Astrophysics and Astronomy.

Proc-2001-Daglis: (2001), *Space Storms and Space Weather Hazards*, NATO Science Series, II. Mathematics, Physics, and Chemistry Vo. 38, Proc. NATO Advanced Study Institute of Space Storms and Space Weather Hazards, held in Hersonissos, Crete, Greece, 19-29 June, 2000, (ed. I.A. Daglis), Kluwer Academic Publishers, Dordrecht.

Proc-2001-Lopez: (2001), *Cool Stars, Stellar Systems and the Sun*, Proc. 11th Cambridge Workshop, (ed. G. Lopez, R. Rebolo, & M.R. Zapatero—Osorio), Astronomical Society of the Pacific Conference Series Vol. 223, ASP, San Francisco.

Proc-2001-Hoshino: (2001) Proc. The Universe of Tokyo Symposium in 2000 on *Magnetic Reconnection in Space and Laboratory Plasmas*, (eds. M. Hoshino, R.L. Stenzel, and Shibata, K.), Earth Planets and Space, Vol. 53.

Proc-2001-Song: (2001) American Geophysical Union, Geophysical Monograph Series Vol. 125, *Space Weather*, Proc. Chapman Conference on Space Weather, Clearwater, Florida, 2000 Mar 20-24, (eds. Song, P., Singer, H.J., and Siscoe, G.L), AGU, Washington DC.

Proc-2001-Wimmer: (2001) *Solar and Galactic Composition*, (ed. R.F. Wimmer—Schwingruber), AIP, New York.

Proc-2002-ESA477: (2002) Proc. Second *Solar Cycle and Space Weather Euroconference*, 24-29 Sept 2001, Vico Equense, Italy. (ed. H. Sawaya—Lacoste), ESA SP-477, Noordwijk: ESA Publications Division.

Proc-2002-ESA505: (2002) European Space Agency Special Publication Vol. 505, *Magnetic Coupling of the Solar Atmosphere*, Proc. SOLMAG 2002 Euroconference and IAU Colloquium 188, held in Santorini, Greece, 2002 June 11-15, (ed. Huguette Sawaya—Lacoste), ESA, ESTEC Noordwijk, The Netherlands.

Proc-2002-ESA506: (2002) European Space Agency Special Publication Vol. 506, *Solar Variability: From Core to outer Frontiers*, Proc. 10th European Solar Physics Meeting, Prague, Czech Republic, 2002 Sept 9-14, (ed. A. Wilson), ESA, ESTEC Noordwijk, The Netherlands.

Proc-2002-Martens: (2002), *Multi-Wavelength Observations of Coronal Structure and Dynamics*, Proc. Yohkoh 10th Anniversary Meeting, held in Kona, Hawaii, 2002 Jan 20-24, (eds. Martens, P.C.H. & Cauffman, D.), COSPAR Colloquia Series, Vol. 13, Pergamon, Elsevier Science, Amsterdam.

Proc-2002-Tromsoe: (2002), *Nonlinear Waver and Chaos in Space Plasmas*, Proc. 4th Internat. Workshop, held in Tromsoe, Norway, 2001.

Proc-2003-Buechner: (2003), *Magnetic Helicity in Space and Laboratory Plasmas: Vistas from X-Ray Observatories*, Advances in Space Research, Vol. 32, Issue 10, Pergamon, Elsevier Science, Amsterdam.

Proc-2003-Dwivedi: (2003), *Dynamic Sun*, (ed. B. Dwivedi), Cambridge University Press, Cambridge.

Proc-2003-Erdélyi: (2003) NATO Science Series: II. Mathematics, Physics, and Chemistry, Vol. 124, *Turbulence, Waves, and Instabilities in the Solar Plasma*, Proc. NATO Advanced Workshop, held in Budapest, Hungary, 2002 Sept 16-20, (eds. R. von Fay—Siebenburgen, K. Petrovay, B. Roberts, & M.J. Aschwanden), Kluwer Academic Publishers, Dordrecht, The Netherlands.

Proc-2003-Klein: (2003), *Energy Conversion and Particle Acceleration in the Solar Corona*, Proc. of CESRA Workshop, held in Ringberg, Germany, 2001 June 2-6, (ed. K.L. Klein), Springer, Berlin, Lecture Notes in Physics, Vol. 612.

Proc-2003-Wilson: (2003), ISCS Symposium *Solar Variability as an Input to the Earth's Environment*, Proc. Workshop held in Tatranksa Lomnica, Slova, (ed. A. Wilson), ESA, ESTEC Noordwijk, The Netherlands.

Proc-2004-Dupree: (2004), *Stars as Suns: Activity, Evolution, and Planets*, IAU General Assembly XXV, Sydney, Australia, 21-25 July 2003, (eds. A.K.Dupree & A.O.Benz), San Francisco, Astronomical Society of Pacific.

Proc-2004-Gary: (2004), *Solar and Space Weather Radiophysics*, AAS/SPD Special Session, Albuquerque, New Mexico, 2002 June 4, (eds. D.E. Gary & C.O. Keller), Astrophysics and Space Science Library, Kluwer Academic Publishers, Dordrecht, The Netherlands.

Proc-2004-Wolf: (2004), Proc. *NIC Symposium 2004*, (eds. D. Wolf, G. Münster, and M. Kremer), John von Neumann Institute of Computing, Jülich, Germany, NIC Series, Vol. 20.

Proc-2004-ESA547: (2004), Proc. 14th SoHO Workshop, Mallorca 2003, European Space Agency Special Publication Vol. 547, ESA, ESTEC Noordwijk, The Netherlands.

Book/Monograph List

Allen, C.W. 1973, *Astrophysical Quantities*, Athlone, London.

Aschwanden, M.J. 2002b, *Particle Acceleration and Kinematics in Solar Flares. A Synthesis of Recent Observations and Theoretical Concepts*, Reprinted from *Space Science Reviews* 101, p.1-227, Kluwer Academic Publishers, Dordrecht, The Netherlands.

Baumjohann, W. & Treumann, R.A. 1997, *Basic Space Plasma Physics*, Imperial College Press, London.

Bellan, P.M. 2002, *Spheromaks. A Practical Application of Magnetohydrodynamic Dynamos and Plasma Self-Organization.* Imperial College Press, London.

Benz, A.O. 1993, (Second edition: 2003), *Plasma Astrophysics. Kinetic Processes in Solar and Stellar Coronae*, Kluwer Academic Publishers, Dordrecht, The Netherlands.

Billings, D.E. 1966, *A Guide to the Solar Corona*, Academic Press, New York.

Bleeker, J., Geiss,J., & Huber,M.C.E. (eds.) 2002, *The Century of Space Science*, Kluwer Academic Publishers, Dordrecht, The Netherlands.

Boyd, T.J.M. & Sanderson,J.J. 2003, *The Physics of Plasmas*, Cambridge University Press, Cambridge.

Bray, R.J., Cram, L.E., Durrant, C.J., & Loughhead, R.E. 1991, *Plasma Loops in the Solar Corona*, Cambridge University Press, Cambridge.

Bruzek, A. & Durrant, C.J. 1977, *Illustrated Glossary for Solar and Solar-Terrestrial Physics*, Astrophysics and Space Science Library Vol. 69, 224p., Reidel, Dordrecht.

Burlaga, L.F. 1995, *Interplanetary Magnetohydrodynamics*, Oxford University Press, Oxford.

Carlowicz, M.J. & Lopez, R.E. 2002, *Storms from the Sun − The emerging science of space weather*, The Joseph Henry Press, Washington DC.

Chandrasekhar, S. 1961, *Hydrodynamic and Hydromagnetic Stability*, Cambridge University Press, Cambridge.

Chen, F.F. 1974, *Introduction to Plasma Physics*, Plenum Press, New York.

Cowling, T.G. 1976, *Magnetohydrodynamics*, 2nd edn., Adam Hilger, Bristol, England.

Cox, A.N. (ed.) 2000, *Allen's Astrophysical Quantities*, AIP Press, 4th edition, Springer−Verlag, New York.

Craig, I.J.D. & Brown, J.C. 1986, *Inverse Problems in Astrophysics*, McGraw−Hill.

Davidson, P.A. 2001, *An Introduction to Magnetohydrodynamics*, Cambridge University Press, Cambridge.

Foukal, P.V. 1990, *Solar Astrophysics*, John Wiley & Sons, New York.

Ginzburg, V.L. 1961, *Propagation of Electromagnetic Waves in a Plasma*, Gordon and Breach, New York.

Ginzburg, V.L. & Syrovatskii,S.I. 1964, *The origin of cosmic rays*, Pergamon Press, New York, 423p.

Golub, L. & Pasachoff, J.M. 1997, *The Solar Corona*, Cambridge University Press, Cambridge.

Goossens, M., 2003, *An Introduction to Plasma Astrophysics and Magnetohydrodynamics*, Astrophysics and Space Science Library Vol. 294, Kluwer Academic Publishers, Dordrecht, The Netherlands.

Guillermier, P. & Koutchmy, S. 1999, *Total Eclipses. Science, Observations, Myths and Legends*, Springer−Praxis Series in Astronomy, Springer−Verlag, Berlin and Praxis Publishing, Chichester, UK.

Hasegawa, A. 1975, *Plasma Instabilities and Nonlinear Effects*, Physics and Chemistry in Space, Vol. 8, 217p., Springer, Berlin.

Hundhausen, A.J. 1972, *Coronal Expansion and Solar Wind*, Springer, New York.

Jackson, J.D. 1962, *Classical Electrodynamics*, John Wiley & Sons, New York.

Jeffrey, A. & Taniuti, T. 1966, *Magnetohydrodynamic Stability and Thermonuclear Containment*, Academic Press, XXX.

Kaplan, S.A. & Tsytovitch, V.N. 1973, *Plasma Astrophysics*, Pergamon Press, London.

Kivelson, M.G. & Russell, C.T. 1995, *Introduction to Space Physics*, Cambridge University Press, Cambridge.

Krueger, A. 1979, *Introduction to Solar Radio Astronomy and Radio Physics*, Geophysics and Astrophysics Monographs, Reidel Publishing Company, Dordrecht, The Netherlands.

Kundu, M.R. 1965, *Solar Radio Astronomy*, Interscience Publishers, New York.

Lamb, H. 1963, *Hydrodynamics*, Dover Publications, New York.

Lang, K.R. 1980, *Astrophysical Formulae, A Compendium for the Physicist and Astrophysicist*, Springer Verlag, Berlin.

Lang, K.R. 1985, *Sun, Earth, and Sky*, Springer Verlag, Berlin.

Lang, K.R. 2000, *The Sun from Space*, Springer Verlag, Berlin.

Lang, K.R. 2001, *The Cambridge Encyclopedia of the Sun*, Cambridge University Press.

Mandelbrot, B.B. 1977, *The Fractal Geometry of Nature*, Freeman, New York.

Mariska, J.T. 1992, *The Solar Transition Region*, Cambridge University Press, Cambridge.

McLean, D.J. & Labrum, N.R. (eds.) 1985, *Solar Radiophysics. Studies of Emission from the Sun at Metre Wavelengths*, Cambridge University Press, Cambridge.

Melrose, D.B. 1980a, *Plasma Astrophysics. Nonthermal Processes in Diffuse Magnetized Plasmas. Volume 1: The Emission, Absorption and Transfer of Waves in Plasmas*, Gordon and Breach Publishers, New York.

Melrose, D.B. 1980b, *Plasma Astrophysics. Nonthermal Processes in Diffuse Magnetized Plasmas. Volume 2: Astrophysical Applications*, Gordon and Breach Publishers, New York.

Melrose, D.B. 1986, *Instabilities in Space and Laboratory Plasmas*, Cambridge University Press, Cambridge.

Milne, E.A. 1930, *Thermodynamics of the Stars*, Springer, Berlin.

Murdin, P. 2000, *Encyclopedia of Astronomy and Astrophysics*, Institute of Physics Publishing, Grove's Dictionaries, Inc., New York.

Parker, E.N. 1963b, Interplanetary Dynamical Processes, Wiley, New York.

Parker, E.N. 1979, *Cosmical Magnetic Fields*, Oxford University Press, Oxford.

Parker, E.N. 1994, *Spontaneous Current Sheets in Magnetic Fields*, Oxford University Press, Oxford.

Planck, M. 1913, *The Theory of Heat Radiation*, reproduced by Dover 1959, New York.

Phillips, K.J.H. 1992, *Guide to the Sun*, Cambridge University Press, Cambridge.

Priest, E.R. 1982, Geophysics and Astrophysics Monographs Volume 21, *Solar Magnetohydro-dynamics*, D.Reidel Publishing Company, Dordrecht, The Netherlands.

Priest, E.R. & Forbes, T. 2000, *Magnetic Reconnection — MHD Theory and Applications*, Cambridge University Press, Cambridge.

Ratcliffe, J.A. 1969, *The Magneto-Ionic Theory and its Applications to the Ionosphere*, Cambridge University Press, Cambridge.

Rybicki, G.B. & Lightman, A.P. 1979, *Radiative Processes in Astrophysics*, Wiley—Interscience & Sons, New York.

Sato, J., Sawa, M., Yoshimura, K., Masuda, S., and Kosugi, T. 2003, *The Yohkoh HXT/SXT Flare Catalogue*, October 1, 1991 - December 14, 2001, Montana State University and Institute of Space and Astronautical Science (ISAS).

Schmidt, G. 1979, *Physics of High Temperature Plasmas*, Academic Press, New York.

Schrijver, C.J. & Zwaan, C. 2000, *Solar and Stellar Magnetic Activity*, Cambridge University Press, Cambridge.

Schroeder, M. 1991, *Fractals, Chaos, Power Laws: Minutes from an Infinite Paradise*, Freeman, New York.

Schüssler, M. & Schmidt, W. 1994, *Solar Magnetic Fields*, Cambridge University Press, Cambridge.

Schuster, H.G. 1988, *Deterministic Chaos: An Introduction*, VCH Verlagsgesellschaft, Weinheim, Germany.

Schwenn, R. & Marsch, E. (eds.) 1991a, *Physics of the Inner Heliosphere: 1. Large-Scale Phenomena*, Physics and Chemistry in Space Vol. 20, Space and Solar Physics, Springer Verlag, Berlin.

Schwenn, R. & Marsch, E. (eds.) 1991b, *Physics of the Inner Heliosphere: 2. Particles, Waves and Turbulence*, Physics and Chemistry in Space Vol. 21, Space and Solar Physics, Springer Verlag, Berlin.

Slottje, C. 1981, *Atlas of Fine Structures of Dynamic Spectra of Solar Type IV-dm and Some Type II Radio Bursts*, Dwingeloo, The Netherlands.

Somov, B.V. 2000, *Cosmic Plasma Physics*, Astrophysics and Space Science Library, Vol. 251, Kluwer Academic Publishers, Dordrecht, The Netherlands.

Spitzer, L. 1967, *The Physics of Fully Ionized Gases*, (2nd edition), Interscience, New York.

Stenflo, J.O. 1994, *Solar Magnetic Fields: Polarized Radiation Diagnostics*, Astrophysics and Space Science Library, Dordrecht: Kluwer Academic Publishers.

Stix, M. 2002, *The Sun. An Introduction*, Berlin: Springer.

Stix, T.H. 1992, *Waves in Plasmas*, New York: AIP Press.

Strong, K.T., Saba, J.L.R., Haisch, B.M., et al. (eds.) 1999, *The Many Faces of the Sun. A Summary of the Results from NASA's Solar Maximum*, Springer, Berlin.

Sturrock, P.A. 1994, *Plasma Physics. An Introduction to the Theory of Astrophysical, Geophysical, and Laboratory Plasmas*, Cambridge: Cambridge University Press.

Svestka, Z. 1976, *Solar Flares*, D.Reidel Publishing Company, Dordrecht.

Tajima, T. & Shibata, K. 2002, *Plasma Astrophysics*, Perseus Publishing, Cambridge, Massachusetts.

Tandberg—Hanssen, E. 1974, *Solar Prominences*, Dordrecht: Reidel.

Tandberg—Hanssen, E. 1995, *The Nature of Solar Prominences*, Dordrecht: Kluwer.

Treumann, R.A. and Baumjohann, W. 1997, *Advanced Space Plasma Physics*, Imperial College Press, London.

White, R.B. 1983, *Handbook of Plasma Physics* (eds. M.N. Rosenbluth and R.Z. Sagdeev), Vol. 1: Basic Plasma Physics I, p.611 (eds. A.A. Galeev and R.N. Sudan).

Zheleznyakov, V.V. 1970, *Radio Emission of the Sun and Planets*, Pergamon Press, Oxford.

Zirin, H. 1988, *Astrophysics of the Sun*, Cambridge University Press, Cambridge.

Zirker, J.B. (ed.) 1977, *Coronal Holes and High Speed Wind Streams*, Monograph from Skylab Solar Workshop I, Colorado Associated University Press, Boulder, Colorado.

Zombek, M.V. 1990, *Handbook of Space Astronomy and Astrophysics*, Second Edition, Cambridge University Press, Cambridge, UK.

PhD Thesis List

Aschwanden, M.J. 1987b, PhD Thesis, *Pulsations of the Radio Emission of the Solar Corona. Analysis of Observations and Theory of the Pulsating Electron-Cyclotron Maser*, ETH Zurich, 173p.

Biesecker, D.A., 1994, PhD Thesis, *On the Occurrence of Solar Flares Observed with the Burst and Transient Source Experiment*, University of New Hampshire.

Falconer, D. 1994, PhD Thesis, *Relative Elemental Abundance and Heating Constraints Determined for the Solar Corona from SERTS Measurements*, NASA Tech.Memo. 104616.

Harvey, J.W. 1969, PhD Thesis, *Magnetic Fields Associated with Solar Active-Region Prominences*, Univ. Colorado, Boulder, Colorado.

Harvey, K.L. 1993, PhD Theseis, *Magnetic Dipoles on the Sun*, Astronomical Institute, Utrecht University.

Sakao, T. 1994, PhD Thesis, *Characteristics of Solar Flare Hard X-ray Sources as Revealed with the Hard X-ray Telescope Aboard the YOHKOH Satellite*, University of Tokyo.

Shimizu, T. 1997, PhD Thesis, *Studies of transient brightenings (microflares) discovered in solar active regions*, School of Science, Univ.Tokyo.

Volwerk, M. 1993, PhD Thesis, *Strong Double Layers in Astrophysical Plasmas*, Utrecht University, The Netherlands.

Reference List

References with the same first author are sorted chronologically.
The author list is limited to the three first authors per reference.
The full references of proceedings, books, and PhD theses are given in previous lists.

Journal Abbreviations

AA	Astronomy and Astrophysics
AASS	Astronomy and Astrophysics Supplement Series
AdSpR	Advances in Space Research
ApJ	The Astrophysical Journal
ApJL	The Astrophysical Journal Letters
ApJS	The Astrophysical Journal Supplement Series
ARAA	Annual Review of Astronomy and Astrophysics
BAAS	Bulletin of the American Astronomical Society
GRL	Geophysics Research Letters
JGR	Journal of Geophysics Research
MNRAS	Monthly Notices of the Royal Astronomical Society
PASA	Publications of the Astronomical Society of Australia
PASJ	Publications of the Astronomical Society of Japan
PASP	Publications of the Astronomical Society of the Pacific
SP	Solar Physics
SPIE	Proc. SPIE (International Society for Optical Engineering)

All References (in compact form)

Abbett, W.P., Fisher, G.H., & Fan, Y. 2000, ApJ 540, 548.

Abbett, W.P., & Fisher, G.H. 2003, ApJ 582, 475.

Abrami, A. 1970, SP 11, 104.

Abrami, A. 1972, Nature 238/80, 25.

Achong, A. 1974, SP 37, 477.

Achour, H., Brekke, P., Kjeldseth−Moe, O., et al. 1995, ApJ 453, 945.

Achterberg, A. 1979, AA 76, 276.

Achterberg, A. & Norman, C.A. 1980, AA 89, 353.

Acton, L.W., Canfield, R.C., Gunkler, T.A., et al. 1982, ApJ 263, 409.

Acton, L.W., Tsuneta, S., Ogawara, Y., et al., 1992, Science, 258, 618.

Akimov, V.V. et al. 1991, Proc-1991-ICRC22, 73.

Akimov, V.V., Belov, A.V., Chertok, I.M., et al. 1994a, Proc-1994-Enome, 371.

Akimov, V.V., Leikov, N.G., Belov, A.V. et al. 1994b, Proc-1994-Ryan, 106.

Akimov, V.V., Leikov, N.G., Kurt, V.G. et al. 1994c, Proc-1994-Ryan, 130.

Akimov, V.V., Ambroz, P., Belov, A.V. et al. 1996, SP 166, 107.

Aletti, V., Velli, M., Bocchialini, K., et al. 2000, ApJ 544, 550.

Alexander, D. 1990, AA 235, 431

Alexander, D. & MacKinnon, A.L. 1993, SP 144, 155.

Alexander, D. & Matthews, S.A. 1994, SP 154, 157.

Alexander, D. & Metcalf, T. 1997, ApJ 489, 442.

Alexander, D., Metcalf, T.R., & Nitta, N.V. 2002, GRL 29/No.10, 41-1.

Alexander, D. & Acton, L.W. 2002, *The Active Sun*, (in Bleeker et al. 2002).

Alexander, R.C., & Brown, J.C. 2002, SP 210, 407.

Alfvén, H. & Carlqvist, P. 1967, SP 1, 220.

Alissandrakis, C.E., Kundu, M.R., & Lantos, P. 1980, AA 82, 30.

Alissandrakis, C.E. 1981, AA 100, 197.

Alissandrakis, C.E. & Kundu, M.R. 1984, AA 139, 271.

Alissandrakis, C.E. & Chiuderi−Drago, F. 1995, SP 160, 171.

Alissandrakis, C.E., Borgioli, F., Chiuderi−Drago, F., et al. 1996, SP 167.

Allen, C.W. 1973, *Astrophysical Quantities*, (see book list).

Altschuler, M.D. & Newkirk,G.Jr. 1969, SP 9, 131.

Aly, J.J. 1984, ApJ 283, 349.

Aly, J.J. 1989, SP 120, 19.

Aly, J.J. & Amari, T. 1997, AA 319, 699.

Amari, T., Luciani, J.F., Aly, J.J., et al. 1966, ApJ 466, L39.

Amari, T., Aly,J.J., Luciani, J.F., et al. 1997, SP 174, 129.

Amari, T., Luciani,J.F., Mikić, Z., & Linker,J. 2000, ApJL 529, L49.

Amari, T., Luciani,J.F., Aly, J.J., et al. 2003a, ApJ 595, 1231.

Amari, T., Luciani,J.F., Aly, J.J., et al. 2003b, ApJ 585, 1073.

Ambrosiano, J., Matthaeus, W.H., Goldstein, M.L., et al. 1988, JGR 93, 14383.

Anastasiadis, A. & Vlahos, L. 1991, AA 245, 271.

Anastasiadis, A. & Vlahos, L. 1994, ApJ 428, 819.

Anderson, K.A., Kane, S.R., Primbsch, J.H., et al. 1978, ITGE GE-16/3, 157.

Anderson, K.A., Lin, R.P., Martel, F., et al. 1979, GRL 6, 401.

Andrews, M.D., Wang, A.H., & Wu, S.T. 1999, SP 187, 427.

Andrews, M.D. 2003, SP 218, 261.

Antiochos, S.K. & Sturrock, P.A. 1978, ApJ 220, 1137.

Antiochos, S.K. 1979, ApJ 232, L125.

Antiochos, S.K. 1980, ApJ 241, 385.

Antiochos, S.K. 1984, ApJ 280, 416.

Antiochos, S.K., Shoub, E.C., An, C.H., et al. 1985, ApJ 298, 876.

Antiochos, S.K. & Noci,G. 1986, ApJ 301, 440.

Antiochos, S.K. 1986, Proc-1986-Poland, 419.

Antiochos, S.K. & Klimchuk, J.A. 1991, ApJ 378, 372.

Antiochos, S.K., Dahlburg, R.B., & Klimchuk, J.A. 1994, ApJ 420, L41.

Antiochos, S.K. 1998, ApJ 502, L181.

Antiochos, S.K., MacNeice, P.J., Spicer, D.S., et al. 1999a, ApJ 512, 985.

Antiochos, S.K., DeVore, C.R., & Klimchuk, J.A. 1999b, ApJ 510, 485.

Antiochos, S.K., MacNeice, P.J., & Spicer, D.S. 2000a, ApJ 536, 494.

Antiochos, S.K., DeLuca, E.E., Golub, L., et al. 2000b, ApJ 542, L151.

Antonucci, E., Gabriel, A.H., Acton, L.W. et al. 1982, SP 78, 107.

Antonucci, E. & Dennis, B.R. 1983, SP 86, 67.

Antonucci, E., Gabriel, A.H., & Patchett, B.E. 1984a, SP 93, 85.

Antonucci, E., Gabriel, A.H., & Dennis, B.R. 1984b, ApJ 287, 917.

Antonucci, E., Dennis, B.R., Gabriel, A.H., et al. 1985, SP 96, 129.

Antonucci, E. 1986, Proc-1986-Swings, 731.

Antonucci, E. and Dodero, M.A. 1986, Proc-1986-Neidig, 363.

Antonucci, E., Rosner, R., & Tsinganos, K. 1986, ApJ 301, 975.

Antonucci, E. 1989, SP 121, 31.

Antonucci, E., Dodero, M.A., & Martin, R. 1990a, ApJS 73, 137.

Antonucci, E. & Somov, B.V. 1992, Proc-1992-ESA348, 293.

Antonucci, E., Dodero, M., & Somov, B.V. 1994, Proc-1993-Uchida, 333.

Antonucci, E., Benna, C., & Somov, B.V. 1996, ApJ 456, 833.

Antonucci, E., Alexander, D., Culhane, J.L. et al. 1999, in Strong et al. (see book list).

Anzer, U. 1968, SP 3, 298.

Anzer, U. 1969, SP 8, 37.

Anzer, U. 1972, SP 24, 324.

Anzer, U. & Heinzel, P. 1999, AA 349, 974.

Anzer, U. 2002, ESA SP-506, 389.

Arnaud, M. & Raymond, J. 1992, ApJ 398, 394.

Asai, A., Shimojo, M., Isobe, H., et al. 2001, ApJ 562, L103.

Asai, A., Ishii, T.T., Kurokawa, H., et al. 2003, ApJ 586, 624.

Aschwanden, M.J. 1986, SP 104, 57.

Aschwanden, M.J. & Benz, A.O. 1986, AA 158, 102.

Aschwanden, M.J. 1987a, SP 111, 113.

Aschwanden, M.J. 1987b, PhD Thesis (see PhD Thesis list).

Aschwanden, M.J. & Benz, A.O. 1988a, ApJ 332, 447.

Aschwanden, M.J. & Benz, A.O. 1988b, ApJ 332, 466.

Aschwanden, M.J. 1990a, AASS 85, 1141.

Aschwanden, M.J. 1990b, AA 237, 512.

Aschwanden, M.J., Benz, A.O., & Kane, S.R. 1990, AA 229, 206.

Aschwanden, M.J. & Güdel, M. 1992, ApJ 401, 736.

Aschwanden, M.J., Bastian, T.S. & Gary, D.E., 1992a, BAAS 24/2, 802.

Aschwanden, M.J., Bastian, T.S., Benz, A.O., & Brosius,J.W. 1992b, ApJ 391, 380.

Aschwanden, M.J., Benz, A.O. & Schwartz, R.A. 1993, ApJ 417, 790.

Aschwanden, M.J. 1994, SP 152, 53.

Aschwanden, M.J. & Bastian, T.S. 1994a, ApJ 426, 425.

Aschwanden, M.J. & Bastian, T.S. 1994b, ApJ 426, 434.

Aschwanden, M.J., Benz, A.O., & Montello, M. 1994a, ApJ 431, 432.

Aschwanden, M.J., Benz, A.O. Dennis, B.R., et al. 1994b, ApJS 90, 631.

Aschwanden, M.J. & Benz, A.O. 1995, ApJ 438, 997.

Aschwanden, M.J. & Schwartz, R.A. 1995, ApJ 455, 699.

Aschwanden, M.J., Lim, J., Gary, D.E., et al. 1995a, ApJ 454, 512.

Aschwanden, M.J., Benz, A.O., Dennis, B.R., et al. 1995b, ApJ 455, 347.

Aschwanden, M.J., Schwartz, R.A., & Alt, D.M. 1995c, ApJ 447, 923.

Aschwanden, M.J. 1996, Proc-1996-Ramaty, 300.

Aschwanden, M.J., & Schwartz, R.A. 1996, ApJ 464, 974.

Aschwanden, M.J., Hudson, H.S., Kosugi, T., et al. 1996a, ApJ 464, 985.

Aschwanden, M.J., Wills, M.J., Hudson, H.S., et al. 1996b, ApJ 468, 398.

Aschwanden, M.J., Kosugi, T., Hudson, H.S., et al. 1996c, ApJ 470, 1198.

Aschwanden, M.J. & Benz, A.O. 1997, ApJ 480, 825.

Aschwanden, M.J. & Treumann, R.A. 1997, Proc-1997-Trottet, 108.

Aschwanden, M.J., Bynum, R.M., Kosugi, T., et al. 1997, ApJ 487, 936.

Aschwanden, M.J. 1998a, ApJ 502, 455.

Aschwanden, M.J. 1998b, Proc-1998-Watanabe, 285.

Aschwanden, M.J., Dennis, B.R., & Benz, A.O. 1998a, ApJ 497, 972.

Aschwanden, M.J., Schwartz, R.A., & Dennis, B.R. 1998b, ApJ 502, 468.

Aschwanden, M.J., Kliem, B., Schwarz, U., et al. 1998c, ApJ 505, 941.

Aschwanden, M.J. 1999a, SP 190, 233.

Aschwanden, M.J. 1999b, chapter 8, p.273, in Strong et al. (see book list).

Aschwanden, M.J. 1999c, ESA SP-448, 1015.

Aschwanden, M.J., Newmark, J.S., Delaboudinière, J.-P., et al. 1999a, ApJ 515, 842.

Aschwanden, M.J., Fletcher, L., Schrijver, C., et al. 1999b, ApJ 520, 880.

Aschwanden, M.J., Kosugi, T., Hanaoka, Y., et al. 1999c, ApJ 526, 1026.

Aschwanden, M.J., Fletcher, L., Sakao, T., et al. 1999d, ApJ 517, 977.

Aschwanden, M.J. 2000, Proc-2000-Ramaty, 197.

Aschwanden, M.J. & Nitta, N. 2000, ApJ 535, L59.

Aschwanden, M.J., Alexander, D., Hurlburt, N., et al. 2000a, ApJ 531, 1129.

Aschwanden, M.J., Nightingale, R., Tarbell, T., et al. 2000b, ApJ 535, 1027.

Aschwanden, M.J., Tarbell, T., Nightingale, R., et al. 2000c, ApJ 535, 1047.

Aschwanden, M.J., Nightingale, R.W., & Alexander, D. 2000d, ApJ 541, 1059.

Aschwanden, M.J. 2001a, ApJ 559, L171.

Aschwanden, M.J. 2001b, ApJ 560, 1035.

Aschwanden, M.J. & Acton, L.W. 2001, ApJ 550, 475.

Aschwanden, M.J. & Alexander, D. 2001, SP 204, 91.

Aschwanden, M.J., Schrijver, C.J., & Alexander, D. 2001, ApJ 550, 1036.

Aschwanden, M.J. 2002a, ApJ 580, L79.

Aschwanden, M.J. 2002b, Space Science Reviews 101, 1.

Aschwanden, M.J. 2002c, COSPAR-CS 13, 57.

Aschwanden, M.J. & Charbonneau, P. 2002, ApJ 566, L59.

Aschwanden, M.J. & Parnell, C.E. 2002, ApJ 572, 1048.

Aschwanden, M.J. & Schrijver, C.J. 2002, ApJS 142, 269.

Aschwanden, M.J., DePontieu, B., Schrijver, C.J., et al. 2002a, SP 206, 99.

Aschwanden, M.J., Brown, J.C., & Kontar, E.P. 2002b, SP 210, 383.

Aschwanden, M.J. 2003, Proc-2003-Erdélyi, 215.

Aschwanden, M.J., Schrijver, C.J., Winebarger, A.R., et al. 2003a, ApJ 588, L49.

Aschwanden, M.J., Nightingale, R.W., Andries, J., et al. 2003b, ApJ 598, 1375.

Aschwanden, M.J. 2004, ApJ 608, 554.

Aschwanden, M.J., Nakariakov, V.M., & Mel'nikov, V.F. 2004a, ApJ 600, 458.

Aschwanden, M.J., Alexander, D., & DeRosa, M. 2004b, Proc-2004-Gary, 243.

Athay, R.G. & Moreton,G.E. 1961, ApJ 133, 935.

Athay, R.G. & White, O.R. 1978, ApJ 226, 1135.

Athay, R.G. & White, O.R. 1979, ApJS 39, 333.

Athay, R.G. 1982, ApJ 263, 982.

Athay, R.G. 1990, ApJ 362, 264.

Arnaud, M. & Raymond, J.C. 1992, ApJ 398, 39.

Aulanier, G. & Démoulin,P. 1998, AA 329, 1125.

Aulanier, G., Démoulin, P., Van Driel−Geztelyi, L., et al. 1998a, AA 335, 309.

Aulanier, G., Démoulin, P., Schmieder, B., et al. 1998b, SP 183, 369.

Aulanier, G., Démoulin, P., Mein,N., et al. 1999, AA 342, 867.

Aulanier, G., Srivastava, N., & Martin,S.F. 2000a, ApJ 543, 447.

Aulanier, G., DeLuca, E.E., Antiochos, S.K., et al. 2000b, ApJ 540, 1126.

Aulanier, G. & Schmieder, B. 2002, AA 386, 1106.

Aulanier, G., DeVore, C.R., & Antiochos, S.K. 2002, ApJ 567, L97.

Aulanier, G. & Démoulin, P. 2003, AA 402, 769.

Aurass, H. & Mann, G. 1987, SP , 112, 359.

Aurass, H., Chernov, G.P., Karlický, M., et al. 1987, SP 112, 347.

Aurass, H., Vrsnak, B., & Mann, G. 2002a, AA 384, 273.

Aurass, H., Shibasaki, K., Reiner, M., et al. 2002b, ApJ 567, 610.

Bagalá, L.G., Mandrini, C.H., Rovira, M.G., et al. 2000, AA 363, 779.

Bai, T. 1982a, ApJ 259, 341

Bai, T. 1982b, Proc-1982-Lingenfelter, 409.

Bai, T., Hudson, H.S., Pelling, R.M., at el. 1983, ApJ 267, 433.

Bai, T. & Sturrock, P. 1989, ARAA 27, 421.

Bai, T. 1993, ApJ 404, 805.

Ballai, I. & Erdélyi, R. 1998, SP 180, 65.

Ballai, I. & Erdélyi, R. 2003, Proc-2003-Erdélyi, 121.

Ballester, J.L. & Priest, E.R. (eds.) 1988, Proc-1988-Ballester.

Ballester, J.L. & Roberts, B. (eds.) 2001, Proc-2001-Ballester.

Balke, A.C., Schrijver, C.J., Zwaan, C., et al. 1993, SP 143, 215.

Balmer, J.J. 1885, Ann.Phys.Chem., 25, 80.

Balogh, A. & Riley, P. 1997, Proc-1997-Jokipii, 359.

Balogh, A. 2001, *Solar Wind: Ulysses*, (in Murdin 2000) .

Balogh, A., Marsden, R.G., & Smith, E.J. 2001, *The Heliosphere − Ulysses*, (see book list).

Balthasar, H., Knoelker, M., Wiehr, E., et al. 1986, AA 163, 343.

Balthasar, H., Wiehr, E., & Stellmacher, G. 1988, AA 204, 286.

Balthasar, H., Wiehr, E., Schleicher, H., et al. 1993, AA 277, 635.

Balthasar, H. & Wiehr, E. 1994, SP 286, 639.

Banaszkiewicz, M., Axford, W.I., & McKenzie, J.F. 1998, AA 337, 940.

Banerjee, D., Teriaca, L., Doyle, J.G., et al. 1998, AA 339, 208.

Banerjee, D., O'Shea, E., Doyle, J.G., et al. 2001, AA 377, 691.

Barat, C., Trottet, G., Vilmer, N., et al. 1994, ApJ 425, L109.

Barbosa, D.D. 1979, ApJ 233, 383.

Baranov, N.V. & Tsvetkov, L.I. 1994, Astronomy Letters 20/3, 327.

Barbosa, D.D. 1979, ApJ 233, 383.

Baring, M.G., Ellison, D.C., & Jones, F.C. 1994, ApJS 90,. 547.

Bashkirtsev, V.S., Kobanov, N.I., & Mashnich, G.P. 1983, SP 82, 443.

Bashkirtsev, V.S. & Mashnich, G.P. 1984, SP 91, 93.

Bastian, T.S. & Gary, D.E. 1992, SP 139, 357.

Bastian, T.S., Ewell, M.W., & Zirin, H. 1993a, ApJ 415, 364.

Bastian, T.S., Ewell, M.W., & Zirin, H. 1993b, ApJ 418, 510.

Bastian, T.S. 1994, ApJ 426, 774.

Bastian, T.S., Dulk, G.A., & Leblanc, Y. 1996, ApJ 473, 539.

Bastian, T.S., Benz, A.O., & Gary, D.E. 1998, ARAA 36, 131.

Bastian, T.S. 2000, Proc-2000-Murdin, Vol.3, 2553.

Batchelor, D. 1989, ApJ 340, 607.

Baumjohann, W. & Treumann, R.A. 1997, *Basic Space Plasma Physics*, (see book list).

Bélien, A.J.C., Martens, P.C.H., & Keppens, R. 1999, ApJ 526, 478.

Bellan, P.M. 2002, *Spheromaks* (see book list).

Bellan, P.M. 2003, AdSpR 32/10, 1923.

Benevolenskaya, E.E., Kosovichev, A.G., Lemen, J.R., et al. 2002, ApJ 571, L181.

Benka, S.G. & Holman, G.D. 1992, ApJ 391, 854.

Benka, S.G. & Holman, G.D. 1994, ApJ 435, 469.

Bentley, R.D. & Mariska, J.T. (eds.) 1996, Proc-1996-Bentley.

Benz, A.O. & Gold, T. 1971, SP 21, 157

Benz, A.O. 1977, ApJ 211, 270.

Benz, A.O. 1980, ApJ 240, 892

Benz, A.O. 1985, SP , 96, 357.

Benz, A.O. 1986, SP 104, 99.

Benz, A.O. 1987a, SP 111, 1.

Benz, A.O. 1987b, Proc-1987-Dennis, 133.

Benz, A.O. & Smith, D.F. 1987, SP 107, 299.

Benz, A.O. & Thejappa, G. 1988, AA 202, 267.

Benz, A.O., Magun,A., Stehling,W., & Su.H. 1992, SP 141, 335.

Benz, A.O. 1993, *Plasma Astrophysics*, (see book list).

Benz, A.O., Krucker,S., Acton, L.W., et al. 1997, AA 320, 993.

Benz, A.O. & Krucker, S. 1998, SP 182, 349.

Benz, A.O. & Krucker, S. 1999, AA 341, 286.

Benz, A.O. 2000, Proc-2000-Murdin, Vol.3, 2529.

Benz, A.O. & Krucker, S. 2002, ApJ 568, 413.

Benz, A.O. & Grigis, P.C. 2002, SP 210, 431.

Benz, A.O., Saint−Hilaire, P., & Vilmer, N.R. 2002, AA 383, 678.

Berger, M.A. 1991, AA 252, 369.

Berger, M.A. 1993, Phys.Rev.Lett. 70/6, 705.

Berger, T.E., DePontieu, B., Fletcher, L., et al. 1999, SP 190, 409.

Berghmans, D. & De Bruyne, P. 1995, ApJ 453, 495.

Berghmans, D., De Bruyne, P., & Goossens, M. 1996, ApJ 472, 398.

Berghmans, D., Clette, F., & Moses, D. 1998, AA 336, 1039.

Berghmans, D. & Clette, F. 1999, SP 186, 207.

Berghmans, D., McKenzie, D., & Clette, F. 2001, AA 369, 291.

Berghmans, D. 2002, ESA SP-506, 501.

Berney, M. & Benz, A.O. 1978, AA 65, 369

Bernold, T.E.X. 1980, AASS 42, 43.

Betta, R.M., Peres, G., Serio, S., et al. 1999a, ESA SP-448, 475.

Betta, R.M., Orlando, S., Giovanni, P., et al. 1999b, Space Sci. Rev. 87, 133.

Bewsher, D., Parnell, C.E., & Harrison, R.A. 2002, SP 206, 21.

Biesecker, D.A. 1994, PhD Thesis (see PhD Thesis list).

Biesecker, D.A., Myers, D.C., Thompson,B.J. et al. 2002, ApJ 569, 1009.

Billings, D.E. 1966, *A Guide to the Solar Corona*, see book list.

Biskamp, D. 1986, Phys. Fluids 29, 1520.

Biskamp, D. & Welter, H. 1979, Phys. Rev. Lett. 44, 1069.

Biskamp, D. & Welter, H. 1989, SP 120, 49.

Biskamp, D. 2003, Proc-2003-Klein, 109.

Birn, J., Gosling, J.T., Hesse, M., et al. 2000, ApJ 541, 1078.

Birn, J., Gorbes, T.G., & Schindler, K. 2003, ApJ 588, 578.

Blackman, E.G. & Brandenburg, A. 2003, ApJ 584, L99.

Blanco, S., Bocchialini, K., Costa, A., et al. 1999, SP 186, 281.

Blake, M.L. & Sturrock, P.A. 1985, ApJ 290, 359.

Bleeker, J., Geiss, J., & Huber, M.C.E. (eds.) 2002, *Century of Space Science*, (see book list).

Block, L.P. 1978, *Astrophys. Space Sci.* 55, 59.

Bloomberg, H.W., Davis, J., & Boris, J.P. 1977, JQSRT 18, 237.

Bocchialini, K., Costa, A., Domenech, G., et al. 2001, SP 199, 133.

Bochsler, P. 2001, *Solar Wind Composition*, (in Murdin 2000) .

Bogdan, T.J. & Low, B.C. 1986, ApJ 306, 271.

Bogod, V.M., & Grebinskij, A.S. 1997, SP 176, 67.

Boltzmann, L. 1884, Ann.Pys. 31, 291.

Bommier, V., Sahal−Bréchot, S., & Leroy, J.L. 1986a, AA 156, 79.

Bommier, V., Leroey, J.L., & Sahal−Bréchot, S. 1986b, AA 156, 90.

Bornmann, P.L. 1999, in Strong et al. (see book list), 301.

Bornmann, P.L., & Lemen, J.R. 1994, Proc-1994-Uchida, 265.

Bougeret, J.L 2000, Proc-2000-Murdin.

Bougeret, J.L. 2001, *Solar Wind: Interplanetary Radio Bursts*, (in Murdin 2000) .

Boyd, T.J.M. & Sanderson, J.J. 2003, *The Physics of Plasmas*, (see book list).

Brandenburg, A. & Zweibel, E.G. 1994, ApJ 427, L91.

Bray, R.J., Cram, L.E., Durrant, C.J., et al. 1991, *Plasma Loops*, (see book list).

Brekke, P. 1993, ApJ 408, 735.

Brekke, P., Kjeldseth−Moe, O., Brynildsen, N., et al. 1997a, SP 170, 163.

Brekke, P., Hassler, D.M., & Wilhelm, K. 1997b, SP 175, 349.

Brekke, P. 1999, SP 190, 379.

Brekke, P., Kjeldseth−Moe, O., Tarbell, T., et al. 1999, Proc-1999-Rimmele, 357.

Brkovic, A., Solanki, S.K., & Ruedi, I. 2001, AA 373, 1056.

Bromund, K.R., McTiernan, J.M., & Kane, S.R. 1995, ApJ 455, 733.

Brosius, J.W. & Holman, G.D. 1988, ApJ 327, 417.

Brosius, J.W. & Holman, G.D. 1989, ApJ 342, 1172.

Brosius, J.W., Willson, R.F., Holman, G.D., et al. 1992, ApJ, 386, 347.

Brosius, J.W., Davila, J.M., Thompson, W.T., et al. 1993, ApJ 411, 410.

Brosius, J.W., Davila, J.M., Thomas, R.J., et al. 1996, ApJS 106, 143.

Brosius, J.W., Davila, J.M., THomas, R.J., et al. 1997, ApJ 488, 488.

Brosius, J.W., Thomas, R.J., Davila, J.M., et al. 2000, ApJ 543, 1016.

Brosius, J.W., Landi. E., Cook, J.W., et al. 2002, ApJ 574, 453.

Brown, D.S. & Priest, E.R. 1999, SP 190, 25.

Brown, D.S. & Priest, E.R. 2001, AA 367, 339.

Brown, J.C. 1971, SP 18, 489.

Brown, J.C. 1972, SP 26, 441.

Brown, J.C. 1974, Proc-1974-Newkirk, 395.

Brown, J.C. & Hoyng, P. 1975, ApJ 200, 734.

Brown, J.C. & McClymont, A.N. 1976, SP 49, 329.

Brown, J.C. & Melrose, D.B. 1977, SP 52, 117.

Brown, J.C., Melrose, D.B., & Spicer, D.S. 1979, ApJ 228, 592.

Brown, J.C. & Smith, D.F. 1980, Rep.Prog.Phys. 43, 125.

Brown, J.C., Carlaw, V.A., Cromwell, D., et al. 1983, SP 88, 281.

Brown, J.C. & Bingham, R. 1984, AA 131, L11.

Brown, J.C. & Loran, J.M. 1985, MNRAS 212, 245.

Brown, J.C. & Emslie, A.G. 1988, ApJ 331, 554.

Brown, J.C. & Emslie, A.G. 1989, ApJ 339, 1123.

Brown, J.C., Karlický, M., MacKinnon, A.L., et al. 1990, ApJS 73, 343.

Brown, J.C. 1991, RSPTA 336, 413.

Brown, J.C., MacKinnon, A.L., VanDenOord, G.H.J., et al. 1991, AA 242, L13.

Brown, J.C., McArthur, G.K., Barrett, R.K., et al. 1998b, SP 179, 379.

Brown, J.C., Krucker, S., Güdel, M., et al. 2000, AA 359, 1185.

Brown, J.C., Aschwanden, M.J., & Kontar, E.P. 2002a, SP 210, 373.

Brown, M.R., Canfield, R.C., & Pevtsov, A.A. (eds.) 1999, Proc-1999-Brown.

Brown, M.R., Cothran, C.D., Landerman, M., et al. 2002b, ApJ 577, L63.

Brown, J.C., Emslie, A.G., & Kontar, E.P. 2003, ApJ 595, L115.

Browning, P.K. & Priest, E.R. 1983, ApJ 266, 848.

Browning, P.K. 1988, J. Plasma Physics 40, 263.

Browning, P.K. 1989, SP 124, 271.

Browning, P.K. & Hood, A.W. 1989, SP 124, 271.

Brueckner, B.E. 1981, Proc-1981-Priest, 113.

Brueckner, G.E. & Bartoe, J.D.F. 1983, ApJ 272, 329.

Brueckner, G.E., Howard, R.A., Koomen, M.J., et al. 1995, SP 162, 357.

Bruzek, A. 1969, Proc-1969-DeJager, 61.

Bruzek, A. & Durrant, C.J. 1977, *Illustrated glossary ...*, (see book list).

Brynildsen, N., Maltby, P., Leifsen, T., et al. 2000, SP 191, 129.

Brynildsen, N., Maltby, P., Fredvik, T., et al. 2002, SP 207, 259.

Buechner, J., & Pevtsov, A.A. (eds.) 2003, Proc-2003-Buechner.

Burgess, D. 1995, in Kivelson & Russell (1995), 129.

Burgess, D. 1997, Proc-1997-Simnett, 117.

Burlaga, L.F. 1995, *Interplanetary Magnetohydrodynamics*, (see book list).

Burlaga, L.F. 2001, *Solar Wind: Magnetic Field*, (in Murdin 2000) .

Cairns, I.H., Robinson, P.A., & Zank, G.P. 2000, Publ. Astron. Soc. Austraila 17, 22.

Caligari, P., Moreno–Insertis, F., & Schüssler,M. 1995, ApJ 441, 886.

Cally, P.S. 1986, SP 103, 27.

Cally, P.S. & Robb, T.D. 1991, ApJ 372, 329.

Cally, P.S. 2003, SP 217, 95.

Cane, H.V., Stone, R.G., Fainberg, J., et al. 1981, JGR 8/12, 1285.

Cane, H.V. 1984, AA 140, 205.

Cane, H.V., Kahler, S.W., & Sheeley, N.R.Jr. 1986, JGR 91, 13321.

Cane, H.V. 1988, JGR 93/A1, 1.

Cane, H.V. & Reames, D.V. 1988a, ApJ 325, 901.

Cane, H.V. & Reames, D.V. 1988b, ApJ 325, 895.

Cane, H.V. & Reames, D.V. 1990, ApJS 73, 253.

Cane, H.V., Reames, D.V., & von Rosenvinge, T.T. 1991, ApJ 373, 675.

Cane, H.V., Richardson, I.G., & St.Cyr, O.C. 2000, GRL 27/21, 3591.

Cane, H.V., Erickson, W.C., & Prestage, N.P. 2002, JGR 107/A10, SSH 14-1.

Canfield, R.C. et al. 1980, Proc-1980-Sturrock, 231.

Canfield, R.C., Gunkler, T.A., Hudson, H.S., et al. 1983, AdSpR 2, 145.

Canfield, R.C., Gunkler, T.A., & Ricchiazzi, P.J. 1984, ApJ 282, 296.

Canfield, R.C. & Gunkler, T.A. 1985, ApJ 288, 353.

Canfield, R.C. et al. 1986, Proc-1986-Kundu, 5-1.

Canfield, R.C. 1986a, Proc-1986-Neidig, 10.

Canfield, R.C. 1986b, Adv. Space Res. 6/6, 167.

Canfield, R.C., Zarro, D.M., Metcalf, T.R. et al. 1990, ApJ 348 333.

Canfield, R.C., Hudson, H.S., & McKenzie, D.E. 1999, GRL 26/6, 627.

Cargill, P.J. & Priest, E.R. 1980, SP 65, 251.

Cargill, P.J. & Priest, E.R. 1983, ApJ 266, 383.

Cargill, P.J., Goodrich, C.C., & Vlahos, L. 1988, AA 189, 254.

Cargill, P.J., Chen, J., & Garren, D.A. 1994, ApJ 423, 854.

Cargill, P.J., Mariska, J.T., & Antiochos, S.K. 1995, ApJ 439, 1034.

Cargill, P.J., 2001, *Solar Flares: Particle Acceleration Mechanisms*, (in Murdin 2000) .

Cargill, P.J. 2001, Proc-2001-Daglis, 177.

Carlquist, P. 1969, SP 7, 377.

Carlowicz, M.J. & Lopez, R.E. 2002, *Storms from the Sun*, (see book list).

Carmichael, H. 1964, Proc-1964-Hess, 451.

Chae, J.C., Yun, H.S., & Poland, A.I. 1997, ApJ 480, 817.

Chae, J.C., Schühle, U. & Lemaire, P. 1998a, ApJ 505, 957.

Chae, J.C., Chae, J., Wang, H., et al. 1998b, ApJ 504, L123.

Chae, J.C., Yun, H.S., & Poland, A.I. 1998c, ApJS 114, 151.

Chae, J.C. 1999, Proc-1999-Rimmele, 375.

Chae, J.C., Giu, J., Wang, H., et al. 1999, ApJ 513, L75.

Chae, J.C. 2000, ApJL 540, L115.

Chae, J.C., Wang, H., Qiu, J., et al. 2000a, ApJ 533, 535.

Chae, J.C., Denker, C., Spirock, T.J., et al. 2000b, SP 195, 333.

Chae, J.C., Wang,H., Goode, P.R., et al. 2000c, ApJ 528, L119.

Chae, J.C., Park,Y.D., Moon, Y.J., et al. 2002a, ApJ 567, L159.

Chae, J.C., Poland, A.I., & Aschwanden, M.J. 2002b, ApJ 581, 726.

Chae, J.C., Moon, Y.J., Wang, H., & Yun,H.S. 2002c, SP 207, 73.

Chae, J.C., Choi, B.K., & Park, M.J. 2002d, J.Korean Astr. Soc. 35, 59.

Chae, J.C, Moon, Y.J., & Park, S.Y. 2003, J.Korean Astr. Soc. 36, 13.

Chandrasekhar, S. 1961, *Hydrodynamic and Hydromagnetic Stability*, (see book list).

Chapman, R.D., Jordan, S.D., Neupert, W.M., et al. 1972, ApJ 174, L97.

Charbonneau, P., McIntosh, S.W., Liu, H.L., et al. 2001, SP 203, 321.

Charikov, Y.E. & Fleishman, G.D. 1991, SP 139, 387.

Charikov, Y.E., Mosunov, A.N., & Prokopjev, A.V. 1993, SP 147, 157.

Chen, F.F. 1974, *Introduction to Plasma Physics*, (see book list).

Chen, J. 1989, ApJ 338, 453.

Chen, J., Burkhart, G.R., & Huang, C.R. 1990, GRL 17, 2237.

Chen, J. 1992, JGR 97, 15011.

Chen, J. 1996, JGR 101, 27499.

Chen, J. 1997a, ApJ 409, L191.

Chen, J. 1997b, Proc-1997-Crooker, 65.

Chen, J. 2000, SSR 95, 165.

Chen, J. & Krall, J. 2003, JGR 108/A11, 1410, SSH 2-1.

Chen, J., Santoro, R.A., Krall, J. et al. 2000, ApJ 533, 481.

Chen, P.F., Fang, C., Tang, Y.H., et al. 1999a, ApJ 513, 516.

Chen, P.F., Fang, C., Ding, M.D., et al. 1999b, ApJ 520, 853.

Chen, P.F. & Shibata, K. 2000, ApJ 545, 524.

Chen, P.F., Wu, S.T., Shibata, K., et al. 2002, ApJ 572, L99.

Cheng, C.C. & Pallavicini, R. 1984, SP 93, 337.

Cheng, C.C., Karpen, J.T., & Doschek, G.A. 1984, ApJ 286, 787.

Cheng, C.C., Vanderveen, K., Orwig, L.E., et al. 1988, ApJ 330, 480.

Cheng, C.C. 1999, in Strong et al., (see booklist), 393.

Cheng, C.Z. & Choe, G.S. 1998, ApJ 505, 376.

Cheng, C.Z., Ren, Y., Choe, G.S. et al. 2003, ApJ 596, 1341.

Chernov, G.P. 1989, Sov.Astron. 33(6), 649

Chernov, G.P. & Kurths, J 1990, Sov. Astron. 34(5), 516.

Chernov, G.P., Markeev, A.K., Poquerusse, M., et al. 1998, AA 334, 314.

Chertok, I.M. & Grechnev, V.V. 2003, Astronomy Reports, 47, 139.

Chertok, I.M. & Grechnev, V.V. 2004, Astronomy Reports, 47/11, 934.

Chiu, Y.T., & Hilton, H.H. 1977, ApJ 212, 873.

Chiuderi, C., Einaudi, G., & Torricelli−Campioni,G. 1981, AA 97, 27.

Chiuderi, C. 1996, Proc-1996-Bentley, 69.

Chiuderi−Drago, F. & Poletto, G. 1977, AA 60, 227.

Chiuderi−Drago, F., Avignon, Y., & Thomas, R.J. 1977, SP 51, 143.

Chiuderi−Drago, F., Landi, E., Fludra, A., et al. 1999, AA 348, 261.

Choe, G.S. & Lee, L.C. 1996, ApJ 472, 372.

Chou, Y.P. & Charbonneau, P. 1996, SP 166, 333.

Chupp, E.L., et al. 1981, ApJ 244, L171.

Chupp, E.L., Forrest, D.J., Vestrand, W.T. et al. 1985, Proc-1985-ICRC19, 126.

Chupp, E.L., Debrunner, H., Flückiger, E. et al. 1987, ApJ 318, 913.

Chupp, E.L., Trottet, G., Marschhäuser, H. et al. 1993, AA 275, 602.

Chupp, E.L. 1995, Nuclear Physics B 39A, 3.

Chupp, E.L. 1996, Proc-1996-Ramaty, 3.

Ciaravella, A., Peres,G., & Serio,S. 1991, SP 132, 279.

Ciaravella, A., Raymond, J.C., Thompson, B.J. et al. 2000, ApJ 529, 575.

Ciaravella, A., Raymond, J.C., Reale, F., et al. 2001, ApJ 557, 351.

Ciaravella, A., Raymond, J.C., VanBallegooijen, A. et al. 2003, ApJ 597, 1118.

Cid, C., Hidalgo, M.A., Sequeiros, J., et al. 2001, SP 198, 169.

Classen, H.T. & Aurass, H. 2002, AA 384, 1098.

Cliver, E.W., Crosby, N.B., & B.R. Dennis 1994, Proc-1994-Ryan, 65.

Cliver, E.W. 2001, *Solar Flare Classification*, (in Murdin 2000) .

Colburn, D.S. & Sonnett, C.P. 1996, SSR 5, 439.

Colgate, S.A. 1978, ApJ 221, 1068.

Conway, A.J., & MacKinnon, A.L. 1998, AA 339, 298.

Conway, A.J., & Willes, A.J. 2000, AA 355, 751.

Conway, A.J., Brown, J.C., Eves, B.A.C., et al. 2003, AA 407, 725.

Coppi, B. & Friedland, A.B. 1971, ApJ 169, 379.

Cook, J.W., Cheng, C.C., Jacobs, V.L., et al. 1989, ApJ 338, 1176.

Cooper, F.C., Nakariakov, V.M., & Tsiklauri, D. 2003, AA 397, 765.

Correia, E. & Kaufmann, P. 1987, SP 111, 143.

Cowley, S.W.H. 1974a, J. Plasma Phys. 12, 319.

Cowley, S.W.H. 1974b, J. Plasma Phys. 12, 341.

Cowling, T.G. 1976, *Magnetohydrodynamics*, (see book list).

Cox, A.N. (ed.) 2000, *Allen's Astrophysical Quantities*, (see book list).

Craig, I.J.D. & Brown, J.C. 1976, AA 49, 239.

Craig, I.J.D., McClymont, A.N., & Underwood, J.H. 1978, AA 70, 1.

Craig, I.J.D., Robb, T.D., & Rollo, M.D. 1982, SP 76, 331.

Craig, I.J.D., MacKinnon, A.L., & Vilmer,N. 1985, Astrophys. Space Sci.116, 377

Craig, I.J.D. & McClymont, A.N. 1986, ApJ 307, 367.

Craig, I.J.D. & Sneyd, A.D. 1986, ApJ 311, 451.

Craig, I.J.D. & Brown, J.C. 1986, *Inverse Problems in Astrophysics*, (see book list).

Craig, I.J.D. & McClymont, A.N. 1991, ApJ 371, L41.

Craig, I.J.D. & Watson, P.G. 1992, ApJ 393, 385.

Craig, I.J.D. & McClymont, A.N. 1993, ApJ 405, 207.

Craig, I.J.D. & Rickard, G.J. 1994, AA 287, 261.

Craig, I,J.D. & Henton, S.M. 1994, ApJ 434, 192.

Craig, I,J.D. & Henton, S.M. 1995, ApJ 450, 280.

Craig, I.J.D. & McClymont, A.N. 1997, ApJ 481, 996.

Craig, I.J.D. & McClymont, A.N. 1999, ApJ 510, 1045.

Craig, I.J.D., Fabling, R.B., Heerikhuisen, J., et al. 1999, ApJ 523, 838.

Craig, I.J.D. & Watson, P.G. 2000a, SP 194, 251.

Craig, I.J.D. & Watson, P.G. 2000b, SP 191, 359.

Craig, I.J.D. & Wheatland, M.S. 2002, SP 211, 275.

Cranmer, S.R., Kohl, J.L., Noci, G., et al. 1999a, ApJ 511, 481.

Cranmer, S.R., Field, G., & Kohl, J.L. 1999b, ApJ 518, 937.

Cranmer, S.R. 2001, *Coronal Holes*, (in Murdin 2000) .

Cranmer, S.R. 2002a, COSPAR-CS 13, 3.

Cranmer, S.R. 2002b, SSR 101, 229.

Crannell, C.J., Joyce, G., Ramaty, R., et al. 1976, ApJ 210, 582.

Cremades, H. & Bothmer, V. 2004, AA, 422, 307.

Crifo, F., Picat, J.P., & Cailloux, M. 1983, SP 83, 143.

Crooker, N., Joselyn, J.A., & Feynmann, J. (eds.) 1997, Proc-1997-Crooker.

Crosby, N.B., Aschwanden, M.J., & Dennis, B.R. 1993, SP 143, 275.

Culhane, J.L., Hiei, E., Doschek, G.A., et al. 1991, SP 136, 89.

Culhane, J.L. & Jordan, C. (eds.) 1991, RSPTA 336, 494p.

Culhane, J.L., Phillips, A.T., Inda−Koide,M., et al. 1994, SP 153, 307.

Cuntz, M. & Suess, S.T. 2001, ApJ 549, L143.

Cuperman, S., Ofman, L., and Semel, M. 1990, AA 230, 193.

Cushman, G.W. & Rense, W.A. 1976, ApJ 207, L61.

Czaykowska, A., DePontieu, B., Alexander, D., et al. 1999, ApJ 521, L75.

Czaykowska, A., Alexander, D., & DePontieu, B. 2001, ApJ 552, 849.

Daglis, I.A. (ed.) 2001, Proc-2001-Daglis.

Dahlburg, R.B., DeVore, C.R., Picone, J.M., et al. 1987 ApJ 315, 385.

Dahlburg, R.B., Antiochos,S.K., & Klimchuk,J.A. 1998, ApJ 495, 485.

Datlowe, D.W. & Lin, R.P. 1973, SP 32, 459.

Datlowe, D.W., Elcan, M.J., & Hudson, H.S. 1974, SP 39, 155.

Davidson, P.A. 2001, *An Introduction to Magnetohydrodynamics*, (see book list).

Davila, J.M. 1987, ApJ 317, 514.

Davila, J.M. 1991, Proc-1991-Ulmschneider, 464.

Decker, R.B. & Vlahos, L. 1986, ApJ 306, 710.

Decker, R.B. 1988, Space Sciecne Reviews 48/3-4, 195.

DeForest, C.E., & Gurman, J.B. 1998, ApJ 501, L217.

DeForest, C.E., Lamy, P. L., & Llebaria, A. 2001, ApJ 560, 490.

De Groot, T. 1970, SP 14, 176.

De Jager, C., Boelee, A., & Rust, D.M. 1984, SP 92, 245.

De Jager, C. & Boelee, A. 1984, SP 92, 227.

De Jager, C. 1985, SP 96, 143.

De Jager, C. 1986, SSR 44, 43.

De Jager, C., Inda−Koide, M., Koide, S. et al. 1995, SP 158, 391.

Delaboudinière, J.P. et al. 1995, SP 162, 291.

Délannée, C. & Aulanier, G. 1999, SP 190, 107.

Délannée, C. 2000, ApJ 545, 512.

Délannée, C., Delaboudiniere, J.P., & Lamy, P. 2000, AA 355, 725.

DelZanna, G., & Bromage, B.J.I. 1999, JGR 104/A5, 9753.

Démoulin, P., Cuperman, S., Semel, M. 1992, AA 263, 351.

Démoulin, P., Hénoux, J.C., Priest, E.R., et al. 1996, AA 308, 643.

Démoulin, P., Bagalá, L.G., Mandrini, C.H., et al. 1997a, AA 325, 305.

Démoulin, P., Hénoux, J.C., Mandrini, C.H., et al. 1997b, SP 174, 73.

Démoulin, P. 2003, AA 402, 769.

Démoulin, P., Van Driel−Gesztelyi, L., Mandrini, C.H., et al. 2003, ApJ 586, 592.

De Moortel, I., Hood, A.W., Ireland, J., et al. 1999, AA, 346, 641.

De Moortel, I., Hood, A.W., & Arber, T.D. 2000a, AA, 354, 334.

De Moortel, I., Ireland, J., & Walsh, R.W. 2000b, AA 355, L23.

De Moortel, I., Ireland, J., Walsh, R.W., et al. 2002a, SP 209, 61.

De Moortel, I., Hood, A.W., Ireland, J., et al. 2002b, SP 209, 89.

De Moortel, I., Ireland, J., Hood, A. W., et al. 2002c, AA 387, L13.

De Moortel, I. & Hood, A.W. 2003, AA 408, 755.

De Moortel, I. & Hood, A.W. 2004, AA 415, 704.

Deng, Y., Wang, J., Yan, Y., et al. 2001, SP 204, 11.

Dennis, B.R. 1985, SP 100, 465.

Dennis, B.R. 1988, SP 118, 49.

Dennis, B.R., Orwig, L.E., & Kiplinger, A.L. (eds.) 1987, Proc-1987-Dennis, 478p.

Dennis, B.R. & Schwartz, R.A. 1989, SP 121, 75.

Dennis, B.R., Orwig, L.E., Kennard, G.S., et al. 1991, NASA TM-4332.

Dennis, B.R. & Zarro, D.M. 1993, SP 146, 177.

Dennis, B.R., Veronig, A., Schwartz, R.A., et al. 2003, AdSpR 32/12, 2459.

DePontieu, B. 1999, AA 347, 696.

DePontieu, B., Berger, T.E., Schrijver, C.J., et al. 1999, SP 190, 419.

DePontieu, B., Martens, P.C.H., & Hudson, H.S. 2001, ApJ 558, 859.

Dere, K.P. & Cook, J.W. 1979, ApJ 229, 772.

Dere, K.P. 1982, SP 75, 189.

Dere, K.P., Bartoe, J.D.F., & Brueckner, G.E. 1989, SP 123, 41.

Dere, K.P., Bartoe, J.D.F., Brueckner, G.E., et al. 1991, JGR 96/A6, 9399.

Dere, K.P. 1996, ApJ 472, 864.

Dere, K.P., Landi, E., Mason, H.E., et al. 1997a, AA 125, 149.

Dere, K.P., Brueckner, G.A., Howard, R.A. et al. 1997b, SP 175, 601.

Dere, K.P., Brueckner, G.E., Howard, R.A. et al. 1999, ApJ 516, 465.

Dere, K.P., Landi, E., Young, P.R., et al. 2001, ApJS 134, 331.

Dermer, C.D. & Ramaty, R. 1986, ApJ 301, 962.

Desai, U.D., Kouveliotou, C., Barat, C., et al. 1987, ApJ 319, 567.

DeToma, G., White, O.R., & Harvey, K.L. 2000, ApJ 529, 1101.

DeVore, C.R. & Antiochos, S.K. 2000, ApJ 539, 954.

Díaz, A.J., Oliver, R., Erdélyi, R., et al. 2001, AA 379, 1083.

Díaz, A.J., Oliver, R., & Ballester, J.L. 2002, ApJ 580, 550.

Díaz, A.J., Oliver, R., & Ballester, J.L. 2003, AA 402, 781.

Ding, M.D. & Fang, C. 1989, AA 225, 204.

Ding, M.D. & Fang, C. 1996, AA 314, 643.

Ding, M.D., Watanabe, T., Shibata, K., et al. 1996, ApJ 458, 391.

Ding, M.D., Fang, C., & Yun, H.S. 1999, ApJ 512, 454.

Dingus, B.L., Sreekumar, P., Bertsch, D.L. et al. 1994, Proc-1994-Ryan, 177.

Dmitruk, P. & Gomez, D.O. 1997, ApJ 484, L83.

Dmitruk, P., Gomez, D.O., & DeLuca, E.E. 1998, ApJ 505, 974.

Dmitruk, P., Milano, L.J., & Matthaeus, W.H. 2001, ApJ 548, 482.

Dmitruk, P., Matthaeus, W.H., Milano, L.J., et al. 2002, ApJ 575, 571.

Dobrzycka,D., Cranmer,S.R., Panasyuk, A.V., et al. 1999, JGR 104/A5, 9791.

Dorch, S.B.F., Archontis, V., & Nordlund, A. 1999, AA 352, L79.

Doschek, G.A., Feldman, U., & Bohlin, J.D. 1976, ApJ 205, L177.

Doschek, G.A., & Feldman, U. 1977, ApJ 212, L143.

Doschek, G.A. et al. 1986, Proc-1986-Kundu, 4-1.

Doschek, G.A. 1990, ApJS 73, 117.

Doschek, G.A. 1991, Proc-1991-Uchida, 121.

Doschek, G.A., Warren, H.P., Laming, J.M., et al. 1997, ApJ 482, L109.

Doschek, G.A., Laming, J.M., Feldman, U., et al. 1998, ApJ 504, 573.

Doschek, G.A., Feldman, U., Laming, J.M., et al. 2001, ApJ 546, 559.

Dowdy, J.F.Jr., Moore, R.L. & Wu, S.T. 1985, SP 99, 79.

Dowdy, J.F.Jr., Rabin, D., & Moore, R.L. 1986, SP 105, 35.

Doyle, J.G. & Phillips, K.J.H. 1992, AA 257, 773.

Doyle, J.G., Banerjee, D., & Perez, M.E. 1998, SP 181, 91.

Doyle, J.G., Teriaca, L., & Banerjee, D. 1999, AA 349, 956.

Drago, F. 1974, Proc-1974-Righini, 120.

Drake, J.F., Biskamp, D., & Zeiler, A. 1997, Geophys. Res. Lett. 24/22, 2921.

Dreicer, H. 1959, Phys.Rev. 115, 238.

Dreicer, H. 1960, Phys.Rev. 117, 329.

Dröge, F. 1967, Z. Astrophys. 66, 200.

Dryer, M. & Maxwell,A. 1979, ApJ 231, 945.

Dryer, M., Wu, S.T., Steinolfson, R.S. et al. 1979, ApJ 227, 1059.

Dryer, M. 1982, SSR 22, 233.

Dryer, M. 1994, SSR 67, 363.

Dryer, M. 1996, SP 169, 421.

Dryer, M., Fry, C.D., Sun, W., et al. 2001, SP 204, 267.

D'Silva, S., & Choudhuri, A. 1993, AA 272, 621.

Duijveman, A., Hoyng, P., & Ionson, J.A. 1981, ApJ 245, 721.

Duijveman, A., Hoyng, P., & Machado, M.E. 1982, SP 81, 137.

Dulk, G.A. & McLean, D.J. 1978, SP 57, 279.

Dulk, G.A. & Dennis, B.R. 1982, ApJ 260, 875.

Dulk, G.A. 1985, ARAA 23, 169.

Dulk, G.A., Sheridan, K.V., Smerd, S.F., et al. 1977, SP 52, 349.

Dulk, G.A. 1990, SP 130, 139.

Dulk, G.A., Kiplinger, A.L., & Winglee, R.M. 1992, ApJ 389, 756.

Dungey, J.W. 1953, Phil. Mag. 44, 725.

Dupree, A.K., Penn, M.J., & Jones, H.P. 1996, ApJ 467, L121.

Edlén, B. 1943, Z.Astrophysik, 22, 30.

Edwin, P.M. & Roberts, B. 1982, SP 76, 239.

Edwin, P.M. & Roberts, B. 1983, SP 88, 179.

Eichler, D. 1979, ApJ 229, 413.

Einaudi, G., Velli, M., Politano, H., et al. 1996a, ApJ 457, L113.

Einaudi, G., Califano, F., & Chiuderi, C. 1996b, ApJ 472, 853.

Elgaroy, Ø. 1980, AA 82, 308.

Ellerman, F. 1917, ApJ 46, 298.

Ellison, D.C. & Ramaty, R. 1985, ApJ 298, 400.

Elwert, G. 1939, Ann.Physik 34, 178.

Elwert, G. & Haug, E. 1971, SP 20, 413.

Emonet, T. & Moreno−Insertis, F. 1996, ApJ 458, 783.

Emslie, A.G. 1978, ApJ 224, 241.

Emslie, A.G., & Rust, D.M. 1980, SP 65, 271.

Emslie, A.G. 1981, ApJ 245, 711.

Emslie, A.G., Brown, J.C., & Machado, M.E. 1981, ApJ 246, 337.

Emslie, A.G. 1983, SP 86, 133.

Emslie, A.G. & Brown, J.C. 1985, ApJ 295, 648.

Emslie, A.G. 1989, SP 121, 105.

Emslie, A.G., Li, P., & Mariska, J.T. 1992, ApJ 399, 714.

Emslie, A.G. & Hénoux, J.C. 1995, ApJ 446, 371.

Emslie, A.G., Hénoux, J.C., Mariska, J.T., et al. 1996, ApJ 470, L131.

Emslie, A.G., Brown, J.C., & MacKinnon, A.L. 1997, ApJ 485, 430.

Emslie, A.G., Mariska, J.T., Montgomery, M.M., et al. 1998, ApJ 498, 441.

Emslie, A.G., Miller, J.A., Vogt, E., et al. 2000, ApJ 542, 513.

Emslie, A.G. 2003, ApJ 595, L119.

Emslie, A.G., Kontar, E.P., Krucker, S., et al. 2003, ApJ 595, L107.

Emslie, A.G., Miller, J.A., & Brown, J.C. 2004, ApJ 602, L69.

Engvold, Ø. & Jensen, E. 1977, SP 52, 37.

Engvold, Ø. 2001a, *Solar Prominence Fine Structure*, (in Murdin 2000) .

Engvold, Ø. 2001b, Proc 2001/Ballester, 123.

Enome, S. 1983, AdSpR 2/11, 201.

Erdélyi, R. & Goossens, M. 1994, Astrophysics and Space Science 213, 273.

Erdélyi, R. & Goossens, M. 1995, SP 294, 575.

Erdélyi, R., Goossens, M., & Ruderman, M.S. 1995, SP 161, 123.

Erdélyi, R. & Goossens, M. 1996, AA 313, 664.

Erdélyi, R., Sarro, L.M., & Doyle, J.G. 1998a, ESA SP-421, 207.

Erdélyi, R., Doyle, J.G., Perez, M.E., et al. 1998b, AA 337, 287.

Erdélyi, R., Ballai, I., & Goossens, M. 2001, AA 368, 662.

Erdoes, G. 2003, Proc-2003-Erdélyi, 367.

Ergun, R.E., Larson, D., Lin, R.P., et al. 1998, ApJ 503, 435.

Erickson, W.C. 1964, ApJ 139, 1290.

Esser, R., Fineschi, S., Dobrzycka, D., et al. 1999, ApJ 510, L63.

Esser, R. 2001, *Solar Wind Acceleration*, (in Murdin 2000) .

Ewell, M.W.Jr., Zirin, H., Jensen, J.B., et al. 1993, ApJ 403, 426.

Falchi, A. & Mauas,J.D. 2002, AA 387, 678.

Falconer, D.A. 1994, PhD Thesis (see PhD Thesis list).

Falconer, D.A., Moore, R.L., Porter, J.G., et al. 1998, ApJ 501, 386.

Falconer, D.A., Moore, R.L., & Gary, G.A. 2001, GRL 106/A11, 25185.

Falconer, D.A., Moore, R.L., & Gary, G.A. 2002, ApJ 569, 1016.

Fan, C.Y., Gloeckler, G., & Simpson, J.A. 1964, Phys. Rev. Lett. 13, 149.

Fan, Y., Fisher, G., & DeLuca, E. 1993, ApJ 405, 390.

Fan, Y., Zweibel, E.G., Linton, M.G., et al. 1999, ApJ 521, 460.

Fan, Y. & Gibson, S.E. 2003, ApJ 589, L105.

Fan, Y. & Gibson, S.E. 2004, ApJ 609, 1123.

Farnik, F., Hudson, H., & Watanabe, T. 1997, AA 320, 620.

Feldman, U. 1992, Physica Scripta, 46, 202.

Feldman, U. & Laming, J.M. 2000, Physica Scripta, 61, 222.

Feldman, U. & Widing, K.G. 2003, Space Sci.Rev. 107, 665.

Fermi, E. 1949, *Phys.Rev.* 75, 1169.

Fermi, E. 1954, ApJ 119, 1.

Fiedler, R.A. 1992, ESA SP-348, 273.

Field, G.B. 1965, ApJ 142, 531.

Filippov, B. 1999, SP 185, 297.

Fisher, G.H., Canfield, R.C., & McClymont, S. 1984, ApJ 281, L79.

Fisher, G.H., Canfield, R.C., & McClymont, S. 1985a, ApJ 289, 414.

Fisher, G.H., Canfield, R.C., & McClymont, S. 1985b, ApJ 289, 425.

Fisher, G.H. 1986, Proc-1986-Mihalas, 53.

Fisher, G.H. 1989, ApJ 346, 1019.

Fisher, G.H. & Hawley, S.L. 1990, ApJ 357, 243.

Fisher, G.H., Fan, Y., & Howard, R.F. 1995, ApJ 438, 463.

Fisher, G.H., Longcope, D.W., Metcalf,T.R., et al. 1998, ApJ 508, 885.

Fisher, R.R. & Musman, S. 1975, ApJ 194, 801.

Fishman, G.J. et al. 1989, Proc-1989-Johnson, 2-39 and 3-47.

Fisk, L.A. 1976, JGR 81, 4633.

Fleck, B., Domingo, V., & Poland, A. (eds.) 1995, SP 162.

Fleck, B. & Svestka, Z. (eds.) 1997, SP 170 and 175.

Fleishman, G.D. & Charikov, Y.E. 1991, Sov.Astron. 35/4, 354.

Fleishman, G.D. & Yastrebov, S.G. 1994a, SP 153, 389.

Fleishman, G.D. & Yastrebov, S.G. 1994b, SP 154, 361.

Fleishman, G.D. & Mel'nikov, V.F. 1998, Physics Uspekhi 41(12), 1157.

Fleishman, G.D. & Arzner, K.J. 2000, AA 358, 776.

Fletcher, L. 1995, AA 303, L9.

Fletcher, L. 1996, AA 310, 661.

Fletcher, L. & Petkaki, P. 1997, SP 172, 267.

Fletcher, L. & Martens, P.C.H. 1998, ApJ 505, 418.

Fletcher, L. 1999, Proc-1999-ESA448, 693.

Fletcher, L., & DePontieu, B. 1999, ApJ 520, L135.

Fletcher, L. & Hudson, H.S. 2001, SP 204, 69.

Fletcher, L., Metcalf, T.R., Alexander, D., et al. 2001, ApJ 554, 451.

Fletcher, L. & Hudson, H.S. 2002, SP 210, 307.

Fludra, A., Lemen, J.R., Jakimiec, J. et al. 1989, ApJ 344, 391.

Fludra, A., DelZanna, G., Alexander, D., et al. 1999, JGR 104/A5, 9709.

Fludra, A. 2001, AA 368, 639.

Foley,C.A., Acton, L.W., Culhane, J.L., et al. 1996, Proc-1996-Uchida, 419.

Foley,C.R., Culhane, J.L., & Acton, L.W. 1997, ApJ 491, 933.

Fontenla,J.M., Tandberg–Hanssen, E., Reichmann, E.J., et al. 1989, ApJ 344, 1034.

Fontenla,J.M., Avrett, E.H., & Loeser, R. 1990, ApJ 355, 700.

Fontenla,J.M., Avrett, E.H., & Loeser, R. 1991, ApJ 377, 712.

Fontenla,J.M., Avrett, E.H., & Loeser, R. 1993, ApJ 406, 319.

Fontenla,J.M., Avrett, E.H., & Loeser, R. 2002, ApJ 572, 636.

Forbes, T.G. & Priest, E.R. 1984, SP 94, 315.

Forbes, T.G. & Malherbe, J.M. 1986a, ApJ 302, L67.

Forbes, T.G. & Malherbe, J.M. 1986b, Proc-1986-Neidig, 443.

Forbes, T.G. & Priest, E.R. 1987, Rev. Geophys. 25, 1583.

Forbes, T.G., Malherbe, J.M., & Priest,E.R. 1989, SP 120, 258.

Forbes, T.G. 1991, Geophys. Astrophys. Fluid Dynamics 62, 15.

Forbes, T.G. & Isenberg, P.A. 1991, ApJ 373, 294.

Forbes, T.G. & Priest, E.R. 1995, ApJ 446, 377.

Forbes, T.G. & Acton, L.W. 1996, ApJ 459, 330.

Forbes, T.G. 1996, Proc-1996-Bentley, 259.

Forbes, T.G. 1997, Proc-1997-Mouradian, 149.

Forbes, T.G. 2000a, Adv.Spaced Res. 26/3, 549.

Forbes, T.G. 2000b, Phil. Trans. Roy. Soc. A, 358, 711.

Forbes, T.G. 2000c, JGR 105/A10, 23153.

Forbes, T.G. 2001, *Solar Flare Models*, (in Murdin 2000) .

Forrest, D.J., Chupp, E.L., Ryan, J.M., et al. 1980, SP 65, 15.

Forrest, D.J. & Chupp, E.L. 1983, Nature 305, 291.

Forrest, D.J., Vestrand, W.T., Chupp, E.L., et al. 1985, Proc-1985-ICRC19, 146.

Forrest, D.J., Vestrand, W.T., Chupp, E.L., et al. 1986, Adv. Space Res. 6, 115.

Foukal, P.V. 1978, ApJ 223, 1046.

Foukal, P.V. 1987, Proc-1987-Athay, 15.

Foukal, P.V. 1990, *Solar Astrophysics*, (see book list).

Fredvik, T., Kjeldseth−Moe, O., Haugan, S.V.H., et al. 2002, Adv. Space Res., 30/3, 635.

Frost, K.J. 1969, ApJ 158, L159.

Frost, K.J., Dennis, B.R. & Lencho, R.J. 1971, Proc-1971-Labuhn, 185.

Fu, Q.J., Gong, Y.F., Jin, S.Z., et al. 1990, SP 130, 161.

Fürst, E. & Hirth, W. 1975, SP 42, 157.

Furth, H.P., Killeen, J., & Rosenbluth, M.N. 1963, *Phys. Fluids* 6, 459.

Furusawa, K. & Sakai, J. 2000, ApJ 540, 1156.

Gabriel, A.H. 1976, Phil.Trans.R.Soc., (London) 281, 399.

Gabriel, A.H., Culhane, J.L., Patchett, B.E., et al. 1995, Adv.Sace Sci., 15/7, 63.

Gaizauskas, V., Zirker, J.B., Sweetland, C., et al. 1997, ApJ 479, 448.

Gaizauskas, V. 2001, *Solar Filament Channels*, (in Murdin 2000) .

Gallagher, P.T., Phillips, K,J.H., Harra−Murnion, L.K., et al. 1999, AA 348, 251.

Gallagher, P.T., Dennis, B.R., Krucker, S., et al. 2002, SP 210, 341.

Gallagher, P.T., Lawrence, G.R., & Dennis, B.R. 2003, ApJ 588, L53.

Galsgaard, K. & Nordlund, A. 1996, J.Geophys. Res. 101, 13,445.

Galsgaard, K. & Nordlund, A. 1997, J.Geophys. Res. 102, 219.

Galsgaard, K., Reddy, R.V., & Rickard, G.J. 1997b, SP 176, 299.

Galsgaard, K., and Longbottom, A.W. 1999, ApJ 510, 444.

Galsgaard, K., Priest, E.R., & Nordlund,A. 2000, SP 193, 1.

Galsgaard, K. & Roussev, I. 2002, AA 383, 685.

Galtier, S. 1999, ApJ 521, 483.

Gan, W.Q., Fang, C., & Zhang, H.Q. 1991, AA 241, 618.

Gan, W.Q., Li, Y.P., Chang, J., et al. 2002, SP 207, 137.

Garcia,H.A. 1998, ApJ 504, 1051.

Garren, D., Chen, J., & Cargill, P. 1993, ApJ 418, 919.

Gary, D.E. & Hurford, G.J. 1987, ApJ 317, 522.

Gary, D.E. & Hurford, G.J. 1989, Proc-1989-Waite, ...

Gary, D.E., Zirin, H., & Wang, H. 1990, ApJ 355, 321.

Gary, D.E., Wang, H., Nitta, N., et al. 1996, ApJ 464, 965.

Gary, D.E. & Hurford, G.J. 1994, ApJ 420, 903.

Gary, D.E., Hartl, M.D., & Shimizu,T. 1997, ApJ 477, 958.

Gary, A., & Démoulin, P. 1995, ApJ 445, 982.

Gary, G.A. 1997, SP 174, 241.

Gary, G.A. 1989, ApJS 69, 323.

Gary, G.A. 2001, SP 203, 71.

Gary, G.A. & Moore, R.L. 2004, ApJ 611, 545.

Gebbie, K.B., Hill, F., Toomre, J., et al. 1981, ApJ 251, L115.

Geiss, J. 1985, Proc-1985-ESA, 37.

Gelfreikh, G.B., Grechnev,V., Kosugi,T., et al. 1999, SP 185, 177.

Georgoulis, M,K., Rust, D.M., Bernasconi, P.N., et al. 2002, ApJ 575, 506.

Giachetti, R., Van Hoven, G., & Chiuderi, C. 1977, SP 55, 731.

Gibson, S.E. & Low,B.C. 1998, ApJ 493, 460.

Ginzburg, V.L. & Zheleznyakov, V.V. 1958, Soviet Astronomy 2, 653.

Ginzburg, V.L. 1961, *Propagation of Electromagnetic Waves in a Plasma*, (see book list).

Ginzburg, V.L. & Syrovatskii, S.I. 1964, *The origin of cosmic rays*, (see book list).

Gleeeson, L.J. & Axford, W.I. 1967, ApJ 149, L115.

Glencross, W.M. 1980, AA 83, 65.

Gold, T. & Hoyle, F. 1960, MNRAS 120, 89.

Goldreich, P. & Julian, W.H. 1969, ApJ 157, 869.

Goldstein, M.L., Roberts, D.A., & Matthaeus, W.H. 1995, ARAA 33, 283.

Goldman, M.V. & Smith, D.F. 1986, Proc-1986-Sturrock, 325.

Goldman, M.V. 1989, Proc-1989-Waite, ...

Golub, L., Krieger, A.S, Silk, J.K., et al. 1974, ApJ 189, L93.

Golub, L., Krieger, A.S., & Vaiana, G.S. 1976a, SP 49, 79.

Golub, L., Krieger, A.S., & Vaiana, G.S. 1976b, SP 50, 311.

Golub, L., Krieger, A.S., Harvey, J.W., et al. 1977, SP 53, 111.

Golub, L., Davis, J.M., & Krieger, A.L. 1979, ApJ 229, L145.

Golub, L., Maxson, C., Rosner, R., et al. 1980, ApJ 238, 343.

Golub, L. 1997, SP 174, 99.

Golub, L. & Pasachoff, J.M. 1997, *The Solar Corona*, (see book list).

Golub, L., Bookbinder, J., DeLuca, E., et al. 1999, Phys.Plasmas 6/5, 2205.

Gomez, D., Schifino, A.S. & Fontan, C.F. 1990, ApJ 352, 318.

Goodrich, C.C. 1985, Proc-1985-Tsurutani, 153.

Goossens, M. 1991, Proc-1991-Priest, 137.

Goossens, M., Hollweg, J.V., & Sakurai, T. 1992, SP 138, 233.

Goossens, M., Ruderman, M.S., & Hollweg, J.V. 1995, SP 157, 75.

Goossens, M., Andries, J., & Aschwanden, M.J. 2002a, AA 394, L39.

Goossens, M., DeGroof, A., & Andries, J. 2002b, ESA-SP 505, 137.

Goossens, M. 2003, *An Introduction to Plasma Astrophysics*, (see book list).

Gopalswamy, N., White, S.M., & Kundu, M.R. 1991, ApJ 379, 366.

Gopalswamy, N. & Kundu, M.R. 1992, ApJ 390, L37.

Gopalswamy, N., Payne, T.E.W., Schmahl, E.J., et al. 1994, ApJ 437, 522.

Gopalswamy, N., Raulin, J.P., Kundu, M.R., et al. 1995, ApJ 455, 715.

Gopalswamy, N., Zhang, J., Kundu, M.R., et al. 1997, ApJ 491, L115.

Gopalswamy, N. & Hanaoka, Y. 1998, ApJ 498, L179.

Gopalswamy, N., Shibasaki, K., Thompson, B.J., et al. 1999, JGR 104/A5, 9767.

Gopalswamy, N., Lara, A., Yashiro,S. et al. 2003, Proc-2003-Wilson, 403.

Gosling, J.T. 1993, JGR 98, 18,937.

Gosling, J.T. 1994, Proc-1994-ESA373, 275.

Gosling, J.T. 1996, ARAA 34, 35.

Gotwols, B.L. 1972, SP 25, 232.

Grad, H., & Rubin, H. 1958, Proc-1958-UN, 190.

Green, R.M. and Sweet, P.A. 1967, ApJ 147, 1153.

Grevesse, N. & Sauval, A.J. 2001, *Solar Abundances*, (in Murdin 2000) .

Grognard, R.J.M. 1985, in *McLean & Labrum 1985*, 253.

Gu, Y.M., Jefferies, J.T., Lindsey, C., et al. 1997, ApJ 484, 960.

Güdel, M. & Benz, A.O. 1988, AASS 75, 243.

Güdel, M. 1990, AA 239, L1.

Güdel, M. & Benz, A.O. 1990, AA 231, 202.

Güdel, M., Aschwanden, M.J., & Benz, A.O. 1991, AA 251, 285.

Güdel, M. & Wentzel, D.G. 1993, ApJ 415, 750.

Güdel, M., Benz, A.O., Schmitt, J.H.M.M., et al. 1996, ApJ 471, 1002.

Güdel, M., Audard, M., Smith, K.W., et al. 2002, ApJ 577, 371.

Gudiksen, B.V. & Nordlund, A. 2002, ApJ 572, L113.

Guhathakurta, M., Fisher, R., & Strong, K. 1996, ApJL 471, L69.

Guhathakurta, M., & Fisher, R. 1998, ApJ 499, L215.

Guillermier, P. & Koutchmy, S. 1999, *Total Eclipses* (see book list).

Guo, W.P. & Wu, S.T. 1998, ApJ 494, 419.

Guo, W.P., Wang, J.F., Liang, B.X. et al. 1992, Proc-1992-Svestka, 381.

Gurnett, D.A. & Anderson, R.R. 1977, JGR 82/1, 632.

Habbal, S.R., Esser, R., & Arndt, M.B. 1993, ApJ 413, 435.

Habbal, S.R. & Woo, R. 2001, *Solar Wind: Coronal Origins*, (in Murdin 2000) .

Haerendel, G. 1992, Nature 360, 241.

Haerendel, G. 1994, ApJS 90, 765.

Hagenaar, H.J. 2001, ApJ 555, 448.

Hagenaar, H.J., Schrijver, C.J., & Title, A.M. 2003, ApJ 584, 1107.

Halberstadt, G. & Goedbloed, J.P. 1995a, AA 301, 559.

Halberstadt, G. & Goedbloed, J.P. 1995b, AA 301, 577.

Hamilton, B., McClements, K.G., Fletcher, L., et al. 2003, SP 214, 339.

Hamilton, R.J. & Petrosian, V. 1990, ApJ 365, 778.

Hamilton, R.J., Lu, E.T., & Petrosian, V. 1990, ApJ 354, 726.

Hamilton, R.J. & Petrosian, V. 1992, ApJ 398, 350.

Hanaoka, Y. 1994, Proc-1994-Enome, 181.

Hanaoka, Y. 1996, SP 165, 275.

Hanaoka, Y. 1997, SP 173, 319.

Hanaoka, Y. 2003, ApJ 596, 1347.

Handy, B.N., Acton, L.W., Kankelborg, C.C. et al. 1999, SP 187, 229.

Hannah, I.G., Fletcher, L., & Hendry, M.A. 2002, ESA SP-506, 295.

Hansen, J.F. & Bellan, P.M. 2001, ApJ 563, L183.

Hansen, R.T., Garcia, C.G., Hansen, S.F. et al. 1974, PASP 86, 500.

Hansteen, V. 2001, *Transition Region Models*, (in Murdin 2000) .

Hara, H. & Ichimoto, K. 1996, Proc-1996-Bentley, 183.

Hara, H. 1997, PASJ 49, 413.

Hara, H., Tsuneta, S., Acton, L.W., et al. 1994, PASJ 46, 493.

Hara, H., Tsuneta, S., Acton, L.W., et al. 1996, Adv. Space Res., 17/4-5, 231.

Hardie, I.S., Hood, A.W., & Allen, H.R. 1991, SP 133, 313.

Hardy, S.J., Melrose, D.B., & Hudson, H.S. 1998, PASA 15, 318.

Harra−Murnion, L.K., Akita, K., & Watanabe, T. 1997, ApJ 479, 464.

Harra, L.K. & Sterling, A.C. 2001, ApJ 561, L215.

Harra, L.K. & Sterling, A.C. 2003, ApJ 587, 429.

Harrington, T.M., Maloy, J.D., McKenzie, D.L., et al. 1972, IEEE Trans.Nucl.Sci. NS-19, 596.

Harrison, R.A. 1986, AA 162, 283.

Harrison, R.A. 1987, AA 182, 337.

Harrison, R.A. et al. 1995, SP 162, 233.

Harrison, R.A. 1997, SP 170, 467.

Harrison, R.A., Lang, J., Brooks, D.H., et al. 1999, AA 351, 1115.

Harrison, R.A., & Lyons, M. 2000, AA 358, 1097.

Harrison, R.A., Bryans, P., & Bingham, R. 2001, AA 379, 324.

Harrison, R.A., Hood, A.W., & Pike, C.D. 2002, AA 392, 319.

Harrison, R.A., Bryans, P., Simnett, G.M., et al. 2003, AA 400, 1071.

Haruki, T. & Sakai, J.I. 2001a, Physics of Plasmas 8/5, 1538.

Haruki, T. & Sakai, J.I. 2001b, ApJ 552, L175.

Harvey, J.W. 1969, PhD Thesis, (see PhD Thesis list).

Harvey, K.L. & Martin, S.F. 1973, SP 32, 389.

Harvey, K.L., Martin, S.F., & Riddle,A.C. 1974, SP 36, 151.

Harvey, K.L. 1993, PhD Thesis, (see PhD Thesis list).

Harvey, K.L., Strong, K.S, Nitta, N., et al. 1994, Proc-1994-Balasubramaniam, 377.

Harvey, K.L. 1996, Proc-1996-Bentley, 9.

Harvey, K.L., Jones, H.P., Schrijver, C.J. et al. 1999, SP 190, 35.

Hasegawa, A. 1975, *Plasma Instabilities and Nonlinear Effects*, (see book list).

Hassler, D.M., Rottman, G.J. & Orrall, Q. 1991, ApJ 372, 710.

Hassler, D.M., Dammasch, I.E., Lemaire, P., et al. 1999, Science, 283, 810.

Haug, E. 1975, Z. Naturforsch. 30a, 1099.

Haug, E. 1997, AA 326, 417.

Hawley, S.L., Fisher, G.H., Simon, T., et al. 1995, ApJ 453, 464.

Hawley, S.L., Allred, J.C., Johns−Krull, C.M. 2003, ApJ 597, 535.

Heasley, J.N. & Mihalas, D. 1976, ApJ 205, 273.

Heerikhuisen, J., Litvinenko, Y.E., & Craig, I.J.D. 2002, ApJ 566, 512.

Heinzel, P. & Anzer, U. 2001, AA 375, 1082.

Hendrix, D.L., VanHoven, G., Mikić, Z., et al. 1996, ApJ 470, 1192.

Hennig, B.S. & Cally, P.S. 2001, SP 201, 289.

Hénoux, J.C., Chambe, G., Smith, D., et al. 1990, ApJS 73, 303.

Heristchi, D. & Boyer, R. 1994, Proc-1994-Ryan, 124.

Hewitt, R.G., Melrose, D.B., & Roennmark, K.G. 1982, Aust.J.Phys. 35, 447.

Hewitt, R.G. & Melrose, D.B. 1983, Aust.J.Phys. 36, 725.

Hewitt, R.G., Melrose, D.B., & Dulk, G.A. 1983, JGR 88, 10065.

Hewitt, R.G. & Melrose, D.B. 1985, SP 96, 157.

Heyvaerts, J., Priest, E.R., & Rust, D.M. 1977, ApJ 216, 123.

Heyvaerts, J. 1981, Proc-1981-Priest, 429.

Heyvaerts, J. & Priest, E.R. 1983, AA 117, 220.

Heyvaerts, J. & Priest, E.R. 1992, ApJ 390, 297.

Heyvaerts, J. 2000, Proc-2000-Rozelot, 1.

Heyvaerts, J. 2001, *Coronal Heating Mechanisms*, (in Murdin 2000) .

Hick, P. & Priest, E.R. 1989, SP 122, 111.

Hiei, E., Hundhausen, A.J., & Sime, D.G. 1993, GRL 20/24, 2785.

Hiei, E., Hundhausen, A.J., & Burkepile, J. 1996, Proc-1996-Bentley, 383.

Hildner, E. 1977, Proc-1977-Shea, 6.

Hillaris, A., Alissandrakis, C.E., & Vlahos, L. 1988, AA 195, 301.

Hillaris, A., Alissandrakis, C.E., Caroubalos, C., et al. 1990a, AA 229, 216.

Hillaris, A., Alissandrakis, C.E., Bougeret, J.L. et al. 1990b, AA 342, 271.

Hinata, S. 1980, ApJ 235, 258.

Hirayama, T. 1974, SP 34, 323.

Hirayama, T. 1985, SP 100, 415.

Hirose, S., Uchida, Y., Uemura, S., et al. 2001, ApJ 551, 586.

Hollweg, J.V. 1971, J.Geophys.Res. 76, 5155.

Hollweg, J.V. 1978, GRL 5, 731.

Hollweg, J.V. 1984a, ApJ 277, 392.

Hollweg, J.V. 1984b, SP 91, 269.

Hollweg, J.V. & Sterling, A.C. 1984, ApJ 282, L31.

Hollweg, J.V. 1985, Proc-1985-Buti, 77.

Hollweg, J.V. 1986, JGR 91, 4111.

Hollweg, J.V. 1987, ApJ 312, 880.

Hollweg, J.V. & Yang, G. 1988, JGR 93/A6, 5423.

Hollweg, J.V. & Johnson, W. 1988, JGR 93, 9547.

Hollweg, J.V. 1991, Proc-1991-Ulmschneider, 423.

Holman, G.D., Eichler, D., & Kundu, M.R. 1980, Proc-1980-Kundu, 457.

Holman, G.D. & Pesses, M.E. 1983, ApJ 267, 837.

Holman, G.D. 1985, ApJ 293, 584.

Holman, G.D., Kundu, M.R., & Kane, S.R. 1989, ApJ 345, 1050.

Holman, G.D. & Benka, S.G. 1992, ApJ 400, L79.

Holman, G.D. 1995, ApJ 452, 451.

Holman, G.D. 1996, Proc-1996-Ramaty, 479.

Holman, G.D. 2000, Proc-2000-Ramaty, 135.

Holman, G.D. 2003, ApJ 586, 606.

Holman, G.D., Sui, L., Schwartz, R.A., et al. 2003, ApJ 595, L97.

Holt, S.S. & Cline, T.L. 1968, ApJ 154, 1027.

Holzer, T.E. 1989, ARAA , 27, 199.

Hood, A.W. & Priest, E.R. 1979a, AA 77, 233.

Hood, A.W. & Priest, E.R. 1979b, SP 64, 303.

Hood, A.W. & Priest, E.R. 1981, SP 73, 289.

Hood, A.W. 1986, SP 103, 329.

Hood, A.W. & Anzer, U. 1990, SP 126, 117.

Hood, A.W., Galsgaard, K., & Parnell, C.E. 2002, ESA SP-505, 285.

Hori, K., Yokoyama, T., Kosugi, T., et al. 1997, ApJ 489, 426.

Hori, K., Yokoyama, T., Kosugi, T., et al. 1998, ApJ 500, 492.

Howard, R.A., Sheeley, N.R.Jr., Koomen, M.J., et al. 1984, Adv. Space Res., 4, 307.

Howard, R.A., Sheeley, N.R.Jr., Koomen, M.J., et al. 1985, JGR 90, 8173.

Howard, R.A., Brueckner, G.E., St.Cyr. O., et al. 1997, Proc-1997-Crooker, 17.

Hoyng, P., Brown, J.C., & VanBeek, H.F. 1976, SP 48, 197.

Hoyng, P., Knight, J.W., & Spicer, D. 1978, SP 58, 139.

Hoyng, P., Duijveman, A., Machado, M.E., et al. 1981a, ApJ 246, L155.

Hoyng, P., Machado, M.E., Duijveman, A., et al. 1981b, ApJ 244, L153.

Hua,X.M. & Lingenfelter, R.E. 1987, ApJ 319, 555.

Huber, M.C.E., Foukal, P.V., Noyes, R.W., et al. 1974, ApJ 194, L115.

Hudson, H.S. 1972, SP 24, 414.

Hudson, H.S. 1978, ApJ 224, 235.

Hudson, H.S., Canfield, R.C., & Kane, S.R. 1978, SP 60, 137.

Hudson, H.S., Bai, T., Gruber, D.E., et al. 1980, ApJ 236, L91.

Hudson, H.S. 1989, Proc-1989-Johnson, ...

Hudson, H.S. 1991a, SP 133, 357.

Hudson, H.S. 1991b, BAAS 23, 1064.

Hudson, H.S., Acton, L.W., Hirayama, T., et al. 1992, PASJ 44, L77.

Hudson, H.S., Strong, K.T., Dennis, B.R., et al. 1994, ApJ 422, L25.

Hudson, H.S. & Ryan, J. 1995, ARAA 33, 239.

Hudson, H.S. & Khan, J.I. 1996, Proc-1996-Bentley, 61.

Hudson, H.S., Acton, L.W., & Freeland,S.L. 1996, ApJ 470, 629.

Hudson, H.S. & Webb, D.F. 1997, Proc-1997-Crooker, 27.

Hudson, H.S. 1999, Proc-1999-Bastian, 159.

Hudson, H.S. & McKenzie, D.E. 2001, Proc-2001-Hoshino, 581.

Hudson, H.S. & Farnik, F. 2002, Proc-2002-ESA506, 261.

Hundhausen, A.J. 1972, *Coronal expansion and solar wind,* (see book list).

Hundhausen, A.J. 1988, Proc-1988-Pizzo, 131.

Hundhausen, A.J. 1993, JGR 98, 13177.

Hundhausen, A.J. 1995, in Kivelson & Russell (1995), 91.

Hundhausen, A.J. 1997, Proc-1997-Crooker, 1.

Hundhausen, A.J. 1999, in Strong et al. (see book list), 143.

Hurford, G.J., Schmahl, E.J., Schwartz, R.A., et al. 2002, SP 210, 61.

Hurford, G.J., Schwartz, R.A., Krucker, S. et al. 2003, ApJ 595, L77.

Hurley, K. & Duprat, G. 1977, SP 52, 107.

Ichimoto, K. & Kurokawa, K. 1984, SP 93, 105.

Illing, R.M.E. & Hundhausen, A.J. 1985, JGR 90, 275.

Illing, R.M.E. & Hundhausen, A.J. 1986, JGR 91, 10951.

Innes, D.E., Inhester, B., Axford, W.I., et al. 1997, Nature 386, 811.

Innes, D.E. & Toth, G. 1999, SP 185, 127.

Innes, D.E. 2001, AA 378, 1067.

Inverarity, G.W., Priest, E.R., & Heyvarts, J. 1995, AA 293, 913.

Inverarity, G.W. & Priest, E.R. 1995a, AA 296, 395.

Inverarity, G.W. & Priest, E.R. 1995b, AA 302, 567.

Inverarity, G.W. & Priest, E.R. 1999, SP 186, 99.

Ionson, J.A. 1978, ApJ 226, 650.

Ionson, J.A. 1982, ApJ 264, 318.

Ionson, J.A. 1983, ApJ 271, 778.

Ionson, J.A. 1984, ApJ 276, 357.

Ireland, J., Walsh, R.W., Harrison, R.A., et al. 1999, AA 347, 355.

Isenberg, P.A. 1990, JGR 95, 6437.

Isenberg, P.A., Forbes,T.G., & Démoulin,P. 1993, ApJ 417, 368.

Isobe, H., Yokoyama, T., Shimojo, M., et al. 2002, ApJ 566, 528.

Jackson, J.D. 1962, *Classical Electrodynamics,* (see book list).

Jain, K., & Tripathy, S.C. 1998, SP 181, 113.

Jakimiec, J. & Jakimiec, M. 1974, AA 34, 415.

Jakimiec, J. Tomczak, M., Falewicz, R., et al. 1998, AA 334, 1112.

Jakimiec, J. 1999, Proc-1999-ESA448, 729.

James,S.P. & Erdélyi, R. 2002, AA 393, L11.

Janssens, T.J. & White, K.P.III, 1969, ApJ 158, L127.

Janssens, T.J., White, K.P.III, & Broussard,R.M. 1973, SP 31, 207.

Jardine, M. & Priest, E.R. 1988a, J. Plasma Phys. 40, 143.

Jardine, M. & Priest, E.R. 1988b, J. Plasma Phys. 40, 505.

Jardine, M. & Priest, E.R. 1988c, Geophys. Astrophys. Fluid Dynamics 42, 163.

Jardine, M. & Priest, E.R. 1989, J. Plasma Phys. 42, 111.

Jardine, M. & Priest, E.R. 1990, J. Plasma Phys. 43, 141.

Jardine, M. 1991, Proc-1991-Ulmschneider, 588.

Jardine, M., & Allen, H.R. 1996, Proc-1996-Bentley, 300.

Jeans, Sir J.H. 1905, Phil.Mag., 10, 91.

Jeans, Sir J.H. 1909, Phil.Mag., 17, 229.

Jensen, E., Maltby, P., & Orrall, F.Q. (eds.) 1979, Proc-1979-Jensen.

Jensen, E. 1986, Proc-1986-Poland, 63.

Jeffrey, A. & Taniuti, T. 1966, *Magnetohydrodynamic Stability* (see book list).

Ji, H.S. & Song, M.T. 2001, ApJ 556, 1017.

Ji, H.S., Song, M.T., & Huang, G.L. 2001, ApJ 548, 1087.

Jiang, Y., Ji, H., Wang, H., et al. 2003, ApJ 597, L161.

Jiao, L., McClymont, A.N., and Mikić, Z. 1997, SP 174, 311.

Jin, S.P. & Ip, W.H. 1991, Phys. Fluids B 3, 1927.

Jin, S.P., Inhester, B., & Innes, D. 1996, SP 168, 279.

Joarder, P.S. & Roberts, B. 1992, AA 256, 264.

Joarder, P.S., Nakariakov, V.M., & Roberts, B. 1997, SP 173, 81.

Jokipii, J.R. 1966, ApJ 143, 961.

Johns, C.M. & Lin, R.P. 1992, SP 137, 121.

Jones, F.C. 1990, ApJ 361, 162.

Jones, F.C. 1994, ApJS 90, 561.

Jordan, C. 1976, Phil.Trans.R.Soc.A 281, 391.

Joshi, A., Chandra, R., & Uddin, W. 2003, SP 217, 173.

Kahler, S.W., Sheeley, N.R.Jr., Howard, R.A., et al. 1984, JGR 89, 9683.

Kahler, S.W. 1987, Rev. Geophysics 25, 663.

Kahler, S.W. 1992, ARAA 30, 113.

Kahler, S.W. 1994, ApJ 428, 837.

Kahler, S.W. 2001, JGR 106, 20,947.

Kahler, S.W. & Reames, D.V. 2003, ApJ 584, 1063.

Kai, K. & Takayanagi, A. 1973, SP 29, 461.

Kaldeich–Schurmann, B. & Vial, J.C. (eds.) 1999, ESA SP-446.

Kamio, S., Kurokawa, H., & Ishii, T.T. 2003, SP 215, 127.

Kanbach, G., Bertsch, D.L., Fichtel, C.E., et al. 1992, Proc-1992-EGRET, 5p.

Kanbach, G., Bertsch, D.L., Fichtel, C.E., et al. 1993, AASS 97, 349.

Kang, H. & Jones, T.W. 1995, ApJ 447, 944.

Kankelborg, C.C., Walker, A.B.C.II, & Hoover, R.B. 1997, ApJ 491, 952.

Kane, S.R., Anderson, K.A., Evans, W.D., et al. 1979, ApJ 233, L151.

Kane, S.R. 1983, SP 86, 355.

Kane, S.R., Fenimore, F.E., Klebesadel, R.W., et al. 1982, ApJ 254, L53.

Kane, S.R., Chupp, E.L., Forrest, D.J., et al. 1986, ApJ 300, L95.

Kane, S.R., McTiernan, J., Loran, J., et al. 1992, ApJ 390, 687.

Kano, R., & Tsuneta, S. 1995, ApJ 454, 934.

Kaplan, S.A. & Tsytovitch, V.N. 1973, *Plasma Astrophysics*, (see book list).

Karimabadi, H., Menyuk, C.R., Sprangle, P., et al. 1987, ApJ 316, 462.

Karlický, M. 1993, SP 145, 137.

Karlický, M. & Odstrčil, D. 1994, SP 155, 171.

Karlický, M., Yan, Y., Fu, Q., et al. 2001, AA 369, 1104.

Karlický, M., Kliem, B., Meszarosova, H., et al. 2002a, Proc-2002-ESA506, 653.

Karlický, M., Farnik, F., & Meszarosova, H. 2002b, AA 395, 677.

Karlický, M. & Farnik, F. 2003, AdSpR 32/12, 2539.

Karpen, J.T. 1982, SP 77, 205.

Karpen, J.T. & Boris, J.P. 1986, ApJ 307, 826.

Karpen, J.T., Antiochos, S,K., & DeVore, C.R. 1995, ApJ 450, 422.

Karpen, J.T., Antiochos, S.K., DeVore, C.R., et al. 1998, ApJ 495, 491.

Karpen, J.T., Antiochos, S.K., Hohensee, M., et al. 2001, ApJ 553, L85.

Katsiyannis, A.C., Williams, D.R., McAteer, R.T.J., et al. 2003, AA 406, 709.

Katsukawa, Y. 2003, PASJ 55, 1025.

Kattenberg, A. & Kuperus, M. 1983, SP 85, 185.

Kaufmann, P. 1972, SP 23, 178.

Kaufmann, P., Costa, J.E.R., Correia, E., et al. 2000, Proc-2000-Ramaty, 318.

Kaufmann, P., Raulin, J.P., Melo, A.M., et al. 2002, ApJ 574, 1059.

Kawamura, K., Omodaka, T., & Suzuki, I. 1981, SP 71, 55

Keppens, R. 2001, *Sunspot Pores*, (in Murdin 2000) .

Kennel, C.F. & Petschek, H.E. 1966, JGR 71(1), 1

Kennel, C.F. 1969, Rev.Geophys. 7, 379

Khodachenko, M., Haerendel, G., & Rucker, H.O. 2003, AA 401, 721.

Kiepenheuer, K.O. (ed.) 1968, Proc-1968-Kiepenheuer.

Kim, E.J. & Diamond, P.H. 2001, ApJ 556, 1052.

King, D.B., Nakariakov, V.M., DeLuca, E.E., et al. 2003, AA 404, L1.

Kiplinger, A.L., Dennis, B.R., Forst, K.J., et al. 1982, Proc-1982-Tanaka, 66.

Kiplinger, A.L., Dennis, B.R., Frost, K.J., et al. 1983, ApJ 273, 783.

Kiplinger, A.L., Dennis, B.R., Emslie, A.G., et al. 1983b, ApJ 265, L99.

Kiplinger, A.L. 1995, ApJ 453, 973.

Kippenhahn, R. & Schlüter, A. 1957, Z.Astrophys. 43, 36.

Kirk, J.G. 1994, Proc-1994-Benz, 225.

Kirk, J.G., Melrose,D.B., Priest,E.R. 1994, Proc-1994-Benz.

Kivelson, M.G. & Russell, C.T. 1995, *Introduction to Space Physics* (see book list).

Klassen, A., Aurass, H., Mann, G., et al. 2000, AASS 141, 357.

Klein, K.L., & Trottet, G. 1994, Proc-1994-Ryan, 187.

Klein, K.L., Chupp, E.L., Trottet, G., et al. 1999a, AA 348, 271.

Klein, K.L., Khan, J.I., Vilmer, N., et al. 1999b, AA 346, L53.

Klein, K.L., Trottet, G., Lantos, P., et al. 2001, AA 373, 1073.

Klein, K.L., & Trottet, G. 2001, SSR 95, 215.

Kliem, B. 1988, ESA SP-285/II, 117.

Kliem, B. 1990, Astron.Nachr. 311/6, 399.

Kliem, B. 1993, Proc-1993-ESA351, 223.

Kliem, B. 1994, ApJS 90, 719.

Kliem, B. 1995, Proc-1995-Benz, 93.

Kliem, B., Schumacher, J. & Shklyar, D.R. 1996, AdSpR 21/4, 563.

Kliem, B. & Schumacher, J. 1997, Proc-1997-ICRC25, 1, 149.

Kliem, B., Karlický, M., & Benz, A.O. 2000, AA 360, 715.

Kliem, B., Dammasch, I.E., Curdt, W., et al. 2002, ApJ 568, L61.

Kliem, B., MacKinnon, A., Trottet, G., et al. 2003, Proc-2003-Klein, 263.

Kliem, B., Titov, V.S., $ Török, T. 2004, AA 413, L23.

Klimchuk, J.A., Antiochos, S.K., & Mariska, J.T. 1987, ApJ 320, 409.

Klimchuk, J.A. & Mariska, J.T. 1988, ApJ 328, 334.

Klimchuk, J.A. 1990, ApJ 354, 745.

Klimchuk, J.A., Lemen, J.R., Feldman, U., et al. 1992, PASJ 44, L181.

Klimchuk, J.A. 1996, Proc-1996-Bentley, 319.

Klimchuk, J.A. 2000, SP 193, 53.

Klimchuk, J.A., Antiochos, S.K., & Norton, D. 2000, ApJ 542, 504.

Klimchuk, J.A. 2001, Proc-2001-Song, 143.

Knight, J.W. & Sturrock, P.A. 1977, ApJ 218, 306.

Knoepfel, H., & Spong, D.A. 1979, Nucl.Fusion, 19/6, 785.

Kobak, T. & Ostrowski, M. 2000, MNRAS 317/4, 973.

Kobrin, M.M., & Korshunov, A.I. 1972, SP 25, 339.

Kocharov, G.E., Chuikin, E.I., Kovaltsov, G.A., et al. 1994, Proc-1994-Ryan, 45.

Koch, H.W. & Motz, J.W. 1959, Rev.Mod.Phys. 31, 920.

Kohl, J.L., Weiser, H., Withbroe, G., et al. 1980, ApJ 241, L117.

Kohl, J.L., Esser, R., Gardner, L.D., et al. 1995, SP 162, 313.

Kohl, J.L. et al. 1997, SP 175, 613.

Kohl, J.L. et al. 1998, ApJ 501, L127.

Kohl, J.L., Esser, R., Cranmer, S.R., et al. 1999, ApJ 510, L59.

Kontar, E.P., Brown, J.C., & McArthur, G.K. 2002, SP 210, 419.

Kontar, E.P., Emslie, A.G., Brown, J.C., et al. 2003, ApJ 595, L123.

Kopp, R.A. & Orrall, F.Q. 1976, AA 53, 363.

Kopp, R.A. & Pneuman, G.W. 1976, SP 50, 85.

Kosugi, T., Makishima, K., Murakami, T., et al. 1991, SP 136, 17.

Kosugi, T. & Somov, B.V. 1998, Proc-1998-Watanabe, 297.

Koutchmy, S., Zhugzhda, Ia.D., & Locans,V. 1983, AA 120, 185.

Kozlovsky, B., Lingenfelter, R.E., & Ramaty, R. 1987, ApJ 316, 801.

Kozlovsky, B., Murphy, R.J., & Ramaty, R. 2002, ApJS 141, 523.

Krall, K.R. & Antiochos, S.K. 1980, ApJ 242, 374.

Krall, J., Chen, J., & Santoro, R. 2000, ApJ 539, 964.

Krauss–Varban, D. and Wu, C.S. 1989, JGR 94, 15367

Krauss–Varban, D. & Burgess, D. 1991, JGR 96, 143.

Krucker, S., Aschwanden, M.J., Bastian, T.S., et al. 1995, AA 302, 551.

Krucker, S., Benz, A.O., & Aschwanden,M.J. 1997a, AA 317, 569.

Krucker, S., Benz, A.O., Bastian,T.S., et al. 1997b, ApJ 488, 499.

Krucker, S. & Benz, A.O. 1998, ApJ 501, L213.

Krucker, S. & Benz, A.O. 1999, Proc-1999-Bastian, 25.

Krucker, S. & Benz, A.O. 2000, SP 191, 341.

Krucker, S., Christe, S., Lin, R.P., et al. 2002, SP 210, 445.

Krueger, A. 1979, *Introduction to Solar Radio Astronomy*, (see book list).

Krueger, A., Hildebrandt, J., & Fuerstenberg, F. 1985, AA 143, 72.

Kruskal, M.D. & Schwarzschild, M. 1954, Proc. Roy. Soc. A223, 348.

Kruskal, M.D., Johnson, J.L, Gottlieb, M.B., et al. 1958, *Phys. Fluids* 1, 421.

Kucera, T.A., Dulk, G.A., Kiplinger, A.L., et al. 1993, ApJ 412, 853.

Kucera, T.A., Love, P.J., Dennis, B.R., et al. 1996, ApJ 466, 1067.

Kucera, T.A., Dennis, B.R., Schwartz, R.A., et al. 1997, ApJ 475, 388.

Kucera, T.A., Aulanier, G., Schmieder, B., et al. 1999, SP 186, 259.

Kucera, T.A. & Antiochos, S.K. 1999, ESA SP-446, 97.

Kuin, N.P.M. & Martens, P.C.H. 1982, AA 108, L1.

Kuin, N.P.M. & Poland, A.I. 1991, ApJ 370, 763.

Kuijpers, J., Van Der Post, P., & Slottje, C. 1981, AA 103, 331.

Kulsrud, R.M. 2001, Earch Planets Space 53, 417.

Kumar, A. & Rust, D.M. 1996, JGR 101, 15667.

Kundu, M.R. 1965, *Solar Radio Astronomy*, (see book list).

Kundu, M.R. & Liu, S.Y. 1976, SP 49, 267.

Kundu, M.R. & Velusamy, T. 1980, ApJ 240, L63.

Kundu, M.R. & Vlahos, L. 1982, Space Sci. Rev. 32, 405.

Kundu, M.R. 1985, SP 100, 491.

Kundu, M.R., Melozzi, M., & Shevgaonkar, R.K. 1986, AA 167, 166.

Kundu, M.R., Schmahl, E.J., Gopalswamy, N., et al. 1989a, Adv.Space.Res., 9/4, 41.

Kundu, M.R., Gopalswamy, N., White, S., et al. 1989b, ApJ 347, 505.

Kundu, M.R., MacDowall, R.J., & Stone, R.G. 1990, Astrophysics and Space Science 165, 101.

Kundu, M.R., White, S.M., & McConnell, D.M. 1991, SP 134, 315.

Kundu, M.R., Shibasaki, K., Enome, S., et al. 1994, ApJ 431, L155.

Kundu, M.R., Nitta, N., White, S.M., et al. 1995, ApJ 454, 522.

Kundu, M.R., White, S.M., Shibasaki, K., et al. 2001, ApJ 547, 1090.

Kunow, H. 2001, *Solar Wind: Co-rotating Interaction Regions*, (in Murdin 2000) .

Kuperus, M. & Tandberg−Hanssen, E. 1967, SP 2, 39.

Kuperus, M. & Raadu, M.A. 1974, AA 31, 189.

Kuperus, M., Ionson, J.A., & Spicer,D. 1981, ARAA 19, 7.

Kuperus, M. & Van Tend, W. 1981, SP 71, 125.

Kurfess, J.D., Bertsch, D.L., Fishman, G.J., et al. 1998, Proc-1998-Dermer, 509.

Kurokawa, H., & Sano, S. 2000, Adv.Space Res. 26/3, 441.

Kurt, V.G., Akimov, V.V., & Leikov, N.G. 1996, Proc-1996-Ramaty, 237.

Kurths, J. & Herzel, H. 1987, Phyica Scripta 25D, 165.

Kurths, J. & Karlický, M. 1989, SP 119, 399.

Kurths, J., Benz, A.O., & Aschwanden, M.J. 1991, AA 248, 270.

Kusano, K., Maeshiro, T., Yokoyama, T., et al. 2002, ApJ 577, 501.

Kusano, K. 2002, ApJ 571, 532.

Laing, G.B. & Edwin, P.M. 1995, SP 161, 269.

Lamb, H. 1963, *Hydrodynamics*, (see book list).

Lampe, M. & Papadopoulos, K. 1977, ApJ 212, 886.

Landi, E., Landini, M., Dere, K.P., et al. 1999, AASS 135, 339.

Landini, M. & Monsignori−Fossi, B.C. 1975, AA 42, 213.

Landini, M. & Monsignori−Fossi, B.C. 1990, AASS 82, 229.

Landman, D.A., Edberg, S.J., & Laney, C.D. 1977, ApJ 218, 888.

Lang, K.R. 1980, *Astrophysical Formulae*, (see book list).

Lang, K.R., Willson, R.F., & Gaizauskas, V. 1983, ApJ 267, 455.

Lang, K.R., Willson, R.F., Smith, K.L. et al. 1987, ApJ 322, 1035.

Lang, K.R. & Willson, R.F. 1989, ApJ 344, L73.

Lantos, P. 1972, SP 22, 387.

Lantos, P. & Avignon, Y. 1975, AA 41, 137.

LaRosa, T.N. 1988, ApJ 335, 425.

LaRosa, T.N. & Emslie, A.G. 1989, SP 120, 343.

LaRosa, T.N. & Moore, R.L. 1993, ApJ 418, 912.

LaRosa, T.N., Moore, R.N., & Shore,S.N. 1994, ApJ 425, 856.

LaRosa, T.N., Moore, R.L., Miller, J.A., et al. 1995, ApJ 467, 454.

Larmor, Sir, J. 1897, Phil.Mag., 44, 503.

Lau, Y.T., Northrop, T.G., & Finn, J.M. 1993, ApJ 414, 908.

Lawrence, J.K. 1991, SP 135, 249.

Lawrence, J.K. & Schrijver, C.J. 1993, ApJ 411, 402.

Leach, J. & Petrosian, V. 1981, ApJ 251, 781.

Leach, J. & Petrosian, V. 1983, ApJ 269, 715.

Leboef, J.N., Tajima, T., & Dawson, J.M. 1982, Phys. Fluids 25, 784.

Ledenev, V.G. 1994, Space Science Rev. 68, 119.

Ledenev, V.G. 1998, SP 179, 405.

Lee, J.W., Hurford, G.J., & Gary, D.E. 1993a, SP 144, 45.

Lee, J.W., Hurford, G.J., & Gary, D.E. 1993b, SP 144, 349.

Lee, J.W., White, S.M., Gopalswamy, N., et al. 1997, SP 174, 175.

Lee, J.W., McClymont, A.N., Mikić, Z., et al. 1998, ApJ 501, 853.

Lee, J.W., White, S.M., Kundu, M.R., et al. 1999, ApJ 510, 413.

Lee, J.W., Gary, D.E., & Shibasaki, K. 2000, ApJ 531, 1109.

Lee, J.W. & Gary, D.E. 2000, ApJ 543, 457.

Lee, J.W., Gary, D.E., Qiu, J. et al. 2002, ApJ 572, 609.

Lee, J.W., Gallagher, P.T., Gary, D.E., et al. 2003, ApJ 585, 524.

Lee, L.C. & Fu, Z.F. 1986, JGR 91, 6807.

Lee, M.A. & Völk, H.J. 1973, Astrophys. Space Sci.24, 31.

Lee, M.A. & Völk, H.J. 1975, ApJ 198, 485.

Lee, M.A. & Roberts, B. 1986, ApJ 301, 430.

Lee, T.T., Petrosian, V., & McTiernan, J.M. 1993c, ApJ 412, 401.

Lee, T.T., Petrosian, V., & McTiernan, J.M. 1995, ApJ 448, 915.

Leer, E. 2001, *Solar Wind: Theory*, (in Murdin 2000) .

Leka, K.D., Canfield, R.C., McClymont, A.N., et al. 1993, ApJ 411, 370.

Leka, K.D., Canfield, R.C., McClymont, A.N., et al. 1996, ApJ 462, 547.

Leka, K.D. & Barnes, G. 2003a, ApJ 595, 1277.

Leka, K.D. & Barnes, G. 2003b, ApJ 595, 1296.

Lemberge, B. 1995, Proc-1995-Benz, 201.

Lenz, D.D., DeLuca, E.E., Golub, L., et al. 1999, ApJ 517, L155.

Lepping, R.P., Jones, J.A. & Burlaga, L.F. 1990, JGR 95, 11957.

Lepping, R.P. 2001, *Solar Wind Shock Waves and Discontinuities*, (in Murdin 2000) .

Lerche, L. & Low, B.C. 1977, SP 53, 385.

Lerche, L. & Low, B.C. 1980, SP 66, 285.

Leroy, J.L., Bommier, V., & Sahal−Bréchot, S. 1984, AA 131, 33.

Leroy, J.L. 1989, Proc-1989-Priest, 77.

Levine, R.H. 1974, ApJ 190, 457.

Levine, R.H., Schulz, M., & Frazier, E.N. 1982, SP 77, 363.

Levinson,A. 1994, ApJ 426, 327.

Li, J., Metcalf, T.R., Canfield, R.C., et al. 1997, ApJ 482, 490.

Li, P., Emslie, A.G., & Mariska, J.T. 1989, ApJ 341, 1075.

Li, P., Emslie, A.G., & Mariska, J.T. 1993, ApJ 417, 313.

Li, P., McTiernan, M., & Emslie, G.A. 1997, ApJ 491, 395.

Li, X., Habbal, S.R., Kohl, J.L., et al. 1998, ApJ 501, L133.

Lin, H., Penn, M.J., & Tomczyk, S. 2000, ApJ 541, L83.

Lin, J., Forbes, T.G., Isenberg, P.A., & Démoulin,P. 1998a, ApJ 504, 1006.

Lin, J., Martin, R., & Wu,N. 1996, AA 311, 1015.

Lin, J., & Forbes, T.G. 2000, JGR , 105/A2, 2375.

Lin, R.P. 1974, SSR 16, 189.

Lin, R.P. & Hudson, H.S. 1976, SP 50, 153.

Lin, R.P., Schwartz, R.A., Pelling, R.M., et al. 1981a, ApJ 251, L109.

Lin, R.P., Potter, D.,W., Gurnett, D.A., et al. 1981b, ApJ 251, 364.

Lin, R.P., Schwartz, R.A., Kane, S.R., et al. 1984, ApJ 283, 421.

Lin, R.P., Levedahl, W.K., Lotko, W., et al. 1986, ApJ 308, 954.

Lin, R.P., & Schwartz, R.A. 1987, ApJ 312, 462.

Lin, R.P. & Johns, C.M. 1993, ApJ 417, L53.

Lin, R.P. & HESSI Team, 1998b, SPIE 3442, 2.

Lin, R.P. 2000, Proc-2000-Martens, 15.

Lin, R.P., Feffer, P.T., & Schwartz, R.A. 2001, ApJ 557, L125.

Lin, R.P., Dennis, B.R., Hurford, G.J., et al. 2002, SP 210, 3.

Lin, Y. 2002, ESA SP-506.

Lingenfelter, R.E., Flamm, E.J., Canfield, E.H. et al. 1965, JGR 70, 4077 and 4087.

Lingenfelter, R.E. & Ramaty, R. 1967, Proc-1967-Shen, 99.

Lingenfelter, R.E. 1994, Proc-1994-Ryan, 77.

Linker, J.A. & Mikić, Z. 1995, ApJ 438, L45.

Linker, J.A., Lionello, R., Mikić, Z. et al. 2001, JGR 106, 25,165.

Lionello, R., Mikić, Z., Linker, J.A., et al. 2002, ApJ 581, 718.

Lipa, B. 1978, SP 57, 191.

Lites, B.W., Bruner, E.C.Jr., Chipman, E.G., et al. 1976, ApJ 210, L111.

Lites, B.W., Low, B.C., Pillet, V.M., et al. 1995, ApJ 446, 877.

Lites, B.W. 2001, *Solar Magnetic Field: Inference by Polarimetry*, (in Murdin 2000) .

Litvinenko, Y.E. & Somov, B.V. 1991, SP 131, 319.

Litvinenko, Y.E. 1995, Astron. Reports 39/1, 99.

Litvinenko, Y.E. & Somov, B.V. 1995, SP 158, 317.

Litvinenko, Y.E. 1996a, SP 167, 321.

Litvinenko, Y.E. 1996b, ApJ 462, 997.

Litvinenko, Y.E. 1997, Phys. Plasmas 4(9), 3439.

Litvinenko, Y.E. & Martin, S.F. 1999, SP 190, 45.

Litvinenko, Y.E. 1999a, ApJ 515, 435.

Litvinenko, Y.E. 1999b, SP 186, 291.

Litvinenko, Y.E. 1999c, AA 349, 685.

Litvinenko, Y.E. 2000a, SP 194, 327.

Litvinenko, Y.E. 2000b, Proc-2000-Ramaty, 167.

Litvinenko, Y.E. & Craig, I,J.D. 2000, ApJ 544, 1101.

Litvinenko, Y.E. & Somov,B.V. 2001, SSR 95, 67.

Litvinenko, Y.E. 2002, Proc-2002-ESA506, 327.

Litvinenko, Y.E. 2003a, Proc-2003-Klein, 213.

Litvinenko, Y.E. 2003b, SP 216, 189.

Litwin, C. & Rosner, R. 1998, ApJ 506, L143.

Liu, Y., Zhao, X.P., Hoeksema, J.T., et al. 2002, SP 206, 333.

Livi, S.H.B., Wang, J., & Martin, S.F. 1985, Australian J. Phys. 38, 855.

Livi, S.H.B., Martin, S., Wang, H., & Ai,G. 1989, SP 121, 197.

Ljepojevic, N.N. & MacNeice, P. 1989, Phys. Rev. A, 40, 981.

Longcope, D.W. & Strauss, H.R. 1994, ApJ 426, 742.

Longcope, D.W. 1996, SP 196, 91.

Longcope, D.W., Fisher, G.H., & Arendt, S. 1996, ApJ 464, 999.

Longcope, D.W. & Klapper, I. 1997, ApJ 488, 443.

Longcope, D.W. & Silva, A.V.R. 1997, SP 179, 349.

Longcope, D.W. 1998, ApJ 507, 433.

Longcope, D.W., Fisher, G.H., & Pevtsov, A.A. 1998, ApJ 507, 417.

Longcope, D.W. & Kankelborg, C.C. 1999, ApJ 524, 483.

Longcope, D.W. & Noonan, E.J. 2000, ApJ 542, 1088.

Longcope, D.W., Kankelborg, C.C., Nelson,J., et al. 2001, ApJ 553, 429.

Longcope, D.W. & Klapper, I. 2002, ApJ 579, 468.

Lothian, R.M. & Hood, A.W. 1989, SP 122, 227.

Lothian, R.M. & Hood, A.W. 1992, SP 137, 105.

Lothian, R.M. & Browning, P.K. 1995, SP 161, 289.

Louarn, P., Roux, A., de Feraudy, H., et al. 1990, JGR 95/A5, 5983.

Loughhead, R.E., Wang, J.L., & Blows, G. 1983, ApJ 274, 883.

Low, B.C. 1975a, ApJ 197, 251.

Low, B.C. 1975b, ApJ 198, 211.

Low, B.C. 1981, ApJ 246, 538.

Low, B.C. 1982, ApJ 263, 952.

Low, B.C., Munro, R.H., & Fisher, R.R. 1982, ApJ 254, 335.

Low, B.C. 1984a, Proc-1984-Hagyard, 49.

Low, B.C. 1984b, ApJ 281, 392.

Low, B.C. 1985, ApJ 293, 31.

Low, B.C. 1991, ApJ 370, 427.

Low, B.C. 1992, ApJ 399, 300.

Low, B.C. 1993a, ApJ 408, 689.

Low, B.C. 1993b, ApJ 408, 693.

Low, B.C. & Smith, D.F. 1993, ApJ 410, 412.

Low, B.C. 1994, Plasma Phys. 1, 1684.

Low, B.C. 1996, SP 167, 217.

Low, B.C. 1999a, Proc-1999-Habbal, 109.

Low, B.C. 1999b, Proc-1999-Brown, 25.

Low, B.C. 2001a, JGR 106, 25141.

Low, B.C. 2001b, *Solar Coronal Mass Ejection: Theory*, (in Murdin 2000) .

Low, B.C. & Zhang, M. 2002, ApJ 564, L53.

Low, B.C., Fong, B., & Fan, Y. 2003, ApJ 594, 1060.

Lu, E.T. & Petrosian, V. 1988, ApJ 327, 405.

Lu, E.T. & Petrosian, V. 1990, ApJ 354, 735.

Lu, E.T. & Hamilton, R.J. 1991, ApJ 380, L89.

Lui, Y., Jiang, Y., Ji, H. et al. 2003, ApJ 593, L140.

MacDowall, R.J., Stone, R.G., & Kundu, M.R. 1987, SP 111, 397.

Machado, M.E. & Moore, R.L 1986, Adv.Space Res., 6/6, 217.

Machado, M.E. & Mauas, P.J. 1987, Proc-1987-Dennis, 271.

Machado, M.E., Moore, R.L., Hernandez, A.M., et al. 1988, ApJ 326, 425.

Machado, M.E. 1991, Proc-1991-Culhane, 425.

Mackay, D.H., Gaizauskas, V., Rickard, G.J., et al. 1997, ApJ 486, 534.

Mackay, D.H., Galsgaard, K., Priest, E.R., et al. 2000a, SP 193, 93.

Mackay, D.H., Gaizauskas, V., & Van Ballegooijen, A.A. 2000b, ApJ 544, 1122.

Mackay, D.H. & Van Ballegooijen, A.A. 2001, ApJ 560, 445.

MacKinnon, A.L., Brown, J.C., Trottet, G., et al. 1983, AA 119, 297

MacKinnon, A.L., Brown, J.C., & Hayward, J. 1985, SP 99, 231.

MacKinnon, A.L. 1986, AA 163, 239

MacKinnon, A.L. 1988, AA 194, 279

MacKinnon, A.L. 1991, AA 242, 256.

MacKinnon, A.L., & Craig, I.J.D. 1991, AA 251, 693.

MacKinnon, A.L. & Toner, M.P. 2003, AA 409, 745.

MacNeice, P., McWhirter, R.W.P., Spicer, D.S., et al. 1984, SP 90, 357.

MacNeice, P. 1986, SP 103, 47.

MacQueen, R.M. 1980, Philos. Trans. Royal Soc. London A297, 605.

MacQueen, R.M. 1993, SP 145, 169.

Madjarska, M.S. & Doyle, J.G. 2002, AA 382, 319.

Madjarska, M.S. & Doyle, J.G. 2003, AA 403, 731.

Magara, T., Mineshige, S., Yokoyama, T., et al. 1996, ApJ 466, 1054.

Magara, T., & Shibata,K. 1999, ApJ 514, 456.

Makhmutov, V.S., Costa, J.E.R., Raulin, J.P., et al. 1998, SP 178, 393.

Makishima, K. 1982, Proc-1982-Tanaka, 120.

Malara, F., Velli, M., & Carbone, V. 1992, Phys. Fluids B4, 3070.

Malherbe, J.M. & Priest, E.R. 1983, AA 123, 80.

Malitson, H.H., Fainberg, J. & Stone,R.G. 1973, ApJ 183, L35.

Maltby, P., Avrett, E.H., Carlsson, M., et al. 1986, ApJ 306, 284.

Maltby, P., Brynildsen, N., Kjeldseth—Moe, O., et al. 2001, AA 373, L1.

Mandelbrot, B.B. 1977, *The Fractal Geometry of Nature*, (see book list).

Mandrini, C.H., Rovira, M.G., Démoulin, P., et al. 1993, AA 272, 609.

Mandrini, C.H., Démoulin, P., Van Driel—Gesztelyi, L., et al. 1996, SP 168, 115.

Mandrini, C.H., Démoulin, P., Bagalá, L.G., et al. 1997, SP 174, 229.

Mandrini, C.H., Démoulin, P. & Klimchuk, J.A. 2000, ApJ , 530, 999.

Mandzhavidze, N. & Ramaty, R. 1992a, ApJ 389, 739.

Mandzhavidze, N. & Ramaty, R. 1992b, ApJ 396, L111.

Mandzhavidze, N. & Ramaty, R. 1993, Nuclear Physics B (Proc.Suppl.) 33A,B, 141.

Mandzhavidze, N., Ramaty, R., Bertsch, D.L., et al. 1996, Proc-1996-Ramaty, 225.

Mandzhavidze, N., Ramaty, R., & Kozlovsky, B. 1999, ApJ 518, 918.

Mandzhavidze, N. & Ramaty, R. 2000, Proc-2000-Ramaty, 64.

Mann, G. 1995, Proc-1995-Benz, 183.

Mann, G., Classen, T., & Aurass, H. 1995, AA 295, 775.

Mann, G., Aurass, H., Klassen, A. et al. 1999, Proc-1999-ESA446, 477.

Mann, G., Classen, H.T., Keppler, E., et al. 2002, AA 391, 749.

Manoharan, P.K., VanDriel—Gesztelyi, L., Pick, M., et al. 1996, ApJ 468, L73.

Mariska, J.T. & Boris, J.P. 1983, ApJ 267, 409.

Mariska, J.T. & Poland, A.I. 1985, SP 96, 317.

Mariska, J.T. 1986, ARAA 24, 23.

Mariska, J.T. 1987, ApJ 319, 465.

Mariska, J.T., Emslie, A.G., & Li,P. 1989, ApJ 341, 1067.

Mariska, J.T. 1992, *The Solar Transition Region*, (see book list).

Mariska, J.T. & Dowdy, J.F.Jr. 1992, ApJ 401, 754.

Mariska, J.T., Doschek, G.A., & Bentley, R.D. 1993, ApJ 419, 418.

Mariska, J.T. 1994, ApJ 434, 756.

Mariska, J.T. 1995, ApJ 444, 478.

Mariska, J.T., Sakao, T., & Bentley, R.D. 1996, ApJ 459, 815.

Mariska, J.T. & McTiernan, J.M. 1999, ApJ 514, 484.

Marik, D. & Erdélyi, R. 2002, AA 393, L73.

Marque, C., Lantos, P., Klein,K.L., et al. 2001, AA 374, 316.

Marsch, E. & Tu, C.Y. 1997a, SP 176, 87.

Marsch, E. & Tu, C.Y. 1997b, AA 319, L17.

Marsch, E. 2001, *Solar Wind: Kinetic Properties*, (in Murdin 2000) .

Marsch, E. & Tu, C.Y. 2001, JGR 106/A1, 227.

Marschhäuser, H., Rieger, E. & Kanbach, G. 1994, Proc-1994-Ryan, 171.

Marsh, K.A. & Hurford, G.J. 1982, ARAA 20, 497.

Marsh, M.S., Walsh,R.W., De Moortel,I. et al. 2003, AA 404, L37.

Martens, P.C.H. 1988, ApJ 330, L131.

Martens, P.C.H. & Kuin, N.P.M. 1989, SP 122, 263.

Martens, P.C.H. & Young, A. 1990, ApJS 73, 333.

Martens, P.C.H., Kankelborg, C.C., & Berger,T.E. 2000, ApJ 537, 471.

Martens, P.C.H. & Cauffman, D. (eds.) 2002, COSPAR-CS 13.

Martin, R.F. 1986, JGR 91, 11985.

Martin, S.F., Bentley, R., Schadee, A., et al. 1984, Adv.Spa.Res. 4/7, 61.

Martin, S.F., Livi, S.H.B. & Wang, J. 1985, Austr.J.Phys. 38, 929.

Martin, S.F. 1988, SP 117, 243.

Martin, S.F. 1990, Proc-1990-Ruzdjak, 1.

Martin, S.F., Billamoria, R., & Tracadas,P.W. 1994, Proc-1994-Rutten, 303.

Martin, S.F. 1998, SP 182, 107.

Martin, S.F. 2001, *Solar Prominence Formation*, (in Murdin 2000) .

Massone, A.M., Piana, M., Conway, A., et al. 2003, AA 405, 325.

Masuda, S., Kosugi, T., Hara, H., et al. 1994, Nature 371, No. 6497, 495.

Masuda, S., Kosugi. T., Hara, H., et al. 1995, PASJ 47, 677.

Masuda, S., Kosugi, T., Tsuneta, S., et al. 1996, Adv.Space Res. 17/4-5, 63.

Masuda, S. 2000, Proc-2000-Martens, 413.

Masuda, S., Kosugi, T., & Hudson, H.S. 2001, SP , 204, 55.

Matsumoto, R., Tajima, T., Shibata, K., et al. 1993, ApJ 414, 357.

Matsumoto, R., Tajima, T., Chou, W., et al. 1996, Proc-1996-Uchida, 355.

Matsumoto, R., Tajima, T., Chou, W., et al. 1998, ApJ 493, L43.

Matthaeus, W.H., Zank, G.P., Oughton, S., et al. 1999, ApJ 523, L93.

Matthaeus, W.H. 2001a, *MHD: Magnetic Reconnection and Turbulence*, (in Murdin 2000) .

Matthaeus, W.H. 2001b, *Solar Wind Turbulence*, (in Murdin 2000) .

Matthaeus, W.H., Mullan, D.J., Dmitruk, P., et al. 2002, Proc-2002-Tromsoe, ...

Matthews, S.A., VanDriel−Gesztelyi, L., Hudson, H.S. et al. 2002, Proc-2002-Martens, 289.

Matsushita, K., Masuda, S., Kosugi, T., et al. 1992, PASJ 44, L89.

Maxwell, A. & Rinehart, R. 1974, SP 37, 437.

Mazzotta, P., Mazzitelli, G., Colafrancesco, S. et al. 1998, AASS 133, 403.

McClymont, A.N. 1989, ApJ 347, L47.

McClymont, A.N. & Canfield, R.C. 1983a, ApJ 265, 497.

McClymont, A.N. & Canfield, R.C. 1983b, ApJ 265, 483.

McClymont, A.N. & Craig, I,J.D. 1985a, ApJ 289, 820.

McClymont, A.N. & Craig, I,J.D. 1985b, ApJ 289, 834.

McClymont, A.N. & Craig, I.J.D. 1987, ApJ 312, 402.

McClymont, A.N., & Mikić, Z. 1994, ApJ 422, 899.

McClymont, A.N., & Craig, I.J.D. 1996, ApJ 466, 487.

McClymont, A.N., Jiao, L., & Mikić, Z. 1997, SP 174, 191.

McClements, K.G. 1990a, AA 230, 213

McClements, K.G. 1990b, AA 234, 487

McClements, K.G., Su, J.J., Bingham, R., et al. 1990, SP 130, 229.

McClements, K.G. & Baynes, N.de B. 1991, AA 245, 262.

McClements, K.G. 1992, AA 258, 542.

McClements, K.G., Bingham, R., Su, J.J., et al. 1993, ApJ 409, 465.

McIntosh, S.W. & Charbonneau, P. 2001, ApJ 563, L165.

McKaig, I. 2001, AA 371, 328.

McKean, M.E., Winglee, R.M., & Dulk, G.A. 1990a, ApJ 364, 295.

McKean, M.E., Winglee, R.M., & Dulk, G.A. 1990b, ApJ 364, 302.

McKenzie, D.E. & Mullan, D.J., 1997, SP 176, 127.

McKenzie, D.E. 2000, SP 195, 381.

McKenzie, D.E. 2002, COSPAR-CS 13, 155.

McKenzie, D.E. & Hudson, H.S. 1999, ApJ 519, L93.

McKenzie, J.F. 1970, *J. Geophys. Res., Space Phys.* 75, 5331.

McLean, D.J, Sheridan, K.V., Steward, R.T., et al. 1971, Nature 234, 140.

McLean, D.J. & Sheridan, K.V. 1973, SP 32, 485.

McLean, D.J. & Labrum, N.R. (eds.) 1985, *Solar Radiophysics*, (see book list).

McTiernan, J.M. & Petrosian, V. 1990, ApJ 359, 524.

McTiernan, J.M. & Petrosian, V. 1991, ApJ 379, 381.

McTiernan, J.M., Fisher,G.H., & Li,P. 1999, ApJ 514, 472.

Meister, C.V. 1995, SP 160, 65.

Melrose, D.B. 1974, SP 37, 353.

Melrose, D.B. & Brown, J.C. 1976, MNRAS 176, 15

Melrose, D.B. & White, S.M. 1979, Astr.Soc.Austr.Proc. 3/56, 369.

Melrose, D.B. 1980a, *Plasma Astrophysics. I.*, (see book list).

Melrose, D.B. 1980b, *Plasma Astrophysics. II.*, (see book list).

Melrose, D.B. & White, S.M. 1981 JGR 86/A4, 2183.

Melrose, D.B. & Dulk, G.A. 1982, ApJ 259, 844.

Melrose, D.B., Hewitt, R.G., & Dulk,G.A. 1984, JGR 89(A2), 897.

Melrose, D.B. 1986, *Instabilities in Space and Laboratory Plasmas*, (see book list).

Melrose, D.B. 1987, SP 111, 89.

Melrose, D.B. & Dulk, G.A. 1987, Physica Scripta T218, 29.

Melrose, D.B. 1990, SP 130, 3.

Melrose, D.B. 1992, ApJ 387, 403.

Melrose, D.B. 1993, Aust.J.Phys. 46, 167.

Melrose, D.B. 1995, ApJ 451, 391.

Melrose, D.B. 1997, ApJ 486, 521.

Melrose, D.B. 2000, Proc-2000-Murdin.

Mendoza−Briceno, C.A., Erdélyi, R., & Sigalotti, L.D.G. 2002, ApJ 579, L49.

Metcalf, T.R., Canfield, R.C., Avrett, E.H. et al. 1990a, ApJ 350, 463.

Metcalf, T.R., Canfield, R.C., & Saba, J.L.R. 1990b, ApJ 365, 391.

Metcalf, T.R., Wülser, J.P., Canfield, R.C. et al. 1992, Proc-1992-Shrader, 536.

Metcalf, T.R., Mickey, D., Canfield, R. 1994, Proc-1994-Ryan, 59.

Metcalf, T.R., Jiao, L., Uitenbroek, H., et al. 1995, ApJ 439, 474.

Metcalf, T.R., & Fisher, G.H. 1996, ApJ 462, 977.

Metcalf, T.R., Alexander, D., Hudson, H.S. et al. 2003, ApJ 595, 483.

Mewe, R. & Gronenschild, E.H.B.M. 1981, AASS 45, 11.

Meyer, J.P. 1985, ApJS 57, 173.

Meyer, J.P. 1996, Proc-1996-Ramaty, 461.

Mikić, Z., Barnes, D.C., & Schnack, D.D. 1988, ApJ 328, 830.

Mikić, Z., Schnack, D.D., and VanHoven, G. 1989, ApJ 338, 1148.

Mikić, Z., Schnack, D.D., and VanHoven, G. 1990, ApJ 361, 690.

Mikić, Z. & McClymont, A.N. 1994, Proc-1994-Balasubramaniam, 225.

Mikić, Z. & Linker, J.A. 1994, ApJ 430, 898.

Mikić, Z. & Linker, J.A. 1999, BAAS 31, 918.

Milano, L.J., Gomez, D.O., & Martens, P.C.H. 1997, ApJ 490, 442.

Milano, L.J., Dmitruk, P., Mandrini, C.H., et al. 1999, ApJ 521, 889.

Miller, J.A. & Ramaty, R. 1987, SP 113, 195.

Miller, J.A., Guessoum, N., & Ramaty, R. 1990, ApJ 361, 701.

Miller, J.A. 1991, ApJ 376, 342.

Miller, J.A. & Ramaty, R. 1992, Proc-1992-Zank, 223.

Miller, J.A. & Viñas, A.F. 1993, ApJ 412, 386.

Miller, J.A. & Roberts, D.A. 1995, ApJ 452, 912.

Miller, J.A. & Reames, D.V. 1996, Proc-1996-Ramaty, 450.

Miller, J.A., LaRosa, T.N., & Moore, R.L. 1996, ApJ 461, 445.

Miller, J.A. 1997, ApJ 491, 939.

Miller, J.A., Cargill, P.J., Emslie, A.G., et al. 1997, JGR 102/A7, 14631.

Miller, J.A. 2000a, Proc-2000-Martens, 277.

Miller, J.A. 2000b, Proc-2000-Ramaty, 145.

Milne, A.M., Priest, E.R., & Roberts, B. 1979, ApJ 232, 304.

Milne, E.A. 1930, *Thermodynamics of the Stars*, (see book list).

Milne, A.M. & Priest, E.R. 1981, SP 73, 157.

Miralles, M.P., Cranmer, S.R., Panasyuk, A.V., et al. 2001, ApJ 549, L257.

Mitchell, H.G.Jr. & Kan, J.R. 1978, J. Plasma Phys. 20, 31.

Miyagoshi, T. & Yokoyama, T. 2003, ApJ 593, L133.

Moghaddam–Taaheri, E. & Goertz, C.K. 1990, ApJ 352, 361.

Mok, Y. 1987, AA 172, 327.

Mok, Y., Schnack,D.D., & VanHoven,G. 1991, SP 132, 95.

Molowny–Horas, R., Oliver, R., Ballester, J.L., & Baudin, F. 1997, SP 172, 181.

Moon, Y.J., Choe, G.S., Yun, H.S., 2002a, ApJ 568, 422.

Moon, Y.J., Chae, J.C., Wang, H., 2002b, ApJ 580, 528.

Moore, R.L. & Fung, P.C.W. 1972, SP 23, 78.

Moore, R.L. 1988, ApJ 324, 1132.

Moore, R.L., LaRosa, T.N., & Orwig, L.E. 1995, ApJ 438, 985.

Moore, R.L., Falconer,D.A., Porter, J.G., et al. 1999, ApJ 526, 505.

Moore, R. et al. 1980, Proc-1980-Sturrock, 341.

Moore, R. 2001, *Solar Prominence Eruption*, (in Murdin 2000) .

Moran, T.G., Gopalswamy, N., Dammasch, I.E., et al. 2001, AA 378, 1037.

Moran, T.G. & Davila, J.M. 2004, Science 305, 66.

Moreton, G.E. & Ramsey, H.E. 1960, PASP 72, 428.

Moreton, G.E. 1961, Sky and Telescope 21, 145.

Moreton, G.E. 1964, Astronom. J. 69, 145.

Mori, K.I., Sakai, J.I., & Zhao, J. 1998, ApJ 494, 430.

Moses, D., Cook, J.W., Bartoe, J.D.F., et al. 1994, ApJ 430, 913.

Moses, R.W., Finn, J.M., & Ling, K.M. 1993, JGR 98, A3, 4013.

Mouschovias, T.C. & Poland, A.I. 1978, ApJ 220, 675.

Mozer, F.S., Carlson, C.W., Hudson, M.K., et al. 1997, Phys. Rev. Lett. 38, 292.

Mrozek,T. & Tomczak, M. 2004, AA 415, 377.

Murawski, K. & Roberts, B. 1993, SP 144, 101.

Murawski, K. & Roberts, B. 1994, SP 151, 305.

Murawski, K., Aschwanden, M.J., & Smith, J.M. 1998, SP 179, 313.

Murdin, P. 2000 *Encyclopedia of Astronomy and Astrophysics* (see book list).

Murphy, R.J., Dermer, C.D., & Ramaty, R. 1987, ApJS 63, 721.a

Murphy, R.J., Kozlovsky, B., & Ramaty, R. 1988, ApJ 331, 1029.

Murphy, R.J., Share, G.H., Grove, J.E. et al. 1994, Proc-1994-Ryan, 15.

Murphy, R.J., Share, G.H., Grove, J.E. et al. 1996, Proc-1996-Ramaty, 184.

Murphy, R.J., Share, G.H., Grove, J.E. et al. 1997, ApJ 490, 883.

Murphy, R.J., Share, G.H., DelSignore, K.W., et al. 1999, ApJ 510, 1011.

Murphy, R.J., Share, G.H., Hua, X.M. et al. 2003, ApJ 595, L93.

Muschietti, L. 1990, SP 130, 201.

Nagai, F. 1980, SP 68, 351.

Nagai, F. & Emslie, A.G. 1984, ApJ 279, 896.

Nakagawa, Y. 1970, SP 12, 419.

Nakagawa, Y., & Raadu,M.A. 1972, SP 25, 127.

Nakagawa, Y., Wu, S.T., & Han, S.M. 1978, ApJ 219, 314.

Nakagawa, Y., Wu, S.T., & Han, S.M. 1981, ApJ 244, 331.

Nakajima, H., Kosugi, T., Kai, K. et al. 1983, Nature, 305, 292.

Nakajima, H., Dennis, B.R., Hoyng, P., et al. 1985, ApJ 288, 806.

Nakakubo, K. & Hara, H. 2000, Adv. Space Res. 25/9, 1905.

Nakariakov, V.M. & Roberts, B. 1995, SP 159, 399.

Nakariakov, V.M., Ofman, L., DeLuca, E., et al. 1999, Science 285, 862.

Nakariakov, V.M., Verwichte, E., Berghmans, D., et al. 2000a, AA 362, 1151.

Nakariakov, V.M., Ofman, L., & Arber, T.D. 2000b, AA 353, 741.
Nakariakov, V.M. & Ofman, L. 2001, AA 372, L53.
Nakariakov, V.M. 2003, Proc-2003-Dwivedi, 314.
Nakariakov, V.M., Mel'nikov, V.F., & Reznikova, V.E. 2003, AA 412, L7.
Nakariakov, V.M., Arber, T.D., Ault, C.E., et al. 2004, MNRAS 349, 705.
Narain, U. & Ulmschneider, P. 1990, Space Science Rev. 54, 377.
Narain, U. & Ulmschneider, P. 1996, Space Science Rev. 75, 453.
Neidig, D.F. 1989, SP 121, 261.
Neidig, D.F. & Kane, S.R. 1993, SP 143, 201.
Nelson, G.J. & Melrose, D.B. 1985, in McLean & Labrum (1985), (see book list).
Nenovski, P., Dermendjiev, V.N., Detchev, M., et al. 2001, AA 375, 1065.
Ness, N.F. 2001, Proc-2001-Daglis, 131.
Neugebauer, M. 2001, Proc-2001-Balogh, 43.
Neukirch, T. 1995, AA 301, 628,
Neukirch, T. 1996, Proc-1996-Bentley, 286.
Neukirch, T. & Rastätter, L. 1999, AA 348, 1000.
Neupert, W.M. 1968, ApJ 153, L59.
Neupert, W.M., Brosius, J.W., Thomas, R.J., et al. 1992, ApJ 392, L95.
Neupert, W.M., Newmark, J., Delaboudinière, J.-P., et al. 1998, SP 183, 305.
Newkirk, G., Altschuler, M.D., & Harvey, J. 1968, Proc-1968-Kiepenheuer, 379.
Newton, E.K., Emslie, A.G., & Mariska, J.T. 1995, ApJ 447, 915.
Nindos, A. & Zirin, H. 1997, SP 182, 381.
Nindos, A., Alissandrakis, C.E., Gelfreikh, G.B., et al. 2002, AA 386, 658.
Nishio, M., Yaji, K., Kosugi, T., et al. 1997, ApJ 489, 976.
Nitta, N., Kiplinger, A., & Kai, K. 1989, ApJ 337, 1003.
Nitta, N., Dennis, B.R., & Kiplinger, A.L. 1990, ApJ 353, 313.
Nitta, N., White, S.M., Kundu, M.R., et al. 1991, ApJ, 374, 374.
Nitta, N., Bastian, T.S., Aschwanden, M.J., et al. 1992, PASJ 44/5, L167.
Nitta, N. 1997, ApJ 491, 402.
Nitta, N., & Yaji, K. 1997, ApJ 484, 927.
Noci, G. 1981, SP 69, 63.
Noci, G. & Zuccarello, F. 1983, SP 88, 193.
Noci, G., Spadaro, D., Zappala, R.A., et al. 1989, ApJ 338, 1131.
Nolte, J.T., Solodyna, C.V., & Gerassimenko, M. 1979, SP 63, 113.
Norman, C.A. & Smith, R.A. 1978, AA 68, 145.
Nordlund, A. & Galsgaard, K. 1997, Proc-1997-Simnett, 179.
Noyes, R.W., Withbroe, G.L., & Kirshner, R.P. 1970, SP 11, 388.
Obridko, V.N. & Staude, J. 1988, AA 189, 232.
Odstrčil, D., Dryer, M., and Smitch, Z. 1996, JGR 101, 19,973.
Odstrčil, D. & Pizzo, V. 1999a, JGR 104, 483.
Odstrčil, D. & Pizzo, V. 1999b, JGR 104, 28,225.
Odstrčil, D., Linker, J.A., Lionello, R., et al. 2002, JGR 107/A12, SSH 14-1.
Ofman, L. & Davila, J.M. 1994, GRL 21/20, 2259.
Ofman, L., Davila, J.M., & Steinolfson, R.S. 1994, ApJ 421, 360.
Ofman, L., Davila, J.M., & Steinolfson, R.S. 1995, ApJ 444, 471.
Ofman, L., Davila, J.M., & Shimizu, T. 1996, ApJ 459, L39.
Ofman, L., Romoli, M., Poletto, G., et al. 1997, ApJ 491, L111.
Ofman, L. & Davila, J.M. 1997, ApJ 476, 357.
Ofman, L. & Davila, J.M. 1998, JGR 103, 23677.
Ofman, L., Klimchuk, J.A. & Davila, J.M. 1998, ApJ 493, 474.

Ofman, L., Nakariakov, V.M., & DeForest, C.E. 1999, ApJ 514, 441.

Ofman, L., Romoli, M., Poletto, G., et al. 2000a, ApJ 529, 529.

Ofman, L., Nakariakov, V.M., & Seghal, N. 2000b, ApJ 533, 1071.

Ofman, L. & Wang, T.J. 2002, ApJ 580, L85.

Ofman, L. & Aschwanden, M.J. 2002, ApJ 576, L153.

Ofman, L. 2002, ApJ 568, L135.

Ofman, L. & Thompson, B.J. 2002, ApJ 574, 440.

Ofman, L. 2003, Proc-2003-Erdélyi, 349.

Ogawara, Y., Takano, T., Kato, T., et al. 1991, SP 136, 1.

Ohsawa, Y. & Sakai, J.I. 1987, ApJ 313, 440.

Ohsawa, Y. & Sakai, J.I. 1988a, SP 116, 157.

Ohsawa, Y. & Sakai, J.I. 1988b, ApJ 332, 439.

Ohyama, M. & Shibata, K. 1996, Proc-1996-Uchida, 525.

Øieroset, M., Phan, T.F., Fujimoto, M., et al. 2001, Nature 412, 6845.

Oliver, R. 2001a, *Solar Prominence Oscillations*, (in Murdin 2000) .

Oliver, R. 2001b, Proc-2001-Ballester, 133.

Oliver, R. & Ballester, J. 2002, SP 206, 45.

Orlando, S., Peres, G., & Serio, S. 1995a, AA 294, 861.

Orlando, S., Peres, G., & Serio, S. 1995b, AA 300, 549.

Orrall, F.Q. & Zirker, J.B. 1961, ApJ 134, 72.

Orrall, F.Q., Rottman, G.J., & Klimchuk, J.A. 1983, ApJ 266, L65.

Orwig, L.E., Frost, K.J., & Dennis, B.R. 1980, SP 65, 25.

Orwig, L.E. & Woodgate, B.E. 1986, Proc-1986-Neidig, 306.

O'Shea, E., Banerjee, D., Doyle, J.G., et al. 2001, AA 368, 1095.

Oughton, S., Matthaeus, W.H., Dmitruk, P., et al. 2001, ApJ 551, 565.

Pallavicini, R., Serio, S., & Vaiana, G.S. 1977, ApJ 216, 108.

Pallavicini, R., Peres, G., Serio, S., et al. 1981, ApJ 247, 692.

Pallavicini, R. & Peres, G. 1983, SP 86, 147.

Papadopoulos, K. 1979, Proc-1979-Akasofu, 289.

Papadopoulos, K. 1979, ASSL 78, 289.

Papagiannis, M.D. & Baker, K.B. 1982, SP 79, 365.

Parenti, S., Bromage, B.J.I., & Bromage, G.E. 2002, AA 384, 303.

Park, B.T. & Petrosian, V. 1995, ApJ 446, 699.

Park, B.T. & Petrosian, V. 1996, ApJS 103, 255.

Park, B.T., Petrosian, V., & Schwartz, R.A. 1997, ApJ 489, 358.

Parker, E.N. 1953, ApJ 117, 431.

Parker, E.N. 1958, ApJ 128, 664.

Parker, E.N. 1963a, ApJS 8, 177.

Parker, E.N. 1963b, Interplanetary Dynamical Processes, (see book list

Parker, E.N. 1965, *Planet. Space Sci.* 13, 9.

Parker, E.N. 1966, ApJ 145, 811.

Parker, E.N. 1969, SSR 9, 651.

Parker, E.N. 1972, ApJ 174, 499.

Parker, E.N. 1977, ARAA 15, 45.

Parker, E.N. 1979, *Cosmical Magnetic Fields*, (see book list).

Parker, E.N. 1983, ApJ 264, 642.

Parker, E.N. 1988, ApJ 330, 474.

Parker, E.N. 1991, ApJ 376, 355.

Parker, E.N. 1994, *Spontaneous Current Sheets in Magnetic Fields*, (see book list).

Parks, G.K. & Winckler, J.R. 1969, ApJ 155, L117.

Parnell, C.E., Priest, E.R., & Golub, L. 1994, SP 151, 57.

Parnell, C.E. 1996, Proc-1996-Bentley, 19.

Parnell, C.E. & Jupp, P.E. 2000, ApJ 529, 554.

Parnell, C.E. 2001, SP 200, 23.

Parnell, C.E. 2002a, COSPAR-CS 13, 47.

Parnell, C.E. 2002b, ESA SP-505, 231.

Parnell, C.E., Bewsher, D., & Harrison, R.A. 2002, SP 206, 249.

Pasachoff, J.M. & Landman, D.A. 1984, SP 90, 325.

Pasachoff, J.M. & Ladd, E.F. 1987, SP 109, 365.

Pasachoff, J.M., Babcock, B.A., Russell, K.D., et al. 2002, SP 207, 241.

Patsourakos, S. & Vial, J.C. 2000, AA 359, L1.

Patsourakos, S. & Vial, J.C. 2002, SP 208, 253.

Pekeris, C.L. 1948, *Geol.Soc.Amer.Mem.* 27, 117.

Pelaez, F., Mandrou, P., Niel, M., et al. 1992, SP 140, 121.

Peres, G. 1997, ESA SP-404, 55.

Peres, G., Rosner, R., Serio, S., et al. 1982, ApJ 252, 791.

Peres, G., Spadaro, D., & Noci, G. 1992, ApJ 389, 777.

Peres, G. & Reale, F. 1993, AA 267, 566.

Peres, G. & Reale, F. 1993, AA 275, L13.

Perez, M.E., Doyle, J.G., Erdélyi, R., et al. 1999, AA 342, 279.

Peter, H. 2001, AA 374, 1108.

Peter, H. & Brkovic, A. 2003, AA 403, 287.

Pesses, M.E. 1979, Proc-1979-ICRC16, 18.

Petrie, G.J.D. & Neukirch, T. 1999, Geophys. Astrophys. Fluid Dyn. 91, 269.

Petrie, G.J.D. & Neukirch, T. 2000, AA 356, 735.

Petrie, G.J.D. & Lothian, R.M. 2003, AA 398, 287.

Petrosian, V. 1973, ApJ 186, 291.

Petrosian, V. 1985, ApJ 299, 987.

Petrosian, V. 1996, Proc-1996-Ramaty, 445.

Petrosian, V. 1999, Proc-1999-Ostrowski, 135.

Petrosian, V. & Donaghy, T.Q. 1999, ApJ 527, 945.

Petrosian, V. & Donaghy, T.Q. 2000, Proc-2000-Ramaty, 215.

Petrosian, V., Donaghy, T.Q., & McTiernan, J.M. 2002, ApJ 569, 459.

Petschek, H.E. 1964, Proc-1964-Hess, 425.

Petschek, H.E. & Thorne, R.M. 1967, ApJ 147, 1157.

Pevtsov, A.A., Canfield, R.C., & Zirin, H. 1996, ApJ 473, 533.

Pevtsov, A.A., Canfield, R.C., & McClymont, A.N. 1997, ApJ 481, 973.

Pevtsov, A.A. 2002, SP 207, 111.

Pevtsov, A.A., Balasubramaniam, K.S., & Rogers, J.W. 2003, ApJ 595, 500.

Phillips, K.J.H. 1992, *Guide to the Sun*, (see book list).

Phillips, K,J.H., Bhatia, A.K., Mason, H.E., et al. 1996, ApJ 466, 549.

Phillips, K.J.H. 2004, ApJ 605, 921.

Piana, M. 1994, AA 288, 949.

Piana, M., Brown, J.C., & Thompson, A.M. 1995, SP 156, 315.

Piana, M., Massone, A.M., Kontar, E.P., et al. 2003, ApJ 595, L127.

Pick, M. & Trottet, G. 1978, SP 60, 353.

Pick, M. & Van den Oord, G.H.J. 1990, SP 130, 83.

Pick, M. 1999, Proc-1999-Bastian, 187.

Pikel'ner, S.B. 1971, SP 17, 44.

Planck, M. 1901, Ann. Physik 4, 553.

Planck, M. 1913, *The Theory of Heat Radiation*, (see book list).

Plunkett, S.P. & Simnett, G.M. 1994, SP 155, 351.

Plunkett, S.P., Vourlidas, A., Simberova, S., et al. 2000, SP 194, 371.

Pneuman, G.W. 1972, ApJ 177, 793.

Pneuman, G.W. & Kopp, R.A. 1977, AA 55, 305.

Pneuman, G.W. 1980, SP 65, 369.

Poedts, S., Goossens, M., & Kerner, W. 1989, SP 123, 83.

Poedts, S., Toth, G., Bélien, A.J.C., et al. 1997, SP 172, 45.

Poedts, S. 1999, ESA SP-448, 167.

Poedts, S. 2002, ESA SP-505, 273.

Poland, A.I., Howard, R.A., Koomen, M.J., et al. 1981, SP 69, 169.

Poland, A.I., Mariska, J.T., & Klimchuk, J.A. 1986, Proc-1986-Poland, 57.

Poland, A.I. & Mariska, J.T. 1986, SP 104, 303.

Poquerusse, M. & McIntosh, P.S. 1995, SP 159, 301.

Porter, L.J., Klimchuk, J.A., & Sturrock, P.A. 1992, ApJ 385, 738.

Porter, L.J. & Klimchuk, J.A. 1995, ApJ 454, 499.

Portier−Fozzani, F., Aschwanden, M.J., Démoulin, P., et al. 2001, SP 203, 289.

Pottasch, S.R. 1964a, MNRAS 128, 73.

Pottasch, S.R. 1964b, Space Sci.Rev., 3, 816.

Priest, E.R. 1972, Quart. Journal Mech. and App. Math. 25, 319.

Priest, E.R. & Cowley, S.W.H. 1975, J.Plasma Phys. 14, 271.

Priest, E.R. & Soward, A.M. 1976, Proc-1976-Bumba, 353.

Priest, E.R. 1978, SP 58, 57.

Priest, E.R. & Smith, D.F. 1979, SP 64, 267.

Priest, E.R. 1981, *Solar Flare Magnetohydrodynamics*, (see book list).

Priest, E.R. 1982, *Solar Magnetohydrodynamics*, (see book list).

Priest, E.R. 1985a, Rep. Prog. Phys. 48, 955.

Priest, E.R. 1985b, Proc-1985-Kundu, 233

Priest, E.R. 1986, Mit.Astron.Ges. 65, 41.

Priest, E.R. & Forbes, T.G. 1986, JGR 91, 5579.

Priest, E.R., Hood, A.W., and Anzer,U. 1989, ApJ 344, 1010.

Priest, E.R. (ed.) 1989, Proc-1989-Priest.

Priest, E.R. & Lee, L.C. 1990, J. Plasma Phys. 44, 337.

Priest, E.R. 1994, *Hydrodynamics*, Proc-1994-Benz, 1.

Priest, E.R., Parnell, C.E., & Martin, S.F. 1994, ApJ 427, 459.

Priest, E.R., Van Ballegooijen, A.A., & Mackay, D.H. 1996, ApJ 460, 530.

Priest, E.R. 1996, Proc-1996-Bentley, 331.

Priest, E.R., Bungey, T.N. & Titov, V.S. 1997, Geophys. Astrophys. Fluid Dyn. 84, 127.

Priest, E.R. & Schrijver, C.J. 1999, SP 190, 1.

Priest, E.R., Foley, C.R., Heyvaerts, J., et al. 1999, Nature 393, 545.

Priest, E.R. 2000, Proc-2000-Ramaty, 13.

Priest, E.R. & Forbes, T. 2000, *Magnetic Reconnection* (see book list).

Priest, E.R., Foley, C.R., Heyvaerts, J., et al. 2000, ApJ 539, 1002.

Priest, E.R., Heyvaerts, J.F., & Title, A.M. 2002, ApJ 576, 533.

Priest, E.R. & Forbes, T.G. 2002, Astron.Astrophys.Rev. 10, 313.

Prince, T.A., Ling, J.C., Mahoney, W.A. et al. 1982, ApJ 255, L81.

Prince, T.A., Forrest, D.J., Chupp, E.L. et al. 1983, Proc-1983-ICRC18, 79.

Pritchett, P.L. & Wu, C.C. 1979, Phys. Fluids 22, 2140.

Pritchett, P.L. 1984, JGR 89(A10), 8957.

Pritchett, P.L. 1986, Phys.Fluids 29(9), 2919.

Pryadko, J.M., & Petrosian, V. 1997, ApJ 482, 774.

Qin, Z.H. & Huang, G.L. 1994, Astrophys. Space Sci. 218, 213.

Qin, Z.H., Li, C., & Fu, Q. 1996, SP 163, 383.

Qiu, J., Wang, H., Chae, J.C., et al. 1999, SP 190, 153.

Qiu, J., Ding, M.D., Wang, H., et al. 2000, ApJ 544, L157.

Qiu, J., Ding, M.D., Wang, H., et al. 2001, ApJ 554, 445.

Qiu, J., Lee, J.W., Gary, D.E., et al. 2002, ApJ 565, 1335.

Querfeld, C.W., Smartt, R.N., Bommier, V., et al. 1985, SP 96, 277.

Raadu, M.A. 1972, SP 22, 425.

Rabin, D. 1991, ApJ 383, 40.

Rabin, D & Moore, R. 1984, ApJ 285, 359.

Rae, I.C. & Roberts, B. 1982, MNRAS 201, 1171.

Ramaty, R. & Kozlovsky, B. 1974, Proc-1974-Leningrad, 25.

Ramaty, R., Kozlovsky, B., & Lingenfelter, R.E. 1975, SSR 18, 341.

Ramaty, R. 1979, Proc-1979-Arons, 135.

Ramaty, R. 1986, Proc-1986-Sturrock, (Chapter 14), 291.

Ramaty, R. & Murphy, R.J. 1987, Space Science Rev. 45, 213.

Ramaty, R. & Mandzhavidze, N. 1994, Proc-1994-Ryan, 26.

Ramaty, R., Mandzhavidze, N., Kozlovsky, B., et al. 1995, ApJ 455, L193.

Ramaty, R. 1996, Proc-1996-Ramaty, 533.

Ramaty, R. & Lingenfelter, R.E. 1996, Proc-1996-Williams, 180.

Ramaty, R., Mandzhavidze, N., & Kozlovsky, B. (eds.) 1996, Proc-1996-Ramaty.

Ramaty, R., Mandzhavidze, N., & Kozlovsky, B. 1996, Proc-1996-Ramaty, 172.

Ramaty, R. & Mandzhavidze, N. (eds.) 2000a, Proc-2000-Ramaty.

Ramaty, R. & Mandzhavidze, N. 2000b, Proc-2000-Martens, 123.

Ramaty, R. & Mandzhavidze, N. 2001, (in Murdin 2000) , p...

Ramsey, H. & Smith, S.F. 1966, Astron. J. 71/3, 197.

Rank, G., Bennett, K., Bloemen, H., et al. 1996, Proc-1996-Ramaty, 219.

Rank, G., Ryan, J., Debrunner, H., et al. 2001, AA 378, 1046.

Ratcliffe, J.A. 1969, *The Magneto-Ionic Theory*, (see book list).

Rayleigh, Lord 1900, Phil.Mag., 49, 539.

Rayleigh, Lord 1905, Nature, 72, 54.

Raymond, J.C. & Doyle, J.G. 1981, ApJ 247, 686.

Raymond, J.C. 1990, ApJ 365, 387.

Reale, F., Betta, R., Peres, G., et al. 1997, AA 325, 782.

Reale, F., Peres, G., Serio, S., et al. 2000a, ApJ 535, 412.

Reale, F., Peres, G., Serio, S., et al. 2000b, ApJ 535, 423.

Reames, D.V., von Rosenvinge, T.T., & Lin, R.P. 1985, ApJ 292, 716.

Reames, D.V. & Stone, R.G. 1986, ApJ 308, 902.

Reames, D.V., Dennis, B.R., Stone, R.G., et al. 1988, ApJ 327, 998.

Reames, D.V. 1990a, ApJ 358, L63.

Reames, D.V. 1990b, ApJS 73, 235.

Reames, D.V., Richardson, I.G., & Barbier, L.M. 1991a, ApJ 382, L43.

Reames, D.V., Kallenrode, M.B. & Stone, R.G. 1991b, ApJ 380, 287.

Reames, D.V. 1992, Proc-1992-Zank 213.

Reames, D.V., Richardson, I.G., & Wenzel, K.P. 1992, ApJ 387, 715.

Reames, D.V., Meyer, J.P., & Voni Rosenvinge, T.T. 1994, ApJS 90, 649.

Reames, D.V. 1995a, Adv. Space Res. 15(7), 41.

Reames, D.V. 1995b, Rev. Geophys. (Suppl), 33, 585.

Reames, D.V., Barbier, L.M., & Ng, C.K. 1996, ApJ 466, 473.

Reames, D.V., Kahler, S.W., & Ng, C.K. 1997, ApJ 491, 414.

Reames, D.V. 1999, SSR 90, 413.

Reames, D.V., Ng, C.K. & Tylka, A.J. 1999, GRL 26, 3585.

Reames, D.V. 2000, Proc-2000-Ramaty, 102.

Reames, D.V. 2001a, *Solar Wind: Energetic Particles*, (in Murdin 2000) .

Reames, D.V. 2001b, Proc-2001-Wimmer, 153.

Reames, D.V., Ng, C.K., & Tylka, A.J. 2001a, ApJ 548, L233.

Reames, D.V., Ng, C.K., & Tylka, A.J. 2001b, ApJ 550, 1064.

Reames, D.V. 2002, ApJ 571, L63.

Reames, D.V. & Tylka, A.J. 2002, ApJ 575, L37.

Reeves, K.K. & Warren, H.P. 2002, ApJ 578, 590.

Régnier, S., Solomon, J., & Vial, J.C. 2001, AA 376, 292.

Reimers, D. 1971a, AA 10, 182.

Reimers, D. 1971b, AA 14, 198.

Reiner, M. 2001, SSR 97, 129.

Ricca, R.L. 1997, SP 172, 241.

Ricchiazzi, P.J. & Canfield, R.C. 1983, ApJ 272, 739.

Rickard, G.J. & E.R. Priest, 1994, SP 151, 107.

Rickard, G.J. & Titov, V.S. 1996, ApJ 472, 840.

Ridgway, C., Priest, E.R., & Amari, T. 1991a, ApJ 367, 321.

Ridgway, C., Amari, T., & Priest, E.R. 1991b, ApJ 378, 773.

Rieger, E. 1989, SP 121, 323.

Rieger, E. & Marschhäuser, H. 1990, Proc-1990-Winglee, 68.

Rieger, E. 1994, ApJS 90, 645.

Rieger, E., Neidig, D.F., Enfger, D.W., et al. 1996, SP 167, 307.

Rieger, E., Gan, W.Q., & Marschhäuser, H. 1998, SP 183, 123.

Riley, P., Gosling, J.T., & Pizzo,V.J. 1997, JGR 192/A7, 14677.

Riley, P. 1999, Proc-1999-Habbal, 131.

Riley, P., Linker, J.A., Mikic, Z., et al. 2003, JGR 108/A7, SSH 2-1, CiteID 1272.

Robb, T.D. & Cally, P.S. 1992, ApJ 397, 329.

Robbrecht, E., Verwichte, E., Berghmans, D., et al. 2001, AA 370, 591.

Roberts, B. & Priest, E.R. 1975, J. Plasma Phys. 14, 417.

Roberts, B. 1981a, SP 69, 27.

Roberts, B. 1981b, SP 69, 39.

Roberts, B., & Frankenthal, S. 1980, SP 68, 103.

Roberts, B., Edwin, P.M., & Benz, A.O. 1983, Nature 305, 688.

Roberts, B., Edwin, P.M., & Benz, A.O. 1984, ApJ 279, 857.

Roberts, B. 1984, ESA SP-20, 137.

Roberts, B. 1985, Proc-1985-Priest, 37.

Roberts, B. 1991a, Proc-1991-Priest, 105.

Roberts, B. 1991b, Geophys. Astrophys. Fluid Dynamics 62, 83.

Roberts, B. & Joarder, P.S. 1994, Proc-1994-Belvedere, 173.

Roberts, B. 2000, SP 193, 139.

Roberts, B. 2001, *Solar Photospheric Magnetic Flux Tubes: Theory*, (in Murdin 2000) .

Roberts, B. 2002, ESA SP-506, 481.

Roberts, B. & Nakariakov, V.M. 2003, Proc-2003-Erdélyi, 167.

Robertson, J.A., Hood, A.W., & Lothian, R.M. 1992, SP 137, 273.

Robinson, P.A. 1986, J. Plasma Physics 36, 63.

Robinson, P.A. 1988, Phys.Fluids 31(3), 525.

Robinson, P.A. 1989, ApJ 341, L99.

Robinson, P.A. 1991a, SP 134, 299.

Robinson, P.A. 1991b, SP 136, 343.

Robinson, P.A. 1996, SP 168, 357.

Robinson, P.A. 1997, Reviews of Modern Physics 69/2, 508.

Rosenberg, H. 1970, AA 9, 159.

Rosenberg, H. 1972, SP 25, 188.

Rosner, R. & Vaiana, G.S. 1977, ApJ 216, 141.

Rosner, R. & Vaiana, G.S. 1978, ApJ 222, 1104.

Rosner, R., Tucker, W.H., & Vaiana, G.S. 1978a, ApJ 220, 643.

Rosner, R., Golub, L., Coppi, B., et al. 1978b, ApJ 222, 317.

Rottman, G.J., Orrall, F.Q., & Klimchuk, J.A. 1982, ApJ 260, 326.

Roumeliotis, G., Sturrock, P., & Antiochos, S.K. 1994, ApJ 423, 847.

Roumeliotis, G. 1996, ApJ 473, 1095.

Roussev, I., Galsgaard, K., Erdélyi, R., et al. 2001a, AA 370, 298.

Roussev, I., Galsgaard, K., Erdélyi, R., et al. 2001b, AA 375, 228.

Roussev, I., Doyle, J.G., Galsgaard, K., et al. 2001c, AA 380, 719.

Roussev, I., Galsgaard, K., & Judge, P.G. 2002, AA 382, 639.

Roussev, I. & Galsgaard, K. 2002, AA 383, 697.

Rudenko, G.V. 2001, SP 198, 5.

Ruderman, M.S., Goossens, M., Ballester, J.L., et al. 1997, AA 328, 361.

Ruderman, M.S., & Roberts, B. 2002, ApJ 577, 475.

Ruderman, M.S. 2003, ApJ 409, 287.

Russell, C.T., Priest, E.R., & Lee, L.C. (eds.) 1990, Proc-1990-Russell.

Russell, C.T. 2001, Proc-2001-Song, 73.

Rust, D.M. 1967, ApJ 150, 313.

Rust, D.M. & Hegwer, F. 1975, SP 40, 141.

Rust, D.M. & Hildner, E. 1978, SP 48, 381.

Rust, D.M. 1983, SSR 34, 21.

Rust, D.M., Simnett, G.M., & Smith, D.F. 1985, ApJ 288, 401.

Rust, D.M. & Kumar, A. 1994, SP 155, 69-97.

Rust, D.M. 1996, Proc-1997-Crooker, 119.

Rust, D.M. & Kumar, A. 1996, ApJ 464, L199.

Rust, D.M. 2001, *Solar Prominences*, (in Murdin 2000) .

Rust, D.M. 2001b, GRL 106/A11, 25075.

Rust, D.M. 2003, Adv. Space Res. 32/10, 1895.

Rutherford, E. 1911, Phil.Mag., 21, 669.

Ryabov, B.I., Pilyeva, N.A., Alissandrakis, C.E., et al. 1999, SP 185, 157.

Ryan, J.M. 1986, SP 105, 365

Ryan, J.M. & Lee, M.A. 1991, ApJ 368, 316.

Ryan, J.M. 1994, Proc-1994-Fichtel, 12.

Ryan, J.M. & Vestrand, W.T. (eds.) 1994, Proc-1994-Ryan.

Ryan, J.M. 2000, SSR 93, 581.

Rybicki, G.B. & Lightman, A.P. 1979, *Radiative Processes in Astrophysics*, (see book list).

Ryutov, D.D. & Sagdeev, R.Z. 1970, Soviet Phys. JETP 31, 396.

Ryutova, M. & Tarbell, T.D. 2000, ApJL 541, L29.

Ryutova, M., Habbal, S., Woo, R., et al. 2001, SP 200, 213.

Saba, J.L.R. & Strong, K.T. 1991, AdSpR 11/1, 117.

Saint—Hilaire, P. & Benz, A.O. 2002, SP 210, 287.

Sakai, J.I. & Tajima, T. 1986, ESA SP-251, 77.

Sakai, J.I. & Ohsawa, Y. 1987, Space Sci.Rev. 46, 113.

Sakai, J.I., Colin, A., & Priest, E.R. 1987, SP 114, 253.

Sakai, J.I. & de Jager, C. 1991, SP 134, 329.

Sakai, J.I. & Koide, S. 1992, SP 142, 399.

Sakai, J.I., Fushiki, T., & Nishikawa, K.I. 1995, SP 158, 301.

Sakai, J.I. & de Jager, C. 1996, Space Sci.Rev. 77, 1.

Sakai, J.I., Kawata, T., Yoshida, K., et al. 2000a, ApJ 537, 1063.

Sakai, J.I., Mizuhata, Y., Kawata, T., et al. 2000b, ApJ 544, 1108.

Sakai, J.I., Minamizuka, R., Kawata, T., et al. 2001a, ApJ , 550, 1075.

Sakai, J.I., Takahata, A., & Sokolov, I.V. 2001b, ApJ 556, 905.

Sakai, J.I. & Furusawa, K. 2002, ApJ 564, 1048.

Sakai, J.I., Nishi, K., & Sokolov, I.V. 2002, ApJ 576, 519.

Sakao, T. 1994, PhD Thesis, (see PhD Thesis list).

Sakao, T., Kosugi,T., & Masuda,S. 1998, Proc-1998-Watanabe, 273.

Sakao, T. 1999, Proc-1999-Bastian, 231.

Sakurai, T. 1976, PASJ 28, 177.

Sakurai, T. & Uchida, Y. 1977, SP 52, 397.

Sakurai, T. 1979, PASJ 31, 209.

Sakurai, T. 1981, SP 69, 343.

Sakurai, T. 1982, SP 76, 301.

Sakurai, T., Makita, M., & Shibasaki, K. 1985, Proc-1985-Schmidt, 312.

Sakurai, T. 1989, SP 121, 347.

Sakurai, T., Goossens, M., & Hollweg, J.V. 1991a, SP 133, 227.

Sakurai, T., Goossens, M., & Hollweg, J.V. 1991b, SP 133, 247.

Sakurai, T., Shibata, K., Ichimoto, K., et al. 1992, PASJ 44, L123.

Sakurai, T., Ichimoto, K., Raju, K.P., et al. 2002, SP 209, 265.

Sarro, L.M., Erdélyi, R., Doyle, J.G., et al. 1999, AA 351, 721.

Sastry, Ch.V., Krishan, V., & Subramanian, K.R. 1981, J. Astrophys. Astron. 2, 59.

Sato, T. 1979, JGR 89, 9761.

Sato, T., Sawa, M., Yoshimura, K. et al. 2003, *Yohkoh Flare Catalogue* (see book list).

Schatzman, E. 1949, Ann. d'Ap., 12, 203.

Schindler, K. & Hornig,G. 2001, *Magnetic Reconnection*, (in Murdin 2000) .

Schlickeiser, R. 2003, Proc-2003-Klein, 230.

Schmahl, E.J., Kundu, M.R., Strong, K.T., et al. 1982, SP 80, 223.

Schmahl, E.J. & Orrall, F.Q. 1979, ApJ 231, L41.

Schmahl, E.J. & Hurford, G.J. 2002, SP 210, 273.

Schmelz, J.T., Holman, G.D,, Brosius, J.W., et al. 1992, ApJ 399, 733.

Schmelz, J.T., Holman, G.D., Brosius, J.W., et al. 1994, ApJ 434, 786.

Schmelz, J.T., Scopes, R.T., Cirtain, J.W., et al. 2001, ApJ 556, 896.

Schmidt, H.U. 1964, Proc-1964-Hess, 107.

Schmidt, G. 1979, *Physics of High Temperature Plasmas*, (see book list).

Schmieder, B., Forbes, T.G., Malherbe, J.M., et al. 1987, ApJ 317, 956.

Schmieder, B., Malherbe, J.M., Simnett, G., et al. 1990, ApJ 356, 720.

Schmieder, B., Démoulin, P., Aulanier, G., et al. 1996, ApJ 467, 881.

Schmieder, B., Aulanier, G., Démoulin, P., et al. 1997a, AA 325, 1213.

Schmieder, B., Démoulin, P., Malherbe, J.M., et al. 1997b, Adv.Space Res. 18/12, 1871.

Scholer, M. 1989, JGR 94, 8805.

Scholer, M. 2003, Proc-2003-Klein, 9.

Schrijver, C.J. et al. 1999, SP 187, 261.

Schrijver, C.J. & Title, A.M. 1999, SP 188, 331.

Schrijver, C.J. & Zwaan, C. 2000, *Solar and Stellar Magnetic Activity*, (see book list).

Schrijver, C.J. 2001a, SP 198, 325.

Schrijver, C.J. 2001b, Proc-2001-Lopez, 131.

Schrijver, C.J. & Aschwanden, M.J. 2002, ApJ 566, 1147.

Schrijver, C.J. & Title, A.M. 2002, SP 207, 223.

Schrijver, C.J., Aschwanden, M.J., & Title, A. 2002, SP 206, 69.

Schroeder, M. 1991, *Fractals, Chaos, Power Laws*, (see book list).

Schroeter, E.H. and Woehl, H. 1976, SP 49, 19.

Schüssler, M. & Schmidt, W. 1994, *Solar Magnetic Fields*, (see book list).

Schüssler, M. 2001, *Solar Magnetic Fields*, (in Murdin 2000) .

Schumacher, J. & Kliem, B. 1996, Phys. Plasmas 3(12), 4703.

Schumacher, J. & Kliem, B. 1997a, Phys. Plasmas 4(10), 3533.

Schumacher, J. & Kliem, B. 1997b, Adv. Space Research 19/12, 1797.

Schumacher, J., Kliem, B., & Seehafer, N. 2000, Phys.Plasmas 7/1, 108.

Schuster, H.G. 1988, *Deterministic Chaos: An Introduction*, (see book list).

Schwadron, N.A., Fisk, L.A., & Zurbuchen, T.H. 1997, SSR 86, 1/4, 51.

Schwartz, R.A., Dennis, B.R., Fishman, G.J., et al. 1992, Proc-1992-Shrader, 457.

Schwenn, R. 1986, SSR 44, 139.

Schwenn, R. & Marsch, E. (eds.) 1991a, *Physics of the Inner Heliosphere. I.*, (see book list).

Schwenn, R. & Marsch, E. (eds.) 1991b, *Physics of the Inner Heliosphere. II.*, (see book list).

Schwenn, R. 2001, *Solar Wind: Global Properties*, (in Murdin 2000) .

Scudder, J.D. 1992a, ApJ 398, 299.

Scudder, J.D. 1992b, ApJ 398, 319.

Scudder, J.D. 1994, ApJ 427, 446.

Sedlacek, Z. 1971, J.Plasma Phys. 5, 239.

Seehafer, N. 1978, SP 58, 215.

Semel, M. 1988, AA 198, 293.

Serio, S., Peres, G., Vaiana, G.S., et al. 1981, ApJ 243, 288.

Sersen, M. 1996, Proc-1996-Bentley, 206.

Shafranov, V.D. 1957, *J. Nucl. Energy II*, 5, 86.

Shapiro, P.R. & Knight, J.W. 1978, ApJ 224, 1028.

Share, G.H. & Murphy, R.J. 1995, ApJ 452, 933.

Share, G.H., Murphy, R.J., & Skibo, J.G. 1996, Proc-1996-Ramaty, 162.

Share, G.H. & Murphy, R.J. 1997, ApJ 485, 409.

Share, G.H., Murphy, R.J., & Ryan, J. 1997, Proc-1997-Dermer, 17.

Share, G.H. & Murphy, R.J. 1998, ApJ 508, 876.

Share, G.H. & Murphy, R.J. 2000, Proc-2000-Ramaty, 377.

Share, G.H., Murphy, R.J., Kiener, J., et al. 2002, ApJ 573, 464.

Share, G.H., Murphy, R.J., Smith, D.M. et al. 2003, ApJ 595, L89.

Share, G.H. & Murphy, R.J. 2004, Proc-2004-Dupree, 133.

Sharma, R.R., Vlahos, L., & Papadopoulos, K. 1982, AA 112, 377.

Sharma, R.R. & Vlahos, L. 1984, ApJ 280, 405.

Sheeley, N.R., Wang, Y.M., Hawley, S.H., et al. 1997, ApJ 484, 472.

Sheeley, N.R., Walters, J.H., Wang, Y.M. et al. 1999, JGR , 104/A11, 24739.

Shemi, A. 1991, MNRAS 251, 221.

Shibasaki, K., Enome, S., Nakajima, H., et al. 1994, PASJ 46, L17.

Shibasaki, K. 2001, ApJ 557, 326.

Shibata, K., Tajima, T., Matsumoto, R., et al. 1989a, ApJ 338, 471.

Shibata, K., Tajima, T., Steinolfson, R.S., et al. 1989b, ApJ 345, 584.

Shibata, K., Nozawa, S., Matsumoto, R., et al. 1990, ApJ 351, L25.

Shibata, K. 1991, Proc-1991-Uchida, 205.

Shibata, K., Ishido, Y., Acton, L.W., et al. 1992a, PASJ 44, L173.

Shibata, K., Nozawa, S., & Matsumoto, R. 1992b, PASJ 44, 265.

Shibata, K. & SXT Team 1993, ESA SP-351, 207.

Shibata, K. 1994, Proc-1994-Pap, 89.

Shibata, K., Nitta, N., Strong, K.T., et al. 1994a, ApJ 431, L51.

Shibata, K., Yokoyama, T., & Shimojo, M. 1994b, Proc-1994-Enome, 75.

Shibata, K. 1995, Proc-1995-Watanabe, 85.

Shibata, K., Masuda, S., Shimojo, M., et al. 1995a, ApJ 451, L83.

Shibata, K. 1996, Proc-1996-Uchida, 13.

Shibata, K., Yokoyama, T., & Shimojo, M. 1996a, Adv.Space Res. 17 No. 4/5, 197.

Shibata, K., Yokoyama, T., & Shimojo, M. 1996b, G.Geomag.Geoelectr. 48, 19.

Shibata, K., Shimojo, M., Yokoyama, T., et al. 1996c, Proc-1996-Bentley, 29.

Shibata, K. 1998, Proc-1998-Watanabe, 187.

Shibata, K. 1999a, Astrophysics and Space Science 264, 129.

Shibata, K. 1999b, Proc-1999-Bastian, 381.

Shibata, K. & Yokoyama, T. 1999, ApJ 526, L49.

Shibata, K. & Tanuma, S. 2001, Earth, Planets and Space 53, 473.

Shimizu, T., Tsuneta, S., Acton, L.W., et al. 1992, PASJ 44, L147.

Shimizu, T., Tsuneta, S., Acton, L.W., et al. 1994, ApJ 422, 906.

Shimizu, T. 1995, PASJ 47, 251.

Shimizu, T. & Tsuneta, S. 1997, ApJ 486, 1045.

Shimizu, T. 1997, PhD Thesis, (see PhD Thesis list).

Shimizu, T. 2002a, COSPAR-CS 13, 29.

Shimizu, T. 2002b, ApJ 574, 1074.

Shimojo, M., Hashimoto, S., Shibata, K., et al. 1996, PASJ 48, 123.

Shimojo, M., Shibata, K., & Harvey, K.L. 1998, SP 178, 379.

Shimojo, M. & Shibata, K. 1999, ApJ 516, 934.

Shimojo, M. & Shibata, K. 2000, ApJ 542, 1100.

Shimojo, M., Shibata, K., Yokoyama, T., et al. 2001, ApJ 550, 1051.

Shrivastava, N. & Ambastha, A. 1998, Astrophys. Space Sci. 262, 29.

Silva, A.V.R., White, S.M., Lin, R.P., et al. 1996 ApJS 106, 621.

Silva, A.V.R., Gary, D.E., White, S.M., et al. 1997a, SP 175, 157.

Silva, A.V.R., Wang, H., Gary, D.E., et al. 1997b, ApJ 481, 978.

Silva, A.V.R., Wang, H., & Gary, D.E. 2000, ApJ 545, 1116.

Simnett,G.M. 1986a, SP 106, 165.

Simnett,G.M. 1986b, SP 104, 67.

Simnett,G.M. 1995, SSR 73, 387.

Simon, M. & Shimabukuro, F.I. 1971, ApJ 168, 525.

Slottje, C. 1981, *Atlas of fine structures* ..., (see book list).

Smith, D.F. 1974, Space Science Reviews, 16, 91.

Smith, D.F. 1977, JGR 82, 704.

Smith, D.F. & Lilliequist, C.G. 1979, ApJ 232, 582.

Smith, D.F., Hildner, E., & Kuin, N.P.M. 1992, SP 137, 317.

Smith, D.F. & Benz, A.O. 1991, SP 131, 351.

Smith, D.F. & Brecht, S.H. 1993, ApJ 406, 298.

Smith, D.M., Lin, R.P., Turin, P., et al. 2000, Proc-2000-Ramaty, 92.

Smith, D.M., Lin, R.P., Turin, P., et al. 2002, SP 210, 33.

Smith, D.M., Share, G.H., Murphy, R.J. et al. 2003, ApJ 595, L81.

Smith, S.F. & Harvey, K.L. 1971, Proc-1971-Macris, 156.

Smith, Z. & Dryer, M. 1990, SP 129, 387.

Solanki, S.K. 2001a, *Solar Photospheric Magnetic Flux Tubes*, (in Murdin 2000) .

Solanki, S.K. 2001b, *Sunspot Magnetic Fields*, (in Murdin 2000) .

Solanki, S.K. 2001c, *Sunspot Models*, (in Murdin 2000) .

Somov, B.V., Syrovatskii, S.I., & Spector, A.R. 1981, SP 73, 145.

Somov, B.V., Syrovatskii, S.I., & Spector, A.R. 1982, SP 81, 281.

Somov, B.V. & Verneta, A.I. 1989, SP 120, 93.

Somov, B.V. 1992, *Physical Processes in Solar Flares*, (see book list).

Somov, B.V. 1996, Proc-1996-Ramaty, 493.

Somov, B.V. & Kosugi, T. 1997, ApJ 485, 859.

Somov, B.V., Kosugi, T., & Sakao, T. 1998, ApJ 497, 943.

Somov, B.V., Kosugi, T., Sakao, T., et al. 1999, Proc-1999-ESA448, 701.

Somov, B.V. 2000, *Cosmic Plasma Physics*, (see book list).

Somov, B.V. & Oreshina, A.V. 2000, AA, 354, 703.

Song, P. et al. (ed.) 2001, Proc-2001-Song.

Sonnerup, B.U.Ö. 1970, J. Plasma Phys. 4, 161.

Soward, A.M. & Priest, E.R. 1977, Phil. Trans. Roy. Soc. Lon., A 284, 369.

Soward, A.M. 1982, J.Plamsa Physics 28/3, 415.

Spadaro, D., Noci, G., Zappala, R.A., et al. 1990a, ApJ 355, 342.

Spadaro, D., Noci, G., Zappala, R.A., et al. 1990b, ApJ 362, 370.

Spadaro, D., Antiochos, S.K., & Mariska, J.T. 1991, ApJ 382, 338.

Spadaro, D., Leto, P., & Antiochos, S.K. 1994, ApJ 427, 453.

Spadaro, D. 1999, ESA SP-448, 157.

Spadaro, D., Lanza, A.F., Lanzafame ,A.C., et al. 2003, ApJ 582, 486.

Spicer, D.S. 1977a, SP 53, 249.

Spicer, D.S. 1977b, SP 53, 305.

Spicer, D.S. 1981a, SP 70, 149.

Spicer, D.S. 1981b, SP 71, 115.

Spicer, D.S. 1982, Space Science Rev., 31, 351.

Spicer, D.S. & Sudan, R.N. 1984, ApJ 280, 448.

Spicer, D.S. & Emslie, A.G. 1988, ApJ 330, 997.

Spitzer, L.Jr. & Härm,R. 1953, Phys. Rev. 89(5), 977.

Spitzer, L. 1967, *The Physics of Fully Ionized Gases* (see book list).

Sprangle, P. & Vlahos, L. 1983, ApJ 273, L95.

Spruit, H.C. 1981, Proc-1981-Jordan, 385.

Spruit, H.C. 1982, SP 75, 3.

Stähli, M., Gary, D.E., & Hurford, G.J. 1989, SP 120, 351.

St.Cyr, O.C., Howard, R.A., Sheeley, N.R., et al. 2000, JGR 105, 169.

Stefan, A.J. 1879, Wien. Ber. 79, 397.

Steinacker, J. & Miller, J.A. 1992, ApJ 393, 764.

Steinacker, J., Jaekel, U., & Schlickeiser, R. 1993, ApJ 415, 342.

Steiner, O., Grossmann−Doerth, U., Knoelker, M., et al. 1998, ApJ 495, 468.

Steiner, O. 2001, *Chromosphere: Magnetic Canopy*, (in Murdin 2000) .

Steinolfson, R.S., Wu, S.T., Dryer, M., et al. 1978, ApJ 225, 259.

Steinolfson, R.S., & Tajima, T. 1987, ApJ 322, 503.

Steinolfson, R.S. & Hundhausen, A.J. 1988, JGR 93, 14269.

Steinolfson, R.S. 1991, ApJ 382, 677.

Steinolfson, R.S. 1992, JGR 97/A7, 10811.

Steinolfson, R.S. & Davila, J.M. 1993, ApJ 415, 354.

Stenflo, J.O. 1994, *Solar Magnetic Fields*, (see book list).

Stenflo, J.O. 2001a, *Solar Magnetic Fields: Zeeman and Hanle Effects*, (in Murdin 2000) .

Stenflo, J.O. 2001b, *Solar Photosphere: Intranetwork ...*, (in Murdin 2000) .

Stepanov, A.V., Urpo, S., & Zaitsev, V.V. 1992, SP 140, 139.

Sterling, A.C., Mariska, J.T., Shibata, K., et al. 1991, ApJ 381, 313.

Sterling, A.C. & Hudson, H.S. 1997, ApJ 491, L55.

Sterling, A.C., Hudson, H.S., Lemen, J.R., et al. 1997, ApJS 110, 115.

Sterling, A.C. & Moore, R.L. 2001, ApJ 560, 1045.

Sterling, A.C., Moore, R.L., & Thompson, B.J. 2001, ApJ 561, L219.

Stern, D.P. 1966, *Space Sci. Rev.* 6, 147.

Stix, M. 2002, *The Sun*, (see book list).

Stix, T.H. 1992, *Waves in Plasmas*, (see book list).

Strachan, N.R. & Priest, E.R. 1994, Geophys. Astrophys. Fluid Dynamics 74, 245.

Strachan, L., Panasyuk, A.V., Dobrzycka, D., et al. 2000, JGR 105, 2345.

Strauss, F.M., Kaufmann, P., & Opher, R. 1980, SP 67, 83

Strong, K.T., Benz, A.O., Dennis, B.R., et al. 1984, SP 91, 325.

Strong, K.T., Harvey, K.L., Hirayama, T., et al. 1992, PASJ 44, L161.

Strong, K.T., Saba, J.L.R., Haisch, B.M., et al. 1999, *The Many Faces of the Sun* (see book list).

Strous, L.H., Scharmer, G., Tarbell, T.D., et al. 1996, AA 306, 947.

Strous, L.H. & Zwaan, C. 1999, ApJ 527, 435.

Stucki, K., Solanki, S.K., Schühle, U., et al. 2000, AA 363, 1145.

Stucki, K., Solanki, S.K., Pike, C.D., et al. 2002, AA 381, 653.

Sturrock, P.A. 1964, Proc-1964-Hess, 357.

Sturrock, P.A. 1966, Nature 5050, 695.

Sturrock, P.A. 1973, Proc-1973-Ramaty, 3.

Sturrock, P.A. (ed.) 1980, Proc-1980-Sturrock.

Sturrock, P.A. & Uchida, Y. 1981, ApJ 246, 331.

Sturrock, P.A., Dixon, W.W., Klimchuk, J.A., et al. 1990, ApJ 356, L31.

Sturrock, P.A. 1991, ApJ 380, 655.

Sturrock, P.A. 1994, *Plasma Physics* (see book list).

Sturrock, P.A., Wheatland, M.S., & Acton, L.W. 1996a, Proc-1996-Uchida, 417.

Sturrock, P.A., Wheatland, M.S., & Acton, L.W. 1996b, ApJ 461, L115.

Sturrock, P.A. 1999, ApJ 521, 451.

Subramanian, P., Dere, K.P., Rich, N.B. et al. 1999, JGR 104, 22321.

Suematsu, Y., Yoshinaga, R., Terao, N., et al. 1990, PASJ 42, 187.

Suess, S.T., Poletto, G., Wang, A.H., et al. 1998, SP 180, 231.

Suetterlin, P., Wiehr, E., Bianda, M., et al. 1997, AA 321, 921.

Sui, L. & Holman, G.D. 2003, ApJ 596, L251.

Suydam, B.R. 1958, Proc-1958-UN, 187.

Svestka, Z. 1976, *Solar Flares*, (see book list).

Svestka, Z., Fontenla, J.M., Machado, M.E., et al. 1987, SP 108, 237.

Svestka, Z. 1994, SP 152, 505.

Sweet, P.A. 1958, Proc-1958-Lehnert, 123.

Sylwester, B. & Sylwester, J. 2000, SP 194, 305.

Syniavskii, D.V. & Zharkova, V.V. 1994, ApJS 90, 729.

Tajima, T., Brunel, F., & Sakai, J. 1982, ApJ 258, L45.

Tajima, T., & Sakai, J. 1986, *IEEE Trans. Plasma Sci.* PS-14, 929.

Tajima, T., Sakai, J., Nakajima, H., et al. 1987, ApJ 321, 1031.

Tajima, T., Benz, A.O., Thaker, M., et al. 1990, ApJ 353, 666.

Tajima, T. & Shibata, K. 2002, *Plasma Astrophysics*, (see book list).

Takakura, T. 1960, PASJ 12, 352.

Takakura, T. 1967, SP 1, 304.

Takakura, T. 1979, SP 62, 383.
Takakura, T., Tsuneta, S., Nitta, N., et al. 1983a, ApJ 270, L83.
Takakura, T., Kaufmann, P., Costa, J.E.R., et al. 1983b, Nature 302, 317.
Takakura, T. 1988, SP 115, 149.
Takeuchi, A. & Shibata, K. 2001, ApJ 546, L73.
Tanaka, K., Akita, K., Watanabe, T., et al. 1982, Annals Tokyo Astron. Obs. 18/4, 237.
Tanaka, K. & Papadopoulos, K. 1983, *Physics of Fluids* 26, 1697.
Tanaka, K. 1987, PASJ 39, 1.
Tandberg–Hanssen, E. 1974, *Solar Prominences*, (see book list).
Tandberg–Hanssen, E.A. 1986, Proc-1986-Poland, 5.
Tandberg–Hanssen, E. 1995, *The Nature of Solar Prominences*, (see book list).
Tandberg–Hanssen, E. 2001, *Solar Prominences: Active*, (in Murdin 2000) .
Tang, Y.H., Li, Y.N., Fang, C., et al. 2000, ApJ 534, 482.
Tapping, K.F. 1978, SP 59, 145.
Tarbell, T.D., Ryutova, M., Covington, J., et al. 1999, ApJ 514, L47.
Tarbell, T.D., Ryutova, M., & Shine, R. 2000, SP 193, 195.
Terekhov, O.V., Shevchenko, A.V., Kuz'min, A.G., et al. 2002, Astronomy Letters 28/6, 397.
Teriaca, L., Doyle, J.G., Erdélyi, R., et al. 1999, AA, 352, L99.
Terradas, J., Oliver, R., & Ballester, J.L. 2001, AA 378, 635.
Teriaca, L., Madjarska, M.S., & Doyle, J.G. 2002, AA 392, 309.
Teriaca, L., Poletto, G., Romoli,M. et al. 2003, ApJ 588, 566.
Terradas, J., Molowny–Horas, R., Wiehr, E., et al. 2002, AA 393, 637.
Thomas, R.J., Neupert, W.M., & Thompson, W.T. 1987, Proc-1987-Dennis, 299.
Thompson, A.M., Brown, J.C., Craig, I.J.D., et al. 1992, AA 265, 278.
Thompson, B.J., Plunkett, S.P., Gurman, J.B., et al. 1998a, GRL 25, 14, 2461.
Thompson, B.J., Gurman, J.B., Neupert, W.M., et al. 1999 ApJ 517, L151.
Thompson, B.J., Reynolds, B., Aurass, H., et al. 2000a, SP 193, 161.
Thompson, B.J., Cliver, E.W., Nitta, N., et al. 2000b, GRL 27/10, 1431.
Thompson, B.J. 2001, *Moreton Waves*, (in Murdin 2000) .
Thompson, W.T. & Schmieder, B. 1991, AA 243, 501.
Timothy, A.F., Krieger, A.S., & Vaiana, G.S. 1975, SP 42, 135.
Title, A.M. & Schrijver, K. 1998, Proc-1998-Donahue, 345.
Todh, G. & Odstrčil, J. 1996, J. Comput. Phys. 182, 82.
Tokman, M. & Bellan, P.M. 2002, ApJ 567, 1202.
Tomczak, M. 1999, AA 342, 583.
Tomczak, M. 2001, AA 366, 294.
Török, T. & Kliem, B. 2003, AA 406, 1043.
Török, T. & Kliem, B. 2004, Proc-2004-Wolf, 25.
Török, T., Kliem, B., & Titov, V.S. 2003, AA 413, L27.
Treumann, R.A. and Baumjohann, W. 1997, *Advanced Space Plasma Physics*, (see book list).
Trottet, G., Pick, M., & Heyvaerts, J. 1979, AA 79, 164.
Trottet, G. & MacQueen, R.M. 1980, SP 68, 177.
Trottet, G., Kerdraon, A., Benz, A.O., et al. 1981, AA 93, 129.
Trottet, G., Vilmer, N., Barat, C. et al. 1993, AASS 97, 337.
Trottet, G. 1994a, Proc-1994-Ryan, 3.
Trottet, G. 1994b, SSR 68, 149.
Trottet, G., Barat, C., Ramaty, R., et al. 1996, Proc-1996-Ramaty, 153.
Trottet, G. & Vilmer, N. 1997, Proc-1997-Simnett, 219.
Trottet, G., Vilmer, N., Barat, C, et al. 1998, AA 334, 1099.
Trottet, G., Schwartz, R.A., Hurley, K., et al. 2003, AA 403, 1157.

Trubnikov, B.A. 1965, Rev. Plasma Phys. 1, 105.

Tskiklauri, D., Aschwanden, M.J., Nakariakov, V.M., et al. 2004, AA ...

Tsinganos, K. 1980, ApJ 239, 746.

Tsubaki, T. & Takeuchi, A. 1986, SP 104, 313.

Tsuneta, S. 1984, Annals Tokoy Astron. Obs.(2nd series), 20/1, 1-50.

Tsuneta, S. 1985, ApJ 290, 353.

Tsuneta, S., Acton, L., Bruner, M., et al. 1991, SP 136, 37.

Tsuneta, S., Hara, H., Shimizu, T., et al. 1992, PASJ 44, L63.

Tsuneta, S., Takahashi, T., Acton, L.W., et al. 1992b, PASJ 44, L211.

Tsuneta, S. 1993a, Proc-1993-Zirin, 239.

Tsuneta, S. 1993b, ESA SP-351, 75.

Tsuneta, S. & Lemen, J.R. 1993, Proc-1993-Linsky, 113.

Tsuneta, S. 1994a, Proc-1994-Balasubramaniam, 338.

Tsuneta, S. 1994b, Proc-1991-Uchida, 115.

Tsuneta, S. 1995a, Proc-1995-Wang, 197.

Tsuneta, S. 1995b, in Proc-1995-Ichimaru, 447.

Tsuneta, S. 1995c, PASJ 47, 691.

Tsuneta, S. 1996a, ApJ 456, 840.

Tsuneta, S. 1996b, Proc-1996-Tsinganos, 85.

Tsuneta, S. 1996c, Proc-1996-Bentley, 409.

Tsuneta, S. 1996d, ApJ 456, L63.

Tsuneta, S. 1997, ApJ 483, 507.

Tsuneta, S., Masuda, S., Kosugi, T., et al. 1997, ApJ 478, 787.

Tsuneta, S. & Naito, T. 1998, ApJ 495, L67.

Tu, C.Y. & Marsch, E. 1997, SP 171, 363.

Tu, C.Y. & Marsch, E. 2001a, AA 368, 1071.

Tu, C.Y. & Marsch, E. 2001b, JGR 106/A5, 8233.

Tucker, W.H. & Koren, M. 1971, ApJ 168, 283.

Tylka, A.J. 2001, JGR 106, 25,333.

Tziotziou, K., Martens, P.C.H., & Hearn, A.G. 1998, AA 340, 203.

Uchida, Y. 1968, SP 4, 30.

Uchida, Y., Altschuler, M.D., & Newkirk, G.Jr. 1973, SP 28, 495.

Uchida, Y. 1974, SP 39, 431.

Uchida, Y. 1980, Proc-1980-Sturrock, 67 and 110.

Uchida, Y., Fujisaki, K., Morita, S., et al. 1996, Proc-1996-Bentley, 347.

Uchida, Y. 1997a, *Physical Review E* 56/2, 2181.

Uchida, Y. 1997b, *Physical Review E* 56/2, 2198.

Uchida, Y., Hirose, S., Cable, S., et al. 1998a, Proc-1998-Webb, 384.

Uchida, Y., Hirose, S., Morita, S., et al. 1998b, Astrophysics and Space Science 264(1/4), 145.

Ugai, M. & Tsuda, M. 1977, J. Plasma Phys. 18, 451.

Ugai, M. 2001, Space Science Reviews 95, 601.

Ulmschneider, P. 1971, AA 12, 297.

Ulmschneider, P., Priest, E.R., and Rosner, R. (eds.) 1991, Proc-1991-Ulmschneider.

Uzdensky, D.A. 2003, ApJ 587, 450.

Van Aalst, M.K., Martens, P.C.H., & Bélien, A.J.C. 1999, ApJL 511, L125.

Van Ballegooijen, A.A. 1986, ApJ 311, 1001.

Van Ballegooijen, A.A., Cartledge, N.P., & Priest,E.R. 1998, ApJ 501, 866.

Van Ballegooijen, A.A., & Martens, P.C.H. 1989, ApJ 343, 971.

Van Ballegooijen, A.A., Priest, E.R., & Mackay,D.H. 2000, ApJ 539, 983.

Van Ballegooijen, A.A.,2001, *Solar Prominence Models*, (in Murdin 2000) .

Van Beek, H.F., Hoyng, P., Lafleur, B., et al. 1980, SP 65, 39.
Van de Hulst, H.C. 1950a, Bull.Astron.Inst.Netherlands 11, 135.
Van de Hulst, H.C. 1950b, Bull.Astron.Inst.Netherlands 11, 150.
Van den Oord, G.H.J. 1990, AA 234, 496.
Van der Linden, R.A.M. & Goossens, M. 1991, SP 134, 247.
Van Doorsselaere, T.V., Andries, J., Poedts, S., et al. 2004, AA 606, 1223.
Van Driel−Gesztelyi, L., Hofmann,A ., Démoulin, P., et al. 1994, SP 149, 309.
Van Driel−Gesztelyi, L., Schmieder, B., Cauzzi, G., et al. 1996, SP 163, 145.
Van Driel−Gesztelyi, L., Wiik, J.E., Schmieder, B., et al. 1997, SP 174, 151.
Van Driel−Gesztelyi, L., Malherbe, J.M., & Démoulin, P. 2000, AA 364, 845.
Van Driel−Gesztelyi, L. 2002, ESA SP-505, 113.
Van Driel−Gesztelyi, L., Démoulin,P., Mandrini,C.H., et al. 2003, ApJ 586, 579.
Van Driel−Gesztelyi, L. 2003, Proc-2003-Erdélyi, 297.
Van Tend, W. & Kuperus, M. 1978, SP 59, 115.
Van Tend, W. 1979, SP 61, 69.
Varvoglis, I. & Papadopoulos, K. 1985, JGR 56, 201.
Vekstein, G.E. & Browning, P.K. 1996, Proc-1996-Bentley, 308.
Vekstein, G.E. & Katsukawa,Y. 2000, ApJ 541, 1096.
Velli, M., Einaudi, G., & Hood, A.W. 1990, ApJ 350, 428.
Vernazza, J.E., Avrett, E.H., & Loeser, R. 1973, ApJ 184, 605.
Vernazza, J.E., Avrett, E.H., & Loeser, R. 1976, ApJS 30, 1.
Vernazza, J.E., Avrett, E.H., & Loeser, R. 1981, ApJS 45, 635.
Veronig, A., Temmer, M., Hanslmeier, A., et al. 2002a, AA 382, 1070.
Veronig, A., Vrsnak, B., Dennis, B.R., et al. 2002b, AA 392, 699.
Veronig, A. 2003, The Observatory 123, 58.
Vesecky, J.F., Antiochos, S.K., & Underwood, J.H. 1979, ApJ 233, 987.
Vestrand, W.T., Forrest, D.J., Chupp, E.L., et al. 1987, ApJ 322, 1010.
Vestrand, W.T. 1988, SP 118, 95.
Vestrand, W.T. & Forrest, D.J. 1993, ApJ 409, L69.
Vestrand, W.T. & Forrest, D.J. 1994, Proc-1994-Ryan, 143.
Vestrand, W.T. & Miller, J.A. 1999, in Strong et al. (1999), chapter 7, 231.
Vestrand, W.T.. Share, G.J., Murphy, R.J. et al. 1999, ApJS 120, 409.
Vilmer, N., Kane, S.R., & Trottet, G. 1982, AA 108, 306.
Vilmer, N., Trottet,G., & MacKinnon,A.L. 1986, AA 156, 64.
Vilmer, N. 1987, SP 111, 207.
Vilmer, N., Trottet, G., Barat, C., et al. 1999, AA 342, 575.
Vilmer, N. & MacKinnon, A.L. 2003, Proc-2003-Klein, 127.
Vlahos, L., Gergely, T.E., & Papadopoulos,K. 1982, ApJ 258, 812.
Vlahos, L., Machado, M.E., Ramaty, R., et al. 1986, Proc-1986-Kundu, 2-1
Vlahos, L. 1987, SP 111, 155.
Vlahos, L. & Sprangle, Ph. 1987, ApJ 322, 463.
Vogt, E., & Hénoux, J.C. 1999, AA 349, 283.
Vogt, E., Sahal−Bréchot,S., & Hénoux, J.C. 2002, Proc-2002-ESA477, 191.
Voitenko, Y.M. 1995, SP 161, 197.
Voitenko, Y.M. 1996, SP 168, 219.
Voitenko, Y.M. & Goossens, M. 1999, Proc-1999-ESA448, 735.
Voitenko, Y.M., Goossens, M., Sirenko, O., et al. 2003, AA 409, 331.
Volwerk, M. & Kuijpers, J. 1994, ApJS 90, 589.
Volwerk, M. 1993, PhD Thesis, (see PhD Thesis List).
Von Steiger, R. 2001, *Transition Region: First Ionization Potential Effect*, (in Murdin 2000) .

Vourlidas, A., Bastian, T.S., & Aschwanden, M.J. 1997, ApJ 489, 403.

Vourlidas, A., Subramanian, P., Dere, K.P., et al. 2000, ApJ 534, 456.

Vrsnak, B., Ruzdjak, V., Messerotti, M., et al. 1987a, SP 111, 23.

Vrsnak, B., Ruzdjak, V., Messerotti, M., et al. 1987b, SP 114, 289.

Vrsnak, B., Ruzdjak, V., & Rompolt,B. 1991, SP 136, 151.

Vrsnak, B. & Lulic, S. 2000a, SP 196, 157.

Vrsnak, B. & Lulic, S. 2000b, SP 196, 181.

Wagner, W.J. 1975, ApJ 198, L141.

Waldmeier, M. 1950, Zeitschrift f. Astrophysik, 27, 24.

Wallenhorst, S.G. 1982, SP 79, 333.

Walsh, R.M., Bell, G.E., & Hood, A.W. 1997, SP 171, 81.

Wang, A.H., Wu, S.T., Suess, S.T., et al. 1995b, SP 161, 365.

Wang, H., Gary, D.E., Lim, J. et al. 1994, ApJ 433, 379.

Wang, H., Gary, D.E., Zirin, H., et al. 1995a, ApJ 453, 505.

Wang, H., Gary, D.E., Zirin, H., et al. 1996, ApJ 456, 403.

Wang, H., Chae, J.C. Qui, J., et al. 1999, SP 188, 365.

Wang, H., Ji, H., Schmahl, E.J., et al. 2002a, ApJ 580, L177.

Wang, H., Spirock, T.J., Qiu, J., et al. 2002b, ApJ 576, 497.

Wang, M. & Xie, R.X. 1997, SP 176, 171.

Wang, T.J., Solanki, S.K., Curdt, W., et al. 2002, ApJ 574, L101.

Wang, T.J., Solanki, S.K., Innes, D.E., et al. 2003a, AA 402, L17.

Wang, T.J., Solanki, S.K., Curdt, W., et al. 2003b, AA, 406, 1105.

Wang, T.J. 2004, Proc-2004-ESA547, 417.

Wang, Y.M., Sheeley, N.R.Jr., Walters, J.H., et al. 1998, ApJ 498, L165.

Wang, Y.M., Sheeley, N.R., Howard, R.A., et al. 1999, GRL 26, No.10, 1349.

Wang, Y.M. 2000, ApJ 543, L89.

Wang, Z., Schmahl, E.J., & Kundu, M.R. 1987, SP 111, 419.

Warmuth, A., Vrsnak, B., Aurass, H., et al. 2001, ApJ 560, L105.

Warren, H.P., Mariska, J.T., & Wilhelm, K. 1997, ApJ 490, L187.

Warren, H.P. & Hassler, D.M. 1999, JGR 104/A5, 9781.

Warren, H.P. & Reeves, K. 2001, ApJ 554, L103.

Warren, H.P., & Warshall, A.D. 2001, ApJ 560, L87-L90.

Warren, H.P., Winebarger, A.R., & Hamilton, P.S. 2002, ApJ 579, L41.

Watanabe, T. 1987, SP 113, 107.

Watanabe, T., Kosugi, T., & Sterling, A.C. (eds.) 1998, Proc-1998-Watanabe.

Watari, S., Kozuka, Y., Ohyama, M., et al. 1995, J.Geomag.Geoelec., 47(11), 1063.

Watko, J.A. & Klimchuk, J.A. 2000, SP 193, 77.

Watson, P.G. & Craig, I.J.D. 1998, ApJ 505, 363.

Watson, P.G. & Craig, I.J.D. 2002, SP 207, 337.

Webb, D.F. 1988, JGR 93, 1749.

Webb, D.F., Holman, G.D., Davis, J.M., et al. 1987, ApJ 315, 716.

Webb, D.F., Rust, D.M., & Schmieder, B. (eds.) 1998, Proc-1998-Webb.

Webb, D.F., Rust, D.M., & Schmieder, B. (eds.) 1998, Proc-1998-Webb, 463.

Webb, D.F. 2001, *Solar Wind: Manifestations of Solar Activity*, (in Murdin 2000) .

Wentzel, D.G. 1961, J.Geophys.Res. 66/2, 359

Wentzel, D.G. 1976, ApJ 208, 595

Wentzel, D.G. 1979, AA 76, 20.

Wentzel, D.G. 1991, ApJ 373, 285.

Whang, Y.C., Zhou, J., Lepping, R.P., et al. 1998, JGR 103, 6513.

Wheatland, M.S. & Melrose, D.B. 1995, SP 158, 283.

Wheatland, M.S. 1999, ApJ 518, 948.

Wheatland, M.S., Sturrock, P.A., & Acton, L.W. 1997, ApJ 482, 510.

Wheatland, M.S., Sturrock, P.A. and Roumeliotis, G. 2000 ApJ 540, 1150.

White, R.B. 1983, *Handbook of Plasma Physics* (see book list).

White, S.M., Melrose, D.B., & Dulk, G.A. 1983, Proc.ASA 5(2), 188.

White, S.M., Kundu, M.R., & Gopalswamy, N. 1991, ApJ 366, L43.

White, S.M. & Kundu, M.R. 1992, SP 141, 347.

White, S.M., Kundu, M.R., Shimizu, T., et al. 1995, ApJ 450, 435.

White, S.M. & Kundu, M.R. 1997, SP 174, 31.

White, S.M., Thomas, R.J., Brosius, J.W., et al. 2000, ApJ 534, L203.

White, S.M., Lee, J.W., Aschwanden, M.J., et al. 2003, SPIE 4853, 531.

Wiegelmann, T. & Neukirch, T. 2002, SP 208, 233.

Wiegelmann, T. & Neukirch, T. 2003, Nonlinear Processes in Geophysics 10, 1.

Wiehl, H.J., Benz, A.O., & Aschwanden, M.J. 1985, SP 95, 167.

Wiehr, E., Balthasar, H., & Stellmacher, G. 1984, SP 94, 285.

Wien, W. 1893, Sitz.Acad.Wiss. Berlin, 1, 551.

Wien, W. 1894, Phil.Mag., 43, 214.

Wild, J.P., Smerd, S.F., & Weiss, A.A. 1963, ARAA 1, 291.

Wild, J.P., & Smerd, S.F. 1972, ARAA 10, 159

Wilhelm, K., Curdt, W., Marsch, E. et al. 1995, SP 162, 189.

Wilhelm, K., Marsch, E., Dwivedi, B.N., et al. 1998, ApJ 500, 1023.

Wilhelm, K., Dammasch, I.E., Marsch, E., et al. 2000, AA 353, 749.

Willes, A.J., & Robinson, P.A. 1996, ApJ 467, 465.

Willes, A.J., Robinson, P.A., & Melrose, D.B. 1996, Phys.Plasmas 3/1, 149.

Williams, D.R., Phillips, K.J.H., Rudaway, P., et al. 2001, MNRAS 326, 428.

Williams, D.R., Mathioudakis, M., Gallagher, P.T., et al. 2002, MNRAS 336, 747.

Willson, R.F., Aschwanden, M.J. & Benz, A.O. 1992, Proc-1992-Shrader, 515.

Wills—Davey, M.J. & Thompson, B.J. 1999, SP 190, 467.

Wilson, P.R. 1980, AA 87, 121.

Winebarger, A.R., Emslie, A.G., Mariska, J.T., et al. 1999, ApJ 526, 471.

Winebarger, A.R., DeLuca, E.E., Golub, L. 2001, ApJ 553, L811

Winebarger, A.R., Warren, H., Van Ballegooijen, A., et al. 2002, ApJ 567, L89.

Winebarger, A.R., Warren, H.P., & Seaton, D.B. 2003a, ApJ , 593, 1164.

Winebarger, A.R., Warren, H.P., & Mariska, J.T. 2003b, ApJ 587, 439.

Winglee, R.M. 1985a, ApJ 291, 160.

Winglee, R.M. 1985b, JGR 90/A10, 9663.

Winglee, R.M. & Dulk, G.A. 1986a, ApJ 307, 808.

Winglee, R.M. & Dulk, G.A. 1986b, ApJ 310, 432.

Winglee, R.M. & Dulk, G.A. 1986c, SP 104, 93.

Winglee, R.M. 1989, ApJ 343, 511.

Winglee, R.M., Kiplinger, A.L., Zarro, D.M. 1991a, ApJ 375, 366.

Winglee, R.M., Kiplinger, A.L., Zarro, D.M. 1991b, ApJ 375, 382.

Winterhalter, D. et al. (eds.) 1996, Proc-1996-Winterhalter.

Withbroe, G.L. & Noyes, R.W. 1977, ARAA 15, 363.

Withbroe, G.L. 1978, ApJ 225, 641.

Withbroe, G.L. 1988, ApJ 325, 442.

Withbroe, G.L. & Gurman, J.B. 1973, ApJ 183, 279.

Withbroe, G.L. & Noyes, R.W. 1977, ARAA 15, 363.

Wolfson, R. & Saran, S. 1998, ApJ 499, 496.

Wolfson, R. & Dlamini, B. 1997, ApJ 483, 961.

Wood, B.E., Karovska, M., Chen, J., et al. 1999, ApJ 512, 484.
Woodgate, B.E., Shine, R.A., Poland, A.I. et al. 1983, ApJ 265, 530.
Woods, D.T., Holzer, T.E., & MacGregaor, K.B. 1990, ApJ 355, 295.
Woods, D.T. & Holzer, T.E. 1991, ApJ 375, 800.
Wragg, M.A. & Priest, E.R. 1981, SP 70, 293.
Wragg, M.A. & Priest, E.R. 1982, SP 80, 309.
Wu, C.S. 1984, JGR 89, 8857.
Wu, C.S. 1985, Space Science Reviews 41, 215.
Wu, S.T., Dryer, M., Nakagawa, Y. et al. 1978, ApJ 219, 324.
Wu, S.T. 1982, SSR 32, 115.
Wu, S.T. et al. 1986, Proc-1986-Kundu, 5-1.
Wu, S.T., Nakagawa, Y., Han, S.M. et al. 1982, ApJ 262, 369.
Wu, S.T., Sun, M.T., Chang, H.M., Hagyard,M.J., & Gary,G.A. 1990, ApJ 362, 698.
Wu, S.T., Guo, W.P., & Wang, J.F. 1995, SP 157, 325.
Wu, S.T. & Guo, W.P. 1997a, Adv. Space Res. 20/12, 2313.
Wu, S.T. & Guo, W.P. 1997b, Proc-1997-Crooker, 83.
Wu, S.T., Guo, W.P., & Dryer, M. 1997c, SP 170, 265.
Wu, S.T., Guo, W.P., Michels, D.J., et al. 1999, JGR 104/A7, 14,789.
Wu, S.T., Wang, A.H., Plunkett, S.P., & Michels,D.J. 2000, ApJ 545, 1101.
Wu, S.T., Andrews, M.D., & Plunkett, S.P. 2001, SSR 95, 191.
Wülser, J.P., Zarro, D.M., & Canfield, R.C. 1992, ApJ 384, 341.
Wülser, J.P., Canfield, R.C., Acton, L.W., et al. 1994, ApJ 424, 459.
Yan, M., Lee, L.C., & Priest, E.R. 1992, JGR 97, 8277.
Yan, M., Lee, L.C., & Priest, E.R. 1993, JGR 98, 7593.
Yan, Y. 1995, SP 159, 97.
Yan, Y., Yu, Q., & Wang, T.J. 1995, SP 159, 115.
Yan, Y. & Sakurai, T. 1997, SP 174, 65.
Yan, Y. & Sakurai, T. 2000, SP 195, 89.
Yan, Y., Aschwanden, M.J., Wang, S.J, et al. 2001, SP 204, 29.
Yang, C.K. & Sonnerup, B.U.O. 1976, ApJ 206, 570.
Yang, W.H., Sturrock, P.A., and Antiochos, S.K. 1986, ApJ 309, 383.
Yeh, T. & Axford, W.I. 1970, J. Plasma Phys. 4, 207.
Yeh, T. & Wu, S.T. 1991, SP 132, 335.
Yi, Z., Engvold, Ø., & Keil, S.L. 1991, SP 132, 63.
Yokoyama, T. & Shibata, K. 1994, ApJ 436, L197.
Yokoyama, T. & Shibata, K. 1995, Nature 375, 6526, 42.
Yokoyama, T. & Shibata, K. 1996, PASJ 48, 353.
Yokoyama, T. & Shibata, K. 1997, ApJ 474, L61.
Yokoyama, T. & Shibata, K. 1998, ApJ 494, L113.
Yokoyama, T. & Shibata, K. 2001, ApJ 549, 1160.
Yokoyama, T., Akita, K., Morimoto, T., et al. 2001, ApJ 546, L69.
Yoshimori, M., Okudaira, K., Hirashima, Y., et al. 1991, SP 136, 69.
Yoshimori, M., Takai, Y., Morimoto, K. et al. 1992, PASJ 44, L107.
Young, P.R., Landi, E., & Thomas, R.J. 1998, AA 329, 291.
Yurchyshyn, V.B. 2002, ApJ 576, 493.
Zaitsev, V.V., Stepanov, A.V., & Chernov, G.P. 1984, SP 93, 363.
Zaitsev, V.V., Stepanov, A.V., Urpo, S., et al. 1998, AA 337, 887.
Zaitsev, V.V., Urpo, S. & Stepanov, A.V. 2000, AA 357, 1105.
Zangrilli, L., Poletto, G., Nicolosi, P., et al. 2002, ApJ 574, 477.
Zank, G.P. & Gaisser, T.K. (eds.) 1992, Proc-1992-Zank.

Zarro, D.M. & Lemen, J.R. 1988, ApJ 329, 456.

Zarro, D.M. & Canfield, R.C. 1989, ApJ 338, L33.

Zarro, D.M., Slater, G.L., & Freeland, S.L. 1988a, ApJ 333, L99.

Zarro, D.M., Canfield, R.C., Strong, K.T., et al. 1988b, ApJ 324, 582.

Zarro, D.M. 1992, Proc-1992-Svestka, 95.

Zarro, D.M., Mariska, J.T., & Dennis, B.R. 1995, ApJ 440, 888.

Zarro, D.M. & Schwartz, R.A. 1996a, Proc-1996-Bentley, 209.

Zarro, D.M., & Schwartz, R.A. 1996b, Proc-1996-Ramaty, 359.

Zarro, D.M., Sterling, A.C., Thompson,B.J., et al. 1999, ApJ 520, L139.

Zhang, H. 2001, AA 372, 676.

Zhang, H.Q., Sakurai, T., Shibata, K., et al. 2000, AA 357, 725.

Zhang, H.Q., Sakurai, T., Shibata, K., et al. 1998, Proc-1998-Watanabe, 391.

Zhang, J., Kundu, M.R., & White, S.M. 2001a, SP 198, 347.

Zhang, J. & Huang, G.L. 2003, ApJ 592, L49.

Zhang, J., Dere, K.P., Howard, R.A., et al. 2001b, ApJ 559, 452.

Zhang, M., Golub, L., DeLuca, E., et al. 2002, ApJ 574, L97.

Zhang, M. & Low, B.C. 2004, ApJ 600, 1043.

Zhao, R.Y., Jin, S.Z., Fu, Q.J., & Li, X.C. 1990, SP 130, 151.

Zhao, X.P., Plunkett, S.P., & Liu, W. 2002, JGR 107, 10.1029/2001JA009143.

Zharkova, V.V. & Syniavskii, D.V. 1997, AA 32, 13.

Zheleznyakov, V.V. 1970, *Radio Emission of the Sun and Planets*, (see book list).

Zheleznyakov, V.V. & Zaitsev, V.V. 1970a, Soviet Astronomy 14(I), 47.

Zheleznyakov, V.V. & Zaitsev, V.V. 1970b, Soviet Astronomy 14(II), 250.

Zhou, Y. & Matthaeus, W.H. 1990, JGR 95, 14881.

Zirin, H. 1988, *Astrophysics of the Sun*, (see book list).

Zirker, J.B. (ed.) 1977, *Coronal holes and high speed wind streams*, (see book list).

Zirker, J.B. 1993, SP 148, 43-60.

Zirker, J.B. & Cleveland, F.M. 1993a, SP 144, 341.

Zirker, J.B. & Cleveland, F.M. 1993b, SP 145, 119.

Zirker, J.B., Engvold, Ø., & Martin,S.F. 1998, Nature 396, 440.

Zirker, J. 2001, *Solar Prominence Chirality*, (in Murdin 2000) .

Zlobec, P., Messerotti, M., Dulk, G.A., et al. 1992, SP 141, 165.

Zlotnik, E.Y. 1968, *Soviet Astron.* 12, 245.

Zodi, A.M., Kaufmann, P., & Zirin, H. 1984, SP 92, 283.

Zombek, M.V. 1990, *Handbook of Space Astronomy and Astrophysics*, (see book list).

Zwaan, C. 1987, ARAA 25, 83.

Zweibel, E.G. 1989, ApJ 340, 550.

Zweibel, E.G. & Haber, D.A. 1983, ApJ 264, 648

Zweibel, E.G. & Boozer, A.H. 1985, ApJ 295, 642.

Index

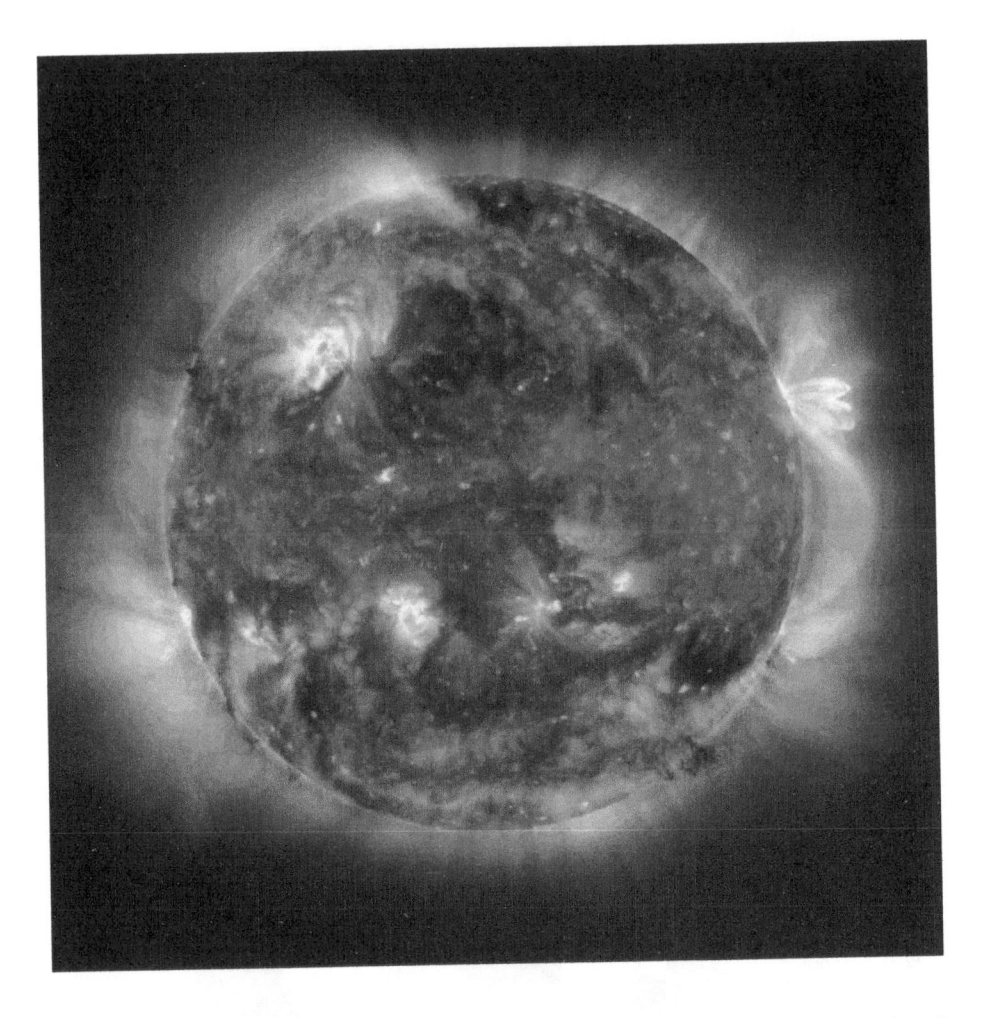

Plate 1. The multi-temperature corona : from the EIT instrument on board the space-based SOHO observatory. This tantalizing picture is a false-color composite of three images all taken in extreme ultraviolet light. Each individual image highlights a different temperature regime in the upper solar atmosphere and was assigned a specific color; red at 2 million, green at 1.5 million, and blue at 1 million degrees K. The combined image shows bright active regions strewn across the solar disk, which would otherwise appear as dark groups of sunspots in visible light images, along with some magnificent plasma loops and an immense prominence at the right-hand solar limb (courtesy of EIT/SoHO).

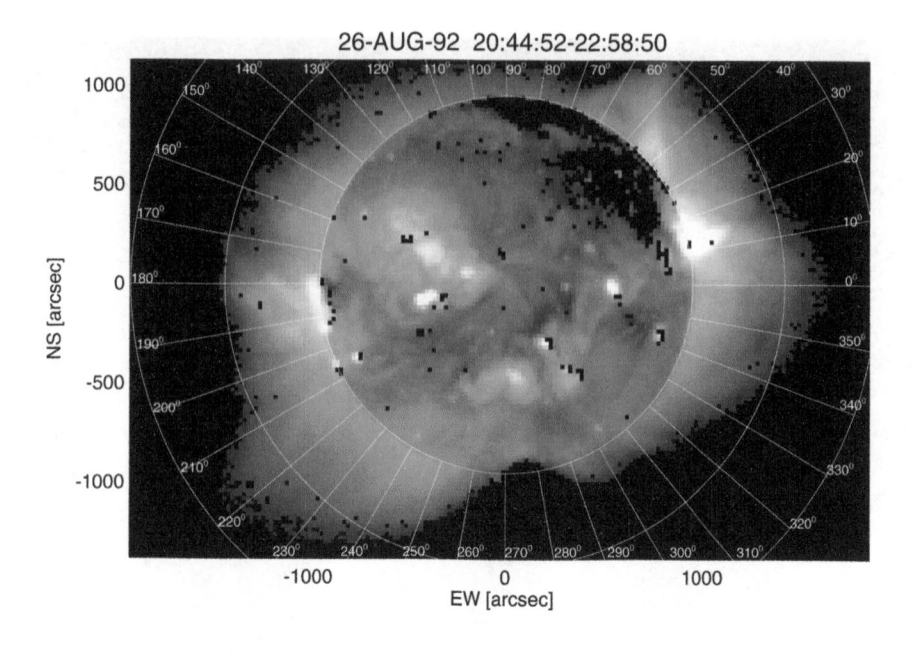

26-AUG-92 20:44:52-22:58:50

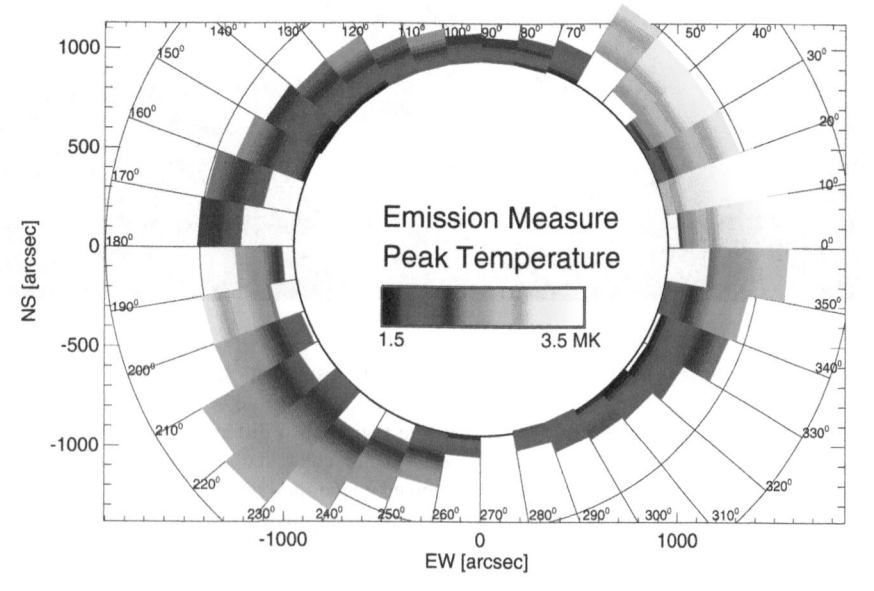

Emission Measure
Peak Temperature

1.5   3.5 MK

Plate 2. *Top:* a composite and large field-of-view soft X-ray (Al.1) map from *Yohkoh /SXT* with subdivision into 36 radial sectors, each 10° wide. The circles indicate the altitude levels of $R = 1.0, 1.5$, and 2.0 solar radii. Note two active regions at the east and west, a coronal streamer in the south−east, and coronal holes in the north and south. *Bottom:* coronal temperature maps are shown for the 36 sectors of the same Yohkoh image. The peak temperature T_0 of the fitted differential emission measure distributions $dEM(T,h)/dT$ is given according to the multi-hydrostatic model defined in Eq. (3.3.7) (Aschwanden & Acton 2001).

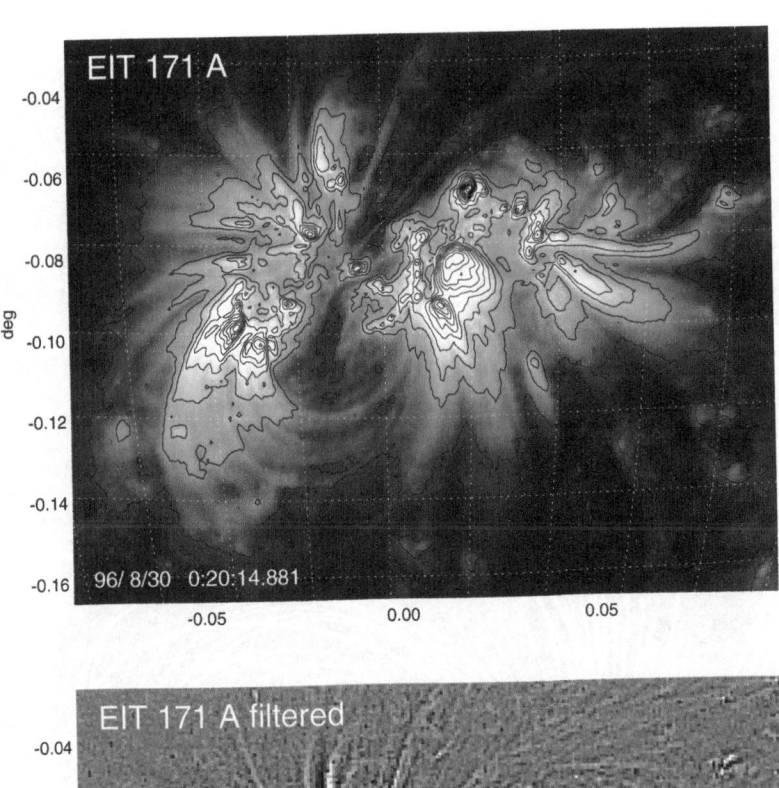

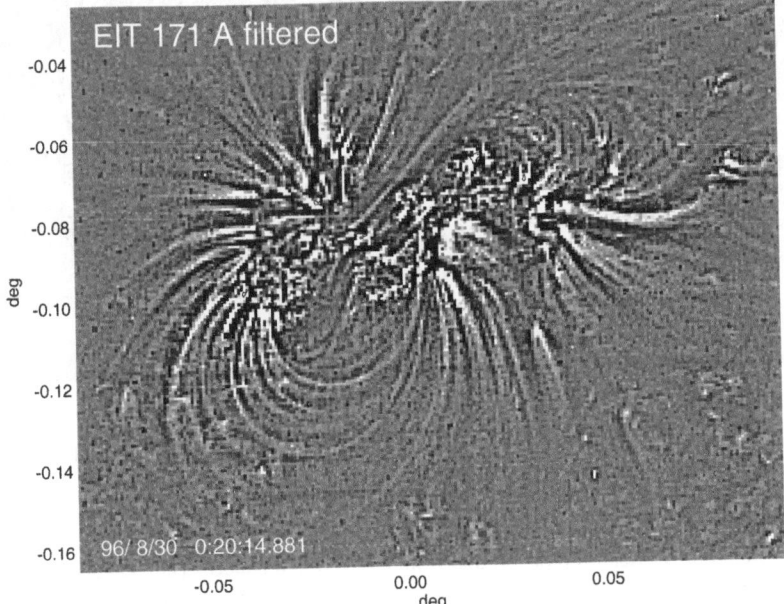

Plate 3: SoHO/EIT Fe IX/X image of AR 7986, recorded on 1996-Aug-30, 0020:14 UT, at a wavelength of 171 Å, sensitive in the temperature range of $T_e = 1.0 - 1.5$ MK (top). The color scale of the image is logarithmic in flux, the contours correspond to increments of 100 DN (data numbers). The heliographic grid has a spacing of $5°$. The filtered image (lower panel) was created by subtracting a smoothed image (using a boxcar of 3×3 pixels) from the original image, in order to enhance the loop fine structure (Aschwanden et al. 1999a).

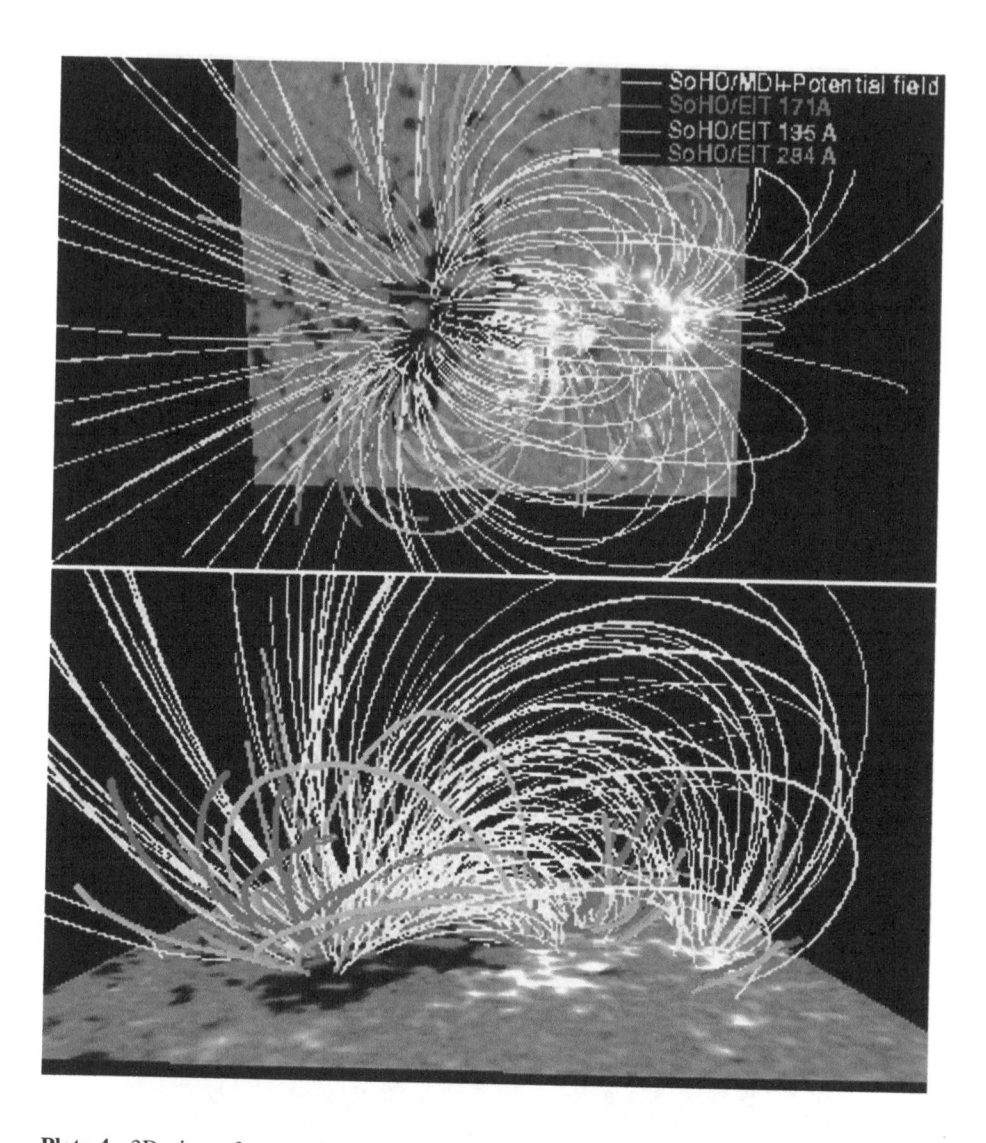

Plate 4. 3D view of magnetic potential field lines (yellow lines) calculated with the Sakurai code by extrapolation of a SoHO/MDI magnetogram recorded on 1996 August 30, 20:48 UT (red surface with white and black polarities), the traced 171 Å loop segments (blue lines), the traced 195 Å loop segments (green lines), and the traced 284 Å loop segments (red lines). The two views are: from vertical (top panel) and from east (bottom panel). The 3D coordinates of the traced EIT loops are based on stereoscopic reconstruction. Note some significant deviations between the observed loops and the potential field model (Aschwanden et al. 2000a).

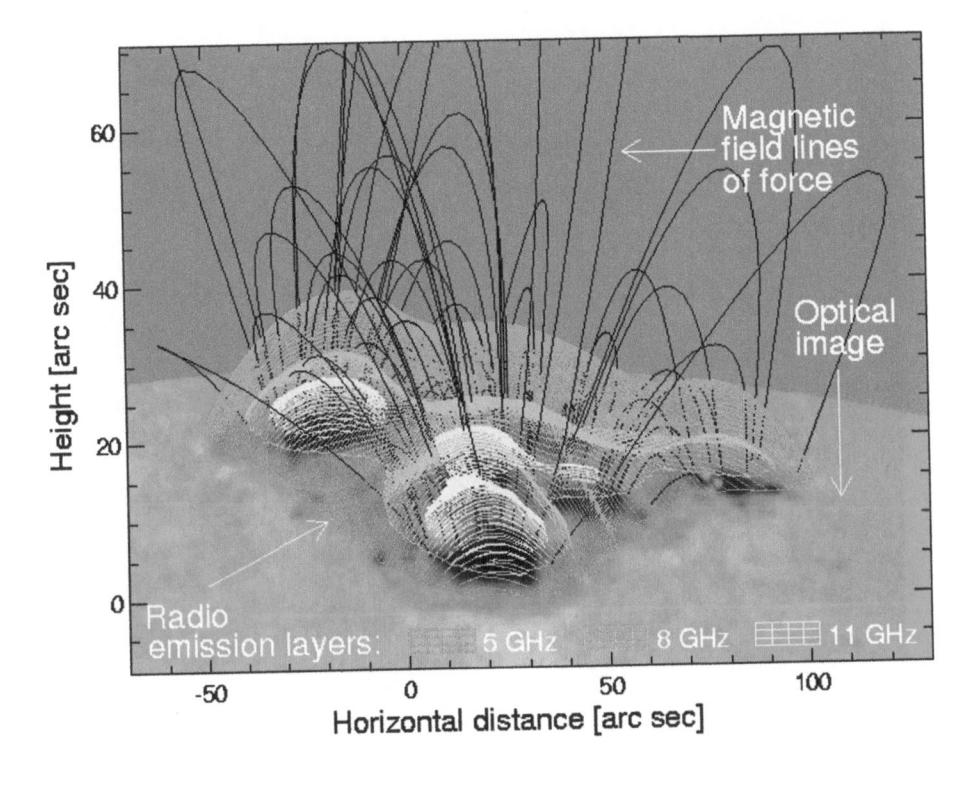

Plate 5. A 3D magnetic field representation is rendered from photospheric magnetograms (optical image in orange), from extrapolated magnetic field lines (black lines), and the iso-gauss contours for three gyroresonant layers that correspond to gyrofrequencies of 5 GHz (green), 8 GHz (blue), and 11 GHz (yellow). The outer contours of each igo-gauss surface demarcate the extent of radio emission at each frequency (Courtesy of Stephen White and Jeong Woo Lee).

Plate 6. *Top:* the small-scale magnetic field connects the network on the spatial scale of supergranulation cells, while large-scale magnetic fields extend up into the corona. This magnetic field extrapolation was computed based on a magnetogram recorded by SoHO/MDI on 1996 Oct 19. The tangled small-scale fields at the bottom of the corona have also been dubbed *"magnetic carpet"*. *Bottom:* horizontal view of the same 3D representation of magnetic field lines (Courtesy of Neal Hurlburt and Karel Schrijver).

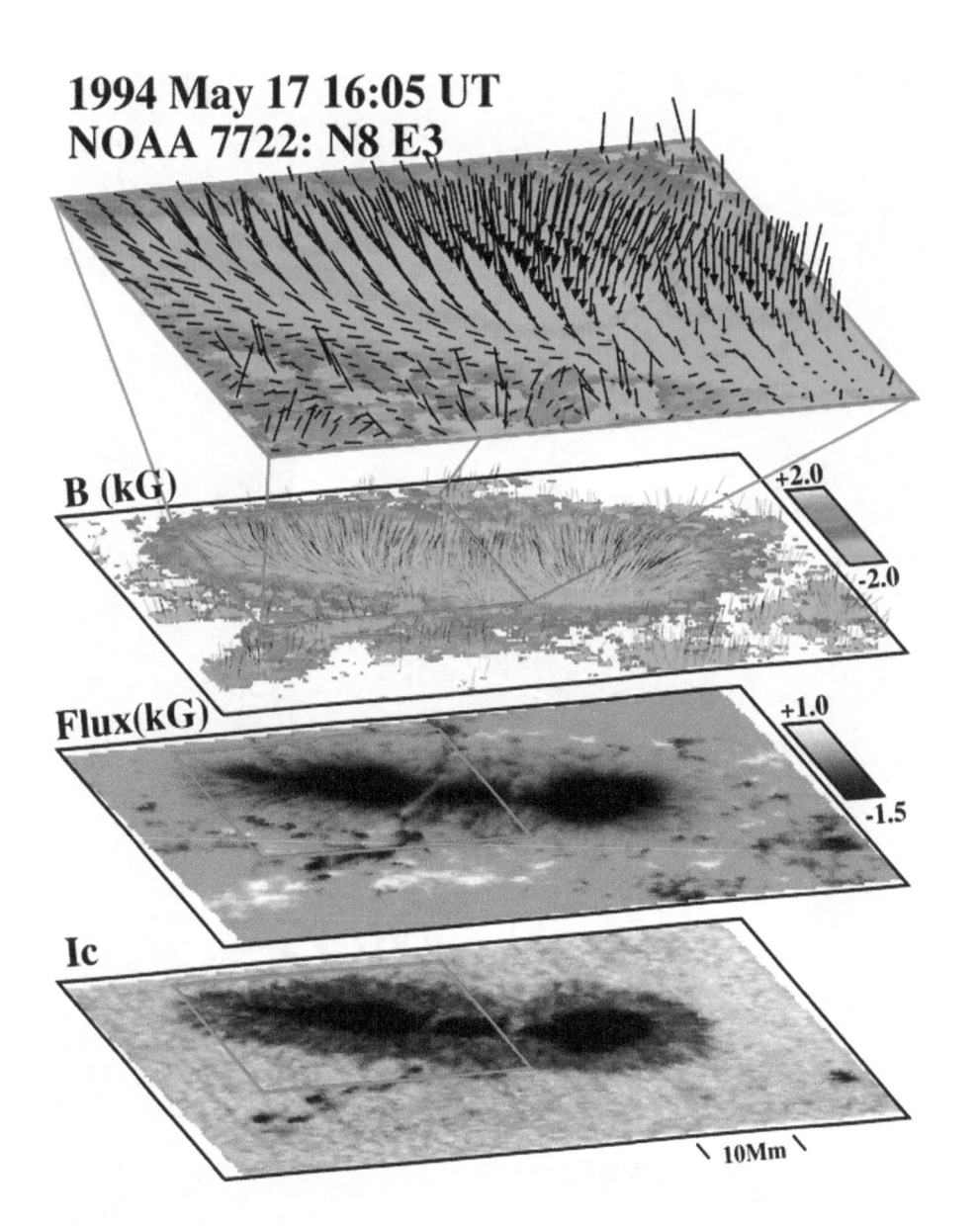

Plate 7. A stack plot of a sunspot magnetogram in white light (bottom level), magnetic field strength (second level), vector field **B** [kG] (third level), with an enlargement of the central sunspot (forth level) of active region NOAA 7722, recorded with the *Advanced Stokes Polarimeter* on 1994 May 17, 16:05 UT (Courtesy of Bruce Lites).

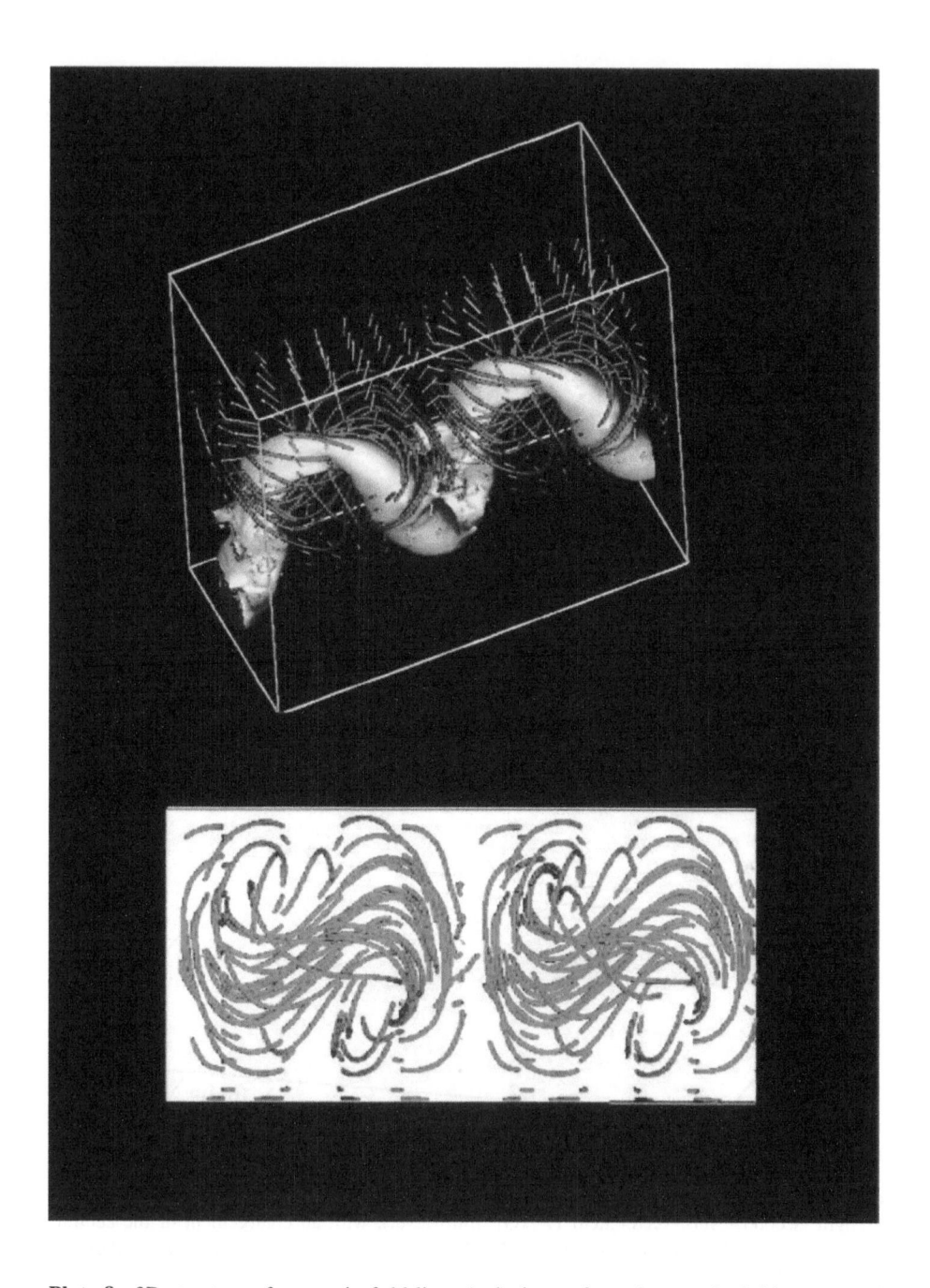

Plate 8. 3D structure of magnetic field lines (red), iso-surface of magnetic field (grey), and velocity vectors (green) of an emerging twisted fluxtube, calculated with a 3D MHD code. The bottom panel shows the projection of magnetic field lines onto the XY-plane. The emergence into the corona leads to a kinked alignment of solar active regions (Matsumoto et al. 1998).

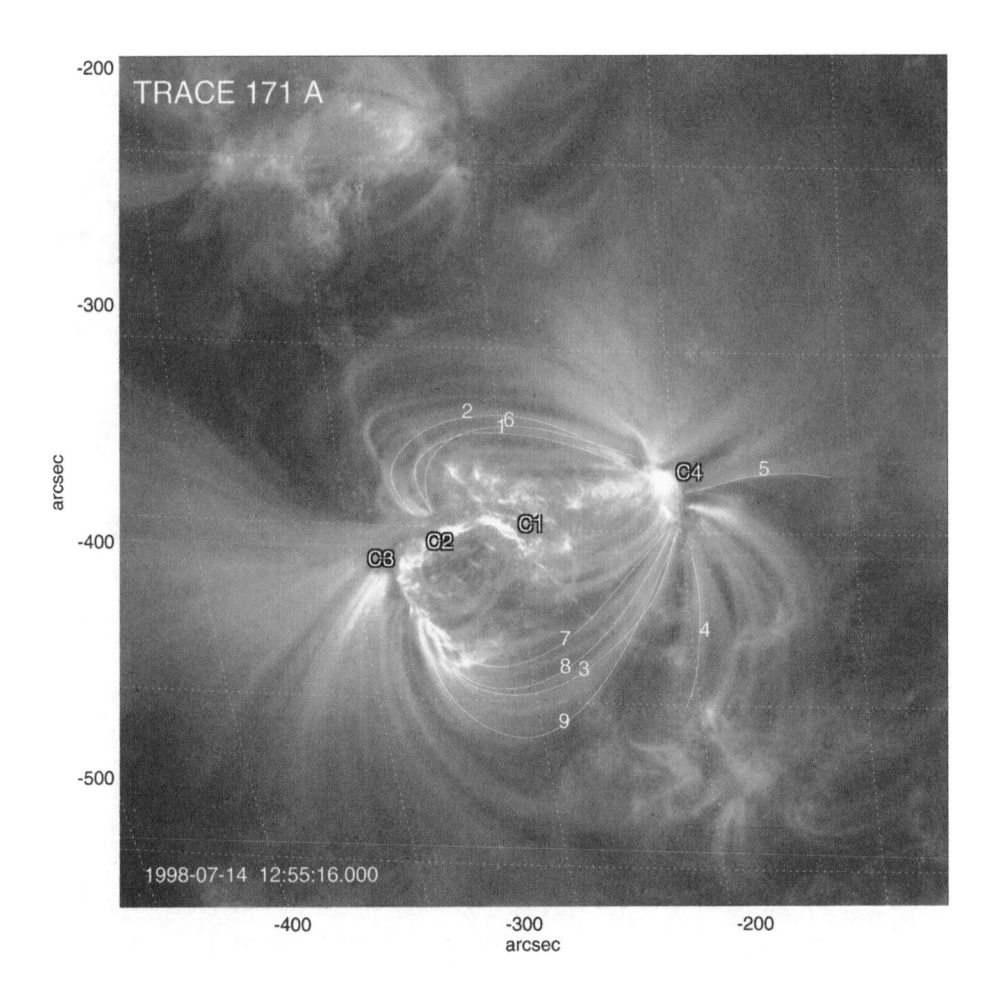

Plate 9. A TRACE 171 Å image is shown at the beginning of the 1998-Jul-14, 12:55:16 UT flare, during which the first kink-mode oscillations were discovered. Kernels of flare emission are located at $C_1 - C4$, which seem to trigger loop oscillations. The diagonal pattern across the brightness maximum at C_1 is a diffraction effect of the telescope. The analyzed loops are outlined with thin lines. Loops #4 and #6 − 9 show pronounced oscillations (Aschwanden et al. 1999b).

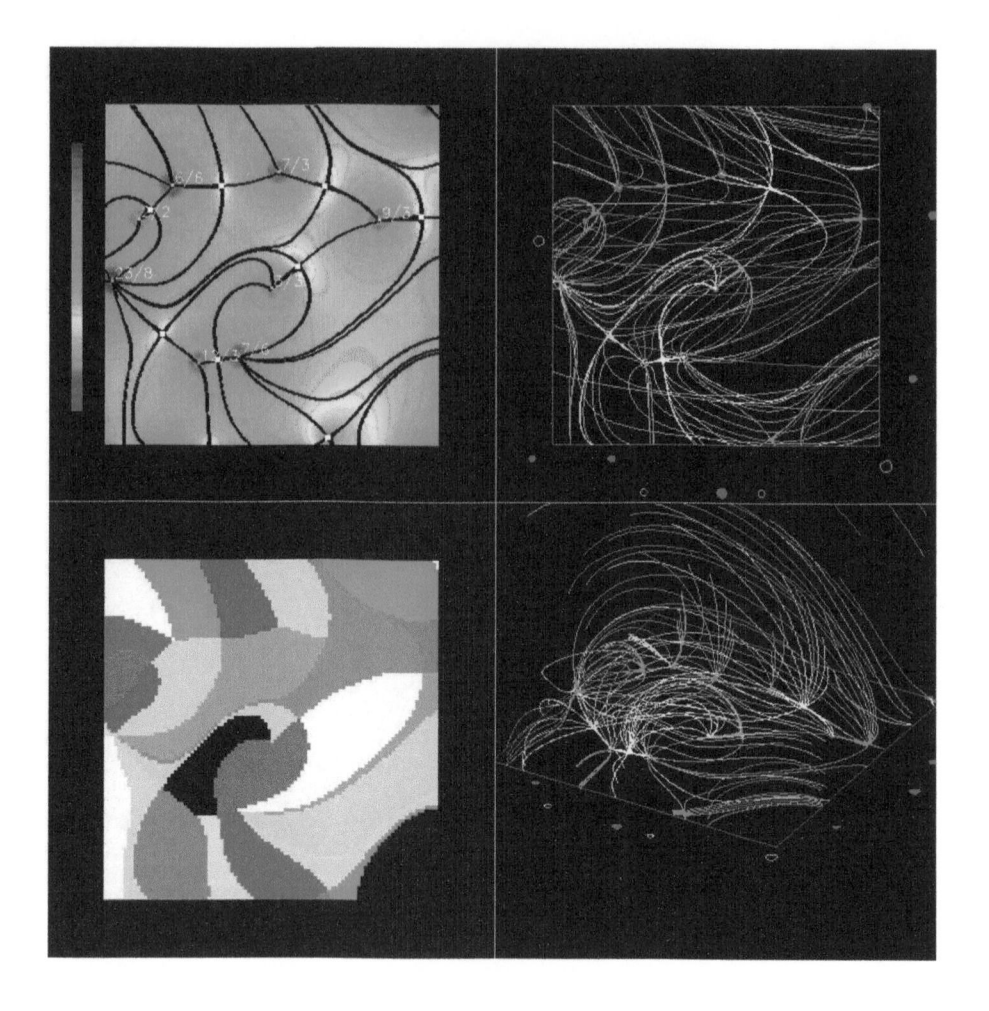

Plate 10. Connectivity domains of a potential magnetic field are visualized by domains with different colors (bottom left). The logarithm of the magnetic field strength is shown with a colored contour map, with nullpoints marked as small white squares and separators marked with black lines (top left). Fan and spine field lines from different perspectives are shown in the right frames. This numerical computation illustrates that most of the low-lying field lines are closed (in the transition region), while only a small fraction of the field lines are open and connect upward to the corona (Schrijver & Title 2002).

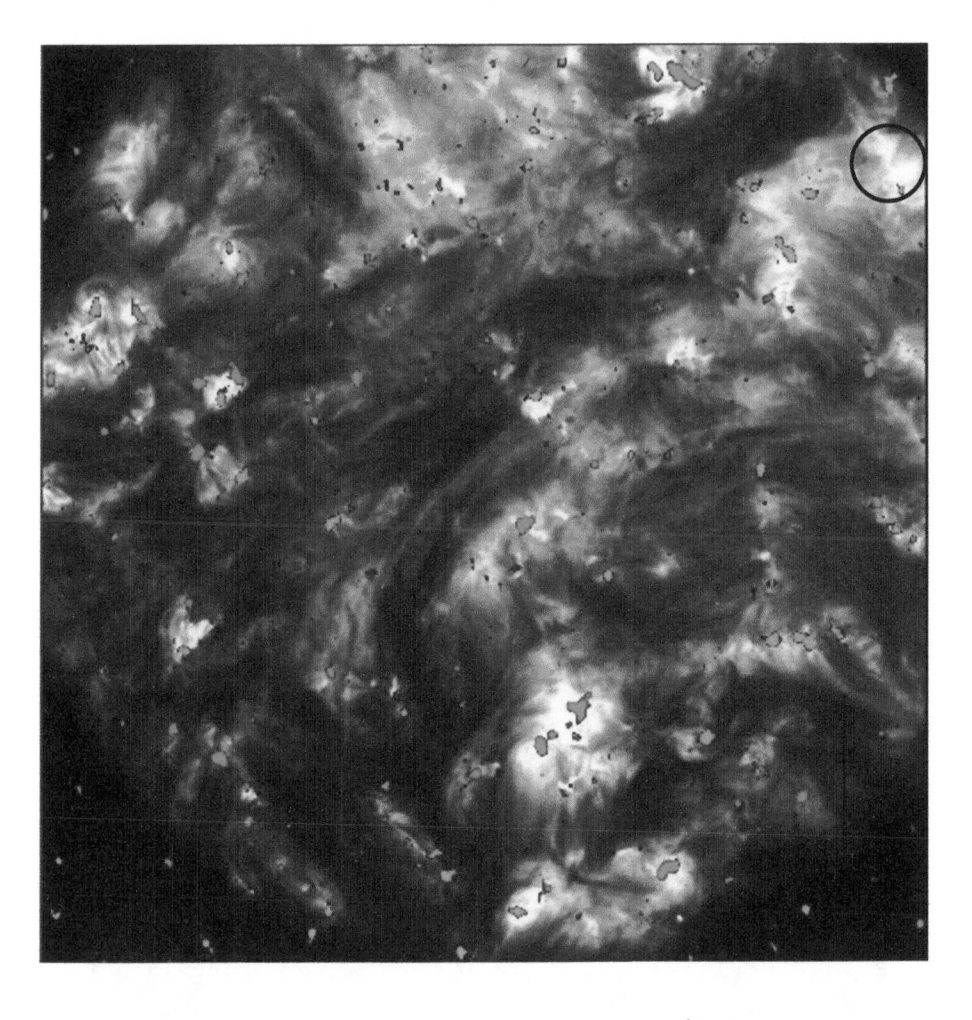

Plate 11. TRACE image of the quiet Sun corona, taken in the 171 Å passband, on 1998-Jun-10, 20:40 UT. The exposure time is 262 s, centered at $(-122, -16)$ arcsec relative to the disk center, displayed with a pixel resolution of $1''$. Superimposed is a threshold SOHO/MDI full-disk magnetogram (with green and red indicating opposite polarities), taken at 20:48 UT, aligned within one arcsec. The color scale saturates at ±15 Mx cm^{-2} and the magnetogram resolution is $1.4''$. Note the detailed correspondence of small magnetic bipoles at the footpoints of coronal nanoflares and large-scale loops (Schrijver & Title 2002).

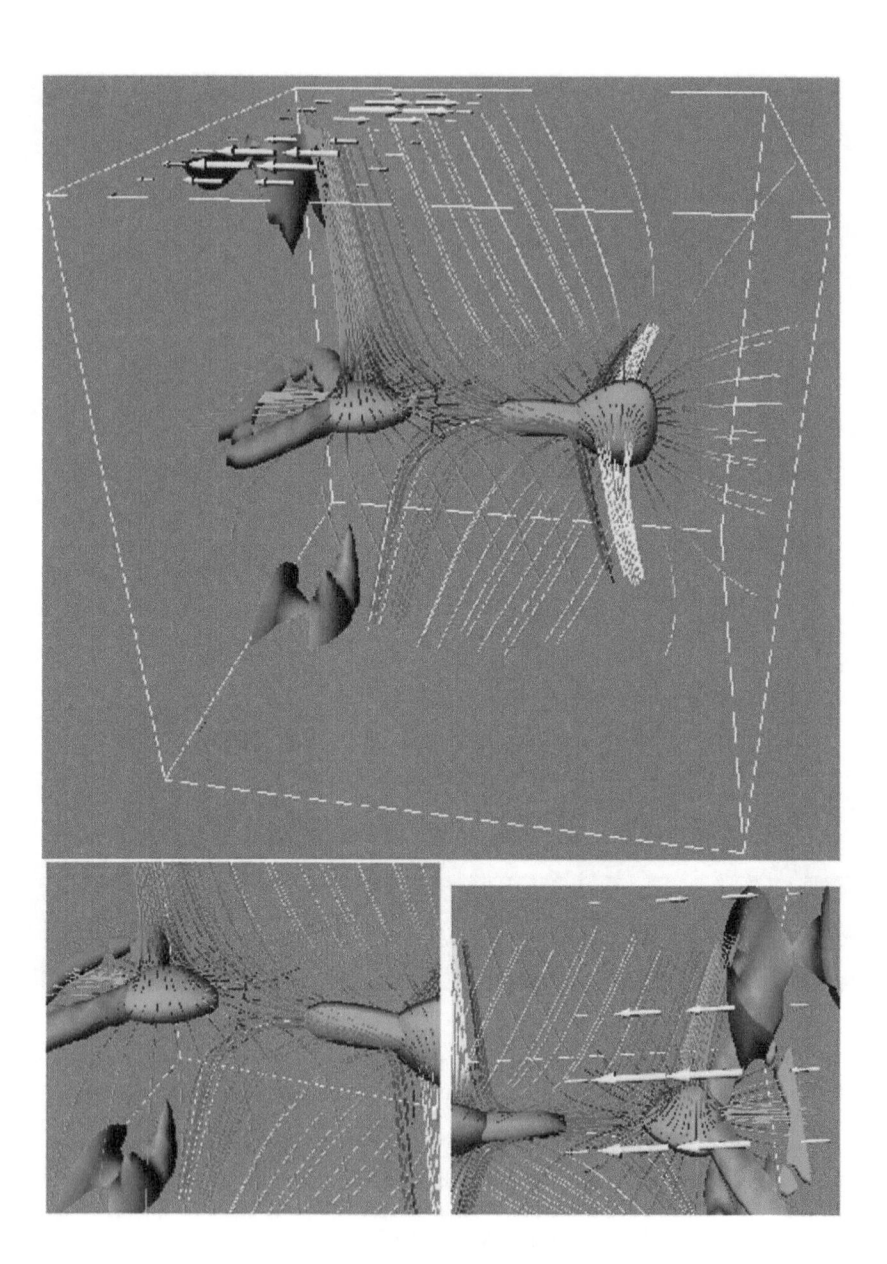

Plate 12. Numerical simulation of 3D separator reconnection. Yellow vectors (bottom right frame) indicate the driving pattern at the top boundary. The two reddish iso-surfaces show the locations of the nullpoints and the purple iso-surfaces the locations of strong current. Red and yellow field lines show the magnetic topology and the two nulls. The green and blue field lines provide the current topology. The bottom right frame is rotated by 180 degrees around the vertical axis. Note that the currents spread along the separators and enable reconnection along the entire separator lines (Galsgaard et al. 1997b).

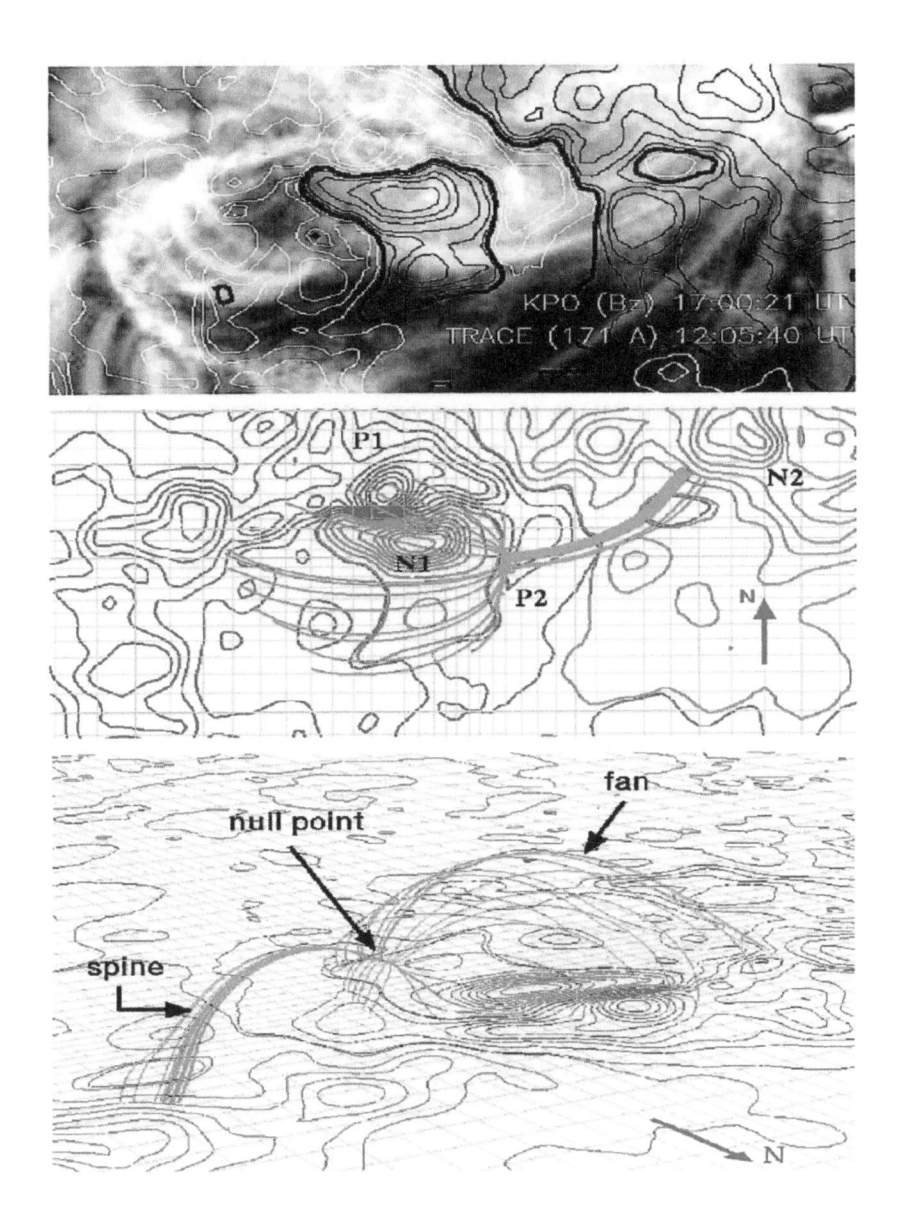

Plate 13. A 3D nullpoint with its associated spine field lines and fan surface is inferred for the first Bastille-Day (1998-Jul-14, 12:55 UT) flare by Aulanier et al. (2000b). *Top:* the magnetic field (with contours at $B = \pm 20, 50, 100, 250, 400, 600$ G) from a KPNO magnetogram is overlaid on a *TRACE* 171 Å image with a FOV of 203×104 Mm. The neutral lines are indicated with thick black lines. *Middle:* extrapolated magnetic field lines are shown that closely trace out a fan-like separatrix surface above the δ-spot (P1-N1) and end in a 3D nullpoint (P2), which is connected through spine field lines to the leading polarity in the west (N2). *Bottom:* a 3D view is shown from a different viewing angle (from north−west) (Aulanier et al. 2000b).

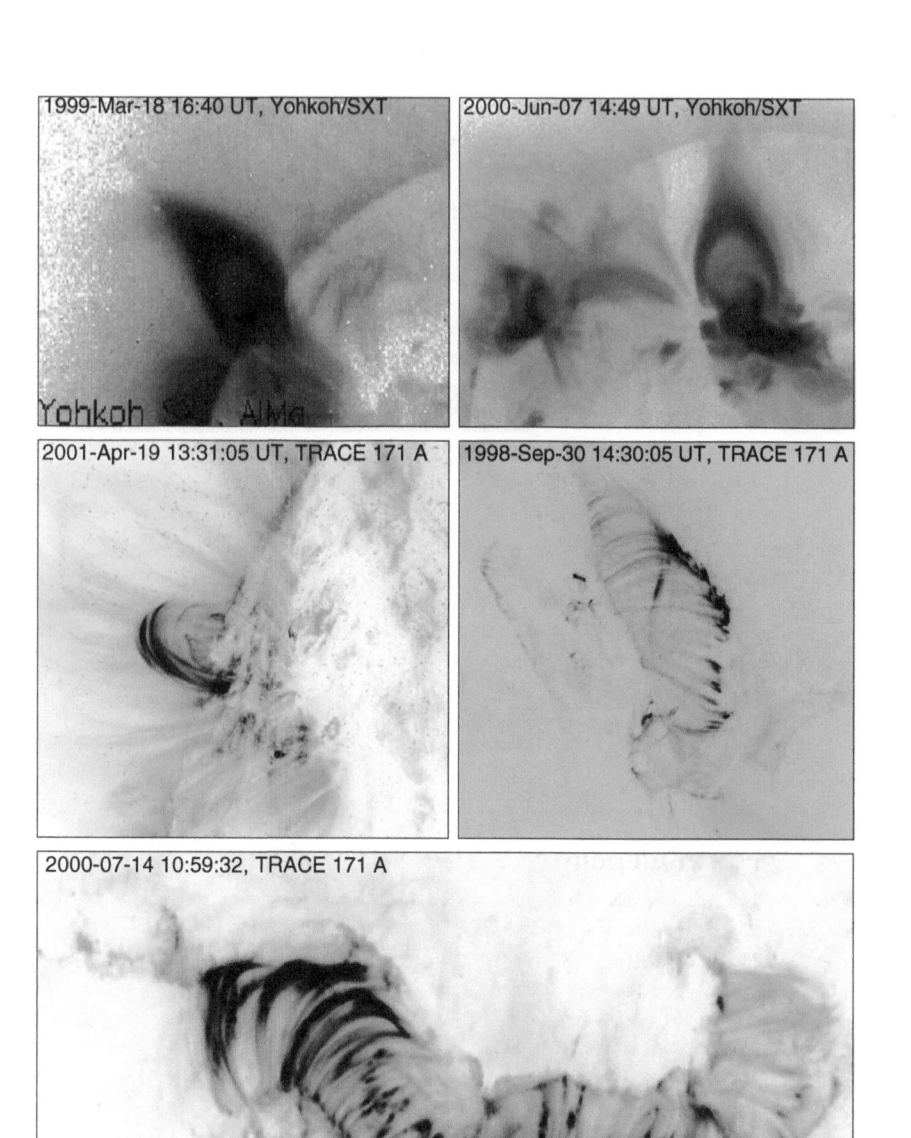

Plate 14. Soft X-ray and EUV images of flare loops and flare arcades with bipolar structure. *Yohkoh/SXT* observed flares (1999-Mar-18, 16:40 UT, and 2000-Jun-07, 14:49 UT) with "candle-flame"-like cusp geometry during ongoing reconnection, while *TRACE* sees postflare loops once they cooled down to $1 - 2$ MK, when they already relaxed into a near-dipolar state. Examples are shown for a small flare (the 2001-Apr-19 flare, 13:31 UT, GOES class M2), and for two large flares with long arcades, seen at the limb (1998-Sep-30, 14:30 UT) and on the disk (the 2000-Jul-14, 10:59 UT, X5.7 flare) (Aschwanden 2002b).

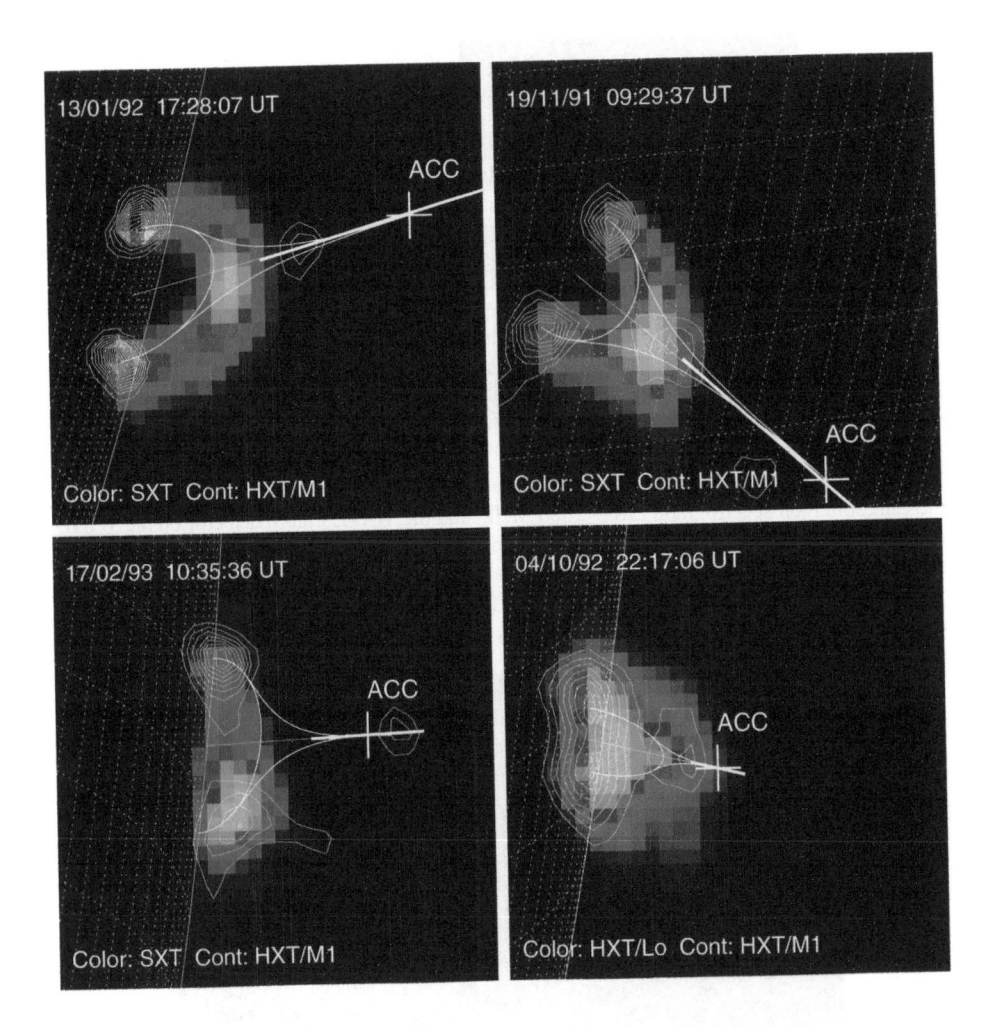

Plate 15. The geometry of the acceleration region inferred from direct detections of above-the-looptop hard X-ray sources with *Yohkoh/HXT* (contours) and simultaneous modeling of electron time-of-flight distances based on energy-dependent time delays of 20 − 200 keV hard X-ray emission measured with *BATSE/CGRO* (crosses marked with ACC). Soft X-rays detected with *Yohkoh/SXT* or thermal hard X-ray emission from the low-energy channel of *Yohkoh/HXT/Lo* are shown in colors, outlining the flare loops (Aschwanden 1999c).

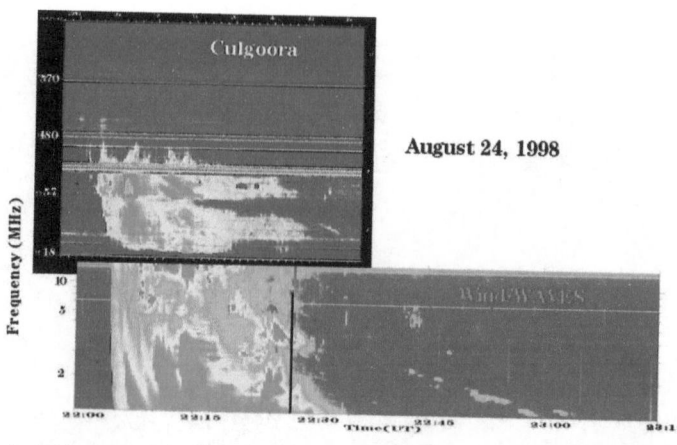

August 24, 1998

Plate 16. A dynamic spectrum of flare/CME-related radio bursts, recorded on 1998-Aug-24 with the Culgoora radiospectrograph (18 − 1000 MHz) and the *WAVES* radio detector on the *WIND* spacecraft (1 − 10 MHz). The composite dynamic spectrum shows type III bursts followed by slower-drifting type II bursts. The frequency axis is shown with increasing frequency in the y-direction, so that electron beams propagating from the Sun away drift in negative y-direction ($d\nu/dt < 0$) (Courtesy of Culgoora Observatory and Wind/WAVES team).

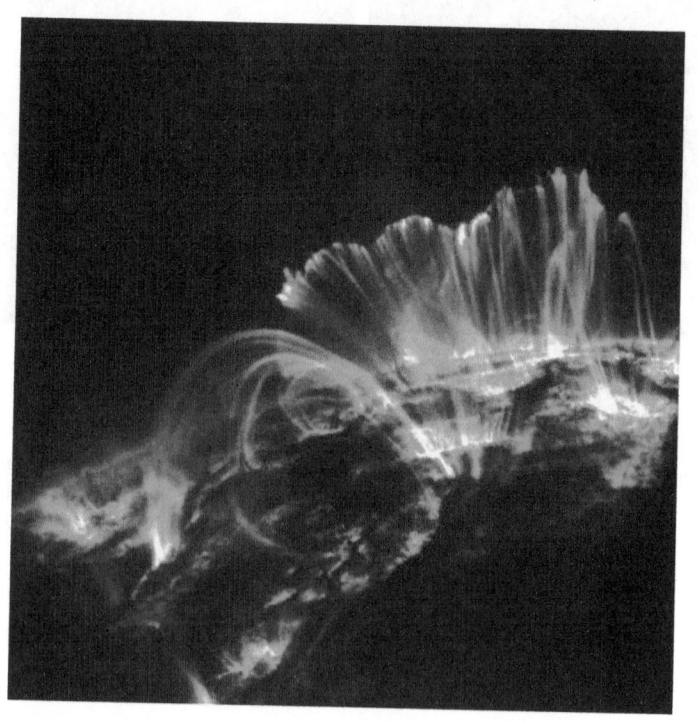

Plate 17. A postflare loop system is imaged with *TRACE* 171 Å, on 2000-Nov-09, 05:03 UT, some 6 hours after a GOES-class M7.4 flare on 2000-Nov-08, 22:42 UT in AR 9213 near the west limb. The flare was accompanied by a coronal mass ejection, observed by SoHO/LASCO. Note the numerous postflare loops which indicate continuous heating over more than 6 hours after the impulsive flare phase (Courtesy of the TRACE team).

Lightning Source UK Ltd.
Milton Keynes UK
UKOW07f1418260416

272995UK00004B/48/P